INTERMEDIATE
algebra
A REAL-WORLD APPROACH

IGNACIO BELLO   FRAN HOPF

INTERMEDIATE
algebra
A REAL-WORLD APPROACH

THIRD EDITION

Ignacio Bello

Hillsborough Community College
Tampa, Florida

Fran Hopf

University of South Florida
Tampa, Florida

 **McGraw-Hill Higher Education**

Boston   Burr Ridge, IL   Dubuque, IA   New York   San Francisco   St. Louis
Bangkok   Bogotá   Caracas   Kuala Lumpur   Lisbon   London   Madrid   Mexico City
Milan   Montreal   New Delhi   Santiago   Seoul   Singapore   Sydney   Taipei   Toronto

INTERMEDIATE ALGEBRA: A REAL-WORLD APPROACH, THIRD EDITION

Published by McGraw-Hill, a business unit of The McGraw-Hill Companies, Inc., 1221 Avenue of the Americas, New York, NY 10020. Copyright © 2009 by The McGraw-Hill Companies, Inc. All rights reserved. Previous edition © 2006. No part of this publication may be reproduced or distributed in any form or by any means, or stored in a database or retrieval system, without the prior written consent of The McGraw-Hill Companies, Inc., including, but not limited to, in any network or other electronic storage or transmission, or broadcast for distance learning.

Some ancillaries, including electronic and print components, may not be available to customers outside the United States.

This book is printed on acid-free paper.

1 2 3 4 5 6 7 8 9 0 QPV/QPV 0 9 8

ISBN 978–0–07–353345–2
MHID 0–07–353345–9

ISBN 978–0–07–320000–2 (Annotated Instructor's Edition)
MHID 0–07–320000–X

Editorial Director: *Stewart K. Mattson*
Senior Sponsoring Editor: *David Millage*
Developmental Editor: *Michelle Driscoll*
Marketing Manager: *Victoria Anderson*
Senior Project Manager: *Vicki Krug*
Senior Production Supervisor: *Sherry L. Kane*
Senior Media Project Manager: *Sandra M. Schnee*
Senior Designer: *David W. Hash*
Cover/Interior Designer: *Asylum Studios*
(USE) Cover Image: ©*Latin Percussion*
Senior Photo Research Coordinator: *Lori Hancock*
Photo Research: *Connie Mueller*
Supplement Coordinator: *Melissa M. Leick*
Compositor: *ICC Macmillan, Inc.*
Typeface: *10/12 Times Roman*
Printer: *Quebecor World Versailles, KY*

The credits section for this book begins on page C-1 and is considered an extension of the copyright page.

www.mhhe.com

About the Authors

Ignacio Bello

attended the University of South Florida (USF), where he earned a B.A. and M.A. in Mathematics. He began teaching at USF in 1967, and in 1971 became a member of the Faculty at Hillsborough Community College (HCC) and Coordinator of the Math and Sciences Department. Professor Bello instituted the USF/HCC remedial program, a program that started with 17 students taking Intermediate Algebra and grew to more than 800 students with courses covering Developmental English, Reading, and Mathematics. Aside from the present series of books *(Basic College Mathematics, Introductory Algebra,* and *Intermediate Algebra),* Professor Bello is the author of more than 40 textbooks including *Topics in Contemporary Mathematics* (ninth edition), *College Algebra, Algebra and Trigonometry,* and *Business Mathematics.* Many of these textbooks have been translated into Spanish. With Professor Fran Hopf, Bello started the Algebra Hotline, the only live, college-level television help program in Florida. Professor Bello is featured in three television programs on the award-winning Education Channel. He has helped create and develop the USF Mathematics Department website (http://mathcenter.usf.edu), which serves as support for the Finite Math, College Algebra, Intermediate Algebra, and Introductory Algebra, and CLAST classes at USF. You can see Professor Bello's presentations and streaming videos at this website, as well as at http://www.ibello.com. Professor Bello is a member of the MAA and AMATYC and has given many presentations regarding the teaching of mathematics at the local, state, and national levels.

Fran Hopf

teaches algebra and finite mathematics both on campus and as distance learning courses for the University of South Florida where she is presently a PhD student in higher education with a cognate in mathematics. She has a bachelor's degree from Florida State University and a master's in education from the University of South Florida. In addition to this textbook, she has authored course supplements for intermediate algebra and has reviewed texts for developmental and liberal arts mathematics. Fran and Ignacio were co-hosts for a live, call-in Algebra Hotline program that aired weekly for three years on the Education Channel in Tampa, Florida which supported a

About the Authors

college algebra telecourse. Fran has developed a curriculum and videos for a math review for the CLAST, a mandatory test given to college students attending state universities in Florida. The CLAST material can be found at http://mathcenter.usf.edu/. She developed and facilitated faculty workshops, "Interactive Online Learning," for the University of South Florida's Center for 21st Century Teaching Excellence. Fran developed and presented, "Just the FAQ's: Algebra 1," video vignettes for the Education Channel in Hillsborough County, which won first place in the 2005 "Southern Sunshine Awards" sponsored by the Southeast Region of the Alliance for Community Media. The series took top honors in the "Instructional Series" category. She is a member of AMATYC and has presented at the AMATYC national convention. Whether in the classroom, on the Internet, or writing curriculum, Fran enjoys sharing her enthusiasm for mathematics.

The Team Bello and Hopf, Hopf and Bello. They have been working together for more than 20 years. As founders of the Algebra Hotline they helped students on live television every Wednesday at 8 P.M. for 3 years. To reach more students, they created and taught telecourses at both HCC and USF. The idea was expanded to cover college algebra and now it is used in intermediate algebra by other instructors. The team has also worked in creating instructor's manuals, solutions manuals, and video tapes as well as spreading the joy of mathematics at local, regional, and national mathematics meetings. Finally, through the auspices of McGraw-Hill, you can see their work, ideas, and techniques right here: *Intermediate Algebra*. The word to students from the team? **Math is fun, and you can do it!**

Contents

Preface xiii

Guided Tour: Features and Supplements xxi

Applications Index xxxi

Chapter one

1 ▸ The Real Numbers

The Human Side of Algebra 1

- 1.1 Numbers and Their Properties 2
- 1.2 Operations and Properties of Real Numbers 17
- 1.3 Properties of Exponents 37
- 1.4 Algebraic Expressions and the Order of Operations 51

Collaborative Learning 1A 64

Collaborative Learning 1B 64

Research Questions 65

Summary 66

Review Exercises 68

Practice Test 1 71

Chapter two

2 ▸ Linear Equations and Inequalities

The Human Side of Algebra 73

- 2.1 Linear Equations in One Variable 74
- 2.2 Formulas, Geometry, and Problem Solving 87
- 2.3 Problem Solving: Integers and Geometry 100
- 2.4 Problem Solving: Percent, Investment, Motion, and Mixture Problems 111
- 2.5 Linear and Compound Inequalities 123
- 2.6 Absolute-Value Equations and Inequalities 142

Collaborative Learning 2A 154

Collaborative Learning 2B 155

Research Questions 155

Summary 156

Review Exercises 158

Practice Test 2 162

Cumulative Review Chapters 1–2 164

vii

Contents

Chapter three

3 ▶ Graphs and Functions

The Human Side of Algebra 165
3.1 Graphs 166
3.2 Using Slopes to Graph Lines 184
3.3 Equations of Lines 199
3.4 Linear Inequalities in Two Variables 210
3.5 Introduction to Functions 223
3.6 Linear Functions 242
Collaborative Learning 3A 262
Collaborative Learning 3B 262
Research Questions 263
Summary 264
Review Exercises 267
Practice Test 3 272
Cumulative Review Chapters 1–3 281

Chapter four

4 ▶ Solving Systems of Linear Equations and Inequalities

The Human Side of Algebra 283
4.1 Systems with Two Variables 284
4.2 Systems with Three Variables 305
4.3 Coin, Distance-Rate-Time, Investment, and Geometry Problems 315
4.4 Systems of Linear Inequalities 327
Collaborative Learning 4A 334
Collaborative Learning 4B 335
Research Questions 336
Summary 336
Review Exercises 338
Practice Test 4 341
Cumulative Review Chapters 1–4 345

Contents

Chapter five

5 ▶ Polynomials

The Human Side of Algebra 347
- **5.1** Polynomials: Addition and Subtraction 348
- **5.2** Multiplication of Polynomials 359
- **5.3** The Greatest Common Factor and Factoring by Grouping 370
- **5.4** Factoring Trinomials 379
- **5.5** Special Factoring 388
- **5.6** General Methods of Factoring 398
- **5.7** Solving Equations by Factoring: Applications 404

Collaborative Learning 5A 417
Collaborative Learning 5B 417
Research Questions 418
Summary 419
Review Exercises 421
Practice Test 5 423
Cumulative Review Chapters 1–5 425

Chapter six

6 ▶ Rational Expressions

The Human Side of Algebra 427
- **6.1** Rational Expressions 428
- **6.2** Multiplication and Division of Rational Expressions 440
- **6.3** Addition and Subtraction of Rational Expressions 448
- **6.4** Complex Fractions 460
- **6.5** Division of Polynomials and Synthetic Division 470
- **6.6** Equations Involving Rational Expressions 481
- **6.7** Applications: Problem Solving 495
- **6.8** Variation 505

Collaborative Learning 6A 513
Collaborative Learning 6B 514
Research Questions 515
Summary 515
Review Exercises 517
Practice Test 6 520
Cumulative Review Chapters 1–6 522

Contents

Chapter seven

7 ▶ Rational Exponents and Radicals

The Human Side of Algebra 525
- **7.1** Rational Exponents and Radicals 526
- **7.2** Simplifying Radicals 537
- **7.3** Operations with Radicals 548
- **7.4** Solving Equations Containing Radicals 559
- **7.5** Complex Numbers 567

Collaborative Learning 7A 578
Collaborative Learning 7B 578
Research Questions 579
Summary 580
Review Exercises 581
Practice Test 7 584
Cumulative Review Chapters 1–7 588

Chapter eight

8 ▶ Quadratic Equations and Inequalities

The Human Side of Algebra 591
- **8.1** Solving Quadratics by Completing the Square 592
- **8.2** The Quadratic Formula: Applications 605
- **8.3** The Discriminant and Its Applications 617
- **8.4** Solving Equations in Quadratic Form 626
- **8.5** Nonlinear Inequalities 635

Collaborative Learning 8A 647
Collaborative Learning 8B 648
Research Questions 648
Summary 649
Review Exercises 650
Practice Test 8 653
Cumulative Review Chapters 1–8 655

Contents

Chapter nine

9 Quadratic Functions and the Conic Sections

The Human Side of Algebra 657
- **9.1** Quadratic Functions (Parabolas) and Their Graphs 658
- **9.2** Circles and Ellipses 680
- **9.3** Hyperbolas and Identification of Conics 698
- **9.4** Nonlinear Systems of Equations 712
- **9.5** Nonlinear Systems of Inequalities 724

Collaborative Learning 9A 731
Collaborative Learning 9B 731
Research Questions 732
Summary 733
Review Exercises 734
Practice Test 9 739
Cumulative Review Chapters 1–9 745

Chapter ten

10 Functions—Inverse, Exponential, and Logarithmic

The Human Side of Algebra 747
- **10.1** The Algebra of Functions 748
- **10.2** Inverse Functions 760
- **10.3** Exponential Functions 773
- **10.4** Logarithmic Functions and Their Properties 784
- **10.5** Common and Natural Logarithms 796
- **10.6** Exponential and Logarithmic Equations and Applications 811

Collaborative Learning 10A 824
Collaborative Learning 10B 825
Research Questions 825
Summary 826
Review Exercises 827
Practice Test 10 831
Cumulative Review Chapters 1–10 836

Contents

Appendix A

A1: Sequences and Series
A2: Arithmetic Sequences and Series
A3: Geometric Sequences and Series
(A1–A3 found online at www.mhhe.com/bello)
A4: Matrices A-2
A5: Determinants and Cramer's Rule A-15
A6: The Binomial Expansion A-28

Selected Answers SA-1
Photo Credits C-1
Index I-1

▶ Preface

▶ From the Authors

The Inspiration for Our Teaching

Bello: I was born in Havana, Cuba and I encountered some of the same challenges in mathematics that many of my current students face, all while attempting to overcome a language barrier. In high school, I failed my freshman math course, which at the time was a complex language for me. However, perseverance being one of my traits, I scored 100% on the final exam the second time around. After working in various jobs (roofer, sheetrock installer, and dock worker), I finished high school and received a college academic scholarship. I enrolled in calculus and made a C. Never one to be discouraged, I became a math major and learned to excel in the courses that had previously frustrated me. While a graduate student at the University of South Florida (USF), I taught at a technical school, Tampa Technical Institute, a decision that contributed to my resolve to teach math and make it come alive for my students the way brilliant instructors such as Jack Britton, Donald Rose, and Frank Cleaver had done for me. My math instructors instilled in me the motivation to be successful. I have learned a great deal about the way in which students learn and how the proper guidance through the developmental mathematics curriculum leads to student success. I believe I have accomplished a strong level of guidance in my textbook series to further explain the language of mathematics carefully to students to help them to reach success as well.

A Lively Approach to Reach Today's Students

Teaching math at the University of South Florida was a great new career for me, but I was disappointed by the materials I had to use. A rather imposing, mathematically correct but boring book was in vogue. Students hated it, professors hated it, and administrators hated it. I took the challenge to write a better book, a book that was not only mathematically correct, but **student-oriented** with **interesting applications**—many suggested by the students themselves—and even, dare we say, entertaining! That book's approach and philosophy proved an instant success and was a precursor to my current series.

Students fondly called my class "The Bello Comedy Hour," but they worked hard, and they performed well. Because my students always ranked among the highest on the common final exam at USF, I knew I had found a way to motivate them through **common-sense language** and humorous, **realistic math applications.** I also wanted to show students they could overcome the same obstacles I had in math and become successful, too. If math has just never been a subject that some of your students have felt comfortable with, then they're not alone! I wrote this book with the **math-anxious** student in mind, so they'll find my tone is jovial, my explanations are patient, and instead of making math seem mysterious, I make it down-to-earth and easily digestible. For example, after I've explained the different methods for simplifying fractions, I speak directly to readers: "Which way should you simplify fractions? The way you understand!" Once students realize that math is within their grasp and not a foreign language, they'll be surprised at how much more confident they feel.

A Real-World Approach: Applications, Student Motivation, and Problem Solving

What is a "real-world approach"? I found that most textbooks put forth "real-world" applications that meant nothing to the real world of my students. How many of my

xiii

Preface

students would really need to calculate the speed of a bullet (unless they are in its way) or cared to know when two trains traveling in different directions would pass by each other (disaster will certainly occur if they are on the same track)? For my students, both traditional and nontraditional, the real world consists of questions such as, "How do I find the best cell phone plan?" and "How will I pay my tuition and fees if they increase by $x\%$?" That is why I introduce mathematical concepts through everyday applications with **real data** and give homework using similar, well-grounded situations (see the Getting Started application that introduces every section's topic and the word problems in every exercise section). Putting math in a real-world context has helped me to overcome one of the problems we all face as math educators: **student motivation.** Seeing math in the real world makes students perk up in a math class in a way I have never seen before, and realism has proven to be the best motivator I've ever used. In addition, the real-world approach has enabled me to enhance students' **problem-solving skills** because they are far more likely to tackle a real-world problem that matters to them than one that seems contrived.

Diverse Students and Multiple Learning Styles

We know we live in a pluralistic society, so how do you write one textbook for everyone? The answer is to build a flexible set of teaching tools that instructors and students can adapt to their own situations. Are any of your students members of a **cultural minority?** So am I! Did they learn **English as a second language?** So did I! You'll find my book speaks directly to them in a way that no other book ever has, and fuzzy explanations in other books will be clear and comprehensible in mine.

Do your students all have the same **learning style?** Of course not! That's why I wrote a book that will help students learn mathematics regardless of their personal learning style. **Visual learners** will benefit from the text's clean page layout, careful use of color highlighting, *"Web Its,"* and the video lectures on the text's website. **Auditory learners** will profit from the audio *e-Professor lectures* on the text's website, and both **auditory** and **social learners** will be aided by the *Collaborative Learning* projects. **Applied** and **pragmatic learners** will find a bonanza of features geared to help them: *Pretests* can be found in MathZone providing practice problems by every example, and *Mastery Tests,* to name just a few. **Spatial learners** will find the chapter *Summary* is designed especially for them, while **creative learners** will find the *Research Questions* to be a natural fit. Finally, **conceptual learners** will feel at home with features like *"The Human Side of Algebra"* and the *"Write On"* exercises. Every student who is accustomed to opening a math book and feeling like they've run into a brick wall will find in my books that a number of doors are standing open and inviting them inside.

Listening to Student and Instructor Concerns

McGraw-Hill has given me a wonderful resource for making my textbook more responsive to the immediate concerns of students and faculty. In addition to sending my manuscript out for review by instructors at many different colleges, several times a year McGraw-Hill holds symposia and focus groups with math instructors where the emphasis is *not* on selling products but instead on the **publisher listening** to the needs of faculty and their students. These encounters have provided me with a wealth of ideas on how to improve my chapter organization, make the page layout of my books more readable, and fine-tune exercises in every chapter so that

Preface

students and faculty will feel comfortable using my book because it incorporates their specific suggestions and anticipates their needs.

R-I-S-E to Success in Math

Why are some students more successful in math than others? Often it is because they know how to manage their time and have a plan for action. Students can use models similar to the following tables to make a weekly schedule of their time (classes, study, work, personal, etc.) and a semester calendar indicating major course events like tests, papers, and so on. Then, try to do as many of the suggestions on the **"R-I-S-E"** list as possible. (Larger, printable versions of these tables can be found in MathZone at www.mhhe.com/bello.)

Weekly Time Schedule

Time	S	M	T	W	R	F	S
8:00							
9:00							
10:00							
11:00							
12:00							
1:00							
2:00							
3:00							
4:00							
5:00							
6:00							
7:00							
8:00							
9:00							
10:00							
11:00							

Semester Calendar

Wk	M	T	W	R	F
1					
2					
3					
4					
5					
6					
7					
8					
9					
10					
11					
12					
13					
14					
15					
16					

R—Read and/or view the material before and after each class. This includes the textbook, the videos that come with the book, the skill checkers from the preceding section, and any special material given to you by your instructor.

I—Interact and/or practice using the CD that comes with the book or the Web exercises suggested in the sections, or seeking tutoring from your school.

S—Study and/or discuss your homework and class notes with a study partner/group, with your instructor, or on a discussion board if available.

E—Evaluate your progress by checking the odd homework questions with the answer key in the back of the book, using the mastery questions in each section of the book as a self-test, and using the Chapter Reviews and Chapter Practice Tests as practice before taking the actual test.

As the items on this list become part of your regular study habits, you will be ready to **"R-I-S-E"** to success in math.

Preface

Improvements in the Third Edition

Based on the valuable feedback of numerous reviewers and users over the years, the following improvements were made to the third edition of *Intermediate Algebra*.

Organizational Changes

- Chapter 3, Graphs and Functions, is reorganized so it develops graphing from lines into linear functions.
- The sections on matrices and determinants have been moved to the appendix, since many instructors do not teach that as part of their regular curriculum.
- A new skill checker set is added at the end of every exercise that practices the skills that will be used in the next section. It can also be used as an in-class warm-up exercise when lecturing on the next section.

Pedagogical Changes

- *Real-World Applications*—many examples, applications, and real-data problems have been added or updated to keep the book's content current.
- *Web Its*—now found in the margin of the Exercises and on MathZone (www.mhhe.com/bello) to encourage students to visit math sites while they're Web surfing and discover the many informative and creative sites that are dedicated to stimulating better education in math.
- *Calculator Corners*—found before the exercise sets, these have been updated with recent information and keystrokes relevant to currently popular calculators.
- *Concept Checkers*—have been added to the end-of-section exercises to help students reinforce key terms and equations.
- *Pretests*—can be found in MathZone providing practice problems for every example. These Pretest results can be compared to the Practice Tests at the end of the chapter to evaluate and analyze student success.
- *The RSTUV approach* to problem solving has been expanded and used throughout this edition as a response to positive comments from both students and users of the previous edition.
- *Translate It*—boxes appear periodically before the word problem exercises to help students turn phrases into equations, reinforcing the RSTUV method.
- *Skill Checker*—now appears at the end of the exercises sets, making sure the students have the necessary skills for the next section.
- *Practice Tests (Diagnostic)*—at the end of every chapter give students immediate feedback and guidance on which section, examples, and pages to review.

▶ Acknowledgments

We would like to thank the following people at McGraw-Hill: David Dietz, who provided the necessary incentives and encouragement for creating this series with the cooperation of Bill Barter; Christien Shangraw, our first developmental editor who worked many hours getting reviewers and gathering responses into concise and usable reports; Randy Welch, who continued and expanded the Christien tradition into a well-honed editing engine with many features, including humor, organization, and very hard work; Liz Haefele, our former editor and publisher, who was encouraging, always on the lookout for new markets; Lori Hancock and her many helpers (LouAnn, Emily, David, and Connie Mueller), who always get the picture; Dr. Tom Porter, of Photos at Your Place, who improved on the pictures provided; Vicki Krug, one of the most exacting persons at McGraw-Hill, who will always give you the time of day and then solve the problem; Hal Whipple, for his help in preparing the answers manuscript; Cindy Trimble, for the accuracy of the text; Jeff Huettman, one of the best 100 producers in the United States, who learned Spanish in anticipation of this project; Marie Bova, for her detective work in tracking down permission rights; and to Professor Nancy Mills, for her expert advice on how my books address multiple learning styles. For the third edition, we thank David Millage, senior sponsoring editor; Michelle Driscoll and Lisa Collette, developmental editors, Torie Anderson, marketing manager; and especially Pat Steele, our very able copy editor. Finally, thanks to our attack secretary, Beverly DeVine, who still managed to send all materials back to the publisher on time, Brad Davis for his exceptional attention to detail in his accuracy check, and Karol McIntosh at Hillsborough Community College for her review comments. To all of them, our many thanks.

We would also like to extend our gratitude to the following reviewers of the Bello series for their many helpful suggestions and insights. They helped us write better textbooks:

Tony Akhlaghi, *Bellevue Community College*

Theresa Allen, *University of Idaho*

John Anderson, *San Jacinto College–South Campus*

Keith A. Austin, *Devry University–Arlington*

Sohrab Bakhtyari, *St. Petersburg College–Clearwater*

Emilie Berglund, *Utah Valley State College*

Fatemah Bicksler, *Delgado Community College*

Brenda Blankenship, *Volunteer State Community College*

Ann Brackebusch, *Olympic College*

Connie Buller, *Metropolitan Community College*

Gail G. Burkett, *Palm Beach Community College*

Linda Burton, *Miami Dade College*

James Butterbach, *Joliet Junior College*

Magdalena Caproiu, *Antelope Valley College*

Judy Carlson, *Indiana University–Purdue University Indianapolis*

Jimmy Chang, *St. Petersburg College*

Ryan Cornett, *San Jacinto College North*

Randall Crist, *Creighton University*

Mark Czerniak, *Moraine Valley Community College*

Acknowledgments

Antonio David, *Del Mar College*

Parsla Dineen, *University of Nebraska–Omaha*

Sue Duff, *Guilford Technical Community College*

Lynda Fish, *St. Louis Community College–Forest Park*

Donna Foster, *Piedmont Technical College*

Deborah D. Fries, *Wor-Wic Community College*

James P. Fulton, *Suffolk County Community College*

Jeanne H. Gagliano, *Delgado Community College*

Debbie Garrison, *Valencia Community College*

Donald K. Gooden, *Northern Virginia Community College–Woodbridge*

Edna Greenwood, *Tarrant County College–Northwest Campus*

David E. Gustafson, *Tarrant County College*

Ken Harrelson, *Oklahoma City Community College*

Joseph Lloyd Harris, *Gulf Coast Community College*

Tony Hartman, *Texarkana College*

Hank Hernandez, *Indiana University Purdue University–Indianapolis*

Susan Hitchcock, *Palm Beach Community College*

Patricia Carey Horacek, *Pensacola Junior College*

Jack Hughes, *St. Petersburg College–Clearwater Campus*

Peter Intarapanich, *Southern Connecticut State University*

Judy Ann Jones, *Madison Area Technical College*

Shelbra Bullock Jones, *Wake Technical Community College*

Rosamma Joseph, *Wilbur Wright College*

Cheryl Kane, *University of Nebraska*

Linda Kass, *Bergen Community College*

Joe Kemble, *Lamar University*

Joanne Kendall, *Blinn College–Brenham*

Gary King, *Ozarks Technical College*

Bernadette Kocyba, *J S Reynolds Community College*

Randa Kress, *Idaho State University*

Carla Kulinsky, *Salt Lake Community College*

Lider-Manuel Lamar, *Seminole Community College*

Marcia J. Lambert, *Pitt Community College*

Marie Agnes Langston, *Palm Beach Community College*

Kathryn Lavelle, *Westchester Community College*

Angela Lawrenz, *Blinn College–Bryan*

Richard Leedy, *Polk Community College*

Barbara Little, *Central Texas College*

Judith L. Maggiore, *Holyoke Community College*

Timothy Magnavita, *Bucks Community College*

Tsun-Zee Mai, *University of Alabama*

Acknowledgments

Harold Mardones, *Community College of Denver*
Lois Martin, *Massasoit Community College*
Mike Martin, *Johnson County Community College*
Gary McCracken, *Shelton State Community College*
Tania McNutt, *Community College of Aurora*
Barbara Miller, *Lexington Community College*
Danielle Morgan, *San Jacinto College–South Campus*
Shauna Mullins, *Murray State University*
Roya Namavar, *Rogers State University*
Deborah A. Okonieski, *The University of Akron*
Diana Orrantia, *El Paso Community College*
Robert Payne, *Stephen F. Austin State University*
Joanne Peeples, *El Paso Community College*
Faith Peters, *Miami Dade College–Wolfson*
Jane Pinnow, *University of Wisconsin–Parkside*
Anne Prial, *Orange County Community College*
Janice F. Rech, *University of Nebraska–Omaha*
Libbie Reeves, *Mitchell Community College*
Karen Roothaan, *Harold Washington College*
Don Rose, *College of the Sequoias*
Pascal Roubides, *Miami Dade College–Wolfson*
Juan Saavedra, *Albuquerque Technical Vocational Institute*
Judith Salmon, *Fitchburg State College*
Mansour Samimi, *Winston-Salem State University*
Susan Santolucito, *Delgado Community College*
Ellen Sawyer, *College of DuPage*
Laura Schaben, *University of Nebraska–Omaha*
Vicki Schell, *Pensacola Junior College*
Sandra Siegrist, *Central Ohio Technical College*
Timothy Siniscalchi, *Palm Beach Community College*
Judith Smalling, *St. Petersburg College–Gibbs*
Carol E. Smith, *Bakersfield College*
Ray Stanton, *Fresno City College*
Patrick Stevens, *Joliet Junior College*
Bryan Stewart, *Tarrant County College–Southwest*
Janet E. Teeguarden, *Ivy Tech Community College of Indiana*
Ann Thrower, *Kilgore College*
Michael Tran, *Antelope Valley College*
Bettie A. Truitt, *Black Hawk College*
Razvan Verzeanu, *Fullerton College*

Acknowledgments

Nguyen Vu, *Rio Hondo College*
James Wang, *University of Alabama*
Betty Weinberger, *Delgado Community College*
Jadwiga Weyant, *Edmonds Community College*
Robert E. White, *Allan Hancock College*
Denise Widup, *University of Wisconsin–Parkside*
Cheryll Wingard, *Community College of Aurora*
Jeff Young, *Delaware Valley College*
Marilyn A. Zopp, *McHenry County College*

Guided Tour

Features and Supplements

Motivation for a Diverse Student Audience

A number of features exist in every chapter to motivate students' interest in the topic and thereby increase their performance in the course:

❯ The Human Side of Algebra

To personalize the subject of mathematics, the origins of numerical notation, concepts, and methods are introduced through the lives of real people solving ordinary problems.

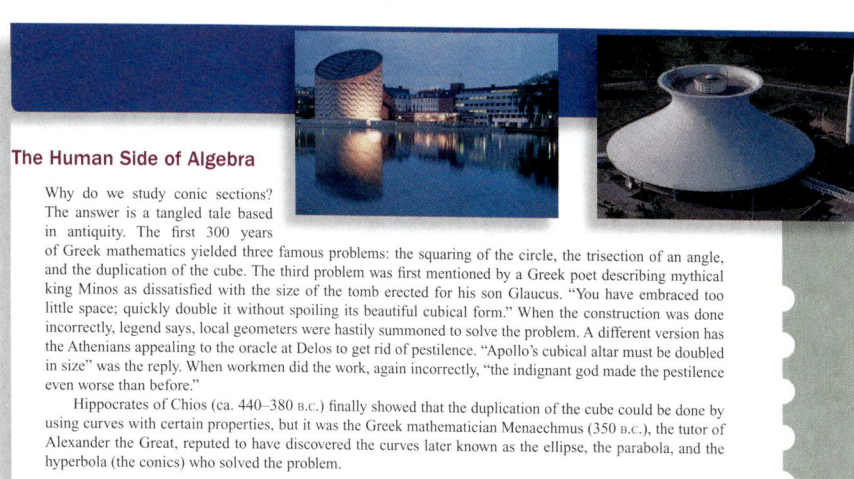

The Human Side of Algebra

Why do we study conic sections? The answer is a tangled tale based in antiquity. The first 300 years of Greek mathematics yielded three famous problems: the squaring of the circle, the trisection of an angle, and the duplication of the cube. The third problem was first mentioned by a Greek poet describing mythical king Minos as dissatisfied with the size of the tomb erected for his son Glaucus. "You have embraced too little space; quickly double it without spoiling its beautiful cubical form." When the construction was done incorrectly, legend says, local geometers were hastily summoned to solve the problem. A different version has the Athenians appealing to the oracle at Delos to get rid of pestilence. "Apollo's cubical altar must be doubled in size" was the reply. When workmen did the work, again incorrectly, "the indignant god made the pestilence even worse than before."

Hippocrates of Chios (ca. 440–380 B.C.) finally showed that the duplication of the cube could be done by using curves with certain properties, but it was the Greek mathematician Menaechmus (350 B.C.), the tutor of Alexander the Great, reputed to have discovered the curves later known as the ellipse, the parabola, and the hyperbola (the conics) who solved the problem.

❯ Getting Started

Each topic is introduced in a setting familiar to students' daily lives, making the subject personally relevant and more easily understood.

▶ Getting Started

Algebra—Wages and Unemployment by Educational Status

Higher Education Means Higher Income and Less Chance of Unemployment The graph here indicates the unemployment rate (pink bars) by percent and the median weekly earnings (green bars) in 2003 by educational attainment.

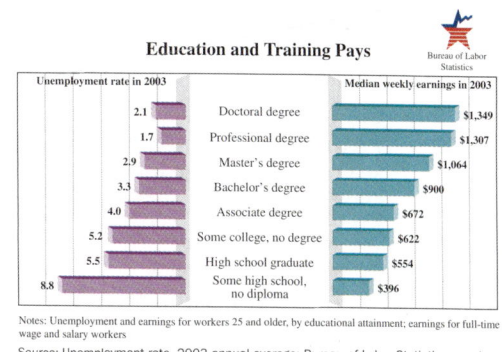

xxi

Guided Tour

› Web It

Appearing in the end of section exercises, this URL refers students to the abundance of resources available on the Web that can show them fun, alternative explanations and demonstrations of important topics.

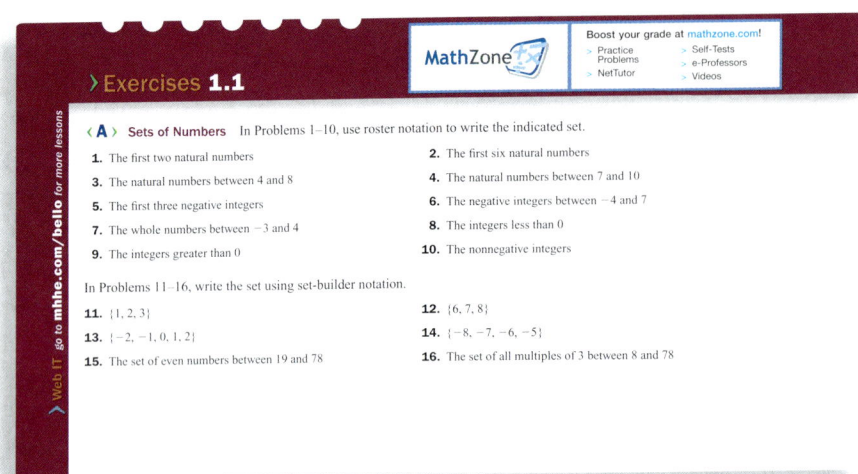

› Write On

Writing exercises give students the opportunity to express mathematical concepts and procedures in their own words, thereby internalizing what they have learned.

› › › Write On

84. How would you know if a system of linear equations is consistent when you are using the specified method?
   a. the graphical method   b. the substitution method
   c. the elimination method

85. How would you know if a system of linear equations is inconsistent when you are using the specified method?
   a. the graphical method   b. the substitution method
   c. the elimination method

86. How would you know if a system of linear equations is dependent when you are using the specified method?
   a. the graphical method   b. the substitution method
   c. the elimination method

87. Write the possible advantages and disadvantages of each of the methods we have studied: graphical, substitution, and elimination.

› Collaborative Learning

Concluding the chapter are exercises for collaborative learning that promote teamwork by students on interesting and enjoyable exploration projects.

› Collaborative Learning 3A

Suppose you interviewed for a job with the two companies shown below and received a job offer from each of the companies. Should you be greedy and take the job with the higher base salary or is there more to it? Let's do some collaboration to help with your decision.

Divide into two groups and have each group select one of the two companies to investigate. To conduct the investigations do the following:

1. Make a table of values for potential sales (x-values) and their respective salary (y-values) based on that company's compensation plan. Let the potential sales begin with $0 and increase by increments of $50,000 up to $500,000.
2. Make a graph plotting the pairs of values from the table.

After each group has completed their company's graph, have someone from each group put their group's graph on the same large poster board for all to view. Have a group discussion about the comparison of these two graphs including the following questions:

1. Does the salary from the $80,000/year job always yield more income?
2. Is there a point where the amount of sales will yield the same income for both jobs? If so, when?
3. Does the income from the $60,000/year job ever yield more income? If so, when?
4. Which job should you take? Explain.

Company:	Sales Consultants of Tacoma		**Job Type:**	Banking
Location:	US-Washington			Business Development
Base Pay:	$80,000.00/Year			Finance
Other Pay:	10% of all sales			Marketing
Employee Type:	Full-Time Employee			Sales
Industry:	Banking–Financial Services		**Req'd Education:**	BS degree
	Consulting		**Req'd Experience:**	At Least 3 Years
	Sales–Marketing		**Req'd Travel:**	Negligible
			Relocation Covered:	No

Guided Tour

> Research Questions

Research questions provide students with additional opportunities to explore interesting areas of math, where they may find the questions can lead to surprising results.

> **Research Questions**

1. There is a charming story about the wooden tally sticks mentioned in the chapter preview. Find out how their disposal literally resulted in the destruction of the old Houses of Parliament in England.
2. Write a paper detailing the Egyptian number system and the base and symbols used, and enumerate the similarities and differences between the Egyptian and our (Hindu-Arabic) system of numeration.
3. Write a paper detailing the Greek number system and the base and symbols used, and enumerate the similarities and differences between the Greek and our system of numeration.
4. Find out about the development of the symbols we use in our present numeration system. Where was the symbol for zero invented and by whom?
5. When were negative numbers introduced, by whom were they introduced, and what were they first called?

Abundant Practice and Problem Solving

Bello and Hopf offers students many opportunities and different skill paths for developing their problem-solving skills.

> Pretest

An optional Pretest can be found in MathZone at www.mhhe.com/bello, and is especially helpful for students taking the course as a review who may remember some concepts but not others. The answer grid is also found online and gives students the page number, section, and example to study in case they missed a question.

> **Pretest Chapter 1**

1. Use roster notation to list the whole numbers between 7 and 12.
2. Write $\frac{1}{6}$ as a decimal.
3. Classify the given number using all the classifications that are true—natural number, whole number, integer, rational number, irrational number, and real number.
   a. $1\frac{1}{2}$ b. -7 c. $\sqrt{11}$
4. Find the additive inverse of -1.2.
5. Find $\left|\frac{-2}{7}\right|$
6. Fill in the blank with $<$, $>$, or $=$ to make the resulting statement true.
   0.7 _____ $\frac{7}{100}$
7. Subtract: $-0.5 - (-1.2)$
8. Subtract: $-18 - 9$
9. Subtract: $\frac{1}{4} - \frac{3}{8}$
10. Multiply: $(-6)(-2.9)$

> **Answers to Pretest Chapter 1**

Answer	If You Missed	Review		
	Question	Section	Examples	Page
1. {8, 9, 10, 11}	1	1.1	1	4
2. 0.1666…	2	1.1	2	5
3. a. Real, rational b. Real, rational, integer c. Real, irrational	3	1.1	3	6
4. 1.2	4	1.1	4	8
5. $\frac{2}{7}$	5	1.1	5	9
6. $>$	6	1.1	6	10–11
7. 0.7	7	1.2	2	20

xxiii

▶ Guided Tour

▶ Paired Examples/Problems

Examples are placed adjacent to similar problems intended for students to obtain immediate reinforcement of the skill they have just observed. These are especially effective for students who learn by doing and who benefit from frequent practice of important methods. Answers to the problems appear at the bottom of the page.

EXAMPLE 1 Factoring trinomials of the form $x^2 + bx + c$
Factor completely:
a. $x^2 + 7x + 12$ b. $x^2 - 6x + 8$ c. $x^2 - 4 - 3x$

SOLUTION

a. To factor $x^2 + 7x + 12$, we need two integers whose product is 12 and whose sum is 7. The numbers are 3 and 4. Thus,
$$x^2 + 7x + 12 = (x + 3)(x + 4)$$

b. To factor $x^2 - 6x + 8$, we need two integers whose product is 8 and whose sum is -6. The product is positive, so both numbers must be negative. They are -2 and -4. [Check: $(-2)(-4) = 8$ and $(-2) + (-4) = -6$.] Thus,
$$x^2 - 6x + 8 = (x - 2)(x - 4)$$

c. First, rewrite $x^2 - 4 - 3x$ in descending order as $x^2 - 3x - 4$. This time, we need integers with product -4 and sum -3. The numbers are 1 and -4. Thus,
$$x^2 - 4 - 3x = x^2 - 3x - 4 = (x + 1)(x - 4)$$
You can check all these results by multiplying. For example,
$$(x + 1)(x - 4) = x^2 - 4x + x - 4 = x^2 - 3x - 4.$$

PROBLEM 1
Factor completely:
a. $x^2 + 7x + 10$
b. $x^2 - 3x - 10$
c. $x^2 - 6 - 5x$

▶ RSTUV Method

The easy-to-remember **"RSTUV"** method gives students a reliable and helpful tool in demystifying word problems so that they can more readily translate them into equations they can recognize and solve.

- **R**ead the problem and decide what is being asked.
- **S**elect a letter or □ to represent this unknown.
- **T**ranslate the problem into an equation.
- **U**se the rules you have studied to solve the resulting equation.
- **V**erify the answer.

EXAMPLE 12 A geometry problem
Two angles are complementary. If the measure of one of the angles is 20° more than the other, what are the measures of these angles?

SOLUTION We use the RSTUV method.
1. **Read the problem.** We are asked to find the measures of these angles.
2. **Select the unknown.** There are two angles involved. Let x be the measure of one of the angles and y the measure of the other angle.
3. **Think of a plan.** Do you know what complementary angles are? They are angles whose sum is 90°. With this information, we have
$x + y = 90$ The sum of the angles is 90°.
$y = x + 20$ One of the angles is 20° more than the other.
4. **Use the substitution method to solve the system.** Substituting $(x + 20)$ for y into $x + y = 90$, we obtain
$$x + (x + 20) = 90$$
$$2x + 20 = 90 \quad \text{Add like terms.}$$
$$2x = 70 \quad \text{Subtract 20.}$$
$$x = 35 \quad \text{Divide by 2.}$$
Thus one angle is 35° and the other 20° more, or 55°.
5. **Verify the answer.** Since the sum of the measures of the two angles is $35° + 55° = 90°$, the angles are complementary and our result is correct.

PROBLEM 12
Two angles are supplementary. If the measure of one of the angles is 40° less than the other, what are the measures of these angles?

TRANSLATE THIS
1. A certain company budgets $3000 a year for temporary office help. They want to stay within a 5% variance. Write an absolute-value inequality statement that will meet that requirement.
2. A family budgets $2500 a year for movies and movie rentals. If their acceptable variance is $200, write an absolute-value inequality statement that will meet that requirement.

The third step in the RSTUV procedure is to TRANSLATE the information into an equation or inequality. In Problems 1–4, TRANSLATE the sentence and match the correct translation with one of the equations A–H.

3. Chip budgets $2800 a year for gasoline. His actual gasoline expense for last year was $3200. If his acceptable variance was $250, write an inequality statement that will check to see if he stayed within the variance.
4. Talitha budgets $10,800 a year for apartment rental. Due to an unexpected raise in the rent she actually spent $11,400 in rent for the year. Write an inequality statement that will check to see if she stayed within her 10% acceptable variance.

A. $-2800 < 3200 - 250 < 2800$
B. $|150 - a| < 3000$
C. $|2500 - a| < 200$
D. $-250 < 2800 - 3200 < 250$
E. $-1140 < 10,800 - 11,400 < 1140$
F. $|200 - a| < 2500$
G. $|3000 - a| < 150$
H. $-1080 < 10,800 - 11,400 < 1080$

▶ Translate This

These boxes appear periodically before the end-of-section exercises to help students turn phrases into equations, reinforcing the RSTUV method.

▶ Exercises

A wealth of exercises for each section are organized according to the learning objectives for that section, giving students a reference to study if they need extra help.

Boost your grade at mathzone.com!
> Practice Problems   > Self-Tests
> NetTutor   > e-Professors
  > Videos

MathZone

> Exercises **5.7**

⟨A⟩ Solving Equations by Factoring   In Problems 1–46, solve the equations.

1. $(x + 1)(x + 2) = 0$
2. $(x + 3)(x + 4) = 0$
3. $(x - 1)(x + 4)(x + 3) = 0$
4. $(x + 5)(x - 3)(x + 2) = 0$
5. $\left(x - \frac{1}{2}\right)\left(x - \frac{1}{3}\right) = 0$
6. $\left(x - \frac{1}{4}\right)\left(x - \frac{1}{7}\right) = 0$
7. $y(y - 3) = 0$
8. $y(y - 4) = 0$
9. $y^2 - 64 = 0$
10. $y^2 - 1 = 0$
11. $y^2 - 81 = 0$
12. $y^2 - 100 = 0$
13. $x^2 + 6x = 0$
14. $x^2 + 2x = 0$
15. $x^2 - 3x = 0$
16. $x^2 - 8x = 0$
17. $y^2 - 12y = -27$
18. $y^2 - 10y = -21$
19. $y^2 = -6y - 5$
20. $y^2 = -3y - 2$
21. $x^2 = 2x + 15$
22. $x^2 = 4x + 12$
23. $3y^2 + 5y + 2 = 0$
24. $3y^2 + 7y + 2 = 0$
25. $2y^2 - 3y + 1 = 0$
26. $2y^2 - 3y - 20 = 0$
27. $2y^2 - y - 1 = 0$
28. $2y^2 - y - 15 = 0$

Hint: For Problems 29–34, multiply each term by the LCD *first*.

29.  $= 0$
30. $\frac{x^2}{2} - \frac{x}{12} - 1 = 0$
31. $\frac{x^2}{3} - \frac{x}{2} = -\frac{1}{6}$
32. $\frac{x^2}{6} + \frac{x}{3} = \frac{1}{2}$

Guided Tour

› Applications

Students will enjoy the exceptionally creative applications in most sections that bring math alive and demonstrate that it can even be performed with a sense of humor.

››› Applications

35. *Supply and demand* When sunglasses are sold at the regular price of $6, a student purchases two pairs. Sold for $4 at the flea market, the student purchases six pairs. Let the ordered pair (d, p) represent the demand and price for sunglasses.
  a. Find the demand equation. (*Hint:* To find the demand equation, use the ordered pairs (2, $6) and (6, $4).)
  b. How many pairs would the student buy if sunglasses are selling for $2?

36. *Supply and demand* When skateboards are sold for $80, 5 of them are sold each day. When they are on sale for $40, 13 of them are sold daily. Let (d, p) represent the demand and price for skateboards.
  a. Find the demand equation.
  b. For 10 skateboards to be sold on a given day, what should the price be?

37. *Supply and demand* A wholesaler will supply 50 sets of video games at $35. If the price drops to $20, she will provide only 20. Let (s, p) represent the supply and the price for video games.
  a. Find the supply equation. (*Hint:* To find the supply equation, use the ordered pairs (50, $35) and (20, $20).)

38. *Supply and demand* When T-shirts are selling for $6, a retailer is willing to supply 6 of them each day. If the price increases to $12, he's willing to supply 12 T-shirts.
  a. Find the supply equation.
  b. If the T-shirts were free, how many would he supply?

› Using Your Knowledge

Optional, extended applications give students an opportunity to practice what they've learned in a multistep problem requiring reasoning skills in addition to numerical operations.

››› Using Your Knowledge

Inequalities and the Environment

84. Do you know why spray bottles use a pump rather than propellants? They do that because some of the propellants have chlorofluorocarbons (CFCs) which deplete the ozone layer. In an international meeting in London, the countries represented agreed to stop producing CFCs by the year 2000. The annual production of CFCs (in thousands of tons) is given by $P = 1260 - 110x$, where x is the number of years after 1980. How many years after 1980 did it take for the production of CFCs to be less than 0? (Answer to the nearest whole number.) What year did that take place?

85. Cigarette smoking also produces air pollution. The annual number of cigarettes consumed per person in the United States is given by $N = 4200 - 70x$, where x is the number of years after 1960. How many years after 1960 was consumption less than 1000 cigarettes per person annually? (Answer to the nearest whole year.) What year did that take place?

Study Aids to Make Math Accessible

Since some students confront math anxiety as soon as they sign up for the course, the Bello/Hopf system provides numerous study aids to make their learning easier.

› Objectives

The objectives for each section not only identify the specific tasks students should be able to perform, but they organize the section itself with letters corresponding to each section heading, making it easy to follow.

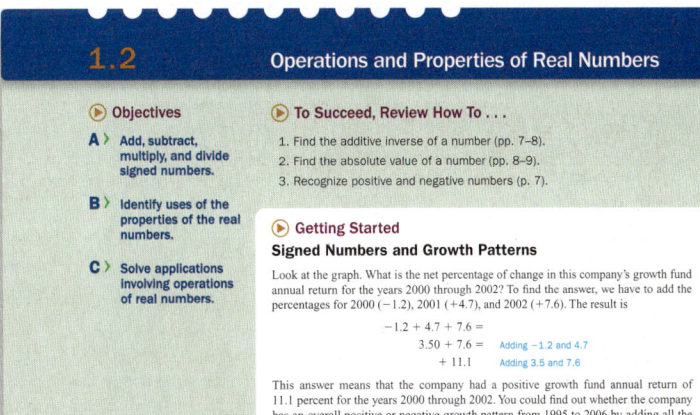

› Reviews

Every section begins with "To succeed, review how to . . . ," which directs students to specific pages to study key topics they need to understand to successfully begin that section.

Guided Tour

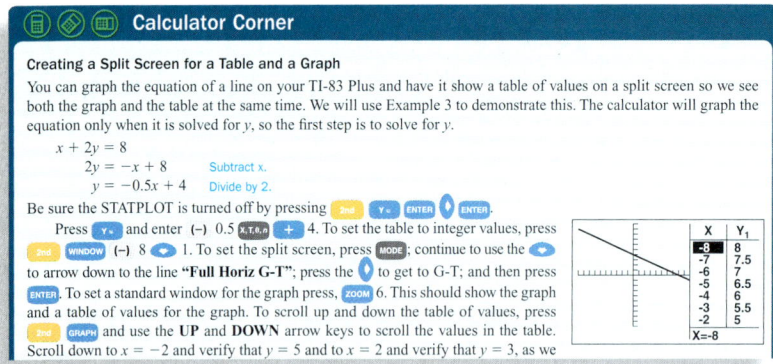

Calculator Corner

When appropriate, optional calculator exercises are included to show students how they can explore concepts through calculators and verify their manual exercises with the aid of technology.

Concept Checker

These boxes have been added to the end-of-section exercises to help students reinforce key terms and equations.

>>> Concept Checker

Fill in the blank(s) with the correct word(s), phrase, or mathematical statement.

115. The number 5 is the _____ of the number $\frac{1}{5}$.
116. Using the _____ property, multiplication of more than two numbers can be done in any grouping and still yield the same product.
117. In the statement $\frac{x}{y} = z$, x is the _____, y is the _____, and z is the _____.
118. Using the _____ property, the expression $a(x + y)$ can be written $ax + ay$.

additive identity	associative
additive inverse	divisor
factors	quotient
product	reciprocal
dividend	multiplicative identity
commutative	

Mastery Tests

>>> Mastery Test

Reduce to lowest terms:

84. $\dfrac{y^3 - x^3}{y - x}$ 85. $\dfrac{x^5 y^7}{xy^3}$ 86. $\dfrac{3y + xy}{y}$

87. $\dfrac{x^2 - y^2}{y^3 - x^3}$ 88. $\dfrac{x^2 - xy}{x^2 - y^2}$ 89. $\dfrac{4y^2 - xy^2}{y^2}$

Brief tests in every section give students a quick checkup to make sure they're ready to go on to the next topic.

Skill Checkers

>>> Skill Checker

Multiply:

103. $6x^2y \cdot 2x$ 104. $6x^2y \cdot 3xy$ 105. $-2x^3y \cdot (-2y^2)$
106. $-2x^3y \cdot (-7xy)$ 107. $(2x)^2$ 108. $3(2 + x)$

These brief exercises help students keep their math skills well honed in preparation for the next section.

Summary

> Summary Chapter 5

Section	Item	Meaning	Example
5.1A	Monomial	A constant times a product of variables with whole-number exponents	$3x^2y$, $-7x$, $0.5x^3$, $\frac{2}{3}x^2yz^4$
	Polynomial	A sum or difference of monomials	$3x^2 - 7x + 8$, $x^2y + y^3$
	Terms	The individual monomials in a polynomial	The terms of $3x^2 - 7x + 8$ are $3x^2$, $-7x$, and 8.
	Coefficient	The numerical factor of a term	The coefficient of $3x^2$ is 3.
	Binomial	A polynomial with two terms	$5x^2 - 7$ is a binomial.
	Trinomial	A polynomial with three terms	$-3 + x^2 + x$ is a trinomial.
5.1B	Degree of a polynomial	Largest sum of the exponents in any term	The degree of $x^3 + 7x$ is 3. The degree of $-2x^3yz^2$ is 6.
	Descending order	Polynomial in one variable ordered from highest exponent to lowest	$7x^3 - 5x^2 + x - 3$ is in descending order.

An easy-to-read grid summarizes the essential chapter information by section, providing an item, its meaning, and an example to help students connect concepts with their concrete occurrences.

Guided Tour

> Review Exercises

Chapter review exercises are coded by section number and give students extra reinforcement and practice to boost their confidence.

> Review Exercises Chapter 5

(If you need help with these exercises, look in the section indicated in brackets.)

1. ⟨**5.1A, B**⟩ Classify as a monomial, binomial, or trinomial and give the degree.
   a. $x^3 + x^2y^3z$
   b. $x^3y^2z^3$
   c. $x^4 - 5x^2y^3 + xyz$

2. ⟨**5.1B**⟩ Write the polynomials in descending order.
   a. $-x^2 + 3x^4 - 5x + 2$
   b. $3x - x^2 + 4x^3$
   c. $6x^2 - 2 + x$

3. ⟨**5.1C, E**⟩ A $50,000 computer depreciates 20% each year and its value $v(t)$ after t years is given by $v(t) = 50,000(1 - 0.20t)$.
   a. Find the value of the computer after 3 yr.
   b. Find the value of the computer after 5 yr.

> Practice Test with Answers

The chapter Practice Test offers students a nonthreatening way to review the material and determine whether they are ready to take a test given by their instructor. The answers to the Practice Test give students immediate feedback on their performance, and the answer grid gives them specific guidance on which section, example, and pages to review for any answers they may have missed.

> Practice Test Chapter 3

(Answers on pages 276–280)
Visit www.mhhe.com/bello to view helpful videos that provide step-by-step solutions to several of the problems below.

1. a. Graph $A(3, 2)$, $B(4, -2)$, $C(-1, -2)$, and $D(-1, 3)$.
   b. Find the coordinates of the points in the figure.

2. Graph the solutions to $3x - y = 3$.

> Answers to Practice Test Chapter 3

Answer	If You Missed		Review	
	Question	Section	Examples	Page
1. a. 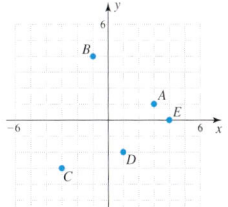	1a	3.1A	1	167

> Cumulative Review

The Cumulative Review covers material from the present chapter and any of the chapters prior to it and can be used for extra homework or for student review to improve their retention of important skills and concepts.

> Cumulative Review Chapters 1–3

1. The number $\sqrt{13}$ belongs to which of these sets? Natural numbers, Whole numbers, Integers, Rational numbers, Irrational numbers, Real numbers. Name all that apply.

2. Graph the additive inverse of $\frac{5}{2}$ on the number line.

3. Divide: $-\frac{1}{6} \div \left(-\frac{7}{12}\right)$

4. Simplify: $(2x^4y^{-4})^3$

5. Evaluate: $-7^3 + \frac{(14 - 10)}{2} + 35 \div 7$

6. Solve: $\frac{3}{7}y - 4 = 2$

7. Solve: $|x - 5| = |x - 9|$

8. Graph: $x \geq 1$

9. Graph: $\{x \mid x > -4 \text{ and } x < 4\}$

10. If $H = 2.45h + 72.98$, find h when $H = 136.68$.

xxvii

Guided Tour

Supplements for Instructors

Annotated Instructor's Edition

This version of the student text contains **answers** to all odd- and even-numbered exercises in addition to helpful **teaching tips.** The answers are printed on the same page as the exercises themselves so that there is no need to consult a separate appendix or answer key.

Computerized Test Bank (CTB) Online

Available through MathZone, this **computerized test bank,** utilizes Brownstone Diploma® algorithm-based testing software to quickly create customized exams. This user-friendly program enables instructors to search for questions by topic, format, or difficulty level; to edit existing questions or to add new ones; and to scramble questions and answer keys for multiple versions of the same test. Hundreds of text-specific open-ended and multiple-choice questions are included in the question bank. Sample chapter tests and final exams in Microsoft Word® and PDF formats are also provided.

Instructor's Solutions Manual

Available on MathZone, the Instructor's Solutions Manual provides comprehensive, **worked-out solutions** to all exercises in the text. The methods used to solve the problems in the manual are the same as those used to solve the examples in the textbook.

McGraw-Hill's MathZone is a complete online tutorial and homework management system for mathematics and statistics, designed for greater ease of use than any other system available. Instructors have the flexibility to create and share courses and assignments with colleagues, adjunct faculty, and teaching assistants with only a few clicks of the mouse. All algorithmic exercises, online tutoring, and a variety of video and animations are directly tied to text-specific materials. Completely customizable, MathZone suits individual instructor and student needs. Exercises can be easily edited, multimedia is assignable, importing additional content is easy, and instructors can even control the level of help available to students while doing their homework. Students have the added benefit of full access to the study tools to individually improve their success without having to be part of a MathZone course. MathZone allows for automatic grading and reporting of easy-to assign algorithmically generated homework, quizzes and tests. Grades are readily accessible through a fully integrated grade book that can be exported in one click to Microsoft Excel, WebCT, or BlackBoard.

MathZone Offers:

- Practice exercises, based on the text's end-of-section material, generated in an unlimited number of variations, for as much practice as needed to master a particular topic.
- Subtitled videos demonstrating text-specific exercises and reinforcing important concepts within a given topic.
- Net Tutor™ integrating online whiteboard technology with live personalized tutoring via the Internet.

Applications Index

bond yields, 115–116, 162, 314, 324, 494, 655
budgets, 142, 152
charity giving, 120
compound interest, 776–777, 782, 789–790, 793, 794, 817, 820, 830, 832
credit card spending, 822
depreciation, 181, 357, 781
equilibrium point, 613
exchange rates, 760
growth funds, 17
hourly wages, 207
interest rates, 468, 603
investment rates, 324, 340
investment returns, 120, 122
investment tracking, 319
minimum wage, 324, 340
national debt, 49, 121
principal of investment, 513
profit, 199–200, 237, 357, 358, 646
property taxes, 750
refinance loans, 118
revenue, 368
revenue *vs.* spending, 321–322
salaries, 184, 262–263, 312, 745
savings account returns, 120
selling price, 119
simple interest, 62, 260, 512, 721
stock prices, 33
tax filing, 497

Food

beef consumption, 613–614
burger fat content, 15
cat ages, 303
catsup, 34
coffee mixing, 305–306
coffee prices, 121
fast food nutrition, 139, 259, 327–328
grill shape, 692
juicing, 310–311
large pancakes, 323
largest shortcakes, 718
poultry consumption, 613–614
recipe proportions, 490, 520
soft drink consumption, 302
sundae size, 323
tea prices, 121

Geometry

angle measures, 301, 302, 312
cone volume, 565
cube surface area, 565
cube volume, 37
pyramid volume, 565
rectangle area, 33, 366–367, 447
rectangle dimensions, 340, 413
rectangle perimeter, 836
sphere curvature, 503
sphere surface area, 565

Home and Garden

apartment size, 722
crop yield, 676
fencing, 670, 676
firewood splitting, 626
furniture shopping, 600
hose flow, 386
lighting, 698
painting, 55, 417–418, 519, 746
picture framing, 718
pool construction, 497–498, 503, 504
property taxes, 750
swimming pool treatment, 603
tree houses, 602

Law Enforcement

accident investigation, 578–579
blood alcohol level, 238, 811
public safety spending, 188
robbery statistics, 238
skid mark measurement, 635–636
speeding, 537

Medicine and Health

AIDS cases, 782
blood alcohol level, 238, 811
blood pressure, 804, 822
body composition, 119
body fat, 808–809
calories consumed, 107
cardiovascular disease, 398
cardiovascular disease in women, 107
cellular mitosis, 773, 825
diabetes incidence, 255
diabetic wound ointment, 822
dieting, 139, 140, 220
doctor salaries, 119
dosages, 493, 503
drug concentration, 808
drug sensitivity, 677
E. coli growth, 782
exercise and pulse, 648
exercise target zones, 181
fat in diet, 154
foot pain, 312
growth charts, 325
height of boys, 793
height of girls, 793
ideal weight, 95, 237
life expectancy, 252–253
Medicare costs, 757
mustache growth, 506
noise threshhold limits, 793–794
prescription benefits, 102
pulse rate, 237
sleep needed by children, 95
smoking and cancer, 248
smoking and life expectancy, 511
target heart rate, 207
threshhold weight, 511
vitamins, 313, 325–326
weights, 511
wound treatment, 822

Records

fattest twins, 326
largest catsup bottle, 34
largest cigar, 30
largest dog, 317
largest flag, 324
largest quilt, 324
largest sundae, 323
shorthand speed, 119
strongest current, 318
tallest buildings, 303
tallest people, 15, 303
tallest President, 326

Science

asteroid belt width, 48
bacteria growth, 781, 818, 821, 824
bacteria population, 807, 837
cephalic index, 512
chlorofluorocarbons, 140
circuit impedance, 576
circuit resistance, 645
comets, 680
computer sizes, 34
current, 446
decibel scale, 796
Doppler effect, 468
Earth core temperature, 33
earthquake magnitude, 784, 789, 793, 795
Earth's orbit, 693
electrical resistance, 467
elevation, 34, 323
energy from Sun, 48
firefighting, 379, 404
flow from hose, 386
focal length, 503
free-falling object, 563–564
gas particle velocity, 546
gas pressure, 447, 519, 546, 746, 836
glycerin solutions, 121
gravity, 510, 520
heat transmission, 369
hydrogen spectrum light, 468
illumination intensity, 510
kinetic energy, 708
leaf shapes, 687
lifting force, 508
loudness and distance, 507
mass of Earth, 48
nuclear reactors, 303
O-ring damage inde, 249
pendulum length, 546
pendulum motion, 458
pH of solution, 803–804, 807, 830
photography solutions, 117–118
planetary motion, 458
planetary orbits, 468, 694
plutonium decay, 781

xxxii

Applications Index

Animals

ant speed, 99, 223–224, 238, 757, 766
baby chick growth, 825
clam catch, 756
crab catch, 756
cricket chirps, 74, 99, 180, 511, 757, 766
elephant weights, 302
fish catch, 756
frog leaps, 676
largest dog, 317
water mites, 645

Business

break even point, 617, 717, 721, 729, 737, 739
candle manufacturing, 208
computer efficiency, 502
corporate travel expenses, 119
cost per unit, 139, 479
customer service costs, 415
demand function, 296, 729
equilibrium point of supply and demand, 301, 302, 613, 713
IT spending, 118
mailing costs, 139
manufacturing, 215–216, 220, 333, 356, 357, 415, 603, 837
maximum revenue, 670, 731
maximum sales, 676
net exports, 14
oil exporters, 14
oil producers, 14
product demand, 438, 446
profit and loss, 724–725
profit calculation, 354, 645, 675
profit function, 753–754, 756
profit margin, 96
profitability, 96
revenue, 357, 368, 370, 675, 679, 729, 827
sales figures, 261, 808
shipping costs, 197
supply and demand, 396, 712–713
supply determination, 207, 296
trade balance, 15
word processor efficiency, 502
work rate, 633
world oil consumption, 11

Construction

awning measurement, 30
bend allowance, 403
bending moments, 388, 642, 645
blueprint scale, 491, 492, 518
bridge arch, 692, 694
building height, 340
cantilever, 458
carpenter efficiency, 502

conic sections in architecture, 680
costs, 510
current, 446
dam weight, 510
deflection of bridge beams, 359, 369, 387, 458
drain pipe, 693
electrical resistance, 467
ellipses in, 731–732
hyperbolas in architecture, 698
ladder heights, 412, 602
parallel resistors, 447
of pools, 497–498, 503, 504
riveting, 502
scale drawings, 64
shear stress, 403
suspension bridge cables, 678

Consumer Issues

air conditioner efficiency, 96
appliance repair costs, 162
bike rental, 731
Blackberry subscribers, 207
book prices, 297
bus rental, 633
cabin rental, 633
candle consumption, 102
car leasing, 97
car loan costs, 103–104
car rental cots, 108, 123, 210–211, 221, 237
catering costs, 514
cell phone call length, 564
cell phone plans, 197
cell phone sales, 781, 823
cigarette consumption, 99, 302
cigarette production, 823
cigarette tax, 110
classified ad costs, 335–336
computer screen aspect ratio, 722
cost of traffic, 325
credit card spending, 822
diamond costs, 242–243
digital camera prices, 118
disc jockey charges, 97
dress sizes, 757, 770, 828
energy prices, 258
ethanol production, 102
film processing, 492
garbage production, 748–749
gas prices, 513–514
high definition TV, 334–335
high heel heights, 417
home prices, 208, 256, 259
home purchase, 112
home sales, 615
long distance costs, 93, 97, 108, 118
mail costs, 99
online shopping, 112
patents, 258
pet-sitting costs, 220
plastic plates, 694

plumber charges, 97
recycling, 238
rent, 415
shipping costs, 197
shoe size, 85–86, 238, 770
skateboard prices, 206, 366, 370
sunglasses price, 206
SUV *vs.* car sales, 284
T-shirt prices, 206
television aspect ratio, 722
television sizes, 600
text messages and age, 282
ticket prices, 415
wholesale prices, 206
world oil demand, 107
wrapping paper size, 62

Education

careers for graduates, 155
college graduate increase, 511
in different countries, 313
grants, 136, 162
Pell Grants, 136, 162
poster dimensions, 105–106
public financing, 120
of Registered Nurses, 119
SAT scores, 677
student numbers, 110, 122
teacher salaries, 184, 194
time taken to graduate, 256
and unemployment, 2
university budgets, 149–150

Entertainment

awards won by Beatles, 108
blogs, 113–115
card games, 440
comedy on television, 119
expenditures, 119
largest painting, 108
new rock songs, 511
radio song play, 460
Super Bowl ratings, 119
types of, 120

Environment

and chlorofluorocarbons, 413
cigarettes and air pollution, 140
garbage production, 748–749
global warming, 511
ocean depth, 512
ozone layer, 239
pollution permits, 438
recycling, 238, 448–449, 781, 821
waste generation, 197

Finances

annual interest, 510
annual percentage rate, 468
billionaires, 313

xxxi

Guided Tour

 www.mathzone.com

McGraw-Hill's MathZone is a complete online tutorial and homework management system for mathematics and statistics, designed for greater ease of use than any other system available. All algorithmic exercises, online tutoring, and a variety of video and animations are directly tied to text-specific materials.

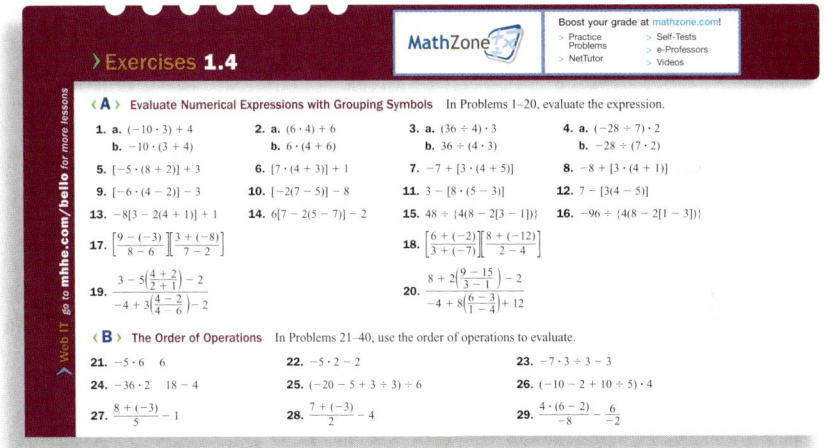

 ALEKS® (www.aleks.com)

ALEKS (**A**ssessment and **LE**arning in **K**nowledge **S**paces) is a dynamic online learning system for mathematics education, available over the Web 24/7. ALEKS assesses students, accurately determines their knowledge, and then guides them to the material that they are most ready to learn. With a variety of reports, Textbook Integration Plus, quizzes, and homework assignment capabilities, ALEKS offers flexibility and ease of use for instructors.

Bello Video Series

The video series is available on DVD and VHS tape and features an instructor introducing topics and working through selected odd-numbered exercises from the text, explaining how to complete them step by step. The DVDs are **closed-captioned** for the hearing impaired and are also **subtitled in Spanish.**

Math for the Anxious: Building Basic Skills, by Rosanne Proga

Math for the Anxious: Building Basic Skills is written to provide a practical approach to the problem of math anxiety. By combining strategies for success with a pain-free introduction to basic math content, students will overcome their anxiety and find greater success in their math courses.

Guided Tour

- Assessment capabilities, powered through ALEKS, which provide students and instructors with the diagnostics to offer a detailed knowledge base through advanced reporting and remediation tools.
- Faculty with the ability to create and share courses and assignments with colleagues and adjuncts, or to build a course from one of the provided course libraries.
- An Assignment Builder that provides the ability to select algorithmically generated exercises from any McGraw-Hill math textbook, edit content, as well as assign a variety of MathZone material including an ALEKS Assessment.
- Accessibility from multiple operating systems and Internet browsers.

 ALEKS® (www.aleks.com)

ALEKS (**A**ssessment and **LE**arning in **K**nowledge **S**paces) is a dynamic online learning system for mathematics education, available over the Web 24/7. ALEKS assesses students, accurately determines their knowledge, and then guides them to the material that they are most ready to learn. With a variety of reports, Textbook Integration Plus, quizzes, and homework assignment capabilities, ALEKS offers flexibility and ease of use for instructors.

- ALEKS uses artificial intelligence to determine exactly what each student knows and is ready to learn. ALEKS remediates student gaps and provides highly efficient learning and improved learning outcomes
- ALEKS is a comprehensive curriculum that aligns with syllabi or specified textbooks. Used in conjunction with a McGraw-Hill texts, students also receive links to text-specific videos, multimedia tutorials, and textbook pages.
- Textbook Integration Plus allows ALEKS to be automatically aligned with syliabi or specified McGraw-Hill textbooks with instructor chosen dates, chapter goals, homework, and quizzes.
- ALEKS with AI-2 gives instructors increased control over the scope and sequence of student learning. Students using ALEKS demonstrate a steadily increasing mastery of the content of the course.
- ALEKS offers a dynamic classroom management system that enables instructors to monitor and direct student progress toward mastery of course objectives.

Supplements for Students

Student's Solutions Manual

This supplement contains complete worked-out solutions to all odd-numbered exercises and all odd- and even-numbered problems in the Review Exercises and Cumulative Reviews in the textbook. The methods used to solve the problems in the manual are the same as those used to solve the examples in the textbook. This tool can be an invaluable aid to students who want to check their work and improve their grades by comparing their own solutions to those found in the manual and finding specific areas where they can do better.

Applications Index

polar moment, 403
radar dish, 678
radioactive decay, 778, 781, 818–819, 821, 829, 831, 832
rockets, 503, 605, 642–643
speed of light, 48
spring tension, 510
stellar magnitude, 808
volcanic eruptions, 794
weight, on moon, 108
wind force, 508, 513
world oil reserves, 48

Social Studies

fear of flying, 821
moods, 35
population demographics, 781, 807
population growth, 197, 824–825
population prediction, 778–779, 781
poverty, 677
rumors, 808
world population, 817

Sports

attendance at games, 822
baseball diamond, 592
baseball height when batted, 676
baseball mound covering, 67
baseball pitching, 416
baseball statistics, 427
baseball trajectory, 676
basketball court size, 302
batting average, 482, 493
billiards, 63
cycling, 493
discus throwing, 260
diving, 348, 512
favorite, among baseball players, 108
free throw statistics, 518
golf handicap, 500–501
Olympics predictions, 770–771
shape of home plate, 411
sumo wrestling, 34
swimming and heart rate, 51
track, 693
weight difference in boxing, 108

Transportation

airport passenger numbers, 113
boat speed, 324, 503, 504, 745
breaking distance, 511, 757
budgeting for, 220
bus fare, 55
bus speeds, 116–117, 122
car speeds, 121, 507
curve speeds, 526, 559
distance traveled, 95, 497–498, 603
driving rate, 493
engine efficiency, 470
flying coordinates, 195
gas mileage, 119, 505, 511, 677
jet speed, 503
map scale, 490
plane speeds, 314, 323, 326
plane wing lifting force, 508
safety of flying, 821
sound barrier, 548
speeding, 537
stopping distance, 645
tolls, 87
train speeds, 121, 281

Weather

air pressure, 237
atmospheric pressure, 781
average temperatures, 181
drought, 647
Fahrenheit/Celsius conversion, 62, 757
global warming and carbon dioxide, 511
global warming and sea levels, 98
heat index, 181–182
high temperature, 33
hurricane intensity, 108
hurricane preparedness, 166
rainfall, 647
tornado frequency, 108
water in snowfall, 509

Chapter 1: The Real Numbers

Section
- 1.1 Numbers and Their Properties
- 1.2 Operations and Properties of Real Numbers
- 1.3 Properties of Exponents
- 1.4 Algebraic Expressions and the Order of Operations

The Human Side of Algebra

The development of the number system used in algebra has been a multicultural undertaking. More than 20,000 years ago, our ancestors needed to count their possessions, their livestock, and the passage of days. Australian aborigines counted to two, South American Indians near the Amazon counted to six, and the Bushmen of South Africa were able to count to ten ($10 = 2 + 2 + 2 + 2 + 2$).

The earliest technique for visibly expressing a number was tallying (from the French verb *tailler* "to cut"). Tallying, a practice that reached its highest level of development in the British Exchequer tallies, used flat pieces of hazelwood about 6–9 inches long and about an inch thick, with notches of varying sizes and types. When a loan was made, the appropriate notches were cut and the stick split into two pieces, one for the debtor, one for the Exchequer. In this manner, transactions could easily be verified by fitting the two halves together and noticing whether the notches coincided, hence the expression "our accounts tallied."

The development of written numbers is due mainly to the Egyptians (about 3000 B.C.), the Babylonians (about 2000 B.C.), the early Greeks (about 400 B.C.), the Hindus (about 250 B.C.), and the Arabs (about 200 B.C.). Here are the numbers three of these civilizations used:

Egyptian, about 3000 B.C.

1	10	100	1000	10,000	100,000	1,000,000

Babylonian, about 2000 B.C.

0	1	10	12	20	60	600

Early Greek, about 400 B.C.

1	5	10	50	100	500	5000

1.1 Numbers and Their Properties

Objectives

A. Write a set of numbers using roster or set-builder notation.

B. Write a rational number as a decimal.

C. Classify a number as natural, whole, integer, rational, irrational, or real.

D. Find the additive inverse of a number.

E. Find the absolute value of a number.

F. Given two numbers, use the correct notation to indicate equality or which is larger.

G. Applications involving real numbers.

To Succeed, Review How To . . .

1. Write the fraction $\frac{a}{b}$ as a decimal by dividing a by b.
2. Distinguish between a positive and a negative number.

Note: The *To Succeed* section tells you what you need to know or review *before* you proceed with this section.

Getting Started

Algebra—Wages and Unemployment by Educational Status

Higher Education Means Higher Income and Less Chance of Unemployment The graph here indicates the unemployment rate (pink bars) by percent and the median weekly earnings (green bars) in 2003 by educational attainment.

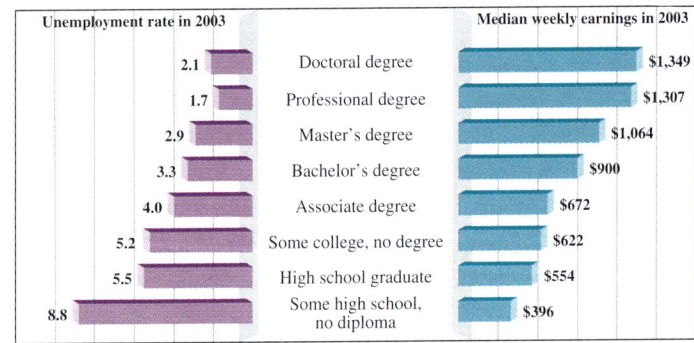

Notes: Unemployment and earnings for workers 25 and older, by educational attainment; earnings for full-time wage and salary workers

Source: Unemployment rate, 2003 annual average: Bureau of Labor Statistics; earnings, March 2003: Bureau of the Census. http://www.bls.gov/.

Look at the graph. What was the difference in the unemployment of a person with a high school diploma as compared to one with a doctoral degree? The answer is $5.5 - 2.1$; or a 3.4% higher rate of unemployment for a person with a high school diploma as compared to a person with a doctoral degree. Now, how many times greater is the salary of a person with a master's degree than that of a person with a high school diploma (green bars). The answer is $1064 \div 554$; or 2 (rounded to the nearest whole number) times greater earnings for a person with a master's degree as compared to a person with a high school diploma.

In arithmetic we are accustomed to expressions such as

$$5.5 - 2.1, \quad 396 + 554, \quad 6 \times 3.3, \quad \text{and} \quad 1064 \div 554$$

In algebra, we use expressions such as

$$5.5 - x, \quad 396 + h, \quad 2\pi r, \quad \text{and} \quad \frac{1064}{s}.$$

The letters x, h, r, and s are variables that stand for different numbers. When a letter stands for *one* number, it's called a constant. For example, if in the graph on the previous page, b were to represent the median weekly earnings for a person with a bachelor's degree in 2003, then b would be a constant. However, earnings change from year to year, so in some other algebra problem we could say the variable s will represent the salary of a person with a bachelor's degree for a given year.

What kind of numbers can we use in place of the variables x, h, r, and s? In algebra, we start with the real numbers and then study the complex numbers. We begin this section by discussing different sets of numbers.

A › Sets of Numbers

The idea of a **set** is familiar in everyday life. You may own a set of dishes, a tool set, or a set of books. In algebra we use sets of numbers.

We use capital letters to denote sets and lowercase letters or numbers to denote **elements** (or **members**) of these sets. If possible, we "list" the elements of a set in braces { } and separate them by commas. Thus, $A = \{1, 2, 3\}$ is the set A that has 1, 2, and 3 for its elements. This notation is known as **roster notation.** If a set has no elements, it is called the **empty,** or **null,** set and is denoted by { } or by the symbol $\varnothing$ (used without braces and read "the empty set" or "the null set").

| NOTE | Do not write {∅} for the empty set. |

Here are three sets of numbers frequently used in algebra.

| NATURAL NUMBERS | The set of numbers used for counting. $$N = \{1, 2, 3, \ldots\}$$ |

The three dots (called an ellipsis) mean that the pattern continues in the indicated direction.

| WHOLE NUMBERS | The set including the natural numbers and zero. $$W = \{0, 1, 2, 3, \ldots\}$$ |

| INTEGERS | The set including the whole numbers and their opposites (negatives). $$I = \{\ldots, -2, -1, 0, 1, 2, \ldots\}$$ |

The Greek letter ∈ (epsilon) is used to indicate that an element *belongs* to a set. Thus, $-5 \in I$ indicates that -5 is in the set I; that is, -5 *is an integer*. On the other hand, $0 \notin N$ indicates that 0 *is not a natural number*.

EXAMPLE 1 Roster notation
Use roster notation to write the following sets:

a. The natural numbers between 2 and 7.
b. The first three whole numbers.
c. The first two negative integers.
d. The only number that is neither positive nor negative.

SOLUTION

a. The set of natural numbers between 2 and 7 is $\{3, 4, 5, 6\}$. (2 and 7 are not included.)
b. The set of the first three whole numbers is $\{0, 1, 2\}$.
c. The set of the first two negative integers is $\{-1, -2\}$.
d. The set containing the only number that is neither positive nor negative is $\{0\}$.

PROBLEM 1
Use roster notation to write the following sets:

a. The natural numbers between 3 and 8.
b. The first four whole numbers.
c. The first four negative integers.
d. The first whole number.

Sets can also be described by a **rule** or statement describing the elements in the set. An example of such a set would be the counting numbers from 1 to 5, which in roster notation would be $\{1, 2, 3, 4, 5\}$.

In the "Getting Started," we mentioned that in algebra we use letters or **variables** to represent different numbers, and when a letter stands for just one number it is called a **constant**. Sometimes a variable and a rule can be used to describe a set, and it is called **set-builder notation**. An example of such a set and the way to read it follows:

$$\{x \mid x \text{ is a counting number from 1 to 5}\}$$

The set of all x such that x is a counting number from 1 to 5.
(variable)   *(rule)*

If a number can be written in the form $\frac{a}{b}$, where a and b are integers and b is not 0 ($b \neq 0$), the number is called a *rational number* (because it is a *ratio* of two integers). Thus,

$$\frac{1}{5}, \quad \frac{-4}{3}, \quad \frac{4}{1} = 4, \quad \text{and} \quad \frac{-7}{1} = -7$$

are rational numbers. There is no obvious pattern with which we can list all the rational numbers, so we use set-builder notation to define this set.

RATIONAL NUMBERS

The set Q of **rational numbers** consists of all the numbers that can be written as the ratio of two integers. Thus,

$$Q = \left\{r \mid r = \frac{a}{b}, a \text{ and } b \text{ integers}, b \neq 0\right\}$$

This is read "Q equals the set of all r such that r equals a divided by b, a and b integers and b not equal to 0."

Answers to PROBLEMS

1. a. $\{4, 5, 6, 7\}$ b. $\{0, 1, 2, 3\}$
   c. $\{-1, -2, -3, -4\}$ d. $\{0\}$

Using set-builder notation, the sets N, W, and I can be written as

$$N = \{x \mid x \text{ is a counting number}\}$$
$$W = \{x \mid x \text{ is a whole number}\}$$
$$I = \{x \mid x \text{ is an integer}\}.$$

B › Writing Rational Numbers as Decimals

The rational number $\frac{a}{b}$ can also be written as a decimal by dividing a by b. The result is either a **terminating decimal** (as in $\frac{1}{2} = 0.5$) or a **nonterminating, repeating decimal** (as in $\frac{1}{3} = 0.333\ldots$). We often place a bar over the repeating digits in a nonterminating, repeating decimal. Thus $\frac{1}{3} = 0.\overline{3}$, and $\frac{2}{11} = 0.181818\ldots = 0.\overline{18}$.

EXAMPLE 2 Writing fractions as decimals
Write as a decimal:

a. $\frac{4}{5}$ b. $\frac{3}{11}$ c. $\frac{95}{30}$

SOLUTION

a. Dividing 4 by 5, we obtain $\frac{4}{5} = 0.8$, a terminating decimal.

b. Dividing 3 by 11, we have

$$\frac{3}{11} = 0.272727\ldots = 0.\overline{27}$$

a nonterminating, repeating decimal.

c. Dividing 95 by 30, we have

$$\frac{95}{30} = 3.1666\ldots = 3.1\overline{6}$$

a nonterminating, repeating decimal.

PROBLEM 2
Write as a decimal:

a. $\frac{3}{8}$ b. $\frac{5}{11}$ c. $\frac{95}{60}$

Calculator Corner

Numerical Calculations

The numerical calculations in this section can be done with a calculator, but be aware that calculator procedures vary. (When in doubt, read the manual.) Start at the "Home Screen" and do Example 2(a) by dividing 4 by 5; that is, find $4 \div 5$. On a TI-83 Plus, press 4 ÷ 5 ENTER. The answer is shown as .8. Do parts b and c of Example 2.

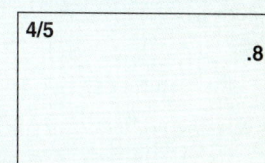

Since any rational number of the form $\frac{a}{b}$ ($b \neq 0$) is either a terminating decimal or a nonterminating, repeating decimal, the set Q of rational numbers can also be defined as follows.

ALTERNATIVE DEFINITION FOR THE SET Q

$Q = \{x \mid x \text{ is a terminating decimal or a nonterminating, repeating decimal}\}$

There are some numbers such as $\sqrt{2}$ (the square root of 2), π, and $\sqrt{10}$ that are *not* rational numbers. These are called *irrational* numbers because they cannot be written as the ratio of two integers. When written as decimals, irrational numbers are nonterminating and nonrepeating. For example, $0.101001000\ldots$ and $3.1234567\ldots$ are irrational. Here is the definition of irrational numbers.

Answers to PROBLEMS

2. a. 0.375
   b. $0.454545\ldots$ or $0.\overline{45}$
   c. $1.58333\ldots$ or $1.58\overline{3}$

IRRATIONAL NUMBERS

Irrational numbers are numbers that *cannot* be written as ratios of two integers. The set of irrational numbers is

$$H = \{x \mid x \text{ is a real number that is not rational}\}$$

ALTERNATIVE DEFINITION FOR THE SET H

$$H = \{x \mid x \text{ is a nonterminating and nonrepeating decimal}\}$$

The **real numbers** include both the rational numbers and the irrational numbers.

REAL NUMBERS

Numbers that are either rational or irrational are called **real numbers.** The set of real numbers R is defined by

$$R = \{x \mid x \text{ is a number that is rational or irrational}\}$$

C › Classifying Numbers

Here are some real numbers:

$$5, 17, -4, -9, 0, \tfrac{3}{5}, 0.6, \tfrac{1}{-10}, -0.\overline{1}, \tfrac{4}{-3}, \sqrt{3}, \pi, 0.345\ldots$$

EXAMPLE 3 Classifying numbers

Classify the given numbers by making a check mark (✓) in the row(s) where the classification is true.

Set	0	$-\tfrac{4}{5}$	-4	$\sqrt{2}$	7	$-\pi$	$0.\overline{8}$	$0.01001000\ldots$
Natural numbers					✓			
Whole numbers	✓				✓			
Integers	✓		✓		✓			
Rational numbers	✓	✓	✓		✓		✓	
Irrational numbers				✓		✓		✓
Real numbers	✓	✓	✓	✓	✓	✓	✓	✓

PROBLEM 3

Classify the given numbers by making a check mark (✓) in the row(s) where the classification is true.

Set	$\tfrac{8}{5}$	π	6	$\sqrt{3}$	-11
Natural numbers					
Whole numbers					
Integers					
Rational numbers					
Irrational numbers					
Real numbers					

Do you have a good idea of the relationship between the sets of numbers we have discussed? We can clarify the situation by using the idea of a *subset*. We say that A is a **subset** of B, denoted by $A \subseteq B$, when all the elements in A are also in B. Thus, because all natural numbers N are whole numbers, $N \subseteq W$ (read "N is a subset of W"). Also, because all whole numbers are integers, $W \subseteq I$. Here is the relationship:

$$N \subseteq W \subseteq I \subseteq Q \subseteq R$$

Answers to PROBLEMS

3.

Set	$\tfrac{8}{5}$	π	6	$\sqrt{3}$	-11
Natural numbers			✓		
Whole numbers			✓		
Integers			✓		✓
Rational numbers	✓		✓		✓
Irrational numbers		✓		✓	
Real numbers	✓	✓	✓	✓	✓

Figure 1.1 shows the sets involved.

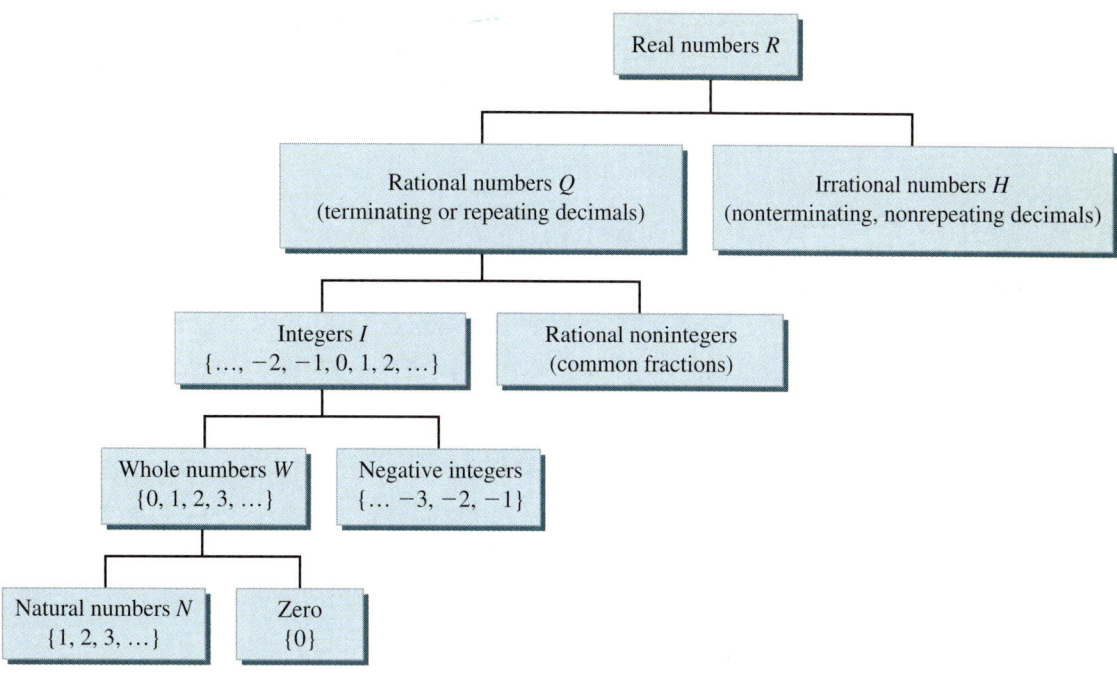

>Figure 1.1

To make a picture or graph of some real numbers, we can think of a number line completely filled with all the real numbers. In this line, the *positive* real numbers are to the *right* of zero; the *negative* real numbers are to the *left* of zero; and zero, in the center of the number line, is called the *origin,* as shown in Figure 1.2.

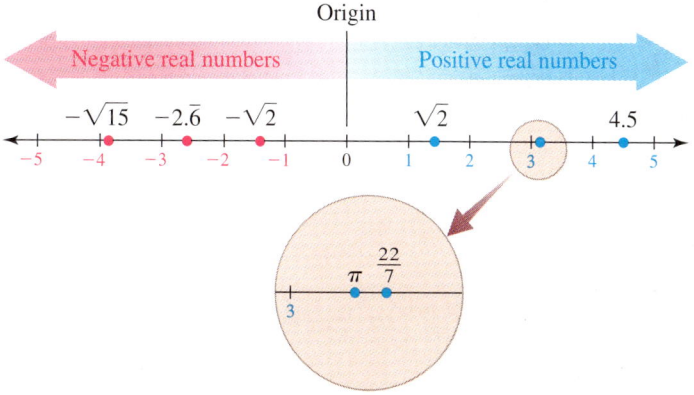

>Figure 1.2

The number corresponding to a point on the number line is the **coordinate** of the point, and the point is called the **graph** of the number. In Figure 1.2, some irrational numbers and some rational numbers are shown. There is exactly one real number for each point graphed on the number line and one point for each real number.

D > Additive Inverses (Opposites)

Each point on the number line has another point *opposite* it with respect to zero. The numbers corresponding to these two points are called the *additive inverses (opposites)* of each other. Thus, 3 and -3 (read "the additive inverse of 3," or "negative 3") are additive inverses, as are -4 and 4. (See Figure 1.3.) A number and its additive inverse are always the same distance from zero.

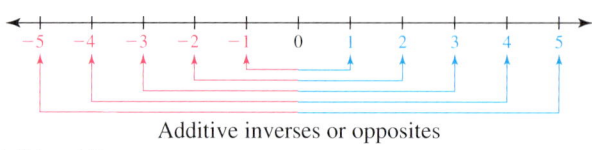

>Figure 1.3

Additive inverses or opposites

Here is the definition.

ADDITIVE INVERSE

The **additive inverse (opposite)** of a is $-a$.

The sum of a number and its additive inverse is always zero—that is, $a + (-a) = (-a) + a = 0$.

EXAMPLE 4 Finding additive inverses
Find the additive inverse:

a. 5   b. -3.5 c. $\frac{2}{3}$ d. $-6x$

SOLUTION

a. The additive inverse of 5 is -5. (See Figure 1.4.)
b. The additive inverse of -3.5 is 3.5. (See Figure 1.4.) In symbols, $-(-3.5) = 3.5$ (read "the additive inverse of negative 3.5 is 3.5," or "the additive inverse of the inverse of 3.5 is 3.5").
c. The additive inverse of $\frac{2}{3}$ is $-\frac{2}{3}$. (See Figure 1.4.)

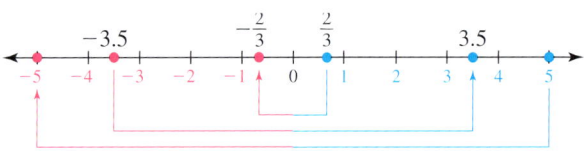

Additive inverses

>Figure 1.4

d. The additive inverse of $-6x$ is $6x$.

PROBLEM 4
Find the additive inverse:

a. 3   b. -2.5 c. $\frac{3}{5}$ d. $9y$

🖩 Calculator Corner

Negative Sign versus Subtraction Key

Texas Instrument (TI) calculators use a (−) key to find the additive inverse of a number and a subtraction key ⊖ to indicate subtraction. To find the *additive inverse* of -3.5, you have to find the additive inverse of a negative number. With a TI calculator, enter (−) (−) 3.5 and then press ENTER to find the answer shown. What does the calculator indicate for ⊖ ⊖ 3.5?

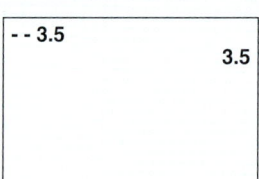

E > Absolute Values

Now, let's go back to the number line. What is the distance between 3 and 0? The answer is 3 units. What about the distance between -3 and 0? The answer is *still* 3 units. The distance between any number a and 0 is called the *absolute value* of a and is denoted by $|a|$. Thus $|-3| = 3$ and $|3| = 3$. In general, we have the following definition.

Answers to PROBLEMS

4. a. -3 b. 2.5 c. $-\frac{3}{5}$ d. $-9y$

1.1 Numbers and Their Properties

> **ABSOLUTE VALUE OF A REAL NUMBER**
>
> The **absolute value** of a real number a, denoted by $|a|$, is defined as the distance between a and 0 on the real-number line. In general,
>
> $$|a| = \begin{cases} a & \text{if } a \text{ is positive} \\ 0 & \text{if } a \text{ is zero} \\ -a & \text{if } a \text{ is negative} \end{cases} \quad \begin{array}{l} |11| = 11 \\ |0| = 0 \\ |-5| = -(-5) = 5 \end{array} \quad \text{and} \quad \begin{array}{l} \left|\frac{1}{2}\right| = \frac{1}{2} \\ \\ \left|-\frac{1}{3}\right| = -\left(-\frac{1}{3}\right) = \frac{1}{3} \end{array}$$

> **CAUTION**
>
> The absolute value of a number represents a distance and a distance is *never* negative, therefore the absolute value of a number is *never* negative. It is always positive or zero. However, if a is not 0, $-|a|$ is *always* negative. Thus,
>
> $$-|-3| = -3, \quad -|4.2| = -4.2, \quad \text{and} \quad -|0.\overline{3}| = -0.\overline{3}.$$

EXAMPLE 5 Finding absolute values

Find the absolute values:

a. $|-8|$ **b.** $\left|\frac{1}{7}\right|$ **c.** $|0|$ **d.** $|4.2|$

e. $|0.\overline{3}|$ **f.** $-|\sqrt{2}|$ **g.** $4|-6| + 7$ **h.** $|18 + 5| - |2|$

PROBLEM 5

Find the absolute value:

a. $|-19|$ **b.** $\left|\frac{1}{6}\right|$

c. $|-0|$ **d.** $|3.1|$

e. $|0.\overline{6}|$ **f.** $-\left|-\frac{5}{7}\right|$

g. $15|3| - 4$ **h.** $|7 - 2| - 5$

SOLUTION

a. $|-8| = 8$ -8 is 8 units from 0. $|-8| = -(-8) = 8.$

b. $\left|\frac{1}{7}\right| = \frac{1}{7}$ $\frac{1}{7}$ is $\frac{1}{7}$ units from 0.

c. $|0| = 0$ 0 is 0 units from 0.

d. $|4.2| = 4.2$ 4.2 is 4.2 units from 0.

e. $|0.\overline{3}| = 0.\overline{3}$ $0.\overline{3}$ is $0.\overline{3}$ units from 0.

f. $-|\sqrt{2}| = -\sqrt{2}$ $-\sqrt{2}$ is $\sqrt{2}$ units from 0 and the opposite of $\sqrt{2}$ is $-\sqrt{2}$

g. This expression involves multiplication and addition along with the absolute value number, so we start by replacing $|-6|$ with 6 and then perform the operations

$$4|-6| + 7 = 4(6) + 7 \quad \text{Replace } |-6| \text{ with 6.}$$
$$= 24 + 7 \quad \text{Multiply.}$$
$$= 31 \quad \text{Add.}$$

h. This expression involves an operation inside of the absolute value symbol so we add $18 + 5$ first.

$$|18 + 5| - |2| = |23| - |2| \quad \text{Add inside the first absolute value symbol.}$$
$$= 23 - 2 \quad \text{Replace } |23| \text{ with 23 and } |2| \text{ with 2.}$$
$$= 21 \quad \text{Subtract.}$$

Answers to PROBLEMS

5. **a.** 19 **b.** $\frac{1}{6}$ **c.** 0 **d.** 3.1 **e.** $0.\overline{6}$ **f.** $-\frac{5}{7}$ **g.** 41 **h.** 0

F › Equality and Inequality

TRICHOTOMY LAW

If you are given any two real numbers a and b, only one of three things can be true:

1. a is **equal to** b, written $a = b$, or
2. a is **less than** b, written $a < b$, or
3. a is **greater than** b, written $a > b$.

On a number line, numbers are shown in order; they *increase* as you move right and *decrease* as you move left. Thus,

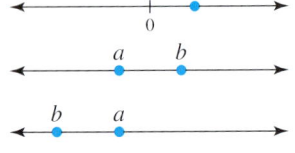

1. $a = b$ means that the graphs of a and b coincide.
2. $a < b$ means that a is to the left of b on the number line.
3. $a > b$ means that a is to the right of b on the number line.

The symbols $<$ and $>$ are called **inequality** signs, and statements such as $a > b$ or $b < a$ are called **inequalities.** For example, $3 < 4$ (and $4 > 3$) because 3 is to the left of 4 on the number line, and $-3 < -2$ (and $-2 > -3$) because -3 is to the left of -2. Similarly, $3.14 > 3.13$ (and $3.13 < 3.14$) because 3.14 is to the right of 3.13 on the number line as shown in Figure 1.5.

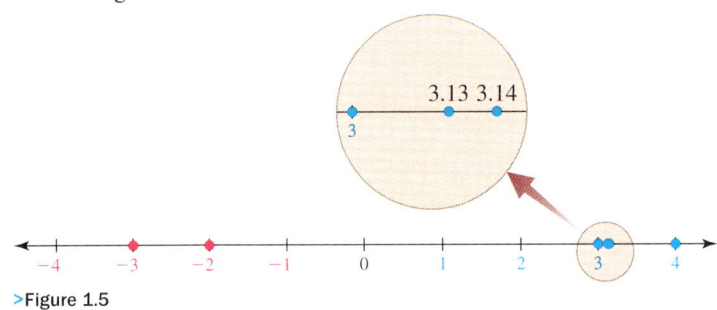

> Figure 1.5

EXAMPLE 6 Determining relationships between numbers

Fill in the blank with $<$, $>$, or $=$ to make the resulting statement true:

a. -3 _____ -1 b. -5 _____ -6 c. $\frac{1}{2}$ _____ 0

d. $\frac{1}{5}$ _____ $\frac{1}{3}$ e. -2.3 _____ -2.2 f. $\frac{1}{3}$ _____ $0.\overline{3}$

SOLUTION

a. $-3 < -1$

b. $-5 > -6$

PROBLEM 6

Fill in the blank with $<$ or $>$ to make the resulting statement true. (*Hint:* The arrow always points to the smaller number.)

a. -8 _____ -4

b. -7 _____ -6

c. 0 _____ $-\frac{1}{3}$

d. $\frac{1}{3}$ _____ $\frac{1}{2}$

e. -2.1 _____ -2.3

Answers to PROBLEMS

6. a. $<$ b. $<$ c. $>$ d. $<$ e. $>$

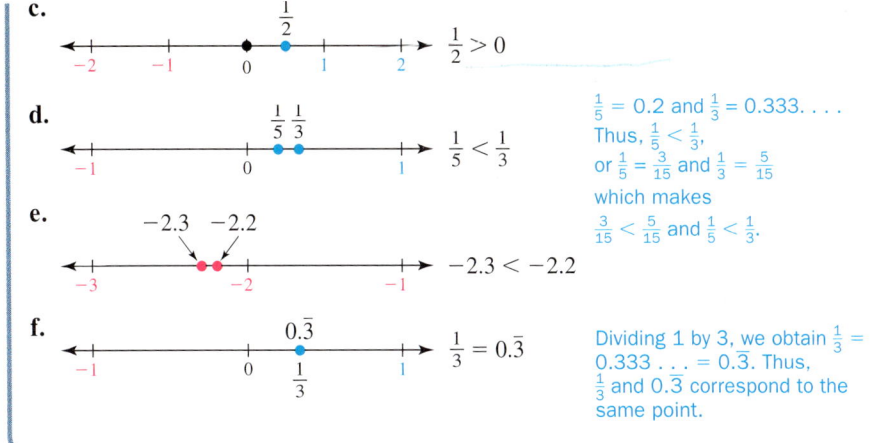

c. $\frac{1}{2} > 0$

d. $\frac{1}{5} = 0.2$ and $\frac{1}{3} = 0.333\ldots$. Thus, $\frac{1}{5} < \frac{1}{3}$, or $\frac{1}{5} = \frac{3}{15}$ and $\frac{1}{3} = \frac{5}{15}$ which makes $\frac{3}{15} < \frac{5}{15}$ and $\frac{1}{5} < \frac{1}{3}$.

e. $-2.3 < -2.2$

f. $\frac{1}{3} = 0.\overline{3}$ Dividing 1 by 3, we obtain $\frac{1}{3} = 0.333\ldots = 0.\overline{3}$. Thus, $\frac{1}{3}$ and $0.\overline{3}$ correspond to the same point.

G ▸ Applications Involving Real Numbers

Now that you have seen some examples of how to write sets with different types of notation and practiced how to compare numbers, let's use this knowledge to solve the application problem in Example 7. Use the graph and table in Figure 1.6 to solve the problem in Example 7.

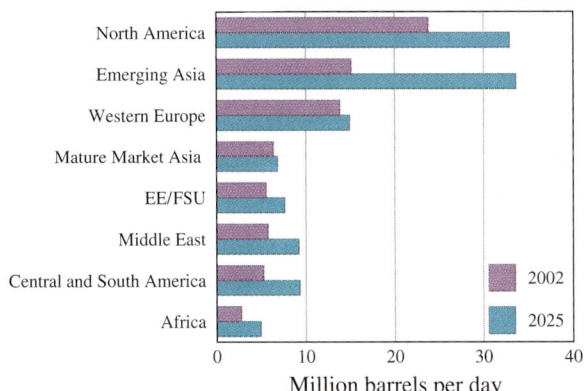

World Oil Consumption by Region and Country Group, 2002 and 2025

	2002	2025
North America	23.8	32.9
Emerging Asia	15.1	33.6
Western Europe	13.8	14.9
Mature Market Asia	6.3	6.8
EE/FSU	5.5	7.6
Middle East	5.7	9.2
Central and South America	5.2	9.3
Africa	2.7	4.9

Source: http://www.eia.doe.gov.

▸ Figure 1.6

Sources: 2002: Data from Energy Information Administration (EIA) *International Energy Annual 2002.* Doe/EIA-0219(2002) (Washington, DC. March 2004). website www.eia.doe.gov/lea/, 2025: Data from EIA, System for the Analysis of Global Energy Markets (2005).

EXAMPLE 7 Using set notation and comparing numbers

The graph in Figure 1.6 and the table include data regarding the millions of barrels of oil being consumed per day by region and country group for 2002 and the projected consumption for 2025.

a. Use the graph and roster notation to make a list of the regions that will have consumed more than 10 million barrels of oil per day for both 2002 and 2025.

b. According to the table, by how many million barrels of oil will the daily consumption of Emerging Asia exceed that of North America in 2025?

c. Use the data to correctly insert <, >, or = symbol for the following statement: In 2025, EE/FSU's consumption of oil _____ Mature Market Asia's consumption of oil.

PROBLEM 7

Use the graph in Figure 1.6 and the table to answer the following questions.

a. Use the graph and roster notation to make a list of the regions that will have consumed less than 10 million barrels of oil per day for both 2002 and 2025.

(continued)

Answers to PROBLEMS

7. a. {Mature Market Asia, EE/FSU, Middle East, Central and South America, Africa} b. 2.2 million barrels per day c. <

SOLUTION

a. On the graph, locate all the regions for which both bars extend to the right of the 10 million barrel mark. There are three regions: {North America, Emerging Asia, Western Europe}

b. The second column contains the data for 2025.

Emerging Asia	33.6 million barrels per day
North America	32.9 million barrels per day
The difference	0.7

The oil consumption of Emerging Asia will exceed that of North America by 0.7 million barrels of oil per day in 2025.

c. To correctly place the $<$, $>$, or $=$ symbol, you must compare the numbers that correspond to each region. According to the table, EE/FSU will consume 7.6 million barrels of oil per day in 2025 and Mature Market Asia will consume 6.8. Since $7.6 > 6.8$, EE/FSU's consumption of oil will be greater than Mature Market Asia's consumption of oil. Insert the $>$ sign.

b. Use the table to find the difference in the amount of barrels of oil Africa consumed in 2002 as compared with its projected consumption for 2025.

c. Use the data to correctly insert $<$, $>$, or $=$ symbol: In 2002, Western Europe's consumption of oil ____ Emerging Asia's consumption of oil.

Calculator Corner

π: Rational?

Let us consider this question: Can you see the *exact* decimal representation of an irrational number like π or $\sqrt{2}$ on your calculator screen? Try entering π as shown. You get 3.141592654, which appears to be rational; however, π is not rational! Can you explain what the problem is? The π symbol is above the ^ key. To engage it, press 2nd and then ^. Your screen should look like the one at the right.

Note: Make sure that your floating decimal (under the MODE menu) is set at "FLOAT."

π

3.141592654

Exercises 1.1

A Sets of Numbers In Problems 1–10, use roster notation to write the indicated set.

1. The first two natural numbers
2. The first six natural numbers
3. The natural numbers between 4 and 8
4. The natural numbers between 7 and 10
5. The first three negative integers
6. The negative integers between -4 and 7
7. The whole numbers between -3 and 4
8. The integers less than 0
9. The integers greater than 0
10. The nonnegative integers

In Problems 11–16, write the set using set-builder notation.

11. $\{1, 2, 3\}$
12. $\{6, 7, 8\}$
13. $\{-2, -1, 0, 1, 2\}$
14. $\{-8, -7, -6, -5\}$
15. The set of even numbers between 19 and 78
16. The set of all multiples of 3 between 8 and 78

In Problems 17–26, classify each statement as true or false. (Recall that N, W, I, Q, H, and R are the sets of natural, whole, integer, rational, irrational, and real numbers, respectively.)

17. $0 \in N$
18. $0 \in W$
19. $-0.3 \notin I$
20. $-0.\overline{8} \in Q$
21. $\sqrt{3} \in H$
22. $8.101001000\ldots \in Q$
23. $8.112233 \notin H$
24. $\sqrt{7} \in H$
25. $\frac{3}{8} \in R$
26. $-0 \in W$

⟨ B ⟩ Writing Rational Numbers as Decimals   In Problems 27–34, write the given number as a decimal.

27. $\frac{2}{3}$
28. $\frac{1}{6}$
29. $\frac{7}{8}$
30. $\frac{5}{6}$
31. $\frac{5}{2}$
32. $\frac{4}{3}$
33. $\frac{7}{6}$
34. $\frac{9}{8}$

⟨ C ⟩ Classifying Numbers   In Problems 35–44, classify the given numbers by placing a check mark in the appropriate row(s).

Set	35. $\frac{-3}{8}$	36. 0	37. $\sqrt{8}$	38. $\frac{3}{7}$	39. $0.\overline{3}$	40. -9	41. 0.9	42. -3.4	43. 3.1416	44. $3.141618\ldots$
Natural numbers										
Whole numbers										
Integers										
Rational numbers										
Irrational numbers										
Real numbers										

In Problems 45–48, classify each statement as true or false.

45. $N \subseteq Q$
46. $I \subseteq W$
47. $R \subseteq W$
48. $Q \subseteq R$

⟨ D ⟩ Additive Inverses (Opposites)   In Problems 49–64, find the additive inverse of the given number.

49. 8
50. -9
51. -7
52. 6
53. $\frac{3}{4}$
54. $-\frac{1}{4}$
55. $-\frac{1}{5}$
56. $\frac{2}{5}$
57. 0.5
58. -0.6
59. $0.\overline{2}$
60. $-0.\overline{3}$
61. $-1.\overline{36}$
62. $2.\overline{38}$
63. π
64. $-\pi$

⟨ E ⟩ Absolute Values   In Problems 65–78, find each value.

65. $|10|$
66. $|-11|$
67. $\left|\frac{3}{5}\right|$
68. $\left|-\frac{5}{7}\right|$
69. $|0.\overline{5}|$
70. $|-0.\overline{7}|$
71. $-|-\pi|$
72. $-|-\sqrt{3}|$
73. $2|-11|$
74. $9|-10|$
75. $12|3|-4$
76. $20|-5|+15$
77. $|7+2|-|6|$
78. $|9-5|+|-15|$

⟨ F ⟩ Equality and Inequality   In Problems 79–88, fill in the blanks with $<$ or $>$ to make the resulting statement true.

79. -5 ___ 2
80. 3 ___ -4
81. -6 ___ -8
82. -7 ___ -5
83. $\frac{1}{2}$ ___ $\frac{1}{4}$
84. $\frac{1}{3}$ ___ $\frac{1}{2}$
85. $-\frac{3}{5}$ ___ $-\frac{1}{4}$
86. $-\frac{1}{3}$ ___ $-\frac{1}{4}$
87. -3.5 ___ -3.4
88. -3.2 ___ -3.1

G › Applications Involving Real Numbers   In Problem 89 use the following table.

Top World Oil Net Exporters, 2004*
(OPEC members in italics)

Country	Net Oil Exports (million barrels per day)
1) Saudi Arabia	8.73
2) Russia	6.67
3) *Norway*	2.91
4) *Iran*	2.55
5) *Venezuela*	2.36
6) *United Arab Emirates*	2.33
7) *Kuwait*	2.20
8) *Nigeria*	2.19
9) Mexico	1.80
10) *Algeria*	1.68
11) *Iraq*	1.48
12) *Libya*	1.34
13) Kazakhstan	1.06
14) *Qatar*	1.02

* Table includes all countries with net exports exceeding 1 million barrels per day in 2004.

Source: Energy Information Administration, http://www.eia.doe.gov/.

89. *Oil exporters*   The table indicates the top world oil exporters by country for 2004, which had net exports that exceeded 1 million barrels of oil per day.

   a. Use roster notation to make a list of all the countries that exported more than $2\frac{1}{2}$ million barrels per day.

   b. Use roster notation to make a list of all the countries that exported less than 1.3 million barrels per day.

In Problem 90 use the following table.

Top World Oil Producers, 2004*
(OPEC members in italics)

Country	Total Oil Production** (million barrels per day)
1) Saudi Arabia	10.37
2) Russia	9.27
3) United States	8.69
4) *Iran*	4.09
5) Mexico	3.83
6) China	3.62
7) *Norway*	3.18
8) Canada	3.14
9) *Venezuela*	2.86
10) *United Arab Emirates*	2.76
11) *Kuwait*	2.51
12) *Nigeria*	2.51
13) United Kingdom	2.08
14) *Iraq*	2.03

* Table includes all countries total oil production exceeding 2 million barrels per day in 2004.

** Total Oil Production includes crude oil, natural gas liquids, condensate, refinery gain, and other liquids.

Source: Energy Information Administration, http://www.eia.doe.gov/.

90. *Oil producers*   The table indicates the top world oil producers by country for 2004, which had a total oil production that exceeded 2 million barrels of oil per day.

   a. Use roster notation to make a list of all the countries that had an oil production of more than $3\frac{1}{2}$ million barrels per day.

   b. Use roster notation to make a list of all the countries that had an oil production of less than 2.3 million barrels per day.

In Problem 91 use the following bar graph.

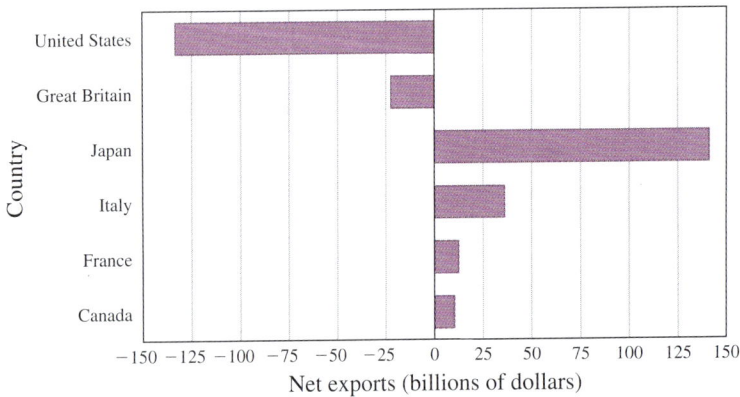

Source: http://cstl.syr.edu/.

91. *Net exports*   The bar graph includes the net exports of selected countries. As you can see by the direction of the bars, some countries have a negative net export, as indicated by negative numbers. This does not mean that those countries do not export any goods, but that they do not export more than they import.

   a. According to the graph, which country has the largest positive net export and approximate that amount in billions of dollars?

   b. According to the graph, which countries have a negative net export and approximate the amounts in billions of dollars?

   c. Insert <, >, or = symbol to make the following a true statement: France's net export ____ Canada's net export.

In Problem 92 use the following bar graph.

America's Trade Balance with Selected Nations, 1993

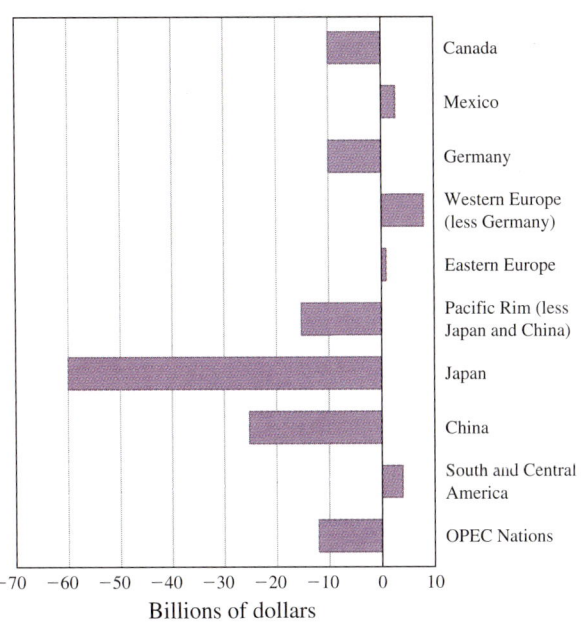

Source: http://cstl.syr.edu/.

92. *America's trade balance* The bar graph indicates America's trade balance with selected nations for 1993. As you can see by the direction of the bars, there were some surpluses or positive trade balances and some deficits or negative trade balances.

   a. According to the graph, with which nation does America have the *least* trade surplus and approximate that amount in billions of dollars?

   b. According to the graph, with which country does America have the largest deficit trade balance and approximate the amount in billions of dollars?

   c. Insert $<$, $>$, or $=$ symbol to make the following a true statement: America's trade balance with Mexico _____ America's trade balance with Western Europe.

>>> Using Your Knowledge

Hamburgers, Lawyers, and Wages In this section we learned how to compare integers and decimals; that is, we learned that $0.33 > 0.32$ and that $\frac{1}{3} < \frac{1}{2}$. To compare 0.33 and $\frac{1}{3}$, we write $\frac{1}{3}$ as a decimal by dividing the numerator by the denominator, obtaining $0.333\ldots$. We then write 0.33 as 0.330 (note the extra zero) and write both numbers in a column with the decimal points aligned.

$$0.333\ldots$$
$$0.330$$

$3 > 0$, so $0.333\ldots > 0.330$

Thus, $\frac{1}{3} > 0.33$.

We can use this knowledge in solving problems.

93. A McDonald's® hamburger weighs 100 grams (g) and contains 11 g of fat; that is, $\frac{11}{100}$ is fat. A Burger King® hamburger is 0.11009 fat. Write $\frac{11}{100}$ as a decimal and determine which restaurant's hamburger has the larger percentage of fat.

94. Lawyers have to know about fractions and decimals, too. In a court case called the *U.S. v. Forty Barrels and Twenty Kegs of Coca-Cola,®* a chemical analysis indicated that $\frac{3}{7}$ of the Coca-Cola was water. A second analysis showed that 0.41 was water. Which of the two analyses indicated more water in the Coke?

95. Using the graph in the *Getting Started*, it can be determined that the percent of increase of the median weekly earnings of a person with a doctoral degree as compared to a person with a bachelor's degree is approximately 50% or 0.50. Writing 164/900 as a decimal will indicate the percent of increase of the median weekly earnings of a person with a master's degree as compared to a person with a bachelor's degree. Determine which of the two degrees had the larger percent of increase in earnings from a bachelor's degree.

Here are the heights of three of the tallest people in the world:

Name	Feet	Inches
Sulaiman Ali Nashnush	8	$\frac{1}{25}$
Gabriel Estavao Monjane	8	$\frac{3}{4}$
Constantine	8	0.8

96. Which one is the tallest of the three?

97. Which one is the second tallest?

98. Which one is the shortest?

>>> Write On

99. Consider the statement, "$5 \subseteq N$, where N is the set of natural numbers." Is this statement true or false? Explain.

100. Consider the statement, "$N \in W$, where W is the set of whole numbers." Is this statement true or false? Explain.

101. Write in your words a definition for the set of rational numbers and the set of irrational numbers using the idea of a fraction.

102. Write in your words a definition of the set of rational numbers and the set of irrational numbers using the idea of a decimal.

103. Explain why every integer is a rational number but not every rational number is an integer.

>>> Concept Checker

Fill in the blank(s) with the correct word(s), phrase, or mathematical statement.

104. The number corresponding to a point on the number line is the _____ of the point, and the point is called the _____ of the number.

105. A _____ is a collection of objects and each member in the collection is called an _____.

106. In algebra when a letter is used to represent a number it is called a _____.

107. Describing a set by making a list of the elements is called _____.

108. A set containing no elements is referred to as the _____.

109. When a letter stands for just *one* number it is called a _____.

set · subset
element · coordinate
variable · graph
constant · additive inverse
roster notation · absolute value
set-builder notation · inequality
empty set

>>> Mastery Test

Write as a decimal:

110. $\frac{5}{7}$ **111.** $\frac{1}{8}$

Find the additive inverse of:

112. 0   **113.** 8

Find the value of each:

114. $-|-\sqrt{17}|$ **115.** $|-2|$

Write the following using set-builder notation:

116. $\{3, 6, 9, \ldots\}$

117. $\{3, 6, 10, 15, \ldots\}$

118. Classify the numbers in the following set as rational, Q, or irrational, H.

$$\left\{\frac{22}{7}, 3.1415, \pi, 3, -5, 0.\overline{77}, 34.010010001\ldots\right\}$$

Fill in the blank with $<$ or $>$ so that the result is a true statement:

119. $\frac{1}{3}$ _____ $0.331332333334\ldots$

120. 3.1416 _____ $\frac{22}{7}$

>>> Skill Checker

Perform the indicated operations.

121. $\frac{3}{5} + \frac{7}{5}$ **122.** $\frac{3}{4} + \frac{1}{4}$ **123.** $\frac{5}{8} - \frac{1}{8}$ **124.** $\frac{4}{7} - \frac{3}{7}$ **125.** $\frac{7}{8} + \frac{3}{4}$ **126.** $\frac{5}{6} + \frac{7}{3}$

127. $\frac{3}{2} - \frac{1}{3}$ **128.** $\frac{3}{4} - \frac{2}{5}$ **129.** $\frac{5}{9} \cdot \frac{2}{7}$ **130.** $\frac{4}{3} \cdot \frac{5}{3}$ **131.** $\frac{5}{7} \cdot \frac{2}{15}$ **132.** $\frac{3}{8} \cdot \frac{2}{5}$

133. $\frac{5}{6} \div \frac{1}{5}$ **134.** $\frac{1}{4} \div \frac{1}{8}$ **135.** $6.1 + 5.48$ **136.** $0.57 + 8.9$

137. $12.68 - 4$ **138.** $75 - 9.4$ **139.** $2.43(1.4)$ **140.** $17.6(0.3)$

141. $87.2 \div 5$ **142.** $0.6 \div 4$ **143.** $18 \div 0.3$ **144.** $16 \div 0.02$

1.2 Operations and Properties of Real Numbers

Objectives

A Add, subtract, multiply, and divide signed numbers.

B Identify uses of the properties of the real numbers.

C Solve applications involving operations of real numbers.

To Succeed, Review How To . . .

1. Find the additive inverse of a number (pp. 7–8).
2. Find the absolute value of a number (pp. 8–9).
3. Recognize positive and negative numbers (p. 7).

Getting Started

Signed Numbers and Growth Patterns

Look at the graph. What is the net percentage of change in this company's growth fund annual return for the years 2000 through 2002? To find the answer, we have to add the percentages for 2000 (-1.2), 2001 ($+4.7$), and 2002 ($+7.6$). The result is

$-1.2 + 4.7 + 7.6 =$
$3.50 + 7.6 =$ Adding -1.2 and 4.7
$+11.1$ Adding 3.5 and 7.6

This answer means that the company had a positive growth fund annual return of 11.1 percent for the years 2000 through 2002. You could find out whether the company has an overall positive or negative growth pattern from 1995 to 2006 by adding all the signed numbers together.

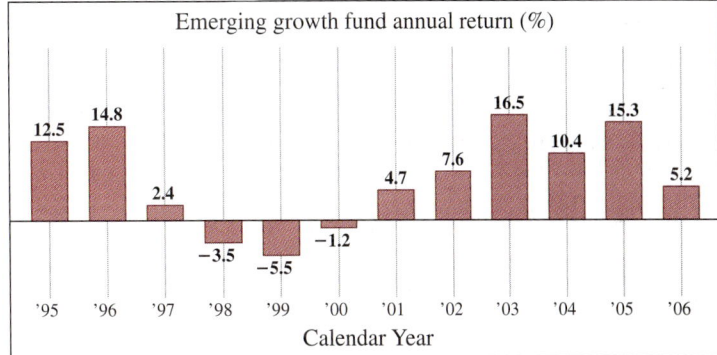

Money Grows on Trees Company

The following are two concepts used when adding and subtracting signed numbers:

1. When adding $-1.2 + 4.7$ it can also be written as $4.7 + (-1.2)$ because of the Commutative Law for addition. Notice the use of parentheses around the negative number in this case. The parentheses makes it less confusing than $4.7 + -1.2$.
2. To add 3.5 and 7.6, we add and keep the sign positive.

In this section, we shall learn to perform additions, subtractions, multiplications, and divisions using real numbers, and we shall study the rules for these four fundamental operations.

A › Operations with Signed Numbers

In algebra, we use the idea of absolute value to define the addition and subtraction of "signed" numbers.

> **PROCEDURE**
> **To Add Two Numbers with the Same Sign**
> **Add** their absolute values and give the sum the common sign (+ if both numbers are positive, − if both numbers are negative).

> **PROCEDURE**
> **To Add Two Numbers with Different Signs**
> 1. Find the absolute value of the numbers.
> 2. Subtract the smaller absolute value from the greater absolute value.
> 3. Use the sign of the number with the greater absolute value for the result obtained in step 2.

If zero is added to any number a, the result is the number a. Thus, $3 + 0 = 3$ and $0 + \frac{1}{2} = \frac{1}{2}$. The number 0 is called the **identity** for addition or the **additive identity.**

ADDITIVE IDENTITY

For any real number a,
$$a + 0 = 0 + a = a$$
(Zero is the identity for addition.)

Note that $-3 + 3 = 0$, $\frac{1}{2} + \left(-\frac{1}{2}\right) = 0$, and $-2.1 + 2.1 = 0$. Thus when **opposites (additive inverses)** are added, their sum is zero.

ADDITIVE INVERSES (OPPOSITES)

For any real number a,
$$a + (-a) = (-a) + a = 0.$$

EXAMPLE 1   Adding signed numbers
Add:

a. $-5 + (-11)$ **b.** $0.8 + (-0.5)$ **c.** $-0.7 + 0.4$

d. $\frac{4}{5} + \left(-\frac{4}{5}\right)$ **e.** $-\frac{3}{8} + \frac{1}{4}$ **f.** $-0.2 + 0$

SOLUTION
a. $-5 + (-11) = -16$ Add 5 and 11 and keep the negative sign.
b. $0.8 + (-0.5) = 0.3$ Subtract 0.5 from 0.8 and use the positive sign.
c. $-0.7 + 0.4 = -0.3$ Subtract 0.4 from 0.7 and use the negative sign.
d. $\frac{4}{5} + \left(-\frac{4}{5}\right) = 0$ Use additive inverses, $a + (-a) = 0$.

PROBLEM 1
Add:

a. $-3 + (-16)$ **b.** $0.7 + (-0.2)$

c. $-0.5 + 0.3$ **d.** $\frac{5}{7} + \left(-\frac{5}{7}\right)$

e. $-\frac{5}{6} + \frac{2}{3}$ **f.** $0 + \left(-\frac{3}{4}\right)$

Answers to PROBLEMS
1. **a.** -19 **b.** 0.5 **c.** -0.2
   d. 0 **e.** $-\frac{1}{6}$ **f.** $-\frac{3}{4}$

e. $\frac{-3}{8} + \left(\frac{1}{4}\right) \cdot \frac{2}{2} = \frac{-3+2}{8} = \frac{-1}{8} = -\frac{1}{8}$

The lowest common denominator (LCD) for the fractions is 8, so we rename $\frac{1}{4}$ by multiplying the numerator and denominator by 2. The numerators have different signs, so subtract 2 from 3 and use the negative sign because $|-3| > |2|$.

f. $-0.2 + 0 = -0.2$ Use the additive identity, $a + 0 = a$.

Calculator Corner

Fractions on the Calculator

Some calculators have special keys to enter fractions. If yours does not, you can always enter fractions as indicated divisions. To do Example 1(e), enter $-\frac{3}{8} + \frac{1}{4}$ as shown on the screen. The answer is given as $-.125$.

Is this the same as the $-\frac{1}{8}$ given in the solution of Example 1(e)? If you divide 1 by 8, you will see that the answers are identical.

With a TI-83 Plus, the decimal $-.125$ can be converted to a fraction by pressing MATH 1 ENTER to obtain $-\frac{1}{8}$ as shown.

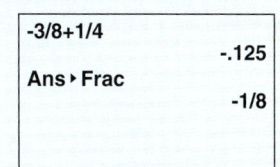

We have developed a procedure for adding real numbers. Next we will use the following example of temperature change to see how subtraction is related to addition.

What is the temperature change from 20° above zero on the Celsius scale to 10° below zero? This means we want to find the difference between 20 and (-10). Using the thermometer in the figure, you can count the number of degrees as designated by the red bracket and see that the difference is 30°. This difference in signed number notation would be written

$$20 - (-10) = 20 + 10 = 30$$

So the procedure for subtracting (-10) requires *adding* the *additive inverse* of (-10) to the first number.

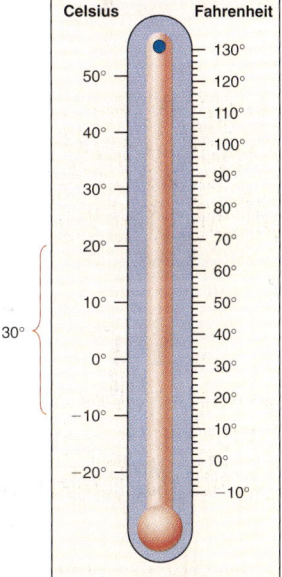

PROCEDURE

To Subtract Two Signed Numbers

Add the additive inverse of the second number to the first number.

If a and b are real numbers, $a - b = a + (-b)$

This means that we can *subtract* by adding the additive inverse or "opposite." Thus,

$$6 - (-3) = 6 + 3 = 9$$

$$-0.7 - 0.2 = -0.7 + (-0.2) = -0.9 \qquad \text{Subtract by adding the opposite.}$$

$$-\frac{1}{5} - \left(-\frac{4}{5}\right) = -\frac{1}{5} + \frac{4}{5} = \frac{3}{5}$$

EXAMPLE 2 Subtracting signed numbers

Subtract:

a. $-20 - 5$ b. $-0.6 - (-0.2)$ c. $-\frac{1}{7} - \left(-\frac{2}{7}\right)$

SOLUTION

a. $-20 - 5 = -20 + (-5) = -25$ Add the opposite of 5.
b. $-0.6 - (-0.2) = -0.6 + 0.2 = -0.4$ Add the opposite of -0.2.
c. $-\frac{1}{7} - \left(-\frac{2}{7}\right) = -\frac{1}{7} + \frac{2}{7} = \frac{1}{7}$ Add the opposite of $-\frac{2}{7}$.

In actual practice most of these operations are carried out mentally.

PROBLEM 2

Subtract:

a. $-15 - 3$ b. $-0.7 - (-0.3)$

c. $-\frac{1}{5} - \left(-\frac{1}{5}\right)$

In arithmetic the product of a and b is written as $a \times b$. In algebra, however, the multiplication sign ($\times$) can be mistaken for the letter x; thus the product of a and b is written in one of the following ways:

NOTATIONS FOR MULTIPLICATION

1. Using a raised dot, · $a \cdot b$
2. Writing a and b next to each other ab
3. Using parentheses $(a)(b)$, $a(b)$, or $(a)b$

The numbers represented by a and b are called **factors**, and the result of the multiplication is called the **product** of a and b.

Before considering multiplication with signed numbers we will state a property of multiplication by zero.

MULTIPLICATION PROPERTY OF ZERO

For all real numbers a,
$$0 \cdot a = a \cdot 0 = 0$$

This means multiplication of any real number by 0 always results in a product of 0. Thus, $3 \cdot 0 = 0$, $-15 \cdot 0 = 0$, and $\frac{2}{7} \cdot 0 = 0$.

Now suppose you own four shares of stock and one day the price is *down* $3, written as -3. Your loss that day would be the product $4 \cdot (-3)$, with factors 4 and -3. What is this product? This multiplication is the repeated addition of -3.

$$4 \cdot (-3) = \underbrace{(-3) + (-3) + (-3) + (-3)}_{\text{4 negative threes}} = -12$$

Next, look at $(-3) \cdot 4$. In Section B, you will see that multiplication of real numbers has the commutative property. Thus,

$$(-3) \cdot 4 = 4 \cdot (-3) = -12$$

We can generalize this idea to show that the product of any two numbers, one *positive* and the other *negative*, is a *negative number*. As in addition, we can state this result in terms of absolute values.

Answers to PROBLEMS

2. a. -18 b. -0.4 c. 0

PROCEDURE

Multiplying Two Numbers with Different Signs
To multiply a *positive* number by a *negative* number, multiply their absolute values. Make the product *negative*.

Thus, $3 \cdot (-2) = -6$, $-4 \cdot 8 = -32$, and $2 \cdot (-3.5) = -7$. The factors in each case, 3 and -2, -4 and 8, and 2 and -3.5, have *different (unlike)* signs.

EXAMPLE 3 Multiplying signed numbers
Multiply:
a. $7 \cdot (-8)$ b. $(-5) \cdot 0$ c. $(-2.5) \cdot 4$

SOLUTION
a. $7 \cdot (-8) = -56$ b. $(-5) \cdot 0 = 0$ c. $-2.5 \cdot 4 = -10.0$ or -10

PROBLEM 3
Multiply:
a. $9 \cdot (-7)$ b. $0 \cdot (-9)$
c. $(-4.5) \cdot 2$

What about the product of two negative integers such as $-2 \cdot (-3)$? First, look for a pattern when multiplying integers by (-3).

This number decreases by 1. This number increases by 3.

$$2 \cdot (-3) = -6$$
$$1 \cdot (-3) = -3$$
$$0 \cdot (-3) = 0$$
$$-1 \cdot (-3) = 3$$
$$-2 \cdot (-3) = 6$$

Thus, $-2 \cdot (-3) = 6$. In this case, -2 and -3 have the *same* sign $(-)$. When we multiply $2 \cdot 3$, 2 and 3 are both positive, so they also have the *same* sign $(+)$. We can summarize this discussion by the following rules.

SIGNS OF MULTIPLICATION

When Multiplying Two Numbers with:	The Product is:
Same (like) signs	Positive $(+)$
Different (unlike) signs	Negative $(-)$

Thus,

$$(-3) \cdot (-4.2) = 12.6 \qquad 4 \cdot (-2.1) = -8.4$$

Same signs → Positive answer   Different signs → Negative answer

Multiplying a number by 1 leaves the number unchanged ($3 \cdot 1 = 3$ and $-7 \cdot 1 = -7$). Thus, 1 is the identity for multiplication.

MULTIPLICATIVE IDENTITY

For any real number a,
$$a \cdot 1 = 1 \cdot a = a.$$
(**1** is the identity element for multiplication.)

Answers to PROBLEMS
3. a. -63 b. 0 c. -9.0 or -9

When multiplying more than two signed numbers, we can use some examples to determine the final sign of the product.

a. $(-3)(2)(-5)$ — two negative factors
$= -6\,(-5)$
$= +30$ — $+$ product

b. $2(-4)(-3)(-1)(-2)$ — four negative factors
$= -8\,(-3)(-1)(-2)$
$= 24\,(-1)(-2)$
$= -24\,(-2)$
$= +48$ — $+$ product

c. $-1(4)(-3)(-2)$ — three negative factors
$= -4\,(-3)(-2)$
$= 12\,(-2)$
$= -24$ — $-$ product

d. $3(-2)(-1)(-5)(-3)(-1)$ — five negative factors
$= -6\,(-1)(-5)(-3)(-1)$
$= 6\,(-5)(-3)(-1)$
$= -30\,(-3)(-1)$
$= 90\,(-1)$
$= -90$ — $-$ product

As you can see from the pattern, the sign of the product is determined by the number of negative factors in the multiplication. The following summarizes the result.

PROCEDURE

When Multiplying More Than Two Signed Numbers

Multiply their absolute values and the sign of the product is:

1. Positive $(+)$ if there are an *even* number of negative factors.
2. Negative $(-)$ if there are an *odd* number of negative factors.

EXAMPLE 4 Multiplying signed numbers
Multiply:

a. $4(-7)(-1)(2)(3)$ **b.** $(-6)(-2)(-3)(1)$

SOLUTION

a. Since there are two negative factors, an *even* number, the product is *positive*. We multiply the absolute value of the numbers.

$$4(-7)(-1)(2)(3) = (+)\,4\,(7)(1)(2)(3)$$
$$= 28(1)(2)(3)$$
$$= 28(2)(3)$$
$$= 56(3)$$
$$= 168$$

PROBLEM 4
Multiply:

a. $-2(-5)(-1)(-3)(-1)$

b. $(-1)(3)(-6)(2)$

Answers to PROBLEMS

4. a. -30 b. 36

b. Since there are three negative factors, an *odd* number, the product is *negative*.
We multiply the absolute value of the numbers.

$$(-6)(-2)(-3)(1) = (-)(6)(2)(3)(1)$$
$$= (-)(12)(3)(1)$$
$$= (-)(36)(1)$$
$$= -36$$

What about fractions? To multiply fractions, we need the following definition.

MULTIPLICATION OF FRACTIONS

If a, b, c, and d are real numbers,

$$\frac{a}{b} \cdot \frac{c}{d} = \frac{a \cdot c}{b \cdot d} \quad (b \neq 0, d \neq 0)$$

The same laws of signs apply. Thus,

$$\left(-\frac{9}{5}\right) \cdot \frac{3}{4} = -\frac{9 \cdot 3}{5 \cdot 4} = -\frac{27}{20}$$

Different signs   Negative answer

When multiplying fractions, it saves time if common factors are divided out before you multiply. Thus to multiply $\frac{5}{7} \cdot \left(-\frac{2}{5}\right)$, we write

$$\frac{\cancel{5}^1}{7} \cdot \left(-\frac{2}{\cancel{5}_1}\right) = -\frac{1 \cdot 2}{7 \cdot 1} = -\frac{2}{7} \quad \frac{5}{5} = 1$$

EXAMPLE 5 Multiplying signed numbers
Multiply:

a. $\left(-\frac{3}{7}\right) \cdot \frac{7}{8}$ **b.** $\left(-\frac{5}{8}\right) \cdot \left(-\frac{4}{15}\right)$

SOLUTION

a. $\left(-\frac{3}{\cancel{7}_1}\right) \cdot \frac{\cancel{7}^1}{8} = -\frac{3 \cdot 1}{1 \cdot 8} = -\frac{3}{8}$ Different (unlike) signs; the answer is negative.

b. $\left(-\frac{\cancel{5}^1}{\cancel{8}_2}\right) \cdot \left(-\frac{\cancel{4}^1}{\cancel{15}_3}\right) = \frac{1 \cdot 1}{2 \cdot 3} = \frac{1}{6}$ Same (like) signs; the answer is positive.

PROBLEM 5
Multiply:

a. $\left(-\frac{1}{3}\right) \cdot \frac{3}{7}$ **b.** $\left(\frac{3}{14}\right) \cdot \left(\frac{7}{6}\right)$

Just as we were able to define subtraction in terms of addition, we can also define division in terms of multiplication.

DIVISION OF REAL NUMBERS

If a and b are real numbers and b is not zero,

$$\frac{a}{b} = q \quad \text{means that} \quad a = b \cdot q$$

where a is called the **dividend**, b is the **divisor**, and q is the **quotient**.

Division is written in one of the following ways, with the first being used more often in algebra.

Answers to PROBLEMS

5. **a.** $-\frac{1}{7}$ **b.** $\frac{1}{4}$

24 Chapter 1 The Real Numbers

> **NOTATIONS FOR DIVISION**
>
> 1. $\dfrac{a}{b}$
> 2. $a \div b$
> 3. $b\overline{)a}$
> 4. $a : b$ (read the ratio of a to b)

We will use the definition of division to explain division with 0. We start with an example of dividing 12 by 4.

$$\frac{12}{4} = ? \quad \text{means} \quad 12 = 4 \cdot ?$$

Using the multiplication facts we know $4 \cdot 3 = 12$. Thus, $\frac{12}{4} = 3$.

Now, let's divide 0 by 4.

$$\frac{0}{4} = ? \quad \text{means} \quad 0 = 4 \cdot ?$$

Using the Multiplication Property of Zero we know that a product of 0 requires one or both of the factors to be 0. Thus, $\frac{0}{4} = 0$.

What about 4 divided by 0? We will try to apply the definition of division and see what happens.

$$\frac{4}{0} = ? \quad \text{means} \quad 4 = 0 \cdot ?$$

Again using the Multiplication Property of Zero, we know multiplication by 0 always results in a product of 0, not 4. In fact, it is impossible to find an answer for this multiplication problem so we say the division problem is impossible as well. The following summarizes division with 0.

> **ZERO IN DIVISION**
>
> For $a \neq 0$, $\dfrac{0}{a} = 0$ and $\dfrac{a}{0}$ is *not* defined. Moreover, $\dfrac{0}{0}$ is indeterminate.

CAUTION
$\dfrac{0}{k}$ is okay but $\dfrac{n}{0}$ is a no-no!

Let's look at some more examples with signed numbers.

$$\frac{48}{-6} = -8 \quad \text{means that} \quad 48 = -6 \cdot (-8)$$

$$\frac{-28}{-7} = 4 \quad \text{means that} \quad -28 = -7 \cdot (4)$$

The same rules of sign that apply to the multiplication of real numbers also apply to the division of real numbers; that is, the quotient of two numbers with the *same* sign is *positive*, and the quotient of two numbers with *different* signs is *negative*. Here are some examples:

$$\left.\begin{array}{ll} \dfrac{24}{6} = 4 & \dfrac{3.2}{1.6} = 2 \\[6pt] \dfrac{-18}{-9} = 2 & \dfrac{-3.3}{-1.1} = 3 \end{array}\right\} \text{Same signs, positive answers}$$

$$\left.\begin{array}{ll} \dfrac{-32}{4} = -8 & \dfrac{-6.3}{0.9} = -7 \\[6pt] \dfrac{35}{-7} = -5 & \dfrac{4.5}{-0.5} = -9 \end{array}\right\} \text{Different signs, negative answers}$$

SIGNS OF DIVISION

When Dividing Two Numbers with	The Quotient Is
Same (like) signs	Positive (+)
Different (unlike) signs	Negative (−)

There are *three* signs associated with every fraction: the sign of the numerator, the sign of the denominator, and the sign of the fraction.
Because

$$\frac{-32}{4} = \frac{32}{-4} = -\frac{32}{4} = -8$$

the following holds true.

SIGNS OF A FRACTION

For any real number a and nonzero real number b, there are two cases of signs of a fraction:

1. $\frac{-a}{b} = \frac{a}{-b} = -\frac{a}{b}$

2. $\frac{-a}{-b} = \frac{a}{b}$

EXAMPLE 6   Dividing signed numbers
Divide:

a. $48 \div 6$ **b.** $\frac{54}{-9}$ **c.** $\frac{-63}{-7}$

d. $-28 \div 4$ **e.** $5 \div 0$ **f.** $3.4 \div 1.7$

g. $\frac{4.8}{-1.2}$ **h.** $\frac{-5.6}{-0.8}$ **i.** $0 \div 3.5$

SOLUTION

a. $48 \div 6 = 8$ 48 and 6 have the same sign; the answer is positive.

b. $\frac{54}{-9} = -6$ 54 and −9 have different signs; the answer is negative.

c. $\frac{-63}{-7} = 9$ −63 and −7 have the same sign; the answer is positive.

d. $-28 \div 4 = -7$ −28 and 4 have different signs; the answer is negative.

e. $5 \div 0$ is not defined. If you make $5 \div 0$ equal any number, say a, you will have

$\frac{5}{0} = a$, which means $5 = a \cdot 0 = 0$ This says that $5 = 0$, which is impossible.

Thus, $\frac{5}{0}$ is not defined.

f. $3.4 \div 1.7 = 2$ 3.4 and 1.7 have the same sign; the answer is positive.

g. $\frac{4.8}{-1.2} = -4$ 4.8 and −1.2 have different signs; the answer is negative.

h. $\frac{-5.6}{-0.8} = 7$ −5.6 and −0.8 have the same sign; the answer is positive.

i. In this case, $0 \div 3.5 = 0$. We can check this using the definition of division:
$0 \div 3.5 = \frac{0}{3.5} = 0$ means $0 = 3.5 \cdot 0$, which is true.

PROBLEM 6
Divide:

a. $60 \div 10$ **b.** $\frac{48}{-3}$

c. $\frac{-18}{-2}$ **d.** $-14 \div 2$

e. $-4 \div 0$ **f.** $4.8 \div 1.6$

g. $\frac{4.2}{-2.1}$ **h.** $\frac{-3.8}{-1.9}$

i. $0 \div 9.2$

Answers to PROBLEMS

6. a. 6   **b.** −16   **c.** 9   **d.** −7
   e. Undefined   **f.** 3   **g.** −2   **h.** 2   **i.** 0

Sometimes simplifying expressions with signed numbers can involve absolute value as shown in Example 7.

EXAMPLE 7   Absolute value and signed numbers
Find:

a. $-3|4-8|-|-2+7|$ **b.** $\dfrac{2|9-12|}{|-5|+|6|}$

SOLUTION
a. $-3|4-8|-|-2+7|$
$= -3|-4|-|5|$ Perform operations inside absolute value symbols, $4-8=-4$ and $-2+7=5$.
$= -3(4)-(5)$ Replace absolute value, $|-4|=(4)$ and $|5|=(5)$.
$= -12-(5)$ Multiply, $-3(4)=-12$.
$= -17$ Subtract.

b. $\dfrac{2|9-12|}{|-5|+|6|}$

$= \dfrac{2|-3|}{|-5|+|6|}$ Perform operation inside absolute value symbol, $9-12=-3$.

$= \dfrac{2(3)}{(5)+(6)}$ Replace absolute value, $|-3|=3$, $|-5|=5$, and $|6|=6$.

$= \dfrac{6}{11}$ Multiply in numerator, add in denominator.

PROBLEM 7
Find:

a. $8|5-10|+|6-12|$

b. $\dfrac{|5|-|14|}{-4|-2+7|}$

Let's look at the division problem $2 \div 5$. We can write $2 \div 5 = \frac{2}{5} = 2 \cdot \frac{1}{5}$. Thus, to divide 2 by 5, we *multiply* 2 by $\frac{1}{5}$. The numbers 5 and $\frac{1}{5}$ are *reciprocals* or *multiplicative inverses*. Here is the definition.

MULTIPLICATIVE INVERSE (RECIPROCAL)   Every nonzero real number a has a **reciprocal (multiplicative inverse)** $\frac{1}{a}$ such that

$$a \cdot \dfrac{1}{a} = 1$$

The reciprocal of 3 is $\frac{1}{3}$, the reciprocal of -6 is $\frac{1}{-6} = -\frac{1}{6}$, and the reciprocal of $\frac{2}{3}$ is $\frac{3}{2}$. The reciprocal of a *positive* number is *positive* and the reciprocal of a *negative* number is *negative*.

EXAMPLE 8   Finding reciprocals
Find the reciprocal:

a. $\dfrac{2}{7}$ **b.** $-\dfrac{4}{5}$ **c.** 0.2   **d.** 0

SOLUTION
a. The reciprocal of $\frac{2}{7}$ is $\frac{7}{2}$.
b. The reciprocal of $-\frac{4}{5}$ is $-\frac{5}{4}$. (Remember that the reciprocal of a negative number is negative.)
c. The reciprocal of 0.2 is obtained by first changing the decimal to its equivalent fraction.

$$0.2 = \dfrac{2}{10} = \dfrac{1}{5}$$

The reciprocal of $\frac{1}{5}$ is 5.

d. The reciprocal of 0 is undefined because $\frac{1}{0}$ or division by 0 is undefined.

PROBLEM 8
Find the reciprocal of:

a. $\dfrac{5}{9}$ **b.** $-\dfrac{7}{8}$ **c.** 0.5   **d.** 8

Answers to PROBLEMS

7. a. 46   **b.** $\dfrac{9}{20}$ **8. a.** $\dfrac{9}{5}$ **b.** $-\dfrac{8}{7}$ **c.** 2   **d.** $\dfrac{1}{8}$

1.2 Operations and Properties of Real Numbers

Division of fractions is done in terms of reciprocals and multiplication. Since

$$\frac{a}{b} = a \cdot \frac{1}{b}$$

to divide by a number (such as b) we multiply by its reciprocal $\frac{1}{b}$. Division is related to multiplication and here is the general definition.

DIVISION OF FRACTIONS

If a, b, c, and d are real numbers,

$$\frac{a}{b} \div \frac{c}{d} = \frac{a}{b} \cdot \frac{d}{c} \quad (b \neq 0, c \neq 0, \text{ and } d \neq 0)$$

Thus to divide by $\frac{c}{d}$, we multiply by the reciprocal $\frac{d}{c}$. For example, to divide $\frac{4}{5}$ by $\frac{2}{3}$, we multiply $\frac{4}{5}$ by the reciprocal of $\frac{2}{3}$, that is, by $\frac{3}{2}$. This is written,

$$\frac{4}{5} \div \frac{2}{3} = \frac{4}{5} \cdot \frac{3}{2}$$

EXAMPLE 9 Dividing fractions

Find the following using reciprocals and multiplication:

a. $\frac{2}{5} \div \left(-\frac{3}{4}\right)$ **b.** $\left(-\frac{5}{6}\right) \div \left(-\frac{7}{2}\right)$ **c.** $\left(-\frac{3}{7}\right) \div \frac{6}{7}$

SOLUTION

a. $\frac{2}{5} \div \left(-\frac{3}{4}\right) = \frac{2}{5} \cdot \left(-\frac{4}{3}\right) = -\frac{8}{15}$

b. $\left(-\frac{5}{6}\right) \div \left(-\frac{7}{2}\right) = \left(-\frac{5}{6}\right) \cdot \left(-\frac{2}{7}\right) = \frac{10}{42} = \frac{5}{21}$

It is easier to "reduce," or divide common factors, before multiplying like this

$$\left(-\frac{5}{\underset{3}{\cancel{6}}}\right) \cdot \left(-\frac{\overset{1}{\cancel{2}}}{7}\right) = \frac{5}{21}$$

c. $\left(-\frac{3}{7}\right) \div \frac{6}{7} = \left(-\frac{3}{7}\right) \cdot \frac{7}{6} = -\frac{21}{42} = -\frac{1}{2}$

You can also "reduce," or divide common factors, before multiplying by writing

$$\left(-\frac{\overset{1}{\cancel{3}}}{\underset{1}{\cancel{7}}}\right) \cdot \frac{\overset{1}{\cancel{7}}}{\underset{2}{\cancel{6}}} = -\frac{1}{2}$$

PROBLEM 9

Divide:

a. $\frac{3}{5} \div \left(-\frac{4}{7}\right)$ **b.** $\left(-\frac{6}{7}\right) \div \left(-\frac{3}{5}\right)$

c. $\left(-\frac{4}{5}\right) \div \frac{8}{5}$

Calculator Corner

Divide Fractions

Your calculator does division of fractions. To do Example 9(c), enter

$$(-3 \div 7) \div (6 \div 7)$$

The result is as shown.

Would you get the same results if the parentheses were not used around one or both of the fractions being divided? Try the problem without the parentheses, and discuss your findings.

```
(-3/7)/(6/7)
                -.5
Ans▶Frac
               -1/2
```

Answers to PROBLEMS

9. **a.** $-\frac{21}{20}$ **b.** $\frac{10}{7}$ **c.** $-\frac{1}{2}$

B ❯ Real-Number Properties

We have already mentioned two properties of the real numbers: zero (0) is the **identity for addition,** and one (1) is the **identity for multiplication.** We continue this section by summarizing some additional real-number properties in Table 1.1 and then examining how these properties can help us in our work with algebra.

The **commutative properties** tell us that the *order* in which we add or multiply two numbers does not affect the result; thus, for the sum

$$\begin{array}{r} 28 \\ +39 \\ \hline \end{array}$$

you can add down (28 + 39) and then check by adding up (39 + 28). Similarly, it is easier to multiply

$$\begin{array}{r} 48 \\ \times 9 \\ \hline \end{array} \quad \text{instead of} \quad \begin{array}{r} 9 \\ \times 48 \\ \hline \end{array}$$

but by the commutative property, the result is the same. On the other hand, the **associative property** tells us that the *grouping* of the numbers when adding or multiplying numbers does not affect the final answer. Thus, if you wish to add 3 + 24 + 6 by adding 24 + 6 first—that is, by finding 3 + (24 + 6)—you will get the same answer as if you had added (3 + 24) + 6.

The **closure property** for addition and multiplication states that if you add or multiply any two real numbers, then the result is a real number. Sometimes we say the operations of addition and multiplication are *closed* over the real numbers.

When multiplying 3 (2 + 4) we can do it two ways and still get the same answer. We can add 2 + 4 to get 6, and then multiply by 3 for a result of *18*. Or, we could multiply 3 times 2, which is 6 and 3 times 4, which is 12, and then add these two results, 6 + 12, to get *18*. We call this property the **distributive property of multiplication over addition.** This property's name is sometimes shortened to just the **distributive property.** The properties of real numbers are summarized in Table 1.1.

Table 1.1 Properties of Real Numbers (*a*, *b*, and *c* represent real numbers)

Property	Addition	Multiplication
Closure	If a and b are real numbers, then $a + b$ is a real number. $-8 + \sqrt{5}$ is a real number.	If a and b are real numbers, then $a \cdot b$ is a real number. $-8 \cdot \sqrt{5}$ is a real number.
Commutative	$a + b = b + a$ $\sqrt{2} + 8 = 8 + \sqrt{2}$	$ab = ba$ $3 \cdot \frac{1}{5} = \frac{1}{5} \cdot 3$
Associative	$a + (b + c) = (a + b) + c$ $\frac{1}{3} + \left(\frac{2}{5} + \frac{3}{4}\right) = \left(\frac{1}{3} + \frac{2}{5}\right) + \frac{3}{4}$	$a(bc) = (ab)c$ $-3 \cdot \left(\sqrt{5} \cdot \frac{1}{8}\right) = (-3 \cdot \sqrt{5}) \cdot \frac{1}{8}$
Identity	$a + 0 = 0 + a = a$ $-\frac{4}{5} + 0 = 0 + \left(-\frac{4}{5}\right) = -\frac{4}{5}$ (0 is called the additive identity.)	$1 \cdot a = a \cdot 1 = a$ $0.\overline{38} \cdot 1 = 1 \cdot 0.\overline{38} = 0.\overline{38}$ (1 is called the multiplicative identity.)
Inverse	$a + (-a) = (-a) + a = 0$ $-a$ is the additive inverse (opposite) of a. $-\frac{1}{5}$ is the opposite of $\frac{1}{5}$. Thus $\frac{1}{5} + \left(-\frac{1}{5}\right) = 0$.	$a \cdot \frac{1}{a} = \frac{1}{a} \cdot a = 1 \quad (a \neq 0)$ $\frac{1}{a}$ is the multiplicative inverse (reciprocal) of a. 8 is the reciprocal of $\frac{1}{8}$. Thus $8 \cdot \frac{1}{8} = 1$.
Distributive	$a(b + c) = ab + ac \quad 3(1 + 6) = 3 \cdot 1 + 3 \cdot 6 \quad a(b - c) = ab - ac \quad 4(1 - 6) = 4 \cdot 1 - 4 \cdot 6$	

EXAMPLE 10 Properties of real numbers

Which property is illustrated in each of the following statements?

a. $(-3) + \frac{7}{5} = \frac{7}{5} + (-3)$

b. $3 \cdot (4 \cdot 7) = (4 \cdot 7) \cdot 3$

c. $(-4) \cdot \left(\frac{1}{2} \cdot \frac{1}{8}\right) = \left(-4 \cdot \frac{1}{2}\right) \cdot \frac{1}{8}$

d. $2 + (3 \cdot 4) = 2 + (4 \cdot 3)$

e. $(-6) + 0 = -6$

f. $(-6) + 6 = 0$

SOLUTION

a. We changed the *order*, so the commutative property of addition applies.
b. Here again we changed the *order* (the 3 changed position). The commutative property of multiplication was used.
c. This time we *grouped* the numbers differently. The associative property of multiplication was used.
d. The commutative property of multiplication applies (the 3 and 4 changed positions).
e. Since 0 is added to the real number (-6), the additive identity property applies.
f. Because we added "opposites," the result is 0. This is an example of the additive inverse property.

PROBLEM 10

Which property is illustrated in each statement?

a. $2 \cdot (3 \cdot 5) = (3 \cdot 5) \cdot 2$

b. $(-4) \cdot \frac{2}{5} = \frac{2}{5} \cdot (-4)$

c. $7 \cdot \left(\frac{1}{4} \cdot \frac{1}{2}\right) = \left(7 \cdot \frac{1}{4}\right) \cdot \frac{1}{2}$

d. $5 + (3 \cdot 9) = 5 + (9 \cdot 3)$

e. $-8 = (-8) + 0$

f. $8 + (-8) = 0$

The associative property can help you find the answer to a problem such as $(3)(-5)(6)$. To use this property, you can write either of the following:

1. $(3)(-5)(6) = (-15)(6)$ Multiply $(3)(-5)$ first and $(-15)(6)$ next.
$= -90$

2. $(3)(-5)(6) = (3)(-30)$ Multiply $(-5)(6)$ first and $(3)(-30)$ next.
$= -90$

The answer is the same in both cases. We shall practice with this type of problem in the exercises at the end of this section.

We have not yet added or subtracted expressions like $2\sqrt{3} + 5\sqrt{3}$ or $8\sqrt{2} - 12\sqrt{2}$. We can do this using the commutative and distributive properties as follows:

$2\sqrt{3} + 5\sqrt{3} = (\sqrt{3} \cdot 2) + (\sqrt{3} \cdot 5)$ By the commutative property of multiplication
$= \sqrt{3}(2 + 5)$ By the distributive property
$= 7\sqrt{3}$ By the commutative property of multiplication

In practice you can see that $2\sqrt{3}$ and $5\sqrt{3}$ are **like** terms that can be combined. Using this idea, $8\sqrt{2} - 12\sqrt{2} = -4\sqrt{2}$. We shall study the distributive property and how to use it to combine like terms in Section 1.4.

C › Applications Involving Operations of Real Numbers

Practicing skill exercises is not all there is to studying mathematics. If you can't use the skills to solve simple day-to-day problems, then you are missing the best part of the lesson. Suppose you have taken a measurement, in inches, for an item you want to buy. However, when you get to the store, the dimensions of the item are given in feet. You will have to know how to convert a measurement in inches to one in feet. That is just what you will do in Example 11. Not only will you need experience with operations in the real numbers, which was learned in this section, but you will need to know that the fraction used in solving these problems is expressed by writing the given measurement over the denominator of 1. For example, when given the measurement 13 centimeters (cm), it

---

Answers to PROBLEMS

10. a. Commutative
   b. Commutative
   c. Associative
   d. Commutative
   e. Additive identity
   f. Additive inverse

can be expressed as $\frac{13 \text{ cm}}{1}$. Also, it will be necessary to know some conversion facts (see Table 1.2). This table includes only a few selected conversions necessary to solve Example 11 and Problems 103–106 in the 1.2 Exercises.

Table 1.2

Convert Within English		Convert Within Metric	
Length	**Weight**	**Length**	**Mass**
1 mi = 5280 ft	1 ton = 2000 lb	1 km = 1000 m	1 kg = 1000 g
1 yd = 3 ft	1 lb = 16 oz	1 m = 100 cm	1 g = 1000 mg
1 ft = 12 in.			

Convert from English to Metric or Metric to English					
English Length	**Metric Length**	**English Capacity**	**Metric Capacity**	**English Weight**	**Metric Mass**
1 mi	1609 m	1 gal	3.785 L	1 lb	0.454 kg
1 yd	0.914 m				
1 ft	0.305 m				
1 in.	2.54 cm				

EXAMPLE 11   Measurement conversions

When measuring for a new awning, Kody concluded that he needed an awning that was 60 inches by 42 inches. However, the dimensions of the awnings at the store were given in feet.

a. Find the dimensions of the awning that Kody needs in feet.
b. Find the dimensions of the awning to the nearest hundredth of a centimeter.

PROBLEM 11

It took Patricio Pena in San Juan, Puerto Rico, about four days to manufacture one of the largest cigars in the world. He used 20 pounds of tobacco and 100 leaves to roll the cigar which cost about $2000. The cigar was 62 feet long.

a. How long was the cigar to the nearest hundredth of a yard?
b. How much did the 20 pounds of tobacco weigh in ounces?
c. How long was the cigar to the nearest hundredth of a meter?

SOLUTION

a. 1) Use Table 1.2 to find the conversion fact: 1 ft = 12 in.

2) Write the first dimension, 60 inches, as a fraction over 1.
$$\frac{60 \text{ in.}}{1}$$

3) Multiply the fraction in step 2 by a fraction using the conversion fact so that the inch unit is in the denominator.
$$\frac{60 \text{ in.}}{1} \cdot \frac{1 \text{ ft}}{12 \text{ in.}}$$

4) Divide out the inch units that appear diagonally.
$$\frac{60 \text{ in.}}{1} \cdot \frac{1 \text{ ft}}{12 \text{ in.}}$$

5) Simplify.
$$\frac{60 \text{ ft}}{12} = 5 \text{ ft}$$

Answers to PROBLEMS

11. a. 20.67 yd   b. 320 oz   c. 18.91 m

The same procedure is used to convert the second dimension, 42 inches.

    1) Use the table to find the conversion fact: 1 ft = 12 in.

    2) Write the second dimension, 42 inches, as a fraction over 1.

$$\frac{42 \text{ in.}}{1}$$

    3) Multiply the fraction in step 2 by a fraction using the conversion fact so that the inch unit is in the denominator.

$$\frac{42 \text{ in.}}{1} \cdot \frac{1 \text{ ft}}{12 \text{ in.}}$$

    4) Divide out the inch units that appear diagonally.

$$\frac{42 \cancel{\text{ in.}}}{1} \cdot \frac{1 \text{ ft}}{12 \cancel{\text{ in.}}}$$

    5) Simplify.

$$\frac{42 \text{ ft}}{12} = 3.5 \text{ ft}$$

The dimensions of the awning are 5 feet by 3.5 feet.

b. 1) Use the table to find the conversion fact: 1 in. = 2.54 cm

    2) Write the first dimension, 60 inches, as a fraction over 1.

$$\frac{60 \text{ in.}}{1}$$

    3) Multiply the fraction in step 2 by a fraction using the conversion fact so that the inch unit is in the denominator.

$$\frac{60 \text{ in.}}{1} \cdot \frac{2.54 \text{ cm}}{1 \text{ in.}}$$

    4) Divide out the inch units that appear diagonally.

$$\frac{60 \cancel{\text{ in}}}{1} \cdot \frac{2.54 \text{ cm}}{1 \cancel{\text{ in.}}}$$

    5) Simplify.

$$60 \cdot 2.54 \text{ cm} = 152.40 \text{ cm}$$

The same procedure is used to convert the second dimension, 42 inches.

    1) Use the table to find the conversion fact: 1 in. = 2.54 cm

    2) Write the second dimension, 42 inches, as a unit fraction:

$$\frac{42 \text{ in.}}{1}$$

    3) Multiply the unit fraction in step 2 by a fraction using the conversion fact so that the inch unit is in the denominator.

$$\frac{42 \text{ in.}}{1} \cdot \frac{2.54 \text{ cm}}{1 \text{ in.}}$$

    4) Divide out the inch units that appear diagonally.

$$\frac{42 \cancel{\text{ in.}}}{1} \cdot \frac{2.54 \text{ cm}}{1 \cancel{\text{ in.}}}$$

$$42 \cdot 2.54 \text{ cm} = 106.68 \text{ cm}$$

    5) Simplify.

The dimensions are 152.40 by 106.68 centimeters.

32 Chapter 1 The Real Numbers

> **Calculator Corner**
>
> 1. What do you have to do so that the answer to Example 1(e) is shown as a fraction on your calculator screen?
> 2. What happens if you try to divide by zero on your calculator?
> 3. If you enter the left side of Example 10(d), 2 + (3 · 4), into your calculator, the result is 14. How would you place the grouping symbols so the result is 20?

Exercises 1.2

Boost your grade at mathzone.com!
> Practice Problems
> NetTutor
> Self-Tests
> e-Professors
> Videos

A Operations with Signed Numbers In Problems 1–86, perform the indicated operations.

1. $\frac{3}{5} + \left(-\frac{1}{5}\right)$
2. $-0.4 + 0.9$
3. $-0.3 + 0.2$
4. $-8 + 5$
5. $-4 + 0$
6. $0 + 0.3$
7. $-0.5 + (-0.3)$
8. $-7 + (-11)$
9. $-\frac{1}{5} + \frac{1}{4} + \frac{3}{20}$
10. $-\frac{4}{7} + \frac{2}{9} + \frac{1}{63}$
11. $6 - 13$
12. $8 - 13$
13. $0.6 - 0.9$
14. $0.3 - 0.8$
15. $\frac{1}{7} - \frac{3}{8}$
16. $\frac{3}{8} - \frac{4}{8}$
17. $-8 - 4 - 2$
18. $-4 - 6 - 3$
19. $-0.4 - 0.2$
20. $-0.3 - 0.5$
21. $-\frac{3}{7} - \frac{2}{9}$
22. $-\frac{4}{9} - \frac{1}{9}$
23. $-6 - (-5)$
24. $-7 - (-9)$
25. $-8 - (-4)$
26. $-9 - (-2)$
27. $-0.7 - (-0.6)$
28. $-0.9 - (-0.3)$
29. $-\frac{2}{7} - \left(-\frac{4}{3}\right)$
30. $-\frac{3}{4} - \left(-\frac{5}{3}\right)$
31. $-5(8)$
32. $-9(6)$
33. $4(-3)$
34. $6(-8)$
35. $-10(0)$
36. $0(-9)$
37. $-3(4)(-5)$
38. $-4(-2)(5)$
39. $-5(2)(3)$
40. $-2(-5)(-9)$
41. $-1.3(-2.2)$
42. $-1.5(-1.1)$
43. $-3(5)(-2)(-1)(4)$
44. $2(-10)(-2)(-3)(-1)$
45. $|7 - 15| - 2|-3 + 6|$
46. $-4|1 - 5| + |2 - 8|$
47. $\frac{5}{6}\left(-\frac{5}{7}\right)$
48. $\frac{3}{8}\left(-\frac{5}{7}\right)$
49. $-\frac{3}{5}\left(-\frac{5}{12}\right)$
50. $-\frac{4}{7}\left(-\frac{21}{8}\right)$
51. $-\frac{6}{7}\left(\frac{35}{8}\right)$
52. $-\frac{7}{5}\left(\frac{15}{28}\right)$
53. $\frac{-18}{9}$
54. $\frac{-32}{16}$
55. $\frac{20}{-5}$
56. $\frac{36}{-3}$
57. $\frac{-14}{-7}$
58. $\frac{-24}{-8}$
59. $\frac{0}{-3}$
60. $\frac{0}{-9}$
61. $\frac{4}{0}$
62. $\frac{-7}{0}$
63. $-\left(\frac{-4}{-2}\right)$

64. $-\left(\dfrac{-10}{-5}\right)$ 65. $-\left(\dfrac{-27}{3}\right)$ 66. $-\left(\dfrac{-9}{3}\right)$ 67. $-\left(\dfrac{15}{-5}\right)$

68. $-\left(\dfrac{18}{-6}\right)$ 69. $\dfrac{-3}{-3}$ 70. $\dfrac{-18}{-9}$ 71. $\dfrac{-16}{4}$

72. $\dfrac{-48}{6}$ 73. $\dfrac{-56}{8}$ 74. $\dfrac{-54}{6}$ 75. $\dfrac{3}{5} \div \left(-\dfrac{4}{7}\right)$

76. $\dfrac{4}{9} \div \left(-\dfrac{1}{7}\right)$ 77. $-\dfrac{2}{3} \div \left(-\dfrac{7}{6}\right)$ 78. $-\dfrac{5}{6} \div \left(-\dfrac{25}{18}\right)$ 79. $-\dfrac{5}{8} \div \dfrac{7}{8}$

80. $-\dfrac{4}{5} \div \dfrac{8}{15}$ 81. $\dfrac{-3.1}{6.2}$ 82. $\dfrac{1.2}{-4.8}$ 83. $\dfrac{-1.6}{-9.6}$

84. $\dfrac{-9.8}{-1.4}$ 85. $\dfrac{|20|-|-6|}{5|4-7|}$ 86. $\dfrac{3|4-3|}{-|2|+|-9|}$

⟨ B ⟩ Real-Number Properties
In Problems 87–94, indicate which property is illustrated in each statement (a and b represent real numbers).

87. $5.6 + 9.2 = 9.2 + 5.6$

88. $-3 \cdot 4 = 4 \cdot (-3)$

89. $-5 \cdot (2 \cdot 7) = (-5 \cdot 2) \cdot 7$

90. $\left(\dfrac{1}{5} + \dfrac{2}{7}\right) + \dfrac{1}{8} = \dfrac{1}{5} + \left(\dfrac{2}{7} + \dfrac{1}{8}\right)$

91. $12 \cdot \dfrac{1}{12} = 1$

92. $3a + (-3a) = 0$

93. $1 \cdot (3 + b) = 3 + b$

94. $(a + b) \cdot 1 = a + b$

95. If the area of a rectangle is found by multiplying the length times the width, express the area of the largest rectangle in the figure in two ways to illustrate the distributive property for $a(b + c)$.

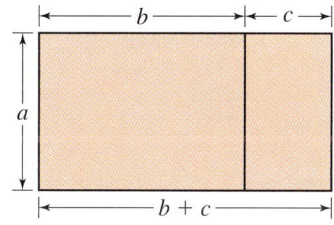

96. Express the shaded area of the rectangle in the figure to illustrate the distributive property for $a(b - c)$.

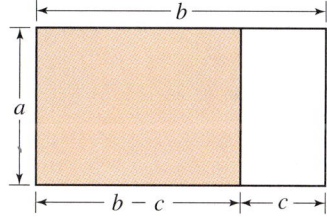

⟨ C ⟩ Applications Involving Operations of Real Numbers

97. *Earth* The temperature in the center core of the Earth reaches $+5000°C$. In the thermosphere (a region in the upper atmosphere), the temperature is $+1500°C$. Find the difference in temperature between the center of the Earth and the thermosphere.

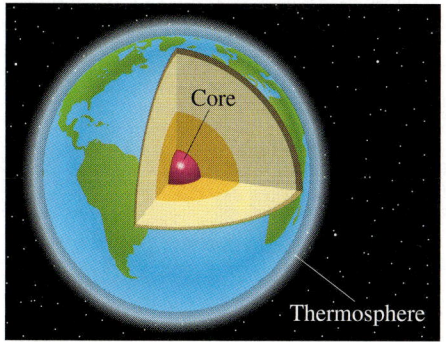

98. The record high temperature in Calgary, Alberta, is $+99°F$. The record low temperature is $-46°F$. Find the difference between these extremes.

99. *Stocks* The price of a certain stock at the beginning of the week was $47. Here are the daily changes in price during the week: $+1, +2, -1, -2, -1$. What was the price of the stock at the end of the week?

100. *Stocks* The price of a stock at the beginning of the week was $37. On Monday, the price went up $2; on Tuesday, it went down $3; and on Wednesday, it went down another $1. What was the price of the stock on Wednesday?

Use for Problems 101–102   *Temperature*   Here are the temperature changes (in degrees Celsius, C) by the hour in a certain city:

$$1\text{ P.M.}, +2 \qquad 2\text{ P.M.}, +1 \qquad 3\text{ P.M.}, -1 \qquad 4\text{ P.M.}, -3$$

101. If the temperature was initially 15°C, what was it at 4 P.M.?

102. If the temperature was initially −15°C, what was it at 4 P.M.?

In Problems 103–106, use the conversion table on page 30.

103. *Elevation*   Mount Everest, the highest mountain on Earth, is on the border between Nepal and China. Its summit is 8848 meters (29,028 feet) above sea level.

   a. What is the height of Mount Everest in kilometers above sea level?

   b. What is the height of Mount Everest to the nearest tenth of a mile above sea level?

104. *Sumo wrestlers*   One of the heaviest professional sumo wrestlers is a Samoan-American named Seleeva Fuali Atisanoe, alias Konishiki. On January 3, 1994, at Tokyo's Ryogoku Kokugikan, he weighed in at 267 kilograms (589 pounds).

   a. What is Konishiki's weight in grams?

   b. What is Konishiki's weight to the nearest tenth of a ton?

105. *Catsup*   The world's largest catsup bottle, which also happens to be a water tower, is in Collinsville, Illinois. It was built by the W. E. Caldwell Company for the G. S. Suppiger catsup bottling plant in 1949. The bottle stands 170 feet high and holds 100,000 gallons of water.

   a. What is the catsup bottle's height to the nearest meter?

   b. What is the catsup bottle's capacity to the nearest liter?

106. *Computers*   One of the smaller PCs, similar in size to a CD player, is the BOLData® Mini PC. Its dimensions are 1.8 × 5.7 × 6.2 inches and it weighs 1.9 pounds.

   a. What are the dimensions of this PC in centimeters?

   b. What does this PC weigh to the nearest tenth of a kilogram?

>>> Using Your Knowledge

Different Moods Have you met anybody *nice* today? Have you had an *unpleasant* experience? Perhaps the person you met was *very nice* or your experience *very unpleasant.* Psychologists and linguists have a numerical way to indicate the difference between nice and very nice or between unpleasant and very unpleasant. Suppose you assign a positive number ($+2$, for example) to the adjective *nice,* and a negative number (say, -2) to *unpleasant,* and a positive number greater than 1 (say, $+1.25$) to *very.* Then, by definition, "**very nice**" means

$$(1.25) \cdot (2) = 2.5$$

(very nice)

and "very unpleasant" means

$$(1.25) \cdot (-2) = -2.5$$

(very unpleasant)

Here are some adverbs and adjectives and their average numerical values as rated by a panel of college students.

Adverbs		Adjectives	
Slightly	0.54	Wicked	−2.5
Rather	0.84	Disgusting	−2.1
Decidedly	0.16	Average	−0.8
Very	1.25	Good	3.1
Extremely	1.45	Lovable	2.4

Find the value of

107. Slightly wicked

108. Decidedly average

109. Extremely disgusting

110. Rather lovable

111. Very good

By the way, if you got all the answers correct, you are 4.495! (Extremely good)

>>> Write On

112. Explain why division by zero is not defined.

113. The full name of the distributive property $a(b + c) = ab + ac$ is the distributive property of multiplication over addition. Is addition distributive over multiplication? Explain and give examples to support your answer.

114. We mentioned that the set of real numbers is *closed* under addition and multiplication. Are the natural numbers closed under addition? What about under subtraction? Explain and give examples to support your answer.

>>> Concept Checker

Fill in the blank(s) with the correct word(s), phrase, or mathematical statement.

115. The number 5 is the _____ of the number $\frac{1}{5}$.

116. Using the _____ property, multiplication of more than two numbers can be done in any grouping and still yield the same product.

117. In the statement $\frac{x}{y} = z$, x is the _____, y is the _____, and z is the _____.

118. Using the _____ property, the expression $a(x + y)$ can be written $ax + ay$.

additive identity
additive inverse
factors
product
dividend
commutative
distributive
associative
divisor
quotient
reciprocal
multiplicative identity
multiplicative inverse

119. The statement $(x)(1) = x$ is an example of the _____ property.

120. In the statement $cd = f$, c and d are the _____ and f is the _____.

>>> Mastery Test

Perform the indicated operation.

121. $-\frac{2}{5} \div \left(-\frac{5}{8}\right)$

122. $-\frac{9}{5} - \left(-\frac{3}{4}\right)$

123. $\left(\frac{3}{5}\right)\left(-\frac{10}{3}\right)$

124. $23.4 + (-29.7)$

125. $-\frac{5}{8} + \left(-\frac{3}{7}\right)$

126. $18.7 - (-13.2)$

127. $3.2(-4)$

128. $-3.4 \div 1.7$

129. $(-3 - 5)(-2) + 8(3 - 7 + 4)$

130. $-4(5 - 7) + (-3 - 5)(-2)$

Name the property used:

131. $(-8) + 8 = 0$

132. $(a + 3) = 1 \cdot (a + 3)$

133. $3(ab) = (3a)b$

134. $8 + (b + c) = (8 + b) + c$

135. $2 \cdot (b + 4) = (b + 4) \cdot 2$

136. $(a + 1) \cdot \frac{1}{a + 1} = 1, a \neq -1$

>>> Skill Checker

Multiply.

137. $(-4)(-4)$

138. $(-6)(-6)(-6)$

139. $-(2)(2)(2)(2)$

140. $-(3)(3)(3)$

In Problems 141–144, fill in the blanks.

	Number	Additive Inverse	Reciprocal
141.	7	_____	_____
142.	-8	_____	_____
143.	0	_____	_____
144.	$-\frac{1}{2}$	_____	_____

145. $2 - 3$

146. $-4 - (-5)$

147. $-7 + 8$

148. $4 + (-7)$

149. $6(-4)$

150. $(-3)(-5)$

151. $0(-2)$

152. $-2(4)$

1.3 Properties of Exponents

Objectives

A Evaluate expressions containing natural numbers as exponents.

B Write an expression containing negative exponents as a fraction.

C Multiply and divide expressions containing exponents.

D Raise a power to a power and a quotient to a power.

E Convert between ordinary decimal notation and scientific notation, and use scientific notation in computations.

To Succeed, Review How To . . .

Perform the four fundamental operations (addition, subtraction, multiplication, and division) using signed numbers (pp. 18–27).

Getting Started

One of the most prominent landmarks in the East Village in Manhattan is a statue of a giant steel black cube. The cube was built at Astor Place in 1968 and has stood there ever since. One day a group of individuals decided to play a prank and turn it into the world's largest Rubik's cube. The cube measures 8 ft by 8 ft by 8 ft. It is standing on one vertex, such that the top reaches around 14 ft in the air. The cube remained painted like the Rubik's cube for an entire day before the New York City maintenance department cleaned it off.

Unlike the "prank" Rubik's cube shown in the picture, the side of each of the squares in a "real" Rubik's cube is 1 centimeter (cm) long. Can you find the **area** of the top of the cube? Can you find the **volume** of the whole cube? As you may recall, to find the area of a square, we multiply the length of two sides of the square. The top of the cube is 3 cm long, therefore the area of the top is

$$(3 \text{ cm})(3 \text{ cm}) = 3^2 \text{ cm}^2 \text{ or } 9 \textbf{ square} \text{ centimeters.}$$

Similarly, the volume of the cube is

$$(3 \text{ cm})(3 \text{ cm})(3 \text{ cm}) = 3^3 \text{ cm}^3 \text{ or } 27 \textbf{ cubic} \text{ centimeters.}$$

$$3^2 = \overbrace{3 \times 3}^{2 \text{ times}} \quad \text{and} \quad 3^3 = \overbrace{3 \times 3 \times 3}^{3 \text{ times}}$$

This is **exponential** notation, one of the topics we shall study in this section.

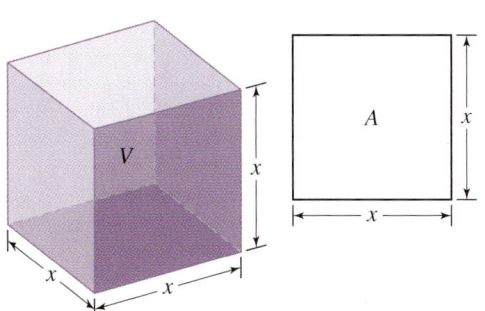

What is the **area** A of the square? The area is x^2 units.

$$A = x \cdot x = x^2 \quad \text{Read "}x\text{ squared or }x\text{ to the second power."}$$

In the expression x^2, the *exponent* 2 indicates that the *base* x is to be used as a factor twice. What about the **volume** V of the cube? It is x^3 units

$$V = x \cdot x \cdot x = x^3 \quad \text{Read "}x\text{ cubed or }x\text{ to the third power."}$$

This time the *exponent* 3 indicates that the *base* x is used as a factor three times. Expressions written as 3^2 or x^3 are in **exponential** notation. In general, we have the following definition.

EXPONENT AND BASE

If a is a real number and n is a natural number,

$$a^n = \underbrace{a \cdot a \cdot a \cdots a}_{n \text{ factors}}$$

where n is called the **exponent** and a is the **base**.

When $n = 1$, the exponent is usually omitted. Thus, $a^1 = a$, $5^1 = 5$, and $c^1 = c$.

A > Natural-Number Exponents

We can **evaluate** (find the value of) expressions containing natural-number exponents. Thus,

$$4^2 = 4 \cdot 4 = 16$$
$$(-3)^2 = (-3)(-3) = 9$$
$$(-2)^3 = (-2)(-2)(-2)$$
$$= 4(-2) \qquad (-2)(-2) = 4$$
$$= -8$$
$$(-4)^4 = (-4)(-4)(-4)(-4)$$
$$= (16)(16)$$
$$= 256$$

Recall the procedure in Section 1.2 for multiplying more than two signed numbers required a positive result if there were an even number of negative factors and a negative result if there were an odd number of negative factors. So, for the **even** exponents 2 and 4 in $(-3)^2 = 9$ and $(-4)^4 = 256$, respectively, we get the **positive** answers 9 and 256. However, for the **odd** exponent 3 in $(-2)^3 = -8$, the answer is **negative.**

> **CAUTION**
> $(-3)^2 \neq -3^2$ because $(-3)^2 = 9$, which is positive, while -3^2 is the additive inverse of 3^2, that is, $-3^2 = -9$. The placement of parentheses when using exponents is extremely important. Always interpret -3^2 as $-(3^2)$.

EXAMPLE 1 Evaluating expressions with natural-number exponents

Evaluate:

a. $(-8)^2$ b. -8^2 c. $(-4)^3$ d. -4^3

SOLUTION

a. $(-8)^2 = (-8)(-8) = 64$ The exponent 2 is applied to the base (-8).
b. $-8^2 = -1 \cdot 8^2$ Here the exponent 2 applies to the base 8 only.
$\quad = -1 \cdot 64$ Since $8^2 = 64$
$\quad = -64$
c. $(-4)^3 = (-4)(-4)(-4) = -64$
d. $-4^3 = -1 \cdot 4^3$
$\quad = -1 \cdot 64$ Since $4^3 = 4 \cdot 4 \cdot 4 = 64$
$\quad = -64$

PROBLEM 1

Evaluate:

a. $(-5)^2$ b. -5^2
c. $(-6)^3$ d. -6^3

Answers to PROBLEMS

1. a. 25   b. -25
   c. -216 d. -216

Calculator Corner

Verify $(-4)^2 \neq -4^2$

We can use a calculator to clarify the concepts in Example 1. Enter $(-4)^2$ and -4^2 in your calculator. Can you see that in $(-4)^2$ we are raising the base -4 to the second power, while in -4^2 we are finding the additive inverse of 4^2? Look at the way you enter each of the expressions. With a TI-83 Plus, $(-4)^2$ is entered as [( ] [(-)] 4 [ )] [x^2], while -4^2 is entered as [(-)] 4 [x^2]. The answers are different. We can see that $(-4)^2 \neq -4^2$.

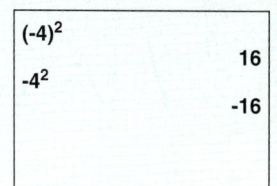

B › Negative and Zero Exponents

In science and technology, negative numbers are also used as exponents. For example, the diameter of a DNA molecule is 10^{-8} m, and the time it takes for an electron to go from source to screen in a TV tube is 10^{-6} sec. What do 10^{-8} and 10^{-6} mean? Look at the pattern obtained by dividing each number on the right side by 10 while the powers of 10 on the left side are decreased by 1.

$$10^3 = 1000$$
$$10^2 = 100$$
$$10^1 = 10$$

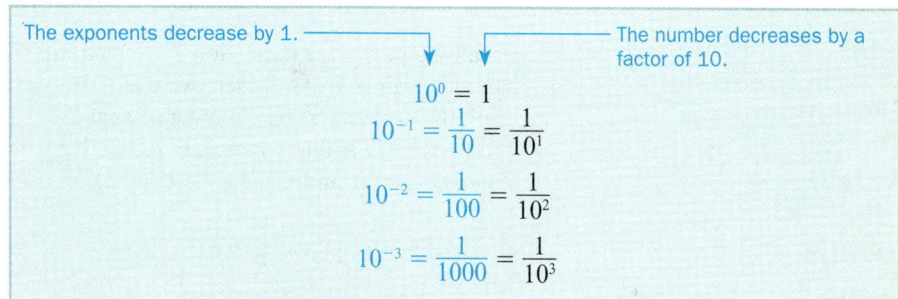

The exponents decrease by 1. ⸻⸻ The number decreases by a factor of 10.

$$10^0 = 1$$
$$10^{-1} = \frac{1}{10} = \frac{1}{10^1}$$
$$10^{-2} = \frac{1}{100} = \frac{1}{10^2}$$
$$10^{-3} = \frac{1}{1000} = \frac{1}{10^3}$$

This procedure yields $10^0 = 1$. In general, we make the following definition.

ZERO EXPONENT

If a is a nonzero real number,
$$a^0 = 1$$
Moreover, 0^0 is *not* defined.

Thus, $5^0 = 1$, $8^0 = 1$, and $9^0 = 1$. 0^0 is **not** defined. Now look again at the numbers in the first box. We obtained

$$10^{-1} = \frac{1}{10}, \quad 10^{-2} = \frac{1}{10^2}, \quad \text{and} \quad 10^{-3} = \frac{1}{10^3}$$

We make the following definition.

NEGATIVE EXPONENTS

If n is a positive integer,
$$a^{-n} = \frac{1}{a^n} \quad (a \neq 0)$$

This definition says that a^{-n} and a^n are reciprocals, provided $a \neq 0$, because

$$a^{-n} \cdot a^n = \frac{1}{a^n} \cdot a^n = 1$$

By definition

$$5^{-2} = \frac{1}{5^2} = \frac{1}{5 \cdot 5} = \frac{1}{25}$$

and

$$(-2)^{-3} = \frac{1}{(-2)^3} = \frac{1}{(-2) \cdot (-2) \cdot (-2)} = \frac{1}{-8} = -\frac{1}{8}$$

Similarly,

$$\frac{1}{4^2} = 4^{-2}$$

and

$$\frac{1}{3^4} = 3^{-4}$$

EXAMPLE 2 Rewriting expressions with negative exponents as fractions

Write the following without negative exponents:

a. 6^{-2} b. $(-4)^{-3}$ c. $x^{-4}y^2$ d. $-3a^{-4}$ e. $5x^{-2}y^3z^{-4}$

SOLUTION

a. $6^{-2} = \frac{1}{6^2} = \frac{1}{6 \cdot 6} = \frac{1}{36}$

b. $(-4)^{-3} = \frac{1}{(-4)^3} = \frac{1}{(-4) \cdot (-4) \cdot (-4)} = \frac{1}{-64} = -\frac{1}{64}$

c. $x^{-4}y^2 = \frac{1}{x^4} \cdot \frac{y^2}{1} = \frac{y^2}{x^4}$

d. $-3a^{-4} = -\frac{3}{1} \cdot \frac{1}{a^4} = -\frac{3}{a^4}$

e. $5x^{-2}y^3z^{-4} = \frac{5}{1} \cdot \frac{1}{x^2} \cdot \frac{y^3}{1} \cdot \frac{1}{z^4} = \frac{5y^3}{x^2z^4}$

PROBLEM 2

Write the following without negative exponents:

a. 5^{-2} b. $(-3)^{-3}$
c. $x^{-5}z^6$ d. $-2b^{-5}$
e. $7x^{-3}y^{-1}z^2$

C › Multiplication and Division with Exponents

Suppose we wish to multiply $x^2 \cdot x^3$. We write

$$\underbrace{\overbrace{x^2}^{} \cdot \overbrace{x^3}^{}}_{x^5}$$
$$x \cdot x \cdot x \cdot x \cdot x$$

We want to know the total number of factors, so we add the exponents 2 and 3 of x^2 and x^3 to find the exponent 5 of the result. Similarly,

$$a^3 \cdot a^4 = a^{3+4} = a^7$$
$$b^2 \cdot b^4 = b^{2+4} = b^6$$
$$c^5 \cdot c^0 = c^{5+0} = c^5$$

Now, let's consider $x^5 \cdot x^{-3}$, where one of the exponents is negative. By the definition of exponents,

$$x^5 \cdot x^{-3} = x \cdot x \cdot x \cdot x \cdot x \cdot \frac{1}{x \cdot x \cdot x}$$
$$= \frac{x \cdot x \cdot \cancel{x} \cdot \cancel{x} \cdot \cancel{x}}{\cancel{x} \cdot \cancel{x} \cdot \cancel{x}}$$
$$= x^2$$

Answers to PROBLEMS

2. a. $\frac{1}{5^2} = \frac{1}{25}$ b. $\frac{1}{(-3)^3} = -\frac{1}{27}$
c. $\frac{z^6}{x^5}$ d. $\frac{-2}{b^5}$ e. $\frac{7z^2}{x^3y}$

Adding exponents,

$$x^5 \cdot x^{-3} = x^{5+(-3)} = x^2 \quad \text{Same answer}$$

Let us try $x^{-2} \cdot x^{-3}$. This time we have

$$x^{-2} \cdot x^{-3} = \frac{1}{x \cdot x} \cdot \frac{1}{x \cdot x \cdot x}$$

$$= \frac{1}{x \cdot x \cdot x \cdot x \cdot x} \quad \text{Multiplying}$$

$$= x^{-5} \quad \text{By the definition of negative exponents}$$

Adding exponents,

$$x^{-2} \cdot x^{-3} = x^{-2+(-3)} = x^{-5} \quad \text{Same answer}$$

This suggests the following rule.

PRODUCT RULE OF EXPONENTS

If a is a real number and m and n are integers,

$$a^m \cdot a^n = a^{m+n} \quad (a \neq 0)$$

This rule tells us that to *multiply* expressions with the *same* base, we *add* the exponents. This rule *does not* apply to expressions such as $x^m \cdot y^n$ because the bases are *different*. $[3^2 \cdot 2^4 = 3 \cdot 3 \cdot 2 \cdot 2 \cdot 2 \cdot 2$, so we have different bases (factors) that cannot be combined.] If the expressions involved have numerical coefficients, we multiply numbers by numbers and letters (variables) by letters using the commutative and associative properties that we have studied. Thus, to multiply $(-3x^2)(5x^3)$, we write

$$(-3x^2)(5x^3) = (-3 \cdot 5)(x^2 \cdot x^3)$$
$$= -15x^{2+3} = -15x^5$$

Similarly,

$$(-4x^{-2})(3x^5) = (-4 \cdot 3)(x^{-2+5})$$
$$= -12x^3 \quad -2 + 5 = 3$$

and

$$(-2x^{-5})(-3x^2) = (-2)(-3)(x^{-5+2})$$
$$= 6x^{-3} \quad -5 + 2 = -3$$
$$= 6 \cdot \frac{1}{x^3}$$
$$= \frac{6}{x^3}$$

NOTE We write the answer without using negative exponents.

EXAMPLE 3 Using the product rule of exponents

Multiply and simplify:

a. $(-4x^2)(5x^5)$ **b.** $(3x^7)(-2x^4)$ **c.** $(4x^3y)(-2x^{-8}y^6)$

SOLUTION

a. $(-4x^2)(5x^5) = (-4 \cdot 5)(x^2 \cdot x^5)$
$= -20x^{2+5}$
$= -20x^7$

b. $(3x^7)(-2x^4) = (3)(-2)(x^7 \cdot x^4)$
$= -6x^{7+4}$
$= -6x^{11}$

PROBLEM 3

Multiply and simplify:

a. $(-3x^{-3})(5x^5)$
b. $(3x^2)(-2x^3)$
c. $(5x^3y^4)(-2x^{-7}y)$

(continued)

Answers to PROBLEMS

3. a. $-15x^2$ **b.** $-6x^5$ **c.** $-\frac{10y^5}{x^4}$

c. $(4x^3y)(-2x^{-8}y^6) = (4)(-2)(x^3 \cdot x^{-8})(y^1 \cdot y^6)$
$= -8(x^{3+(-8)})(y^{1+6})$
$= -8(x^{-5})(y^7)$
$= -8\left(\frac{1}{x^5}\right)(y^7)$
$= -\frac{8y^7}{x^5}$

Now consider the following division:

$$\frac{x^5}{x^3} = \frac{x \cdot x \cdot x \cdot x \cdot x}{x \cdot x \cdot x}$$
$$= x \cdot x \cdot \frac{\cancel{x} \cdot \cancel{x} \cdot \cancel{x}}{\cancel{x} \cdot \cancel{x} \cdot \cancel{x}}$$
$$= x^2 \cdot 1$$
$$= x^2$$

The exponent 2 in the answer can be obtained by *subtracting* $5 - 3 = 2$. Similarly,

$$\frac{x^5}{x^{-3}} = x^{5-(-3)} = x^8$$

This is because

$$\frac{x^5}{x^{-3}} = \frac{x^5}{\frac{1}{x^3}} = x^5 \div \frac{1}{x^3} = x^5 \cdot x^3 = x^8$$

This line of reasoning suggests the following law.

> **QUOTIENT RULE OF EXPONENTS**
>
> If a is a real number and m and n are integers,
> $$\frac{a^m}{a^n} = a^{m-n} \quad (a \neq 0)$$

This law tells us that to *divide* expressions with the same base, we *subtract* the exponent of the denominator from that of the numerator. Thus,

$$\frac{x^8}{x^5} = x^{8-5} = x^3$$

$$\frac{x^3}{x^8} = x^{3-8} = x^{-5} = \frac{1}{x^5}$$

$$\frac{x^{-3}}{x^{-8}} = x^{-3-(-8)} = x^{-3+8} = x^5$$

$$\frac{x^8}{x^8} = x^{8-8} = x^0 = 1$$

EXAMPLE 4 Using the quotient rule of exponents
Divide and simplify:

a. $\dfrac{4x^8}{2x^5}$ **b.** $\dfrac{-12x^4}{-3x^6}$ **c.** $\dfrac{5x^{-4}}{-15x^{-6}}$ **d.** $\dfrac{30x^3}{-15x^{-6}}$

PROBLEM 4
Divide and simplify:

a. $\dfrac{6x^9}{12x^3}$ **b.** $\dfrac{-18x^3}{-9x^8}$

c. $\dfrac{10x^{-6}}{-20x^{-8}}$ **d.** $\dfrac{45x^4}{-15x^{-6}}$

SOLUTION

a. $\dfrac{4x^8}{2x^5} = \dfrac{4}{2} \cdot \dfrac{x^8}{x^5}$
$= 2x^{8-5}$
$= 2x^3$

b. $\dfrac{-12x^4}{-3x^6} = \dfrac{-12}{-3} \cdot \dfrac{x^4}{x^6}$
$= 4x^{4-6}$
$= 4x^{-2}$ $4 - 6 = -2$
$= \dfrac{4}{x^2}$

Answers to PROBLEMS

4. **a.** $\dfrac{x^6}{2}$ **b.** $\dfrac{2}{x^5}$ **c.** $-\dfrac{x^2}{2}$ **d.** $-3x^{10}$

c. $\dfrac{5x^{-4}}{-15x^{-6}} = \dfrac{5}{-15} \cdot \dfrac{x^{-4}}{x^{-6}}$

$= -\dfrac{1}{3} \cdot x^{-4-(-6)}$

$= -\dfrac{1}{3} \cdot x^{-4+6}$ $\quad -4-(-6)$
$\quad\quad\quad\quad\quad\quad\quad = -4+6$

$= -\dfrac{1}{3} \cdot x^2$

$= -\dfrac{x^2}{3}$

d. $\dfrac{30x^3}{-15x^{-6}} = \dfrac{30}{-15} \cdot \dfrac{x^3}{x^{-6}}$

$= -2x^{3-(-6)}$

$= -2x^9$ $\quad 3-(-6)$
$\quad\quad\quad\quad = 3+6 = 9$

NOTE A **negative coefficient**, like -15 in the denominator of Example 4(c), does not follow the same rule as the definition of a **negative exponent** like x^{-2} in Example 4(b). The -15 stays in the denominator, but the x^{-2} moves to the denominator and becomes x^2.

D › Raising a Product and a Quotient to a Power

Now let us consider $(5^3)^2$. By definition,

$$(5^3)^2 = (5^3)(5^3) = 5^{3+3} = 5^6$$

The exponent is $6 = 3 \cdot 2$, so we could have obtained the answer by multiplying exponents in the expression $(5^3)^2$. Similarly,

$$(3^{-2})^3 = \left(\dfrac{1}{3^2}\right)\left(\dfrac{1}{3^2}\right)\left(\dfrac{1}{3^2}\right) = \dfrac{1}{3^6} = 3^{-6}$$

Again, the exponent -6 could be obtained by multiplying the original exponents, -2 and 3. Generalizing this result, we have the following rule.

POWER RULE OF EXPONENTS

If a is a real number and m and n are integers,

$$(a^m)^n = a^{m \cdot n} \quad (a \neq 0)$$

To raise a power to another power, we multiply the exponents,

$$(x^3)^4 = x^{3 \cdot 4} = x^{12}$$

$$(y^6)^{-3} = y^{6 \cdot (-3)} = y^{-18} = \dfrac{1}{y^{18}}$$

$$(z^{-5})^4 = z^{-5 \cdot 4} = z^{-20} = \dfrac{1}{z^{20}}$$

Let's consider $(5x^4)^3$. By definition,

$$(5x^4)^3 = (5x^4)(5x^4)(5x^4)$$
$$= (5^1 \cdot 5^1 \cdot 5^1)(x^4 \cdot x^4 \cdot x^4)$$
$$= 5^{3 \cdot 1} \cdot x^{3 \cdot 4}$$
$$= 5^3 \cdot x^{12}$$

The exponent 3 is applied to the 5 and the x^4; that is, if we raise several factors inside parentheses to a power, we raise each factor to the given power. In general, we have the following rule.

| RAISING A PRODUCT TO A POWER | If a and b are real numbers and m, n, and k are integers, $$(a^m b^n)^k = a^{m \cdot k} b^{n \cdot k} \quad (a \neq 0, b \neq 0)$$ |

This result can be generalized further to apply to any number of factors inside the parentheses.

EXAMPLE 5 Raising a product to a power
Simplify:
a. $(3x^5 y^3)^{-2}$ **b.** $(-2x^5 y^{-4} z)^3$

SOLUTION
a. $(3x^5 y^3)^{-2} = (3)^{-2} (x^5)^{-2} (y^3)^{-2}$

$= \dfrac{1}{3^2} \cdot x^{5 \cdot (-2)} \cdot y^{3 \cdot (-2)}$ $(3)^{-2} = \dfrac{1}{3^2}$

$= \dfrac{1}{3^2} \cdot x^{-10} \cdot y^{-6}$

$= \dfrac{1}{3^2} \cdot \dfrac{1}{x^{10}} \cdot \dfrac{1}{y^6}$ $x^{-10} = \dfrac{1}{x^{10}}, y^{-6} = \dfrac{1}{y^6}$

$= \dfrac{1}{9 x^{10} y^6}$

b. $(-2x^5 y^{-4} z)^3 = (-2)^3 (x^5)^3 (y^{-4})^3 (z)^3$

$= -8 x^{5 \cdot 3} y^{-4 \cdot 3} \cdot z^{1 \cdot 3}$

$= -8 x^{15} y^{-12} z^3$

$= -\dfrac{8 x^{15} z^3}{y^{12}}$ $y^{-12} = \dfrac{1}{y^{12}}$

PROBLEM 5
Simplify:
a. $(4x^3 y^4)^{-2}$ **b.** $(-3x^2 y^{-5} z)^3$

We have already raised a product to a power. Can we raise a quotient to a power? Let us try $(2^3/3^4)^2$. By the definition of exponents,

$$\left(\dfrac{2^3}{3^4}\right)^2 = \dfrac{2^3}{3^4} \cdot \dfrac{2^3}{3^4} = \dfrac{2^{3+3}}{3^{4+4}} = \dfrac{2^6}{3^8}$$

The same answer is obtained by multiplying each of the exponents in the numerator and denominator by 2. Here is the general rule.

| RAISING A QUOTIENT TO A POWER | If a and b are real numbers and m, n, and k are integers, $$\left(\dfrac{a^m}{b^n}\right)^k = \dfrac{a^{m \cdot k}}{b^{n \cdot k}} \quad (a \neq 0, b \neq 0)$$ |

EXAMPLE 6 Raising a quotient to a power
Simplify:
a. $\left(\dfrac{x^4}{y^{-3}}\right)^{-2}$ **b.** $\left(\dfrac{3x^{-3} y^2}{2y^3}\right)^3$

SOLUTION
a. $\left(\dfrac{x^4}{y^{-3}}\right)^{-2} = \dfrac{(x^4)^{-2}}{(y^{-3})^{-2}} = \dfrac{x^{4 \cdot (-2)}}{y^{-3 \cdot (-2)}} = \dfrac{x^{-8}}{y^6}$ $x^{-8} = \dfrac{1}{x^8}$

$= \dfrac{1}{x^8 y^6}$

PROBLEM 6
Simplify:
a. $\left(\dfrac{x^5}{y^{-3}}\right)^{-4}$ **b.** $\left(\dfrac{2x^{-3} y^4}{3y^3}\right)^2$

Answers to PROBLEMS

5. **a.** $\dfrac{1}{16 x^6 y^8}$ **b.** $-\dfrac{27 x^6 z^3}{y^{15}}$ 6. **a.** $\dfrac{1}{x^{20} y^{12}}$ **b.** $\dfrac{4y^2}{9x^6}$

b. In this case, it is easier to do the operations inside the parentheses first.

$$\left(\frac{3x^{-3}y^2}{2y^3}\right)^3 = \left(\frac{3}{2} \cdot x^{-3} \cdot y^{2-3}\right)^3 \qquad \frac{y^2}{y^3} = y^{2-3}$$

$$= \left(\frac{3y^{-1}}{2x^3}\right)^3 \qquad x^{-3} = \frac{1}{x^3}$$

$$= \left(\frac{3}{2x^3 y}\right)^3 \qquad y^{-1} = \frac{1}{y}$$

$$= \frac{3^3}{(2x^3 y)^3}$$

$$= \frac{27}{8x^9 y^3}$$

Since the reciprocal of $\frac{a}{b}$ is $\frac{b}{a}$,

$$\left(\frac{a}{b}\right)^{-1} = \left(\frac{b}{a}\right)$$

Thus,

$$\left[\left(\frac{a}{b}\right)^{-1}\right]^n = \left(\frac{b}{a}\right)^n$$

RAISING A QUOTIENT TO A NEGATIVE POWER	If a and b are real numbers and n is an integer, $$\left(\frac{a}{b}\right)^{-n} = \left(\frac{b}{a}\right)^n \qquad (a \neq 0, b \neq 0)$$

This means that a fraction raised to the $-n$th power is equivalent to its *reciprocal* raised to the nth power. In Example 6(a), we could write

$$\left(\frac{x^4}{y^{-3}}\right)^{-2} = \left(\frac{y^{-3}}{x^4}\right)^2 = \frac{y^{-6}}{x^8} = \frac{1}{x^8 y^6}$$

You can use this method when you work the exercise set.

E ▸ Scientific Notation

In science and other areas of endeavor, very large or very small numbers occur frequently. For example, a red cell of human blood contains 270,000,000 hemoglobin molecules, and the mass of a single carbon atom is 0.000 000 000 000 000 000 000 019 9 gram. Numbers in this form are difficult to write and to work with, so they are written in scientific notation for which we have the following definition.

> **SCIENTIFIC NOTATION**
> A number is said to be in **scientific notation** if it is written in the form
> $$m \times 10^n$$
> where m is a number greater than or equal to 1 and less than 10 ($1 \leq m < 10$) and n is an integer.

For any given number, the m is obtained by placing the decimal point so that there is exactly one nonzero digit to its left. The integer n is then the number of places that the decimal point must be moved from its position in m to its original position; it is positive if the point must be moved to the right and negative if the point must be moved to the left. Thus

$5.3 = 5.3 \times 10^0$ Decimal point in 5.3 must be moved 0 places.

$87 = 8.7 \times 10^1 = 8.7 \times 10$ Decimal point in 8.7 must be moved 1 place to the right to get 87.

$68{,}000 = 6.8 \times 10^4$ Decimal point in 6.8 must be moved 4 places to the *right* to get 68,000.

$0.49 = 4.9 \times 10^{-1}$ Decimal point in 4.9 must be moved 1 place to the *left* to get 0.49.

$0.072 = 7.2 \times 10^{-2}$ Decimal point in 7.2 must be moved 2 places to the *left* to get 0.072.

$0.0003875 = 3.875 \times 10^{-4}$ Decimal point in 3.875 must be moved 4 places to the *left* to get 0.0003875.

EXAMPLE 7 Writing numbers in scientific notation
Write in scientific notation:

a. 270,000,000 **b.** 0.000 000 000 000 000 000 019 9

SOLUTION

a. $270{,}000{,}000 = 2.7 \times 10^8$
b. $0.000\,000\,000\,000\,000\,000\,019\,9 = 1.99 \times 10^{-23}$

PROBLEM 7
Write in scientific notation:

a. 350,000
b. 0.000000378

EXAMPLE 8 Scientific notation to standard decimal notation
Write in standard decimal notation:

a. 2.5×10^{10} **b.** 7.4×10^{-6}

SOLUTION

a. $2.5 \times 10^{10} = 25{,}000{,}000{,}000$ **b.** $7.4 \times 10^{-6} = 0.0000074$

PROBLEM 8
Write in standard decimal notation:

a. 3.5×10^5
b. 8.2×10^{-3}

We can use the laws of exponents when working with numbers in scientific notation. Example 9 shows you how.

EXAMPLE 9 Calculations in scientific notation
Do the following calculations, and write the answers in scientific notation:

a. $(5 \times 10^4) \times (9 \times 10^{-7})$ **b.** $\dfrac{6 \times 10^5}{3 \times 10^{-4}}$

SOLUTION

a. $(5 \times 10^4) \times (9 \times 10^{-7}) = (5 \times 9) \times (10^4 \times 10^{-7})$
$= 45 \times 10^{4-7}$
$= 45 \times 10^{-3}$

Remember in scientific notation, $m \times 10^n$, m must be less than 10, which 45 is not. The next step shows how to write 45 in scientific notation.

$= \overbrace{4.5 \times 10^1} \times 10^{-3}$
$= 4.5 \times 10^{1-3}$
$= 4.5 \times 10^{-2}$

b. $\dfrac{6 \times 10^5}{3 \times 10^{-4}} = \dfrac{6}{3} \times \dfrac{10^5}{10^{-4}}$
$= 2 \times 10^{5-(-4)}$
$= 2 \times 10^9$ or 2.0×10^9

PROBLEM 9
Do the calculations and write the answers in scientific notation:

a. $(3 \times 10^3)(8 \times 10^{-7})$
b. $\dfrac{4 \times 10^6}{2 \times 10^{-5}}$

Answers to PROBLEMS

7. a. 3.5×10^5 **b.** 3.78×10^{-7}
8. a. 350,000 **b.** 0.0082
9. a. 2.4×10^{-3}
   b. 2.0×10^{11} or 2×10^{11}

Calculator Corner

Scientific Notation

Consult the *Keystroke Guide* or your calculator manual to determine how to enter a number written in scientific notation. To do Example 7(a) with a TI-83 Plus, key in 270 000 000 **MODE** and use your ▶ button to select **Sci**. Press **ENTER**. Go back to the Home screen **2nd** **MODE** and press **ENTER**. The answer is 2.7E8, which means 2.7×10^8.

Write the given calculator answers in scientific notation.

1. 3.8 E 5 **2.** 6.4 E 2 **3.** 1.56 E −2 **4.** 2.45 E −5

```
270000000
            2.7E8
```

Exercises 1.3

⟨A⟩ Natural-Number Exponents In Problems 1–10, evaluate.

1. -4^2 **2.** $(-4)^2$ **3.** $(-5)^2$ **4.** -5^2 **5.** -5^3

6. $(-5)^3$ **7.** $(-6)^4$ **8.** -6^4 **9.** -2^5 **10.** $(-2)^5$

⟨B⟩ Negative and Zero Exponents In Problems 11–20, write the expression given as a fraction in simplified form and without negative exponents.

11. 4^{-2} **12.** $(-2)^{-3}$ **13.** x^{-5} **14.** a^{-8} **15.** $2x^{-6}$

16. $-5y^{-7}$ **17.** $m^2 n^{-1}$ **18.** $b^{-3} c$ **19.** $6x^2 y^{-1} z^{-3}$ **20.** $4a^{-5} b c^{-2}$

⟨C⟩ Multiplication and Division with Exponents In Problems 21–50, perform the indicated operations and simplify.

21. $2^{-4} \cdot 2^{-2}$ **22.** $4^{-1} \cdot 4^{-2}$ **23.** $(3x^6) \cdot (4x^{-4})$

24. $(4y^7) \cdot (5y^{-3})$ **25.** $(-3y^{-3}) \cdot (5y^5)$ **26.** $(-5x^{-7}) \cdot (4x^8)$

27. $(-4a^3) \cdot (-5a^{-8})$ **28.** $(-2b^4) \cdot (-3b^{-7})$ **29.** $(3x^{-5}) \cdot (5x^2 y)(-2xy^2)$

30. $(4y^{-6}) \cdot (5xy^4)(-2x^2 y)$ **31.** $(-2x^{-3} y^2)(3x^{-2} y^3)(4xy)$ **32.** $(-3xy^{-5})(4x^2 y)(2x^3 y^2)$

33. $(4a^{-2} \cdot b^{-3})(5a^{-1} b^{-1})(-2ab)$ **34.** $(2a^{-5} \cdot b^{-2})(3a^{-1} b^{-1})(5ab)$ **35.** $(6a^{-3} \cdot b^3)(5a^2 b^2)(-ab^{-5})$

36. $(7a^6 \cdot b^{-6})(2ab^5)(-a^{-7} b)$ **37.** $\dfrac{8x^7}{4x^3}$ **38.** $\dfrac{8a^3}{4a^2}$

39. $\dfrac{-8a^4}{-16a^2}$ **40.** $\dfrac{-9y^5}{-18y^2}$ **41.** $\dfrac{12x^5 y^3}{-6x^2 y}$ **42.** $\dfrac{18x^6 y^2}{-9xy}$

43. $\dfrac{-6x^{-4}}{12x^{-5}}$ **44.** $\dfrac{8x^{-3}}{4x^{-4}}$ **45.** $\dfrac{-14a^{-5}}{-21a^{-2}}$ **46.** $\dfrac{-2a^{-6}}{-6a^{-3}}$

47. $\dfrac{-27a^{-4}}{-36a^{-4}}$ **48.** $\dfrac{-5x^{-3}}{10x^{-3}}$ **49.** $\dfrac{3a^{-2} \cdot b^5}{2a^4 b^2}$ **50.** $\dfrac{x^{-3} \cdot y^6}{x^4 \cdot y^3}$

Answers to CALCULATOR CORNER

1. 3.8×10^5 **2.** 6.4×10^2

3. 1.56×10^{-2} **4.** 2.45×10^{-5}

D Raising a Product and a Quotient to a Power

In Problems 51–60, simplify the expression given and write your answer without negative exponents.

51. $(2x^3y^{-2})^3$
52. $(3x^2y^{-3})^2$
53. $(2x^{-2}y^3)^2$
54. $(3x^{-4}y^4)^3$
55. $(-3x^3y^2)^{-3}$
56. $(-2x^5y^4)^{-4}$
57. $(x^{-6}y^{-3})^2$
58. $(y^{-4}z^{-3})^5$
59. $(x^{-4}y^{-4})^{-3}$
60. $(y^{-5}z^{-3})^{-4}$

In Problems 61–70, simplify.

61. $\left(\dfrac{a}{b^3}\right)^2$
62. $\left(\dfrac{a^2}{b}\right)^3$
63. $\left(\dfrac{-3a}{2b^2}\right)^{-3}$
64. $\left(\dfrac{-2a^2}{3b^0}\right)^{-2}$
65. $\left(\dfrac{a^{-4}}{b^2}\right)^{-2}$
66. $\left(\dfrac{a^{-2}}{b^3}\right)^{-3}$
67. $\left(\dfrac{x^5}{y^{-2}}\right)^{-3}$
68. $\left(\dfrac{x^6}{y^{-3}}\right)^{-2}$
69. $\left(\dfrac{x^{-4}y^3}{x^5y^5}\right)^{-3}$
70. $\left(\dfrac{x^{-2}y^0}{x^7y^2}\right)^{-2}$

E Scientific Notation

In Problems 71–74, write the numbers in scientific notation.

71. 268,000,000 (U.S. population in the year 2000)
72. 1,900,000,000 (dollars spent on waterbeds and accessories in 1 year)
73. 0.00024 (probability of four of a kind in poker)
74. 0.00000009 (wavelength of an X-ray in centimeters)

In Problems 75–78, write the numbers in decimal notation.

75. 8×10^6 (bagels eaten per day in the United States)
76. $\$6.85 \times 10^9$ (estimated combined wealth of the five wealthiest women)
77. 2.3×10^{-1} (kilowatt per hour used by your TV)
78. 4×10^{-11} (joule of energy released by splitting one uranium atom)

❯❯❯ Applications

In Problems 79–81, write your answer in scientific notation.

79. *Astronomy* The width of the asteroid belt is 2.8×10^8 kilometers (km). The speed of *Pioneer 10* in passing through this belt was 1.4×10^5 km/hr. Thus, *Pioneer 10* took
$$\frac{2.8 \times 10^8}{1.4 \times 10^5} \text{ hr}$$
to go through the belt. How many hours was that?

80. *Earth's mass* The mass of Earth is 6×10^{21} tons. The sun is about 300,000 times as massive. Thus, the mass of the sun is $(6 \times 10^{21}) \times 300{,}000$ tons. How many tons is that?

81. *Speed of light* The velocity of light can be measured by knowing the distance from the sun to Earth (1.47×10^{11} meter, m) and the time it takes for sunlight to reach Earth (490 sec). Thus, the velocity of light is
$$\frac{1.47 \times 10^{11}}{490} \text{ m/sec}$$
How many meters per second is that?

82. *Oil reserves* Oil reserves in the United States are estimated to be 3.5×10^{10} barrels. Production amounts to 3.2×10^9 barrels per year. At this rate, how long would U.S. oil reserves last? (Give your answer to the nearest year.)

83. *Oil reserves* The world's oil reserves are estimated to be 6.28×10^{11} barrels. Production is 2.0×10^{10} barrels per year. At this rate, how long would the world's oil reserves last? (Give your answer to the nearest year.)

84. *Energy from the sun* Scientists have estimated that the total energy received from the sun each minute is 1.02×10^{19} calories. The surface area of the Earth is 5.1×10^8 km² (square kilometers), so the average amount of energy received per square centimeter of Earth's surface per minute (the solar constant) is
$$\frac{1.02 \times 10^{19}}{(5.1 \times 10^8) \times 10^{10}} \qquad (1 \text{ km}^2 = 10^{10} \text{ cm}^2)$$
How many calories per square centimeter is that?

85. *U.S. debt* The outstanding historical debt for the United States in 1940 was approximately $43,000,000,000. By the year 2005 that debt had grown to $7,900,000,000,000. To the nearest hundred, how many times greater is the debt in 2005 as compared to 1940?

86. *Debt* The Bureau of the Public Debt may accept gifts donated to the United States Government to reduce debt held by the public. The total gifts given for the fiscal years 1997 and 2005, were approximately $956,000 and $1,460,000, respectively. To the nearest tenth, how many times greater were the gifts donated in 2005 as compared to those in 1997?

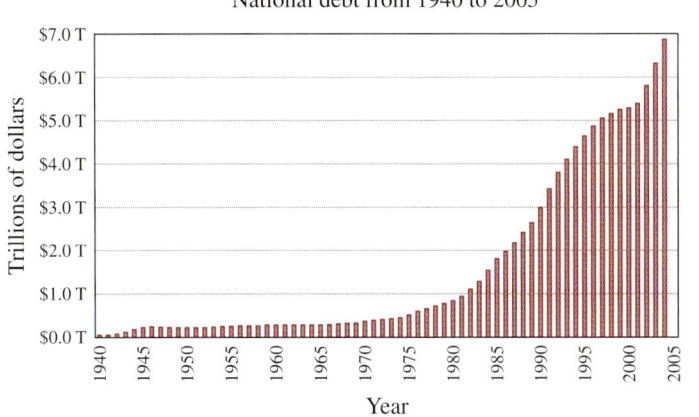

National debt from 1940 to 2005

Source: http://tinyurl.com/2czpzv

>>> Using Your Knowledge

Scientific Calculators If you have a scientific calculator, and you multiply 9,800,000 by 4,500,000, the display may show

$$\boxed{4.41 \quad 13}$$

This means that the answer is 4.41×10^{13}.

87. The display on a calculator shows

$$\boxed{3.34 \quad 5}$$

Write this number in scientific notation.

88. The display on a calculator shows

$$\boxed{-9.97 \quad -6}$$

Write this number in scientific notation.

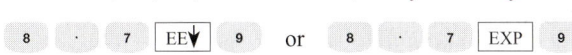

89. To enter large or small numbers in a calculator with scientific notation, you must write the number using this notation first. To enter the number 8,700,000,000 in the calculator, you must know that 8,700,000,000 is 8.7×10^9, *then* you can key in

8 · 7 [EE▼] 9 or 8 · 7 [EXP] 9

The calculator displays

$$\boxed{8.7 \quad 09}$$

a. What would the display read when you enter the number 73,000,000,000?

b. What would the display read when you enter the number 0.000000123?

>>> Write On

90. Why is a^0 defined as 1?

91. Why is $0^n = 0$ for "n" a natural number?

92. Why is 0^0 not defined?

93. Why is $x^m x^n = x^{m+n}$, using the concept of a factor?

94. Why is $\frac{x^m}{x^n} = x^{m-n}$, using the concept of a factor?

95. Why is $(x^m)^n = x^{m \cdot n}$, using the concept of a factor?

⟩⟩⟩ Concept Checker

Fill in the blank(s) with the correct word(s), phrase, or mathematical statement.

96. When multiplying two numbers with the same base, the _____ rule says to keep the base and _____ the exponents.

97. A number written as 5.4×10^{-3} is said to be in _____.

98. When simplifying $(x^{-2})^{-5}$, use the _____ rule which says to keep the base and _____ the exponents.

99. In the number 6^4, 6 is referred to as the _____ and 4 is the _____.

100. When dividing two numbers with the same base, the _____ rule says to keep the base and _____ the exponents.

base	quotient
exponent	power
exponential notation	add
zero exponent	subtract
negative exponent	multiply
product	scientific notation

⟩⟩⟩ Mastery Test

Multiply and simplify:

101. $(-3x^4y^2)(3x^{-7}y)$

102. $(-2x^{-5}y^{-3})(-4xy)$

Divide and simplify:

103. $\dfrac{45y^4}{-15y^{-7}}$

104. $\dfrac{-6x^{-8}}{30x^6}$

Evaluate:

105. $(-5)^{-4}$

106. -2^6

Simplify:

107. $(-2x^4y^{-5})^3$

108. $(-3x^{-4}y^5)^{-2}$

Write as a fraction:

109. $(-3)^{-5}$

110. $(-x)^{-5}$

Simplify:

111. $\left(\dfrac{2x^{-4}y^3}{3y^5}\right)^{-2}$

112. $\left(-\dfrac{5x^{-5}y^7}{7y^3}\right)^{-2}$

Write in scientific notation:

113. 387,000,000

114. $(4 \times 10^{-5}) \times (6 \times 10^2)$

⟩⟩⟩ Skill Checker

Perform the indicated operations.

115. $(-4)12$

116. $-11(-9)$

117. -6^2

118. $(-6)^2$

119. $2 - 13$

120. $-4 + (-15)$

121. $\dfrac{-19}{-19}$

122. $\dfrac{0}{-17}$

123. $-20 - 5 + 4$

124. $3 + (-5) - 2$

125. $(-3)(0)2$

126. $-2(4)10$

1.4 Algebraic Expressions and the Order of Operations

Objectives

A Evaluate numerical expressions with grouping symbols.

B Evaluate expressions using the correct order of operations.

C Evaluate algebraic expressions.

D Use the distributive property to simplify expressions.

E Simplify expressions by combining like terms.

F Simplify expressions by removing grouping symbols and combining like terms.

To Succeed, Review How To . . .

1. Perform the four fundamental operations using signed numbers (pp. 18–27).
2. Calculate powers of integers (pp. 38–40).
3. Use the identity for multiplication (p. 28).

Getting Started

Swimming and Your Heart Rate

How do you calculate your ideal heart rate when swimming? One way is to subtract your age from 205 and multiply by 0.70.

This means that if you are a years old, you would subtract your age a from 205 and multiply by 0.70.

Should you subtract your age from 205 first and then multiply by 0.70, or multiply your age by 0.70 first, and then subtract from 205? In algebra we can make our meaning clear by using parentheses. The formula is written as $(205 - a) \cdot 0.70$ to indicate that the subtraction should be done first. (Try it by substituting your age for a.)

In the "Getting Started" in Section 1.1, it was mentioned that in algebra we use letters or **variables** to represent different numbers, and when a letter stands for just one number it is called a **constant**. If an expression consists only of numbers with operations and grouping symbols, it is called a **numerical expression**. However, when an expression consists of variables, numbers, and operation symbols it is called an **algebraic expression**. **Terms** in an algebraic expression are separated by + or − signs. In the algebraic expression "$4x^2 - 2x + 8$", there are three terms; $4x^2$, $-2x$, and the constant 8.

In Sections A and B, we will learn how to evaluate *numerical expressions* with grouping symbols and the correct order of operations. In Section C, we will use those skills to help us evaluate an *algebraic expression*.

A > Evaluate Numerical Expressions with Grouping Symbols

When an expression consists only of numbers with operations and grouping symbols, it is called a **numerical expression**. Some examples of numerical expressions are

$$4 + 3 - 2 \qquad 5(9 - 6) \qquad 12 - \frac{6}{3}$$

In algebra and arithmetic, *parentheses* ( ) are grouping symbols used to indicate which operations are to be performed first. Square brackets [ ] and braces { } are also used as grouping symbols; they can be used in the same manner as parentheses.

$$4 \cdot (3 + 2), \qquad 4 \cdot [3 + 2], \qquad \text{and} \qquad 4 \cdot \{3 + 2\}$$

all mean that we must first add 3 and 2 and then multiply this sum by 4. The expressions $4 \cdot (3 + 2)$ and $(4 \cdot 3) + 2$ have *different* meanings. In the first expression we add 3 and 2 first, while in the second we multiply 4 by 3 first. Thus $4 \cdot (3 + 2) = 4 \cdot (5) = 20$, but $(4 \cdot 3) + 2 = (12) + 2 = 14$. Hence, $4 \cdot (3 + 2) \neq (4 \cdot 3) + 2$.

EXAMPLE 1 Evaluating expressions containing parentheses
Evaluate:
a. $(-4 \cdot 5) + 6$
b. $-4 \cdot (5 + 6)$
c. $-48 \div (4 \cdot 3)$
d. $(-48 \div 4) \cdot 3$

SOLUTION Perform the operations inside the parentheses *first*.

a. $(-4 \cdot 5) + 6 = -20 + 6$ Multiply -4 and 5 first.
$= -14$

b. $-4 \cdot (5 + 6) = -4 \cdot 11$ Add 5 and 6 first.
$= -44$

c. $-48 \div (4 \cdot 3) = -48 \div 12$ Multiply 4 and 3 first.
$= -4$

d. $(-48 \div 4) \cdot 3 = -12 \cdot 3$ Divide -48 by 4 first.
$= -36$

PROBLEM 1
Evaluate:
a. $(-6 \cdot 4) + 7$ b. $-6 \cdot (4 + 7)$
c. $-36 \div (3 \cdot 4)$ d. $(-36 \div 3) \cdot 4$

If we have more than one set of grouping symbols, we perform the operations inside the innermost grouping symbols first.

$[2 \cdot (88 + 14)] + 12 = [2 \cdot (102)] + 12$ Add 88 and 14 inside the parentheses first.
$= 204 + 12$
$= 216$

EXAMPLE 2 Evaluating expressions containing grouping symbols
Evaluate:
a. $[-5 \cdot (6 + 4)] + 9$
b. $[-10 \cdot (8 - 3)] - 9$

SOLUTION

a. $[-5 \cdot (6 + 4)] + 9 = [-5 \cdot 10] + 9$ Add 6 and 4 first.
$= -50 + 9$
$= -41$

b. $[-10 \cdot (8 - 3)] - 9 = [-10 \cdot 5] - 9$ Subtract 3 from 8 first.
$= -50 - 9$
$= -59$

PROBLEM 2
Evaluate:
a. $[-8 \cdot (7 + 2)] + 8$
b. $[-10 \cdot (9 - 6)] - 9$

In some cases we have to evaluate expressions in which a bar is used to indicate division. For instance, if we wish to convert a temperature given in degrees Fahrenheit to degrees Celsius, we have to evaluate the expression

$$\frac{5 \cdot (F - 32)}{9}$$

where F represents the temperature in degrees Fahrenheit. If the temperature is 77°F, the corresponding Celsius temperature is calculated as follows:

$$\frac{5 \cdot (77 - 32)}{9} = \frac{5 \cdot (45)}{9}$$
$$= \frac{225}{9}$$
$$= 25$$

You can also do this by dividing 45 by 9 first and then multiplying the result, 5, by 5, to obtain 25.

The corresponding temperature is 25°C.

Answers to PROBLEMS

1. a. -17 b. -66 c. -3 d. -48
2. a. -64 b. -39

EXAMPLE 3 An application using grouping symbols

In September 1933, a freak heat flash struck the city of Coimbra in Portugal. On this day, the temperature rose to 158°F for 120 sec. How many degrees Celsius is that?

SOLUTION   In this case, $F = 158$; thus, the Celsius temperature is given by

$$\frac{5 \cdot (158 - 32)}{9} = \frac{5 \cdot (126)}{9}$$
$$= \frac{630}{9}$$
$$= 70$$

You can also do this by dividing 126 by 9 first and then multiplying the result, 14, by 5, to obtain 70.

The corresponding Celsius temperature is 70°C.

PROBLEM 3

Evaluate:
$$\frac{5 \cdot (149 - 32)}{9}$$

B > The Order of Operations

If an expression does not contain parentheses or brackets, we must establish the order in which operations are to be performed. For example, the expression $6 + 9 \div 3$ might be evaluated in two ways:

$6 + 9 \div 3$	Divide 9 by 3.	$6 + 9 \div 3$	Add $6 + 9$.
$6 + 3$	Add.	$15 \div 3$	Divide by 3.
9		5	

The first way resulted in a value of 9, and the second way resulted in 5.

To avoid this ambiguity, we agree to perform any sequence of operations from left to right and in the following order.

PROCEDURE

Order of Operations

- **P**   1. Do the operations inside parentheses (or other grouping symbols) starting with the innermost grouping symbols. When simplifying fractions, do the operations above and below the fraction bar and then simplify the fraction, if possible.
- **E**   2. Evaluate all exponential expressions.
- **(MD)** 3. Perform multiplications and divisions as they occur from left to right.
- **(AS)** 4. Perform additions and subtractions as they occur from left to right.

The letters to the left in the procedure are used to help remember this order of operations. One sentence used to remember the letters is, "**P**lease **E**xcuse **M**y **D**ear **A**unt **S**ally." The multiplications and divisions are equally important so they are done as they appear from left to right in the expression. Thus, to evaluate $24 \div 6 \cdot 2 + (7 - 9) - 1 + 2^3$ we let **PE(MD)(AS)** guide us.

$$24 \div 6 \cdot 2 + (7 - 9) - 1 + 2^3 = 24 \div 6 \cdot 2 + (-2) - 1 + 2^3 \quad \text{Parentheses } (7 - 9) = -2$$
$$= 24 \div 6 \cdot 2 + (-2) - 1 + 8 \quad \text{Exponents } 2^3 = 8$$
$$= 4 \cdot 2 + (-2) - 1 + 8 \quad \text{(MD) Divide } 24 \div 6 = 4.$$
$$= 8 + (-2) - 1 + 8 \quad \text{(MD) Multiply } 4 \cdot 2 = 8.$$
$$= 6 - 1 + 8 \quad \text{(AS) Add } 8 + (-2) = 6.$$
$$= 5 + 8 \quad \text{(AS) Subtract } 6 + (-1) = 5.$$
$$= 13 \quad \text{(AS) Add } 5 + 8 = 13.$$

Answers to PROBLEMS

3. 65

EXAMPLE 4 Order of operations

Evaluate:

a. $25 \div 5 - (6 - 10) - 8^2$ **b.** $\dfrac{3(7-9)}{-2} - \dfrac{18}{-6}$ **c.** $-6^2 + \dfrac{(4-8)}{2} + 10 \div 5$

SOLUTION

a. $25 \div 5 - (6 - 10) - 8^2$
$= 25 \div 5 - (-4) - 8^2$ **P** Perform the operation inside the parentheses.
$= 25 \div 5 - (-4) - 64$ **E** Exponent $8^2 = 64$.
$= 5 - (-4) - 64$ **(MD)** Perform multiplications and divisions from left to right.
$= 9 - 64$ **(AS)** Perform addition and subtraction from left to right. $5 - (-4) = 9$
$= -55$

b. $\dfrac{3(7-9)}{-2} - \dfrac{18}{-6}$ Simplify the fractions first.

$= \dfrac{3(-2)}{-2} - \dfrac{18}{-6}$ **P** Perform the operation inside the parentheses. (no exponents)

$= \dfrac{-6}{-2} - \dfrac{18}{-6}$ **(MD)** Perform multiplication and division from left to right. $\dfrac{-6}{-2} = 3$ and $\dfrac{18}{-6} = -3$

$= 3 - (-3)$ **(AS)** Perform addition and subtraction from left to right.
$= 6$

c. $-6^2 + \dfrac{4-8}{2} + 10 \div 5$

$= -6^2 + \dfrac{-4}{2} + 10 \div 5$ **P** Perform the operation inside the parentheses.

$= -36 + \dfrac{-4}{2} + 10 \div 5$ **E** Exponent $-6^2 = -36$.

$= -36 + (-2) + 2$ **(MD)** Perform multiplications and divisions from left to right.

$= -38 + 2$ **(AS)** Perform additions and subtractions from left to right. $-36 + (-2) = -38$

$= -36$

PROBLEM 4

Evaluate:

a. $14 \cdot 3 - (5 + 4) - 2^3$

b. $\dfrac{-24}{3} - \dfrac{2(8-12)}{-4}$

c. $-7^2 + \dfrac{2-8}{2} + 20 \div 5$

Calculator Corner

Calculator Does Order of Operations

Some calculators automatically follow the order of operations. Thus if you key in 3 [×] 4 [+] 5, the calculator multiplies 3 by 4 *first,* and then adds 5 to obtain 17 as shown. On the other hand, if you enter 3 [+] 4 [×] 5, the calculator does **not** add 3 and 4 first, but rather *multiplies* 4 by 5 first and then adds 3 to obtain 23.

As before, if you want to add 3 and 4 first, you have to use parentheses and key in [(] 3 [+] 4 [)] [×] 5. The result is 35 as shown.

```
3*4+5
           17
3+4*5
           23
(3+4)*5
           35
```

C › Evaluating Algebraic Expressions

When an expression consists of variables, numbers, operations, and grouping symbols it is called an **algebraic expression**. Terms in an algebraic expression are separated by $+$ and $-$ signs. Some examples are

$$2x + 3 \qquad \dfrac{7}{5} - \dfrac{6z}{5} \qquad 4x^2 - 2x + 8 \qquad \sqrt{y} + 9.$$

Answers to PROBLEMS

4. **a.** 25   **b.** -10 **c.** -48

Without explanation the algebraic expression $5x$ has no real application. However, if we know that we have to ride the same bus to work for 5 days and the cost of the round trip, (x), will vary depending on whether we ride the regular bus or the express bus, then 5 times x or $5x$ could represent the cost of the bus fare for the 5 days.

Suppose we decide to ride the regular bus, which costs $2 for one round trip. Then we could evaluate the cost of the bus fare for the week by using the algebraic expression $5x$ and replacing the x with $2. Thus, the cost of the bus fare for the 5 days would be $5(\$2)$ or $10. $10 is called the **value** of the algebraic expression. What if we choose the express bus, which costs $3 for a round trip? Then the cost of the 5-day bus fare becomes $5(\$3)$ or $15. You can see that the value of the expression $5x$ varies depending on the number used to replace x. This is a very simple example of how to **evaluate an algebraic expression.** To evaluate the expressions in Example 5, we need some of the skills learned in simplifying with grouping symbols and the correct order of operations.

EXAMPLE 5 Evaluating an algebraic expression

a. $(n - 2)180$ gives the sum of the measures of the angles in a polygon of n sides. Find the sum of the measures of the angles if a polygon has 5 sides.

b. Evaluate the algebraic expression
$$2x - 3(y^2 + 1)$$
if $x = -4$ and $y = 5$.

SOLUTION

a. Replace n with 5 and perform the correct order of operations.
$$(n - 2)180 = (5 - 2)180$$
$$= (3)180 \quad \text{Simplify inside parentheses.}$$
$$= 540 \quad \text{Multiply.}$$
Thus, the sum of the measures of the angles in a polygon with 5 sides is 540°.

b. Replace x with -4 and y with 5 and perform the correct order of operations.
$$2x - 3(y^2 + 1) = 2(-4) - 3(5^2 + 1)$$
$$= 2(-4) - 3(25 + 1) \quad \text{Evaluate exponent inside parentheses.}$$
$$= 2(-4) - 3(26) \quad \text{Add inside parentheses.}$$
$$= -8 - 78 \quad \text{Multiply.}$$
$$= -86 \quad \text{Subtract.}$$
The value of the algebraic expression is -86.

PROBLEM 5

a. $\dfrac{5(F - 32)}{9}$ gives the Celsius temperature when the Fahrenheit temperature, F, is known. Find the Celsius temperature when $F = 50$.

b. Evaluate the algebraic expression
$$3x^2 - (y + 11) + 10 \div 2$$
if $x = -1$ and $y = 7$.

D › Using the Distributive Property to Simplify

In algebra the distributive property, $a(b + c) = ab + ac$, is used to remove parentheses in expressions such as $3(x + 5)$ or $4(x - 7)$, where x is a real number. Thus,
$$3(x + 5) = 3x + 3 \cdot 5 = 3x + 15$$
and
$$4(x - 7) = 4x - 4 \cdot 7 = 4x - 28$$

> **DISTRIBUTIVE PROPERTY OF MULTIPLICATION OVER ADDITION**
> If a, b, and c are real numbers, then
> $$a(b + c) = ab + ac.$$

Answers to PROBLEMS

5. a. 10°C   b. -10

EXAMPLE 6 Removing parentheses

Remove the parentheses (simplify) using the distributive property.

a. $-2(x + 8)$
b. $0.5(7 - y)$

SOLUTION

a. $-2(x + 8) = (-2 \cdot x) + (-2 \cdot 8)$
$= -2x + (-16)$
$= -2x - 16$ *Recall that $a + (-b) = a - b$.*

b. $0.5(7 - y) = 0.5 \cdot 7 - 0.5y$
$= 3.5 - 0.5y$

PROBLEM 6

Simplify:

a. $-3(a + b)$
b. $0.2(6 - b)$

Expressions of the form $-(a + b)$ or $-(a - b)$, require special consideration. We first recall the following.

MULTIPLICATIVE IDENTITY For any real number a, $a = 1 \cdot a$.

Any real number has an additive inverse and the additive inverse of a is $-a$, so the additive inverse of $1 \cdot a$ is $-1 \cdot a$.

MULTIPLICATION BY -1 For any real number a, $-a = -1 \cdot a$.

Hence,
$$-(a + b) = -1 \cdot (a + b)$$
$$= (-1)(a) + (-1)(b)$$
$$= -a - b$$

PROCEDURE
Additive Inverse of a Sum
$$-(a + b) = -a - b$$

Similarly,
$$-(a - b) = -1 \cdot (a - b)$$
$$= -1 \cdot [a + (-b)]$$
$$= (-1)(a) + (-1)(-b)$$
$$= -a + b$$

PROCEDURE
Additive Inverse of a Difference
$$-(a - b) = -a + b$$

These rules tell us that to remove the parentheses in an expression preceded by a negative sign, we *change the sign of every term inside the parentheses* (find each term's additive inverse) or, equivalently, *multiply each term inside the parentheses by -1*.

Answers to PROBLEMS

6. **a.** $-3a - 3b$ **b.** $1.2 - 0.2b$

EXAMPLE 7 Removing parentheses

Remove the parentheses (simplify) using the distributive property.

a. $-(x - 2)$ **b.** $-(ab + 3)$

SOLUTION

a. $-(x - 2) = -1 \cdot (x - 2)$
$= -1 \cdot x + (-1)(-2)$
$= -x + 2$

Changing the signs inside the parentheses in $-(x - 2)$ will immediately yield

$-(x - 2) = -x + 2$
(change sign)

b. $-(ab + 3) = -1 \cdot (ab + 3) = -1 \cdot ab + (-1)(3)$
$= -ab + (-3)$
$= -ab - 3$

Changing signs inside the parentheses will immediately yield the answer $-ab - 3$.

PROBLEM 7

Simplify:

a. $-(y - 6)$ **b.** $-(xy + 7)$

We can summarize this discussion by the following two facts.

PROCEDURE

Removing Parentheses Using the Distributive Property

1. If the factor in front of the parentheses has no written sign, multiply each term inside the parentheses by this factor; that is,
$$a(b - c + d - e) = ab - ac + ad - ae$$

2. If the factor in front of the parentheses is preceded by a negative sign, multiply each term inside the parentheses by this factor and change the sign of each of these terms; that is,
$$-a(b - c + d - e) = -ab + ac - ad + ae$$

EXAMPLE 8 Removing parentheses

Remove the parentheses (simplify) using the distributive property.

a. $4(x - 2y + 3)$ **b.** $-5x(2x + y - z)$
c. $0.4(-3x + 2y - 7z - 8)$ **d.** $-0.5x(y + 3z - 5)$

SOLUTION

a. $4(x - 2y + 3) = 4x - 8y + 12$
b. $-5x(2x + y - z) = -10x^2 - 5xy + 5xz$
c. $0.4(-3x + 2y - 7z - 8) = -1.2x + 0.8y - 2.8z - 3.2$
d. $-0.5x(y + 3z - 5) = -0.5xy - 1.5xz + 2.5x$

PROBLEM 8

Simplify:

a. $5(a - 3b + 4)$

b. $-3(4x + y - z)$

c. $0.5(-2x + 3y - 6z - 4)$

d. $-0.3x(y + 2z - 4)$

Answers to PROBLEMS

7. **a.** $-y + 6$ **b.** $-xy - 7$
8. **a.** $5a - 15b + 20$ **b.** $-12x - 3y + 3z$ **c.** $-x + 1.5y - 3z - 2$ **d.** $-0.3xy - 0.6xz + 1.2x$

NOTE

There are four cases of removing parentheses to remember.

1. $(x + 2) = x + 2$ Drop ( ).
2. $-(x + 2) = -x - 2$ Change signs.
3. $3(x + 2) = 3x + 6$ Distribute 3.
4. $-3(x + 2) = -3x - 6$ Distribute -3.

E › Combining Like Terms

Earlier in the section it was mentioned that **terms** in an algebraic expression are separated by $+$ or $-$ signs. In the algebraic expression "$4x^2 - 2x + 8$" there are three terms; $4x^2$, $-2x$, and the constant 8.

Suppose we wish to simplify $3x + 2(x + 5)$. We start by using the distributive property to simplify $2(x + 5)$ and obtain

$$3x + 2(x + 5) = 3x + 2x + 10$$

The terms $3x$ and $2x$ are called *like terms*. They differ only in their numerical factors **(coefficients)**. Similarly, $-3y$ and $5y$ are like terms, and $9z^2$ and $-3z^2$ are like terms. In general, we have the following definition.

LIKE TERMS

Constant terms or terms with exactly the same variable factors are called **similar** or **like** terms.

NOTE

Like terms differ only in their *numerical* coefficients (the numbers being multiplied by the variables).

We can *combine* like terms by using a variation of the distributive property. As you recall, the distributive property states that, for any real numbers a, b, and c, $a(b + c) = ab + ac$ and $a(b - c) = ab - ac$.

Using the commutative property of multiplication, we can rewrite the two distributive properties as follows.

PROCEDURE

Distributive Properties for Like Terms

$$ba + ca = (b + c)a$$
$$ba - ca = (b - c)a$$

Now

$$3x + 2x = (3 + 2)x = 5x$$

Similarly,

$$7z^2 - 5z^2 = (7 - 5)z^2 = 2z^2$$

and

$$8xy + 3xy - 2xy = (8 + 3 - 2)xy$$
$$= 9xy$$

Let's consider adding $x + x$. The coefficient of x is understood to be 1 so we can write, $x + x = 1 \cdot x + 1 \cdot x = (1 + 1)x = 2x$. Also, if a set of parentheses is preceded by a plus sign, we can remove the parentheses and combine any like terms. Using the commutative and associative properties,

$$3x + (2 + 5x) = 3x + 2 + 5x$$
$$= 8x + 2$$

NOTE You can combine like terms by adding or subtracting their coefficients.

We use this idea in Example 9.

EXAMPLE 9 Combining like terms
Simplify:

a. $5x + 2(x - 4)$ b. $-3(x^2 + 5) - 2x + x^2$ c. $5xy - y(x + 1) + (y + 3)$

SOLUTION

a. $5x + 2(x - 4) = 5x + 2x - 8$ Distributive property
$ = (5x + 2x) - 8$ Associative property
$ = 7x - 8$ Combine like terms.

b. $-3(x^2 + 5) - 2x + x^2 = -3x^2 - 15 - 2x + x^2$ Distributive property
$ = -3x^2 + x^2 - 2x - 15$ Commutative property
$ = (-3x^2 + x^2) - 2x - 15$ Group like terms.
$ = -2x^2 - 2x - 15$ Combine like terms.

c. $5xy - y(x + 1) + (y + 3) = 5xy - xy - y + y + 3$ Distributive property
$ = (5xy - xy) + (-y + y) + 3$ Group like terms.
$ = 4xy + 3$ Combine like terms.

PROBLEM 9
Simplify:

a. $8y + 3(y - 4)$

b. $-4(y^2 + 2) - 3y + 5y^2$

c. $6xy - 3x(y + 2) + 6(x + 8)$

F> Removing Other Grouping Symbols

To avoid confusion when grouping symbols occur within other grouping symbols, we do not write $((x + 5) + 3)$. Instead, we may use a different grouping symbol, the brackets [ ], and write $[(x + 5) + 3]$. To simplify (combine like terms) in such expressions, the innermost grouping symbols are removed first. This procedure is illustrated in the next example.

EXAMPLE 10 Removing other grouping symbols
Remove the grouping symbols and simplify:

a. $[(4x^2 - 1) + (2x + 5)] - [(x - 2) + (3x^2 - 3)]$
b. $[4(x + 5) - 15] + 5[3 + 2(6 - x)]$

SOLUTION
We first remove the innermost parentheses and then add like terms. Thus

a. $[(4x^2 - 1) + (2x + 5)] - [(x - 2) + (3x^2 - 3)]$
$= [4x^2 - 1 + 2x + 5] - [x - 2 + 3x^2 - 3]$ Remove parentheses.
$= [4x^2 + 2x + 4] - [3x^2 + x - 5]$ Combine like terms.
$= 4x^2 + 2x + 4 - 3x^2 - x + 5$ Multiply by -1 and remove brackets.
$= x^2 + x + 9$ Combine like terms.

PROBLEM 10
Remove the grouping symbols and simplify:

a. $[(2x^2 - 3) + (3x + 1)]$
$ - [(x + 1) + (x^2 - 2)]$

b. $[-7 + 3(a - 2)]$
$ + [-4(9 - a) + 8]$

(continued)

Answers to PROBLEMS

9. a. $11y - 12$ b. $y^2 - 3y - 8$ c. $3xy + 48$ 10. a. $x^2 + 2x - 1$ b. $7a - 41$

b. $[4(x + 5) - 15] + 5[3 + 2(6 - x)]$
$= [4x + 20 - 15] + 5[3 + 12 - 2x]$ Remove parentheses.
$= [4x + 5] + 5[15 - 2x]$ Combine like terms.
$= 4x + 5 + 75 - 10x$ Remove brackets.
$= -6x + 80$ Combine like terms.

Calculator Corner

Check the results of the following examples with your calculator:
1. Example 1(a).
2. Example 2(a).
3. Example 3
4. Example 4(a).
5. Why can't we check the results of Example 7(b) with a calculator?

Exercises 1.4

> Boost your grade at mathzone.com!
> Practice Problems
> NetTutor
> Self-Tests
> e-Professors
> Videos

‹ A › **Evaluate Numerical Expressions with Grouping Symbols**   In Problems 1–20, evaluate the expression.

1. **a.** $(-10 \cdot 3) + 4$
   b. $-10 \cdot (3 + 4)$
2. **a.** $(6 \cdot 4) + 6$
   b. $6 \cdot (4 + 6)$
3. **a.** $(36 \div 4) \cdot 3$
   b. $36 \div (4 \cdot 3)$
4. **a.** $(-28 \div 7) \cdot 2$
   b. $-28 \div (7 \cdot 2)$
5. $[-5 \cdot (8 + 2)] + 3$
6. $[7 \cdot (4 + 3)] + 1$
7. $-7 + [3 \cdot (4 + 5)]$
8. $-8 + [3 \cdot (4 + 1)]$
9. $[-6 \cdot (4 - 2)] - 3$
10. $[-2(7 - 5)] - 8$
11. $3 - [8 \cdot (5 - 3)]$
12. $7 - [3(4 - 5)]$
13. $-8[3 - 2(4 + 1)] + 1$
14. $6[7 - 2(5 - 7)] - 2$
15. $48 \div \{4(8 - 2[3 - 1])\}$
16. $-96 \div \{4(8 - 2[1 - 3])\}$
17. $\left[\dfrac{9 - (-3)}{8 - 6}\right]\left[\dfrac{3 + (-8)}{7 - 2}\right]$
18. $\left[\dfrac{6 + (-2)}{3 + (-7)}\right]\left[\dfrac{8 + (-12)}{2 - 4}\right]$
19. $\dfrac{3 - 5\left(\dfrac{4 + 2}{2 + 1}\right) - 2}{-4 + 3\left(\dfrac{4 - 2}{4 - 6}\right) - 2}$
20. $\dfrac{8 + 2\left(\dfrac{9 - 15}{3 - 1}\right) - 2}{-4 + 8\left(\dfrac{6 - 3}{1 - 4}\right) + 12}$

‹ B › **The Order of Operations**   In Problems 21–40, use the order of operations to evaluate.

21. $-5 \cdot 6 \quad 6$
22. $-5 \cdot 2 - 2$
23. $-7 \cdot 3 \div 3 - 3$
24. $-36 \cdot 2 \quad 18 - 4$
25. $(-20 - 5 + 3 \div 3) \div 6$
26. $(-10 - 2 + 10 \div 5) \cdot 4$
27. $\dfrac{8 + (-3)}{5} - 1$
28. $\dfrac{7 + (-3)}{2} - 4$
29. $\dfrac{4 \cdot (6 - 2)}{-8} - \dfrac{6}{-2}$
30. $\dfrac{5 \cdot (6 - 2)}{-4} - \dfrac{16}{-4}$
31. $4 \div 2 + 3 - 5^2$
32. $8 \div 4 + 7 - 2^2$
33. $4 + 6 \cdot 4 \div 2 - 2^3$
34. $6 + 6 \div 3 - 3^3$
35. $-5^2 + \dfrac{2 - 10}{4} + 12 \div 4$
36. $-4^2 + \dfrac{3 - 7}{2} + 18 \div 9$
37. $-3^3 + 4 - 6 \cdot 8 \div 4 - \dfrac{8 - 2}{-3}$
38. $-2^3 + 6 - 6 \div 3 \cdot 2 - \dfrac{9 - 3}{-6}$
39. $4 \cdot 9 \div 3 \cdot 10^3 - 2 \cdot 10^2$
40. $5 \cdot 8 \div 4 \cdot 10^3 - 2 \cdot 10^2$

⟨C⟩ Evaluating Algebraic Expressions In Problems 41–44, evaluate the algebraic expressions.

41. $2(l + w)$ gives the perimeter of a rectangle. Find the perimeter if $l = 12\frac{1}{2}$ and $w = 6$.

42. $1.8C + 32$ gives the temperature in degrees Fahrenheit. Find the temperature when $C = 15$.

43. $P(1 + r)$ gives the amount compounded annually for one year. Find the value of the amount if $P = \$1000$ and $r = 5\%$.

44. Pe^{rt} gives the amount when continuous interest is computed. Find the amount if $P = \$2000$, $e = 3$, $r = 100\%$, and $t = 2$.

In Problems 45–50, evaluate the algebraic expressions for the given values of the variables.

45. $2x - y$; for $x = 4$, $y = -5$

46. $a - 3b$; for $a = -7$, $b = 20$

47. $t^2 - 5t + 8$; for $t = -3$

48. $-z^2 + z - 1$; for $z = 2$

49. $5n^2 + 2(n - m)$; for $n = -4$, $m = 6$

50. $5(p^2 - r) + p \div 5$; for $p = 10$, $r = 50$

⟨D⟩ Using the Distributive Property to Simplify In Problems 51–80, remove the parentheses (simplify).

51. $4(x - y)$

52. $3(a - b)$

53. $-9(a - b)$

54. $-6(x - y)$

55. $0.3(4x - 2)$

56. $0.2(3a - 9)$

57. $-\left(\frac{3a}{2} - \frac{6}{7}\right)$

58. $-\left(\frac{2x}{3} - \frac{1}{5}\right)$

59. $-(2x - 6y)$

60. $-(3a - 6b)$

61. $-(2.1 + 3y)$

62. $-(5.4 + 4b)$

63. $-4ab(a + 5)$

64. $-6xy(x + 8)$

65. $-x(6 + y)$

66. $-y(2x + 3)$

67. $8y(x - y)$

68. $9b(a - b)$

69. $a^2(2a - 7b)$

70. $x^2(3x - 9y)$

71. $0.5(x + y - 2)$

72. $0.8(a + b - 6)$

73. $-\frac{6}{5}(a - b + 5)$

74. $-\frac{2}{3}(x - y + 4)$

75. $-2x(x - y + 3z + 5)$

76. $-4a(a - b + 2c + 8)$

77. $-0.3(x + y - 2z - 6)$

78. $-0.2(a + b - 3c - 4)$

79. $-\frac{5}{2}(a - 2b + c + 2d - 2)$

80. $-\frac{4}{7}(2a - b + 3c + 7d - 7)$

⟨E⟩ Combining Like Terms In Problems 81–95, remove the parentheses and combine like terms.

81. $6x + 3(x - 2)$

82. $8y + 6(y - 3)$

83. $-4(x + 2) - 5x$

84. $-5(x + 3) - 6x$

85. $(5L - 3W) - (W - 6L)$

86. $(2ab - 2ac) - (ab - 4ac)$

87. $5xy - y(8x + 1) + (y + 1)$

88. $3xy - y(7x + 2) + 2(y + 2)$

89. $\frac{2x}{9} - \left(\frac{x}{9} - 2\right)$

90. $\frac{5x}{7} - \left(\frac{2x}{7} - 3\right)$

91. $4a - (a + b) + 3(b + a)$

92. $8x - 3(x + y) - (x - y)$

93. $7x^2 - 3x(x + y) - x(x + y)$

94. $4a(b - a) + 3b(b + a) - 2a(a + b)$

95. $-(x + y - 2) + 3(x - y + 6) - (x + y - 16)$

⟨F⟩ Removing Other Grouping Symbols In Problems 96–105, remove the grouping symbols and simplify.

96. $[(a^2 - 4) + (2a^3 - 5)] + [(4a^3 + a) + (a^2 + 9)]$

97. $(x^2 + 7 - x) + [-2x^3 + (8x^2 - 2x) + 5]$

98. $[(0.4x - 7) + 0.6x^2] - [(0.3x^2 - 2) - 0.8x]$

99. $\left[\left(\frac{5}{7}x^2 + \frac{1}{5}x\right) - \frac{1}{8}\right] - \left[\left(\frac{3}{7}x^2 - \frac{3}{5}x\right) + \frac{5}{8}\right]$

100. $[3(x + 2) - 10] + [5 + 2(5 + x)]$

101. $[3(2a - 4) + 5] - [2(a - 1) + 6]$

102. $[6(a - b) + 2a] - [3b - 4(a - b)]$

103. $[4a - (3 + 2b)] - [6(a - 2b) + 5a]$

104. $-[-(x + y) + 3(x - y)] - [4(x + y) - (3x - 5y)]$

105. $-[-(0.2x + y) + 3(x - y)] - [2(x + 0.3y) - 5]$

〉〉〉 Applications

In Problems 106–110, use your skills of evaluating an expression along with the given formulas to solve the problems.

106. *Area* Sergio wants to paint the back of his house. The total area can be found by adding the areas of the white triangle and the yellow rectangle as shown in Figure 1.7.

  a. Find the area of the triangle by evaluating the formula $A = \frac{1}{2}bh$.

  b. Find the area of the rectangle by evaluating the formula $A = bh$.

  c. Find the total area.

107. *Temperature* Formulas can be used to change Fahrenheit temperatures to Celsius and vice versa. When given a temperature in Fahrenheit degrees, use $C = \frac{5}{9}(F - 32)$ to find the temperature in Celsius. However, to find the temperature in Fahrenheit when given the degrees in Celsius, use $F = \frac{9}{5}C + 32$.

  a. If the temperature is 28°C, find the temperature in Fahrenheit to the nearest degree.

  b. If a cake must be baked at 375°F, find the temperature in Celsius to the nearest degree.

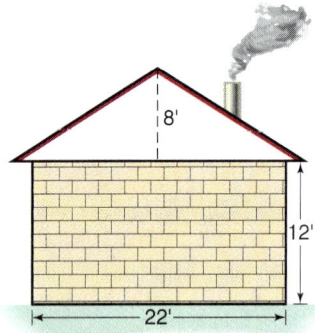

\> Figure 1.7

108. *Simple interest* Tran wants to save his money in a bank where he will receive simple interest for his investment. To calculate simple interest, multiply the principal amount *of the investment* (p) times the rate of interest (r) times the amount of time (t) in years ($I = prt$).

  a. Use the simple interest formula to evaluate Tran's interest after investing $3400 at 4% interest for 3 years.

  b. How much money will he have altogether (principal and interest) at the end of 3 years?

109. *Simple interest* When money is borrowed from a bank, a fee is charged called the interest. Kyrstin borrowed from a bank that will charge her simple interest ($I = prt$).

  a. Use the simple interest formula to evaluate Kyrstin's interest if she borrows $4600 at 7.5% interest for 4 years.

  b. What is the total amount she will owe the bank?

110. *Surface area* Sheena has to buy wrapping paper to cover the surface area of a box. To find the surface area of the box, she will use the formula, $SA = 2lw + 2wh + 2lh$, to get the area in square units where l is the length, w is the width and h is the height. If the gift box has a length of 2.5 feet, a width of 1.25 feet, and a height of 1.5 feet, find the surface area of the gift box.

>>> Using Your Knowledge

Average Velocity, Momentum, and Kinetic Energy The distributive property is helpful in solving problems in many areas. Use your knowledge of the distributive property to remove the parentheses in Problems 111–114.

111. If your car is accelerating at a constant rate, and v_1 is the initial velocity and v_2 is the final velocity, the *average* velocity is
$$v_a = \frac{1}{2}(v_1 + v_2)$$

112. The momentum M of a billiard ball is the product of its mass m and its velocity v. If two billiard balls of equal mass m and moving in the same straight line, collide with velocities v_1 and v_2, respectively, the total momentum M is given by
$$M = m(v_1 + v_2)$$

113. The total kinetic energy (KE) of the billiard balls in Problem 112 is given by
$$KE = \frac{1}{2}m(v_1^2 + v_2^2)$$

114. The length of a belt L needed to connect two pulleys of radius r_1 and r_2, respectively, with centers d units apart is
$$L = \pi(r_1 + r_2) + 2d$$

>>> Write On

115. Explain why $(32 \div 4) \cdot 2$ is different from $32 \div (4 \cdot 2)$.

116. Write the definition for "like terms" in your own words.

>>> Concept Checker

Fill in the blank(s) with the correct word(s), phrase, or mathematical statement.

117. In the expression, $-5x^3$, -5 is the _____.

118. Terms with exactly the same variable factors are said to be _____.

119. $3 \cdot 5 + (-7) - 12 \div 4$ is an example of a(n) _____.

120. A letter that represents a number is called a _____.

121. _____ are separated by $+$ and $-$ signs.

122. An expression that consists of variables, numbers, and operation symbols is called a(n) _____.

variable terms
constant like terms
numerical expression numerical coefficient
algebraic expression

>>> Mastery Test

Evaluate:

123. $-3^2 + \left(\frac{4-8}{2}\right) + 48 \div 6$

124. $\left(\frac{2-14}{6}\right) + 16 \div 4 - 5^2$

Simplify:

125. $-2(x - 3y + 2z - 4)$

126. $-\frac{2}{3}(a + 6b - 9c - 12)$

127. $\frac{3}{8}x - \left(\frac{x}{8} - 5\right)$

128. $\frac{5}{7}x - \left(4 - \frac{5}{7}x\right)$

129. $[(a^2 - 5) + (2a^3 - 3)] - [(4a^3 + a) - (a^2 - 9)]$

130. $\left[\left(\frac{5}{9}x^2 + \frac{1}{5}x\right) - \frac{1}{3}\right] - \left[\left(\frac{2}{9}x^2 - \frac{2}{5}x\right) - \frac{4}{3}\right]$

Evaluate:

131. $\left[\dfrac{8-(-4)}{8-10}\right]\left[\dfrac{5+(-9)}{7-3}\right]$

132. $\left[\dfrac{7+(-4)}{4-7}\right]\left[\dfrac{9+(-14)}{3-5}\right]$

››› Skill Checker

133. $(-4)+(-4)$ **134.** $13-15$ **135.** $x+(-x)$ **136.** $-8x+0$

137. $12(-3)$ **138.** $-4(-15)$ **139.** $\dfrac{-24}{8}$ **140.** $\dfrac{2}{-10}$

141. $\dfrac{-10}{-10}$ **142.** $\dfrac{3x}{3x}$ **143.** $0(-2)$ **144.** $\dfrac{0}{-4}$

145. $3(2x-7)$ **146.** $-(8-5x)$ **147.** $9x-4(x+2)$ **148.** $-7x+6(8-2x)$

› Collaborative Learning 1A

The real numbers we have studied in this chapter are used in many day-to-day applications. For instance, when deciding how much furniture to buy for a room and how to arrange it, it would be useful to be able to make a scale drawing. A scale drawing is a drawing that represents a real object. The scale of the drawing is the ratio of the size of the drawing to the actual size of the object. For example, you might let 1 inch = 1 foot. It is helpful to use grid paper when making a scale drawing. Measurements can be in the English system or in the metric system. The United States has not completely changed to the metric system, so it is important to be able to measure in both systems.

Divide into groups and perform the following tasks. Have one person from each group present the results to the class.

1. Measure the dimensions of your classroom to the nearest inch and then measure it again to the nearest centimeter.
2. Measure each desk and cabinet in your classroom first to the nearest inch and then again to the nearest centimeter.
3. Using grid paper, make two scale drawings of your classroom, one in the English system and one in the metric system. The drawing should include the placement of the desks and any cabinets in the room.

After listening to each group's results and viewing their scale drawings, determine the following.

1. Were there any differences in the measurements when comparing each group's drawings? If so, discuss how the differences might occur.
2. Of the English and metric scales used, which scale did you like the best and why?

› Collaborative Learning 1B

Even pranksters have to know their math. The giant Rubik's cube in the "Getting Started" from Section 1.3 took a lot of planning and math to accomplish. Let's see if you can do the math. Remember the cube is 8 feet by 8 feet by 8 feet. The colors of the Rubik's cube are red, orange, yellow, green, blue, and white. Each face of the cube should be one color and contain nine squares with a 2-inch black strip separating the squares. It was decided to use cardboard to cover each face of the cube and duct tape to make the black strip separating the squares. Working in groups, divide the following questions so that each group answers a share of the questions. Then the groups should collaborate to see how much the prank will cost.

‹CONTINUED›

1. Find the cost of the cardboard.
   a. How much cardboard is necessary to cover the six faces of the cube?
   b. Research the price and sizes of cardboard sheets. Make a table of the results.
   c. Use the results from **a** and **b** to choose the most cost-effective size cardboard sheets to buy and determine how many sheets will be required. Then, compute the total cost for the cardboard. Remember to include shipping if you decide to purchase online.
2. Find the cost of the paint.
   a. How many square feet of paint will be needed to cover each face of the cube?
   b. Research the price of paint by the various size containers (pints, quarts, gallons) and how many square feet each size will cover. Make a table of the results.
   c. Use the results from **a** and **b** to choose the most economical size to buy and determine how many containers of each paint color are necessary. Then, compute the total cost for the paint.
3. Find the cost of the duct tape.
   a. Two-inch duct tape can be used to make the strips separating the nine squares on each face. How many linear feet of duct tape are necessary for the strips on the entire cube?
   b. Research the price of 2-inch black duct tape and the various lengths contained on the roll. Make a table of the results.
   c. Use the results from **a** and **b** to decide what size roll of duct tape to buy and how many rolls are needed. Then, compute the total cost for the duct tape.
4. What is the most economical cost of the prank?

Research Questions

1. There is a charming story about the wooden tally sticks mentioned in the chapter preview. Find out how their disposal literally resulted in the destruction of the old Houses of Parliament in England.
2. Write a paper detailing the Egyptian number system and the base and symbols used, and enumerate the similarities and differences between the Egyptian and our (Hindu-Arabic) system of numeration.
3. Write a paper detailing the Greek number system and the base and symbols used, and enumerate the similarities and differences between the Greek and our system of numeration.
4. Find out about the development of the symbols we use in our present numeration system. Where was the symbol for zero invented and by whom?
5. When were negative numbers introduced, by whom were they introduced, and what were they first called?
6. Write a short paper about the Rhind, or Ahmes, papyrus. What is the significance of each of the names *Rhind* and *Ahmes*? What is the content of the papyrus and who discovered it?
7. Find out what "gematria" (not geometry!) is, the significance of 666, and the reason why many old editions of the Bible substitute the number 99 for *amen* at the end of a prayer.

Summary Chapter 1

Section	Item	Meaning	Example
1.1A	Empty or null set, $\emptyset$	The set containing no elements	The set of natural numbers between 5 and 6 is $\emptyset$.
	Natural numbers	$N = \{1, 2, 3, \ldots\}$	2, 76, and 308 are natural numbers.
	Whole numbers	$W = \{0, 1, 2, \ldots\}$	0, 8, and 93 are whole numbers.
	Integers	$I = \{\ldots, -2, -1, 0, 1, 2, \ldots\}$	-7 and 23 are integers.
	Rational numbers	$Q = \{r \mid r = \frac{a}{b}, a \text{ and } b \text{ are integers and } b \neq 0\}$	$\frac{1}{5}, -\frac{2}{3}, 0, 9, 1.4,$ and $0.\overline{3}$ are rational numbers.
1.1B	Rational numbers	The set Q of rational numbers is the same as the set of terminating or nonterminating, repeating decimals.	0.345 and $0.\overline{3}$ are rational numbers.
	Irrational numbers	$H = \{x \mid x \text{ is not rational}\}$	$\sqrt{2}$ and π are irrational numbers.
	Real numbers (R)	The set of all rationals and irrationals	$\frac{2}{7}, -\frac{2}{3}, 0, 9, 1.4, 0.\overline{3}, \sqrt{2},$ and π are real numbers.
1.1C	$N \subseteq W \subseteq I \subseteq Q \subseteq R$	N is a subset of W, W is a subset of I, and so on.	Every natural number is a whole number, every whole number is an integer, and so on.
1.1D	Additive inverses (opposites)	a and $-a$ are additive inverses.	8 and -8 are additive inverses.
1.1E	Absolute value $\lvert a \rvert$	The distance from 0 to a on the number line, $\lvert a \rvert = \begin{cases} a & \text{when } a \geq 0 \\ -a & \text{when } a < 0 \end{cases}$	$\lvert -8 \rvert = 8, \left\lvert \frac{2}{3} \right\rvert = \frac{2}{3},$ and $\lvert -0.4 \rvert = 0.4$
1.1F	Trichotomy law	If a and b are real numbers, then 1. $a = b$, or 2. $a < b$, or 3. $a > b$	
1.2A	Adding signed numbers with the same sign	Add their absolute values and give the sum the common sign.	$-3 + (-7) = -(\lvert -3 \rvert + \lvert -7 \rvert) = -10$
	Adding signed numbers with different signs	Subtract the smaller absolute value from the greater absolute value and use the sign of the number with the greater absolute value.	$3 + (-5) = -(5 - 3) = -2$ $-7 + 9 = +(9 - 7) = 2$
	Subtraction	If a and b are real numbers, $a - b = a + (-b)$.	$3 - (-4) = 3 + 4 = 7$
	$a \cdot b, ab, (a)(b), a(b), (a)b$	The product of a and b	$5 \cdot 2 = (5)(2) = 5(2) = (5)2 = 10$
	Multiplying signed numbers with different signs	Multiply their absolute values; the product is negative.	$3 \cdot (-4) = -12$ $-7 \cdot 2 = -14$

Section	Item	Meaning	Example
1.2A	Multiplying signed numbers with the same sign	Multiply their absolute values; the product is positive.	$3 \cdot 8 = 24$ and $(-9)(-2) = 18$
	Multiplication of fractions	$\dfrac{a}{b} \cdot \dfrac{c}{d} = \dfrac{a \cdot c}{b \cdot d}$ $(b \neq 0, d \neq 0)$	$\left(-\dfrac{3}{4}\right) \cdot \dfrac{2}{7} = -\dfrac{3}{14}$
	Division	If a, b, c are real numbers, $\dfrac{a}{b} = c$ means $a = bc$. $(b \neq 0)$	$\dfrac{6}{3} = 2$ means $6 = 3 \cdot 2$.
	Dividing signed numbers	The quotient of two real numbers with the same sign is positive, with different signs, negative.	$\dfrac{6}{-3} = -2, \dfrac{-6}{3} = -2,$ and $\dfrac{-8}{-2} = 4$
	Zero in division problems	$\dfrac{0}{a} = 0$ $(a \neq 0)$ $\dfrac{a}{0}$ is not defined. $\dfrac{0}{0}$ is indeterminate.	$\dfrac{0}{9} = 0, \dfrac{0}{-8} = 0,$ and $\dfrac{0}{-2.4} = 0$ $\dfrac{9}{0}, \dfrac{-8}{0},$ and $\dfrac{-2.4}{0}$ are not defined.
	Reciprocal	The reciprocal of a is $\dfrac{1}{a}$. $(a \neq 0)$	The reciprocal of -7 is $-\dfrac{1}{7}$.
	Division of fractions	$\dfrac{a}{b} \div \dfrac{c}{d} = \dfrac{a}{b} \cdot \dfrac{d}{c}$ $(b \neq 0, c \neq 0, d \neq 0)$	$\dfrac{3}{4} \div \dfrac{9}{2} = \dfrac{3}{4} \cdot \dfrac{2}{9} = \dfrac{1}{6}$

1.2B Real-Number Properties

If $a, b,$ and c are real numbers:

Name	Addition	Multiplication
Closure	$a + b$ is a real number.	$a \cdot b$ is a real number.
Commutative	$a + b = b + a$	$a \cdot b = b \cdot a$
Associative	$a + (b + c) = (a + b) + c$	$a \cdot (b \cdot c) = (a \cdot b) \cdot c$
Identity	$a + 0 = 0 + a = a$ (0 is the identity.)	$1 \cdot a = a \cdot 1 = a$ (1 is the identity.)
Inverse	For each real number a, there is a unique real number $-a$ such that $a + (-a) = -a + a = 0$. ($-a$ is called the opposite of a.)	For each nonzero real number a, there is a unique real number $\dfrac{1}{a}$ such that $a \cdot \dfrac{1}{a} = \dfrac{1}{a} \cdot a = 1$. $\left(\dfrac{1}{a}\right.$ is called the reciprocal of $a.\Big)$

Distributive property of multiplication over addition $a(b + c) = ab + ac$
Distributive property of multiplication over subtraction $a(b - c) = ab - ac$

Section	Item	Meaning	Example
1.3A	Exponent	$a^n = a \cdot a \cdot a \cdots a$, where $n \in N$ (n factors) n is called the exponent.	$3^4 = 3 \cdot 3 \cdot 3 \cdot 3 = 81$
	Base	In the expression a^n, a is called the base.	In the expression 2^5, 2 is the base and 5 is the exponent.
1.3B	Negative exponent	$a^{-n} = \dfrac{1}{a^n}$ $(a \neq 0)$	$5^{-2} = \dfrac{1}{5^2} = \dfrac{1}{25}$
	Zero exponent	$a^0 = 1$ $(a \neq 0)$	$2^0 = 1$ and $\left(\dfrac{-1}{4}\right)^0 = 1$
1.3C	Product rule of exponents	$a^m \cdot a^n = a^{m+n}$ $(a \neq 0)$	$x^5 \cdot x^4 = x^9$
	Quotient rule	$\dfrac{a^m}{a^n} = a^{m-n}$ $(a \neq 0)$	$\dfrac{x^8}{x^3} = x^5$

(continued)

Section	Item	Meaning	Example
1.3D	Power rule of exponents Raising a product to a power Raising a quotient to a power	$(a^m)^n = a^{m \cdot n}$ $(a \neq 0)$ $(a^m b^n)^k = a^{m \cdot k} b^{n \cdot k}$ $(a \neq 0, b \neq 0)$ $\left(\dfrac{a^m}{b^n}\right)^k = \dfrac{a^{m \cdot k}}{b^{n \cdot k}}$ $(a \neq 0, b \neq 0)$	$(x^2)^5 = x^{10}$ $(x^3 y^5)^6 = x^{3 \cdot 6} y^{5 \cdot 6} = x^{18} y^{30}$ $\left(\dfrac{x^5}{y^3}\right)^4 = \dfrac{x^{5 \cdot 4}}{y^{3 \cdot 4}} = \dfrac{x^{20}}{y^{12}}$
1.3E	Scientific notation	A number is in scientific notation when it is written in the form $m \times 10^n$, where m is greater than or equal to 1 and less than 10 and n is an integer.	$352 = 3.52 \times 10^2$ is in scientific notation.
1.4B	Order of operations (from left to right)	**P** Parentheses (or other grouping symbols, including division bars) **E** Exponentiation **(MD)** Multiplication / Division **(AS)** Addition / Subtraction	$12 \div 6 \cdot 2 + 3^2 - (4 + 5)$ $= 12 \div 6 \cdot 2 + 3^2 - 9$ **Parentheses** $= 12 \div 6 \cdot 2 + 9 - 9$ **Exponents** $= 2 \cdot 2 + 9 - 9$ **(MD)** Divide $= 4 + 9 - 9$ Multiply $= 13 - 9$ **(AS)** Add $= 4$ Subtract
1.4D	$-1 \cdot a$ $-(a + b)$ $-(a - b)$	$-1 \cdot a = -a$ $-(a + b) = -a - b$ $-(a - b) = -a + b$	$-1 \cdot 4 = -4$ $-(x + 7) = -x - 7$ $-(x - 3) = -x + 3$
1.4E	Similar or like terms	Two or more terms that differ only in their numerical coefficients	$-3a$ and $7a$ are similar or like terms. $5x^2y$ and $-2x^2y$ are like terms.

Review Exercises Chapter 1

(If you need help with these exercises, look in the section indicated in brackets.)

1. ⟨ **1.1A** ⟩ Use roster notation to list the natural numbers that fall between
   a. 3 and 9   b. 4 and 8

2. ⟨ **1.1B** ⟩ Write as a decimal:
   a. $\dfrac{1}{5}$ b. $\dfrac{2}{5}$

3. ⟨ **1.1B** ⟩ Write as a decimal:
   a. $\dfrac{1}{9}$ b. $\dfrac{2}{9}$

4. ⟨ **1.1C** ⟩ Classify the given number by making a check mark (✓) in the appropriate row(s).

Set	0.3	0	$\dfrac{-3}{4}$	-5	$\sqrt{3}$
Natural numbers	__	__	__	__	__
Whole numbers	__	✓	__	__	__
Integers	__	__	__	✓	__
Rational numbers	__	__	__	__	__
Irrational numbers	__	__	__	__	__
Real numbers	__	__	__	__	__

5. ⟨ **1.1D** ⟩ Find the additive inverse:
   a. -3.5 b. $\dfrac{3}{4}$

6. ⟨ **1.1E** ⟩ Find:
   a. $|-9|$ b. $|4.2|$

7. ⟨**1.1E**⟩ Find:
   a. $3|-7|$
   b. $|8| - |-5|$

8. ⟨**1.1F**⟩ Fill in the blank with $<, >,$ or $=$ to make the resulting statement true:
   a. -8 _____ -7
   b. -4 _____ -3

9. ⟨**1.1F**⟩ Fill in the blank with $<, >,$ or $=$ to make the resulting statement true:
   a. $\frac{1}{4}$ _____ $\frac{1}{5}$
   b. $\frac{3}{4}$ _____ 0.75

10. ⟨**1.1F**⟩ Fill in the blank with $<, >,$ or $=$ to make the resulting statement true:
    a. $\frac{1}{5}$ _____ 0.25
    b. $0.\overline{6}$ _____ $\frac{2}{3}$

11. ⟨**1.2A**⟩ Add:
    a. $-3 + (-8)$
    b. $-5 + 2$

12. ⟨**1.2A**⟩ Subtract:
    a. $\frac{1}{7} - \frac{3}{7}$
    b. $-0.2 - 0.4$

13. ⟨**1.2A**⟩ Subtract:
    a. $8 - (-4)$
    b. $-3 - (-7)$

14. ⟨**1.2A**⟩ Subtract:
    a. $\frac{3}{4} - \left(-\frac{1}{5}\right)$
    b. $\frac{5}{6} - \left(-\frac{1}{4}\right)$

15. ⟨**1.2A**⟩ Multiply:
    a. $9 \cdot (-4)$
    b. $-2.4 \cdot 6$

16. ⟨**1.2A**⟩ Multiply:
    a. $\left(-\frac{3}{4}\right) \cdot \frac{7}{8}$
    b. $\left(-\frac{5}{6}\right) \cdot \left(-\frac{2}{7}\right)$

17. ⟨**1.2A**⟩ Divide (if possible):
    a. $\frac{0}{7}$
    b. $\frac{8}{0}$

18. ⟨**1.2A**⟩ Find the reciprocal:
    a. $-\frac{3}{5}$
    b. 0.3

19. ⟨**1.2A**⟩ Divide:
    a. $-\frac{3}{5} \div \frac{4}{15}$
    b. $\frac{3.6}{-1.2}$

20. ⟨**1.2B**⟩ Name the property illustrated in the statement:
    a. $(4 + 9) + 5 = 5 + (4 + 9)$
    b. $(3 + 5) + 8 = 3 + (5 + 8)$

21. ⟨**1.3A**⟩ Evaluate:
    a. $(-3)^4$
    b. -3^4

22. ⟨**1.3B**⟩ Evaluate or simplify:
    a. 9^0
    b. $\left(\frac{1}{7}\right)^0$
    c. $(-8)^{-3}$
    d. x^{-10}

23. ⟨**1.3B**⟩ Write as a fraction:
    a. $12x^{-2}$
    b. $-4x^3y^{-1}$

24. ⟨**1.3C**⟩ Multiply and simplify:
    a. $(3x^4y)(-5x^{-8}y^9)$
    b. $(4x^{-3}y^{-1})(-6x^{-8}y^{-7})$

25. ⟨**1.3C**⟩ Divide and simplify:
    a. $\frac{48x^4}{16x^6}$
    b. $\frac{8x^5}{-2x^{-6}}$

26. ⟨**1.3C**⟩ Divide and simplify:
    a. $\frac{-5x^{-3}}{15x^{-4}}$
    b. $\frac{8x^{-4}}{-4x^7}$

27. ⟨**1.3D**⟩ Simplify:
    a. $(-2x^7y^{-6})^3$
    b. $(-2x^{-6}y^{-6})^4$

28. ⟨**1.3D**⟩ Simplify:
    a. $\left(\frac{x^6}{y^{-3}}\right)^{-4}$
    b. $\left(\frac{x^{-5}}{y^3}\right)^{-5}$

29. ⟨**1.3E**⟩ *Write in scientific notation:*
   a. 340,000
   b. 0.000047

30. ⟨**1.3E**⟩ *Write in decimal notation:*
   a. 3.7×10^4
   b. 7.8×10^{-3}

31. ⟨**1.4A**⟩ *Evaluate:*
   a. $[-8 \cdot (9 + 2)] + 13$
   b. $[-7(3 - 8)] + 15$

32. ⟨**1.4B**⟩ *Evaluate:*
   a. $6^2 \div 3 - 9 \cdot 2 \div 3 + 3$
   b. $\dfrac{5 \cdot (68 - 32)}{9}$

33. ⟨**1.4B**⟩ *Evaluate:*
   a. $-3^2 + \dfrac{4 - 10}{2} + 15 \div 3$
   b. $-4^3 + \dfrac{2 - 10}{2} - 25 \div 5$

34. ⟨**1.4C**⟩ *Evaluate the algebraic expression:*
   a. $2(lw + lh + wh)$ gives the surface area of a rectangular box. Find the area if $l = 12$ in., $w = 8$ in., and $h = 4$ in.
   b. $-2x^2 - (4 - y)5$ if $x = -2$ and $y = -1$

35. ⟨**1.4D**⟩ *Remove the parentheses (simplify):*
   a. $-3(x - 7)$
   b. $3(x + 8) - (x + 7)$

36. ⟨**1.4E, F**⟩ *Simplify:*
   a. $3(x + 2y - 2) - 2(2x - 2y + 5)$
   b. $[(5x^2 - 3) + (4x + 5)] - [(x - 4) + (2x^2 - 2)]$

Practice Test Chapter 1

(Answers on page 72)

Visit www.mhhe.com/bello to view helpful videos that provide step-by-step solutions to several of the problems below.

1. Use roster notation to list the natural numbers between 5 and 9.

2. Write as a decimal:
   a. $\dfrac{3}{8}$
   b. $\dfrac{2}{3}$

3. Classify the given number by making a check mark (✓) in the appropriate row(s).

Set	0.5	0	−6	$\dfrac{-2}{7}$	$\sqrt{5}$
Natural numbers					
Whole numbers					
Integers					
Rational numbers					
Irrational numbers					
Real numbers					

4. Find the additive inverse of $\dfrac{4}{5}$.

5. Find:
   a. $|-9|$
   b. $|4| - 2\,|-5|$

6. Fill in the blank with $<$, $>$, or $=$ to make the resulting statement true:
   a. $-\dfrac{1}{4}$ _____ $-\dfrac{1}{3}$
   b. 0.4 _____ $\dfrac{2}{5}$

7. Add:
   a. $-9 + 5$
   b. $-0.8 + (-0.7)$

8. Subtract:
   a. $-16 - 7$
   b. $-0.6 - (-0.4)$

9. Subtract:
   a. $-\dfrac{1}{8} - \dfrac{3}{4}$
   b. $-\dfrac{3}{4} - \left(-\dfrac{5}{6}\right)$

10. Multiply:
    a. $6 \cdot (-9)$
    b. $-4 \cdot (-1.2)$

11. Multiply or divide:
    a. $-\dfrac{1}{2} \cdot \dfrac{2}{9}$
    b. $-\dfrac{3}{2} \div \dfrac{9}{8}$

12. Name the property illustrated in the statement:
    a. $(7 + 3) + 6 = (3 + 7) + 6$
    b. $(2 + 9) + 4 = 2 + (9 + 4)$

13. Name the property illustrated in the statement:
    a. $3 \cdot \dfrac{1}{3} = 1$
    b. $0.3 + (-0.3) = 0$

14. Evaluate:
    a. $(-3)^4$
    b. -3^4

15. Write without negative exponents:
    a. 7^{-2}
    b. x^{-8}
    c. $5a^{-4}b$

16. Perform the indicated operation and simplify:
    a. $(3x^4 y)(-4x^{-8} y^8)$
    b. $\dfrac{48x^4}{16x^{-8}}$

17. Simplify:
    a. $(-2x^8 y^{-2})^3$
    b. $\left(\dfrac{x^5}{y^{-3}}\right)^{-3}$

18. Write 6.5×10^{-3} in standard decimal notation.

19. Write 8.5×10^5 in standard decimal notation.

20. Perform the calculation and write your answer in scientific notation:
    $$(7.1 \times 10^5) \times (4 \times 10^{-7})$$

21. Evaluate:
    a. $[-7(4 + 3)] + 9$
    b. $\dfrac{5 \cdot (131 - 32)}{9}$

22. Evaluate: $-4^3 + \dfrac{6 - 12}{2} + 15 \div 3$.

23. Evaluate:
    a. $\dfrac{1}{2}(b_1 + b_2)h$ gives the area of a trapezoid. Find the area of the trapezoid if $b_1 = 8$, $b_2 = 3$, and $h = 6$.
    b. $7 - x^2 + (20 \div y) - xy$; if $x = -2$ and $y = 4$.

24. Simplify:
    a. $-5(x + 7)$
    b. $7x - (3x + 1) + (2x + 2)$

25. Simplify:
    $$[(5x^2 - 3) + (3x + 7)] - [(x - 3) + (2x^2 - 2)]$$

Answers to Practice Test Chapter 1

Answer	If You Missed Question	Review Section	Review Examples	Review Page
1. $\{6, 7, 8\}$	1	1.1A	1	4
2. a. 0.375 b. $0.\overline{6}$	2	1.1B	2	5
3. (see table below)	3	1.1C	3	6
4. $-\dfrac{4}{5}$	4	1.1D	4	8
5. a. 9 b. -6	5	1.1E	5	9
6. a. $>$ b. $=$	6	1.1F	6	10–11
7. a. -4 b. -1.5	7	1.2A	1	18–19
8. a. -23 b. -0.2	8	1.2A	2a, b	20
9. a. $-\dfrac{7}{8}$ b. $\dfrac{1}{12}$	9	1.2A	2c	20
10. a. -54 b. 4.8	10	1.2A	3	21
11. a. $-\dfrac{1}{9}$ b. $-\dfrac{4}{3}$	11	1.2A	5, 6, 8, 9	23, 25–27
12. a. Commutative property of addition b. Associative property of addition	12	1.2B	10	29
13. a. Inverse (reciprocal) property for multiplication b. Inverse (opposite) property for addition	13	1.2B	10	29
14. a. 81 b. -81	14	1.3A	1	38
15. a. $\dfrac{1}{7^2} = \dfrac{1}{49}$ b. $\dfrac{1}{x^8}$ c. $\dfrac{5b}{a^4}$	15	1.3B	2	40
16. a. $-\dfrac{12y^9}{x^4}$ b. $3x^{12}$	16	1.3C	3, 4	41–43
17. a. $-\dfrac{8x^{24}}{y^6}$ b. $\dfrac{1}{x^{15}y^9}$	17	1.3D	5, 6	44–45
18. 0.0065	18	1.3E	8	46
19. $850,000$	19	1.3E	8	46
20. 2.84×10^{-1}	20	1.3E	7, 8, 9	46
21. a. -40 b. 55	21	1.4A	1, 2, 3	52–53
22. -62	22	1.4B	4	54
23. a. 33 b. 16	23	1.4C	5	55
24. a. $-5x - 35$ b. $6x + 1$	24	1.4D, E	6, 7, 8, 9	56, 57, 59
25. $3x^2 + 2x + 9$	25	1.4F	10	59–60

Problem 3:

Set	0.5	0	−6	$\dfrac{-2}{7}$	$\sqrt{5}$
N					
W		✓			
I		✓	✓		
Q	✓	✓	✓	✓	
H					✓
R	✓	✓	✓	✓	✓

Chapter 2: Linear Equations and Inequalities

Section

- **2.1** Linear Equations in One Variable
- **2.2** Formulas, Geometry, and Problem Solving
- **2.3** Problem Solving: Integers and Geometry
- **2.4** Problem Solving: Percent, Investment, Motion, and Mixture Problems
- **2.5** Linear and Compound Inequalities
- **2.6** Absolute-Value Equations and Inequalities

The Human Side of Algebra

Who invented algebra, anyway? One of the earliest accounts of an algebra problem is in the Rhind papyrus, an ancient Egyptian document written by an Egyptian priest named Ahmes (ca. 1620 B.C.) and purchased by Henry Rhind. When Ahmes wished to find a number such that the number added to its seventh made 19, he symbolized the number by the sign we translate as "heap." Today, we write the problem as

$$x + \frac{x}{7} = 19$$

Can you find the answer? Ahmes says it is $16 + \frac{1}{2} + \frac{1}{8}$.

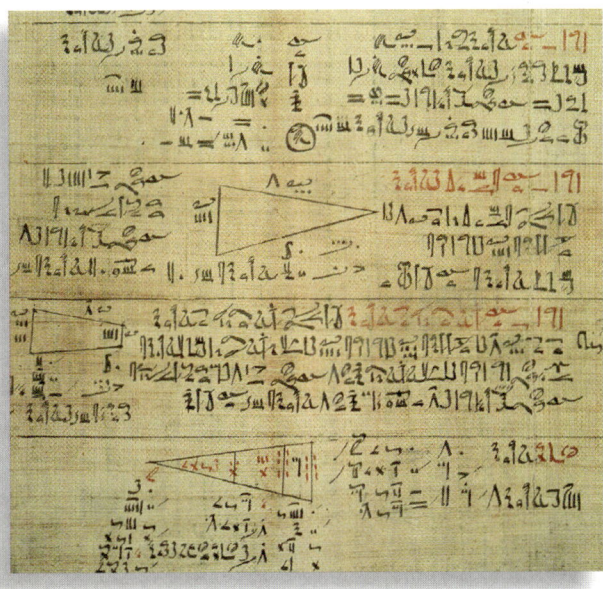

Rhind Papyrus

Western Europeans, however, learned their algebra from the works of Mohammed ibn Musa al-Khowarizmi (this translates as Mohammed the son of Moses of Khowarizmi; ca. A.D. 820), an astronomer and mathematician of Baghdad and author of the treatise *Hisab al-jabr w'al muqabalah,* the science of restoring (placing variables on one side of an equation) and reduction (collecting like terms). With the passage of time, the word *al-jabr* evolved into our present word *algebra,* a subject that we shall continue to study now.

2.1 Linear Equations in One Variable

Objectives

A Determine whether a number is a solution of a given equation.

B Solve linear equations using the properties of equality.

C Solve linear equations in one variable involving fractions or decimals.

D Solve linear equations with one solution, no solutions, or infinitely many solutions.

To Succeed, Review How To . . .

1. Add, subtract, multiply, and divide positive and negative numbers (pp. 18–27).
2. Use the commutative, associative, and distributive properties (p. 28).
3. Find the sum of opposites (additive inverses) (p. 18).
4. Find the product of a non-zero number and its reciprocal (p. 26).

Getting Started

Crickets, Ants, and Temperatures

Does temperature affect animal behavior? You certainly know about bears hibernating in the winter and languid students in the spring. But what about crickets and ants? Farmers claim that they can tell the temperature F in degrees Fahrenheit by listening to the number of chirps N a certain type of cricket makes in 1 minute! How? By using the formula

$$F = \frac{N}{4} + 40$$

Thus, if a cricket is chirping 80 times a minute, the temperature is $F = \frac{80}{4} + 40 = 60°F$ (60 degrees Fahrenheit). Now suppose the temperature is 90°F. How fast is the cricket chirping? To find the answer, you must know how to solve the equation

$$90 = \frac{N}{4} + 40$$

Similarly, the speed S of an ant, in centimeters per second (cm/sec), is given by $S = \frac{1}{6}(C - 4)$, where C is the temperature in degrees Celsius. If an ant is moving at 4 centimeters per second, what is the temperature? Here, you have to solve the equation $4 = \frac{1}{6}(C - 4)$. You will learn how to solve these and other similar equations in this chapter.

Finally, if you know the relationship between Fahrenheit and Celsius temperatures and you look at these two formulas, can you tell, as the weather gets cooler, whether the crickets stop chirping before the ants stop crawling?

How do we express our ideas in algebra? We start with a collection of letters (**variables**) and real numbers (**constants**) and then perform the basic operations of addition, subtraction, multiplication, and division. The result is an **algebraic expression** such as

$$2x + 7, \quad x^2 - 3x + 5, \quad \text{or} \quad \frac{x^5 y^7}{z^3}$$

CAUTION

In expressions such as

$$\frac{x^5 y^7}{z^3}$$

with variables in the denominator, the denominator cannot be zero.

An **equation** is a sentence stating that two algebraic expressions are equal. Some examples are

$$8 = 9 - 3x \qquad \frac{2}{y} = 6 - y \qquad 8 = x^2 + 2x$$

Notice: equations will always contain an *equal* sign.

There are three important properties of equality: the *reflexive, symmetric,* and *transitive* properties.

PROPERTIES OF EQUALITY

For all real numbers a, b, and c,

1. $a = a$ — Reflexive Property
2. If $a = b$, then $b = a$. — Symmetric Property
3. If $a = b$ and $b = c$, then $a = c$. — Transitive Property

Study Skills Hint

To practice the language of mathematics, from each section list the new vocabulary words and their meanings on index cards. These cards should be reviewed before and after each class and each homework assignment. Begin with these words from this section:

1. variables
2. constants
3. algebraic expressions
4. equation

Examples of the Reflexive Property:
$$0.5 = 0.5$$
$$x + 7 = x + 7$$
$$x^2 + 5x - 7 = x^2 + 5x - 7$$

Examples of the Symmetric Property:

If $x = 7$, then $7 = x$.
If $C = 2\pi r$, then $2\pi r = C$.
If $y = x^2 + 3x - 7$, then $x^2 + 3x - 7 = y$.

Examples of the Transitive Property:

If $x = y$ and $y = 2a$, then $x = 2a$.
If $C = 2\pi r$ and $2\pi r = \pi d$, then $C = \pi d$.
If $x = \frac{a}{b}$ and $\frac{a}{b} = \frac{ac}{bc}$, then $x = \frac{ac}{bc}$.

In this section, we consider *linear equations* involving only real numbers and *one* variable. Here are some examples of linear equations in one variable:

$$x = 8, \quad 2x - 5 = 7, \quad 3(y - 1) = 2y + 8, \quad \text{and} \quad 3k + 7 = 10$$

In general, we have the following definition.

LINEAR EQUATIONS

A **linear equation** in one variable is an equation that can be written in the form

$$ax + b = c$$

where a, b, and c are real numbers and $a \neq 0$.

The highest power of the variable in a linear equation is 1, so linear equations are also called **first-degree equations.**

A › Solutions of an Equation

Some equations are always *true* (identities: $2 + 2 = 4$, $5 - 3 = 2$), some are always *false* (contradictions: $2 + 2 = 22$, $5 - 3 = -2$), and some are neither true nor false. For example, the equation $x + 1 = 5$ is neither true nor false. It is a *conditional* equation, and its truth or falsity depends on the value of x.

In the conditional equation $x + 1 = 5$, the **variable** x can be replaced by many numbers, but only one number will make the resulting statement true. This number is called the *solution* of the equation.

> **SOLUTIONS OF AN EQUATION**
>
> The **solutions (roots)** of an equation are the replacement values of the variable that make the equation a true statement. To **solve** an equation is to find all its solutions.

To determine whether a number is a solution of an equation, we replace the variable by the number. For example, 4 is a solution of $x + 1 = 5$ because replacing x with 4 in the equation yields $4 + 1 = 5$, a true statement, but -6 is *not* a solution because $-6 + 1 \neq 5$. Since 4 is the *only* number that yields a true statement, the solution set of $x + 1 = 5$ is $\{4\}$.

EXAMPLE 1 Determining when a number is a solution
Determine whether:

a. 8 is a solution of $x - 5 = 3$
b. 5 is a solution of $3 = 2 - y$
c. 6 is a solution of $\frac{1}{3}z - 4 = 2z - 14$

SOLUTION

a. Substituting 8 for x in $x - 5 = 3$, we have $8 - 5 = 3$, a true statement. Thus, 8 is a solution of $x - 5 = 3$.
b. Substituting 5 for y in $3 = 2 - y$, we obtain $3 = 2 - 5$, a false statement. Hence, 5 is not a solution of $3 = 2 - y$.
c. If we replace z by 6 in $\frac{1}{3}z - 4 = 2z - 14$, we obtain

$$\frac{1}{3}(6) - 4 = 2(6) - 14$$
$$2 - 4 = 12 - 14$$
$$-2 = -2$$

a true statement. Thus, 6 is a solution of $\frac{1}{3}z - 4 = 2z - 14$.

PROBLEM 1
Determine whether:

a. 6 is a solution of $x - 2 = 4$
b. 7 is a solution of $4 = 11 - y$
c. 4 is a solution of $\frac{1}{2}z - 2 = z - 2$

B › Solving Equations Using the Properties of Equality

We have learned how to determine whether a number is a solution of an equation; now we will learn how to find these solutions—that is, how to *solve* the equation. The procedure is to find an *equivalent* equation whose solution is obvious. For example, the equations $x = 2$ and $x + 3 = 5$ are equivalent because $x = 2$ has the obvious solution 2, which is the only solution of the equation $x + 3 = 5$.

> **EQUIVALENT EQUATIONS**
>
> Two or more equations are **equivalent** if they have the same solution set.

Answers to PROBLEMS

1. a. Yes: $6 - 2 = 4$ b. Yes: $4 = 11 - 7$ c. No: $\frac{1}{2}(4) - 2 \neq 4 - 2$

2.1 Linear Equations in One Variable

To solve an equation, we use the idea that adding or subtracting the same number on both sides of the equation and multiplying or dividing both sides of an equation by the same nonzero number produces an equivalent equation. Here is the principle.

PROPERTIES OF EQUALITY

If c is a real number, then the following equations are equivalent.

$$a = b$$

Add c. $a + c = b + c$

Subtract c. $a - c = b - c$

Multiply by c. $a \cdot c = b \cdot c$ $\quad (c \neq 0)$

Divide by c. $\dfrac{a}{c} = \dfrac{b}{c}$ $\quad (c \neq 0)$

We will use the properties of equality to write a series of equivalent equations until we reach the obvious solution. Suppose we want to solve $x + 5 = 7$. We are trying to find a value of x that will satisfy the equation, so we try to get x by itself on one side of the equation. To do this, we "undo" the addition of 5 by *adding* the inverse of 5 on both sides.

$x + 5 = 7$	Given
$x + 5 + (-5) = 7 + (-5)$	Add (-5) to both sides.
$x + 0 = 2$	$5 + (-5) = 0$, additive inverses
$x = 2$	$x + 0 = x$, additive identity

The solution is 2, and the solution set is $\{2\}$. We can also solve the equation by subtracting 5 from both sides:

$x + 5 = 7$	Given
$x + 5 - 5 = 7 - 5$	Subtract 5 from both sides.
$x + 0 = 2$	$5 - 5 = 0$
$x = 2$	

Using the same idea, we can solve $8 = 3x - 7$ by first adding 7 to both sides.

$8 = 3x - 7$	Given
$8 + 7 = 3x - 7 + 7$	Add 7 to both sides.
$15 = 3x$	
$\dfrac{15}{3} = \dfrac{3x}{3}$	Divide both sides by 3.
$5 = x$	

The solution is 5, and the solution set is $\{5\}$.

EXAMPLE 2 Solving linear equations using the equality properties

Solve:
a. $2x - 4 = 6$ **b.** $\dfrac{2}{3}y - 3 = 9$

SOLUTION

a. To solve this equation, we want the variable x by itself on one side of the equation. We start by adding 4 to both sides.

$2x - 4 = 6$	Given
$2x - 4 + 4 = 6 + 4$	Add 4 to both sides.
$2x = 10$	
$\dfrac{1}{2} \cdot 2x = \dfrac{1}{2} \cdot 10$	Multiply both sides by $\dfrac{1}{2}$, the reciprocal of 2.
$x = 5$	$\dfrac{1}{2} \cdot 2 = 1$ and $\dfrac{1}{2} \cdot 10 = 5$

PROBLEM 2

Solve:
a. $3x - 8 = 4$ **b.** $\dfrac{3}{4}x - 5 = 10$

(continued)

Answers to PROBLEMS

2. **a.** 4 **b.** 20

The solution set is $\{5\}$.
Instead of multiplying both sides by $\frac{1}{2}$ to solve $2x = 10$, we could have divided both sides by 2

$$\frac{2x}{2} = \frac{10}{2}$$

to obtain the same result, $x = 5$. To check this solution, we substitute 5 for x in the original equation and use the following diagram:

$$\begin{array}{c|c} 2x - 4 \stackrel{?}{=} 6 \\ \hline 2(5) - 4 & 6 \\ 10 - 4 & \\ 6 & \end{array}$$

Both sides yield 6, so we have a true statement and our result is correct.

b.
$\frac{2}{3}y - 3 = 9$ Given

$\frac{2}{3}y - 3 + 3 = 9 + 3$ Add 3 to both sides.

$\frac{2}{3}y = 12$

$\frac{3}{2} \cdot \frac{2}{3}y = \frac{3}{2} \cdot 12$ Multiply both sides by $\frac{3}{2}$, the reciprocal of $\frac{2}{3}$.

$y = 18$ $\frac{3}{2} \cdot 12 = \frac{3}{2} \cdot \frac{12}{1} = 18$

The solution set is $\{18\}$. You can check this answer by substituting 18 for y in $\frac{2}{3}y - 3 = 9$ to obtain

$$\begin{array}{c|c} \frac{2}{3}y - 3 \stackrel{?}{=} 9 \\ \hline \frac{2}{3} \cdot 18 - 3 & 9 \\ 12 - 3 & \\ 9 & \end{array}$$

Both sides yield 9, so we have a true statement and our result is correct.

EXAMPLE 3   Solving linear equations using the equality properties
Solve: $4a - 7 = a + 4$

SOLUTION   We start by adding the inverse of a on both sides so that all variables will be on the left. Here are the steps:

$4a - 7 = a + 4$ Given

$4a + (-a) - 7 = a + (-a) + 4$ Add $(-a)$ to both sides so that all variables will be on the left.

$3a - 7 = 4$

$3a - 7 + 7 = 4 + 7$ Add 7 to both sides.

$3a = 11$

$a = \frac{11}{3}$ Divide both sides by 3 (or multiply by $\frac{1}{3}$).

The solution is $\frac{11}{3}$, and the solution set is $\{\frac{11}{3}\}$, as can be checked by substituting $\frac{11}{3}$ for a in $4a - 7 = a + 4$. Remember $\frac{11}{3}$ can also be written as $3\frac{2}{3}$ or $3.\overline{6}$.

PROBLEM 3
Solve: $5a - 8 = a + 5$

Sometimes we need to simplify an equation before applying the properties of equality. For instance, to solve $x + 6 = 3(2x - 2)$, we use the distributive property to simplify the right-hand side of the equation and then solve for x. We do this next.

EXAMPLE 4   Simplifying equations before solving
Solve: $x + 6 = 3(2x - 2)$

PROBLEM 4
Solve: $3 + x = 2(2x - 3)$

Answers to PROBLEMS
3. $\frac{13}{4}$ 4. 3

SOLUTION

$$x + 6 = 3(2x - 2) \quad \text{Given}$$
$$x + 6 = 6x - 6 \quad \text{Simplify the right-hand side (distributive property).}$$

We have two choices; we can isolate the variables on the left or on the right. To avoid negative expressions, this time we keep them on the right by adding $(-x)$ to both sides. We have

$$x + (-x) + 6 = 6x + (-x) - 6 \quad \text{Add } (-x) \text{ so all variables will be on the right.}$$
$$6 = 5x - 6$$
$$6 + 6 = 5x - 6 + 6 \quad \text{Add 6 to both sides.}$$
$$12 = 5x$$
$$\frac{12}{5} = x \quad \text{Divide by 5 (or multiply by } \tfrac{1}{5}\text{).}$$

The solution is $\frac{12}{5}$, and the solution set is $\left\{\frac{12}{5}\right\}$. This solution can also be written as $2\frac{2}{5}$ or 2.4.

C ❯ Solving Linear Equations Involving Fractions or Decimals

All the equations we have solved in this section can be written in the form $ax + b = c$, where x is the variable and a, b, and c are real numbers. Such equations are called **linear equations**. (You will see in Chapter 3 that the graph of $ax + b = y$ is a *straight line*.)

$$2x - 4 = 6 \quad \text{and} \quad \frac{x}{10} + \frac{x}{8} = 9$$

are linear equations. In the equation first, $2x$, -4, and 6 are called **terms:** $2x$ is a variable term and -4 and 6 are constant terms.

This terminology along with the properties we have studied thus far will help us develop a general procedure for solving linear equations in one variable.

To solve the equation $3(x - 2) + 20 = -2x - 2(8 - 6x) + 9$, we would like to have the variable on one side of the equal sign and a constant on the other side ($x = $ constant). The given equation has the variable terms on both sides and included within parentheses. First, use the distributive property to remove parentheses:

$$3x - 6 + 20 = -2x - 16 + 12x + 9$$

Second, simplify on each side of the equal sign by combining any like terms:

$$3x + 14 = 10x - 7$$

Since the variable term appears on both sides of the equal sign, we will subtract $3x$ from both sides:

$$14 = 7x - 7$$

Now there is a constant term on both sides of the equal sign, so we add 7 to both sides.

$$21 = 7x$$

The final step is to divide both sides by 7, the coefficient of the variable term. This can also be accomplished by multiplying both sides by the reciprocal of 7 which is $\frac{1}{7}$. The result will be $3 = x$, that is, $x = 3$.

Some equations will contain fractions and decimals. The following is a procedure that can be used to solve any linear equation in one variable.

PROCEDURE

Procedure for Solving Linear Equations

1. Clear fractions or decimals, if necessary, by multiplying both sides of the equation by the LCD of the fractions or decimals.
2. Remove parentheses and collect like terms on each side (simplify), if necessary.

(continued)

3. Add or subtract the same quantity on both sides of the equation so that one side has only the terms containing variables.
4. Add or subtract the same quantity on both sides of the equation so that the other side has only a constant.
5. If the coefficient of the variable is not 1, multiply both sides by the reciprocal of the coefficient (or equivalently divide both sides by the coefficient).
6. Be sure to check your answer in the original equation.

You can remember the first five steps as "CRAM":

Clear fractions/decimals.
Remove parentheses/simplify.
Add/subtract to get the variable isolated.
Multiply/divide to make the coefficient a 1.

If an equation involves fractions, we "clear" them by multiplying both sides of the equation by the *smallest* number that is a multiple of each denominator—the least common denominator (LCD). Thus, to solve

$$\frac{x}{6} + \frac{x}{4} = 10$$

we have to find the LCD of $\frac{x}{6}$ and $\frac{x}{4}$. One way is to pick the larger of the two denominators and double it, triple it, and so on, until the other number divides into the result. Using this idea, we find that 12 is the LCD. Multiplying both sides of the equation by 12, we have

$$12 \cdot \left(\frac{x}{6} + \frac{x}{4}\right) = 12 \cdot (10)$$

$12 \cdot \frac{x}{6} + 12 \cdot \frac{x}{4} = 12 \cdot 10$ Use the distributive property.
$2x + 3x = 120$ Simplify.
$5x = 120$ Combine like terms.
$x = 24$ Divide both sides by 5.

The solution is 24, and the solution set is {24}. Here is the check:

$$\frac{x}{6} + \frac{x}{4} \stackrel{?}{=} 10$$
$$\frac{24}{6} + \frac{24}{4} \mid 10$$
$$4 + 6$$
$$10$$

NOTE

When finding the LCD be sure to consider all denominators that appear in the equation (that includes looking on both sides of the equal sign).

EXAMPLE 5   Solving linear equations using the CRAM procedure

Solve:

a. $\frac{7}{24} = \frac{x}{8} + \frac{1}{6}$ **b.** $\frac{1}{5} - \frac{x}{4} = \frac{7(x+3)}{10}$

SOLUTION   We use the six steps as follows.

a. Given: $\frac{7}{24} = \frac{x}{8} + \frac{1}{6}$ Remember CRAM.

1. $24 \cdot \frac{7}{24} = 24\left(\frac{x}{8} + \frac{1}{6}\right)$ Clear fractions; multiply by 24, the LCD.

$\overset{1}{\cancel{24}} \cdot \frac{7}{\cancel{24}} = \overset{3}{\cancel{24}} \cdot \frac{x}{\cancel{8}} + \overset{4}{\cancel{24}} \cdot \frac{1}{\cancel{6}}$ Remove parentheses/simplify.

PROBLEM 5

Solve:

a. $\frac{7}{12} = \frac{x}{4} + \frac{1}{3}$

b. $\frac{1}{3} - \frac{x}{5} = \frac{8(x+2)}{15}$

Answers to PROBLEMS

5. **a.** 1   **b.** −1

2. $\quad 7 = 3x + 4$

3. $\quad 7 - 4 = 3x + 4 - 4 \quad$ Add -4 to both sides.

4. $\quad 3 = 3x$

5. $\quad \dfrac{3}{3} = \dfrac{3x}{3} \quad$ Multiply by $\frac{1}{3}$ (or divide by 3).

   $\quad 1 = x$

   The solution set is $\{1\}$.

6. **CHECK** Since $\dfrac{1}{8} = \dfrac{3}{24}$ and $\dfrac{1}{6} = \dfrac{4}{24}$,

   $$\dfrac{7}{24} = \dfrac{1}{8} + \dfrac{1}{6} = \dfrac{3}{24} + \dfrac{4}{24}$$

   which is true.

b. Given: $\quad \dfrac{1}{5} - \dfrac{x}{4} = \dfrac{7(x + 3)}{10}$

1. $\quad 20\left(\dfrac{1}{5} - \dfrac{x}{4}\right) = 20 \cdot \left[\dfrac{7(x + 3)}{10}\right] \quad$ Clear fractions; multiply by 20, the LCD.

2. $\quad \overset{4}{20} \cdot \dfrac{1}{5} - \overset{5}{20} \cdot \dfrac{x}{4} = \overset{2}{20} \cdot \dfrac{7(x + 3)}{10} \quad$ Remove parentheses/simplify.

   $\quad 4 - 5x = 14(x + 3)$
   $\quad 4 - 5x = 14x + 42$

3. $\quad 4 - 5x + 5x = 14x + 5x + 42 \quad$ Add 5x to both sides.

   $\quad 4 = 19x + 42$

4. $\quad 4 - 42 = 19x + 42 - 42 \quad$ Add -42 to both sides (or subtract 42).

   $\quad -38 = 19x$

5. $\quad -\dfrac{38}{19} = \dfrac{19x}{19} \quad$ Multiply by $\frac{1}{19}$ (or divide by 19).

   $\quad -2 = x$

   The solution set is $\{-2\}$.

6. **CHECK** $\quad \dfrac{1}{5} - \dfrac{x}{4} \overset{?}{=} \dfrac{7(x + 3)}{10} \quad$ for $x = -2$

   $$\begin{array}{c|c} \dfrac{1}{5} - \dfrac{(-2)}{4} & \dfrac{7(-2 + 3)}{10} \\ \dfrac{1}{5} + \dfrac{1}{2} & \dfrac{7(1)}{10} \\ \dfrac{7}{10} & \dfrac{7}{10} \end{array}$$

EXAMPLE 6 Clearing fractions in linear equations

Solve:

a. $\dfrac{x + 1}{3} + \dfrac{x - 1}{10} = 5$

b. $\dfrac{x + 1}{3} - \dfrac{x - 1}{8} = 4$

SOLUTION

a. The LCD of

$$\dfrac{x + 1}{3} \quad \text{and} \quad \dfrac{x - 1}{10}$$

PROBLEM 6

Solve:

a. $\dfrac{x + 2}{4} + \dfrac{x - 1}{5} = 3$

b. $\dfrac{x + 3}{2} - \dfrac{x - 2}{3} = 5$

(continued)

Answers to PROBLEMS

6. a. 6 **b.** 17

is $3 \cdot 10 = 30$, since 3 and 10 do not have any common factors. Multiplying both sides by 30, we have

$$30\left(\frac{x+1}{3} + \frac{x-1}{10}\right) = 30 \cdot (5)$$

$$30\left(\frac{x+1}{3}\right) + 30\left(\frac{x-1}{10}\right) = 30 \cdot 5 \qquad \text{Use the distributive property.}$$

$$10(x+1) + 3(x-1) = 150 \qquad \text{Since } \overset{10}{\cancel{30}}\left(\frac{x+1}{\cancel{3}}\right) = 10(x+1) \text{ and}$$
$$\overset{3}{\cancel{30}}\left(\frac{x-1}{\cancel{10}}\right) = 3(x-1).$$

$$10x + 10 + 3x - 3 = 150 \qquad \text{Use the distributive property.}$$
$$13x + 7 = 150 \qquad \text{Add like terms.}$$
$$13x = 143 \qquad \text{Subtract 7.}$$
$$x = 11 \qquad \text{Divide by 13.}$$

The solution set is $\{11\}$. The check is left to you.

b. Here the LCD is $3 \cdot 8 = 24$. Multiplying both sides by 24, we obtain

$$24\left(\frac{x+1}{3} - \frac{x-1}{8}\right) = 24 \cdot (4)$$

$$24\left(\frac{x+1}{3}\right) - 24\left(\frac{x-1}{8}\right) = 24 \cdot 4 \qquad \text{Use the distributive property.}$$

$$8(x+1) - 3(x-1) = 96 \qquad \text{Since } \overset{8}{\cancel{24}}\left(\frac{x+1}{\cancel{3}}\right) = 8(x+1) \text{ and}$$
$$\overset{3}{\cancel{24}}\left(\frac{x-1}{\cancel{8}}\right) = 3(x-1).$$

$$8x + 8 - 3x + 3 = 96 \qquad \text{Use the distributive property.}$$
$$5x + 11 = 96 \qquad \text{Add like terms.}$$
$$5x = 85 \qquad \text{Subtract 11.}$$
$$x = 17 \qquad \text{Divide by 5.}$$

The solution set is $\{17\}$. Check this answer in the original equation.

If an equation has fractions, we "clear" the fractions by multiplying both sides of the equation by the LCD. If an equation has decimals, we will "clear" the decimals by multiplying both sides by a power of 10. The power of 10 should have as many zeros as the number of decimal places in the coefficient with the greatest number of decimal places. In Example 7, the coefficient 3.15 has the greatest number of decimal places (2), so we multiply both sides by 100.

EXAMPLE 7 Solving linear equations involving decimals
Solve: $14.5 - 3.15x = 5.5$ Remember CRAM.

SOLUTION
$$100 \cdot (14.5 - 3.15x) = 100 \cdot (5.5) \qquad \text{Clear decimals; multiply by 100.}$$
$$100(14.5) - 100(3.15x) = 100(5.5) \qquad \text{Remove parentheses/simplify.}$$
$$1450 - 315x = 550$$
$$1450 - 1450 - 315x = 550 - 1450 \qquad \text{Add } -1450 \text{ to both sides.}$$
$$\frac{-315x}{-315} = \frac{-900}{-315} \qquad \text{Multiply by } -\frac{1}{315} \text{ (or divide by } -315\text{).}$$
$$x = \frac{900}{315} \qquad \text{Reduce the fraction by dividing numerator and denominator by 45.}$$
$$x = \frac{20}{7}$$

The solution is $\frac{20}{7}$, and the solution set is $\left\{\frac{20}{7}\right\}$, which you can check by substitution in the original equation. Also, the solution can be written as a repeating decimal or as 2.86 rounded to the nearest hundredth.

PROBLEM 7
Solve: $12.5 - 1.25x = 6.5$

Answers to PROBLEMS

7. $\frac{24}{5}$ or 4.8

Later in this chapter we shall translate word problems into equations such as $1.30P + 1.50(50 - P) = 72.50$. But, first let's solve this equation.

EXAMPLE 8 Solving linear equations involving decimals

Solve: $1.30P + 1.50(50 - P) = 72.50$

SOLUTION

$$1.3P + 1.5(50 - P) = 72.5$$ Rewrite the equation by dropping the zero in the hundredths place for each coefficient.

$$10[1.3P] + 10[1.5(50 - P)] = 10[72.5]$$ Clear decimals; multiply by 10.
$$10(1.3P) + 10(1.5)(50 - P) = 10(72.5)$$ Remove parentheses/simplify.
$$13P + 15(50 - P) = 725$$
$$13P + 750 - 15P = 725$$
$$-2P + 750 = 725$$
$$-2P + 750 - 750 = 725 - 750$$ Add -750 to both sides.
$$\frac{-2P}{-2} = \frac{-25}{-2}$$ Multiply by $-\frac{1}{2}$ (or divide by -2).
$$P = \frac{25}{2} = 12.5$$ Simplify the answer.

The solution is 12.5, and the solution set is $\{12.5\}$. The solution set can also be written as $\{12\frac{1}{2}\}$. Make sure you check this!

PROBLEM 8

Solve: $3.30d + 1.30(30 - d) = 42.50$

D ▶ Solving Linear Equations with One Solution, No Solutions, or Infinitely Many Solutions

So far all our equations have had one solution. They are called **conditional equations**. There are two other possibilities.

EQUATIONS WITH NO SOLUTIONS OR INFINITELY MANY SOLUTIONS

1. Equations with *no* solutions are *contradictions*. $x + 4 = x - 2$
2. Equations with all real numbers as solutions are *identities*. $2x + 8 = 2(x + 4)$

Consider the following example:

$$-4x + 6 = 2(1 - 2x) + 3$$ Clear fractions: none.
$$-4x + 6 = 2 - 4x + 3$$ Remove parentheses/simplify.
$$-4x + 6 = -4x + 5$$
$$-4x + 4x + 6 = -4x + 4x + 5$$ Add $4x$ to both sides.
$$6 = 5 \quad \text{False!}$$

We cannot find a replacement value for x to satisfy this equation. This equation is a **contradiction**; it has no solution. Its solution set is $\varnothing$ (the empty set). On the other hand, consider the following equation:

$$-4x + 6 = 2(1 - 2x) + 4$$ Clear fractions: none.
$$-4x + 6 = 2 - 4x + 4$$ Remove parentheses/simplify.
$$-4x + 6 = -4x + 6$$ Add $4x$ to both sides.
$$6 = 6 \quad \text{True!}$$

Here, any x will be a solution (try 0, 1, or any other number in the original equation). This equation is an **identity**. Any real number is a solution. Its solution set is ***R*, the set of real numbers**.

Answers to PROBLEMS

8. $\frac{7}{4}$ or 1.75

84 Chapter 2 Linear Equations and Inequalities

EXAMPLE 9 Solving linear equations with one solution, no solutions, or infinitely many solutions

Solve and identify as a contradiction, an identity, or a conditional equation.

a. $3x + 8 = 3(x + 1) + 5$ **b.** $3x + 8 = 3(x + 1) - 3x + 5$
c. $3x + 8 = 3(x + 1) + 2$

SOLUTION We use the six-step procedure we studied.

a. Given: $3x + 8 = 3(x + 1) + 5$ Clear fractions: none.
$3x + 8 = 3x + 3 + 5$ Remove parentheses/simplify.
$3x + 8 = 3x + 8$

We can stop here. This equation is always true. It is an identity, and its solution set is R.

b. Given: $3x + 8 = 3(x + 1) - 3x + 5$ Clear fractions: none.
$3x + 8 = 3x + 3 - 3x + 5$ Remove parentheses/simplify.
$3x + 8 = 8$
$3x = 0$ Add -8.
$x = 0$ Multiply by $\frac{1}{3}$ (or divide by 3).

This equation has a solution, $x = 0$, and it is called a conditional equation. This conditional equation is true only under the condition that the variable is replaced with the value "0."

c. Given: $3x + 8 = 3(x + 1) + 2$ Clear fractions: none.
$3x + 8 = 3x + 3 + 2$ Remove parentheses/simplify.
$3x + 8 = 3x + 5$

We can stop here. This equation is a contradiction; it is always false. (Try subtracting $3x$ from both sides.) It has no solutions, and its solution set is $\emptyset$.

PROBLEM 9
Solve and identify as a contradiction, an identity, or a conditional equation.

a. $4(x + 1) + 3 = 4x + 7$
b. $5(x + 2) + 1 = 3 + 5x$
c. $6(x + 3) = 6 - 6x$

CAUTION
Do not write $\{\emptyset\}$ to represent the empty set because the set $\{\emptyset\}$ is **not** empty, it has one element, $\emptyset$! Always use $\emptyset$ or $\{\ \}$ to represent the empty set.

Boost your grade at mathzone.com!
> Practice Problems
> NetTutor
> Self-Tests
> e-Professors
> Videos

› Exercises 2.1

‹A› Solutions of an Equation   In Problems 1–10, determine if the number following the equation is a solution of the equation.

1. $2x + 8 = 14$; 3
2. $5x + 5 = 10$; 2
3. $-2x + 1 = 3$; -1
4. $-3x + 4 = 10$; -2
5. $2y - 5 = y - 2$; 3
6. $3y - 7 = y + 1$; 4
7. $\frac{4}{5}t - 1 = 5t$; $-\frac{1}{5}$
8. $6t + 1 = t + \frac{2}{3}$; $-\frac{1}{3}$
9. $\frac{1}{2}x + 5 = 5 - \frac{1}{3}x$; 3
10. $-\frac{1}{3}x + 1 = x - 3$; 3

‹B› Solving Equations Using the Properties of Equality   In Problems 11–31, solve the equation.

11. $3x - 4 = 8$
12. $5a + 16 = 6$
13. $2y + 8 = 10$
14. $4b - 6 = 2$
15. $-3z - 6 = -12$
16. $-4r - 3 = 5$

Answers to PROBLEMS

9. **a.** R; identity **b.** $\emptyset$; contradiction **c.** -1; conditional

17. $-5y + 2 = -8$

18. $-3x + 2 = -10$

19. $3x + 5 = x + 19$

20. $4x + 6 = x + 9$

21. $7(x - 1) - 3 + 5x = 3(4x - 3) + x$

22. $5(x + 1) + 3x + 2 = 8(x + 2) + 2x + 1$

23. $6v - 8 = 8v + 8$

24. $8t + 3 = 15t - 11$

25. $7m - 4m + 12 = 0$

26. $10k + 25 - 5k = 35$

27. $4(2 - z) + 8 = 8(2 - z)$

28. $4(3 - y) + 8 = 12(3 - y)$

29. $5(x + 3) = 3(x + 3) + 6$

30. $y - (5 - 2y) = 7(y - 1) - 2$

31. $5(4 - 3a) = 7(3 - 4a)$

⟨C⟩ Solving Linear Equations Involving Fractions or Decimals In Problems 32–70, solve the equation.

32. $\frac{3}{4}y - 4 = \frac{1}{4}y - 2$

33. $-\frac{7}{8}c + 5 = -\frac{5}{8}c + 3$

34. $x + \frac{2}{3}x = 10$

35. $-2x + \frac{1}{4} = 2x + \frac{4}{5}$

36. $6x + \frac{1}{7} = 2x - \frac{2}{7}$

37. $\frac{t}{6} + \frac{t}{8} = 7$

38. $\frac{f}{9} + \frac{f}{12} = 14$

39. $\frac{x}{2} + \frac{x}{5} = \frac{7}{10}$

40. $\frac{a}{3} + \frac{a}{7} = \frac{20}{21}$

41. $\frac{c}{3} - \frac{c}{5} = 2$

42. $\frac{F}{4} - \frac{F}{7} = 3$

43. $\frac{W}{6} - \frac{W}{8} = \frac{5}{12}$

44. $\frac{m}{6} - \frac{m}{10} = \frac{4}{3}$

45. $\frac{x}{5} - \frac{3}{10} = \frac{1}{2}$

46. $\frac{3y}{7} - \frac{1}{14} = \frac{1}{14}$

47. $\frac{x + 4}{4} - \frac{x + 2}{3} = -\frac{1}{2}$

48. $\frac{w - 1}{2} + \frac{w}{8} = \frac{7w + 1}{16}$

49. $\frac{x + 1}{4} - \frac{2x - 2}{3} = 3$

50. $\frac{z + 4}{3} = \frac{z + 6}{4}$

51. $\frac{2h - 1}{3} = \frac{h - 4}{12}$

52. $\frac{5 - 6y}{7} - \frac{-7 - 4y}{3} = 2$

53. $\frac{2w + 3}{2} - \frac{3w + 1}{4} = 1$

54. $\frac{8x - 23}{6} + \frac{1}{3} = \frac{5}{2}x$

55. $\frac{7r + 2}{6} + \frac{1}{2} = \frac{r}{4}$

56. $\frac{x + 1}{2} + \frac{x + 2}{3} + \frac{x + 4}{4} = -8$

57. $6.3x - 8.4 = 16.8$

58. $15.5a + 49.6 = 18.6$

59. $-12.6y - 25.2 = 50.4$

60. $6.4y - 19.2 = 32$

61. $2.1y + 3.5 = 0.7y + 83.3$

62. $2.4x + 3.6 = 0.6x + 5.4$

63. $3.5(x + 3) = 2.1(x + 3) + 4.2$

64. $7.2(3 - t) = 2.4(3 - t) + 4.8$

65. $0.40y + 0.20(32 - y) = 9.60$

66. $0.30x + 0.35(50 - x) = 16$

67. $0.65x + 0.40(50 - x) = 25.375$

68. $0.09y + 0.12(200 - y) = 19.20$

69. $0.06P + 0.08(2000 - P) = 130$

70. $0.15P + 0.10(6000 - P) = 660$

⟨D⟩ Solving Linear Equations with One Solution, No Solutions, or Infinitely Many Solutions In Problems 71–76, solve the equation and identify each as a contradiction, an identity, or a conditional statement.

71. $4(x - 2) + 4 = 4x - 4$

72. $8 - (2 - 3x) = 3(x + 2)$

73. $29 - 10x = 4 + 5(2x - 3)$

74. $1 + (2x - 6x) = 2x - 4(x + 2)$

75. $7 + (9 - 5x) = 5(2 - x)$

76. $-(4 - 6x) = 2x + 4(x + 2)$

⟩⟩⟩ Using Your Knowledge

If the Shoe Fits Use the knowledge you have gained to solve the following problems about shoes. The relationship between your shoe size S and the length of your foot L (in inches) is given by

$$S = 3L - 22 \quad \text{(for men)}$$
$$S = 3L - 21 \quad \text{(for women)}$$

77. If Tyrone wears a size 11, what is the length L of his foot?

78. If Maria wears a size 7, what is the length L of her foot?

79. Sam's size 7 tennis shoes fit Sue perfectly! What size women's tennis shoe does Sue wear?

80. The largest shoes ever sold was a pair of size 42 built for the giant Harley Davidson of Avon Park, Florida. How long is Mr. Davidson's foot?

81. How long of a foot requires a size 14, the largest standard shoe size for men?

82. How long is your foot when your shoe size is the same as the length of your foot and you are
   a. A man?   b. A woman?

83. In 1951, Eric Shipton photographed a 23-in. footprint believed to be that of the Abominable Snowman.
   a. What size shoe does the Abominable Snowman need?
   b. If the Abominable Snowman turned out to be a woman, what size shoe would she need?

❯❯❯ Write On

Write a paragraph.

84. Explain the difference between a conditional equation and an identity.

85. Define a contradictory equation.

86. Detail the terminology used for equations that have no solutions, exactly one solution, and infinitely many solutions, and give examples.

❯❯❯ Concept Checker

Fill in the blank(s) with the correct word(s), phrase, or mathematical statement.

87. An equation with no solution is called a(n) _____.

88. An example of a(n) _____ equation is $2x + 5 = 4$.

89. $2x + 4 = 6$ and $2x = 2$ are called _____ equations because they have the same solution.

90. The number or numbers that make an equation true are called the _____.

91. An identity will have the _____ as its solution set.

92. $2x + 4 - 6 + 9x$ is an example of a(n) _____.

algebraic expression **conditional**
linear **identity**
solutions **contradiction**
real numbers **∅**
equivalent

❯❯❯ Mastery Test

Solve:

93. $\dfrac{x+4}{8} - \dfrac{x+2}{6} = -\dfrac{1}{4}$

94. $\dfrac{y}{8} - \dfrac{1}{4} = \dfrac{7y+2}{12}$

95. $0.8t + 0.4(32 - t) = 19.2$

96. $0.60x + 3.6 = 0.40(x + 12)$

97. $7(x - 1) - x = 3 - 5x + 3(4x - 3)$

98. $8(x + 2) + 4x + 3 = 5x + 4 + 5(x + 1)$

99. Is -5 a solution of $-\dfrac{1}{5}x + 2 = -x + 4$?

100. Is $-\dfrac{1}{4}$ a solution of $4t - 1 = t + \dfrac{1}{4}$?

❯❯❯ Skill Checker

Simplify:

101. $3x - (4x + 7) - 2x$

102. $5x - 3 - (7 + 4x)$

103. $6x - 2(3x - 1)$

104. $8x - 4(3 - 2x) + 5x$

Evaluate:

105. $\dfrac{400 - 2 \cdot 150}{2}$

106. $\dfrac{300.48 - 2 \cdot 100}{2}$

107. $\dfrac{5}{9}(F - 32)$ when $F = 41$

108. $\dfrac{5}{9}(F - 32)$ when $F = 32$

Solve:

109. $x - 4 = -5$

110. $10 = x - 12$

111. $3x - 1 = 7$

112. $18 = 4x - 6$

2.2 Formulas, Geometry, and Problem Solving

Objectives

A > Solve a formula or literal equation for a specified variable and then evaluate the answer for given values of the variables.

B > Translate from word expressions into mathematical expressions.

C > Write a formula for a given situation that has been described in words.

D > Solve problems about angle measures.

To Succeed, Review How To . . .

1. Solve linear equations (pp. 76–83).
2. Evaluate expressions using the correct order of operations (pp. 53–54).

Getting Started
Trucks, Axles, and Tolls

You are a truck driver traveling on a toll road in Florida. If your truck has five axles, how much do you have to pay? You can use the toll schedule and a linear equation to figure this out when there are more than three axles. The cost C for a truck with n axles, where n is greater than 3, is given by

$$C = 0.50 + 0.25(n - 3)$$

For a truck with five axles, $n = 5$ and the answer is

$$C = 0.50 + 0.25(5 - 3)$$
$$= 0.50 + 0.25(2)$$
$$= 0.50 + 0.50$$
$$= \$1.00$$

Many problems in algebra can be solved when the correct formula is used. Suppose the cost of the toll is known and we want to know how many axles the truck has. Then we would need a formula for finding the number of axles given the cost of the toll. Since we are interested in n in the original formula, we could use our algebra skills to solve the formula for n. Thus, we are solving for a *specified variable*. To remind you that we are solving for n, the n is in color. Here is the procedure.

$C = 0.50 + 0.25(n - 3)$	Given
$100C = 50 + 25(n - 3)$	Clear decimals; multiply by 100.
$100C = 50 + 25n - 75$	Remove parentheses; simplify.
$100C = 25n - 25$	
$100C + 25 = 25n$	Add 25 to both sides.
$\dfrac{100C + 25}{25} = n$	Multiply by $\frac{1}{25}$ (or divide by 25).

In this section we shall discuss writing formulas, evaluating formulas, and solving for a specified variable in a formula.

A > Solving Formulas or Literal Equations for a Specified Variable

Remember, the equations we studied in Section 2.1 are called **linear equations in *one variable*** ($2x + 3 = 7x - 5$) because they involve only one variable, "x." However, equations may have two, three, four, or more variables. Equations that have more than one variable are called **literal equations**. An example of a literal equation is $3a + b = 8$.

A **formula** is a type of literal equation in which a relationship is described by using variables. An example of a formula is $P = 2l + 2w$, where P is the perimeter of a rectangle and $2l + 2w$ describes the relationship of how to use the length and width of the rectangle to calculate the perimeter.

Sometimes we are asked to solve for one of the variables in a literal equation or formula. When solving for a *specified variable,* the goal is much like solving linear equations in one variable. That is, we want the specified variable on one side of the equal sign and all the other variables and constants on the other side. Here is a procedure to follow when solving for a specified variable.

PROCEDURE

Procedure for Solving for a Specified Variable in a Formula or Literal Equation

1. Clear fractions or decimals, if necessary, by multiplying both sides of the equation by the LCD of the fractions or decimals.
2. Remove parentheses and simplify, if necessary.
3. Add or subtract the same quantity on both sides of the equation so that one side has only the terms containing the *specified variable* and the other side contains all the other variables and constants.
4. Multiply by the reciprocal of the coefficient of the *specified variable* if it is not a "1."

To apply this procedure, we parallel the solving of an equation with one variable to solving an equation with two variables for a specified variable.

Solve Equation with One Variable (Solve Linear Equation in One Variable)		Solve Equation with Two Variables (Solve Literal Equation for Specified Variable)	
$3x + 7 = -5$	Solve for x.	$3a + b = 8$	Solve for a.
$3x + 7 - 7 = -5 - 7$	Subtract 7.	$3a + b - b = 8 - b$	Subtract b.
$3x = -12$	Simplify.	$3a = 8 - b$	Simplify.
$x = -4$	Divide by 3.	$a = \dfrac{8 - b}{3}$	Divide by 3.

EXAMPLE 1   Solving for a specified variable in a literal equation

Anthropologists know how to estimate the height of a man (in centimeters) using only one bone. They use the formula

$$\underbrace{H}_{\text{Height of the man}} = 2.89 \underbrace{h}_{\text{Length of the humerus}} + 70.64$$

a. Solve for h.
b. A man is 157.34 centimeters tall. How long is his humerus?

SOLUTION

a. We have to solve for h; that is, we isolate h on one side of the equation. It is not necessary to clear decimals because we are rewriting the formula for "h."

$H = 2.89h + 70.64$ Clear decimals (not necessary).
    Remove parentheses (not necessary).

$H - 70.64 = 2.89h + 70.64 - 70.64$ Add -70.64 to both sides.

$H - 70.64 = 2.89h$

$\dfrac{H - 70.64}{2.89} = h$ Multiply by $\tfrac{1}{2.89}$ (or divide by 2.89).

Now we can find h for any given value of H.

PROBLEM 1

The comparable formula for a woman is $H = 2.75h + 71.48$.

a. Solve for h.
b. A woman is 126.48 centimeters tall. How long is her humerus?

Answers to PROBLEMS

1. a. $h = \dfrac{H - 71.48}{2.75}$ **b.** 20 cm

b. Substitute 157.34 for H to get

$$h = \frac{157.34 - 70.64}{2.89} = \frac{86.7}{2.89} = 30$$

The man's humerus is 30 centimeters long.

EXAMPLE 2 Solving for a specified variable in a formula

The formula for converting degrees Fahrenheit (°F) to degrees Celsius (°C) is

$$C = \frac{5}{9}(F - 32)$$

a. Solve for F.
b. If the temperature is 35°C, what is the equivalent Fahrenheit temperature?

SOLUTION

a. We use CRAM for solving linear equations.

Given: $C = \frac{5}{9}(F - 32)$

1. $9 \cdot C = \overset{1}{\cancel{9}} \cdot \frac{5}{\cancel{9}}(F - 32)$ Clear fractions; multiply by 9.

    $9C = 5(F - 32)$

2. $9C = 5F - 160$ Remove parentheses.

3. $9C + 160 = 5F$ Add 160 to both sides.

4. $\frac{9C + 160}{5} = F$ Multiply by $\frac{1}{5}$ (or divide by 5).

b. Substitute 35 for C. $F = \frac{9 \cdot 35 + 160}{5} = \frac{475}{5} = 95$

The equivalent Fahrenheit temperature is 95°F.

PROBLEM 2

a. Solve for a in the formula $A = \frac{1}{2}h(a + b)$.

b. If $A = 40$, $h = 10$, and $b = 5$, what is a?

Many of the formulas we encounter in algebra come from geometry. In geometry, the distance around a polygon is called the **perimeter,** and the number of square units occupied by the figure is its **area.** Table 2.1 shows the formulas for the perimeters and areas of some common geometric figures.

Table 2.1 Perimeters and Areas

Name	Shape	Perimeter	Area
Square		$P = 4s$	$A = s^2$
Rectangle		$P = 2l + 2w$	$A = lw$
Triangle		$P = s_1 + s_2 + b$	$A = \frac{1}{2}bh$

(continued)

Answers to PROBLEMS

2. a. $a = \frac{2A}{h} - b$ or $a = \frac{2A - bh}{h}$

b. 3

Table 2.1 (continued)

Name	Shape	Perimeter	Area
Circle		$C = 2\pi r = \pi d$ (Circumference, Radius, Diameter)	$A = \pi r^2$

$2r = d$ and $\pi \approx 3.14$

EXAMPLE 3 Solving geometric formulas for a specified variable

The perimeter P of a rectangle is given by

$$P = 2l + 2w$$

a. Solve for w.

b. One of the largest posters ever made was a rectangular greeting card 166 feet long with a perimeter of 458.50 feet. How wide was this poster?

SOLUTION

a.
$$P = 2l + 2w \quad \text{Clear fractions: none.}$$
$$\text{Remove parentheses: none.}$$
$$P - 2l = 2l - 2l + 2w \quad \text{Add } -2l \text{ to both sides.}$$
$$P - 2l = 2w$$
$$\frac{P - 2l}{2} = w \quad \text{Multiply by } \tfrac{1}{2} \text{ (or divide by 2).}$$

b. The perimeter P was 458.50 feet and the length l was 166 feet, so we substitute 458.50 for P and 166 for l to obtain

$$w = \frac{458.50 - 2 \cdot 166}{2}$$
$$= \frac{458.50 - 332}{2} = 63.25$$

The poster was 63.25 feet wide.

PROBLEM 3

a. Solve for l in the formula $P = 2l + 2w$.

b. The Mona Lisa, a painting by Leonardo da Vinci, has a perimeter of 102.8 inches and is 20.9 inches wide. What is the length of this painting?

EXAMPLE 4 Solving interest rate problems

If you invest P dollars at the simple interest rate r, the amount of money A you will receive at the end of t years is $A = P + Prt$.

a. Solve for P.

b. At the end of 5 years, an investor receives \$2250 on her investment at 10% simple interest. How much money did she invest?

SOLUTION

a. This time P appears twice on the right-hand side of the equation. Using the distributive property, we write $A = P(1 + rt)$. To get P by itself on the right, we need only divide by $(1 + rt)$. Here are all the steps:

$$A = P + Prt \quad \text{Given}$$
$$A = P(1 + rt) \quad \text{Use the distributive property.}$$
$$\frac{A}{1 + rt} = \frac{P(1 + rt)}{1 + rt} \quad \text{Divide by } 1 + rt.$$
$$P = \frac{A}{1 + rt}$$

PROBLEM 4

If $2A = ah + bh$,

a. Solve for h.

b. If A is 150, $a = 10$, and $b = 20$, what is h?

Answers to PROBLEMS

3. a. $L = \frac{P - 2W}{2}$ **b.** 30.5 in. **4. a.** $h = \frac{2A}{a + b}$ **b.** 10

b. We substitute 2250 for A, 10% = 0.10 for r, and 5 for t to obtain

$$P = \frac{2250}{1 + 0.10 \cdot 5} = \frac{2250}{1.5} = 1500$$

She invested $1500.

EXAMPLE 5 Solving for a specified variable

Do you know how to add $a_1 + a_2 + a_3 + \cdots + a_n$? If the difference between successive terms is a constant, the sum is

$$S = \frac{n(a_1 + a_n)}{2}$$

where n is the number of terms to be added. Note that a_1 is the first term and the subscript n in a_n indicates the number of the term.

a. Solve for n.

b. $S = 2 + 4 + 6 + \cdots + 100 = 2550$. Thus, $a_1 = 2$, $a_n = 100$, and $S = 2550$. Verify that we have added 50 terms.

SOLUTION

a. We want n by itself on the right, so we multiply by 2 and divide by $a_1 + a_n$. Here are the steps:

$$S = \frac{n(a_1 + a_n)}{2}$$

$$2 \cdot S = \cancel{2} \cdot \frac{n(a_1 + a_n)}{\cancel{2}} \quad \text{Clear fractions; multiply by 2.}$$

$$2S = n(a_1 + a_n)$$

$$\frac{2S}{a_1 + a_n} = \frac{n(a_1 + a_n)}{a_1 + a_n} \quad \text{Multiply by } \frac{1}{a_1 + a_n} \text{ [or divide by } (a_1 + a_n)\text{].}$$

$$\frac{2S}{a_1 + a_n} = n$$

b. We substitute $a_1 = 2$, $a_n = 100$, and $S = 2550$ in the formula

$$n = \frac{2S}{a_1 + a_n} \quad \text{to get} \quad n = \frac{2 \cdot 2550}{2 + 100} = \frac{5100}{102} = 50$$

PROBLEM 5

If $A = \dfrac{n(P_1 + P_n)}{2}$,

a. Solve for n.

b. If $A = 50$, $P_1 = 32$, and $P_n = 18$, find n.

B › Translating Words into Mathematical Expressions

You have probably noticed the frequent occurrence of certain words in the statements of word problems. Table 2.2 presents a brief mathematics dictionary to help you translate these words properly and consistently.

Table 2.2 Mathematics Dictionary

Words	Translation	Example	Translation
Add	+	Add n to 7	$7 + n$
More than		7 more than n	
Sum		The sum of n and 7	
Increased by		n increased by 7	
Added to		7 added to n	

(continued)

Answers to PROBLEMS

5. a. $n = \dfrac{2A}{P_1 + P_n}$ **b.** 2

Table 2.2 (continued)

Words	Translation	Example	Translation
Subtract Less than Minus Difference Decreased by Subtracted from	−	Subtract 9 from x 9 less than x x minus 9 Difference of x and 9 x decreased by 9 9 subtracted from x	$x - 9$
Of Product Times Multiply by	×	$\frac{1}{2}$ of a number x The product of $\frac{1}{2}$ and x $\frac{1}{2}$ times a number x Multiply $\frac{1}{2}$ by x	$\frac{1}{2}x$
Divide Ratio Divided by Divide into Quotient	÷	Divide 10 by x Ratio of 10 to x 10 divided by x x divided into 10 The quotient of 10 and x	$\frac{10}{x}$
Is the same as, yields, gives, is, equals	=	The sum of x and 10 is the same as the sum of 10 and x.	$x + 10 = 10 + x$

Study Skills Hint

To learn the terms in this mathematics dictionary, write them and their translation on an index card. These cards should be reviewed before and after each class and each homework assignment.

CAUTION

Phrases to look for:

1. "7 less than x" compared to "7 *is* less than x"
   $$x - 7 \qquad\qquad 7 < x$$

2. A number x subtracted *from* y
   $$y - x$$

3. x divided by y can also be stated y divided into x
   $$\frac{x}{y}$$

We are now ready to use our mathematics dictionary to translate words into mathematical expressions.

EXAMPLE 6 Translating words into mathematical expressions

Cover the "Translation" column below and see if you can translate the words into mathematical expressions.

Words	Translation
8 more than a number z	$z + 8$
The quotient of twice a number x and 9	$\frac{2x}{9}$
5 subtracted from 3 times a number m	$3m - 5$
The product of a number t and 4 minus 6	$4t - 6$
The sum of the consecutive integers n and $(n + 1)$	$n + (n + 1)$

PROBLEM 6

Translate the words into mathematical expressions.

a. A number x increased by 5
b. The quotient of a number m and 4
c. Twice a number t decreased by 7
d. The product of a number y and 6 subtracted from 9
e. Ten more than three-fourths of a number n

Answers to PROBLEMS

6. a. $x + 5$ b. $\frac{m}{4}$ c. $2t - 7$
   d. $9 - 6y$ e. $\frac{3}{4}n + 10$

C › Writing and Evaluating Formulas

In Section A you were asked to solve for a specified variable in a formula. There are many applications in which a formula is given in words and you have to translate it into algebra. We will use the translation skills learned in Section B to translate statements into an algebra equation or formula. For example, anthropologists know that a female's height H (in centimeters) can be found by multiplying the length L of her femur by 1.95 and adding 28.68.

Algebraically this is $H = 1.95L + 28.68$
(Multiplying f by 1.95 and adding 28.68)

See how your translation skills work in Example 7.

EXAMPLE 7   Writing formulas from words

The cost of a long distance call anywhere, anytime in the United States is advertised as 39¢ for the first minute and 6¢ for each additional minute or fraction of a minute thereafter.

a. Write a formula for the cost C of a call from Miami to Dallas with time t minutes.
b. Solve the formula in part **a** for the time t minutes.
c. Roberto's telephone bill showed a $1.59 charge for a Miami-to-Dallas call. Use the formula found in part **b** to find the number of minutes for which he was charged.

SOLUTION   First we write the words relating the information. Use the variable "C" to represent the cost of the call and the variable "t" to represent the total time of the call in minutes. Let's look at the costs for 1, 2, 3, and 4 minutes and see whether we can notice a pattern.

	Cost of the call	is	first minute charge	plus	$0.06	times	number of minutes after the first
For 1 minute:	C	=	$0.39	+	$0.06	·	$(1 - 1)$
For 2 minutes:	C	=	$0.39	+	$0.06	·	$(2 - 1)$
For 3 minutes:	C	=	$0.39	+	$0.06	·	$(3 - 1)$
For 4 minutes:	C	=	$0.39	+	$0.06	·	$(4 - 1)$
For t minutes:	C	=	$0.39	+	$0.06	·	$(t - 1)$

You pay 6¢ for each minute after the first, so you have to multiply 0.06 by $(t - 1)$.

a. In general, the formula for C is $C = 0.39 + 0.06(t - 1)$.
b. To solve for t, think of t as the unknown and C as a number, then follow the procedure for solving a linear equation. (It is not necessary to clear the decimals.)

$C = 0.39 + 0.06(t - 1)$ Clear decimals; not necessary.
$C = 0.39 + 0.06t - 0.06$ Remove parentheses/simplify.
$C = 0.33 + 0.06t$
$C - 0.33 = 0.06t$ Add -0.33 to both sides.
$\dfrac{C - 0.33}{0.06} = t$ Multiply by $\dfrac{1}{0.06}$ (or divide by 0.06).

c. Substituting 1.59 for C in the preceding equation gives us

$$\dfrac{1.59 - 0.33}{0.06} = \dfrac{1.26}{0.06} = 21$$

Roberto was charged for 21 minutes.

PROBLEM 7

The cost of a long distance call is advertised as 25¢ for the first minute and 10¢ for each additional minute or fraction of a minute thereafter.

a. Write a formula for the cost C of a call with time t minutes.
b. Solve the formula in part **a** for the time t minutes.
c. If one long distance charge on Nancy's bill was $2.75, for how many minutes was she charged?

Answers to PROBLEMS

7. a. $C = 0.25 + 0.10(t - 1), t \geq 1$
   b. $t = \dfrac{C - 0.15}{0.1}$ **c.** 26 minutes

D › Angle Measures

In geometry, angles are measured in **degrees,** and 1° is defined as $\frac{1}{360}$ of a complete revolution. Certain relationships between angles and their measures are shown in Table 2.3.

Table 2.3

Angle Relationships	Examples
If L_1 and L_2 are parallel lines crossed by a **transversal,** t, as shown to the right, angles 1, 2, 7, and 8 are **exterior angles,** and angles 3, 4, 5, and 6 are **interior angles.** Angles 1 and 5 are **corresponding angles,** as are angles 3 and 7. Corresponding angles are equal. The measures of angles 1 and 5 are equal and the measures of angles 3 and 7 are equal.	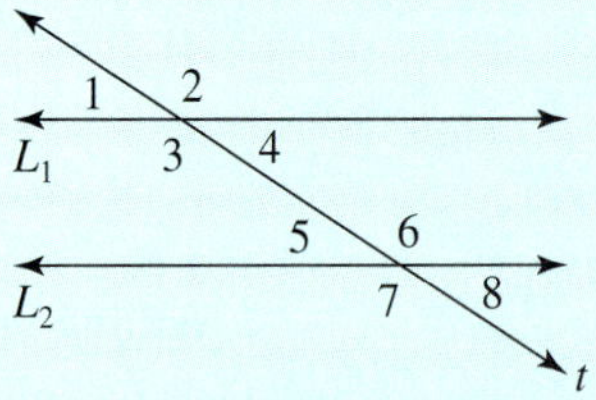
Alternate interior angles (the pairs of angles *between* the parallel lines and on *opposite* sides of the transversal) are equal; that is, the measures of angles 3 and 6 are equal. Similarly, the measures of angles 4 and 5 are equal.	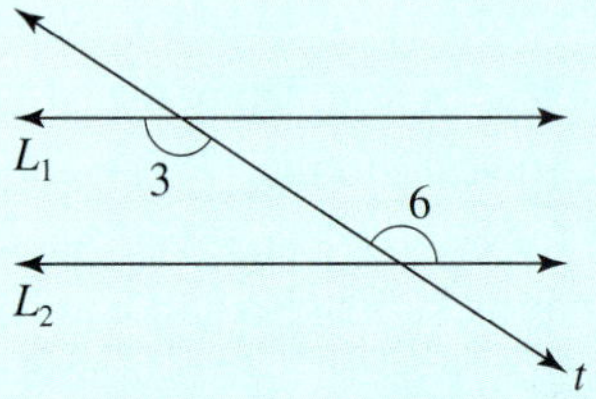 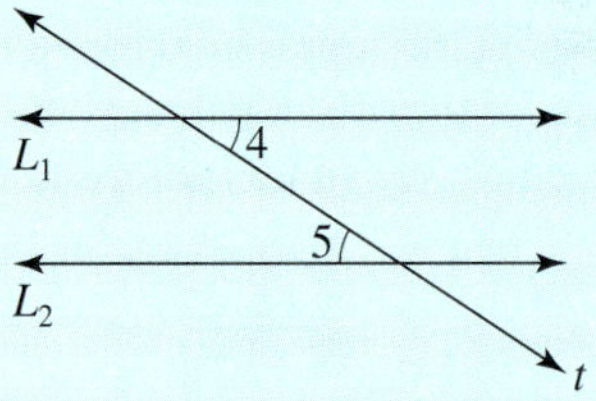
Vertical angles (angles *opposite* each other) are equal. That is, the following pairs of angles are equal: 1 and 4; 2 and 3; 5 and 8; and 6 and 7.	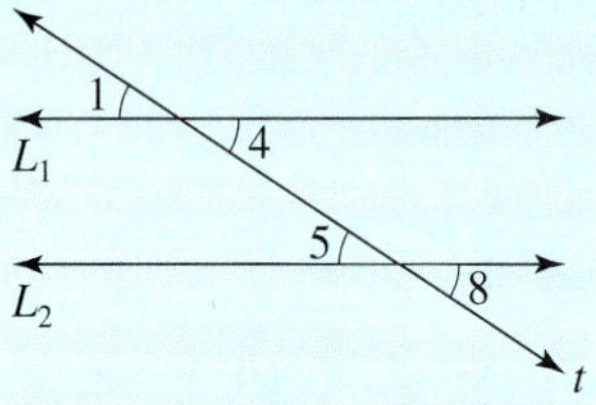 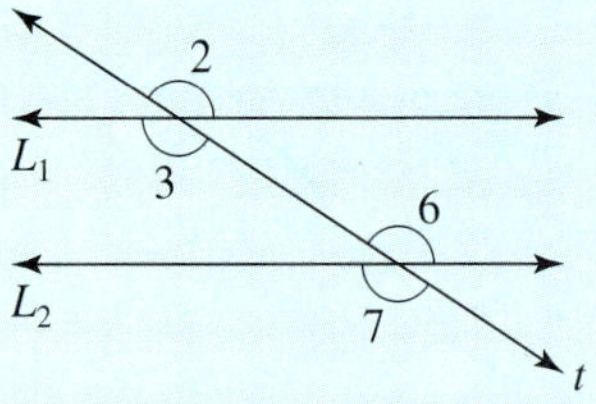

Here is a quick way of remembering all these facts: Look at angles 1 and 2. Angle 1 is *acute* (less than 90°), and angle 2 is *obtuse* (greater than 90°). Think of the measure of angle 1 as "small" and that of angle 2 as "big."

Remember:

All the acute ("small") angles in the figure have the same measure. Thus, angles 1, 4, 5, and 8 have the same measure.

All the obtuse ("big") angles in the figure have the same measure. Thus, angles 2, 3, 6, and 7 have the same measure.

EXAMPLE 8 Finding the measures of angles

If lines L_1 and L_2 are parallel, find x and the measures of the two unknown angles.

SOLUTION The angles whose measures are given are both "small" (alternate exterior angles); hence, they have equal measures. Thus

$6x - 5 = 4x + 15$ Clear fractions: none. Remove parentheses: none.
$6x - 4x - 5 = 4x - 4x + 15$ Add $-4x$ to both sides.
$2x - 5 = 15$
$2x - 5 + 5 = 15 + 5$ Add 5 to both sides.
$2x = 20$ Multiply by $\frac{1}{2}$ (or divide by 2).
$x = 10$

Substituting $x = 10$ in $6x - 5$, we find that the measure of one angle is $6 \cdot 10 - 5 = 55°$. The angles are equal, so the other angle also measures $55°$.

PROBLEM 8

If lines L_1 and L_2 are parallel, find x and the measures of the two unknown angles.

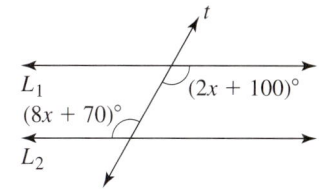

Boost your grade at mathzone.com!
> Practice Problems
> NetTutor
> Self-Tests
> e-Professors
> Videos

Exercises 2.2

<A> **Solving Formulas or Literal Equations for a Specified Variable** In Problems 1–12, solve the given literal equation for the indicated variable.

1. $V = \pi r^2 h$ for h
2. $V = \frac{1}{3}\pi r^2 h$ for h
3. $V = lwh$ for w
4. $V = lwh$ for h
5. $P = s_1 + s_2 + b$ for b
6. $P = s_1 + s_2 + b$ for s_2
7. $A = \pi(r^2 + rs)$ for s
8. $T = 2\pi(r^2 + rh)$ for h
9. $\frac{V_2}{V_1} = \frac{P_1}{P_2}$ for V_2
10. $\frac{V_2}{V_1} = \frac{P_1}{P_2}$ for P_1
11. $2x + 3y = 12$ for y
12. $2x + 3y = 12$ for x

13. The distance D traveled in time T by an object moving at rate R is given by $D = RT$.
    a. Solve for T.
    b. The distance between two cities A and B is 220 miles (mi). How long would it take a driver traveling at 55 miles per hour to go from A to B?

14. The ideal weight W (in pounds) of a man is related to his height H (in inches) by the formula $W = 5H - 190$.
    a. Solve for H.
    b. If a man weighs 160 pounds, how tall should he be?

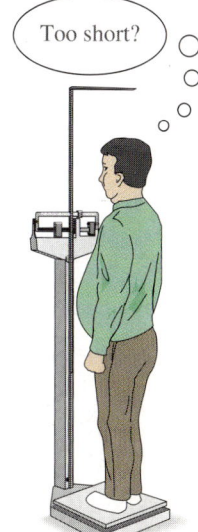

Too short?

15. The number of hours H a growing child should sleep is $H = 17 - \frac{A}{2}$, where A is the age of the child in years.
    a. Solve for A.
    b. The parents of an infant cannot wait until the child sleeps just 8 hours a day. At what age will that happen?

Answers to PROBLEMS

8. $x = 5$; both angles measure $110°$.

16. The efficiency energy rating E of an air conditioner is given by $E = \frac{B}{W}$, where B is the cooling capacity (per hour) in British thermal units and W is the watts of energy consumed.

a. Solve for W.

b. How many watts of electricity would an air conditioner with $E = 9$ and $B = 9000$ consume in 1 hr?

17. The operating profit margin o for a business is

$$o = \frac{c + e}{n}$$

where c is the cost of goods sold, e is the operating expense, and n is the net sales.

a. Solve for c.

b. If the operating expenses of a business amounted to $18,500, the operating profit margin was 96% = 0.96, and the net sales were $50,000, what was the cost of the goods sold?

18. The acid-test ratio A for a business is given by

$$A = \frac{C + R}{L}$$

where C is the amount of cash on hand, R is the amount of receivables, and L is the current liability.

a. Solve for R.

b. If the A ratio for a business is 1, its current liability is $7800, and the business has $1200 cash on hand, what should the accounts receivable be?

19. The sum of the measures of the three angles of a triangle is 180°. The formula is

$$A + B + C = 180$$

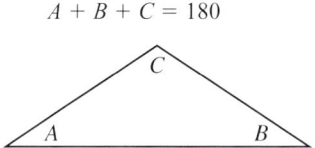

where A, B, and C represent the measures of the three angles.

a. Solve for B.

b. If the measure of angle A is 47° and the measure of angle C is 119°, find the measure of angle B.

20. The area A of a trapezoid is given by $A = \frac{1}{2}h(a + b)$.

a. Solve for b.

b. If the area of a trapezoid is 60 square units, its height h is 10 units, and side a is 7 units, what is the length of side b?

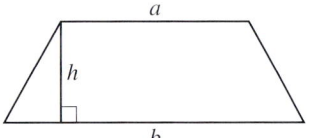

Use the following information in Problems 21–22.

The relationship between the length L of the femur bone of a man and his height H is given by

$$H = 1.88L + 32 \quad (1)$$

where H and L are both measured in inches. The corresponding equation for a woman is

$$H = 1.95L + 29 \quad (2)$$

21. a. Solve equation (1) for L.

b. Can a 20-in. femur belong to a man whose height was 6 ft?

c. Can the bone belong to a man whose height was 69.6 in.?

d. Solve equation (2) for L.

e. Can a femur measuring 20 in. in length belong to a woman whose height was 5 ft, 8 in.?

22. A police pathologist wants to check the accuracy of equations (1) and (2). Use your calculator to find L (to one decimal place) for:

a. A man 5 ft, 6 in. in height.

b. A man 5 ft, 8 in. in height.

c. A woman 5 ft, 6 in. in height.

d. A woman 5 ft, 10 in. in height.

‹ B › Translating Words Into Mathematical Expressions In Problems 23–30, translate the words into mathematical expressions.

23. The product of 4 and a number m increased by 18

24. 25 more than twice a number n

25. The quotient of a number t and 3

26. The quotient of a number p and the number plus 2

27. The difference of $\frac{1}{2}$ a number b and 7

28. $\frac{2}{3}$ of a number c subtracted from 12

29. The product of the sum of x and y and the difference of x and y

30. The quotient of the sum of x and y and the difference of x and y

⟨ C ⟩ Writing and Evaluating Formulas

31. A new Acura can be leased for $2000 down and $309 per month.
   a. Find a formula for the total cost C of the car after m months.
   b. Use the formula from part **a** to find the total cost C after leasing the car for the required 39 months.

32. A plumber charges $25 per hour ($h$) plus $30 for the service call.
   a. Find a formula for C if C is the total cost for the call.
   b. Use the formula from part **a** to find the number of hours worked by a plumber who charged $142.50.

33. A disc jockey charges $150 per hour with a set up fee of $75.
   a. Write a formula for the cost C of the disc jockey for h hours.
   b. If the cost of the DJ for Cindy's birthday was $675, how many hours did he work?

34. A finance company will lend you $1000 for a finance charge of $20 plus simple interest at 1% per month. This means that you will pay interest of 1% of $1000—that is, $10 per month in addition to the $20 finance charge.
   a. Find a formula for F if F is the finance charge and m is the number of months before you repay the loan.
   b. Use the formula from part **a** to find the number of months outstanding on a $1000 loan that has already cost $140 in finance charges.

35. The cost C of a Tampa-to-Los Angeles call during business hours is 36¢ for the initial minute and 8¢ for each additional minute or fraction thereof.
   a. Write a formula for the cost C of a Tampa-to-Los Angeles call if t is the total time of the call in minutes.
   b. Use the formula from part **a** to find the time of a phone call from Tampa to Los Angeles if the cost was $4.04.

⟨ D ⟩ Angle Measures
In Problems 36–43, find x and then find the measure of each marked angle. In Problems 40–43, assume that L_1 and L_2 are parallel.

36.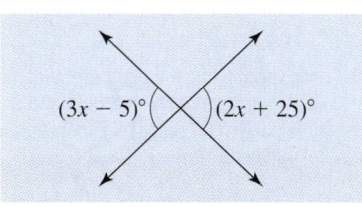

$3x - 5 = 2x + 25$
$1x = 30$

37.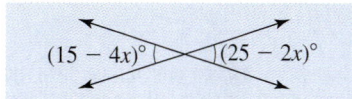

38.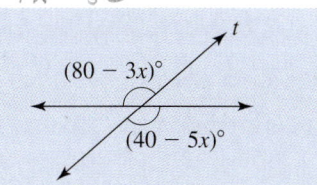

$4x18 = 16x + 2$

39.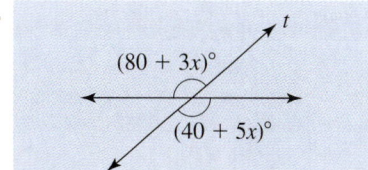

40.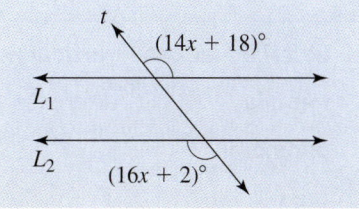

41.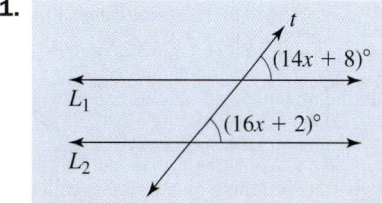

42.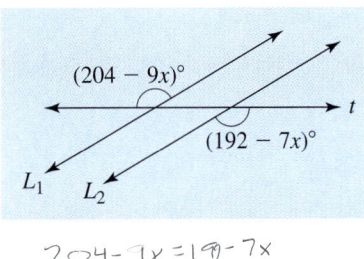

$204 - 9x = 192 - 7x$
$12 = 2x$

43.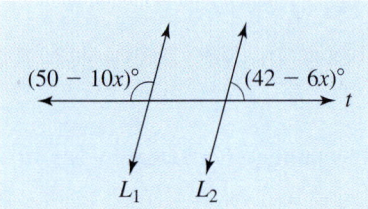

(*Hint:* The two marked angles are supplementary. Their sum is 180°.)

> > > **Using Your Knowledge**

Global Warming and Sea Levels Do you know what global warming is? It is the theory that over the past half-century, the earth has begun to warm significantly due to factors like pollution (from your car and power plants); deforestation (fewer trees, less shade, less rain, more heat); and the depletion of the ozone layer (ozone absorbs ultraviolet radiation; less ozone, more sun and more heat). The average global temperature C (in degrees Celsius) can be approximated by

where t is the number of years after 1960.

44. Solve for t in the equation.

45. In what year will the average global temperature be 20°C? Give your answer to the nearest year.

46. In what year will the average global temperature be 25°C? Give your answer to the nearest year.

47. The average global temperature in 1962 was about 15°C. An increase of 1°C causes a 1-foot rise in the sea level. What was the rise in sea level from 1962 to the year 2000?

> > > **Write On**

48. In this section you have solved formulas for specified variables. Write a short paragraph explaining what that means.

49. A linear equation in one variable is an equation that can be written as $ax + b = c$ (a, b, and c are real numbers and $a \neq 0$). Write an explanation of how you would solve for x in this equation. Use your explanation to solve the equation $-3x + 10 = 16$ for x.

50. The definition of linear equation in one variable (Problem 49) states that a cannot be zero ($a \neq 0$). Explain what happens when a is 0. Can $c = 0$?

> > > **Concept Checker**

Fill in the blank(s) with the correct word(s), phrase, or mathematical statement.

51. A _____ is a type of literal equation that describes a relationship using variables.

52. If two lines intersect, then the opposite angles are equal and are called _____.

53. Equations with two or more variables are called _____.

54. The number of square units occupied by a figure is called its _____.

literal equations

formula

perimeter

area

vertical angles

alternate interior angles

>>> Mastery Test

In Problems 55 and 56, find x and the measures of the marked angles.

55.

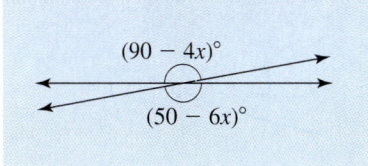

56. Assume L_1 and L_2 are parallel lines.

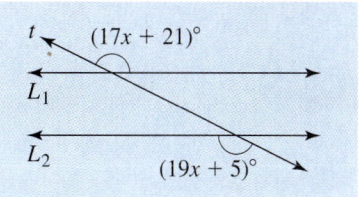

57. Suppose the cost C of mailing a first class letter is 39¢ for the first ounce (oz) and 24¢ for each additional ounce or fraction thereof. If w is the weight of a letter in ounces:

  a. Write a formula for the cost C of mailing the letter in dollars.

  b. Solve for w in the formula.

  c. What is the cost of mailing a letter weighing 10 oz?

58. The annual number N of cigarettes per person consumed in the United States is the difference between 4200 and the product of 70 and t, where t is the number of years after 1960.

  a. Write a formula for N.

  b. In what year did consumption drop to 2100 cigarettes per person?

59. The speed S (in centimeters per second) for a certain type of ant is $S = \frac{1}{6}(C - 4)$, where C is the temperature in degrees Celsius.

  a. Solve for C.

  b. If the ant is moving at a rate of 2 centimeters per second, what is the temperature?

60. The number of chirps N a certain type of cricket makes per minute is $N = 4(F - 40)$, where F is the temperature in degrees Fahrenheit.

  a. Solve for F.

  b. If the cricket is chirping 80 times per minute, what is the temperature?

>>> Skill Checker

Simplify:

61. $n + (n + 1) + (n + 2)$

62. $n + (n + 2) + (n + 4)$

Solve:

63. $(p + 50) + p = 110$

64. $p + (p + 60{,}000) = 110{,}000$

65. $0.30m + 25 = 52$

66. $0.25m + 20 = 37.50$

67. $x + (90 - x) + 3x = 180$

68. $(90 - x) + x + 5x = 180$

2.3 Problem Solving: Integers and Geometry

Objectives

A ▸ Translate a word statement into a mathematical equation.

B ▸ Solve word problems of a general nature.

C ▸ Solve word problems about integers.

D ▸ Solve word problems about geometric formulas and angles.

▸ To Succeed, Review How To . . .

1. Simplify expressions (pp. 55–60).
2. Solve linear equations involving decimals (pp. 82–83).

▸ Getting Started
RSTUV Procedure

In the preceding sections we learned how to solve certain types of equations. Now we are ready to apply this knowledge to solve problems. These problems will be stated in words and are consequently called **word** or **story problems.** Word problems frighten many students, but do not panic. We have a surefire method for tackling *any* word problem. Here is our five-step procedure:

1. **R**ead the problem. Not once or twice, but until you understand it.
2. **S**elect the unknown; that is, find out what the problem asks for.
3. **T**hink of a plan to solve the problem.
4. **U**se the techniques you are studying to carry out the plan.
5. **V**erify the answer.

Look at the first letter in each sentence. To help you remember the steps, we call it the **RSTUV** method. Later in this section we have additional tips on how to use this method, but first we have to discuss the terminology we shall use and practice translating word statements into mathematical equations.

A ▸ Translating Words into Mathematical Equations

In Section 2.2B, we translated words into mathematical expressions. An important part of learning how to solve word problems is to be able to translate word statements into mathematical **equations.** Translating mathematics is much like translating a foreign language. We will consider one word or phrase at a time and its mathematical translation. Remember to refer to the Mathematics Dictionary in Table 2.2 when necessary. We will use the statement, "Three more than 5 times a number results in 13," to demonstrate the steps for translating words into a mathematical equation. We start by writing the word statement on one line.

2.3 Problem Solving: Integers and Geometry

Three more than 5 times a number results in 13.   Given word statement

Three more than 5 times a number results in 13.   Identify constants and phrases.

   3   5x   13   Translate.

Three *more than* 5 times a number *results in* 13.   Translate operations and equal sign.

   3   +   5x   =   13   Mathematical equation

EXAMPLE 1 Translating words into mathematical equations

Translate the statements into mathematical equations. **Do not solve.**

a. A number increased by 7 yields 3.
b. Nine more than twice a number is 18.
c. 45 is the result when the product of a number and 10 is subtracted from 15.
d. The difference of two consecutive integers, n and $(n + 1)$, is always -1.

SOLUTION

a. A number increased by 7 yields 3.   Given word statement
    A number increased by 7 yields 3.   Identify constants and phrases.
       x 7   3   Translate.
    A number *increased by* 7 *yields* 3.   Translate operations and equal sign.
       x +   7   =   3   Mathematical equation

b. Nine more than twice a number is 18.   Given word statement
    Nine more than twice a number is 18.   Identify constants and phrases.
       9           $2x$ 18   Translate.
    Nine *more than* twice a number *is* 18.   Translate operations and equal sign.
       9   + $2x$ =   18   Mathematical equation

c. 45 is the result when the product of a number and 10 is subtracted from 15.   Given word statement
    45 is the result when the product of a number and 10 is subtracted from 15.   Identify constants and phrases.
       45              $10x$ 15   Translate.
    45 *is the result* when the product of a number and 10 *is subtracted from* 15.   Translate operations and equal sign.
       45   =   15   − $10x$ Mathematical equation

d. The difference of the consecutive integers, n and $(n + 1)$, is always -1.   Given word statement
    The difference of the consecutive integers, n and $(n + 1)$, is always -1.   Identify constants and phrases.
                n $(n + 1)$ -1 Translate.
    The *difference* of two consecutive integers, n and $(n + 1)$, *is always* -1.   Translate operations and equal sign.
       n − $(n + 1)$ = -1 Mathematical equation

PROBLEM 1

Translate the statements into mathematical equations.
Do not solve.

a. Six more than 4 times a number is 14.
b. Twice a number plus 12 results in 4.
c. 12 is the result when 8 is subtracted from the product of a number and 5.
d. The difference of two consecutive even integers, n and $(n + 2)$, is always -2.

Now that we have practiced the skills for translating statements into mathematical equations, it is time to practice translating real-life statements into mathematical equations. In Example 2, we have written some statements from Internet research on current facts and trends. Let's translate them into mathematical equations.

Answers to PROBLEMS

1. a. $6 + 4x = 14$ or $4x + 6 = 14$
   b. $2x + 12 = 4$
   c. $12 = 5x - 8$
   d. $n - (n + 2) = -2$

EXAMPLE 2 Translating real-life statements into mathematical equations

Translate the statements into mathematical equations. **Do not solve.**

a. 45% of the world's ethanol production, p, amounts to 16.5 million liters.
b. The amount of money spent by women for the purchase of candles, A, is 96% of the $2 billion spent annually on U.S. retail sales of candles.
c. Three-fifths of Medicare managed-care drug benefit plans, m, are reporting that they will cap prescription drug benefits bringing the number of plans to 6.2 million.

SOLUTION

a. 45% of the world's ethanol production, p, amounts to 16.5 million liters.   *Given statement*

45% of	production, p,	amounts to	16.5 million liters.	Identify constants and phrases.
0.45	p		16.5	Translate.

45% *of* production, p, *amounts to* 16.5 million liters.   *Translate operations and equal sign.*

$$0.45 \cdot p = 16{,}500{,}000$$ *Mathematical equation*

b. Money spent by women for candles, A, is 96% of the $2 billion spent annually.   *Given statement*

Money spent, A, was	96%	of	$2 billion	Identify constants and phrases.
A	0.96		2,000,000,000	Translate.

Money spent, A, is 96% *of* the $2 billion spent annually.   *Translate operations and equal sign.*

$$A = 0.96 \cdot 2{,}000{,}000{,}000$$ *Mathematical equation.*

c. Three-fifths of plans, m, are reporting they will cap prescription drug benefits to 6.2 million.   *Given statement*

$\tfrac{3}{5}$ of	plans, m,	will cap to	6.2 million	Identify constants and phrases.
$\tfrac{3}{5}$	m		6,200,000	Translate.

$\tfrac{3}{5}$ *of* plans, m, *will cap to* 6.2 million   *Translate operations and equal sign.*

$$\tfrac{3}{5} \cdot m = 6{,}200{,}000$$ *Mathematical equation*

PROBLEM 2

Translate the statements into mathematical equations. Do not solve.

a. World biofuel production reached 670,000 barrels per day, which is the equivalent of about 1% of the global transport fuel market, g.
b. The number of hungry people, h, in developing countries in the second half of the 1990s increased by 18 million to 800 million.
c. If the total amount ($1.6 billion) spent for inpatient hospital costs for cardiovascular disease patients in Minnesota is divided by the 70,000 hospitalizations, the result is the average cost per hospital stay, C.

B > General Word Problems

The mathematics dictionary in Section 2.2 and the previous practice translating statements into mathematical equations will play an important role in the solution of word problems. You will find that the words contained in the dictionary are often key words when you translate a problem. But there are other details that can help you as well.

The **RSTUV** method for solving word problems was introduced in the "Getting Started." We will use the following word problem to develop the hints and tips for applying the **RSTUV** procedure in solving word problems.

Harold purchased a new ink cartridge for his printer. The cost of the ink cartridge, $34.99, plus the sales tax came to a total of $37.44. What was the amount of the sales tax?

Hints and Tips Mathematics is a language. As such you have to learn how to read it. You may not understand or even get through reading the problem the first time. Read it

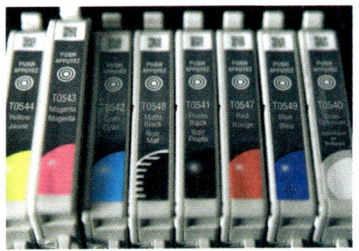

Answers to PROBLEMS

2. a. $670{,}000 = 0.01g$
b. $h + 18{,}000{,}000 = 800{,}000{,}000$
c. $\dfrac{1{,}600{,}000{,}000}{70{,}000} = C$

again and as you do, pay attention to key words or instructions such as *compute, draw, write, construct, make, show, identify, state, simplify, solve,* and *graph*. (Can you think of others?)

1. **Read the problem.** Using a complete sentence, identify what you are being asked by writing a "Find" statement. "Find the sales tax."

   How can you answer a question if you don't know what the question is? One way to look for the unknown is to look for the question mark (?) and read the material to its left. Not all problems will contain a question mark. Try to determine what is given and what is missing.

2. **Select the unknown.** Using a "Let" statement, represent the unknown(s) as a variable expression. "Let x = the sales tax."

   Problem solving requires many skills and strategies: Some of them are *look for a pattern; examine a related problem; make tables, pictures, diagrams; write an equation; work backwards;* and *make a guess*. (Can you think of others?)

3. **Think of a plan.** Using the information, translate it first to a word equation and then to an algebra equation.

$$\text{Cost} + \text{tax} = \text{total}$$
$$\$34.99 + x = \$37.44$$

If you are studying a mathematical technique, it is almost certain that you will have to use it to solve the given problem. Look for specific procedures given to solve certain problems.

4. **Use the techniques you are studying to carry out the plan.** Using your algebra skills, solve the equation.

$$\begin{array}{r} 34.99 + x = 37.44 \\ -34.99 \quad\quad -34.99 \\ \hline x = \$2.45 \end{array}$$

Look back and check the results of the original problem. Is your answer reasonable? Can you find it some other way?

5. **Verify the answer.** Replace the result into the word problem and check.

$$\$34.99 + \$2.45 = \$37.44$$

Now let's try Example 3 using the **RSTUV** method.

EXAMPLE 3 Car loan costs

J. R. Clementi bought a $20,000 car. He put $2500 down, and his monthly payment is $335. He is told that when the car loan is paid off he will have paid a total of $26,620 for the car. How many months will it take J. R. to pay off the car?

SOLUTION We use the RSTUV method.

1. **Read and write a "Find" statement.** Find how many months it will take J. R. to pay off the car.

2. **Select the unknown using a "Let" statement.** Let m = the number of months to pay off the car.

   The monthly payment is $335 and he will make "$m$" payments, so we can use the following variable expression to represent the amount spent on monthly payments.

$$\$335 \, (m) = \text{money spent on monthly payments}$$

(*Hint:* Often numbers that are not needed are mentioned in a problem, like the "$20,000 car" in this problem.)

PROBLEM 3

Kyra rented an intermediate sedan at $25 per day plus $0.20 per mile. How many miles can Kyra travel in one day for $50?

(continued)

Answers to PROBLEMS

3. Kyra can travel 125 miles.

3. **Think of a word equation and translate it to an algebra equation.**

   Down payment plus the monthly payments = total paid for the car.
   $2500 + \$335\,(m) = \$26{,}620$

4. **Use your algebra skills to solve the equation.**

   $335\,m = 24{,}120$ Subtract 2500.
   $m = 72$ Divide by 335.

5. **Verify the results.**

   $2500 + 335(72) = 26{,}620$

It will take J. R. 72 months (6 years) to pay off the car.

C > Integer Word Problems

Many popular algebra problems deal with integers. If you are given an integer, can you find the integer that follows it? For example, the integer that follows 7 is $7 + 1 = 8$, and the one that follows -6 is $-6 + 1 = -5$.

CONSECUTIVE INTEGERS

If n is any integer, the next **consecutive** integer is $n + 1$.

On the other hand, if you are given an *even* integer such as 6, the next *even* integer is 8 (add 2 this time). The even integer after -4 is $-4 + 2 = -2$.

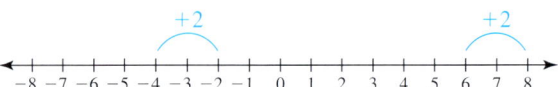

CONSECUTIVE EVEN (OR ODD) INTEGERS

If n is an *even* (or *odd*) integer, the next *even* (or *odd*) integer is $n + 2$.

If $n = 34$, the next even integer is $n + 2 = 34 + 2 = 36$. Similarly, if $n = 21$, the next odd integer is $n + 2 = 21 + 2 = 23$.

We use this idea in the next example. (Don't forget to consult the mathematics dictionary when necessary.)

EXAMPLE 4 Integer problems
The sum of three consecutive odd integers is 129. Find the integers.

SOLUTION Use the RSTUV method. We are asking for three consecutive odd integers. Look at the following number line and notice how far apart the odd integers are.

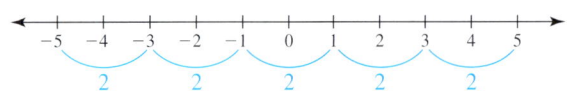

1. **Read and write a "Find" statement.** Find three consecutive odd integers.
2. **Select the unknown using a "Let" statement.** If we let n be the first of the odd integers, then we can represent the other two with variable expressions by using n and the spacing on the number line. We want three consecutive odd integers, so we need to find the next two consecutive odd integers. After noting on the graph that the odd integers are "2" apart, the odd integer after n is $(n + 2)$ and the one after $(n + 2)$ is $(n + 4)$.

PROBLEM 4
The sum of three consecutive odd integers is 249. Find the integers.

Answers to PROBLEMS

4. The three consecutive odd integers are 81, 83, and 85.

Let n = the first odd integer
$(n + 2)$ = the second odd integer
$(n + 4)$ = the third odd integer

3. Think of a word equation and translate it to an algebra equation.

First odd integer plus second plus third = the sum of the three
$$n + (n + 2) + (n + 4) = 129$$

4. Use your algebra skills to solve the equation.

$3n + 6 = 129$ Remove parentheses; simplify.
$3n = 123$ Add -6.
$n = 41$ Divide by 3 for the first odd integer.
$(n + 2) = 43$ The second odd integer
$(n + 4) = 45$ The third odd integer

5. Verify the results.

$$41 + 43 + 45 = 129$$

Thus, the three consecutive odd integers are 41, 43, and 45.

D › Geometry Problems

As you recall, the *perimeter* of a rectangle (distance around) is $P = 2l + 2w$, where l is the length and w is the width of the rectangle. Here's a problem using this formula.

EXAMPLE 5   A geometry word problem

The students at Osaka Gakun University made a rectangular poster whose length was 130 feet more than its width. If the perimeter of the poster was 416 feet, find its dimensions.

PROBLEM 5

A rectangular poster is 100 inches longer than it is wide. If its perimeter is 360 inches, find its dimensions.

SOLUTION   Use the RSTUV method. The key words are "rectangular" and "perimeter." Drawing the geometric shape and labeling it will help.

1. Read and write a "Find" statement. Find the dimensions, which are the length and width.

2. Select the unknown using a "Let" statement. If we let "w" represent the measure of the width, we can represent the measure of the length with a variable expression using "w."

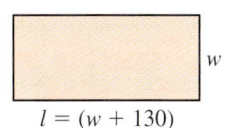

The length is 130 feet more than the width so we can represent the length as $(w + 130)$.

Let w = the measure of the width
$(w + 130)$ = the measure of the length

3. Think of a word equation and translate it to an algebra equation. We use the formula for the perimeter of the rectangle to write the word equation and translate it to an algebra equation.

$$2l + 2w = P$$
2(length) + 2(width) = the perimeter of the rectangle
$$2(w + 130) + 2(w) = 416$$

4. Use your algebra skills to solve the equation.

$2w + 260 + 2w = 416$ Remove parentheses.
$4w + 260 = 416$ Simplify.
$4w = 156$ Add -260.
$w = 39$ Divide by 4 for the width.
$(w + 130) = 169$ The length

Answers to PROBLEMS

5. The width is 40 in. and the length is 140 in.

(continued)

5. Verify the results.
$$2(169) + 2(39) = 416$$
Thus, the dimensions of the rectangle are width of 39 ft and length of 169 ft.

In Section 2.2, we studied vertical angles and angles formed by transversals that intersect parallel lines. Table 2.4 shows some other types of angles and their relationships.

Table 2.4

Angle	Relationship
A **right angle** is an angle whose measure is 90°. If the sum of the measures of two angles is 90°, the angles are **complementary angles** and they are **complements** of each other.	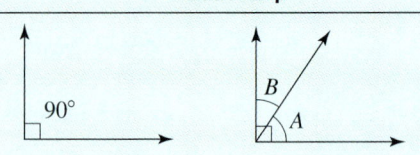 Right angle Complementary angles If the measure of one of the angles is $x°$, the measure of its complement is $(90 - x)°$.
A **straight angle** is an angle whose measure is 180°. If the sum of the measures of two angles is 180°, the angles are **supplementary angles** and they are **supplements** of each other.	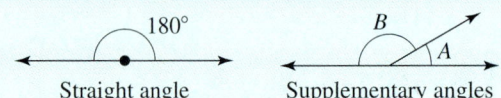 Straight angle Supplementary angles If the measure of one of the angles is $x°$, the measure of its supplement is $(180 - x)°$.
The **sum** of the measures of the angles of a triangle is 180°. You can use this fact to find the measure of the third angle when the measures of the other two angles are given.	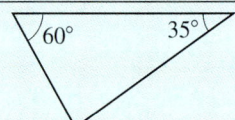 The measure of the other angle, call it $x°$, is such that $$60 + 35 + x = 180$$ $$95 + x = 180$$ $$x = 85$$ The third angle measures 85°.

EXAMPLE 6   A geometry problem involving angles

Two of the angles in a triangle are complementary. The third angle is twice the measure of one of the complementary angles. What is the measure of each of the angles?

SOLUTION   Use the RSTUV method. The key words are "complementary angles" and "angles in a triangle." You can use Table 2.4 to help with this problem. You should also draw the geometric shape.

1. Read and write a "Find" statement. Find the measures of the three angles of the triangle.

2. Select the unknown using a "Let" statement.
If we let x be the measure of one of the complementary angles, then we can represent the measure of the other complementary angle as $(90 - x)$. Well, that takes care of two of the three angles in the triangle, but what about the third angle? It says that the third angle is twice the measure of one of the complementary angles, so we represent its measure with $2(x)$.

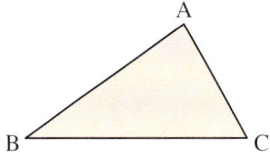

Let x = the measure of one of the complementary angles
$(90 - x)$ = the measure of the other complementary angle
$2(x)$ = the measure of the third angle of the triangle

PROBLEM 6

If the measures of the three angles of a triangle are related such that the second angle is twice the first angle and the third angle is 20° more than the first angle, find the measure of each of the angles.

Answers to PROBLEMS

6. The three angle measures are 40°, 80°, and 60°.

3. **Think of a word equation and translate it to an algebra equation.** In Table 2.4, it states that the sum of the measures of the angles of a triangle is 180°. We use this fact to write a word equation that we will translate into an algebra equation.

$$\underbrace{\text{Measure of the first angle}}_{x} + \underbrace{\text{measure of the second angle}}_{(90 - x)} + \underbrace{\text{measure of the third angle}}_{2(x)} \text{ is } 180$$
$$x + (90 - x) + 2(x) = 180$$

4. **Use your algebra skills to solve the equation.**

$$2x + 90 = 180 \quad \text{Remove parentheses; simplify.}$$
$$2x = 90 \quad \text{Add } -90.$$
$$x = 45 \quad \text{Divide by 2 for the first angle measure.}$$
$$(90 - x) = 45 \quad \text{For the second angle measure}$$
$$2(x) = 90 \quad \text{For the third angle measure}$$

5. **Verify the results.**

$$45 + 45 + 90 = 180$$

The measures of the three angles of the triangle are 45°, 45°, and 90°.

> Exercises 2.3

Boost your grade at **mathzone.com**!
> Practice Problems
> NetTutor
> Self-Tests
> e-Professors
> Videos

< A > **Translating Words into Mathematical Equations** In Problems 1–8, translate the sentences into mathematical equations. Do not solve.

1. The product of 12 and a number (m) is the number increased by 9.

2. Fifteen more than three times a number (n) results in -18.

3. One half of a number (x) less 12 is the same as two times the number divided by 3.

4. The square of a number (y), decreased by twice the number itself is 10 more than the number.

5. The 500,000 deaths by cardiovascular disease for women together with the deaths for men (M) account for the approximately 950,000 deaths each year attributed to this disease.

6. Approximately 53% of the world's poorest countries (P) will yield the 25 poor countries who import all of their oil.

7. The increase in average daily calories consumed between 1985 and 2000 (I) was 12% of the 2500 average daily calories consumed in 1985.

8. From 2002 to 2004, the world oil demand increased. If the United States' increase of 4.9% in demand is subtracted from China's increase (C), the result is 21.5%.

< B > **General Word Problems** In Problems 9–18, write the given statement as an equation and then solve the equation.

9. If 4 times a number is increased by 5, the result is 29. Find the number.

10. Eleven more than twice a number is 19. Find the number.

11. The sum of 3 times a number and 8 is 35. Find the number.

12. If 6 is added to 7 times a number, the result is 76. Find the number.

13. If the product of 3 and a number is decreased by 2, the result is 16. Find the number.

14. Five times a certain number is 9 less than twice the number. What is the number?

15. Five times a certain number is the same as 12 increased by twice the number. What is the number?

16. If 5 is subtracted from half a number, the result is 1 less than the number itself. Find the number.

17. One-third of a number decreased by 2 yields 10. Find the number.

18. One-fifth of a certain number plus 2 times the number is 11. What is the number?

In Problems 19–24, use the RSTUV method to obtain the solution.

19. According to a *Sports Illustrated* 2003 survey of major league baseball players, golf was voted the favorite nonbaseball activity. The other four activities on the list were hunting, fishing, time with the family, and movies, in that order. Golf received approximately three times as many votes as the other four combined. Using 100% as the total amount of votes, find the percent of the votes received by golf.

20. According to that same *Sports Illustrated* 2003 survey, Barry Bonds was voted the greatest living baseball player. The other five players on the list were Alex Rodriguez, Willie Mays, Nolan Ryan, Hank Aaron, and Pete Rose, in that order. Bonds received approximately $1\frac{1}{2}$ times as many votes as the other five combined. Using 100% as the total amount of votes, find the percent of the votes received by Barry Bonds.

21. Alex is 12 years old and his sister Ramie is 2 years old. In how many years will Alex be exactly twice as old as Ramie?

22. Morgan is 19 years old and her sister Erin is 3 years old. In how many years will Morgan be exactly three times as old as Erin?

23. The cost of renting a car is $18 per day plus $0.20 per mile traveled. Margie rented a car and paid $44 at the end of the day. How many miles did Margie travel?

24. The Zone LD company offers a long distance phone rate plan that requires a monthly fee of $2 and an interstate rate of $0.04 per minute. If Kyra's long distance phone bill for one month is $13.32, how many long distance minutes did she use?

⟨ C ⟩ Integer Word Problems
In Problems 25–36, consult your mathematics dictionary, if necessary, to solve the integer problems.

25. The sum of three consecutive even integers is 138. Find the integers.

26. The sum of three consecutive odd integers is 135. Find the integers.

27. The sum of three consecutive odd integers is −27. Find the integers.

28. The sum of three consecutive even integers is −24. Find the integers.

29. The sum of two numbers is 179, and one of them is 5 more than the other. Find the numbers.

30. The larger of two numbers is 6 times the smaller. Their sum is 147. Find the numbers.

31. The Saffir-Simpson scale is used to measure the intensity of a hurricane. Hurricanes are measured in categories from 1 to 5 according to their wind rates in miles per hour. The wind speed of a category 4 hurricane exceeds that of the wind speed of a category 1 hurricane by 60 miles per hour. If the two wind rates were added, they would total 250.
   a. Find the wind rate for a category 1 hurricane.
   b. Find the wind rate for a category 4 hurricane.

32. Would you believe Florida gets more tornadoes than Kansas or Oklahoma? The tornadoes in Florida are rated F-0 and F-1 as compared to the F-5 rated storms of the Midwest. The Fujita scale uses the damage a tornado causes to gauge the storm's wind speed in mi/hr. An F-5 rated (incredible) tornado can have winds that exceed three times an F-0 rated (minor damage) tornado by 100 miles per hour. If the two wind rates were added, they would total 388.
   a. Find the wind rate for an F-0 rated tornado.
   b. Find the wind rate for an F-5 rated tornado.

33. To find the weight of an object on the moon, you can divide its weight on Earth by 6. The crew of Apollo 16 collected lunar rocks and soil weighing 35.5 pounds on the moon. What is their weight on Earth?

34. The weight of an object on the moon is obtained by dividing its Earth weight by 6. If an astronaut weighs 28 pounds on the moon, what is the corresponding Earth weight?

35. The Beatles have 20 more Recording Industry Association of America awards than Paul McCartney. If the number of awards received by the Beatles and McCartney total 74, how many awards do the Beatles have?

36. The greatest weight difference ever recorded in a major boxing bout was 140 pounds in a match between John Fitzsimmons and Ed Punkhorst. If the combined weight of the contestants was 484 pounds, find Fitzsimmons' weight. (He was the lighter of the two.)

⟨ D ⟩ Geometry Problems
In Problems 37–50, solve the geometry problems.

37. The largest painting in the world used to be the *Panorama of the Mississippi*, by John Banvard. If the length of this painting was 4988 feet more than its width and its perimeter was 10,024 feet, find the dimensions of this rectangular painting.

38. In 2001, *Hero: The World's Largest Painting by One Artist*, painted by Eric Waugh, was recorded in the *Guinness Book of World Records* as having a perimeter of 820 feet. If the length of this painting exceeds its width by 50 feet, find its dimensions.

39. The scientific building with the greatest capacity is the Vehicle Assembly Building at the John F. Kennedy Space Center. The width of this building is 198 feet less than its length. If the perimeter of the rectangular building is 2468 feet, find its dimensions.

40. The largest fair hall is located in Hanover, Germany. The length of this hall exceeds its width by 295 feet. If the perimeter of this rectangular hall is 4130 feet, find its dimensions.

41. An angle has four times the measure of its complement. What is the measure of the angle?

42. The sum of the measures of an angle and one third a second angle is 32°. If the angles are complementary, what are their measures?

43. An angle has three times the measure of its supplement. What is the measure of the angle?

44. An angle is 5° less than twice another angle. Find their measures if they are supplementary angles.

45. Find x and the measure of each of the complementary angles.

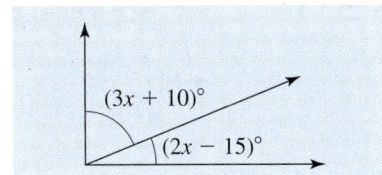

46. Find x and the measure of each of the complementary angles.

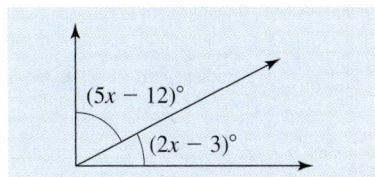

47. Find x and the measure of each of the supplementary angles.

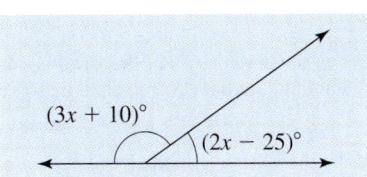

48. Find x and the measure of each of the supplementary angles.

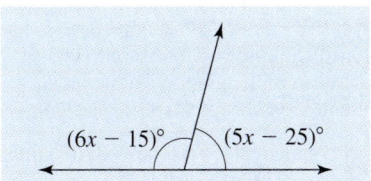

49. A triangle has two complementary angles. If the measure of the third angle is three times that of the first, what are the measures of the angles in this triangle?

50. A triangle has two complementary angles. If the measure of the third angle is five times that of the first, what are the measures of the angles in this triangle?

››› Using Your Knowledge

Puzzles and Riddles Here is a puzzle and a riddle that you can solve using the knowledge you've gained in these sections.

51. "What number," asked Professor Bourbaki, absentmindedly trying to find his algebra book, "must be added to the numerator and to the denominator of the fraction $\frac{1}{4}$ to obtain the fraction $\frac{2}{3}$?" "Well," he said after he found his book. Well?

52. Not much is known about Diophantus, sometimes called the father of algebra, except that he lived between A.D. 100 and 400. His age at death, however, is well known, because one of his admirers described his life using this algebraic riddle:

> Diophantus' youth lasted $\frac{1}{6}$ of his life. He grew a beard after $\frac{1}{12}$ more of his life. After $\frac{1}{7}$ more of his life, he married. Five years later, he had a son. The son lived exactly $\frac{1}{2}$ as long as his father, and Diophantus died just 4 years after his son's death.

How many years did Diophantus live?

››› Write On

53. When reading a word problem, what is the first thing you should try to determine?

54. How can you verify the answer in a word problem?

55. Make your own integer word problem and show a detailed solution of it using the RSTUV method.

56. Make your own geometry word problem and show a detailed solution of it using the RSTUV method.

57. Step 3 of the RSTUV method calls for you to "Think of a plan" to solve the problem. Some strategies you can use include *looking for a pattern* and *making a picture*. Can you think of three other strategies?

⟩⟩⟩ Concept Checker

Fill in the blank(s) with the correct word(s), phrase, or mathematical statement.

58. If an angle has a measure of 90° it is called a _____.

59. If the sum of the measures of two angles is 90°, the angles are called _____.

60. To help solve word problems, the _____ method is recommended to guide your thinking process.

61. An angle with a measure of 180° is called a _____.

RSTUV

right angle

straight angle

complementary angles

supplementary angles

⟩⟩⟩ Mastery Test

62. The sum of three consecutive integers is 27. What are the integers?

63. The sum of three consecutive odd integers is 99. What are the three integers?

64. The square of a number x, increased by 4 times the number itself, is 2 less than the number. Write an equation expressing this statement.

65. If a number is decreased by 12, the result is the same as the product of 7 and the number. Write an equation expressing this statement.

66. The measure of an angle is 8 times that of another.

   a. Find their measures if the angles are complementary.

   b. Find their measures if the angles are supplementary.

67. Find x and the measure of each marked angle.

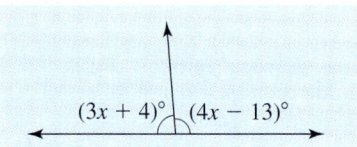

68. Denmark's cigarette tax tops the world. The tax on a pack of 20 cigarettes exceeds the price of the cigarettes by $3.03! If the total price of a pack of cigarettes in Denmark is $4.33, what is the tax?

69. The total number of students in U.S. schools and colleges in 2004 was about 71 million. The approximate number of students in secondary schools was one million less than that of the number of students in colleges, while in elementary schools there were 2.5 times as many students as in college. What is the number in (millions) of students enrolled in college in 2004? Give your answer to nearest million.

Source: http://nces.ed.gov/.

⟩⟩⟩ Skill Checker

Simplify:

70. $100(4.5)$ **71.** $1000(0.05)$ **72.** $10(3.4)(x + 2)$ **73.** $100(0.5)(x - 3)$

Solve:

74. $n + 0.30n = 26$

75. $n - 0.60n = 80$

76. $220 = 2(W + 70) + 2W$

77. $300 = 2(W + 60) + 2W$

78. $0.05x + 0.10(800 - x) = 20$

79. $0.10s + 0.20(6000 - s) = 1000$

80. $110T = 80(T + 3)$

81. $80T = 50(T + 3)$

2.4 Problem Solving: Percent, Investment, Motion, and Mixture Problems

Objectives

A Solve percent problems.

B Solve investment problems.

C Solve uniform motion problems.

D Solve mixture problems.

To Succeed, Review How To . . .

1. Perform the four fundamental operations using decimals (pp. 18–27).
2. Solve linear equations involving decimals (pp. 82–83).

Getting Started
Web Shopping Costs

List Price: $1,299.00   List Price: $399.00   List Price: ?

Sale Price: ?   Sale Price: $139.65   Sale Price: $20.99

You Save: 63%   You Save: ?%   You Save: 30%

These items can be purchased on the Internet. Underneath the items are the online savings. To supply the answers for the question marks requires knowing how to solve percent problems. Suppose you wanted to find out what the sale price of the laptop computer would be with a 63% discount. Knowing the percent of discount allows you to find the dollar amount of the discount.

$$\text{percent of discount times list price} = \text{discount dollar amount}$$
$$(0.63) \times \$1299 = \$818.37$$

Remember $818.37 is the discount. We use the discount to find the sale price.

$$\text{list price minus discount} = \text{sale price}$$
$$\$1299 - \$818.37 = \$480.63$$

Those two equations could be combined to make one formula:

$$\text{list price} - (\text{percent of discount times list price}) = \text{sale price}$$
$$\$1299 - (0.63)(\$1299) = \$480.63$$

To find the percent when buying the camera, solve $399.00 - (x\%)(399.00) = 139.65$ to get 65%. To find the list price for the vacuum, solve $x - 0.30(x) = 20.99$ to get $29.99. These types of equations are explained on p. 112.

Don't forget to read the fine print for the cost of shipping and handling. Some companies charge by the shipping weight and some charge a flat fee like $9.95. These were examples of solving percent problems. In this section, we will also study how to solve investment, motion, and mixture problems.

A › Percent Problems

In algebra, rates, increases, decreases, and discounts are often written in terms of percents (%). *Percent* means "by the hundred." Thus, 28% means 28 parts of 100 or $\frac{28}{100}$. In most applications, however, percents are written as decimals. Thus,

$$28\% = \frac{28}{100} = 0.28$$

$$30\% = \frac{30}{100} = 0.30$$

$$120\% = \frac{120}{100} = 1.2$$

and

$$34.8\% = \frac{34.8}{100} = \frac{348}{1000} = 0.348$$

There are three basic types of percent problems. All of these can be translated into algebra equations.

TYPE 1 asks to find a percent of a number.

> PROBLEM   30% of 80 is what number?
> TRANSLATION $0.30 \times 80 = n$
> SOLUTION $24 = n$

TYPE 2 asks what percent of a number is another given number.

> PROBLEM   What percent of 40 is 8? Or 8 is what percent of 40?
> TRANSLATION $n \times 40 = 8$ or $8 = n \times 40$
> SOLUTION $n = \frac{8}{40} = \frac{1}{5} = 20\%$

TYPE 3 asks to find a number when a percent of the number is given.

> PROBLEM   10 is 40% of what number?
> TRANSLATION $10 = 0.40 \times n$
> SOLUTION $\frac{10}{0.40} = \frac{1000}{40} = 25 = n$

EXAMPLE 1   Percent problems

When purchasing their first home the Gonzalez family were prequalified to purchase a home up to $110,000. They had saved $6000 for the down payment. When they found their dream home, it was priced at $102,650 with a 5% down payment. Obviously the house was in their price range.

a. Did they have enough for the down payment?

b. If they did purchase the house, what would be the amount of the mortgage?

SOLUTION   Use the RSTUV method.

a. Is $6000 enough for their down payment?

1. Read and write a "Find" statement. Find the amount of the down payment and compare that amount to $6000.

2. Select the unknown using a "Let" statement. Let x = the amount of the down payment.

3. Think of a word equation and translate it to an algebra equation.

> Percent times cost of house = down payment
> (0.05) $(\$102{,}650) =$ x

4. Use your algebra skills to solve the equation.

> $\$5133 = x$ Rounded to the nearest dollar

Comparison: $5133 is less than $6000.

PROBLEM 1

The Tyler family want to purchase a home for $122,850 with a 10 percent down payment.

a. What is the down payment (to the nearest dollar)?

b. What is the amount of the mortgage?

Answers to PROBLEMS

1. a. $12,285   **b.** $110,565

5. **Verify the results.** The verification is left to you.
Yes, the Gonzalez family have enough for the down payment.

b. What is the amount of the mortgage?

Cost of the house − the down payment = amount of mortgage
$102,650 − $5133 = $97,517

The Gonzalez family will have a mortgage of $97,517.

EXAMPLE 2 Solving percent increase problems

According to O'Hare Airport statistics, the total number of passengers using the airport in January 2003 was approximately 5 million. This was a 12% increase over the total for January 2002. Approximately how many passengers used O'Hare Airport in January 2002? (Round the answer to the nearest thousand.)

SOLUTION Use the RSTUV method. The key words are "12% increase." The problem will be solved like one in the *Getting Started*.

1. **Read and write a "Find" statement.** Find the number of passengers using O'Hare Airport in January 2002.

2. **Select the unknown using a "Let" statement.** Let x = the number of passengers in January 2002.

3. **Think of a word equation and translate it to an algebra equation.** If we add the increase to the number of passengers in January 2002, that should equal the number of passengers in January 2003. To find the increase we have to multiply 12% times the number of passengers in January 2002 $[(0.12)(x)]$.

number in January 2002 + increase = number in January 2003
x + $[(0.12)(x)]$ = 5,000,000

4. **Use your algebra skills to solve the equation.**

$1x + [(0.12)(x)] = 5,000,000$ Remember $x = 1x$.
$1.12x = 5,000,000$ Simplify.
$x = 4,464,286$ Divide by 1.12.
$x = 4,464,000$ (Rounded to nearest thousand)

5. **Verify the results.** The verification is left for you.

The number of passengers using O'Hare Airport in January 2002 was approximately 4,464,000.

PROBLEM 2

The number of passengers using a certain airport for the year 2004 was approximately 67 million. If that was down by 1.3% from 2003, how many passengers used the airport in 2003? (Round the answer to the nearest thousand.)

Now we pose the question, "Does blogging do the mind good?" as we study bar graphs. A **blog** (short for weblog) is a personal online journal that is frequently updated and intended for general public consumption. We will use the results from a 2005 blogger survey to construct bar graphs representing the data sets for two of the questions. Sometimes it is necessary to use percents to help in the construction of bar graphs as we will see in Example 3 and Problem 3.

EXAMPLE 3 A bar graph involving percents

How often does a blogger blog? Use the results of the 2005 blogger survey of 821 people to fill in the last column of the table on the next page and construct a bar graph. (Round answers to the nearest whole number.)

PROBLEM 3

Why do bloggers blog? Use the results of the 2005 blogger survey of 821 people to fill in the last column of the table on the next page and construct a bar graph. (Round answers to the nearest whole number.)

(continued)
(Answer on page 114)

Answers to PROBLEMS

2. The number of passengers for the year 2003 was approximately 67,882,000.

Frequency of Blogs	Percent of People	Number of People
Daily	25.7%	
Every Few Days	37.9%	
Multiple Times Per Day	18.4%	
Weekly	10.6%	
Other	7.4%	

Source: https://extranet.edelman.com/.bloggerstudy/.

Reasons for Blogging	Percent of People	Number of People
Create a record of my thoughts	31.5%	
Other	9.6%	
Connect with others	20.3%	
Visibility as an authority	33.9%	
Generate revenue	4.6%	

SOLUTION This is a two-step problem. We will use the RSTUV method for step 1—finding the number of people in each category. In step 2, we will construct the bar graph.

STEP 1

1. **Read and write a "Find" statement.** Find the number of people for each category in the chart.
2. **Select the unknown using a "Let" statement.**
   Let D = number of people who blog daily.
   Let E = number of people who blog every few days.
   Let M = number of people who blog multiple times/day.
   Let W = number of people who blog weekly.
   Let O = number of people who gave other answers.
3. **Think of a word equation and translate it to an algebra equation.** The previous work involving percent will enable us to make the generalized statement,

$$\text{Percent of the Total is the Part}$$
$$\% \cdot Total = Part$$

4. **Use your algebra skills to solve the equations.** We are given the total number of people surveyed as 821 and the percents for each category are in the second column.
   Substitute data for each category into the formula and solve.

$0.257\,(821) = D$ Multiply and round to nearest whole number.
$211 = D$ **(25.7% = 0.257)**

$0.379\,(821) = E$ Multiply and round to nearest whole number.
$311 = E$ **(37.9% = 0.379)**

$0.184\,(821) = M$ Multiply and round to nearest whole number.
$151 = M$ **(18.4% = 0.184)**

$0.106\,(821) = W$ Multiply and round to nearest whole number.
$87 = W$ **(10.6% = 0.106)**

$0.074\,(821) = O$ Multiply and round to nearest whole number.
$61 = O$ **(7.4% = 0.074)**

Answers to PROBLEMS

3. 259, 79, 167, 278, 38

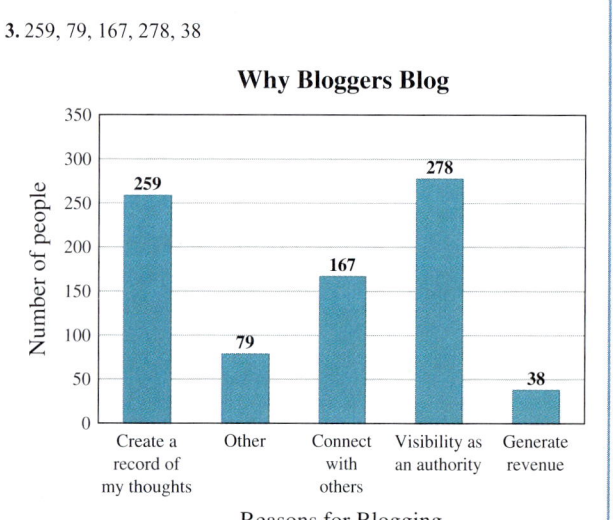

Why Bloggers Blog

Frequency of Blogs	Percent of People	Number of People
Daily	25.7%	211
Every Few Days	37.9%	311
Multiple Times Per Day	18.4%	151
Weekly	10.6%	87
Other	7.4%	61

5. Verify the results. Left to you.

STEP 2

Now we will construct the bar graph with a horizontal axis labeled "Frequency of blogs" and a vertical axis labeled "Number of users." Looking at column 3, we can see the data ranges from 61 to 311 so the vertical scale will be 0 to 400, using increments of 100. Use the results from step 1 to draw and label the appropriate length bar for each category.

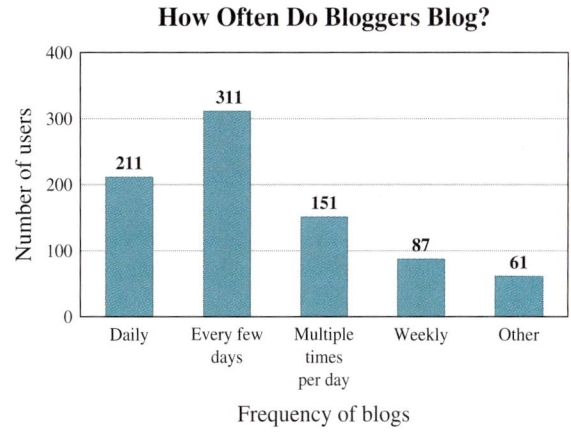

B › Investment Problems

Investment problems also use percents. If you invest P dollars at a rate r, your **annual interest** is

$$I = Pr$$

When working this type of problem, it's helpful to organize the information in a table. We do this in Example 4.

EXAMPLE 4 Solving investment problems

A woman has some stocks yielding 5% annually and some bonds that yield 10%. If her investments total $6000 and her annual income from the investments is $500, how much does she have invested in stocks and how much in bonds?

SOLUTION Use the RSTUV method.

1. Read and write a "Find" statement. Find how much she has invested in stocks and how much in bonds.

2. Select the unknown using a "Let" statement. Let $s =$ the amount she has invested in stocks. This makes the amount invested in bonds $(6000 - s)$.

PROBLEM 4

A man has two investments totaling $8000. One investment yields 7% and the other 10%. If the total annual interest is $710, how much is invested at each rate?

(continued)

Answers to PROBLEMS

4. $3000 at 7%; $5000 at 10%

3. **Think of a word equation and translate it to an algebra equation.** Use the following to help visualize the problem. Enter the information in a table.

	Principal	×	Rate	=	Interest
Stocks	s		0.05		$0.05s$
Bonds	$6000 - s$		0.10		$0.10(6000 - s)$

The total interest is the sum of the entries in the last column, that is, $0.05s$ and $0.10(6000 - s)$. This amount must be $500.

$$\underbrace{0.05s}_{\text{interest from stocks}} + \underbrace{0.10(6000 - s)}_{\text{interest from bonds}} = \underbrace{500}_{\text{total interest}}$$

4. **Use your algebra skills to solve the equation.**

$$0.05s + 0.10(6000 - s) = 500 \quad \text{Given}$$
$$0.05s + 600 - 0.10s = 500 \quad \text{Remove parentheses; simplify.}$$
$$600 - 0.05s = 500$$
$$600 - 600 - 0.05s = 500 - 600 \quad \text{Add } -600.$$
$$-0.05s = -100$$
$$\frac{-0.05s}{-0.05} = \frac{-100}{-0.05} \quad \text{Divide by } -0.05.$$
$$s = 2000$$

The woman has $2000 in stocks and the rest, $4000, in bonds.

5. **Verify the results.** Five % of $2000 = $100 and 10% of $4000 = $400, so the total interest is $500.

C > Uniform Motion Problems

When traveling at a constant rate R, the distance D traveled in time T is given by

$$D = RT$$

This formula is similar to the one used for interest problems, so we again use a table to organize the information. Here's how we do it.

EXAMPLE 5 Solving distance problems

A Supercruiser bus leaves Miami for San Francisco traveling at the rate of 40 mi/hr. Three hours later a car leaves Miami for San Francisco traveling at 55 mi/hr on the same route as the bus. How far from Miami does the car overtake the bus (assume they both travel at constant rates)?

SOLUTION Use the RSTUV method.

1. **Read and write a "Find" statement.** To find the distance, we need to know the time it takes the car to overtake the bus.
2. **Select the unknown using a "Let" statement.** Let T = time it takes the car to overtake the bus.

PROBLEM 5

A bus leaves Los Angeles traveling at 50 miles per hour. An hour later a car leaves at 60 miles per hour to try and catch the bus. How far from Los Angeles does the car catch the bus?

Answers to PROBLEMS

5. 300 miles

3. Think of a word equation and translate it to an algebra equation. We translate the given information and write it in a table. If the car travels for T hours, the bus travels for $(T + 3)$ hours because it left 3 hours earlier.

	Rate	×	Time	=	Distance
Car	55		T		$55T$
Bus	40		$(T+3)$		$40(T+3)$

When the car overtakes the bus, they will have traveled the same distance. According to the table, the car has traveled $55T$ miles in distance and the bus $40(T + 3)$ miles. Thus,

$$\underbrace{55T}_{\text{distance of car}} = \underbrace{40(T + 3)}_{\text{distance of bus}}$$

4. Use your algebra skills to solve the equation.

$55T = 40(T + 3)$	Given
$55T = 40T + 120$	Remove parentheses.
$55T - 40T = 40T - 40T + 120$	Add $-40T$.
$15T = 120$	Simplify.
$\dfrac{15T}{15} = \dfrac{120}{15}$	Divide by 15.
$T = 8$	

It takes the car 8 hr to overtake the bus. Since $R \cdot T = D$, in 8 hr the car travels $8 \times 55 = 440$ mi, and overtakes the bus 440 mi from Miami.

5. Verify the results. The distances are equal so we can use either one. The car has traveled for 8 hr at 55 mi/hr; thus it travels $55 \times 8 = 440$ mi, whereas the bus has traveled 11 hr at 40 mi/hr, a total of $40 \times 11 = 440$ mi.

D > Mixture Problems

The last type of problem we discuss is the **mixture problem,** a type of problem in which two or more things are put together to form a mixture. Again, we use a table to organize the information.

EXAMPLE 6   Solving mixture problems

How many ounces of a 50% acetic acid solution should a photographer add to 32 ounces of a 5% acetic acid solution to obtain a 10% acetic acid solution?

SOLUTION   Use the RSTUV method.

1. Read and write a "Find" statement. Find the number of ounces of the 50% solution that should be added.

2. Select the unknown using a "Let" statement. Let $x =$ the number of ounces of 50% solution to be added.

3. Think of a word equation and translate it to an algebra equation. To translate the problem, we use a table. In this case, the headings for the table contain the percent of acetic acid and the amount to be mixed. The product of these two numbers will give us the amount of pure acetic acid.

PROBLEM 6

How many gallons of a 10% salt solution should be added to 15 gallons of a 20% salt solution to obtain an 18% solution?

(continued)

Answers to PROBLEMS

6. 3.75 gallons

Percent of Acetic Acid	×	Amount to be mixed	=	Amount of Pure Acid
50% solution or 0.50		x oz		$0.50x = 0.5x$
5% solution or 0.05		32 oz		$0.05(32) = 1.6$
10% solution or 0.10 (Final mixture)		$(x + 32)$ oz		$0.10(x + 32) = 0.1(x + 32)$

The percents have been converted to decimals. You *should not* add the percents in this column.

We have x oz of one and 32 oz of the other, so we have $(x + 32)$ ounces of the mixture.

The sum of the total amounts of pure acetic acid should be the same as the amount of pure acetic acid in the final mixture, so we have

amt. in 50% + amt. in 5% = amt. of acid in final mixture

$$0.5x + 1.6 = 0.1(x + 32)$$

4. Use your algebra skills to solve the equation.

$5x + 16 = x + 32$ Clear decimals; multiply by 10.

$5x + 16 - 16 = x + 32 - 16$ Add -16.

$5x = x + 16$

$5x - x = x - x + 16$ Add $-x$.

$4x = 16$

$\dfrac{4x}{4} = \dfrac{16}{4}$ Divide by 4.

$x = 4$

The photographer must add 4 ounces of the 50% solution.

5. Verify the results. The verification of this fact is left to you.

TRANSLATE THIS

1. A certain company sells its 5-megapixel digital camera for $255. The newer model, a 6-megapixel camera, sells for $360. Write an equation that will find the percent of increase in price (I) for the newer model?

2. A survey of 1300 executives indicated Information Technology (IT) spending in 2005 (N) showed a 7% increase over the amount spent in 2004 (P). Write an equation that will find the IT amount of spending in 2005 (N)?

Source: http://www.line56.com/.

3. Unlimited local and long-distance calls are advertised by one company for a monthly fee (F) plus the $49.99 cost of a router. Write an equation that will find the total cost of the phone service (C) for *1 year*?

Source: http://www.vonage.com/.

The third step in the RSTUV procedure is to TRANSLATE the information into an equation. In Problems 1–6 TRANSLATE the sentence and match the correct translation with one of the equations **A–L.**

A. $65\% = 13P$

B. $N = P + 0.07P$

C. $120 = 3.36S$

D. $I = \dfrac{225}{360}$

E. $I = 110{,}000(0.38)(30)$

F. $C = 49.99 + 12F$

G. $N = 0.07P$

H. $I = \dfrac{360 - 255}{255}$

I. $65\% = P + 13\%$

J. $I = 110{,}000(0.038)(30)$

K. $C = 12F$

L. $120 = S + 3.36S$

4. Research indicates that some refinance loans can be obtained at 3.8% interest. Assuming a $110,000 loan is paid off in 30 years at this rate, write an equation that will find the simple interest for that loan.

5. The sales made by small online companies grew 336% to $120 billion from 2000 to 2002. Write an equation that will find the online sales for 2000 (S).

Source: http://www.bigstep.com/.

6. In a survey, participants were asked their preference on cleaning up spills. The results indicated the percent who preferred dish towels exceeded the percent who preferred paper towels by 13%. If 65% preferred using dish towels, write an equation to find the percent that preferred paper towels (P).

Source: http://yahoo.usatoday.com/.

C. Uniform Motion Problems

In Problems 30–35, use the RSTUV method to solve the motion problems.

30. A car leaves a town traveling at an average speed of 60 km/hr. Two hours later a highway patrol officer leaves from the same starting point to overtake the car. If the average speed of the officer is 90 km/hr, how far from town does the officer overtake the car?

31. A group of smugglers crosses the border in a car traveling in a straight line at 96 km/hr. An hour later, the border patrol starts after them in a light plane traveling 144 km/hr.
    a. How long will it be before the border patrol reaches the smugglers?
    b. At what distance from the border will the border patrol overtake the smugglers?

32. A freight train leaves the station traveling at 30 mi/hr. One hour later, a passenger train leaves the same station on a parallel track traveling at 60 mi/hr. How far from the station does the passenger train overtake the freight train?

33. A bus leaves the station traveling at 55 mph. One hour later, the wife of one of the passengers shows up at the station with a briefcase belonging to an absent-minded professor riding the bus. If she immediately starts after the bus at an astonishing 90 mph, how far, to the nearest mile, from the station is the briefcase reunited with the professor?

34. An accountant and her boss have to travel to a nearby town. The accountant catches a train traveling at 50 mi/hr while the boss leaves 1 hr later in a car traveling at 60 mi/hr. They have decided to meet at the train station and, strangely enough, they get there at exactly the same time! If the train and the car traveled in straight lines on parallel paths, how far is it from one town to the other?

35. The basketball coach at a local high school left for work on her bicycle, traveling at 15 mi/hr. Thirty minutes later, her husband noticed that she had left her lunch. He got in his car and traveled 60 mi/hr to take her lunch to her. Luckily, he got to school at exactly the same time as his wife. How far is it from the house to the school?

D. Mixture Problems

In Problems 36–40, use the RSTUV method to solve the mixture problems.

36. How many liters (L) of a 40% glycerin solution must be mixed with 10 L of an 80% glycerin solution to obtain a 65% solution?

37. How many parts of glacial acetic acid (99.5%) must be added to 100 parts of a 10% solution of acetic acid to give a 28% solution?

38. If the price of copper is 65¢ per pound and the price of zinc is 30¢ per pound, how many pounds of copper and zinc should be mixed to make 70 lb of brass selling for 45¢ per pound?

39. Oolong tea sells for $19 per pound. How many pounds of Oolong should be mixed with regular tea selling at $4 per pound to produce 50 lb of tea selling for $7 per pound?

40. You think the prices of coffee are high? You haven't seen anything yet! Jamaican Blue coffee sells for about $20 per pound! How many pounds of Jamaican Blue should be mixed with 80 lb of regular coffee selling at $8 per pound so that the result is a mixture selling for $10.40 per pound? (You can cleverly advertise this as "Containing the incomparable Jamaican Blue coffee.")

>>> Using Your Knowledge

We Owe, We Owe, and to the Deficit We Go   The last page of one year's IRS instructions for Form 1040 indicated that the federal income that year was $1090.5 billion, while outlays (expenses) amounted to $1380.9 billion. (By the way, 1090.5 billion = 1.0905 *trillion*.)

41. What was the difference between income and expenses?

42. The pie chart indicates that 21% of the federal income was borrowed to cover the deficit. How much money was borrowed to cover the deficit?

43. The pie chart indicates that 35% of the federal income comes from personal taxes. How much money comes from personal taxes?

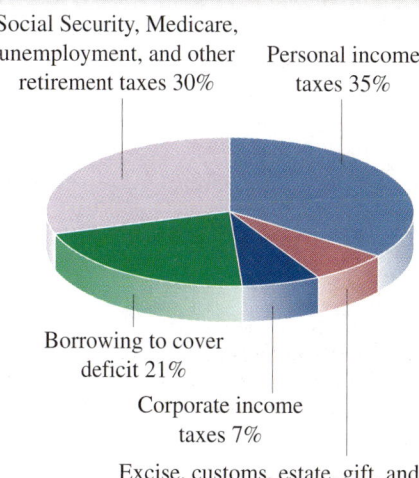

Social Security, Medicare, unemployment, and other retirement taxes 30%
Personal income taxes 35%
Borrowing to cover deficit 21%
Corporate income taxes 7%
Excise, customs, estate, gift, and miscellaneous taxes 7%

>>> Write On

44. Formulate three different problems involving percents, and show a detailed solution using the RSTUV method.

45. Formulate your own investment word problem, and show a detailed solution using the RSTUV method.

46. Formulate your own motion word problem, and show a detailed solution using the RSTUV method.

47. Most of the problems in this text have precisely the information you need to solve them. In real life, however, some of the information given may be irrelevant. Such irrelevant information is called a *red herring*. Find some problems with red herrings and point them out.

48. Ask your pharmacist or your chemistry instructor if he or she mixes products of different concentrations to make new mixtures. Write a paragraph on your findings.

>>> Mastery Test

49. A man has two investments totaling $8000. One investment yields 5% and the other 10%. If the total annual interest is $650, how much is invested at each rate?

50. How many gallons of a 10% salt solution must be added to 15 gallons of a 20% salt solution to obtain a 16% solution?

51. The number of passengers using the Atlanta airport in the year 2000 was 31.2 million, a 30% increase over the 1999 figure. How many passengers used the Atlanta airport in 1999?

52. A bus leaves Los Angeles traveling at 50 mi/hr. An hour later a car traveling on the same road leaves at 70 miles per hour to try and overtake the bus. How far from Los Angeles does the car overtake the bus?

53. The number of students enrolled in higher education in the United States increased from 12 million in 1980 to 14.4 million in 1990. What was the percent increase?

>>> Skill Checker

Simplify:

54. $-9 + 7$
55. $12 - 18$
56. $x - 5x$
57. $-4y + 8y$
58. $\frac{1}{2}(-22)$
59. $-3(-13)$
60. $3(x - 7)$
61. $-(2x + 1)$

Fill in the blank with < or > to make the resulting statement true.

62. -3 ___ -1
63. -1.3 ___ -1.4
64. $\frac{1}{3}$ ___ $\frac{1}{2}$
65. $-\frac{1}{3}$ ___ $-\frac{1}{2}$

Solve:

66. $-14 = 5x - 4$
67. $2x + 3 = x - 4$
68. $7x = 4x - 2(3x + 9)$

2.5 Linear and Compound Inequalities

Objectives

A Graph linear inequalities.

B Solve and graph linear inequalities.

C Solve and graph compound inequalities.

D Translate sentences and solve applications involving inequalities.

To Succeed, Review How To . . .

1. Solve linear equations (pp. 76–83).
2. Add, subtract, multiply, and divide integers (pp. 18–27).
3. Properly use the symbols $>$ and $<$ when comparing numbers (p. 10).

Getting Started
Inequalities and Rental Cars

Suppose a rental car costs $25 each day and $0.20 per mile. If you drive m miles, the cost C for the day is

$$C = \underbrace{0.20m}_{\text{Cost for } m \text{ miles}} + \underbrace{25}_{\text{Cost per day}}$$

Now suppose your daily cost must be under $50. This means that $0.20m + 25 < 50$. How many miles can you drive? The expression $0.20m + 25 < 50$ is an example of a linear inequality and can be solved using the same techniques as those used to solve linear equations. Thus,

$0.20m + 25 < 50$	Given
$0.20m + 25 - 25 < 50 - 25$	Add -25.
$0.20m < 25$	
$m < \dfrac{25}{0.20}$	Divide by 0.20.
$m < 125$	

Hence if you drive less than 125 mi, your cost will be less than $50 per day.

An **inequality** is a statement that two expressions are *not* equal. Some examples of linear inequalities are $x + 2 < 5$, $x - 7 < 3$, and $3x - 8 > 15$.

LINEAR INEQUALITIES

> A **linear inequality** in one variable is an inequality that can be written in one of the following forms
>
> $$ax + b < c \quad \text{or} \quad ax + b > c$$
>
> where a, b, and c are real numbers, and $a \neq 0$.

Note: Similar statements can be written using $\leq$ and $\geq$.

As with linear equations, we *solve* an inequality by finding all the replacement values of the variable that make the inequality a true statement. This is done by finding an *equivalent* inequality whose solution is obvious. For example, the inequalities

$x + 2 > 5$ and $x > 3$ are equivalent; they are both satisfied by all real numbers greater than 3. We write the *solution set* of $x + 2 > 5$ in set-builder notation as

$$\{x | x > 3\}$$

This is read "the set of all x's such that x is greater than 3."

Let's consider another form of equivalent inequalities. If we compare 5 with 3, we can write "$5 > 3$." However, we can also compare them correctly by writing "$3 < 5$." This can be generalized as follows.

EQUIVALENT INEQUALITIES

If $a > c$, then $c < a$, where a and c are two non-equal real numbers.

This means when *a* is greater than *c*, then *c* must be less than *a*.

A ❯ Graphing Linear Inequalities

There are infinitely many numbers in the solution set of the inequality $x > 3$ (4, 7.5, $\sqrt{10}$, and $\frac{10}{3}$ are a few of them). We cannot list all these numbers, so we show all solutions of $x > 3$ **graphically** by using a number line. This type of representation is called the **graph** of the inequality. The graph is a picture of all the solutions to a statement. The heavy line in the figure below is the graph of the solution set for $x > 3$.

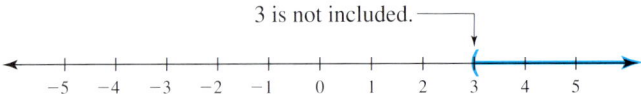

The number 3 is *excluded* from the graph. This is shown by drawing a parenthesis at the point 3.

This graph can also be drawn with an open circle at 3.

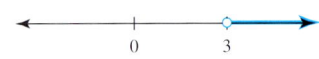

EXAMPLE 1 Graphing linear inequalities
Graph:

a. $x \geq -1$ **b.** $x < -2$ **c.** $0 < x$

PROBLEM 1
Graph:

a. $x \geq -2$ **b.** $x < 1$
c. $-3 \leq x$

SOLUTION

a. The numbers that satisfy the inequality $x \geq -1$ are the numbers *greater than or equal to* -1, that is, the number -1 and all the numbers to the *right* of -1 ($\geq$ points to the right). The graph is shown in the following figure. That -1 is included is shown by drawing a bracket at the point -1. (*Hint:* When graphing a linear inequality, the arrow will point in the direction to shade on the number line as long as the variable is on the left side of the inequality symbol.)

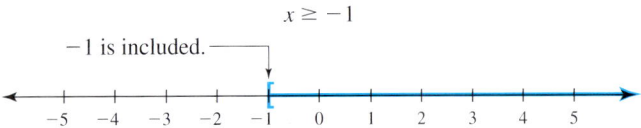

This graph can also be drawn with a closed circle at -1.

Answers to PROBLEMS

1. a.    **b.**    **c.**

b. The numbers that satisfy the inequality $x < -2$ are the numbers *less than* -2, that is, the numbers to the *left* of but *not including* -2 ($<$ points to the left). The graph of these points is

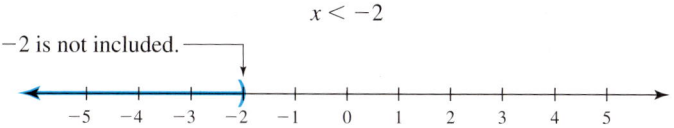

c. Since the statement $0 < x$ is equivalent to $x > 0$, we will use $x > 0$ to graph the solutions. The numbers that satisfy the inequality $x > 0$ are the numbers to the right of but not including 0. The graph of these points is as shown.

$$0 < x \quad (x > 0)$$

The inequalities $x > 3$, $x \geq -1$, and $x < -2$ resulted in graphs that did *not* have a finite length; they are called **unbounded** (or **infinite**) **intervals**. The four basic types of unbounded intervals are summarized in Table 2.5, where we use a parenthesis (or open circle) to denote the inequality symbol $<$ or $>$ and a bracket (or closed circle) to indicate $\leq$ or $\geq$. An **open** interval does not contain either of its end points.

Study Skills Hint

To practice how to write your answer in interval notation, it would be a good idea to write your own examples of the four cases in Table 2.5 on an index card. Practice reading and writing this notation before and after each class and each homework assignment. For example, you might begin with:

1. $\{x | x > 2\}$ The set of all "x" such that x is greater than 2.
2. $(2, \infty)$ The open interval from 2 to positive infinity.

Table 2.5 Unbounded Interval Notation

Set Notation	Interval Notation	Type of Interval	Graph	Inequality
$\{x \mid x > a\}$	$(a, +\infty)$	Open		$x > a$
$\{x \mid x < b\}$	$(-\infty, b)$	Open		$x < b$
$\{x \mid x \geq a\}$	$[a, +\infty)$	Half-open		$x \geq a$
$\{x \mid x \leq b\}$	$(-\infty, b]$	Half-open		$x \leq b$

For example, to name the interval on the graph

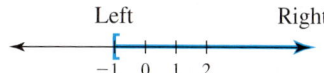

we write the left-end number, -1, and then the infinity symbol, with the appropriate enclosures. The interval is written $[-1, \infty)$.

In all these intervals, a parenthesis is always used with the symbols ∞ (**positive** infinity) and $-\infty$ (**negative** infinity). The symbols ∞ and $-\infty$ do not represent real numbers; they are used to indicate that the interval is unbounded.

B › Solving Linear Inequalities

The inequalities $x + 2 > 5$ and $x > 3$ are equivalent. This is because we can add -2 to both sides of $x + 2 > 5$ to obtain the equivalent inequality

$$x + 2 + (-2) > 5 + (-2) \quad \text{or} \quad x > 3$$

We used the first of the following properties.

> **PROPERTIES OF INEQUALITIES: ADDITION AND SUBTRACTION**
>
> If c is a real number, then the following inequalities are all *equivalent*.
>
> $$a < b$$
> Add c. $a + c < b + c$
> Subtract c. $a - c < b - c$
>
> Similar statements hold for $>$, $\leq$, and $\geq$.

Now consider the inequality $2x - 3 < x + 1$. To solve this inequality, we need all the variables by themselves on one side. We can use the steps learned in Section 2.1 for solving linear equations (CRAM, page 79–80). Thus we proceed as follows:

$2x - 3 < x + 1$ Clear fractions: none.
                    Remove parentheses: none.
$2x - x - 3 < x - x + 1$ Add $-x$ to both sides.
$x - 3 < 1$
$x - 3 + 3 < 1 + 3$ Add 3 to both sides.
$x < 4$

The solution set is $\{x \mid x < 4\}$ or, in interval notation, $(-\infty, 4)$.

The graph of this inequality is shown in the figure. You can check that this solution is correct by selecting a number from the graph (say 0) and then replacing x with the number 0 in the original inequality to obtain $2(0) - 3 < 0 + 1$ or $-3 < 1$, a true statement. This is only a "partial" check because we did not try *all* the numbers in the graph.

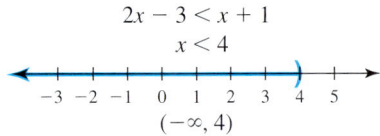

EXAMPLE 2   Solving and graphing inequalities

Solve, graph, and write the solution set in interval notation.

a. $3x - 2 < 2(x - 2)$ **b.** $4(x + 1) \geq 3x + 7$

SOLUTION

a.
$3x - 2 < 2(x - 2)$ Clear fractions: none.
$3x - 2 < 2x - 4$ Remove parentheses.
$3x - 2x - 2 < 2x - 2x - 4$ Add $-2x$ to both sides.
$x - 2 < -4$
$x - 2 + 2 < -4 + 2$ Add 2 to both sides.
$x < -2$

The solution set is $\{x \mid x < -2\}$ or, in interval notation, $(-\infty, -2)$. The graph is

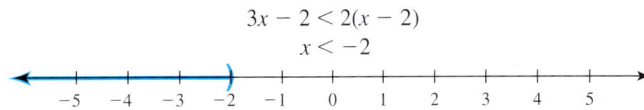

b.
$4(x + 1) \geq 3x + 7$ Clear fractions: none.
$4x + 4 \geq 3x + 7$ Remove parentheses.
$4x - 3x + 4 \geq 3x - 3x + 7$ Add $-3x$ to both sides.
$x + 4 \geq 7$
$x + 4 - 4 \geq 7 - 4$ Add -4 to both sides.
$x \geq 3$

PROBLEM 2

Solve, graph, and write the solution set in interval notation.

a. $4x - 7 < 3(x - 2)$

b. $3(x + 1) \geq 2x + 5$

Answers to PROBLEMS

2. a. $x < 1$; $(-\infty, 1)$ 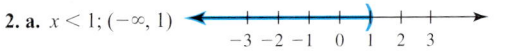   **b.** $x \geq 2$; $[2, \infty)$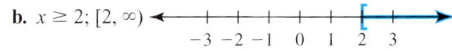

The solution set is $\{x|x \geq 3\}$ or, in interval notation, $[3, \infty)$. The graph of this inequality is

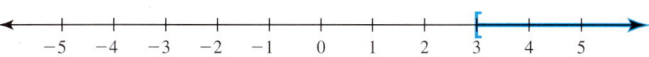

Before we state the multiplication and division properties of inequalities, let's see what happens if we multiply both sides of an inequality by a positive number. Consider these true inequalities:

$$1 < 3 \qquad -1 < 3 \qquad -3 < -1$$
$$4 \cdot 1 \; ? \; 4 \cdot 3 \qquad 4 \cdot (-1) \; ? \; 4 \cdot 3 \qquad 4 \cdot (-3) \; ? \; 4 \cdot (-1) \qquad \text{Multiply by 4.}$$
$$4 < 12 \qquad -4 < 12 \qquad -12 < -4$$

The resulting inequalities are all *true*, and the inequality symbol points in the *same* direction as in the original. Multiplying both sides of an inequality by a *positive* number *preserves the sense,* or direction, of the inequality.

Now, let's multiply both sides of the original inequality by a *negative* number, say -4. We have

$$1 < 3 \qquad -1 < 3 \qquad -3 < -1$$
$$-4 \cdot 1 \; ? \; -4 \cdot 3 \qquad -4 \cdot (-1) \; ? \; -4 \cdot 3 \qquad -4 \cdot (-3) \; ? \; -4 \cdot (-1) \qquad \text{Multiply by } -4.$$
$$-4 > -12 \qquad 4 > -12 \qquad 12 > 4$$

This time, however, we had to reverse the direction of the inequalities to maintain a true statement. Multiplying both sides of an inequality by a *negative* number *reverses the sense* of the inequality. Division is defined in terms of multiplication, so these two properties apply to division as well. They can be stated as follows.

PROPERTIES OF INEQUALITIES: MULTIPLICATION AND DIVISION

If c is a real number, then the following inequalities are all equivalent.

$$a < b$$
$$a \cdot c < b \cdot c \qquad \text{if c is positive } (c > 0)$$
$$\frac{a}{c} < \frac{b}{c} \qquad \text{if c is positive } (c > 0)$$

$$a \cdot c > b \cdot c \qquad \text{if c is negative } (c < 0)$$
$$\frac{a}{c} > \frac{b}{c} \qquad \text{if c is negative } (c < 0)$$

Similar statements hold for $>$, $\leq$, and $\geq$.

CAUTION

We can still multiply or divide both sides of an inequality by any nonzero number *as long as we remember to reverse the sense (direction) of the inequality if the number is negative.*

To solve $-2x > 4$, we must divide both sides by -2 (or multiply by $-\frac{1}{2}$). When doing this, remember to reverse the sense (direction) of the inequality.

$$-2x > 4 \qquad \text{Given}$$
$$\frac{-2x}{-2} < \frac{4}{-2} \qquad \text{Multiply by } -\frac{1}{2} \text{ (or divide by } -2\text{) and reverse the sign.}$$
$$x < -2 \qquad \text{Simplify.}$$

EXAMPLE 3 Solving and graphing inequalities

Solve, graph, and write the solution set in interval notation.

a. $4(x - 1) \leq 6x + 2$ **b.** $2x + 9 \geq 5x + 3$

SOLUTION

a. We follow the same procedure (CRAM) as that for solving linear equations given on page 79–80.

Given:	$4(x - 1) \leq 6x + 2$	Clear fractions: none.
	$4x - 4 \leq 6x + 2$	Remove parentheses.
	$4x - 6x - 4 \leq 6x - 6x + 2$	Add $-6x$ to both sides.
	$-2x - 4 \leq 2$	
	$-2x - 4 + 4 \leq 2 + 4$	Add 4 to both sides.
	$-2x \leq 6$	
	$\dfrac{-2x}{-2} \geq \dfrac{6}{-2}$	Multiply by $-\tfrac{1}{2}$ (or divide by -2).
	$x \geq -3$	Remember to reverse the sense of the inequality.

The solution set is $\{x \mid x \geq -3\}$ or, in interval notation, $[-3, \infty)$. The graph is

If we wish to avoid multiplying (or dividing) by negative numbers, we can add $-4x$ and -2 to obtain $-6 \leq 2x$. Then, dividing by 2, we get $-3 \leq x$, equivalent to $x \geq -3$.

b. Again, we isolate the x's on one side. This time, avoid multiplying or dividing by negative numbers. Do this by noting that there are more x's on the right-hand side of the inequality. Isolate the x's on the right.

Given:	$2x + 9 \geq 5x + 3$	Clear fractions: none. Remove parentheses: none.
	$2x - 2x + 9 \geq 5x - 2x + 3$	Add $-2x$ to both sides.
	$9 \geq 3x + 3$	
	$9 - 3 \geq 3x + 3 - 3$	Add -3 to both sides.
	$6 \geq 3x$	
	$\dfrac{6}{3} \geq \dfrac{3x}{3}$	Multiply by $\tfrac{1}{3}$ (or divide by 3).
	$2 \geq x$	

The inequality $x \leq 2$ is equivalent to $2 \geq x$. So, the solution set is $\{x \mid x \leq 2\}$ or, in interval notation $(-\infty, 2]$. The graph is

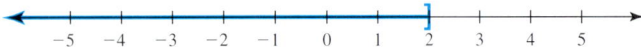

This could also be solved as follows:

$$2x + 9 \geq 5x + 3$$
$$2x - 5x + 9 \geq 5x - 5x + 3$$
$$-3x + 9 \geq 3$$
$$-3x + 9 - 9 \geq 3 - 9$$
$$\dfrac{-3x}{-3} \leq \dfrac{-6}{-3}$$
$$x \leq 2$$

PROBLEM 3

Solve, graph, and write the solution set in interval notation.

a. $3(x - 1) \leq 5x + 1$

b. $3x + 11 \geq 5x + 5$

Answers to PROBLEMS

3. a. $x \geq -2$; $[-2, \infty)$ 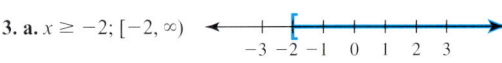 **b.** $x \leq 3$; $(-\infty, 3]$

2.5 Linear and Compound Inequalities 129

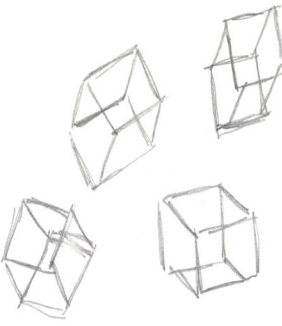

The inequality $x \leq 2$, in Example 3b, can be equivalently expressed as $2 \geq x$. To maintain equivalent inequality statements, if you exchange sides in an inequality you must reverse the inequality sign.

The two inequalities in Example 3 did not contain fractions. When fractions are present, we clear them by multiplying both sides of the inequality by the LCD of the fractions involved, as Example 4 shows.

EXAMPLE 4 Solving and graphing inequalities with fractions

Solve, graph, and write the solution set in interval notation.

$$\frac{x}{4} - \frac{x}{6} > \frac{x-3}{6}$$

SOLUTION

Given: $\frac{x}{4} - \frac{x}{6} > \frac{x-3}{6}$

$12\left(\frac{x}{4} - \frac{x}{6}\right) > 12\left(\frac{x-3}{6}\right)$ Clear fractions: multiply by 12.

$\overset{3}{\cancel{12}}\left(\frac{x}{\cancel{4}}\right) - \overset{2}{\cancel{12}}\left(\frac{x}{\cancel{6}}\right) > \overset{2}{\cancel{12}}\left(\frac{x-3}{\cancel{6}}\right)$

$3x - 2x > 2(x - 3)$ Remove parentheses/simplify.

$x > 2x - 6$

$x - 2x > 2x - 2x - 6$ Add $-2x$ to both sides.

$-x > -6$

$\frac{-x}{-1} < \frac{-6}{-1}$ Multiply by -1 (or divide by -1).

$x < 6$ Remember to reverse the sense of the inequality.

The solution set is $\{x \mid x < 6\}$ or, in interval notation, $(-\infty, 6)$, and the graph is

PROBLEM 4

Solve, graph, and write the solution set in interval notation.

$$\frac{x}{3} - \frac{x}{4} > \frac{x-4}{4}$$

C › Solving Compound Inequalities

Sometimes we connect inequalities by using the word *or* as in

$$x < -2 \text{ or } x > 1$$

The resulting inequality is a **compound** inequality and its graph is based on the idea of the *union* of two sets.

UNION OF TWO SETS (A or B) (A ∪ B)	If A and B are sets, the **union** of A and B, denoted by A ∪ B, is the set of elements in either A or B.

To graph $\{x \mid x < -2 \text{ or } x > 1\}$, we want to find solutions that will satisfy one or the other or both of the two inequality statements. First graph $x < -2$, then graph $x > 1$, and finally graph the *union*. The graph of $x < -2$ is

$(-\infty, -2)$

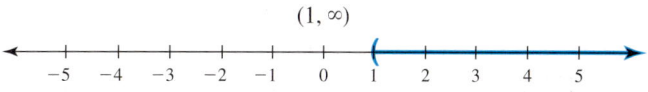

The graph of $x > 1$ is

$(1, \infty)$

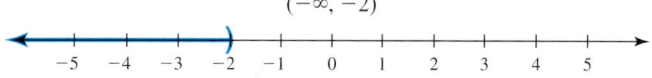

Answers to PROBLEMS

4. $x < 6$; $(-\infty, 6)$

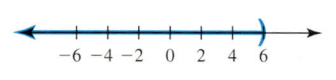

The union is

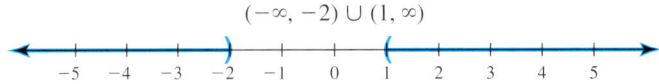
$(-\infty, -2) \cup (1, \infty)$

The solution set consists of all numbers less than -2 **or** greater than 1 or, in interval notation, $(-\infty, -2) \cup (1, \infty)$.

EXAMPLE 5 Solving a compound inequality with "or"

Graph and write the solution set in interval notation.

$$\{x \mid x < -3 \text{ or } x \geq 1\}$$

SOLUTION We do each of the graphs separately and then take the union of the two graphs. The graph of $x < -3$ is

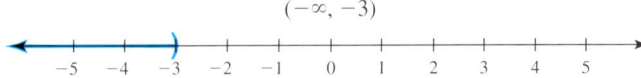
$(-\infty, -3)$

The graph of $x \geq 1$ is

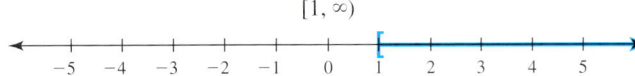
$[1, \infty)$

The union is

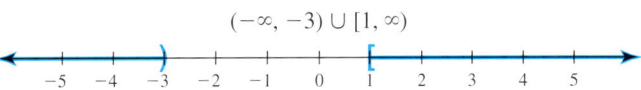
$(-\infty, -3) \cup [1, \infty)$

PROBLEM 5

Graph and write the solution set in interval notation.

$$\{x \mid x \leq 1 \text{ or } x > 4\}$$

The inequality $3 \leq s \leq 5$ is also a compound inequality because it is equivalent to two other inequalities:

$$3 \leq s \quad \text{and} \quad s \leq 5$$

These inequalities use "and" as their connective. We can graph these inequalities using the idea of *intersection*.

INTERSECTION OF TWO SETS (A and B) (A ∩ B) If A and B are sets, the **intersection** of A and B, denoted by A ∩ B, is the set of elements in both A and B.

To graph $\{s \mid 3 \leq s \text{ and } s \leq 5\}$, we want to find solutions that will satisfy *both* statements of inequality. First graph $3 \leq s$. (Remember $3 \leq s$ is the same as $s \geq 3$.)

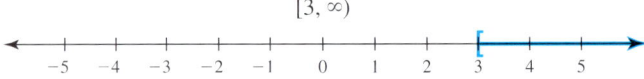
$[3, \infty)$

The graph of $s \leq 5$ is

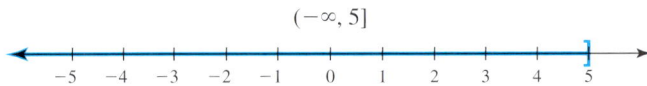
$(-\infty, 5]$

Answers to PROBLEMS

5. $(-\infty, 1] \cup (4, \infty)$

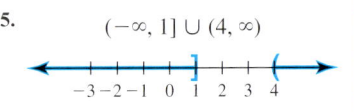

The intersection A ∩ B is the set of numbers the two graphs have in common, that is, from 3 to 5, including 3 and 5.

$$[3, \infty) \cap (-\infty, 5] = [3, 5]$$

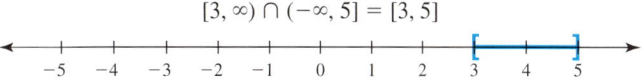

The intersection, [3, 5], is the graph of $\{s \mid 3 \leq s \text{ and } s \leq 5\}$ or $\{s \mid 3 \leq s \leq 5\}$.

The interval [3, 5] has a finite length, $5 - 3 = 2$. Because of this, [3, 5] is called a **bounded** interval. In general, if a and b are real numbers and $a < b$, the intervals shown in Table 2.6 are bounded intervals and a and b are called the **endpoints** of each interval.

Table 2.6 Bounded Interval Notation

Set Notation	Interval Notation	Type of Interval	Graph	Inequality
$\{x \mid a \leq x \leq b\}$	$[a, b]$	Closed	[a ———— b]	$a \leq x \leq b$
$\{x \mid a < x < b\}$	(a, b)	Open	(a ———— b)	$a < x < b$
$\{x \mid a \leq x < b\}$	$[a, b)$	Half-open	[a ———— b)	$a \leq x < b$
$\{x \mid a < x \leq b\}$	$(a, b]$	Half-open	(a ———— b]	$a < x \leq b$

EXAMPLE 6 Solving a compound inequality with "and"

Graph and write the solution set in interval notation.

$$\{x \mid x > -2 \text{ and } x < 1\}$$

SOLUTION The graph of $x > -2$ is

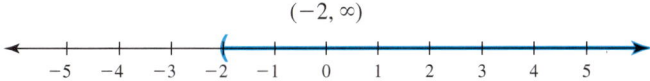

The graph of $x < 1$ is

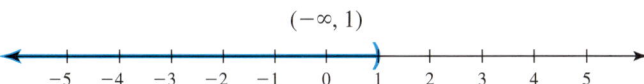

The numbers common to both graphs are in the open interval from -2 to 1. Thus, the intersection is $(-2, 1)$ or

$$(-2, \infty) \cap (-\infty, 1) = (-2, 1)$$

PROBLEM 6

Graph and write the solution set in interval notation:

$$\{x \mid x > -1 \text{ and } x < 2\}$$

Answers to PROBLEMS

6. $(-1, 2)$

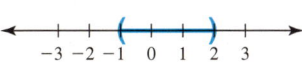

Now suppose you wish to solve the compound inequality $2x + 7 < 9$ *and* $x + 3 \geq -1$. We solve each inequality, obtaining

$2x + 7 < 9$ Given   and $x + 3 \geq -1$
$2x < 2$ Add -7. $x \geq -4$ Add -3.
$x < 1$ Divide by 2.

The graph of $x < 1$ is

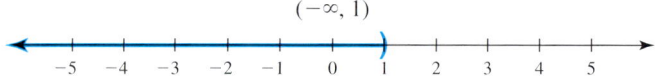

The graph of $x \geq -4$ is

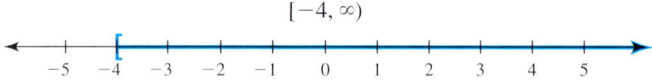

The numbers common to both graphs are in the half-open interval from -4 to 1; including -4 but not 1. The intersection is $[-4, 1)$ or

$$(-\infty, 1) \cap [-4, \infty) = [-4, 1)$$

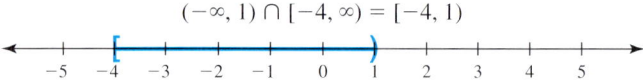

The intersection consists of all points such that $-4 \leq x$ *and* $x < 1$, which we can write as $-4 \leq x < 1$. A compound inequality that can be written in the form $a < x$ and $x < b$ (a and b real numbers) can be expressed more concisely as $a < x < b$.

EQUIVALENT STATEMENTS FOR "AND"

$a < x$ and $x < b$ is equivalent to $a < x < b$ (where a and b are real numbers with $a < b$).

The graph of $a < x < b$ is the solution set of $a < x$ and $x < b$, that is, the interval (a, b). Also be aware that inequality statements may be written with or without set builder notation. Graphing $\{x | 3 < x + 2 \text{ and } x < 3\}$ means the same as graphing $3 < x + 2$ and $x < 3$. We use these ideas to solve compound inequalities in Example 7.

EXAMPLE 7 Solving a compound inequality with "and"

Solve, graph, and write the solution set in interval notation.

a. $3 < x + 2$ and $x < 3$ **b.** $5 \geq -x$ and $x - 1 \leq -4$
c. $x + 1 \leq 5$ and $-2x < 6$

PROBLEM 7

Solve, graph, and write the solution set in interval notation.

a. $1 < x - 1$ and $x < 4$
b. $3 \geq -x$ and $x + 6 \leq 5$
c. $x + 2 \leq 6$ and $-3x < 12$

SOLUTION

a. We wish to write $3 < x + 2$ and $x < 3$ in the form $a < x < b$, so we add -2 to both sides of the first inequality to obtain

$$1 < x \quad \text{and} \quad x < 3$$

which is equivalent to

$$1 < x < 3 \quad \text{or} \quad (1, 3)$$

The graph is

Answers to PROBLEMS

7. a. $2 < x < 4$; $(2, 4)$ **b.** $-3 \leq x \leq -1$; $[-3, -1]$

c. $-4 < x \leq 4$; $(-4, 4]$

b. We wish to write $5 \geq -x$ and $x - 1 \leq -4$ in the form $a < x < b$, so we multiply the first inequality by -1 and add 1 to both sides of the second inequality to obtain

$$-5 \leq x \quad \text{and} \quad x \leq -3$$

which is equivalent to

$$-5 \leq x \leq -3 \quad \text{or} \quad [-5, -3]$$

The graph is

c. We solve the first inequality, $x + 1 \leq 5$, by subtracting 1 from both sides. We then have

$$x \leq 4 \quad \text{and} \quad -2x < 6$$

We then divide the second inequality by -2. (Remember to reverse the inequality sign.)

$$x \leq 4 \quad \text{and} \quad x > -3$$

Rearranging these inequalities gives

$$-3 < x \quad \text{and} \quad x \leq 4$$

which is equivalent to

$$-3 < x \leq 4 \quad \text{or} \quad (-3, 4]$$

The graph is

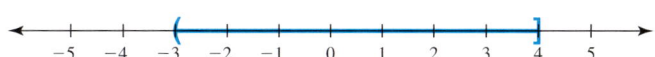

Suppose we want to solve $-5 < 2x + 3 \leq 9$. This inequality is equivalent to $-5 < 2x + 3$ and $2x + 3 \leq 9$, so it is also a compound inequality. We can solve this inequality by using the inequality properties we studied, keeping in mind that if we do any operations on the center expression ($2x + 3$), we must do the same operations on the outside expressions. As before, we wish to isolate x (this time in the middle). We start by adding (-3) to all three parts. We then have

$$
\begin{array}{rcll}
-5 < & 2x + 3 & \leq 9 & \text{Given} \\
-5 - 3 < & 2x + 3 - 3 & \leq 9 - 3 & \text{Add } -3. \\
-8 < & 2x & \leq 6 & \text{Simplify.} \\
\dfrac{-8}{2} < & \dfrac{2x}{2} & \leq \dfrac{6}{2} & \text{Multiply by } \tfrac{1}{2} \text{ (or divide by 2).} \\
-4 < & x & \leq 3 & \text{Simplify.}
\end{array}
$$

The solution is $-4 < x \leq 3$ or $(-4, 3]$.
The graph is

EXAMPLE 8 Solving and graphing compound inequalities

Solve, graph, and write the solution set in interval notation.

a. $6 < 4x + 6 \leq 10$ **b.** $-2 \leq -3x - 5 < 4$

SOLUTION

a. We start by adding -6 to each part.

$$6 - 6 < 4x + 6 - 6 \leq 10 - 6 \quad \text{Add } -6.$$
$$0 < 4x \leq 4$$
$$\frac{0}{4} < \frac{4x}{4} \leq \frac{4}{4} \quad \text{Divide by 4.}$$
$$0 < x \leq 1$$

The solution is $0 < x \leq 1$ or $(0, 1]$.

The graph is

b. We start by adding 5 to each part.

$$-2 + 5 \leq -3x - 5 + 5 < 4 + 5 \quad \text{Add 5.}$$
$$3 \leq -3x < 9$$
$$\frac{3}{-3} \geq \frac{-3x}{-3} > \frac{9}{-3} \quad \text{Divide by } -3 \text{ and } \textit{reverse} \text{ the sense (direction).}$$
$$-1 \geq x > -3$$
$$-3 < x \leq -1 \quad \text{Rewrite equivalently.}$$

The solution is $-3 < x \leq -1$ or $(-3, -1]$.

The graph is

Check the answer by substituting a number from the interval $(-3, -1]$, say -2, into the original inequality.

PROBLEM 8

Solve, graph, and write the solution set in interval notation.

a. $-7 < 5x + 3 < 7$

b. $-4 \leq -2x - 2 < 6$

CAUTION

When writing a compound inequality such as $-2 < -3x - 5 \leq 4$, make sure that the numbers are in the correct position. If you write $4 \leq -3x - 5 < -2$, you are implying that $4 < -2$, which is *wrong*. Compound inequalities must be written with the symbols pointing in the same direction toward the smaller number.

No matter how hard you try to find one, the inequality $\{x \mid x > 5 \text{ \textbf{and} } x < -5\}$ has no solutions. There is no number that can be greater than $+5$ and at the same time be less than -5.

You may have noticed that all linear inequalities we have solved have *unbounded* intervals for their graphs, while all compound inequalities have *bounded* intervals or *unbounded* intervals for their graphs. Table 2.7 shows what happens in general.

Answers to PROBLEMS

8. a. $-2 < x < \frac{4}{5}$; $\left(-2, \frac{4}{5}\right)$ **b.** $-4 < x \leq 1$; $(-4, 1]$

Table 2.7 Equations, Inequalities, and Their Solution Sets

Type	Solution Set	Graph
Linear equation $ax + b = c$	$\{p\}$	
Linear inequality $ax + b < c$	$(-\infty, p)$ or (p, ∞)	
Compound inequality "or" $x < p$ or $x > q$	$(-\infty, p) \cup (q, \infty)$	
Compound inequality "and" $c < ax + b < d$ $c < ax + b$ and $ax + b < d$	(p, q)	

D › Translating Sentences into Inequalities and Solving Applications Using Inequalities

At the beginning of this section, we mentioned that the cost of renting a car must be under $50. We then translated this phrase by writing "< 50."

Here are some other phrases and their translations using inequalities.

Words	Translation	In Symbols
x is at least 10	x is 10 or more	$x \geq 10$
x is at most 20	x is 20 or less	$x \leq 20$
x is no more than 30	x is 30 or less	$x \leq 30$
x is no less than 40	x is 40 or more	$x \geq 40$

EXAMPLE 9 Translating sentences involving inequalities

Translate into an inequality.

a. The height h of a human (in feet) has never been known to exceed 9 feet.
b. The weight w of a human is at most 1400 pounds.
c. The number n of puppies born in a single litter is no more than 23.
d. The cat population p in the United States is at least 76 million.

SOLUTION

a. $h \leq 9$ (Robert Wadlow was the tallest at 8 ft, 11.1 in.)
b. $w \leq 1400$ (Jon Browner Minnoch weighed 1400 lb.)
c. $n \leq 23$ (Lena, a foxhound, had 23 live puppies June 9, 1944.)
d. $p \geq 76$ million (according to the American Pet Association, 2006)

PROBLEM 9

Translate into an inequality.

a. x does not exceed 23.
b. y is at most 180.
c. z is no more than 10.
d. p is at least 45.

In Example 9, we practiced translating sentences involving inequalities. Now in Example 10, we will use those skills to solve an application involving inequalities.

Answers to PROBLEMS

9. a. $x \leq 23$ b. $y \leq 180$
   c. $z \leq 10$ d. $p \geq 45$

EXAMPLE 10 Solving applications involving inequalities

Federal Pell Grant awards for the academic year 2005–2006 at Morehead State University ranged from $400 to $4050. Part-time undergraduate tuition was $180 per hour. Relonda, a part-time undergraduate student, took as many hours as her Pell Grant allowed. What range of hours could she consider taking for the year? (Round to the nearest hour.)

Source: http://www.morehead-st.edu/.

SOLUTION
We will use the RSTUV method for solving word problems to set up the solution.

1. **Read and write a "Find" statement.** Find the range of hours she could consider taking for the year.
2. **Select the unknown using a "Let" statement.** Let x = the number of hours she could consider taking

   Since the tuition was $180 per hour, $180x$ = the amount of tuition she would pay.
3. **Think of a word sentence and translate it to an inequality.** The amount of tuition paid must be between $400 and $4050 inclusive.

$$400 \leq 180x \leq 4050$$

4. **Use your algebra skills to solve the inequality.**

$$400 \leq 180x \leq 4050$$
$$\frac{400}{180} \leq \frac{180}{180}x \leq \frac{4050}{180}$$ Divide by 180; keep the whole number and drop the remainder.
$$2 \leq x \leq 22$$

5. **Verify the result.** We leave that to you.

   Relonda could consider taking from 2 to 22 hours paid by the Pell Grant.

PROBLEM 10
Federal Supplemental Education Opportunity Grant (FSEOG) awards may range from $100 to $4000 per academic year, depending on a student's financial need, the other financial aid funds received by the student, and the availability of funds at the school. Wallace is a nonresident, undergraduate, and his tuition is $480 per hour. If the money he gets from this grant is all he has to pay for his tuition, what range of hours can he consider taking for the year? (Round to the nearest hour.)

Boost your grade at **mathzone.com**!
> Practice Problems
> NetTutor
> Self-Tests
> e-Professors
> Videos

> Exercises 2.5

< A > Graphing Linear Inequalities In Problems 1–4, graph the inequality and write the solution set using interval notaion.

1. $x > 3$
2. $x > 4$
3. $x \leq -3$
4. $x \leq -4$

< B > Solving Linear Inequalities In Problems 5–30, solve the inequality, graph the solution set, and write the solution set in interval notation.

5. $2x \geq 6$
6. $3x \geq 8$
7. $-3x \leq 3$
8. $-5x \leq 10$
9. $-4x > -8$
10. $-6x > -12$

Answers to PROBLEMS

10. $0 \leq x \leq 8$;   Wallace may be able to take from 0 to 8 hours.

11. $3x + 6 \leq 9$

12. $4y - 9 \leq 3$

13. $-2y - 4 \geq -10$

14. $-2z - 2 \geq -10$

15. $-3x + 1 < -14$

16. $-3x + 4 < -8$

17. $3a + 6 \leq a + 10$

18. $4b + 4 \leq b + 7$

19. $7z - 12 > 8z - 8$

20. $3z + 7 > 5z + 19$

21. $10 - 5x \leq 7 - 8x$

22. $6 - 4y \leq -14 + 6y$

23. $5(x + 2) \leq 3(x + 3) + 1$

24. $5(4 - 3x) < 7(3 - 4x) + 12$

25. $-4x + \frac{1}{2} > 4x + \frac{8}{5}$

26. $12x + \frac{2}{7} > 4x - \frac{4}{7}$

27. $\frac{x}{5} - \frac{x}{4} \leq 1$

28. $\frac{x}{3} - \frac{x}{2} \leq 1$

29. $\frac{7x + 2}{6} + \frac{1}{2} \geq \frac{3}{4}x$

30. $\frac{8x - 23}{6} + \frac{1}{3} \geq \frac{5}{2}x$

‹ C › **Solving Compound Inequalities** In Problems 31–70, solve, graph, and write the solution set in interval notation. If the solution set is empty, write $\varnothing$ for the answer. (Remember, inequality statements may be written with or without set builder notation.)

31. $\{x \mid x + 4 > 7 \text{ or } x - 2 < -6\}$

32. $\{x \mid x - 4 > 1 \text{ or } x + 2 < 1\}$

33. $3x + 4 > 10 \text{ or } 3x - 1 < 2$

34. $2x - 5 > 5 \text{ or } 2x < -4$

35. $2x \leq x + 4 \text{ or } x - 2 > 3$

36. $3x - 2 > 7 \text{ or } -5x \geq -5$

37. $\{x \mid -3x + 2 < -4 \text{ or } 2x - 1 < 3\}$

38. $\{x \mid -5x + 1 < -4 \text{ or } -2x + 1 > 5\}$

39. $-6x - 2 \geq -14 \text{ or } -7x + 2 < -19$

40. $-3x - 5 < 7$ or $-2x + 1 > -5$

41. $\{x \mid x \leq 4 \text{ and } x \geq -2\}$

42. $\{x \mid x > 0 \text{ and } x \leq 5\}$

43. $x + 1 \leq 7$ and $x > 2$

44. $x > -5$ and $x - 1 < 0$

45. $\{x \mid 2x - 1 > 1 \text{ and } x + 1 < 4\}$

46. $\{x \mid x - 1 < 1 \text{ and } 3x - 1 > 11\}$

47. $x < -5$ and $x > 5$

48. $x \geq 0$ and $x < -2$

49. $\{x \mid x + 1 \geq 2 \text{ and } x \leq 4\}$

50. $\{x \mid x \leq 5 \text{ and } x > -1\}$

51. $\{x \mid x < 3 \text{ and } -x < -2\}$

52. $\{x \mid -x < 5 \text{ and } x < 2\}$

53. $\{x \mid x + 1 < 4 \text{ and } -x < -1\}$

54. $\{x \mid x - 2 < 1 \text{ and } -x < 2\}$

55. $\{x \mid x - 2 < 3 \text{ and } 2 > -x\}$

56. $\{x \mid x - 3 < 1 \text{ and } 1 > -x\}$

57. $\{x \mid x + 2 < 3 \text{ and } -4 < x + 1\}$

58. $\{x \mid x + 4 < 5 \text{ and } -1 < x + 2\}$

59. $\{x \mid x - 1 > 2 \text{ and } x + 7 < 11\}$

60. $\{x \mid x - 2 > 1 \text{ and } -x > -5\}$

61. $-3 < x - 1 < 3$

62. $-4 < x + 1 < 4$

63. $-8 < 2y + 4 < 6$

64. $-2 \leq 3x + 1 \leq 7$

65. $4 \leq 3y - 8 \leq 10$

66. $3 \leq 4z + 3 \leq 5$

67. $-1 < \dfrac{x}{2} < 2$

68. $-2 < \dfrac{y}{2} < 1$

69. $2 < 4 + \dfrac{2}{3}a < 6$

70. $1 < 5 + \dfrac{4}{5}b < 9$

⟨ D ⟩ Translating Sentences and Solving Applications Using Inequalities   In Problems 71–76, translate the statement into an inequality.

71. The height h (in feet) of any mountain does not exceed that of Mt. Everest, 29,028 feet.

72. The number e of possible eclipses in a year is at most 7.

73. The number e of possible eclipses in a year is at least 2.

74. The altitude h (in feet) attained by the first liquid-fueled rocket was no more than 41 feet.

75. There are no less than 4×10^{25} nematode sea worms in the world. (Let n be the number of nematodes.)

76. There are at least 713 million people (p) who speak Mandarin Chinese.

77. When the variable cost per unit is $12 and the fixed cost is $160,000, the total cost for a certain product is $C = 12n + 160,000$ (n is the number of units sold). If the unit price is $20, the revenue R is $20n$. What is the minimum number of units that must be sold to make a profit? (You need $R > C$ to make a profit.)

78. Using a mailing company the cost of first-class mail is 41¢ for the first ounce and 17¢ for each additional ounce. A delivery company will charge $4.00 for delivering a package weighing up to 2 pounds (32 oz). When would the mailing company's price, $P = 0.41 + 0.17(x - 1)$ (where x is the weight of the package in ounces), be cheaper than the delivery company's price? (Round to smaller whole number)

79. The parking cost at a garage is $C = 1 + 0.75(h - 1)$, where h is the number of hours you park and C is the cost in dollars. When is the cost C less than $10?

In Problems 80–83, use the table below and the RSTUV method to help solve the application.

Fast Food	Protein Grams (Multiply by 4 to change to calories)	Carb Grams (Multiply by 4 to change to calories)	Total Fat Grams (Multiply by 9 to change to calories)
McDonald's Egg McMuffin	18 g	29 g	12 g
McDonald's Sausage Biscuit	10 g	30 g	28 g
Subway 6" Cheese Steak	24 g	47 g	10 g
Subway 6" Turkey & Ham	20 g	47 g	5 g
Boston Market Three Piece Dark Meat	49 g	4 g	24 g
Boston Market Chicken Pot Pie	29 g	59 g	49 g

80. Gina is maintaining an 1800 calorie diet each day. It has been recommended that no more than 15% of her daily caloric intake (1800 calories) be protein. However, nutrition labels give the protein content in grams and must be changed to calories. To find the amount of protein calories, multiply the amount of protein in grams by 4. For example, if a food item contains 15 grams of protein, then it has 4×15 or 60 calories of protein.

  a. Up to how many grams of protein can Gina eat per day to meet this 15% requirement?

  b. Using the table, can she have a McDonald's Egg McMuffin, a Subway 6-inch Turkey and Ham, and a Boston Market Three Piece Dark Meat Meal and still be within her recommended daily protein requirement?

81. Derek wants to maintain a 2200 calorie diet each day. He wants $\frac{1}{2}$ or less of the daily calories (2200 calories) to be in carbs (carbohydrates). However, nutrition labels give the carb content in grams and must be changed to calories. To change the grams to calories, multiply the amount of carbs in grams by 4. For example, if a food item contains 20 grams of carbs, then it has 4×20 or 80 calories of carbs.

  a. Up to how many grams of carbs can Derek eat per day to maintain that $\frac{1}{2}$ or less of his daily caloric intake meets this carb requirement?

  b. Using the table, can he have two McDonald's Sausage Biscuits, two Subway 6-inch Cheese Steaks, and two Boston Market Chicken Pot Pies and still be within his daily carb requirement?

82. Tyler is an active young man and wants to maintain a 2800 calorie diet each day. He wants the amount of fat he eats to be at most 30% of the daily calories (2800 calories). However, nutrition labels give the fat content in grams and must be changed to calories. To change the grams to calories, multiply the amount of fat in grams by 9. For example, if a food item contains 25 grams of fat, then it has 9×25 or 225 calories of fat.
   a. Up to how many grams of fat can Tyler eat per day so that no more than 30% of his daily caloric intake meets this fat requirement?
   b. Using the table, can he have two McDonald's Egg McMuffins, one Subway 6-inch Cheese Steak, one Subway 6-inch Turkey and Ham, and one Boston Market Chicken Pot Pie and still be within his daily fat requirement?

83. Kyrstin maintains a daily intake of 1600 calories. She is strict about keeping 15% of her daily intake in protein calories and 30% in fat calories.
   a. Up to how many carb calories can she eat and not exceed her 1600 calorie limit for the day?
   b. If it is recommended that her carb intake not exceed more than $\frac{1}{2}$ the total daily caloric intake, will she be able to have the maximum amount of carb calories as indicated in answer **a**?

〉〉〉 Using Your Knowledge

Inequalities and the Environment

84. Do you know why spray bottles use a pump rather than propellants? They do that because some of the propellants have chlorofluorocarbons (CFCs) which deplete the ozone layer. In an international meeting in London, the countries represented agreed to stop producing CFCs by the year 2000. The annual production of CFCs (in thousands of tons) is given by $P = 1260 - 110x$, where x is the number of years after 1980. How many years after 1980 did it take for the production of CFCs to be less than 0? (Answer to the nearest whole number.) What year did that take place?

85. Cigarette smoking also produces air pollution. The annual number of cigarettes consumed per person in the United States is given by $N = 4200 - 70x$, where x is the number of years after 1960. How many years after 1960 was consumption less than 1000 cigarettes per person annually? (Answer to the nearest whole year.) What year did that take place?

〉〉〉 Write On

86. What is the main difference in the techniques used when solving equations as opposed to inequalities?

In Problems 87–90, describe in words the real numbers in the intervals.

87. $(1, \infty)$
88. $(-\infty, 3]$
89. $(4, \infty)$
90. $[-2, \infty)$

91. Look at the following argument, where it is assumed that $a > b$.

$2 > 1$	Known.
$2(b - a) > 1(b - a)$	Multiply by $(b - a)$.
$2b - 2a > b - a$	Simplify.
$2b > b + a$	Add 2a.
$b > a$	Subtract b.

But we assumed that $a > b$. Write a paragraph explaining what went wrong.

〉〉〉 Concept Checker

Fill in the blank(s) with the correct word(s), phrase, or mathematical statement.

92. A visual way to represent the solutions to a statement is called a _____ .

93. If a compound inequality statement has the connective "and," then use _____ to graph the solution set.

94. A(n) _____ is a statement that two expressions are not equal.

95. Included in the graph of the solution to the statement "$x < 3$" will be the values from _____ to 3.

inequality
compound inequality
graph
positive infinity
negative infinity
union
intersection

>>> Mastery Test

Translate into an inequality:

96. x does not exceed 23.

97. p is at least 45.

Graph:

98. $x \leq -3$

99. $x > 4$

Solve and graph:

100. $-3 < -3x - 6 < 3$

101. $-2 \leq 2 + \frac{4}{5}x < 6$

102. $2(x - 1) \geq 3x + 1$

103. $3x + 10 < 6x + 4$

104. $\frac{1}{4}(x - 5) < \frac{1}{3}(x - 4)$

105. $2x + 3 > 1$ or $5x - 3 < -13$

106. $\{x \mid 2x + 9 > 11 \text{ or } x + 3 > -1\}$

107. $\{x \mid 2x + 3 < 5x - 3 \text{ and } 3x - 2 < 3 + 2x\}$

108. $\{x \mid -3x + 5 < 9 \text{ and } 3x - 7 < -3 + 2x\}$

>>> Skill Checker

Fill in the blank with $<$ or $>$ to make the result a true statement.

109. $-7 \underline{} -8$

110. $\frac{1}{3} \underline{} \frac{1}{2}$

111. $0.34 \underline{} 0.342$

112. $-0.234 \underline{} -0.233$

Find:

113. $|-9|$

114. $\left|-\frac{1}{5}\right|$

115. $|-0.34|$

116. $|\sqrt{2}|$

Graph:

117. $x < 5$

118. $x \geq -3$

119. $x > 1$ or $x < 0$

120. $-4 \leq x \leq 1$

2.6 Absolute-Value Equations and Inequalities

Objectives

A ▸ Solve absolute-value equations.

B ▸ Solve absolute-value inequalities of the form $|ax + b| < c$ or $|ax + b| > c$, where $c > 0$.

C ▸ Solve applications involving absolute-value inequalities.

To Succeed, Review How To . . .

1. Find the absolute value of a number (p. 9).
2. Graph an inequality (pp. 124–125).

▸ Getting Started

Budget Variances

Businesses and individuals usually try to predict how much money will be spent on certain items over a certain time. If you budget $120 for a month's utilities and a heat wave or cold snap makes your actual expenses jump to $150, the $30 difference might be an acceptable **variance**. Suppose b represents the budgeted amount for an item, a represents the actual expense, and you want to be within $10 of your estimate. The item will pass the variance test if the actual expenses a are within $10 of the budgeted amount b; that is, $b - a$ is between -10 and 10. In symbols, $-10 < b - a < 10$ or equivalently $|b - a| < 10$.

In general, if a and b are as before, a certain item will pass the variance test if $|b - a| < c$, where c is the acceptable variance. The quantity c can be a definite amount or a percent of the budget. For example, if you budget $50 for gas and you want to be within 10% of your budget, how much gas money can you spend and still be within your variance? Since 10% of $50 = (0.10)(50) = 5$, you can see that if you spend between $45 and $55, you will

be within your 10% variance. Here $|b - a| < c$ becomes $|50 - a| < 5$ or equivalently $-5 < 50 - a < 5$.

$$-55 < -a < -45 \quad \text{Add } -50.$$
$$55 > a > 45 \quad \text{Multiply by } -1.$$
$$45 < a < 55 \text{ (as expected)} \quad \text{Rewrite.}$$

$|50 - a| < 5$ is equivalent to $-5 < 50 - a < 5$; we shall develop the rule for this concept in this section.

A ▸ Absolute-Value Equations

The absolute value of a, denoted by $|a|$, is the distance between a and zero on a number line. Thus, $|3| = 3$, $|-5| = 5$, and $|-0.7| = 0.7$. The equation $|x| = 2$ is read as "the distance between x and zero on the number line is 2." x is 2 units from zero, so x can only be $+2$ or -2; that is, $x = \pm 2$.

To solve absolute-value equations, we need the following definition.

> **THE SOLUTIONS OF $|x| = a$ WHERE $a \geq 0$**
>
> If $a \geq 0$, the solutions of $|x| = a$ are
>
> $$x = a \text{ and } x = -a.$$

> **CAUTION**
>
> If a is negative ($a < 0$), $|x| = a$ has no solution. An example of an absolute value equation with no solution is $|3x - 4| = -7$.

EXAMPLE 1 Solving absolute-value equations

Solve:

a. $|x| = 8$ b. $|y| = 2.5$ c. $|z| = -6$ d. $|w| = 0$

SOLUTION

a. Using a number line, we can locate those points on the line that are 8 units in distance from 0.

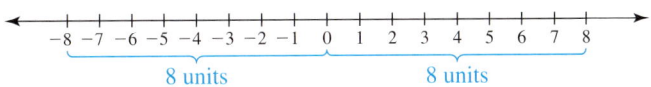

8 and -8 are 8 units from 0, so the solutions of $|x| = 8$ are 8 and -8. The solution set of $|x| = 8$ is $\{-8, 8\}$.

b. Using a number line, we can locate those points on the line 2.5 units in distance from 0.

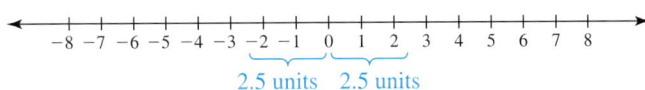

2.5 and -2.5 are 2.5 units from 0, so the solutions of $|y| = 2.5$ are 2.5 and -2.5. The solution set of $|y| = 2.5$ is $\{-2.5, 2.5\}$.

c. A number line can't help us solve $|z| = -6$ because there is no such thing as a -6 distance from 0. Distance is always expressed as a positive number or 0. Thus, the equation $|z| = -6$ has no solutions. Its solution set is $\emptyset$ (the empty set).

d. Using a number line we can see that there is only one point 0 units from 0 and that is the number 0.

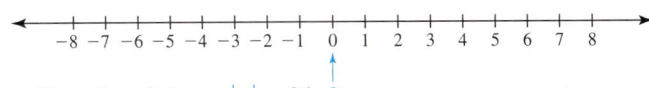

The only solution to $|w| = 0$ is 0.

PROBLEM 1

Solve:

a. $|x| = 7$ b. $|y| = 3.4$
c. $|z| = -9$

The ideas of Example 1 can be generalized to more complicated equations. Look at the pattern:

$\|x\| = 8$ has solutions	$x = 8$	and	$x = -8$
$\|y\| = 2.5$ has solutions	$y = 2.5$	and	$y = -2.5$
$\|x + 1\| = 8$ has solutions	$x + 1 = 8$	and	$x + 1 = -8$
(Solve each equation by adding (-1) on each side.)	$x = 7$	and	$x = -9$

Answers to PROBLEMS

1. a. $-7; 7$ b. $-3.4; 3.4$
   c. No solutions

144   Chapter 2   Linear Equations and Inequalities

We can check the solutions by substituting them into the equation $|x + 1| = 8$. Substituting 7 for x, we obtain $|7 + 1| = |8| = 8$, a true statement. If we substitute -9 for x, we have $|-9 + 1| = |-8| = 8$, also true.

STATEMENT TRANSLATION FOR ABSOLUTE-VALUE EQUATIONS

If $|\text{expression}| = a$, where $a \geq 0$, then
$$\text{expression} = a \quad \text{or} \quad \text{expression} = -a$$

If $|\text{expression 1}| = |\text{expression 2}|$, then

Equal   Opposites
expression 1 = expression 2   expression 1 = $-$(expression 2)

Be sure $|\text{expression}|$ is isolated on one side before applying the translation statement. Example: $|x + 3| - 4 = 6$ cannot be translated until written as $|x + 3| = 10$.

Remember if the $|\text{expression}|$ equals a negative number, the equation has no solutions.

EXAMPLE 2   Solving absolute-value equations

Solve:

a. $|2x + 1| = 6$ **b.** $\left|\frac{3}{4}x + 1\right| + 5 = 11$

SOLUTION

a. The solutions of $|2x + 1| = 6$ are

$2x + 1 = 6$ and $2x + 1 = -6$
$2x = 5$ and $2x = -7$ Add (-1) to both sides.
$x = \frac{5}{2}$ and $x = -\frac{7}{2}$ Multiply both sides by $\frac{1}{2}$.

The solution set of $|2x + 1| = 6$ is $\left\{-\frac{7}{2}, \frac{5}{2}\right\}$.

Check the solutions by substituting them, one at a time, into the original equation.

b. To use the definition of absolute value, we must isolate $\left|\frac{3}{4}x + 1\right|$. We first add (-5) to both sides.

$\left|\frac{3}{4}x + 1\right| + 5 = 11$ Given

$\left|\frac{3}{4}x + 1\right| + 5 + (-5) = 11 + (-5)$ Add -5.

$\left|\frac{3}{4}x + 1\right| = 6$ Simplify.

The solutions of this equation are

$\frac{3}{4}x + 1 = 6$ and $\frac{3}{4}x + 1 = -6$

$\frac{3}{4}x = 5$ and $\frac{3}{4}x = -7$ Add -1.

$\frac{4}{3} \cdot \frac{3}{4}x = \frac{4}{3} \cdot 5$ and $\frac{4}{3} \cdot \frac{3}{4}x = \frac{4}{3} \cdot (-7)$ Multiply by $\frac{4}{3}$.

$x = \frac{20}{3}$ and $x = -\frac{28}{3}$ Simplify.

The solution set of $\left|\frac{3}{4}x + 1\right| + 5 = 11$ is $\left\{-\frac{28}{3}, \frac{20}{3}\right\}$.

Check the solutions by substituting them, one at a time, into the original statement.

PROBLEM 2

Solve:

a. $|3x + 1| = 5$

b. $\left|\frac{3}{2}x + 5\right| - 4 = 6$

Answers to PROBLEMS

2. a. $-2; \frac{4}{3}$ **b.** $-10; \frac{10}{3}$

EXAMPLE 3 Solving absolute-value equations

Solve:

a. $|x - 3| = |x - 5|$ **b.** $|4x + 5| = |7 - 2x|$

SOLUTION

a. The equation $|x - 3| = |x - 5|$ is true if $x - 3$ and $x - 5$ are equal to each other or if they are opposites. The opposite of $(x - 5)$ is $-(x - 5)$. Here are the two cases.

Equal		**Opposites**	
$x - 3 = x - 5$		$x - 3 = -(x - 5)$	
$-x + x - 3 = -x + x - 5$	Add $-x$.	$x - 3 = -x + 5$	Remove parentheses.
$-3 = -5$		$x = -x + 8$	Add 3.
This is a contradiction. There is no solution for this case.		$2x = 8$	Add x.
		$x = 4$	Divide by 2.

The solution set is $\{4\}$. We can verify this by replacing x by 4 in $|x - 3| = |x - 5|$ to obtain

$$|4 - 3| = |4 - 5|$$
$$|1| = |-1|$$
$$1 = 1$$

This is a true statement, so our solution is correct.

b. The equation $|4x + 5| = |7 - 2x|$ is true if $(4x + 5)$ and $(7 - 2x)$ are equal to each other or if they are opposites. The opposite of $(7 - 2x)$ is $-(7 - 2x)$. Here are the two cases.

Equal		**Opposites**	
$4x + 5 = 7 - 2x$		$4x + 5 = -(7 - 2x)$	Remove parentheses.
$4x + 2x + 5 = 7 - 2x + 2x$	Add $2x$.	$4x + 5 = -7 + 2x$	
$6x + 5 = 7$		$4x - 2x + 5 = -7 + 2x - 2x$	Add $-2x$.
$6x + 5 - 5 = 7 - 5$	Add -5.	$2x + 5 = -7$	
$6x = 2$		$2x + 5 - 5 = -7 - 5$	Add -5.
$\dfrac{6x}{6} = \dfrac{2}{6}$	Divide by 6.	$\dfrac{2x}{2} = \dfrac{-12}{2}$	Divide by 2.
$x = \dfrac{1}{3}$		$x = -6$	

The solution set is $\left\{\dfrac{1}{3}, -6\right\}$.

The verification of the solutions is left to you.

PROBLEM 3

Solve:

a. $|x - 5| = |x - 8|$

b. $|3 - 5x| = |x + 9|$

B › Absolute-Value Inequalities

As you recall, $|x| = 2$ means that the distance from 0 to x on the number line is 2 units. What would $|x| < 2$ and $|x| > 2$ mean? Here are the statements, their translations, their graphs, and their meanings in terms of the compound inequalities we just studied.

Answers to PROBLEMS

3. **a.** $\dfrac{13}{2}$ **b.** $-1, 3$

STATEMENT	TRANSLATION
$\|x\| = 2$	x is exactly 2 units from 0. 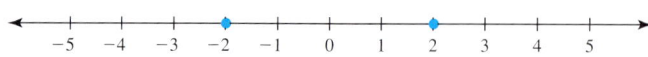
$\|x\| < 2$	x is less than 2 units from 0. 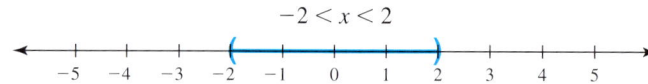
$\|x\| > 2$	x is more than 2 units from 0. 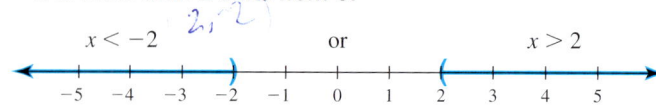

In general, the following holds.

$\|x\| < a$ WHERE a IS POSITIVE ($a > 0$) IS EQUIVALENT TO $-a < x < a$

The statement $|x| < a$ is equivalent to $-a < x < a$, where $a > 0$. A similar relationship exists if $\leq$ replaces $<$.

CAUTION

If $a < 0$, $|x| < a$ has no solutions. Think about it. The $|x|$ represents a positive number or 0. Can a positive number or 0 be less than a negative number? Of course not! An example of an absolute value inequality with no solutions is $|2x - 5| < -3$.

$$|x| < 4 \quad \text{is equivalent to} \quad -4 < x < 4$$
$$|y| \leq 2.5 \quad \text{is equivalent to} \quad -2.5 \leq y \leq 2.5$$
$$|z| < \frac{3}{4} \quad \text{is equivalent to} \quad -\frac{3}{4} < z < \frac{3}{4}$$

What does $|x + 1| < 2$ mean? $|x| < a$ is equivalent to $-a < x < a$, so $|x + 1| < 2$ is equivalent to $-2 < x + 1 < 2$.

To solve $-2 < x + 1 < 2$, add -1, and simplify like this:

$$-2 - 1 < x + 1 - 1 < 2 - 1 \quad \text{Add } -1.$$
$$-3 < \quad x \quad < 1 \quad \text{Simplify.}$$

The solution is $(-3, 1)$, and the graph is

In general, $|ax + b| < c$, where $c > 0$, is equivalent to $-c < ax + b < c$.

EXAMPLE 4 Absolute-value inequalities of the form $|ax + b| \leq c$

Solve, graph, and write the solution set in interval notation.

$$|3x + 4| \leq 8$$

SOLUTION $|3x + 4| \leq 8$ is equivalent to $-8 \leq 3x + 4 \leq 8$. *Translation*
$$-8 - 4 \leq 3x + 4 - 4 \leq 8 - 4 \quad \text{Add } -4.$$
$$-12 \leq \quad 3x \quad \leq 4$$

PROBLEM 4

Solve, graph, and write the solution set in interval notation.

$$|3x - 2| \leq 5$$

Answers to PROBLEMS

4. $-1 \leq x \leq \frac{7}{3}$; $\left[-1, \frac{7}{3}\right]$

$$\frac{-12}{3} \leq \frac{3x}{3} \leq \frac{4}{3} \quad \text{Divide by 3.}$$

$$-4 \leq x \leq \frac{4}{3}$$

The solution set is $\left[-4, \frac{4}{3}\right]$, and the graph is

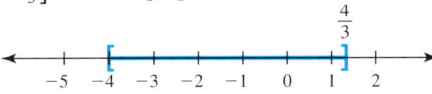

Let's look at $|x| > 2$. Its graph consists of all points such that $x < -2$ or $x > 2$. $|x| > 2$ is equivalent to $x < -2$ or $x > 2$. If we generalize this result, we obtain the following.

$|x| > a$ WHERE $a > 0$ IS EQUIVALENT TO $x < -a$ OR $x > a$

The statement $|x| > a$ is equivalent to

$$x < -a \text{ or } x > a, \text{ where } a > 0.$$

A similar relationship applies if $\geq$ replaces $>$.

CAUTION

If $a < 0$ (a is negative), $|x| > a$ is always true and has infinitely many solutions—the set of real numbers (R). The $|x|$ represents a positive number. Will a positive number such as $|x|$ always be greater than a negative number? Of course it will! An example of an absolute value inequality with the real numbers (R) as its solution set is $|3x + 4| > -5$.

$|x| > 5$	is equivalent to	$x < -5$ or $x > 5$
$|x| > 3.4$	is equivalent to	$x < -3.4$ or $x > 3.4$
$|x| \geq \frac{1}{5}$	is equivalent to	$x \leq -\frac{1}{5}$ or $x \geq \frac{1}{5}$

What does $|x - 1| > 3$ mean? $|x| > a$ is equivalent to $x < -a$ or $x > a$, so $|x - 1| > 3$ means

$$x - 1 < -3 \quad \text{or} \quad x - 1 > 3$$
$$x < -2 \quad \text{or} \quad x > 4 \quad \text{Solve each inequality by adding 1 to both sides.}$$

The solution set is $(-\infty, -2) \cup (4, \infty)$, and the graph of the solution set is

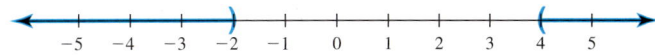

In general, $|ax + b| > c$, where $c > 0$, is equivalent to $ax + b < -c$ or $ax + b > c$.

STATEMENT TRANSLATION FOR ABSOLUTE-VALUE INEQUALITIES

If $|\text{expression}| > a$, where $a > 0$, then

$$\text{expression} < -a \quad \text{or} \quad \text{expression} > a$$

If $|\text{expression}| < a$, where $a > 0$, then

$$-a < \text{expression} < a$$

In the previous translations, the $|\text{expression}|$ must be isolated on the left side of the inequality sign before the equivalent translation can be made.

EXAMPLE 5 Absolute-value inequalities of the form $|ax + b| > c$

Solve, graph, and write the solution set in interval notation.

$$|-2x + 3| > 1$$

SOLUTION The inequality $|-2x + 3| > 1$ is equivalent to

$$-2x + 3 < -1 \quad \text{or} \quad -2x + 3 > 1 \quad \text{Translation}$$

Add (-3) on both sides of each inequality.

$$-2x + 3 + (-3) < -1 + (-3) \quad \text{or} \quad -2x + 3 + (-3) > 1 + (-3)$$
$$-2x < -4 \quad \text{or} \quad -2x > -2$$

Divide by -2 and reverse the inequality signs.

$$\frac{-2x}{-2} > \frac{-4}{-2} \quad \text{or} \quad \frac{-2x}{-2} < \frac{-2}{-2}$$
$$x > 2 \quad \text{or} \quad x < 1$$

The solution set is $(-\infty, 1) \cup (2, \infty)$, and the graph is

PROBLEM 5

Solve, graph, and write the solution set in interval notation.

$$|-2x + 1| > 5$$

EXAMPLE 6 More complex absolute-value inequalities

Solve, graph, and write the solution set in interval notation.

$$10 \geq 4|5x - 3| + 2$$

SOLUTION Before making the equivalent translation statement, the |expression| must be isolated on the left side of the inequality sign. We will subtract 2, divide by 4, and rewrite with |expression| on the left side.

$$10 \geq 4|5x - 3| + 2$$
$$8 \geq 4|5x - 3| \quad \text{Subtract 2.}$$
$$2 \geq |5x - 3| \quad \text{Divide by 4.}$$
$$|5x - 3| \leq 2 \quad \text{Rewrite with |expression| on left side.}$$

The statement is ready for the equivalent translation.

$$-2 \leq 5x - 3 \leq 2 \quad \text{Translation}$$
$$1 \leq 5x \leq 5 \quad \text{Add 3.}$$
$$\frac{1}{5} \leq x \leq 1 \quad \text{Divide by 5.}$$

The solution set is $\left[\frac{1}{5}, 1\right]$, and the graph is

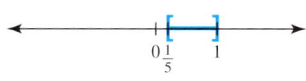

PROBLEM 6

Solve, graph, and write the solution set in interval notation:

$$5 \geq 2|2x - 6| - 3$$

To find the solution sets to absolute value equations and inequalities you must be able to write the appropriate translation statement, Table 2.8 is a summary of this information.

Table 2.8 Summary for Solving Absolute-Value Equations and Inequalities

1. The solutions of the equation $|ax + b| = c, c \geq 0$, are obtained by solving the two equations

$$ax + b = c \quad \text{and} \quad ax + b = -c$$

Note: If c is negative ($c < 0$), there is no solution.

Answers to PROBLEMS

5. $x < -2$ or $x > 3$; $(-\infty, -2) \cup (3, \infty)$ 6. $1 \leq x \leq 5$; $[1, 5]$

> **Table 2.8** *(continued)*
>
> 2. The solutions of the equation $|ax + b| = |cx + d|$ are obtained by solving the equations
>
> $$ax + b = cx + d \quad \text{and} \quad ax + b = -(cx + d)$$
>
> 3. The solution set of the inequality $|ax + b| < c,\ c > 0$, is obtained by solving the inequality $-c < ax + b < c$. The graph is *bounded*. (If c is negative, there is no solution.)
> 4. The solution set of the inequality $|ax + b| > c,\ c > 0$, is obtained by solving the inequalities $ax + b > c$ or $ax + b < -c$. The graph is the union of two *unbounded* intervals. (If c is negative, $|ax + b| > c$ is always true, and the solution set is the set of real numbers, R.)

C ⟩ Applications Involving Absolute-Value Inequalities

In the *Getting Started*, we discussed how much money was budgeted for certain items over a certain time. The difference between the budgeted amount and the actual amount spent is called the **variance**. In general, if a is the actual expense and b is the budgeted amount, then a certain item will pass the variance test if $|b - a| < c$, where c is the preset variance.

EXAMPLE 7 Solving applications involving absolute-value inequalities

The following table is an excerpt from the University of Alberta's expense budget in thousands of dollars for the year ended March 31, 2003.
Source: http://www.uofaweb.ualberta.ca/.

Faculty Support	Budgeted	Actual	Variance
Indirect Instructional	17,262	14,923	
Misc. Recurring Transfers	3,106	3,106	
Libraries	33,249	32,505	
Physical Facilities	22,891	20,554	
Utilities	24,561	28,490	
Student Services	6,362	6,218	
Administration	74,511	66,224	

a. If we use a 10% acceptable variance for the budgeted expense of indirect instructional, did the actual amount stay within the variance?

b. The following year's budget kept the same amount budgeted for administration with a 10% acceptable variance. In what range would the actual amount needed to have been to stay within the acceptable variance?

c. Which one of the line items had a "0" variance?

SOLUTION

a. The general rule for variance is $|b - a| < c$, where b is the budgeted amount, a is the actual amount, and c is the variance. For the line item, indirect instructional, $b = 17,262$, $a = 14,923$, and the variance c is 10% of the budgeted amount, 17,262. We substitute these values into the rule and simplify.

PROBLEM 7
Use the table from Example 7 to solve the following.

a. If we use a 10% acceptable variance for the budgeted expense of libraries, did the actual amount stay within the acceptable variance?

b. The following year's budget kept the same amount budgeted for utilities with a 10% acceptable variance. In what range would the actual amount needed to have been to stay within the acceptable variance?

(continued)

Answers to PROBLEMS

7. a. Yes, $-3324.9 < 744 < 3324.9$
b. $22,104.9 < a < 27,017.1$

$$|b - a| < c$$
$$|17{,}262 - 14{,}923| < 0.10(17{,}262) \quad \text{Substitution}$$
$$|17{,}262 - 14{,}923| < 1726.2 \quad \text{Simplify the right side.}$$
$$-1726.2 < 17{,}262 - 14{,}923 < 1726.2 \quad \text{Translate the absolute value.}$$
$$-1726.2 < 2339 < 1726.2 \quad \text{Simplify the middle.}$$

Since this is a false statement, 2339 is not between -1726.2 and $+1726.2$, we conclude that the actual amount did not stay within the 10% variance.

b. The general rule for variance is $|b - a| < c$, where b is the budgeted amount, a is the actual amount, and c is the variance. For the line item, Administration, $b = 74{,}511$, $a = $ actual amount, and the variance c is 10% of the budgeted amount, 74,511. We substitute these values into the rule and simplify.

$$|b - a| < c$$
$$|74{,}511 - a| < 0.10(74{,}511) \quad \text{Substitution}$$
$$|74{,}511 - a| < 7451.1 \quad \text{Simplify the right side.}$$
$$-7451.1 < 74{,}511 - a < 7451.1 \quad \text{Translate the absolute value.}$$
$$-81{,}962.1 < -a < -67{,}059.9 \quad \text{Add } -74{,}511.$$
$$81{,}962.1 > a > 67{,}059.9 \quad \text{Divide by } -1 \text{ (Reverse the inequality sign).}$$
$$67{,}059.9 < a < 81{,}962.1 \quad \text{Rewrite equivalently.}$$

This means that to stay within 10% of the amount budgeted for Administration, the actual amount needed to be between \$67,059.9 and \$81,962.1 thousands of dollars.

c. The variance is obtained by subtracting the actual amount from the budgeted amount. To get a result of "0" in subtraction, the two numbers being subtracted must be the same. Looking down the "Budgeted" and "Actual" columns on the table, we see there is one line item with the same numbers for both and that is "Misc. Recurring Transfers."

TRANSLATE THIS

The third step in the RSTUV procedure is to **TRANSLATE** the information into an equation or inequality. In Problems 1–4, **TRANSLATE** the sentence and match the correct translation with one of the equations A–H.

1. A certain company budgets \$3000 a year for temporary office help. They want to stay within a 5% variance. Write an absolute-value inequality statement that will meet that requirement.

2. A family budgets \$2500 a year for movies and movie rentals. If their acceptable variance is \$200, write an absolute-value inequality statement that will meet that requirement.

3. Chip budgets \$2800 a year for gasoline. His actual gasoline expense for last year was \$3200. If his acceptable variance was \$250, write an inequality statement that will check to see if he stayed within the variance.

4. Talitha budgets \$10,800 a year for apartment rental. Due to an unexpected raise in the rent she actually spent \$11,400 in rent for the year. Write an inequality statement that will check to see if she stayed within her 10% acceptable variance.

A. $-2800 < 3200 - 250 < 2800$
B. $|150 - a| < 3000$
C. $|2500 - a| < 200$
D. $-250 < 2800 - 3200 < 250$
E. $-1140 < 10{,}800 - 11{,}400 < 1140$
F. $|200 - a| < 2500$
G. $|3000 - a| < 150$
H. $-1080 < 10{,}800 - 11{,}400 < 1080$

> Exercises **2.6**

< A > Absolute-Value Equations   In Problems 1–30, solve the equation.

1. $|x| = 13$
2. $|y| = 17$
3. $|y| - 2.3 = 0$
4. $|x| - 3.7 = 0$
5. $|x| = 0$
6. $|y| = -3$
7. $|z| = -4$
8. $|x + 1| = 10$
9. $|x + 7| = 2$
10. $|x + 9| = 3$
11. $|2x - 4| = 8$
12. $|3x - 6| = 9$
13. $|5a - 2| - 8 = 0$
14. $|6b - 3| - 9 = 0$
15. $\left|\frac{1}{2}x + 4\right| = 6$
16. $\left|\frac{1}{3}x + 2\right| = 7$
17. $\left|\frac{2}{3}z - 3\right| = 9$
18. $\left|\frac{2}{5}x - 6\right| = 4$
19. $|x + 2| = |x + 4|$
20. $|y + 6| = |y + 2|$
21. $|2y + 4| = |4y + 6|$
22. $|3x - 2| = |6x + 4|$
23. $2|a + 1| - 3 = 9$
24. $2|3a + 1| + 5 = 13$
25. $3|2x + 1| - 4 = -6$
26. $5|x - 1| - 6 = -8$
27. $|x - 4| = |4 - x|$
28. $|2x - 2| = |2 - 2x|$
29. $|5x + 10| = |10 - 5x|$
30. $|6x - 3| = |3 - 6x|$

< B > Absolute-Value Inequalities   In Problems 31–68 solve, graph, and write the solution set in interval notation.

31. $|x| < 4$
32. $|y| \le 1.5$
33. $|z| + 4 \le 6$
34. $|x| + 3 < 7$
35. $|a| - 2 \le 2$
36. $|b| + 2 \le 5$
37. $|x - 1| < 2$
38. $|x - 3| \le 1$
39. $|x + 3| < -2$
40. $|2x - 3| < -4$
41. $|2x + 3| \le 1$
42. $|3x + 2| < 5$
43. $|4x + 2| - 4 < 2$
44. $|3x + 3| - 2 < 4$
45. $|x| > 2$
46. $|y| \ge 2.5$
47. $|z| + 5 \ge 6$
48. $|x| + 2 > 5$
49. $|a| - 1 \ge 2$
50. $|b| + 1 \ge 5$

51. $|x - 1| > 1$

52. $|x - 3| \geq 2$

53. $|x + 3| > -1$

54. $|2x - 3| > -4$

55. $|2x + 3| \geq 1$

56. $|3x + 2| > 5$

57. $|3 - 4x| > 7$

58. $|5 - 3x| > 11$

59. $|4x + 2| - 4 \geq 2$

60. $|3x + 3| - 2 \geq 4$

61. $\left|2 - \frac{1}{2}a\right| > 1$

62. $\left|1 + \frac{2}{3}b\right| > 2$

63. $-2 < |2x + 4|$

64. $-3 \leq |-2x - 4|$

65. $-3|x - 5| > 6$

66. $-4|x + 3| > 12$

67. $9 \geq 5|3x - 2| + 4$

68. $3 < 4|-2x + 3| - 5$

‹ C › **Applications Involving Absolute-Value Inequalities** In Problems 69–72, use $|b - a| < c$, where b is the budgeted amount, a is the actual expense, and c is the acceptable variance.

69. *Budgets* A company budgets $500 for office supplies. If their acceptable variance is $50, write an inequality giving the amounts between which the actual expense a must fall.

70. *Budgets* A company budgets $800 for maintenance. If their acceptable variance is 5% of their budgeted amount, write an inequality giving the amounts between which the actual expense a must fall.

71. *Budgets* George budgets $300 for miscellaneous monthly expenses. His actual expenses for 1 month amounted to $290. Was he within a 5% budget variance?

72. *Budgets* If George from Problem 71 spent $310, was he within a 5% budget variance?

››› Using Your Knowledge

The Boundaries of Inequalities The inequality $|-2x + 3| > 1$ can be solved by finding its *boundary numbers,* the solutions resulting when the inequality sign (> in this case) is replaced by an equal sign. We write $|-2x + 3| = 1$ with solutions

$$-2x + 3 = 1 \quad \text{or} \quad -2x + 3 = -1$$
$$-2x = -2 \quad \quad\quad -2x = -4 \quad \text{Add } -3.$$
$$x = 1 \quad\quad\quad\quad x = 2 \quad\quad \text{Multiply by } -\tfrac{1}{2}.$$

The boundary numbers are 1 and 2.

The graph of an inequality of the form $|ax + b| > c$ is the union of two unbounded intervals, so the graph is

as in Example 5. Try doing Problems 37–42 and 51–60 using this technique.

››› Write On

73. Using the idea of distance between two numbers, write your own definition for the absolute value of a number.

74. Write your own explanation of why the equation $|x| = a$ has no solution if a is negative.

75. Explain how to set up the translation statements that can be used to solve the equation $|ax + b| = |cx + d|$.

76. Write in your own words the reasons why:
    a. The inequality $|x| < a\ (a < 0)$ has no solution.
    b. The inequality $|x| > a\ (a < 0)$ is always true.

››› Concept Checker

Fill in the blank(s) with the correct word(s), phrase, or mathematical statement.

77. $3 < x$ and $x > 3$ are called _____ inequalities.

78. If $x + 5 < -8$, then $x < -13$ is possible because of the _____ Property of Inequality.

79. $x > 2$ is an example of a(n) _____ interval.

80. The _____ of two sets A and B is the set of elements in either A or B.

equivalent   multiplication
open   intersection
half-open   union
subtraction

››› Mastery Test

Solve:

81. $|x - 1| = |x - 8|$

82. $|4 - x| = |7 + 2x|$

83. $|3x + 1| = 5$

84. $|z| - \frac{3}{4} = 0$

85. $\left|\frac{2}{3}x + 1\right| + 3 = 9$

86. $|y| = 3.4$

Solve, graph, and write the solution set in interval notation.

87. $|-2x + 1| - 1 > 4$

88. $\left|x - \frac{2}{3}\right| \leq 5$

89. $|-2 + 3x| \geq 5$

90. $|2x - 3| - 5 < 0$

››› Skill Checker

Evaluate:

91. $-16t^2 + 10t - 15$ when $t = 2$

92. $x^2 - 3x$ when $x = -3$

Simplify:

93. $(-6x^2 + 3x + 5) + (7x^2 + 8x + 2)$

94. $(3x - 2x^2 + 4) + (5 - 7x + 4x^2)$

95. $(3x - 7x^2 + 4) - (3x^2 + 5x + 1)$

96. $(5 - 3x^2 - 2x) - (8x^2 + 4x - 2)$

Solve:

97. $2x + 3 = 5x - 6$

98. $y - 7 = 3(y + 1)$

Solve for y:

99. $2x + y = 8$

100. $3x - 4y = 8$

Collaborative Learning 2A

How much is "too much" when it comes to calories from fat in your daily diet? Most dieticians recommend that the proportion of calories from fat in your daily diet should be 30% or less, although many people consume more than 50% of their total calories as fat. Usually nutrition facts are based on a daily 2000 calorie diet and that fat contains 9 calories per gram of food energy.

To take a closer look at this we will divide into groups, plan two menus for a day, and then collaborate the findings. The first menu will be any items of your choosing as long as it totals 2000 calories for the day. The second menu will not only have to total 2000 calories, but it should not exceed 30% in fat calories. To plan the menus you may use the following charts or any other resource, including books or Internet sites.

1. Use two poster boards, one for each menu, to make a table that includes columns with all the food items for the day, their calorie contents, fat grams, and fat percents. To find the percent of fat calories you will have to answer the following questions:

   a. What is the total number of fat calories allowed in a 2000 calorie diet if 30% is allowed for fat calories?

   b. If fat contains 9 calories per gram and you know how many fat grams there are in a food, what would you do to find the number of fat calories in that food?

   c. Knowing the number of fat calories in a food and the total number of fat calories allowed, what would you do to find the percent of fat calories?

2. After the groups have completed their menus on poster board, have a class discussion about your findings. Be sure to include the following questions.

   a. What kinds of food items have a higher percent of fat calories? What kinds have a lower percent?

   b. How does the percent of fat calories from fast food compare with other types of food?

   c. Which meal (breakfast, lunch, or dinner) contains most of the percent of fat calories?

Item	Calorie content	Fat grams
Bread and Bakery		
Bagel (85g)	216	1.4
Biscuit (15g)	74	3.3
Bread, white (slice, 37g)	84	0.6
Danish pastry (67g)	287	17.4
Doughnut (49g)	140	2
Dairy		
Butter (10g)	74	8.2
Cheddar cheese (40g)	172	14.8
Cream cheese (34g)	58	4.8
Eggs, 3 (57g)	84	6.2
Milk, semi-skim (200ml)	96	3.2
Fruit		
Apple (112g)	53	0.1
Banana (150g)	143	0.5
Grapes (50g)	30	0.1
Melon (1oz/28g)	7	0.1
Orange (160g)	59	0
Strawberries (1oz/28g)	7	0
Drinks		
Coffee (1 cup/220ml)	15.4	0.9
Coke, can (330ml)	139	0
Orange juice (1 glass/200ml)	88	0
Tea (1 mug/270ml)	29	0.5

Item	Calorie content	Fat grams
Vegetables		
Carrots (60g)	13	0.2
Celery (40g)	2	0.1
Chips (100g)	253	9.9
Peas (60g)	32	0.4
Potato, baked (180g)	245	0.4
Salad (100g)	19	0.3
Meats		
Bacon (25g)	64	4
Beef in gravy (83ml)	45	2.7
Chicken breast (200g)	342	13
Ham slice (30g)	35	1
Kebob (168g)	429	28.6
Fast Food		
Big Mac (215g)	492	23
Cheeseburger (148g)	379	18.9
Chicken, KFC (67g)	195	12
Hamburger (108g)	254	7.7
Pizza (1/2 pizza/135g)	263	4.9
Pizza Deluxe (1 slice/66g)	171	6.7
Potato Wedges (135g)	279	13

› Collaborative Learning 2B

Do you know the 10 hottest careers for college grads? Government economists have estimated which occupations would grow the fastest between 2002 and 2012, and they are listed in the following table. The numbers are in thousands. It seems these careers could be listed in one of two categories—the technology field or the medical/health field. Maybe you are trying to decide on one of these two areas to pursue as a career. Knowing which ones will have more opportunities could help you make that choice. It is said that a picture is worth a thousand words, so that is what we will use to help you consider a career choice. Divide into groups and do the following.

1. Separate these 10 careers into two categories—Technical or Medical/Health. Make a chart for each field that will include a column for the percent of increase for each career. Percent increase is found by dividing the difference of the two numbers by the original number.
2. Make a bar graph for each of the two categories so that each bar represents the percent of increase for a career in that field.
3. Have each group present their bar graphs and discuss the following.
   a. Of the two categories, which field seems to have a higher percentage increase?
   b. Which career has the highest percent of increase in the technology field? The lowest?
   c. Which career has the highest percent of increase in the medical/health field? The lowest?
   d. What other information would you want to know about a career besides the availability of jobs?

10 Fastest Growing Occupations for College Grads		
Occupation	2002	2012
Network systems and data communications analysts	186	292
Physician assistants	63	94
Medical records and health information technicians	147	216
Computer software engineers, applications	394	573
Computer software engineers, systems software	281	409
Physical therapist assistants	50	73
Fitness trainers and aerobics instructors	183	264
Database administrators	110	159
Veterinary technologists and technicians	53	76
Dental hygienists	148	212

Source: United States Bureau of Labor Statistics.

› Research Questions

1. Write a short paragraph about the contents of the Rhind papyrus with special emphasis on the algebraic method called "the method of false position."
2. Try to find three problems that appeared on the Rhind papyrus and explain the techniques used to solve them. Then try to solve them yourself using the techniques you've learned in this chapter. (*Hint:* Problems 25, 26, and 27 are easy to solve.)
3. Find out about al-Khowarizmi and then write a few paragraphs about him and his life. (Include what academy he belonged to, the titles of the books he wrote, and the types of problems that appeared in these books.)
4. A Latin corruption of the name "al-Khowarizmi" meant "the art of computing with Hindu Arabic numerals." What is this word and what does it mean today?

(continued)

5. The traditional explanation of the word *jabr* is "the setting of a broken bone." Write a paragraph relating how this word reached Spain as *algebrista* and the context in which it was used.

6. Diophantus of Alexandria wrote a book described "as the earliest treatise devoted to algebra." Write a few paragraphs about Diophantus and, in particular, the problems contained in his book.

7. Write a short paragraph detailing the types of symbols used in Diophantus' book.

Summary Chapter 2

Section	Item	Meaning	Example
2.1	Equation	A sentence using = as its verb	$x + 1 = 5$ and $3 - y = 7$ are equations.
2.1A	Solutions of an equation	The replacement values of the variable that make the equation a true statement	5 is a solution of $x + 3 = 8$.
2.1B	Equivalent equations	Two or more equations are equivalent if they have the same solution set.	$x + 2 = 5$ and $x = 3$ are equivalent.
	Properties of equality	You can *add* or *subtract* the same quantity on both sides and *multiply* or *divide* both sides of an equation by the same nonzero quantity, and the result will be an equivalent equation.	If C is a real number and $A = B$, $A + C = B + C$ $A - C = B - C$ $A \cdot C = B \cdot C \, (C \neq 0)$ $\frac{A}{C} = \frac{B}{C} \, (C \neq 0)$ are equivalent equations.
2.1C	Linear equation	An equation that can be written in the form $ax + b = c$, where a, b, and c are real numbers and $a \neq 0$.	$3x + 7 = 9$ and $-x - 3 = \frac{2}{3}$ are linear equations.
	CRAM Procedure	Clear fractions/decimals. Remove parentheses/simplify. Add/subtract to get variable isolated. Multiply/divide to make coefficient a 1.	
2.2A	Perimeter	The distance around a polygon (figure)	The perimeter P of a rectangle is $P = 2L + 2W$, where L is the length and W is the width.
2.2D	Angles and transversals	All the acute angles (1, 4, 5, and 8) are equal. All the obtuse angles (2, 3, 6, and 7) are equal.	L_1 and L_2 are parallel.

Section	Item	Meaning	Example
2.3B	Word problem	The RSTUV procedure to solve a word problem is **R**ead **S**elect a variable **T**hink of a plan **U**se algebra **V**erify	
2.3C	Consecutive integers	If n is an integer, n and $n + 1$ are consecutive integers.	8 and 9; 28 and 29 are consecutive integers.
2.3D	Right angle	An angle whose measure is 90°	90°
	Complementary angles	Two angles whose measures sum to 90°	A and B are complementary
	Straight angle	An angle whose measure is 180°	180°
	Supplementary angles	Two angles whose measures sum to 180° (A and B are supplementary.)	A and B are supplementary
2.4B	$I = Pr$	The annual interest I is the product of the principal P and the rate r.	The annual interest on a $3000 investment (principal) at a 5% rate is $I = 3000 \cdot 0.05 = \$150$.
2.4C	$D = RT$	When moving at a constant rate R, the distance D traveled in time T is $D = RT$.	A car moving at 55 mi/hr for 3 hr will travel $D = 55 \cdot 3 = 165$ mi.
2.5A	Linear inequality	An inequality that can be written in the form $ax + b > c$, where a, b, and c are real numbers	$-2x + 3 > 5$ and $3 - 2x > 7$ are linear inequalities.
2.5B	Properties of inequalities	If $A < B$, then $A + C < B + C$.	The number C can be added to both sides to obtain an equivalent inequality.
		If $A < B$, then $A - C < B - C$.	The number C can be *subtracted* from both sides to obtain an equivalent inequality.
		If $A < B$, then $A \cdot C < B \cdot C$ when $C > 0$.	Both sides can be *multiplied* by $C > 0$ to obtain an equivalent inequality.
		If $A < B$, then $A \cdot C > B \cdot C$ when $C < 0$.	Both sides can be *multiplied* by $C < 0$ to obtain an equivalent inequality, provided you change the sense (direction) of the inequality,
		If $A < B$, then $\frac{A}{C} < \frac{B}{C}$ when $C > 0$.	Both sides can be *divided* by $C > 0$ to obtain an equivalent inequality.
		If $A < B$, then $\frac{A}{C} > \frac{B}{C}$ when $C < 0$.	Both sides can be *divided* by $C < 0$ to obtain an equivalent inequality, provided you change the sense (direction) of the inequality.

(continued)

Section	Item	Meaning	Example
2.5C	Union of sets	If A and B are sets, the union of A and B, denoted by $A \cup B$, is the set of elements that are in either A or B.	
	Intersection of sets	If A and B are sets, the intersection of A and B, denoted by $A \cap B$, is the set of elements that are in both A and B.	
2.6A	Absolute-value equation	The solutions of $|x| = a$ are $x = a$ and $x = -a$ for $a \geq 0$.	The solutions of $|x| = 2$ are 2 and -2.
2.6B	$|x| < a$	For $a \geq 0$, $|x| < a$ is equivalent to $-a < x < a$.	$|x| < 3$ is equivalent to $-3 < x < 3$.
	$|x| > a$	For $a \geq 0$, $|x| > a$ is equivalent to $x > a$ or $x < -a$.	$|x| > 3$ is equivalent to $x > 3$ or $x < -3$.

> Review Exercises Chapter 2

(If you need help with these exercises, look in the section indicated in brackets.)

1. ⟨**2.1A**⟩ *Does -3 satisfy the equation?*
  a. $7 = 8 - x$
  b. $9 = 8 + x$
  c. $4 = 1 - x$

2. ⟨**2.1B**⟩ *Solve:*
  a. $\frac{2}{3}y - 3 = 5$
  b. $\frac{2}{3}y - 5 = 5$
  c. $\frac{2}{3}y - 7 = 5$

3. ⟨**2.1B**⟩ *Solve:*
  a. $x + 2 = 2(2x - 2)$
  b. $x + 3 = 3(2x - 4)$
  c. $x + 4 = 4(2x - 6)$

4. ⟨**2.1C**⟩ *Solve:*
  a. $\frac{x+4}{3} - \frac{x-4}{5} = 4$
  b. $\frac{x+9}{4} - \frac{x-12}{5} = 6$
  c. $\frac{x+8}{3} - \frac{x-8}{5} = 8$

5. ⟨**2.1C**⟩ *Solve:*
  a. $\frac{x}{4} - \frac{x}{3} = \frac{x-4}{4}$
  b. $\frac{x}{5} - \frac{x}{4} = \frac{x-5}{5}$
  c. $\frac{x}{7} - \frac{x}{3} = \frac{x-7}{7}$

6. ⟨**2.1C**⟩ *Solve:*
  a. $0.05P + 0.10(2000 - P) = 175$
  b. $0.08P + 0.10(5000 - P) = 460$
  c. $0.06P + 0.10(10{,}000 - P) = 840$

7. ⟨**2.2A**⟩ *Solve for h and evaluate when $H = 82.48$.*
  a. $H = 2.5h + 72.48$
  b. $H = 2.5h + 77.48$
  c. $H = 2.5h + 84.98$

8. ⟨**2.2A**⟩ *Solve for A.*
  a. $B = \frac{2}{7}(A - 7)$
  b. $B = \frac{3}{7}(A - 5)$
  c. $B = \frac{4}{5}(A - 7)$

9. ⟨**2.2A**⟩ The perimeter of a rectangle is $P = 2L + 2W$, where L is the length and W is the width. Solve for L and give the dimensions when:

   a. The perimeter is 180 ft and the length is 10 ft more than the width.

   b. The perimeter is 220 ft and the length is 10 ft more than the width.

   c. The perimeter is 260 ft and the length is 10 ft more than the width.

10. ⟨**2.2D**⟩

   a. If L_1 and L_2 are parallel lines, find x and the measure of the unknown angles.

   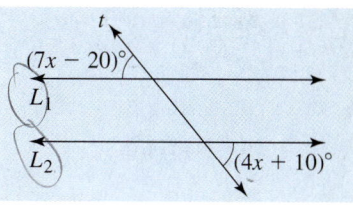

   b. Find x and the measure of the unknown angles if the measure of one angle is $(3x + 12)°$ and the measure of the other angle is $(4x - 3)°$.

   c. Find x and the measure of the unknown angles if the measure of one angle is $(2x + 18)°$ and the measure of the other angle is $(3x - 2)°$.

11. ⟨**2.3B**⟩ Juan rented a car for $30 per day plus 15¢ per mile. How many miles did Juan travel if he paid:

   a. $45 for 1 day?   b. $52.50 for 1 day?
   c. $60 for 1 day?

12. ⟨**2.3C**⟩ Find three consecutive odd integers whose sum is:

   a. 153   b. 159
   c. 207

13. ⟨**2.4A**⟩ A woman's salary is increased by 20%. What was her salary before the increase if her new salary is:

   a. $24,000   b. $36,000
   c. $18,000

14. ⟨**2.4A**⟩ In a recent survey, 380 new home buyers were asked how they found the homes they bought. Use the results to the survey to fill in the last column of the table and construct the bar graph.

Real Estate Agent	49%	
Yard Sign	21%	
Friend or Relative	12%	
Internet Listing	8%	
Newspaper Ad	6%	
Open House	4%	

Source: http://recenter.tamu.edu.

Home Buyer Survey

15. ⟨**2.4B**⟩ An investor bought some municipal bonds yielding 5% annually and some certificates of deposit (CDs) yielding 10% annually. If his total investment amounts to $20,000, find how much money is invested in bonds and how much in certificates of deposit given his annual interest is:

   a. $1750
   b. $1150
   c. $1500

16. ⟨**2.4C**⟩ A car leaves a town traveling at 40 mi/hr. An hour later another car leaves the same town in the same direction traveling at 50 mi/hr.

   a. How far from town does the second car overtake the first?

   b. Repeat the problem where the first car is traveling at 50 mi/hr and the second one at 60 mi/hr.

   c. Repeat the problem where the first car is traveling at 40 mi/hr and the second one at 60 mi/hr.

17. ⟨**2.4D**⟩ *How many liters of a 40% salt solution must be mixed with 50 L of a 10% salt solution to obtain:*

   a. A 30% solution?
   b. A 20% solution?
   c. A 10% solution?

18. ⟨**2.5A**⟩ *Graph:*

   a. $x \geq -2$
   b. $x \geq -3$
   c. $x \geq -4$

19. ⟨**2.5B**⟩ *Solve, graph, and write in interval notation:*

   a. $2(x - 1) \leq 4x + 4$
   b. $3(x - 1) \leq 6x + 3$
   c. $4(x - 1) \leq 8x + 4$

20. ⟨**2.5B**⟩ *Solve, graph, and write in interval notation:*

   a. $\frac{x}{4} - \frac{x}{3} < \frac{x-4}{4}$
   b. $\frac{x}{5} - \frac{x}{3} < \frac{x-5}{5}$
   c. $\frac{x}{7} - \frac{x}{3} < \frac{x-7}{7}$

21. ⟨**2.5C**⟩ *Graph and write in interval notation:*

   a. $\{x \mid x < -2 \text{ or } x \geq 3\}$
   b. $\{x \mid x < -3 \text{ or } x \geq 2\}$
   c. $\{x \mid x < -4 \text{ or } x \geq 1\}$

22. ⟨**2.5C**⟩ *Solve, graph, and write in interval notation:*

   a. $x + 1 \leq 3 \text{ and } -4x < 8$
   b. $x + 1 \leq 4 \text{ and } -3x < 9$
   c. $x + 1 \leq 5 \text{ and } -2x < 4$

23. ⟨**2.5C**⟩ *Solve, graph, and write in interval notation:*

   a. $-4 \leq -2x - 6 < 4$
   b. $-3 \leq -2x - 5 < 3$
   c. $-6 \leq -2x - 4 < 2$

24. ⟨**2.5D**⟩ *Brenda is maintaining a 2000 calorie diet each day. It has been recommended that no more than 15% of her daily caloric intake (2000 calories) be protein. However, nutrition labels give the protein content in grams and must be changed to calories. To find the number of protein calories, multiply the number of protein grams by 4. Up to how many grams of protein can Brenda eat per day to meet this 15% requirement?*

25. ⟨**2.6A**⟩ *Solve:*

  a. $\left|\frac{2}{7}x + 2\right| + 5 = 9$

  b. $\left|\frac{2}{7}x + 2\right| + 3 = 9$

  c. $\left|\frac{2}{7}x + 2\right| + 1 = 9$

26. ⟨**2.6A**⟩ *Solve:*

  a. $|x - 1| = |x - 3|$

  b. $|x - 3| = |x - 5|$

  c. $|x - 5| = |x - 7|$

27. ⟨**2.6B**⟩ *Solve, graph, and write in interval notation:*

  a. $|3x - 1| \leq 2$

  b. $|4x - 1| \leq 3$

  c. $|5x - 1| \leq 4$

28. ⟨**2.6B**⟩ *Solve, graph, and write in interval notation:*

  a. $|3x - 1| \geq 2$

  +1) 3x ≥ 2 +1   +1 3x-1 ≤ -2+1)
  1/3) 7x ≥ 1   1/3 < x ≤ -1/3

  b. $|4x - 1| \geq 3$

  c. $|5x - 1| \geq 4$

Practice Test Chapter 2

(Answers on page 163)

Visit www.mhhe.com/bello to view helpful videos that provide step-by-step solutions to several of the problems below.

1. Does 4 satisfy the equation $5 = 9 - x$?

2. Solve: $4y + 3 = -9$.

3. Solve: $-z + 8 = 5z - 6$.

4. Solve: $4(5 - x) + 6 = 2x - (x + 4)$.

5. Solve: $\frac{6}{5} + \frac{3x}{15} = \frac{x+4}{10}$.

6. Solve: $0.06P + 0.07(1500 - P) = 96$.

7. $H = 2.75h + 71.48$
   a. Solve for h. b. Find h if $H = 140.23$.

8. Solve for A in $B = \frac{3}{4}(A - 8)$.

9. The perimeter of a rectangle is $P = 2l + 2w$, where l is the length and w is the width.
   a. Solve for l.
   b. If the perimeter is 100 ft and the length is 20 ft more than the width, what are the dimensions of the rectangle?

10. If L_1 and L_2 are parallel lines, find x and the measure of the unknown angles.

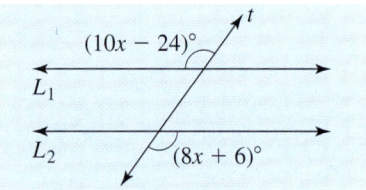

11. The bill for repairing an appliance totaled $72.50. If the repair shop charges $35 for the service call, plus $25 for each hour of labor, how many hours of labor did the repair take?

12. The sum of three consecutive odd integers is 117. What are the three integers?

13. A woman's salary is increased by 20% to $48,000. What was her salary before the increase?

14. An investor bought some municipal bonds yielding 5% annually and some certificates of deposit yielding 7%. If his total investment amounts to $20,000 and his annual interest is $1100, how much money is invested in bonds and how much in certificates of deposit?

15. A freight train leaves a station traveling at 40 mi/hr. Two hours later, a passenger train leaves the same station traveling in the same direction at 60 mi/hr. How far from the station does the passenger train overtake the freight train?

16. How many gallons of a 30% salt solution must be mixed with 40 gal of a 12% salt solution to obtain a 20% solution?

17. Solve, graph, and write the solution set in interval notation of $4(x - 1) \leq 8x + 4$.

18. Solve, graph, and write the solution set in interval notation of $\{x \mid x < -1 \text{ or } x \geq 2\}$.

19. Solve, graph, and write the solution set in interval notation of $x + 1 \leq 4$ and $-2x < 6$.

20. Solve, graph, and write the solution set in interval notation of $-4 \leq -2x - 6 < 0$.

21. Federal Pell Grant awards can range from $300 to $4000. At a certain Florida university the in-state tuition is $110 per hour. Kevin attends that university and will only be able to take as many hours as his Pell Grant allows. What range of hours can he consider taking? (Round to the lower whole number hour.)

22. Solve: $\left|\frac{3}{4}x + 2\right| + 4 = 9$.

23. Solve: $|x - 3| = |x - 7|$.

24. Solve and graph $|2x - 1| \leq 5$.

25. Solve and graph $|2x + 1| > 3$.

Answers to Practice Test Chapter 2

Answer	If You Missed Question	Section	Review Examples	Page
1. Yes	1	2.1A	1	76
2. -3	2	2.1B	2	77–78
3. $\frac{7}{3}$	3	2.1B	3	78
4. 6	4	2.1B	4	78–79
5. -8	5	2.1C	5, 6	80–82
6. 900	6	2.1C	7, 8	82–83
7. a. $h = \frac{H - 71.48}{2.75}$ b. 25	7	2.2A	1	88–89
8. $A = \frac{4B + 24}{3}$	8	2.2A	2	89
9. a. $l = \frac{P - 2w}{2}$ b. 15 ft by 35 ft	9	2.2A	3	90
10. $x = 15$; $126°$	10	2.2D	8	95
11. 1.5 hr	11	2.3B	3	103–104
12. 37; 39; 41	12	2.3C	4	104–105
13. $40,000	13	2.4A	2	113
14. $15,000 bonds; $5000 CD	14	2.4B	4	115–116
15. 240 mi	15	2.4C	5	116–117
16. 32	16	2.4D	6	117–118
17. $x \geq -2$; $[-2, \infty)$	17	2.5B	2, 3	126–128
18. $x < -1$ or $x \geq 2$; $(-\infty, -1) \cup [2, \infty)$	18	2.5C	5	130
19. $-3 < x \leq 3$; $(-3, 3]$	19	2.5C	6, 7	131–133
20. $-3 < x \leq -1$; $(-3, -1]$	20	2.5C	8	134
21. $300 < 110x < 4000$ or He can take from 2 to 36 hours.	21	2.5D	10	136
22. 4; $-\frac{28}{3}$	22	2.6A	1, 2	143–144
23. 5	23	2.6A	3	145
24. $-2 \leq x \leq 3$	24	2.6B	4	146–147
25. $-2 < x$ or $x > 1$	25	2.6B	5	148

Cumulative Review Chapters 1–2

1. Use roster notation to list the elements of the set of even natural numbers less than 8.

2. Write 0.19 as a fraction.

3. Add: $-5 + (-6)$.

4. Multiply: $5 \cdot (-4)$.

5. Simplify:
$[(3x^2 - 3) + (7x + 1)] - [(x - 2) + (8x^2 - 5)]$.

6. Perform the indicated operation and simplify:
$\dfrac{80x^8}{16x^{-6}}$.

7. Evaluate: $[-2(4 + 4)] + 7$.

8. Does the number -4 satisfy the equation $5 = 9 - x$?

9. Solve: $x + 5 = 3(2x - 1)$.

10. Solve: $\dfrac{x + 2}{5} - \dfrac{x - 2}{7} = 4$.

11. Solve: $0.04P + 0.06(1700 - P) = 65$.

12. Solve: $\left|\dfrac{3}{7}x + 8\right| + 3 = 9$.

13. Graph: $3(x + 1) \leq 4x - 3$.

14. Graph: $\{x \mid x < -1 \text{ or } x \geq 2\}$.

15. Graph: $-4 \leq -4x - 8 < 4$.

16. Graph: $|2x + 3| > 5$.

17. Solve for A in $B = \dfrac{6}{7}(A - 11)$.

18. The perimeter of a rectangle is $P = 2l + 2w$, where l is the length and w is the width. If the perimeter is 100 feet and the length is 10 feet more than the width, what are the dimensions?

19. A woman's salary was increased by 30% to $29,900. What was her salary before the increase?

20. How many gallons of a 60% salt solution must be mixed with 60 gallons of a 21% salt solution to obtain a solution that is 50% salt?

Section

- 3.1 Graphs
- 3.2 Using Slopes to Graph Lines
- 3.3 Equations of Lines
- 3.4 Linear Inequalities in Two Variables
- 3.5 Introduction to Functions
- 3.6 Linear Functions

Chapter 3 *three*

▶ Graphs and Functions

The Human Side of Algebra

The inventor of the Cartesian coordinate system, René Descartes, was born March 31, 1596, near Tours, France. Trained as a gentleman and educated in Latin, Greek, and rhetoric, he maintained a healthy skepticism toward all he was taught. Disenchanted with these studies, he took an interlude of pleasure in Paris, later moving to the suburb of St. Germain, where he worked on mathematics for 2 years. In 1637, at age 41, his friends persuaded him to print his masterpiece, known as the *Discourse on Method,* which included an essay on geometry that is probably the most important thing he ever did. His *Analytic Geometry,* a combination of algebra and geometry, revolutionized the study of geometry and made much of modern mathematics possible.

In 1649, Queen Christine of Sweden chose Descartes as her private philosophy teacher. Unfortunately, she insisted that her lessons begin promptly at five o'clock in the morning in the ice-cold library of her palace. This proved to be too much for the frail Descartes, who soon caught "inflammation of the lungs," from which he died on February 11, 1650, at the age of 54. What price for fame!

3.1 Graphs

Objectives

A Given an ordered pair of numbers, find its graph, and vice versa.

B Graph linear equations by finding solutions satisfying the equation.

C Graph lines by finding the *x*- and *y*-intercepts.

D Graph horizontal and vertical lines.

E Graph nonlinear equations by finding the solutions satisfying the equation.

To Succeed, Review How To . . .

1. Evaluate an expression (pp. 51–55).
2. Solve linear equations (pp. 76–83).

Getting Started
Hurricane Preparedness

The map shows the **coordinates** of the hurricane to be near the vertical line indicating 90° of longitude and the horizontal line indicating 25° of latitude. If we agree to list the longitude first, we can identify this point by using the ordered pair

(90, 25)

This is the longitude. ↑   ↑ This is the latitude.

So if we know the *coordinates* of the hurricane, we can find it on the map, and if we know the *point* where the hurricane is located on the map, we can find its coordinates. This is an example of a rectangular coordinate system, which we shall study in this section.

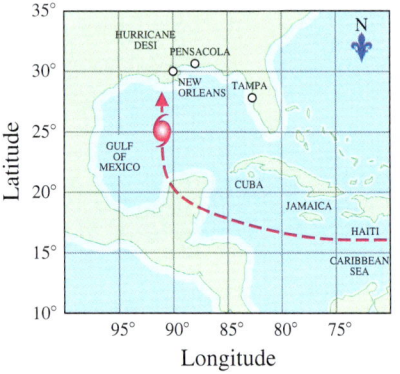

A ▸ Graphing and Finding Ordered Pairs

Graphs such as bar graphs and circle graphs can be very useful to display data sets as we did in some of the exercises in Chapter 2. In algebra, we use a rectangular coordinate plane to display sets of points much like the graphing described in the *Getting Started* regarding the location of a hurricane on a map. We use two real numbers called an **ordered pair** to locate a point on the rectangular coordinate system. This graphing system is sometimes called a **coordinate plane** or a **Cartesian coordinate system** named for the French mathematician René Descartes. We draw two perpendicular number lines called the *x*-axis and the *y*-axis, intersecting at a point (0, 0) called the **origin**. (See Figure 3.1.) On the *x*-axis, positive is to the right, whereas on the *y*-axis, *positive* is up. The two axes divide the plane into four regions called **quadrants.** These quadrants are numbered counterclockwise using Roman numerals and starting in the upper right-hand region, as shown.

Every point *P* in the plane can be associated with an ordered pair of real numbers (*x*, *y*), and every ordered pair

"Beadwork and graphs"
Patterns of beauty require knowing about coordinates and the four quadrants. (See Section 3.1 Exercises 9–12.)

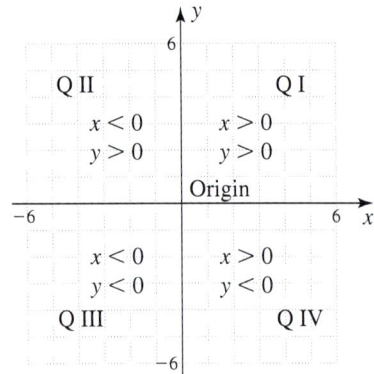

> Figure 3.1

3.1 Graphs

(x, y) can be associated with a point P in the plane. For example, in Figure 3.2 the point A can be associated with the ordered pair $(1, 3)$, the point B with the ordered pair $(-3, 2)$, and the point C with the ordered pair $(0, -2)$. The **graphs** of the points $A(1, 3)$, $B(-3, 2)$, and $C(0, -2)$ are indicated by the blue dots.

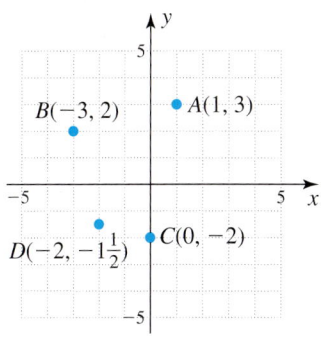

> Figure 3.2

A NOTE ABOUT COORDINATES

If the *x*-coordinate is *positive*, the point is to the *right* of the vertical axis.
If the *x*-coordinate is *negative*, the point is to the *left* of the vertical axis.
If the *x*-coordinate is *zero*, the point is *on* the vertical axis.
If the *y*-coordinate is *positive*, the point is *above* the horizontal axis.
If the *y*-coordinate is *negative*, the point is *below* the horizontal axis.
If the *y*-coordinate is *zero*, the point is *on* the horizontal axis.

You can use the following diagram to help identify the placement of a point according to the signs of the ordered pair.

$$\begin{array}{c|c} (-, +) & (+, +) \\ \hline (-, -) & (+, -) \end{array}$$

Since the coordinates of the points on the graph are real numbers, some coordinates may contain fractions and decimals as well as integers. Such an example would be point D in Figure 3.2, whose coordinates are $\left(-2, -1\frac{1}{2}\right)$.

EXAMPLE 1 Graphing ordered pairs

Graph the ordered pairs $A(2, 3)$, $B(-1, 2)$, $C(-2, -1)$, and $D(0, -3)$.

SOLUTION To graph the ordered pair $(2, 3)$, we start at the origin and move 2 units to the *right* and then 3 units *up*, reaching the point whose coordinates are $(2, 3)$. The other three pairs are graphed in a similar manner and are shown in Figure 3.3.

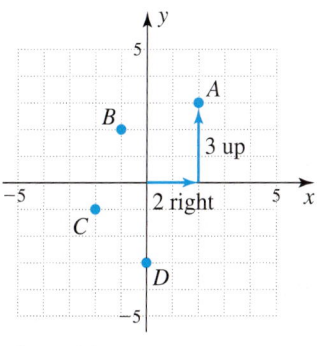

> Figure 3.3

PROBLEM 1

Graph the ordered pairs.
$A(3, 2), B(-1, 4), C(-2, 0),$
$D(3, -1)$

Every point P in the plane is associated with an ordered pair (x, y), so we should be able to find the coordinates of any point, as shown in Example 2.

Answers to PROBLEMS
1.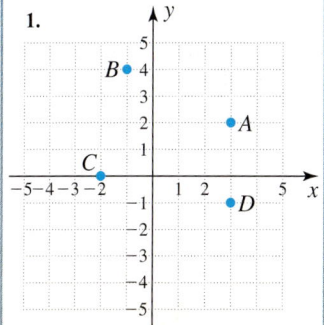

EXAMPLE 2 Finding coordinates

Determine the coordinates of each of the points shown in Figure 3.4.

SOLUTION Point A is 4 units to the *right* of the origin and 1 unit *above* the horizontal axis. Thus, the ordered pair corresponding to A is (4, 1). The coordinates of the other four points can be found in a similar manner. The summary appears in the table.

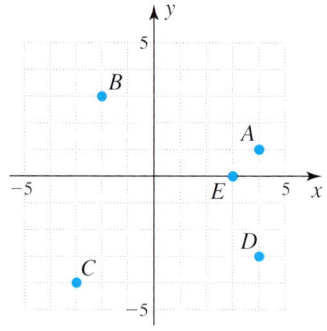

>Figure 3.4

Point	Start at the origin, move	Coordinates
A	4 units *right*, 1 unit *up*	(4, 1)
B	2 units *left*, 3 units *up*	(−2, 3)
C	3 units *left*, 4 units *down*	(−3, −4)
D	4 units *right*, 3 units *down*	(4, −3)
E	3 units *right*, 0 units *up*	(3, 0)

PROBLEM 2

Determine the coordinates of each of the points shown.

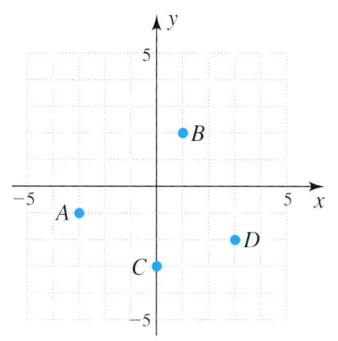

B › Graphing Linear Equations

Some information is best described graphically by displaying it as ordered pairs. The total cost of a long-distance call with a flat rate of 7 cents per minute or portion thereof and ranging from 1 to 5 minutes is shown in Table 3.1.

Figure 3.5 represents these points when plotted.

Consider the pattern involved for finding the total cost C.

Minutes	Total Cost	Ordered Pair
1	7	(1, 7)
2	14	(2, 14)
3	21	(3, 21)
4	28	(4, 28)
5	35	(5, 35)

Table 3.1

$C = 1$ minute times (7)
$C = 2$ minutes times (7)
$C = 3$ minutes times (7)
$C = 4$ minutes times (7)
$C = 5$ minutes times (7)

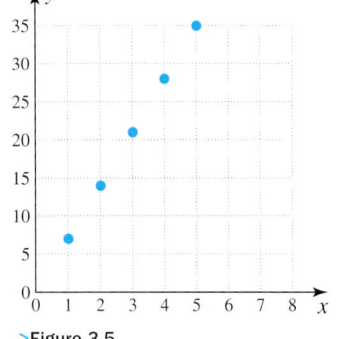

>Figure 3.5

In general, for x minutes, the total cost can be found by $C = x$ times (7) or $C = 7x$.

We used the variable C because it related to the context of the problem, but we could replace the variable C with a y and write the equation as $y = 7x$. We call this an equation in two variables, x and y.

If in this equation x is replaced with 3 and y by 21, we obtain the true statement $21 = 7(3)$. Thus, we say that the ordered pair (3, 21) is a **solution** of the equation $y = 7x$ or that it satisfies the equation. On the other hand, (21, 3) is *not* a solution of $y = 7x$ because $3 \neq 7(21)$.

All the ordered pairs in the table are solutions to this equation. The x is called the independent variable and the y is the dependent variable. There are infinitely many ordered pairs of real numbers that will solve the equation $y = 7x$ and when graphed (see Figure 3.6) will appear to be on the same straight line.

It can be proved that every solution of $y = 7x$ corresponds to a point on the line, and vice versa. The line obtained by joining the solution points is the **graph** of all the solutions to the equation $y = 7x$. In fact, it can be shown that any equation in two variables x and y of the form $y = mx + b$ or $Ax + By = C$ (where m, b, A, B, and C are constants; A and B are not both 0.) has a straight line for its graph. These equations are called **linear equations.** Here is the definition.

Answers to PROBLEMS

2. $A(-3, -1)$; $B(1, 2)$; $C(0, -3)$; $D(3, -2)$

STANDARD FORM OF LINEAR EQUATIONS

Any equation that can be written in the form
$$Ax + By = C$$
is a **linear equation** in two variables, and the graph of its solutions is a *straight line*. The form $Ax + By = C$ is called the **standard form** of the equation, and A, B, and C are integers, where A and B are not both 0, and A is nonnegative.

NOTE
We prefer to write an equation in the form $3x + 2y = 6$ instead of $\frac{1}{2}x + \frac{1}{3}y = 1$; that is, we use the standard form with integer coefficients for A, B, and C with A nonnegative.

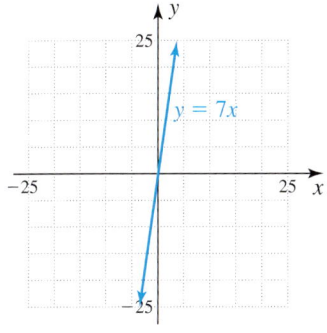

> Figure 3.6

The **solution set** of an equation in two variables can be written using set notation. For example, the solution set of $y = 7x$ can be written as $\{(x, y) \mid y = 7x\}$. The solution set consists of infinitely many points, so it would be impossible to list all these points. However, we can find some of these points as we did in the table, graph them, and then connect the points with a line. The line in Figure 3.6 represents only a *part* of the complete graph, which continues without end in both directions, as indicated by the arrows in the figure. Although it takes only *two* points to determine a straight line, we shall include a third point here in our solutions to make sure that we have drawn the correct line.

Calculator Corner

Make a Table of Values for a Given Equation

You can use your TI-83 Plus or TI-84 Plus to produce a table of values for any given linear equation and set of x-values. Suppose in the example of the long distance phone call the company offers the same 7 cents per minute but will time the call to the nearest half-minute. Then we would want a table of values that would display the minutes (x) as 1, 1.5, 2, 2.5, ..., and so on. Remember to clear any old lists.

X	Y
1	7
1.5	10.5
2	14
2.5	17.5
3	21
3.5	24.5
4	28

MODE ▼ ▼ ▼ ▼ ENTER   (function mode)

Y= 7 X,T,θ,n   (enters equation)

2nd WINDOW 1 ENTER .5 ENTER

2nd GRAPH

So for a $2\frac{1}{2}$ (2.5) minute phone call we would owe 17.5 cents. By using the down arrow  you can scroll through a continuation of the table. How much would a $20\frac{1}{2}$ minute phone call cost?

EXAMPLE 3   Graphing lines
Graph the solutions to $x + 2y = 8$.

SOLUTION   To graph the solutions to $x + 2y = 8$, we find three ordered pairs that solve the equation and then plot (graph) the points on a coordinate plane. To find solutions, we arbitrarily assign a value for x, the *independent variable*. We have chosen -2, 0, and 2. One at a time, these values are substituted for x in the equation, and we solve for the related y-value, the *dependent variable*. We use a *table* to help facilitate the process.

For $x = -2$	For $x = 0$	For $x = 2$
$x + 2y = 8$	$x + 2y = 8$	$x + 2y = 8$
$-2 + 2y = 8$	$0 + 2y = 8$	$2 + 2y = 8$
$2y = 10$	$2y = 8$	$2y = 6$
$y = 5$	$y = 4$	$y = 3$

x	y
-2	5
0	4
2	3

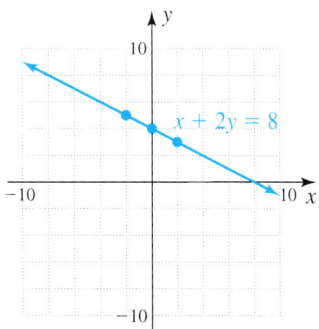

>Figure 3.7

Each row of the table represents a solution to the equation, $x + 2y = 8$, and can be written as the three ordered pair, $(-2, 5)$, $(0, 4)$, and $(2, 3)$. We graph these points and then draw a line through them, as shown in Figure 3.7.

PROBLEM 3
Graph the solutions to $3x - y = 3$.

EXAMPLE 4   Graphing lines
Graph the solutions to $y = \frac{1}{2}x - 1$.

SOLUTION   To graph the solutions to $y = \frac{1}{2}x - 1$, we find three ordered pairs that solve the equation, plot the corresponding points on a coordinate plane, and then connect the points with a straight line representing all the solutions to the equation. We have arbitrarily chosen the values -2, 0, and 4 for the independent variable x. One at a time, these values are substituted for x in the equation, and we solve for the related dependent variable, y. We use a table to help facilitate the process.

PROBLEM 4
Graph the solutions to

$$y = -\frac{1}{3}x + 2$$

Answers to PROBLEMS

3.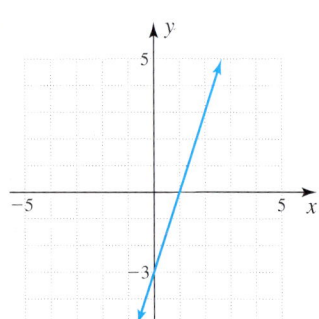

4.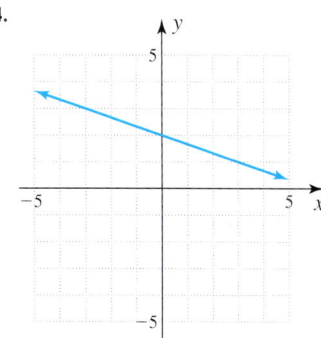

For $x = -2$ For $x = 0$ For $x = 4$

$y = \frac{1}{2}x - 1$ $y = \frac{1}{2}x - 1$ $y = \frac{1}{2}x - 1$

$y = \frac{1}{2}(-2) - 1$ $y = \frac{1}{2}(0) - 1$ $y = \frac{1}{2}(4) - 1$

$y = -1 - 1$ $y = 0 - 1$ $y = 2 - 1$

$y = -2$ $y = -1$ $y = 1$

x	y
-2	-2
0	-1
4	1

We graph the points $(-2, -2)$, $(0, -1)$, and $(4, 1)$ and then draw a line through them, as shown in Figure 3.8.

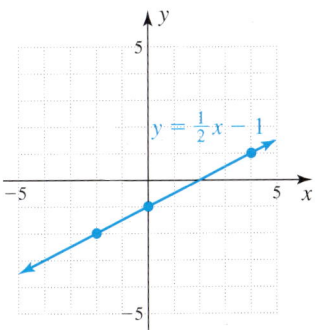

> Figure 3.8

Calculator Corner

Creating a Split Screen for a Table and a Graph

You can graph the equation of a line on your TI-83 Plus and have it show a table of values on a split screen so we see both the graph and the table at the same time. We will use Example 3 to demonstrate this. The calculator will graph the equation only when it is solved for y, so the first step is to solve for y.

$x + 2y = 8$
$2y = -x + 8$ Subtract x.
$y = -0.5x + 4$ Divide by 2.

Be sure the STATPLOT is turned off by pressing [2nd] [Y=] [ENTER] [▼] [ENTER].
  Press [Y=] and enter (−) 0.5 [X,T,θ,n] [+] 4. To set the table to integer values, press [2nd] [WINDOW] (−) 8 [▼] 1. To set the split screen, press [MODE]; continue to use the [▼] to arrow down to the line "**Full Horiz G-T**"; press the [▶] to get to G-T; and then press [ENTER]. To set a standard window for the graph press, [ZOOM] 6. This should show the graph and a table of values for the graph. To scroll up and down the table of values, press [2nd] [GRAPH] and use the **UP** and **DOWN** arrow keys to scroll the values in the table. Scroll down to $x = -2$ and verify that $y = 5$ and to $x = 2$ and verify that $y = 3$, as we indicated in our solution to Example 3.

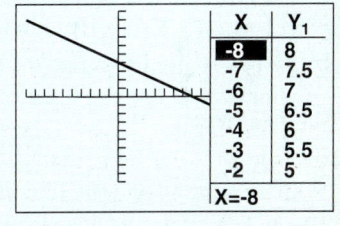

C ❯ Graphing Lines Using Intercepts

The equation $x + 2y = 8$ has a *straight* line as its graph of solutions. We have shown that any equation in two variables x and y of the form $Ax + By = C$ (where A and B are not both zero) has a straight line for its graph. This is why $Ax + By = C$ is called a *linear equation*.

  Two points determine a straight line, so it is sufficient to locate two points to graph all solutions to a linear equation. The easiest points to compute are those involving zeros—points of the form $(x, 0)$ and $(0, y)$. For example, if we let $x = 0$ in the equation $x + 2y = 8$, we obtain $2y = 8$, or $y = 4$. Thus, $(0, 4)$ is a point on the graph. Similarly, if we let $y = 0$ in the equation $x + 2y = 8$, we have $x = 8$. Therefore, $(8, 0)$ is also on the graph. The points $(8, 0)$ and $(0, 4)$ are the points at which the line crosses

172   Chapter 3   Graphs and Functions

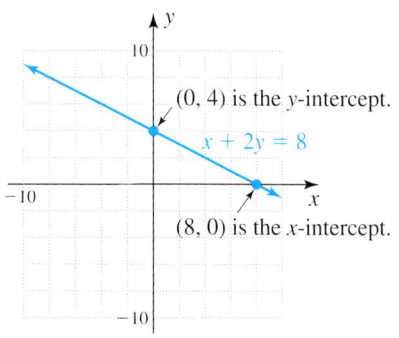

> Figure 3.9

the *x*- and *y*-axes, respectively, so they are called the **x-** and **y-intercepts.** These two intercepts, as well as the line determined by them (the graph of $x + 2y = 8$), are shown in Figure 3.9. Here is the procedure we used to find the intercepts.

PROCEDURE
Finding the Intercepts

To find the *x*-intercept, let $y = 0$ and find x: $(x, 0)$ is the *x*-intercept.

To find the *y*-intercept, let $x = 0$ and find y: $(0, y)$ is the *y*-intercept.

EXAMPLE 5   Graphing lines by finding intercepts

Find the *x*- and *y*-intercepts and then graph the lines:

a. $y = 3x + 6$ **b.** $2x + 3y = 0$

SOLUTION

a. We first find the *x*- and *y*-intercepts. For $x = 0$, $y = 3x + 6$ becomes

$$y = 3(0) + 6 = 6$$

$(0, 6)$ is the *y*-intercept. For $y = 0$, $y = 3x + 6$ becomes

$$0 = 3x + 6$$
$$-6 = 3x \quad \text{Subtract 6.}$$
$$x = -2 \quad \text{Divide by 3.}$$

$(-2, 0)$ is the *x*-intercept. We join the points $(-2, 0)$ and $(0, 6)$ with a line and obtain the graph of all solutions to $y = 3x + 6$ shown in Figure 3.10.

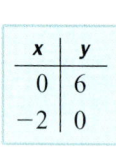

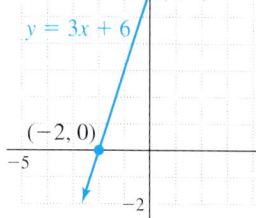

> Figure 3.10

b. As before, we try to find the *x*- and *y*-intercepts. For $x = 0$, $2x + 3y = 0$ becomes

$$2(0) + 3y = 0 \quad \text{or} \quad y = 0$$

$(0, 0)$ is the *y*-intercept. If we now let $y = 0$, we get the ordered pair $(0, 0)$ again. This is because any line of the form $Ax + By = 0$ (where A and B are not both zero)

PROBLEM 5

Find the *x*- and *y*-intercepts and then graph the lines:

a. $y = x - 2$ **b.** $3x + y = 0$

Answers to PROBLEMS

5. a. x: $(2, 0)$ and y: $(0, -2)$ **b.** $(0, 0)$ is both the *x*- and *y*-intercept.

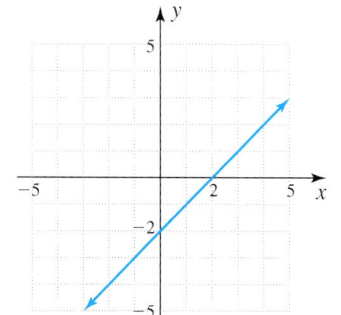 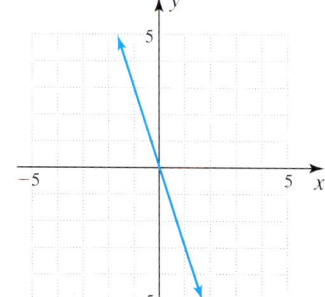

goes through the origin. In these cases, we select a different value for x, say $x = 3$. Then $2x + 3y = 0$ becomes

$2(3) + 3y = 0$
$3y = -6$ Subtract 6.
$y = -2$ Divide by 3.

x	y
0	0
3	-2

$(3, -2)$ is another point on the line. We join the points $(0, 0)$ and $(3, -2)$ with a line and obtain the graph of all solutions to $2x + 3y = 0$ shown in Figure 3.11.

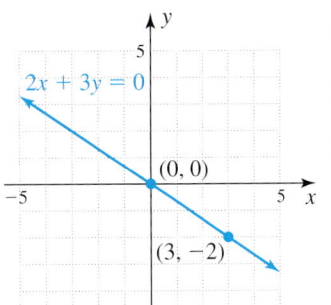

>Figure 3.11

Calculator Corner

Finding the x- and y-intercepts

You can find the x- and y-intercepts of a line on your TI-83 Plus by graphing it using an integer window and the trace feature. To find the x- and y-intercepts of $x + 2y = 8$, solve for y.

$$y = -0.5x + 4$$

Then use the following steps:

  0.5 X,T,θ,n  4 ZOOM 8 ENTER TRACE 0 ENTER

That should indicate "$y = 4$," the y-intercept.

To find the x-intercept, continue to press the right arrow ▶ until the screen displays "$y = 0$" on the right. On the left it says "$x = 8$."

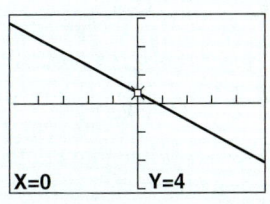

D › Graphing Horizontal and Vertical Lines

The procedure we have discussed works *only* for equations that can be written in the form $Ax + By = C$, where both A and B are not zero. The equation $2y = 6$ can be written in the form $0 \cdot x + 2y = 6$, so we *cannot* graph $2y = 6$ by finding its intercepts. The equation $2y = 6$ assigns to every value of x a y-value of 3. If we solve for y in $2y = 6$, we get $y = 3$, which has no specific x-coordinate. Thus, for $x = 1$, $y = 3$, and for $x = 2$, $y = 3$ (y is always 3). If we graph the points $(1, 3)$ and $(2, 3)$ and connect them with a straight line, we see that the result is a horizontal line, as shown in Figure 3.12. Similarly, the equation $2x = 6$ assigns an x-value of 3 to every y. For $y = 1$, $x = 3$, and for $y = 5$, $x = 3$. If we graph the points $(3, 1)$ and $(3, 5)$ and draw a straight line through them, we see that the result is a vertical line, as shown in Figure 3.13. In general, we have the following.

>Figure 3.12

>Figure 3.13

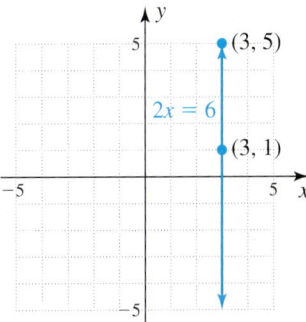

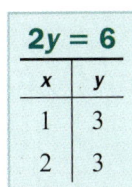

2y = 6	
x	y
1	3
2	3

2x = 6	
x	y
3	1
3	5

HORIZONTAL AND VERTICAL LINES

The graph of all solutions to $y = C$ is a *horizontal* line.

The graph of all solutions to $x = C$ is a *vertical* line.

EXAMPLE 6 Graphing horizontal and vertical lines

Graph the solutions to:

a. $2x = 8$ b. $5y = -10$

SOLUTION

a. $2x = 8$ is equivalent to $x = 4$, so the graph of $2x = 8$ is a vertical line for which $x = 4$. If we choose the solutions $(4, 1)$ and $(4, 2)$ and draw a straight line through them, we obtain the graph of all solutions to $2x = 8$, shown in Figure 3.14.

b. $5y = -10$ is equivalent to $y = -2$, so $5y = -10$ is a horizontal line for which $y = -2$. If we choose the solutions $(1, -2)$ and $(2, -2)$ and draw a straight line through them, we obtain the graph of all solutions to $5y = -10$, shown in Figure 3.15.

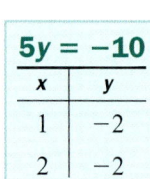

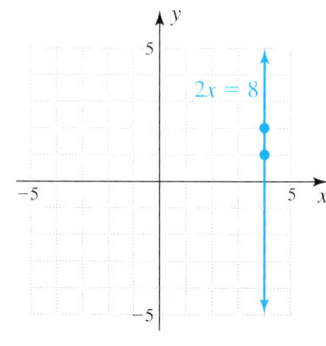

>Figure 3.14

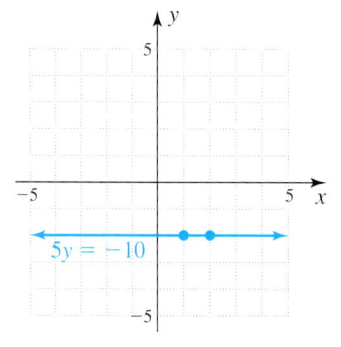

>Figure 3.15

PROBLEM 6

Graph the solutions to:

a. $3x = -9$ b. $-2y = -4$

E> Graphing Nonlinear Equations

Not all equations have graphs whose solution sets make a straight line. When the graph of the solution set of an equation is not a straight line, then the equation is called a **nonlinear equation**. We will need to graph several points to be able to sketch the graph of these nonlinear equations, as demonstrated in Examples 7, 8, and 9.

Answers to PROBLEMS

6. a. 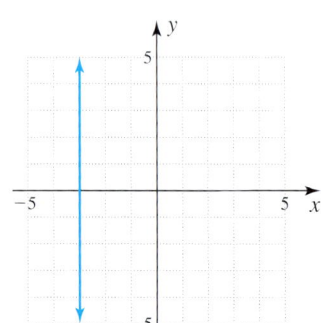   b.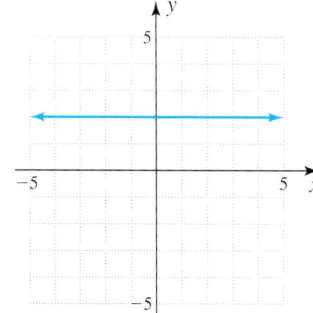

EXAMPLE 7 Graphing nonlinear equations

Graph the solutions to $y = x^2 + 3$.

SOLUTION To graph the solutions to $y = x^2 + 3$, first we will find several ordered pairs that solve the equation. We have arbitrarily chosen the values $-2, -1, 0, 1,$ and 2 for the independent variable x. One at a time, these values are substituted for x in the equation, and we solve for the related dependent variable, y. The table lists the solutions. In Figure 3.16 we graph the ordered pair solutions and sketch the curve.

x	y
-2	7
-1	4
0	3
1	4
2	7

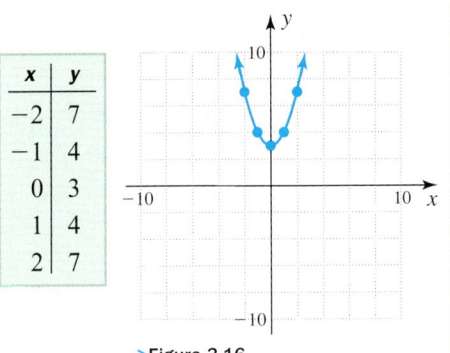

>Figure 3.16

PROBLEM 7

Graph the solutions to $y = -x^2 - 1$.

EXAMPLE 8 Graphing nonlinear equations

Graph the solutions to $y = -|x|$.

SOLUTION To graph the solutions to $y = -|x|$, first we will find several ordered pairs that solve the equation. We have arbitrarily chosen the values $-2, -1, 0, 1,$ and 2 for the independent variable x. One at a time, these values are substituted for x in the equation, and we solve for the related dependent variable, y. The table lists the solutions. In Figure 3.17 we graph the ordered pair solutions and sketch the curve.

x	y
-2	-2
-1	-1
0	0
1	-1
2	-2

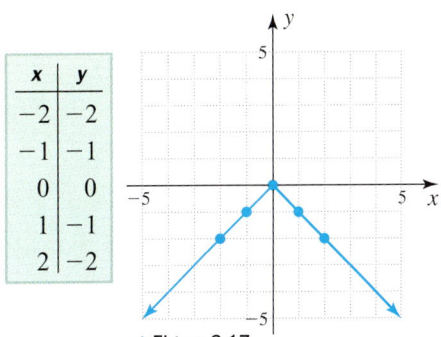

>Figure 3.17

PROBLEM 8

Graph the solutions to $y = |x| - 2$.

EXAMPLE 9 Graphing nonlinear equations

Graph the solutions to $y = \frac{2}{x}$.

SOLUTION To graph the solutions to $y = \frac{2}{x}$, first we will find several ordered pairs that solve the equation. We have arbitrarily chosen the values $-3, -2, -1, 0, 1, 2,$ and 3 for the independent variable x. One at a time, these values are substituted for x in the equation, and we solve for the related dependent variable, y. The table lists the solutions. In Figure 3.18 we graph the ordered pair solutions and sketch the curve.

x	y
-3	$-\frac{2}{3}$
-2	-1
-1	-2
0	undef.
1	2
2	1
3	$\frac{2}{3}$

>Figure 3.18

PROBLEM 9

Graph the solutions to $y = -\frac{2}{x}$.

Answers to PROBLEMS

7.

8.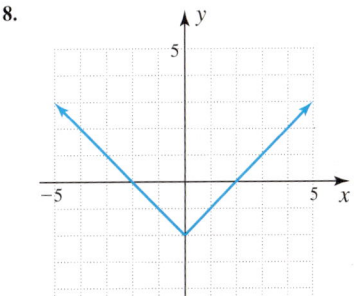

9.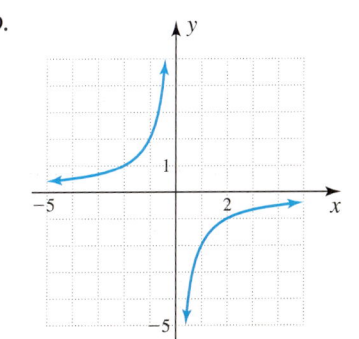

Exercises 3.1

⟨ A ⟩ Graphing and Finding Ordered Pairs   In Problems 1–8, graph the ordered pair.

1. $(-4, 3)$
2. $(-3, 4)$
3. $(-3, -2)$
4. $(-2, -3)$
5. $(0, -2)$
6. $(2, 0)$
7. $\left(\frac{1}{2}, 3\right)$
8. $\left(3, -\frac{1}{2}\right)$

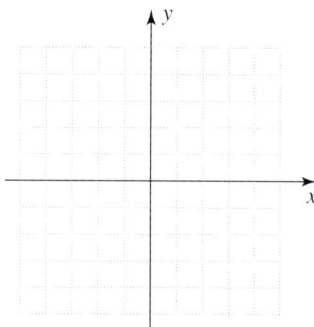

In Problems 9–12, Figure 3.19 depicts the colors of a beadwork pattern. Use the graph to answer the following questions.

9. What quadrant(s) will contain red beads?

10. What quadrant(s) will contain blue beads?

11. What quadrant(s) will not contain red beads?

12. What quadrant(s) will not contain pink beads?

>Figure 3.19

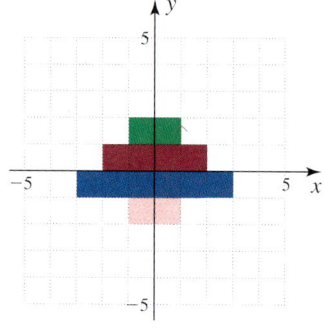

In Problems 13–20, determine the coordinates of each of the points in Figure 3.20.

13. C
14. D
15. E
16. F
17. G
18. H
19. I
20. J

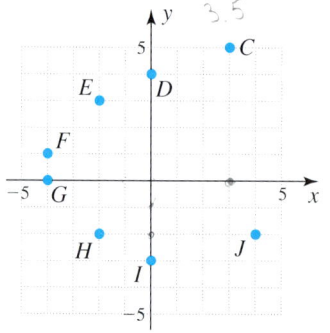

>Figure 3.20

⟨ **B** ⟩ **Graphing Linear Equations** In Problems 21–26, complete the ordered pairs to satisfy the equation and then graph the line of all solutions.

21. $y = x + 3$ $(-2, \ )$, $(-1, \ )$, $(0, \ )$, $(1, \ )$, $(2, \ )$

22. $y = 2x + 1$ $(-1, \ )$, $(0, \ )$, $(1, \ )$

23. $x - y = 4$ $(-1, \ )$, $(0, \ )$, $(1, \ )$

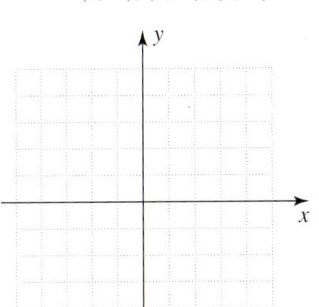
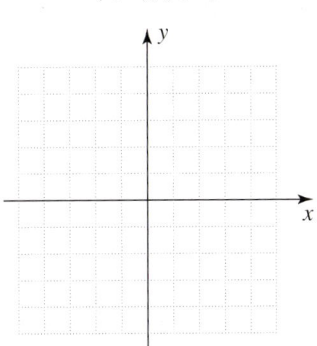
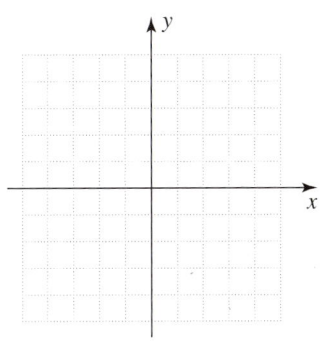

24. $x - 3y = 6$ $(0, \ )$, $(3, \ )$, $(-3, \ )$

25. $2x - y - 3 = 0$ $(-1, \ )$, $(0, \ )$, $(1, \ )$

26. $2x + y + 2 = 0$ $(-1, \ )$, $(0, \ )$, $(1, \ )$

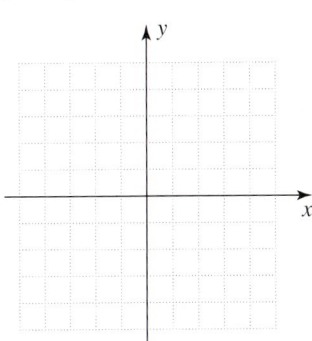
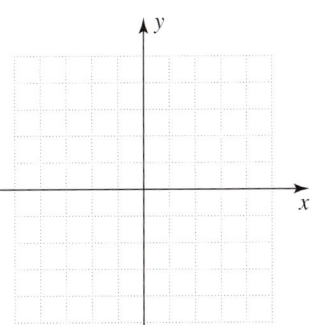
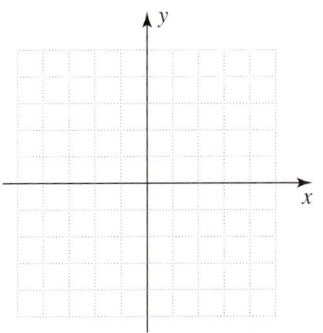

⟨ **C** ⟩ **Graphing Lines Using Intercepts** In Problems 27–40, find the x- and y-intercepts and then graph all solutions to the equation.

27. $y = x - 5$

28. $2y = 4x - 2$

29. $2x + 3y = 6$

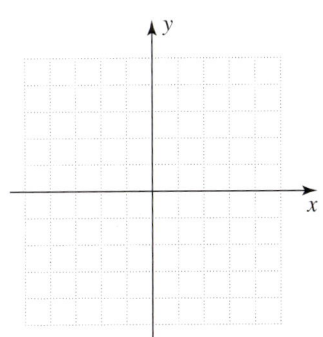
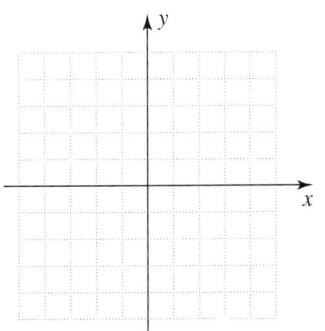
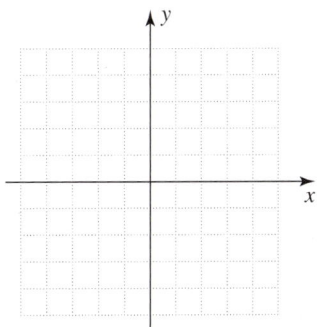

30. $3x + 2y = 6$

31. $2x - y = 4$

32. $3x - y = 3$

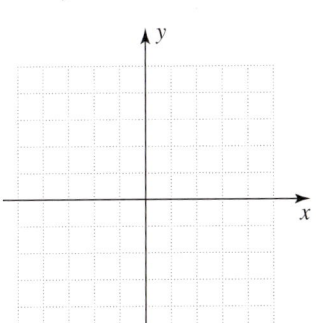
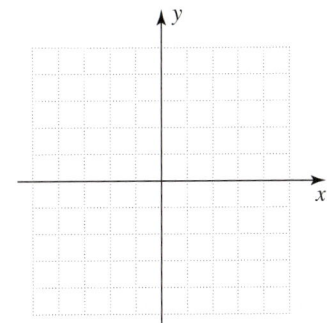
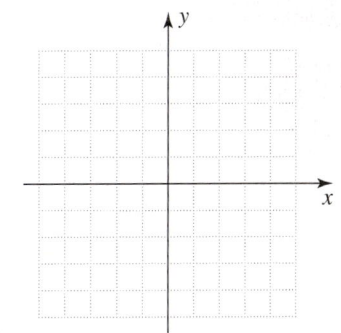

33. $2x + y - 4 = 0$

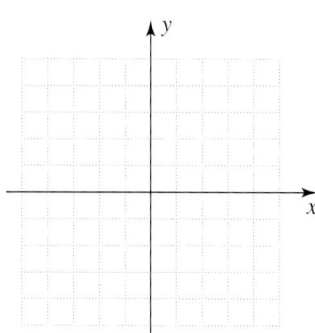

34. $3x + y - 3 = 0$

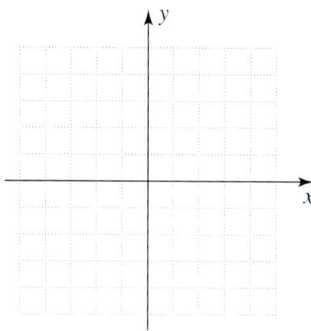

35. $y + 4x = 0$

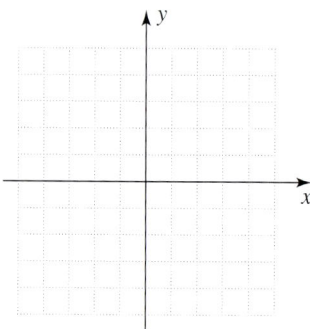

36. $y + 3x = 0$

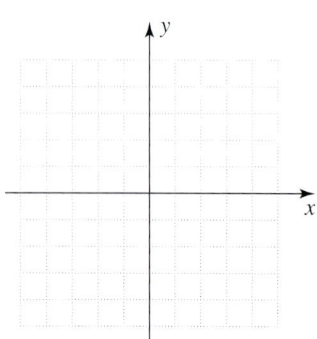

37. $2x - 5y = -10$

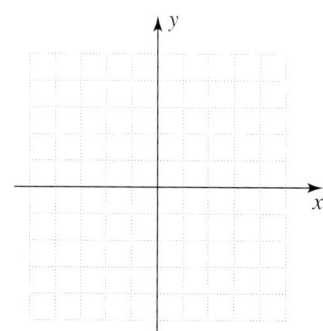

38. $2x - 3y = -6$

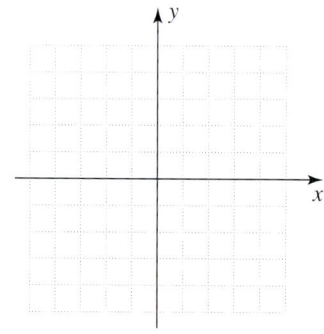

39. $x - y - 5 = 0$

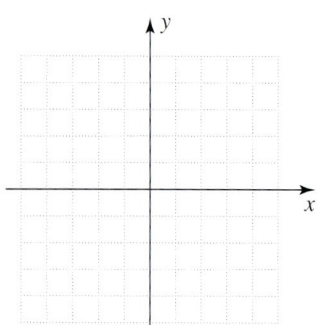

40. $2x - 3y - 12 = 0$

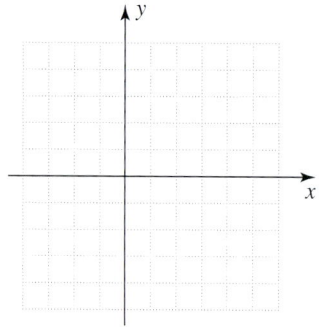

< D > **Graphing Horizontal and Vertical Lines** In Problems 41–50, determine whether the given line is horizontal or vertical and then graph all the solutions.

41. $-\dfrac{7}{2}x = 14$

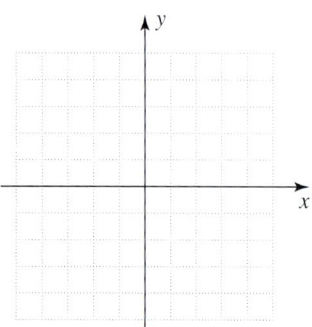

42. $-\dfrac{1}{2}x = 2$

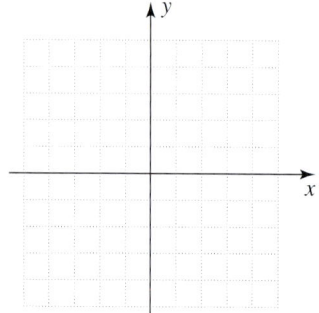

43. $\dfrac{3}{2}x = 6$

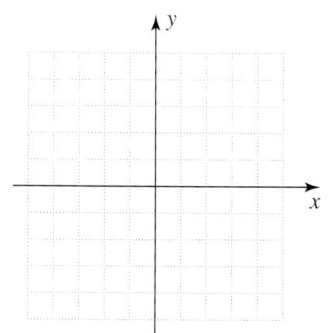

44. $-\frac{5}{2}y = 10$

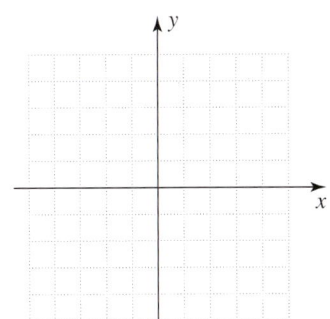

45. $-\frac{3}{4}x = 3$

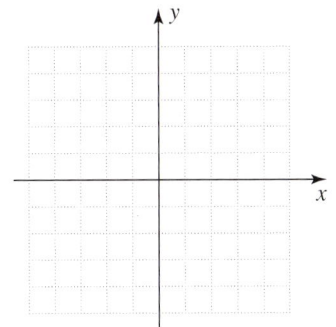

46. $-\frac{3}{7}y = -\frac{6}{7}$

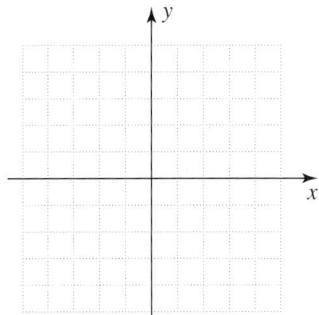

47. $-\frac{1}{3} + y = \frac{2}{3}$

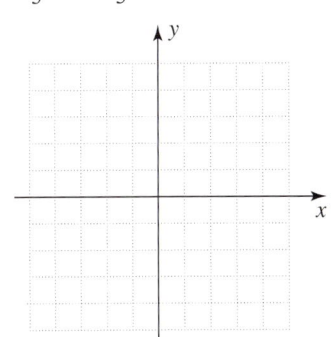

48. $-\frac{1}{5} + y = \frac{4}{5}$

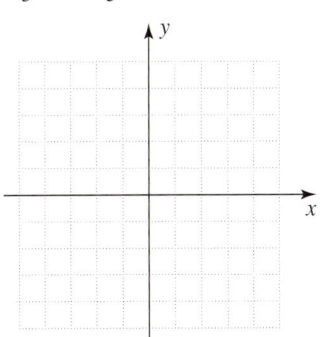

49. $\frac{2}{3} = x - \frac{4}{3}$

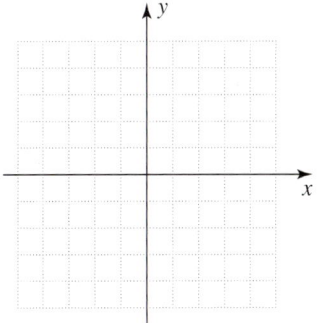

50. $\frac{3}{4} = x - \frac{5}{4}$

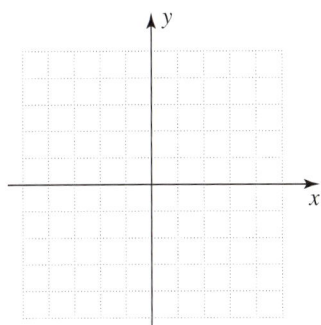

< E > **Graphing Nonlinear Equations** In Problems 51–60, graph all solutions to the equations.

51. $y = x^2 + 2$

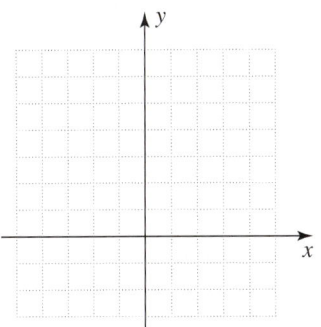

52. $y = x^2 - 1$

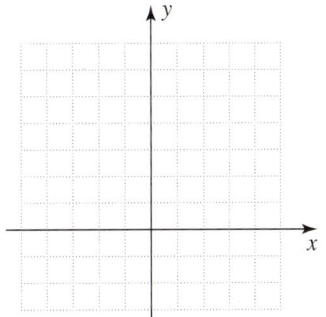

53. $y = -x^2$

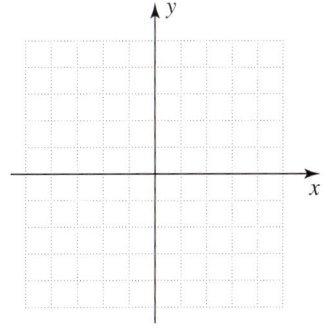

54. $y = -x^2 + 3$

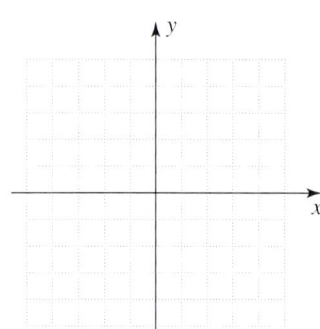

55. $y = 1 + x^2$

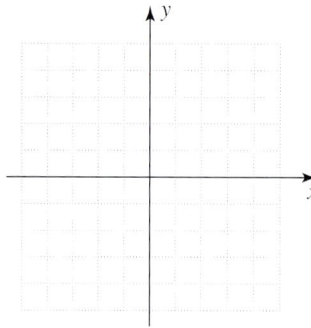

56. $y = 2 - x^2$

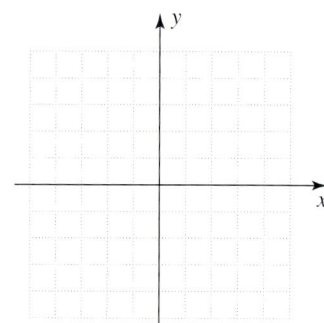

57. $y = |x|$

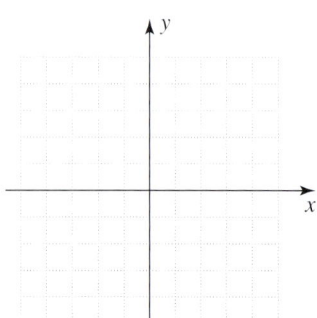

58. $y = |x| - 2$

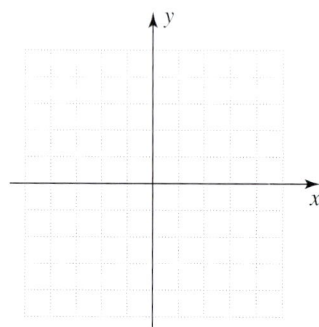

59. $y = \frac{1}{x}$

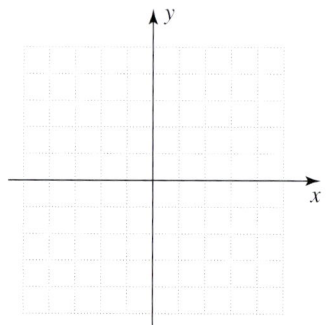

60. $y = \frac{-1}{x}$

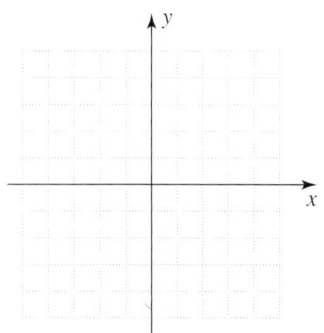

〉〉〉 Applications

61. *Cricket chirps* The effect of temperature on cricket chirps was discussed on page 74. Now we look at the formula for finding the number of chirps given the temperature. The number N of chirps a certain type of cricket makes per minute is given by

$$N = 4(T - 40)$$

where T is the temperature in degrees Fahrenheit. The ordered pairs corresponding to this equation are of the form (T, N).

a. Does the ordered pair (60, 80) satisfy the equation?

b. Based on your answer to part **a**, how many chirps does a cricket make when the temperature is 60°F?

c. How many chirps does a cricket make when the temperature is 80°F?

62. *Cricket chirps* Based on the formula in Problem 61, crickets stop chirping when $N = 0$. The corresponding ordered pair will be $(T, 0)$.

a. Find T.

b. At what temperature will crickets stop chirping?

63. *Depreciation* The depreciation (decrease in value) of a new car each year is approximately 15% to 20% and can be found by the equation

$$V = -3000t + 23{,}000$$

where t is the number of years after the purchase and V is the value of the car for that year.

a. Find the purchase value of the car. (*Hint:* $t = 0$)
b. Find the value of the car 1 year from the purchase.
c. Find the value of the car 5 years from the purchase.
d. According to this formula when will the value of the car become 0? (Round the answer to the nearest year.)
e. Explain how a car's value may or may not be 0.

64. *Temperatures* The average monthly temperatures, in degrees Fahrenheit, for Phoenix from January to June can be approximated by the equation

$$T = 7.5m + 56$$

where m is the number of the month (1–6) and T is the average temperature for that month.

a. Find the temperature in January. (*Hint:* $m = 1$)
b. Find the temperature in April.
c. Find the temperature in June.
d. Do you think that the equation will accurately predict the temperature for November? Explain your answer.

Exercise Are you exercising too hard? Your target zone can tell you. It works like this. Take your pulse after exercising, find your age on the *x*-axis in Figure 3.21, and follow a vertical line up to the lower edge of the shaded area; then go across to the number on the *y*-axis at the left. That pulse rate is the lower limit for your target zone. To find the upper limit, continue vertically on your age line to the top of the shaded area and then go across to the *y*-axis to locate the pulse rate. Use this information and Figure 3.21 to solve Problems 65–68.

65. What is the lower limit pulse rate for a 20-year-old person? Write the answer as an ordered pair (age, limit).

66. What is the upper limit pulse rate for a 20-year-old person? Write the answer as an ordered pair.

67. What is the upper limit pulse rate for a 45-year-old person? Write the answer as an ordered pair.

68. What is the lower limit pulse rate for a 50-year-old person? Write the answer as an ordered pair.

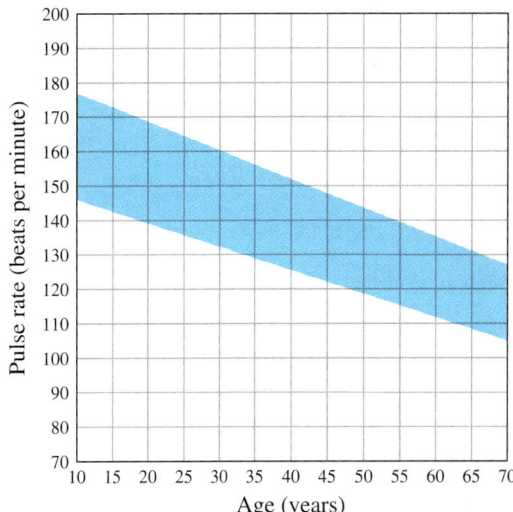

>Figure 3.21

>>> Using Your Knowledge

The Long, Hot Summer Exercises The ideas presented in this section are vital for understanding graphs. For example, do you exercise in the summer? To determine the risk of exercising in the heat, you must know how to read the graph. To do this, first find the temperature on the *y*-axis, then read across to the right, stopping at the vertical line representing the relative humidity. For example, on a 90°F day, if the humidity is less than 30% the weather is in the safe zone.

69. If the humidity is 50%, how high can the temperature rise and still be in the safe zone for exercising? (Answer to the nearest degree.)

70. If the humidity is 70%, at what temperature will the danger zone start?

71. If the temperature is 100°F, what does the humidity have to be so that it is safe to exercise?

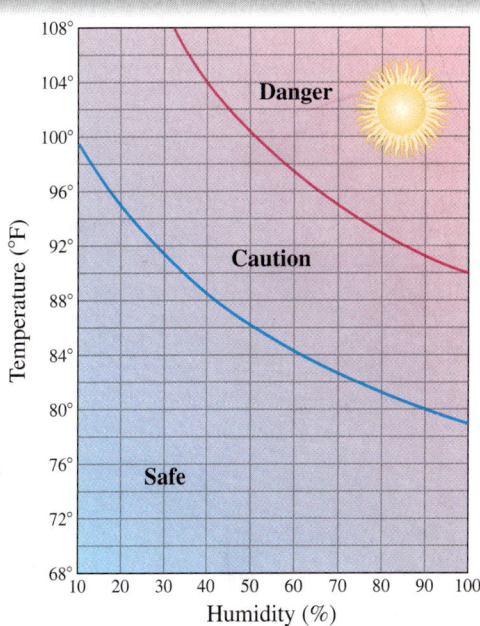

72. Between what temperatures should you use caution when exercising if the humidity is 80%?

73. If you start jogging at 1 P.M. when the temperature is 86°F and the humidity is 60%, how many degrees can the temperature rise before you get to the danger zone?

〉〉〉 Write On

74. What does the graph of an equation represent to you? Can you ever draw the *complete* graph of a line?

75. Why are *linear equations* named that way?

76. What happens when $A = 0$ in the equation $Ax + By = C$?

77. What happens when $B = 0$ in the equation $Ax + By = C$?

78. Why does setting $x = 0$ give the *y*-intercept for a line?

79. Why does setting $y = 0$ give the *x*-intercept for a line?

80. If two points determine a line, why did we use three points when graphing lines?

81. Explain in your words the procedure you use to graph lines by (a) graphing points and (b) using the intercepts.

〉〉〉 Concept Checker

Fill in the blank(s) with the correct word(s), phrase, or mathematical statement.

82. Two real numbers used to locate a point on a rectangular coordinate system are called a(n) _____.

83. The axes on a rectangular coordinate system separate the plane into four _____.

84. The *x*- and *y*-axes intersect at the _____.

85. If the solution set to an equation forms a line when graphed, then the equation is called a _____.

coordinate system

ordered pair

origin

quadrants

linear equation

nonlinear equation

〉〉〉 Mastery Test

Graph each equation.

86. $y = x - 3$

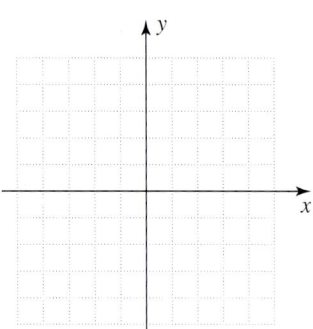

87. $2y = -6$

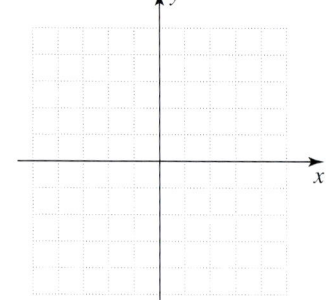

88. $2x - y = 6$

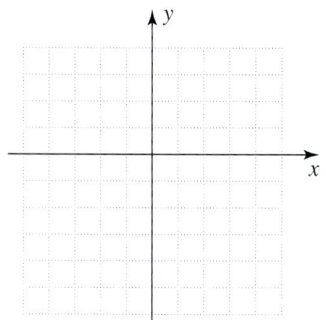

89. $-3x - y = -6$

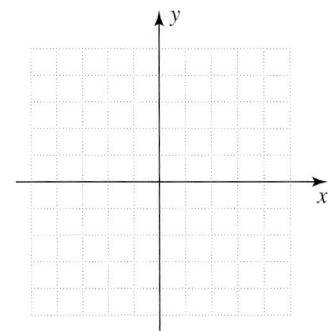

90. $y = 1 - x^2$

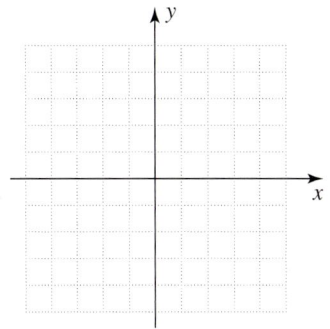

91. $y = x^2 - 4$

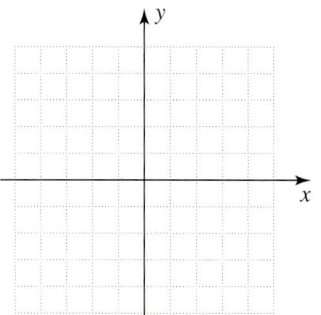

92. Graph the points $A(3, 2)$, $B(-1, 4)$, $C(0, -2)$, $D(-2, 1)$, and $E(2, -1)$.

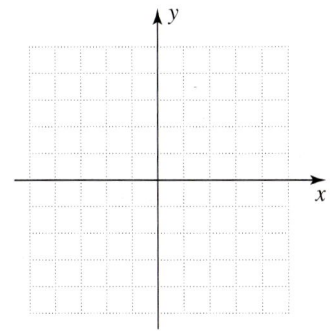

93. What are the coordinates of the points A, B, and C in Figure 3.22?

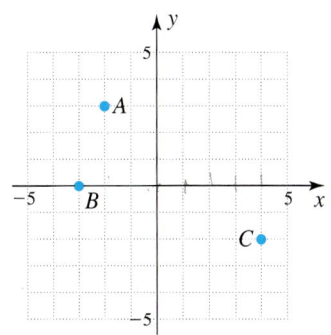

> Figure 3.22

》》 Skill Checker

Find y for the indicated value of x.

94. $y = 3x + 6$, $x = 2$

95. $7y = 3x + 6$, $x = -2$

96. $y = -\frac{2}{3}x + 4$, $x = 3$

97. $y = -\frac{2}{3}x + 4$, $x = -3$

98. $y = x^2 + 2x - 3$, $x = -3$

99. $y = -x^2 + x - 2$, $x = -3$

100. $y = \frac{1}{2x+5}$, $x = 5$

101. $y = \frac{2}{x-5}$, $x = 5$

Find x for the indicated value of y.

102. $y = 3x + 6$, $y = 3$

103. $y = 2x + 8$, $y = 0$

104. $y = -\frac{2}{3}x + 4$, $y = 8$

105. $y = -\frac{2}{3}x + 4$, $y = -8$

3.2 Using Slopes to Graph Lines

Objectives

A Find the slope of a line passing through two given points.

B Use the definition of slope to decide whether two lines are perpendicular, parallel, or neither.

C Graph a line given its slope and a point on the line.

D Find the slope and *y*-intercept given the equation of a line.

To Succeed, Review How To . . .

1. Add, subtract, multiply, and divide integers (pp. 18–26).
2. Find the reciprocal of a number (p. 26).

Getting Started
Sloping Salaries

Sources: Consumer Price Index (all urban consumers, not seasonally adjusted), Bureau of Labor Statistics. http://www.bls.gov/cpi/home.htm. American Federation of Teachers, annual survey of state departments of education, various years.

As you can see by looking at the graph, the average salaries for beginning teachers have been fluctuating over the period of time from 1964 to 2004. The salaries from 1994 seem to be on an upward incline. Let's see how we can use the graph to describe the change in the yearly average beginning salary from 1994 to 2004.

A line segment joining two points in a plane has a certain "steepness" or inclination. According to the graph, the average beginning salary in 1994 was about $45,000; in 2004, it was about $47,000. The change is

$$47,000 - 45,000 = 2000$$

during the 10-year period.

The average yearly raise for beginning teacher salaries is given by the ratio

$$\frac{\textit{Increase} \text{ in 10 yr}}{\textit{Number} \text{ of years}} = \frac{47,000 - 45,000}{10} = \frac{2000}{10} = 200$$

Thus, the teachers' yearly average beginning salary went up about $200 each year.

The steepness of the line segment we just described algebraically is called the **slope** of the line, and in this section, we shall show you how to determine the slope of a line given various situations.

A › Finding the Slope of a Line

In the *Getting Started* we considered the "sloping salaries" for beginning teachers. The actual calculation was done by *dividing* the difference in the salaries at the start of the 10-year period and at the end, *by* the difference in time lapsed. This can also be called a ratio of the differences or the **rate of change.** The value of the ratio calculated from the rate of change will be called the **slope** of the line. This will enable us to assign a value to the **steepness** of a line.

In mathematics the ratio of the change (difference) in y to the change in x between two points on a line is called the *slope* of the line segment and is denoted by the letter m. Thus,

$$m = \frac{\text{change in } y}{\text{change in } x} = \frac{\text{difference in } y\text{'s}}{\text{difference in } x\text{'s}} = \frac{\text{rise}(\uparrow)}{\text{run}(\rightarrow)}$$

We summarize this discussion with the following definition and formula.

DEFINITION OF SLOPE

If $A(x_1, y_1)$ and $B(x_2, y_2)$ are any two distinct points on a line L (that is not parallel to the y-axis), then the **slope** of L, denoted by m, is

$$m = \frac{y_2 - y_1}{x_2 - x_1}$$

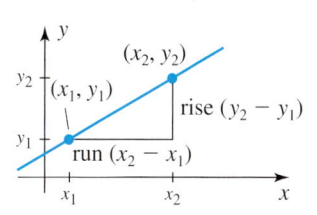

NOTE
The slope is a number that measures the "steepness" of a line. For positive numbers, the larger the number, the steeper the line.

When using this definition, it doesn't matter which point is taken for A and which for B. For example, the slope of the line passing through the points $A(0, -6)$ and $B(3, 3)$ is

$$m = \frac{3 - (-6)}{3 - 0} = \frac{9}{3} = 3$$

If we choose A to be $(3, 3)$ and B to be $(0, -6)$, the slope is the same.

$$m = \frac{-6 - 3}{0 - 3} = \frac{-9}{-3} = 3$$

Thus, A and B can be interchanged without changing the value of the resulting slope.

On the left side (x_2, y_2) is the first point and (x_1, y_1) is the second point.

On the right side (x_1, y_1) is the first point and (x_2, y_2) is the second point.

$$\frac{y_2 - y_1}{x_2 - x_1} = \frac{y_1 - y_2}{x_1 - x_2}$$

EXAMPLE 1 Finding slopes given two points on a line
Find the slope of the line passing through the given points:

a. $A(-3, 1), B(-1, -2)$
b. $A(-1, 5), B(-1, 6)$
c. $A(1, 1), B(2, 3)$
d. $A(3, 4), B(1, 4)$

PROBLEM 1
Find the slope of the line passing through the given points:

a. $A(-2, 3), B(1, 5)$
b. $A(4, -5), B(4, 3)$
c. $A(2, 2), B(4, -4)$
d. $A(6, -1), B(7, -1)$

(continued)

Answers to PROBLEMS

1. a. $\frac{2}{3}$ **b.** Undefined
 c. -3 **d.** 0

SOLUTION

a. The slope, as shown in Figure 3.23, is

$$m = \frac{-2-1}{-1-(-3)} = \frac{-3}{2} \quad \begin{array}{l}\leftarrow \text{rise} \\ \leftarrow \text{run}\end{array}$$

$\frac{-3}{2}$ is in fraction form, so if we start at A, we move 3 units *down* (the change in y) and 2 units *right* (the change in x) ending at B.

b. The line passing through $A(-1, 5)$ and $B(-1, 6)$ in Figure 3.24 is a line *parallel* to the y-axis. The fact that the change in x is zero in lines parallel to the y-axis (that is, *vertical* lines) requires that they be excluded from the definition; their slope is *undefined*. If we tried to apply the definition, we would get

$$m = \frac{6-5}{-1-(-1)} = \frac{1}{0}$$

which is *undefined*.

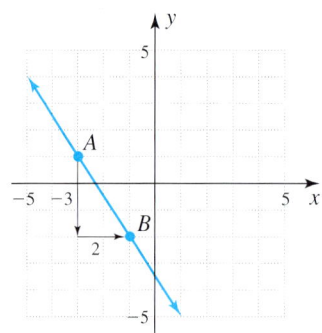
>Figure 3.23

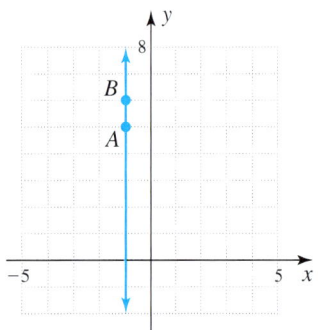
>Figure 3.24

NOTE

If the slope is a number that measures the "steepness" of a line, then the vertical line, being the steepest, should have a slope that names "the largest number." Is there a "largest number"? Of course not, so we can say there is "no slope" because there is not a "largest number." *Beware that "no slope" is not the same as "0" slope.*

c. The slope, as shown in Figure 3.25, is

$$m = \frac{3-1}{2-1} = \frac{2}{1} = 2$$

The change in y (rise) is 2 units up, the change in x (run) is 1 unit right.

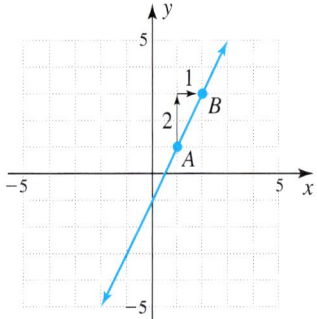
>Figure 3.25

d. The slope, as shown in Figure 3.26, is

$$m = \frac{4 - 4}{1 - 3} = \frac{0}{-2}$$

Since

$$\frac{0}{-2} = 0$$

the slope is zero. The line is parallel to the *x*-axis and is a horizontal line. Horizontal lines have zero slope.

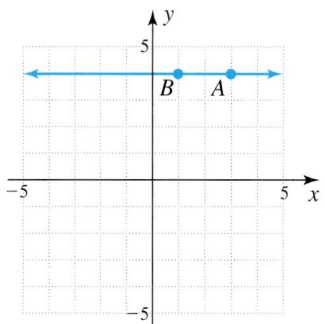

> Figure 3.26

Table 3.2 summarizes the information in Example 1.

Table 3.2 Slope Summary

A line that *rises* from left to right has a *positive* slope.

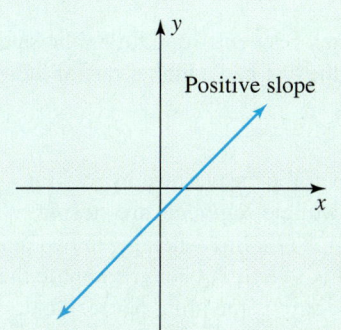

A horizontal line has zero slope. Since $y_2 - y_1 = 0$,

$$m = \frac{y_2 - y_1}{x_2 - x_1} = \frac{0}{x_2 - x_1}$$

$$m = 0$$

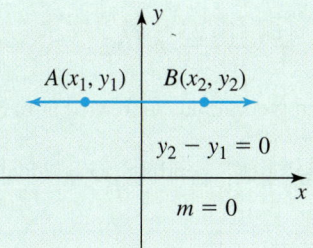

A line that *falls* from left to right has a *negative* slope.

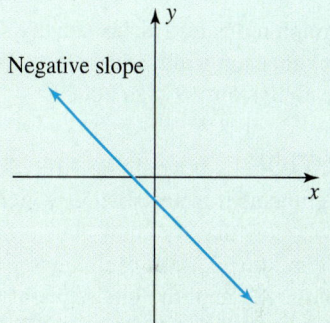

The slope of a *vertical* line is undefined. Since $x_2 - x_1 = 0$,

$$m = \frac{y_2 - y_1}{x_2 - x_1} = \frac{y_2 - y_1}{0}$$

so *m* is *undefined*.

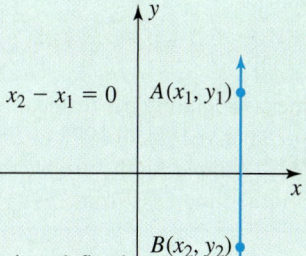

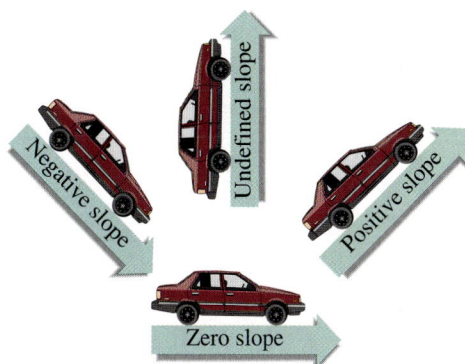

EXAMPLE 2 Application using slope

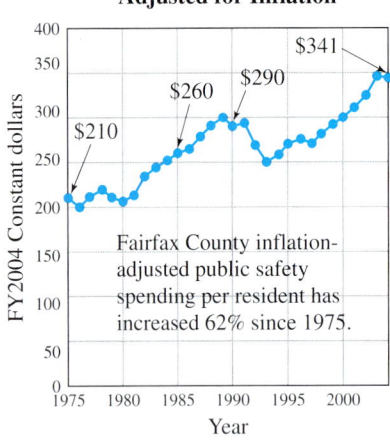

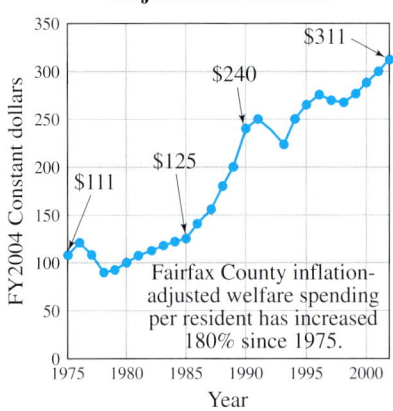

Fairfax County Taxpayers Alliance
Analysis based on Fairfax County Adopted budgets, FY1977–2003, and FY04 Advertised budget 3/10/03

Source: Fairfax County Taxpayer Alliance.

Each graph indicates Fairfax County spending per resident—one for public safety and the other for health and welfare. Which of the two had a higher rate of increase for spending from 1998 to 2002?

SOLUTION It would be difficult to discover the answer just by looking at the graph as it is not obvious which of the two line segments are steeper. However, now that we know how to find the slope of a line when given two points, we can find the slope of each line segment from 1998 to 2002 and compare them. To find the slope of the line segment for public safety spending per resident, we will use $260 for 1998 and $325 for 2002.

$$m = \frac{y_2 - y_1}{x_2 - x_1} = \frac{325 - 260}{2002 - 1998} = \frac{65}{4} = \frac{16.25}{1} = 16.25$$

This means spending per resident for public safety has increased at the rate of $16.25 per year from 1998 to 2002.

To find the slope of the line segment for health and welfare spending per resident, we will use $270 for 1998 and $311 for 2002.

$$m = \frac{y_2 - y_1}{x_2 - x_1} = \frac{311 - 270}{2002 - 1998} = \frac{41}{4} = \frac{10.25}{1} = 10.25$$

This means spending per resident for health and welfare has increased at the rate of $10.25 per year from 1998 to 2002.

Thus, comparing the two we see that public safety spending per resident had a higher rate of increase from 1998 to 2002.

PROBLEM 2
Using the graphs from Example 2, what was the rate of increase for each of the two types of spending from 1985 to 1990?

Answers to PROBLEMS

2. $6 per year increase for public safety spending and $23 per year for health and welfare spending

B › Parallel and Perpendicular Lines

The definition of slope can be used to determine whether two lines are parallel. Two parallel lines have the same steepness (inclination) and thus the same slope, so we have the following.

SLOPES OF PARALLEL LINES

Two nonvertical lines L_1 and L_2 with slopes m_1 and m_2 are parallel if and only if $m_1 = m_2$.

NOTE

If L_1 and L_2 are both vertical lines, they are parallel.

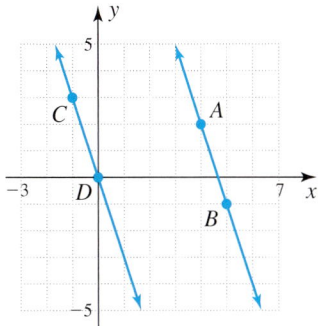

> Figure 3.27

For example, consider the two lines, one passing through points $A(4, 2)$ and $B(5, -1)$ and the other through points $C(-1, 3)$ and $D(0, 0)$, as shown in Figure 3.27. The slope of AB is

$$m_1 = \frac{-1 - 2}{5 - 4} = \frac{-3}{1} = -3$$

and the slope of CD is

$$m_2 = \frac{0 - 3}{0 - (-1)} = \frac{-3}{1} = -3$$

Thus, $m_1 = m_2$, so both lines have the same slope and are parallel.

It is shown in more advanced courses that two lines with slopes m_1 and m_2 are perpendicular (meet at a 90° angle) if

$$m_1 = \frac{-1}{m_2}$$

that is, if their slopes are *negative reciprocals*. This means that

$$\text{if} \quad m_1 = \frac{-1}{m_2}, \quad \text{then} \quad m_1 \cdot m_2 = \frac{-1}{m_2} \cdot m_2 = -1$$

SLOPES OF PERPENDICULAR LINES

The lines L_1 and L_2 with slopes m_1 and m_2, respectively, are perpendicular if and only if the slopes are **negative reciprocals**; that is, $m_1 \cdot m_2 = -1$ ($m_1, m_2 \neq 0$).

We can show that the line passing through $A(4, 2)$ and $B(5, -1)$ is perpendicular to the line passing through $C(2, 1)$ and $D(5, 2)$. Since the slope of AB is

$$m_1 = \frac{-1 - 2}{5 - 4} = \frac{-3}{1} = -3$$

and that of CD is

$$m_2 = \frac{2 - 1}{5 - 2} = \frac{1}{3}$$

we have $m_1 \cdot m_2 = -3 \cdot \frac{1}{3} = -1$. These perpendicular lines can be graphed, as shown in Figure 3.28.

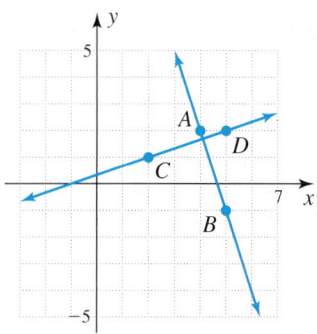

> Figure 3.28

EXAMPLE 3 Finding whether lines are parallel or perpendicular

A line L_1 has slope $\frac{2}{3}$; find:

a. Whether another line passing through $A(5, 1)$ and $B(8, 3)$ is parallel or perpendicular to L_1
b. Whether the line passing through $C(-1, -4)$ and $D(-3, -1)$ is parallel or perpendicular to L_1

SOLUTION

a. The slope of AB is

$$m = \frac{3-1}{8-5} = \frac{2}{3}$$

The slope of L_1 is also $\frac{2}{3}$, so line AB is parallel to L_1.

b. The slope of CD is

$$m = \frac{-1-(-4)}{-3-(-1)} = \frac{3}{-2} = \frac{-3}{2}$$

The slope of L_1 is $\frac{2}{3}$ and

$$\frac{2}{3} \cdot \left(\frac{-3}{2}\right) = -1$$

so line CD is perpendicular to L_1.

PROBLEM 3

A line L_1 has slope $\frac{3}{4}$; find:

a. Whether another line passing through $A(2, 6)$ and $B(6, 9)$ is parallel or perpendicular to L_1
b. Whether the line passing through $C(-9, -1)$ and $D(-6, -5)$ is parallel or perpendicular to L_1

EXAMPLE 4 Finding y when lines are perpendicular

If the line through $A(4, y)$ and $B(-2, -5)$ is perpendicular to a line whose slope is $-\frac{2}{3}$, find y.

SOLUTION The slope of the line through points A and B is

$$\frac{y-(-5)}{4-(-2)} = \frac{y+5}{6}$$

If the line through A and B is perpendicular to a line whose slope is $-\frac{2}{3}$, the slope of the line through A and B must be $\frac{3}{2}$ (the negative reciprocal of $-\frac{2}{3}$). Thus

$$\frac{y+5}{6} = \frac{3}{2}$$

$$\frac{6 \cdot (y+5)}{6} = \frac{6 \cdot 3}{2} \quad \text{Multiply both sides by 6.}$$

$$y + 5 = 9 \quad \text{Simplify.}$$

$$y = 4 \quad \text{Subtract 5 from both sides.}$$

The line through $A(4, 4)$ and $B(-2, -5)$ has slope

$$\frac{4-(-5)}{4-(-2)} = \frac{9}{6} = \frac{3}{2}$$

making it perpendicular to a line whose slope is $\frac{-2}{3}$, and our result is correct.

PROBLEM 4

The line through $A(5, y)$ and $B(-4, 8)$ is perpendicular to a line whose slope is $\frac{-3}{4}$. Find y.

C > Graphing Lines Using the Slope and a Point

We can graph a line if we know its slope and a point on the line. For convenience we will use the y-intercept as the point but the procedure could be done using any point on the line. For example, suppose a line has a y-intercept of -2 and has a slope of $\frac{2}{3}$. The y-intercept of -2 means that one point on the line is $(0, -2)$. To graph the line, we recall that the slope of a line is the ratio

$$m = \frac{\text{change in } y}{\text{change in } x} = \frac{2}{3}$$

Answers to PROBLEMS

3. a. Parallel b. Perpendicular
4. 20

We start at the point (0, −2) and go 2 units up (change in y is 2) and 3 units right (change in x is 3), ending at (3, 0). We then draw a line through the points (0, −2) and (3, 0) to obtain the graph shown in blue in Figure 3.29. If the slope of the line had been $\frac{-2}{3}$, we would move 2 units down (change in y) and then go 3 units right (change in x), ending at (3, −4). This line with the negative slope is shown in red.

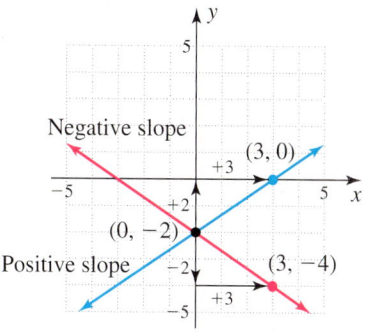

> Figure 3.29

EXAMPLE 5 Graphing lines using the slope and a point

a. Graph a line that has y-intercept 2 and has slope $-\frac{3}{4}$.

b. Graph a line that has slope 3 and passes through the point (−2, 1).

SOLUTION

a. $-\frac{3}{4}$ can be written as $\frac{-3}{4}$.

We start at the y-intercept point (0, 2) and go 3 units down (the change in y). We then go right 4 units (the change in x), ending at (4, −1), as shown in Figure 3.30. The graph is obtained by drawing a line through the points (0, 2) and (4, −1).

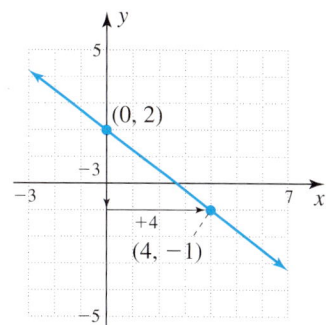

> Figure 3.30

b. The slope 3 can be written as $\frac{3}{1}$. We start by plotting the point (−2, 1). From that point, (−2, 1) we go 3 units up (change in y). We then go right 1 unit (the change in x), ending at (−1, 4), as shown in Figure 3.31.

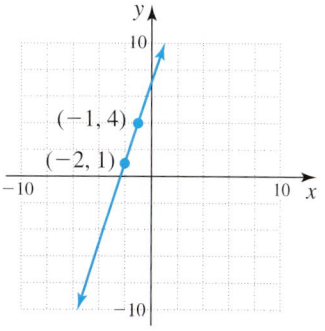

> Figure 3.31

PROBLEM 5

a. Graph a line that has y-intercept −1 and has slope $-\frac{1}{2}$.

b. Graph a line that has slope $\frac{1}{2}$ and passes through the point (1, 2).

Answers to PROBLEMS

5. a. 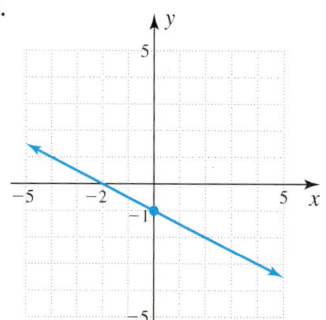   b.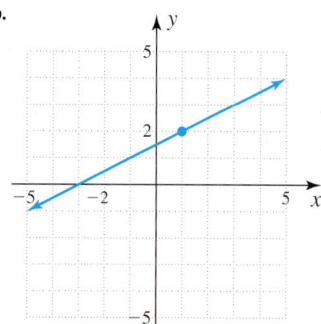

D ❯ Slope-Intercept Form for a Linear Equation

Can the coefficients and constant in a linear equation be related to the graph of the line? Let's graph the equation $-x + 2y = 4$. Using a table, we get the following points to plot.

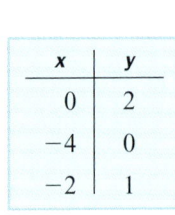
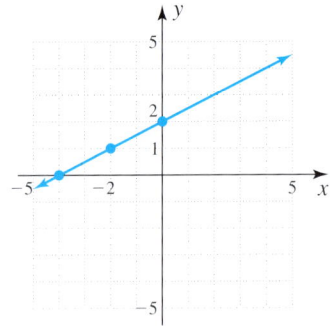

The y-intercept is 2 and the slope is $m = \frac{2 - 0}{0 - (-4)} = \frac{2}{4} = \frac{1}{2}$.

The numbers in the equation $-x + 2y = 4$ don't reflect that information, but solve the equation for y and see what happens.

$$-x + 2y = 4$$
$$2y = x + 4 \qquad \text{Add x to both sides.}$$
$$y = \frac{1}{2}x + 2 \qquad \text{Divide both sides by 2.}$$

The coefficient of x, $\frac{1}{2}$, is the same value we obtained when we calculated the slope of the line. Also, the constant, 2, in the equation is the same as the y-intercept we observed on our graph. We can use this example to generalize that when the equation is solved for y, then the equation, $y = mx + b$, has m as its **slope** and b as its **y-intercept.**

THE SLOPE-INTERCEPT FORM OF A LINE

$$y = mx + b$$

slope ↑   ↑ y-intercept

Calculator Corner

Finding the Slope Given 2 Points

Use your TI-83 Plus to find the slope of a line. To do this, you need two points. If you have an equation written in the form $y = ax + b$, a is the slope of the line. (The calculator uses a instead of m.) Let's do Example 1(a). The two given points are $A(-3, 1)$ and $B(-1, -2)$. We will make two lists on the calculator, but we must first be sure the lists are clear. Press [2nd] [+] 4 [ENTER] to ClrAllLists. Press [STAT] 1, and under L_1 press (-) 3 [ENTER] and (-) 1 [ENTER]. Press the right arrow to get to L_2 and then press 1 [ENTER] and (-) 2 [ENTER]. The strokes needed to give you the equation of a line of the form $y = ax + b$ using a technique called "linear regression" are [STAT] ▶ 4 [ENTER]. We see that the slope $a = -1.5 = -\frac{3}{2}$, as shown in the screen. Try Example 1(b) and see what happens.

```
LinReg
y=ax+b
a=-1.5
b=-3.5
```

EXAMPLE 6 Finding the slope and y-intercept given the equation of a line

Given the following linear equations, find the slope and y-intercept of each.

a. $y = 5x - 3$
b. $-y = \frac{1}{4}x + 2$
c. $-3x + 2y = 7$

SOLUTION

a. The equation is solved for y so the coefficient of x and the constant are the slope and y-intercept, respectively.

$$y = 5x - 3$$

slope = 5, y-intercept = -3

b. We start by solving the equation for y.

$$-y = \frac{1}{4}x + 2$$

$$y = \frac{-1}{4}x - 2 \quad \text{Divide both sides by } -1.$$

slope is $\frac{-1}{4}$ and y-intercept is -2

c. We solve the equation for y.

$$-3x + 2y = 7$$
$$2y = 3x + 7 \quad \text{Add 3x to both sides.}$$
$$y = \frac{3}{2}x + \frac{7}{2} \quad \text{Divide both sides by 2.}$$

slope is $\frac{3}{2}$ and y-intercept is $\frac{7}{2}$

PROBLEM 6

Given the following linear equations, find the slope and y-intercept of each.

a. $y = -4x + 1$
b. $-y = \frac{1}{2}x - 3$
c. $2x - 3y = 5$

Exercises 3.2

Boost your grade at mathzone.com!
> Practice Problems
> NetTutor
> Self-Tests
> e-Professors
> Videos

⟨ A ⟩ Finding the Slope of a Line   In Problems 1–7, find the slope of the line passing through the given points and graph the line.

1. $A(2, 4)$, $B(-1, 0)$

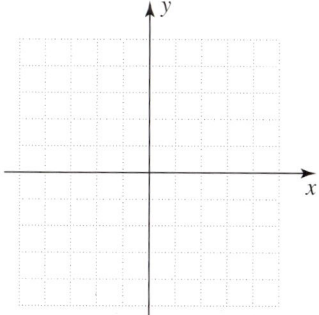

2. $A(3, -2)$, $B(8, 4)$

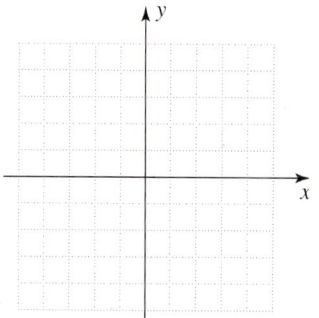

Answers to PROBLEMS

6. a. Slope -4; y-intercept 1
   b. Slope $-\frac{1}{2}$; y-intercept 3
   c. Slope $\frac{2}{3}$; y-intercept $-\frac{5}{3}$

3. $C(-4, -5), D(-1, 3)$

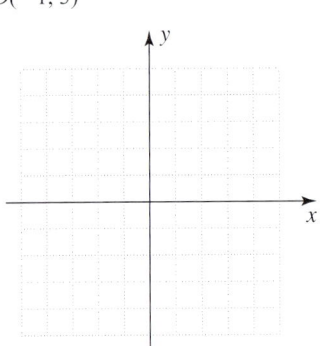

4. $C(5, 7), D(-2, 3)$

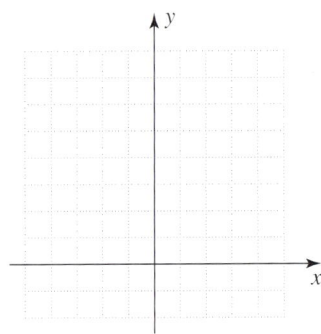

5. $E(4, 8), G(1, -1)$

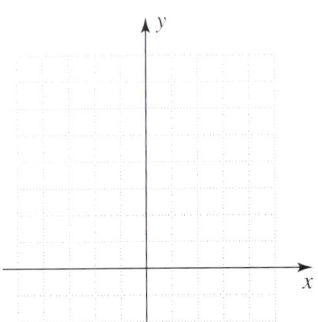

6. $H(-2, -2), I(-2, 4)$

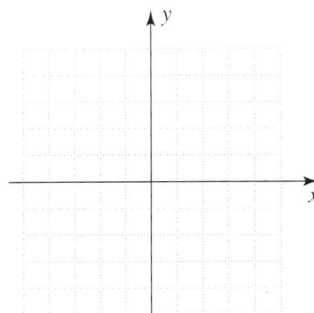

7. $A(3, -1), B(-2, -1)$

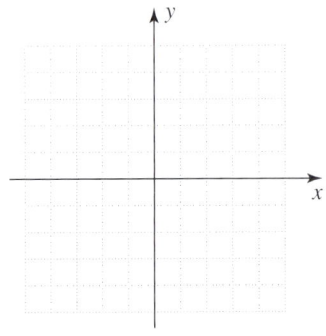

In Problems 8–10, use the graph.

The graph indicates the salary trends for the Fairfax County government and school system. Both the schools' average salaries and the county government's average salaries show a steady increase from 1998 to 2004.

8. a. Find the slope of the line segment for the county average salary from 1998 to 2004. Use $50,000 for the 1998 salary and $58,000 for 2004.

   b. Using the slope, describe the rate of change of the salary increase per year for the county average salary (to the nearest dollar).

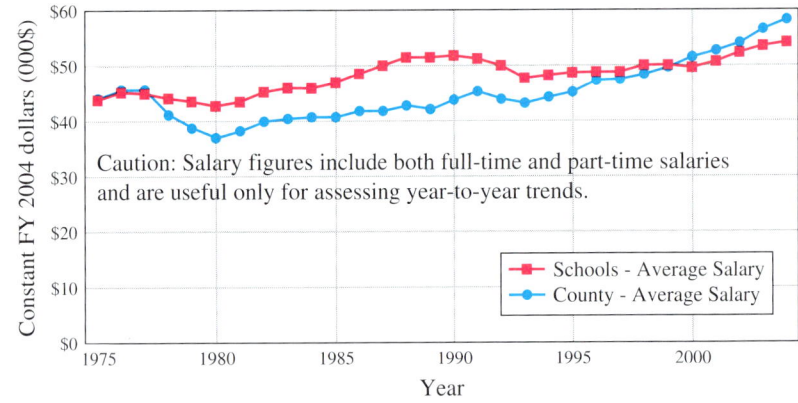

Fairfax County Government and School System Salary Trends, FY 1975–2004 (Inflation Adjusted)

Fairfax County Taxpayers Alliance analysis based on Fairfax County and Fairfax County Public Schools budgets. FY 1975–FY 2004 and data from the Fairfax County Office of Management and Budget. 4/6/03

Source: Fairfax County Taxpayers Alliance.

9. a. Find the slope of the line segment for the schools' average salary from 1998 to 2004. Use $50,000 for the 1998 salary and $54,000 for 2004.

   b. Using the slope, describe the rate of change of the salary increase per year for the schools' average salary (to the nearest dollar).

10. Comparing the results from Problems 8 and 9, which average salary is increasing at a higher rate? Explain how the graph confirms the answer.

⟨ B ⟩ **Parallel and Perpendicular Lines** In Problems 11–20, determine whether lines AB and CD are parallel, perpendicular, or neither.

11. $A(1, 6)$, $B(-1, 4)$ and $C(1, -2)$, $D(2, -1)$
12. $A(0, 4)$, $B(1, -1)$ and $C(0, 1)$, $D(-5, -1)$
13. $A(2, 0)$, $B(4, 5)$ and $C(8, 0)$, $D(3, 2)$
14. $A(1, 1)$, $B(-1, 2)$ and $C(1, -1)$, $D(0, -3)$
15. $A(-1, 1)$, $B(1, 2)$ and $C(1, -1)$, $D(0, -1)$
16. $A(1, 1)$, $B(3, 3)$ and $C(1, -1)$, $D(0, 2)$
17. $A(1, -1)$, $B(2, -\frac{1}{2})$ and $C(2, -2)$, $D(1, 0)$
18. $A(1, 1)$, $B(\frac{1}{5}, 0)$ and $C(1, 1)$, $D(0, \frac{9}{5})$
19. $A(0, 1)$, $B(14, -1)$ and $C(0, 2)$, $D(7, 1)$
20. $A(2, -2)$, $B(1, -7)$ and $C(1, -3)$, $D(0, -8)$

In Problems 21–30, find x or y.

21. The line through $A(x, 4)$ and $B(6, 8)$ is parallel to a line whose slope is 1.
22. The line through $A(x, 5)$ and $B(-2, 3)$ is parallel to a line whose slope is $\frac{2}{5}$.
23. The line through $A(x, 2)$ and $B(2, 6)$ is perpendicular to a line whose slope is $\frac{1}{2}$.
24. The line through $A(x, 4)$ and $B(-3, -4)$ is perpendicular to a line whose slope is $\frac{2}{5}$.
25. The line through $A(x, -6)$ and $B(-2, -1)$ is perpendicular to a line whose slope is $-\frac{2}{3}$.
26. The line through $A(2, y)$ and $B(3, 4)$ is parallel to a line whose slope is 2.
27. The line through $A(3, y)$ and $B(1, -2)$ is parallel to a line whose slope is -3.
28. The line through $A(2, y)$ and $B(1, -4)$ is perpendicular to a line whose slope is $\frac{1}{3}$.
29. The line through $A(x, 4)$ and $B(3, 5)$ is perpendicular to the horizontal line $y = 5$.
30. The line through $A(-4, 2)$ and $B(x, 7)$ is perpendicular to the horizontal line $y = -3$.

⟨ C ⟩ **Graphing Lines Using the Slope and a Point** In Problems 31–40, graph the line with the indicated slope and passing through the given point.

31. Slope 2; y-intercept $(0, -1)$

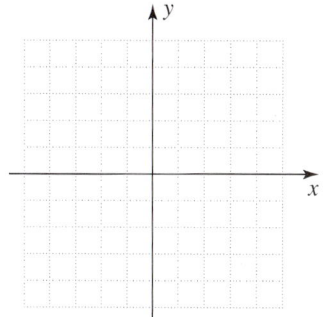

32. Slope $\frac{2}{3}$; y-intercept $(0, 1)$

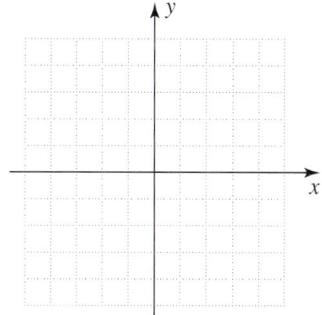

33. Slope $-\frac{2}{3}$ through $(1, 1)$

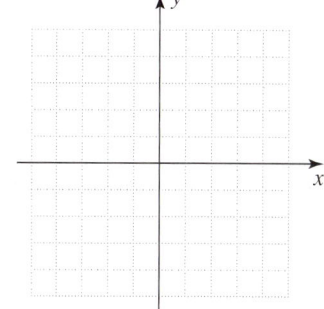

34. Slope $-\frac{3}{4}$ through $(-1, -1)$

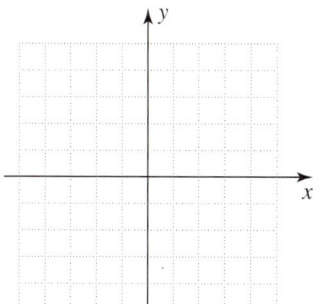

35. Slope -2 through $(2, 3)$

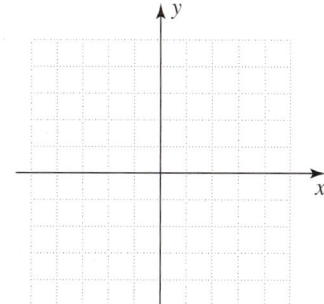

36. Slope 1 through $(3, 2)$

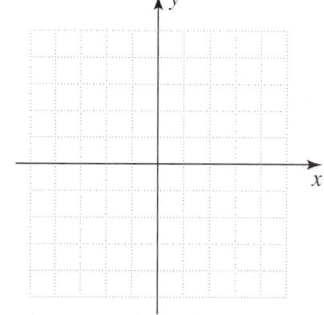

37. Slope 0 through (0, 2)

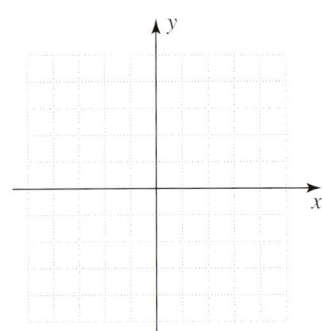

38. Slope undefined through (−1, 2)

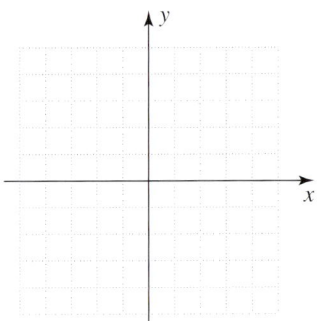

39. Slope 0; y-intercept (0, −3)

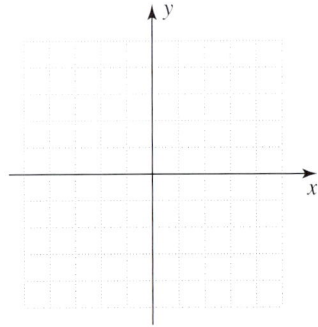

40. Slope −1 through (0, 0)

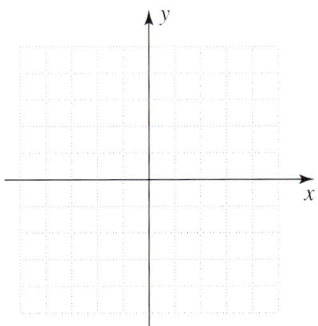

The midpoint of the line segment AB joining the points $A(x_1, y_1)$ and $B(x_2, y_2)$ is the point (x_m, y_m), where

$$x_m = \frac{x_1 + x_2}{2} \quad \text{and} \quad y_m = \frac{y_1 + y_2}{2}$$

In Problems 41–46, find the midpoint of the line segment AB.

41. $A(3, 4)$ and $B(7, 2)$

42. $A(2, 5)$ and $B(-6, 3)$

43. $A(0, -8)$ and $B(-8, 0)$

44. $A(-3, -5)$ and $B(1, 3)$

45. $A(-5, -2)$ and $B(-6, 2)$

46. $A(0, -4)$ and $B(-5, -2)$

‹ D › Slope-Intercept Form for a Linear Equation   In Problems 47–52, find the slope and y-intercept for the given linear equation.

47. $y = x - 4$

48. $y = -x + 2$

49. $-3y = x + 6$

50. $-2y = -x - 8$

51. $4x + 2y = 8$

52. $15x + 3y = 9$

››› Applications

53. *Flying* A pilot is flying from New Orleans, with coordinates (90, 30), to Philadelphia, with coordinates (76, 40). What are the coordinates of the point exactly halfway between the two cities?

54. *Flying* A pilot is flying from Buenos Aires, with coordinates (58, 36), to Sao Paulo, with coordinates (46, 24). What are the coordinates of the point exactly halfway between the two cities?

Triangles In Problems 55–60, determine whether the three given points form the vertices (corners) of a right triangle. (*Hint*: A triangle is a right triangle if two of the sides are perpendicular. What do you know about the slopes of two perpendicular lines?)

55. $A(2, 2), B(0, 5), C(-20, 12)$

56. $A(0, 6), B(-3, 0), C(9, -6)$

57. $A(2, 2), B(0, 5), C(-19, -12)$

58. $A(0, 0), B(6, 0), C(3, 3)$

59. $A(2, 2), B(-4, -14), C(-20, -8)$

60. $A(3, 2), B(0, -4), C(12, -10)$

61. *U.S. Population* The U.S. population has been growing according to the equation

$$y = 2.2x + 180$$

where y is the approximate population (in millions) and x is the number of years after 1960. For example, in 1970, $x = 10$ and $y = 2.2 \cdot 10 + 180 = 202$. This means that the U.S. population was about 202 million in 1970. Use the equation $y = 2.2x + 180$ to find the approximate population in each of the following years:

a. 1990
b. 2000
c. 2010
d. In what year will the U.S. population reach approximately 312 million?
e. Graph the equation $y = 2.2x + 180$.

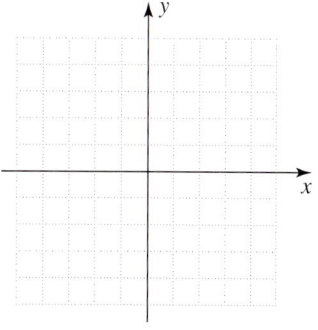
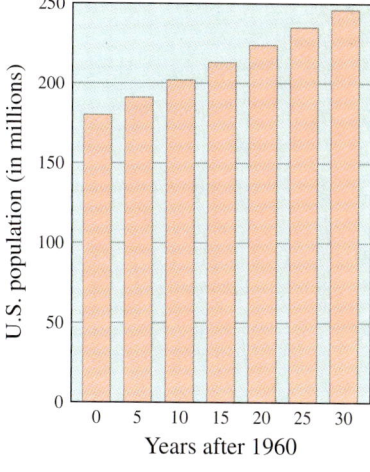

62. *Waste* The municipal waste y (in millions of tons) generated in the United States is approximately $y = 3.4x + 88$, where x is the number of years after 1960. Use the equation $y = 3.4x + 88$ to find the approximate amount of waste generated in each of the following years:

a. 1990
b. 2000
c. 2010
d. In what year did the amount of waste generated reach approximately 241 million tons?
e. Graph the equation $y = 3.4x + 88$.

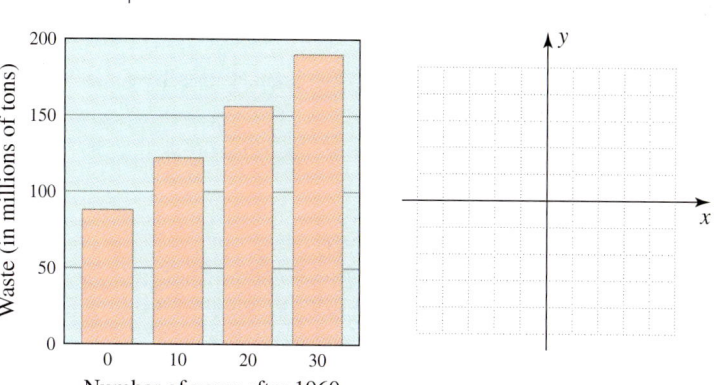

63. *Shipping costs* For shipping packages, The Speedy Delivery Company charges a basic fee of $4.00 plus $0.40 per pound. The total cost for shipping a package can be found by the equation, $C = \$4.00 + \$0.40p$, where p represents the weight of the package in pounds. Use the equation to find the cost of shipping a package that weighs the specified amount.

a. 5
b. 15.5
c. If the cost of the shipping was $19.40, how much did the package weigh?
d. Graph the equation, $C = 4 + 0.4p$. Use "p" for the x-axis and "C" for the y-axis.

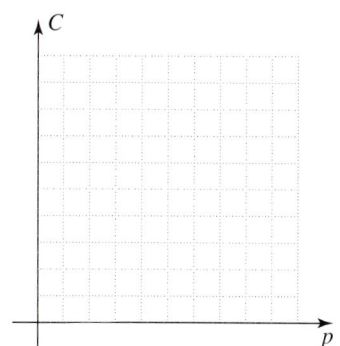

64. *Cell phone service* The Small Talk cell phone service company charges their customers $0.35 per minute for their phone usage with no other fees. The equation that would find the cost of m minutes of "small talk" is given by $C = 0.35m$. Find the cost for the following minutes of phone usage.

a. 122 minutes
b. 458 minutes
c. If the cost was $89.25, how many minutes were used?
d. Graph the equation, $C = 0.35m$.

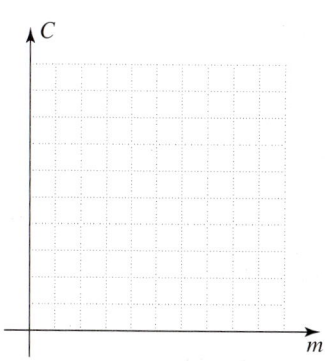

>>> Using Your Knowledge

Distances and Slopes   The following problems require using your knowledge about a topic previously discussed.

65. A line has *x*-intercept 2 and *y*-intercept -3. What is the slope of the line?

66. A line has *x*-intercept -3 and *y*-intercept 2. What is the slope of the line?

67. Find the slope of a line parallel to the line through the points $(3, 1.3)$ and $(2, 3.6)$.

68. Find the slope of a line parallel to the line through the points $(5, 1.8)$ and $(6, \frac{-3}{4})$.

69. Find the slope of a line perpendicular to a second line passing through the points $(3, 0.\overline{6})$ and $(2, 0.\overline{3})$.

70. Find the slope of a line perpendicular to a second line passing through the points $(5, 2\pi)$ and $(4, -\pi)$.

71. The line through the points $(1, c)$ and $(2, 3c)$ is parallel to another line whose slope is 2. What are the possible values for c?

72. The line through the points $(6, 2c)$ and $(3, \frac{1}{2}c)$ is perpendicular to another line whose slope is 2. What is the value of c?

>>> Write On

73. Explain in your own words why the slope of a horizontal line is zero.

74. Explain in your own words why the slope of a vertical line is undefined.

75. Explain in your own words why two parallel lines have the same slope.

76. What does it mean graphically if the slope of a line is positive?

77. What does it mean graphically if the slope of a line is negative?

>>> Concept Checker

Fill in the blank(s) with the correct word(s), phrase, or mathematical statement.

78. If the slope of two lines are opposite in signs and reciprocals, then the lines are _____.

79. The slope of a line represents the _____.

80. The slope of a vertical line is _____.

81. When looking at the graph of a line if it falls from left to right, then its slope is _____.

82. The numerator of the ratio for finding the slope of a line given two points is the _____.

rate of change   negative
difference in *y*-values   zero
difference in *x*-values   undefined
slope   parallel
slope-intercept form   perpendicular
positive

>>> Mastery Test

Graph:

83. The line through the point $(2, 1)$ with slope $-\frac{1}{3}$.

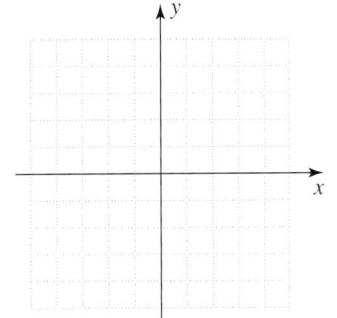

84. The line through the point $(-2, 1)$ with slope $\frac{1}{4}$.

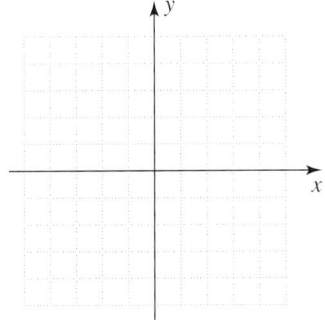

85. The line through $A(5, y)$ and $B(-1, -4)$ is perpendicular to a line whose slope is $-\frac{3}{4}$. Find y.

86. The line through $A(x, 2)$ and $B(3, 4)$ is parallel to a line whose slope is 1. Find x.

87. Line L_1 has slope $\frac{3}{2}$. Find whether the line passing

  a. through $A(6, 3)$ and $B(3, 5)$ is parallel or perpendicular to L_1.

  b. through $A(-6, -8)$ and $B(-4, -5)$ is parallel or perpendicular to L_1.

88. Find the slope of the line passing through the points.

  a. $A(-4, 2)$ and $B(-2, -3)$
  b. $A(-2, 4)$ and $B(-2, 7)$
  c. $A(2, 1)$ and $B(4, 5)$
  d. $A(2, 3)$ and $B(1, 3)$

⟩⟩⟩ Skill Checker

In Problems 89–94, solve for y.

89. $y - 4 = 3(x + 2)$

90. $y + 6 = -1(x - 5)$

91. $y + 6 = \frac{2}{3}(x + 3)$

92. $y - (-5) = -\frac{1}{2}(x + 3)$

93. $y - (-7) = 2[x - (-1)]$

94. $y + 9 = -3[x - (-8)]$

In Problems 95–100, rewrite the equations in standard form ($Ax + By = C$, where A, B, and C are integers and A is positive).

95. $y = 2x + 1$

96. $-x + 3y = -12$

97. $\frac{1}{3}x + y = \frac{5}{3}$

98. $\frac{y}{2} - \frac{7}{2}x = 6$

99. $y + 2 = \frac{3}{4}(x - 5)$

100. $y - (-8) = -\frac{2}{5}[x - (-1)]$

3.3 Equations of Lines

▶ Objectives

Find the equation and the graph of a line given

A⟩ Two points.

B⟩ One point and the slope.

C⟩ The slope and the y-intercept.

D⟩ One point and the fact that the line is parallel or perpendicular to a given line.

E⟩ The slope is that of a horizontal or vertical line.

▶ To Succeed, Review How To . . .

1. Graph lines when two points are given (pp. 168–174).
2. Write a linear equation in standard form (p. 169).
3. Solve an equation for a specified variable (pp. 87–89).

▶ Getting Started
How's the Profit?

Suppose a company wants to examine its profit over a six-year period beginning in 2000. They might start with a table listing the yearly profits.

Year	2000	2002	2004	2006
Profit in Millions	$4.2	$5.6	$7.0	$8.4

After comparing the profits, you see that for each 2-year period, profit has increased by 1.4 million dollars. So the annual rate of change is 0.7 million dollars, which can be found using the ratio

$$\frac{\text{change in profit}}{\text{change in time}} = \frac{5.6 - 4.2}{2} = 0.7$$

Here's another way to create the table.
Suppose we let $t = 0$ in 2000. We are now creating a new time line. It is *relative* to the first year in our data. Our table looks like this:

Now we can describe the data set with the ordered pairs (0, 4.2), (2, 5.6), and (4, 7.0). Let's use the *x-y* coordinate system to graph this data, where the *x*-axis will represent the time in years after 2000 and the *y*-axis will represent the profit in millions of dollars. Here the annual rate of change is the *slope* of the linear relationship between profit and time. In Section 3.2, we learned that when an equation is written as $y = mx + b$, the *m* is the slope of the line and the *b* is the *y*-intercept. The *y*-intercept is (0, 4.2) and the slope is the annual rate of change, 0.7, so we can use the *slope-intercept* form, $y = mx + b$, to write the equation of this profit line. Thus, the equation of the profit line is $y = 0.7x + 4.2$ where *y* is the profit in millions of dollars and *x* is the time in years after 2000.

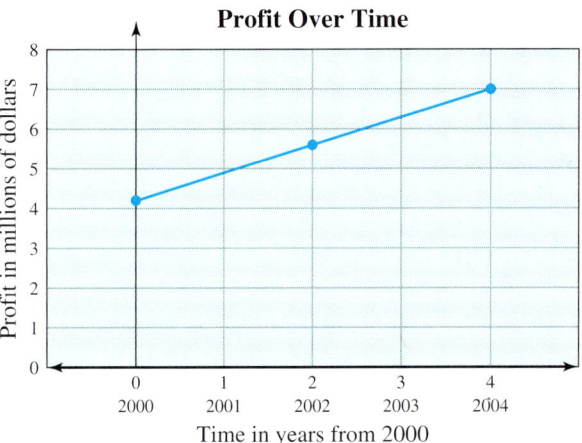

Years after 2000	0	2	4
Profit in Millions	$4.2	$5.6	$7.0

Assuming the profit continues to increase at the same rate, we can use this equation to predict our future profits. For example, let's calculate the year 2010 profit. The year 2010 corresponds to $x = 10$. This gives us $y = 0.7(10) + 4.2 = 11.2$ million dollars. There are no guarantees in the business world, but having information like this profit equation could help a business when planning for their future. As we study more about writing equations of lines in this section, remember it could mean a profit for your future.

A › Finding Equations Given Two Points

We are now ready to work with the graph of a line to find its equation. In Section 3.1, we defined the standard form of an equation as $Ax + By = C$, where *A*, *B*, and *C* are integers and *A* is nonnegative. We worked from an equation of a line to draw its graph. In general, if a line goes through two points $P_1(x_1, y_1)$ and $P_2(x_2, y_2)$, as shown in Figure 3.32, an equation for this line can be found as follows:

1. Select a general point $P(x, y)$ on the line.
2. The slope of the line P_1P_2 is

$$m = \frac{y_2 - y_1}{x_2 - x_1} \quad (x_2 \neq x_1)$$

3. The slope of the line P_1P is

$$m = \frac{y - y_1}{x - x_1} \quad (x \neq x_1)$$

4. Since the slopes are equal,

$$\frac{y - y_1}{x - x_1} = \frac{y_2 - y_1}{x_2 - x_1}$$

or $\quad y - y_1 = \frac{y_2 - y_1}{x_2 - x_1} \cdot (x - x_1) \quad$ Multiply both sides by $(x - x_1)$.

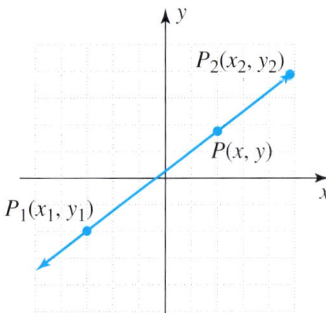

›Figure 3.32

3.3 Equations of Lines

NOTE
Since
$$\frac{y_2 - y_1}{x_2 - x_1} = m$$
it's easier to remember this equation as $y - y_1 = m(x - x_1)$.

We summarize this discussion as follows.

THE POINT-SLOPE FORM OF A LINE

An equation of a line passing through the points (x_1, y_1) and (x_2, y_2) is given by
$$y - y_1 = m(x - x_1)$$
where $m = \dfrac{y_2 - y_1}{x_2 - x_1}$ and $(x_2 \neq x_1)$.

EXAMPLE 1 Finding an equation given two points

Find, write in standard form, and graph an equation of the line passing through the points $(5, 2)$ and $(6, 4)$.

SOLUTION To use the point-slope form of the line we must first find the slope m. Letting $(x_1, y_1) = (5, 2)$ and $(x_2, y_2) = (6, 4)$, we substitute into the slope formula.

$$m = \frac{4 - 2}{6 - 5} = \frac{2}{1} = 2$$

Now we use the point $(5, 2)$ and $m = 2$ to substitute into the point-slope form of the line.

$y - y_1 = m(x - x_1)$ Point-slope form
$y - 2 = 2(x - 5)$ Distributive property
$y - 2 = 2x - 10$ Subtract y and add 10
$8 = 2x - y$

In standard form, $Ax + By = C$, the equation of this line is $2x - y = 8$. The graph is shown in Figure 3.33.

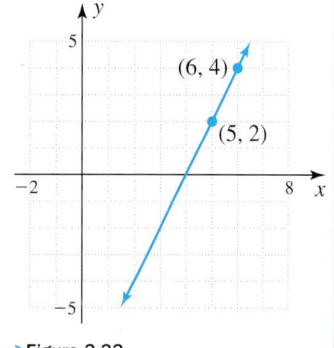
> Figure 3.33

PROBLEM 1
Find, write in standard form, and graph an equation of the line going through the points $(3, 1)$ and $(4, 3)$.

Answers to PROBLEMS
1. $2x - y = 5$

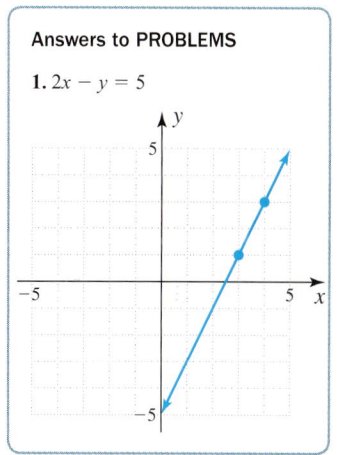

Calculator Corner

Find the Equation Given Two Points

As you recall from Section 3.2 we can use the TI-83 Plus to find the equation of a line given two points. Before we rework Example 1, let's clear any old lists by pressing [2nd] [+] [4] [ENTER]. Also, let's set the window to a standard window by pressing [ZOOM] [6]. Now, press [STAT] [1] and enter 5 and 6 under L_1 and enter 2 and 4 under L_2. Press [STAT] [▶] [4] [ENTER] to get the equation

$$y = ax + b, \text{ with } a = 2 \text{ and } b = -8, \text{ that is, the line}$$
$$y = 2x - 8, \text{ equivalent to } 2x - y = 8.$$

Now graph the two points we entered in the table by using [2nd] [Y=] [ENTER]. Remember to turn on the stat plot by highlighting [ON] and pressing [ENTER]; then choose the type by highlighting the scattergram and pressing [ENTER]. Pressing [GRAPH] will show the two points.

To verify that the line $y = 2x - 8$ does contain these two points, press [Y=] and enter $2x - 8$ on the first line and then press [GRAPH]. Does the line pass through the two points?

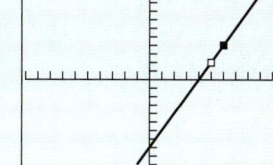

B › Finding Equations Given a Point and the Slope

The point-slope form of a line enables us to find an equation of a line when a point $P(x_1, y_1)$ and the slope m are given. We use this equation in Example 2.

USING THE POINT-SLOPE FORM OF THE LINE

$$y - y_1 = m(x - x_1)$$

EXAMPLE 2 Finding an equation given a point and the slope

Find, write in standard form, and graph an equation of the line with slope $m = -2$ passing through the point (3, 5).

SOLUTION Here $m = -2$, $(x_1, y_1) = (3, 5)$. Substituting into the point-slope form of the line, we get

$y - y_1 = m(x - x_1)$
$y - 5 = -2(x - 3)$
$y - 5 = -2x + 6$ Distributive property
$2x + y = 11$ Add 2x and 5 to both sides.

Standard form of the equation is $2x + y = 11$.
To graph the equation, we start at (3, 5). Since

$$m = -2 = -\frac{2}{1}$$

go 2 units down (the change in y) and 1 unit right (the change in x), ending at (4, 3). The graph is the line through the points (3, 5) and (4, 3), as shown in Figure 3.34.

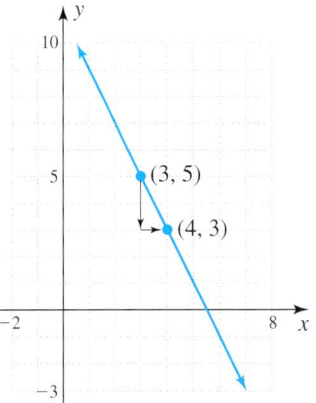
> Figure 3.34

PROBLEM 2

Find, write in standard form, and graph an equation of the line with slope $m = -3$ passing through the point (1, 2).

Answers to PROBLEMS

2. $3x + y = 5$

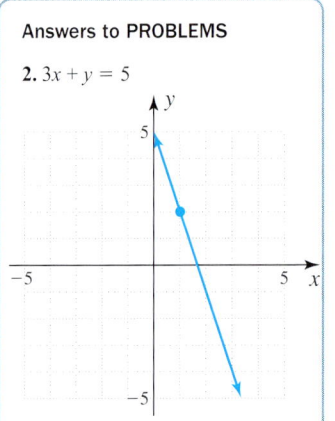

C › Finding Equations Given the Slope and the y-Intercept

If, in $y - y_1 = m(x - x_1)$, the point $P(x_1, y_1)$ is on the y-axis, then $x_1 = 0$. If we let $y_1 = b$, then $P(x_1, y_1) = P(0, b)$. The point-slope form of the equation is $y - b = m(x - 0) = mx$. Solving for y by adding b to both sides, we have the following equivalent equation.

THE SLOPE-INTERCEPT FORM OF A LINE

EXAMPLE 3 Finding an equation given the slope and the y-intercept

Find the slope-intercept form and graph the equation of the line with slope 5 and y-intercept 3.

PROBLEM 3

Find the slope-intercept form and graph the equation of the line with slope 3 and y-intercept 2.

(Answer on page 203)

SOLUTION Using the slope-intercept form of a line, $y = mx + b$, with $m = 5$ and $b = 3$, the equation is $y = 5x + 3$. To graph this line, we start at the y-intercept, $(0, 3)$. The slope is $5 = \frac{5}{1}$, so we go 5 units up and 1 unit right, ending at $(1, 8)$. The graph is the line drawn through the points $(0, 3)$ and $(1, 8)$, as shown in Figure 3.35

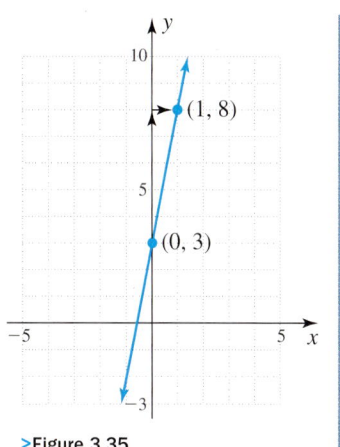

>Figure 3.35

Answers to PROBLEMS

3. $y = 3x + 2$

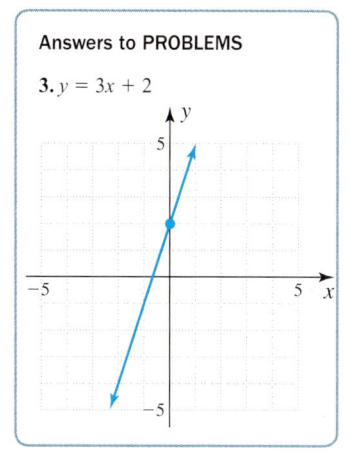

D › Finding the Equation of a Line Through a Given Point and Parallel or Perpendicular to a Given Line

We are already familiar with particular aspects of the slopes of parallel lines and perpendicular lines. We can use the fact that two parallel lines have identical slopes to find an equation of a line parallel to a given line. Likewise, the fact that two perpendicular lines have slopes that are negative reciprocals of each other can be used to find an equation of a line perpendicular to a given line. We illustrate these ideas next.

EXAMPLE 4 Finding an equation of a line through a given point and parallel or perpendicular to a given line

Find an equation of the line passing through the point $(6, 1)$ and

a. Parallel to the line $y - 3x = 1$. Write the final equation in slope-intercept form.
b. Perpendicular to the line $y - 3x = 1$. Write the final equation in slope-intercept form.

SOLUTION

a. We first write the equation $y - 3x = 1$ in the slope-intercept form: $y = 3x + 1$. The slope of this line is 3. If we wish to construct another line parallel to $y = 3x + 1$ passing through $(6, 1)$, we use the point-slope form, with $(x_1, y_1) = (6, 1)$ and $m = 3$, the same slope as that of $y = 3x + 1$. Thus,

$$y - y_1 = m(x - x_1)$$
$$y - 1 = 3(x - 6)$$
$$y - 1 = 3x - 18$$
$$y = 3x - 17$$

b. The line $y - 3x = 1$ has slope 3. A line perpendicular to this line must have a slope of $-\frac{1}{3}$. Using the point-slope form again, we find that an equation of the line perpendicular to $y - 3x = 1$ and passing through $(6, 1)$ is

$$y - y_1 = m(x - x_1)$$
$$y - 1 = -\frac{1}{3}(x - 6)$$

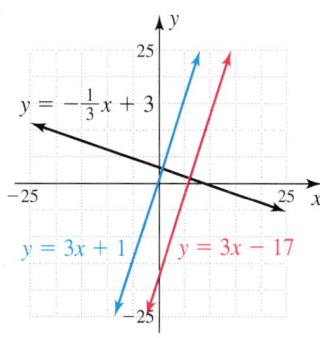

>Figure 3.36

PROBLEM 4

Find an equation of the line passing through the point $(4, 3)$ and

a. Parallel to the line $y + 2x = 7$. Write the final equation in slope-intercept form.
b. Perpendicular to the line $y + 2x = 7$. Write the final equation in slope-intercept form.

(continued)

Answers to PROBLEMS

4. a. $y = -2x + 11$ b. $y = \frac{1}{2}x + 1$

$$y - 1 = -\frac{1}{3}x + 2$$

$$y = -\frac{1}{3}x + 3$$

The graphs of all three lines are shown in Figure 3.36.

E > Finding Equations of Horizontal and Vertical Lines

In Section 3.2 we studied lines that have slope $= 0$, which are horizontal lines with equation $y = C$ (where C is some constant). Lines whose slopes are undefined are vertical lines and have equations of the form $x = C$ (where C is some constant). The next example will use this information to illustrate how to write the equations of horizontal and vertical lines when given a point on the line.

EXAMPLE 5 Finding the equations of horizontal and vertical lines

Given that a line passes through the point $(-4, 2)$, find and graph the equation of the line if:

a. the slope is 0. **b.** the slope is undefined.

SOLUTION

a. Given the slope $= 0$, we know that it must be a horizontal line with equation $y = C$. To find C, we use the fact that the line passes through the point $(-4, 2)$, which indicates $C = 2$ because that is the y-value of the point on the line. Thus, the equation of the line is $y = 2$. The graph is the blue horizontal line shown in Figure 3.37.

b. Given the slope is undefined we know that the line must be vertical with equation $x = C$. To find C, we use the fact that the line passes through the point $(-4, 2)$, which indicates $C = -4$ because that is the x-value of the point on the line. Thus, the equation of the line is $x = -4$. The graph is the red vertical line shown in Figure 3.37.

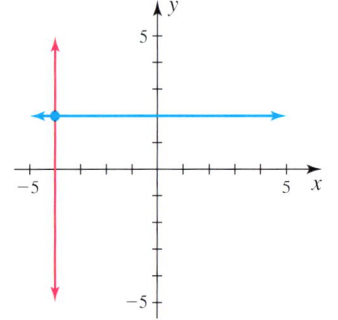

> Figure 3.37

PROBLEM 5

Given that a line passes through the point $(1, -3)$, find and graph the equation of the line if:

a. the slope is 0.

b. the slope is undefined.

Answers to PROBLEMS

5. **a.** $y = -3$ **b.** $x = 1$

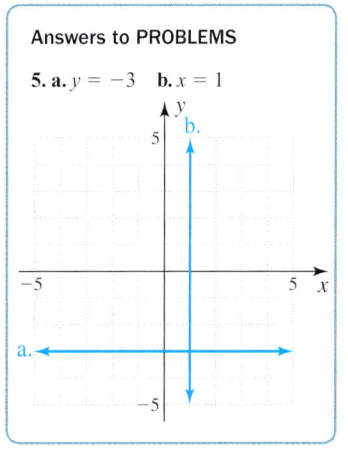

Table 3.3 gives you a summary of most of the formulas dealing with linear equations to which you can refer.

Table 3.3 Linear Equations

If You Want an Equation	Write
In standard form	$Ax + By = C$
For a vertical line	$x = C$ (C a constant)
For a horizontal line	$y = C$ (C a constant)
For a line going through the two points (x_1, y_1) and (x_2, y_2)	$y - y_1 = m(x - x_1)$, where $m = \dfrac{y_2 - y_1}{x_2 - x_1}$
For a line with slope m going through a point (x_1, y_1)	$y - y_1 = m(x - x_1)$
For a line with slope m and y-intercept b	$y = mx + b$

Table 3.3 (continued)

If You Want the Formula	Write
For the slope m of a line passing through two points (x_1, y_1) and (x_2, y_2)	$m = \dfrac{y_2 - y_1}{x_2 - x_1}$
For the slope of a vertical line	Undefined
For the slope of a horizontal line	$m = 0$
For the slope of a line parallel to the line $y = mx + b$	m
For the slope of a line perpendicular to the line $y = mx + b$	$-\dfrac{1}{m}$

Calculator Corner

Be sure to clear all lists and clear any equations in **Y=** before you begin. Given the point (2, 1), we want a line parallel to $y - x = 1$, that is, $y = x + 1$, which has slope $1 = \frac{1}{1}$. This means that the change in x is 1 and the change in y is 1. Starting at the point (2, 1), go 1 unit right and 1 unit up, ending at (3, 2). We now have two points, (2, 1) and (3, 2), so we can follow our previous steps to make the two lists and get an equation of the form $y = ax + b$. For (2, 1) and (3, 2), $a = 1$ and $b = -1$, that is $y = x - 1$. Now graph the line perpendicular to $y - x = 1$ that goes through the point (2, 1).

A warning about your calculator: Look closely at the graph of the two lines. The lines $y = x - 1$ and $y = -x + 3$ are perpendicular. Turn off the Stat Plot, and graph these two lines using your calculator and a standard window (see Window 1). They don't look perpendicular. Why? Because the standard window is not square (the units on the x-axis are larger than the units on the y-axis). To remedy this, change to a square window by pressing **ZOOM** 5 (see Window 2). Your faith in your calculator should now be restored.

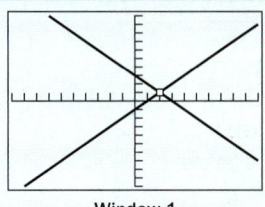
Window 1
$Y_1 = x - 1$
$Y_2 = -x + 3$

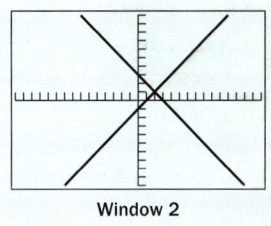
Window 2

Exercises 3.3

Boost your grade at mathzone.com!
> Practice Problems   > Self-Tests
> NetTutor   > e-Professors
                > Videos

< A > Finding Equations Given Two Points In Problems 1–6, find the equation of the line passing through the given points. Write the final equation in standard form.

1. $(1, -1)$ and $(2, 2)$
2. $(-3, -4)$ and $(-2, 0)$
3. $(3, 2)$ and $(2, 3)$
4. $(3, 0)$ and $(0, 5)$
5. The line with x-intercept 2 and y-intercept 4.
6. The line with x-intercept -3 and y-intercept -1.

< B > Finding Equations Given a Point and the Slope In Problems 7–14, find the equation of the line with the given slope and passing through the given point. Write the final equation in standard form.

7. Slope 2, point $(-3, 5)$
8. Slope $\frac{1}{2}$, point $(2, 3)$
9. Slope -3, point $(-1, -2)$
10. Slope $-\frac{1}{3}$, point $(2, -4)$
11. Slope 1, point $(0, 0)$
12. Slope -1, point origin
13. Slope 4, point $(0, 6)$
14. Slope -2, point $(5, 0)$

⟨ C ⟩ Finding Equations Given the Slope and the y-Intercept

In Problems 15–18, find the slope-intercept form of the equation of the line with the given slope and y-intercept. Then graph the line.

15. Slope 5, y-intercept 2
16. Slope $\frac{1}{4}$, y-intercept 2
17. Slope $-\frac{1}{5}$, y-intercept $-\frac{1}{3}$
18. Slope $\frac{2}{3}$, y-intercept $\frac{1}{2}$

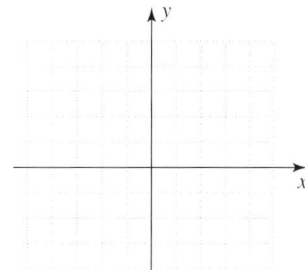
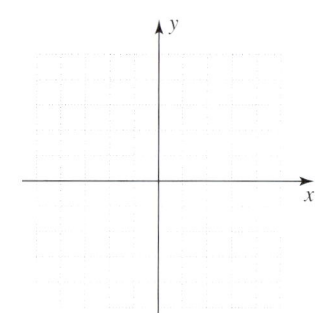
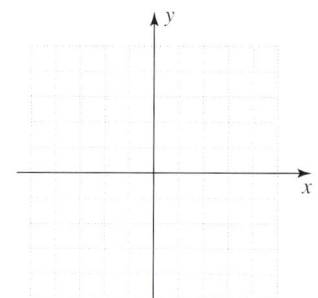
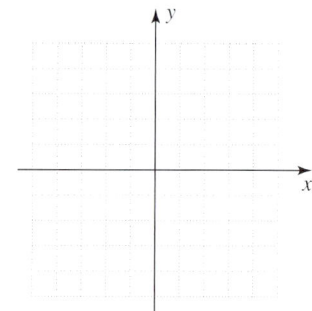

⟨ D ⟩ Finding the Equation of a Line Through a Given Point and Parallel or Perpendicular to a Given Line

In Problems 19–26, find the slope-intercept form of the equation of the line passing:

19. Through $(1, -2)$ and parallel to $y = 2x + 1$
20. Through $(-1, -2)$ and parallel to $2y = -4x + 5$
21. Through $(-5, 3)$ and parallel to $2y + 6x = 8$
22. Through $(-3, -5)$ and parallel to $3y - 6x = 12$
23. Through $(1, 1)$ and perpendicular to $2y = x + 6$
24. Through $(2, 3)$ and perpendicular to $3y = -x + 5$
25. Through $(-2, -4)$ and perpendicular to $y - x = 3$
26. Through $(-3, 5)$ and perpendicular to $2y - x = 5$

⟨ E ⟩ Finding Equations of Horizontal and Vertical Lines

In Problems 27–34, find an equation of the line described.

27. A line passing through the point $(3, 4)$ and with slope 0.
28. A line passing through the point $(-2, -4)$ and with slope 0.
29. The slope of the line is undefined, and it passes through the point $(-2, 4)$.
30. The slope of the line is undefined, and it passes through the point $(-4, -5)$.
31. A vertical line passing through $(-2, 3)$.
32. A vertical line passing through $(-3, -1)$.
33. A horizontal line passing through $(3, 2)$.
34. A horizontal line passing through $(-3, -4)$.

⟩⟩⟩ Applications

35. *Supply and demand* When sunglasses are sold at the regular price of $6, a student purchases two pairs. Sold for $4 at the flea market, the student purchases six pairs. Let the ordered pair (d, p) represent the demand and price for sunglasses.

    a. Find the demand equation. (*Hint:* To find the demand equation, use the ordered pairs (2, $6) and (6, $4).)

    b. How many pairs would the student buy if sunglasses are selling for $2?

36. *Supply and demand* When skateboards are sold for $80, 5 of them are sold each day. When they are on sale for $40, 13 of them are sold daily. Let (d, p) represent the demand and price for skateboards.

    a. Find the demand equation.

    b. For 10 skateboards to be sold on a given day, what should the price be?

37. *Supply and demand* A wholesaler will supply 50 sets of video games at $35. If the price drops to $20, she will provide only 20. Let (s, p) represent the supply and the price for video games.

    a. Find the supply equation. (*Hint:* To find the supply equation, use the ordered pairs (50, $35) and (20, $20).)

    b. At what price would the wholesaler stop selling the video games; in other words, when is the supply zero?

    c. If the price went up to $40 per game, how many games would she be willing to supply?

38. *Supply and demand* When T-shirts are selling for $6, a retailer is willing to supply 6 of them each day. If the price increases to $12, he's willing to supply 12 T-shirts.

    a. Find the supply equation.

    b. If the T-shirts were free, how many would he supply?

39. *Supply and demand* The supply s of a product is given by $s = 3p - 6$, while the demand d is $d = -2p + 14$. What will the price p be when the supply s equals the demand d? (This price is called the *equilibrium price*.)

40. *Supply and demand* The supply and demand for a product are given by $s = 3p - 10$ and $d = -2p + 40$, respectively. At what price p will the supply equal the demand?

41. *E-mail device* The Blackberry e-mail device is becoming quite popular. The number of subscribers in the fiscal year ending 2005 was 2.5 million. The number of subscribers for the fiscal year 2006 was 5 million. Let the ordered pairs (5, 2.5) and (6, 5) represent the year and the number of subscribers in millions, respectively, for the years 2005 and 2006. Graph the points (5, 2.5) and (6, 5), connect them with a line, and write the equation of the line using those two points.
Source: http://yahoo.usatoday.com.

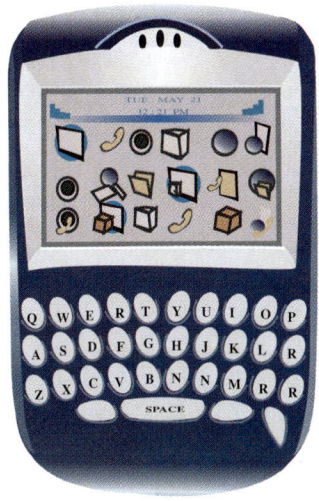

42. *Hourly wages* According to the U.S. Department of Labor, statistics regarding the average hourly earnings of production workers show a steady increase over the years 1997 through 2002 with $12.27 as the average for January 1997 and $14.77 as the average for January 2002. Let the ordered pairs (1, 12.27) and (6, 14.77) represent the year and the average hourly earnings of production workers, respectively, for the years 1997 and 2006. Graph the points (1, 12.27) and (6, 14.77), connect them with a line, and write the equation of the line using those two points.
Source: http://data.bls.gov.

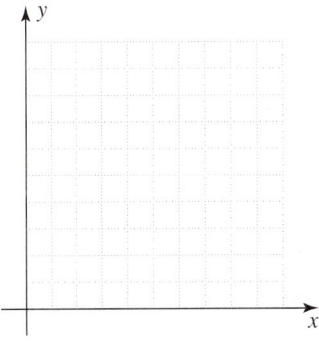

43. *Target heart rate* Often people will take their pulse rate during or right after exercise, to check their heart rate. This helps determine whether they are exercising at a healthy pace. Your heart rate (and pulse) during and after exercise will be higher than your resting heart rate. Your level of fitness can be determined by knowing your heart rate during peak exercise, and how quickly your heart rate increases as you exercise. The table indicates a person's age in years associated with a *target* heart beat per minute (BPM).

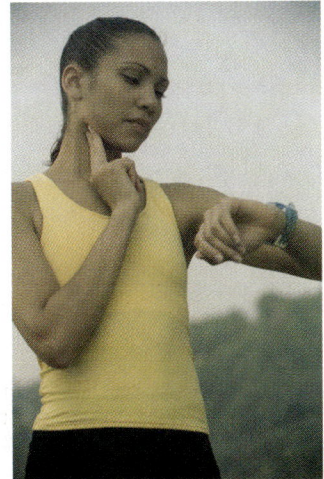

Age in Years (x)	Target Heart Beat per Minute for Exercising (y)
15	143
20	140
25	137
30	134

a. Using the data in the table, write the equation of the line that would give the target heart beat per minute (y) given a person's age (x).

b. Using the equation found in part **a**, predict the target heart beat for a person doing exercise who is 33 years old. (Round answer to nearest whole number.)

Source: http://www.americanheart.org.

44. Real estate The median sale price of single-family homes in Estes Park, Colorado, had a steady rise from 2001 to 2005. If the median sale price of homes in that area continues at that rate, could you predict what a home would cost in 2012? Use the information in the table to answer the questions.

Year (x)	Median Sale Price of Single-Family Homes in Estes Park (y)
(2001) 1	$257,500
(2002) 2	$265,700
(2003) 3	$273,900
(2004) 4	$282,100
(2005) 5	$290,300

a. Using the data in the table, write the equation of the line that would give the median sale price of a single-family home (y) given the number of years after 2000 (x).

b. Using the equation found in part **a**, predict what a home would cost in 2012. (*Hint:* Remember x is the number of years after 2000.)

Source: http://www.co.larimer.co.us.

❯❯❯ Using Your Knowledge

Business by Candle Lights In economics and business, the slope, m, and the y-intercept, b, of an equation play an important role. Let's see how.

Suppose you wish to go into the business of manufacturing fancy candles. First, you have to buy some supplies such as wax, paint, and so on. Assume these supplies cost you $100. This is the *fixed cost*. Now suppose it costs $2 to manufacture each candle. This is the *marginal cost*. What would be the total cost y if the marginal cost is $2, x units are produced, and the fixed cost is $100? The answer is

$$y = \underbrace{2x}_{\text{Cost for } x \text{ units}} + \underbrace{100}_{\text{Fixed cost}}$$
(Total cost)

In general, an equation of the form

$$y = mx + b$$

gives the total cost y of producing x units, where m is the marginal cost of producing 1 unit and b is the fixed cost.

45. Find the total cost y of producing x units of a product costing $2 per unit if the fixed cost is $50.

46. Find the total cost y of producing x units of a product whose production cost is $7 per unit if the fixed cost is $300.

47. The total cost y of producing x units of a certain product is given by

$$y = 2x + 75$$

a. What is the production cost for each unit?

b. What is the fixed cost?

❯❯❯ Write On

48. How do you decide what formula to use when you're asked to find an equation of a line?

In Problems 49–51, write the procedure you use to draw the graph of a line:

49. When a point and the slope are given.

50. When the y-intercept and the slope are given.

51. When the x- and y-intercepts are given.

❯❯❯ Concept Checker

Fill in the blank(s) with the correct word(s), phrase, or mathematical statement.

52. What is the form of the equation, $2x - y = 8$? _____.

53. In the equation, $y = \frac{1}{2}x + \frac{1}{4}$, the number $\frac{1}{4}$ represents the _____ of the line.

54. $y = \frac{1}{3}$ is the equation of a _____ line.

point-slope form   slope

~~slope-intercept~~   ~~y-intercept~~
~~form~~

standard form   horizontal

vertical

55. What is the form of the equation $3 - x = y$? _____.

56. What form would be used to write the equation of a line given the slope is 4 and the line goes through the point (2, 5)? _____.

>>> Mastery Test

57. Find an equation of the line passing through the point (1, 1) and:
  a. Parallel to $2y - 6x = 5$.
  b. Perpendicular to $2y - 6x = 5$.

58. Find the slope and the y-intercept of the line $8x + 4y = 16$.

59. Find the slope m and the y-intercept b of the line $y = 3$.

60. A line has slope 3 and y-intercept 2. Find the slope-intercept form of the equation of the line and then graph it.

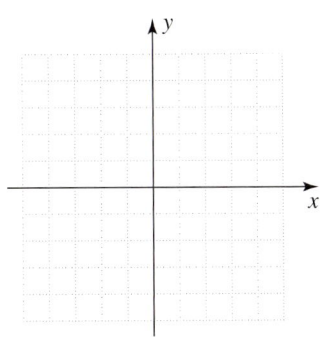

61. The slope of a line is undefined and it passes through the point $(-2, 0)$. What is the equation of this line?

62. Find an equation of the line with slope -3 and passing through the point (1, 2) and then graph it.

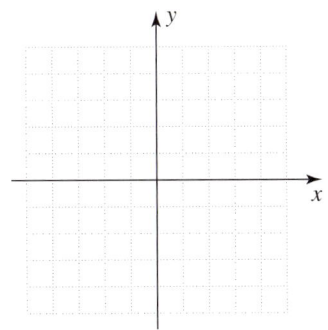

63. Find an equation of the line passing through points (3, 1) and (4, 3), write it in standard form, and then graph it.

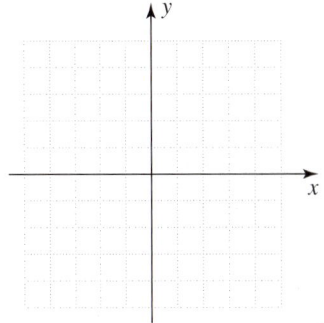

64. Find an equation of the line passing through points (3, 2) and (4, 2) and then graph it.

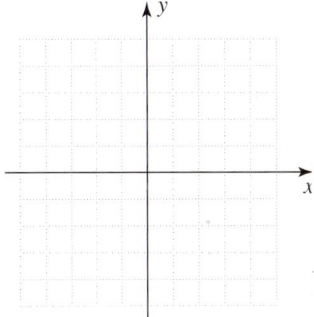

>>> Skill Checker

In Problems 65–68, find the x- and y-intercepts.

65. $y = 3x + 2$ **66.** $-x + 4y = 12$ **67.** $y = -3$ **68.** $7 = x$

In Problems 69–70, complete the ordered pairs so they are solutions to the given equations.

69. $y = -4x + 1; (\_, -3), (0, \_)$

70. $3x - 2y = -18; (\_, 0), (-4, \_)$

In Problems 71–74, solve the absolute-value inequality statements.

71. $|x| < 3$ **72.** $|x| \geq 1$ **73.** $|y| \geq 5$ **74.** $|y| < 4$

3.4 Linear Inequalities in Two Variables

Objectives

A Graph linear inequalities.

B Graph inequalities involving absolute values.

C Solve applications involving linear inequalities.

To Succeed, Review How To . . .

1. Find the x- and y-intercepts of a line (pp. 171–179).
2. Solve linear equations (pp. 76–83).
3. Graph lines (pp. 168–174).
4. Solve inequalities involving absolute values (pp. 145–149).

Getting Started
Renting Cars

Suppose you want to rent a car for a few days. Here are some prices for an intermediate car rental.

    Rental A: $36 per day, $0.15 per mile
    Rental B: $49 per day, $0.33 per mile

The total cost C for the Rental A car is

$$C = \underbrace{36d}_{\text{Cost for } d \text{ days}} + \underbrace{0.15m}_{\text{Cost for } m \text{ miles}}$$

Now suppose you want the cost C to be $180. Then $180 = 36d + 0.15m$. We graph this equation by finding the intercepts. When $d = 0$, $180 = 0.15m$, or

$$m = \frac{180}{0.15} = 1200$$

When $m = 0$, $180 = 36d$, or

$$d = \frac{180}{36} = 5$$

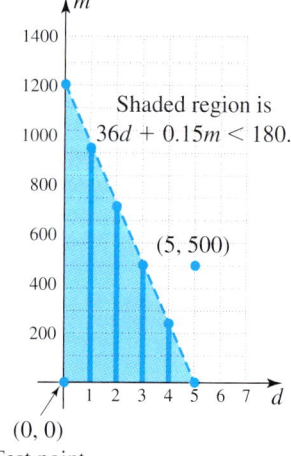

Shaded region is $36d + 0.15m < 180$.

(5, 500)

(0, 0) Test point

We join (0, 1200) and (5, 0) with a line and then graph the discrete points corresponding to 1, 2, 3, or 4 days.

But what if you want the cost to be less than $180? We then have

$$36d + 0.15m < 180$$

We have graphed the points on the line $36d + 0.15m = 180$. Where are the points for which $36d + 0.15m < 180$? As the graph shows, the line $36d + 0.15m = 180$ divides the plane into three parts:

1. The points *below* the line
2. The points *on* the line
3. The points *above* the line

It can be shown that if any point on one side of the line $Ax + By = C$ satisfies the inequality $Ax + By < C$, then all points on that side satisfy the inequality and no point on the other side of the line does. Let's select $(0, 0)$ as a test point. Since

$$36 \cdot 0 + 0.15 \cdot 0 < 180$$
$$0 < 180$$

is true, all points below the line (shown shaded) satisfy the inequality. [As a check, note that $(5, 500)$, above the line, doesn't satisfy the inequality, because $36 \cdot 5 + 0.15 \cdot 500 < 180$ is a false statement.] The line $36d + 0.15m = 180$ is *not* part of the graph and is shown dashed. To apply this result to our rental-car problem, d must be an integer. The solution to our problem consists of the points on the heavy-line segments at $d = 1$, 2, 3, and 4. The graph shows, for instance, that you can rent a car for 2 days and go about 700 miles at a cost less than $180. What we have just set up is called a *linear inequality*, which we shall now examine in more detail.

A › Graphing Linear Inequalities

LINEAR INEQUALITY IN TWO VARIABLES

A **linear inequality** is a statement that can be written in the form
$$Ax + By \leq C, \quad Ax + By \geq C, \quad Ax + By < C, \quad \text{or } Ax + By > C$$
where *A* and *B* not both zero.

The graph of a linear inequality should reflect all the solutions to the statement. This will require a **boundary line** that separates the plane so that the solutions, which will either be above or below that line, can be shaded. If the boundary line is part of the solution set then it will be a solid line and if it is not, then it will be a dashed line. In the *Getting Started*, we graphed a linear inequality and the following is a summary of the procedure we used.

PROCEDURE
Graphing a Linear Inequality
1. **Graph the line.** Graph the line associated with the inequality. This line is called the boundary line. If the inequality involves $\leq$ or $\geq$, draw a **solid line**; this means the line is included in the solution. If the inequality involves $<$ or $>$, draw the **line dashed**, which means the line is not part of the solution.
2. **Test a point.** Choose a test point not on the line. [$(0, 0)$ if possible.]
3. **Shade the region.** If the test point satisfies the inequality, shade the region containing the test point; otherwise, shade the region on the other side of the line. The shaded region represents all the solutions to the inequality.

EXAMPLE 1 Graphing a linear inequality
Graph all the solutions to $x - 2y < -4$.

SOLUTION We follow the three-step procedure.
1. We first graph the boundary line $x - 2y = -4$ by finding the intercepts.

x	y	
0	2	When $x = 0$, $-2y = -4$ and $y = 2$.
−4	0	When $y = 0$, $x - 2 \cdot 0 = -4$ and $x = -4$.

The boundary line is shown dashed to indicate that it isn't part of the solution (Figure 3.38).

PROBLEM 1
Graph all the solutions to
$$3x - 2y < -6.$$

(Answer on page 212)

(continued)

2. We select an easy test point and see if it satisfies the inequality. If it does, the solution lies on the same side of the boundary line as the test point; otherwise, the solution is on the other side of the line. An easy point is (0, 0), *below* the line. If we substitute $x = 0$ and $y = 0$ in the inequality $x - 2y < -4$, we obtain

$$0 - 2 \cdot 0 < -4$$
$$0 < -4$$

which is false.

3. The point (0, 0) is not part of the solution. Because of this, the solution consists of the points *above* (on the other side of) the boundary line $x - 2y = -4$ and is shown shaded in Figure 3.38.

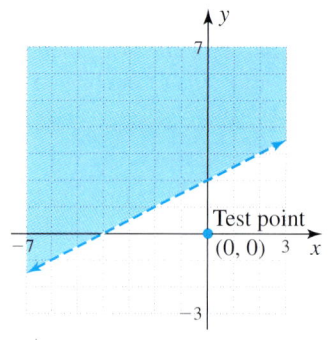

>Figure 3.38

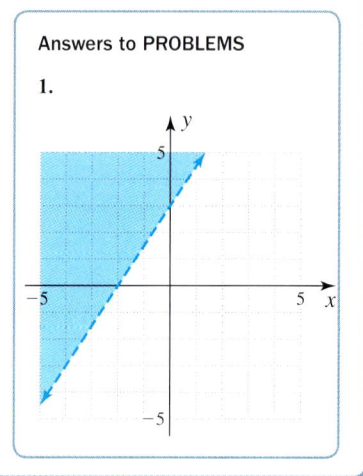

Answers to PROBLEMS

1.

In Example 1, the test point was not part of the solution for the given inequality. Now we give an example in which the test point is part of the solution for the inequality.

EXAMPLE 2 Graphing an inequality where the test point is part of the solution

Graph all the solutions to $y \leq -2x + 4$.

SOLUTION We use our three-step procedure.

1. As usual, we first graph the boundary line $y = -2x + 4$.

x	y	
0	4	When $x = 0$, $y = 4$.
2	0	When $y = 0$, $0 = -2x + 4$, or $x = 2$.

The graph of the boundary line is shown in Figure 3.39.

2. Now we select the point (0, 0) as a test point. When $x = 0$ and $y = 0$, we have

$$y \leq -2x + 4$$
$$0 \leq -2 \cdot 0 + 4$$
$$0 \leq 4$$

which is true.

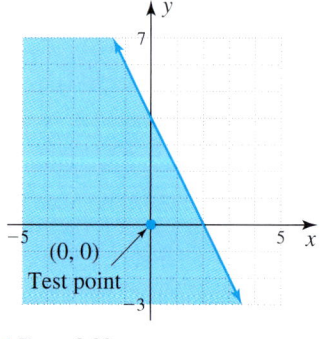

>Figure 3.39

3. Thus all the points on the same side of the boundary line as (0, 0)—that is, the points *below* the line—are solutions of $y \leq -2x + 4$. These solutions are shown shaded in Figure 3.39. This time, the line is *solid* because it's part of the solution.

PROBLEM 2

Graph all the solutions to $y \geq -x - 3$.

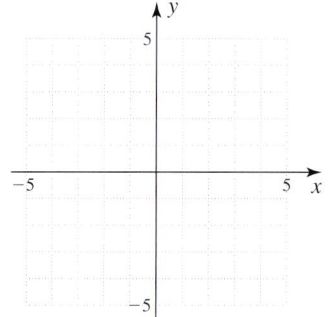

Answers to PROBLEMS

2.

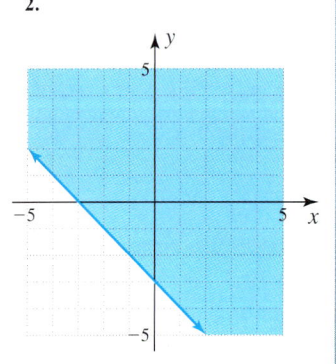

NOTE

If you solve an inequality for y obtaining:

(1) $y > ax + b$, the solution set consists of all the points *above* $y = ax + b$

(2) $y < ax + b$, the solution set consists of all the points *below* $y = ax + b$

Using the information from the note on the previous page, rework Example 1. First, solve for y so that it is on the left of the inequality sign.

$$x - 2y < -4$$
$$-2y < -x - 4 \quad \text{Subtract } x.$$
$$y > \tfrac{1}{2}x + 2 \quad \text{Divide by } -2 \text{ (reverse the inequality sign when dividing by a negative number).}$$

1. This time to graph the boundary line we use the y-intercept, 2, and the slope, $\tfrac{1}{2}$. We will get the same line as shown in Figure 3.38.
2. We don't have to use a test point when the inequality is in this form. Instead, we know the solutions are above the boundary line because the inequality sign is "greater than."
3. Thus, we shade above the boundary line.

We will use this method to graph the linear inequality in Example 3.

EXAMPLE 3   Graphing a linear inequality without using a test point

Graph all the solutions to $3x + 2y < 4$.

SOLUTION   First, solve for y so that y is on the left of the inequality sign.

$$3x + 2y < 4$$
$$2y < -3x + 4 \quad \text{Subtract } 3x.$$
$$y < -\tfrac{3}{2}x + 2 \quad \text{Divide by } 2.$$

1. Graph the boundary line by plotting the y-intercept, 2, and then the slope $-\tfrac{3}{2}$. The line will be dashed since the inequality sign does not include the equal sign. See Figure 3.40 for the graph of the boundary line.
2. We know the solutions are below the boundary line because the inequality sign is "less than."
3. So, we shade below the line. See Figure 3.41 for the graph of all the solutions to $3x + 2y < 4$.

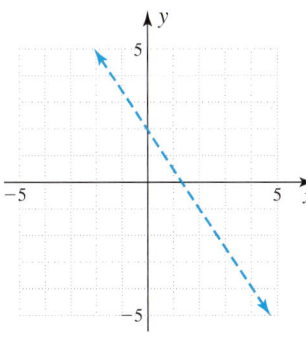

>Figure 3.40

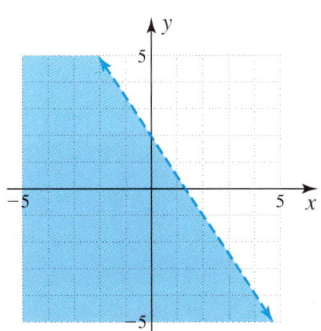

>Figure 3.41

PROBLEM 3
Graph all the solutions to $-x + 4y > -12$.

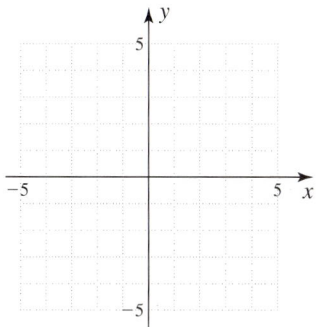

Answers to PROBLEMS

3.

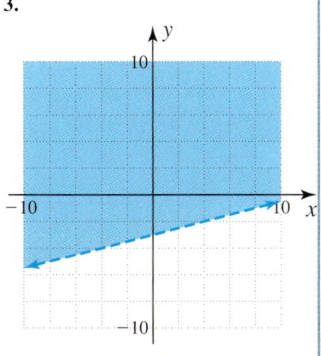

There is another case to consider where graphing a linear inequality can be done without using a test point: when the boundary line is either horizontal or vertical. If it is a vertical boundary line and the inequality sign is *greater than*, $>$, then shade to the *right* of the line. If the sign is *less than*, $<$, then shade to the *left*. Similarly, when the boundary line is horizontal and the inequality sign is *greater than*, $>$, then shade *above* the line. If the sign is *less than*, $<$, then shade *below* the line.

EXAMPLE 4 Graphing a linear inequality without using a test point

Graph all the solutions to $x \geq -1$.

SOLUTION We first graph the vertical boundary line $x = -1$. This time, we don't even need a test point! All points to the *right* of this line have x-coordinates greater than -1 (points to the left have x-coordinates less than -1). The graph of all the solutions, which includes the boundary line $x = -1$, is shown in Figure 3.42.

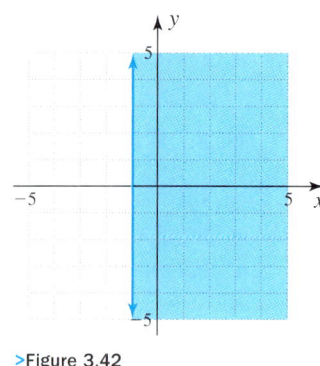

>Figure 3.42

PROBLEM 4

Graph all the solutions to $y < 2$.

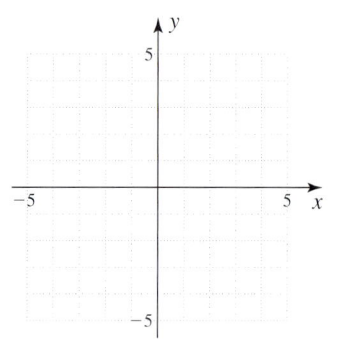

B > Graphing Absolute-Value Inequalities

As you recall from Section 2.6,

$$|x| \leq a \text{ is equivalent to } -a \leq x \leq a$$

If we graph the solutions to $|x| \leq 1$, we must graph all the points satisfying the inequality $-1 \leq x \leq 1$ (that is, the points between -1 and 1) as well as the boundary lines $x = -1$ and $x = 1$. These points are shown in Figure 3.43.

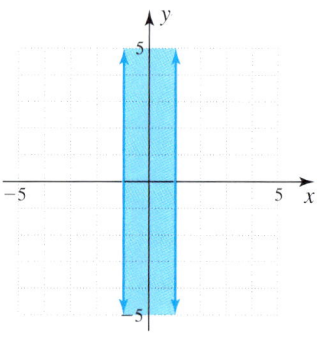

>Figure 3.43

EXAMPLE 5 Graphing an absolute-value inequality

Graph all the solutions to $|y| \leq 2$.

SOLUTION $|y| \leq 2$ is equivalent to $-2 \leq y \leq 2$, so the graph of all the solutions, shown in Figure 3.44, consists of all points bounded by the horizontal lines $y = -2$ and $y = 2$, as well as these two boundary lines; these lines are therefore shown as solid lines.

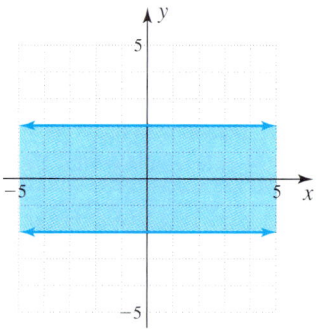

>Figure 3.44

PROBLEM 5

Graph all the solutions to $|x - 3| \leq 4$.

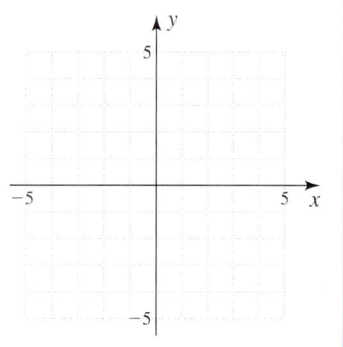

Answers to PROBLEMS

4.
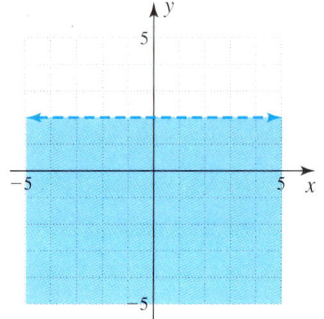

5.
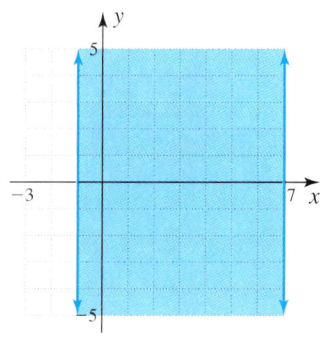

EXAMPLE 6 Graphing an absolute-value inequality

Graph all the solutions to $|x + 1| > 2$.

SOLUTION The inequality $|x + 1| > 2$ is equivalent to

$$x + 1 > 2 \quad \text{or} \quad x + 1 < -2$$
$$x > 1 \quad \text{or} \quad x < -3$$

The graph of all the solutions to $|x + 1| > 2$ consists of all points to the *right* of the vertical boundary line $x = 1$ and all points to the *left* of the boundary line $x = -3$. The boundary lines $x = 1$ and $x = -3$ are *not* part of the graph. They are therefore shown as dashed lines in Figure 3.45.

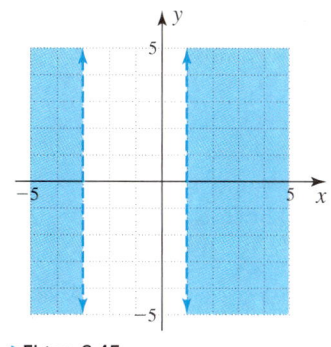

> Figure 3.45

PROBLEM 6

Graph all the solutions to $|y| \geq 1$.

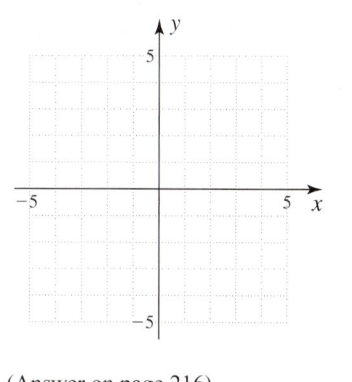

(Answer on page 216)

C > Applications Involving Linear Inequalities

Knowing how to interpret the graph of a linear inequality can be useful when answering questions in an application. Remember if a point lies in the shaded region, then it is a solution to the inequality statement. That is, the ordered pair naming that point will make the inequality statement true.

In the *Getting Started*, we saw how the graph of the linear inequality, $36d + 0.15m < 180$, could help give a visual interpretation of how many days and miles we could rent the car and have less than \$180 in cost. The shaded region represents the ordered pairs, (miles, days) that we can rent the car and have less than \$180 in cost. If we rented the car for 2 days and drove 400 miles, would we have less than \$180 in car rental costs? All we have to do is locate the point on the graph and see if it lies in the shaded region. If it does, then we know it must be less than \$180 in cost. Point A in the figure represents the graph of (2, 400). Since it is in the shaded region we know we have spent less than the \$180 cost.

What if we rented the car for 4 days and drove 800 miles, would we have less than \$180 in rental costs? Again, locate the point and you will know the answer. This time the graph of (4, 800) represented by the point B in the figure lies above the shaded region and cannot be a solution. Then the answer is no, we would not have less than \$180 in rental costs.

EXAMPLE 7 Using graphs of linear inequalities to solve applications

In a certain manufacturing plant, some parts can be made by machine X while other parts must be made by machine Y. It takes 3 hours for machine X to make its part and 5 hours for machine Y to make its part. These machines are permitted to run at most a combined total of 35 hours per week.

a. Write the inequality statement that represents how many of each type of parts machine X and machine Y can make and operate at most 35 total hours a week.

b. Graph the inequality from part **a**; letting the *x*-axis represent the number of parts made by machine X and the *y*-axis the number of parts made by machine Y.

c. Name one possible ordered pair in quadrant I that lies in the shaded region and explain what that ordered pair represents.

PROBLEM 7

Suppose you are on a diet that allows only calories from protein (p) and fat (f) and you must not eat more than 1800 calories a day. If 1 gram of protein is 4 calories and 1 gram of fat is 9 calories, then answer the following questions.

(continued)

SOLUTION

a. Using machine X we let $3x$ represent the hours needed to make x parts. Using machine Y we let $5y$ represent the hours needed to make y parts. The sum of the hours worked is not to exceed 35 hours, so we can write the inequality as $3x + 5y \le 35$.

b. Graph the x-intercept, $11\frac{2}{3}$, and the y-intercept, 7. Draw a solid line connecting the two points and shade below the line. Since the number of parts each machine makes will be a positive number we use only quadrant I for the graph.

c. Point A, (6, 2), on the graph in part **b** lies in the shaded region and means machine X can produce six parts while machine Y produces two parts and they won't exceed the 35 hours a week.

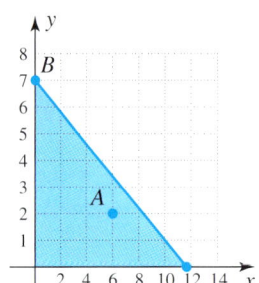

a. Write the inequality statement that represents how many grams of protein and how many grams of fat can be eaten and the calorie intake still not exceed 1800 calories.

b. Graph the inequality from part **a** plotting grams of protein (p) on the x-axis and grams of fat (f) on the y-axis.

c. Name one possible ordered pair in quadrant I outside of the shaded region and explain what that order pair represents.

Calculator Corner

Graphing Linear Inequalities

You can use your calculator to graph linear inequalities, but before you can use it properly, you need to do some work on your own. In Example 1, you graphed $x - 2y = -4$ and used a test point to find the region that satisfied $x - 2y < -4$. If you want to use your calculator to do this problem, you first have to solve for y in $x - 2y < -4$ (add $2y + 4$ to both sides and then divide each term by 2) to obtain

$$\frac{x}{2} + 2 < y \quad \text{or equivalently,} \quad y > \frac{x}{2} + 2$$

Now graph the associated equation,

$$y = \frac{x}{2} + 2$$

shown in Window 1.

The solution set of

$$y > \frac{x}{2} + 2$$

(equivalent to the original inequality $x - 2y < -4$) consists of the points *above* the graph of the line, as shown in Window 2. (We want the y's that are strictly "greater than" so we do not include the points on the line itself in the solution set.) To shade the region above, press $Y_1 =$, two left arrows, and toggle the enter key until the appropriate shading appears, then press GRAPH (see Window 2). You try Example 2 (see Window 3).

Window 1

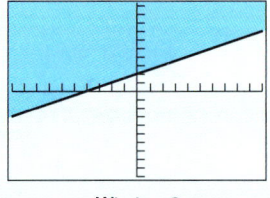

$x - 2y < -4$ or $y > \frac{x}{2} + 2$

Window 2

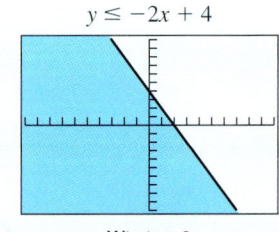

$y \le -2x + 4$

Window 3

Answers to PROBLEMS

6.

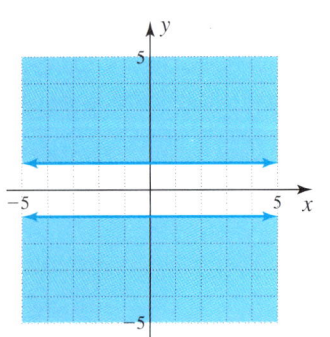

7. a. $4p + 9f \le 1800$

b.

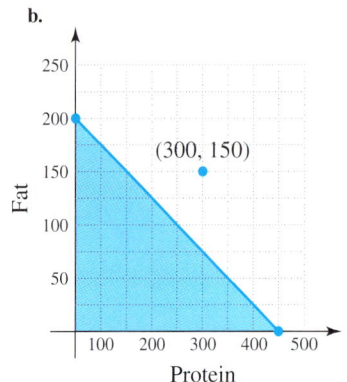

c. (300, 150) means eating 300 g of protein and 150 g of fat will exceed the 1800 calorie limit.

Exercises 3.4

⟨ A ⟩ Graphing Linear Inequalities In Problems 1–20, graph the solutions to the inequalities.

1. $x + 2y > 4$

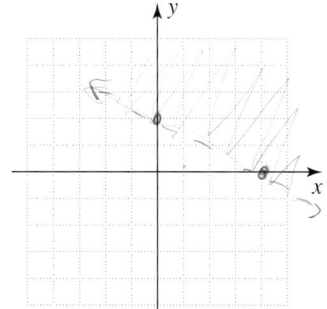

2. $x + 3y > 3$

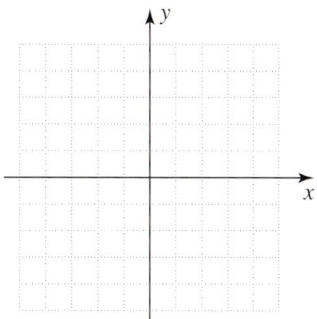

3. $-2x - 5y \leq -10$

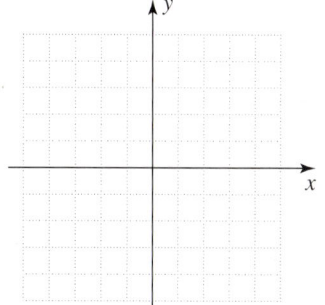

4. $-3x - 2y \leq -6$

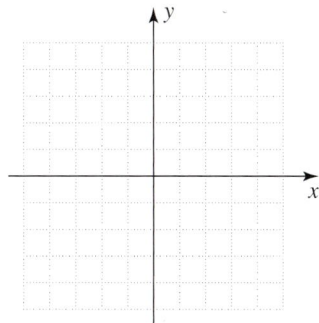

5. $y \geq 2x - 2$

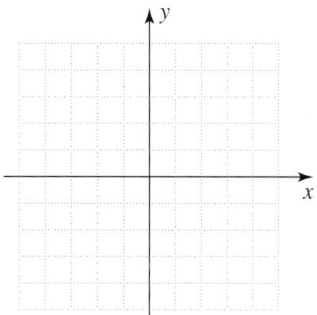

6. $y \geq -2x + 4$

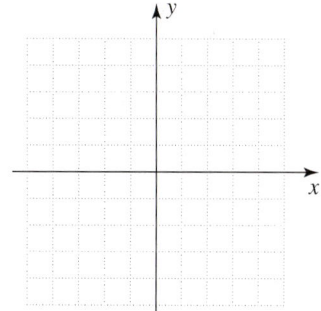

7. $6 < 3x - 2y$

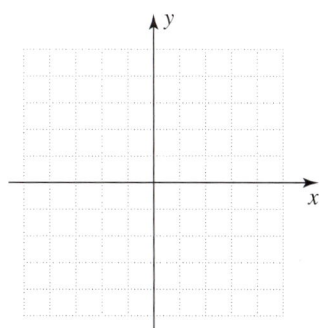

8. $6 < 2x - 3y$

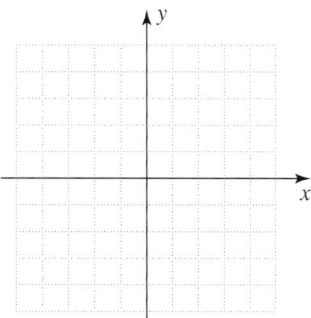

9. $4x + 3y \geq 12$

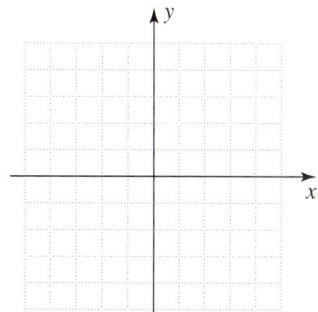

10. $-3y \geq 6x + 6$

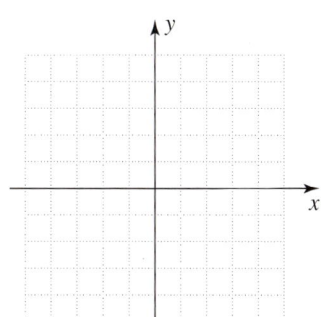

11. $10 < -5x + 2y$

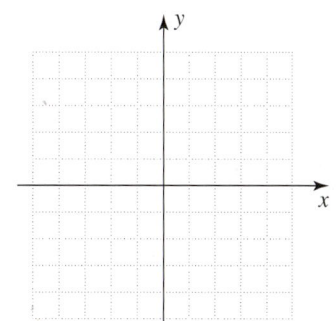

12. $4 < -2x - 4y$

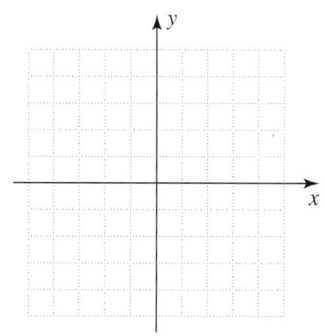

13. $2x \geq 2y - 4$

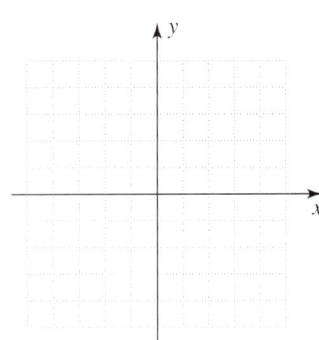

14. $2x \geq 4y + 2$

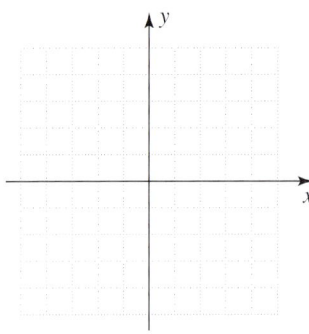

15. $2y < -4x + 8$

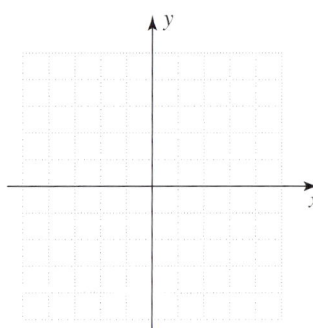

16. $3y < -6x + 9$

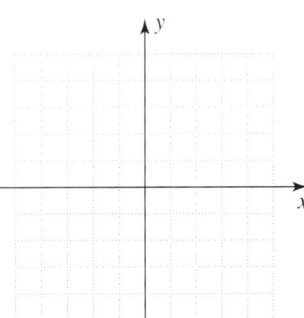

17. $x \geq -3$

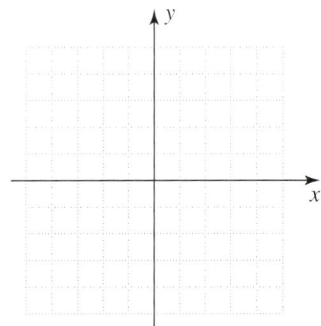

18. $x \geq -4$

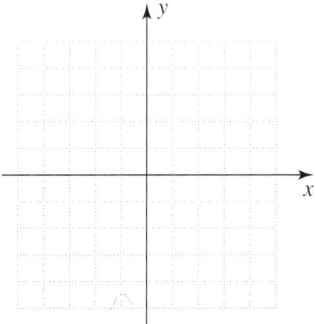

19. $y < 3$

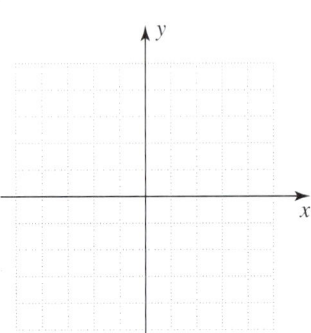

20. $y < -2$

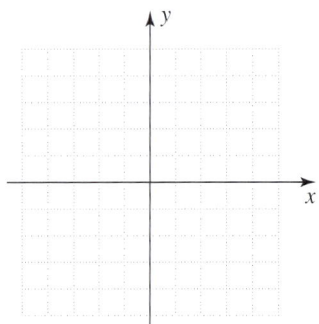

⟨ B ⟩ Graphing Absolute-Value Inequalities In Problems 21–38, graph the solutions to the inequalities.

21. $|x| < 1$

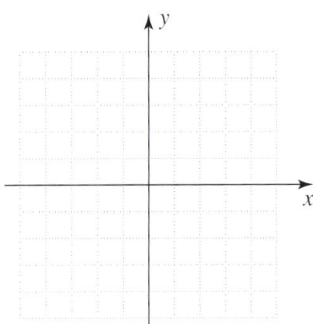

22. $|x| < 3$

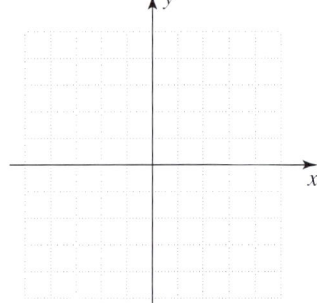

23. $|y| < 4$

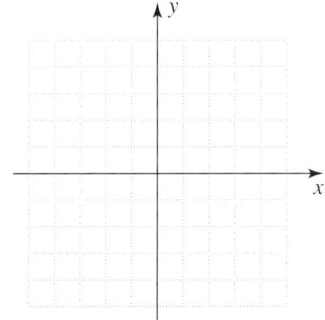

24. $|y| < 3$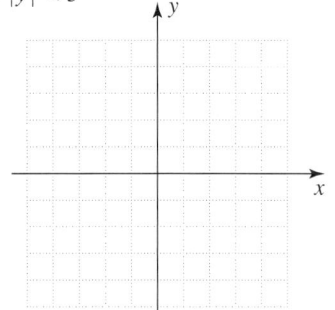

25. $|x| \geq 1$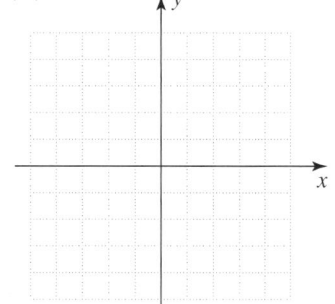

26. $|x| \geq 2$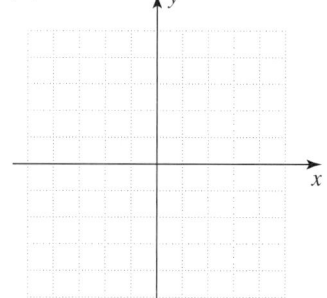

27. $|y| \geq 2$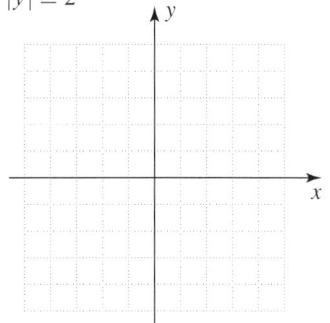

28. $|y| \geq 4$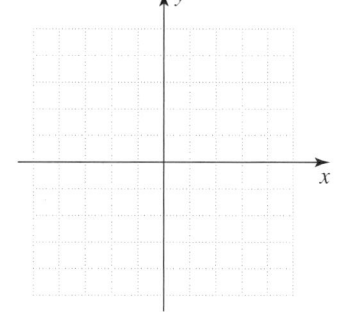

29. $|x + 2| < 1$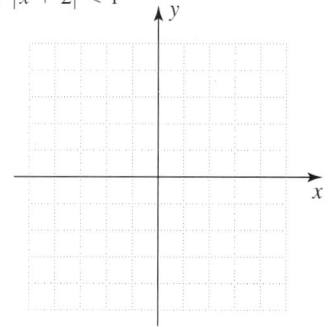

30. $|x + 3| < 1$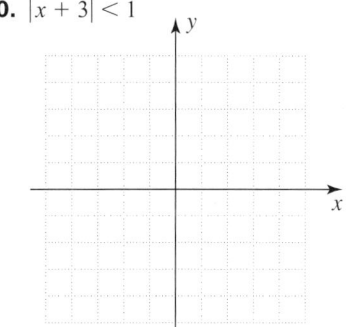

31. $|y + 2| < 1$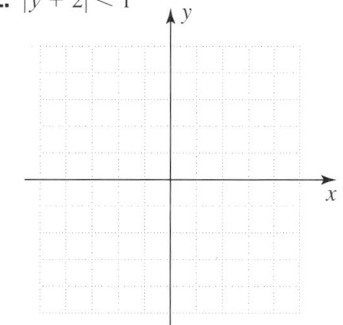

32. $|y + 2| < 2$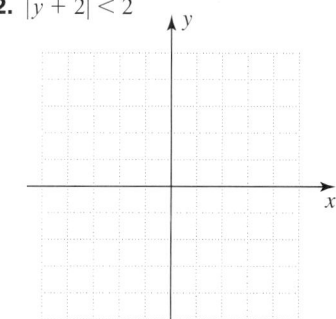

33. $|x + 1| \geq 3$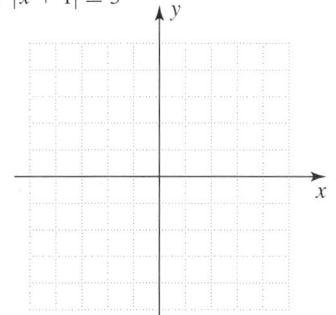

34. $|x + 2| \geq 1$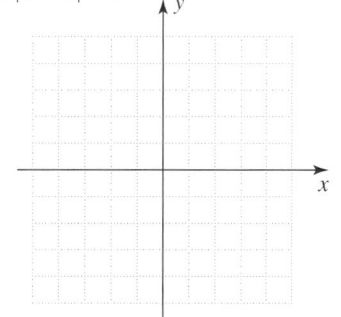

35. $|x - 1| \leq 2$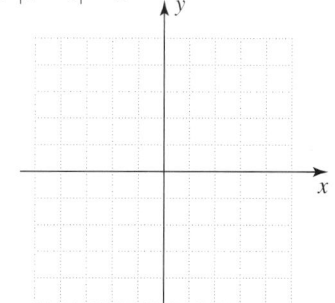

36. $|x - 2| \leq 1$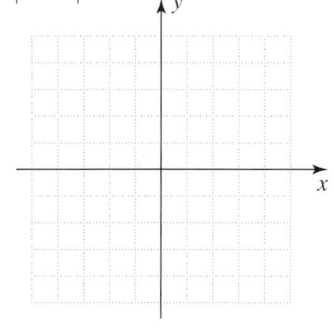

37. $|y - 2| < 1$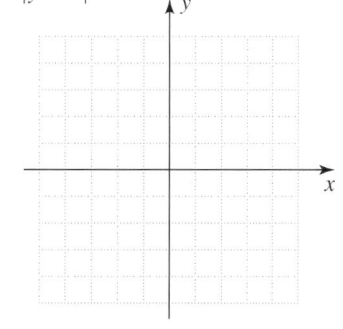

38. $|y - 3| < 1$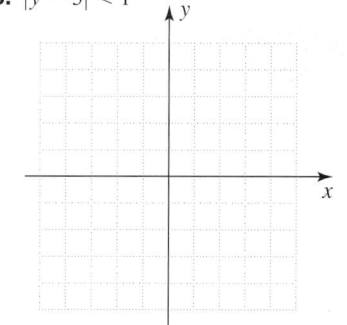

‹C› Applications Involving Linear Inequalities

39. *Budgets* Francesca has budgeted $250 for a road trip with her new car. If gas costs $2.50 per gallon and her car gets 20 miles per gallon, then dividing $2.50 by 20 gives a cost of $0.125 per mile ($m$). Her hotel accommodations are $75 per day ($d$). The equation representing the total cost of her trip (C) is $0.125m + 75d = C$.

a. Write the inequality statement that represents her staying within her budget.

b. Graph the inequality from part **a** using miles (m) as the x-axis and days (d) as the y-axis.

c. Use the graph to decide if she could drive 1000 miles over 4 days and be within her budget.

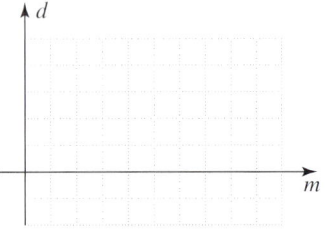

d. Use the graph to decide if she were gone for 2 days and drove 400 miles, would she be within her budget?

40. *Earnings* Travis has a pet-sitting service and charges $15 per day ($d$) to take care of the basic needs of the pet. In addition to that, he will spend time playing with the pet at the rate of $6 per hour ($h$). The total pet-sitting cost (C) to his clients is represented by the equation $15d + 6h = C$. Travis would like to take the money from his job and buy an MP3 player. He is considering the models that start at $120.

a. Write the inequality statement that represents Travis earning enough to be able to afford one of the MP3 players that he likes.

b. Graph the inequality from part **a** using days (d) as the x-axis and hours (h) as the y-axis.

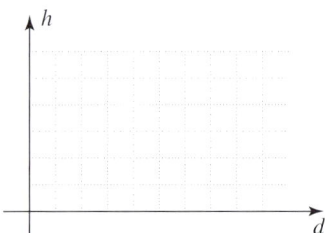

c. Use the graph to decide if he could pet-sit for 6 days, spend 12 hours of pet play time, and be able to make enough money to buy one of the MP3 player models he likes.

d. Use the graph to decide if he could pet-sit for 4 days, spend 8 hours of pet play time, and be able to make enough money to buy one of the MP3 player models he likes.

41. *Manufacturing* In a certain manufacturing plant some parts can be made by machinery while other parts must be made by hand. Usually it takes a worker 2 hours to make the part by machinery (M) and 6 hours to make the other part by hand (H). Employees are allowed to work at most 40 hours per week.

a. Write the inequality statement that represents how many of each type of part a worker can make and work at most 40 hours a week.

b. Graph the inequality from part **a** using how many parts made by machinery (M) as the x-axis and how many parts made by hand (H) as the y-axis.

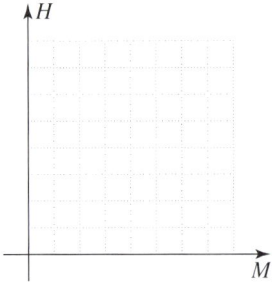

c. Name one possible ordered pair that lies in the shaded region and explain what that ordered pair represents.

42. *Diets* Suppose you are on a diet that allows only calories from protein (p) and carbohydrates (c) and you must not eat more than 1600 calories a day. If 1 gram of protein is 4 calories and 1 gram of carbohydrates is 4 calories, then answer the following:

a. Write the inequality statement that represents how many grams of protein and how many grams of carbohydrates can be eaten and the calorie intake still does not exceed 1600 calories.

b. Graph the inequality from part **a** using how many grams of protein (p) as the x-axis and how many grams of carbohydrates (c) as the y-axis.

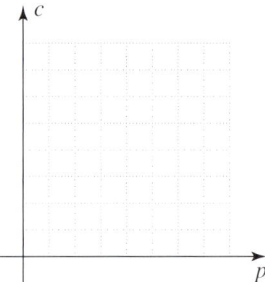

c. Name one possible ordered pair in quadrant 1 outside of the shaded region and explain what that order pair represents.

>>> Using Your Knowledge

The More You Drive, the Costlier It Is You can use what you know about linear inequalities to save money when you rent a car. Here are the rental prices for an intermediate car obtained from a telephone survey.

Rental A: $41 a day, unlimited mileage included
Rental B: $36 a day, $0.15 per mile

43. If you compare the Rental A price ($41) with the Rental B price, you see that Rental B appears to be cheaper.

 a. How far can you drive a Rental B car in one day if you wish to spend exactly $41? (Answer to the nearest mile.)

 b. If you are planning on driving 100 mi in one day, which car would you rent, Rental A or Rental B?

44. How far can you drive a Rental B car in 1 day if you wish to spend exactly $42? (Answer to the nearest mile.)

45. Based on your answers to Problems 43 and 44, which is the cheaper rental price? (*Hint:* It has to do with the miles you drive.)

>>> Write On

46. Write the procedure you would use to graph the solution set of the inequality $ax + by > c$.

47. How would you describe the graph of the inequality $x \geq k$ (where k is a constant)?

48. Explain what it means when a dashed line is used as the boundary in the graph of the solution set of a linear inequality.

49. Explain what it means when a solid line is used as the boundary in the graph of the solution set of a linear inequality.

>>> Concept Checker

Fill in the blank(s) with the correct word(s), phrase, or mathematical statement.

50. The equation $y = 3$ is the _____ used when graphing the inequality, $2y < 6$.

51. In graphing the inequality, $y > \frac{1}{2}x + 4$, the shading of the solution would be _____ the boundary line.

52. The statement $-x + 2y < 5$ is called a _____.

53. In graphing the inequality, $3 > x + y$, the boundary line would be a _____ line.

54. In graphing the inequality $x < 5$, the boundary line would be a _____ line.

linear inequality	vertical
boundary line	horizontal
shading	above
solid	below
dashed	

>>> Mastery Test

Graph the solutions to each inequality.

55. $|x + 2| > 3$

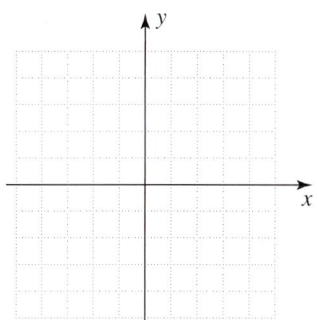

56. $|x - 1| < 4$

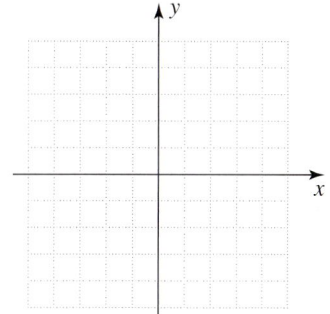

57. $|y| \leq 1$

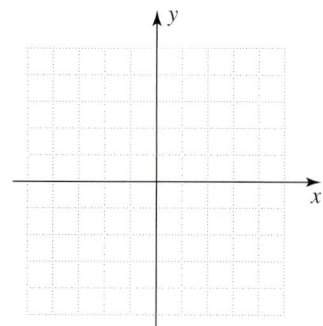

58. $|y| > 3$

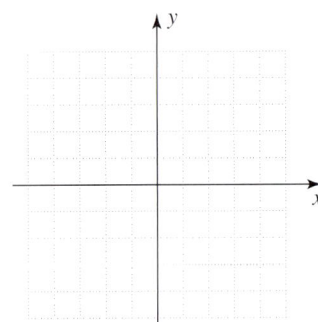

59. $x \geq -2$

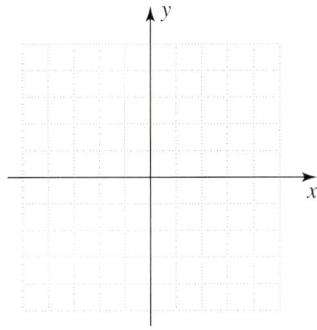

60. $x < 3$

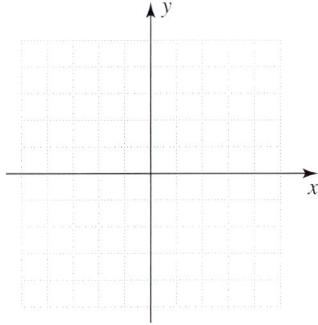

61. $3x - 2y < -6$

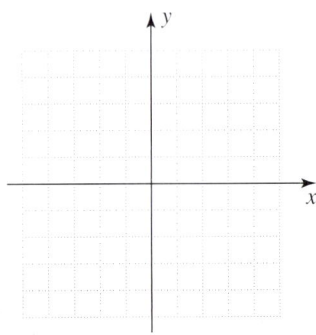

62. $-2x - 3y \geq -6$

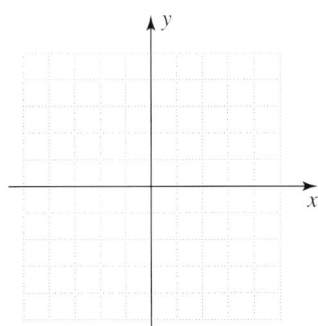

63. Suppose you rented a car for a few days and the cost was $48 per day ($d$) plus $0.32 per mile ($m$). Write and graph the statement of inequality that would represent the cost being less than $160.

›>> Skill Checker

In Problems 64–67, use the graph to answer the following questions for each problem.

a. Does the vertical line intersect the curve in more than one point?

b. For those cases where the vertical line intersects the graph in two or more points, would the x- or y-value be repeated when naming the ordered pairs of the intersection points?

64.

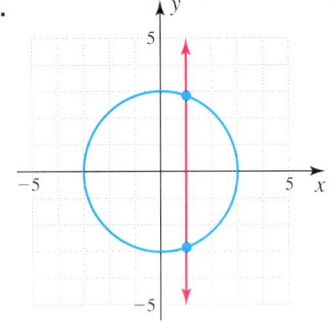

65.

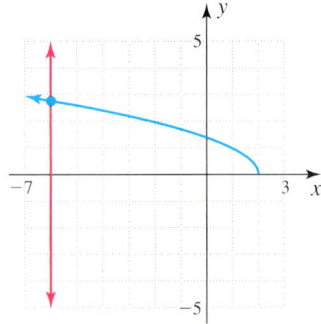

66.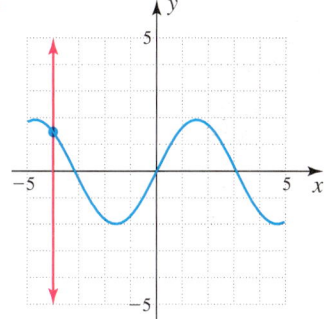

67.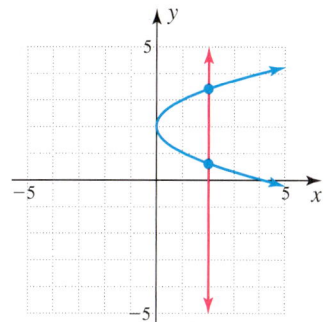

In Problems 68–75, evaluate the expression. If there is not a real number answer, then state "undefined in the reals."

68. $6\left(\frac{1}{2}\right) - 4$

69. $-15\left(\frac{2}{3}\right) + 8$

70. $\frac{4(-5) + 2}{-9 - (-9)}$

71. $\frac{6 - 5(-2)}{-12 - 12}$

72. $7(-1)^3 - 5(-1) - 14$

73. $3(-2)^2 + 5(-2) + 8$

74. $\sqrt{4 - 8(5)}$

75. $\sqrt{4 + 2(6)}$

3.5 Introduction to Functions

▶ Objectives

A ▶ Find the domain and range of a relation and determine whether the relation is a function.

B ▶ Use the vertical line test to determine if a relation is a function.

C ▶ Find the domain of a function defined by an equation.

D ▶ Find the value of a function.

▶ To Succeed, Review How To . . .

1. Graph a line (pp. 168–174).
2. Evaluate an expression (pp. 51–55).

▶ Getting Started

Traveling Ants

In Section 2.1 on page 74 we discussed the claim that farmers have relating temperatures with the number of chirps of a cricket. Here we will discuss the linear relationship between the temperature and the rate of travel (speed) of certain ants. If y is the speed in centimeters per second and x is the temperature in degrees Celsius, the relationship is

$$y = \frac{1}{6}(x - 4)$$

Thus if the temperature is 10°C, the speed is

$$y = \frac{1}{6}(10 - 4) = \frac{1}{6} \cdot 6 = 1 \text{ cm/sec}$$

If the temperature is 16°C, the speed is

$$y = \frac{1}{6}(16 - 4) = \frac{1}{6} \cdot 12 = 2 \text{ cm/sec}$$

We can make a table showing two related sets of numbers, one for the temperature x and the other for the speed y, as shown. (By the way, x has to be greater than 4°C and less than 35°C. Why?)

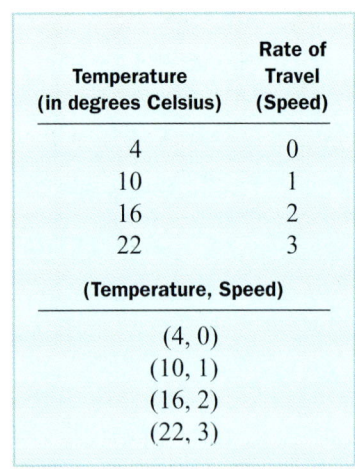

The numbers in the first column are called values of the **independent variable** because they are chosen *independently* of the second number. The numbers in the second column are called values of the **dependent variable** because they *depend* on the values of the numbers in the first column. The numbers in our table are written as ordered pairs, and these ordered pairs can also be written as the set

$$\{(4, 0), (10, 1), (16, 2), (22, 3)\}$$

The concept of ordered pairs begins our investigation into an important idea in algebra: the concept of a function.

A › Finding the Domain and Range and Determining Whether a Relation Is a Function

To study about a function we first describe a **relation** as a correspondence between two sets. Examples of relations are

- The set of books in the library corresponding to the set of last page numbers in each book.
- The set of university e-mail usernames corresponding to the set of passwords for each.
- The set of counting numbers corresponding to the set of squares of each counting number.

The first set in the correspondence is called the **domain,** and the second set is the **range.** In the first example, the set of books would be the domain, and the set of last page numbers would be the range.

Sometimes we use a **mapping diagram** as a visual way to indicate the related correspondence. Figure 3.46 is a mapping diagram of the correspondence between the set of counting numbers, $\{1, 2, 3, 4\}$, and the set of the squares of each, $\{1, 4, 9, 16\}$.

›Figure 3.46

The equation in the *Getting Started*, $y = \frac{1}{6}(x - 4)$, is an example of linear equations we studied in Section 3.1. The set of ordered pairs, $\{(4, 0), (10, 1), (16, 2), (22, 3)\}$, is a partial solution set to this equation. A relation is a *set of ordered pairs,* so this equation with its ordered pair solution set would be an example of a relation. The first members of the ordered pairs (x-values) form the domain of the relation, and the second members of the ordered pairs (y-values) form the range of the relation.

3.5 Introduction to Functions

Any set of ordered pairs is a relation, which we describe as follows.

RELATION, DOMAIN, AND RANGE

A **relation** is a correspondence between two sets, such as a set of ordered pairs. The set of all first coordinates is the **domain** of the relation, and the set of all second coordinates is the **range** of the relation.

Now that we have described a relation, we will introduce the concept of a function as a special case of a relation where each domain member is related to *exactly one* range member. We will use the three relations described in Figure 3.47 to investigate how to determine if a relation is a function. This will be done by comparing the way the members of the domain are related to the members of the range.

Relation A		Relation B		Relation C	
Domain	Range	Domain	Range	Domain	Range
Canine	Bull dog, Chow, Poodle	17, 27, 38, 47	7, 8	coffee	$1.50
Feline	Manx			soda	$1.75
				milkshake	$3.50

> Figure 3.47

In *Relation A*, there is one domain member, Canine, that is related to more than one range member. That is, when considering the set of ordered pairs describing this relation, the domain member "Canine" has been repeated with different range members, {(Canine, Bull dog), (Canine, Chow), (Canine, Poodle)}. *Relation A is not a function*.

In *Relation B*, each domain member is related to exactly one range member. When considering the set of ordered pairs describing this relation, no domain member repeats itself with a different range member, {(17, 7), (27, 7), (38, 8), (47, 7)}. Notice it is acceptable for a range member, in this case "7," to repeat itself, as long as the domain members are different. *Relation B is a function*.

In *Relation C*, each domain member is related to exactly one range member. In the set of ordered pairs describing this relation, no domain member repeats itself with a different range member, {(coffee, $1.50), (soda, $1.75), (milkshake, $3.50)}. *Relation C is a function*.

Here is the definition of a function.

FUNCTION

A **function** is a relation in which each domain member is related to *exactly one* range member. That is, no two different ordered pairs have the same first coordinate.

EXAMPLE 1 Finding the domain and range and determining if a relation is a function

Find the domain and range and determine if the relation is a function.

a. $A = \{(1, 2), (2, 3), (3, 4)\}$

b.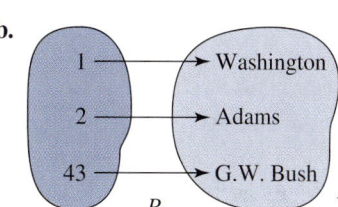

c. $C = \{$(coffee, Starbucks), (burgers, Burger King), (coffee, City Brew), (fries, McDonald's)$\}$

PROBLEM 1

Find the domain and range and determine if the relation is a function.

a. $C = \{(-1, 0), (1, 0), (2, 1)\}$

b.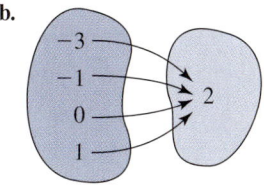

c. $D = \{(f, \$15), (g, \$20), (h, \$25), (f, \$30)\}$

(continued)

Answers to PROBLEMS

1. a. Domain: $\{-1, 1, 2\}$; Range: $\{0, 1\}$; Function   **b.** Domain: $\{-3, -1, 0, 1\}$; Range: $\{2\}$; Function
c. Domain: $\{f, g, h\}$; Range: $\{\$15, \$20, \$25, \$30\}$; Not a function

SOLUTION

a. Domain: {1, 2, 3}

Range: {2, 3, 4}

The relation A is a function since each domain member is related to exactly one range member. Another way to decide if a set of ordered pairs is a function is to be sure no "x"-value repeats itself with a different y-value.

b. Domain: {1, 2, 43}

Range: {Washington, Adams, G.W. Bush}

The set of ordered pairs for this relation is {(1, Washington), (2, Adams), (43, G.W. Bush)}. Since no "x" repeats itself with a different y-value, this relation is a function.

c. Domain: {coffee, burgers, fries}

Range: {Starbucks, Burger King, City Brew, McDonald's}

As we compare x's in this set of ordered pairs, we see that coffee has repeated itself with different y-values, (coffee, Starbucks) and (coffee, City Brew). Thus, relation C is *not* a function.

A relation is a set of ordered pairs, so relations can be graphed in the Cartesian plane. The numbers used when graphing in this plane are the real numbers. Let's consider the equation $y = 3x + 2$ and find some of the ordered pairs that solve the equation. When no domain is specified, the domain is assumed to be the set of all real numbers for which the relation is defined. This relation is defined as multiplying the real number x by 3 and then adding 2. Since there are no restrictions on multiplying and adding real numbers, the domain is the set of real numbers. To find solutions, we arbitrarily assign values for x from the set of real numbers. This means x is the **independent variable.** We choose -2, 0, and 1. We can call these the "input" values. One at a time these values are substituted for x in the equation and we solve for the related y-value. The outcome of the y-variable, sometimes called the "output" value, depends on the value of x so it is called the **dependent variable.**

Input	Relation	Output
x	$y = 3x + 2$	y
-2	$3(-2) + 2$	-4
0	$3(0) + 2$	2
1	$3(1) + 2$	5

These ordered pairs, $(-2, -4)$, $(0, 2)$, and $(1, 5)$, are three of infinitely many solutions to the relation $y = 3x + 2$. When inputting any real number into this relation, we see that we will always get a real number, so we can say the range is the set of real numbers. As you can see, if we were to continue this table of inputs and outputs, for each value in the domain there would be exactly one value in the range. That establishes this relation as a function. In Example 2, we will find the domain and range and the graph of a relation given a set of ordered pairs or an equation.

EXAMPLE 2 Finding the domain, range, and graph of a relation

a. Find the domain and range and the graph of the relation $A = \{(1, 2), (2, 3), (3, 4)\}$.

b. Find the domain and range and the graph of the relation $\{(x, y) \mid y = 3x - 4\}$.

PROBLEM 2

Find the domain, range, and graph of the relation:

a. $\{(2, 1), (7, 5), (9, 8)\}$

b. $\{(x, y) \mid y = -2x + 1\}$

(Answers on page 227)

SOLUTION

a. The domain of A is the set of first coordinates, so $D = \{1, 2, 3\}$. The range of A is the set of second coordinates, so $R = \{2, 3, 4\}$. The graph of relation A is shown in Figure 3.48.

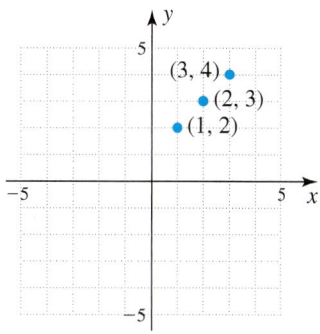

> Figure 3.48

b. The domain is the set of all real numbers since any real number x can be used as the input value. Similarly, the range of y is the set of all real numbers.

The graph of the relation is the graph of the equation $y = 3x - 4$. The following table will help graph the relation as shown in Figure 3.49.

Input x	Relation y = 3x − 4	Output y
−1	3(−1) − 4	−7
0	3(0) − 4	−4
2	3(2) − 4	2

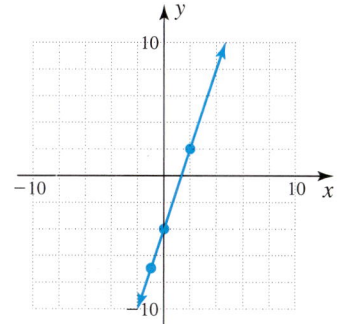

> Figure 3.49

Answers to PROBLEMS

2. a. $D = \{2, 7, 9\}$ and $R = \{1, 5, 8\}$

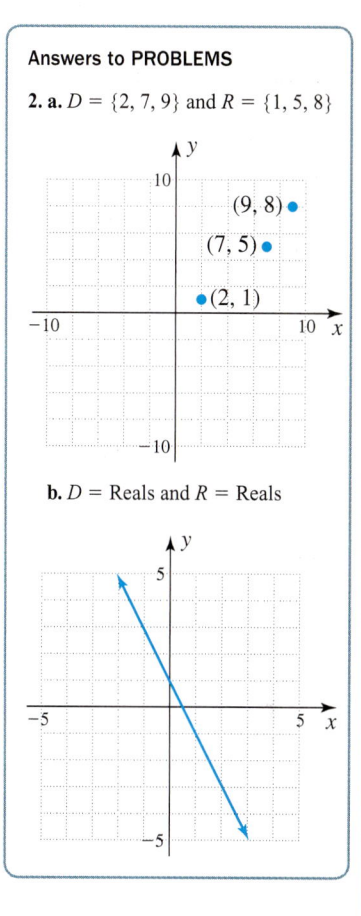

b. $D = $ Reals and $R = $ Reals

In Example 2b, we named the domain and range of the given equation and then graphed the equation. Suppose we were given the graph of a relation, as shown in Figure 3.50, and asked to find the domain and range.

One way to consider finding the domain is to find the smallest and largest values of x on the graph as indicated by the horizontal line. Since the graph stops on the left at $x = -2$ and x continues to positive infinity on the right, we say the domain goes from $x = -2$ to positive infinity. This can also be written in interval notation as $D = [-2, \infty)$. Similarly, to find the range we must find the smallest and largest values of y on the graph as indicated by the vertical line. The smallest y-value on the graph is -4, and the y-values continue to positive infinity. The range values go from $y = -4$ to positive infinity and can be written in interval notation as $R = [-4, \infty)$.

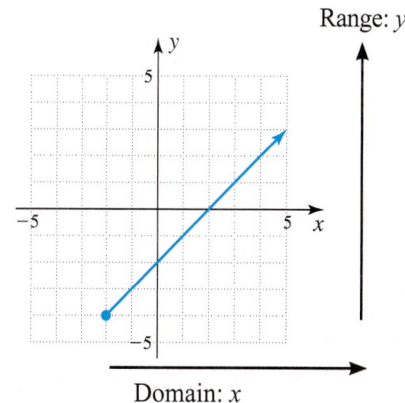

> Figure 3.50

228   Chapter 3   Graphs and Functions

In Example 3, we will be given the graphs of some relations and asked to name the domain and range of each.

EXAMPLE 3 Finding the domain and range given a graph
Find the domain and range of each graph.

a.
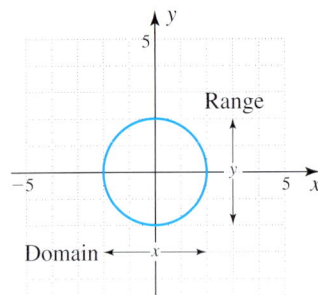

b.
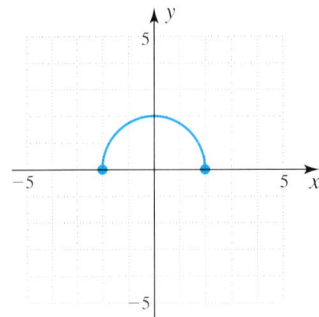

c.
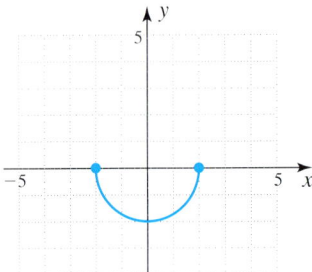

d.
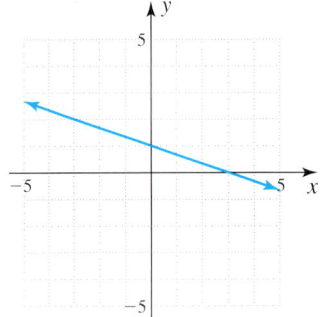

PROBLEM 3
Find the domain and range of each graph.

a.
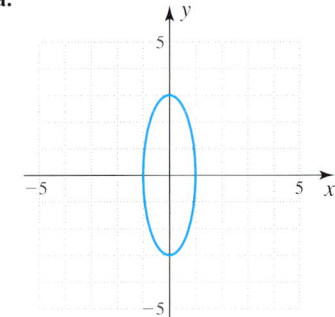

b.
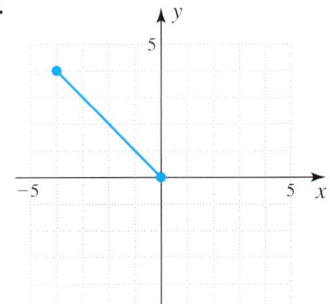

c.
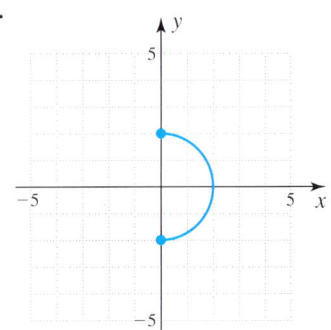

d.
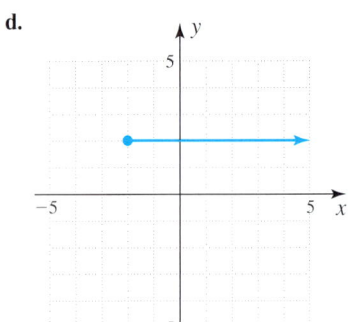

SOLUTION

a. The graph is a circle of radius 2 centered at the origin. From the graph, it is clear that x and y can be any real numbers between -2 and 2, inclusive. The domain is $D = \{x \mid -2 \leq x \leq 2\}$, and the range is $R = \{y \mid -2 \leq y \leq 2\}$.

b. The graph is the top half of the circle from part **a**. The domain is $D = \{x \mid -2 \leq x \leq 2\}$, and the range is $R = \{y \mid 0 \leq y \leq 2\}$.

c. The graph is the bottom half of the circle from part **a**. The domain is $D = \{x \mid -2 \leq x \leq 2\}$, and the range is $R = \{y \mid -2 \leq y \leq 0\}$.

d. The graph is a line. The domain is $D = \{x \mid x \text{ is a real number}\}$, and the range is $R = \{y \mid y \text{ is a real number}\}$.

Answers to PROBLEMS

3. a. $D = \{x \mid -1 \leq x \leq 1\}$ and
   $R = \{y \mid -3 \leq y \leq 3\}$

   b. $D = \{-4 \leq x \leq 0\}$ and
   $R = \{0 \leq y \leq 4\}$

   c. $D = \{0 \leq x \leq 2\}$ and
   $R = \{-2 \leq y \leq 2\}$

   d. $D = \{x \geq -2\}$ and
   $R = \{2\}$

B › Functions and the Vertical Line Test

The graph of a relation can be used to determine whether the relation is a function. Any two points with the same first coordinate will be on a vertical line parallel to the y-axis, so if any vertical line intersects the graph more than once, the relation is *not* a function. Testing to see if a relation is a function by determining whether a vertical line crosses the graph more than once is called the **vertical line test.**

VERTICAL LINE TEST If a vertical line parallel to the y-axis intersects the graph of a relation more than once, the relation is *not* a function.

Using this test, we can see that the relation in Example 3, part **a**, is *not* a function. (A vertical line crosses the graph in more than one place.) On the other hand, the graphs in parts **b, c,** and **d** are functions.

EXAMPLE 4   Using the vertical line test

Use the vertical line test to determine whether the graph of the relation defines a function:

a.

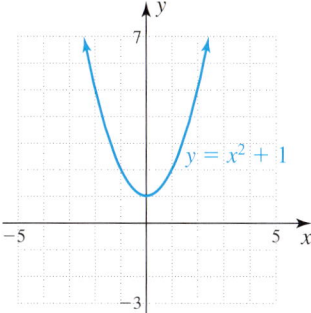

b.

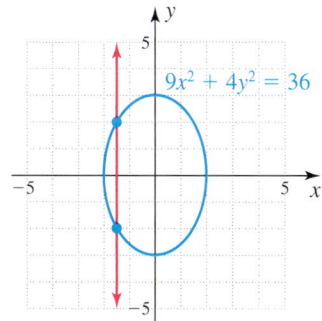

The vertical line shown in red crosses the graph in *two* points.

SOLUTION
a. The relation is a function because no vertical line crosses the graph more than once.
b. The graph is an **ellipse.** We can draw a vertical line that crosses the graph in more than one point, so the relation is **not** a function.

PROBLEM 4
Use the vertical line test to determine whether the graph of the relation defines a function:

a.

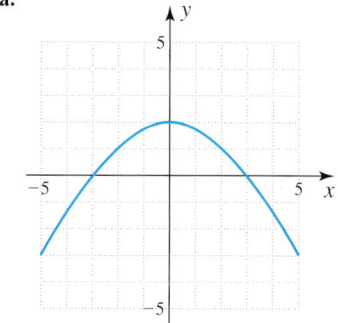

b.

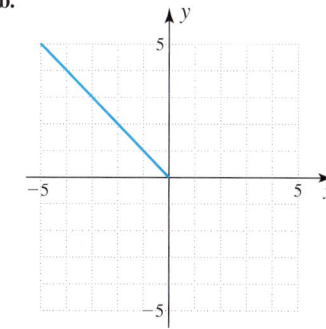

c.

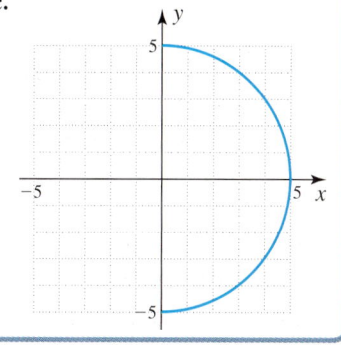

Answers to PROBLEMS

4. a. Yes **b.** Yes **c.** No

C ▸ Finding the Domain of a Function

When relations are defined by means of an equation, the domain is the set of all possible replacements for the variable x that result in real numbers for y. We cannot replace x by values that will produce zero in the denominator of a fraction or produce the square root of a negative number. For example, the domain of $y = \frac{1}{x}$ is the set of all real numbers *except* 0 and the domain of $y = \sqrt{x}$ is the set of all real numbers $x \geq 0$.

EXAMPLE 5 Finding the domain of a function given an equation

Find the domain of the function defined by:

a. $y = \dfrac{1}{x+3}$ **b.** $y = \sqrt{x-3}$ **c.** $y = \dfrac{1}{3}x + 2$

SOLUTION

a. We cannot replace x by values that will produce zero in the denominator, so we must avoid the case in which

$$x + 3 = 0$$
$$x = -3$$

The domain of

$$y = \dfrac{1}{x+3}$$

is the set of all real numbers except -3 or $\{x \mid x \in \text{Reals, where } x \neq -3\}$.

b. The square root of a negative number is not a real number, thus we must make the expression under the radical, $x - 3$, nonnegative, that is,

$$x - 3 \geq 0$$

or $x \geq 3$

The domain is $\{x \mid x \geq 3\}$.

c. Multiplication and addition have no restrictions in the real numbers, so we can use any real number for x. The domain is the set of real numbers or

$$\{x \mid x \in R\}$$

PROBLEM 5

Find the domain of the function defined by:

a. $f(x) = \dfrac{1}{x-4}$

b. $y = \sqrt{x+2}$

c. $y = |x| + 3$

Calculator Corner

Finding Domain and Range

To find the domain and range of

$$y = \dfrac{1}{(x-2)(x-3)}$$

using your TI-83 Plus: First graph y using a decimal window:

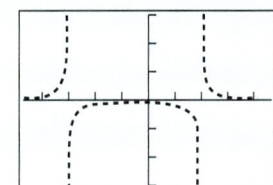

Sometimes the dot mode of your calculator gives a better view of points that may not be in the domain.

Press MODE, use arrows to get to DOT, then press ENTER GRAPH

Do you see that $x = -3$ and $x = 2$ are not part of the graph?

If you use the TRACE key to move your cursor to $x = 2$, what value do you get for y?

Try graphing $y = \sqrt{x-3}$. What do you get for the domain?

Answers to PROBLEMS

5. a. $\{x \mid x \in \text{Reals, where } x \neq 4\}$ **b.** $\{x \mid x \geq -2\}$ **c.** $\{x \mid x \in \text{Reals}\}$

D › Finding the Value of a Function

Water pressure is a *function* of depth, as you can see in the photo. The higher pressure at the lower holes of the bucket makes water squirt out in a nearly flat trajectory, whereas the lower pressure at the upper holes produces only a weak stream. The pressure y of the water (in pounds per square foot) at a depth of x feet (x is a positive number) is

$$y = 62.5x$$

If we wish to emphasize that the pressure y is a function of the depth x, we can use the **function notation** $f(x)$ (read "f of x") and write

$$f(x) = 62.5x$$

To find the pressure at 2 ft below the surface, we find the value of y when $x = 2$. If $x = 2$, then $y = 62.5(2) = 125$. Using function notation, we would write $f(2) = 62.5(2) = 125$. This table illustrates both notations:

y in terms of x $y = 62.5x$	Function Notation $f(x) = 62.5x$
If $x = 2$, then $y = 62.5(2) = 125$.	$f(2) = 62.5(2) = 125$
If $x = 4$, then $y = 62.5(4) = 250$.	$f(4) = 62.5(4) = 250$
If $x = 5$, then $y = 62.5(5) = 312.5$.	$f(5) = 62.5(5) = 312.5$

If $y = f(x)$, the symbols $f(x)$ and y are interchangeable; they both represent the value of the function for the given value of x. If

$$y = f(x) = 2x + 3$$

then

Function Notation	Ordered Pair
$f(1) = 2(1) + 3 = 5$	$(1, 5)$
$f(0) = 2(0) + 3 = 3$	$(0, 3)$
$f(-6) = 2(-6) + 3 = -9$	$(-6, -9)$
$f(4) = 2(4) + 3 = 11$	$(4, 11)$
$f(a) = 2(a) + 3 = 2a + 3$	$(a, 2a + 3)$
$f(w + 2) = 2(w + 2) + 3 = 2w + 7$	$(w + 2, 2w + 7)$

These ordered pairs are graphed as shown in Figure 3.51.

Whatever appears between the parentheses in $f(\ )$ is to be substituted for x in the rule that defines $f(x)$.

Instead of describing a function in set notation, we frequently say "the function defined by $f(x) = \ldots$," where the ellipsis dots are to be replaced by the expression for the value of the function. For instance, "the function defined by $f(x) = 2x + 3$" has the same meaning as "the function $f = \{(x, y) | y = 2x + 3\}$." The function $f(x) = 2x + 3$ is an example of a linear function because its graph is a straight line. In general, we have the following definition.

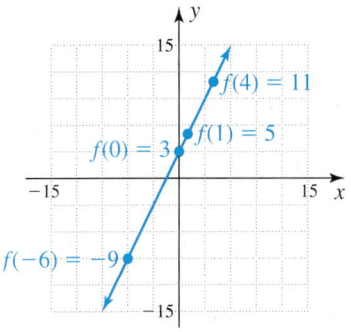

›Figure 3.51

LINEAR FUNCTION

A **linear function** is a function that can be written in the form

$$f(x) = mx + b$$

where m and b are real numbers.

EXAMPLE 6 Finding the values of a function given the rule

Let $f(x) = 3x + 5$ and find: (for **a** and **b** write the results as ordered pairs)

a. $f(4)$ **b.** $f(2)$ **c.** $f(2) + f(4)$ **d.** $f(x + 1)$

SOLUTION

a. $f(x) = 3x + 5$, so $f(4) = 3 \cdot 4 + 5 = 12 + 5 = 17$; $(4, 17)$

b. $f(2) = 3 \cdot 2 + 5 = 6 + 5 = 11$; $(2, 11)$

c. $f(2) = 11$ and $f(4) = 17$, so $f(2) + f(4) = 11 + 17 = 28$

d. $f(x + 1) = 3(x + 1) + 5 = 3x + 8$

PROBLEM 6

Let $f(x) = 2x - 3$ and find: (for **a** and **b** write the results as ordered pairs)

a. $f(5)$
b. $f(1)$
c. $f(1) - f(5)$
d. $f(x - 1)$

EXAMPLE 7 Finding the values of a function given ordered pairs

Let $f = \{(3, 2), (4, -2), (5, 0)\}$ and find:

a. $f(3)$ **b.** $f(4)$ **c.** $f(3) + f(4)$

SOLUTION

a. In the ordered pair $(3, 2)$, $x = 3$ and $y = 2$. Thus $f(3) = 2$.

b. In the ordered pair $(4, -2)$, $x = 4$ and $y = -2$. Thus $f(4) = -2$.

c. $f(3) + f(4) = 2 + (-2) = 0$

PROBLEM 7

Let $f = \{(2, 3), (-1, 4), (0, 6)\}$ and find:

a. $f(2)$ **b.** $f(-1)$ **c.** $f(2) + f(-1)$

DEFINITIONS OF A FUNCTION

1. A **function** is a rule or a correspondence that assigns exactly **one** *range* value to each *domain* value (as in Example 6).

2. A **function** is a relation in which **no two** ordered pairs have the same first coordinate and different second coordinates (as in Example 7).

3. A **function** is a *mapping* that assigns exactly **one** range element to each domain element.

Calculator Corner

Evaluate a Function by Storing a Value

Use your TI-83 Plus to evaluate the function of Example 6:
Store a value for x, say, 4

4

Now tell the calculator you want to evaluate the function by entering:

Try evaluating the function for -4.

Answers to PROBLEMS

6. a. 7; $(5, 7)$ **b.** -1; $(1, -1)$ **c.** -8 **d.** $2x - 5$ **7. a.** 3   **b.** 4   **c.** 7

Exercises 3.5

> Practice Problems > Self-Tests
> NetTutor > e-Professors
>   > Videos

Boost your grade at mathzone.com!

A **Finding the Domain and Range and Determining If a Relation Is a Function** In Problems 1–12, find the domain and range, and determine whether the relation is a function.

1. $\{(-3, 0), (-2, 1), (-1, 2)\}$
2. $\{(-1, -2), (0, -1), (1, 0)\}$
3. $\{(3, 0), (4, 0), (5, 0)\}$

4. $\{(0, 1), (0, 2), (0, 3)\}$
5. $\{(1, 2), (1, 3), (2, 2), (2, 3)\}$
6. $\{(2, 1), (1, 2), (3, 4), (4, 3)\}$

7.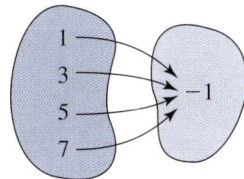
8.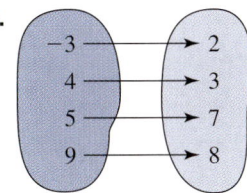
9.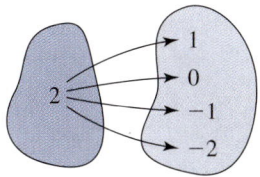

10.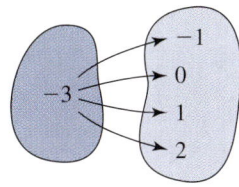
11. $\{(2, \text{February}), (6, \text{June}), (11, \text{November})\}$
12. $\{(\text{fish, trout}), (\text{dog, poodle}), (\text{cat, manx}), (\text{fish, salmon})\}$

In Problems 13–16, find the domain and range and then graph the relation.

13. $B = \{(-2, 1), (0, -3), (1, 4), (-2, 2)\}$

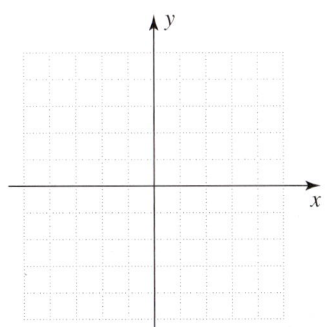

14. $C = \{(0, 0), (2, 0), (-1, 3), (3, -4)\}$

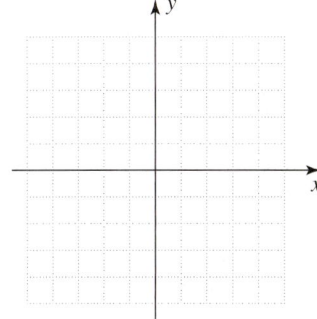

15. $\{(x, y) \mid y = -2x + 1\}$

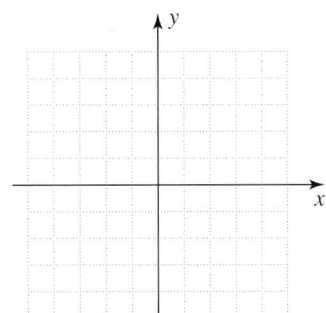

16. $\{(x, y) \mid y = 3x - 1\}$

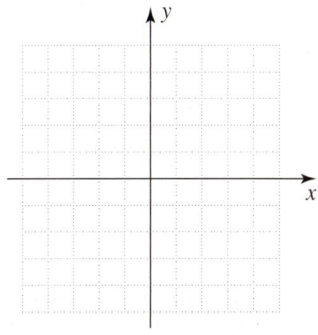

B Functions and the Vertical Line Test
In Problems 17–36, give the domain and range, and use the vertical line test to determine whether the relation is a function.

17.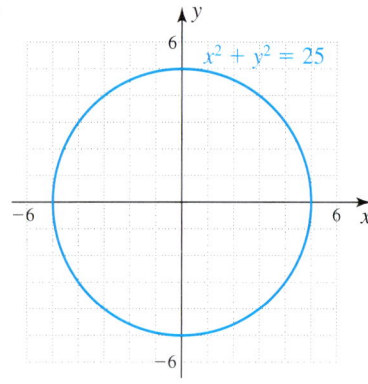

18.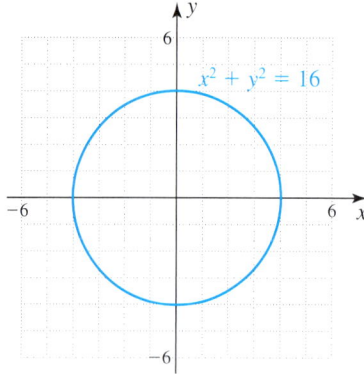

19.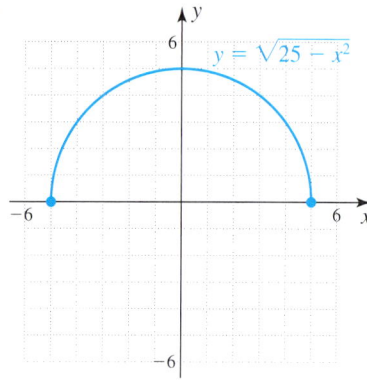

20.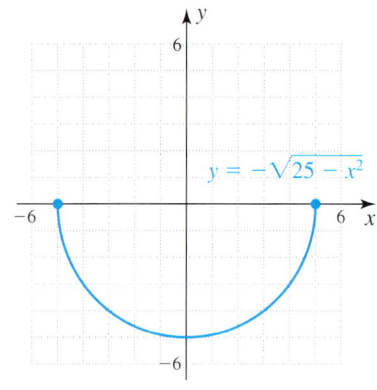

21.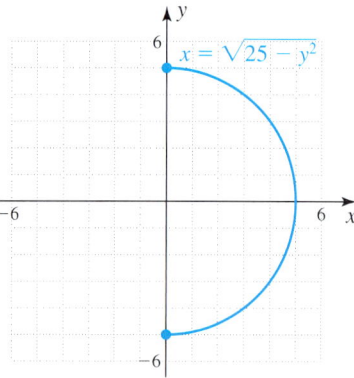

22.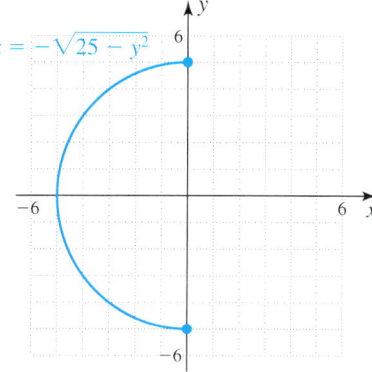

23.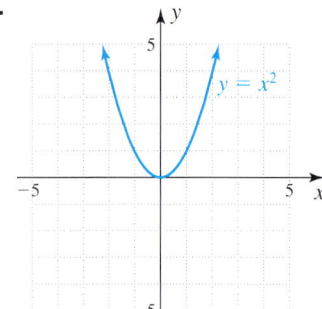

24.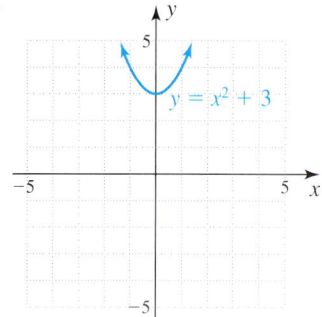

25.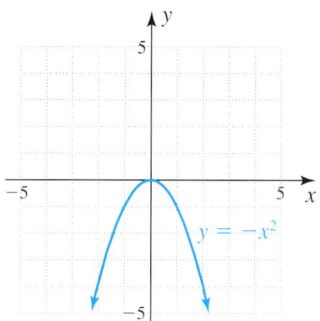

26.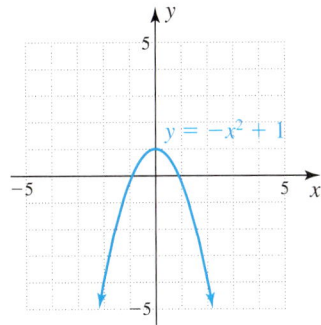

27.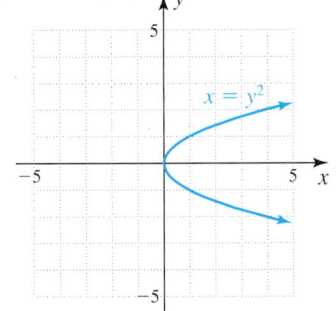

28.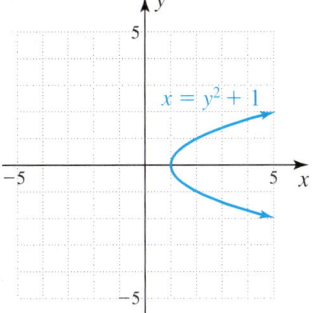

29.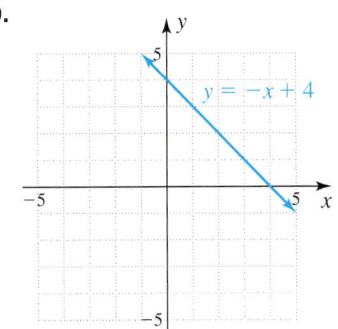

30.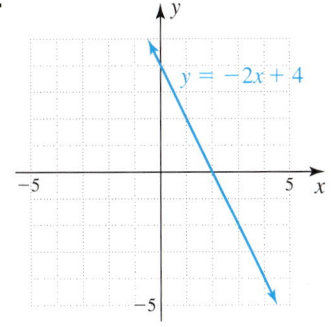

31.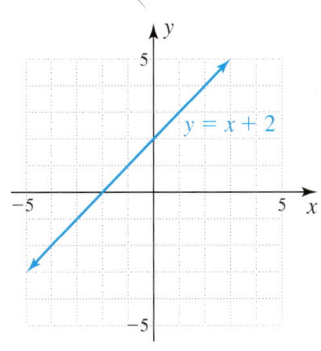

32.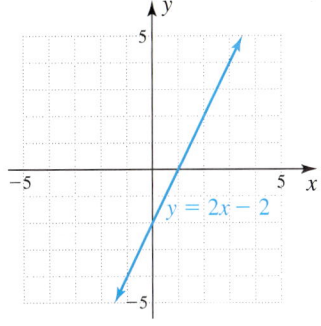

33.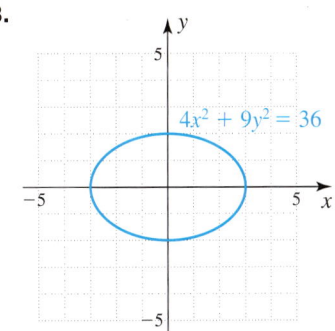

34.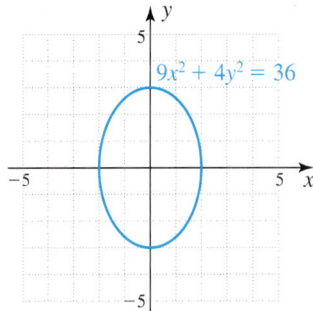

35.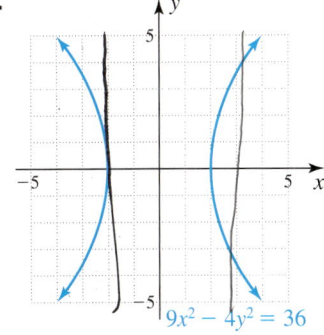

36.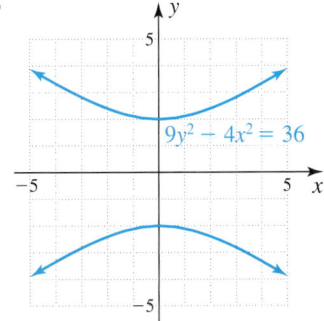

In Problems 37–42, use the vertical line test to determine whether the graphs represent functions.

37.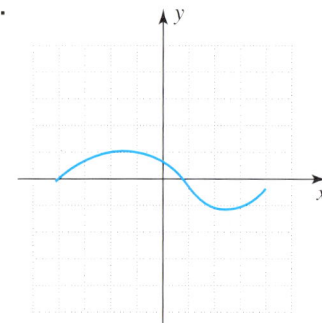

38.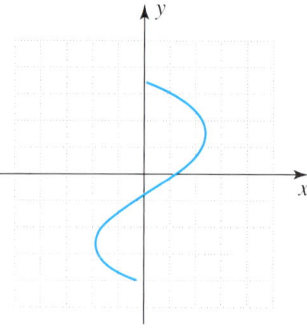

39.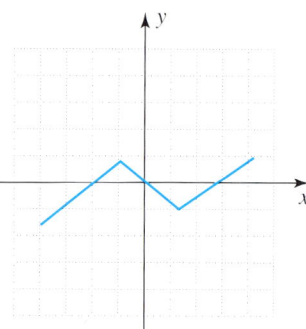

40.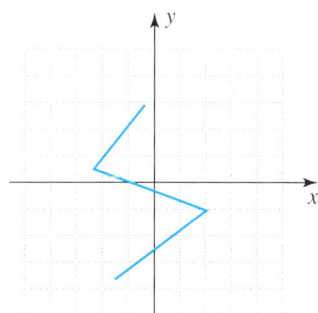

41.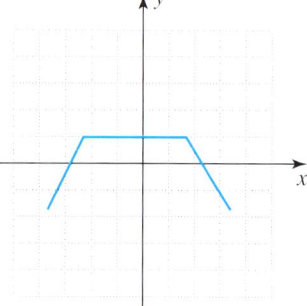

42.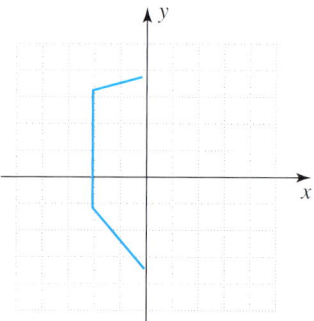

⟨ C ⟩ Finding the Domain of a Function In Problems 43–54, find the domain of the function defined by the given equation.

43. $y = \sqrt{x - 5}$

44. $y = \sqrt{x + 5}$

45. $y = \sqrt{4 - 2x}$

46. $y = \sqrt{6 + 3x}$

47. $y = \sqrt{x + 1}$

48. $y = \dfrac{1}{2x}$

49. $y = \dfrac{1}{x - 5}$

50. $y = \dfrac{1}{x + 5}$

51. $y = \dfrac{x + 2}{x + 5}$

52. $y = \dfrac{x + 3}{x - 6}$

53. $y = \dfrac{x}{(x + 2)(x + 1)}$

54. $y = \dfrac{x}{(x + 3)(x + 2)}$

⟨ D ⟩ Finding the Value of a Function In Problems 55–60, the rule defining a function is given.

55. Given: $f(x) = 3x + 1$. Find:
   a. $f(0)$ **b.** $f(2)$ **c.** $f(-2)$

56. Given: $g(x) = -2x + 1$. Find:
   a. $g(0)$ **b.** $g(1)$ **c.** $g(-1)$

57. Given: $F(x) = \sqrt{x - 1}$. Find:
   a. $F(1)$ **b.** $F(5)$ **c.** $F(26)$

58. Given: $G(x) = x^2 + 2x - 1$. Find:
   a. $G(0)$ **b.** $G(2)$ **c.** $G(-2)$

59. Given: $f(x) = \dfrac{1}{3x + 1}$ Find:
   a. $f(1)$ **b.** $f(1) - f(2)$ **c.** $\dfrac{f(1) - f(2)}{3}$

60. Given: $f(x) = \dfrac{x - 2}{x + 3}$ Find:
   a. $f(2)$ **b.** $f(3)$ **c.** $f(3) - f(2)$

The functions defined by $f(x) = 3x - 4$ and $g(x) = x^2 + 2x + 4$ are used in Problems 61–64.

61. Find:
   a. $f(3)$
   b. $g(3)$
   c. $f(3) + g(3)$

62. Find:
   a. $f(4)$
   b. $g(4)$
   c. $f(4) - g(4)$

63. Find:
   a. $f(-2)$
   b. $g(-3)$
   c. $f(-2) \cdot g(-3)$

64. Find:
   a. $f(-1)$
   b. $g(-2)$
   c. $\dfrac{f(-1)}{g(-2)}$

The functions $f = \{(1, 3), (-1, 5), (-3, 7), (-5, 9)\}$ and $g = \{(-2, 4), (0, 6), (2, 8), (4, 10)\}$ are used in Problems 65–68.

65. Find:
  a. $f(1)$ **b.** $g(-2)$ **c.** $f(1) + g(-2)$

66. Find:
  a. $f(-1)$ **b.** $g(0)$ **c.** $f(-1) - g(0)$

67. Find:
  a. $f(-3)$ **b.** $g(2)$ **c.** $f(-3) \cdot g(2)$

68. Find:
  a. $f(-5)$ **b.** $g(4)$ **c.** $\dfrac{f(-5)}{g(4)}$

>>> Applications

69. *Profit* The revenue obtained from selling x textbooks is given by $R(x) = 30x - 0.0005x^2$. The cost of producing the books is $C(x) = 100{,}000 + 6x$.
  a. Find the profit function $P(x) = R(x) - C(x)$.
  b. Find the profit when 10,000 books are sold.

70. *Temperature* The Fahrenheit temperature reading F is a function of the Celsius temperature reading C. This function is given by
$$F(C) = \frac{9}{5}C + 32$$
  a. If the temperature is 15°C, what is the Fahrenheit temperature?
  b. Water boils at 100°C. What is the corresponding Fahrenheit temperature?
  c. The freezing point of water is 0°C or 32°F. How many Fahrenheit degrees below freezing is a temperature of $-10°C$?
  d. The lowest temperature attainable is $-273°C$; this is the zero point on the absolute temperature scale. What is the corresponding Fahrenheit temperature?

71. *Pulse rate* When you exercise, your pulse rate should be within a certain *target zone*. The *upper limit* U of your target zone when exercising is a function of your age a (in years) and is given by
$$U(a) = -a + 190 \quad \text{(your pulse or heart rate)}$$
Find the highest safe heart rate for a person who is
  a. 50 years old
  b. 60 years old

72. *Pulse rate* The lower limit L of your target zone when exercising is a function of your age a (in years) and is given by
$$L(a) = -\frac{2}{3}a + 150$$
The target zone for a person a years old consists of all the heart rates between $L(a)$ and $U(a)$, inclusive. (See Problem 71.) If a person's heart rate is R, that person's target zone is described by $L(a) \leq R \leq U(a)$.
  a. Find the target zone for a person who is 30 years old.
  b. Find the target zone for a person who is 45 years old.

73. *Weight* The ideal weight w (in pounds) of a man is a function of his height h (in inches). This function is defined by
$$w(h) = 5h - 190$$
  a. If a man is 70 inches tall, how much should he weigh?
  b. If a man weighs 200 pounds, what should his height be?

74. *Renting a car* The cost C in dollars of renting a car for 1 day is a function of the number of miles traveled, m. For a car renting for $20 per day and 20¢ per mile, this function is given by
$$C(m) = 0.20m + 20$$
  a. Find the cost of renting a car for 1 day and driving 290 miles.
  b. An executive is given a $60-a-day expense budget. How many miles can she drive?

75. *Below sea level* The pressure P (in pounds per square foot) at a depth of d feet below the surface of the ocean is a function of the depth. This function is given by
$$P(d) = 63.9d$$
What is the pressure on a submarine at a depth of
  a. 10 ft?
  b. 100 ft?

76. *Distance* If a ball is dropped from a point above the surface of the earth, the distance s (in meters) that the ball falls in t seconds is a function of t. This function is given by
$$s(t) = 4.9t^2$$
Find the distance that the ball falls in
  a. 2 sec
  b. 5 sec

77. *Shoe size* Your shoe size S is a *linear* function of the length L (in inches) of your foot and is given by

$$S = m(L) = 3L - 22 \quad \text{(for men)}$$
$$S = f(L) = 3L - 21 \quad \text{(for women)}$$

a. What is the independent variable?

b. What is the dependent variable?

c. If the length of a man's foot is 11 inches, what is his shoe size?

d. If the length of a woman's foot is 11 inches, what is her shoe size?

78. *Ants* The speed S (in centimeters per second) of an ant is a *linear* function of the temperature C (in degrees Celsius) and is given by

$$S = f(C) = \frac{1}{6}(C - 4)$$

a. What is the independent variable?

b. What is the dependent variable?

c. What is the speed of an ant on a hot day when the temperature is 28°C?

d. What is the speed of an ant on a cold day when the temperature is 10°C?

79. *Robberies* According to FBI data, the number of robberies (per 100,000 population) is a *quadratic* function of t, the number of years after 1980, and is given by

$$R(t) = 1.85t^2 - 19.14t + 262$$

a. What was the number of robberies (per 100,000) in 2000?

b. What do you predict that the number of robberies would be in the year 2010?

80. *Recycling* Do you recycle glass, newspaper, and plastic? The waste recovered (in millions of tons) is a *quadratic* function of t, the number of years after 1960, and is given by

$$G(t) = 0.04t^2 - 0.59t + 7.42$$

a. How many million tons of waste were recovered in 1960?

b. How many million tons of waste would you expect to be recovered in the year 2010?

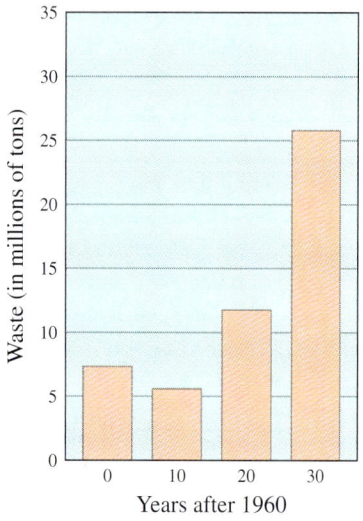

81. *Blood alcohol level* The graphs show that your blood alcohol level L is related to the number of drinks D you consume, your weight, sex, and the time period in which the drinks were consumed.

a. Do the two graphs indicating the effects of alcohol for males and females represent functions?

b. What are the domain and range for the graph representing males?

c. What are the domain and range for the graph representing females?

Blood alcohol levels

Fatalities

Most states consider a driver drunk with a blood alcohol concentration (BAC) of 0.08%. Approximately 40% of all fatal accidents involving drivers and pedestrians in 2004 were alcohol related.

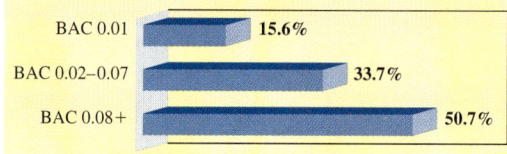

Effects of alcohol

How blood alcohol levels are affected by each drink over a 2-hour period:

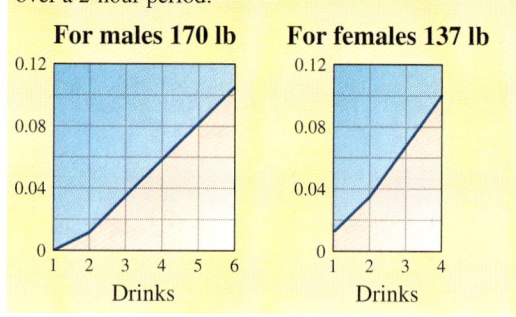

82. *Emissions* The graphs show the worldwide emissions (in parts per trillion, ppt) of two ozone-destroying chemicals: CFC-12 and CFC-11.

   a. Do both graphs represent functions?

   b. What is the common domain of the functions?

   c. What is the range for CFC-11?

   d. What is the range for CFC-12?

   e. From 1980 to 1992, CFC-12 emissions can be modeled by the linear function $f(t) = 20t + 300$, where t is the number of years after 1980. What is $f(0)$, and is it close to the value for 1980 shown on the graph?

   f. What is $f(10)$, and is it close to the corresponding value on the graph?

   g. If you use this same linear model, what would be the CFC-12 emissions in the year 2000? Is this prediction consistent with the value represented on the graph? What is the difference in parts per trillion between the linear model and the value represented on the graph?

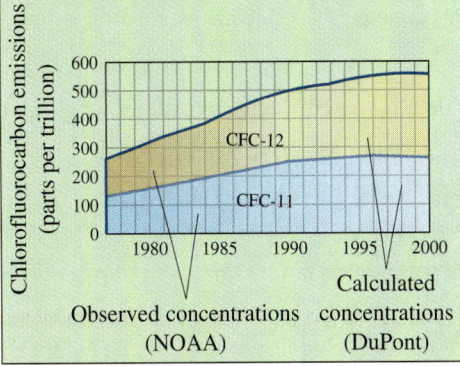

Source: Data from AP/Wide World Photos.

>>> Write On

In Problems 83–84, what is the difference between:

83. A function and a relation when they are written as a set of ordered pairs?

84. The graph of a function and the graph of a relation?

In Problems 85–86, write your own definition of:

85. The domain of a function.

86. The range of a function.

In Problems 87–88, answer true or false. Explain your answer.

87. Every relation is a function.

88. Every function is a relation.

>>> Concept Checker

Fill in the blank(s) with the correct word(s), phrase, or mathematical statement.

89. The set of all second coordinates of a relation is called the _____.

90. A relation in which each domain member is related to exactly one range member is called a _____.

91. The variable "x" in a relation can also be called the _____.

92. A correspondence between two sets can be called a _____.

relation **independent variable**
domain
range **dependent variable**
function
mapping diagram

>>> Using Your Knowledge

Everything You Always Wanted to Know about Relations (in Mathematics) A special kind of relation that's important in mathematics is called an **equivalence relation.** A relation R is an equivalence relation if it has the following three properties:

a. Reflexive: If a is an element of the domain of R, then (a, a) is an element of R.

b. Symmetric: If (a, b) is an element of R, then (b, a) is an element of R.

c. Transitive: If (a, b) and (b, c) are both elements of R, then (a, c) is an element of R.

A simple example of an equivalence relation is

$$R = \{(x, y) | y = x, x \text{ an integer}\}$$

To show that R is an equivalence relation, we check the three properties:

a. Reflexive: If a is an integer, then $a = a$, so (a, a) is an element of R.

b. Symmetric: Suppose that (a, b) belongs to R. Then, by the definition of R, a and b are integers and $b = a$. But if $b = a$, then $a = b$, so (b, a) also belongs to R.

c. Transitive: Suppose that (a, b) and (b, c) both belong to R. Then $b = a$ and $c = b$, so $c = a$. Hence (a, c) also belongs to R.

Because R has all three properties, it is an equivalence relation.

The pairs in a relation don't have to be numbers, and some interesting relations occur outside the realm of numbers. For example,

$$R = \{(x, y) | y \text{ is a member of the same immediate family as } x, x \text{ is a person}\}$$

is a relation. Is R an equivalence relation?

We check the three properties as before:

a. Reflexive: Given a person A, is (A, A) an element of R? Yes, A is obviously a member of the same immediate family as A.

b. Symmetric: Suppose that (A, B) is an element of R. Then B is a member of the same immediate family as A. But then A is a member of the same immediate family as B, so (B, A) is an element of R.

c. Transitive: Suppose that (A, B) and (B, C) both belong to R. Then A, B, and C are all members of the same immediate family. Thus, (A, C) is an element of R.

Again, we see that R has all three properties, so it is an equivalence relation.
Can you determine which of the following are equivalence relations?

93. $R = \{(x, y) | x \text{ and } y \text{ are triangles and } y \text{ is similar to } x\}$ (*Note:* Here, "is similar to" means "has the same shape as.")

94. $R = \{(x, y) | x \text{ and } y \text{ are integers and } y > x\}$

95. $R = \{(x, y) | x \text{ and } y \text{ are positive integers and } y \text{ has the same parity as } x\}$; that is, y is odd if x is odd and y is even if x is even.

96. $R = \{(x, y) | x \text{ and } y \text{ are boys and } y \text{ is the brother of } x\}$

97. $R = \{(x, y) | x \text{ and } y \text{ are positive integers and when } x \text{ and } y \text{ are divided by 3, } y \text{ leaves the same remainder as } x\}$

98. $R = \{(x, y) | x \text{ is a fraction } \frac{a}{b} \text{ where } a \text{ and } b \text{ are integers } (b \neq 0), \text{ and } y \text{ is an equivalent fraction } \frac{ma}{mb} \text{ where } m \text{ is a nonzero integer}\}$

>>> Mastery Test

99. If $f = \{(4, 3), (5, -1), (6, 0)\}$, find:

  a. $f(4)$ **b.** $f(5)$ **c.** $f(4) - f(5)$

100. If $f(x) = 2x - 3$, find:

  a. $f(4)$ **b.** $f(2)$ **c.** $f(x + 1)$

Find the domain of each function:

101. $f(x) = \dfrac{1}{x - 2}$

102. $y = \sqrt{x - 3}$

Use the vertical line test to determine whether the relation shown is a function:

103.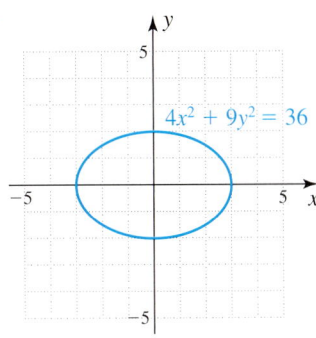

104.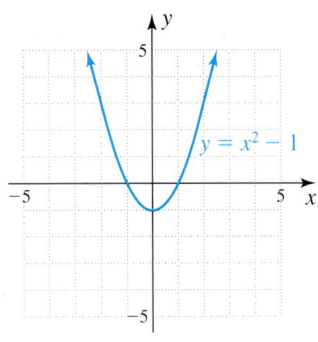

Find the domain and range of the relations:

105.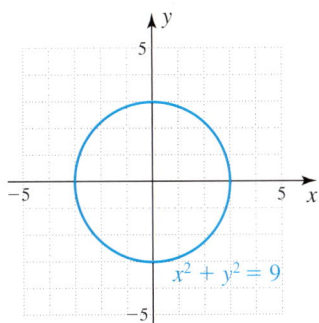

106.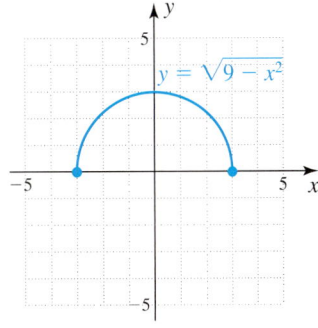

107.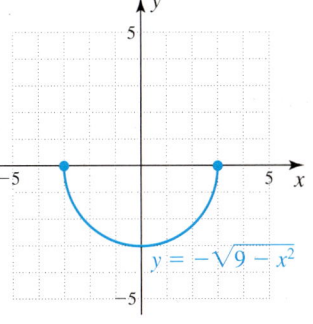

108. Find the domain and range of the relation $\{(x, y) \mid y = 2x + 4\}$.

109. Find the domain and range of the relation $B = \{(7, 8), (8, 9), (9, 10)\}$.

》》》 Skill Checker

110. Graph the following ordered pairs and name the axis where each point is graphed.

A: $(0, -3)$
B: $\left(0, 4\frac{1}{2}\right)$
C: $(2, 0)$
D: $\left(-\frac{1}{2}, 0\right)$

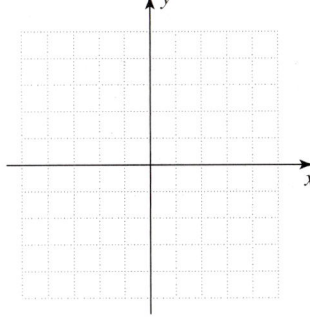

111. Find the reciprocal of each number:

A: $\frac{2}{3}$ B: $\frac{1}{5}$
C: -8 D: $-\frac{3}{4}$

In Problems 112–118, simplify.

112. $\dfrac{1 - 6}{3 - (-2)}$

113. $\dfrac{10 - 6}{14 - 2}$

114. $\dfrac{-7 - 5}{6 - 6}$

115. $\dfrac{-8 + 8}{-3 - 3}$

116. $\dfrac{1 - \frac{3}{5}}{-1 - 9}$

117. $\dfrac{\frac{1}{2} - (-1)}{1 - \frac{1}{4}}$

118. $\dfrac{1 - 0.7}{-0.2 - 0.3}$

3.6 Linear Functions

Objectives

A Identify linear and nonlinear functions from graphs or equations.

B Find the equation of a linear function from a graph.

C Use mathematical modeling with linearly related data.

To Succeed, Review How To . . .

1. Plot points (pp. 166–168).
2. Graph lines (pp. 168–174).
3. Write equations of lines (pp. 200–205).

Getting Started
Diamonds Are Forever, But Can We Afford Them?

Everyone knows that as the diamond gets larger and the clarity gets better, the price goes up. Examining the data for the weight of some diamonds and their costs may give the information needed to predict the cost of any weight diamond. The weight of a diamond is measured in carats. Let's get started by plotting the points from the data set in Table 3.4 on a graph called a *scatter plot,* Figure 3.52.

Carats	Price	Carats	Price
0.17	355	0.18	462
0.16	328	0.28	823
0.17	350	0.16	336
0.18	325	0.20	498
0.25	642	0.23	595
0.16	342	0.29	860
0.15	322	0.12	223
0.19	485	0.26	663
0.21	483	0.25	750
0.15	323	0.27	720

Table 3.4

The points on the scatter plot closely resemble a line and so we can say this data is *linearly related*.

We could use a graphing utility to find the *line of best fit,* or line of *least squares,* by finding the linear regression equation, $y = ax + b$. However, we will use the skills we learned in Section 3.3 for how to write the equation of a line given two points. But which two points do we choose? The line of best fit should be drawn as near to as many of the points as possible to more clearly show the linear trend. We will choose points (0.16, 342) and (0.27, 720). We find the slope of the line first, using the slope formula.

$$m = \frac{y_2 - y_1}{x_2 - x_1} = \frac{720 - 342}{0.27 - 0.16} = \frac{378}{0.11} = 3436 \quad \text{(Rounded to the nearest whole number)}$$

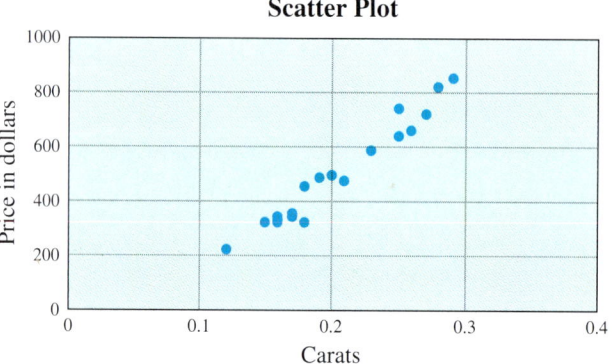

▶Figure 3.52

To find the equation of the line, we use the point-slope formula.

$$y - y_1 = m(x - x_1)$$
$$y - 342 = 3436(x - 0.16)$$
$$y - 342 = 3436x - 550 \quad \text{(550 rounded)}$$
$$y = 3436x - 208 \quad \text{or} \quad f(x) = 3436x - 208$$

In this equation, x is the weight of the diamond in carats and y is the price in dollars.

Diamonds are forever, but can we afford them? Now we have an equation to use to predict the price of a diamond given the weight in carats. Can you afford to buy a one-carat diamond? Substitute 1 in for x and solve for y.

$$y = 3436(1) - 208$$
$$y = 3228$$

If your budget allows for $3228, then you can afford a one-carat diamond.

This is an example of modeling linearly-related data, which we will study in this section.

A > Identifying Linear Functions

It is not difficult to recognize the graph of a line, but are all lines necessarily a function? Recall from Section 3.5 that a function is a relation in which each domain member is related to exactly one range member. When examining the graph of a function, it must pass the "vertical line test." Examine the graphs of the lines in Figure 3.53. All the lines except for the one in Graph C pass the vertical line test, so we can see that all lines are functions with the exception of vertical lines.

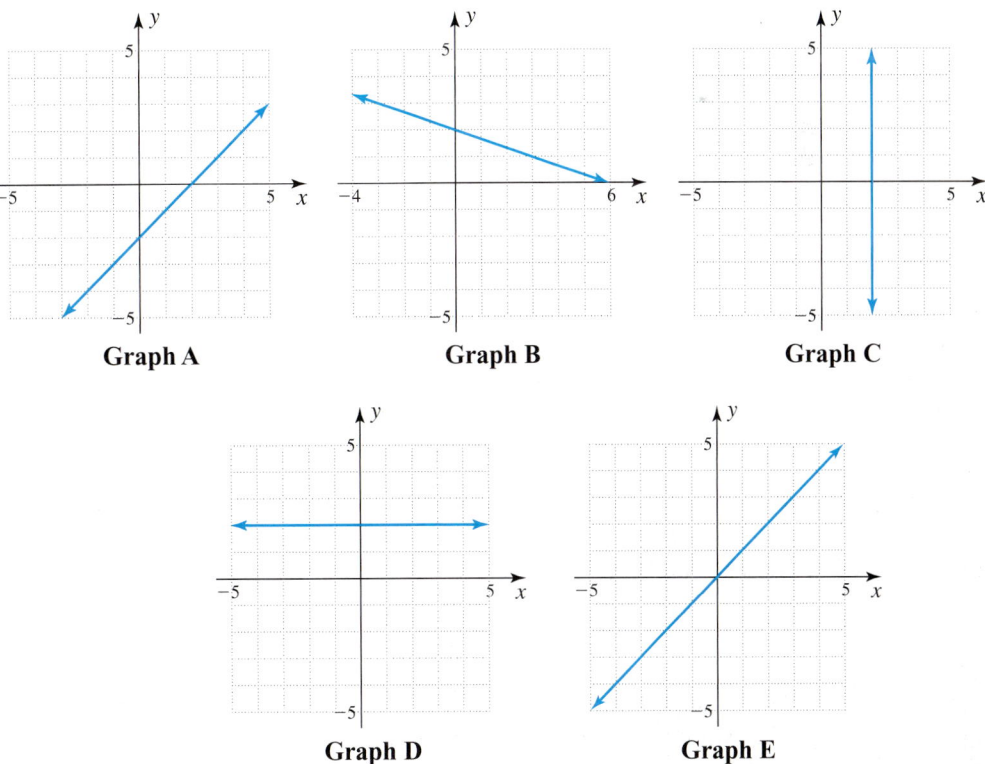

>Figure 3.53

If a function is not linear, then it is said to be a **nonlinear function.** Examples of graphs that are nonlinear functions are shown in Figure 3.54.

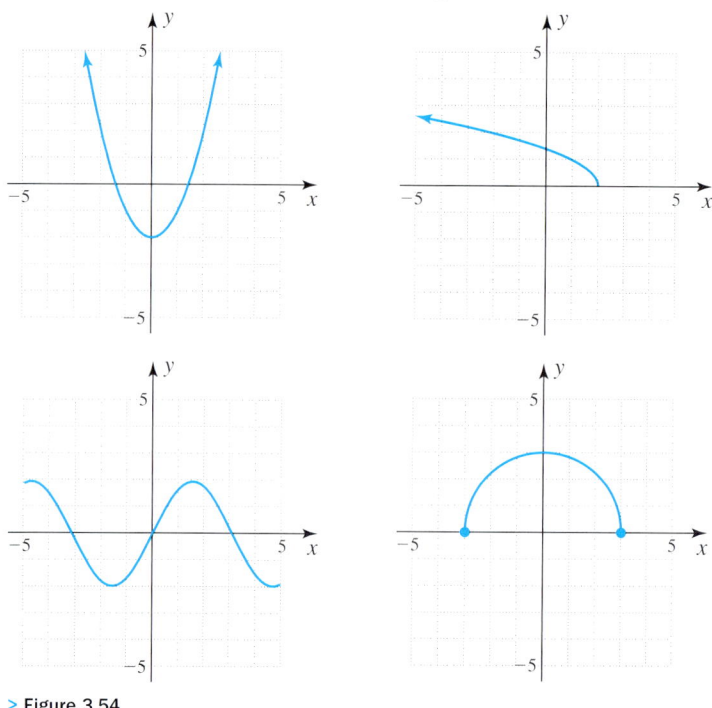

> Figure 3.54

We have studied two forms of a linear equation—standard form ($Ax + By = C$) and slope-intercept form ($y = mx + b$). When referring to the equation of a line as a **linear function,** we will use the $y = mx + b$ form and write it as follows.

LINEAR FUNCTIONS	A linear function *f* is any function that can be expressed by $f(x) = mx + b$

In Figure 3.53, Graphs A, B, and E are linear functions with domain and range being the set of all real numbers. When observing the graph of a line from left to right, if it *rises,* then the slope is positive as in the case of Graphs A and E. This is called an **increasing function.** However, in Graph B we notice the line *falls* as we observe it from left to right, so it has a negative slope. It is called a **decreasing function.** Graph D, the horizontal line, is also a linear function with equation $y = 2$. As a function, it can be expressed $f(x) = 2$. This means the domain can be any real number but y is *constantly* 2. This is a special case of the linear function and is called the **constant function.** Special cases of linear functions are summarized as follows.

SPECIAL CASES OF LINEAR FUNCTIONS

Increasing Function
$f(x) = mx + b$
m is positive

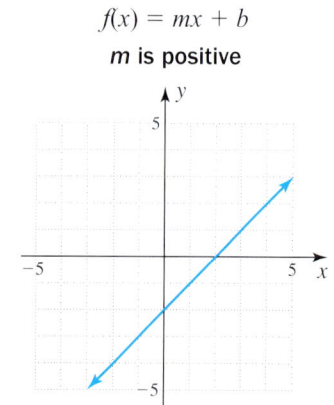

Decreasing Function
$f(x) = mx + b$
m is negative

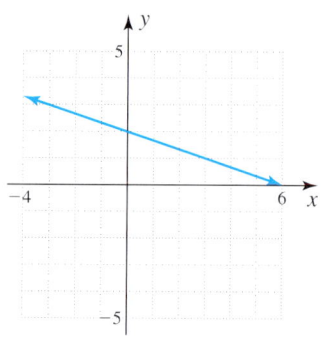

Constant Function

$f(x) = b$

$m = 0$

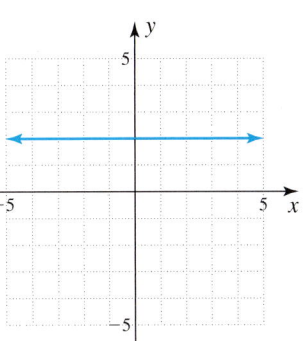

In studying functions, it is important to be able to identify linear and nonlinear functions as well as special cases of linear functions by observing their graphs, as we will do in Example 1.

EXAMPLE 1 Identifying special cases of linear functions from their graphs

Which of the following graphs are linear functions and which are nonlinear functions? Of those that are linear functions, identify them as increasing, decreasing, or constant functions.

a.
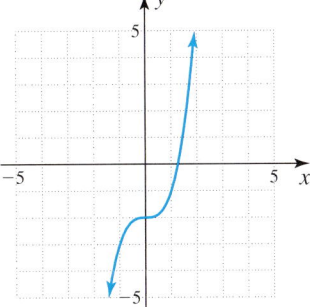

b.
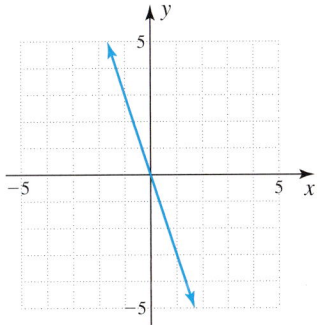

c.
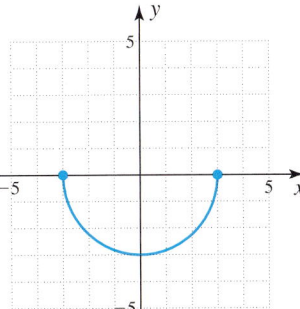

d.
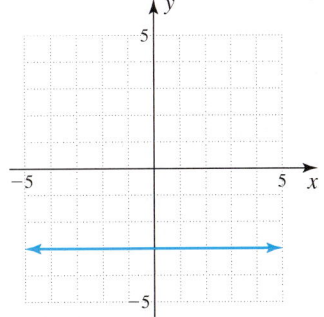

e.
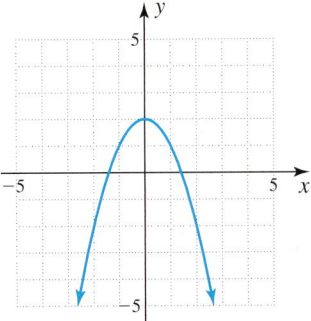

f.
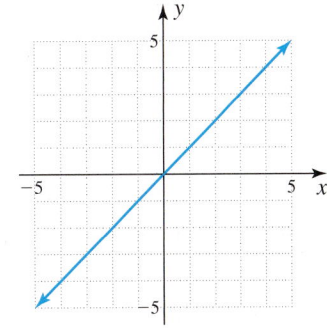

PROBLEM 1

Which of the following graphs are linear functions and which are nonlinear functions? Of those that are linear functions, identify them as increasing, decreasing, or constant functions.

a.
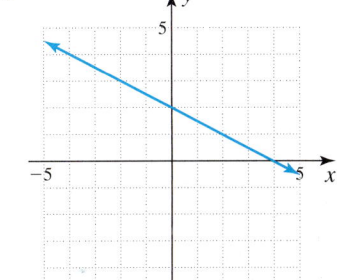

b.
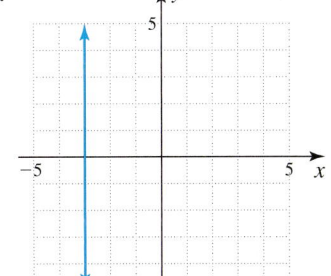

c.
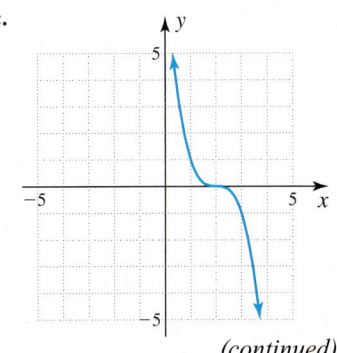

(continued)

Answers to PROBLEMS

1. **a.** Decreasing linear function   **b.** Straight vertical line but not a function   **c.** Nonlinear function

SOLUTION

a. The graph is not a straight line. Therefore, this is a nonlinear function.
b. The graph is a nonvertical straight line, and so it is a linear function. Since the line *falls* from left to right, the slope is negative, and thus, the graph is a decreasing linear function.
c. This is not the graph of a straight line, so it is a nonlinear function.
d. This graph is a horizontal line with slope 0, so it is a constant function.
e. This is a nonlinear function since it is not the graph of a straight line.
f. The graph of this nonvertical straight line *rises* from left to right, so we say it is an increasing linear function.

d.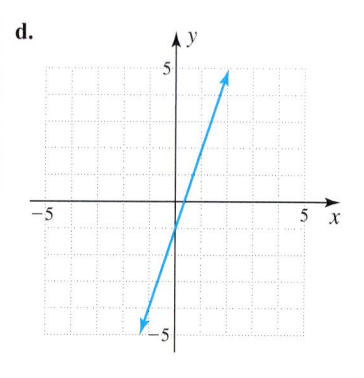

We have discussed how to identify linear functions from their graphs, but how do we recognize when an equation is that of a linear function? The equation will be a linear function if the exponents of the two variables x and y are positive 1 and it can be expressed in the form, $y = mx + b$ or $f(x) = mx + b$. For example, $2x + y = 5$ is an example of a linear function since both the exponents of x and y are understood to be positive 1 and it can be written as $y = -2x + 5$ or $f(x) = -2x + 5$. Another example to consider is $3x^2 + y = 4$, which is a case where *both* variables are *not* raised to the first power. Even though the equation can be written $y = -3x^2 + 4$, it is not a linear function. One example that could be confusing at first glance would be the equation, $y = \frac{2}{x} + 6$. In the fraction $\frac{2}{x}$, the exponent of x is positive one, but we know from the study of exponents in Section 1.3 that $\frac{2}{x}$ means $2x^{-1}$. Since the exponent of the variable x is not positive 1, we say this equation is not a linear function.

EXAMPLE 2 Identifying special cases of linear functions from their equations

Which of the following equations are linear functions and which are nonlinear functions? If they are linear functions, then write them using function notation and identify them as increasing, decreasing, or constant functions.

a. $-6x + 2y = -4$ b. $g(x) = \sqrt{x}$ c. $y = 5x^3 + 4$
d. $p(x) = 6.5$ e. $h(x) = \frac{1}{x}$ f. $x + y = 0$

SOLUTION

a. This is a linear function since both variables have exponents of positive 1 and it can be written as $f(x) = 3x - 2$. The slope is $+3$, so it is increasing.
b. Although the variable x is raised to the first power, the fact that it is the square root of x makes it a nonlinear function.
c. The exponent of the variable y is one but the exponent of x is 3, making it a nonlinear function.
d. This equation can also be written as $y = 6.5$, which is the special case of a linear function called a constant function.
e. Since $\frac{1}{x}$ by definition means x^{-1}, it is a nonlinear function.
f. This equation can be written as $y = -x$ and has positive 1 as the exponent of both variables. This is a linear function, $f(x) = -x$, with a negative slope, so it is decreasing.

PROBLEM 2

Which of the following equations are linear functions and which are nonlinear functions? Of those that are linear functions, identify them as increasing, decreasing, or constant functions.

a. $f(x) = -x + 1$
b. $g(x) = -x^2$
c. $12 = x - 4y$
d. $y = -\frac{1}{2}$

B ▶ Finding the Equation of a Linear Function from a Graph

In the *Getting Started* section, we demonstrated how being able to write the equation of a line for a given data set could be helpful in predicting values. In Section 3.3, we studied in detail how to write the equation of a line given certain information. We will apply it in Example 3.

Answers to PROBLEMS

1. d. Increasing linear function
2. a. Linear decreasing function
    b. Nonlinear function
    c. Linear increasing function
    d. Linear constant function

EXAMPLE 3   Writing the equation of a linear function

The graph of this linear function describes the number C of calories burned when running t minutes.

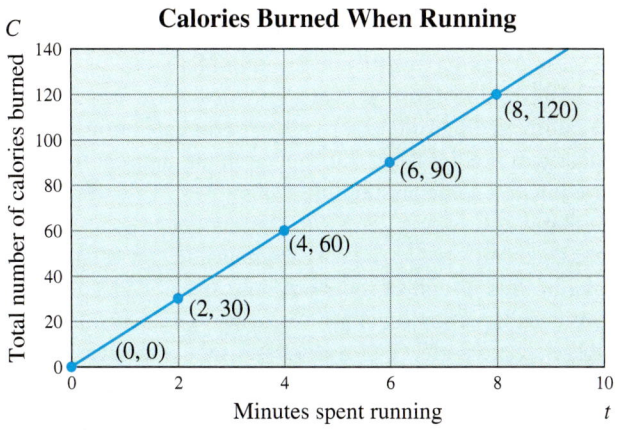

a. What is the rate of change of the number of calories burned?
b. Write the equation of the line where C is the calories burned and t is time in minutes. Then write the equation in function notation.
c. How many calories would the runner burn if he ran 1 hour?
d. How long would it take the runner to burn 630 calories?

SOLUTION

a. The rate of change can be found by comparing the change in calories to the change in minutes. We use the time interval from 0 to 2 minutes and get the ratio

$$\frac{\text{Change in calories}}{\text{Change in minutes}} = \frac{30-0}{2-0} = \frac{30}{2} = \frac{15}{1}$$

The rate of change is $\frac{15}{1}$ or we say the runner will burn 15 calories per minute. The rate of change is the same as the slope of the line.

b. Since the rate of change found in part **a** is the same as the slope of the line, we know $m = 15$. The graph indicates the y-intercept is 0. So we use the slope and y-intercept form of the equation, $y = mx + b$, to write $C = 15t + 0$. Thus, the equation of the line is $C = 15t$ or $f(t) = 15t$.

c. The function describing this linear relationship is written for t minutes, so we rewrite 1 hour as 60 minutes. Using the equation $f(t) = 15t$, we replace t with 60.

$$f(t) = 15(60) = 900$$

The runner will burn 900 calories if he runs for one hour.

d. The equation describing this linear relationship is $f(t) = 15t$ or $C = 15t$, so we replace C with 630 and solve for t.

$$630 = 15t \quad \text{Divide both sides by 15.}$$
$$42 = t$$

It will take the runner 42 minutes to burn 630 calories.

PROBLEM 3

The following line graph describes the possible linear relationship between a salesman's monthly gross pay and his sales for his first 2 months.

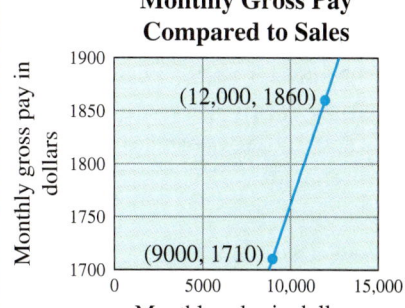

a. Write the equation of the line describing this linear relationship where P is the monthly gross pay and s is the monthly sales. Then write the equation in function notation.
b. What is the rate of change of the monthly gross pay?
c. Assuming his sales continue to increase at this same rate, what would his monthly gross pay be if his monthly sales were $20,000?

Answers to PROBLEMS

3. a. $P = 0.05s + 1260$;
$f(s) = 0.05s + 1260$ b. $1 gross pay per $20 in sales or $\frac{1}{20}$
c. $2260

C ❯ Mathematical Modeling Using Linearly Related Data

The study of linear functions would not be complete without including applications that demonstrate how mathematics can model some aspects of the world. **Mathematical modeling** is the process of taking a verbal description of a problem, assigning variables to the unknown quantities, and forming a mathematical model or algebraic equation. Gathering data in real life settings either through experiments or observations can often lead to a function that will predict what will happen in the future. As we learned earlier, not all functions are linear—some are nonlinear. However, we will restrict our study in this section to data that will lead to a linear function model.

Before we can make a linear model, we have to know if the set of data is linearly related. To begin the process, we find or collect some ordered pairs of data and graph the ordered pairs on a graph that is called a **scatter plot.** It is called a scatter plot because the points rarely fall in a straight line or a perfect curve. What we want to look for in the scatter plot is whether or not the points tend to gather in a "linear" fashion. In Tables 3.5 and 3.6, there are two data sets—one regarding a person's smoking index and the incidence of lung cancer and the other is about the length of an alligator compared to its weight. Figures 3.55 and 3.56 are the corresponding scatter plots made by plotting the ordered pairs of points from the tables.

Smoking Index	Incidence of Cancer	Smoking Index	Incidence of Cancer
77	84	107	86
137	116	112	96
117	123	113	144
94	128	110	139
116	155	125	113
102	101	133	125
111	118	115	146
93	113	105	115
88	104	87	79
102	88	91	85
91	104	100	120
104	129	76	60

Table 3.5

Length	Weight	Length	Weight
58	28	85	84
61	44	86	83
63	33	86	80
68	39	86	90
69	36	88	70
72	38	89	84
72	61	90	106
74	54	90	102
74	51	94	110
76	42	94	130
78	57	114	197
82	80	128	366

Table 3.6

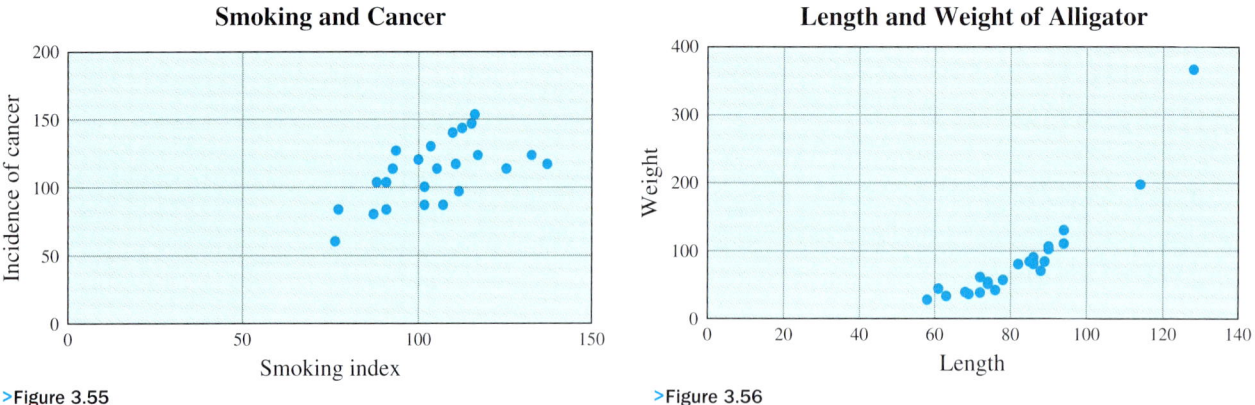

❯Figure 3.55 ❯Figure 3.56

Comparing the two scatter plots, we see that the one in Figure 3.55 tends to resemble a line in the way the points are scattered; the one in Figure 3.56 tends to be more of a curve. We can conclude that the data set regarding the smoking index compared to the incidence of lung cancer, Table 3.5, is linearly related. We could make a linear function from that data and use it to predict the incidence of lung cancer from any given smoking index. But first we must practice making scatter plots and deciding if the data is linearly related as we will do in Example 4.

EXAMPLE 4 Identifying linearly related data sets

Make a scatter plot of the given data sets and decide if the data is linearly related.

a. Rubber O-rings, such as those used on the space shuttle Challenger, can lose their resiliency due to low temperatures. This could result in damage to the aircraft and loss of life, which was the case in the 1986 launch. Studying this data set could lead to safer air shuttle take-offs.

Temp. °C	Damage Index	Temp. °C	Damage Index
12	11	21	4
14	4	21	0
14	4	21	4
17	2	21	0
19	0	21	0
19	0	22	0
19	0	23	0
19	0	24	4
19	0	24	0
20	0	24	0

b. The stride rate, in steps per second, of a runner is related to speed; the greater the stride rate the greater the speed. Studying this data could help a runner be more efficient by getting the stride rate to its optimum.

Stride Rate	Speed
3.05	4.83
3.12	5.15
3.17	5.33
3.25	5.68
3.36	6.09
3.46	6.42
3.55	6.74

PROBLEM 4

Make a scatter plot of the given data sets and decide if the data is linearly related.

a. These data examine the trends in carbon dioxide concentration in parts per million (ppm) in the atmosphere over a period of 16 years.

Year	CO_2 (ppm)	Year	CO_2 (ppm)
1	338.5	9	351.4
2	339.8	10	352.7
3	341	11	354
4	342.6	12	355.5
5	344.3	13	356.2
6	345.7	14	357
7	347	15	358.8
8	348.8	16	360.7

(continued)

Answers to PROBLEMS

4. a. Linearly related data

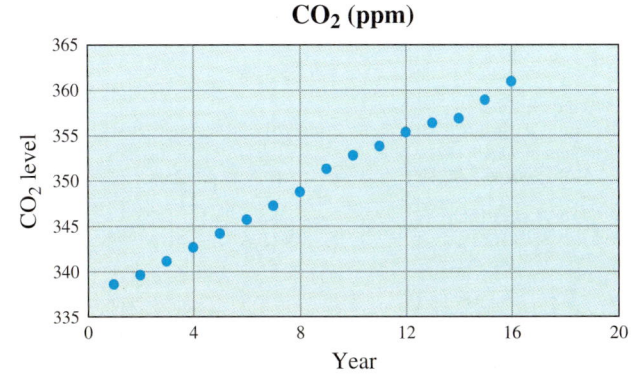

SOLUTION

a. Plotting the ordered pairs from the data table gives the scatter plot in Figure 3.57. The points seem to have a curved shape but definitely not a trend toward being in the shape of a line. So we say these data are not linearly related. To pursue a function model for these data that best captures this curved pattern, we would have to consider some nonlinear functions we haven't studied yet.

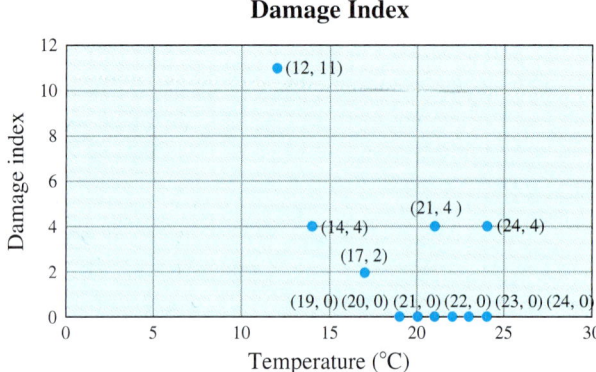

>Figure 3.57

b. Plotting the ordered pairs from the data table, we get the scatter plot in Figure 3.58. The points do not form a perfect straight line but there is a definite trend towards the data being linearly related. Knowing the linear function related to this data could help a runner find his optimum stride rate.

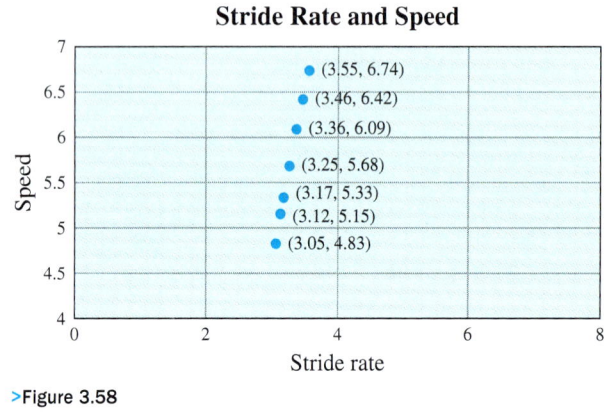

>Figure 3.58

b. This data examines the trend in the world population from the year 0 to 2005.

World Population

Year	Population (billions)
0	0.3
1000	0.32
1750	0.8
1800	1
1930	2
1960	3
1974	4
1987	5
1990	5.2
1995	5.7
2005	6.5

In the prior discussion of the data concerning the smoking index compared to the incidence of lung cancer it was established that there was a linear relationship with that data (Table 3.5). We can approximate the *best line* that will fit the data by connecting the

---

Answers to PROBLEMS

4. b. Nonlinearly related data

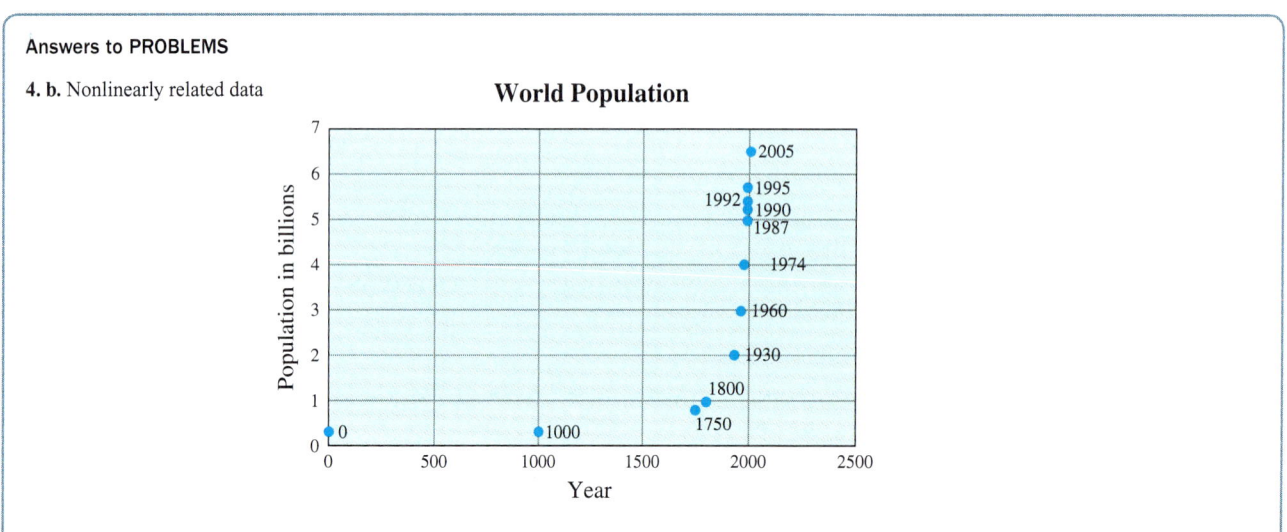

3.6 Linear Functions

Smoking Index	Incidence of Cancer	Smoking Index	Incidence of Cancer
77	84	107	86
137	116	112	96
117	**123**	113	144
94	128	110	139
116	155	125	113
102	101	133	125
111	118	115	146
93	113	105	115
88	104	87	79
102	88	91	85
91	104	100	120
104	129	76	60
107	86		

Table 3.5

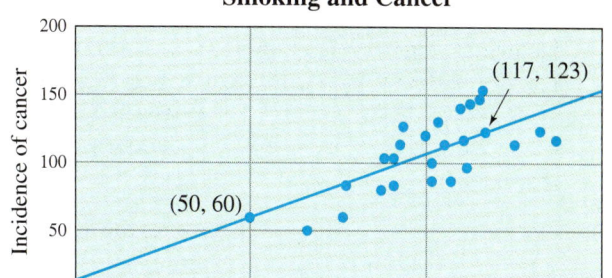

> Figure 3.59

two points that will have most of the points on the line, or close to it. This is called the **line of best fit.** The two points that we connect do not have to be in our data set. The two points we have chosen to use for the line of best fit are (117, 123), which is highlighted in our data table and another point (50, 60). The line is drawn in Figure 3.59. Given those two points we will write the equation of that line by first finding the slope and then using the point-slope formula.

Slope formula:

$$m = \frac{y_2 - y_1}{x_2 - x_1} = \frac{123 - 60}{117 - 50} = \frac{63}{67} = 0.94$$

Point-slope formula:

$$y - y_1 = m(x - x_1)$$
$$y - 60 = 0.94(x - 50)$$
$$y - 60 = 0.94x - 47$$
$$y = 0.94x + 13$$

Now we have a linear function, $f(x) = 0.94x + 13$, which models this data. A more accurate way to find the line of best fit is to use a graphing utility. Putting this data into a graphing utility, the line of best fit is the function, $f(x) = 0.89x + 17.9$. The two functions are very similar and should be able to adequately predict the risk of lung cancer based on a person's smoking index.

The smoking index is based on the number of cigarettes smoked per day, and the incidence of cancer is based on the mortality records for lung cancer deaths. So if a person knows their smoking index, they can substitute that value for x in the function and predict their incidence of lung cancer.

Calculator Corner

Entering Lists and Making a Scatter Plot

We are going to plot data using two features: plotting data points and making a scatter plot *or* scattergram.

First, erase any old data. With a TI-83 Plus, clear any old lists by pressing [2nd] [+] [4] [ENTER]; then press [STAT] [1]. Enter the *x*-coordinates under L1, (2, −1, −2, 0), by entering each number followed by [ENTER]. To enter the *y*-coordinates, press the right arrow to get to L2 and enter 3, 2, −1, and −3. Now press [2nd] [Y=] [ENTER] [ENTER] to turn the plot feature ON. Go to "**TYPE**" (line 2) and select the first type of graph, called a scattergram, by pressing [ENTER]. Finally, press [ZOOM] [9] and the scattergam of the points appears, as shown in the window. Pressing [ZOOM] [9] selects the correct window for you!

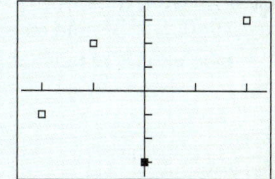

STEPS TO WRITING A LINEAR FUNCTION THAT MODELS LINEARLY RELATED DATA

1. First decide if the data is linearly related by making a scatter plot of the points in the data table. If it is linear, then follow steps 2–5.
2. Find two points that, when connected, will give the line of best fit.
3. Use those two points to find the slope of the line.
$$m = \frac{y_2 - y_1}{x_2 - x_1}$$
4. Use the slope and one point in the point-slope formula to find the equation of the line.
$$(y - y_1) = m(x - x_1)$$
5. Once the equation is written as $y = mx + b$, it can be rewritten as $f(x) = mx + b$.

We've heard it said that people are living longer these days than they did 50 or 100 years ago. How much longer are people living? We will examine the data in Example 5 and answer that question.

EXAMPLE 5 How long can you expect to live?

According to the U.S. National Center for Health Statistics, life expectancy has been steadily increasing since 1950. Table 3.7 indicates the life expectancy from birth of a person born in each of the years listed.

How can we use this information to find the approximate life expectancy for the year 2010?

Year	Life Expectancy
1950	68.2
1960	69.7
1970	70.8
1980	73.7
1990	75.4
2000	76.9
2010	?

Table 3.7

PROBLEM 5

Using the table in Example 5, find the equation of the line using $t = 0$, $y = 68.2$ and $t = 50$, $y = 76.9$. How does that equation compare to the result in Example 5?

SOLUTION If we could develop an algebraic equation using this information that would model the life expectancy based on the number of years after 1950 the person was born, then the solution would be easy to obtain.

We start with step 1 of the five steps to writing a linear function that models linearly related data.

1. Make a scatter plot of the points in the data table. The time column is in 10-year periods, so we will rename the years 0, 10, 20, 30, 40, 50 as our t values, representing the 10-year periods after 1950. For 1950, $t = 0$, for 1960, $t = 10$, and so on. The second column, "Life Expectancy," will be labeled y. The new table and its graph shown here indicate that the scatter of

---

Answers to PROBLEMS

5. $y = 0.174t + 68.2$
This equation is similar to the solution of Example 5.

the points closely resembles a line, so we will follow steps 2–5 to write the linear function.

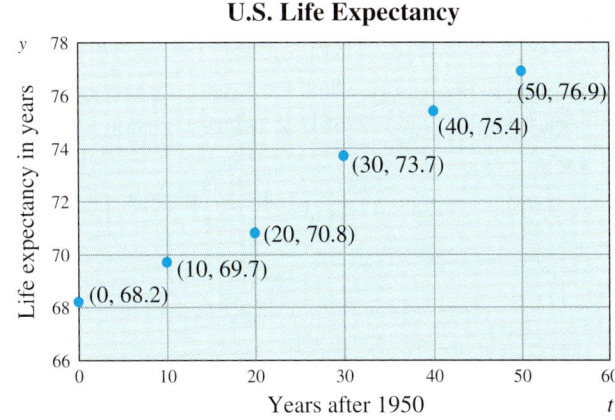

t	y
0	68.2
10	69.7
20	70.8
30	73.7
40	75.4
50	76.9
60	?

U.S. Life Expectancy

2. The two points we will use to draw the line of best fit are highlighted in the data table.
3. We use those two points, (10, 69.7) and (40, 75.4), to find the slope.

$$m = \frac{y_2 - y_1}{x_2 - x_1} = \frac{75.4 - 69.7}{40 - 10} = \frac{5.7}{30} = 0.19$$

4. The point-slope formula with $m = 0.19$ and the point (10, 69.7) is used to find the equation.

$$y - y_1 = m(x - x_1)$$
$$y - 69.7 = 0.19(t - 10)$$
$$y - 69.7 = 0.19t - 1.9$$
$$y = 0.19t + 67.8$$

5. The function that will approximate the life expectancy for the year 2010 is

$$f(x) = 0.19t + 67.8$$

To be more precise, we could use a graphing utility to find the line of best fit, or line of **least squares**, by finding the **linear regression equation**, $y = ax + b$. (See the Calculator Corner below.)

Since $t = 60$ for the year 2010, we substitute 60 for t in the function.

$$f(60) = 0.19(60) + 67.8 = 79.2$$

A person born in 2010 can expect to live about 79.2 years.

Calculator Corner

Write the Equation of a Line

As you recall from Section 3.3, we can use the TI-83 Plus to find the equation of a line given two points. Before we begin Example 5, let's clear any old lists by pressing [2nd] [+] [4] [ENTER]. Now, press [STAT] [1] and enter 0, 10, 20, 30, 40, and 50 under L1 and enter the y-values under L2. Press [STAT] [▶] [4] [ENTER] to get the equation $y = ax + b$, with $a \approx 0.18$ and $b \approx 67.9$, that is, the line with equation

$$y = 0.18x + 67.9$$

As you can see, we came close to this equation when we approximated the line of best fit in Example 5, where we got

$$y = 0.19t + 67.8$$

Exercises 3.6

A **Identifying Linear Functions** In Problems 1–12, identify whether the graphs are linear functions, nonlinear functions, or not a function. If the graph is a linear function, then identify it as an increasing, decreasing, or constant function.

1.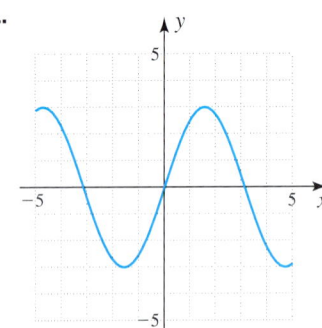

2.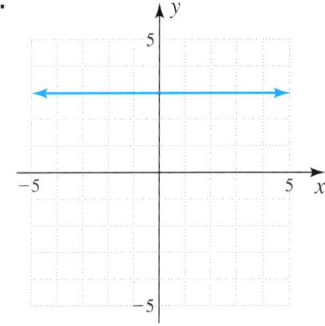

3.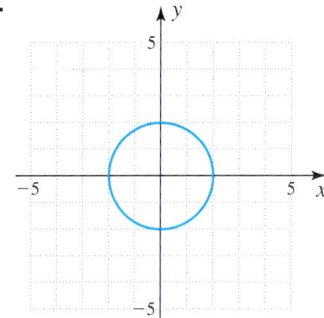

4.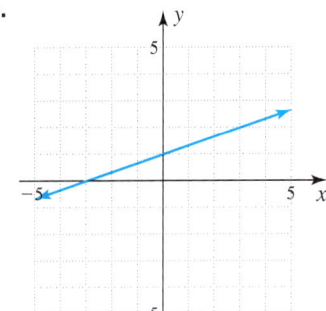

5.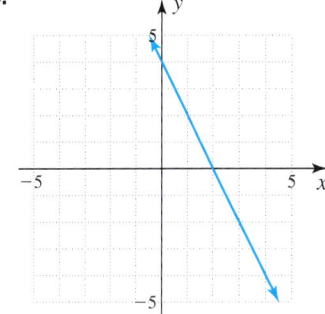

6.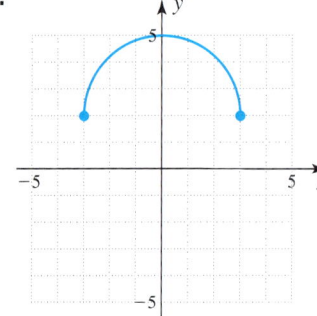

7.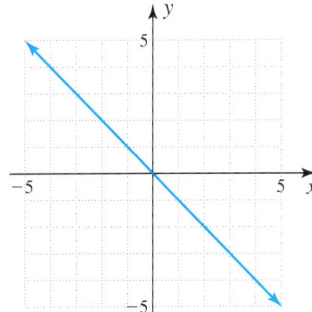

8.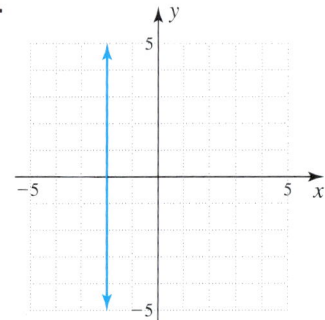

9.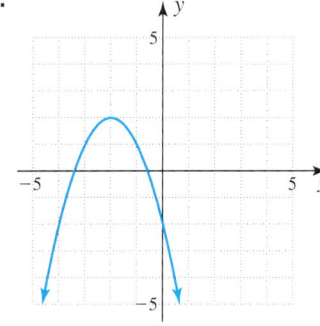

10.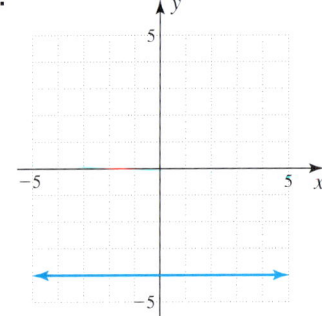

11.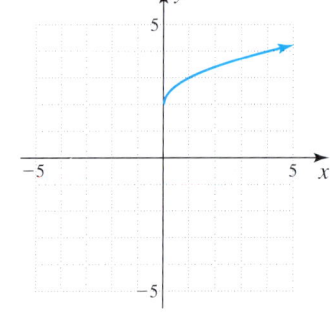

12.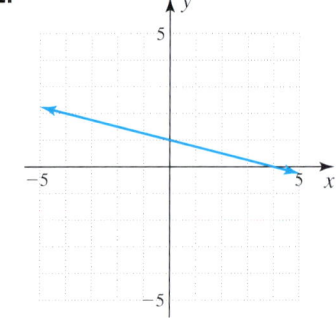

In Problems 13–24, identify whether the equation is a linear function or a nonlinear function. If it is a linear function, write it using function notation and identify it as an increasing, decreasing, or constant function.

13. $10 = x - 5y$

14. $f(x) = -x^3 + 4x$

15. $g(x) = -4$

16. $y = -\frac{3}{2}x$

17. $x = 3$

18. $-y = -2x + 6$

19. $p(x) = 5 - 3x$

20. $y = \frac{7}{x} + 3$

21. $h(x) = 4 + x^2$

22. $2.6 = y$

23. $x = 8 + y$

24. $f(x) = \sqrt{x + 1}$

< **B** > **Finding the Equation of a Linear Function from a Graph** In Problems 25–28, find an equation for the specified line shown in the graph. Write the equation in function notation.

25. L_1

26. L_2

27. L_3

28. L_4

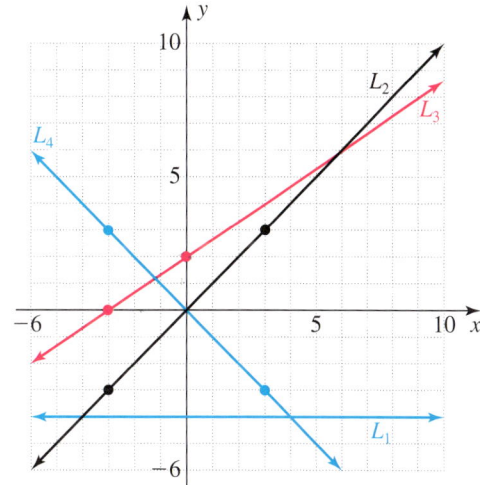

In Problems 29–32, use each graph to solve the application.

29. The graph of the linear function to the right describes the number N, in millions, of cases of diabetes in the country in years x since 1983.

a. Find the equation of the line where N is the number of cases and x is the time in years since 1983. Write the equation in function notation.

b. What is the rate of change of the number of cases of diabetes?

c. What is x when the year is 2008?

d. Use the equation of the function to predict the number of cases of diabetes in 2008.

30. After walking into a cold room (45°F), the heat is turned on high. The graph of the linear function to the right describes the change in temperature F over the time t in minutes.

a. Find the equation of the line where F is Fahrenheit degrees and t is the time in minutes. Write the equation in function notation.

b. What is the rate of change of the temperature over time?

c. Use the equation of the function to predict the temperature in 10 minutes.

d. The heat should be turned down after the room reaches 80°F. How many minutes will it take for the room to become 80°F?

31. The graph of the linear function to the right describes the cost of homes in a new housing development. The cost is based on the square footage of living space in the home.

  a. Find the equation of the line where C is the cost and x is the square footage. Write the equation in function notation.

  b. What is the cost of a home per square foot?

  c. Use the equation of the function to predict the cost of a home with 1475 square footage.

  d. If your bank approved you for a home mortgage loan of $138,000, how big of a house could you buy? (Round the answer to the nearest square foot.)

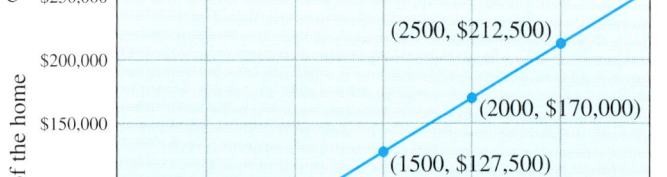
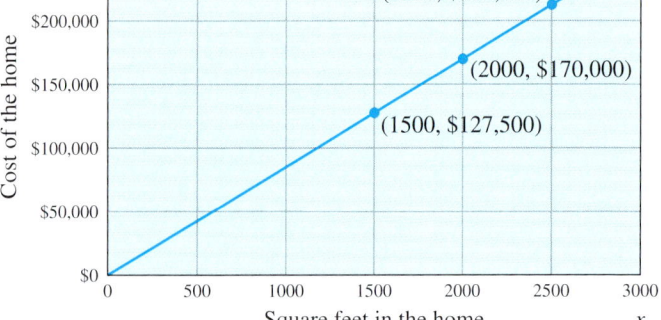

32. Studies show that college students are taking longer to graduate from college. The graph of the linear function approximates the percentage of college students graduating in less than 5 years between 1988 and 2002.

  a. Find the equation of the line where P is the percentage and x is the years after 1988. Write the equation in function notation.

  b. What is the rate of change in the percentage of college students graduating in less than 5 years?

  c. Use the equation of the function to estimate the percentage of college students who graduated in 2000 after less than 5 years in school.

  d. Use the equation to predict the percentage of college students who will graduate in 2012 after less than 5 years in school.

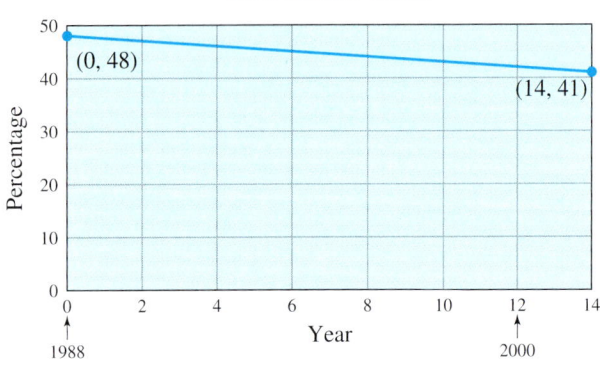
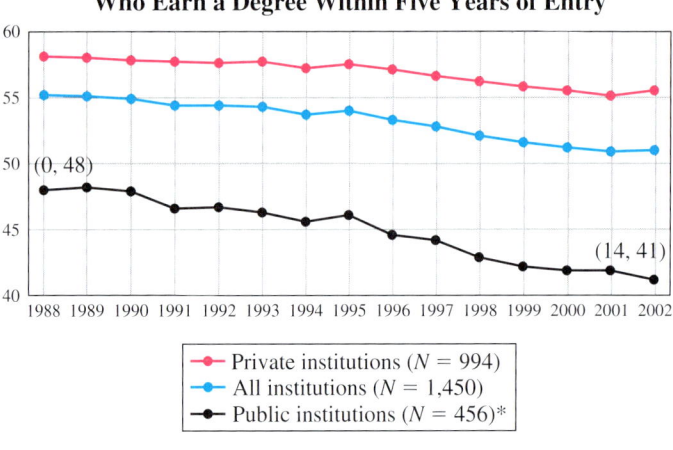

‹ C › Mathematical Modeling Using Linearly Related Data In Problems 33–36, identify which scatter plots display data that is linearly related.

33.

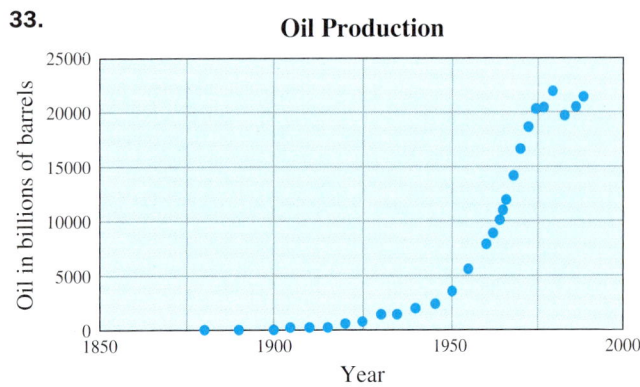

34.

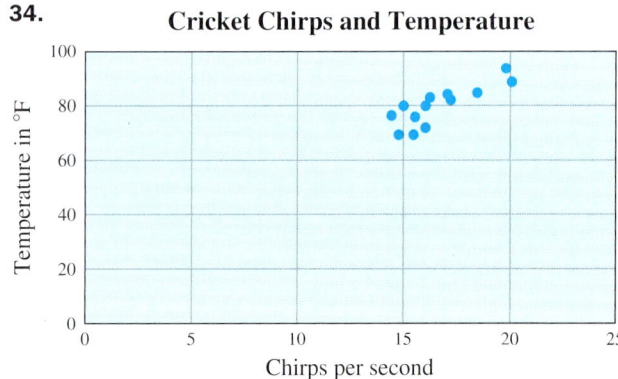

35.

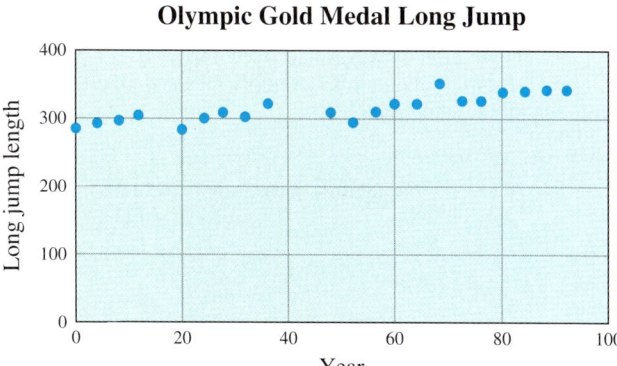

36.

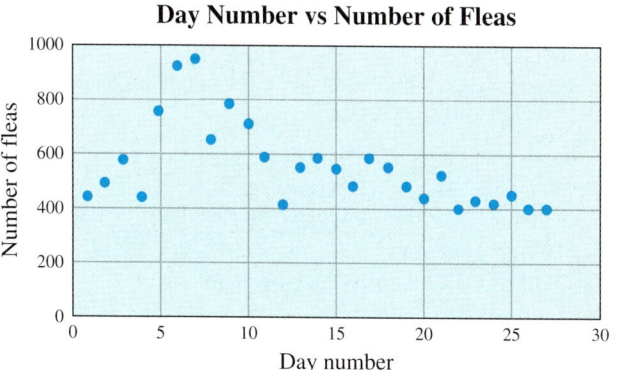

In Problems 37–38, make a scatter plot of the given data sets and decide if the data is linearly related.

37.

Year Since 1900	High Jump	Year Since 1900	High Jump
−4	71.25	52	80.32
0	74.8	56	83.25
4	71	60	85
8	75	64	85.75
12	76	68	88.25
20	76.25	72	87.75
24	78	76	88.5
28	76.375	80	92.75
32	77.625	84	92.5
36	79.9375	88	93.5
48	78	92	92

Olympic Gold and The High Jump

38.

Day No.	No. of Fleas
1	436
2	495
3	575
4	444
5	754
6	915
7	945
8	655
9	782
10	704

Fleas

In Problems 39–42, use mathematical modeling.

39. According to the U.S. Energy Information Administration, the energy prices in the following table indicate the amount of money spent on electricity in dollars per million Btu by the residential consumer in the United States for each of the years listed in the first column.

Year	Dollars per Million Btu Spent on Electricity
2000	$24.50
2001	$25.40
2002	$23.90
2003	$23.10
2004	$22.90
2008	?

a. Rename the table by assigning variable names to the columns, adjusting the numbers in the first column so 2000 corresponds to $x = 0$, and then graph the data.

Dollars per Million BTu Spent on Electricity

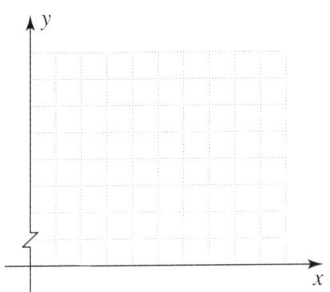

b. Use the coordinates (2, 23.9) and (4, 22.9) to draw the line of best fit.

c. Use the two points named in part **b** to write the equation of the line of best fit.

d. Use the equation in part **c** to predict the amount of money that will be spent on electricity in dollars per million Btu in the year 2008.

e. What seems to be the trend in the dollars per million Btu spent on electricity for the future in the United States?

40. Have you ever had a great idea you thought should be patented? Thousands of people in the United States get patents each year. The data in the following table is from the U.S. Patent Office, and it indicates the number of utility patents in thousands for each of the years listed below.

Year	Number of Utility Patents in Thousands
2002	167.3
2003	169.0
2004	164.3
2005	143.8

a. Rename the table by assigning variable names to the columns, adjust the numbers in the first column so 2002 corresponds to $x = 0$, and then graph the data.

b. Use the coordinates (0, 167.3) and (3, 143.8) to draw the line of best fit.

c. Use the two points named in part **b** to write the equation of the line of best fit.

d. Use the equation in part **c** to predict the number of utility patents there will be in the year 2012.

Source: http://www.uspto.gov.

41. Marco and Jenna just got married and have decided to save their money so they can buy their first home. They plan to buy the home in the year 2014. They researched the median cost for a single-family home in their area and came up with the data listed in Table 3.8. Answer the following questions to predict how much they can expect to pay for a home in 2014.

Quarter	Median Cost for a Single-Family Home
2004 q3	231.6
2004 q4	286.4
2005 q1	370.1
2005 q2	368.9
2005 q3	386.6
2005 q4	391.2
2006 q1	377
2006 q2	376.2

Table 3.8

a. Make a scatter plot of the data in Table 3.8.

Median Cost for a Single-Family Home

b. Assuming the data is linearly related, use the coordinates (2, 286.4) and (7, 377) to draw the line of best fit.

Median Cost for a Single-Family Home

c. Use the two points named in part **b** to write the equation of the line.

d. Use the equation in part **c** to predict the median cost of a home 34 quarters from q3 in 2004.

42. Could the number of fat grams in fast food predict how many calories you will eat? The following table is a list of the fat grams in certain types of fast food and how many calories are associated with that item. Complete the following to find the answer to the question.

Fast Food Item	Total Fat (g)	Total Calories
Hamburger	9	260
Cheeseburger	13	320
$\frac{1}{4}$-lb Burger	21	420
$\frac{1}{4}$-lb Burger/cheese	30	530
Large Burger	31	560
Special Burger	31	550
Special Burger/Bacon	34	590
Fried Chicken	25	500
Fried Fish	28	560
Grilled Chicken	20	440
Lite Grilled Chicken	5	300

a. Make a scatter plot of the data in the table.

Fat Grams vs Calories

b. Does the scatter plot indicate that the data is linearly related? If so, then write the equation of the line of best fit and predict what the calories would be for a fast food item that has 24 grams of fat.

›› Write On

In Exercises 43–44, determine if the statement is true or false. Explain your answer.

43. Every line is a linear function.

44. Every constant function is a line.

Write your own definition of the following phrases.

45. Line of best fit

46. Linearly related data

›› Concept Checker

Fill in the blank(s) with the correct word(s), phrase, or mathematical statement.

47. Graph the points from the _____ to make a scatter plot.

48. If the points on the scatter plot resemble a line, then it models a(n) _____.

49. If the slope of the line of the linear function is positive, then the function is called a(n) _____.

50. The linear function $f(x) = -3$ is called a(n) _____.

linear function	constant function
nonlinear function	line of best fit
increasing function	scatter plot
decreasing function	data set

›› Mastery Test

51. Write the equation of the function in the graph. Identify whether the function is increasing, decreasing, or constant.

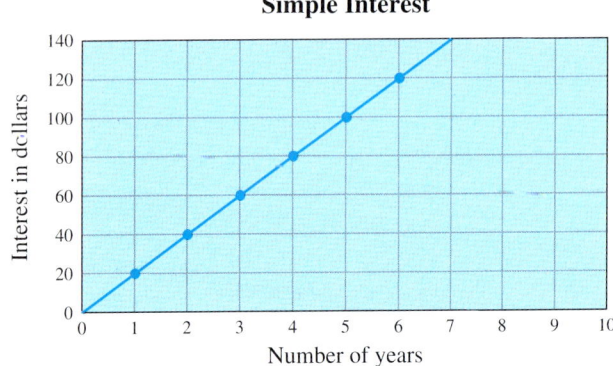

Simple Interest

52. Write the linear equation $4x - 2y = 6$ as a linear function.

53. If a linear function has a slope of 0, is it an increasing, decreasing, or constant function?

54. What type of function could be written to model the data in the scatter plot?

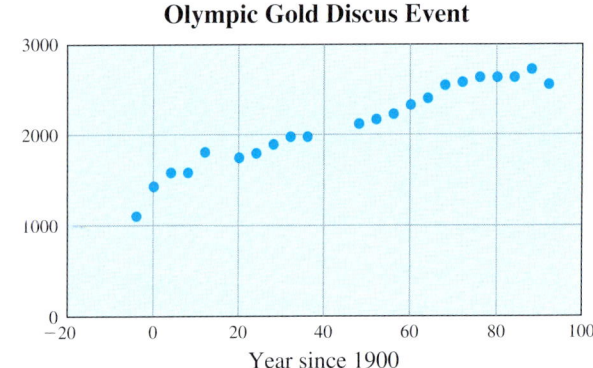

Olympic Gold Discus Event

55. Given the function $f(x) = \frac{1}{2}x - 4$, identify whether it is increasing, decreasing, or constant.

56. Given the graph of the function at the right, identify whether it is increasing, decreasing, or constant.

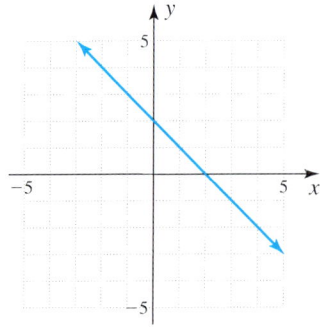

57. Make a scatter plot of the data in the table. Identify whether it models a linear or a nonlinear function.

Sales Report

Week	Sales in Thousands
1	20
2	35
3	28
4	50
5	44

Sales Report

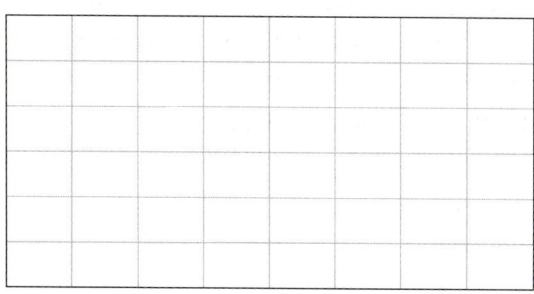

58. Given the linear equation $3y = 15$, write it as a function.

>>> Skill Checker

In Problems 59–62, find the x- and y-intercepts.

59. $3x + 2y = 9$

60. $-14 = 7x - 5y$

61. $4y = -32$

62. $x = 6$

In Problems 63–66, find the slope and y-intercept of the line.

63. $10 = -3x + 5y$

64. $\frac{1}{2}y + 2 = 4x$

65. $50 = -10y$

66. $y - 2 = -1$

In Problems 67–70, graph the lines.

67. $y = -x + 2$

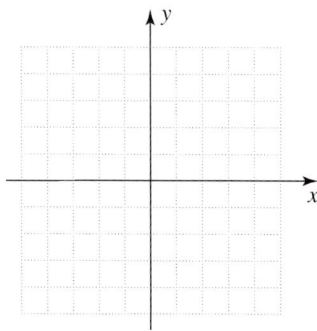

68. $1 + 2x = y$

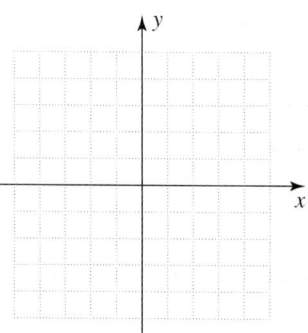

69. $y = 2.4$

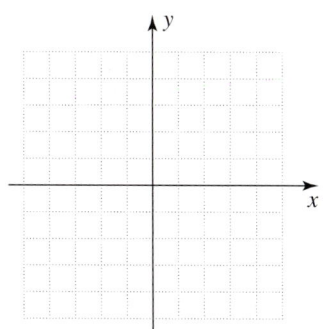

70. $x = -3.5$

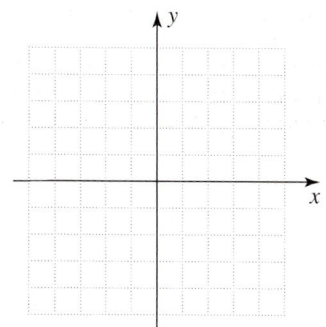

> Collaborative Learning 3A

Suppose you interviewed for a job with the two companies shown below and received a job offer from each of the companies. Should you be greedy and take the job with the higher base salary or is there more to it? Let's do some collaboration to help with your decision.

Divide into two groups and have each group select one of the two companies to investigate. To conduct the investigations do the following:

1. Make a table of values for potential sales (*x*-values) and their respective salary (*y*-values) based on that company's compensation plan. Let the potential sales begin with $0 and increase by increments of $50,000 up to $500,000.
2. Make a graph plotting the pairs of values from the table.

After each group has completed their company's graph, have someone from each group put their group's graph on the same large poster board for all to view. Have a group discussion about the comparison of these two graphs including the following questions:

1. Does the salary from the $80,000/year job always yield more income?
2. Is there a point where the amount of sales will yield the same income for both jobs? If so, when?
3. Does the income from the $60,000/year job ever yield more income? If so, when?
4. Which job should you take? Explain.

Company:	Sales Consultants of Tacoma	Job Type:	Banking
Location:	US-Washington		Business Development
Base Pay:	$80,000.00/Year		Finance
Other Pay:	10% of all sales		Marketing
Employee Type:	Full-Time Employee		Sales
Industry:	Banking–Financial Services	Req'd Education:	BS degree
	Consulting	Req'd Experience:	At Least 3 Years
	Sales–Marketing	Req'd Travel:	Negligible
		Relocation Covered:	No

Company:	Sales Reps of Syracuse	Job Type:	Banking
Location:	US-New York		Business Development
Base Pay:	$60,000.00/Year		Finance
Other Pay:	20% of all sales		Marketing
Employee Type:	Full-Time Employee		Sales
Industry:	Banking–Financial Services	Req'd Education:	BS degree
	Consulting	Req'd Experience:	At Least 3 Years
	Sales–Marketing	Req'd Travel:	Negligible
		Relocation Covered:	No

> Collaborative Learning 3B

Do you know the annual average salary offered by U.S. employers to graduates of your college or university? The following tables indicate the annual salaries offered to MIT classes of 2001, 2002, 2003, and 2004. We will use some collaborative learning to study these numbers so we can predict the annual salary for a specific year in the future. We will assume this data is linearly related. Divide into groups and complete the following.

1. Make a graph using the average base salary for a bachelors degree graduate from each of the four years. Let the *x*-axis represent the year (for 2001 let $x = 1$, for 2002 let $x = 2$, etc.) and the *y*-axis represent the average salary.
2. Name the two points on the graph that can be connected to make the line of best fit. Use those ordered pairs to write the equation of the line in slope-intercept form. (This could also be done using a graphing calculator.)

3. This equation should be a predictor of salaries for the future. Let's see how close the equation value comes to the actual value for the first 4 years ($x = 1$, $x = 2$, $x = 3$, and $x = 4$). Make a table to display those comparisons and then add $x = 5$, $x = 6$, and $x = 7$.

4. Research your own college or university for this data, and answer the first three questions again. How does your school's annual salaries offered to bachelors degree graduates compare with those at MIT? Discuss some of the things that might make a difference in comparing these salaries.

Annual Salaries Offered By U.S. Employers to MIT Class of 2004 Graduates

Degree	Base Yearly Salary			# of Reports
	Low	Average	High	
Bachelors	$30,000	$56,211	$90,000	73
Masters	$35,000	$71,632	$105,000	55
M. Eng	$43,000	$71,010	$90,000	49
Doctoral	$50,000	$84,081	$110,000	31
			Total	208

Annual Salaries Offered By U.S. Employers to MIT Class of 2003 Graduates

Degree	Base Yearly Salary			# of Reports
	Low	Average	High	
Bachelors	$30,000	$54,761	$94,000	111
Masters	$40,000	$66,367	$94,000	68
M. Eng	$40,000	$73,514	$150,000	88
Doctoral	$42,000	$81,867	$105,000	45
			Total	312

Annual Salaries Offered By U.S. Employers to MIT Class of 2002 Graduates

Degree	Base Yearly Salary			# of Reports
	Low	Average	High	
Bachelors	$30,000	$53,497	$90,000	145
Masters	$38,000	$73,914	$187,000	99
M. Eng	$40,000	$72,032	$110,000	78
Doctoral	$42,000	$84,020	$120,000	94
			Total	416

Annual Salaries Offered By U.S. Employers to MIT Class of 2001 Graduates

Degree	Base Yearly Salary			# of Reports
	Low	Average	High	
Bachelors	$40,000	$53,000	$100,000	64
Masters	$38,000	$66,500	$120,000	39
M. Eng	$30,000	$74,000	$102,500	44
Doctoral	$30,000	$79,900	$110,000	30
			Total	177

Source: Data from Massachusetts Institute of Technology (MIT), Cambridge, MA.

> Research Questions

1. Some historians claim that the official birthday of analytic geometry is November 10, 1619. Investigate and write a report on why this is so and the events that led Descartes to the discovery of analytic geometry.

2. Find out what led Descartes to make his famous pronouncement, "*Je pense, donc je suis*" (I think, therefore I am), and write a report about the contents of one of his works, *La Géométrie*.

3. She was 19, a capable ruler, a good classicist, a remarkable athlete, and an expert hunter and horsewoman. Find out who this queen was and what connections she had with Descartes.

4. Upon her arrival at the University of Stockholm, one newspaper reporter wrote "Today we do not herald the arrival of some vulgar, insignificant prince of noble blood. No, the Princess of Science has honored our city with her arrival." Write a report identifying this woman and discussing the circumstances leading to her arrival in Sweden.

5. When she was 6 years old, the "princess's" room was decorated with a unique type of wallpaper. Write a paragraph about this wallpaper and its influence on her career.

6. In 1888, the "princess" won the Prix Bordin offered by the French Academy of Sciences. Write a report about the contents of her prizewinning essay and the motto that accompanied it.

Summary Chapter 3

Section	Item	Meaning	Example
3.1A	Origin	The origin is the point where the x- and y-axes intersect.	The coordinates of the origin are (0, 0).
	Quadrant	The x- and y-axes divide the plane into four quadrants.	QII, QI, QIII, QIV
3.1B	Solution	An ordered pair (a, b) is a solution of an equation in two variables x and y if a true statement results when a and b are substituted for x and y.	(3, 4) is a solution of $x + y = 7$.
	Linear equation in standard form	Any equation that can be written in the form $Ax + By = C$ (A, B, C are integers, A and B not both zero, and A is nonnegative)	$7x + 8y = 3$ is a linear equation in standard form.
3.1C	x-intercept	The x-coordinate of a point at which the graph crosses the x-axis	The x-intercept of $2x + y = 6$ is $x = 3$.
	y-intercept	The y-coordinate of a point at which the graph crosses the y-axis	The y-intercept of $2x + y = 6$ is $y = 6$.
3.1D	Horizontal line	A line whose equation can be written in the form $y = C$, C a constant	$y = 4$, $y = -3$, $y = 0.25$, and $2y = 6$ are equations of horizontal lines.
	Vertical line	A line whose equation can be written in the form $x = C$, C a constant	$x = 4$, $x = -3$, $x = 0.25$, and $2x = 6$ are equations of vertical lines.
3.1E	Nonlinear equation	The graph of the solution set of a nonlinear equation is not a straight line	$y = -x^2 + 1$
3.2A	Slope	The slope of the line through (x_1, y_1) and (x_2, y_2) is $m = \frac{y_2 - y_1}{x_2 - x_1}$.	The slope of the line through (3, 5) and (9, 7) is $m = \frac{7 - 5}{9 - 3} = \frac{1}{3}$.
	Slope of a vertical line	The slope of a vertical line is undefined.	The slope of the line $x = 2$ is undefined.
	Slope of a horizontal line	The slope of a horizontal line is zero.	The slope of the line $y = -3$ is zero.

Section	Item	Meaning	Example
3.2B	Slopes of parallel lines	Two different lines L_1 and L_2 with slopes m_1 and m_2 are parallel if and only if $m_1 = m_2$.	The lines $y = -2x + 3$ and $y = -2x + 9$ are parallel.
	Slopes of perpendicular lines	Two lines L_1 and L_2 with slopes m_1 and m_2 are perpendicular if and only if $m_1 = -\frac{1}{m_2}$.	A line perpendicular to the line $y = -2x + 3$ will have a slope of $\frac{1}{2}$, because the slope of $y = -2x + 3$ is -2.
3.2D	Slope-intercept form of a line	Given the equation $y = mx + b$, m = slope, and b = y-intercept	Given the equation $y = 2x + 3$, the slope = 2 and the y-intercept is 3.
3.3A	Finding an equation given two points	An equation of the line going through the points (x_1, y_1) and (x_2, y_2) is $y - y_1 = m(x - x_1)$ where $m = \frac{y_2 - y_1}{x_2 - x_1}$.	An equation of the line through the points (3, 1) and (4, 3) is $y - 1 = 2(x - 3)$ since $m = \frac{3-1}{4-3} = \frac{2}{1} = 2$.
3.3B	Finding an equation given a point and the slope	An equation of the line going through the point (x_1, y_1) and with slope m is $y - y_1 = m(x - x_1)$.	An equation of the line going through the point (2, 5) and with slope -3 is $y - 5 = -3(x - 2)$.
3.3C	Finding an equation given the slope and y-intercept	An equation of the line with slope m and y-intercept b is $y = mx + b$.	An equation of the line with slope 3 and y-intercept -4 is $y = 3x - 4$.
3.3E	Equations of horizontal and vertical lines	$y = c$, $m = 0$, horizontal $x = c$, m is undefined, vertical	An equation of the line with $m = 0$ and passing through the point (2, 5) is $y = 5$.
3.4A	Linear inequality	A statement that can be written in the form $Ax + By \leq C$ or $Ax + By \geq C$ where A and B are not both zero (You can also use the symbols $<$ and $>$.)	$5x - 2y \geq 10$ and $y \leq 3x + 1$ are linear inequalities. Graph the boundary line and then shade the side where the solutions occur.
3.5A	Relation	A set of ordered pairs	$\{(2, -1), (4, 6), (5, 2)\}$
	Domain	The set of first coordinates of a relation	$\{2, 4, 5\}$ is the domain of the preceding relation.
	Range	The set of second coordinates of a relation	$\{-1, 2, 6\}$ is the range of the preceding relation.
	Function	A relation in which no two different ordered pairs have the same first coordinate	$\{(1, 2), (2, 4), (3, 6)\}$ is a function.
3.5B	Vertical line test	If any vertical line intersects the graph of a relation more than once, the relation is *not* a function.	Not a function
3.5D	Function notation	Use of a letter such as f to denote a function and $f(x)$ to mean the value of the function for the given value of x	$f = \{(x, y) \mid y = x^2\}$. For this function, $f(x) = x^2$, $f(2) = 4$, and $f(-3) = 9$.
	Linear function	A function that can be written as $f(x) = mx + b$, where m and b are real numbers	$f(x) = 3x + 5$ and $g(x) = -2x - 7$ are linear functions, but $h(x) = x^2 - 1$ and $i(x) = \frac{1}{x}$ are not.

(continued)

Section	Item	Meaning	Example
3.6A	Increasing function	When the graph of a line rises from left to right, the slope is positive.	$f(x) = \frac{1}{4}x + 1$
	Decreasing function	When the graph of a line falls from left to right, the slope is negative.	$f(x) = -2x - 3$
	Constant function	The graph is a horizontal line. The slope is 0.	$f(x) = 4$
3.6C	Mathematical modeling	Process of taking a verbal description of a problem, assigning variables to the unknown quantities, and forming an algebra equation	The table indicates the yearly salary increases for Sue's job. Write the equation that will predict her salary in the ninth year. It is $y = 2000x + 1000$. Year | Increase 1 | $3000 2 | $5000 3 | $7000 4 | $9000

Review Exercises Chapter 3

(If you need help with these exercises, look in the section indicated in brackets.)

1. ⟨**3.1A**⟩ *Graph.*

   a. $A(1, 4)$, $B(-3, 1)$, $C(-3, -2)$, and $D(3, -1)$

   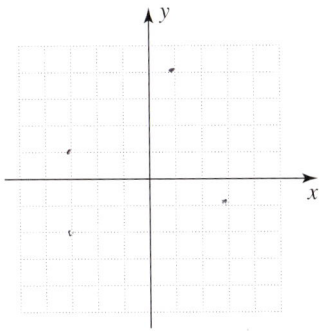

   b. $A(4, 1)$, $B(-1, 3)$, $C(-2, -3)$, and $D(1, -3)$

   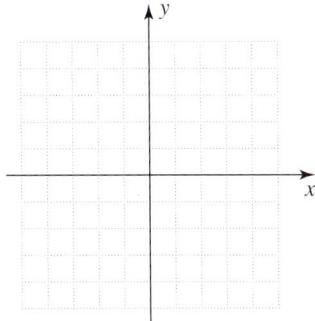

2. ⟨**3.1A**⟩ *Determine the coordinates of each of the points.*

   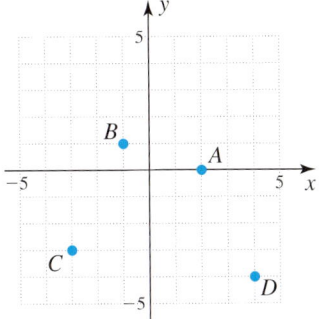

3. ⟨**3.1A**⟩ *Determine the coordinates of each of the points.*

   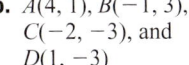

4. ⟨**3.1B**⟩ *Graph the solutions to each equation.*

   a. $x + 2y = 4$

   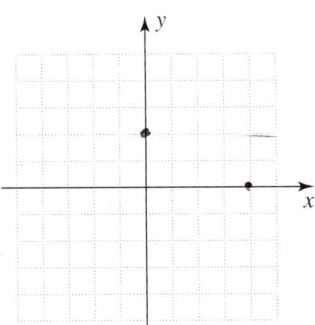

   b. $2x - y = 2$

   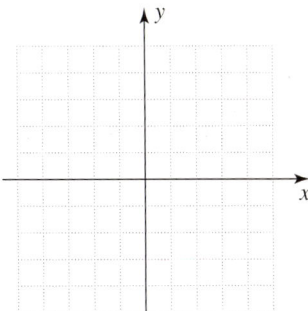

(continued)

5. ⟨**3.1C**⟩ *Find the x- and y-intercepts and graph the solutions to the line.*

   a. $y = 3x + 3$ 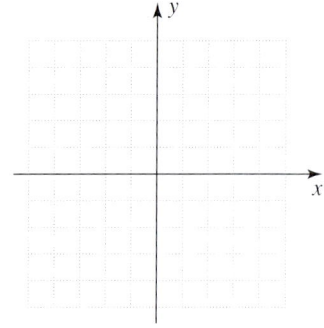   **b.** $y = 2x - 4$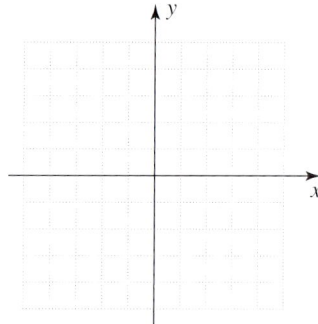

6. ⟨**3.1D**⟩ *Graph the solutions to each equation on the same coordinate system.*

   a. $2x = 6$ **b.** $3y = 6$

   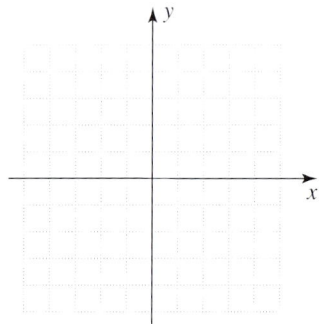

7. ⟨**3.1D**⟩ *Graph the solutions to each equation on the same coordinate system.*

   a. $2x = -6$ **b.** $3y = -6$

   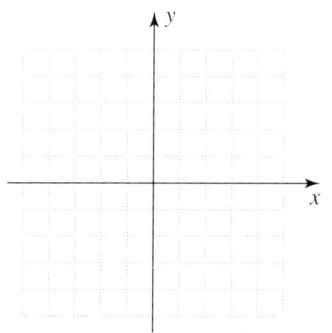

8. ⟨**3.2A**⟩ *Find the slope of the line that passes through the given points.*

   a. $A(-3, 2)$ and $B(1, 0)$
   b. $A(4, -2)$ and $B(4, -7)$

9. ⟨**3.2A**⟩ *Find the slope of the line that passes through the given points.*

   a. $A(3, 4)$ and $B(4, 3)$
   b. $A(1, -1)$ and $B(-3, 5)$

10. ⟨**3.2B**⟩ *A line L has slope $\frac{3}{4}$. Determine whether the line passing through the two given points is parallel or perpendicular to L.*

    a. $A(1, 3)$ and $B(-2, 7)$
    b. $A(1, 3)$ and $B(5, 6)$

11. ⟨**3.2B**⟩ *A line L has slope -2. Determine whether the line through the two given points is parallel or perpendicular to L.*

    a. $A(2, -1)$ and $B(1, 1)$
    b. $A(3, -1)$ and $B(2, -3)$

12. ⟨**3.2B**⟩ *The line through $(2, 4)$ and $(5, y)$ is perpendicular to a line with the given slope. Find y.*

    a. $m = \frac{3}{2}$ **b.** $m = -2$

13. ⟨**3.2C**⟩ *A line passes through the point $(-1, 2)$ and has the given slope. Graph the line.*

    a. $m = -\frac{1}{2}$ **b.** $m = \frac{2}{3}$

    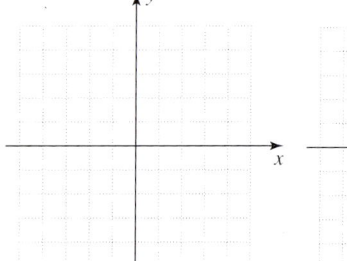   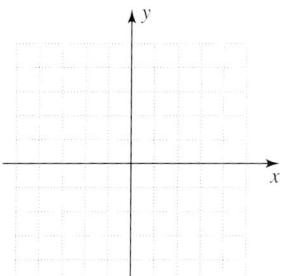

14. ⟨**3.2D**⟩ *Find the slope and y-intercept for the given linear equation.*

    a. $6x - y = 2$
    b. $4y = 2x - 8$

15. ⟨**3.3A**⟩ *Find an equation of the line through the two given points, and write the equation in standard form.*

    a. $A(2, 5)$ and $B(-1, 2)$
    b. $A(-4, 3)$ and $B(-2, -2)$

16. ⟨**3.3B**⟩ *Find an equation of the line with slope 2 and passing through the given point, and write the equation in standard form.*

   a. $A(-1, -3)$
   b. $A(5, 0)$

17. ⟨**3.3C**⟩

   a. A line has slope 3 and *y*-intercept 2. Find the slope-intercept equation of this line.
   b. Repeat part **a** if the slope is -3 and the *y*-intercept is 4.

18. ⟨**3.3D**⟩ *Find an equation of the line passing through the point (2, 1) and parallel to the given line.*

   a. $2x + y = 7$
   b. $3x - y = 4$

19. ⟨**3.3D**⟩ *Find an equation of the line passing through the point (2, 1) and perpendicular to the line.*

   a. $2x + 3y = 7$
   b. $3x - 2y = 4$

20. ⟨**3.3E**⟩ *Find the equation of the line passing through the point $(-3, 7)$ with:*

   a. slope 0
   b. undefined slope

21. ⟨**3.4A**⟩ *Graph:*

   a. $2x - y < 1$
   b. $y \geq x - 3$
   c. $x \geq 4$
   d. $2y < -6$

22. ⟨**3.4B**⟩ *Graph:*

   a. $|y| > 2$
   b. $|x + 2| \leq 3$

23. ⟨**3.5A**⟩ *Find the domain and the range of each relation.*

   a. $\{(0, 5), (2, 9), (3, 10), (5, 8)\}$
   b. $\{(0, 6), (2, 10), (3, 11), (5, 9)\}$

24. ⟨**3.5A**⟩ *Find the domain and the range and graph the relation.*

   a. $\{(x, y) \mid y = 1 - x\}$
   b. $\{(x, y) \mid y = 2 - x\}$

(continued)

25. ⟨**3.5A**⟩ *Find the domain and the range.*

a.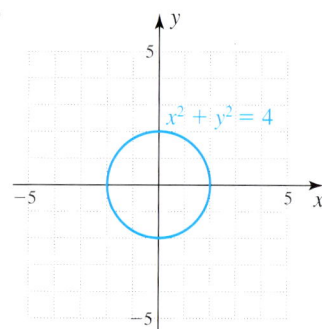

b.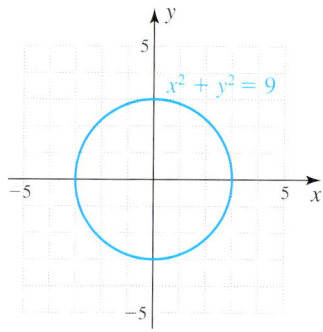

26. ⟨**3.5A**⟩ *Find the domain and the range.*

a.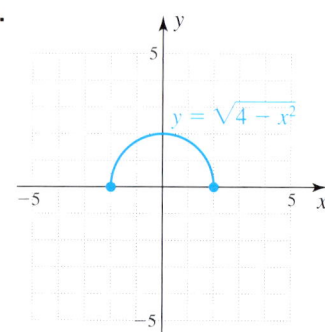

b.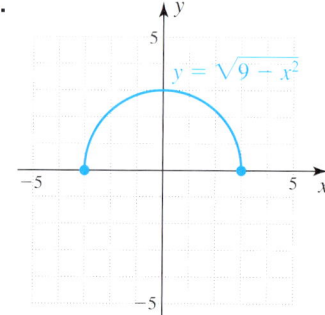

27. ⟨**3.5B**⟩ *Use the vertical line test to determine whether the relation defines a function.*

a.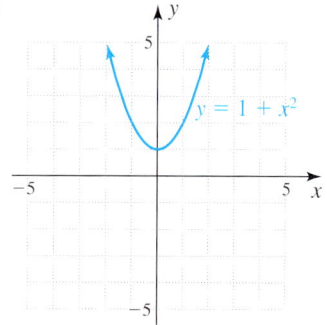

b.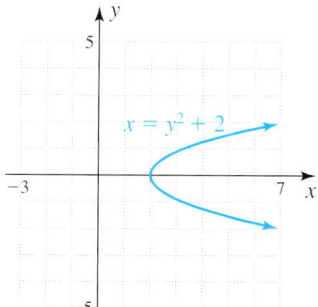

28. ⟨**3.5C**⟩ *Find the domain of the function.*

  a. $y = \dfrac{2}{x-1}$

  b. $y = \dfrac{2}{x-2}$

29. ⟨**3.5C**⟩ *Find the domain of the function.*

  a. $y = \sqrt{x-3}$

  b. $y = \sqrt{x-4}$

30. ⟨**3.5D**⟩ *Let* $f(x) = x - 4$. *Find:*

  a. $f(2)$ **b.** $f(1)$ **c.** $f(2) - f(1)$

31. ⟨**3.5D**⟩ *Let* $f(x) = x^2 - 3$. *Find:*

  a. $f(2)$ **b.** $f(1)$ **c.** $f(2) - f(1)$

32. ⟨**3.5D**⟩ *Let* $f = \{(2, 0), (3, 3), (1, -1)\}$. *Find:*

  a. $f(2)$ **b.** $f(1)$ **c.** $f(2) - f(1)$

33. ⟨**3.5D**⟩ *Let* $f = \{(2, 1), (3, 4), (1, 0)\}$. *Find:*

  a. $f(2)$ **b.** $f(1)$ **c.** $f(2) - f(1)$

34. ⟨**3.6A**⟩ *Identify which of the following equations are linear functions. If they are linear functions, then write them using function notation and identify them as increasing, decreasing, or constant functions.*

  a. $y = 2 - 4x + x^2$
  b. $5x = 3 - y$
  c. $13 = y + 12$

35. ⟨**3.6B**⟩ *Find the equation of the line shown in the graph.*

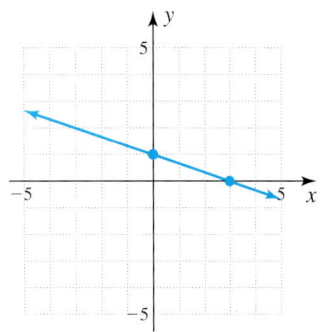

36. ⟨**3.6C**⟩ *The points in the following table can be represented by a linear model. Plot the points on a graph, draw the line of best fit, and write the equation of that line in slope-intercept form.*

x	1	2	3	4	5
y	1	3	4	3	4

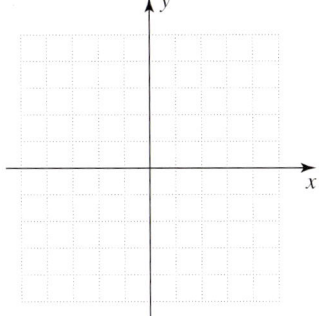

Practice Test Chapter 3

(Answers on pages 276–280)

Visit www.mhhe.com/bello to view helpful videos that provide step-by-step solutions to several of the problems below.

1. **a.** Graph $A(3, 2)$, $B(4, -2)$, $C(-1, -2)$, and $D(-1, 3)$.
   b. Find the coordinates of the points in the figure.

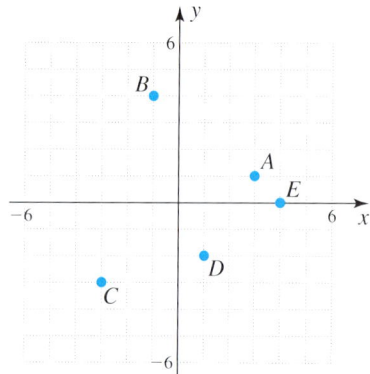

2. Graph the solutions to $3x - y = 3$.

3. Find the x- and y-intercepts of $y = 3x + 2$ and then graph the solutions to the equation.

4. Graph the solutions to each equation.
   a. $3x = -9$ **b.** $2y = -4$

5. Find the slope of the line passing through the two given points.
   a. $A(-2, 2)$ and $B(1, 1)$
   b. $A(-2, 4)$ and $B(-2, 8)$
   c. $A(5, 6)$ and $B(-5, -2)$

6. A line L_1 has slope $\frac{3}{2}$. Determine whether the line passing through the two given points is parallel or perpendicular to L_1.
   a. $A(4, 2)$ and $B(1, 4)$
   b. $A(-1, -4)$ and $B(-4, -6)$

7. The line through $A(1, -2)$ and $B(-1, y)$ is perpendicular to a line with slope $-\frac{2}{3}$. Find y.

8. Graph the line that has a y-intercept of 1 and slope $-\frac{1}{2}$.

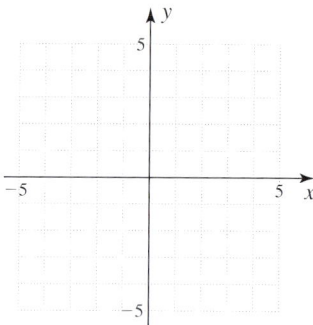

9. Find the slope and y-intercept of $4x - y = 3$.

10. Find an equation of the line that passes through $(4, 3)$ and $(2, 4)$. Then write the equation in standard form and graph the line.

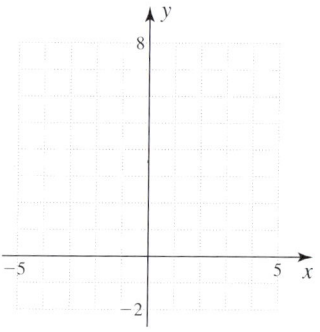

11. Find an equation of the line with slope -2 and passing through the point $(2, -3)$. Then graph the line.

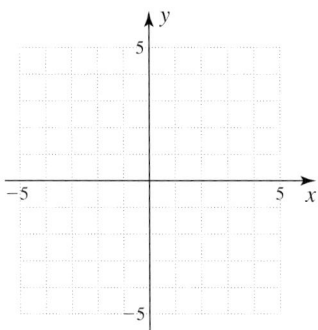

12. A line has slope 3 and y-intercept 2. Find the slope intercept equation of this line and graph the line.

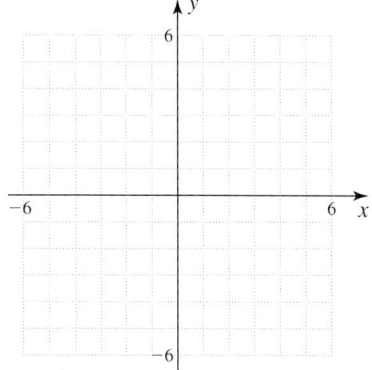

13. Find an equation of the line that passes through the point $(1, 2)$ and:

 a. Parallel to the line $2x - 3y = 5$. Write the equation in standard form.

 b. Perpendicular to the line $2x - 3y = 5$. Write the equation in standard form.

14. a. Given that a line passes through the point $(6, -5)$, find the equation of the line if the slope is 0.

 b. Given that a line passes through the point $(-2, 4)$, find the equation of the line if it is a vertical line.

15. Graph:

 a. $x + 4y < 4$

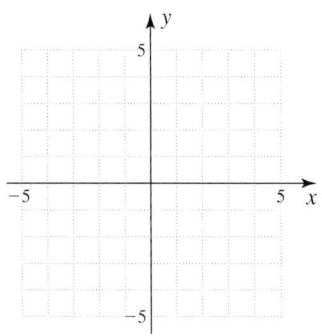

 b. $y + 2 \geq 1$

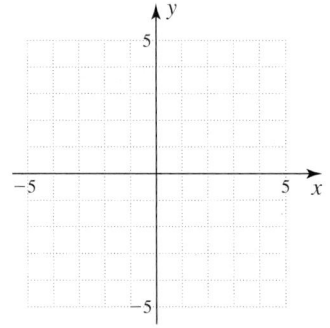

16. Graph:

 a. $|x + 1| \geq 5$

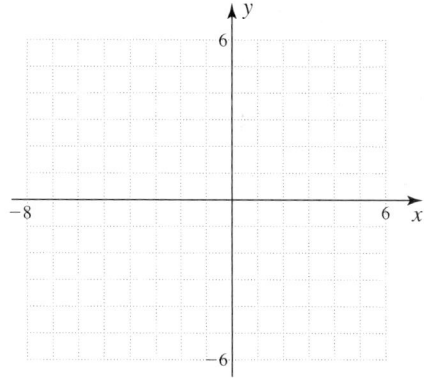

 b. $|y| < \dfrac{1}{2}$

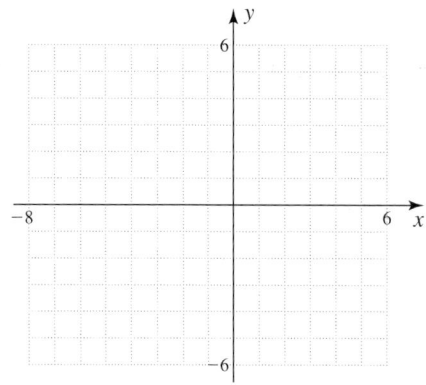

17. Find the domain and range of the relation {(1, 3), (2, 5), (3, 7), (4, 9)}.

18. Find the domain and the range of the relation $y = 4 + x$ shown in the graph.

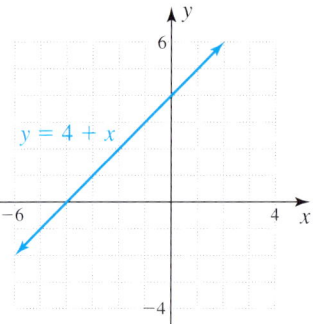

19. Find the domain and the range of the relation.

a.
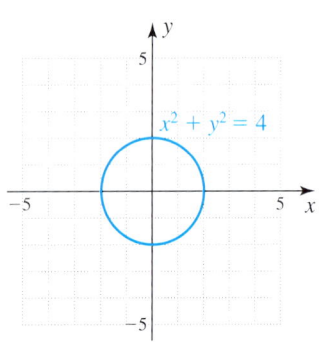

b.
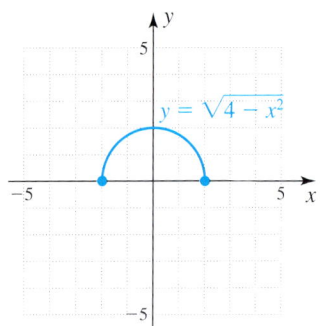

c.
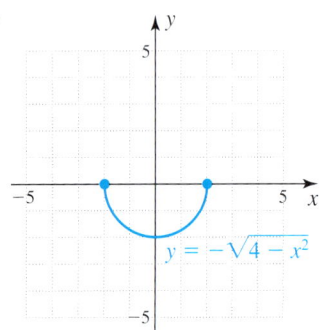

20. Use the vertical line test to determine whether the graph of the given relation defines a function.

a.
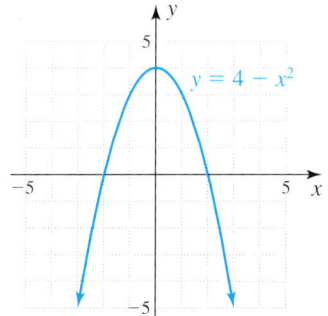

b.
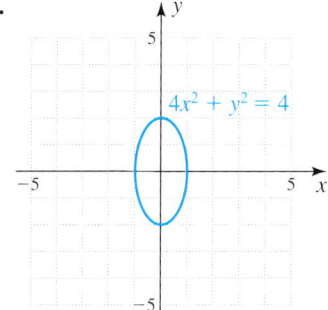

21. Find the domain of the function.
  a. $y = \dfrac{2}{x+3}$
  b. $y = \sqrt{x+2}$

22. Let $f(x) = 4x - 3$. Find: (For *a* and *b* write the results as ordered pairs.)
  a. $f(2)$
  b. $f(1)$
  c. $f(2) - f(1)$

23. Let $f = \{(1, 4), (2, -1), (3, 2)\}$. Find:
  a. $f(2)$
  b. $f(1)$
  c. $f(2) - f(1)$

24. Identify which of the following equations are linear functions. If they are linear functions, then write them using function notation and identify them as increasing, decreasing, or constant functions.
  a. $2 + 3y = 11$
  b. $-4 + x = -2y$
  c. $x^3 + 2x + 1 = y$

25. The points in the table can be represented by a linear model. Plot the points on a graph, draw the line of best fit, and write the equation of that line in slope-intercept form.

x	y
5	3
6	2
7	5
8	4
9	6

Answers to Practice Test Chapter 3

Answer		If You Missed	Review		
		Question	Section	Examples	Page
1. a. [graph with points D, A, C, B]		1a	3.1A	1	167
b. $A(3, 1)$; $B(-1, 4)$; $C(-3, -3)$; $D(1, -2)$; $E(4, 0)$		1b	3.1A	2	168
2. [graph of $3x - y = 3$]		2	3.1B	3	170
3. [graph of $y = 3x + 2$]	$x: \left(-\frac{2}{3}, 0\right)$; $y: (0, 2)$	3	3.1C	5	172–173
4. a. [graph of $3x = -9$]		4a	3.1D	6	174

Answer	If You Missed	Review		
	Question	Section	Examples	Page
b. *(graph of $2y = -4$)*	4b	3.1D	6	174
5. a. $-\dfrac{1}{3}$ b. Undefined c. $\dfrac{4}{5}$	5	3.2A	1	185–187
6. a. Perpendicular b. Neither	6	3.2B	3	190
7. $y = -5$	7	3.2B	4	190
8. *(graph)*	8	3.2C	5	191
9. Slope = 4; y-intercept = -3	9	3.2D	6	193
10. $x + 2y = 10$ *(graph)*	10	3.3A	1	201

Answer	If You Missed		Review	
	Question	Section	Examples	Page
11. $2x + y = 1$	11	3.3B	2	202
12. $y = 3x + 2$	12	3.3C	3	202–203
13. a. $2x - 3y = -4$ **b.** $3x + 2y = 7$	13	3.3D	4	203–204
14. a. $y = -5$ **b.** $x = -2$	14	3.3E	5	204
15. a.	15a	3.4A	1	211–212
b.	15b	3.4A	4	214

(continued)

Answer	If You Missed	Review		
	Question	Section	Examples	Page
16. a.	16a	3.4B	6	215
b.	16b	3.4B	5	214
17. $D = \{1, 2, 3, 4\}$; $R = \{3, 5, 7, 9\}$	17	3.5A	1	225–226
18. The domain and range are the set of real numbers.	18	3.5A	2	226–227
19. a. $D = \{x \mid -2 \leq x \leq 2\}$; $R = \{y \mid -2 \leq y \leq 2\}$ **b.** $D = \{x \mid -2 \leq x \leq 2\}$; $R = \{y \mid 0 \leq y \leq 2\}$ **c.** $D = \{x \mid -2 \leq x \leq 2\}$; $R = \{y \mid -2 \leq y \leq 0\}$	19	3.5A	3	228
20. a. A function **b.** Not a function	20	3.5B	4	229
21. a. All real numbers except -3 **b.** All real numbers greater than or equal to -2	21	3.5C	5	230
22. a. $(2, 5)$ **b.** $(1, 1)$ **c.** 4	22	3.5D	6	232
23. a. -1 **b.** 4 **c.** -5	23	3.5D	7	232

Answer	If You Missed		Review	
	Question	Section	Examples	Page
24. a. linear function; $f(x) = 3$; constant function b. linear function; $f(x) = -\frac{1}{2}x + 2$; decreasing function c. nonlinear function	24	3.6A	2	246
25.	25	3.6C	5	252–253

$y = 0.8x - 1.6$
(Answers will vary.)

Cumulative Review Chapters 1–3

1. The number $\sqrt{13}$ belongs to which of these sets? Natural numbers, Whole numbers, Integers, Rational numbers, Irrational numbers, Real numbers. Name all that apply.

2. Graph the additive inverse of $\frac{5}{2}$ on the number line.

3. Divide: $-\frac{1}{6} \div \left(-\frac{7}{12}\right)$

4. Simplify: $(2x^4y^{-4})^3$

5. Evaluate: $-7^3 + \frac{(14-10)}{2} + 35 \div 7$

6. Solve: $\frac{3}{7}y - 4 = 2$

7. Solve: $|x - 5| = |x - 9|$

8. Graph: $x \geq 1$

9. Graph: $\{x \mid x > -4 \text{ and } x < 4\}$

10. If $H = 2.45h + 72.98$, find h when $H = 136.68$.

11. The sum of three consecutive odd integers is 75. What are the three integers?

12. A freight train leaves a station traveling at 30 miles per hour. Two hours later, a passenger train leaves the same station in the same direction at 40 miles per hour. How far from the station does the passenger train overtake the freight train?

13. Find the range of the relation $\{(-2, -3), (5, -2), (-3, -4)\}$.

14. Find the domain and range of $\{(x, y) \mid y = 6 + x\}$.

15. Find the domain of $y = \sqrt{x + 25}$.

16. Let $f = \{(-2, 1), (-3, 2), (1, 4)\}$. Find $f(-3) - f(-2)$.

17. The lower limit L (heartbeats per minute) of your target zone is given by $L_1(a) = -\frac{2}{3}a + 150$, where a is your age in years. Find L for a person who is 24 years old.

18. Graph: $x - y = 2$

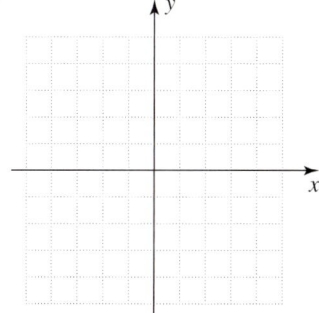

19. Find the x- and y-intercepts of $y = -5x + 2$.

20. Graph: $2x = -2$

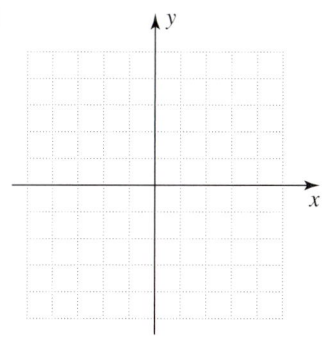

21. A line L_1 has slope 1. Find whether the line through (4, 6) and (7, 3) is parallel or perpendicular to line L_1.

22. Find an equation of the line with slope -2 and passing through the point $(-2, 1)$. Write the answer in standard form.

23. Find the slope and the y-intercept of the line $6x - 3y = 36$.

24. The graph shows that the percentage of teens sending text messages increases with age.

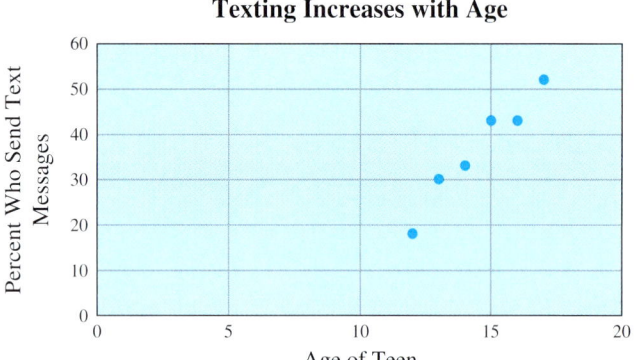

Texting Increases with Age

a. Draw the "line of best fit" and then use two points from this data, (12, 18) and (16, 43), to write an equation of the linear function, $P(a)$, that models this data.

b. Use the equation from part **a** to estimate the percent of 19-year-olds that use text messaging.

c. According to this model, by what percent does the number of teens sending text messages increase with every year of age?

25. Graph: $y \leq 6x - 6$

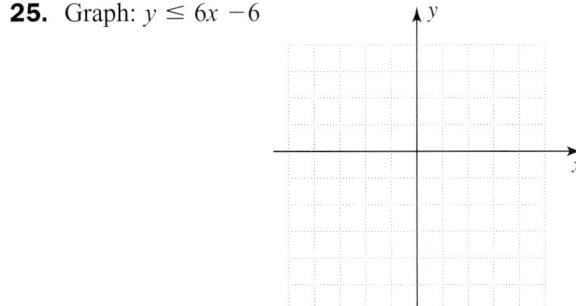

Chapter 4: Solving Systems of Linear Equations and Inequalities

Section

- **4.1** Systems with Two Variables
- **4.2** Systems with Three Variables
- **4.3** Coin, Distance-Rate-Time, Investment, and Geometry Problems
- **4.4** Systems of Linear Inequalities

The Human Side of Algebra

The study of systems of linear equations in the Western world was initiated by Gottfried Wilhelm Leibniz. In 1693, Leibniz solved a system of three equations by eliminating two of the unknowns and using a determinant to obtain the solution. It was Colin Maclaurin, however, who used determinants to solve simultaneous linear equations in two, three, and four unknowns. Maclaurin was born in Scotland and educated at the University of Glasgow, which he entered at the incredible age of 11! He became professor of mathematics at Aberdeen at 19 and taught at the prestigious University of Edinburgh at 25.

Who was the real author of Cramer's rule?

Ironically, Maclaurin's name is associated with a portion of analysis (the Maclaurin series) discovered by Brook Taylor. The general Taylor series, in turn, had been known long before to James Gregory and Jean Bernoulli. If Maclaurin's name is recalled in connection with a series he did not first discover, this is compensated by a contribution he did make that bears the name of someone else who discovered and printed it later: Cramer's rule, published in 1750, was probably known to Maclaurin as early as 1729.

4.1 Systems with Two Variables

Objectives

Find the solution of a system of two linear equations using:

A The graphical method

B The substitution method

C The elimination method

D Solve applications involving systems of equations

To Succeed, Review How To . . .

1. Find the *x*- and *y*-intercepts of a line (pp. 171–173).
2. Graph a line (pp. 168–174).
3. Find the slope of a line given its equation (pp. 192–193).

Getting Started

How Do Yearly Sales of SUVs Compare to Passenger Car Sales?

Sport utility vehicles (SUVs) are becoming more popular among new vehicle buyers. The number of yearly sales of SUVs in North America in millions and the number of yearly sales of passenger cars in millions can be approximated by

$$\text{SUVs sold:} \quad y = 100x + 2500$$
$$\text{Passenger cars sold:} \quad y = 200x + 500$$

where *x* represents the year, with $x = 0$ corresponding to 1999. This *system* of *two* equations can be graphed as shown.

There are circumstances, like increased gas prices, that can affect the sales of SUVs. However, if this rate continues, in which year will there be as many SUVs sold as passenger cars? The graph indicates it will occur in 2019—when $x = 20$.

Using the first equation from the system and replacing *x* with 20, the number of vehicles (in millions) sold was

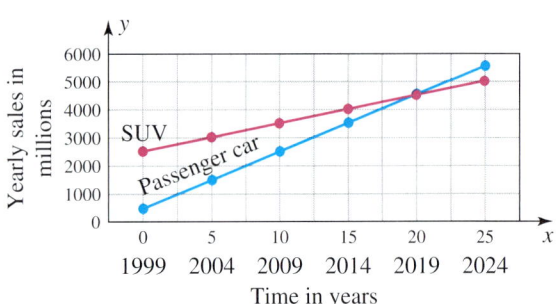

Projected Yearly SUV Sales Versus Passenger Car Sales in North America

$$y = 100(20) + 2500 = 4500$$

If we denote the year by *x* and the number of vehicles sold by *y*, the point $(x, y) = (20, 4500)$ represents the point at which the lines *intersect*. This point, (20, 4500), is the *solution* of the system of equations. The point (20, 4500) *satisfies* both equations.

We can also solve this system of equations by substitution. The variable *y* is the same in both equations. We substitute $100x + 2500$ for *y* in the second equation to obtain

$$100x + 2500 = 200x + 500 \qquad \text{There are no fractions to clear or parentheses to remove, so we can solve this by}$$

$$2500 = 100x + 500 \qquad \text{Adding } (-100x) \text{ on both sides and then } (-500) \text{ to both sides}$$

$$2000 = 100x$$

$$20 = x \qquad \text{and multiplying both sides by } \tfrac{1}{100}.$$

If you substitute 20 for *x* in the first equation (as we did before), you obtain 4500 for the value of *y*.

These are the graphical and substitution methods for solving a system of equations; we will study these and other methods in this section.

A › Finding Solutions Using the Graphical Method

When comparing two sets of data that are linearly related it is important to know when the two equations will have the same solution. A more simplistic view of this concept might be explained with the following question. What two numbers have a sum of 6 and a difference of 4? It sounds easy enough to do without algebra, but let's use our skills and develop the two equations that will answer this question. Using x to represent one number and y for the other, we get the following equations.

Sum is 6: $\quad x + y = 6$
Difference is 4: $\quad x - y = 4$

This is called a **system of equations** with two unknowns. We are looking for an ordered pair that will solve both equations, called the **solution to the system.**

There are many solutions to the sum equation, $x + y = 6$, and those solutions are represented by the red line in the graph in Figure 4.1. Likewise, there are many solutions to the difference equation, $x - y = 4$, and they are represented by the blue line in the graph in Figure 4.1. It is their point of intersection, $(5, 1)$, that solves both equations and is the solution to the system.

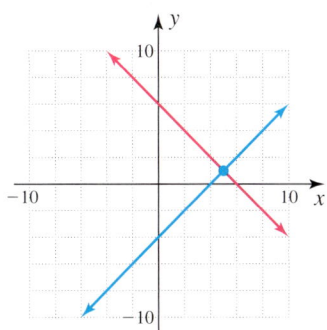
>Figure 4.1

We can verify that $(5, 1)$ is the solution to the system by substituting the value $x = 5$ and $y = 1$ into both equations to see if the values satisfy both equations.

Sum is 6: $\quad x + y = 6 \quad 5 + 1 = 6 \quad$ True
Difference is 4: $\quad x - y = 4 \quad 5 - 1 = 4 \quad$ True

In Example 1, we will identify the solution of a system of equations by observing the graph and then verifying that the proposed solution solves both equations.

EXAMPLE 1   Identifying the solution of a system of equations from the graph

The system of equations is graphed. Use the graph to identify the solution to the system. Use the system of equations to verify the solution.

a. $2x - y = 5$
$\ y = -x + 1$

b. $x - 3y = 6$
$\ 6x - 18y = -18$

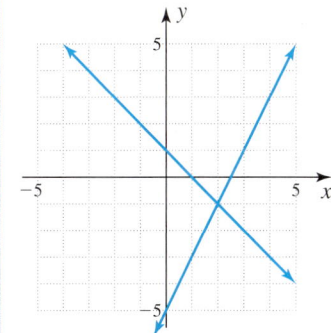
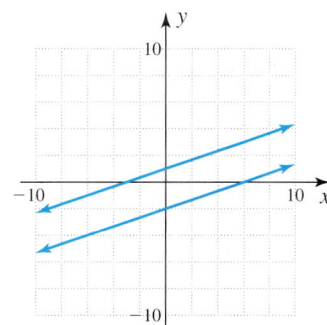

PROBLEM 1

The system of equations is graphed. Use the graph to identify the solution to the system. Use the system of equations to verify the solution.

a. $x + y = 4$
$\ 2x + 2y = -6$

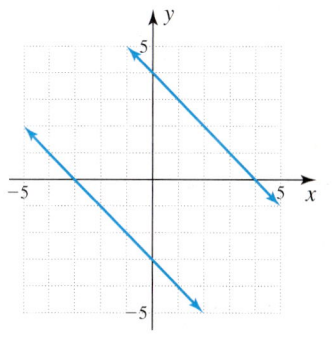

(continued)

Answers to PROBLEMS

1. a. Parallel lines; no solution

SOLUTION

a. Observing the graph, we see that the lines intersect at the point $(2, -1)$, so $(2, -1)$ is the solution.

We verify the solution by substituting $x = 2$ and $y = -1$ into each equation to see if it makes a true statement. Substituting into the first equation,

$$2x - y = 5$$
$$2(2) - (-1) = 5$$
$$4 + 1 = 5 \quad \text{True}$$

Substituting into the second equation,

$$y = -x + 1$$
$$(-1) = -(2) + 1$$
$$-1 = -1 \quad \text{True}$$

This verifies that $(2, -1)$ is the solution to the system.

b. The graph of this system indicates that the lines are parallel. Since they are parallel, they will not intersect. Since the lines don't intersect, the system has *no solution*.

b. $x + 2y = 2$
$-x + y = 4$

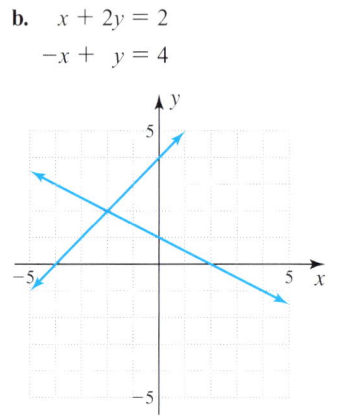

As we have seen, the solution set of a system of linear equations can be estimated by graphing the equations on the same axes and determining the coordinates of any points of intersection. This is called the **graphical method,** and we use it in Example 2. Remember, the *solution* of the system is a point (an ordered pair) that satisfies both equations. When a system has one or more solutions it is called a **consistent** system.

EXAMPLE 2 Using the graphical method to solve a system with one solution

Use the graphical method to find the solution of the system:

$$2x - y = 2$$
$$y = x - 1$$

PROBLEM 2

Use the graphical method to find the solution of the system:

$$x - y = -1$$
$$y = -x - 1$$

SOLUTION We first graph the equation $2x - y = 2$ using the x- and y-intercepts shown in the table.

When $x = 0$, $2x - y = 2$ becomes $2(0) - y = 2$, or $y = -2$.
When $y = 0$, $2x - y = 2$ becomes $2x - 0 = 2$, or $x = 1$.

x	y
0	-2
1	0

The points $(0, -2)$ and $(1, 0)$ are on the graph of the equation. We draw a line through these two points (the blue line in Figure 4.2).

We now graph $y = x - 1$. The equation is in slope-intercept form ($y = mx + b$). We can see the y-intercept is -1 (the constant) and the slope of the line is 1 (the coefficient of x). We use this information to graph the line like we did in Section 3.2. We plot -1 on the y-axis and use the slope ($1 = \frac{1}{1}$) to plot the point $(1, 0)$ so we can

---

Answers to PROBLEMS

1. b. Intersecting lines; solution $(-2, 2)$;
$(-2) + 2(2) = 2$
$-2 + 4 = 2$
$2 = 2$ True
$-(-2) + (2) = 4$
$2 + 2 = 4$
$4 = 4$ True

2.

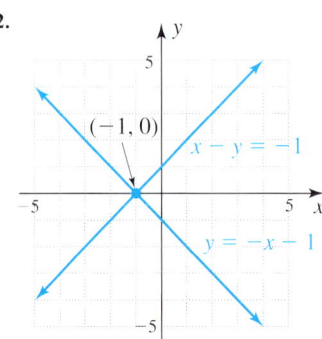

Solution: $(-1, 0)$

draw the line through the two points. The graph for $y = x - 1$ is shown in red in Figure 4.2.

The lines intersect at $(1, 0)$. The point $(1, 0)$ is the solution of the system of equations, and the system is said to be **consistent** because it has *one* solution.

CHECK For $x = 1$, $y = 0$, $2x - y = 2$ becomes $2(1) - 0 = 2$ (true). For $x = 1$, $y = 0$, $y = x - 1$ becomes $0 = 1 - 1$ (true). Thus, $(1, 0)$ is the correct solution for the system; the solution set is $\{(1, 0)\}$.

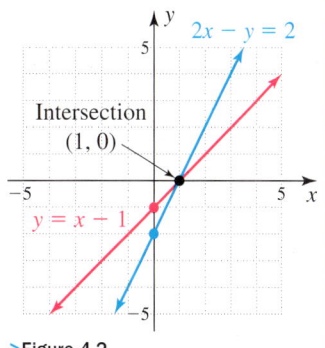

> Figure 4.2

 Calculator Corner

Finding a Solution of a Linear System by Graphing

In Example 2, rewrite the first equation, $2x - y = 2$, so that it is solved for y ($y = 2x - 2$). Then graph $Y_1 = 2x - 2$ and $Y_2 = x - 1$. To verify that the solution is $(1, 0)$, use a decimal window by pressing ZOOM 4, use the TRACE key and arrows until you find the point of intersection or, with a TI-83 Plus; use the intersect feature by pressing 2nd TRACE 5 ENTER ENTER ENTER to find the intersection, as shown in the window. Since the system has a solution, it is a *consistent* system.

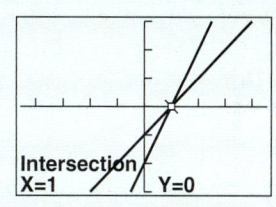

If a system has no solution, it is called an **inconsistent** system. This will be obvious when solving by the graphing method because the lines will not intersect, as we will see in Example 3.

EXAMPLE 3 Using the graphical method to solve a system with no solution

Use the graphical method to find the solution of the system:

$$-2x + y = 4$$
$$-6x + 3y = 18$$

SOLUTION We first graph the equation $-2x + y = 4$ by solving the equation for y so it will be in slope-intercept form ($y = mx + b$). That will make it easy to name the y-intercept and slope of the line. We will use the y-intercept and slope to graph the line.

$-2x + y = 4$ To solve for y, add $2x$ to both sides.
$y = 2x + 4$ The y-intercept is 4 and the slope is 2.

We plot 4 on the y-axis and use the slope $(2 = \frac{2}{1})$ to plot the point $(1, 6)$ so we can draw the line through the two points. The two points, as well as the completed graph, are shown in red in Figure 4.3.

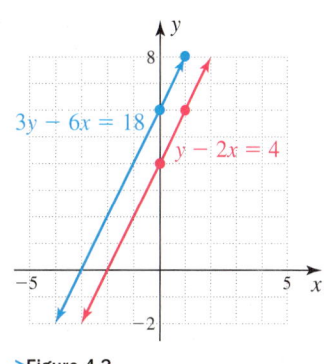

> Figure 4.3

PROBLEM 3

Use the graphical method to find the solution of the system:

$$y - 3x = 3$$
$$2y - 6x = 12$$

(continued)

Answers to PROBLEMS

3.

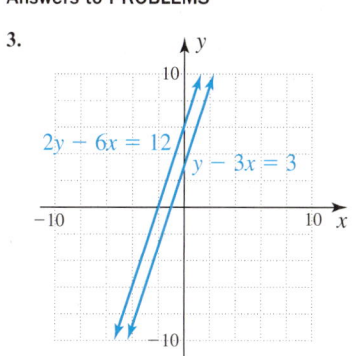

No solution, inconsistent

We now graph $-6x + 3y = 18$. We solve the equation for y by adding $6x$ to both sides of the equation and then dividing by 3.

$$-6x + 3y = 18$$
$$3y = 6x + 18$$
$$y = 2x + 6 \quad \text{The y-intercept is 6 and the slope is } \tfrac{2}{1}.$$

We plot 6 on the y-axis and use the slope, $\tfrac{2}{1}$, to plot the point $(1, 8)$ so we can draw the blue line in Figure 4.3 through the two points.

The graphs of the two lines we have drawn in Figure 4.3 look parallel. However, we must be mathematically sure that they are parallel lines. What do we notice when we compare the slopes of these two lines? Yes, they are the same—both have a slope of 2. In Chapter 3 we learned that if lines have the same slopes then they are parallel. So, not only do the lines look parallel, but we can conclude that they are parallel. Notice however that they have different y-intercepts. The lines are parallel so their graphs cannot intersect. There is *no solution* for this system. The solution set is $\emptyset$ and the system is said to be **inconsistent.**

If a system has coinciding lines, then it is called a **dependent** system and has infinitely many solutions. Such a system will be noticeable when solving by the graphing method because the two equations will graph as the same line. We will use the graphing method to solve the dependent system in Example 4.

EXAMPLE 4   Using the graphical method to solve a system with infinitely many solutions

Use the graphical method to find the solution of the system:

$$2x + \tfrac{1}{2}y = 2$$
$$y = -4x + 4$$

SOLUTION   To write the first equation in slope-intercept form, we solve for y by adding $(-2x)$ to both sides and then multiplying both sides by 2.

$$2x + \tfrac{1}{2}y = 2$$
$$\tfrac{1}{2}y = -2x + 2$$
$$y = -4x + 4 \quad \text{y-intercept 4 and slope } -4$$

The graph is shown in Figure 4.4. The second equation is already solved for y, and it is exactly the same as the equation we got for the first equation when we solved for y. What does this mean? It means that the graphs of the lines $2x + \tfrac{1}{2}y = 2$ and $y = -4x + 4$ coincide (are the same). Thus a solution of one equation is automatically a solution of the other. This solution set is $\{(x, y) \mid y = -4x + 4\}$. There are *infinitely* many solutions. Such a system is called **dependent.**

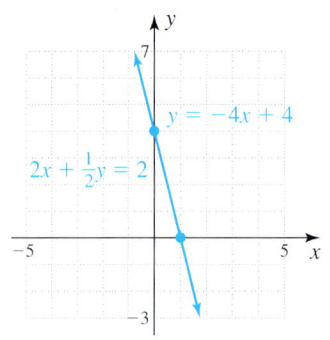

>Figure 4.4

PROBLEM 4

Use the graphical method to find the solution of the system:

$$x + \tfrac{1}{2}y = -2$$
$$y = -2x - 4$$

Answers to PROBLEMS

4.

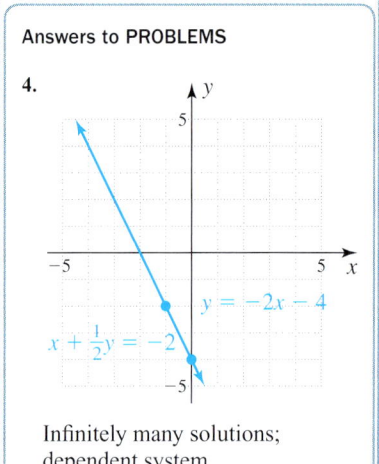

Infinitely many solutions; dependent system

A system of two linear equations in two variables can have one of three possible solution sets. They are summarized in the following table followed by their graphs in Figure 4.5. For ease of classification, we will use the name "consistent" for a system with one solution, even though a dependent system is also consistent.

4.1 Systems with Two Variables

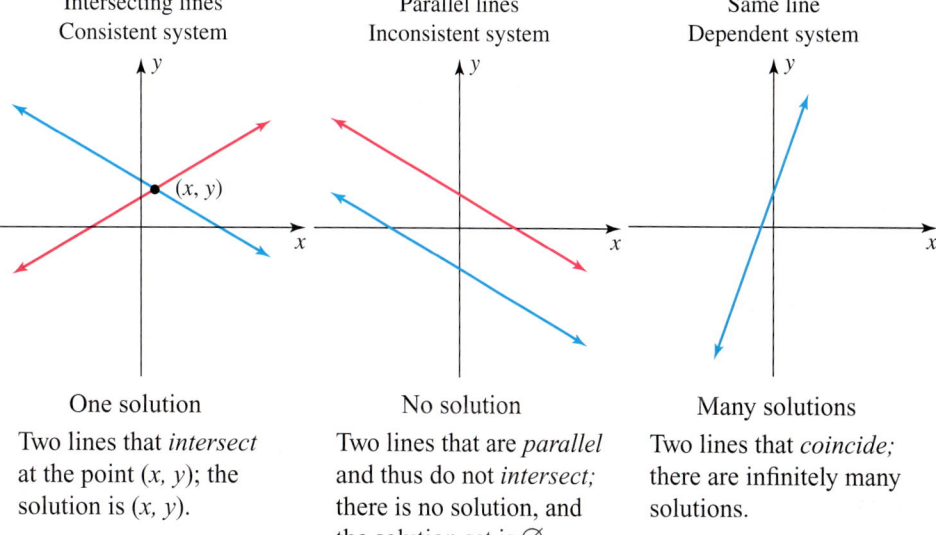

Summary Table for Solving a System by Graphing

Solution	Name of System	Slope Comparison	y-intercept Comparison
One Solution Lines intersect (Example 2)	Consistent	Different slopes	
No Solution Lines parallel (Example 3)	Inconsistent	Same slopes	Different y-intercepts
Infinitely Many Solutions Lines coincide (Example 4)	Dependent	Same slopes	Same y-intercepts

Intersecting lines
Consistent system

Parallel lines
Inconsistent system

Same line
Dependent system

One solution

Two lines that *intersect* at the point (x, y); the solution is (x, y).

No solution

Two lines that are *parallel* and thus do not *intersect*; there is no solution, and the solution set is $\emptyset$.

Many solutions

Two lines that *coincide;* there are infinitely many solutions.

> Figure 4.5

B › Finding Solutions Using the Substitution Method

Sometimes the graphical method isn't accurate enough to determine the exact solutions of a system of equations. For example, to solve the system

$$y = 2.5x + 72$$
$$y = 3x + 70$$

we can start by letting $x = 0$ in the first equation to obtain $y = 2.5(0) + 72$, or $y = 72$. To graph this point, we need a piece of graph paper with 72 units, or we have to make each division on the graph represent multiple units. But there is a way out. Keep in mind that we are looking for a point that satisfies *both* equations and notice that the second equation tells us that y is $3x + 70$. Thus we can *substitute* $3x + 70$ for y in the equation $y = 2.5x + 72$ to obtain

$$3x + 70 = 2.5x + 72$$
$$3x = 2.5x + 2 \quad \text{Subtract 70.}$$
$$0.5x = 2 \quad \text{Subtract 2.5x.}$$
$$x = \frac{2}{0.5} \quad \text{Divide by 0.5.}$$
$$x = 4$$

Now substitute 4 for x in the equation $y = 3x + 70$ to obtain $y = 3(4) + 70 = 82$.

The solution of the system is $(4, 82)$, and the method we used is called the **substitution method.**

> **NOTE**
>
> The substitution method is most useful when both equations are solved for one of the variables in terms of the other (as in the system we just solved) or when at least one of the equations is in this form, as in Example 2.

To solve the system of Example 2 by substitution, we write

$$2x - y = 2$$
$$y = x - 1$$

Since $y = x - 1$, we substitute $(x - 1)$ for y in the first equation, $2x - y = 2$, which becomes

$$2x - y = 2$$
$$2x - (x - 1) = 2$$
$$2x - x + 1 = 2$$
$$x + 1 = 2$$
$$x = 1$$

Substituting 1 for x in the second equation, $y = x - 1$, gives

$$y = 1 - 1 = 0$$

The solution is $(1, 0)$, and the solution set is $\{(1, 0)\}$ as shown in Example 2.

EXAMPLE 5 Using the substitution method to solve a system with one solution

Use the substitution method to solve the system:

$$3y + x = 9 \quad \textbf{(1)}$$
$$x - 3y = 0 \quad \textbf{(2)}$$

SOLUTION We solve equation (2) for x to obtain $x = 3y$. Substituting $3y$ for x in equation (1),

$$x = 3y$$

$3y + x = 9$ becomes $3y + 3y = 9$

$$6y = 9$$
$$y = \frac{3}{2}$$

Since $x = 3y$ and $y = \frac{3}{2}$,

$$x = 3\left(\frac{3}{2}\right) = \frac{9}{2}$$

The solution of the system is $\left(\frac{9}{2}, \frac{3}{2}\right)$, and the solution set is $\left\{\left(\frac{9}{2}, \frac{3}{2}\right)\right\}$ or $\left\{\left(4\frac{1}{2}, 1\frac{1}{2}\right)\right\}$.

CHECK $3\left(\frac{3}{2}\right) + \frac{9}{2} = 9$ and $\frac{9}{2} - 3 \cdot \left(\frac{3}{2}\right) = 0$

The system is consistent. You can verify this by graphing $3y + x = 9$ and $x - 3y = 0$.

PROBLEM 5

Use the substitution method to solve the system:

$$x + 2y = 3$$
$$x - 3y = 0$$

Answers to PROBLEMS

5. $\left(\frac{9}{5}, \frac{3}{5}\right)$ or $\left(1\frac{4}{5}, \frac{3}{5}\right)$

Without graphing, how do we recognize that a system such as $x + y = 3$ and $y = -x - 3$ will have no solution? We can do this with the substitution method as follows. The system is

$$x + y = 3 \quad (1)$$
$$y = -x - 3 \quad (2)$$

We substitute $(-x - 3)$ for y in equation (1), and

$$y = (-x - 3)$$

$x + y = 3 \quad$ becomes $\quad x + (-x - 3) = 3$
$$-3 = 3$$

Since $-3 = 3$ is a contradiction, the system $x + y = 3$ and $y = -x - 3$ is inconsistent; it has *no* solution. Graphically, the lines are parallel, so the solution set is $\varnothing$.

EXAMPLE 6 Using the substitution method to solve a system with no solution

Use the substitution method to solve the system:

$$x + y = 5 \quad (1)$$
$$y = -x \quad (2)$$

SOLUTION Substituting $(-x)$ in equation (1) for y, we get

$$x + (-x) = 5$$
$$0 = 5$$

This is a contradiction, so the system $x + y = 5$ and $y = -x$ has *no* solution. The graph of this system would be parallel lines, and the solution set is $\varnothing$.

PROBLEM 6

Use the substitution method to solve the system:

$$x - y = 2$$
$$y = x$$

Calculator Corner

Find the Solution for an Inconsistent System

To verify the result of Example 6, solve each equation for y and graph

$$Y_1 = -x + 5$$

and

$$Y_2 = -x$$

The slopes are the same, so the lines are parallel, as shown in the window, and there is no solution. The system is **inconsistent** and the solution set is $\varnothing$.

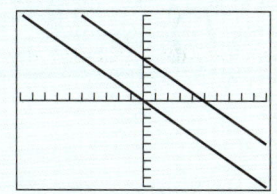

EXAMPLE 7 Using the substitution method to solve a system with infinitely many solutions

Use the substitution method to solve the system:

$$x + 2y = 4 \quad (1)$$
$$2x = 8 - 4y \quad (2)$$

SOLUTION Solving equation (1) for x, we obtain $x = (4 - 2y)$. We now substitute $(4 - 2y)$ for x in equation (2).

$$2x = 8 - 4y$$
$$2(4 - 2y) = 8 - 4y$$
$$8 - 4y = 8 - 4y$$

PROBLEM 7

Use the substitution method to solve the system:

$$x - 3y = 4$$
$$2x = 8 + 6y$$

(continued)

Answers to PROBLEMS

6. There is no solution; the system is inconsistent, and the solution set is $\varnothing$.

7. There are infinitely many solutions; the system is dependent. $\{(x, y) \mid y = \frac{1}{3}x - \frac{4}{3}\}$

This equation is an identity, so any value of y will make it true. There are infinitely many solutions. The graph of this system would be coinciding lines. For example, if $x = 0$ in equation (1),

$$2y = 4 \quad \text{and} \quad y = 2$$

$(0, 2)$ is a solution of the system. If we let $y = 0$ in equation (2),

$$2x = 8 - 4(0) \quad \text{and} \quad x = 4$$

$(4, 0)$ is another solution. You can continue to obtain solutions by assigning numbers to one of the variables in either equation and solving for the other variable. The solution set is obtained by solving $x + 2y = 4$ for y and writing

$$\left\{(x, y) \mid y = -\frac{x}{2} + 2\right\}$$

Calculator Corner

Find the Solution for a Dependent System

To verify the result of Example 7, first solve each equation for y

$$Y_1 = -\frac{x}{2} + 2$$

and

$$Y_2 = -\frac{x}{2} + 2$$

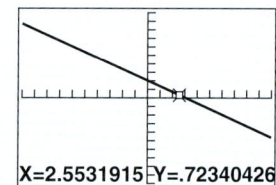

You can stop here; you don't need a calculator! The equations are equivalent, the system is dependent, and there are infinitely many solutions. You can find many of these solutions using the TRACE key. One such solution, point (2.5531915, 0.72340426), is shown in the window. Can you verify that this is a solution? If we had set the window with ZOOM 4, the TRACE key would have given solutions that would be easier to verify! You can also use 2nd GRAPH to find more solutions.

C > Solving by Elimination

A third method for solving a system of linear equations is called the **elimination method**. The key step in this method is *to obtain coefficients that differ only in sign* for one of the variables so that this variable is eliminated by adding the two equations. The procedure is given next.

PROCEDURE

Procedure for Solving a System of Two Equations in Two Unknowns by Elimination

1. Clear any fractions or decimals that may be present.
2. Multiply both sides of the equations (as needed) by numbers that make the coefficients of one of the variables additive inverses (opposites).
3. Add the two equations.
4. Solve for the remaining variable.
5. Substitute this solution into one of the original equations and solve for the second variable.
6. Check the solution.

EXAMPLE 8 Using the elimination method to solve a system with one solution

Use the elimination method to solve the system:

$$0.2x + 0.3y = 1 \quad (1)$$
$$x - \frac{1}{2}y = 3 \quad (2)$$

SOLUTION We will eliminate x by the following steps.

1. Clear the fractions and decimals.

   $0.2x + 0.3y = 1 \quad\quad 2x + 3y = 10$ Multiply both sides of equation (1) by 10. (3)

   $x - \frac{1}{2}y = 3 \quad\quad 2x - y = 6$ Multiply both sides of equation (2) by 2. (4)

2. Make the x coefficients, additive inverses.

   $2x + 3y = 10 \quad -2x - 3y = -10$ Multiply both sides of equation (3) by -1. (5)

   $ 2x - y = 6$ Copy equation (4). (6)

3. Add the two equations (5) and (6). $-4y = -4$
4. Solve for y by dividing by -4. $y = 1$
5. Substitute $y = 1$ into equation (4). $2x - y = 6$ (4)

   $2x - 1 = 6$

   $2x = 7$ Add 1 to both sides.

   $x = 3.5$ Divide both sides by 2.

The solution is $(3.5, 1)$ and the system is **consistent**.

6. The check is left for you.

PROBLEM 8

Use the elimination method to solve the system:

$$x + 0.2y = 4$$
$$4x + \frac{2}{3}y = 2$$

Calculator Corner

Find the Solution for a Consistent System

We can verify the results of Example 8 by solving each equation for y and graphing

$$Y_1 = -\frac{2}{3}x + \frac{10}{3}$$

and

$$Y_2 = 2x - 6$$

Using an integer or decimal window and the TRACE key or the intersect (2nd TRACE 5) feature, we can verify that the system is **consistent** and the solution is $(3.5, 1)$ as shown in the window.

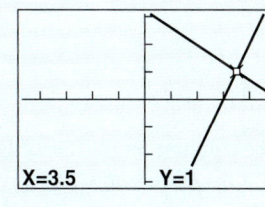

In many cases, it is not enough to multiply *one* of the equations by a number to eliminate one of the variables. For example, to find the common solution of

$$2x + 3y = -1 \quad (1)$$
$$3x + 2y = -4 \quad (2)$$

we multiply both equations by numbers chosen so that the coefficients of one of the variables become opposites in the resulting equations. We will eliminate the y variable by using the 6 steps from the procedure we used in Example 8.

Answers to PROBLEMS

8. $(-17, 105)$

1. Clear the fractions and decimals. There are none.
2. Make the y coefficients additive inverses by multiplying equation (1) by the y coefficient of equation (2) and then multiplying equation (2) by the negative of the y coefficient of equation (1).

$$2x + 3y = -1 \qquad 4x + 6y = -2 \quad \text{Multiply both sides of equation (1) by 2.} \quad \textbf{(3)}$$
$$3x + 2y = -4 \qquad -9x - 6y = 12 \quad \text{Multiply both sides of equation (2) by } -3. \quad \textbf{(4)}$$

3. Add the two equations (3) and (4). $-5x = 10$
4. Solve for x by dividing by -5. $x = -2$
5. Substitute $x = -2$ into equation (1).

$$2x + 3y = -1$$
$$2(-2) + 3y = -1$$
$$-4 + 3y = -1 \quad \text{Simplify.}$$
$$3y = 3 \quad \text{Add 4 to both sides.}$$
$$y = 1 \quad \text{Divide both sides by 3.}$$

The solution is $(-2, 1)$ and the system is *consistent*, as shown in the *Calculator Corner*.

6. The check is left for you.

EXAMPLE 9 Using the elimination method to solve a system with no solutions

Use the elimination method to solve the system:

$$3x - 6y = 2 \quad \textbf{(1)}$$
$$-8x + 16y = 1 \quad \textbf{(2)}$$

SOLUTION We proceed as before.

1. Clear the fractions and decimals. There are none.
2. Make the x coefficients additive inverses by multiplying equation (1) by the negative of the x coefficient of equation (2) and then multiplying equation (2) by the x coefficient of equation (1).

$$3x - 6y = 2 \qquad 24x - 48y = 16 \quad \text{Multiply both sides of equation (1) by 8.} \quad \textbf{(3)}$$
$$-8x + 16y = 1 \qquad -24x + 48y = 3 \quad \text{Multiply both sides of equation (2) by 3.} \quad \textbf{(4)}$$

3. Add the two equations (3) and (4). $0 = 19$
4. This is a contradiction, so the system has no solution.

The solution set is $\emptyset$, and the system is *inconsistent*.

If we were to write the equations in slope-intercept form, we would notice that the slopes are the same and the y-intercepts are different. This means the lines are parallel.

PROBLEM 9

Use the elimination method to solve the system.

$$-2x + 6y = 1$$
$$5x - 15y = 2$$

EXAMPLE 10 Using the elimination method to solve a system with infinitely many solutions

Use the elimination method to solve the system:

$$\frac{x}{6} + \frac{y}{2} = 1 \quad \textbf{(1)}$$
$$\frac{x}{9} + \frac{y}{3} = \frac{2}{3} \quad \textbf{(2)}$$

PROBLEM 10

Use the elimination method to solve the system:

$$\frac{x}{20} + \frac{y}{12} = \frac{1}{2}$$
$$\frac{x}{5} + \frac{y}{3} = 2$$

Answers to PROBLEMS

9. No solution. The system is inconsistent, and the solution set is $\emptyset$. **10.** Infinitely many solutions; the system is dependent; $\{(x, y) \mid y = -\frac{3}{5}x + 6\}$

SOLUTION We proceed as before.

1. Clear the fractions and decimals.

$$x + 3y = 6 \quad \text{Multiply both sides of equation (1) by 6.} \quad (3)$$
$$x + 3y = 6 \quad \text{Multiply both sides of equation (2) by 9.} \quad (4)$$

2. Make the x coefficients additive inverses.

$$-x - 3y = -6 \quad \text{Multiply both sides of equation (3) by } -1. \quad (5)$$
$$x + 3y = 6 \quad \text{Copy equation (4).} \quad (6)$$

3. Add the two equations (5) and (6). $0 = 0$

We obtain the true statement $0 = 0$, so the equations are dependent. This system has infinitely many solutions. The two equations in this system represent coinciding lines.

The solution set is obtained by solving $x + 3y = 6$ for y and is written as $\{(x, y) \mid y = -\frac{x}{3} + 2\}$.

> **NOTE**
>
> In Step 1, we came up with identical equations. Any time we have two equivalent (or identical) equations, the system is dependent, and we write the solution set as $\{(x, y) \mid y = mx + b\}$.

Which Method Shall I Use? We have discussed three methods for solving systems of equations: graphical, substitution, and elimination. If you are wondering how to decide which method to use, Table 4.1 gives you some guidelines.

Table 4.1 Solving Systems of Equations: A Summary

Method	Suggested Use	Disadvantages
Graphical	When the coefficients of the variables and the solutions are integers. You get a picture of the situation.	If the solutions are not integers, they are hard to read on the graph.
Substitution	When one of the variables is isolated (alone) on one side of the equation.	If fractions are involved, you may have much computation.
Elimination	When variables with coefficients that are the same or additive inverses of each other ($2x$ and $-2x$, for example) are present.	You may have lots of computations involving signed numbers.

D › Solving Applications Involving Systems of Equations

Mercy Gonzalez is a buyer for the cosmetic counters of the BestDeal drug store chain. BestDeal has decided to have a fragrance promotion and sell one of their more popular fragrances at a premium price. But which one? She can choose from several different items that have been sold at different prices in past promotions. Her problem is to decide which selling price will best suit the needs of both the customers and the stores. Mercy has data from previous similar promotions to help her make a decision.

Supply Function, S

Selling Price of Each Item	Number Supplied Per Week Per Store
$15	55
$25	145
$35	235

Based on these data, the equation of the supply function is $S = 9x - 80$, where x is the selling price of the item. The graph is shown in Figure 4.6.

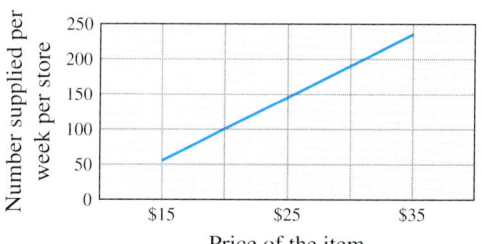

> Figure 4.6

Demand Function, D

Selling Price of Each Item	Number Requested (Demand) Per Week Per Store
$15	295
$25	225
$35	155

The equation of the demand function is $D = -7x + 400$, where x is the selling price of the item. The graph is shown in Figure 4.7.

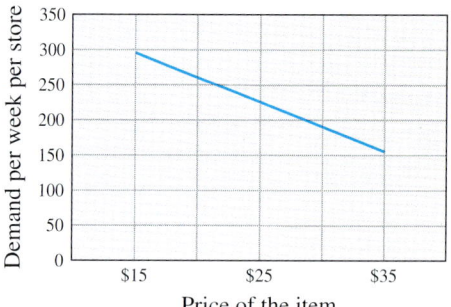

> Figure 4.7

Now let's graph both functions on the same axes. As price increases, supply increases. As price increases, demand decreases. The point at which the graphs intersect is the **equilibrium point;** see Figure 4.8. That is, the price of the item at this point will indicate that the number of items the customers are willing to buy (demand) is the same as the number the store is willing to supply. By definition, if D is the demand function, S is the supply function, and p is the price, the equilibrium point is where the demand equals supply: $D(p) = S(p)$.

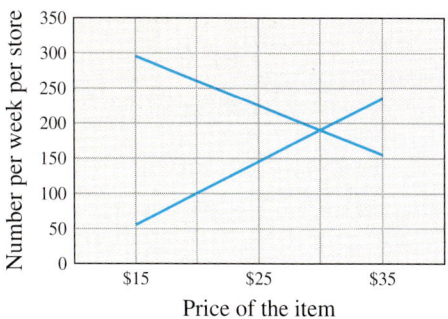

> Figure 4.8

We can see by the graph in Figure 4.8 that Mercy should price the item at $30. Let's verify this by using the supply and demand equations to find the equilibrium point.

$$D(p) = S(p)$$
$$-7x + 400 = 9x - 80 \quad \text{Substitute } -7x + 400 \text{ for } D(p) \text{ and } 9x - 80 \text{ for } S(p).$$
$$-7x + 480 = 9x \quad \text{Add 80 to both sides.}$$
$$480 = 16x \quad \text{Add 7x to both sides.}$$
$$\$30 = x \quad \text{Divide both sides by 16.}$$

There is no doubt about it. Mercy makes the decision to go with the fragrance priced at $30.

EXAMPLE 11 A supply and demand problem

Booksellers supply fewer books when prices go down and more books when prices go up (because the profit margin is larger with the higher price). If a store knows the supply function is $S(p) = 20p + 200$ and the demand function is $D(p) = 740 - 100p$ for their art booklets, find the price of the book at the equilibrium point.

SOLUTION We use the RSTUV method from Section 2.3.

1. Read the problem. We are asked to find the price of the book at the equilibrium point.

2. Select the unknown. Since the equations are written, p is the price of the book, and S and D stand for supply and demand.

3. Think of a plan. At the equilibrium point $S = D$, so we will rename both of those variables with y. This gives the following system of equations

$$y = 20p + 200 \qquad (1)$$
$$y = 740 - 100p \qquad (2)$$

4. Use the substitution method to solve.

$$(740 - 100p) = 20p + 200 \quad \text{Substitute } (740 - 100p) \text{ for } y \text{ in equation (1).}$$
$$740 = 120p + 200 \quad \text{Add 100p to both sides.}$$
$$540 = 120p \quad \text{Subtract 200 from both sides.}$$
$$\frac{540}{120} = p \quad \text{Divide both sides by 120.}$$
$$\$4.50 = p$$

The price at which the supply equals the demand is $4.50. At this point, the number of books suppliers are willing to offer is

$$S(4.5) = 20(4.5) + 200$$
$$= 90 + 200$$
$$= 290 \text{ books}$$

5. Verify the answer. The answer can be verified by showing that the supply and the demand are both 290 when the price is $4.50.

PROBLEM 11

Find the equilibrium point if the supply and demand functions are given by

$$S(p) = 0.4p + 3$$

and

$$D(p) = -0.2p + 27$$

Answers to PROBLEMS

11. $p = 40$

EXAMPLE 12   A geometry problem

Two angles are complementary. If the measure of one of the angles is 20° more than the other, what are the measures of these angles?

SOLUTION   We use the RSTUV method.

1. **Read the problem.** We are asked to find the measures of these angles.
2. **Select the unknown.** There are two angles involved. Let x be the measure of one of the angles and y the measure of the other angle.
3. **Think of a plan.** Do you know what complementary angles are? They are angles whose sum is 90°. With this information, we have

$$x + y = 90 \quad \text{The sum of the angles is 90°.}$$
$$y = x + 20 \quad \text{One of the angles is 20° more than the other.}$$

4. **Use the substitution method to solve the system.** Substituting $(x + 20)$ for y into $x + y = 90$, we obtain

$$x + (x + 20) = 90$$
$$2x + 20 = 90 \quad \text{Add like terms.}$$
$$2x = 70 \quad \text{Subtract 20.}$$
$$x = 35 \quad \text{Divide by 2.}$$

Thus one angle is 35° and the other 20° more, or 55°.

5. **Verify the answer.** Since the sum of the measures of the two angles is $35° + 55° = 90°$, the angles are complementary and our result is correct.

PROBLEM 12

Two angles are supplementary. If the measure of one of the angles is 40° less than the other, what are the measures of these angles?

Calculator Corner

Problems 1–3 refer to Window 1.

1. How many solutions does the system have?
2. Do the lines have the same slope?
3. Is the system consistent or inconsistent?

Problems 4–6 refer to Window 2.

4. How many solutions does the system have?
5. Do the lines have the same slope?
6. Is the system consistent or inconsistent?

Problems 7–9 refer to Window 3.

7. How many solutions does the system have? (Assume a pair of equations, Y_1 and Y_2, have been graphed.)
8. Is the system dependent?
9. Do the lines have the same slope?

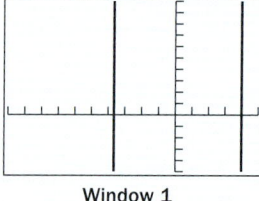
Window 1

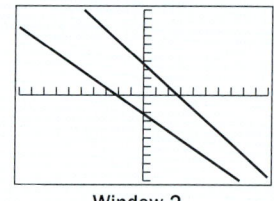
Window 2

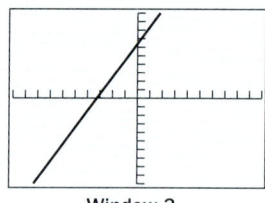
Window 3

Answers to PROBLEMS

12. 70°, 110°

> Exercises 4.1

Boost your grade at mathzone.com!
> Practice Problems
> NetTutor
> Self-Tests
> e-Professors
> Videos

⟨ A ⟩ **Finding Solutions Using the Graphical Method** In Problems 1–4, use the graph of the system to identify the solution to the system.

1. $x - 2y = 1$
   $4x + 3y = 15$

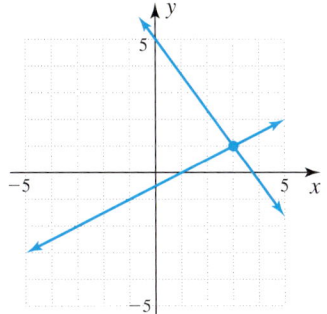

2. $y = -x + 3$
   $y - 2x = 6$

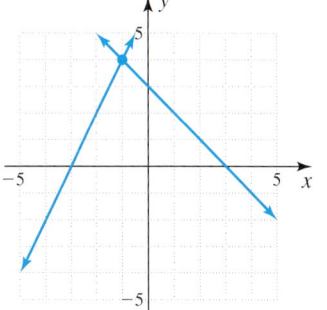

3. $-x + y = 2$
   $2y = 2x - 6$

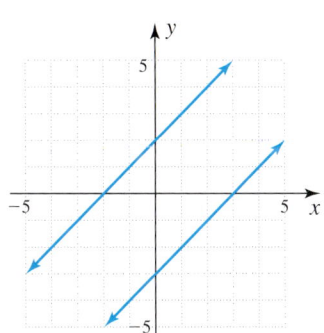

4. $x + 2y = 0$
   $2x + 4y = 12$

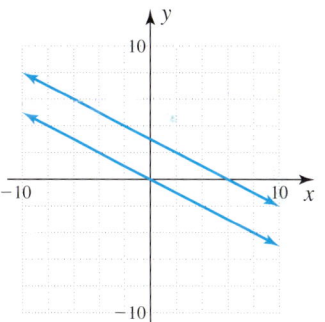

In Problems 5–14, graph the system of equations. Identify the lines as intersecting, parallel, or coinciding. Name the solution to the system.

5. $x - 2y = 6$
   $y = 2x$

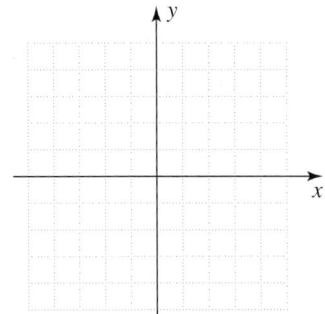

6. $2x + y = 4$
   $y = 2x$

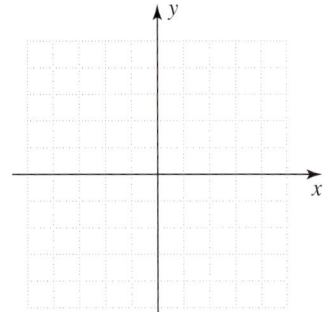

7. $y = x - 3$
   $y = 2x - 4$

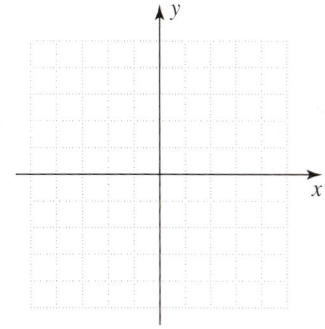

8. $y = x - 1$
$y = 3x - 3$

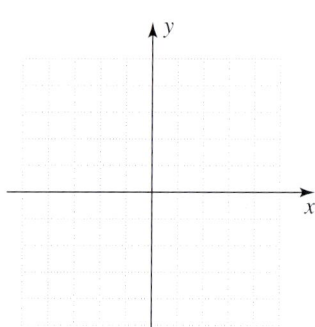

9. $2y = -x + 4$
$y = -2x + 4$

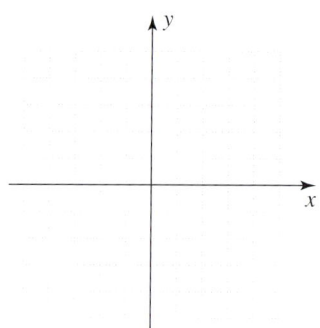

10. $2y = -x + 2$
$y = -2x + 2$

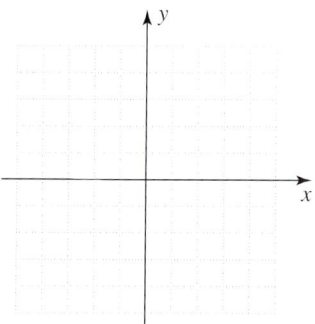

11. $2x - y = -2$
$y = 2x + 4$

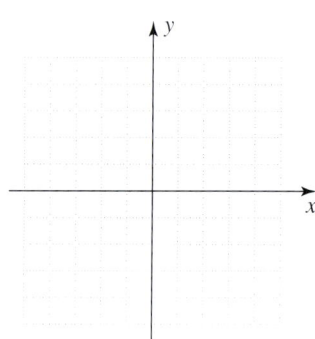

12. $2x + y = -2$
$y = -2x + 4$

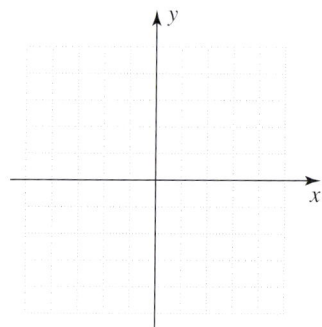

13. $3x + 4y = 12$
$8y = 24 - 6x$

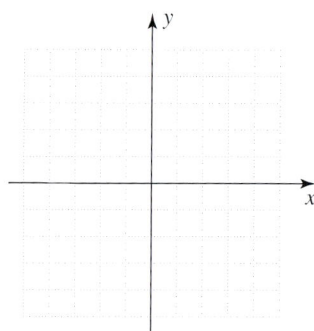

14. $2x - 3y = 6$
$6x = 18 + 9y$

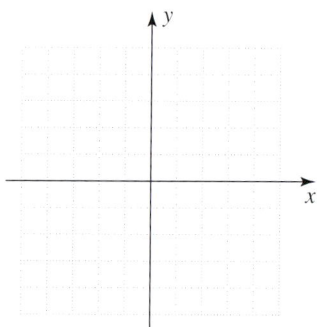

⟨ B ⟩ Finding Solutions Using the Substitution Method In Problems 15–30, solve by the substitution method. Label each system as consistent (one solution), inconsistent (no solution), or dependent (many solutions).

15. $y = 2x - 4$
$-2x = y - 4$

16. $y = 2x + 2$
$-x = y + 1$

17. $x + y = \dfrac{5}{2}$
$3x + y = \dfrac{9}{2}$

18. $x + y = \dfrac{5}{2}$
$3x + y = \dfrac{3}{2}$

19. $y - 4 = 2x$
$y = 2x + 2$

20. $y + 5 = 4x$
$y = 4x + 7$

21. $x = 8 - 2y$
$-x + 2y = 2$

22. $x = 4 - 2y$
$x - 2y = 0$

23. $x + 2y = 4$
$x = -2y + 4$

24. $x + 3y = 6$
$x = -3y + 6$

25. $x = 2y + 1$
$y = 2x + 1$

26. $y = 3x + 2$
$x = 3y + 2$

27. $2x - y = -4$
$-2x = 2y$

28. $5x + y = 5$
$5x = 15 - 3y$

29. $x = 5 - y$
$0 = x - 4y$

30. $x = 3 - y$
$0 = 2x - y$

⟨ C ⟩ Solving by Elimination In Problems 31–58, solve each system by elimination.

31. $x + y = 8$
$x - y = 2$

32. $x + y = 3$
$x - y = 1$

33. $x + 4y = 2$
$x - 4y = -2$

34. $x - 5y = 15$
$x + 5y = -5$

35. $-x - 2y = -2$
$x - 2y = -2$

36. $x + 3y = -7$
$-x + 2y = -3$

37. $2x + y = 7$
$3x - 2y = 0$

38. $2x + y = 4$
$3x - 2y = -1$

39. $2x - 2y = 6$
$x + y = 2$

40. $3x + 5y = 26$
$5x + 3y = 22$

41. $3x + 5y = 1$
$-6x - 10y = 2$

42. $5x - 2y = 4$
$-10x + 4y = 1$

43. $2x + y = 8$
$3x - y = 7$

44. $x - 3y = -2$
$x + 3y = 4$

45. $2x - 5y = 9$
$4x - 10y = 18$

46. $3x - 3y = -15$
$x - y = -5$

47. $6x + 5y = 12$
$9x - 4y = -5$

48. $5x + 4y = 6$
$4x - 3y = 11$

49. $2x - 3y = 16$
$x - y = 7$

50. $3x - 2y = 35$
$x - 5y = 42$

51. $18x - 15y = 1$
$10x - 12y = 3$

52. $6x - 9y = -2$
$3x - 5y = -6$

53. $\frac{x}{3} + \frac{y}{6} = \frac{2}{3}$
$\frac{2}{5}x + \frac{y}{4} = \frac{1}{5}$

54. $\frac{x}{6} + \frac{y}{3} = \frac{1}{2}$
$\frac{3}{5}x + \frac{y}{4} = \frac{17}{20}$

55. $\frac{5}{6}x + \frac{y}{4} = 7$
$\frac{2}{3}x - \frac{y}{8} = 3$

56. $\frac{1}{5}x + \frac{2}{5}y = 1$
$\frac{1}{4}x - \frac{1}{3}y = -\frac{5}{12}$

57. $\frac{2}{x} + \frac{3}{y} = -\frac{1}{2}$
$\frac{3}{x} - \frac{2}{y} = \frac{17}{12}$

Hint: Multiply the first equation by 2 and the second by 3 and add, or let $a = \frac{1}{x}$, $b = \frac{1}{y}$ and solve the system:
$2a + 3b = -\frac{1}{2}$
$3a - 2b = \frac{17}{12}$

58. $\frac{4}{x} + \frac{2}{y} = \frac{26}{21}$
$\frac{2}{x} - \frac{1}{y} = -\frac{1}{21}$

Hint: Multiply the second equation by 2 and add, or let $a = \frac{1}{x}$, $b = \frac{1}{y}$ and solve the resulting system.

⟨ D ⟩ Solving Applications Involving Systems of Equations In Problems 59–63, find the price p at the equilibrium point for each pair of demand and supply functions.

59. $D(p) = 620 - 10p$
$S(p) = 20p + 200$

60. $D(p) = 200 - 2p$
$S(p) = 50 + p$

61. $D(p) = 1500 - 5p$
$S(p) = 300 + p$

62. $D(p) = 1100 - 8p$
$S(p) = 200 + 17p$

63. $D(p) = 450 - 5p$
$S(p) = 2p + 170$

64. *Equilibrium point* The demand function for a certain product is $D(p) = 500 - 5p$, and the supply function is $S(p) = 8p + 110$. What will the price of this product be at the equilibrium point? What will the demand for the product be at the equilibrium point?

65. *Equilibrium point* The demand function for skateboards is $D(p) = 960 - 16p$, where p is the price in dollars. If the manufacturer is willing to supply $S(p) = 14p + 120$ boards when the price is p, what is the price of each skateboard when the equilibrium point is reached? What is the demand at that price; that is, how many skateboards sell at that price?

66. *Equilibrium point* The demand function for sunglasses is $D(p) = 24 - 0.4p$, where p is the price in dollars. If the supply function is $S(p) = p + 10$, what will the price of sunglasses be when the equilibrium point is reached? How many sunglasses would suppliers be willing to offer at that price?

67. *Equilibrium point* Find the price p (in dollars) and the supply and the demand at the equilibrium point when the demand function is given by $D(p) = 480 - 3p$ and the supply function is $S(p) = 25p + 60$.

68. *Equilibrium point* A store will buy 80 digital cellular phones if the price is $350 and 120 if the price is $300. The wholesaler is willing to supply 60 phones if the price is $280 and 140 if the price is $370. If the supply and demand functions are linear, write the equations of the functions, make a graph, and find the equilibrium point.

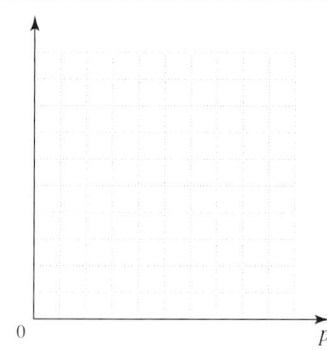

In Problems 69–72, (a) write a system of two equations describing the situation, and (b) find the measure of each angle.

69. *Angles* Two angles are complementary, and the measure of one of the angles is 15° more than the other.

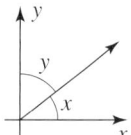

The sum of the measures of the two angles is 90°.

70. *Angles* Two angles are complementary, and the measure of one of the angles is twice the measure of the other angle.

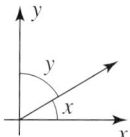

The sum of the measures of the two angles is 90°.

71. *Angles* Two angles are supplementary, and the measure of one of the angles is four times the measure of the other.

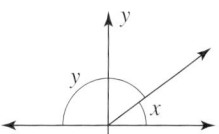

The sum of the measures of the two angles is 180°.

72. *Angles* Two angles are supplementary, and the measure of one of the angles is 30° more than the other.

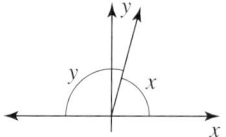

The sum of the measures of the two angles is 180°.

In Problems 73–82, use the RSTUV method to solve.

73. *Elephants* The heaviest pair of African elephant tusks on record weighed a total of 465 pounds. One of the tusks weighed 15 pounds more than the other.

  a. Write a system of two equations representing this situation.

  b. What does each tusk weigh?

74. *Basketball court* Do you know who invented basketball? It was Dr. James Naismith who in 1921 created a game that could be played indoors in Massachusetts. Although the height of the players has changed dramatically, the size of the court hasn't. The perimeter (distance around) of the maximum-size basketball court is 288 feet, and the length L of the court is 44 feet more than the width W.

  a. Write a system of two equations representing this situation.

  b. What are the dimensions of the maximum-size basketball court?

75. *Cigarettes* Do you smoke? How many cigarettes a year? Together, smokers in the United States and Japan consume 4637 cigarettes per person each year. If the average Japanese smoker consumes 437 cigarettes a year more than the average U.S. smoker, what is the annual consumption of cigarettes per person in each country?

76. *Soft drink consumption* The average American drinks 29 more Cokes® per year than the average Mexican. If together they consume 555 Cokes per year, how many Cokes are consumed each year by the average American and how many by the average Mexican?

77. *Ancient Babylonian problem*   Find the solution to this ancient Babylonian problem:

   There are two silver rings; $\frac{1}{7}$ of the first ring and $\frac{1}{11}$ of the second ring are broken off so that what is broken off weighs 1 shekel. The first, diminished by $\frac{1}{7}$, weighs as much as the second diminished by $\frac{1}{11}$. What did the silver rings (together) originally weigh?

   (*Hint:* Consider the system of equations

   $$\frac{x}{7} + \frac{y}{11} = 1 \quad \text{and} \quad \frac{6x}{7} = \frac{10y}{11}$$

   Clear denominators by multiplying by the LCD, and then use the substitution method to solve the system.)

78. *Nuclear reactors*   Is there a nuclear reactor near your city? The total number of reactors in the United States is 160, but not all of them are in operation. The number of operable reactors is four more than double the number of inoperable reactors. How many U.S. reactors are operable and how many are inoperable?

79. *Cats*   According to the *Guinness Book of World Records*, the sum of the ages of the two oldest cats is 70 years. The difference of their ages is 2 years. What are their ages?

80. *Height*   According to the *Guinness Book of World Records*, the sum of the heights of the shortest and tallest persons on record is 130 inches. If the difference in their heights was 84 inches, how tall was each?

81. *Tallest buildings in the world*   The combined height of two of the tallest buildings in the world, Taipei 101 in Taiwan and Petronas Tower 1 in Malaysia, is 3,150 feet. The difference in height between the two buildings is 184 feet. How tall is each building?

82. *Weight*   At one time, the combined weight of the McCreary brothers was 1300 pounds. Their weight difference was 20 pounds. What was the weight of each brother?

〉〉〉 Using Your Knowledge

Tweedledee and Tweedledum   Have you read *Alice in Wonderland*? Do you know who the author of this book is? The answer is Lewis Carroll. Did you also know that he was an accomplished mathematician and logician? These interests show up occasionally in his children's books. Take, for example, *Through the Looking Glass,* where one of the characters, Tweedledee, is talking to Tweedledum:

   Tweedledee: The sum of your weight and twice mine is 361 pounds.

   Tweedledum: Contrariwise, the sum of your weight and twice mine is 360 pounds.

83. If Tweedledee weighs x pounds and Tweedledum weighs y pounds, can you use the knowledge gained in this section to find their weights?

〉〉〉 Write On

84. How would you know if a system of linear equations is consistent when you are using the specified method?

   a. the graphical method   **b.** the substitution method
   c. the elimination method

85. How would you know if a system of linear equations is inconsistent when you are using the specified method?

   a. the graphical method   **b.** the substitution method
   c. the elimination method

86. How would you know if a system of linear equations is dependent when you are using the specified method?

   a. the graphical method   **b.** the substitution method
   c. the elimination method

87. Write the possible advantages and disadvantages of each of the methods we have studied: graphical, substitution, and elimination.

〉〉〉 Concept Checker

Fill in the blank(s) with the correct word(s), phrase, or mathematical statement.

88. If two lines of a system intersect, then the system has _____ solution(s) and is said to be a(n) _____ system.

89. When solved by the graphical method, systems of equations that have infinitely many solutions have _____ lines.

90. When solving a system of equations by the substitution method, if both variables cancel out and the resulting statement is a contradiction, then the system has _____ solution(s) and is called a(n) _____ system.

system of equations
consistent
inconsistent
dependent
one

no
infinitely many
intersecting
parallel
coinciding

Mastery Test

Solve the system by elimination.

91. $5x = 6 - 4y$
$3y = 4x - 11$

92. $3x = 6 - 3y$
$2y = 5x + 11$

93. $\dfrac{x}{3} + \dfrac{y}{9} = \dfrac{2}{3}$
$\dfrac{x}{2} + \dfrac{y}{6} = 1$

94. $\dfrac{x}{3} + \dfrac{y}{4} = \dfrac{11}{6}$
$\dfrac{x}{6} + \dfrac{y}{8} = \dfrac{7}{12}$

95. $3x + 2y = 8$
$x - y = 1$

96. $2x + 3y = 9$
$x - y = 2$

Solve the system by substitution.

97. $3x - 3y = -4$
$2x = 8 + 6y$

98. $x - y = 2$
$y = x$

99. $2y + x = 8$
$x - 3y = 0$

Use the graphical method to solve the system and classify the system as consistent, inconsistent, or dependent.

100. $2x + 2y = 8$
$2x - y = 2$

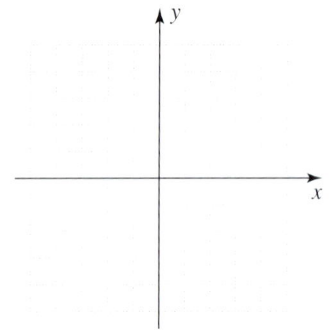

101. $2x + y = 4$
$2y + 4x = 6$

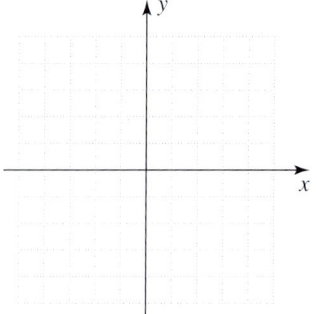

102. $2x + y = 5$
$2x + 4y = 8$

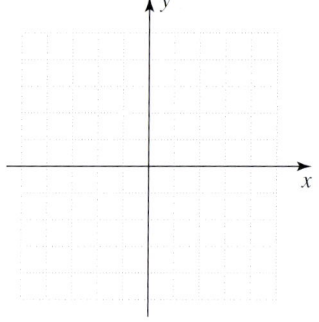

103. $x + \dfrac{1}{2}y = -2$
$y = -2x - 4$

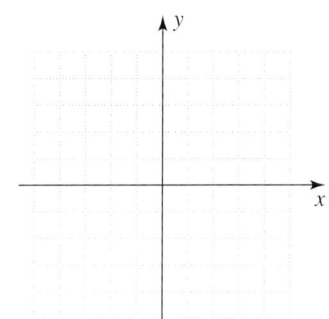

104. $y - 3x = 3$
$2y - 6x = 12$

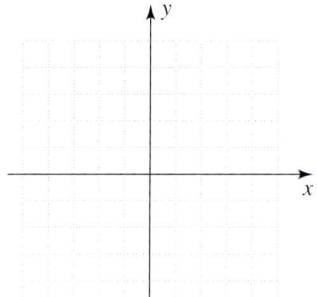

105. $x - y = -1$
$y = -x - 1$

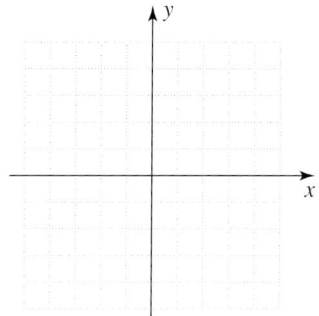

106. The height of the Eiffel Tower and its antenna is 1052 feet. The difference in height between the tower and the antenna is 920 feet. How tall is the antenna and how tall is the tower?

>>> Skill Checker

Find y for the indicated value of x.

107. $4y = 3x + 6, x = -3$ **108.** $5x - 3y = 8, x = 3$ **109.** $y = -8x + 2, x = \frac{1}{4}$ **110.** $y = 6x - 3, x = -\frac{1}{2}$

Solve the equation for y.

111. $6x - y = 3$ **112.** $3x = 2y - 8$ **113.** $-5 = -\frac{2}{3}x + y$ **114.** $-y = x + \frac{1}{4}$

4.2 Systems with Three Variables

▶ Objectives

A▶ Solve a system of three equations and three unknowns by the elimination method.

B▶ Determine whether a system of three equations in three unknowns is consistent, inconsistent, or dependent.

C▶ Solve applications involving systems of three equations.

▶ To Succeed, Review How To . . .

1. Solve linear equations (pp. 76–83).
2. Evaluate an expression (pp. 51–55).
3. Solve a system of two equations in two unknowns (pp. 285–295).

▶ Getting Started
Coffee Mixing: An Application

The man has three coffees that sell for $13.00, $14.00, and $15.00 per pound. If he calls these coffees A, B, and C and decides to make 50 pounds of a mixture containing x pounds of A, y pounds of B, and z pounds of C, then

$$x + y + z = 50 \quad (1)$$

In this mixture, he will have twice as much of brand B as of brand C. Thus,

$$y = 2z \quad (2)$$

Finally, if he sells the 50 lb at $14.20 per pound, the total price will be $710.00 so that

$$\$13x + \$14y + \$15z = \$710 \quad (3)$$

Now we rewrite equations (1), (2), and (3) in *standard form*:

$$x + y + z = 50 \quad (4)$$
$$y - 2z = 0 \quad \text{Subtract } 2z \text{ from both} \quad (5)$$
$$13x + 14y + 15z = 710 \quad \text{sides of equation (2).} \quad (6)$$

We have obtained a system of linear equations in three unknowns. There are many ways to solve this system. If we use the elimination method, we take two *different* pairs of equations and eliminate the same variable from each pair. Equation (5) *does not* contain x. Because of this, we select equations (4) and (6) and eliminate x by multiplying equation (4) by -13 and adding equation (6), as follows:

$$-13x - 13y - 13z = -650 \quad (7)$$
$$13x + 14y + 15z = 710 \quad (6)$$
$$\overline{}$$
$$y + 2z = 60 \quad \text{Add.} \quad (8)$$

The new system, consisting of equations (5) and (8), is a system of two equations in two unknowns. We solve it using the techniques of Section 4.1. To eliminate z from this system, we add equations (5) and (8), obtaining

$$y - 2z = 0 \quad (5)$$
$$y + 2z = 60 \quad (8)$$
$$\overline{}$$
$$2y = 60 \quad \text{Add.} \quad (9)$$
$$y = 30$$

Substituting $y = 30$ in equation (5) gives

$$y - 2z = 0$$
$$30 - 2z = 0$$
$$z = 15$$

Using $y = 30$ and $z = 15$ in equation (4) gives

$$x + 30 + 15 = 50$$
$$x + 45 = 50$$
$$x = 5$$

The solution for the system is the *ordered triple* (5, 30, 15). You can check this by substituting 5 for x, 30 for y, and 15 for z in each of the original equations. This means the man mixed 5 pounds of the $13.00 coffee, 30 pounds of the $14.00 coffee, and 15 pounds of the $15.00 coffee.

A › Solving Equations by Elimination

In Section 4.1, we used the graphical, substitution, and elimination methods to solve systems of equations involving two variables in two equations. When we have three variables, the graphical method isn't used because a three-dimensional coordinate system is required. However, we can use the elimination method to solve a system of three linear equations in three unknowns, as we illustrated in the *Getting Started*. The solution to the system of three equations is the values of x, y, and z that will solve all three equations. The solution is written as an **ordered triple** (x, y, z). Now let's examine this method in more detail. To solve a system of three linear equations in three unknowns by elimination, we use the following procedure.

PROCEDURE

Solving a Three-Equation System with Three Unknowns by Elimination

1. Select a pair of equations and eliminate one variable from this pair.
2. Select a *different* pair of equations and eliminate the same variable as in step 1.
3. Solve the pair of equations resulting from steps 1 and 2. (Use the procedure outlined in Section 4.1.)
4. Substitute the values found in step 3 in the simplest of the original equations, and then solve for the third variable.
5. Check by substituting the values in each of the original equations.

EXAMPLE 1 Using the elimination method to solve a system of three equations with one solution

Solve the system:

$$x + y + z = 12 \quad (1)$$
$$2x - y + z = 7 \quad (2)$$
$$x + 2y - z = 6 \quad (3)$$

If we add these two, z is eliminated.
If we add these two, z is eliminated.

SOLUTION It's easiest to eliminate z from the two pairs of equations, (1) and (3), and (2) and (3).

1. Adding (1) and (3), we obtain
$$2x + 3y = 18 \quad (4)$$

2. Adding (2) and (3), we have
$$3x + y = 13 \quad (5)$$

3. We now have the system
$$2x + 3y = 18 \quad (4)$$
$$3x + y = 13 \quad (5)$$

   Multiplying equation (5) by -3, we have
$$2x + 3y = 18 \quad (4)$$
$$-9x - 3y = -39 \quad (6)$$
$$-7x = -21 \quad \text{Add.}$$
$$x = 3$$

   Substituting 3 for x in equation (5), we get $9 + y = 13$, or $y = 4$.

4. We know now that x is 3 and y is 4. We can substitute these values in equation (1) to obtain $3 + 4 + z = 12$. Solving, we find $z = 5$.

5. The solution of the system is $(3, 4, 5)$, as can easily be verified:
$$3 + 4 + 5 = 12 \quad (1)$$
$$2(3) - 4 + 5 = 7 \quad (2)$$
$$3 + 2(4) - 5 = 6 \quad (3)$$

The system is *consistent*.

PROBLEM 1
Solve the system:

$$x + y + z = 4$$
$$x - y + z = 2$$
$$2x + 2y - z = -4$$

Calculator Corner

Solving a System with Three Equations

To solve systems of three equations with your calculator, select two pairs of equations and eliminate the same variable from each pair; then graph the resulting pair of equations.

In Example 1, z was eliminated from each pair. We then graph

$$Y_1 = -\tfrac{2}{3}x + 6 \quad (4)$$
$$Y_2 = -3x + 13 \quad (5)$$

and find the point of intersection, as shown in the window, by using [2nd] [TRACE] [5] [ENTER] [ENTER] [ENTER].

The result is $x = 3$ and $y = 4$. Now substitute $x = 3$ and $y = 4$ in the first equation and solve for z.

$$x + y + z = 12$$
$$3 + 4 + z = 12$$
$$z = 5$$

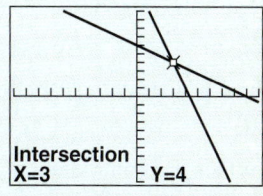

Answers to PROBLEMS

1. $(-1, 1, 4)$; consistent system

Sometimes in a system of three equations and three unknowns, one or two of the equations will be missing one of the variables. The solution to such a system will be developed in Example 2.

EXAMPLE 2 Using the elimination method to solve a system of three equations when one or more equations has a missing variable

Solve the system:
$$2x - y = 4 \quad (1)$$
$$y - z = 6 \quad (2)$$
$$x - y + 2z = 8 \quad (3)$$

Adding will eliminate y. (between (1) and (2))
Adding will eliminate y. (between (2) and (3))

SOLUTION Eliminate y from the two pairs of equations, (1) and (2), and (2) and (3).

1. Adding equations (1) and (2), we obtain
$$2x - z = 10 \quad (4)$$
2. Adding equations (2) and (3), we obtain
$$x + z = 14 \quad (5)$$
3. We now have the system
$$2x - z = 10 \quad (4)$$
$$x + z = 14 \quad (5)$$

Adding equations (4) and (5), we have
$$3x = 24$$
$$x = 8$$

Substituting 8 for x in equation (5), we get
$$8 + z = 14$$
$$z = 6$$

4. We know now that $x = 8$ and $z = 6$. We can substitute these values in equation (3) to obtain $8 - y + 2(6) = 8$. Solving, we find $y = 12$.
5. The solution of the system is (8, 12, 6) and can be verified:
$$2(8) - 12 = 4$$
$$12 - 6 = 6$$
$$8 - 12 + 2(6) = 8$$

PROBLEM 2
Solve the system:
$$3x + y - 2z = 2$$
$$y + z = 5$$
$$-x + 2z = 6$$

EXAMPLE 3 Using the elimination method to solve a system of three equations with no solution

Solve the system:
$$x + 3y - z = 1 \quad (1)$$
$$x - y + z = 4 \quad (2)$$
$$3x + y + z = 3 \quad (3)$$

If we add these two, z is eliminated. (between (1) and (2))
If we add these two, z is eliminated. (between (1) and (3))

SOLUTION Adding first (1) and (2) and then (1) and (3), we obtain
$$2x + 2y = 5 \quad \text{The sum of (1) and (2)} \quad (4)$$
$$4x + 4y = 4 \quad \text{The sum of (1) and (3)} \quad (5)$$

Multiplying equation (4) by -2 to eliminate x, we have
$$-4x - 4y = -10 \quad \text{This is } -2(2x + 2y) = -2 \cdot 5. \quad (6)$$
$$4x + 4y = 4 \quad (5)$$
$$0 = -6 \quad \text{Add.} \quad (7)$$

This is a contradiction. It's impossible for 0 to be equal to -6. This system has *no solution*; it is an *inconsistent system*.

PROBLEM 3
Solve the system:
$$x + 8y - z = 8$$
$$-x + 2y - z = 4$$
$$2x + y + z = 2$$

Answers to PROBLEMS

2. (4, 0, 5)
3. No solution; inconsistent system

EXAMPLE 4 Using the elimination method to solve a system of three equations with infinitely many solutions

Solve the system:

$$x - 2y + 3z = 4 \quad (1)$$
$$2x - y + z = 1 \quad (2)$$
$$x + y - 2z = -3 \quad (3)$$

SOLUTION We use the five-step procedure.

1. Multiplying equation (2) by -2 and adding to equation (1) (to eliminate y), we get

$$x - 2y + 3z = 4 \quad (1)$$
$$-4x + 2y - 2z = -2 \quad (4)$$
$$\overline{-3x + z = 2} \quad (5)$$

2. Adding equations (2) and (3) to eliminate y, we obtain

$$2x - y + z = 1 \quad (2)$$
$$x + y - 2z = -3 \quad (3)$$
$$\overline{3x - z = -2} \quad (6)$$

3. We now have the system

$$-3x + z = 2 \quad (5)$$
$$3x - z = -2 \quad (6)$$
$$\overline{0 = 0} \quad \text{Add.} \quad (7)$$

The system is *dependent* and has infinitely many solutions of the form (x, y, z). One such solution is obtained if we let $x = 0$ in equation (6), which gives $z = 2$.

4. Substituting $x = 0$ and $z = 2$ in equation (2), we have

$$2 \cdot 0 - y + 2 = 1 \quad (2)$$
$$-y + 2 = 1$$
$$y = 1$$

5. So $(0, 1, 2)$ is *one* of the solutions for the system, as can be easily checked by substituting $x = 0$, $y = 1$, and $z = 2$ in the original system. Here are some more solutions you can check: $(1, 6, 5)$ and $(-1, -4, -1)$.

PROBLEM 4

Solve the system:

$$x + 2y + z = -10$$
$$x + y - z = -3$$
$$5x + 7y - z = -29$$

B › Consistent, Inconsistent, and Dependent Systems

As was the case with systems of two equations in two unknowns, the solution of three equations in three unknowns always produces one of three different possibilities.

> **POSSIBLE SOLUTIONS OF THREE EQUATIONS IN THREE UNKNOWNS**
>
> The system is *consistent* and *independent*, as in Example 1; it has one solution consisting of an ordered triple (x, y, z).
> The system is *inconsistent*, as in Example 3. It has no solution.
> The system is *consistent* and *dependent*, as in Example 4. It has infinitely many solutions.

Answers to PROBLEMS

4. Infinitely many solutions; dependent system

In the case of two unknowns, that a system is *consistent* tells us that if we graph the lines associated with the system, the lines *intersect*. If we graph a linear equation in three unknowns, the graph is a plane. If a system of three linear equations in three unknowns has a single solution, it means that the three planes corresponding to their equations intersect at a point, as shown in Figure 4.9.

If the equations are *inconsistent,* the planes do not intersect at a common point, as shown in Figures 4.10 and 4.11. Finally, if the equations are *dependent,* the three planes can intersect in a line, as in Figure 4.12 (or the three planes can be coincident). Any point on the intersection is a solution; consequently, there are infinitely many solutions.

>Figure 4.9 One solution: Planes intersect in exactly one point.

>Figure 4.10 No solution: Planes intersect two at a time but have no common point of intersection. The equations are *inconsistent,* and the lines of intersection are parallel.

>Figure 4.11 No solution: Three parallel planes with no point of intersection. The equations are *inconsistent.*

>Figure 4.12 Infinitely many solutions: The three planes intersect along a common line. The equations are *dependent.*

C › Solving Applications Involving Systems of Three Equations

There are problems to solve that would be best accomplished by writing a system of three equations with three unknowns as we will demonstrate in Cindy's dilemma in Example 5.

EXAMPLE 5 Solving an application involving three equations

Cindy has been searching for a juicing recipe that will help her immune system and is told that it should contain vitamin C, vitamin B, and beta-carotene. She chooses three vegetables (V_1, V_2, V_3) known to contain these three vitamins.

She finds a table that gives the units per ounce of each vitamin that is in each of the three vegetables (see Table 4.2). Her dilemma: how many ounces of each vegetable should she juice to obtain 260 units of vitamin C, 140 units of vitamin B, and 90 units of beta-carotene in her juice drink?

PROBLEM 5
Cindy is told that during the flu season she should increase the amount of vitamin C to 520 units, the amount of vitamin B to 260, and the amount of beta-carotene to 170 units. How many ounces of each vegetable should she juice for this recipe?

Table 4.2

	V_1	V_2	V_3
Vitamin C	50	60	
Vitamin B	20		20
Beta-carotene	10	20	10

Answers to PROBLEMS

5. 8 oz of V_1; 2 oz of V_2; 5 oz of V_3

SOLUTION We are asked to find how many ounces of each vegetable should be used. We use the following variables.

Let $x = $ the number of ounces of V_1
$y = $ the number of ounces of V_2
$z = $ the number of ounces of V_3

Using Table 4.2, we notice that only V_1 and V_2 contain vitamin C and because the numbers given represent the units of vitamin C per ounce of each vegetable, we must multiply those numbers by the respective ounces for those two vegetables, namely x and y ounces.

This means: $\quad\quad\quad\quad 50x + 60y$ represents the total units of vitamin C.
The total units of vitamin C is 260, so

$$50x + 60y = 260 \quad\quad (1)$$

The table also indicates that only V_1 and V_3 contain vitamin B. To represent the units of vitamin B from those vegetables, we use the numbers from the table and the variables representing the ounces of those vegetables, x and z.

This means: $\quad\quad\quad\quad 20x + 20z$ represents the total units of vitamin B.
The total units of vitamin B is 140, so

$$20x + 20z = 140 \quad\quad (2)$$

Beta-carotene is found in all three vegetables so we multiply the respective numbers from the table by x, y, and z ounces.

This means: $\quad\quad\quad 10x + 20y + 10z$ represents the total units of beta-carotene.

The total units of beta-carotene is 90, so

$$10x + 20y + 10z = 90 \quad\quad (3)$$

Using equations (1), (2), and (3), we can solve the system to find the answer.
  By dividing each equation in the system by 10 we get,

$$5x + 6y = 26 \quad\quad (1)$$
$$2x + 2z = 14 \quad\quad (2)$$
$$x + 2y + z = 9 \quad\quad (3)$$

Adding equation (2) with the result of multiplying equation (3) by -2 we get

$$\begin{array}{rl} 2x + 2z = 14 & (2) \\ -2x - 4y - 2z = -18 & \text{-2 times equation (3)} \quad (4) \\ \hline -4y = -4 & \\ y = 1 & \end{array}$$

We substitute $y = 1$ into equation (1).

$$5x + 6(1) = 26 \quad\quad (1)$$
$$5x + 6 = 26$$
$$5x = 20 \quad\text{Subtract 6 from both sides.}$$
$$x = 4 \quad\text{Divide both sides by 5.}$$

Now we substitute $x = 4$ and $y = 1$ into equation (3) to find z.

$$4 + 2(1) + z = 9 \quad\quad (3)$$
$$6 + z = 9$$
$$z = 3 \quad\text{Subtract 6 from both sides.}$$

Cindy's dilemma is solved. To have the proper amount of each vitamin in her juice drink, she must juice 4 oz of V_1, 1 oz of V_2, and 3 oz of V_3.

Exercises 4.2

⟨A⟩ Solving Equations by Elimination
⟨B⟩ Consistent, Inconsistent, and Dependent Systems

In Problems 1–20, solve the system. Label the system as consistent, inconsistent, or dependent.

1. $\begin{aligned} x + y + z &= 12 \\ x - y + z &= 6 \\ x + 2y - z &= 7 \end{aligned}$

2. $\begin{aligned} x + y + z &= 13 \\ x - 2y + 4z &= 10 \\ 3x + y - 3z &= 5 \end{aligned}$

3. $\begin{aligned} x + y + z &= 4 \\ 2x + 2y - z &= -4 \\ x - y + z &= 2 \end{aligned}$

4. $\begin{aligned} 2x - y + z &= 3 \\ x + 4y - z &= 6 \\ 3x + 2y + 3z &= 16 \end{aligned}$

5. $\begin{aligned} 2x - y + z &= 3 \\ x + 2y + z &= 12 \\ 4x - 3y + z &= 1 \end{aligned}$

6. $\begin{aligned} x - 3y - 2z &= -12 \\ 2x + y - 3z &= -1 \\ 3x - 2y - z &= -5 \end{aligned}$

7. $\begin{aligned} x - 2y - 3z &= 2 \\ x - 4y - 13z &= 14 \\ -3x + 5y + 4z &= 2 \end{aligned}$

8. $\begin{aligned} 2x + 2y + z &= 3 \\ -x + y - z &= 5 \\ 3x + 5y + z &= 8 \end{aligned}$

9. $\begin{aligned} 2x + 4y + 3z &= 3 \\ 10x - 8y - 9z &= 0 \\ 4x + 4y - 3z &= 2 \end{aligned}$

10. $\begin{aligned} 9x + 4y - 10z &= 6 \\ 6x - 8y + 5z &= -1 \\ 12x + 12y - 15z &= 10 \end{aligned}$

11. $\begin{aligned} x - 2y - z &= 3 \\ 2x - 5y + z &= -1 \\ x - 2y - z &= -3 \end{aligned}$

12. $\begin{aligned} x - 3y + 6z &= -8 \\ 3x - 2y - 10z &= 11 \\ 5x - 6y - 2z &= 7 \end{aligned}$

13. $\begin{aligned} 2x + y + z &= 5 \\ -x + 2y - z &= 3 \\ 3x + 4y + z &= 10 \end{aligned}$

14. $\begin{aligned} x - 3y + z &= 2 \\ x + 2y - z &= 1 \\ -7x + y + z &= -10 \end{aligned}$

15. $\begin{aligned} x + y &= 5 \\ y + z &= 3 \\ x + z &= 7 \end{aligned}$

16. $\begin{aligned} x + 2y &= -1 \\ 2y + z &= 0 \\ x + 2z &= 11 \end{aligned}$

17. $\begin{aligned} x - 2y &= 0 \\ y - 2z &= 5 \\ x + y + z &= 8 \end{aligned}$

18. $\begin{aligned} 2y + z &= 9 \\ z - 2y &= 1 \\ x + y + z &= 1 \end{aligned}$

19. $\begin{aligned} 5x - 3z &= 2 \\ 2z - y &= -5 \\ x + 2y - 4z &= 8 \end{aligned}$

20. $\begin{aligned} 5x - 2z &= 1 \\ 3z - y &= 6 \\ x + 2y - z &= -1 \end{aligned}$

⟨C⟩ Solving Applications Involving Systems of Three Equations   In Problems 21–31, use a system of three equations in three unknowns to solve the problem.

21. *Numbers*   The sum of three numbers is 48. The second number is double the first, and the third number is triple the first. What are the numbers?

22. *Numbers*   The sum of three numbers is 49. The second number is 8 more than the first, and the third is 9 more than the second. Find the numbers.

23. *Angles*   The sum of the measures of the three angles in a triangle is 180°. The measure of one of the angles is twice that of the smallest angle, and the measure of the largest angle is three times that of the smallest. Find the measure of each angle.

24. *Angles*   The sum of the measures of three angles in a triangle is 180°. The sum of the measures of the first and the second angle is 20° less than that of the third angle, and the measure of the second angle is 20° more than that of the first. Find the measure of each angle.

25. *Salaries*   In 2004, the sum of the salaries of the president, vice president, and chief justice of the United States was $805,800. The president made $197,100 more than the vice president. If the vice president and the chief justice made the same salary, how much did each of these people make?

26. *Foot pain*   In a survey of 1000 people, the number of respondents complaining about corns, heel pain, or ingrown toenails was 150. Two more people suffered heel pain than ingrown toenails, and the number of respondents with either heel pain or ingrown toenails exceeded the number of those with corns by 10. Write a system of three equations and three unknowns, solve it, and find the number of people complaining about corns only, heel pain only, and ingrown toenails only.

27. *Vitamins* Do you take vitamins? The annual amount spent on vitamins B, C, and E in the United States has reached $885 million. Vitamin C is the most popular of the three; $90 million more was spent on it than on vitamin B, and $15 million more was spent on vitamin E than on B. Find the amount of money spent on each of these vitamins.

28. *Pizza* Who makes all the dough? Pizza Hut, Domino's, and Papa John's account for 72% of the total pizza market. Pizza Hut sells 40% more than Domino's, while Domino's sells 4% more than Papa John's. Find each company's market share (%).

29. *Musical instruments* More than 57 million adults play a musical instrument. The number of piano, guitar, and organ players combined is 46 million. The number of guitar and organ players exceeds the number of piano players by 4 million, but the number of guitar players is only 2 million less than those playing piano. Find out how many adults play each of these instruments.

30. *Billionaires* In a recent year, Bill Gates (computer software), John Kluge (media), and the Walton family (retailing) had combined fortunes totaling $16.9 billion. Kluge's fortune was $0.4 billion more than the Walton family's and $0.8 billion less than Gates's. Find out just how rich these people were during this particular year.

31. *Schools* If you went to school in Japan one year, then moved to Israel the next, and finally ended up in the United States the third year, you would have attended school for a total of 639 days. In Israel, you would have attended 36 more days than in the United States. In Japan, you would have attended even longer: 27 more days than in Israel. How long is the school year in each of these countries?

❯❯❯ Using Your Knowledge

Crickets, Ants, and the Weather Can you tell how hot it is by listening to the crickets? You can if you know the formula. The number of chirps n a certain cricket makes per minute is related to the temperature F, in degrees Fahrenheit, by

$$F = \frac{n}{4} + 40 \tag{1}$$

Ants are also affected by temperature changes. The crawling speed d (in centimeters per second) of a certain ant is related to the temperature C, in degrees Celsius, by

$$C = 6d + 4 \tag{2}$$

The relationship between C and F is given by

$$C = \frac{5}{9}(F - 32) \tag{3}$$

32. Use the substitution method with equations (2) and (3) to solve for F in terms of d.

33. Substitute the expression obtained for F in Problem 32 in equation (1) to find the relationship between d and n.

34. Now use your answer from Problem 33 to solve this: If the cricket is chirping 112 times a minute, how fast is the ant crawling?

❯❯❯ Write On

35. Why is the graphical method difficult to use when solving a system of equations in three unknowns?

36. When would you use the substitution method when solving a system of three equations in three unknowns?

37. If you graph the linear equation $2x + y = 4$, what is the resulting graph? What do you think the graph will be if you graph the equation $2x + y + z = 4$?

38. How do you know if a system of three equations in three unknowns is inconsistent?

39. How do you know if a system of three equations in three unknowns is dependent?

〉〉〉 Concept Checker

Fill in the blank(s) with the correct word(s), phrase, or mathematical statement.

40. The solution to a consistent system of three equations with three unknowns is called a(n) _____.

41. When solving a system of three equations with three unknowns, if the result is a contradiction, then the system has _____ solution(s).

42. When solving a system of three equations, if the variables cancel out and the result is two equal numbers, then it is called a _____ system.

system of equations	one
consistent	no
inconsistent	infinitely many
dependent	ordered triple

〉〉〉 Mastery Test

Solve each system of equations.

43. $x + 2y + z = -10$
$x + y - z = -3$
$5x + 7y - z = -29$

44. $x + y + z = 4$
$x - 2y - z = 1$
$2x - y - 2z = -1$

45. $x + 8y - z = 8$
$-x + 2y - z = 4$
$2x + y + z = 2$

46. $2x + 2y - 6z = 5$
$-x - y + 3z = 4$
$3x - y + z = 2$

47. $x + y + z = 4$
$x - y + z = 2$
$2x + 2y - z = -4$

48. $x + 2y + z = 4$
$-3x + 4y - z = -4$
$-2x - 4y - 2z = -8$

〉〉〉 Skill Checker

Solve.

49. A person bought some bonds yielding 5% annually and some certificates yielding 7%. If the total investment amounts to $20,000 and the interest received is $1160, how much is invested in bonds and how much in certificates?

50. A car leaves a town going north at 30 miles per hour. Two hours later, another car leaves the same town traveling on the same road in the same direction at 40 miles per hour. How far from the town does the second car overtake the first one?

51. How many gallons of a 30% solution must be mixed with 40 gallons of a 12% solution to obtain a 20% solution?

52. Kiran can complete a typing job in 3 hours. Veronica does it in 2 hours. How long would it take to complete the job if they work together?

53. A plane traveled 840 miles with a 30-mile/hour tail wind in the same time it took to travel 660 miles against the wind. What is the plane's speed in still air?

4.3 Coin, Distance-Rate-Time, Investment, and Geometry Problems

Objectives

A Solve coin problems with two or more unknowns.

B Solve general problems with two or more unknowns.

C Solve rate, time, and distance (motion) problems with two or more unknowns.

D Solve investment problems with two or more unknowns.

E Solve geometry problems with two or more unknowns.

F Solve problems that can be modeled with a system.

To Succeed, Review How To . . .

1. Use the RSTUV method (pp. 100–102).
2. Solve a system of linear equations in two or three unknowns (pp. 285–295, 305–309).

Getting Started

Money Problems!

The pile of money contains $3.25 in nickels and dimes. There are five more nickels than dimes. How many nickels and how many dimes are in the pile?

We use the RSTUV method to solve this problem.

1. Read the problem. We are asked to find the number of nickels and dimes.

2. Select the unknowns. Let n be the number of nickels and d the number of dimes.

3. Think of a plan. We have to translate two statements, yielding two equations and two unknowns. First note that:

If you have 1 nickel, you have $0.05(1).
If you have 2 nickels, you have $0.05(2).
If you have n nickels, you have $0.05(n)$.

Similarly,

If you have d dimes, you have $0.10(d)$.

Now we can translate the first statement.

The pile contains $3.25 {Total amount} = {amount in nickels} in nickels and + {amount in dimes} dimes.

$$3.25 = 0.05n + 0.10d$$
$$325 = 5n + 10d \quad \text{Multiply by 100.}$$
$$65 = n + 2d \quad \text{Divide by 5.}$$

The statement "there are five more nickels than dimes" means that

{The number of nickels} = {5 more than} + {the number of dimes}

$$n = 5 + d$$

The complete problem can be reduced to the system of equations

$$65 = n + 2d$$
$$n = 5 + d$$

4. Use the substitution method or your calculator to solve this system.

Substituting $5 + d$ for n in the first equation gives

$$65 = (5 + d) + 2d$$
$$65 = 5 + d + 2d \quad \text{Remove parentheses.}$$
$$65 = 5 + 3d \quad \text{Combine like terms.}$$
$$60 = 3d \quad \text{Subtract 5.}$$
$$20 = d \quad \text{Divide by 3.}$$

$n = 5 + d$, so we substitute 20 for d:
$$n = 5 + 20 = 25$$

Hence we have 25 nickels and 20 dimes.

5. Verify the answer. We do have 5 more nickels (25 in total) than dimes (20 in total) and $3.25 ($1.25 in nickels and $2.00 in dimes).

In Chapter 2, we learned how to develop our problem-solving skills and use our knowledge of solving equations to find the answers to certain types of applications. There are some applications that are easier to solve when using two or more equations with two or more unknowns, or what we refer to as systems of equations. In this section, we will start with some examples of the more traditional word problems dealing with coins, motion, investment, and geometry. We will conclude with the type of problems that can be modeled by a system of two linear functions.

A > Solving Coin Problems

Coin problems are a classic way to practice setting up systems of equations.

EXAMPLE 1 Nickels and dimes

Jack has $3 in nickels and dimes. He has twice as many nickels as he has dimes. How many nickels and how many dimes does he have?

SOLUTION As usual, we use the RSTUV method.

1. Read the problem. We are asked to find the number of nickels and dimes, so we need two variables.

2. Select the unknowns. Let n be the number of nickels and d the number of dimes.

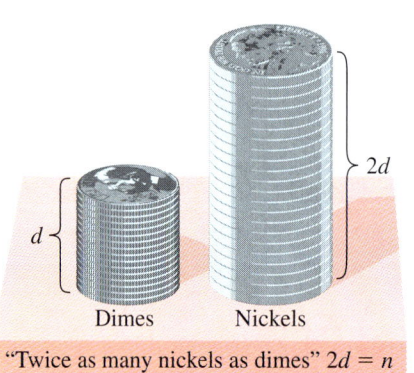

"Twice as many nickels as dimes" $2d = n$

3. Think of a plan. Translate the problem. Jack has $3 (300 cents) in nickels and dimes:
$$300 = 5n + 10d$$

He has twice as many nickels as he has dimes:
$$n = 2d$$

We then have the system
$$5n + 10d = 300$$
$$n = 2d$$

4. Use the substitution method or your calculator to solve the system. This time it's easy to use the substitution method.

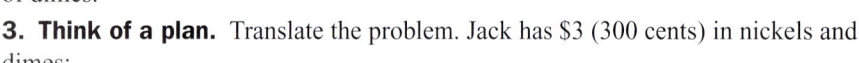

$5n + 10d = 300 \rightarrow 5(2d) + 10d = 300$ Let $n = 2d$.
$\qquad\qquad\qquad\quad 10d + 10d = 300$ Simplify.

PROBLEM 1

Jill has $9.90 in dimes and quarters. She has 3 times as many dimes as quarters. How many dimes and how many quarters does she have?

Answers to PROBLEMS

1. 54 dimes; 18 quarters

$$20d = 300 \quad \text{Combine like terms.}$$
$$d = 15 \quad \text{Divide by 20.}$$
$$n = 2(15) = 30 \quad \text{Substitute } d = 15 \text{ in } n = 2d.$$

Thus, Jack has 15 dimes ($1.50) and 30 nickels ($1.50).

5. Verify the answer. It's easy to verify this answer because we know Jack has $3 and twice as many nickels as dimes.

B > Solving General Problems

We can also use systems of equations to solve other problems. Here is an interesting one.

EXAMPLE 2   A whopper of a dog

JR thought his dog Bleau, an American bull dog with a 22-inch neck, was a big dog. Then he compared him to the *Guinness Book of World Records'* largest dog, "Hercules," an English mastiff with a 38-inch neck.

If the sum of their weights is 357 pounds and the difference is 207 pounds, find the weight of each dog.

PROBLEM 2

The total combined height of a redwood tree and a silver maple is 436 feet. If the height differential is 292 feet and the redwood tree is taller, what is the height of each tree?

SOLUTION

1. Read the problem. We are asked to find the weight of each dog so we need two variables.

2. Select the unknowns. Let b = the weight of Bleau in pounds and h = the weight of Hercules in pounds.

3. Think of a plan. We need two facts to translate into two algebraic equations.

Fact 1: The sum of the weights is 357. $h + b = 357$
Fact 2: The difference of the weights is 207. $h - b = 207$

Now we use our skills to solve the system.

4. Use the elimination method or your calculator to solve the system. We use the elimination method since the b coefficients are additive inverses (opposites) and by adding the equations together they will be eliminated.

$$
\begin{array}{r}
h + b = 357 \\
\underline{h - b = 207} \\
2h = 564 \quad \text{Add equations.} \\
h = 282 \quad \text{Divide both sides by 2.}
\end{array}
$$

Substitute $h = 282$ into $h + b = 357$
$$282 + b = 357 \quad \text{Subtract 282.}$$
$$b = 75$$

So Hercules is a whopper of a dog as he weighs 282 pounds, while Bleau weighs a mere 75 pounds.

5. Verify the answer. You can verify Hercules' weight by going online to the *Guinness Book of World Records*.

Answers to PROBLEMS

2. Redwood: 364 ft; silver maple: 72 ft

C › Solving Motion Problems

Remember the motion problems we solved in Chapter 2? They can also be solved using two variables. The procedure is almost the same. We write the given information in a chart labeled $R \times T = D$, and use the RSTUV method.

EXAMPLE 3   The current in Norway

The world's strongest current is the Saltstraumen in Norway. The current is so strong that a boat, which can go 48 miles downstream (with the current) in 1 hour, takes 4 hours to go the same 48 miles upstream (against the current). How fast is the current flowing?

SOLUTION

1. Read the problem. We are asked to find the rate of speed of the current, so we need two variables: the rate of speed of the boat and the rate of speed of the current.

2. Select the unknowns. Let x be the rate of speed of the boat in still water and y be the rate of speed of the current. Then $(x + y)$ is the rate of speed of the boat downstream; $(x - y)$ is the rate of speed of the boat upstream.

3. Think of a plan. We enter this information in a chart:

	Rate of Speed (mi/hr)	Time (hr)	Distance (mi)	$R \times T = D$
Downstream	$(x + y)$	1	48	$x + y = 48$
Upstream	$(x - y)$	4	48	$(x - y)4 = 48$

4. Use the elimination method or your calculator to solve the system. Our system of equations is simplified as follows:

$x + y = 48$ →leave as is→ $x + y = 48$
$(x - y)4 = 48$ →divide by 4→ $x - y = 12$
$2x = 60$ Add.
$x = 30$ Divide by 2.
$30 + y = 48$ Substitute $x = 30$ in $x + y = 48$.
$y = 18$ Subtract 30.

The speed of the boat in still water is $x = 30$ mi/hr and the speed of the current is 18 miles per hour.

5. Verify the answer. The verification is left to you. You may wish to compare how we solved this type of problem using one variable in Section 2.4 with the method used here.

PROBLEM 3

A plane goes 1200 miles with a tail wind in 3 hours. It takes 4 hours to travel the same distance against the wind. Find the velocity of the wind and the velocity of the plane in still air.

D › Solving Investment Problems

The investment problems we studied in Section 2.4 are easier to solve if we use more than one variable. We illustrate their solution next.

Answers to PROBLEMS

3. Wind velocity 50 mi/hr; plane velocity 350 mi/hr

EXAMPLE 4 Tracking investments

An investor divides $20,000 among three investments at interest rates of 6%, 8%, and 10%. If the total annual income is $1700 and the income of the 10% investment exceeds the income from the 6% and 8% investments by $300, how much was invested at each rate?

SOLUTION

1. Read the problem. We are asked to find how much was invested at each rate, so we need three variables.

2. Select the unknowns. Let x be the amount invested at 6%, y the amount invested at 8%, and z the amount invested at 10%.

3. Think of a plan. Let's use a chart with the heading:

$$\text{Principal} \times \text{Rate} = \text{Annual Interest}$$

We enter the information in the chart:

	Principal	×	Rate	=	Annual Interest
1st investment	x		6%		$0.06x$
2nd investment	y		8%		$0.08y$
3rd investment	z		10%		$0.10z$
Total	$20,000				$1700

Looking at the column labeled "Principal," we see that

$$x + y + z = 20{,}000 \quad (1)$$

From the column labeled "Annual Interest," we can see that the total interest earned is $1700; thus

$$0.06x + 0.08y + 0.10z = 1700$$

Multiplying by 100 gives

$$6x + 8y + 10z = 170{,}000 \quad (2)$$

We also know that

{The income from the 10% investment} {exceeds the income from the 6% and 8% investments} {by $300}

$$0.10z = 0.06x + 0.08y + 300$$
$$10z = 6x + 8y + 30{,}000 \quad \text{Multiply by 100.}$$
$$-6x - 8y + 10z = 30{,}000 \quad (3)$$

4. Use the elimination method or your calculator to solve the system.
We now have the system

$$x + y + z = 20{,}000 \quad (1)$$
$$6x + 8y + 10z = 170{,}000 \quad (2)$$
$$-6x - 8y + 10z = 30{,}000 \quad (3)$$

Adding equations (2) and (3), we have

$$20z = 200{,}000 \quad (4)$$
$$z = 10{,}000$$

Now multiply equation (1) by 6 and add it to equation (3):

$$6x + 6y + 6z = 120{,}000 \quad (5)$$
$$\underline{-6x - 8y + 10z = 30{,}000} \quad (3)$$
$$-2y + 16z = 150{,}000 \quad (6)$$

(continued)

PROBLEM 4

Spencer received an inheritance. He wants to invest in three accounts—the first at 4%, the second at 6%, and the third at 8%. The second investment is three times the first and the third investment is $7000 more than the second. If the annual interest from all three is $2860, find the amount of Spencer's inheritance.

Answers to PROBLEMS

4. 1st = $5000; 2nd = $15,000; 3rd = $22,000; thus, the inheritance was $42,000.

Substitute 10,000 for z in equation (6):

$$-2y + 16(10{,}000) = 150{,}000 \quad (7)$$
$$-2y + 160{,}000 = 150{,}000$$
$$-2y = -10{,}000$$
$$y = 5000$$

Finally, substitute 10,000 for z and 5000 for y in equation (1):

$$x + 5000 + 10{,}000 = 20{,}000$$
$$x + 15{,}000 = 20{,}000$$
$$x = 5000$$

Thus, $x = 5000$, $y = 5000$, and $z = 10{,}000$.

5. Verify the answer. You can verify that if we invest $5000 at 6%, $5000 at 8%, and $10,000 at 10%, the conditions of the problem are satisfied.

E › Solving Geometry Problems

Problems involving the perimeter of a rectangle also involve two unknowns. Example 5 shows how to set up a system of two equations to solve such problems.

EXAMPLE 5 A "check" using geometry and perimeter

The check is in the mail? Certainly not one of the physically largest checks ever written, which had a perimeter of 202 feet and was 39 feet longer than wide! Can you find the dimensions of this check given by InterMortgage of Leeds, England, to a Yorkshire TV telephone appeal?

PROBLEM 5

The perimeter of the base of the Washington Monument is $183\frac{5}{6}$ feet. If the length is $18\frac{1}{4}$ feet longer than the width, find the dimensions of the base.

SOLUTION We continue to use the RSTUV procedure.

1. Read the problem. We are asked to find the dimensions of the check. Even though the problem doesn't specify it, we assume that the check is rectangular.

2. Select the unknowns. We let W be the check's width and L its length.

3. Think of a plan. First, translate the problem. A picture helps.

The perimeter P is 202 ft and the perimeter is also $2L + 2W$, so we have the equation

$$2L + 2W = 202 \quad (1)$$

Also, "the check was 39 feet longer than wide" means

$$L = W + 39 \quad (2)$$

Equation (1) can be simplified if we divide each of its terms by 2. Since $L = W + 39$, it will be easy to use the substitution method.

We have the system

$$2L + 2W = 202 \quad (1)$$
$$L = W + 39 \quad (2)$$

First, divide each term in equation (1) by 2 to obtain

$$L + W = 101 \quad (3)$$
$$L = W + 39 \quad (2)$$

Answers to PROBLEMS

5. Width = $36\frac{5}{6}$ ft or $36.8\overline{3}$ ft;

length = $55\frac{1}{12}$ ft or $55.08\overline{3}$ ft

4. Use the substitution method or your calculator to solve the system.
Substituting $W + 39$ for L in equation (3), we have

$$W + (W + 39) = 101$$
$$2W + 39 = 101 \quad \text{Simplify.}$$
$$2W = 62 \quad \text{Subtract 39 from both sides.}$$
$$W = 31 \quad \text{Divide both sides by 2.}$$

From equation (2), $L = W + 39$ and we know that $W = 31$, so
$$L = 31 + 39 = 70$$

Thus, the check is 31 feet wide by 70 feet long.

5. Verify the answer. The verification that the perimeter is 202 feet and that the check is 39 feet longer than wide is left to you.

F › Solving Problems That Can Be Modeled with a System

In Section 3.6 we learned how to recognize certain data as being linearly related and how to write the linear function that models it. Next we will use that skill along with our knowledge of systems of equations to solve Example 6.

EXAMPLE 6 Revenue versus spending—When will the two ever meet?

Since 1962 federal tax revenue and federal spending have generally been increasing as shown by the graph in Figure 4.13. The government spent more than it took in during most of the years shown. However, there is an increasing linear trend for both the spending and the revenue since 2003. Using the information from the graph, the linear function that models the federal spending is $y = 100x + 2000$ and the linear function that models the federal revenue is $y = 133x + 1400$ where x represents the years since 2000. Assuming the data continues to increase at this rate, when will the federal revenue be the same as the federal spending?
Source: http://www.heritage.org.

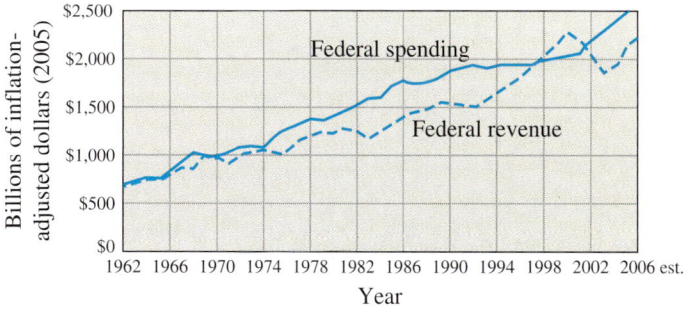

› Figure 4.13

PROBLEM 6

Data was collected that compares the average monthly temperatures in degrees Celsius of New York with Sydney for the first 6 months of the year. Observing the data, it is noted that New York's temperatures model an increasing linear function while Sydney's model a decreasing linear function. If we let x represent each of the months (1–6) and y represent the average monthly temperature in degrees Celsius, then the function that approximates New York's temperatures is $y = 4.5x - 5.5$ and the function that approximates Sydney's is $y = -1.3x + 25.3$. What month will the two cities have the same temperature?
Source: http://www.coolantarctica.com.

SOLUTION

1. Read the problem. We are asked to find when the federal revenue will be the same as the federal spending. This means we want to know how many years after 2000 it will take for the two to be the same amount of money.

(continued)

Answers to PROBLEMS

6. Month 5, May

2. Select the unknowns.

Let x = the time in years after 2000
and y = the amount of money.

3. Think of a plan. We use the two given linear functions to create the system.

$$y = 100x + 2000$$
$$y = 133x + 1400$$

Now we use our skills to solve the system.

4. Use the substitution method or your calculator to solve the system. We use the substitution method since both equations are solved for y, and set the right sides equal to each other.

$$100x + 2000 = 133x + 1400$$
$$2000 = 33x + 1400 \quad \text{Subtract 100x.}$$
$$600 = 33x \quad \text{Subtract 1400.}$$
$$18 = x \quad \text{(rounded to nearest whole number)} \quad \text{Divide by 33.}$$

That means it will take approximately 18 years from 2000 for the federal revenue to equal the federal spending. By adding 18 to 2000 we get 2018, when we can expect the amount on federal spending to be the same as the amount of federal revenue.

5. Verify the answer. You can verify this by extending the graph.

Calculator Corner

Summary of Solving a Word Problem by Graphing a System

You can use your calculator as a tool to assist you in solving the problems in this section. There are two important considerations:

1. Which variable will you designate as the independent variable (the one that is easier to solve for)?
2. What size window will let you see the part of the graph at which the given lines intersect? (Use a window that contains the x- and y-intercepts of the lines involved.)

For instance, in the *Getting Started,* go through the first three steps of the RSTUV method (your calculator can't do this for you) to obtain the system

$$65 = n + 2d \quad \text{and} \quad n = 5 + d$$

It's easier to solve for n, so we have the equivalent system

$$n = 65 - 2d \quad \text{and} \quad n = 5 + d$$

Graph

$$Y_1 = 65 - 2X \quad \text{and} \quad Y_2 = 5 + X$$

In the first equation, when $X = 0$, $Y_1 = 65$ and when $Y_1 = 0$, $X = 32.5$, so we use a [0, 35] by [0, 65] window, with a scale of 5 for X and Y (Xscl = 5, Yscl = 5). Using your intersection feature (2nd TRACE 5 and ENTER three times; see the Window), we get $X = 20$ and $Y = 25$. Thus, $d = 20$ and $n = 25$ as before; that is, we have 20 dimes and 25 nickels. Now you can verify that these two numbers satisfy the conditions of the original problem.

It is easier to do problems involving three variables algebraically.

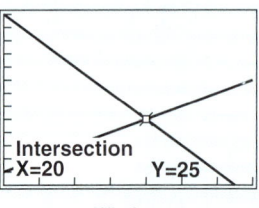

Window

4.3 Coin, Distance-Rate-Time, Investment, and Geometry Problems

> Exercises 4.3

Boost your grade at mathzone.com!
> Practice Problems
> NetTutor
> Self-Tests
> e-Professors
> Videos

⟨ A ⟩ Solving Coin Problems   In Problems 1–10, use two or more unknowns to solve these coin problems.

1. Natasha has $6.25 in nickels and dimes. If she has twice as many dimes as she has nickels, how many dimes and how many nickels does she have?

2. Mida has $2.25 in nickels and dimes. She has four times as many dimes as nickels. How many dimes and how many nickels does she have?

3. Dora has $5.50 in nickels and quarters. She has twice as many quarters as she has nickels. How many of each coin does she have?

4. Mongo has 20 coins consisting of nickels and dimes. If the nickels were dimes and the dimes were nickels, he would have 50¢ more than he now has. How many nickels and how many dimes does he have?

5. Desi has 10 coins consisting of pennies and nickels. Strangely enough, if the nickels were pennies and the pennies were nickels, she would have the same amount of money as she now has. How many pennies and nickels does she have?

6. Don has $26 in his pocket. If he has only 1-dollar bills and 5-dollar bills, and he has a total of 10 bills, how many of each bill does he have?

7. A person went to the bank to deposit $300. The money was in 10- and 20-dollar bills, 25 bills in all. How many of each did the person have?

8. A woman has $5.95 in nickels and dimes. If she has a total of 75 coins, how many nickels and how many dimes does she have?

9. A man has $7.05 in nickels, dimes, and quarters. The quarters are worth $4.60 more than the dimes and the dimes are worth 25¢ more than the nickels. How many nickels, dimes, and quarters does the man have?

10. Amy has $2.50 consisting of nickels, dimes, and quarters in her piggy bank. She has the same amount in nickels and dimes, and twice as much in nickels as she has in quarters. How many nickels, dimes, and quarters does Amy have?

⟨ B ⟩ Solving General Problems   In Problems 11–20, use two or more unknowns to solve these general problems.

11. The sum of two numbers is 102. Their difference is 16. What are the numbers?

12. The difference between two numbers is 28. Their sum is 82. What are the numbers?

13. The sum of two integers is 126. If one of the integers is 5 times the other, what are the integers?

14. The difference between two integers is 245. If one of the integers is 8 times the other, find the integers.

15. The difference between two numbers is 16. One of the numbers is 5 times the other. What are the numbers?

16. The sum of two numbers is 116. One of the numbers is 50 less than the other. What are the numbers?

17. Longs Peak is 145 feet higher than Pikes Peak. If you were to put these two peaks one on top of the other, you would still be 637 feet short of reaching the elevation of Mt. Everest, 29,002 feet. Find the elevations of Longs Peak and Pikes Peak.

18. The height of the Empire State building and its antenna (Figure 4.14) is 1472 feet. The difference in height between the building and the antenna is 1028 feet. How tall is the antenna and how tall is the building?

> Figure 4.14 The Empire State building and its antenna.

19. One of the largest sundaes ever made contained about 6700 pounds of topping. The topping flavors were chocolate, butterscotch, and caramel. There was the same amount of butterscotch as caramel but 600 pounds more of chocolate than butterscotch. How many pounds of each were included in the topping?

20. One of the largest pancakes ever made used buckwheat flour, Puritan mix, and 15 gallons of syrup. The flour and mix weighed 100 pounds more than the 15 gallons of syrup. What was the weight of the syrup if the whole pancake weighed 4100 pounds? By the way, 68 pounds of butter were added before it was consumed!

⟨ C ⟩ Solving Motion Problems   In Problems 21–25, use two unknowns to solve these motion problems.

21. A plane flies 540 miles with a tail wind in $2\frac{1}{4}$ hours. The plane makes the return trip against the same wind and takes 3 hours. Find the speed of the plane in still air and the speed of the wind.

22. A motorboat runs 45 miles downstream in $2\frac{1}{2}$ hours and 39 miles upstream in $3\frac{1}{4}$ hours. Find the speed of the boat in still water and the speed of the current.

23. A motorboat can travel 15 miles per hour downstream and 9 miles per hour upstream on a certain river. Find the rate of the current and the rate at which the boat can travel in still water.

24. It takes a motorboat $1\frac{1}{3}$ hours to go 20 miles downstream and $2\frac{2}{9}$ hours to return. Find the rate of the current and the rates at which the boat can travel in still water.

25. A plane flying with the wind took 2 hours for a 1000 mile flight and $2\frac{1}{2}$ hours for the return flight. Find the wind velocity and the speed of the plane in still air.

⟨ D ⟩ Solving Investment Problems In Problems 26–30, use two or more unknowns to solve these investment problems.

26. Two sums of money totaling $20,000 earn 8% and 10% annual interest. If the interest from both investments amounts to $1900, how much is invested at each rate?

27. An investor invested $10,000, part at 6% and the rest at 8%. Find the amount invested at each rate if the annual income from the two investments is $720.

28. Andy Cabazos has $30,000 in three investments paying 4%, 6%, and 10%. The total interest on the 4% and 6% investments is $280 less than that obtained from the 10% investment. If his annual income from these investments is $2,120, how much does he have invested at each rate?

29. Marlene McGuire invested $25,000 in municipal bonds. The first investment paid 6%, the second 8%, and the third, 10%. If her annual income from these bonds was $2000 and the interest she received on the combined 6% and 8% investments equaled the interest on the 10% investment, how much money did she have in each category?

30. Marc Goldstein divided $20,000 into three parts. One part yielded 4%, another 8%, and the third one, 6%. If his total return was $1080 and he made $40 less on his 8% investment than on his 4% investment, what amount did he invest in each category?

⟨ E ⟩ Solving Geometry Problems In Problems 31–34, use two unknowns to solve these geometry problems.

31. The perimeter of the SuperFlag (Figure 4.15) is 1520 feet. If the length of the flag exceeds the width by 250 feet, what are the dimensions of the flag?

32. One of the largest flags *actually flown* from a flagpole was 98 feet longer than it was wide. If its perimeter was 1016 feet, what were the dimensions of the flag? (It was a Brazilian flag, flown in Brazil.)

33. One of the world's largest quilts boasted a 438 foot perimeter with its length being 49 feet more than its width. What were the dimensions of the quilt? It took 7000 North Dakotans to make it.

>Figure 4.15 The SuperFlag, shown hanging on Hoover Dam, is the World's Largest Flag according to the *Guinness Book of World Records*.

34. If you walked around one of the largest rectangular swimming pools in the world, located in Morocco, you would end up walking 3640 feet. If the pool is 1328 feet longer than it is wide, what are the dimensions of the pool?

⟨ F ⟩ Solving Problems That Can Be Modeled with a System In Problems 35–38, use a system of linear functions that models the data to solve these problems.

35. A comparative study examining the minimum wages from 1995 to 2004 for some European countries was completed by a European industrial relations observatory online. In 1995 the minimum wage for Spain was 377 Euros while Greece was 350 Euros. The study indicated Greece had one of the largest increases in wages over that period of time. The linear function that approximates the wage for Greece is $y = 21.9x + 329$, and $y = 10.9x + 363$ is the linear function that approximates the wage for Spain where x is the years after 1995 and y is the wage in Euros. In what year was the minimum wage for Greece the same as Spain? What was that wage to the nearest Euro?

Source: http://www.eiro.eurofound.eu.int.

36. Using standard growth charts, this mom and dad can make some predictions about baby Kyrstin's growth. The standard growth for height in inches at the 50th percentile for a baby girl from birth to 36 months is given by a curve. However, from 12 months to 36 months the curve approximates the linear function, $y = 0.38x + 17$, where x is the month and y is the weight in pounds. If $y = 0.5x + 14$ approximates baby Kyrstin's growth rate for 12 months to 36 months, when will her weight be equal to the 50th percentile standard weight? How much will she weigh?

Source: http://www.kidsgrowth.com.

37. There is a claim that maintaining a compact region when developing urban areas can cut back on the vehicle miles traveled (VMT). The graph compares the cost of congestion per capita for Portland-Vancouver, Atlanta, and Miami in thousands of dollars. The linear functions that approximate the cost for each are Portland-Vancouver, $y = 60x + 300$; Atlanta, $y = 190x + 683$; Miami, $y = 60x + 900$. In these functions, x is the number of years after 1990 and y is the costs of congestion per capita in thousands of dollars. In what year were the costs of congestion per capita for Atlanta the same as Miami? What was that cost to the nearest thousand?

Source: http://www.cnu.org.

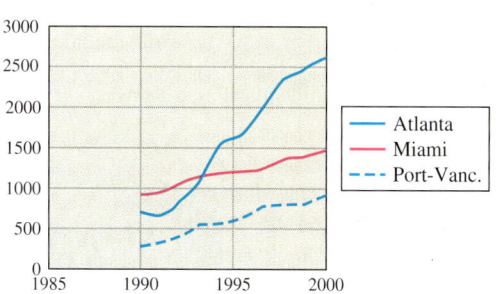

Costs of Congestion Per Capita for Portland-Vancouver, Atlanta, and Miami

38. Using the linear function models and the graph from Problem 37, how does the rate of increase of the costs of congestion per capita of Miami compare with Portland-Vancouver? At that rate compare their costs for 2005.

>>> Using Your Knowledge

The A, B, C's of Vitamins In this section, we solved coin, general, distance, investment, and geometry problems. Now we will discuss mixture problems. Use your knowledge to solve these mixture problems.

A dietician wants to arrange a diet composed of three basic foods A, B, and C. The diet must include 170 units of calcium, 90 units of iron, and 110 units of vitamin B. The table gives the number of units per ounce of each of the needed ingredients contained in each of the basic foods.

	Units per Ounce		
Nutrient	**Food A**	**Food B**	**Food C**
Calcium	15	5	20
Iron	5	5	10
Vitamin B	10	15	10

If a, b, and c are the number of ounces of basic foods A, B, and C eaten by an individual, write an equation indicating the specified amount of each nutrient.

39. The amount of calcium needed

40. The amount of iron needed

41. The amount of vitamin B needed

>>> Write On

42. State the advantages and disadvantages of solving word problems using systems of equations instead of the techniques we studied in Chapter 2.

43. Write a word problem that uses the system
$$x + y = 10$$
$$x - y = 2$$
to obtain its solution.

>>> Mastery Test

44. The perimeter of a rectangle is 170 centimeters. If its length is 15 centimeters more than its width, what are the dimensions of the rectangle?

45. An investor divides $20,000 among three investments at 6%, 8%, and 10%. If her total income is $1740 and the income from the 10% investment exceeds the income from the 6% and 8% investments by $260, how much did she invest at each rate?

46. A plane flies 1200 mi with a tail wind in 3 hr. It takes 4 hr to fly the same distance against the wind. Find the wind velocity and the velocity of the plane in still air.

47. A change machine gave Jill $2 in nickels and dimes. She had twice as many nickels as dimes. How many nickels and how many dimes did the machine give Jill?

48. The McGuire twins, Benny and Billy, weighed a total of 1598 pounds, which is a record in the *Guinness Book of World Records*. Besides their heavy weights they are most famous for their cross-country picture on motorcycles. If their weight differential was 30 pounds, what was the weight of each of the twins?

49. The tallest president of the United States was Abraham Lincoln and the shortest was James Madison. Their height differential was $1\frac{1}{12}$ feet. If their combined height was $11\frac{7}{12}$ feet, how tall was each president?

>>> Skill Checker

In Problems 50–51, use the linear function, $y = \frac{1}{2}x - 4$.

50. Find y given $x = 15$.

51. Find x, given $y = 24$.

In Problems 52–53, graph the linear inequality.

52. $y > 3x + 2$

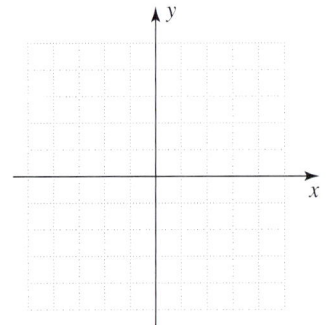

53. $5x + 3y \leq 6$

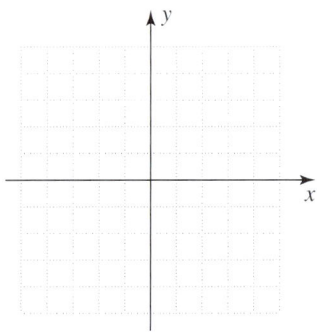

In Problems 54–55, solve the system by graphing and by substitution.

54. $x + 2y = -4$
$x = 2$

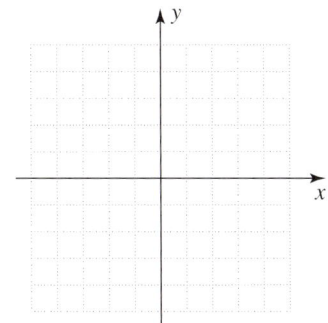

55. $y - x = 2$
$y = -3$

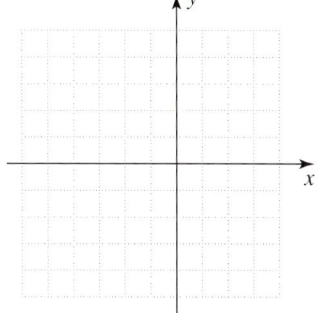

4.4 Systems of Linear Inequalities

Objectives

A Graph systems of two linear inequalities.

B Graph systems of inequalities.

To Succeed, Review How To . . .

1. Evaluate functions (pp. 231–232).
2. Solve systems of equations (pp. 285–295, 305–309).

Getting Started
Nutrition: You Want a Coke with That?

Are burgers and fries your menu of choice but you've got a budget problem and you're concerned about your fat and protein intake? A regular hamburger contains about 11 grams of fat and 12 grams of protein while French fries have 12 grams of fat and 3 grams of protein. Your daily consumption of fat and protein should be about 56 and 51 grams, respectively. (You can check this out in the *Fast Food Guide* by Jacobson and Fritschner.) Suppose that a regular hamburger costs $0.79 and regular fries are $1.09. How many of each can you eat so that you meet the recommended fat and protein intake and, at the same time, minimize your cost? Let's approach this problem by constructing a table where h and f represent the number of hamburgers and fries, respectively.

	Hamburgers	French Fries	Total
Fat (g)	11	12	$11h + 12f$
Protein (g)	12	3	$12h + 3f$
Cost ($)	0.79	1.09	$0.79h + 1.09f$

The recommended amounts of fat and protein are 56 and 51 grams, respectively, so we have the following system of equations.

Fat: $\quad 11h + 12f = 56$

Protein: $\quad 12h + 3f = 51$

When $h = 0$ in the first equation, $f = \frac{56}{12} = \frac{14}{3}$, so we graph $(0, \frac{14}{3})$.

When $f = 0$, $h = \frac{56}{11}$, and we graph $(\frac{56}{11}, 0)$ and join the two points with a line. We graph the second equation accordingly.

Now suppose you want to consume no more than 56 grams of fat and at least 51 grams of protein *and* you want to do this as cheaply as possible. The total cost function C is given by $C = 0.79h + 1.09f$, where h is the number of hamburgers and f the number of fries you buy. The shaded triangle in the graph is called the **feasible region**, and the minimum or maximum for a linear function such as C occurs at the corner points (**vertices**) or $(\frac{17}{4}, 0), (\frac{56}{11}, 0),$ or $(4, 1)$. To find the minimum or maximum of C, evaluate $C = 0.79h + 1.09f$ at each of the vertices.

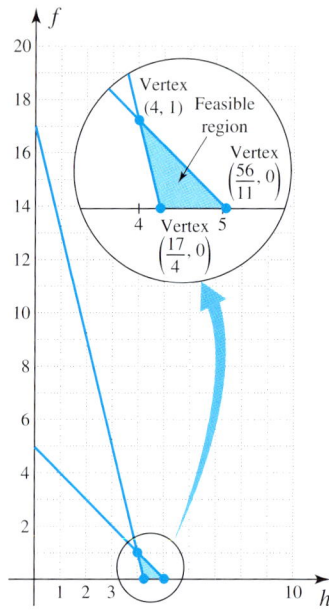

For $(\frac{17}{4}, 0), \quad C = 0.79 \cdot \frac{17}{4} + 1.09 \cdot 0 = \$3.36$

For $(\frac{56}{11}, 0) \quad C = 0.79 \cdot \frac{56}{11} + 1.09 \cdot 0 = \$4.02$

For $(4, 1) \quad C = 0.79 \cdot 4 + 1.09 \cdot 1 = \$4.25$

The ordered pair $(\frac{17}{4}, 0)$ means that you eat $4\frac{1}{4}$ hamburgers with no French fries. This produced the smallest cost, $3.36. Thus, the most economical way to meet your daily fat and protein requirements is to eat approximately 4 hamburgers and no fries. We leave the decision about the Coke up to you!

This problem is an example of an optimization problem and linear programming, which we shall discuss in *Using Your Knowledge*. First we learn how to graph solutions to systems with two linear inequalities.

A > Graphing Systems of Two Linear Inequalities

If we graph two or more linear inequalities **on the same coordinate system,** we have a *system of linear inequalities.* The solution set of the system is the set of points that satisfy *all* the inequalities in the system—the region that is *common* to every graph in the system. We illustrate how to find this solution set in Examples 1 and 2 for systems of two linear inequalities.

EXAMPLE 1 Graphing systems of two linear inequalities

Graph the solution set of the system of inequalities:

$$x \leq 0 \quad \text{and} \quad y \geq 2$$

PROBLEM 1

Graph:

$$y \geq 0$$
$$x < 3$$

Answer on page 329

SOLUTION $x = 0$ is a vertical line corresponding to the y-axis, so $x \leq 0$ consists of the points to the *left* of or *on* the line $x = 0$. Remember $\geq$ and $\leq$ symbols include solid lines in their graphs, while $>$ and $<$ are dashed lines. The condition $y \geq 2$ defines all points *above* or *on* the line $y = 2$. The solution set is the set satisfying both conditions $x \leq 0$ and $y \geq 2$; this is the **intersection** of the two solution sets and is the region common to both graphs, as shown by the darker region in Figure 4.16.

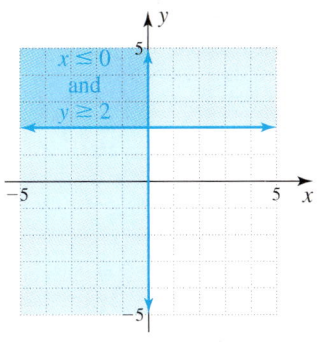

> Figure 4.16

EXAMPLE 2 Graphing systems of two linear inequalities

Graph the solution set of the system of inequalities:

$$y + x \geq 2 \quad \text{and} \quad y - x \leq 2$$

SOLUTION First, we graph the line $y + x = 2$. When $x = 0$, $y = 2$, so we graph $(0, 2)$. When $y = 0$, $x = 2$, and we graph $(2, 0)$. Joining the points $(0, 2)$ and $(2, 0)$ with a line, we have the graph of $y + x = 2$. Use $(0, 0)$ as a test point. Does $(0, 0)$ satisfy $y + x \geq 2$? Letting $x = 0$ and $y = 0$ we obtain $0 + 0 \geq 2$, which is not true. The solution set consists of all points *above* or *on* the line $y + x = 2$.

Now graph the line $y - x = 2$. When $x = 0$, $y = 2$, so we graph $(0, 2)$. When $y = 0$, $x = -2$, so we graph $(-2, 0)$ and join it to $(0, 2)$ with a line, the graph of $y - x = 2$. Using $(0, 0)$ as a test point for $y - x \leq 2$, we let $x = 0$ and $y = 0$ to obtain $0 - 0 \leq 2$, which is true, so we shade all the points *below* or *on* the line $y - x = 2$.

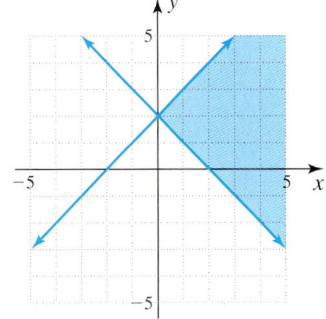

> Figure 4.17

The solution set of the system is the **intersection** of the solution sets of $y + x \geq 2$ and $y - x \leq 2$, as shown in Figure 4.17.

PROBLEM 2
Graph:

$$y > x + 2$$
$$x + 2y < 4$$

B > Graphing Systems of Inequalities

Many problems in business, economics, and the social sciences involve solving systems of linear inequalities arising from certain restrictions or conditions (*constraints*) such as "less than," "more than," "at least," "no more than," "a minimum of," and "a maximum of." As you recall, a **solution** of a system of linear inequalities is a point (x, y) that

Answers to PROBLEMS

1.
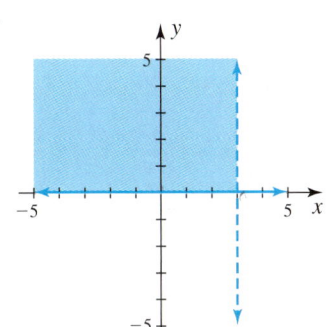

2.
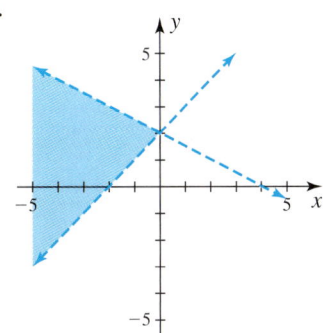

satisfies each inequality in the system. For example, in our fat/protein intake problem in *Getting Started*, the point (4, 1) satisfies the system of inequalities:

$$11h + 12f \leq 56 \quad \text{No more than 56 grams of fat.}$$
$$12h + 3f \geq 51 \quad \text{At least 51 grams of protein.}$$
$$f \geq 0 \quad \text{Some fries.}$$
$$h \geq 0 \quad \text{Some hamburgers.}$$

We then use this information and the following procedure.

PROCEDURE

Graphing a System of Inequalities

1. Sketch the line corresponding to each inequality using dashed lines for inequalities involving $<$ or $>$ and solid lines for inequalities with $\leq$ or $\geq$.
2. Use a test point to shade the half-plane that is the graph of each linear inequality. (If the test point satisfies the inequality, shade all points on the same side of the line as the test point.)
3. The graph of the system is the intersection of the half-planes, that is, the region consisting of the points satisfying *all* the inequalities.

EXAMPLE 3   Graphing a system of inequalities

Graph the system of inequalities and label the vertices:

$$x + y \leq 5$$
$$2x + 3y < 12$$
$$x \geq 0$$
$$y \geq 0$$

SOLUTION Sketch the line $x + y = 5$, as shown by the solid line in Figure 4.18. Select the test point (0, 0). Since $0 + 0 \leq 5$ is true, select all the points *below* or *on* the line $x + y = 5$ as the solution set. Now sketch the line $2x + 3y = 12$, as shown by the dashed line in Figure 4.18. Again, select the test point (0, 0). $2 \cdot 0 + 3 \cdot 0 < 12$, so the solution set consists of the points *below* the line $2x + 3y = 12$. The graph of $x \geq 0$ consists of the points *on* and to the *right* of the y-axis, and the graph of $y \geq 0$ consists of all points *on* or *above* the x-axis. The region common to all these graphs (shown shaded) is the solution set for the system of inequalities. The vertices are (0, 0), (0, 4), (5, 0), and (3, 2). (This last vertex is obtained by solving the system of equations $x + y = 5$ and $2x + 3y = 12$.)

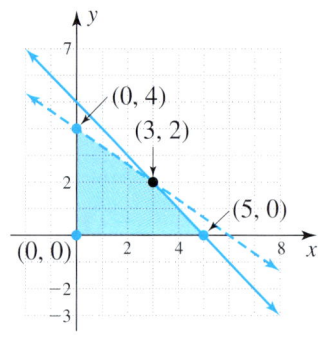

>Figure 4.18

PROBLEM 3

Graph:

$$x \leq 0$$
$$y \leq 0$$
$$y > -x - 2$$

Answers to PROBLEMS

3.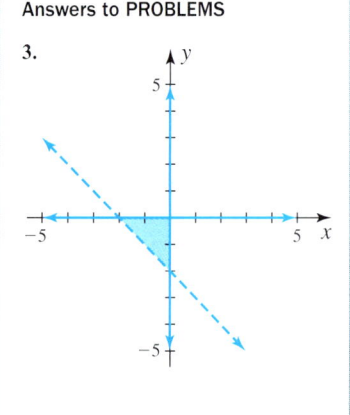

Exercises 4.4

> Boost your grade at mathzone.com!
> Practice Problems
> NetTutor
> Self-Tests
> e-Professors
> Videos

A **Graphing Systems of Two Linear Inequalities** In Problems 1–10, graph the solution set of the given system of inequalities.

1. $x - y \geq 2$ and $x + y \leq 6$

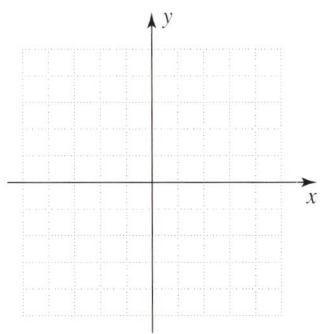

2. $x + 2y \leq 3$ and $x \leq y$

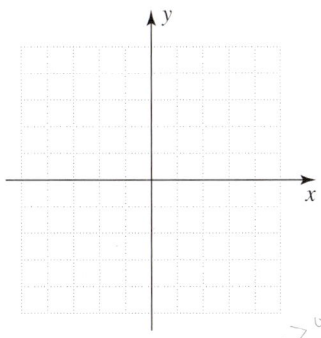

3. $2x - 3y \leq 6$ and $4x - 3y \geq 12$

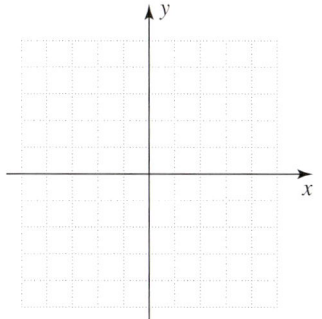

4. $2x - 5y \leq 10$ and $3x + 2y \leq 6$

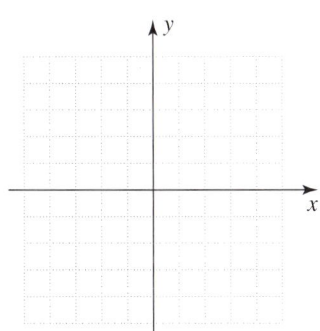

5. $2x - 3y \leq 5$ and $x \geq y$

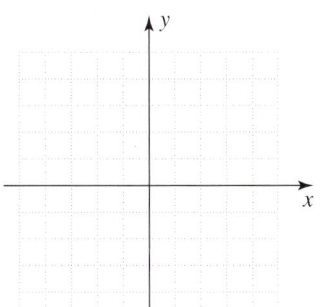

6. $x \leq 2y$ and $x + y < 4$

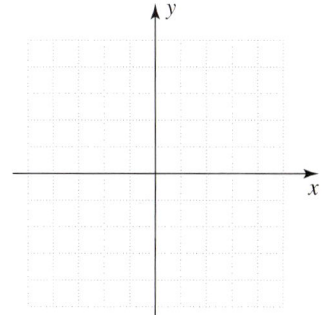

7. $x + 3y \leq 6$ and $x \geq y$

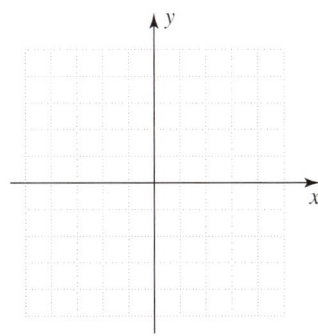

8. $2x - y \leq 2$ and $x \leq y$

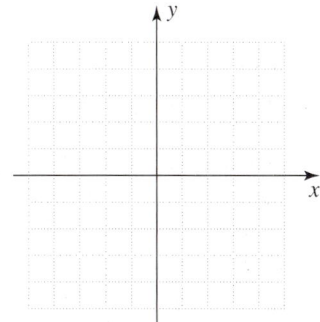

9. $x - y \leq 1$ and $3x - y < 3$

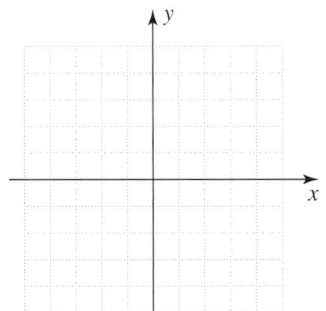

10. $x - y \geq -2$ and $x + y \leq 6$

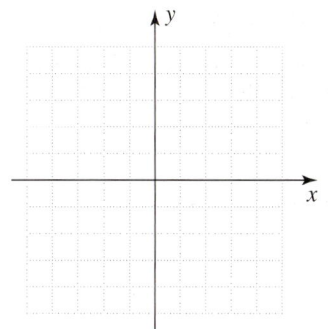

‹ B › Graphing Systems of Inequalities   In Problems 11–16, graph the solution set of the system of inequalities.

11. $x \geq 1$
     $x \leq 4$
     $y \leq 4$
     $x - 3y \leq -2$

12. $y - x \leq 0$
     $x \leq 4$
     $y \geq 0$
     $x + 2y \leq 6$

13. $x + y \geq 1$
     $2y - x \leq 1$
     $x \leq 1$

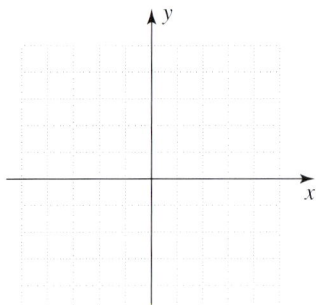
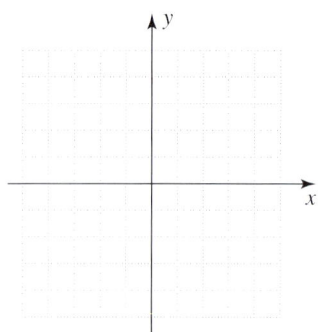
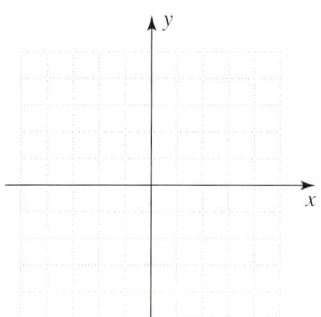

14. $2x + y \geq 18$
     $x + y \geq 12$
     $2y \leq 30$

15. $x \geq 1$
     $y \geq 2$
     $4 \leq 2x + y$
     $2x + y \leq 6$

16. $2x + y \geq 6$
     $0 \leq y \leq 4$
     $0 \leq x \leq 3$

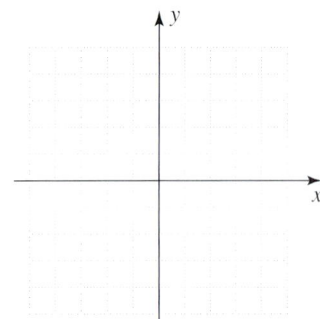
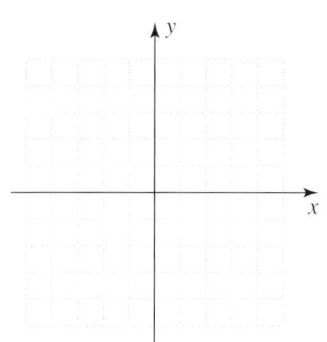
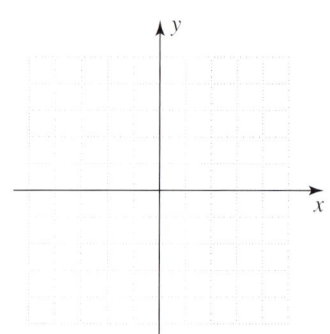

››› Using Your Knowledge

Linear Programming   As we demonstrated in the *Getting Started,* linear inequalities can be used to solve **optimization problems,** problems in which we find the greatest or the least value of a function. The technique used to solve such problems is called **linear programming.** A two-variable linear programming problem consists of two parts:

1. An **objective function** giving the quantity we wish to maximize or minimize. (In the *Getting Started,* $C = 0.79h + 1.09f$ is the objective function.)

2. A system of **constraints** (linear inequalities) whose solution set is called the set of **feasible solutions.** (As in the *Getting Started,* the system of inequalities is the set of constraints, and the shaded triangle in the graph is the set of feasible solutions.) Sometimes there is a set of **implied constraints** that state that the variables *cannot* be negative.

As mentioned, if there is an optimal solution, it must occur at one of the vertices of the set of feasible solutions. This solution can be found by testing the function at each of the vertices. Let's clarify this with an example of finding the optimal solution.

Find the maximum value of:

$$C = 2x + 3y \quad \text{Objective function}$$
$$\left.\begin{array}{r}2x + y \leq 6 \\ x - y \leq 3\end{array}\right\} \text{Constraints on } C$$
$$\left.\begin{array}{r}x \geq 0 \\ y \geq 0\end{array}\right\} \text{Implied constraints}$$

SOLUTION As before, graph the line $2x + y = 6$ (Figure 4.19) and use the test point $(0, 0)$. Since $0 + 0 \le 6$, the points *below* or *on* the line $2x + y = 6$ are in the solution set of $2x + y \le 6$. Next, graph the line $x - y = 3$ and select the test point $(0, 0)$. Again, $0 - 0 \le 3$ so the solution set consists of the points *above* or *on* the line $x - y = 3$. The lines $2x + y = 6$ and $x - y = 3$ intersect at the point $(3, 0)$. The points satisfying $x \ge 0$ are the points *on* or to the *right* of the y-axis, and the points satisfying $y \ge 0$ are *on* or *above* the x-axis. The vertices of the feasible region are $(0, 0)$, $(0, 6)$, and $(3, 0)$. At these three vertices, the objective function C has the following values:

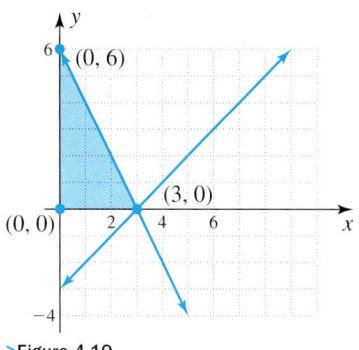

>Figure 4.19

At $(0, 0)$, $C = 2 \cdot 0 + 3 \cdot 0 = 0$
At $(0, 6)$, $C = 2 \cdot 0 + 3 \cdot 6 = 18$ ← Maximum
At $(3, 0)$, $C = 2 \cdot 3 + 3 \cdot 0 = 6$

The maximum value of C is 18 and occurs when $x = 0$ and $y = 6$.

For Problems 17–18, use linear programming to solve.

17. The E-Z-Park storage lot can hold at most 100 cars and trucks. A car occupies 100 square feet and a truck 200 square feet, and the lot has a usable area of 12,000 square feet. The storage charge is $20 per month for a car and $35 per month for a truck. How many of each should be stored to bring E-Z-Park the maximum revenue?

18. The Zig-Zag Manufacturing Company produces two products, zigs and zags. Each of these products has to be processed through all three machines, as shown in the table. If Zig-Zag makes $12 profit on each zig and $8 profit on each zag, find the number of each that the company should produce to maximize its profit.

Zig-Zag Company data

Machine	Machine Time Available (hours)	Production Time (hours) Zigs	Zags
I	Up to 100	4	12
II	Up to 120	8	8
III	Up to 84	6	0

〉〉〉 Write On

19. Is it possible for a system of linear inequalities to have no solution? If so, find an example of such a system.

20. Would it be possible to list all the solutions in the solution set of a system of linear inequalities? Explain.

〉〉〉 Concept Checker

Fill in the blank(s) with the correct word(s), phrase, or mathematical statement.

21. The set of points on the graph that satisfies all the inequalities in the system is called the _____ .

22. If an inequality statement in a system has a "$>$" symbol, then the boundary line representing that inequality is a _____ line.

23. The boundary line for the inequality $x \le 0$ is a solid _____ line through the origin.

system of linear inequalities

solution set

solid

dashed

vertical

horizontal

slanted

>>> Mastery Test

24. Graph the solution set to the system of linear inequalities:

$$y > 2x + 1$$
$$x \leq -1$$

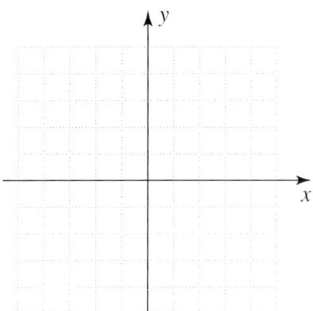

25. Graph the solution set to the system of linear inequalities:

$$y \leq -2$$
$$x - y < 4$$

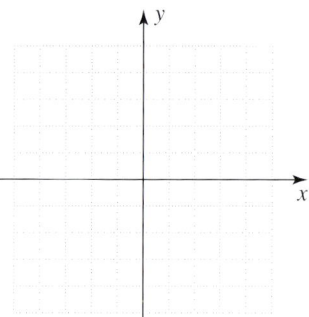

26. Graph the solution set to the system of linear inequalities:

$$x + y \leq 5$$
$$2x - y \leq 4$$
$$x \geq 0$$
$$y \geq 0$$

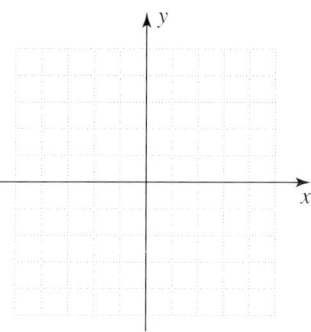

>>> Skill Checker

In Problems 27–28, name the base and exponent.

27. $3x^5$

28. $(3x + 2)^5$

In Problems 29–30, evaluate.

29. $3x^5$, if $x = -2$

30. $(3x + 2)^5$, if $x = -1$

In Problems 31–32, combine like terms.

31. $3x - 7 + 2x^2 - 8x - 10 - 2x^2$

32. $24 - 7x + x^2 + 18x - 10 + x - x^2$

In Problems 33–34, simplify by removing the parentheses.

33. $-(5x - 12)$

34. $a - (-b + c)$

> Collaborative Learning 4A

The transition to digital and high-definition television (HDTV) is driving up the sales for digital displays as shown in the first bar graph. How does that compare to the United States' investment in upgrading to HDTV as shown in the second bar graph? Let's do some collaboration and find out. Divide into two groups and answer the following questions.

Group 1

1. Using the bar graph for the projected sales of HDTVs, write the equation of the line of best fit.
2. Graph the equation of the line found in number 1 on a coordinate system on a large poster board. Label it the "supply" equation.

Group 2

3. Using the bar graph for the "United States Investment Needed for Upgrading to HDTV," write the equation of the line of best fit.
4. Graph the equation of the line found in number 3 on the same poster board graph as Group 1. Label it the "demand" equation.

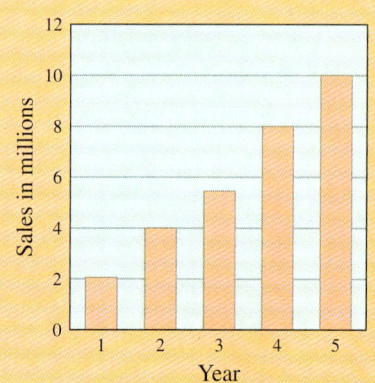

Projected Sales of HDTVs

The two lines graphed on the poster board represent a system of equations where one line represents the supply (projected sales) of HDTVs and the other represents the demand for investment by the United States for upgrading to HDTVs. As a class, discuss the following questions.

5. Use the graph to estimate the year at which the projected sales of HDTVs will equal the United States investment for upgrading to HDTVs. What will the amount of sales for HDTVs be for that point?

6. What do you call the point at which the supply equation intersects the demand equation?

7. Discuss other real world data that could be considered as a supply and demand situation.

8. Choose one or two of the topics discussed in number 7, research the data, analyze it, and find the equilibrium point.

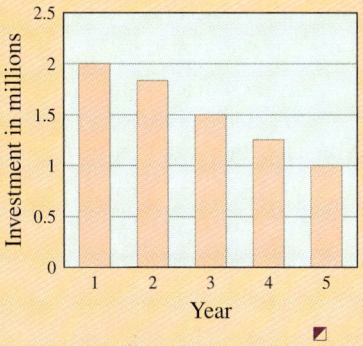

United States Investment Needed for Upgrading to HDTV

> Collaborative Learning 4B

Have you ever considered selling an item by advertising it in the classified ads? One way to advertise is in your local newspaper classified ads while another might be to use the Web.

Here is one newspaper's pricing for classified ads.

Our classified line ads traditionally do not have artwork and are charged out by the number of lines of text. There are approximately 18 characters (including spaces) per line. The charge for an ad depends on the number of lines and the number of days the ad runs. The minimum number of times an ad can be scheduled is three days. We have price breaks for ads running in 6-, 12-, and 24-day increments. However, there will be a set fee of $2 for all 6-day ads plus line cost and a set fee of $1 for all 12-day ads plus line cost.

For example: A three-line ad for three days costs $8.74; a three-line ad for 6 days costs $15.04; and a three-line ad for 12 days costs $24.32.

Divide into three groups. Have the first group complete the table using the information from the ad above (based on a three-line ad). Then use the information from the table to graph each of the three rate lines (3-day, 6-day, and 12-day rate) on a large poster board graph.

Days	Cost at 3-day Rate	Cost at 6-day Rate	Cost at 12-day Rate
3	$8.74		
6	17.48	$15.04 + $2 = ?	
9	26.22		
12	?	$30.08 + $2 = ?	$24.32 + $1 = ?
15	?		
18	?	?	
21	?		
24	?	?	?

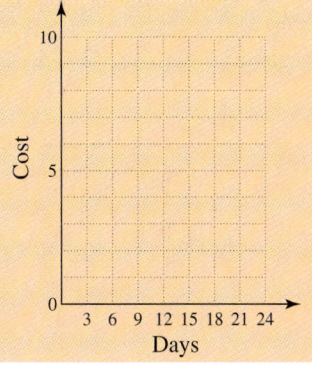

<CONTINUED>

The second group should research the pricing to advertise in your local newspaper and the third group should research the pricing to advertise on the Web. These groups should construct tables and graphs as well. When all groups have completed their work, discuss the following.

1. Using the graph from group 1, will the three lines ever intersect at the same point? What is the name of that type of system of equations?
2. In the graph from group 1, what is the cost at the point of intersection of the line representing the three-day rate and the line representing the six-day rate? Which of the two was the better rate before the intersection point and which is the better rate after that point?
3. How does the information from group 2 and group 3 compare to group 1? Discuss some of the reasons why one form of advertising may or may not be cheaper than another.

> Research Questions

1. Leibniz was probably the first person to use elimination to solve a system of three equations in three unknowns. But the evidence of a systematic method of solving a system of three equations appeared in China, probably in about 250 B.C. Here is the problem:

    There are three grades of corn. After threshing, three bundles of top grade, two bundles of medium grade, and one bundle of low grade make 39 dou (a measure of volume). Two bundles of top grade, three bundles of medium grade, and one bundle of low grade will produce 34 dou. The yield of one bundle of top grade, two bundles of medium grade, and three bundles of low grade is 26 dou. How many dou are contained in each bundle of each grade?

    a. Write a system of three equations in three unknowns representing this situation.
    b. Find out the name of the book where this problem originated and write a short paragraph detailing the contents of the book.
    c. Write a short paragraph explaining how the Chinese solved this problem.

2. Another method used for solving systems of equations that wasn't mentioned in this chapter but is included in the appendix is solving by determinants. This method is also known as Cramer's rule.

    a. What is a square determinant?
    b. Use the system $2x + 3y = -6$ and $x - 2y = 4$ to set up the determinants that would solve the system.
    c. Discuss how Cramer and Maclaurin contributed to solving systems of equations by determinants.

> Summary Chapter 4

Section	Item	Meaning	Example
4.1A	Consistent system	Graphs intersect at one point. There is one solution.	$2x - y = 2$ and $y = x - 1$ form a consistent system intersecting at $(1, 0)$.
	Inconsistent system	Graphs are parallel lines. There is no solution.	$y - 2x = 4$ and $3y - 6x = 18$ form an inconsistent system.
	Dependent system	Graphs coincide. There are infinitely many solutions.	$2x + \frac{1}{2}y = 2$ and $y = -4x + 4$ form a dependent system.

Section	Item	Meaning	Example
4.1B	Substitution method (two variables)	A method where one equation is solved for a variable that is substituted into the other equation.	$y = 2x$ and $2x + y = 4$ can be solved by substituting $y = 2x$ into $2x + y = 4$ to obtain $$2x + 2x = 4 \text{ or}$$ $$4x = 4$$ $$x = 1$$ Thus $y = 2 \cdot 1 = 2$.
4.1C	Elimination method (two variables)	A method where equations are multiplied by suitable numbers so that addition eliminates one of the variables.	For the system $$x - 2y = 4$$ $$x + y = 6$$ multiplying the second equation by 2 yields $$x - 2y = 4$$ $$2x + 2y = 12$$ so that addition eliminates the y, leaving $3x = 16$ or $x = \frac{16}{3}$.
4.2A	Elimination method (three variables)	A method in which two equations are selected and one variable is eliminated; then a different pair of equations is selected, and the same variable is eliminated. The system is then solved as in 4.1C.	Consider the system $$2x - y + z = 4 \quad (1)$$ $$-x - y - z = 0 \quad (2)$$ $$-x + 2y - z = 2 \quad (3)$$ Add (1) and (2), then add (1) and (3). We get $$x - 2y = 4$$ $$x + y = 6$$ Solve this system as in 4.1C, then substitute the values of x and y into (1), (2), or (3) to find the value of z.
4.2B	Consistent system	A system with one solution consisting of an ordered triple of the form (x, y, z)	The system $$x + y + z = 6$$ $$x - y - z = -4$$ $$x + y - z = 0$$ is consistent. The solution is $(1, 2, 3)$.
	Inconsistent system	A system with no solution	The system $$x + y + z = 4$$ $$-x - y - z = 3$$ $$x + 2y + z = 5$$ is inconsistent. The solution set is $\emptyset$.
	Dependent system	A system with infinitely many solutions	The system $$x + y + z = 4$$ $$-x - y - z = -4$$ $$x + 2y - z = 5$$ is dependent.

(continued)

Section	Item	Meaning	Example
4.4B	System of linear inequalities	The solution set is the set of all points that satisfy all the inequalities of the system.	$2x + y < 1$ $x \leq 0$ $y \geq 0$ Solution:

Review Exercises Chapter 4

(If you need help with these exercises, look in the section indicated in brackets.)

1. ⟨**4.1A**⟩ *Use the graph to identify the solution to the system if possible. If the system has a solution, use the system of equations to verify the solution.*

    a. $5x + y = 5$
    $-6x + y = -6$

    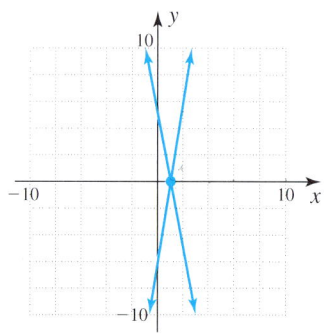

2. ⟨**4.1A**⟩ *Use the graphical method to solve the system.*

    a. $2x - y = 2$
    $y = 3x - 4$

    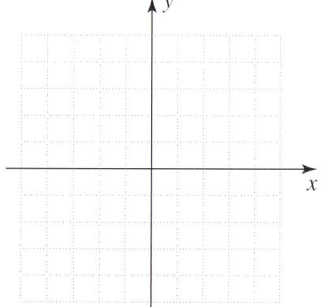

    b. $y = -\dfrac{1}{4}x + 2$
    $x + 4y = -12$

    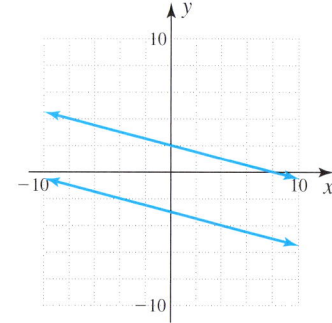

    b. $x - 2y = 0$
    $y = x - 2$

    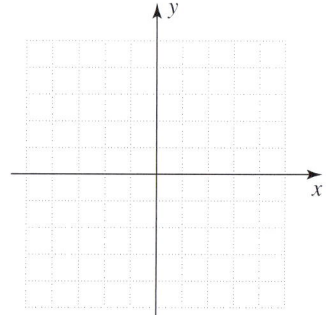

3. ⟨**4.1A**⟩ *Use the graphical method to solve the system.*
   a. $2y - x = 3$
      $4y = 2x + 8$

   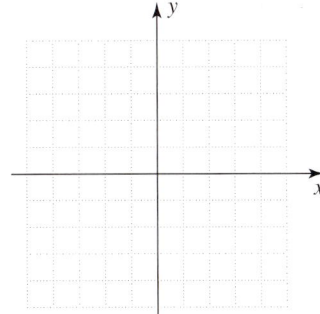

   b. $3y + x = 5$
      $2x = 8 - 6y$

   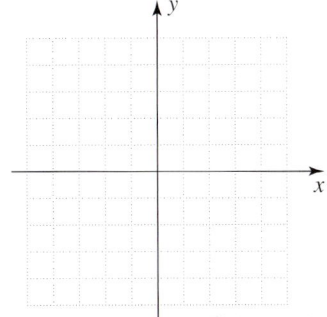

4. ⟨**4.1A**⟩ *Use the graphical method to solve the system.*
   a. $3x + 2y = 6$
      $y = 3 - \frac{3}{2}x$

   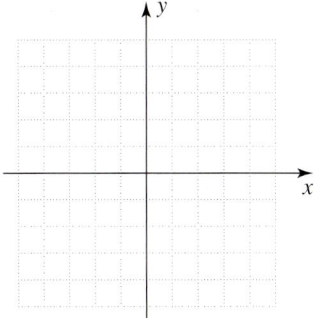

   b. $x + 2y = 4$
      $2x = 8 - 4y$

   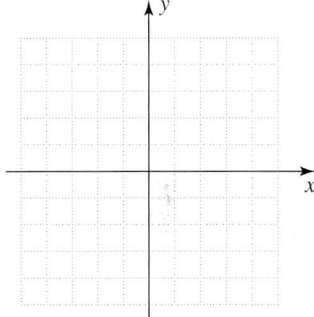

5. ⟨**4.1B**⟩ *Solve by the substitution method.*
   a. $2x - y = 4$
      $x + y = 5$
   b. $2x + 3y = 10$
      $x - y = -1$

6. ⟨**4.1B**⟩ *Solve by the substitution method.*
   a. $2x + 4y = 7$
      $x = -2y - 1$
   b. $2y + x = 5$
      $3x = 10 - 6y$

7. ⟨**4.1B**⟩ *Solve by the substitution method.*
   a. $2y - x = 5$
      $2x = 4y - 10$

   b. $x + 5y = 5$
      $y = 1 - \frac{x}{5}$

8. ⟨**4.1C**⟩ *Solve by the elimination method.*
   a. $x - 3y = 7$
      $2x - y = 9$
   b. $2x + 3y = 4$
      $x + y = 1$

9. ⟨**4.1C**⟩ *Solve by the elimination method.*
   a. $2x + 3y = 7$
      $6x + 9y = 14$
   b. $3x - 4y = 5$
      $6x - 8y = 15$

10. ⟨**4.1C**⟩ *Solve by the elimination method.*
    a. $\frac{x}{5} + \frac{y}{2} = \frac{1}{5}$
       $2x + 5y = 2$
    b. $\frac{x}{3} - \frac{y}{4} = 2$
       $\frac{x}{6} - \frac{y}{8} = 1$

11. ⟨**4.1D**⟩ The demand function for a certain product is $D(p) = 1595 - 5p$, where p is the price in dollars, and the supply function is $S(p) = 3p + 850$. What will the price of this product be at the equilibrium point?

12. ⟨**4.2A**⟩ *Solve the system.*
    a. $x - y + z = 1$
       $x + y - z = 7$
       $2x + y + z = 9$
    b. $2x - 3y + z = -4$
       $2x + y + z = 4$
       $4x + 9y + 3z = 10$

13. ⟨4.2A⟩ *Solve the system.*
   a. $2x + y - 2z = 4$
      $-x + y + 2z = 2$
      $3y + 2z = 12$
   b. $x + 2y = 4$
      $y + 2z = 6$
      $2x + 2y - 4z = -4$

14. ⟨4.3A⟩ *How many of each coin do Joey and Alice have?*
   a. Joey has $4 in nickels and dimes, and he has five more nickels than dimes.
   b. Alice has $2 in nickels and dimes, and she has five fewer nickels than dimes.

15. ⟨4.3B⟩ *Determine the height of each building.*
   a. The total height of a building and a flagpole on the roof is 200 feet. The building is nine times as high as the flagpole.
   b. The total height of a building and a flagpole on the roof is 180 feet. The building is eight times as high as the flagpole.

16. ⟨4.3C⟩ *Find the speed of each current.*
   a. A motorboat can go 12 miles downstream on a river in 20 minutes. It takes this boat 30 minutes to go upstream the same 12 miles.
   b. A motorboat can go 6 miles downstream on a river in 15 minutes. It takes this boat 20 minutes to go upstream the same 6 miles.

17. ⟨4.3D⟩ *Find out how much Bill and Betty have invested at each rate.*
   a. Bill has three investments totaling $40,000. These investments earn interest at 4%, 6%, and 8%. Bill's annual income from these investments is $2600. The income from the 8% investment exceeds the total income from the other two investments by $600.
   b. Betty has three investments totaling $45,000. These investments earn interest at 4%, 6%, and 8%. Betty's annual income from these investments is $2900. The income from the 8% investment exceeds the total income from the other two investments by $300.

18. ⟨4.3E⟩ *What are the dimensions of each rectangle?*
   a. The perimeter of a rectangle is 100 inches and the length is 30 inches more than the width.
   b. The perimeter of a rectangle is 80 inches and the length is three times the width.

19. ⟨4.3F⟩ *Find minimum wage for the country.* In a study of minimum wages in some European countries completed by a European industrial relations observatory online, the data for each country were modeled by linear functions. The linear function that approximates the minimum wage for Poland is $y = 12.2x + 83.9$ and for Malta it is $y = 3.6x + 90.2$, where x is the number of years after 1995 and y is the wage in Euros. In what year was the minimum wage for Poland the same as that for Malta? What was the wage to the nearest whole number?

20. ⟨4.4A, B⟩ *Graph the solution set.*
   a. $y \geq x$
      $x > -2$
   b. $3x - 4y \geq -12$
      $x < 1$
      $y \geq 0$

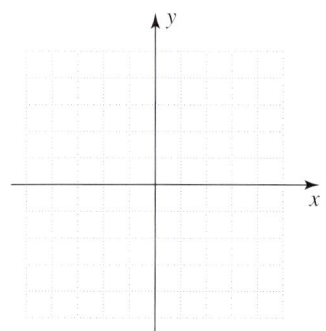

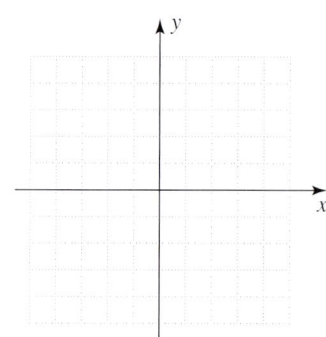

Practice Test Chapter 4

(Answers are on pages 343–344)

Visit www.mhhe.com/bello to view helpful videos that provide step-by-step solutions to several of the problems below.

In Problems 1–2, the system of equations is graphed. Use the graph to identify the solution to the system, if possible. If the system has a solution, use the system of equations to verify the solution.

1. $2x - y = 12$
$4x + y = 6$

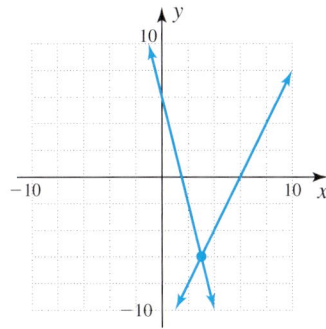

2. $-x + 3y = 6$
$y = \frac{1}{3}x - 1$

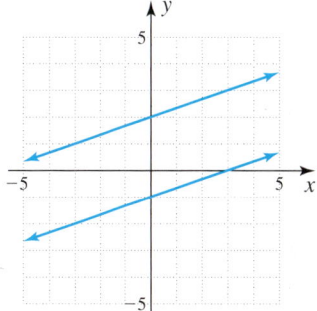

3. Use the graphical method to solve the system.
$$x - 3y = 3$$
$$y = x - 1$$

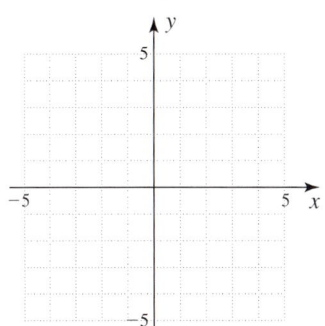

4. Use the graphical method to solve the system.
$$y - 3x = -3$$
$$3y = 9x + 9$$

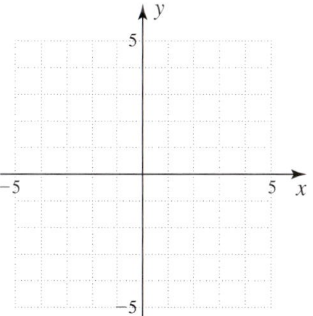

5. Use the substitution method to solve the system.
$$x - 2y = 4$$
$$x = 1 + y$$

6. Use the substitution method to solve the system.
$$2x - y = 6$$
$$2y = 4x - 12$$

7. Solve by the elimination method.
$$3x + 4y = 5$$
$$x + y = 1$$

8. Solve by the elimination method.
$$3x + 4y = 5$$
$$6x + 8y = 9$$

9. Solve the system.
$$\frac{x}{2} + \frac{y}{3} = 2$$
$$\frac{x}{4} + \frac{y}{8} = 1$$

10. Solve the system.
$$2x = 3y - 10$$
$$2y = 3x + 10$$

11. The demand function for a certain product is $D(p) = 1050 - 7p$, where p is the price in dollars, and the supply function is $S(p) = 8p + 975$. What will the price of this product be at the equilibrium point?

12. Solve the system.
$$x + y + z = 2$$
$$2x + y - z = 5$$
$$x + y - z = 4$$

13. Solve the system.
$$2x + y - 2z = 7$$
$$-x + y + 2z = -2$$
$$-3y + 2z = 9$$

14. Pedro has $3.50 in nickels and dimes. He has 10 more nickels than dimes. How many of each coin does he have?

15. The total height of a building and a flagpole on the roof is 240 feet. The building is nine times as high as the flagpole. How high is the building?

16. A motorboat can go 10 miles downstream on a river in 20 minutes. It takes 30 minutes for this boat to go back upstream the same 10 miles. Find the speed of the current.

17. Annie has three investments totaling $60,000. These investments earn interest at 4%, 6%, and 8%. Annie's total annual income from these investments is $4000. The income from the 8% investment exceeds the total income from the other two investments by $800. Find how much she has invested at each rate.

18. The profits of two companies were compared from 2004 to 2007. The linear function that approximates the profits of company A is $y = 150x + 150$, where x represents the years from 2004 and y is the profit in millions of dollars. For company B, the linear function that approximates its profit is $y = 100x + 250$. When did company A and company B have the same profit?

19. Graph the solution set.
$$x > -1$$
$$y \geq x$$

20. Graph the solution set.
$$x + y \leq 2$$
$$y \leq 2$$
$$x \leq 2$$

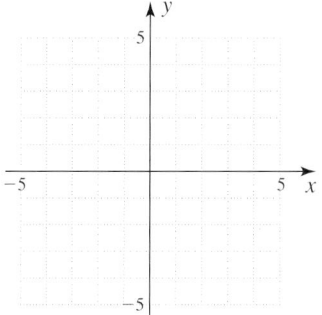

Answers to Practice Test Chapter 4

Answer	If You Missed Question	Section	Review Examples	Page
1. Solution: $(3, -6)$ $2(3) - (-6) = 12 \quad 4(3) + (-6) = 6$ $6 + 6 = 12 \quad 12 - 6 = 6$ $12 = 12 \quad 6 = 6$	1	4.1	1	285–286
2. No solution; parallel lines	2	4.1	1	285–286
3. The solution is $(0, -1)$.	3	4.1	2	286–287
4. Parallel lines; no solution	4	4.1	3	287–288
5. $(-2, -3)$	5	4.1	5	290
6. Dependent, consistent; infinitely many solutions	6	4.1	7	291–292
7. $(-1, 2)$	7	4.1	8	293
8. Inconsistent; no solution	8	4.1	9	294
9. $(4, 0)$	9	4.1	10	294–295
10. $(-2, 2)$	10	4.1	8–9	293–294
11. $5	11	4.1	11	297
12. $(1, 2, -1)$	12	4.2	1	307
13. $(7, -1, 3)$	13	4.2	2	308

Answer	If You Missed		Review	
	Question	Section	Examples	Page
14. 30 nickels; 20 dimes	14	4.3	1	316–317
15. 216 ft	15	4.3	2	317
16. 5 mi/hr	16	4.3	3	318
17. $10,000 at 4%; $20,000 at 6%; $30,000 at 8%	17	4.3	4	319–320
18. In 2 yr or 2006	18	4.3	6	321–322
19.	19	4.4	1, 2	328–329
20.	20	4.4	3	330

Cumulative Review Chapters 1–4

1. The number $\sqrt{13}$ belongs to which of these sets? Natural numbers, whole numbers, integers, rational numbers, irrational numbers, and real numbers. Name all that apply.

2. Simplify:

   $[(2x^2 - 6) + (6x + 6)] - [(x - 2) + (4x^2 - 5)]$

3. Perform the indicated operation and simplify:

   $\dfrac{36x^6}{9x^{-7}}$

4. Simplify: $(3x^3y^{-4})^2$

5. Evaluate: $[-7(2 + 6)] + 6$

6. Solve: $x + 6 = 3(5x - 2)$

7. Solve: $0.03P + 0.08(1300 - P) = 75$

8. Solve: $|x - 6| = |x - 8|$

9. Graph: $2(x - 3) \leq 3x - 5$

   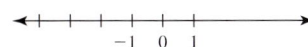

10. Graph: $-5 \leq -5x - 15 < 5$

    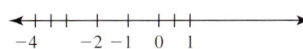

11. Graph: $|2x + 3| > 4$

12. If $H = 2.85h + 73.82$, find h when $H = 139.37$.

13. A woman's salary was increased by 10% to $29,700. What was her salary before the increase?

14. How many gallons of a 40% acid solution must be mixed with 30 gallons of a 16% acid solution to obtain a solution that is 30% acid?

15. Graph: $x - y = 3$

    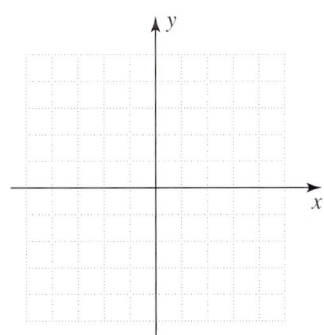

16. Find the slope of the line passing through the points $A(-7, -7)$ and $B(0, 8)$.

17. A line passes through the point $(5, -5)$ and has slope 2. Graph this line.

    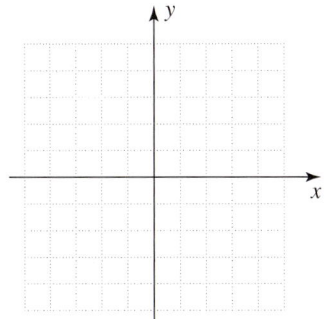

18. Find the slope and the y-intercept of the line $9x + 3y = -54$.

19. Find an equation of the line that passes through the point (3, 6) and is parallel to the line $8x + 6y = -2$. Write the equation in standard form.

20. Graph: $y \leq x + 1$

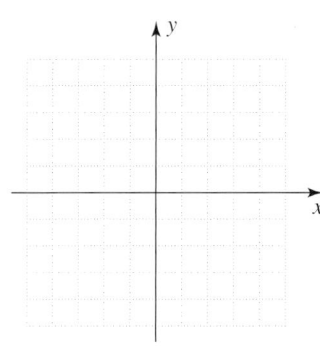

21. Graph: $|y| \leq 4$

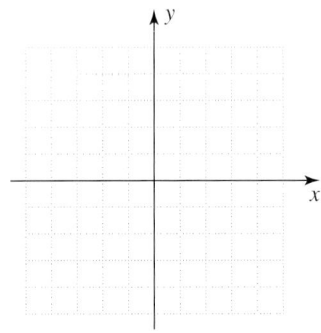

22. Find the domain and range of $\{(x, y) | y = 2 + x\}$.

23. Find $f(2)$ given $f(x) = 3x - 2$.

24. Use the graphical method to solve the system:

$$x - 3y = -3$$
$$x = -3 - 3y$$

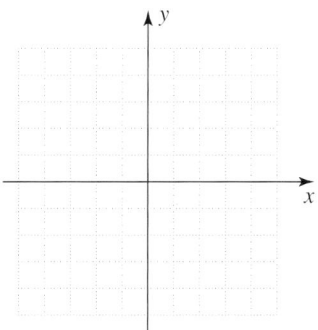

25. Use the substitution method to solve the system:

$$x - 2y = 3$$
$$6y = 3x - 9$$

26. Solve the system:

$$\frac{x}{2} + \frac{y}{5} = 6$$
$$\frac{x}{6} + \frac{y}{3} = 6$$

27. Solve the system:

$$2x + y - 3z = 7$$
$$-4x - y - 2z = 5$$
$$x - y - z = -2$$

28. A motorboat can go 12 miles downstream on a river in 20 minutes. It takes 30 minutes for this boat to go back upstream the same 12 miles. Find the speed of the boat.

Section

- **5.1** Polynomials: Addition and Subtraction
- **5.2** Multiplication of Polynomials
- **5.3** The Greatest Common Factor and Factoring by Grouping
- **5.4** Factoring Trinomials
- **5.5** Special Factoring
- **5.6** General Methods of Factoring
- **5.7** Solving Equations by Factoring: Applications

Chapter

5 five

▶ Polynomials

The Human Side of Algebra

Algebra has gone through three stages: rhetorical, in which statements and equations were written in ordinary language; syncopated, in which familiar terms were abbreviated; and symbolic, in which every part of an expression is written in symbols. At the time of Euclid, letters were used to represent quantities entered into equations until Diophantus introduced "the syncopation of algebra," using his own shorthand to express quantities and operations. (For example, Diophantus wrote the square of the unknown as Δ^Y.)

François Vieta (1540–1603), a French lawyer and member of parliament, used vowels to designate unknown quantities and consonants to represent constants. Vieta, however, retained part of the verbal algebra by writing *A quadratus* for x^2, *A cubus* for x^3, and so on. Vieta's contribution was a significant step toward a more abstract mathematics, but one of his most interesting contributions was his motto, *Leave no problem unsolved*. See if you can apply this motto to your study of algebra.

5.1 Polynomials: Addition and Subtraction

Objectives

A ▶ Classify polynomials.

B ▶ Find the degree of a polynomial and write in descending order.

C ▶ Evaluate a polynomial function.

D ▶ Add and subtract polynomials.

E ▶ Solve applications involving sums or differences of polynomials.

To Succeed, Review How To . . .

1. Define base and exponent (p. 38).
2. Evaluate expressions involving exponents (pp. 38–39).
3. Use the properties of real numbers (pp. 28–29).
4. Combine like terms (pp. 58–59).
5. Remove parentheses in an expression preceded by a minus sign using the distributive property (pp. 56–58).

▶ Getting Started

Diving

A man dives from an altitude of 118 feet into water 12 feet deep. Do you know how high above sea level he will be after falling for t seconds? This height is given by

$$H = -16t^2 + 118$$

The right-hand side of this formula is an algebraic expression called a *polynomial*.

Polynomials have many applications. However, in this section we will begin by introducing some concepts and definitions before solving applications involving polynomials.

A ▶ Classifying Polynomials—Monomials, Binomials, and Trinomials

An expression consisting of a constant or a constant times a product of variables with *whole-number exponents* is called a *monomial*. This means the exponents must come from the set $\{0, 1, 2, 3,\ldots\}$. For example,

$$2x, \quad -5x^2y, \quad 10x^5yz^2, \quad -\frac{2}{3}x^4, \quad \text{and} \quad -7y^3$$

are all monomials.

A **polynomial** is a sum or difference of monomials. Thus,

$$2x^2 + xy - 7y^3$$

is a polynomial. The individual monomials in a polynomial are called the **terms** of the polynomial. The terms are separated by $+$ and $-$ signs. In the term ax^k, where

x is the only variable, a is the **coefficient** and k is the **degree** of the term. In the polynomial $2x^2 + xy - 7y^3$, the terms are $2x^2$, with coefficient 2; xy, with coefficient 1 (since $xy = 1xy$); and $-7y^3$, with coefficient -7.

Notice,

$$5x^{-1} \quad \text{or} \quad \frac{5}{x}, \quad \frac{x+y}{z}, \quad \text{and} \quad 2 + \sqrt{x} \quad \text{or} \quad 2 + x^{1/2}$$

are *not* polynomials. In each case not all variables have whole number exponents. (In the first two expressions we are dividing by a variable. The third expression involves taking the square root of the variable.)

Polynomials can be classified according to the number of terms they have. Polynomials with one term are called **monomials,** polynomials with two *unlike* terms are called **binomials,** and polynomials with three *unlike* terms are called **trinomials.** Polynomials with more than three unlike terms are just called polynomials, as shown in the table.

Type		Examples	
Monomials	$-8x$,	$3x^2y$,	$9m^{10}$, -3
Binomials	$x - y$,	$-3x^2 + xy$,	$-16t^2 + 118$
Trinomials	$x + y - z^5$,	$2x^2 - 3x + \sqrt{2}$,	$z^3 + 2z - 12$
Polynomials	$t^5 + 3t^3 - 2t^2 - 8$,	$-3m^4 + 3n^2 + mn - 4m^3 + 2$	

EXAMPLE 1   Classifying polynomials
Classify each of the following polynomials as a monomial, binomial, trinomial, or polynomial if more than 3 terms:

a. $x + x^2$ **b.** $-9x$ **c.** $3x^2 - y + 3xyz$ **d.** $x^5 - 3x^3 + 5x^2 - 7$

SOLUTION

a. $x + x^2$ has two terms. It is a binomial.
b. $-9x$ has one term. It is a monomial.
c. $3x^2 - y + 3xyz$ has three terms. It is a trinomial.
d. $x^5 - 3x^3 + 5x^2 - 7$ has more than three terms. It is a polynomial.

PROBLEM 1
Classify each of the following polynomials as a monomial, binomial, trinomial, or polynomial:

a. $-12xy$
b. $7 + x + x^3$
c. $z^6 - 7z^4 - 8z^2 + 5z - 4$
d. $25x^2 - 16y^2$

B › Degree of a Polynomial and Writing in Descending Order

In Example 1, the polynomials $(x + x^2)$, $(-9x)$, and $(x^5 - 3x^3 + 5x^2 - 7)$ contain only one variable. The second polynomial is written in *descending order,* from the highest exponent on the variable to the lowest exponent on the variable. The first polynomial written in descending order would be, $x^2 + x$.

Single-variable polynomials can also be classified according to the *greatest exponent* on the variable. This number is called the *degree* of the polynomial. (Recall that in the monomial ax^k, k is the degree.) In general, we have the following definition.

DEGREE OF A POLYNOMIAL IN ONE VARIABLE

The **degree** of a polynomial in one variable is the *greatest* exponent on that variable.

Thus, $-8x^5$ is of the *fifth* degree, $-3x^2 + 8x^4 - 2x$ is of the *fourth* degree, and $0.5x$ is of the *first* degree. (Note that $x = x^1$.) Since $x^0 = 1$, $-3 = -3 \cdot 1 = -3x^0$. Similarly, $9 = 9x^0$. Thus, the degree of nonzero numbers such as -3 and 9 is 0. The number 0 is

Answers to PROBLEMS

1. **a.** monomial   **b.** trinomial   **c.** polynomial   **d.** binomial

called the **zero polynomial** and is not assigned a degree. These ideas can be extended to include polynomials in more than one variable. For example, the expression $3x^2 - y$ is a polynomial in *two* variables (x and y), and $3x^2 - 2xy - 3xyz^2$ is a polynomial in *three* variables (x, y, and z). To find the degree of these polynomials, we look at the degree of each term. Here is the definition.

DEGREE OF A POLYNOMIAL IN SEVERAL VARIABLES

The degree of a polynomial in several variables is the greatest sum of the exponents of the variables in any one term of the polynomial.

Thus, the degree of $3x^2 - y$ is 2 (the degree of the first term, $3x^2$). To find the degree of $3x^2 - 2xy - 3xyz^2$, find the degree of each term:

$$3x^2 \quad - \quad 2x^1y^1 \quad - \quad 3x^1y^1z^2$$

Degree: 2 $1 + 1 = 2$ $1 + 1 + 2 = 4$

The greatest sum of exponents in any one term is 4, so the degree of $3x^2 - 2xy - 3xyz^2$ is 4.

EXAMPLE 2   Finding the degree of a polynomial and writing in descending order

Find the degree of the given polynomials:

a. $-5x^2 + 3x^5 + 9$ **b.** $-x^2 + xy^2z^3 - x^5$ **c.** 3   **d.** 0

e. Write Example 2(a) in descending order.

SOLUTION

a. The degree of $-5x^2 + 3x^5 + 9$ is 5.
b. The degree of $-x^2 + x^1y^2z^3 - x^5$ is $1 + 2 + 3 = 6$, the sum of the exponents in $x^1y^2z^3$.
c. The degree of 3 is 0.
d. 0 has no degree.
e. Descending order means highest exponent on the variable to lowest. Thus, $3x^5 - 5x^2 + 9$ is in descending order.

PROBLEM 2

Find the degree of the given polynomials:

a. $-2x$
b. $4 - 7x + 9x^4$
c. $3xz^6 - 7z^4 - z^2 + 5z - 8$
d. 25
e. Write Problem 2(b) in descending order.

C › Evaluating Polynomial Functions

In mathematics, polynomials in one variable are also called **polynomial functions** and are represented by using symbols such as $P(t)$ (read "P of t"), $Q(x)$, and $D(y)$, *where the symbol in parentheses indicates the variable being used.* For example, we may have

$P(t) = -16t^2 + 10t - 15$ t is the variable.

$Q(x) = x^2 - 3x$ x is the variable.

$D(y) = -3y - 9$ y is the variable.

With this notation, it's easy to indicate the value of a polynomial for specific values of the variable, as was done in Chapter 3. Thus, $P(2)$ represents the value of the polynomial $P(t)$ when 2 is substituted for t in the polynomial. Similarly, $Q(3)$ represents the value of $Q(x)$ for $x = 3$. Thus,

$P(t) = -16t^2 + 10t - 15$

$P(2) = -16(2)^2 + 10(2) - 15$ $P(2)$ represents the value of the polynomial function $P(t)$ when $t = 2$. We also say that we are *evaluating* $P(t)$ at $t = 2$.

$ = -64 + 20 - 15 = -59$

Similarly, to evaluate $Q(x)$ at $x = 3$ in

$Q(x) = x^2 - 3x$

Answers to PROBLEMS

2. **a.** 1   **b.** 4   **c.** 7   **d.** 0
    e. $9x^4 - 7x + 4$

we find
$$Q(3) = 3^2 - 3(3) = 0$$

> **CAUTION**
> Always enclose the substituted number in parentheses.

Using these ideas, we can find the height above sea level of the diver in the *Getting Started* section. As you recall, his altitude after t seconds was given as $H = -16t^2 + 118$. This can also be written as $H(t) = -16t^2 + 118$. Thus, after 1 sec, his altitude will be

$$H(1) = -16(1)^2 + 118 = 102 \text{ ft.}$$ *Here we are evaluating $H(t)$ when $t = 1$.*

After 2 sec, it will be

$$H(2) = -16(2)^2 + 118 = -64 + 118 = 54 \text{ ft}$$ *Here we are evaluating $H(t)$ when $t = 2$.*

EXAMPLE 3   Evaluating polynomials using a formula

Find the altitude of the diver in the *Getting Started* after 3 seconds. Recall the height is given by $H(t) = -16t^2 + 118$. (Here $H = H(t)$)

SOLUTION   After 3 seconds, the formula tells us that his altitude will be $H(3) = -16(3)^2 + 118 = -26$ feet— that is, 26 feet below sea level. The water is only 12 feet deep at this point, so this is impossible. Moreover, divers cannot continue to free-fall after they hit the surface. In other words, the formula doesn't apply when $t = 3$.

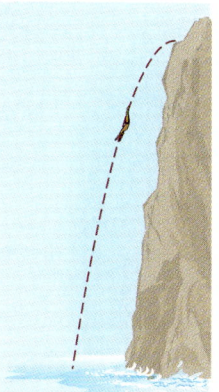

PROBLEM 3

Find the altitude of the diver in the *Getting Started* after 2.5 seconds.

EXAMPLE 4   Evaluating polynomials

Let $P(x) = x^2 - 2x + 3$ and $Q(x) = x^2 + 3x - 5$. Find:

a. $P(0)$ **b.** $Q(-1)$ **c.** $P(0) + Q(-1)$

SOLUTION

a. To find $P(0)$, we substitute 0 for x in $P(x)$.

$$P(x) = x^2 - 2x + 3$$
$$P(0) = (0)^2 - 2(0) + 3$$
$$= 0 - 0 + 3$$
$$= 3$$

Hence, $P(0) = 3$.

b. To find $Q(-1)$, we substitute -1 for x in $Q(x)$.

$$Q(x) = x^2 + 3x - 5$$
$$Q(-1) = (-1)^2 + 3(-1) - 5$$
$$= 1 - 3 - 5$$
$$= -7$$

Hence, $Q(-1) = -7$.

c. Since $P(0) = 3$ and $Q(-1) = -7$,

$$P(0) + Q(-1) = 3 + (-7) = -4$$

PROBLEM 4

If $P(x)$ and $Q(x)$ are as in Example 4, find:

a. $P(1)$
b. $Q(-2)$
c. $P(1) - Q(-2)$

Answers to PROBLEMS

3. 18 ft   **4. a.** 2   **b.** -7 **c.** 9

D › Adding and Subtracting Polynomials

The graph in Figure 5.1 illustrates the federal revenue from 1960 to 2005. Generally the graph is increasing, but it does indicate some peaks and valleys. From 2000 to 2003 the federal revenue had a decline and can be approximated by the linear model $R_1(x) = -133x + 7633$, where x represents the years after 1960 (1960 = 0, 1965 = 5, etc.).

From 2003 to 2005 the federal revenue had an increase and can be approximated by the linear model $R_2(x) = 150x - 4550$, where x represents the years after 1960. We can find the difference between these two polynomials, which we will do in Example 7.

Source: http://www.heritage.org.

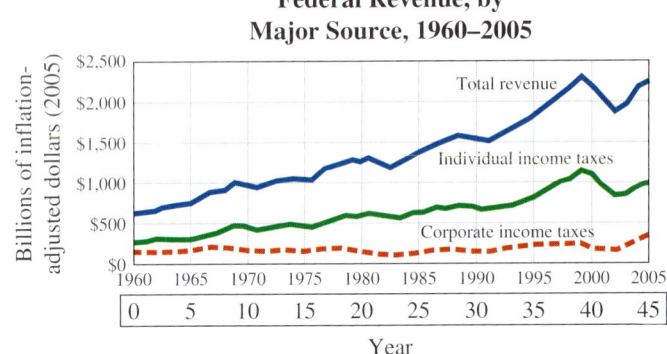

>Figure 5.1

Source: Receipts from FY 2007 Historical Tables, Budget of the United States Government, Table 2.1 and Table 2.3.

The procedure we use to add polynomials is dependent on the fact that the same properties used in the addition of numbers also apply to polynomials. We list these properties next.

PROPERTIES FOR ADDING POLYNOMIALS

If P, Q, and R are polynomials,

$P + Q = Q + P$ Commutative property of addition
$P + (Q + R) = (P + Q) + R$ Associative property of addition
$P(Q + R) = PQ + PR$
$(Q + R)P = QP + RP$ } Distributive property

Now suppose we wish to add $(5x^2 + 3x + 9) + (7x^2 + 2x + 1)$. For our solution, we use the commutative, associative, and distributive properties just mentioned. With these facts, the terms in the expression $(5x^2 + 3x + 9) + (7x^2 + 2x + 1)$ can be added as follows:

$(5x^2 + 3x + 9) + (7x^2 + 2x + 1)$ Given
$= (5x^2 + 7x^2) + (3x + 2x) + (9 + 1)$ Group like terms.
$= (5 + 7)x^2 + (3 + 2)x + (9 + 1)$ Use the distributive property.
$= 12x^2 + 5x + 10$

This addition is done more efficiently by writing the terms of the polynomials in order of descending (or ascending) degree and then placing like terms in a column:

$$\begin{array}{r} 5x^2 + 3x + 9 \\ (+)\ \underline{7x^2 + 2x + 1} \\ 12x^2 + 5x + 10 \end{array}$$

As we have already seen, $a - b = a + (-b)$. The signs in a polynomial are always taken to indicate positive or negative coefficients, and the operation involved is assumed to be addition. Thus,

$$(6x^2 - 9x - 8) + (x^2 + 3x - 1)$$
$$= [6x^2 + (-9x) + (-8)] + [x^2 + 3x + (-1)]$$
$$= (6x^2 + x^2) + (-9x + 3x) + [-8 + (-1)]$$
$$= (6 + 1)x^2 + (-9 + 3)x + [-8 + (-1)]$$
$$= 7x^2 + (-6)x + (-9)$$
$$= 7x^2 - 6x - 9$$

Using the column method, this problem can be done more efficiently by writing

$$\begin{array}{r} 6x^2 - 9x - 8 \\ (+)\ x^2 + 3x - 1 \\ \hline 7x^2 - 6x - 9 \end{array}$$

EXAMPLE 5 Adding polynomials using the column method
Add: $10x^3 + 8x^2 - 7x - 3$ and $9 - 4x + x^2 - 5x^3$

SOLUTION The first polynomial is in descending order. We've written $9 - 4x + x^2 - 5x^3$ in descending order: $-5x^3 + x^2 - 4x + 9$. We then place like terms in a column and add.

$$\begin{array}{r} 10x^3 + 8x^2 - 7x - 3 \\ (+)\ -5x^3 + x^2 - 4x + 9 \\ \hline 5x^3 + 9x^2 - 11x + 6 \end{array}$$

PROBLEM 5
Add the polynomials:

$(9z^4 - 6z^3 - z^2 + 12z - 8)$ and $(5 + z - 7z^2 + 4z^3 - 8z^4)$

The polynomial $2x^3 + 4x - 7$ is written in descending order and we note there is an x^3-term and an x-term, but there is no x^2-term. When this occurs, we say the polynomial is missing a term. In the polynomial $4x^5 - 3x^3 + 5x - 8$, there are two terms missing—the x^4- and the x^2-terms. In Example 6, we will add polynomials with missing terms.

EXAMPLE 6 Adding polynomials with missing terms
Add: $P(x) = 2x^3 - 1.5x + 6.1$ and $H(x) = -x^2 + 4x - 3.8$

SOLUTION As before, we place like terms in a column, leaving space for any missing terms, and then add as follows:

$$\begin{array}{r} 2x^3 - 1.5x + 6.1 \\ (+)\ -x^2 + 4.0x - 3.8 \\ \hline 2x^3 - x^2 + 2.5x + 2.3 \end{array}$$

PROBLEM 6
Add the polynomials:

$P(t) = -1.2t^3 - 4.5t^2 + 3.6t - 15$ and $Q(t) = 9.18t^2 + 1.4t - 7.5$

To subtract polynomials, we first recall the following.

SUBTRACTING POLYNOMIALS

$a - (b + c) = a - b - c$ By the distributive property

For example, the difference between the revenue $R(p) = 60 - 0.3p^2$ and the cost $C(p) = 4000 - 20p$ is

$$(60p - 0.3p^2) - (4000 - 20p) = 60p - 0.3p^2 - 4000 + 20p$$
$$= -0.3p^2 + 80p - 4000$$

Similarly,

$$(3x^2 + 4x - 5) - (5x^2 + 2x + 3) = 3x^2 + 4x - 5 - 5x^2 - 2x - 3$$
$$= -2x^2 + 2x - 8$$

Answers to PROBLEMS

5. $z^4 - 2z^3 - 8z^2 + 13z - 3$
6. $-1.2t^3 + 4.68t^2 + 5t - 22.5$

354 Chapter 5 Polynomials

To subtract $(5x^2 + 2x + 3)$ from $(3x^2 + 4x - 5)$, we changed the sign of each term in $(5x^2 + 2x + 3)$ and then added. This procedure can also be done in columns:

$$\begin{array}{r} 3x^2 + 4x - 5 \\ (-)\ \underline{(5x^2 + 2x + 3)} \end{array} \quad \text{is written} \quad \begin{array}{r} 3x^2 + 4x - 5 \\ (+)\ \underline{-5x^2 - 2x - 3} \\ -2x^2 + 2x - 8 \end{array}$$

> **NOTE**
> Subtracting *a from b* means to find $b - a$.

When using the column method, the polynomial we are "subtracting from" must be written on top.

EXAMPLE 7 Subtracting polynomials
Subtract: $(-133x + 7633)$ from $(150x - 4550)$

SOLUTION

$$\begin{array}{r} 150x - 4550 \\ (-)\ \underline{-133x + 7633} \end{array} \quad \text{is written} \quad \begin{array}{r} 150x - 4550 \\ (+)\ \underline{133x - 7633} \\ 283x - 12{,}183 \end{array}$$

The same result can be obtained by combining like terms, using the usual rules of signs. Thus,

$$(150x - 4550) - (-133x + 7633)$$
$$= 150x - 4550 + 133x - 7633$$
$$= 283x - 12{,}183$$

PROBLEM 7
Subtract: $(8y^3 - 6y^2 + 5y - 3)$ from $(4y^3 + 4y^2 - 3)$

E ▸ Applications Involving Polynomials

The profit P derived from selling x units of a product is related to the cost C and the revenue R, and is given by $P = R - C$. (Thus, the profit P is the revenue R minus the cost C.)

EXAMPLE 8 Solving a word problem with polynomials
A company produces x DVDs at a weekly cost $C = 2x + 1000$ (dollars). What is the company's weekly profit P if the revenue R is given by $R = 50x - 0.1x^2$ and it produces and sells 300 DVDs a week?

SOLUTION The weekly profit can be found by subtracting the cost from the revenue, so we need to find

$$P = R - C$$
$$= (50x - 0.1x^2) - (2x + 1000)$$
$$= 50x - 0.1x^2 - 2x - 1000$$
$$= -0.1x^2 + 48x - 1000$$

If the company produces and sells 300 DVDs, $x = 300$ and

$$P(300) = -0.1(300)^2 + 48(300) - 1000$$
$$= -9000 + 14{,}400 - 1000$$
$$= 4400 \text{ (dollars)}$$

The profit P when the company sells 300 DVDs is $4400.

PROBLEM 8
Find the profit if cost C and revenue R are as follows and the company produces and sells 200 items, x:

$C = 5x + 400$ (dollars) and
$R = 50x - 0.2x^2$ (dollars)

Answers to PROBLEMS

7. $-4y^3 + 10y^2 - 5y$ 8. $600

Exercises 5.1

‹A› Classify Polynomials—Monomials, Binomials, and Trinomials
‹B› Degree of a Polynomial and Writing in Descending Order

In Problems 1–10, classify the given polynomial as a monomial, binomial, or trinomial and give its degree.

1. xyz^2
2. u^2vw^3
3. $x^2 + yz^2$
4. $x^2y + z^3 - x^6$
5. $x + y^2 + z^3$
6. $xy + y^3$
7. $x^2yz - xy^3 - u^2v^3$
8. 8
9. 0
10. $3xyz - uv^2 + v^7$

‹B› Degree of a Polynomial and Writing in Descending Order   In Problems 11–16, write the polynomials in descending order.

11. $3x^2 + 5x - x^4 + 7$
12. $-6 + x - x^5 + x^3$
13. $9x - 8x^2 + 20$
14. $114x^2 + 12x^4 - 5 + x^6$
15. $2 + x - 5x^2$
16. $-11x + x^3 - 7 + x^4$

‹C› Evaluating Polynomial Functions   In Problems 17–20, evaluate the polynomial for the specified values of the variables.

17. z^3 for $z = -2$
18. $(xy)^3$ for $x = 2, y = -1$
19. $(x - 2y + z)^2$ for $x = 2, y = -1, z = -2$
20. $(x - y)(x - z)$ for $x = 2, y = -1, z = -2$

In Problems 21–30, find the specified function values.

21. If $P(x) = 4x^2 + 4x - 1$, find $P(-1)$.
22. If $P(x) = -3x^2 + 3x + 2$, find $P(0)$.
23. If $Q(y) = y^2 - 7y - 2$, find $Q(-2)$.
24. If $R(t) = t^2 - 2t + 7$, find $R(-2)$.
25. If $S(u) = -16u^2 + 120$, find $S(4)$.
26. If $V(t) = -16t^2 + 80t$, find $V(3)$.
27. $P(x) = 2x^2 + 3x$ and $Q(y) = -3y^2 - 7y + 1$
    a. Find $P(0)$.
    b. Find $Q(1)$.
    c. Find $P(0) + Q(1)$.
28. $P(x) = 3x^2 - 2x + 5$ and $Q(y) = -2y^2 + 3y - 1$
    a. Find $P(-2)$.
    b. Find $Q(0)$.
    c. Find $P(-2) + Q(0)$.
29. $P(x) = x^2 - 2x + 5$ and $Q(y) = -2y^2 + 5y - 1$
    a. Find $P(-1)$.
    b. Find $Q(1)$.
    c. Find $P(-1) - Q(1)$.
30. $P(x) = x^2 - 3x + 5$ and $Q(y) = -2y^2 + 5y - 1$
    a. Find $P(-2)$.
    b. Find $Q(2)$.
    c. Find $P(-2) - Q(2)$.

‹D› Adding and Subtracting Polynomials   In Problems 31–55, perform the indicated operations.

31. $(x^2 + 4x - 8) + (5x^2 - 4x + 3)$
32. $(3x^2 + 2x + 1) + (8x^2 - 7x + 5)$
33. $5x^2 + 3x + 4$
    $(+)\ -4x^2 - 5x - 8$
34. $-5x^2 + 4x - 3$
    $(+)\ 6x^2 + 4x - 7$
35. $(4x^2 + 7x - 5) - (3x + x^2 + 4)$
36. $(8x^2 - 6x + 3) - (4x + 2x^2 - 6)$

37. $-3y^2 + 6y - 5$
    $(-)\ \underline{8y^2 + 7y - 2}$

38. $-4y^2 + 5y - 2$
    $(-)\ \underline{5y^2 - 3y + 6}$

39. $(x^3 - 6x^2 + 4x - 2) + (3x^3 - 6x^2 + 5x - 4)$

40. $(-6x^3 - 3x + 2x^2 + 2) + (2x^3 - 6x^2 + 8x - 4)$

41. Add $(-8y^3 + 5y + 7y^2 - 5)$ and $(8y^3 + 7y - 6)$.

42. Add $(5y^3 + 3y - 8)$ and $(-9y^3 - 6y + 2y^2 + 3)$.

43. Subtract $(3v^3 + v - v^2 + 2)$ from $(6v^3 - 3v^2 + 2v - 5)$.

44. Subtract $(5v^3 + 3v^2 - 6v + 4)$ from $(3v^3 - 7v^2 + 3v - 1)$.

45. $(4u^3 - 5u^2 - u + 3) + (2u + 9u^3 + 7)$

46. $(x^3 + y^3 - 8xy + 3) + (10xy - y^3 + 2x^3 - 6)$

47. $(x^3 + y^3 - 6xy + 7) + (3x^3 - y^3 + 8xy - 8)$

48. $(2x^3 - y^3 + 3xy - 5) - (x^3 + y^3 - 3xy + 9)$

49. $(x^3 - y^3 + 5xy - 2) + (x^2 + y^3 + 5xy + 2)$

50. $(4x^2 + y^2 - 3x^2y^2) - (x^3 + 3y^2 - 3x^2y^2)$

51. $(a + a^2) + (9a - 4a^2) + (a^2 - 5a)$

52. Subtract $(x + x^2)$ from $(2x - 5x^2) + (7x - x^2)$.

53. $2y + (x + 3y) - (x + y)$

54. $8y - (y + 3x) + 7y$

55. $(3x^2 + y) - (x^2 - 3y) + (3y + x^2)$

In Problems 56–60, let $P(x) = x^2 - 2x + 3$ and $Q(x) = 2x^2 + 3x - 1$. Simplify:

56. $P(x) - Q(x)$

57. $P(0) + Q(0)$

58. $P(1) - Q(-1)$

59. $P(x) - P(x)$

60. $[P(x) + Q(x)] + P(x)$

In Problems 61–70, justify each of the equalities by using one of the three properties given on page 352 of the text.

61. $3x^2 + 9x = 9x + 3x^2$

62. $(-y^3) + 7y = 7y + (-y^3)$

63. $(8x + 9)4 = 32x + 36$

64. $(7 - 2x)(-3) = -21 + 6x$

65. $x^2 + (x + 5) = (x^2 + x) + 5$

66. $(y^3 + y^2) + 4 = y^3 + (y^2 + 4)$

67. $x^7 + (3 + x) = x^7 + (x + 3)$

68. $(y + 7) + y^3 = y^3 + (y + 7)$

69. $3(x^2 + 5) = 3x^2 + 15$

70. $8(x^3 - 5x) = 8x^3 - 40x$

⟨ E ⟩ Applications Involving Polynomials

71. *Height of an object* The height $H(t)$ (in feet) of an object t seconds (sec) after being thrown straight up with an initial velocity of 64 feet per second is given by

    $$H(t) = -16t^2 + 64t$$

    Find the height of the object after:

    a. 1 sec.
    b. 2 sec.

72. *Manufacturing* The total dollar cost $C(x)$ of manufacturing x units of a certain product each week is given by

    $$C(x) = 10x + 400$$

    Find the cost of manufacturing:

    a. 500 units.
    b. 1000 units.

73. *Revenue* The dollar revenue obtained by selling x units of a certain product each week is given by

$$R(x) = 100x - 0.03x^2$$

Find the revenue when:

a. 500 units are sold.
b. 1000 units are sold.

74. *Profit* In business, you can calculate your gross profit by subtracting the cost from the revenue. If the cost C and revenue R (in dollars) when x units are sold are given by

$$C(x) = 10x + 400 \quad \text{and} \quad R(x) = 100x - 0.03x^2$$

Find the gross profit when:

a. 500 units are sold.
b. 1000 units are sold.

75. *Depreciation* If a $100,000 computer depreciates 10% each year, its value $V(t)$ after t years is given by

$$V(t) = 100{,}000 - 0.10t(100{,}000)$$
$$= 100{,}000(1 - 0.10t)$$

Find the value of the computer after:

a. 5 years.
b. 10 years.

76. *Manufacturing* The total number N of units of output per day when the number of employees is m given by $N = 20m - \frac{1}{2}m^2$. Find how many units are produced per day in a company with

a. 10 employees.
b. 20 employees.

In Problems 77–81, $P = R - C$, where P is the profit, R is the revenue, and C is the cost.

77. *Profit* The cost C in dollars of producing x items is given by $C = 100 + 0.3x$. If the revenue is given by $R = 1.50x$, find the profit P when 100 items are produced and sold.

78. *Profit* The cost C in dollars of producing x pairs of jogging shoes is given by $C = 2000 + 60x$. If the revenue is given by $R = 90x$, find the profit when 200 pairs are produced and sold.

79. *Profit* The cost C in dollars of producing x pairs of jeans is given by $C = 1500 + 20x$. If the revenue is given by

$$R = 50x - \frac{x^2}{20}$$

find the profit when 100 pairs of jeans are produced and sold.

80. *Profit* The cost C in dollars of producing x pairs of sunglasses is given by $C = 30{,}000 + 60x$. Find the profit when 300 pairs of sunglasses are manufactured and sold if the revenue is given by

$$R = 200x - \frac{x^2}{30}$$

81. *Profit* The cost C in dollars of producing x pairs of shoes is given by $C = 100{,}000 + 30x$. Find the profit when 500 pairs of shoes are manufactured and sold if the revenue is given by

$$R = 300x - \frac{x^2}{50}$$

⟩⟩⟩ Using Your Knowledge

Sums of Areas The addition of polynomials can be used to find the sum of the areas of several rectangles. To find the total area of the rectangles, add the individual areas as shown.

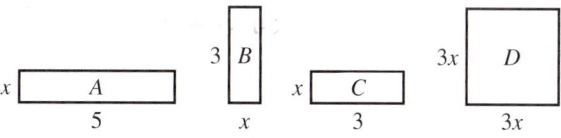

The total area is

$$\underbrace{5x}_{\text{Area of A}} + \underbrace{3x}_{\text{area of B}} + \underbrace{3x}_{\text{area of C}} + \underbrace{(3x)^2}_{\text{area of D}}$$
$$\underbrace{}_{11x} + 9x^2$$

or, in descending order, $9x^2 + 11x$.

Find the sum of the areas of the rectangles.

82.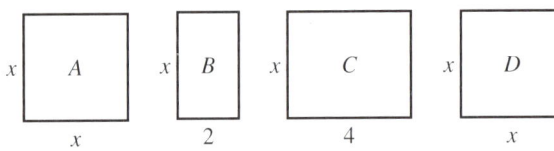

83.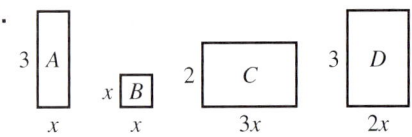

84.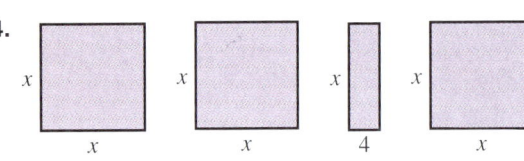

85.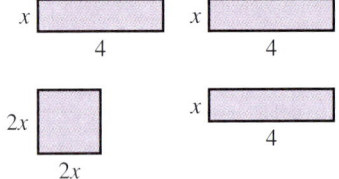

86.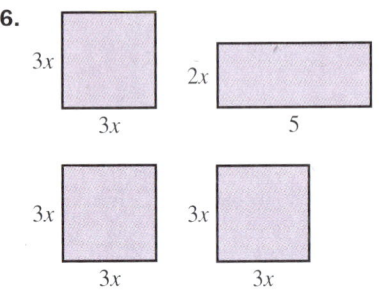

⟩⟩⟩ Write On

87. Write your own definition of a polynomial.

88. Describe the procedure you use to find the degree of a polynomial.

89. If $P(x)$ and $Q(x)$ are polynomials and you subtract $Q(x)$ from $P(x)$, what happens to the signs of all the terms of $Q(x)$?

⟩⟩⟩ Concept Checker

Fill in the blank(s) with the correct word(s), phrase, or mathematical statement.

90. A _____ is a sum or difference of monomials.

91. The polynomial $3x^2 + 4$ has two _____.

92. Three is the _____ of the polynomial $5x^3 + x - 7$.

polynomial **descending order**
monomial
binomial **coefficient**
trinomial ~~terms~~
degree

⟩⟩⟩ Mastery Test

93. A company produces x units of a product at a cost $C = 3x + 100$ (dollars).
    a. If the revenue R is given by $R = 50x - 0.2x^2$ and the profit is $P = R - C$, find P.
    b. What is the company's profit when it produces and sells 100 units?

94. The height of an object t seconds after being thrown straight up with an initial velocity of 96 feet per second is given by
$$H(t) = -16t^2 + 96t$$
What is the height of the object after 3 seconds?

95. Find the degree of $-2x^2 + 9 - 7x^4$.

96. Find the degree of $-3xy^2z + x^2 - x^3$.

97. Classify as monomials, binomials, trinomials, or polynomials if more than 3 terms:
   a. $x^2 - x$
   b. $t^5 - 4t^2 + 3t - 2t^3 + 5$
   c. $x^2y^3z^5$
   d. $\frac{2}{3}x^2 - 8 - 3x$

If $P(x) = x^2 - 3x + 2$ and $Q(x) = x^2 + 2x - 5$, find:

98. $P(2)$ and $Q(2)$

99. $P(2) - Q(2)$

100. $P(x) + Q(x)$

101. $P(x) - Q(x)$

102. Subtract $(8x^3 - 6x^2 + 5x - 3)$ from $(4x^2 + 4x + 3)$.

》》》 Skill Checker

Multiply:

103. $6x^2y \cdot 2x$

104. $6x^2y \cdot 3xy$

105. $-2x^3y \cdot (-2y^2)$

106. $-2x^3y \cdot (-7xy)$

107. $(2x)^2$

108. $3(2 + x)$

Simplify:

109. $6x - 3 + 2x - 10x + 1$

110. $6x^2 + 8x - 6x^2 + 5x - 9$

111. $5 - (3x + 4)$

112. $-4(x + 2) + 7$

5.2 Multiplication of Polynomials

▶ Objectives

A ▶ Multiply a monomial by a polynomial.

B ▶ Multiply two polynomials.

C ▶ Use the FOIL method to multiply two binomials.

D ▶ Square a binomial sum or difference.

E ▶ Find the product of the sum and the difference of two terms.

F ▶ Use the ideas discussed to solve applications.

▶ To Succeed, Review How To . . .

1. Use the distributive property to simplify expressions (pp. 55–58).
2. Use the properties of exponents (pp. 40–45).
3. Combine like terms (pp. 58–59).

▶ Getting Started

Building Bridges with Algebra

How much does a bridge beam bend (deflect) when a car or truck goes over the bridge? There's a formula that can tell us. For a certain beam of length L at a distance x from one end, the deflection is given by

$$(x - L)(x - 2L)$$

To multiply these two binomials, we first learn how to do several types of related multiplications. We show you how to multiply $(x - L)(x - 2L)$ in Example 3 of this section.

A › Multiplying Monomials by Polynomials

When multiplying polynomials such as $6x^2y$ and $(2x + 3xy)$, we use the commutative, associative, and distributive properties of multiplication for polynomials. These properties are generalizations of the properties discussed in Chapter 1. We state them here for polynomials.

MULTIPLICATION PROPERTIES FOR POLYNOMIALS

If P, Q, and R are polynomials,

$P \cdot Q = Q \cdot P$ Commutative property of multiplication

$P \cdot (Q \cdot R) = (P \cdot Q) \cdot R$ Associative property of multiplication

$\left. \begin{array}{l} P(Q + R) = PQ + PR \\ (Q + R)P = QP + RP \end{array} \right\}$ Distributive property

To multiply $6x^2y(2x + 3xy)$, we proceed as follows:

$6x^2y(2x + 3xy) = 6x^2y(2x) + 6x^2y(3xy)$ Use the distributive property.

$= 12x^3y + 18x^3y^2$ Multiply and rearrange using the commutative and associative properties.

We multiplied the coefficients but added the exponents. We can multiply $4x^2(3x^3 + 7x^2 - 4x + 8)$ in a similar way. Thus,

$4x^2(3x^3 + 7x^2 - 4x + 8) = 4x^2(3x^3) + 4x^2(7x^2) + 4x^2(-4x) + 4x^2(8)$

$= 12x^5 + 28x^4 - 16x^3 + 32x^2$

RULE TO MULTIPLY A MONOMIAL BY A POLYNOMIAL

To multiply a monomial by a polynomial, use the distributive property to multiply each term of the polynomial by the monomial. Remember to multiply coefficients and add exponents.

EXAMPLE 1   Multiplying a monomial by a polynomial

Multiply:

a. $5x^2(3x^3 + 3x^2 - 2x - 3)$ b. $-2x^3y(x^2 + 7xy - 2y^2)$

SOLUTION

a. $5x^2(3x^3 + 3x^2 - 2x - 3) = 15x^5 + 15x^4 - 10x^3 - 15x^2$

b. $-2x^3y(x^2 + 7xy - 2y^2) = -2x^5y - 14x^4y^2 + 4x^3y^3$

PROBLEM 1

Multiply:

a. $4x^3(5x^4 - 7x^3 + x^2 + 2x - 1)$

b. $-3xy^2(9x^2 - 6xy + 10y^2)$

Calculator Corner

Using Graphs to Check Polynomial Multiplication

A calculator is the perfect tool to verify products of polynomials in one variable. The procedure is as follows: Graph the original problem as Y_1 and the answer (the product) as Y_2. If the two graphs are identical, you have done the multiplication correctly. Before you do this, however, find out how exponents are entered in your calculator. With a TI-83 Plus, press the exponent key ^ and then enter the exponent. (Some calculators also have x^2 and x^3 keys.)

To verify the results in Example 1(a), use a standard window (ZOOM 6) and graph

$Y_1 = 5x^2(3x^3 + 3x^2 - 2x - 3)$
$Y_2 = 15x^5 + 15x^4 - 10x^3 - 15x^2$

Answers to PROBLEMS

1. a. $20x^7 - 28x^6 + 4x^5 + 8x^4 - 4x^3$ b. $-27x^3y^2 + 18x^2y^3 - 30xy^4$

The result is shown in Window 1. What happens when you encounter a wrong answer? For example, suppose you obtained $(x + 2)^2 = x^2 + 4$. If you let $Y_1 = (x + 2)^2$ and $Y_2 = x^2 + 4$, you get the two graphs in Window 2. Since the graphs are different, it indicates that the multiplication is incorrect.

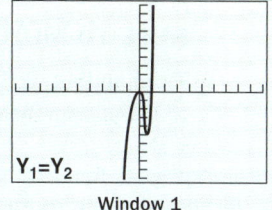
Window 1

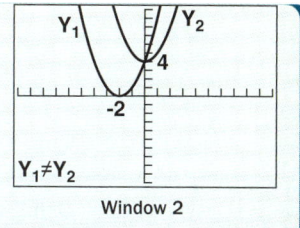
Window 2

B › Multiplying Two Polynomials

To multiply $(3x + 2)(x + 3)$, we use the distributive property

$$a(b + c) = ab + ac$$

and multiply $(3x + 2)$ by each term of $(x + 3)$ to obtain

$$\underbrace{(3x + 2)}_{a}\underbrace{(x + 3)}_{(b + c)} = \underbrace{(3x + 2)}_{a}\underbrace{(x)}_{b} + \underbrace{(3x + 2)}_{a}\underbrace{(3)}_{c}$$
$$= 3x^2 + 2x + 9x + 6$$
$$= 3x^2 + 11x + 6$$

This multiplication can also be done by arranging the work as in ordinary multiplication and placing the resulting like terms of the products in the same column. The procedure looks like this:

$$\begin{array}{r} x + 3 \\ 3x + 2 \\ \hline 2x + 6 \\ 3x^2 + 9x \\ \hline 3x^2 + 11x + 6 \end{array}$$

Multiply by 2: $2(x + 3) = 2x + 6$.
Multiply by 3x: $3x(x + 3) = 3x^2 + 9x$.
Add like terms: $2x + 9x = 11x$.

Now let's use this same technique to multiply two polynomials like $(x + 5)$ and $(x^2 + x - 2)$. We call this technique the **vertical scheme.**

PROCEDURE

Vertical Scheme for Multiplying Polynomials

Step 1

$$\begin{array}{r} x^2 + x - 2 \\ x + 5 \\ \hline 5x^2 + 5x - 10 \end{array}$$

Multiply.
$5(x^2 + x - 2)$
$= 5x^2 + 5x - 10$

Step 2

$$\begin{array}{r} x^2 + x - 2 \\ x + 5 \\ \hline 5x^2 + 5x - 10 \\ x^3 + x^2 - 2x \end{array}$$

Multiply.
$x(x^2 + x - 2)$
$= x^3 + x^2 - 2x$

Step 3

$$\begin{array}{r} x^2 + x - 2 \\ x + 5 \\ \hline 5x^2 + 5x - 10 \\ x^3 + x^2 - 2x \\ \hline x^3 + 6x^2 + 3x - 10 \end{array}$$

Add like terms.

We could have obtained the same result by using the distributive property and multiplying $(x + 5)$ by x^2, $(x + 5)$ by x, and $(x + 5)$ by -2. The result would look like this:

$$(x + 5)(x^2 + x - 2) = (x + 5)(x^2) + (x + 5)x + (x + 5)(-2)$$
$$= x^3 + 5x^2 + x^2 + 5x - 2x - 10$$
$$= x^3 + (5x^2 + x^2) + (5x - 2x) - 10$$
$$= x^3 + 6x^2 + 3x - 10$$

RULE TO MULTIPLY ANY TWO POLYNOMIALS
To multiply two polynomials, multiply each term of one polynomial by every term of the other polynomial and combine like terms.

EXAMPLE 2 Using the vertical scheme to multiply polynomials

Multiply:

a. $(x - 3)(x^2 - 2x - 4)$ b. $(x + 2)(x^2 - 2x + 4)$

SOLUTION We use the vertical scheme.

a.
$$\begin{array}{r} x^2 - 2x - 4 \\ x - 3 \\ \hline -3x^2 + 6x + 12 \\ x^3 - 2x^2 - 4x \\ \hline x^3 - 5x^2 + 2x + 12 \end{array}$$
Multiply by -3.
Multiply by x.
Add like terms.

b.
$$\begin{array}{r} x^2 - 2x + 4 \\ x + 2 \\ \hline 2x^2 - 4x + 8 \\ x^3 - 2x^2 + 4x \\ \hline x^3 + 8 \end{array}$$
Multiply by 2.
Multiply by x.
Add like terms.

PROBLEM 2

Multiply:

a. $(x + 4)(3x^2 + x - 5)$
b. $(3x - 1)(9x^2 - 6x + 2)$

In Example 2a the result of multiplying $(x - 3)(x^2 - 2x - 4)$ was $x^3 - 5x^2 + 2x + 12$. Since $(b - c)a = ba - ca$, you can also do this problem by multiplying $x(x^2 - 2x - 4)$ first and then multiplying $-3(x^2 - 2x - 4)$ to obtain the following result:

$$\overbrace{(x - 3)}^{(b - c)} \overbrace{(x^2 - 2x - 4)}^{a} = \overbrace{x}^{b}\overbrace{(x^2 - 2x - 4)}^{a} \overbrace{- 3}^{-c} \overbrace{(x^2 - 2x - 4)}^{a}$$
$$= x^3 - 2x^2 - 4x - 3x^2 + 6x + 12$$
$$= x^3 + (-2x^2 - 3x^2) + (-4x + 6x) + 12$$
$$= x^3 - 5x^2 + 2x + 12$$

C ▸ Multiplying Two Binomials

Let's use the vertical scheme to multiply $(x + 4)$ and $(x - 7)$.

$$\begin{array}{r} x + 4 \\ x - 7 \\ \hline -7x - 28 \\ x^2 + 4x \\ \hline x^2 - 7x + 4x - 28 \end{array}$$
Multiply by -7.
Multiply by x.
Add.

If you look at the four terms in the last line before adding like terms, you can see the procedure for multiplying two binomials.

The first term, x^2, is the product of the first terms.

$(x + 4)(x - 7) \quad x^2 \quad$ First terms $\quad$ **F**

The second term, $-7x$, is the product of the outside terms.

$(x + 4)(x - 7) \quad -7x \quad$ Outside terms $\quad$ **O**

The third term, $+4x$, is the product of the inside terms.

$(x + 4)(x - 7) \quad 4x \quad$ Inside terms $\quad$ **I**

The last term, -28, is the product of the last two terms.

$(x + 4)(x - 7) \quad -28 \quad$ Last terms $\quad$ **L**

Answers to PROBLEMS

2. a. $3x^3 + 13x^2 - x - 20$
   b. $27x^3 - 27x^2 + 12x - 2$

To multiply $(x + 4)(x - 7)$, we write

$$
\begin{array}{cccc}
\text{First} & \text{Outside} & \text{Inside} & \text{Last} \\
\text{F} & \text{O} & \text{I} & \text{L}
\end{array}
$$

$$(x + 4)(x - 7) = x^2 - 7x + 4x - 28$$
$$= x^2 - 3x - 28$$

Here is the procedure for multiplying two binomials using the FOIL method. This is the first of several special products.

PROCEDURE

Using FOIL to Multiply Two Binomials $(x + a)(x + b)$

To find the product of two binomials, multiply the terms in this order:

$$
\begin{array}{cccc}
\text{First} & \text{Outside} & \text{Inside} & \text{Last} \\
\text{F} & \text{O} & \text{I} & \text{L}
\end{array}
$$

$$(x + a)(x + b) = x^2 + bx + ax + ab$$
$$= x^2 + (b + a)x + ab$$

Let's do one more example, step by step, to give you more practice.

F $(x + 7)(x - 4)$ x^2

O $(x + 7)(x - 4)$ $x^2 - 4x$

I $(x + 7)(x - 4)$ $x^2 - 4x + 7x$

L $(x + 7)(x - 4) = x^2 - 4x + 7x - 28$

Thus,

$$(x + 7)(x - 4) = x^2 + 3x - 28$$

EXAMPLE 3 Using the FOIL method to multiply two binomials

Multiply:

a. $(5x + 2)(2x + 3)$ **b.** $(3x - y)(4x - 3y)$ **c.** $(x - L)(x - 2L)$

SOLUTION

$$
\begin{array}{cccc}
\text{F} & \text{O} & \text{I} & \text{L}
\end{array}
$$

a. $(5x + 2)(2x + 3) = (5x)(2x) + (5x)(3) + (2)(2x) + (2)(3)$
$$= 10x^2 + 15x + 4x + 6$$
$$= 10x^2 + 19x + 6$$

$$
\begin{array}{cccc}
\text{F} & \text{O} & \text{I} & \text{L}
\end{array}
$$

b. $(3x - y)(4x - 3y) = (3x)(4x) + (3x)(-3y) + (-y)(4x) + (-y)(-3y)$
$$= 12x^2 - 9xy - 4xy + 3y^2$$
$$= 12x^2 - 13xy + 3y^2$$

$$
\begin{array}{cccc}
\text{F} & \text{O} & \text{I} & \text{L}
\end{array}
$$

c. $(x - L)(x - 2L) = x \cdot x + (x)(-2L) + (-L)(x) + (-L)(-2L)$
$$= x^2 - 2Lx - Lx + 2L^2$$
$$= x^2 - 3Lx + 2L^2$$

PROBLEM 3

Multiply:

a. $(x + 4)(7x - 5)$
b. $(3x - y)(9x + 2y)$
c. $(B - 3A)(B - 8A)$

Answers to PROBLEMS

3. **a.** $7x^2 + 23x - 20$
   b. $27x^2 - 3xy - 2y^2$
   c. $B^2 - 11AB + 24A^2$

D › Squaring Binomial Sums or Differences

Now suppose we want to find $(x + 7)^2$. The exponent 2 means that we must multiply $(x + 7)(x + 7)$. Using FOIL, we write

$$(x + 7)(x + 7) = x^2 + 7x + 7x + 7 \cdot 7$$
$$= x^2 + 14x + 49$$

The result, $x^2 + 14x + 49$, is called a perfect square trinomial. Did you see how the middle term was calculated? It's the sum of $7x$ and $7x$, that is, $2 \cdot 7x$. Also, the last term is 7^2. In general, we have the following.

PROCEDURE
Square of a Binomial Sum $(x + a)^2$

$$\overset{\text{F} \qquad \text{O} \qquad \text{I} \qquad \text{L}}{(x + a)^2 = (x + a)(x + a) = x^2 + ax + ax + a \cdot a}$$
$$= x^2 + 2ax + a^2$$

The same pattern applies to the square of a binomial difference.

PROCEDURE
Square of a Binomial Difference $(x - a)^2$

$$\overset{\text{F} \qquad \text{O} \qquad \text{I} \qquad \text{L}}{(x - a)^2 = (x - a)(x - a) = x^2 - ax - ax + a \cdot a}$$
$$= x^2 - 2ax + a^2$$

To find the square of a binomial, add the square of the first term, twice the product of the two terms, and the square of the second term. The sign of the middle term is $+$ for binomial sums and $-$ for binomial differences. The sign of the last term is always $+$. You can use this arrow diagram to remember the steps.

$$(x + 7)^2$$
Square $\quad (2 \cdot 7 \cdot x) \quad$ Square
$$x^2 + 14x + 49$$

EXAMPLE 4 Squaring a binomial
Multiply:

a. $(2x + 3)^2$
b. $(5x - 2y)^2$
c. $-3x(2x - 3)^2$
d. $[(2x + 1) + y]^2$

SOLUTION

a. $(2x + 3)^2 = (2x)^2 + 2 \cdot 3 \cdot 2x + (3)^2$ or
$= 4x^2 + 12x + 9$

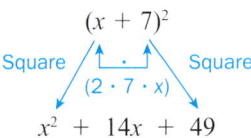

b. $(5x - 2y)^2 = (5x)^2 - 2 \cdot 2y \cdot 5x + (-2y)^2$ or
$= 25x^2 - 20xy + 4y^2$

$(5x - 2y)^2$
Square $\quad (2 \cdot -2y \cdot 5x) \quad$ Square
$25x^2 - 20xy + 4y^2$

PROBLEM 4
Multiply:

a. $(x + 4)^2$
b. $(7x - 5y)^2$
c. $-6x(2x + 5)^2$
d. $[(x - 8) + y]^2$

Answers to PROBLEMS

4. a. $x^2 + 8x + 16$ **b.** $49x^2 - 70xy + 25y^2$
c. $-24x^3 - 120x^2 - 150x$ **d.** $x^2 + 2xy + y^2 - 16x - 16y + 64$

c. $-3x(2x - 3)^2 = -3x[(2x)^2 - 2 \cdot (3)(2x) + (3)^2]$
$= -3x[4x^2 - 12x + 9]$
$= -12x^3 + 36x^2 - 27x$

d. $[(2x + 1) + y]^2 = (2x + 1)^2 + 2 \cdot y \cdot (2x + 1) + y^2$
$= (4x^2 + 4x + 1) + 4xy + 2y + y^2$
$= 4x^2 + 4xy + y^2 + 4x + 2y + 1$

E › Product of a Sum and a Difference

Suppose we multiply the sum of two terms by the difference of the same two terms; that is, suppose we want to find the product of

sum · difference
$(x + 7)(x - 7)$

Using FOIL, we get

$$\overset{F\quad O\quad I\quad L}{(x + 7)(x - 7) = x^2 - 7x + 7x - 7^2}$$
$$= x^2 + 0x - 7^2$$
$$= x^2 - 49$$

Since multiplication is commutative, $(x - 7)(x + 7) = (x + 7)(x - 7)$; thus,

$$(x - 7)(x + 7) = (x + 7)(x - 7) = x^2 - 49$$

PROCEDURE

Product of the Sum and Difference of the Same Two Monomials

$$(x - a)(x + a) = x^2 - a^2$$
$$(x + a)(x - a) = x^2 - a^2$$

The FOIL method is still operative here.

EXAMPLE 5 Finding the product of the sum and difference of the same two monomials

Multiply:

a. $(x + 10)(x - 10)$ b. $(2x + y)(2x - y)$
c. $-(3x - 5y)(3x + 5y)$ d. $[2x - (3y + 1)][2x + (3y + 1)]$

SOLUTION

a. $(x + 10)(x - 10) = x^2 - 10^2$
$= x^2 - 100$

b. $(2x + y)(2x - y) = (2x)^2 - y^2$
$= 4x^2 - y^2$

c. $-(3x - 5y)(3x + 5y) = -[(3x)^2 - (5y)^2]$
$= -9x^2 + 25y^2$ Since $-(a - b) = -a + b$
$= 25y^2 - 9x^2$ By the commutative property

d. $[2x - (3y + 1)][2x + (3y + 1)] = (2x)^2 - (3y + 1)^2$
$= 4x^2 - (9y^2 + 6y + 1)$
$= 4x^2 - 9y^2 - 6y - 1$

PROBLEM 5

Multiply:

a. $(z - 9)(z + 9)$
b. $(y + 3x)(y - 3x)$
c. $-2(4x + 3y)(4x - 3y)$
d. $[6x + (2y - 5)][6x - (2y - 5)]$

Answers to PROBLEMS

5. a. $z^2 - 81$ b. $y^2 - 9x^2$ c. $18y^2 - 32x^2$ d. $36x^2 - 4y^2 + 20y - 25$

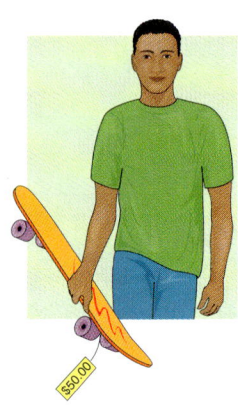

F › Applications Involving Multiplication of Polynomials

The profit P is the revenue R minus the cost C; that is, $P = R - C$. Do you know how revenue is calculated? Suppose you have 10 skateboards and sell them for $50 each. Your revenue is $R = 10 \cdot 50 = \$500$.

$$R = \underbrace{\begin{pmatrix}\text{number of}\\ \text{items sold}\end{pmatrix}}_{x} \cdot \underbrace{\begin{pmatrix}\text{price of}\\ \text{each item}\end{pmatrix}}_{p}$$

or

$$R = xp$$

EXAMPLE 6 Supply, demand, and skateboards

The research department of a company determines that the number of skateboards that are selling (the demand) is given by $x = 1000 - 10p$, where p is the price of each skateboard.

a. Write a formula for the revenue R.
b. Find the revenue obtained from selling the skateboards for $50 each.

SOLUTION

a. The revenue is $R = \begin{pmatrix}\text{number of}\\ \text{items sold}\end{pmatrix} \cdot \begin{pmatrix}\text{price of}\\ \text{each item}\end{pmatrix}$

$$R = \quad x \quad \cdot p$$
$$R = (1000 - 10p)p \quad \text{Substitute } x = 1000 - 10p.$$
$$= 1000p - 10p^2$$

b. When the skateboards are sold for $50 each, then $p = 50$,

$$R = 1000(50) - 10(50)^2$$
$$= 50{,}000 - 10(2500)$$
$$= 50{,}000 - 25{,}000$$
$$= 25{,}000$$

The revenue is $25,000.

PROBLEM 6

Using the formula for Example 6(a) find the revenue obtained from selling the skateboards for $75 each.

When setting up geometry problems involving the area of a rectangle, it might be necessary to use the skills of multiplying polynomials, as we will do in Example 7.

EXAMPLE 7 Writing a formula for the area of a rectangle

a. Write the formula for the area of a rectangle whose width is $(x + 2)$ and whose length is $(4x + 3)$.
b. Find the area if $x = 3$.

SOLUTION

a. The formula for finding the area of a rectangle is

$$A = (L)(W)$$

$W = (x + 2)$
$L = (4x + 3)$

Substitute $L = (4x + 3)$ and $W = (x + 2)$ into the formula.

$$A = (4x + 3)(x + 2)$$
$$A = 4x^2 + 11x + 6 \quad \text{Multiply the two binomials.}$$

PROBLEM 7

a. Write the formula for the area of a rectangle whose width is $(x + 1)$ and whose length is $(3x + 5)$.
b. Find the area if $x = 4$.

Answers to PROBLEMS

6. $18,750   7. a. $A = 3x^2 + 8x + 5$ b. 85 square units

b. Substitute $x = 3$ into the formula from part **a.**

$$A = 4x^2 + 11x + 6$$
$$A = 4 \cdot 3^2 + 11 \cdot 3 + 6$$
$$A = 36 + 33 + 6$$
$$A = 75 \text{ square units}$$

Calculator Corner

Equivalent Expressions

In Example 1(a), we discovered that:

$$5x^2(3x^3 + 3x^2 - 2x - 3) = 15x^5 + 15x^4 - 10x^3 - 15x^2$$

Thus, $5x^2(3x^3 + 3x^2 - 2x - 3)$ and $15x^5 + 15x^4 - 10x^3 - 15x^2$ are equivalent expressions. This means that for any replacement value of x, we obtain the same number for both $5x^2(3x^3 + 3x^2 - 2x - 3)$ and $15x^5 + 15x^4 - 10x^3 - 15x^2$. For example, when $x = 1$, the value of each expression is 5.

If we view the polynomials as functions,

$$f(x) = 5x^2(3x^3 + 3x^2 - 2x - 3)$$

and

$$g(x) = 15x^5 + 15x^4 - 10x^3 - 15x^2$$

For any given x, $f(x)$ and $g(x)$ are identical so their graphs are also identical. This can be verified by entering $f(x)$ as Y_1 and $g(x)$ as Y_2. The graphs are the same. Most calculators can numerically verify these equivalencies. To show that $f(x) = g(x)$, store 1 into X by pressing 1 STO▶ ALPHA STO▶ ENTER, and then enter $f(x)$ and ENTER. The answer will be 5 as shown in Window 1. Now, enter $g(x)$ and press ENTER. You get 5 again.

```
                    1
5X^2(3X^3+3X^2-2
X-3)
                    5
15X^5+15X^4-10X^
3-15X^2
                    5
```
Window 1

Exercises 5.2

Boost your grade at mathzone.com!
> Practice Problems
> NetTutor
> Self-Tests
> e-Professors
> Videos

⟨A⟩ Multiplying Monomials by Polynomials In Problems 1–10, perform the indicated multiplications.

1. $3x(4x - 2)$
2. $4x(x - 6)$
3. $-3x^2(x - 3)$
4. $-5x^3(x^2 - 8)$
5. $-8x(3x^2 - 2x + 1)$
6. $-4x^2(3x^2 - 5x - 1)$
7. $-3xy^2(6x^2 + 3y^2 - 7)$
8. $-2x^2y^3(6xy^3 - 2x^2y + 9)$
9. $2xy^3(3x^2y^3 - 5xy^2 + xy)$
10. $3x^4y(6x^3y^2 - 10x^2y + xy)$

⟨B⟩ Multiplying Two Polynomials In Problems 11–20, perform the indicated multiplications.

11. $(x + 3)(x^2 + x + 5)$
12. $(x + 2)(x^2 + 5x + 6)$
13. $(x + 4)(x^2 - x + 3)$
14. $(x + 5)(x^2 - x + 2)$

15. $\quad x^2 - x - 2$
    $\quad \underline{x + 3}$

16. $\quad x^2 - x - 3$
    $\quad \underline{x + 4}$

17. $(x - 2)(x^2 + 2x + 4)$
18. $(x - 3)(x^2 + x + 1)$

19. $\quad x^2 - x + 2$
    $\quad \underline{x^2 - 1}$

20. $\quad x^2 - 2x + 1$
    $\quad \underline{x^2 - 2}$

‹C› Multiplying Two Binomials   In Problems 21–38, perform the indicated multiplications.

21. $(3x + 2)(3x + 1)$
22. $(x + 5)(2x + 7)$
23. $(5x - 4)(x + 3)$
24. $(2x - 1)(x + 5)$

25. $(3a - 1)(a + 5)$
26. $(3a - 2)(a + 7)$
27. $(y + 5)(2y - 3)$
28. $(y + 1)(5y - 1)$

29. $(x - 3)(x - 5)$
30. $(x - 6)(x - 1)$
31. $(2x - 1)(3x - 2)$
32. $(3x - 5)(x - 1)$

33. $(2x - 3a)(2x + 5a)$
34. $(5x - 2a)(x + 5a)$
35. $(x + 7)(x + 8)$
36. $(x + 1)(x + 9)$

37. $(2a + b)(2a + 4b)$
38. $(3a + 2b)(3a + 5b)$

‹D› Squaring Binomial Sums or Differences
‹E› Product of a Sum and a Difference

In Problems 39–74, use the special product rules to perform the indicated multiplications.

39. $(4u + v)^2$
40. $(3u + 2v)^2$
41. $(2y + z)^2$
42. $(4y + 3z)^2$

43. $(3a - b)^2$
44. $(4a - 3b)^2$
45. $(a + b)(a - b)$
46. $(a + 4)(a - 4)$

47. $(5x - 2y)(5x + 2y)$
48. $(2x - 7y)(2x + 7y)$
49. $-(3a - b)(3a + b)$
50. $-(2a - 5b)(2a + 5b)$

51. $3x(x + 1)(x + 2)$
52. $3x(x + 2)(x + 3)$
53. $-3x(x - 1)(x - 3)$
54. $-2x(x - 5)(x - 1)$

55. $x(x + 3)^2$
56. $3x(x + 7)^2$
57. $-2x(x - 1)^2$
58. $-5x(x - 3)^2$

59. $(2x + y)(2x - y)y^2$
60. $(3x + y)(3x - y)x^2$
61. $\left(x + \frac{3}{4}\right)^2$
62. $\left(x + \frac{2}{5}\right)^2$

63. $\left(2y - \frac{1}{5}\right)^2$
64. $\left(3y - \frac{3}{4}\right)^2$
65. $\left(\frac{3}{4}p + \frac{1}{5}q\right)^2$

66. $\left(\frac{2}{5}p + \frac{1}{4}q\right)^2$
67. $[(3x + 1) + 4y]^2$
68. $[(2x + 1) + 3y]^2$

69. $[(3x - 1) - 4y]^2$
70. $[(2x - 1) - 4y]^2$
71. $[2y + (3x - 1)]^2$

72. $[3y + (2x - 1)]^2$
73. $[4p - (3q - 1)]^2$
74. $[2p - (3q - 1)]^2$

‹F› Applications Involving Multiplication of Polynomials   In Problems 75 and 76, $R = xp$, where x is the number of items sold and p is the price of the item.

75. *Revenue*   The demand x for a certain product is given by $x = 1000 - 30p$.
    a. Write a formula for the revenue R.
    b. What is the revenue when the price is $20?

76. *Revenue*   A company manufactures and sells x jogging suits at p dollars every day. If $x = 3000 - 30p$, write a formula for the daily revenue R and use it to find the revenue on a day in which the suits were selling for $40.

77. Area Write the formula for the area of a rectangle whose width is $(x + 6)$ and whose length is $(2x + 1)$. Find the area if $x = 2$.

78. Area Write the formula for the area of a rectangle whose width is $(x - 4)$ and whose length is $(5x + 2)$. Find the area if $x = 5$.

In Problems 79–80, multiply the expression given.

79. Heat transmission The heat transmission between two objects of temperature T_1 and T_2 involves the expression

$$(T_1^2 + T_2^2)(T_1^2 - T_2^2)$$

where T_1^2 means the square of T_1 and T_2^2 means the square of T_2.

80. Deflection of a beam The deflection of a certain beam involves the expression

$$w(l^2 - x^2)^2$$

Multiply this expression.

❯❯❯ Using Your Knowledge

Avoiding Multiplication Mistakes A common fallacy (mistake) when multiplying binomials is to assume that

$$(x + y)^2 = x^2 + y^2$$

Here are some arguments that should convince you otherwise.

81. Let $x = 1, y = 2$.
  a. What is $(x + y)^2$?
  b. What is $x^2 + y^2$?
  c. Does $(x + y)^2 = x^2 + y^2$?

82. Let $x = 2, y = 1$.
  a. What is $(x - y)^2$?
  b. What is $x^2 - y^2$?
  c. Does $(x - y)^2 = x^2 - y^2$?

83. Look at the large square. Its area is $(x + y)^2$. The square is divided into four smaller regions numbered 1, 2, 3, and 4.
  a. What is the area of square 1?
  b. What is the area of rectangle 2?
  c. What is the area of square 4?
  d. What is the area of rectangle 3?

84. The total area of the square is $(x + y)^2$. It's also the sum of the areas of the regions numbered 1, 2, 3, and 4. What is the sum of these four areas? (Simplify your answer.)

85. From your answer to Problem 84, what can you say about $x^2 + 2xy + y^2$ and $(x + y)^2$?

86. If $x^2 + y^2$ is the sum of the areas of squares 1 and 4, does $x^2 + y^2 = (x + y)^2$?

❯❯❯ Write On

87. Write in your own words the procedure for multiplying two monomials with integer coefficients.

88. Explain why $(x - y)^2$ is different from $x^2 - y^2$.

89. Describe in your own words the procedure for multiplying two polynomials.

❯❯❯ Concept Checker

Fill in the blank(s) with the correct word(s), phrase, or mathematical statement.

90. A perfect square trinomial such as $x^2 + 12x + 36$ is the result of _____.

91. When multiplying two binomials, use _____.

92. The difference of squares such as $x^2 - 25$ is the result of the _____.

93. When multiplying $(x + 3)(x^2 + 2x + 6)$, you can use _____.

distribute

vertical scheme

FOIL

product of sum and difference of same two monomials

squaring a binomial

>>> Mastery Test

Multiply:

94. $(5x - y)(2x - 3y)$

95. $(4x + 3y)(3x + 2y)$

96. $(3x + 5y)^2$

97. $(5x - 3y)^2$

98. $(5x - 3y)(5x + 3y)$

99. $(3x + y)(3x - y)$

100. $(2x - 3)(x^2 - 3x + 5)$

101. $(x - 2)(x^2 - 4x - 3)$

102. $-3xy^2(x^3 - 7xy - 2x^2)$

103. $4x^3(5x^3 + 3x^2 - 2x - 5)$

104. $[3x - (2y - 1)]^2$

105. $[(3x + 1) - 4y]^2$

106. The research department of a company determined that the number of skateboards that will be sold is given by $x = 1000 - 20p$, where p is the price of each skateboard.

    a. If the revenue is the number of skateboards to be sold times the price of each skateboard, write a formula for the revenue R.

    b. Find the revenue obtained by selling the skateboards for $50 each.

>>> Skill Checker

The distributive property can be written as $ab + ac = a(b + c)$. Use this rule to rewrite each expression.

107. $3x + 3y$

108. $5x + 5y$

109. $2xz + 2xy$

110. $4mn + 4mp$

111. $3ab - 3ac$

112. $6st - 6sw$

Simplify using the properties of exponents.

113. $3x^2 \cdot y + 3x^2 \cdot x^5$

114. $6x \cdot 2y + 6x \cdot 9x$

115. $\dfrac{12x^5 y}{4x^2}$

116. $\dfrac{-18x^4 y^6}{3xy^2}$

5.3 The Greatest Common Factor and Factoring by Grouping

▶ Objectives

A ▶ Factor out the greatest common factor of a polynomial.

B ▶ Factor a polynomial with four terms by grouping.

▶ To Succeed, Review How To . . .

1. Use the distributive property to multiply expressions (pp. 55–58).
2. Use the properties of exponents (pp. 40–45).

▶ Getting Started
Factoring with Interest

Fixed Rate Callable CDs

Noncallable CD Term	Possible CD Term	Current CD Rates	Theoretical APY	Minimum Deposit
0.5 Yr	6.0 Yr	4.00%	4.07%	$25,000
0.5 Yr	10.0 Yr	5.25%	5.32%	$25,000
1.0 Yr	15.0 Yr	5.50%	5.64%	$25,000

Source: Federally Insured Savings Network.

If you buy a 1.0-year, $25,000 certificate of deposit, what amount A would you get at the end of the year? Assuming the interest will be calculated once at the end of the year, you will get $25,000 plus interest. The annual interest is $25,000 · 0.055, so the amount will be $A = 25{,}000 + 25{,}000 \cdot 0.055$.

If you invest P dollars at an annual rate r, the amount A, in dollars, that you get is

$$A = P + Pr$$

The expression $P + Pr$ can be written in a simpler way if we **factor** it, that is, if we write it as the product of its factors. Factoring is the reverse of multiplication. If you multiply P by $1 + r$, you will write $P(1 + r) = P + Pr$. If you factor $P + Pr$, you should write

$$P + Pr = P(1 + r) \quad \text{Use the distributive property to factor } P + Pr.$$

Using the distributive property to factor an expression changes a *sum of terms* into a *product of factors*. Here are some examples of factoring. Check the results by multiplying the *factors* on the right-hand side of the equation to obtain the *terms* on the left-hand side.

$$5x + 15 = 5(x + 3)$$
$$6x^2 + 6xy = 6x(x + y)$$
$$-12x^3 - 42x^2y = -6x^2(2x + 7y)$$

A > The Greatest Common Factor

To factor the preceding expressions, you must know how to find the **greatest common factor (GCF)** of their terms. In arithmetic, we factor integers into a product of **primes**. (A **prime number** is a number that has exactly two factors: itself and 1.) The prime numbers are 2, 3, 5, 7, 11, 13, 17, and so on. We can write 18 and 12 as a product of primes by using successive divisions by primes:

$$\begin{array}{c|c} 2 & 18 \\ 3 & 9 \\ 3 & 3 \\ & 1 \end{array} \qquad \begin{array}{c|c} 2 & 12 \\ 2 & 6 \\ 3 & 3 \\ & 1 \end{array}$$

Thus, $18 = 2 \cdot 3 \cdot 3 = 2 \cdot 3^2$ and $12 = 2 \cdot 2 \cdot 3 = 2^2 \cdot 3$. To obtain the GCF, we write all the primes in columns as shown below, and then we select the prime with the smaller exponent from each column.

1. In the first column, we have 2 and 2^2; select 2.
2. In the second column, we have 3^2 and 3; select 3.
3. The GCF is the product of the numbers selected: $2 \cdot 3 = 6$.

$$18 = \boxed{2} \cdot \boxed{3^2}$$
$$12 = \boxed{2^2} \cdot \boxed{3}$$
$$\text{GCF} = 2 \cdot 3 = 6$$

Pick the factor with the smaller exponent in each column.

The common factors in $18 = 2 \cdot 3 \cdot 3$ and $12 = 2 \cdot 2 \cdot 3$ are 2 and 3. The GCF of 18 and 12 is $2 \cdot 3 = 6$.

Similarly, to find the greatest common factor of two terms, such as xy^2 and x^2y, we compare the factors of each term to see if they have any in common. If they do, then the GCF is the product of all the common factors with the smallest exponent.

$$xy^2 = \boxed{x} \cdot \boxed{y^2}$$
$$x^2y = \boxed{x^2} \cdot \boxed{y}$$

$$\text{GCF} = x \cdot y = xy$$

To find the GCF of x^3y^5 and x^2y^7,

$$x^3y^5 = \boxed{x^3} \cdot \boxed{y^5}$$
$$x^2y^7 = \boxed{x^2} \cdot \boxed{y^7}$$

$$\text{GCF} = x^2 \cdot y^5 = x^2y^5$$

In general, we have the following.

> **GREATEST COMMON FACTOR**
>
> The term ax^n is the **greatest common monomial factor** (GCF) of a polynomial in x (with integer coefficients) if
>
> 1. *a* is the *greatest* integer that divides each of the coefficients in the polynomial.
> 2. *n* is the *smallest* exponent of x in all the terms of the polynomial.

To factor $12x^3 + 18x^5$, we could write

$$12x^3 + 18x^5 = 2x(6x^2 + 9x^4) \qquad \text{Check by multiplying.}$$

But this is *not* completely factored. The greatest number that divides 12 and 18 is 6, and the power of x with the smaller exponent in the terms of the polynomial is x^3. Thus, the GCF is $6x^3$. The complete factorization is

$$12x^3 + 18x^5 = 6x^3(2 + 3x^2)$$

It may help your accuracy and understanding if you write an intermediate step showing the greatest common factor present in each term. When factoring $12x^3 + 18x^5$, you can write

$$12x^3 + 18x^5 = 6x^3 \cdot 2 + 6x^3 \cdot 3x^2$$
$$= 6x^3(2 + 3x^2)$$

In addition, when factoring expressions such as $-6x + 18$, even though 6 is the greatest integer that divides each coefficient, we will use -6 as the GCF because the first term is negative.

$$-6(x - 3)$$

This is the *preferred* choice because then the first term of the binomial $x - 3$ has a positive sign.

5.3 The Greatest Common Factor and Factoring by Grouping

EXAMPLE 1 Factoring binomials
Factor completely:

a. $8x - 24$ **b.** $-6y^2 + 12y$ **c.** $10x^2 - 25x^3$

SOLUTION

a. $8x - 24 = 8 \cdot x - 8 \cdot 3$ 8 is the GCF of $8x$ and 24.
$ = 8(x - 3)$

b. $-6y^2 + 12y = -6y \cdot y - 6y \cdot (-2)$ $-6y$ is the GCF of $-6y^2$ and $12y$.
$ = -6y(y - 2)$

c. $10x^2 - 25x^3 = 5x^2 \cdot 2 - 5x^2 \cdot 5x$ $5x^2$ is the GCF of $10x^2$ and $25x^3$.
$ = 5x^2(2 - 5x)$

Check each answer by multiplying the expression to the right of the equals sign.

PROBLEM 1
Factor completely:
a. $6x + 48$
b. $-3y^2 + 21y$
c. $4x^2 - 32x^3$

You can also factor polynomials with more than two terms, as shown in Example 2.

EXAMPLE 2 Factoring polynomials
Factor completely:

a. $6x^3 + 12x^4 + 18x^2$ **b.** $10x^6 - 15x^5 + 20x^7 + 30x^2$

SOLUTION

a. $6x^3 + 12x^4 + 18x^2 = 6x^2 \cdot x + 6x^2 \cdot 2x^2 + 6x^2 \cdot 3$
$ = 6x^2(x + 2x^2 + 3) = 6x^2(2x^2 + x + 3)$

b. $10x^6 - 15x^5 + 20x^7 + 30x^2 = 5x^2 \cdot 2x^4 - 5x^2 \cdot 3x^3 + 5x^2 \cdot 4x^5 + 5x^2 \cdot 6$
$ = 5x^2(2x^4 - 3x^3 + 4x^5 + 6)$
$ = 5x^2(4x^5 + 2x^4 - 3x^3 + 6)$

Check by multiplying.

PROBLEM 2
Factor completely:
a. $7x^3 + 14x^4 - 49x^2$
b. $3x^6 - 6x^5 + 12x^7 + 27x^2$

Calculator Corner

Adjusting the Window to Check Factoring Polynomials
To check the factorization in Example 2(b), graph

$$Y_1 = 10x^6 - 15x^5 + 20x^7 + 30x^2$$

and

$$Y_2 = 5x^2(4x^5 + 2x^4 - 3x^3 + 6)$$

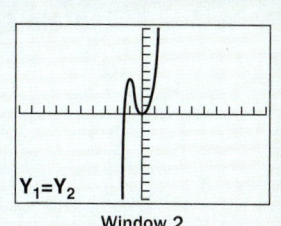

Window 1   Window 2

Window 1 shows the same graph for Y_1 and Y_2.
But how do we know that portions of the graph that may not be shown are the same? What we need is a more complete graph. To do this, adjust the vertical scale by selecting Ymin = -100, Ymax = 100, and Yscl = 10. A more complete graph is shown in Window 2 and Y_1 and Y_2 appear to be identical, so the factorization is correct.

Answers to PROBLEMS
1. **a.** $6(x + 8)$ **b.** $-3y(y - 7)$
   c. $4x^2(1 - 8x)$
2. **a.** $7x^2(2x^2 + x - 7)$
   b. $3x^2(4x^5 + x^4 - 2x^3 + 9)$

EXAMPLE 3 Factoring polynomials with fractional coefficients

Factor completely:

$$\frac{3}{4}x^2 - \frac{1}{4}x^4 + \frac{5}{4}x^5$$

SOLUTION The procedure for finding the GCF only works with polynomials with integer coefficients. However, since all the coefficients have a denominator of 4, the greatest common factor is $\frac{1}{4}x^2$.

$$\frac{3}{4}x^2 - \frac{1}{4}x^4 + \frac{5}{4}x^5 = \frac{1}{4}x^2 \cdot 3 - \frac{1}{4}x^2 \cdot x^2 + \frac{1}{4}x^2 \cdot 5x^3$$

$$= \frac{1}{4}x^2(3 - x^2 + 5x^3)$$

Check by multiplying.

PROBLEM 3

Factor completely:

$$\frac{2}{5}x^2 - \frac{3}{5}x^4 + \frac{4}{5}x^5$$

So far we have factored polynomials whose terms have a greatest common monomial factor. However, sometimes the polynomial will have a common binomial factor, as we illustrate in Example 4.

EXAMPLE 4 Factoring polynomials with a common binomial factor

Factor:

a. $3x^2(a - b) + 5(a - b)$ **b.** $2x(x^3 + 4y) - 7(x^3 + 4y)$

SOLUTION

a. Step one: Determine the common binomial factor: $(a - b)$.

$$3x^2(a - b) + 5(a - b) = 3x^2(a - b) + 5(a - b)$$

Step two: Factor $(a - b)$ out of each term.

$$= (a - b)(3x^2 + 5)$$

b. Step one: Determine the common binomial factor: $(x^3 + 4y)$.

$$2x(x^3 + 4y) - 7(x^3 + 4y) = 2x(x^3 + 4y) - 7(x^3 + 4y)$$

Step two: Factor $(x^3 + 4y)$ out of each term.

$$= (x^3 + 4y)(2x - 7)$$

PROBLEM 4

Factor:

a. $x^2(y + z) - 7(y + z)$

b. $4x^2(6x^3 + 1) + 5y(6x^3 + 1)$

B > Factoring by Grouping

How do we factor expressions with four terms when there seems to be no common factor except 1? One way is to group and factor the first two terms and also the last two terms and then look for a common binomial factor. Here is the three-step procedure.

> **PROCEDURE**
>
> **Factoring by Grouping (Four Terms)**
> 1. Group terms with common factors using the commutative and associative properties.
> 2. Factor each resulting group.
> 3. Factor out the common binomial (GCF), it one exists, by the distributive property.

Answers to PROBLEMS

3. $\frac{1}{5}x^2(2 - 3x^2 + 4x^3)$

4. **a.** $(y + z)(x^2 - 7)$
   b. $(6x^3 + 1)(4x^2 + 5y)$

Let's factor $x^3 + 2x^2 + 3x + 6$. As you can see, this expression doesn't appear to have a common factor other than 1. We use our three-step procedure:

1. Group terms. $\quad x^3 + 2x^2 + 3x + 6 = (x^3 + 2x^2) + (3x + 6)$
2. Factor each group. $\quad = x^2(x + 2) + 3(x + 2)$
3. Factor out the (binomial) GCF. $\quad = (x + 2)(x^2 + 3)$

Thus, $x^3 + 2x^2 + 3x + 6 = (x + 2)(x^2 + 3)$. By the commutative property of multiplication, either $(x + 2)(x^2 + 3)$ or $(x^2 + 3)(x + 2)$ is a correct factorization for $x^3 + 2x^2 + 3x + 6$. Check by multiplying.

EXAMPLE 5 Factoring by grouping
Factor completely:

a. $3x^3 + 6x^2 + 4x + 8$
b. $6x^3 - 3x^2 - 4x + 2$

SOLUTION

a. Check to be sure that all terms do not have a GCF other than 1. Then proceed by steps.
1. Group terms. $\quad 3x^3 + 6x^2 + 4x + 8 = (3x^3 + 6x^2) + (4x + 8)$
2. Factor each group. $\quad = 3x^2(x + 2) + 4(x + 2)$
3. Factor out the (binomial) GCF. $\quad = (x + 2)(3x^2 + 4)$

You can write $3x^3 + 6x^2 + 4x + 8$ as $(3x^2 + 4)(x + 2)$ in Step 3. Since $(x + 2)(3x^2 + 4) = (3x^2 + 4)(x + 2)$, both answers are correct! Do you know why?

b. Check to be sure the terms do not have a GCF other than 1. Then proceed by steps.
1. Group terms. $\quad 6x^3 - 3x^2 - 4x + 2 = (6x^3 - 3x^2) + (-4x + 2)$
2. Factor each group. $\quad = 3x^2(2x - 1) + (-2)(2x - 1)$
3. Factor out the (binomial) GCF. $\quad = (2x - 1)(3x^2 - 2)$

Thus, $6x^3 - 3x^2 - 4x + 2 = (2x - 1)(3x^2 - 2)$. We leave the verification to you.

You might save some time if in Step 1 you realize that $-4x + 2 = -(4x - 2)$ and write

$$6x^3 - 3x^2 - 4x + 2 = (6x^3 - 3x^2) - (4x - 2)$$
$$= 3x^2(2x - 1) - 2(2x - 1)$$
$$= (2x - 1)(3x^2 - 2)$$

If the third term is preceded by a minus sign, write $-(\ )$ with the appropriate signs for the expression inside the parentheses.

PROBLEM 5
Factor completely:

a. $2x^3 - 2x^2 + 3x - 3$
b. $6x^3 - 9x^2 - 2x + 3$

Calculator Corner

Adjusting the Window to Check Factoring Polynomials

To verify the results of Example 5(a), we graph

$$Y_1 = 3x^3 + 6x^2 + 4x + 8$$

and

$$Y_2 = (x + 2)(3x^2 + 4)$$

Using a standard window (ZOOM 6), we get part of the graph as shown in Window 1.

But we need a more complete graph, so in WINDOW we change the vertical (Y) scale to $[-100, 100]$ with Yscl = 10. Window 2 shows that both graphs are identical and our factorization is probably correct.

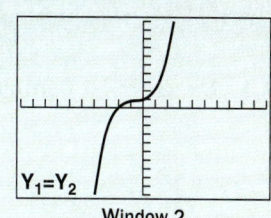

Window 1   Window 2

Answers to PROBLEMS

5. a. $(x - 1)(2x^2 + 3)$
   b. $(2x - 3)(3x^2 - 1)$

Example 6 will require two types of factoring. This will be done in two steps where the first step should be to factor out the GCF and the second step to factor by grouping. The polynomial would not be completely factored if you just did one or the other of these steps. For this reason, we shall continue to use the term "factor completely" when giving factoring directions.

EXAMPLE 6 Factoring out the common factor and factoring by grouping
Factor completely:

a. $2x^4 - 4x^3 - x^2 + 2x$

b. $6x^6 - 9x^4 + 4x^3 - 6x$

PROBLEM 6
Factor completely:

a. $3x^4 - 6x^3 - x^2 + 2x$

b. $6x^6 - 9x^4 + 2x^3 - 3x$

SOLUTION
a. First, factor out the common factor x and write

$$2x^4 - 4x^3 - x^2 + 2x = x(2x^3 - 4x^2 - x + 2)$$

Then factor $2x^3 - 4x^2 - x + 2$.

Use the three-step procedure for grouping. Can you tell what is happening without the words? If not, check the procedure box on page 374.

1. $2x^4 - 4x^3 - x^2 + 2x = x[(2x^3 - 4x^2) - (x - 2)]$
2. $\qquad\qquad\qquad\qquad = x[2x^2(x - 2) - 1(x - 2)]$
3. $\qquad\qquad\qquad\qquad = x[(x - 2)(2x^2 - 1)]$

Thus, $2x^4 - 4x^3 - x^2 + 2x = x(x - 2)(2x^2 - 1)$.

b. Factor out x to obtain

$$6x^6 - 9x^4 + 4x^3 - 6x = x(6x^5 - 9x^3 + 4x^2 - 6)$$

Then factor $6x^5 - 9x^3 + 4x^2 - 6$.

Now use the procedure for grouping.

1. $6x^6 - 9x^4 + 4x^3 - 6x = x[(6x^5 - 9x^3) + (4x^2 - 6)]$
2. $\qquad\qquad\qquad\qquad = x[3x^3(2x^2 - 3) + 2(2x^2 - 3)]$
3. $\qquad\qquad\qquad\qquad = x(2x^2 - 3)(3x^3 + 2)$

Thus,

$$6x^6 - 9x^4 + 4x^3 - 6x = x(2x^2 - 3)(3x^3 + 2)$$

Don't forget the x!

We leave the verification to you.

> Exercises 5.3

Boost your grade at mathzone.com!
> Practice Problems
> NetTutor
> Self-Tests
> e-Professors
> Videos

< A > The Greatest Common Factor In Problems 1–26, factor completely.

1. $8x + 16$
2. $15x + 45$
3. $-5y + 25$
4. $-4y + 28$
5. $-8x - 24$
6. $-6x - 36$
7. $4x^2 + 36x$
8. $5x^3 + 20x$
9. $-5x^2 - 35x^4$
10. $-8x^3 - 16x^6$
11. $3x^3 + 6x^2 + 39x$
12. $8x^3 + 4x^2 - 36x$
13. $63y^3 - 18y^2 + 27y$
14. $10y^3 - 5y^2 + 20y$
15. $36x^6 + 12x^5 - 18x^4 + 30x^2$

Answers to PROBLEMS

6. **a.** $x(x - 2)(3x^2 - 1)$
   b. $x(3x^3 + 1)(2x^2 - 3)$

16. $15x^7 - 15x^6 + 30x^3 - 20x^2$

17. $48y^8 + 16y^5 - 24y^4 + 16y^3$

18. $12y^9 - 4y^6 + 6y^5 + 8y^4$

19. $\frac{4}{7}x^3 + \frac{3}{7}x^2 - \frac{9}{7}x + \frac{3}{7}$

20. $\frac{2}{5}x^3 + \frac{3}{5}x^2 - \frac{2}{5}x + \frac{4}{5}$

21. $\frac{7}{8}y^9 + \frac{3}{8}y^6 - \frac{5}{8}y^4 + \frac{5}{8}y^2$

22. $\frac{4}{3}y^7 - \frac{1}{3}y^5 + \frac{2}{3}y^4 - \frac{7}{3}y^3$

23. $x^2(x + 5) + 4(x + 5)$

24. $2x^3(y - 7) - 3(y - 7)$

25. $z(4x + 9) - 1(4x + 9)$

26. $6x^2(3x - 1) + 7(3x - 1)$

⟨ B ⟩ Factoring by Grouping In Problems 27–52, factor completely.

27. $x^3 + 2x^2 + 3x + 6$

28. $x^3 + 3x^2 + 4x + 12$

29. $y^3 - 3y^2 + y - 3$

30. $y^3 - 5y^2 + y - 5$

31. $4x^3 + 6x^2 + 10x + 15$

32. $6x^3 + 3x^2 + 8x + 4$

33. $6x^3 - 2x^2 + 3x - 1$

34. $6x^3 - 9x^2 + 2x - 3$

35. $4y^3 + 8y^2 + 3y + 6$

36. $2y^3 + 6y^2 + 7y + 21$

37. $2a^6 + 3a^4 + 2a^2 + 3$

38. $3a^6 + 2a^4 + 3a^2 + 2$

39. $3x^5 + 12x^3 + 5x^2 + 20$

40. $2x^5 + 2x^3 + 3x^2 + 3$

41. $6y^5 + 9y^3 + 2y^2 + 3$

42. $12y^5 + 8y^3 + 3y^2 + 2$

43. $4y^7 + 12y^5 + y^4 + 3y^2$

44. $2y^7 + 2y^5 + y^4 + y^2$

45. $3a^7 - 6a^5 - 2a^4 + 4a^2$

46. $4a^7 - 12a^5 - 3a^4 + 9a^2$

47. $8a^5 - 12a^4 - 10a^3 + 15a^2$

48. $3y^7 - 21y^4 + y^5 - 7y^2$

49. $x^6 - 2x^5 + 2x^4 - 4x^3$

50. $v^6 + 3v^5 - 7v^4 - 21v^3$

51. $(x - 4)(x + 2) + (x - 4)(x + 3)$

52. $(y + 2)(y - 7) + (2y + 3)(y - 7)$

53. Factor completely $\alpha L t_2 - \alpha L t_1$, where α is the coefficient of linear expansion, L is the length of the material, and t_2 and t_1 are the temperatures in degrees Celsius.

54. Factor the expression $-kx - kl$ completely, which represents the restoring force of a spring stretched an amount l from its equilibrium position and then an additional x units.

55. When solving for the equivalent resistance of two circuits, we have to factor the expression $R^2 - R - R + 1$. Factor this expression completely by grouping.

56. The bending moment of a cantilever beam of length L, at x inches from its support, involves the expression $L^2 - Lx - Lx + x^2$. Factor this expression completely by grouping.

⟩⟩⟩ Using Your Knowledge

General Formulas There are many formulas that can be simplified by factoring completely. Here are a few.

57. The vertical shear at any section of a cantilever beam of uniform cross section is

$$-wl + wz$$

Factor this expression completely.

58. The bending moment of any section of a cantilever beam of uniform cross section is

$$-Pl + Px$$

Factor this expression completely.

59. The surface area of a square pyramid is

$$a^2 + 2as$$

Factor this expression completely.

60. The energy of a moving object is given by

$$800m - mv^2$$

Factor this expression completely.

61. The height (in feet after t seconds) of a rock thrown from the roof of a certain building is given by

$$-16t^2 + 80t + 240$$

Factor this expression completely. (*Hint:* -16 is a common factor.)

〉〉〉 Write On

62. What should the first step be in factoring a polynomial?

63. Write your own definition for the greatest common factor of a list of integers.

64. The procedure we outlined to find the greatest common monomial factor of a polynomial doesn't apply to Example 3. Explain why.

65. Describe the relationship between multiplying and factoring a polynomial.

〉〉〉 Concept Checker

Fill in the blank(s) with the correct word(s), phrase, or mathematical statement.

66. _____ are numbers that have exactly two factors: itself and 1.

67. If asked to factor a four-term polynomial and there is no greatest common factor then try to _____.

68. $3x$ is called the _____ of $6xy$ and $9xz$.

factor

greatest common factor (GCF)

primes

factor out the GCF

factor by grouping

composites

〉〉〉 Mastery Test

Factor completely:

69. $3x^4 - 6x^3 - x^2 + 2x$

70. $6x^6 - 9x^4 + 2x^3 - 3x$

71. $2x^3 + 2x^2 + 3x + 3$

72. $6x^3 - 9x^2 - 2x + 3$

73. $7x^3 + 14x^4 - 49x^2$

74. $3x^6 - 6x^5 + 12x^7 + 27x^2$

75. $4x^2 - 32x^3$

76. $6x - 48$

77. $-3y^2 + 21y$

78. $\frac{2}{5}x^2 - \frac{4}{5}x^4 - \frac{1}{5}x^5$

〉〉〉 Skill Checker

Multiply:

79. $(x + 3)(x + 4)$

80. $(x + 7)(x + 2)$

81. $(x + 5)(x - 2)$

82. $(x - 5)(x + 3)$

83. $(5x + 2y)^2$

84. $(3x + 4y)^2$

85. $(5x - 2y)^2$

86. $(3x - 4y)^2$

87. $(u + 6)(u - 6)$

88. $(2a + 7b)(2a - 7b)$

5.4 Factoring Trinomials

Objectives

A Factor a trinomial of the form $x^2 + bx + c$ (b and c are integers).

B Factor a trinomial of the form $ax^2 + bx + c$ using trial and error.

C Factor a trinomial of the form $ax^2 + bx + c$ using the *ac* test.

To Succeed, Review How To . . .

1. Multiply two binomials using the FOIL method (p. 363).
2. Add and multiply signed numbers (pp. 18–27).

Getting Started
Applying Factoring to Fire Fighting

How many hundred gallons of water per minute can this fire truck pump? If it has a hose 100 feet long, you can find the answer by solving $2g^2 + g - 36 = 0$. The expression $2g^2 + g - 36$ is factorable. How do we factor it? By using reverse multiplication, but let's start with a simpler problem.

A Factoring Trinomials of the Form $x^2 + bx + c$

In Section 5.2, we multiplied two binomials using the FOIL method:

$$\overset{\text{F}\quad\text{O}\quad\text{I}\quad\text{L}}{(x + 4)(x - 7) = x^2 - 7x + 4x - 28 = x^2 - 3x - 28}$$

and

$$(x + 7)(x + 4) = x^2 + 4x + 7x + 28 = x^2 + 11x + 28$$

Both of the binomial products result in a trinomial. This means if we are given a trinomial of the form $x^2 + bx + c$ and asked to factor it, we would anticipate the strong possibility that the factors should be two binomials.

To factor $x^2 + 11x + 28$, we recall that

$$(x + a)(x + b) = x^2 + bx + ax + ab$$
$$= x^2 + (b + a)x + ab$$

Rewriting this trinomial, we have the following factoring form.

> **PROCEDURE**
> **Factoring Trinomials of the Form $x^2 + (b + a)x + ab$**
> $$x^2 + (b + a)x + ab = (x + a)(x + b)$$

How to Determine the Signs When Factoring Trinomials

If ab is $+$ then both have same sign	If ab is $-$ then one is $+$ and one is $-$
If middle term is $+$ then both are $+$ $$(x + a)(x + b)$$	If middle term is $-$ then the number with larger absolute value is $-$ $$(x - a)(x + b)$$
If middle term is $-$ then both are $-$ $$(x - a)(x - b)$$	If middle term is $+$ then the number with larger absolute value is $+$ $$(x + a)(x - b)$$

The *leading coefficient* of a polynomial in one variable is the coefficient of the term with the highest exponent on the variable. When factoring a trinomial with a leading coefficient of 1, we need two integers a and b whose product is the last term and whose sum is the coefficient of the middle term. Since

$$x^2 + (b + a)x + ab = (x + a)(x + b)$$

we write

$$x^2 + 11x + 28 = (x + a)(x + b)$$

We need two integers a and b whose product ab is 28 and whose sum is 11. The product is positive, so both numbers must have the same sign. Since the middle term of the trinomial is positive, then both numbers are positive. The numbers are 7 and 4. [Check: $7 \cdot 4 = 28$ and $7 + 4 = 11$, the coefficient of the middle term.] Thus,

$$x^2 + 11x + 28 = (x + 7)(x + 4)$$

Since the multiplication of polynomials is commutative,

$$x^2 + 11x + 28 = (x + 4)(x + 7)$$

Now suppose we want to factor $x^2 - 8x + 12$. This time, we need two integers whose product is 12 and whose sum is -8 (the coefficient of the middle term). The numbers are -6 and -2. [Check: $(-6)(-2) = 12$ and $(-6) + (-2) = -8$.] Thus,

$$x^2 - 8x + 12 = (x - 6)(x - 2)$$

You can check this by multiplying $(x - 6)$ by $(x - 2)$.

EXAMPLE 1 Factoring trinomials of the form $x^2 + bx + c$

Factor completely:

a. $x^2 + 7x + 12$ **b.** $x^2 - 6x + 8$ **c.** $x^2 - 4 - 3x$

SOLUTION

a. To factor $x^2 + 7x + 12$, we need two integers whose product is 12 and whose sum is 7. The numbers are 3 and 4. Thus,

$$x^2 + 7x + 12 = (x + 3)(x + 4)$$

b. To factor $x^2 - 6x + 8$, we need two integers whose product is 8 and whose sum is -6. The product is positive, so both numbers must be negative. They are -2 and -4. [Check: $(-2)(-4) = 8$ and $(-2) + (-4) = -6$.] Thus,

$$x^2 - 6x + 8 = (x - 2)(x - 4)$$

c. First, rewrite $x^2 - 4 - 3x$ in descending order as $x^2 - 3x - 4$. This time, we need integers with product -4 and sum -3. The numbers are 1 and -4. Thus,

$$x^2 - 4 - 3x = x^2 - 3x - 4 = (x + 1)(x - 4)$$

You can check all these results by multiplying. For example,

$$(x + 1)(x - 4) = x^2 - 4x + x - 4 = x^2 - 3x - 4.$$

PROBLEM 1
Factor completely:

a. $x^2 + 7x + 10$

b. $x^2 - 3x - 10$

c. $x^2 - 6 - 5x$

Answers to PROBLEMS

1. a. $(x + 2)(x + 5)$ **b.** $(x + 2)(x - 5)$ **c.** $(x + 1)(x - 6)$

We used the example of the product of the factors $(x + 4)(x + 7)$ or $x^2 + 11x + 28$ to illustrate the form of a trinomial referred to as $x^2 + bx + c$. We can build on that form by looking at the product of the factors $(x + 4y)(x + 7y)$, which is $x^2 + 11xy + 28y^2$. The factoring of this trinomial form follows.

> **PROCEDURE**
>
> **Factoring Trinomials of the Form $x^2 + (b + a)xy + cy^2$**
>
> $$x^2 + (b + a)xy + cy^2 = (x + ay)(x + by)$$

Example 2 will illustrate how to factor trinomials of the form $x^2 + (b + a)xy + cy^2$.

EXAMPLE 2 Factoring trinomials with two variables of the form $x^2 + bxy + cy^2$

Factor completely:

a. $x^2 - 8xy + 7y^2$ b. $x^2 + 3xy + 7y^2$

SOLUTION

a. We need two integers whose product is 7 and whose sum is -8. The product is positive, so both numbers must be negative. The numbers are -1 and -7. Thus,

$x^2 - 8xy + 7y^2 = (x - 1y)(x - 7y)$ (The y variable is in the 2nd term of each binomial.)

$= (x - y)(x - 7y)$ Check by multiplication.

b. We need two integers whose product is 7 and whose sum is 3. There are no such numbers. The polynomial $x^2 + 3xy + 7y^2$ is not factorable using integer coefficients.

A polynomial not factorable using integer coefficients is called a **prime polynomial**.

PROBLEM 2

Factor completely:

a. $x^2 - 2xy + 5y^2$

b. $x^2 - 6xy - 16y^2$

B › Factoring Trinomials of the Form $ax^2 + bx + c$

One way to factor polynomials when the coefficient of the squared term is greater than 1 is to use trial and error. To factor the polynomial $2g^2 + g - 36$ mentioned in the *Getting Started* at the beginning of this section, we can rely on our experience with FOIL. To obtain $2g^2 + 1g - 36$, multiply

$$(2g + \underline{})(g - \underline{})$$

Fill the blanks with two integers that have a product of -36 and check whether the middle term is $1g$. The possibilities are as follows.

Trial Factors	Middle Term
$(2g + \underline{1})(g - \underline{36})$	$-71g$
$(2g + \underline{2})(g - \underline{18})$	$-34g$
$(2g + 2) = 2(g + 1)$, so we won't use even numbers in both terms of a factor.	
$(2g + \underline{3})(g - \underline{12})$	$-21g$
$(2g + \underline{4})(g - \underline{9})$	$-1g$

Thus, $2g^2 + g - 36 = (2g + 9)(g - 4)$.

Answers to PROBLEMS

2. a. Prime   b. $(x + 2y)(x - 8y)$

EXAMPLE 3 Factoring trinomials of the form $ax^2 + bx + c$

Factor completely: $6x^2 + 17x + 12$

SOLUTION The factors of 6 are 6 and 1 or 3 and 2. The possible combinations are

$(6x + \underline{})(x + \underline{})$ or $(3x + \underline{})(2x + \underline{})$

Trial Factors	
$(6x + 12)(x + 1)$	$(6x + 1)(x + 12)$
$(6x + 6)(x + 2)$	$(6x + 2)(x + 6)$
$(6x + 4)(x + 3)$	$(6x + 3)(x + 4)$
$(3x + 12)(2x + 1)$	$(3x + 1)(2x + 12)$
$(3x + 6)(2x + 2)$	$(3x + 2)(2x + 6)$
$(3x + 4)(2x + 3)$	$(3x + 3)(2x + 4)$

The only combination yielding the correct middle term is

$(3x + 4)(2x + 3) = 6x^2 + 9x + 8x + 12 = 6x^2 + 17x + 12$

PROBLEM 3

Factor completely:

$6x^2 + 13x + 6$

C > The ac Test

The method used in Example 3 is not efficient when the coefficients are large. Moreover, we don't even know whether the polynomial is factorable. We remedy this situation with the following test.

> **THE ac TEST**
>
> The polynomial $ax^2 + bx + c$ is factorable only if there are two integers whose product is ac and whose sum is b.

To find if $3x^2 + 2x + 5$ is factorable, we first look at $3 \cdot 5 = 15$. If we can find two integers whose product is 15 and whose sum is 2, we can factor the polynomial. No such integers exist, so $3x^2 + 2x + 5$ is *prime*. On the other hand, the polynomial $5x^2 + 11x + 2$ is factorable because we can find two integers (10 and 1) whose product, ac, is $5 \cdot 2 = 10$ and whose sum is 11. The number ac plays such an important part in the procedure used to factor $ax^2 + bx + c$ that we call it the **key number.**

We now give you a method that uses the key number to factor such trinomials. For example, suppose we want to factor $2x^2 - 7x - 4$. We proceed as follows.

1. Find the key number, ac, $2 \cdot (-4) = -8$.
   $2x^2 - 7x - 4 \quad \boxed{-8}$
2. Find the factors of the key number and use the appropriate ones to rewrite the middle term as a sum that equals $-7x$.
   $2x^2 - 8x + 1x - 4 \quad -8, 1$
3. Group the terms (as in Section 5.3).
   $(2x^2 - 8x) + (1x - 4)$
4. Factor each group.
   $2x(x - 4) + 1(x - 4)$
5. The binomial $(x - 4)$ is the GCF.
   $(x - 4)(2x + 1)$

Thus, $2x^2 - 7x - 4 = (x - 4)(2x + 1)$. You can check that this is the correct factorization by multiplying $(x - 4)$ by $(2x + 1)$.

Answers to PROBLEMS

3. $(3x + 2)(2x + 3)$

Suppose you want to factor the trinomial $5x^2 + 7x + 2$. Here is one way of doing it.

1. Find the key number ($5 \cdot 2 = 10$). $5x^2 + \underline{7x} + 2$ ⑩
2. Find the factors of the key number that sum to 7 and use them to rewrite the middle term. $5x^2 + \underline{5x + 2x} + 2$ 5, 2
3. Group the terms. $(5x^2 + 5x) + (2x + 2)$
4. Factor each group. $5x(x + 1) + 2(x + 1)$

5. $(x + 1)$ is the GCF. $(x + 1)(5x + 2)$

Thus, $5x^2 + 7x + 2 = (x + 1)(5x + 2)$.

Here is another way to proceed.

1. Find the key number. $5x^2 + \underline{7x} + 2$ ⑩
2. Find the factors of the key number that sum to 7 and use them to rewrite the middle term. $5x^2 + \underline{2x + 5x} + 2$ 2, 5
3. Group the terms. $(5x^2 + 2x) + (5x + 2)$
4. Factor each group. $x(5x + 2) + 1(5x + 2)$

5. $(5x + 2)$ is the GCF. $(5x + 2)(x + 1)$

In this case, we found that

$$5x^2 + 7x + 2 = (5x + 2)(x + 1)$$

Which is the correct factorization, $(5x + 2)(x + 1)$ or $(x + 1)(5x + 2)$? Both are correct. The multiplication of polynomials is commutative, so the order in which the product is written makes no difference. You can write the factorization of $ax^2 + bx + c$ in *two* ways.

EXAMPLE 4   Factoring trinomials using the key number

Factor completely:

a. $6x^2 - 3x + 4$ **b.** $4x^2 - 3 - 4x$

SOLUTION

a. We proceed by steps:
1. Find the key number ($6 \cdot 4 = 24$). $6x^2 - 3x + 4$ ㉔
2. Find factors of the key number that sum to -3 and use them to rewrite the middle term. Unfortunately, it is impossible to find two integers with product 24 and sum -3. This trinomial is *not* factorable. It is a prime polynomial.

b. We first rewrite the polynomial (in descending order) as $4x^2 - 4x - 3$ and then proceed by steps.
1. Find the key number $4x^2 - \underline{4x} - 3$ ⊖12
   $[4 \cdot (-3) = -12]$.
2. Find factors of the key number that sum to -4 and use them to rewrite the middle term. $4x^2 \underline{- 6x + 2x} - 3$ $-6, 2$
3. Group the terms. $(4x^2 - 6x) + (2x - 3)$
4. Factor each group. $2x(2x - 3) + 1(2x - 3)$
5. $(2x - 3)$ is the GCF. $(2x - 3)(2x + 1)$

$4x^2 - 4x - 3 = (2x - 3)(2x + 1)$ can be easily verified by multiplication.

PROBLEM 4

Factor completely:

a. $5x^2 - 2x + 2$

b. $3x^2 - 4 - 4x$

Answers to PROBLEMS

4. a. Prime   **b.** $(3x + 2)(x - 2)$

EXAMPLE 5 — Factoring trinomials with two variables using the key number

Factor completely: $6x^2 + xy - y^2$

SOLUTION

1. Find the key number
   $[6 \cdot (-1) = -6]$. $\qquad 6x^2 + \underline{xy} - y^2 \quad \boxed{-6}$
2. Find factors of the key number that sum to 1 and use them to rewrite the middle term. $\qquad 6x^2 + \underline{3xy - 2xy} - y^2 \quad 3, -2$
3. Group the terms. $\qquad (6x^2 + 3xy) - (2xy + y^2) \quad -(2xy + y^2) = -2xy - y^2$
4. Factor each group. $\qquad 3x(2x + y) - y(2x + y)$
5. $(2x + y)$ is the GCF. $\qquad (\underline{2x + y})(3x - y)$

Thus, $6x^2 + xy - y^2 = (2x + y)(3x - y)$.

PROBLEM 5

Factor completely:

$$2x^2 + xy - 3y^2$$

If the terms of the trinomial have a common factor, we factor it out first, as in the next example.

EXAMPLE 6 — Factoring out common factors

Factor completely: $12x^3y^2 + 14x^2y^3 - 6xy^4$

SOLUTION The greatest common factor of these three terms is $2xy^2$. Thus, $12x^3y^2 + 14x^2y^3 - 6xy^4 = 2xy^2(6x^2 + 7xy - 3y^2)$. We then factor $6x^2 + 7xy - 3y^2$.

1. The key number is -18. $\qquad 6x^2 + \underline{7xy} - 3y^2$
2. The factors of -18 with a sum of 7 are 9 and -2. Rewrite the middle term. $\qquad 6x^2 + \underline{9xy - 2xy} - 3y^2$
3. Group the terms. $\qquad (6x^2 + 9xy) - (2xy + 3y^2)$
4. Factor each group. $\qquad 3x(\underline{2x + 3y}) - y(\underline{2x + 3y})$
5. $(2x + 3y)$ is the GCF. $\qquad (\underline{2x + 3y})(3x - y)$

Thus,
$$12x^3y^2 + 14x^2y^3 - 6xy^4 = 2xy^2(6x^2 + 7xy - 3y^2)$$
$$= 2xy^2(2x + 3y)(3x - y)$$

You can check this by multiplying all the factors on the right-hand side of the equation.

PROBLEM 6

Factor completely:

$$12x^4y + 2x^3y^2 - 4x^2y^3$$

We have factored polynomials of the form $ax^2 + bx + c$, where $a > 0$. If a is negative, it's helpful to factor out -1 first and proceed as before.

EXAMPLE 7 — Factoring $ax^2 + bx + c$ where a is negative

Factor completely: $-2x^2 - 7x - 5$

SOLUTION

$$-2x^2 - 7x - 5 = -1(2x^2 + 7x + 5) \qquad \text{Factor out } -1.$$

PROBLEM 7

Factor completely:

$$-3x^2 + 14x + 24$$

Answers to PROBLEMS

5. $(2x + 3y)(x - y)$
6. $2x^2y(3x + 2y)(2x - y)$
7. $-(3x + 4)(x - 6)$

1. Find the key number [2 · (5) = 10].   [$2x^2 + \underline{7x} + 5$]
2. Find the factors of the key number (5 and 2) that add up to the coefficient of the middle term and rewrite the middle term. $-1[2x^2 + \underline{5x + 2x} + 5]$
3. Group the terms. $-1[(2x^2 + 5x) + (2x + 5)]$
4. Factor each group. $-1[x(2x + 5) + 1(2x + 5)]$
5. $(2x + 5)$ is the GCF. $-1[(2x + 5)(x + 1)]$

Since $-1 \cdot a = -a$,

$$-1(2x^2 + 7x + 5) = -(2x + 5)(x + 1)$$

Finally, some polynomials that look complicated can be factored if we make a substitution. For example, to factor $6(y - 1)^2 - 5(y - 1) - 4$, we can think of $(y - 1)$ as A, and write $6A^2 - 5A - 4$, factor this polynomial, and substitute $(y - 1)$ for A in the final answer. We illustrate the procedure in the next example.

EXAMPLE 8   Factoring by substitution
Factor completely: $6(y - 1)^2 - 5(y - 1) - 4$

SOLUTION   Let $A = (y - 1)$; that is,

$$6(y - 1)^2 - 5(y - 1) - 4 = 6A^2 - 5A - 4.$$

1. The key number is $6(-4) = -24$. $6A^2 - \underline{5A} - 4$
2. The factors of -24 adding to -5 are -8 and 3. $6A^2 - \underline{8A + 3A} - 4$
3. Group the terms. $(6A^2 - 8A) + (3A - 4)$
4. Factor each group. $2A(3A - 4) + 1(3A - 4)$
5. Factor out the GCF, $(3A - 4)$. $(3A - 4)(2A + 1)$

Substitute $(y - 1)$ for A and simplify:

$$\begin{aligned}6A^2 - 5A - 4 &= (3A - 4)(2A + 1) \\ &= [3(y - 1) - 4][2(y - 1) + 1] \\ &= [3y - 3 - 4][2y - 2 + 1] \\ &= (3y - 7)(2y - 1)\end{aligned}$$

PROBLEM 8
Factor completely:

$7(z - 3)^2 - 2(z - 3) - 5$

Calculator Corner

1. Can you factor $x^2 - 2$? The graph of $Y_1 = x^2 - 2$ is shown in Window 1. It crosses the horizontal axis, but $x^2 - 2$ is *not* factorable using integer coefficients. Can you explain what is wrong?
2. To find the positive x-intercept at which $x^2 - 2$ crosses the x-axis, let $Y_1 = x^2 - 2$ and $Y_2 = 0$, then press [2nd] [TRACE] [5] and [ENTER] three times to find the point of intersection. To find the negative x-intercept, use the [TRACE] key and arrows to move the cursor near the negative intercept. Press [2nd] [TRACE] [5] and [ENTER] three times.
3. One of the x-intercepts of $x^2 - 2$ is $x = 1.4142136$. Do you recognize this number? If not, go to the home screen and press $\sqrt{2}$.
4. The x-intercepts of $x^2 - 4$ are $x = 2$ and $x = -2$. Also, $x^2 - 4 = (x + 2)(x - 2)$. If the x-intercepts of $x^2 - 2$ are $x = -\sqrt{2}$ and $x = \sqrt{2}$, what is the factorization of $x^2 - 2$?
5. Reexamine Exercise 1. Do you know what is wrong now?

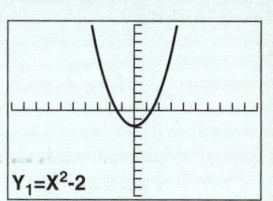

Window 1

Answers to PROBLEMS

8. $(7z - 16)(z - 4)$

Exercises 5.4

⟨A⟩ Factoring Trinomials of the Form $x^2 + bx + c$ In Problems 1–16, factor completely.

1. $x^2 + 5x + 6$
2. $x^2 + 15x + 56$
3. $a^2 + 7a + 10$
4. $a^2 + 10a + 24$
5. $x^2 + x - 12$
6. $x^2 + 5x - 6$
7. $x^2 - 2 + x$
8. $x^2 - 18 + 7x$
9. $x^2 - x - 2$
10. $x^2 - 5x - 14$
11. $x^2 - 3x - 10$
12. $x^2 - 4x - 21$
13. $a^2 - 16a + 63$
14. $a^2 - 4a + 3$
15. $y^2 + 22 - 13y$
16. $y^2 + 11 - 12y$

⟨B⟩ Factoring Trinomials of the Form $ax^2 + bx + c$
⟨C⟩ The ac Test

In Problems 17–46, factor completely.

17. $9x^2 + 37x + 4$
18. $2x^2 + 5x + 2$
19. $3a^2 - 5a - 2$
20. $8a^2 - 2a - 21$
21. $2y^2 - 3y - 20$
22. $6y^2 - 13y - 5$
23. $4x^2 - 11x + 6$
24. $16x^2 - 16x + 3$
25. $6x^2 + x - 12$
26. $20y^2 + y - 1$
27. $21a^2 + 11a - 2$
28. $18x^2 - 3x - 10$
29. $6x^2 + 7xy - 3y^2$
30. $3x^2 + 13xy - 10y^2$
31. $7x^4 - 10x^3y + 3x^2y^2$
32. $6x^4 - 17x^3y + 5x^2y^2$
33. $15x^2y^3 - xy^4 - 2y^5$
34. $5x^2y^3 - 6xy^4 - 8y^5$
35. $15x^3y^2 - 2x^2y^3 - 2xy^4$
36. $4x^3y^2 - 13x^2y^3 - 3xy^4$
37. $-2b^2 + 13b - 20$
38. $-3b^2 - 16b + 12$
39. $-12y^2 - 7y + 12$
40. $-12y^2 - 8y + 15$
41. $2(y + 2)^2 + (y + 2) - 3$
42. $3(y + 3)^2 - 11(y + 3) + 6$
43. $2(x + 1)^2 - 13(x + 1) + 20$
44. $3(u - 1)^2 + 16(u - 1) - 12$
45. $-(a^2 + 2a)^2 - 2(a^2 + 2a) - 1$
46. $-(y^2 - 6y)^2 - 18(y^2 - 6y) - 81$

In Problems 47–50, factor the given expression completely.

47. The flow g (in hundreds of gallons per minute) in 100 feet of $2\tfrac{1}{2}$-inch rubber-lined hose when the friction loss is 21 pounds per square inch is given by

$$2g^2 + g - 21$$

48. The flow g (in hundreds of gallons per minute) in 100 feet of $2\tfrac{1}{2}$-inch rubber-lined hose when the friction loss is 55 pounds per square inch is given by

$$2g^2 + g - 55$$

49. The equivalent resistance R of two electric circuits is given by

$$2R^2 - 3R + 1$$

50. The time t at which an object thrown upward at 12 meters per second will be 4 meters above the ground is given by

$$5t^2 - 12t + 4$$

❯❯❯ Using Your Knowledge

Factoring Applications The ideas presented in this section are important in many fields. Use your knowledge to factor the expressions given in Problems 51 and 52 completely.

51. The deflection of a beam of length L at a distance of 3 feet from its end is given by

$$2L^2 - 9L + 9$$

52. If the distance from the end of the beam in Problem 51 is x feet, then the deflection is given by

$$2L^2 - 3xL + x^2$$

53. The distance (in meters) traveled in t seconds by an object thrown upward at 12 meters per second is

$$-5t^2 + 12t$$

To determine the time at which the object will be 7 meters above the ground, we must solve the equation

$$5t^2 - 12t + 7 = 0$$

Factor the trinomial on the left-hand side of this equation completely.

❯❯❯ Write On

54. The factors of a polynomial are $3x^3$, $x + 5$, and $2x - 1$. Write a procedure that can be used to find the polynomial.

55. If you multiply $(2x - 2)(x - 1)$, you get $2x^2 - 4x + 2$. However, $(2x - 2)(x - 1)$ is *not* the complete factorization for $2x^2 - 4x + 2$. Why not?

56. Without using trial and error, explain why the polynomial $x^2 + 3x - 17$ is prime.

❯❯❯ Concept Checker

Fill in the blank(s) with the correct word(s), phrase, or mathematical statement.

57. When using the _____, the polynomial $ax^2 + bx + c$ is factorable only if there are two integers whose product is ac and whose sum is b.

58. When factoring the polynomial $x^2 + 6x + 8$ both middle signs of the binomial factors must be _____.

59. The trinomial $x^2 + x + 5$ is _____.

ac test	negative
factor tree	prime
positive	factorable

❯❯❯ Mastery Test

Factor completely if possible:

60. $6x^2 + 13x + 6$
61. $6x^2 - 17x + 5$
62. $x^2 - 3x - 10$
63. $x^2 - 5x - 6$
64. $x^2 - 2xy + 5y^2$
65. $x^2 - 7xy + 10y^2$
66. $12x^4y + 2x^3y^2 - 4x^2y^3$
67. $2x^2 + xy - 3y^2$
68. $5x^2 - 2x + 2$

69. $3x^2 - 4 - 4x$
70. $-8x^2 + 2x + 21$
71. $-21x^2 - 11x + 2$
72. $2(y - 3)^2 + 5(y - 3) + 2$
73. $-9(y - 2)^2 - 37(y - 2) - 4$

❯❯❯ Skill Checker

Multiply:

74. $(2a + b)^2$
75. $(3a + 2b)^2$
76. $(a - 2b)^2$
77. $(2a - 3b)^2$
78. $(a + b)(a - b)$
79. $(9a - 2b)(9a + 2b)$
80. $(2x - 3y)(2x + 3y)$
81. $(5x + 7y)(5x - 7y)$
82. $(x + 3)(x^2 - 3x + 9)$
83. $(x + 4)(x^2 - 4x + 16)$

5.5 Special Factoring

Objectives

A Factor a perfect square trinomial.

B Factor the difference of two squares.

C Factor the sum or difference of two cubes.

To Succeed, Review How To . . .

1. Square a binomial (pp. 364–365).
2. Multiply a binomial sum by a binomial difference (p. 365).
3. Multiply a binomial and a trinomial (pp. 361–362).

Getting Started
Applying Factoring to Bending Moments

At x feet from its support, the bending moment (the product of a quantity and the distance from its perpendicular axis) for the crane in the photograph involves the expression

$$\frac{w}{2}(x^2 - 20x + 100)$$

where w is the weight of the crane in pounds per foot. The expression $x^2 - 20x + 100$ is the result of expanding the square of the binomial $(x - 10)^2$ and is called a *perfect square trinomial*. We learned that

$(a + b)^2 = a^2 + 2ab + b^2$ Perfect square trinomial

and

$(a - b)^2 = a^2 - 2ab + b^2$ Perfect square trinomial

The trinomials on the right-hand sides of the equations are perfect square trinomials.

The expression $x^2 - 20x + 100$ is also a perfect square trinomial. Let's factor it using the fact that $(a - b)^2 = a^2 - 2ab + b^2$. Write $x^2 - 20x + 100$ as $x^2 - 2 \cdot x \cdot 10 + 10^2$, where $a = x$ and $b = 10$. We then have $x^2 - 2 \cdot x \cdot 10 + 10^2 = (x - 10)^2$.

A › Factoring Perfect Square Trinomials

To factor perfect square trinomials, we rewrite $(x + a)^2 = x^2 + 2ax + a^2$ and $(x - a)^2 = x^2 - 2ax + a^2$ as follows.

> **PROCEDURE**
>
> **Factoring Perfect Square Trinomials**
>
> $$x^2 + 2ax + a^2 = (x + a)^2$$
> $$x^2 - 2ax + a^2 = (x - a)^2$$

In a perfect square trinomial, the following applies:

1. The first and last terms (x^2 and a^2) are perfect squares and positive.
2. The middle term is twice the product of the two terms in the binomial being squared ($2ax$), or it is the additive inverse of this product ($-2ax$).

If you can write the trinomial in the form shown on the left below, then you may factor it as shown on the right. To factor $x^2 + 6x + 9$, note that the first and last terms are perfect squares $[(x)^2 = x^2$ and $3^2 = 9]$ and positive and the middle term is $2 \cdot 3 \cdot x = 6x$. Hence,

$$x^2 + 6x + 9 = x^2 + 2 \cdot 3x + 3^2 = (x + 3)^2$$

In trying to factor $x^2 - 8x + 16$, we notice the first and last terms are perfect squares and positive.

$$x^2 - 8x + 16$$
$$(x)^2 \qquad (4)^2$$
$$(x - 4)^2$$

Middle term check: $\overbrace{-2 \cdot x \cdot 4}^{\text{The additive inverse of twice the product of the terms of the binomial}} = -8x$

The sign between the terms in the binomial is the same as the middle sign of the trinomial.

You can also factor $x^2 - 8x + 16$ by finding two factors whose product is 16 and whose sum is -8, as we did in the previous section. The factors are -4 and -4. Thus, $x^2 - 8x + 16 = (x - 4)(x - 4) = (x - 4)^2$.

EXAMPLE 1 Factoring perfect square trinomials in one variable
Factor completely:

a. $x^2 - 10x + 25$ **b.** $36 + 12x + x^2$ **c.** $x^2 + 7x + 49$

SOLUTION
a. $x^2 - 10x + 25 = x^2 - 2 \cdot 5 \cdot x + 5^2 = (x - 5)^2$
b. $36 + 12x + x^2 = 6^2 + 2 \cdot 6 \cdot x + x^2 = (6 + x)^2$
You can verify that parts **a** and **b** are correct by multiplying $(x - 5)^2$ and $(6 + x)^2$.
c. $x^2 + 7x + 49$ has positive perfect squares for the first (x^2) and last (7^2) terms. However, the middle term is *not* $2 \cdot 7 \cdot x$. Thus, $x^2 + 7x + 49$ is not a perfect square; it isn't even factorable. We cannot find two integers whose product is 49 and whose sum is 7, so, by the *ac* test, $x^2 + 7x + 49$ is a prime polynomial.

PROBLEM 1
Factor completely:

a. $x^2 - 16x + 64$
b. $x^2 + 8x + 64$
c. $x^2 + 18x + 81$

We can use the same idea to factor perfect square trinomials in two variables. To factor $25x^2 + 20xy + 4y^2$, we write

$$25x^2 + 20xy + 4y^2 = (5x)^2 + 2 \cdot (5x)(2y) + (2y)^2$$
$$= (5x + 2y)^2$$

Calculator Corner

Factor Perfect Square Trinomials by Graphing
The graph of $Y_1 = x^2 - 10x + 25$ in Example 1(a) touches the horizontal axis at $x = 5$ (Window 1), and its factorization is $(x - 5)^2$. Graph $Y_1 = x^2 + 10x + 25$. Where does it touch the horizontal axis?
Thus, if the graph of a perfect square trinomial just touches the horizontal axis at $x = a$, the factorization is $(x - a)^2$. If you know your algebra, you noticed that when the middle term of a perfect square trinomial is preceded by a *minus* sign, the factorization is of the form $(x - a)^2$.
Can you use your calculator to help you factor $x^2 + 7x + 49$ in Example 1(c)? No! The graph does not touch or cut the horizontal axis (Window 2).
To graph this polynomial, you need to set your viewing window so that Ymin $= 0$ and Ymax $= 100$. Then you have to know the *ac* test to discover that $x^2 + 7x + 49$ is prime.

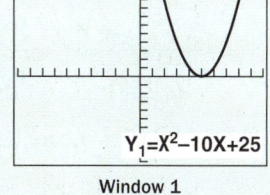
Window 1

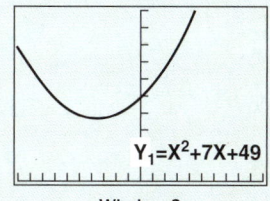
Window 2

Answers to PROBLEMS
1. a. $(x - 8)^2$ **b.** Prime
   c. $(x + 9)^2$

EXAMPLE 2 Factoring perfect square trinomials in two variables
Factor completely:

a. $9x^2 - 12xy + 4y^2$ b. $4x^2 - 10xy + 25y^2$ c. $16x^2 + 24xy + 9y^2$

SOLUTION

a. $9x^2 - 12xy + 4y^2 = (3x)^2 - 2 \cdot 3x \cdot 2y + (2y)^2 = (3x - 2y)^2$

b. Even though the first term, $(2x)^2$, and last term, $(5y)^2$, are perfect squares, the middle term is *not* $2 \cdot 2x \cdot 5y$. Thus, $4x^2 - 10xy + 25y^2$ is not a perfect square. The *ac* test shows that it is a prime polynomial.

c. $16x^2 + 24xy + 9y^2 = (4x)^2 + 2 \cdot 4x \cdot 3y + (3y)^2 = (4x + 3y)^2$

PROBLEM 2
Factor completely:

a. $6x^2 + 30xy + 25y^2$

b. $4x^2 - 12xy + 9y^2$

B › Factoring the Difference of Two Squares

As you recall from Section 5.2,

$$(x + a)(x - a) = x^2 - a^2$$

where $x^2 - a^2$ is called the **difference of two squares.**

An algebraic term is a perfect square if the numerical coefficient is a perfect square and the exponent of each variable is divisible by 2. Table 5.1 illustrates some of these perfect squares. An example of a perfect square term is $121x^4y^8$.

Table 5.1

Number	Perfect Square Coefficients	Perfect Square Variables
1	1	x^2
2	4	x^4
3	9	x^6
4	16	x^8
5	25	x^{10}
6	36	x^{12}
7	49	x^{14}
8	64	x^{16}
9	81	x^{18}
10	100	x^{20}
11	121	x^{22}
12	144	x^{24}

NOTE
A coefficient of 4 is a perfect square and x^4 is a perfect square.
However, a coefficient of 8 is not a perfect square, but x^8 is a perfect square.

The corresponding factoring formula follows.

Answers to PROBLEMS

2. a. Prime   b. $(2x - 3y)^2$

PROCEDURE

Factoring the Difference of Two Squares

$$x^2 - a^2 = (x + a)(x - a)$$

NOTE

In general, $x^2 + a^2$ is not factorable using real numbers.

To factor $x^2 - 9$, we write

$$x^2 - 9 = x^2 - 3^2 = (x + 3)(x - 3)$$

Similarly,

$$25x^2 - 16y^2 = (5x)^2 - (4y)^2 = (5x + 4y)(5x - 4y)$$

EXAMPLE 3 Factoring the difference of two squares

Factor completely:

a. $9x^2 - 1$ **b.** $81x^4 - 16y^4$ **c.** $x^2 + 49$

SOLUTION

a. $9x^2 - 1 = (3x)^2 - 1^2 = (3x + 1)(3x - 1)$

b. $81x^4 - 16y^4 = (9x^2)^2 - (4y^2)^2$ ⟵ Difference of two squares

$= (9x^2 + 4y^2)(9x^2 - 4y^2)$

$= \underbrace{(9x^2 + 4y^2)}_{\text{Not factorable}}\underbrace{(3x + 2y)(3x - 2y)}_{\text{Factored}}$

c. $x^2 + 49$ is not a difference of squares. It is a prime polynomial.

PROBLEM 3

Factor completely:

a. $4x^2 - 25$

b. $x^2 + 64$

c. $16x^4 - 81y^4$

The polynomial $(x + y)^2 - 4$ is also the difference of two squares. If you think of $(x + y)$ as A and 4 as 2^2, you are factoring

$$A^2 - 2^2 = (A + 2)(A - 2)$$

or equivalently

$$(x + y)^2 - 2^2 = [(x + y) + 2][(x + y) - 2]$$
$$= (x + y + 2)(x + y - 2)$$

If $(x + y)^2$ appears as $x^2 + 2xy + y^2$, factor it first. The procedure is

$x^2 + 2xy + y^2 - 4$ Given

$= (x^2 + 2xy + y^2) - 4$ Group the first three terms.

$= (x + y)^2 - 2^2$ Factor the trinomial.

$= [(x + y) + 2][(x + y) - 2]$ Factor the difference of two squares.

$= (x + y + 2)(x + y - 2)$

Answers to PROBLEMS

3. **a.** $(2x + 5)(2x - 5)$ **b.** Prime **c.** $(4x^2 + 9y^2)(2x + 3y)(2x - 3y)$

EXAMPLE 4 Factoring by grouping

Factor completely:

a. $x^2 - 6x + 9 - y^2$ **b.** $x^2 + 6xy + 9y^2 - 4$

PROBLEM 4

Factor completely:

a. $x^2 + 2xy + y^2 - z^2$

b. $x^2 + 10xy + 25y^2 - 16$

SOLUTION

a. If you try to factor by grouping the four terms into pairs as $(x^2 - 6x) + (9 - y^2)$ and factor each binomial, you get $x(x - 6) + (3 + y)(3 - y)$ but there is no common binomial factor. However, the first three terms are a perfect square trinomial. Thus,

$$x^2 - 6x + 9 = (x - 3)^2$$

and

$$\begin{aligned}
x^2 - 6x + 9 - y^2 &= (x^2 - 6x + 9) - y^2 && \text{Group the first three terms.} \\
&= (x - 3)^2 - y^2 && \text{The difference of two squares} \\
&= [(x - 3) + y][(x - 3) - y] \\
&= (x + y - 3)(x - y - 3)
\end{aligned}$$

b. Since $x^2 + 6xy + 9y^2 = (x + 3y)^2$,

$$\begin{aligned}
x^2 + 6xy + 9y^2 - 4 &= (x^2 + 6xy + 9y^2) - 4 && \text{Group the first three terms.} \\
&= (x + 3y)^2 - 2^2 && \text{The difference of two squares} \\
&= [(x + 3y) + 2][(x + 3y) - 2] \\
&= (x + 3y + 2)(x + 3y - 2)
\end{aligned}$$

C › Factoring the Sum or Difference of Two Cubes

We know how to factor the difference of two squares. Can we factor the difference of two cubes—that is, $x^3 - a^3$? Yes! We can also factor $x^3 + a^3$, the sum of two cubes.

An algebraic term is a perfect cube if the numerical coefficient is a perfect cube and the exponent of each variable is divisible by 3. Table 5.2 illustrates some of these perfect cubes. An example of a perfect cube term is $64x^3y^6$.

Table 5.2

Number	Perfect Cube Coefficients	Perfect Cube Variables
1	1	x^3
2	8	x^6
3	27	x^9
4	64	x^{12}
5	125	x^{15}
6	216	x^{18}

NOTE

In the example, $64x^3y^6$, the coefficient 64 is both a perfect square and a perfect cube. Since the exponents on both variables are divisible by 3, in this example we use 64 as a perfect cube.

The corresponding factoring formulas follow.

PROCEDURE

Factoring the Sum or Difference of Two Cubes

$$x^3 + a^3 = (x + a)(x^2 - ax + a^2)$$

$$x^3 - a^3 = (x - a)(x^2 + ax + a^2)$$

Answers to PROBLEMS

4. a. $(x + y + z)(x + y - z)$

b. $(x + 5y + 4)(x + 5y - 4)$

We did not give corresponding product formulas, so we verify these results:

$$
\begin{array}{r}
x^2 - ax + a^2 \\
x + a \\
\hline
ax^2 - a^2x + a^3 \quad \text{Multiply } a(x^2 - ax + a^2). \\
x^3 - ax^2 + a^2x \quad \text{Multiply } x(x^2 - ax + a^2). \\
\hline
x^3 + a^3
\end{array}
$$

Similarly,

$$
\begin{array}{r}
x^2 + ax + a^2 \\
x - a \\
\hline
-ax^2 - a^2x - a^3 \quad \text{Multiply } -a(x^2 + ax + a^2). \\
x^3 + ax^2 + a^2x \quad \text{Multiply } x(x^2 + ax + a^2). \\
\hline
x^3 - a^3
\end{array}
$$

Now that we have verified the formulas, we can factor sums and differences of cubes. For example,

$$8 - x^3 = 2^3 - x^3$$
$$= (2 - x)(2^2 + 2x + x^2)$$
$$= (2 - x)\underbrace{(4 + 2x + x^2)}_{\text{Not factorable}}$$

Middle term of the trinomial factor is the product of the two terms of the binomial factor with the opposite sign.

and

$$x^3 + 125 = x^3 + 5^3$$
$$= (x + 5)(x^2 - 5x + 5^2)$$
$$= (x + 5)\underbrace{(x^2 - 5x + 25)}_{\text{Not factorable}}$$

Notice that when factoring the sum or difference of two cubes, the factors are always a binomial and a trinomial.

Calculator Corner

Using Graphs to Check the Factorization of Cubes

The graph of $x^3 + 125$ cuts the horizontal axis at $x = -5$, so you know that $x - (-5) = x + 5$ is a factor of $x^3 + 125$.

You can verify that $x^3 + 125 = (x + 5)(x^2 - 5x + 25)$ by graphing $Y_1 = x^3 + 125$ and $Y_2 = (x + 5)(x^2 - 5x + 25)$ and making sure you get the same graph. However, only algebra can help you find the other factor, $x^2 - 5x + 25$. If you divide $x^3 + 125$ by $x + 5$, you get $x^2 - 5x + 25$.

EXAMPLE 5 Factoring the sum and difference of two cubes

Factor completely:

a. $27 - 8x^3$ **b.** $125x^3 + 64y^3$ **c.** $(x + y)^3 - 8$

SOLUTION

a. $27 - 8x^3 = 3^3 - (2x)^3$ Remember the factors are a binomial and a trinomial.

These will always be opposite signs.

$$= (3 - 2x)[3^2 + 3 \cdot 2x + (2x)^2]$$
$$= (3 - 2x)(9 + 6x + 4x^2)$$

b. $125x^3 + 64y^3 = (5x)^3 + (4y)^3$
$$= (5x + 4y)[(5x)^2 - 5x \cdot 4y + (4y)^2]$$
$$= (5x + 4y)(25x^2 - 20xy + 16y^2)$$

PROBLEM 5

Factor completely:

a. $64 + \frac{1}{27}x^3$

b. $8x^3 - y^3$

c. $(a - b)^3 + 1$

(continued)

Answers to PROBLEMS

5. a. $\left(4 + \frac{1}{3}x\right)\left(16 - \frac{4}{3}x + \frac{1}{9}x^2\right)$ **b.** $(2x - y)(4x^2 + 2xy + y^2)$ **c.** $(a - b + 1)(a^2 - 2ab + b^2 - a + b + 1)$

c. $(x + y)^3 - 8$ is the difference of two cubes. As before, think of $(x + y)$ as A and 8 as 2^3. Thus, you are factoring

$$A^3 - 2^3 = (A - 2)(A^2 + 2A + 2^2)$$

then replace A with $(x + y)$ to get

$$[(x + y)^3 - 2^3] = [(x + y) - 2][(x + y)^2 + 2(x + y) + 4] \quad \text{Simplify.}$$
$$= (x + y - 2)(x^2 + 2xy + y^2 + 2x + 2y + 4)$$

EXAMPLE 6 Factoring the difference of squares and cubes
Factor completely: $x^6 - 64$

SOLUTION x^6 and 64 are both perfect squares and perfect cubes. Factor first as the difference of two squares and then as the sum and difference of cubes.

$$\begin{aligned}
x^6 - 64 &= x^6 - 2^6 \\
&= (x^3)^2 - (2^3)^2 &&\text{Difference of squares} \\
&= (x^3 + 2^3)(x^3 - 2^3) &&\text{Sum/difference of cubes} \\
&= [(x + 2)(x^2 - 2x + 4)][(x - 2)(x^2 + 2x + 4)] \\
&= (x + 2)(x - 2)(x^2 + 2x + 4)(x^2 - 2x + 4)
\end{aligned}$$

You can also factor $x^6 - 64$ by writing it as the difference of two cubes.

$$\begin{aligned}
x^6 - 64 &= (x^2)^3 - (2^2)^3 \\
&= (x^2 - 2^2)[(x^2)^2 + 2^2 x^2 + (2^2)^2] \\
&= (x^2 - 4)(x^4 + 4x^2 + 16) \\
&= (x + 2)(x - 2)(x^4 + 4x^2 + 16)
\end{aligned}$$

But the expression $(x^4 + 4x^2 + 16)$ is **not** completely factored. You can factor it by first *adding* and *subtracting* $4x^2$ to obtain

$$\begin{aligned}
x^4 + 4x^2 + 16 &= x^4 + 4x^2 + 4x^2 + 16 - 4x^2 \\
&= (x^4 + 8x^2 + 16) - 4x^2 &&\text{By the associative property} \\
&= (x^2 + 4)^2 - 4x^2 &&\text{Factor } x^4 + 8x^2 + 16. \\
&= [(x^2 + 4) + 2x][(x^2 + 4) - 2x] &&\text{Factor the difference of two squares.} \\
&= (x^2 + 2x + 4)(x^2 - 2x + 4) &&\text{By the commutative property}
\end{aligned}$$

Thus, as before,

$$x^6 - 64 = (x^2)^3 - (2^2)^3 = (x + 2)(x - 2)(x^2 + 2x + 4)(x^2 - 2x + 4)$$

PROBLEM 6
Factor completely: $x^6 - 1$

NOTE
As you can see, it's easier to factor $x^6 - 64$ when it is *first* written as the difference of two squares as we did first.

Answers to PROBLEMS

6. $(x + 1)(x - 1)(x^2 + x + 1)(x^2 - x + 1)$

Calculator Corner

1. The graph of the quadratic function $f(x)$ in Window 1 touches the x-axis at $x = 3$. What is the y-intercept? What is $f(x)$?
2. If the graph of the quadratic function $g(x)$ touches the x-axis at $x = -2$ only and intersects the y-axis at a distance of 4 units from the origin, there are two possible expressions for $g(x)$. What are these two expressions?
3. The function $h(x)$ in Window 2 is a cubic function and has only one x-intercept $x = 2$ and a y-intercept of -8. What is $h(x)$?
4. If there is another cubic function $i(x)$ with only one x-intercept $x = -2$ and a y-intercept of -8, what is $i(x)$?
5. Use your calculator to show that $x^2 + 9 \neq (x + 3)^2$.
6. Use your calculator to show that $x^2 - 8x - 16 \neq (x - 4)^2$.
7. Given a polynomial function $f(x)$, how can you tell, using a graphing calculator, if $f(x)$ cannot be factored as a product of linear factors with integer coefficients?

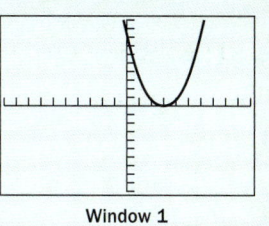
Window 1

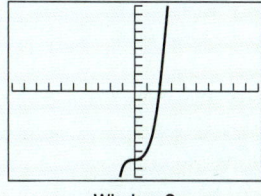
Window 2

> Exercises 5.5

Boost your grade at mathzone.com!
> Practice Problems > Self-Tests
> NetTutor > e-Professors
 > Videos

< A > **Factoring Perfect Square Trinomials** In Problems 1–24, factor completely.

1. $x^2 + 2x + 1$
2. $x^2 + 20x + 100$
3. $y^2 + 22y + 121$
4. $y^2 + 14x + 49$

5. $1 + 4x + 4x^2$
6. $1 + 6x + 9x^2$
7. $9x^2 + 30xy + 25y^2$
8. $25x^2 + 30xy + 9y^2$

9. $36a^2 + 48a + 16$
10. $9a^2 + 60a + 100$
11. $y^2 - 2y + 1$
12. $25 - 10y + y^2$

13. $49 - 14x + x^2$
14. $x^2 - 100x + 2500$
15. $49a^2 - 28ax + 4x^2$
16. $4a^2 - 12ax + 9x^2$

17. $16x^2 - 24xy + 9y^2$
18. $9x^2 - 42xy + 49y^2$
19. $9x^4 + 12x^2 + 4$
20. $25y^4 + 20y^2 + 4$

21. $16x^4 - 24x^2 + 9$
22. $4y^4 - 20y^2 + 25$
23. $1 + 2x^2 + x^4$
24. $4 + 12x^2 + 9x^4$

< B > **Factoring the Difference of Two Squares** In Problems 25–44, factor completely.

25. $y^2 - 64$
26. $y^2 - 121$
27. $a^2 - \frac{1}{9}$

28. $x^2 - \frac{1}{16}$
29. $64 - b^2$
30. $81 - b^2$

31. $36a^2 - 49b^2$
32. $36a^2 - 25b^2$
33. $\frac{x^2}{9} - \frac{y^2}{16}$

34. $\frac{x^2}{16} - \frac{9y^2}{25}$
35. $a^2 + 4ab + 4b^2 - c^2$
36. $9a^2 + 6ab + b^2 - 1$

37. $4x^2 - 4xy + y^2 - 1$
38. $9x^2 - 30xy + 25y^2 - 9$
39. $9y^2 - 12xy + 4x^2 - 25$

40. $16y^2 - 40xy + 25x^2 - 36$
41. $16a^2 - (x^2 + 6xy + 9y^2)$
42. $25a^2 - (4x^2 - 4xy + y^2)$

43. $y^2 - a^2 + 2ab - b^2$
44. $9y^2 - 9x^2 + 6xz - z^2$

⟨ C ⟩ Factoring the Sum or Difference of Two Cubes In Problems 45–70, factor completely.

45. $x^3 + 125$

46. $x^3 + 64$

47. $1 + a^3$

48. $343 + a^3$

49. $8x^3 + y^3$

50. $125x^3 + 8y^3$

51. $x^3 - 1$

52. $x^3 - 216$

53. $125a^3 - 8b^3$

54. $216a^3 - 125b^3$

55. $x^6 - 64$

56. $y^6 - 1$

57. $x^6 - \dfrac{1}{64}$

58. $y^6 - 729$

59. $\dfrac{x^6}{64} - 1$

60. $\dfrac{y^6}{729} - 1$

61. $(x - y)^3 + 1$

62. $(x + 2y)^3 + 8$

63. $1 + (x + 2y)^3$

64. $27 + (x + y)^3$

65. $(y - 2x)^3 - 1$

66. $(y - 4x)^3 - 1$

67. $27 - (x + 2y)^3$

68. $8 - (y - 4x)^3$

69. $64 + (x^2 - y^2)^3$

70. $27 + (y^2 - x^2)^3$

⟩⟩⟩ Using Your Knowledge

Is There Demand for the Supply? Have you heard of supply and demand? In business the supply and demand of a product can be expressed by using polynomials. In Problems 71–73, factor the given expression completely.

71. A business finds that when x units of an item are demanded by consumers, the price per unit is given by

$$D(x) = 100 - x^2$$

Factor $100 - x^2$ completely.

72. When x units are supplied by sellers, the price per unit of an item is given by

$$S(x) = x^3 + 216$$

Factor $x^3 + 216$ completely.

73. When x units of a certain item are produced, the cost is given by

$$C(x) = 8x^3 + 1$$

Factor completely $8x^3 + 1$.

⟩⟩⟩ Write On

74. In Example 6, $(x + 2)(x - 2)(x^2 - 2x + 4)(x^2 + 2x + 4)$ is the "preferred" factorization of $x^6 - 64$. Explain why you think this is true.

75. Write a procedure to determine whether a trinomial of the form $ax^2 + bx + c$ is a perfect square trinomial.

76. Find two different values of k that will make $9x^2 + kx + 4$ a perfect square trinomial. Explain the procedure you used to find your answer.

77. Find a value for k that will make $4x^2 + 12x + k$ a perfect square trinomial. Describe the procedure you used to find your answer.

78. Explain why $a^2 + b^2$ cannot be factored.

79. Can $a^4 + 64$ be factored? (*Hint:* Add and subtract $16a^2$.)

$$a^4 + 64 = a^4 + 16a^2 + 64 - 16a^2$$
$$= (a^4 + 16a^2 + 64) - 16a^2$$

Can you factor the expression now? If so, what is the factorization?

⟩⟩⟩ Concept Checker

Fill in the blank(s) with the correct word(s), phrase, or mathematical statement.

80. The factorization of the _____ is always a binomial factor times a trinomial factor.

81. An example of a _____ is $x^2 + 49$.

82. One of the requirements for a trinomial to be a _____ is that the first and last terms have to be perfect squares and positive.

perfect square trinomial

difference of squares

sum or difference of cubes

prime polynomial

⟩⟩⟩ Mastery Test

Factor completely if possible.

83. $64 - \dfrac{1}{27}x^3$

84. $8x^3 + 27y^3$

85. $x^6 - 1$

86. $x^6 + 1$

87. $16x^2 - 1$

88. $16x^4 - 81y^4$

89. $x^2 + 2xy + y^2 + 16$

90. $x^2 + 10xy + 25y^2 - 16$

91. $4x^2 - 4xy + 9y^2$

92. $4x^2 - 12xy + 9y^2$

93. $x^2 - 16x + 64$

94. $x^2 + 18x + 81$

95. $(x - y)^4 - (x - y)^2$

96. $(x + y)^2 - (y - x)^2$

97. $(x - y)^3 + 125$

98. $125 + (x + y)^3$

⟩⟩⟩ Skill Checker

Factor out the greatest common factor.

99. $3x^2y + 6x^4z - 9x$

100. $-5xy^2 + 7wy^5$

Factor by grouping.

101. $3x^3 + 21x^2 + 4x + 28$

102. $x^3 + x^2 + 2x + 2$

Factor the trinomial completely.

103. $x^2 - 2xy - 15y^2$

104. $6x^2 - 5x - 4$

5.6 General Methods of Factoring

Objective

A ▶ Factor a polynomial using the procedures given in the text.

To Succeed, Review How To . . .

1. Use the factoring procedures (pp. 371–376, 379–385, 388–394).

▶ Getting Started

Applying Factoring to Your Arteries

Have you heard of blocked arteries? The velocity of the blood inside a partially blocked artery (right) depends on the thickness of the inside wall (r) and the diameter of the artery to the outside wall (R) and is given by

$$CR^2 - Cr^2$$

where C is a constant. How do we factor this expression? We shall follow a general pattern that uses one or more of the techniques we've learned.

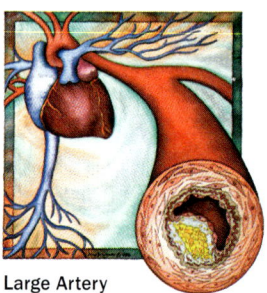

Large Artery

A ▶ Factoring Polynomials

One of the most important skills when factoring polynomials is to know when the polynomial is completely factored. Here are the guidelines you need.

COMPLETELY FACTORED POLYNOMIAL

A polynomial is **completely factored** when

1. The polynomial is written as the product of prime polynomials with *integer* coefficients.
2. The monomial factors need not be factored completely. $6x^3(2x + 1)$ is the factored form of $12x^4 + 6x^3$. You don't need to write $6x^3$ as $2 \cdot 3 \cdot x \cdot x \cdot x$.

PROCEDURE

A General Factoring Strategy

1. Factor out the GCF, if there is one.
2. Count the number of terms in the given polynomial (or inside the parentheses if the GCF was factored out).

   A. If there are *two terms*, check for
   - Difference of two squares

   $$x^2 - a^2 = (x + a)(x - a)$$

   - Difference of two cubes

   $$x^3 - a^3 = (x - a)(x^2 + ax + a^2)$$

- Sum of two cubes

$$x^3 + a^3 = (x + a)(x^2 - ax + a^2)$$

The sum of two squares, $x^2 + a^2$, is not factorable.

B. If there are *three terms*, check for
- Perfect square trinomials

$$x^2 + 2ax + a^2 = (x + a)^2$$
$$x^2 - 2ax + a^2 = (x - a)^2$$

- Trinomials of the form

$$ax^2 + bx + c \quad (a > 0)$$

Use the *ac* method or trial and error. If $a < 0$, factor out -1 first.

C. If there are *four terms*, factor by grouping.

3. Check the result by multiplying the factors.

Factoring Strategy Flow Chart

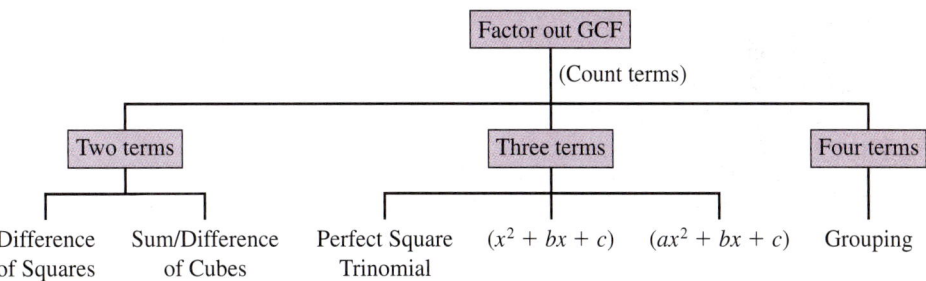

To factor $CR^2 - Cr^2$ (see *Getting Started*) we follow these steps.

1. Factor out the GCF, C. $\quad CR^2 - Cr^2 = C(R^2 - r^2)$
2. Factor the difference of two squares inside the parentheses. $\quad = C(R + r)(R - r)$
3. Now check.

If you were the doctor, you could tell the patient that when R and r are very close ($R - r$ is close to 0) the artery will be almost blocked.

EXAMPLE 1 Using the general factoring strategy to factor a binomial

Factor completely:

a. $8x^5 - x^2y^3$ **b.** $6x^5 + 24x^3$

SOLUTION We use the steps in our general factoring strategy.

a. $8x^5 - x^2y^3$
1. Factor out the GCF, x^2. $\quad 8x^5 - x^2y^3 = x^2(8x^3 - y^3)$
2. Factor the difference of two cubes inside the parentheses. $\quad = x^2(2x - y)(4x^2 + 2xy + y^2)$
3. Now check.

b. $6x^5 + 24x^3$
1. Factor out $6x^3$, the GCF: $\quad 6x^5 + 24x^3 = 6x^3(x^2 + 4)$.
2. Since $x^2 + 4$ is the sum of two squares, it is not factorable.
3. Check this.

The complete factorization of $6x^5 + 24x^3$ is $6x^3(x^2 + 4)$.

PROBLEM 1

Factor completely:

a. $27x^5 - x^2y^3$

b. $8x^5 + 72x^3$

Answers to PROBLEMS

1. **a.** $x^2(3x - y)(9x^2 + 3xy + y^2)$
   b. $8x^3(x^2 + 9)$

EXAMPLE 2 Using the general factoring strategy to factor a trinomial

Factor completely:

a. $12x^5 + 12x^4y + 3x^3y^2$

b. $36x^2y^2 - 24xy^3 + 4y^4$

SOLUTION As usual, factor out the GCFs, $3x^3$ in part **a** and $4y^2$ in part **b,** first. We then have perfect square trinomials inside the parentheses. Here are the steps.

a. $12x^5 + 12x^4y + 3x^3y^2$
   1. Factor out the GCF, $3x^3$. $\quad 12x^5 + 12x^4y + 3x^3y^2 = 3x^3(4x^2 + 4xy + y^2)$
   2. Factor the perfect square trinomial. $\quad = 3x^3 \; (2x + y)^2$
   3. Now check.

b. $36x^2y^2 - 24xy^3 + 4y^4$
   1. Factor out the GCF, $4y^2$. $\quad 36x^2y^2 - 24xy^3 + 4y^4 = 4y^2(9x^2 - 6xy + y^2)$
   2. Factor the perfect square trinomial inside the parentheses. $\quad = 4y^2 \; (3x - y)^2$
   3. Don't forget to check this.

PROBLEM 2

Factor completely:

a. $36x^5 + 24x^4y + 4x^3y^2$

b. $12x^2y^2 - 12xy^3 + 3y^4$

EXAMPLE 3 Using the general factoring strategy to factor a polynomial

Factor completely:

a. $4x^3y - 10x^2y^2 - 6xy^3$

b. $2x^5 + x^4y + x^3y^2$

SOLUTION

a. The GCF is $2xy$. After factoring out this GCF, we have three terms inside the parentheses. We can use the *ac* method or trial and error to finish the problem. The steps are as follows.
   1. Factor out the GCF, $2xy$. $\quad 4x^3y - 10x^2y^2 - 6xy^3 = 2xy(2x^2 - 5xy - 3y^2)$
   2. Use the *ac* method or trial and error to factor $2x^2 - 5xy - 3y^2$. $\quad = 2xy(2x + y)(x - 3y)$
   3. Check the answer by multiplying the factors. $\quad 2xy(2x + y)(x - 3y) = 4x^3y - 10x^2y^2 - 6xy^3$

b. Factor out the GCF, x^3. $\quad 2x^5 + x^4y + x^3y^2 = x^3(2x^2 + xy + y^2)$

   The expression inside the parentheses is *not* factorable. According to the *ac* method, the key number is 2. We need two integers whose product is 2 and whose sum is 1; no such integers exist.

PROBLEM 3

Factor completely:

a. $9x^3y - 15x^2y^2 - 6xy^3$

b. $24x^4 + x^3y + x^2y^2$

EXAMPLE 4 Using the general factoring strategy to factor a polynomial

Factor completely:

a. $2x^5 - x^4y + 4x^3y^2$

b. $2y^2 - 32$

SOLUTION

a. We start by factoring out the GCF, x^3.

   $$2x^5 - x^4y + 4x^3y^2 = x^3(2x^2 - xy + 4y^2)$$

   Although $2x^2 - xy + 4y^2$ is a trinomial, it is *not* factorable. (The *ac* method gives $2 \cdot 4 = 8$, and there are no factors whose product is 8 and whose sum is -1.) Thus, the factorization shown is the complete factorization.

PROBLEM 4

Factor completely:

a. $2x^4 + x^3y + 2x^2y^2$

b. $3x^2 - 75$

Answers to PROBLEMS

2. a. $4x^3(3x + y)^2$ b. $3y^2(2x - y)^2$ 3. a. $3xy(3x + y)(x - 2y)$ b. $x^2(24x^2 + xy + y^2)$
4. a. $x^2(2x^2 + xy + 2y^2)$ b. $3(x + 5)(x - 5)$

b. We start by factoring out the GCF, 2.

$$2y^4 - 32 = 2(y^4 - 16)$$
$$= 2(y^2 + 4)(y^2 - 4) \quad \text{Factor the difference of two squares, } (y^4 - 16).$$
$$= 2(y^2 + 4)(y + 2)(y - 2) \quad \text{Factor the difference of two squares, } (y^2 - 4).$$

Now all factors are prime polynomials, and the factorization is complete.

EXAMPLE 5 Factoring by grouping
Factor completely: $4x^3 - 12x^2 - x + 3$

SOLUTION In this case, there is no common factor other than 1. The polynomial has four terms, so we factor by grouping.

$$4x^3 - 12x^2 - x + 3 = (4x^3 - 12x^2) - (x - 3)$$
$$= 4x^2(x - 3) - 1(x - 3) \quad \text{Group by twos and factor.}$$
$$= (x - 3)(4x^2 - 1) \quad \text{Factor out the GCF, } (x - 3).$$
$$= (x - 3)(2x + 1)(2x - 1) \quad \text{Factor the difference of two squares, } (4x^2 - 1).$$

PROBLEM 5
Factor completely:

$9x^3 + 18x^2 - 25x - 50$

EXAMPLE 6 Factoring by grouping
Factor completely: $x^2 - 6x + 9 - 9y^2$

SOLUTION The polynomial has four terms, so you may try to factor it by grouping. If we try groups of two, we have

$$x^2 - 6x + 9 - 9y^2 = (x^2 - 6x) + (9 - 9y^2)$$
$$= x(x - 6) + 9(1 - y^2)$$

But there is no GCF, so we can't factor any further using this technique. However, the first three terms form a perfect square trinomial, so we write

$$x^2 - 6x + 9 - 9y^2 = (x^2 - 6x + 9) - 9y^2 \quad \text{Group the first 3 terms.}$$
$$= (x - 3)^2 - (3y)^2 \quad \text{The difference of two squares}$$
$$= [(x - 3) + 3y][(x - 3) - 3y] \quad \text{Simplify.}$$
$$= (x + 3y - 3)(x - 3y - 3)$$

PROBLEM 6
Factor completely:

$x^2 + 8x + 16 - 81y^2$

EXAMPLE 7 General factoring strategies
Factor completely: $-2x^2 - 9x + 5$

SOLUTION This is a polynomial of the form $ax^2 + bx + c$ with $a = -2$, which is negative, so we first factor out -1 to obtain

$$-2x^2 - 9x + 5 = -1(2x^2 + 9x - 5)$$

1. The key number is $2(-5) = -10$.
2. We write the middle term as $10x - 1x$. $= -1(2x^2 + 10x - 1x - 5)$
3. Grouping by twos. $= -1[(2x^2 + 10x) + (-1x - 5)]$
4. Factoring each group. $= -1[2x(x + 5) - 1(x + 5)]$
5. Factoring the common binomial. $= -1(x + 5)(2x - 1)$

Since $-1 \cdot a = -a,$ $-2x^2 - 9x + 5 = -(x + 5)(2x - 1)$

PROBLEM 7
Factor completely:

$-3x^2 + 17x + 28$

Answers to PROBLEMS

5. $(x + 2)(3x + 5)(3x - 5)$ **6.** $(x + 9y + 4)(x - 9y + 4)$ or $(x - 9y + 4)(x + 9y + 4)$
7. $-(x - 7)(3x + 4)$

Exercises 5.6

A Factoring Polynomials   In Problems 1–78, factor completely.

1. $3x^4 - 3x^3 - 18x^2$
2. $4x^5 - 12x^4 - 16x^3$
3. $5x^4 + 10x^3y - 40x^2y^2$
4. $6x^7 + 18x^6y - 60x^5y^2$
5. $-3x^6 - 6x^5 - 21x^4$
6. $-6x^5 - 18x^4 - 12x^3$
7. $2x^6y - 4x^5y^2 - 10x^4y^3$
8. $3x^8y - 12x^7y^2 - 9x^6y^3$
9. $-4x^6 + 12x^5y - 18x^4y^2$
10. $-5x^6 - 25x^5 - 30x^4$
11. $6x^3y^2 + 12x^2y^2 + 2xy^2 + 4y^2$
12. $6x^3y^2 + 24x^2y^2 + 3xy^2 + 12y^2$
13. $-9x^4y - 9x^3y - 6x^2y - 6xy$
14. $-8x^4y - 16x^3y - 6x^2y - 12xy$
15. $-4x^4 - 4x^3y + 2x^2y + 2xy^2$
16. $-9x^4 - 18x^3y + 3x^2y + 6xy^2$
17. $3x^2y^2 + 24xy^3 + 48y^4$
18. $8x^2y^2 + 24xy^3 + 18y^4$
19. $-18kx^2 - 24kxy - 8ky^2$
20. $-12kx^2 - 60kxy - 75ky^2$
21. $16x^3y^2 - 48x^2y^3 + 36xy^4$
22. $45x^3y^2 - 60x^2y^3 + 20xy^4$
23. $kx^2 - 12kx + 36$
24. $kx^2 - 20kx + 25$
25. $3x^5 + 12x^4y + 12x^3y^2$
26. $2x^5 + 16x^4y + 32x^3y^2$
27. $18x^6 + 12x^5y + 2x^4y^2$
28. $12x^6 + 12x^5y + 3x^4y^2$
29. $12x^4y^2 - 36x^3y^3 + 27x^2y^4$
30. $18x^4y^2 - 24x^3y^3 + 8x^2y^4$
31. $6x^3 + 12x^2 - 6x - 12$
32. $4x^3 + 16x^2 - 16x - 64$
33. $7x^4 - 7y^4$
34. $9x^4 - 9z^4$
35. $2x^6 - 32x^2y^4$
36. $x^7 - 81x^3y^4$
37. $-2x^2 - 12x - 18$
38. $-2x^2 - 20x - 50$
39. $-3x^2 - 12x - 12$
40. $-4x^2 - 24x - 36$
41. $-4x^4 - 4x^3y - x^2y^2$
42. $-9x^4 - 6x^3y - x^2y^2$
43. $-9x^2y^2 - 12xy^3 - 4y^4$
44. $-4x^2y^2 - 12xy^3 - 9y^4$
45. $-8x^2y^2 + 24xy^3 - 18y^4$
46. $-18x^4 + 24x^3y - 8x^2y^2$
47. $-18x^3 - 24x^2y - 8xy^2$
48. $-12x^3 - 36x^2y - 27xy^2$
49. $-18x^3 - 60x^2y - 50xy^2$
50. $-12x^3 - 60x^2y - 75xy^2$
51. $-x^3 + xy^2$
52. $-x^3 + 9xy^2$
53. $-x^4 + 4x^2y^2$
54. $-x^4 + 16x^2y^2$
55. $-4x^4 + 9x^2y^2$
56. $-9x^4 + 4x^2y^2$
57. $-8x^3 + 18xy^2$
58. $-12x^3 + 3x$
59. $-18x^4 + 8x^2y^2$
60. $-12x^4 + 27x^2y^2$

61. $27x^2 - x^5$ **62.** $64x^3 - x^6$ **63.** $x^7 - 8x^4$

64. $8x^{10} - \dfrac{1}{27}x^7$ **65.** $27x^4 + 8x^7$ **66.** $8x^5 + 27x^8$

67. $27x^7 + 64x^4y^3$ **68.** $8x^8 + 27x^5y^3$ **69.** $x^2 + 4x + 4 - y^2$

70. $x^2 + 8x + 16 - y^2$ **71.** $x^2 + y^2 - 6y + 9$ **72.** $x^2 + y^2 - 8x + 16$

73. $x^2 - y^2 - 4y - 4$ **74.** $x^2 - y^2 - 8y - 16$ **75.** $-9x^2 + 30xy - 25y^2$

76. $-9x^2 + 12xy - 4y^2$ **77.** $18x^3 - 60x^2y + 50xy^2$ **78.** $-12x^3 + 60x^2y - 72xy^2$

〉〉〉 Using Your Knowledge

Technical Applications Many of the ideas presented in this section are used by engineers and technicians. Use your knowledge to factor the expressions in Problems 79–82 completely.

79. The bend allowance needed to bend a piece of metal of thickness t through an angle A when the inside radius of the bend is R is given by

$$\dfrac{2\pi A}{360}R + \dfrac{2\pi A}{360}Kt \quad \text{K is a constant.}$$

Factor this expression completely.

80. The change in kinetic energy of a moving object of mass m with initial velocity v_1 and terminal velocity v_2 is given by

$$\tfrac{1}{2}mv_1^2 - \tfrac{1}{2}mv_2^2$$

Factor this expression completely.

81. The parabolic distribution of shear stress on the cross section of a certain beam is given by

$$\dfrac{3Sd^2}{2bd^3} - \dfrac{12Sz^2}{2bd^3}$$

Factor this expression completely.

82. The polar moment of inertia of a hollow round shaft of inner diameter d_1 and outer diameter d is given by

$$\dfrac{\pi d^4}{32} - \dfrac{\pi d_1^4}{32}$$

Factor this expression completely.

〉〉〉 Write On

83. Explain why the expression $x^2(x^2 - 5) + y^2(x^2 - 5)$ is not completely factored.

84. Is the statement "$9(x^2 - x)$ is factored" true or false? Explain.

85. Explain the difference between the statements "a polynomial is factored" and "a polynomial is completely factored." Give examples of polynomials that are factored but are not completely factored.

86. Suppose you factor $x^2 - 2$ as $(x + \sqrt{2})(x - \sqrt{2})$. Is $x^2 - 2$ completely factored according to the guidelines we mentioned at the beginning of this section? Explain.

〉〉〉 Mastery Test

Factor completely if possible.

87. $9x^3 - 18x^2 - x + 2$ **88.** $4x^3 - 12x^2 - x + 3$ **89.** $2x^4 + x^3y + 2x^2y^2$

90. $3x^5 + x^4y + x^3y^2$ **91.** $x^2 - 10x + 25 - y^2$ **92.** $y^2 - 9x^2 + 12x - 4$

93. $-8y^2 + 2y + 21$

94. $-2 - 5y - 2y^2$

95. $36x^5 + 24x^4y + 4x^3y^2$

96. $12x^2y^2 - 12xy^3 + 3y^4$

97. $8x^5 + 72x^3$

98. $27x^5 - x^2y^3$

❯❯❯ Skill Checker

Solve.

99. $3x - 7 = 0$

100. $5x + 4 = 0$

Evaluate.

101. $3x^3 + 4x - 5$; if $x = -2$

102. $x^3 - x^2 - 2x - 2$; if $x = 3$

Multiply.

103. $(x - 4)^2$

104. $(2x + 3)^2$

5.7 Solving Equations by Factoring: Applications

▶ Objectives

A ❯ Solve equations by factoring.

B ❯ Use the Pythagorean theorem to find the length of one side of a right triangle.

C ❯ Solve applications involving quadratic equations.

▶ To Succeed, Review How To . . .

1. Factor polynomials (pp. 398–401).
2. Solve linear equations (pp. 76–83).
3. Evaluate expressions (pp. 51–55).

▶ Getting Started
Quadratic Equations in Fire Fighting

How much water is the fire truck pumping if the friction loss is 36 pounds per square inch? You can find out by solving the equation

$2g^2 + g - 36 = 0$ (g in hundreds of gallons per minute)

This equation is a *quadratic equation in standard form*. In this section we shall study how to solve equations like this one.

We have already studied linear equations, equations that can be written in the form $ax + b = c$, where a, b, and c are real numbers and $a \neq 0$. We are now ready to study *quadratic equations*. These equations can be written in standard form and then some of them can be solved by the factoring methods we have studied. Here is the definition we need.

| **STANDARD FORM OF A QUADRATIC EQUATION** | An equation that can be written in the **standard form** $$ax^2 + bx + c = 0$$ where a, b, and c are constants and $a \neq 0$ is a **quadratic equation.** |

Here are some quadratic equations:
$$x^2 = 5, \quad 3x^2 - 8x + 7 = 0, \quad \text{and} \quad x^2 - 2x = 4$$

Of these, only $3x^2 - 8x + 7 = 0$ is in standard form, with $a = 3$, $b = -8$, and $c = 7$. *Note:* In a quadratic equation, the highest exponent on the variable is 2.

A › Solving Equations by Factoring

The equation $2g^2 + g - 36 = 0$ can be solved by factoring. As you recall from Section 5.4, $2g^2 + g - 36$ can be factored as $(2g + 9)(g - 4)$. We then write

$$2g^2 + g - 36 = 0 \quad \text{Given}$$
$$(2g + 9)(g - 4) = 0 \quad \text{Factor.}$$

The product of the factors is zero. The only way this can happen is if at least one of the factors is zero. (Try getting zero for a product without having any zero factors.) Here is the property we need.

| **ZERO-PRODUCT PROPERTY** | For all real numbers a and b, $a \cdot b = 0$ means that $a = 0$ or $b = 0$ (or both). |

Thus,
$$(2g + 9)(g - 4) = 0$$
means that
$$2g + 9 = 0 \quad \text{or} \quad g - 4 = 0$$
$$2g = -9 \quad \text{or} \quad g = 4 \quad \text{Solve the linear equations.}$$
$$g = -\frac{9}{2} \quad \text{or} \quad g = 4$$

The two possible solutions are $g = -\frac{9}{2}$ and $g = 4$. Since g is the flow of water in hundreds of gallons per minute, g must be positive, so we discard the negative solution $g = -\frac{9}{2}$. Thus, the fire truck can pump 400 gal/min.

PROCEDURE

Solving a Quadratic Equation

To solve a quadratic equation, follow these three steps using the suggested acronym, "OFF" to help remember the steps.

1. Set the equation equal to **0**.   0
2. **F**actor completely.   F
3. Set each linear **F**actor equal to 0 and solve.   F

EXAMPLE 1 Solving equations using the zero-product property

Solve by factoring:

a. $x^2 - 9 = 0$ **b.** $x^2 + 8x = 0$

SOLUTION

a.
$x^2 - 9 = 0$	0
$(x + 3)(x - 3) = 0$	F Factor.
$x + 3 = 0$ or $x - 3 = 0$	F Factors = 0. Use the zero-product property, with $a = x + 3$ and $b = x - 3$.
$x = -3$ or $x = 3$	Solve the equations $x + 3 = 0$ and $x - 3 = 0$.

We can check the solutions by substituting in the original equation, $x^2 - 9 = 0$.

CHECK

$x^2 - 9 = 0$		$x^2 - 9 = 0$	
$(-3)^2 - 9$	0	$3^2 - 9$	0
$9 - 9$		$9 - 9$	
0		0	

In both cases, the result is true. The solution set is $\{3, -3\}$.

b.
$x^2 + 8x = 0$	0
$x(x + 8) = 0$	F Factor.
$x = 0$ or $x + 8 = 0$	F Factors = 0. Use the zero-product property, with $a = x$ and $b = x + 8$.
$x = 0$ or $x = -8$	Solve the equation $x + 8 = 0$.

CHECK

$x^2 + 8x = 0$		$x^2 + 8x = 0$	
$(0)^2 + 8(0)$	0	$(-8)^2 + 8(-8)$	0
$0 + 0$		$64 - 64$	
0		0	

Since both results check, the solution set is $\{0, -8\}$.

PROBLEM 1

Solve by factoring:

a. $x^2 - 36 = 0$

b. $2x^2 - 50x = 0$

Calculator Corner

Using a Graph to Find Solutions to a Quadratic Equation

To solve $x^2 - 9 = 0$ in Example l(a), graph $Y_1 = x^2 - 9$ with a standard window. The points at which the graph cuts the horizontal axis, 3 and -3, are the points at which $Y_1 = x^2 - 9 = 0$. Thus, 3 and -3 are the solutions of $x^2 - 9 = 0$.

You can verify this by letting $Y_2 = 0$ and pressing [2nd] [TRACE] 5, then [ENTER] three times.

The equation $x^2 = -x + 6$ is *not* in standard form. To write it in standard form, we add x and subtract 6 on both sides to obtain $x^2 + x - 6 = 0$, which can be solved by factoring. It is easier to factor a quadratic expression when the ax^2 term is positive. That is why we left the x^2 on the left side and moved the other two terms to the left side by using the addition property of equality. As you recall, to factor $x^2 + x - 6$, we need to find integers whose product is -6 and whose sum is 1. These integers are 3 and -2. Thus, we have

---

Answers to PROBLEMS

1. a. $6, -6$ **b.** $0, 25$

$$x^2 = -x + 6 \qquad \text{Given}$$
$$x^2 + x - 6 = 0 \qquad \text{0 Write in standard form (add x and subtract 6).}$$
$$(x + 3)(x - 2) = 0 \qquad \text{F Factor.}$$
$$x + 3 = 0 \quad \text{or} \quad x - 2 = 0 \qquad \text{F Factors = 0. Use the zero-product property.}$$
$$x = -3 \quad \text{or} \quad x = 2 \qquad \text{Solve } x + 3 = 0 \text{ and } x - 2 = 0.$$

CHECK

$x^2 = -x + 6$		$x^2 = -x + 6$	
$(2)^2$	$-(2) + 6$	$(-3)^2$	$-(-3) + 6$
4	4	9	$3 + 6$
			9

The solution set is $\{2, -3\}$.

EXAMPLE 2 Writing and solving equations in standard form

Solve by factoring:

a. $x^2 = 6x - 8$ **b.** $x^2 + x = 2$

SOLUTION

a. We first write the equation in standard form by subtracting $6x$ and adding 8 to obtain $x^2 - 6x + 8 = 0$. To factor $x^2 - 6x + 8$, we must find two integers whose product is 8 and whose sum is -6. These numbers are -4 and -2. Here is the procedure:

$$x^2 = 6x - 8 \qquad \text{Given}$$
$$x^2 - 6x + 8 = 0 \qquad \text{0 Subtract 6x and add 8.}$$
$$(x - 4)(x - 2) = 0 \qquad \text{F Factor.}$$
$$x - 4 = 0 \quad \text{or} \quad x - 2 = 0 \qquad \text{F Factors = 0. Use the zero-product property.}$$
$$x = 4 \quad \text{or} \quad x = 2 \qquad \text{Solve } x - 4 = 0 \text{ and } x - 2 = 0.$$

The solution set is $\{2, 4\}$. Check this.

b. The equation $x^2 + x = 2$ is *not* in standard form. To solve an equation by factoring, we must first write the equation in the standard form $ax^2 + bx + c = 0$. Thus, subtracting 2 from both sides of $x^2 + x = 2$, we have

$$x^2 + x - 2 = 0 \qquad \text{0}$$
$$(x + 2)(x - 1) = 0 \qquad \text{F Factor.}$$
$$x + 2 = 0 \quad \text{or} \quad x - 1 = 0 \qquad \text{F Factors = 0. Use the zero-product property.}$$
$$x = -2 \quad \text{or} \quad x = 1 \qquad \text{Solve } x + 2 = 0 \text{ and } x - 1 = 0.$$

The solution set is $\{1, -2\}$. The check is left to you.

PROBLEM 2

Solve by factoring:

a. $x^2 = 6x - 5$

b. $x^2 = 12 - x$

Calculator Corner

Using a Graph to Find Solutions to a Quadratic Equation

To do Example 2(a), write the equation in the standard form $x^2 - 6x + 8 = 0$, and graph $Y_1 = x^2 - 6x + 8$. Since the graph cuts the horizontal axis at $x = 2$ and $x = 4$, these are the solutions. Graph $Y_1 = x^2 + x - 2$ to verify the solutions to Example 2(b).

Answers to PROBLEMS

2. **a.** 1, 5 **b.** $-4, 3$

To solve the equation $6x^2 - x - 2 = 0$, we first factor $6x^2 - x - 2$ by trial and error or by the *ac* method (shown). For $6x^2 - x - 2$, the key number is $6 \cdot (-2) = -12$. Thus, we have to find integers whose product is -12 and whose sum is -1 (that is, -4 and 3) and use these numbers to rewrite the middle term $-x$. We have

$6x^2 - x - 2 = 0$	Given
$6x^2 - 4x + 3x - 2 = 0$	Write the middle term $-x$ as $-4x + 3x$.
$2x(3x - 2) + 1(3x - 2) = 0$	Factor the first and last pairs of terms.
$(3x - 2)(2x + 1) = 0$	Factor out the common factor, $3x - 2$.
$3x - 2 = 0$ or $2x + 1 = 0$	Factors = 0. Use the zero-product property.
$3x = 2$ or $2x = -1$	Solve $3x - 2 = 0$ and $2x + 1 = 0$.
$x = \frac{2}{3}$ or $x = \frac{-1}{2}$	

The solution set is $\{\frac{2}{3}, -\frac{1}{2}\}$. We check this for $x = \frac{2}{3}$.

CHECK

$$\begin{array}{c|c}
6x^2 - x - 2 = 0 & \\
6\left(\frac{2}{3}\right)^2 - \frac{2}{3} - 2 & 0 \\
6\left(\frac{4}{9}\right) - \frac{2}{3} - 2 & \\
\frac{8}{3} - \frac{2}{3} - 2 & \\
2 - 2 & \\
0 &
\end{array}$$

The check that $x = -\frac{1}{2}$ is a solution is left to you.

EXAMPLE 3 Solving quadratic equations by factoring using the *ac* method

Solve by factoring:

a. $12x^2 + 5x - 3 = 0$
b. $6x^2 - x = 1$
c. $x(x + 2) = (3x - 1)x + 1$

SOLUTION

a. You can factor $12x^2 + 5x - 3$ by trial and error or note that the key number for $12x^2 + 5x - 3$ is $(12)(-3) = -36$. Thus,

$12x^2 + 5x - 3 = 0$	Given
$12x^2 + 9x - 4x - 3 = 0$	Write $5x$ using coefficients whose sum is 5 and whose product is -36; that is, $5x = 9x - 4x$.
$3x(4x + 3) - 1(4x + 3) = 0$	Factor the first and last pairs of terms.
$(4x + 3)(3x - 1) = 0$	Factor out the common binomial factor, $4x + 3$.
$4x + 3 = 0$ or $3x - 1 = 0$	Factors = 0. Use the zero-product property.
$4x = -3$ or $3x = 1$	Solve $4x + 3 = 0$ and $3x - 1 = 0$.
$x = \frac{-3}{4}$ or $x = \frac{1}{3}$	

The solution set is $\{\frac{1}{3}, -\frac{3}{4}\}$. Check this.

PROBLEM 3

Solve by factoring:

a. $12x^2 + 13x - 4 = 0$
b. $8x^2 - 2x = 1$
c. $x(x - 3) = (4x + 7)x - 8$

Answers to PROBLEMS

3. a. $-\frac{4}{3}, \frac{1}{4}$ b. $-\frac{1}{4}, \frac{1}{2}$ c. $-4, \frac{2}{3}$

b. $6x^2 - x = 1$ is *not* in standard form. Subtracting 1 from both sides of the equation, we have

$6x^2 - x - 1 = 0$ O   The key number is -6.

$6x^2 - 3x + 2x - 1 = 0$ The integers whose product is -6 with sum -1 are -3 and 2.

$3x(2x - 1) + (2x - 1) = 0$ F   Factor the first and last pairs of terms.

$(2x - 1)(3x + 1) = 0$ Factor out $(2x - 1)$.

$2x - 1 = 0$ or $3x + 1 = 0$ F   Factors = 0. Use the zero-product property.

$2x = 1$ or $3x = -1$ Solve the equations $2x - 1 = 0$ and $3x + 1 = 0$.

$x = \dfrac{1}{2}$ or $x = \dfrac{-1}{3}$

The solution set is $\left\{\dfrac{1}{2}, -\dfrac{1}{3}\right\}$. Check this.

c. $x(x + 2) = (3x - 1)x + 1$ is *not* in standard form. First simplify both sides and then write the result in standard form.

$x(x + 2) = (3x - 1)x + 1$ Given

$x^2 + 2x = 3x^2 - x + 1$ Distributive property

$0 = 3x^2 - x^2 - 2x - x + 1$ Subtract $x^2 + 2x$.

$0 = 2x^2 - 3x + 1$ O   Simplify.

$2x^2 - 3x + 1 = 0$ Symmetric property

$2x^2 - 2x - x + 1 = 0$ Since the key number is 2, rewrite $-3x = -2x - x$.

$(2x^2 - 2x) + (-x + 1) = 0$ Group in pairs.

$2x(x - 1) - 1(x - 1) = 0$ F   Factor the GCF from each pair.

$(x - 1)(2x - 1) = 0$ The GCF is $(x - 1)$.

$x - 1 = 0$ or $2x - 1 = 0$ F   Factors = 0. Zero-product property

$x = 1$ or $x = \dfrac{1}{2}$ Solve $x - 1 = 0$ and $2x - 1 = 0$.

The solution set is $\left\{1, \dfrac{1}{2}\right\}$. Check this.

Calculator Corner

Using a Graph to Find Solutions to a Quadratic Equation

To do Example 3(c), first rewrite the equation as

$Y_1 = x(x + 2) - (3x - 1)x - 1,$
$Y_2 = 0$

and graph (you don't have to do the simplification). Using a decimal window, `ZOOM` `4`, and* `2nd` `TRACE` `5` with `ENTER` three times, you will find the solution 0.5 as shown. (Verify that the other solution is 1 by tracing to the other *x*-intercept.)

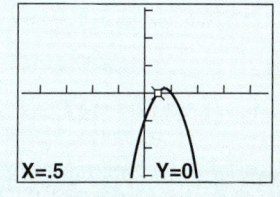

*Some calculators have an **intersect** and a **solve** feature. To find the point at which the curve intersects the horizontal axis, graph the horizontal axis $Y_2 = 0$. Get the cursor to a point near the intersection you want to find. Press `2nd` `TRACE` `5`. The calculator asks "First Curve?" "Second Curve?" and "Guess?" Press `ENTER` each time, and the calculator gives the intersection $x = 0.5, y = 0$.

In Chapter 2, we learned to clear the fractions when solving linear equations. In Example 4, we will use the same process when solving quadratic equations with fractions.

EXAMPLE 4 Solving quadratic equations with fractions

Solve:
$$\frac{x^2}{8} + \frac{x}{4} = 1$$

SOLUTION Multiply each term of $\frac{x^2}{8} + \frac{x}{4} = 1$ by the LCD, 8, to clear the fractions.

$$8 \cdot \frac{x^2}{8} + 8 \cdot \frac{x}{4} = 8 \cdot 1 \quad \text{Multiply by 8.}$$

$$8 \cdot \frac{x^2}{8} + \overset{2}{\cancel{8}} \cdot \frac{x}{\cancel{4}} = 8 \cdot 1 \quad \text{Divide out common factors.}$$

$$x^2 + 2x = 8 \quad \text{Simplify.}$$

$$x^2 + 2x - 8 = 0 \quad \text{Set equation equal to 0.}$$

$$(x + 4)(x - 2) = 0 \quad \text{Factor.}$$

$$x + 4 = 0 \quad x - 2 = 0 \quad \text{Set each factor = 0 and solve.}$$

$$x = -4 \quad x = 2$$

PROBLEM 4

Solve:
$$\frac{x^2}{5} - \frac{x}{3} = \frac{4}{5}$$

We have seen how solving quadratic equations, $ax^2 + bx + c = 0$, can be done by using the zero-product property. We don't have to limit our use of this property to solving just quadratic equations. In fact, we can use it to solve a polynomial equation of any degree that can be set equal to zero and then factored. We will illustrate this in Example 5 with an equation whose highest exponent on the variable is 3.

EXAMPLE 5 Solving polynomial equations by grouping and the zero-product property

Solve: $x^3 + 3x^2 - 4x - 12 = 0$

SOLUTION Since the polynomial has four terms, try to factor it by grouping. Here are the steps.

$$x^3 + 3x^2 - 4x - 12 = 0 \quad \text{Given}$$
$$x^2(x + 3) - 4(x + 3) = 0 \quad \text{Group for factoring.}$$
$$(x + 3)(x^2 - 4) = 0 \quad \text{Factor out the GCF.}$$
$$(x + 3)(x + 2)(x - 2) = 0 \quad \text{Factor } x^2 - 4 \text{ into } (x + 2)(x - 2).$$
$$\text{Factors = 0.}$$
$$x + 3 = 0 \quad \text{or} \quad x + 2 = 0 \quad \text{or} \quad x - 2 = 0$$
$$x = -3 \quad \text{or} \quad x = -2 \quad \text{or} \quad x = 2$$

The solution set is $\{2, -2, -3\}$. Check this.

PROBLEM 5

Solve:
$$x^3 + 2x^2 - 9x - 18 = 0$$

Calculator Corner

Using a Graph to Find Solutions to a Polynomial Equation

Let's find the solutions of $x^3 + 3x^2 - 4x - 12 = 0$ in Example 5 using a calculator. First graph $Y_1 = x^3 + 3x^2 - 4x - 12$ and $Y_2 = 0$ in a standard window. The points at which the graph cuts the horizontal axis, $x = -3, x = -2$, and $x = 2$ are the solutions of the equation as shown in Window 1 and can be verified using 2nd TRACE 5 with ENTER three times.

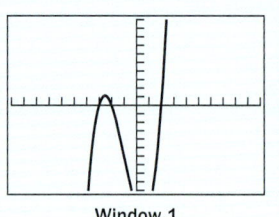

Window 1

Answers to PROBLEMS

4. $-\frac{4}{3}, 3$ 5. $-2, -3, 3$

B ▸ Using the Pythagorean Theorem

Quadratic equations can be used to find the lengths of the sides of right triangles using the **Pythagorean theorem**.

THE PYTHAGOREAN THEOREM

In any right triangle (a triangle with a 90° angle), the square of the longest side (hypotenuse) is equal to the sum of the squares of the other two sides (the legs). In symbols, this is

$$c^2 = a^2 + b^2$$

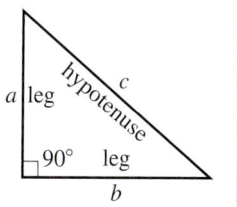

We can find the missing length of the right triangle by using the Pythagorean theorem, as shown in Example 6.

EXAMPLE 6 Using the Pythagorean theorem to solve word problems

The lengths of the three sides of a right triangle are consecutive integers. What are these lengths?

SOLUTION   We use the RSTUV method.

1. Read the problem. We need to find the lengths of the sides of the right triangle.

2. Select the unknowns. If x is an integer, what are the next two consecutive integers?

Let the length of the shortest side be x. Since the lengths of the sides are consecutive integers, we have

x	Length of the shortest side
$x + 1$	Length of the next side
$x + 2$	Length of the hypotenuse (the longest side)

3. Think of a plan. Make a diagram and use the Pythagorean theorem to obtain the equation

$$c^2 = a^2 + b^2$$
$$(x + 2)^2 = (x + 1)^2 + x^2$$

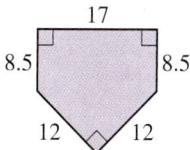

4. Use algebra to solve the resulting equation.

$x^2 + 4x + 4 = x^2 + 2x + 1 + x^2$	Multiply.
$x^2 + 4x + 4 = 2x^2 + 2x + 1$	Simplify.
$0 = x^2 - 2x - 3$	Subtract x^2, $4x$, and 4 from both sides.
$x^2 - 2x - 3 = 0$	Write in standard form.
$(x - 3)(x + 1) = 0$	Factor.
$x - 3 = 0$ or $x + 1 = 0$	Factors = 0. Use the zero-product property.
$x = 3$ or $x = -1$	Solve $x - 3 = 0$ and $x + 1 = 0$.

The lengths of the sides must be positive, so we discard the negative answer, -1. Thus, the shortest side is 3 units, so the other two sides are 4 and 5 units.

5. Verify the solution.   The verification is left to you.

PROBLEM 6

According to Kreutzer and Kerley (1990), the Little League rule book's specification of the shape of home plate is a pentagon and looks like this:

Is this physically possible? Explain.

Answers to PROBLEMS

6. No, this is not possible because

$$12^2 + 12^2 = 17^2$$
$$288 \neq 289$$

EXAMPLE 7 An application of the Pythagorean theorem

A ladder is leaning against the side of a house with its bottom x feet away from the wall and resting against the wall of the house $x + 7$ feet above the ground. If the length of the ladder is 1 foot more than its height above the ground, how long is the ladder?

SOLUTION As before, we use the RSTUV method.

1. **Read the problem.** Form a mental picture of the dimensions.
2. **Select the unknown.** Let x feet be the distance from the foot of the ladder to the wall as stated in the problem.
3. **Think of a plan.** Make a diagram of the situation.

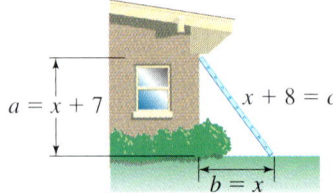

$a = x + 7$, $x + 8 = c$, $b = x$

The distance from the bottom of the ladder to the house is x feet. The ladder is resting on the side of the house $x + 7$ feet from the ground, and the length of the ladder is 1 foot more than the height from the ground. Thus, the length of the ladder is $x + 7 + 1$ or $x + 8$ feet. By the Pythagorean theorem, we have the equation

$$c^2 = a^2 + b^2$$
$$(x + 8)^2 = (x + 7)^2 + x^2$$

4. **Use algebra to solve the resulting equation.**

$(x + 8)^2 = (x + 7)^2 + x^2$	Given
$x^2 + 16x + 64 = x^2 + 14x + 49 + x^2$	Simplify.
$0 = x^2 - 2x - 15$	Subtract $x^2 + 16x + 64$ from both sides.
$x^2 - 2x - 15 = 0$	By the symmetric property
$(x - 5)(x + 3) = 0$	F Factor. We need two numbers whose product is -15 and whose sum is -2 (-5 and 3).
$x - 5 = 0$ or $x + 3 = 0$	F Factors = 0. Use the zero-product property.
$x = 5$ or $x = -3$	Solve each equation.

x represents a length, so $x = -3$ must be discarded as an answer. The foot of the ladder is 5 feet from the wall and leans against the wall $x + 7 = 5 + 7 = 12$ feet from the ground. The ladder's length is $x + 8 = 5 + 8 = 13$ feet.

5. **Verify the solution.** We note that $13^2 = 5^2 + 12^2$.

PROBLEM 7

Gina swims diagonally across her rectangular pool every day. If her pool is 7 meters longer than it is wide and it is 17 meters diagonally across, how long is her pool?

C > Applications Involving Quadratic Equations

Many problems can be studied using quadratic equations. For example, do you use hair spray containing chlorofluorocarbons (CFCs) for propellants? A U.N.-sponsored conference negotiated an agreement to stop producing CFCs by the year 2000 because they harm the ozone layer. In Example 8, we will check to see if the goal was reached.

Answers to PROBLEMS

7. 15 m

EXAMPLE 8 An application of quadratic equations

The production of CFCs for use as aerosol propellants (in thousands of tons) can be approximately represented by $P(t) = -0.4t^2 + 22t + 120$, where t is the number of years after 1960. When will production be stopped?

SOLUTION Production will be stopped when $P(t) = 0$; we need to solve the equation

$P(t) = -0.4t^2 + 22t + 120 = 0$

$-4t^2 + 220t + 1200 = 0$ O Multiply by 10 (to clear decimals).

$t^2 - 55t - 300 = 0$ Divide by -4 (to obtain t^2).

$(t - 60)(t + 5) = 0$ F Factor. We need two numbers whose product is -300 and whose sum is -55 (-60 and 5).

$t - 60 = 0$ or $t + 5 = 0$ F Factors = 0. Use the zero-product property.

$t = 60$ or $t = -5$ Solve each equation.

Since t represents the number of years *after* 1960, production will be zero (stopped) 60 years after 1960 or in 2020 (not in the year 2000 as promised!). The answer $t = -5$ has to be discarded because it represents 5 years *before* 1960, but the equation applies only to years *after* 1960.

PROBLEM 8

John's courtyard measures 6 meters by 8 meters. He plans a garden inside, with a grass border of uniform width on all four sides. The garden must contain exactly 15 square meters of area to meet the plans. How wide a border should John create? He comes up with the equation $(6 - 2x)(8 - 2x) = 15$ to solve the problem. (*To see how he arrived at that equation go to http://math.uww.edu/~mcfarlat/141/story11j.htm.*) Solve the quadratic equation to find the width of the border.

To find the dimensions of a rectangle knowing the area use the formula $A = lw$, as shown in Example 9.

EXAMPLE 9 Finding the dimensions of a rectangle using quadratic equations

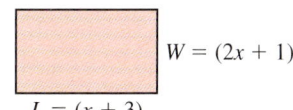

The expressions for the width and length of a rectangle are marked in the diagram. If the area is 7 square feet, find the dimensions of the rectangle.

SOLUTION The formula for finding the area of a rectangle is

$A = (L)(W)$

Substitute $A = 7$, $L = (x + 3)$, and $W = (2x + 1)$ into the formula

$7 = (x + 3)(2x + 1)$

$7 = 2x^2 + 7x + 3$ Multiply.

$0 = 2x^2 + 7x - 4$ O Set equation equal to 0.

$0 = (2x - 1)(x + 4)$ F Factor.

$2x - 1 = 0$ $x + 4 = 0$ F Set each factor = 0 and solve.

$2x = 1$ $x = -4$

$x = \frac{1}{2}$

To find the dimensions of the rectangle, we must replace each value of x into the expressions for width and length.

First, use $x = \frac{1}{2}$:

Width $= (2x + 1) = \left[2\left(\frac{1}{2}\right) + 1\right] = (1 + 1) = 2$ feet

Length $= (x + 3) + \left(\frac{1}{2} + 3\right) = 3\frac{1}{2}$ feet

Next use $x = -4$:

Length $= (x + 3) = (-4 + 3) = -1$

Since it is not possible to have a negative number for a dimension, we discard $x = -4$ and conclude that the dimensions of this rectangle are 2 feet by $3\frac{1}{2}$ feet.

PROBLEM 9

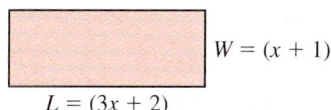

The expressions for the width and length of a rectangle are marked in the diagram. If the area is 10 square feet, find the dimensions of the rectangle.

Answers to PROBLEMS

8. 1.5 m   9. 2 ft by 5 ft

Calculator Corner

1. A polynomial function $P(x)$ has the graph shown in Window 1. The equation $P(x) = 0$ has two solutions.
   a. What is the degree of $P(x)$?
   b. What is the equation for $P(x)$?

2. A polynomial function $Q(x)$ has the graph shown in Window 2. The equation $Q(x) = 0$ has three solutions.
   a. What is the degree of $Q(x)$?
   b. What is the equation for $Q(x)$?

3. Suppose the graph of a polynomial function $R(x)$ crosses the x-axis n times.
   a. At most, how many solutions does $R(x)$ have?
   b. What is the degree of $R(x)$?
   c. If the solutions of $R(x)$ are $r_1, r_2, r_3, \ldots r_n$, can you write an equation for $R(x)$?

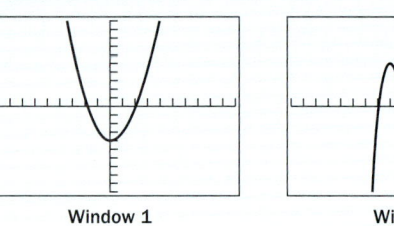

Window 1   Window 2

Exercises 5.7

Boost your grade at mathzone.com!
> Practice Problems
> NetTutor
> Self-Tests
> e-Professors
> Videos

< A > Solving Equations by Factoring   In Problems 1–46, solve the equations.

1. $(x + 1)(x + 2) = 0$
2. $(x + 3)(x + 4) = 0$
3. $(x - 1)(x + 4)(x + 3) = 0$
4. $(x + 5)(x - 3)(x + 2) = 0$
5. $\left(x - \frac{1}{2}\right)\left(x - \frac{1}{3}\right) = 0$
6. $\left(x - \frac{1}{4}\right)\left(x - \frac{1}{7}\right) = 0$
7. $y(y - 3) = 0$
8. $y(y - 4) = 0$
9. $y^2 - 64 = 0$
10. $y^2 - 1 = 0$
11. $y^2 - 81 = 0$
12. $y^2 - 100 = 0$
13. $x^2 + 6x = 0$
14. $x^2 + 2x = 0$
15. $x^2 - 3x = 0$
16. $x^2 - 8x = 0$
17. $y^2 - 12y = -27$
18. $y^2 - 10y = -21$
19. $y^2 = -6y - 5$
20. $y^2 = -3y - 2$
21. $x^2 = 2x + 15$
22. $x^2 = 4x + 12$
23. $3y^2 + 5y + 2 = 0$
24. $3y^2 + 7y + 2 = 0$
25. $2y^2 - 3y + 1 = 0$
26. $2y^2 - 3y - 20 = 0$
27. $2y^2 - y - 1 = 0$
28. $2y^2 - y - 15 = 0$

Hint: For Problems 29–34, multiply each term by the LCD *first*.

29. $\frac{x^2}{12} + \frac{x}{3} - 1 = 0$
30. $\frac{x^2}{2} - \frac{x}{12} - 1 = 0$
31. $\frac{x^2}{3} - \frac{x}{2} = -\frac{1}{6}$
32. $\frac{x^2}{6} + \frac{x}{3} = \frac{1}{2}$
33. $\frac{x^2}{12} + \frac{x}{2} = -\frac{2}{3}$
34. $\frac{x^2}{3} + \frac{x}{3} = \frac{1}{4}$
35. $(2x - 1)(x - 3) = 3x - 5$
36. $(3x + 1)(x - 2) = x + 7$

37. $(2x + 3)(x + 4) = 2(x - 1) + 4$
38. $(5x - 2)(x + 2) = 3(x + 1) - 7$
39. $(2x - 1)(x - 1) = x - 1$
40. $(3x - 2)(3x - 1) = 1 - 3x$
41. $x^3 + 4x^2 - 4x - 16 = 0$
42. $x^3 - 4x^2 - 4x + 16 = 0$
43. $x^3 - 5x^2 - 9x + 45 = 0$
44. $x^3 + 5x^2 - 9x - 45 = 0$
45. $3x^3 + 3x^2 = 12x + 12$
46. $2x^3 - 2x^2 - 18x + 18 = 0$

⟨B⟩ Using the Pythagorean Theorem   Use the Pythagorean theorem to solve Problems 47–50.

47. The sides of a right triangle are consecutive even integers. Find their lengths.

48. The hypotenuse of a right triangle is 4 centimeters longer than the shortest side and 2 centimeters longer than the remaining side. Find the lengths of the sides of the triangle.

49. The hypotenuse of a right triangle is 16 inches longer than the shortest side and 2 inches longer than the remaining side. Find the lengths of the sides of the triangle.

50. The hypotenuse of a right triangle is 8 inches longer than the shortest side and 1 inch longer than the remaining side. Find the lengths of the sides of the triangle.

⟨C⟩ Applications Involving Quadratic Equations   In Problems 51–54, use

$$d = 5t^2 + V_0 t$$

where d is the distance (in meters) traveled in t seconds by an object thrown downward with an initial velocity V_0 (in meters per second).

51. *Falling object*   An object is thrown downward with an initial velocity of 5 meters per second from a height of 10 meters. How long does it take the object to hit the ground?

52. *Falling object*   An object is thrown downward from a height of 28 meters with an initial velocity of 4 meters per second. How long does it take the object to reach the ground?

53. *Falling object*   An object is thrown downward from a building 15 meters tall with an initial velocity of 10 meters per second. How long does it take the object to hit the ground?

54. *Falling object*   How long would it take a package thrown downward from a plane with an initial velocity of 10 meters per second to hit the ground 175 meters below?

55. *Business*   It costs a business $(0.1x^2 + x + 50)$ dollars to serve x customers. How many customers can be served for a cost of $250?

56. *Business*   The cost of serving x customers is given by $(x^2 + 10x + 100)$ dollars. If $1300 is spent serving customers, how many customers are served?

57. *Manufacturing*   A manufacturer will produce x units of a product when its price is $(x^2 + 25x)$ dollars per unit. How many units will be produced when the price is $350 per unit?

58. *Manufacturing*   When the price of a ton of raw materials is $(0.01x^2 + 5x)$ dollars, a supplier will produce x tons of it. How many tons will be produced when the price is $5000 per ton?

59. *Ticket prices*   To attract more students, the campus theater decides to reduce ticket prices by x dollars from the current $5.50 price.

  a. What is the new price after the x dollar reduction?

  b. If the number of tickets sold is $100 + 100x$ and the revenue is the number of tickets sold times the price of each ticket, what is the revenue?

  c. If the theater wishes to have $750 in revenue, how much is the price reduction?

  d. If the reduction must be less than $1, what is the reduction?

60. *Rent*   An apartment owner wants to increase the monthly rent from the current $250 in n increases of $10.

  a. What is the new price after the n increases of $10?

  b. If the number of apartments rented is $70 - 2n$ and the revenue is the number of apartments rented times the rent, what is the revenue?

  c. If the owner wants to receive $17,980 per month, how many $10 increases can the owner make?

  d. What will be the monthly rent?

61. *Rectangle*   The expressions for the width and length of a rectangle are marked in the diagram. If the area is 11 square feet, find the dimensions of the rectangle.

62. *Rectangle*   If the length of a rectangle is 3 more than twice the width and the area is 14 square centimeters, find the dimensions of the rectangle.

$W = (x + 2)$

$L = (5x + 4)$

>>> Using Your Knowledge

Play Ball! Have you been to a baseball game lately? Did anybody hit a home run? The trajectory of a baseball is usually complicated, but we can get help from *The Physics of Baseball,* by Robert Adair. According to Mr. Adair, after t seconds, starting 1 second after the ball leaves the bat, the height of a ball hit at a 35° angle rotating with an initial backspin of 2000 revolutions per minute (rpm) and hit at about 110 miles per hour is given by

$$H(t) = -80t^2 + 239t + 3 \text{ (in feet)}$$

63. How many seconds will it be before the ball hits the ground?

The distance traveled by the ball is given by

$$D(t) = -5t^2 + 115t - 110 \text{ (in feet)}$$

64. How far will the ball travel before it hits the ground?

65. How far will the ball travel in 6 seconds, the time it takes a high fly ball to hit the ground?

>>> Write On

66. Write an explanation of the difference between quadratic and linear equations.

67. Explain why the zero-factor property works for more than two numbers whose product is zero.

68. Explain the differences in the procedure for solving $3(x-1)(x+4) = 0$ and $3x(x-1)(x+4) = 0$.

69. Write a word problem that uses the Pythagorean theorem in its solution.

>>> Concept Checker

Fill in the blank(s) with the correct word(s), phrase, or mathematical statement.

70. In a right triangle, the _____ says the hypotenuse squared is equal to the sum of the squares of the sides.

71. An example of a _____ is $x^2 + 2x = 8$.

72. If $(a)(b) = 0$, then either $a = 0$, $b = 0$, or both is illustrating the _____.

quadratic equation

standard form

zero-product property

Pythagorean theorem

>>> Mastery Test

Solve.

73. $x^3 + 2x^2 - 9x - 18 = 0$

74. $8x^2 - 2x = 1$

75. $12x^2 + 13x - 4 = 0$

76. $x^2 = 6x - 5$

77. $x^2 + x = 6$

78. $x^2 - 16 = 0$

79. $x^2 + 9x = 0$

80. The lengths of one leg of a right triangle and its hypotenuse are consecutive integers. If the shortest leg is 7 units shorter than the longer leg, find the lengths of the three sides.

>>> Skill Checker

Simplify.

81. $\dfrac{3x^5}{15x^7}$

82. $\dfrac{10x^{-3}}{5x^6}$

83. $\dfrac{20x^7}{10x^{-3}}$

84. $\dfrac{8x^{-4}}{16x^{-5}}$

85. $\dfrac{18x^{-10}}{9x^2}$

86. $\dfrac{4x^{-5}}{2x^{-5}}$

Factor.

87. $7x^2y - 7y^2$

88. $7x^2 - 7y^2$

89. $25x^2 - 9$

90. $x^2 + 5x - 14$

91. $2x^2 - 7x - 15$

92. $x^3 - 8$

Collaborative Learning 5A

How high could the heel of a shoe measure before you risk taking a tumble to the floor? At the Institute of Physics in London, scientists have come up with a polynomial formula that will tell you how high the heels can be without the risk of falling.

$$h = Q\left(12 + \frac{3s}{8}\right)$$

In this formula, h is the maximum height of the heel in centimeters, Q is a sociological factor with a value between 0 and 1 (explained later), and s is the size of the shoe in British measurement. According to the physicist at the University of Surrey who led the research, Paul Stevenson, it is based on the Pythagorean formula. Using this formula it was calculated that, when sober, Carrie Bradshaw of the television show *Sex and the City*, could wear a heel of just over 5 inches.

The formula for Q:

$$Q = \frac{p(y + 9)L}{(t + 1)(A + 1)(y + 10)(L + £20)}$$

The variables for this formula are as follows: p is the probability—from 0 to 1, with 1 being very probable—that the shoes will attract a mate for their wearer; y is the number of years' experience in high heels; L is the shoes' price in British pounds; t is the time in months since the shoes were fashionable (with 0 being right now); and A is the units of alcohol that the wearer plans to consume while in the shoes.

To use this formula in the United States, some of the measurements must be converted to U.S. measurements.

1. Find the conversion equivalences for changing from British women shoe sizes to U.S. women sizes.
2. Find the conversion equivalence for changing from the British pound to the U.S. dollar.
3. Find the conversion equivalence for changing from centimeters to inches.
4. Complete the table below to find the height of the heel in inches.

Size (U.S.)	s (Brit)	p (0–1)	y (yr)	Price (US $)	L (Brit £)	A (units)	t (mo)	Height of heel (cm)	Height of heel (in.)
6	?	0.5	10	$60	?	2	4		
7	?	1	20	$100	?	0	0		
8	?	0.8	15	$30	?	1	10		

5. Interview five people to get the necessary measurements so you can determine the appropriate height of heel. Study the results and have a discussion to see if there are any conclusions that can be drawn about what variables may or may not lead to a "safe" height of the heel.
6. Rewrite the formula so that the conversions are in the formula.

Collaborative Learning 5B

Did Picasso know about polynomials? Who knows, but if he did, it would have helped him paint his home! Here is how: suppose there are four walls and opposite sides have the same dimensions, then the total area of the four walls can be found by the polynomial, $A = 2(l_1 w_1) + 2(l_2 w_2)$, where l_1 and w_1 are the dimensions of one wall and l_2 and w_2 are the dimensions of an adjacent wall. Divide into groups and prepare a bid to paint your classroom. Along with this polynomial for finding the total area, you will need paint store information (from fliers or Internet research), measuring tape, calculators, and a chart (provided here) to itemize your bid.

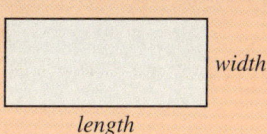

<CONTINUED>

Start by measuring the walls to find how many square feet you will have to paint. Remember to take into consideration things like whether you will paint around the bulletin boards and whether you will subtract the space taken up by the windows and doors when calculating total area.

Each gallon of paint could cover approximately 400–500 square feet of wall space, depending on the brand and grade of paint. Supplies will include brushes, rollers, drop cloths, and so on.

Determine how many painters will be painting, the number of hours they will work, and the hourly wage. Then complete the information on the chart. Transfer the information to a poster board to present to the class. Have a class discussion about your results addressing the following questions.

1. Was the estimate low enough to be in the running for the job, but high enough to give you some profit?
2. How important was it to prepare a precise estimate?
3. Does the size of the job have anything to do with how precise an estimate should be?
4. What happens if the estimate is too low or too high?
5. What other kinds of work would involve an understanding of area?
6. After listening to all the presenters, what would you do differently if you were asked to make another painting bid?

Items			Total
Number of Painters			
Hours to Complete the Job			
Area to be Painted			
	How Many	Cost Each	
Cost of Paint			
Cost of Total Hours of Wages			
Cost of Supplies			
Profit			
Final Bid			

› Research Questions

1. Who was the first person to use the notation $x, xx, x^3, x^4, \ldots$ for exponents, and when did this notation first occur?

2. In Section 5.1, we mentioned that the altitude of a diver above the water is given by $H(t) = -16t^2 + 118$. An eminent mathematician, born in Pisa in 1564, performed experiments from the Leaning Tower of Pisa. Find out the name of this mathematician, the nature of his experiments, and his conclusions.

3. How old do you think quadratic equations are? There are clay tablets indicating that the Babylonians of 2000 B.C. were familiar with formulas for solving quadratic equations. Write a paper about Babylonian mathematics with special emphasis on the solution of quadratic equations.

4. About 4000 years ago, the Egyptians used trained surveyors, the *harpedonaptae*. Find out what the word *harpedonaptae* means. Then write a paragraph explaining what these surveyors did and the ways in which they used the Pythagorean theorem in their work.

5. If you are looking for multicultural discovery, the Pythagorean theorem is it! Write a paper about the proofs of the Pythagorean theorem by
    a. The ancient Chinese
    b. Bhaskara
    c. The early Greeks
    d. Euclid
    e. President Garfield (not the cat, the twentieth U.S. president!)
    f. Pappus

Summary Chapter 5

Section	Item	Meaning	Example
5.1A	Monomial	A constant times a product of variables with whole-number exponents	$3x^2y$, $-7x$, $0.5x^3$, $\frac{2}{3}x^2yz^4$
	Polynomial	A sum or difference of monomials	$3x^2 - 7x + 8$, $x^2y + y^3$
	Terms	The individual monomials in a polynomial	The terms of $3x^2 - 7x + 8$ are $3x^2$, $-7x$, and 8.
	Coefficient	The numerical factor of a term	The coefficient of $3x^2$ is 3.
	Binomial	A polynomial with two terms	$5x^2 - 7$ is a binomial.
	Trinomial	A polynomial with three terms	$-3 + x^2 + x$ is a trinomial.
5.1B	Degree of a polynomial	Largest sum of the exponents in any term	The degree of $x^3 + 7x$ is 3. The degree of $-2x^3yz^2$ is 6.
	Descending order	Polynomial in one variable ordered from highest exponent to lowest	$7x^3 - 5x^2 + x - 3$ is in descending order.
5.1C	Evaluate a polynomial	Indicate the value of a polynomial by replacing the variable with specific values.	If $P(x) = x^2 + 3x - 1$, then $P(2) = (2)^2 + 3(2) - 1$ $= 4 + 6 - 1$ $= 9$
5.1D	Commutative property of addition	If P and Q are polynomials, $P + Q = Q + P$.	$x + 3x^2 = 3x^2 + x$
	Associative property of addition	If P, Q, and R are polynomials, $P + (Q + R) = (P + Q) + R$.	$x^2 + (3 + 7x) = (x^2 + 3) + 7x$
	Distributive property	If P, Q, and R are polynomials, $P(Q + R) = PQ + PR$ $(Q + R)P = QP + RP$.	$2x(x^2 + 3) = 2x^3 + 6x$
5.2A	Commutative property of multiplication	If P and Q are polynomials, $P \cdot Q = Q \cdot P$.	$x \cdot 3x^2 = 3x^2 \cdot x$
	Associative property of multiplication	If P, Q, and R are polynomials, $P \cdot (Q \cdot R) = (P \cdot Q) \cdot R$.	$x^2 \cdot (3 \cdot 7x) = (x^2 \cdot 3) \cdot 7x$
5.2C	Product of two binomials	$(x + a)(x + b) = x^2 + (b + a)x + ab$	$(x + 5)(x - 7) = x^2 - 2x - 35$
5.2D	Square of a binomial sum	$(x + a)^2 = x^2 + 2ax + a^2$	$(x + 5y)^2 = x^2 + 10xy + 25y^2$
	Square of a binomial difference	$(x - a)^2 = x^2 - 2ax + a^2$	$(x - 5y)^2 = x^2 - 10xy + 25y^2$
5.2E	Product of a sum and a difference	$(x + a)(x - a) = x^2 - a^2$	$(x + 2y)(x - 2y) = x^2 - 4y^2$
5.3A	Greatest common factor (GCF) of a polynomial in x (ax^n)	1. a is the greatest integer that divides each of the coefficients in the polynomial. 2. n is the smallest exponent of x in all terms of the polynomial.	The GCF of $3x^6 + 6x^3$ is $3x^3$.

(continued)

Section	Item	Meaning	Example
5.3B	Factoring by grouping (four terms)	1. Group terms with common factors. 2. Factor each group. 3. Factor out the (binomial) GCF.	$2x^3 + 4x^2 - 3x - 6$ $= (2x^3 + 4x^2) - (3x + 6)$ $= 2x^2(x + 2) - 3(x + 2)$ $= (x + 2)(2x^2 - 3)$
5.4A	Factoring $x^2 + (b + a)x + ab$	$x^2 + (b + a)x + ab = (x + a)(x + b)$	$x^2 + 5x + 4 = (x + 1)(x + 4)$
5.4C	The ac test	$ax^2 + bx + c$ is factorable only if there are two integers whose product is ac and whose sum is b.	$3x^2 + 5x + 2$ is factorable, since there are two integers with product 6 and sum 5.
5.5A	Factoring perfect square trinomials	$x^2 + 2ax + a^2 = (x + a)^2$ $x^2 - 2ax + a^2 = (x - a)^2$	$x^2 + 10x + 25 = (x + 5)^2$ $x^2 - 14x + 49 = (x - 7)^2$
5.5B	Factoring the difference of two squares	$x^2 - a^2 = (x + a)(x - a)$	$16x^2 - 9y^2 = (4x + 3y)(4x - 3y)$
5.5C	Factoring the sum or difference of two cubes	$x^3 + a^3 = (x + a)(x^2 - ax + a^2)$ $x^3 - a^3 = (x - a)(x^2 + ax + a^2)$	$8x^3 + 27 = (2x + 3)(4x^2 - 6x + 9)$ $8x^3 - 27 = (2x - 3)(4x^2 + 6x + 9)$

5.6A	General factoring strategy	1. Factor out the GCF.
		2. Check for:
		    Difference of two squares
		    Difference of two cubes
		    Sum of two cubes
		    Perfect square trinomials
		    Trinomials of the form $ax^2 + bx + c$
		    Four terms (grouping)
		3. Check by multiplying factors.	

5.7	Quadratic equation	An equation that can be written in the form $ax^2 + bx + c = 0$ $(a \neq 0)$	$3x^2 + 5x = -6$ is a quadratic equation.
5.7A	Zero-product property	For all real numbers a and b, $a \cdot b = 0$ means that $a = 0$ or $b = 0$, or both.	$(x + 1)(x + 2) = 0$ means $x + 1 = 0$ or $x + 2 = 0$.
5.7B	Pythagorean theorem	In any right triangle, the square of the longest side is equal to the sum of the squares of the other two sides: $c^2 = a^2 + b^2$.	If the sides of a right triangle are of lengths 3 and 4 and the hypotenuse is 5, $3^2 + 4^2 = 5^2$.

Review Exercises Chapter 5

(If you need help with these exercises, look in the section indicated in brackets.)

1. ⟨**5.1A, B**⟩ Classify as a monomial, binomial, or trinomial and give the degree.
   a. $x^3 + x^2y^3z$
   b. $x^3y^2z^3$
   c. $x^4 - 5x^2y^3 + xyz$

2. ⟨**5.1B**⟩ Write the polynomials in descending order.
   a. $-x^2 + 3x^4 - 5x + 2$
   b. $3x - x^2 + 4x^3$
   c. $6x^2 - 2 + x$

3. ⟨**5.1C, E**⟩ A $50,000 computer depreciates 20% each year and its value v(t) after t years is given by $v(t) = 50{,}000(1 - 0.20t)$.
   a. Find the value of the computer after 3 yr.
   b. Find the value of the computer after 5 yr.

4. ⟨**5.1C**⟩ Let $P(x) = x^2 - 3x + 5$ and find:
   a. $P(-1)$
   b. $P(2)$
   c. $P(-3)$

5. ⟨**5.1D**⟩ Add $2x^3 + 5x^2 - 3x - 1$ and:
   a. $8 - 7x + 3x^2 - x^3$
   b. $9 - 8x^2 + 3x^3$
   c. $7 - 4x + x^3$

6. ⟨**5.1D**⟩ Subtract $7x^3 - 5x^2 + 3x - 1$ from:
   a. $4x^3 + 2x^2 + 2$
   b. $7x^3 + 5x^2 + 4x - 7$
   c. $8x^2 - 9x + 3$

7. ⟨**5.2A**⟩ Multiply.
   a. $-2x^2y(x^2 + 3xy - 2y^3)$
   b. $-3x^2y^2(x^2 + 3xy - 2y^3)$
   c. $-4xy^2(x^2 + 3xy - 2y^3)$

8. ⟨**5.2B**⟩ Multiply.
   a. $(x - 1)(x^2 - 3x - 2)$
   b. $(x - 2)(x^2 - 3x - 2)$
   c. $(x + 3)(x^2 - 3x - 2)$

9. ⟨**5.2C**⟩ Multiply.
   a. $(2x + 3y)(4x - 5y)$
   b. $(2x + 3y)(3x + 2y)$
   c. $(2x + 3y)(5x - 3y)$

10. ⟨**5.2D**⟩ Multiply.
    a. $(2x + 5y)^2$
    b. $(3x + 7y)^2$
    c. $(4x + 9y)^2$

11. ⟨**5.2D**⟩ Multiply.
    a. $(3x - 2y)^2$
    b. $(4x - 7y)^2$
    c. $(5x - 6y)^2$

12. ⟨**5.2E**⟩ Multiply.
    a. $(3x + 2y)(3x - 2y)$
    b. $(4x + 3y)(4x - 3y)$
    c. $(5x + 3y)(5x - 3y)$

13. ⟨**5.3A**⟩ Factor completely.
    a. $15x^5 - 20x^4 + 10x^3 + 25x^2$
    b. $9x^5 - 12x^4 + 6x^3 + 15x^2$
    c. $6x^5 - 8x^4 + 4x^3 + 10x^2$

14. ⟨**5.3B**⟩ Factor completely.
    a. $6x^6 - 2x^4 + 15x^3 - 5x$
    b. $6x^6 - 8x^4 + 15x^3 - 20x$
    c. $6x^6 - 4x^4 + 9x^3 - 6x$

15. ⟨**5.4A**⟩ Factor completely.
    a. $x^2 - 3xy - 18y^2$
    b. $x^2 - 4xy - 12y^2$
    c. $x^2 - 5xy - 6y^2$

16. ‹ **5.4B** › *Factor completely.*
  a. $2x^2 - 7xy - 30y^2$
  b. $2x^2 - 3xy - 20y^2$
  c. $2x^2 - 5xy - 25y^2$

17. ‹ **5.4C** › *Factor completely.*
  a. $-18x^4y - 3x^3y^2 + 6x^2y^3$
  b. $-30x^4y - 35x^3y^2 - 10x^2y^3$
  c. $36x^4y - 30x^3y^2 - 36x^2y^3$

18. ‹ **5.5A** › *Factor completely.*
  a. $4x^2 - 28xy + 49y^2$
  b. $9x^2 - 42xy + 49y^2$
  c. $16x^2 - 56xy + 49y^2$

19. ‹ **5.5A** › *Factor completely.*
  a. $9x^2 + 24xy + 16y^2$
  b. $9x^2 + 30xy + 25y^2$
  c. $9x^2 + 36xy + 36y^2$

20. ‹ **5.5B** › *Factor completely.*
  a. $81x^4 - y^4$
  b. $x^4 - 16y^4$
  c. $81x^4 - 16y^4$

21. ‹ **5.5B** › *Factor completely.*
  a. $x^2 - 4x + 4 - y^2$
  b. $x^2 - 6x + 9 - y^2$
  c. $x^2 + 8x + 16 - y^2$

22. ‹ **5.5C** › *Factor completely.*
  a. $27x^3 + 8y^3$
  b. $27x^3 + 64y^3$
  c. $64x^3 + 27y^3$

23. ‹ **5.5C** › *Factor completely.*
  a. $27x^3 - 8y^3$
  b. $27x^3 - 64y^3$
  c. $64x^3 - 27y^3$

24. ‹ **5.6A** › *Factor completely.*
  a. $27x^6 - 8x^3y^3$
  b. $27x^7 - 64x^4y^3$
  c. $64x^8 - 27x^5y^3$

25. ‹ **5.6A** › *Factor completely.*
  a. $27x^6 + 3x^4$
  b. $4x^6 + 64x^4$
  c. $2x^6 - 18x^4$

26. ‹ **5.6A** › *Factor completely.*
  a. $27x^4 + 36x^3y + 12x^2y^2$
  b. $36x^4 - 48x^3y + 16x^2y^2$
  c. $45x^4 + 30x^3y + 5x^2y^2$

27. ‹ **5.6A** › *Factor completely.*
  a. $27x^4 - 18x^3y + 3x^2y^2$
  b. $36x^4 - 48x^3y + 16x^2y^2$
  c. $45x^4 + 60x^3y + 20x^2y^2$

28. ‹ **5.6A** › *Factor completely.*
  a. $12x^3y - 44x^2y^2 - 16xy^3$
  b. $15x^3y + 65x^2y^2 + 20xy^3$
  c. $18x^3y - 60x^2y^2 - 48xy^3$

29. ‹ **5.6A** › *Factor completely.*
  a. $2x^3 - x^2 - 2x + 1$
  b. $18x^3 - 9x^2 - 2x + 1$
  c. $32x^3 - 16x^2 - 2x + 1$

30. ‹ **5.7A** › *Solve.*
  a. $x^2 = -x + 12$
  b. $x^2 = -x + 20$
  c. $x^2 = -2x + 24$

31. ‹ **5.7A** › *Solve.*
  a. $6x^2 + x = 1$
  b. $8x^2 + 2x = 1$
  c. $10x^2 + 3x = 1$

32. ‹ **5.7A** › *Solve.*
  a. $x^3 + 2x^2 - x - 2 = 0$
  b. $x^3 + 4x^2 - x - 4 = 0$
  c. $x^3 + 2x^2 - 9x - 18 = 0$

33. ‹ **5.7B** › *Find the dimensions of a right triangle whose sides have the given dimensions.*
  a. x, $x + 7$, and $x + 8$ units long
  b. x, $x + 3$, and $x + 6$ units long
  c. x, $x + 4$, and $x + 8$ units long

34. ‹ **5.7C** › *The expressions for the width and length of a rectangle are marked in the diagram. If the area is 6 square feet, find the dimensions of the rectangle.*

$W = (x - 1)$
$L = (3x + 4)$

Practice Test Chapter 5

(Answers on page 424)

Visit www.mhhe.com/bello to view helpful videos that provide step-by-step solutions to several of the problems below.

1. Classify as a monomial, binomial, or trinomial and give the degree of $xy^3z^4 - x^7$.

2. Write in descending order: $-4 + 3x^2 - x^4 + 2x^3$

3. The total dollar cost $C(x)$ of manufacturing x units of a product each week is given by $C(x) = 15x + 300$.
   a. Find the cost of manufacturing 400 units.
   b. Find the cost of manufacturing 1500 units.

4. Let $P(x) = x^2 - 3x + 2$. Find $P(-2)$.

5. Add $6x^3 + 8x^2 - 6x - 4$ and $6 - 3x + x^2 - 3x^3$.

6. Subtract $8x^3 - 6x^2 + 5x - 3$ from $5x^3 + 3x^2 + 3$.

7. Multiply $-3x^2y(x^2 + 5xy - 3y^3)$.

8. Multiply.
   a. $(x - 2)(x^2 - 4x - 5)$
   b. $(3x + 5y)(4x - 7y)$

9. Multiply.
   a. $(2x + 3y)^2$
   b. $(3x - 4y)^2$

10. Multiply $(3x + 4y)(3x - 4y)$.

11. Factor completely: $12x^6 - 16x^5 + 8x^4 + 20x^3$.

12. Factor completely: $6x^7 + 6x^5 + 15x^4 + 15x^2$.

13. Factor completely:
    a. $x^2 - 3xy - 18y^2$
    b. $2x^2 + xy - 10y^2$

14. Factor completely: $36x^4y + 12x^3y^2 - 8x^2y^3$.

15. Factor completely:
    a. $16x^2 - 24xy + 9y^2$
    b. $9x^2 + 30xy + 25y^2$

16. Factor completely: $x^4 - 16y^4$.

17. Factor completely: $x^2 - 10x + 25 - y^2$.

18. Factor completely:
    a. $27x^3 + 8y^3$
    b. $8y^3 - 27x^3$

19. Factor completely:
    a. $8x^7 - x^4y^3$
    b. $6x^8 + 24x^6$

20. Factor completely:
    a. $8x^4 + 24x^3y + 18x^2y^2$
    b. $48x^2y^2 - 72xy^3 + 27y^4$

21. Factor completely: $9x^3y - 33x^2y^2 - 12xy^3$.

22. Solve.
    a. $x^2 = -3x + 10$
    b. $6x^2 + 7x = 3$

23. Solve $x^3 - x^2 - 4x + 4 = 0$.

24. The sides of a right triangle are x, $x + 2$, and $x + 4$ units long. Find the lengths of the sides of the triangle.

25. The length of a rectangle is 1 more than twice the width and the area is 10 square meters. Find the dimensions of the rectangle.

Answers to Practice Test Chapter 5

Answer	If You Missed Question	Section	Review Examples	Page
1. Binomial; 8	1	5.1A, B	1, 2	349, 350
2. $-x^4 + 2x^3 + 3x^2 - 4$	2	5.1B	2	350
3. a. $6300 b. $22,800	3	5.1C, E	3, 8	351, 354
4. 12	4	5.1C	3	351
5. $3x^3 + 9x^2 - 9x + 2$	5	5.1D	5, 6	353
6. $-3x^3 + 9x^2 - 5x + 6$	6	5.1D	7	354
7. $-3x^4y - 15x^3y^2 + 9x^2y^4$	7	5.2A	1	360
8. a. $x^3 - 6x^2 + 3x + 10$ b. $12x^2 - xy - 35y^2$	8	5.2B, C	2, 3	362, 363
9. a. $4x^2 + 12xy + 9y^2$ b. $9x^2 - 24xy + 16y^2$	9	5.2D	4	364–365
10. $9x^2 - 16y^2$	10	5.2E	5	365
11. $4x^3(3x^3 - 4x^2 + 2x + 5)$	11	5.3A	1, 2, 3	373, 374
12. $3x^2(x^2 + 1)(2x^3 + 5)$	12	5.3B	5, 6	375, 376
13. a. $(x + 3y)(x - 6y)$ b. $(x - 2y)(2x + 5y)$	13	5.4A, B, C	1, 2, 3, 4, 5	380–384
14. $4x^2y(3x + 2y)(3x - y)$	14	5.4C	6	384
15. a. $(4x - 3y)^2$ b. $(3x + 5y)^2$	15	5.5A	1, 2	389, 390
16. $(x^2 + 4y^2)(x + 2y)(x - 2y)$	16	5.5B	3	391
17. $(x - 5 + y)(x - 5 - y)$ or $(x + y - 5)(x - y - 5)$	17	5.5B	4	392
18. a. $(3x + 2y)(9x^2 - 6xy + 4y^2)$ b. $(2y - 3x)(4y^2 + 6xy + 9x^2)$	18	5.5C	5	393–394
19. a. $x^4(2x - y)(4x^2 + 2xy + y^2)$ b. $6x^6(x^2 + 4)$	19	5.6A	1	399
20. a. $2x^2(2x + 3y)^2$ b. $3y^2(4x - 3y)^2$	20	5.6A	2	400
21. $3xy(3x + y)(x - 4y)$	21	5.6A	3	400
22. a. $2, -5$ b. $\frac{1}{3}, -\frac{3}{2}$	22	5.7A	1, 2, 3	406–409
23. $1, 2, -2$	23	5.7A	5	410
24. 6, 8, and 10 units	24	5.7B	6	411
25. 2 m by 5 m	25	5.7C	9	413

Cumulative Review Chapters 1–5

1. Use braces to list the elements of the set of even natural numbers less than 8.

2. Write 0.19 as a fraction.

3. Divide: $-\dfrac{5}{9} \div \left(-\dfrac{1}{27}\right)$

4. Perform the calculation and write the answer in scientific notation: $(7.5 \times 10^3) \times (8 \times 10^{-7})$

5. Evaluate: $-4^3 + \dfrac{(4-12)}{2} + 6 \div 3$

6. Solve: $\dfrac{3}{7}y - 4 = 2$

7. Solve: $\left|\dfrac{4}{3}x + 6\right| + 8 = 15$

8. Graph: $\dfrac{x}{8} - \dfrac{x}{3} < \dfrac{x-8}{8}$

9. Graph: $\{x \mid x < -2 \text{ or } x \geq 3\}$

10. Graph: $|2x - 3| \leq 1$

11. Solve for A in $B = \dfrac{2}{5}(A - 11)$.

12. A freight train leaves a station traveling at 45 mi/hr. Two hours later, a passenger train leaves the same station in the same direction at 55 mi/hr. How far from the station does the passenger train overtake the freight train?

13. Find the x- and y-intercepts of $y = -3x - 5$.

14. Graph: $4x = -16$

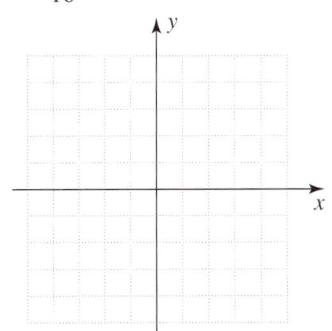

15. A line L_1 has slope $\dfrac{1}{2}$. Find whether the line passing through $(-3, -7)$ and $(3, -4)$ is parallel or perpendicular to line L_1.

16. Find the standard form of the equation of the line that passes through the points $(3, -1)$ and $(7, 4)$. Find slope

17. A line has slope -3 and y-intercept -1. Find the slope-intercept equation of the line.

18. Graph: $x \geq 3$

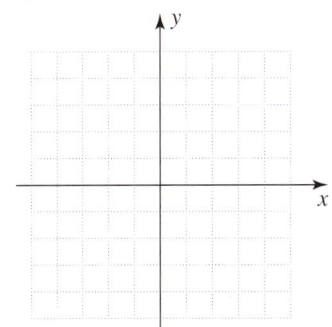

19. Solve by the elimination method:
$3x + 4y = -13$
$x + y = -4$

20. Solve the system:
$3x - 2y + z = 5$
$2x - y - 3z = 0$
$3x - y - z = 4$

21. Bert has $3.75 in nickels and dimes. He has 12 more nickels than dimes. How many of each coin does he have?

22. If $R(x) = x^2 - 5x - 5$, find $R(-4)$.

23. Multiply: $(2x - 7y)^2$

24. Factor completely: $8c^3 + 125$

25. Solve for x: $2x^2 + 9x = -9$

Chapter 6: Rational Expressions

Section
- 6.1 Rational Expressions
- 6.2 Multiplication and Division of Rational Expressions
- 6.3 Addition and Subtraction of Rational Expressions
- 6.4 Complex Fractions
- 6.5 Division of Polynomials and Synthetic Division
- 6.6 Equations Involving Rational Expressions
- 6.7 Applications: Problem Solving
- 6.8 Variation

The Human Side of Algebra

One of the most spectacular mathematical achievements of the sixteenth century was the discovery of an algebraic solution to cubic and quartic equations. The story of the discovery rivals the plots of contemporary novels. It starts with Scipione del Ferro's formula for solving the cubic $x^3 + mx = n$, a formula he passed on to his pupil Antonio Maria Fior. Enter Nicolo de Brescia, cruelly known as Tartaglia, "the stammerer," because of a speech impediment acquired during the sacking of Brescia by the French, which left his father dead and Nicolo with a saber cut that cleft his jaw and palate. So poor was his mother that she could only pay his tutor for a meager 15 days, and even then he was relegated to using tombstones as slates on which to work his exercises. Tartaglia announced in 1535 the discovery of a more general formula to solve $x^3 + mx^2 = n$. Fior, believing the announcement to be a bluff, challenged him to a problem-solving contest. Who won? Tartaglia and his formula handily defeated Fior, who failed to solve a single problem, thus entering the annals of mathematical history in ignominious defeat.

6.1 Rational Expressions

Objectives

A Find the values that make a rational expression undefined.

B Write an equivalent rational expression with the indicated denominator.

C Write a rational expression in one of the standard forms.

D Reduce a rational expression to lowest terms.

To Succeed, Review How To...

1. Find the GCF of two expressions (pp. 371–372).
2. Use the properties of exponents (pp. 38–45).
3. Factor polynomials (pp. 398–401).

Getting Started

A championship baseball team such as the one shown who went 54–6 for its season, must have players who can hit the ball well. Batting average, on-base percentage, slugging percentage, and total bases are a few of the statistics used to describe each player's performance. But what about some of the more advanced statistics used to measure a player's worth like the Isolated Power or ISO?

ISO attempts to separate and measure a player's ability to hit for power as exhibited in extra base hits (doubles, triples, and homeruns). This statistic measures just the player's extra base abilities. The formula is

$$ISO = \frac{\text{Total bases} - \text{Hits}}{\text{At bats}} = \frac{T - H}{B}$$

In 2005, Alex Rodriguez won the American League MVP and had an ISO of 0.289. David Ortiz was the runner up for MVP and had an ISO of 0.304. Even though Ortiz had a slightly higher ISO, it doesn't mean he had a better year but that he exhibited more "raw power."

Source: http://www.associatedcontent.com.

The *fraction*, $\frac{T-H}{B}$, is referred to as a rational expression, which we will study in this chapter. The word *fraction* is derived from the Latin word *fractio*, which means "to break" or "to divide." Any **fraction** of the form $\frac{a}{b}$, where a and b are integers and $b \neq 0$, is a rational number. We extend this idea to polynomial expressions as follows.

RATIONAL EXPRESSION

If P and Q are polynomials, then the algebraic expression

$$\frac{P}{Q} \quad \text{where} \quad Q \neq 0$$

is a **rational expression.** If $Q = 0$, the expression is undefined.

Some examples of rational expressions are

$$\frac{x^2 - 2x + 1}{x + 2}, \frac{1}{y^2 - 1}, \frac{x - 2}{3}$$

Studying these rational expressions in their abstract form will become more meaningful as we apply them to real-life experiences. For example, if you take three tests and want to find your mean average, the statistics formula is

$$\text{Mean} = \frac{\Sigma x}{n} = \frac{\text{sum of the test scores}}{\text{number of tests taken}} = \frac{t_1 + t_2 + t_3}{3}$$

where $\frac{t_1 + t_2 + t_3}{3}$ is a rational expression. If the test scores are 80, 75, and 93, then we can **evaluate** the rational expression to find the mean average. That is, we substitute the three scores into the rational expression and simplify.

$$\frac{t_1 + t_2 + t_3}{3} = \frac{80 + 75 + 94}{3} = \frac{249}{3} = 83$$

The value of the rational expression, 83, is the mean average of the test scores.

Let's evaluate the rational expression $\frac{x^2 - 2x}{x + 2}$ for three different values of x in the real numbers.

For $x = 1$: $\quad \dfrac{1^2 - 2(1)}{1 + 2} = \dfrac{1 - 2}{3} = \dfrac{-1}{3}$

For $x = 0$: $\quad \dfrac{0^2 - 2(0)}{0 + 2} = \dfrac{0 - 0}{2} = \dfrac{0}{2} = 0$

For $x = -2$: $\quad \dfrac{(-2)^2 - 2(-2)}{-2 + 2} = \dfrac{4 + 4}{0} = \dfrac{8}{0} \quad$ Undefined

The last evaluation illustrates the need to be cautious about the values used to substitute in for the variables in a rational expression. When $x = 1$, the rational expression was defined as $-\frac{1}{3}$, and when $x = 0$, it was defined as 0. However, when $x = -2$, the rational expression was **undefined**. This is the topic we will study first—finding the values that make a rational expression *undefined*.

A › Finding the Values That Make a Rational Expression Undefined

As we begin the study of rational expressions, it is important to note the difference between a rational *expression* and a rational *equation*. Some examples follow.

Rational *Expressions*	Rational *Equations*
$\dfrac{2x^2y^4z}{3ab}$	$\dfrac{x}{3} + \dfrac{x}{5} = 2$
$\dfrac{x - 4y}{5}$	$\dfrac{3}{x} = \dfrac{5}{x + 2}$
$\dfrac{6x}{3x + 1}$	$\dfrac{x}{x - 4} - \dfrac{x - 5}{x - 3} = \dfrac{3x + 16}{x^2 - x - 12}$

The visual distinction between the two is that rational *equations* will have an equal sign. Some of the actions involved with studying rational *expressions,* as we will do in Sections 6.1 through 6.5, include evaluating, simplifying, adding, subtracting, multiplying, and dividing. However, in Section 6.6 when we study rational *equations,* we will be *solving* them.

There may be infinitely many real numbers that make a rational expression defined, but if any real numbers make it undefined, then we must *not* use them as replacements for the variable. The values that make a rational expression, or rational function, undefined are the values that make the denominator of the fraction zero.

We are looking for the domain of the expression. For instance,

$$\frac{1}{x+3} \text{ is undefined for } x = -3$$

The domain of $\frac{1}{x+3}$ is all real numbers where $x \neq -3$.

EXAMPLE 1 Finding values that make a rational expression undefined

For what values are the following rational expressions undefined?

a. $\frac{x+1}{3x-4}$ b. $\frac{8x}{x^2+9}$ c. $\frac{x^2+3x+8}{x^2+2x-3}$

SOLUTION A rational expression is undefined for values of x that make the denominator zero. We find these values by setting the denominator equal to zero and solving the resulting equation.

a.
$$3x - 4 = 0 \quad \text{Set the denominator equal to zero.}$$
$$3x = 4 \quad \text{Add 4 to both sides.}$$
$$x = \frac{4}{3} \quad \text{Divide both sides by 3.}$$

The number $\frac{4}{3}$ will make the rational expression undefined.

b.
$$x^2 + 9 = 0 \quad \text{Set the denominator equal to zero.}$$

Since $x^2 = -9$ has no real number solution, we conclude that there aren't any real numbers that will make the rational expression undefined.

c. Since this is a quadratic equation, we use "OFF" as our guide.

$$x^2 + 2x - 3 = 0 \quad \text{O Set the denominator equal to zero.}$$
$$(x-1)(x+3) = 0 \quad \text{F Factor.}$$
$$x - 1 = 0 \text{ or } x + 3 = 0 \quad \text{F Factors = 0. Use the zero-factor property.}$$
$$x = 1 \text{ or } \quad x = -3 \quad \text{Solve each equation.}$$

The numbers 1 and -3 make the rational expression undefined.

PROBLEM 1

For what values are the following rational expressions undefined?

a. $\frac{-5}{2x}$ b. $\frac{x-5}{2x+3}$

c. $\frac{x^2-6x}{x^2-6x-7}$

To avoid mentioning over and over that the denominators of algebraic fractions must not be zero, we make the following rule.

UNDEFINED RATIONAL EXPRESSIONS

The variables in a rational expression may not be replaced by values that will make the denominator zero.

B ▶ Writing Equivalent Rational Expressions

The fraction $\frac{4}{5}$ can be written as an equivalent fraction with a denominator of 10 by multiplying both numerator and denominator by 1 in the form of $\frac{2}{2}$ to obtain

$$\frac{4}{5} \cdot 1 = \frac{4 \cdot 2}{5 \cdot 2} = \frac{8}{10}$$

Answers to PROBLEMS

1. a. 0   b. $-\frac{3}{2}$ c. $-1, 7$

FUNDAMENTAL PROPERTY OF RATIONAL EXPRESSIONS

If P, Q, and K are polynomials, then

$$\frac{P}{Q} = \frac{P \cdot K}{Q \cdot K}$$

for all values for which the denominators are not zero.

In arithmetic and algebra, it is sometimes necessary to write a fraction or rational expression equivalently with a new denominator. Sometimes this is called *building up* the fraction or rational expression. Understanding the procedure in arithmetic will facilitate learning to do it with rational expressions, so we will do an arithmetic example beside one with a rational expression.

Steps	Arithmetic Example of Building Up a Rational Number	Algebra Example of Building Up a Rational Expression
	Write $\frac{4}{7}$ equivalently with a denominator of 42.	Write $\frac{5x^3}{3y^2}$ equivalently with a denominator of $6y^7$.
1. Factor the new denominator with the old denominator factored out to get the *multiplier*.	New denominator, $42 = 7(6)$ So, "6" is the multiplier.	New denominator, $6y^7 = 3y^2(2y^5)$ So, "$2y^5$" is the multiplier.
2. Multiply numerator and denominator by the multiplier.	$\frac{4}{7} \cdot \frac{6}{6} = \frac{24}{42}$	$\frac{5x^3}{3y^2} \cdot \frac{2y^5}{2y^5} = \frac{10x^3y^5}{6y^7}$

In the last step, we used the Fundamental Property of Rational Expressions when we multiplied the numerator and denominator by the same number. Remember $\frac{6}{6}$ and $\frac{2y^5}{2y^5}$ both equal 1, so we are multiplying the original number by 1. That is why the result will be equivalent. This skill will be used in Section 6.3 when adding or subtracting rational expressions with unlike denominators.

EXAMPLE 2 Writing equivalent rational expressions with a specified denominator

Write:

a. $\frac{5}{8a^2}$ with a denominator of $16a^5$

b. $\frac{x+7}{9y^3}$ with a denominator of $18y^4$

c. $\frac{3x+1}{x-1}$ with a denominator of $x^2 + 2x - 3$

PROBLEM 2

Write:

a. $\frac{3}{7x}$ with a denominator of $28x^4$

b. $\frac{4-x}{5y^2}$ with a denominator of $40y^2$

c. $\frac{x+1}{2x+1}$ with a denominator of $2x^2 - 5x - 3$

(continued)

Answers to PROBLEMS

2. **a.** $\frac{12x^3}{28x^4}$ **b.** $\frac{32-8x}{40y^2}$

   c. $\frac{x^2 - 2x - 3}{(2x+1)(x-3)}$

   or $\frac{x^2 - 2x - 3}{2x^2 - 5x - 3}$

SOLUTION First find the multiplier and then multiply the numerator and denominator by it.

a.

Write $\frac{5}{8a^2}$ equivalently with a denominator of $16a^5$.

1. Factor the new denominator with the old denominator factored out to get the *multiplier*.

   New denominator, $16a^5 = 8a^2(2a^3)$ So, "$2a^3$" is the multiplier.

2. Multiply numerator and denominator by the multiplier.

   $\frac{5}{8a^2} \cdot \frac{2a^3}{2a^3} = \frac{10a^3}{16a^5}$

This means $\frac{5}{8a^2} = \frac{10a^3}{16a^5}$. This is one time we *do not* reduce our answer.

b.

Write $\frac{x+7}{9y^3}$ equivalently with a denominator of $18y^4$.

1. Factor the new denominator with the old denominator factored out to get the *multiplier*.

   New denominator, $18y^4 = 9y^3(2y)$ So, "$2y$" is the multiplier.

2. Multiply numerator and denominator by the multiplier.

   $\frac{x+7}{9y^3} \cdot \frac{2y}{2y} = \frac{(x+7)2y}{18y^4} = \frac{2xy + 14y}{18y^4}$

This means $\frac{x+7}{9y^3} = \frac{2xy + 14y}{18y^4}$. When *building up* a fraction, *do not* reduce.

c.

Write $\frac{3x+1}{x-1}$ equivalently with a denominator of $x^2 + 2x - 3$.

1. Factor the new denominator with the old denominator factored out to get the *multiplier*.

   New denominator, $x^2 + 2x - 3 = (x-1)(x+3)$ So, "$(x+3)$" is the multiplier.

2. Multiply numerator and denominator by the multiplier.

   $\frac{3x+1}{x-1} \cdot \frac{(x+3)}{(x+3)}$
   $= \frac{(3x+1)(x+3)}{(x-1)(x+3)} = \frac{3x^2 + 10x + 3}{x^2 + 2x - 3}$

This means $\frac{3x+1}{x-1} = \frac{3x^2 + 10x + 3}{x^2 + 2x - 3}$. Remember, *do not* reduce when *building up*.

C › Writing a Rational Expression in Standard Form

There are three signs associated with a fraction:

1. The sign in front of the fraction
2. The sign of the numerator
3. The sign of the denominator

Using our definition of a quotient and the fundamental property of rational expressions, we can conclude that

$$\frac{-a}{b} = \frac{a}{-b} = -\frac{a}{b} = -\frac{-a}{-b} \quad \text{and} \quad \frac{a}{b} = \frac{-a}{-b} = -\frac{a}{-b} = -\frac{-a}{b}$$

This means

$$-\frac{-12}{-6} = \frac{-12}{6} = \frac{12}{-6} = -\frac{12}{6} = -2 \quad \text{and} \quad \frac{5}{15} = \frac{-5}{-15} = -\frac{5}{-15} = -\frac{-5}{15} = \frac{1}{3}$$

The forms $\frac{-a}{b}$ and $\frac{a}{b}$, in which the sign of the fraction and that of the denominator are positive, are called the **standard forms** of the fractions. Thus, $\frac{-2}{9}$ and $\frac{4}{7}$ are in standard form, but $\frac{2}{-9}$ and $\frac{-4}{-7}$ are not.

STANDARD FORM OF A FRACTION

$$-\frac{a}{b} = \frac{-a}{b} \qquad \frac{-a}{-b} = \frac{a}{b}$$

$$\frac{a}{-b} = \frac{-a}{b} \qquad -\frac{-a}{b} = \frac{a}{b}$$

$$-\frac{-a}{-b} = \frac{-a}{b} \qquad -\frac{a}{-b} = \frac{a}{b}$$

In expressions with more than one term in the numerator or denominator, there are alternative standard forms. For example,

$$\underbrace{\frac{-1}{x-y}} = \underbrace{\frac{-1}{-(y-x)}} = \frac{1}{y-x} \qquad \text{Recall that } x - y = -(y - x), \text{ since } -(y - x) = -y + x = x - y.$$

same

Either

$$\frac{-1}{x-y} \quad \text{or} \quad \frac{1}{y-x}$$

can be used as the standard form. We prefer $\frac{1}{y-x}$

because it has only one minus or negative sign, whereas $\frac{-1}{x-y}$

has two.

EXAMPLE 3 Writing rational expressions in standard form

Write in standard form:

a. $\dfrac{x}{-2}$ **b.** $-\dfrac{-3}{y}$ **c.** $-\dfrac{x-y}{5}$

SOLUTION

a. $\dfrac{x}{-2} = \dfrac{-x}{2}$ **b.** $-\dfrac{-3}{y} = \dfrac{3}{y}$ **c.** $-\dfrac{x-y}{5} = \dfrac{-(x-y)}{5}$, or $\dfrac{y-x}{5}$

PROBLEM 3

Write in standard form:

a. $-\dfrac{7}{y}$ **b.** $-\dfrac{-x}{4}$

c. $-\dfrac{3a-b}{8}$

D › Reducing Rational Expressions to Lowest Terms

The fundamental property of rational expressions can also be used to **simplify (reduce)** fractions—to write fractions as equivalent ones in which no integers other than 1 can be divided exactly into both the numerator and denominator. For example, the fraction $\frac{14}{21}$ can be simplified by writing the numerator and denominator in factored form and using the fundamental principle of rational expressions. Thus,

$$\frac{14}{21} = \frac{2 \cdot \overset{1}{\cancel{7}}}{3 \cdot \underset{1}{\cancel{7}}} = \frac{2}{3}$$

Answers to PROBLEMS

3. **a.** $\dfrac{-7}{y}$ **b.** $\dfrac{x}{4}$ **c.** $\dfrac{b-3a}{8}$

Here we are dividing the numerator and denominator by the common factor 7. We usually write

$$\frac{\cancel{14}^2}{\cancel{21}_3}$$

and say that the fraction $\frac{2}{3}$ is in **lowest terms**.

The rational expression

$$\frac{(x+3)(x^2-4)}{3(x+2)(x^2+x-6)}$$

can also be written in lowest terms by using the fundamental property of rational expressions.

PROCEDURE

Procedure for Reducing Rational Expressions

1. Write the numerator and denominator of the rational expression in completely factored form.
2. Find the greatest common factor (GCF) of the numerator and denominator and replace the quotient of the common factors by the number 1, since $\frac{a}{a} = 1$.
3. Rewrite the rational expression in lowest terms.

We are now ready to simplify $\frac{6x-8}{9x^2-4}$.

Here are the steps:

1. Write the numerator and denominator in completely factored form.

$$\frac{2(3x-4)}{(3x+4)(3x-4)} \quad \begin{array}{l}\text{Factor out the GCF of 2.}\\ \text{Factor the difference of two squares.}\end{array}$$

2. Find the GCF of the numerator and denominator and replace the quotient of the common factors by the number 1.

$$\frac{2\overset{1}{\cancel{(3x-4)}}}{(3x+4)\cancel{(3x-4)}}$$

3. Rewrite the rational expression in lowest terms.

$$\frac{2}{(3x+4)}$$

Let's try another example. Simplify $\frac{(x+3)(x^2-4)}{3(x+2)(x^2+x-6)}$.

Here are the steps:

1. Write the numerator and denominator in completely factored form.

$$\frac{(x+3)(x+2)(x-2)}{3(x+2)(x+3)(x-2)}$$

2. Find the GCF of the numerator and denominator (factors are rearranged to be in the same order) and replace the quotient of the common factors by the number 1.

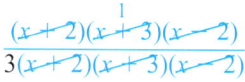

3. Rewrite the rational expression in lowest terms.

$$\frac{1}{3}$$

The whole procedure can be written as

$$\frac{(x+3)(x^2-4)}{3(x+2)(x^2+x-6)} = \frac{\overset{1}{\cancel{(x+3)}}\overset{1}{\cancel{(x+2)}}\overset{1}{\cancel{(x-2)}}}{3\cancel{(x+2)}\cancel{(x+3)}\cancel{(x-2)}} = \frac{1}{3}$$

CAUTION
Only common factors can be divided out.

$$\frac{\cancel{y}(x+8)}{\cancel{y}(x+4)} = \frac{x+8}{x+4}$$

Rational expressions like $\frac{x+8}{x+4}$ cannot be simplified since the numerator and denominator are sums, *not* products, and the x is a term, *not* a factor.

EXAMPLE 4 Reducing rational expressions to lowest terms
Reduce each rational expression to lowest terms:

a. $\dfrac{x^3 y^4}{xy^6}$ b. $\dfrac{xy - y^2}{x^2 - y^2}$ c. $\dfrac{2x + xy}{x}$

SOLUTION

a. $\dfrac{x^3 y^4}{xy^6} = \dfrac{x^2 \cdot \overset{1}{\cancel{xy^4}}}{y^2 \cdot \cancel{xy^4}}$ Factor the numerator and denominator using the GCF as a factor.

$= \dfrac{x^2}{y^2}$ Divide out xy^4, the GCF.

b. $\dfrac{xy - y^2}{x^2 - y^2} = \dfrac{y \cdot \overset{1}{\cancel{(x-y)}}}{(x+y)\cancel{(x-y)}}$ Factor the numerator and denominator completely.

$= \dfrac{y}{x+y}$ Divide out $(x-y)$, the GCF.

c. $\dfrac{2x + xy}{x} = \dfrac{\overset{1}{\cancel{x}}(2+y)}{\cancel{x}}$ Factor the numerator completely.

$= 2 + y$ Divide out x, the GCF.

PROBLEM 4
Reduce to lowest terms:

a. $\dfrac{3x^5 y^7}{6x^2 y^3}$

b. $\dfrac{2x^2 + 4xy}{x^2 - 4y^2}$

c. $\dfrac{3y + xy}{y^2}$

Now, consider the case where additive inverses are divided, such as $\frac{2}{-2}$ or $\frac{-9}{9}$. We can see the result will always be -1. That is easy to recognize in arithmetic but what about in rational expressions?

Using the rational expression $\dfrac{x-5}{5-x}$, if $x = 7$, the expression becomes

$$\frac{7-5}{5-7} = \frac{2}{-2} = -1$$

In the same expression, if $x = -4$, the expression becomes

$$\frac{-4-5}{5-(-4)} = \frac{-9}{9} = -1$$

In general, $\dfrac{a-b}{b-a}$ will yield the quotient of additive inverses. The result will always equal -1.

Answers to PROBLEMS

4. a. 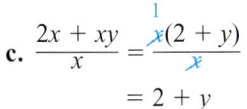   b. $\dfrac{2x}{x-2y}$ c. $\dfrac{3+x}{y}$

QUOTIENT OF ADDITIVE INVERSES

$$\frac{a-b}{b-a} = -1$$

We use this idea in Example 5.

EXAMPLE 5 Reducing quotients involving additive inverses

Reduce to lowest terms:

a. $\dfrac{x^3 - y^3}{y - x}$ b. $\dfrac{x^2 - y^2}{y^3 - x^3}$

SOLUTION

a. $\dfrac{x^3 - y^3}{y - x} = \dfrac{\overset{-1}{(x - y)}(x^2 + xy + y^2)}{y - x}$ Factor the numerator as the difference of perfect cubes.

$= -1(x^2 + xy + y^2)$ Quotient of additive inverses: $\dfrac{x - y}{y - x} = -1$

$= -(x^2 + xy + y^2)$

b. $\dfrac{x^2 - y^2}{y^3 - x^3} = \dfrac{(x + y)(x - y)}{(y - x)(y^2 + xy + x^2)}$ Factor the numerator as the difference of perfect squares.

Factor the denominator as the difference of perfect cubes.

$= \dfrac{\overset{-1}{(x - y)}(x + y)}{(y - x)(y^2 + xy + x^2)}$ Quotient of additive inverses: $\dfrac{x - y}{y - x} = -1$

$= \dfrac{-(x + y)}{y^2 + xy + x^2}$

PROBLEM 5

Reduce to lowest terms:

a. $\dfrac{x^3 - 8}{2 - x}$ b. $\dfrac{1 - 9x^2}{27x^3 - 1}$

Boost your grade at **mathzone.com**!
- Practice Problems
- NetTutor
- Self-Tests
- e-Professors
- Videos

> Exercises 6.1

< A > **Finding Values That Make a Rational Expression Undefined** In Problems 1–12, find the values (if any exist) that make the rational expression undefined.

1. $\dfrac{5}{3x}$
2. $\dfrac{x + 1}{2x}$
3. $\dfrac{x}{x + 3}$
4. $\dfrac{y}{y - 4}$
5. $\dfrac{5m - 5}{5m + 10}$
6. $\dfrac{4q + 4}{4q - 4}$
7. $\dfrac{m + 7}{m^2 - m - 2}$
8. $\dfrac{u - 9}{u^2 + 6u + 5}$
9. $\dfrac{p^2 - 2p}{p^2 - 9}$
10. $\dfrac{z^2 - 3z - 4}{9z^2 - 4}$
11. $\dfrac{4v^2 - 9}{4v^2 + 9}$
12. $\dfrac{9y^2 - 16}{16 + y^2}$

< B > **Writing Equivalent Rational Expressions** In Problems 13–30, write the given rational expression with the indicated denominator.

13. $\dfrac{2x}{3y}$; denominator $6y^3$
14. $\dfrac{-3y}{2x}$; denominator $8x^2$
15. $\dfrac{5}{3(x - 7)}$; denominator $18(x - 7)$
16. $\dfrac{4x}{2(x + 3)}$; denominator $12(x + 3)$

Answers to PROBLEMS

5. **a.** $-(x^2 + 2x + 4)$ or $-x^2 - 2x - 4$ **b.** $\dfrac{-(1 + 3x)}{9x^2 + 3x + 1}$

17. $\dfrac{6}{(x+y)}$; denominator $(x+y)^2$

18. $\dfrac{3x}{(x+8)}$; denominator $(x+8)^2$

19. $\dfrac{x}{x+y}$; denominator x^2-y^2

20. $\dfrac{-y}{x-y}$; denominator x^2-y^2

21. $\dfrac{-x}{2x-3y}$; denominator $4x^2-9y^2$

22. $\dfrac{-x}{2x-y}$; denominator $4x^2-y^2$

23. $\dfrac{4x}{x+1} = \dfrac{?}{x^2-x-2}$

24. $\dfrac{5y}{y-1} = \dfrac{?}{y^2+2y-3}$

25. $\dfrac{-5x}{x+3}$; denominator x^2+x-6

26. $\dfrac{-3y}{y-4}$; denominator y^2-2y-8

27. $\dfrac{3}{x+y}$; denominator x^3+y^3

28. $\dfrac{-4}{x+y}$; denominator x^3+y^3

29. $\dfrac{x}{x-y} = \dfrac{?}{x^3-y^3}$

30. $\dfrac{-y}{x-y} = \dfrac{?}{x^3-y^3}$

⟨C⟩ **Writing a Rational Expression in Standard Form** In Problems 31–40, write each fraction in standard form.

31. $-\dfrac{y}{-2}$

32. $-\dfrac{x-3}{y}$

33. $-\dfrac{x}{x-5}$

34. $\dfrac{2x-y}{-x}$

35. $-\dfrac{-2x}{-5y}$

36. $\dfrac{-x}{-y}$

37. $\dfrac{-(x+y)}{-(x-y)}$

38. $\dfrac{-(3x+y)}{-(x-5y)}$

39. $\dfrac{-1}{-(x-2)}$

40. $\dfrac{-y}{-(x+1)}$

⟨D⟩ **Reducing Rational Expressions to Lowest Terms** In Problems 41–70, simplify each fraction.

41. $\dfrac{x^4 y^2}{xy^5}$

42. $\dfrac{x^5 y^3 z^2}{x^2 y^6 z^4}$

43. $\dfrac{3x-3y}{x-y}$

44. $\dfrac{4x^2}{4x-4y}$

45. $\dfrac{3x-2y}{9x^2-4y^2}$

46. $\dfrac{4x^2-9y^2}{2x+3y}$

47. $\dfrac{(x-y)^3}{x^2-y^2}$

48. $\dfrac{x^2-y^2}{(x+y)^3}$

49. $\dfrac{ay^2-ay}{ay}$

50. $\dfrac{a^3+2a^2+a}{a}$

51. $\dfrac{x^2+2xy+y^2}{x^2-y^2}$

52. $\dfrac{x^2+3x+2}{x^2+2x+1}$

53. $\dfrac{y^2-8y+15}{y^2+3y-18}$

54. $\dfrac{y^2+7y-18}{y^2-3y+2}$

55. $\dfrac{2-y}{y-2}$

56. $\dfrac{3(x-y)}{4(y-x)}$

57. $\dfrac{9-x^2}{x-3}$

58. $\dfrac{25-9x^2}{3x-5}$

59. $\dfrac{y^3-8}{2-y}$

60. $\dfrac{2+x}{x^3+8}$

61. $\dfrac{3x-2y}{2y-3x}$

62. $\dfrac{5y - 2x}{2x - 5y}$

63. $\dfrac{x^2 + 4x - 5}{1 - x}$

64. $\dfrac{x^2 - 2x - 15}{5 - x}$

65. $\dfrac{x^2 - 6x + 8}{4 - x}$

66. $\dfrac{x^2 - 8x + 15}{3 - x}$

67. $\dfrac{2 - x}{x^2 + 4x - 12}$

68. $\dfrac{3 - x}{x^2 + 3x - 18}$

69. $-\dfrac{3 - x}{x^2 - 5x + 6}$

70. $-\dfrac{4 - x}{x^2 - 3x - 4}$

》》》 Using Your Knowledge

Buying Pollution Permits

71. Can you buy a permit to pollute the air? Unfortunately, yes. On March 29, 1993, the Chicago Board of Trade sold a permit for about $21 million to emit 150,000 tons of sulfur dioxide. (This amount represents 1% of the yearly sulfur dioxide emissions.) If the price (in millions of dollars) for removing $p\%$ of the sulfur dioxide is given by

$$\dfrac{2100\,p}{100 - p}$$

find the price that should be paid to remove

  a. 20%
  b. 40%
  c. 60%
  d. Can we afford to remove 100% of the sulfur dioxide? Explain.

72. The demand for a product is given by

$$N(p) = \dfrac{2p + 100}{10p + 10}$$

where $N(p)$ represents the number of units people are willing to buy when the price is p dollars ($1 \leq p \leq 10$).

  a. Reduce this expression to lowest terms.
  b. Find the demand when the price is $3.
  c. What happens to the demand as the price increases?

73. A vendor's profit (in dollars) for the sale of x pairs of sunglasses is

$$\dfrac{5x^2 - 5}{x + 1} \quad (1 \leq x \leq 100)$$

  a. Reduce this expression to lowest terms.
  b. What is the profit when 10 sunglasses are sold?
  c. What is the maximum profit possible?

》》》 Write On

We have already mentioned that when reducing rational expressions, you may only divide out factors. Explain what is wrong with the simplifications in Problems 74 and 75.

74. $\dfrac{x + y}{x} = 1 + y$

75. $\dfrac{y}{x + y} = \dfrac{1}{x + 1}$

76. In this section, we used the fundamental property of rational expressions in two different ways. Write a paragraph explaining these two ways.

77. Just before Example 4, we showed that

$$\dfrac{(x + 3)(x^2 - 4)}{3(x + 2)(x^2 + x - 6)} = \dfrac{1}{3}$$

Is this equation always true? Explain why or why not.

78. Explain that since

$$\dfrac{a - b}{b - a} = -1, \quad \text{then} \quad \dfrac{a - b}{-a + b} = -1$$

79. Explain how you would reduce

$$\dfrac{4x^2 - 7x + 1}{-4x^2 + 7x - 1}$$

and then explain how to reduce a rational expression where the numerator and denominator differ only in sign.

››› Concept Checker

Fill in the blank(s) with the correct word(s), phrase, or mathematical statement.

80. Writing $\frac{14x^2}{28x}$ as $\frac{x}{2}$ is called _____.

81. $\frac{3}{x} + 5 = \frac{1}{2x}$ is an example of a _____.

82. Writing $\frac{5}{7x}$ as $\frac{15x^2}{21x^3}$ is called _____.

83. When evaluating $\frac{x+6}{5x-5}$ for $x = 1$, the result is _____.

fraction

rational expression

rational equation

building up

reducing

undefined

0

››› Mastery Test

Reduce to lowest terms:

84. $\frac{y^3 - x^3}{y - x}$

85. $\frac{x^5 y^7}{xy^3}$

86. $\frac{3y + xy}{y}$

87. $\frac{x^2 - y^2}{y^3 - x^3}$

88. $\frac{x^2 - xy}{x^2 - y^2}$

89. $\frac{4y^2 - xy^2}{y^2}$

90. For what values of x is $\frac{x-1}{x^2 + 2x - 3}$ undefined?

91. For what values of x is $\frac{x+1}{x^2 - 1}$ undefined?

Write in standard form:

92. $\frac{x}{-7}$

93. $-\frac{-4}{x}$

94. $-\frac{a-b}{8}$

Write:

95. $\frac{7}{8}$ with a denominator of 16.

96. $\frac{3x^2}{8y^3}$ with a denominator of $24y^6$.

97. $\frac{4x+1}{x+2}$ with a denominator of $x^2 - x - 6$.

››› Skill Checker

Simplify:

98. $\frac{4x^3}{2x}$

99. $\frac{8x^4}{2x}$

100. $\frac{-16x^4}{8x^2}$

101. $\frac{-48x^5}{12x^3}$

Factor completely:

102. $x^2 + x - 12$

103. $x^2 - 2x - 15$

104. $9x^2 - 4y^2$

105. $4x^2 - 9y^2$

106. $x^3 - 1$

107. $x^3 - 8$

108. $8x^3 + 1$

109. $27x^3 + 8$

Multiply:

110. $\frac{3}{7} \cdot \frac{14}{9}$

111. $-\frac{5}{8} \cdot \frac{16}{15}$

112. $\left(-\frac{4}{9}\right)\left(-\frac{27}{8}\right)$

Divide:

113. $-\frac{3}{4} \div \left(-\frac{3}{8}\right)$

114. $-\frac{4}{5} \div \frac{8}{15}$

115. $\frac{6}{7} \div \left(-\frac{3}{14}\right)$

6.2 Multiplication and Division of Rational Expressions

Objectives

A Multiply rational expressions.

B Divide rational expressions.

C Use multiplication and division together to simplify rational expressions.

To Succeed, Review How To . . .

1. Simplify quotients using the properties of exponents (pp. 42–43).
2. Multiply polynomials (pp. 361–362).

Getting Started

It's All in the Cards

A regular deck of 52 cards has four kings. The probability that you pick two kings when drawing two cards is

$$\frac{4}{52} \cdot \frac{3}{51} = \frac{4}{4 \cdot 13} \cdot \frac{3}{3 \cdot 17} \quad \text{Factor the denominators.}$$

$$= \frac{\overset{1}{\cancel{4}}}{\cancel{4} \cdot 13} \cdot \frac{\overset{1}{\cancel{3}}}{\cancel{3} \cdot 17} \quad \text{Simplify.}$$

$$= \frac{1}{221} \quad \text{Multiply.}$$

Save time by simplifying in the first step:

$$\frac{4}{52} \cdot \frac{3}{51} = \frac{\overset{1}{\cancel{4}}}{\underset{13}{\cancel{52}}} \cdot \frac{\overset{1}{\cancel{3}}}{\underset{17}{\cancel{51}}}$$

In this section, you will learn how to multiply rational expressions using rules involving these three steps: factor, simplify, and multiply.

A Multiplying Rational Expressions

The multiplication of rational expressions follows the same procedure as the multiplication of rational numbers. In Section 1.2 we defined multiplication of fractions as $\frac{a}{b} \cdot \frac{c}{d} = \frac{a \cdot c}{b \cdot d}$, where $b, d \neq 0$. Next, we rewrite that definition for the multiplication of rational expressions.

MULTIPLICATION OF RATIONAL EXPRESSIONS

If P, Q, S, and T are polynomials, then

$$\frac{P}{Q} \cdot \frac{S}{T} = \frac{P \cdot S}{Q \cdot T} \quad (Q \neq 0, T \neq 0)$$

PROCEDURE

Procedure to Multiply Rational Expressions
1. **Factor** the numerators and denominators completely.
2. **Simplify** each rational expression completely.
3. **Multiply** the remaining factors in the numerator and denominator.
4. Make sure the final product is in **lowest terms**.

It is not necessary to factor if the numerators and denominators are monomials as in the case of this next example.

$$\frac{6x^3}{2y} \cdot \frac{4y^5}{9x^2}$$

Both numerators and denominators are to be simplified; proceed as follows:

$$\frac{6x^3}{2y} \cdot \frac{4y^5}{9x^2} = \frac{24x^3y^5}{18x^2y} = \frac{4xy^4 \cdot \cancel{6x^2y}}{3 \cdot \cancel{6x^2y}} = \frac{4xy^4}{3}$$

As in *Getting Started,* the common factors in the numerator and denominator can be divided out *before* doing the multiplications in the numerator and denominator. The procedure goes like this:

$$\frac{\cancel{6}x^3}{\cancel{2}y} \cdot \frac{\cancel{4}y^5}{\cancel{9}x^2} = \frac{2 \cdot x \cdot 2 \cdot y^4}{1 \cdot 1 \cdot 3 \cdot 1} = \frac{4xy^4}{3}$$

Now let's multiply when the numerators and denominators are not all monomials. Remember to factor and then divide out common factors before you do the multiplications.

$$\frac{2x^4}{x^2 - 4} \cdot \frac{x^2 - x - 6}{4x^4} = \frac{2x^4}{(x+2)(x-2)} \cdot \frac{(x+2)(x-3)}{4x^4} \quad \text{Factor.}$$

$$= \frac{\cancel{2x^4}}{(\cancel{x+2})(x-2)} \cdot \frac{(\cancel{x+2})(x-3)}{\cancel{4x^4}} \quad \text{Simplify.}$$

$$= \frac{x-3}{2(x-2)} \quad \text{or} \quad \frac{x-3}{2x-4}$$

EXAMPLE 1 Multiplying rational expressions

Multiply:

a. $\dfrac{4x}{4x^2 - 9} \cdot \dfrac{6x - 9}{12x}$ **b.** $\dfrac{3x + 15}{x + 1} \cdot \dfrac{2x^2 + 2x}{x^2 + 7x + 10}$

SOLUTION

a. Remember the key steps are factor, simplify, and multiply.

$$\frac{4x}{4x^2 - 9} \cdot \frac{6x - 9}{12x} = \frac{4x}{(2x+3)(2x-3)} \cdot \frac{3(2x-3)}{12x} \quad \text{Factor.}$$

$$= \frac{\cancel{4x}}{(2x+3)(\cancel{2x-3})} \cdot \frac{\cancel{3}(\cancel{2x-3})}{\cancel{12x}} \quad \text{Simplify.}$$

$$= \frac{1}{2x+3} \quad \text{Multiply.}$$

b. Again the key steps are factor, simplify, and multiply.

$$\frac{3x + 15}{x + 1} \cdot \frac{2x^2 + 2x}{x^2 + 7x + 10} = \frac{3(x+5)}{(x+1)} \cdot \frac{2x(x+1)}{(x+2)(x+5)} \quad \text{Factor.}$$

$$= \frac{3\cancel{(x+5)}}{\cancel{(x+1)}} \cdot \frac{2x\cancel{(x+1)}}{(x+2)\cancel{(x+5)}} \quad \text{Simplify.}$$

$$= \frac{6x}{x + 2} \quad \text{Multiply.}$$

PROBLEM 1

Multiply:

a. $\dfrac{3x + 1}{5x} \cdot \dfrac{10x^2 + 5x}{9x^2 - 1}$

b. $\dfrac{x - 6}{x^2 - 11x + 30} \cdot \dfrac{2x - 10}{x^2 + x}$

Answers to PROBLEMS

1. **a.** $\dfrac{2x + 1}{3x - 1}$

   b. $\dfrac{2}{x(x + 1)}$ or $\dfrac{2}{x^2 + x}$

EXAMPLE 2 Multiplying rational expressions

Multiply:

a. $\dfrac{2-x}{x+1} \cdot \dfrac{x^2+3x+2}{x^2-4}$

b. $\dfrac{x^2+3x+2}{x^2+5x+4} \cdot \dfrac{x^2+2x-3}{x^2+x-2}$

SOLUTION

a. $\dfrac{2-x}{x+1} \cdot \dfrac{x^2+3x+2}{x^2-4} = \dfrac{(2-x)}{(x+1)} \cdot \dfrac{(x+1)(x+2)}{(x+2)(x-2)}$ Factor.

$= \dfrac{\overset{-1}{\cancel{(2-x)}}}{\cancel{(x+1)}} \cdot \dfrac{\cancel{(x+1)}\cancel{(x+2)}}{\cancel{(x+2)}\cancel{(x-2)}}$ Simplify. Recall that $\dfrac{2-x}{x-2} = -1$.

$= -1$

b. $\dfrac{x^2+3x+2}{x^2+5x+4} \cdot \dfrac{x^2+2x-3}{x^2+x-2} = \dfrac{(x+2)(x+1)}{(x+4)(x+1)} \cdot \dfrac{(x-1)(x+3)}{(x-1)(x+2)}$ Factor.

$= \dfrac{\cancel{(x+2)}\cancel{(x+1)}}{(x+4)\cancel{(x+1)}} \cdot \dfrac{\cancel{(x-1)}(x+3)}{\cancel{(x-1)}\cancel{(x+2)}}$ Simplify.

$= \dfrac{x+3}{x+4}$

PROBLEM 2

Multiply:

a. $\dfrac{3-x}{2x+4} \cdot \dfrac{x^2+5x+6}{x^2-9}$

b. $\dfrac{x^2+4x+3}{x^2+5x+6} \cdot \dfrac{x^2-2x-8}{x^2-2x-3}$

B ▶ Dividing Rational Expressions

To divide fractions in Section 1.2, we used the definition that said $\dfrac{a}{b} \div \dfrac{c}{d} = \dfrac{a}{b} \cdot \dfrac{d}{c}$, where b, c, and d are not zero. We rewrite that definition for the division of rational expressions.

DIVISION OF RATIONAL EXPRESSIONS

If P, Q, S, and T are polynomials, then

$$\dfrac{P}{Q} \div \dfrac{S}{T} = \dfrac{P}{Q} \cdot \dfrac{T}{S} \qquad (Q, S, \text{ and } T \neq 0)$$

The quotient of two rational expressions is the product of the first and the reciprocal of the second. For example,

$$\dfrac{2}{5} \div \dfrac{3}{8} = \dfrac{2}{5} \cdot \dfrac{8}{3} = \dfrac{16}{15}$$
   ↑ Reciprocal

The reciprocal of $\dfrac{3}{8}$ is $\dfrac{8}{3}$.

and

$$\dfrac{3x^2}{2y} \div \dfrac{6x}{4y^2} = \dfrac{3x^2}{2y} \cdot \dfrac{4y^2}{6x}$$
   ↑ Reciprocal

$$= \dfrac{\overset{1 \cdot xy}{\cancel{12x^2 y^2}}}{\cancel{12xy}}$$

$$= xy$$

Answers should be given in simplified form.

Answers to PROBLEMS

2. **a.** $\dfrac{-1}{2}$ **b.** $\dfrac{x-4}{x-3}$

6.2 Multiplication and Division of Rational Expressions

EXAMPLE 3 Dividing rational expressions

Divide:

a. $\dfrac{x^3y}{z} \div \dfrac{x^2y}{z^4}$ **b.** $\dfrac{5x^2 - 5}{3x + 6} \div \dfrac{x + 1}{3}$ **c.** $\dfrac{x + 3}{x - 3} \div (x^2 + 6x + 9)$

SOLUTION

a. $\dfrac{x^3y}{z} \div \dfrac{x^2y}{z^4} = \dfrac{x^3y}{z} \cdot \underbrace{\dfrac{z^4}{x^2y}}_{\text{Reciprocal}}$

$$= \dfrac{x^3 y z^4}{x^2 y z} $$

$$= xz^3$$

b. $\dfrac{5x^2 - 5}{3x + 6} \div \dfrac{x + 1}{3} = \dfrac{5(x + 1)(x - 1)}{3(x + 2)} \cdot \underbrace{\dfrac{3}{x + 1}}_{\text{Reciprocal}}$ Factor.

$$= \dfrac{5(x + 1)(x - 1)}{3(x + 2)} \cdot \dfrac{3}{x + 1} \quad \text{Simplify.}$$

$$= \dfrac{5(x - 1)}{x + 2} \quad \text{or} \quad \dfrac{5x - 5}{x + 2} \quad \text{Multiply.}$$

c. We first write $x^2 + 6x + 9$ as $\dfrac{x^2 + 6x + 9}{1}$.

$$\dfrac{x + 3}{x - 3} \div (x^2 + 6x + 9) = \dfrac{x + 3}{x - 3} \div \dfrac{x^2 + 6x + 9}{1}$$

$$= \dfrac{x + 3}{x - 3} \cdot \dfrac{1}{x^2 + 6x + 9} \quad \text{Multiply by reciprocal.}$$

$$= \dfrac{(x + 3)}{(x - 3)} \cdot \dfrac{1}{(x + 3)(x + 3)} \quad \text{Factor.}$$

$$= \dfrac{(x + 3)}{(x - 3)} \cdot \dfrac{1}{(x + 3)(x + 3)} \quad \text{Simplify.}$$

$$= \dfrac{1}{(x - 3)(x + 3)} \quad \text{or} \quad \dfrac{1}{x^2 - 9} \quad \text{Multiply.}$$

PROBLEM 3

Divide:

a. $\dfrac{x^5y}{z} \div \dfrac{x^3y}{z^6}$

b. $\dfrac{3x^2 - 3}{5x + 10} \div \dfrac{x + 1}{5}$

c. $\dfrac{x + 2}{x - 2} \div (x^2 + 4x + 4)$

C › Using Multiplication and Division to Simplify Rational Expressions

Sometimes both multiplication and division are involved, as shown in Examples 4 and 5. The rational expressions following any division sign should be inverted. Then proceed with multiplication. Let's try an arithmetic example first to get the idea.

$$\dfrac{2}{7} \div \dfrac{4}{15} \cdot \dfrac{7}{3} = \dfrac{2}{7} \cdot \dfrac{15}{4} \cdot \dfrac{7}{3} \quad \text{Invert fraction after division sign.}$$

$$= \dfrac{2}{7} \cdot \dfrac{15}{4} \cdot \dfrac{7}{3} \quad \text{Simplify.}$$

$$= \dfrac{5}{2} \quad \text{Multiply.}$$

Answers to PROBLEMS

3. **a.** x^2z^5

b. $\dfrac{3(x - 1)}{x + 2}$ or $\dfrac{3x - 3}{x + 2}$

c. $\dfrac{1}{(x + 2)(x - 2)}$ or $\dfrac{1}{x^2 - 4}$

EXAMPLE 4 Operations involving multiplication and division of rational expressions

Perform the indicated operations:

$$\frac{3-x}{x+2} \div \frac{x^3-27}{x+4} \cdot \frac{x^3+8}{x+4}$$

SOLUTION First rewrite the division as multiplication by inverting the rational expression following the division sign.

$$\frac{3-x}{x+2} \div \frac{x^3-27}{x+4} \cdot \frac{x^3+8}{x+4}$$

$$= \frac{3-x}{x+2} \cdot \frac{x+4}{x^3-27} \cdot \frac{x^3+8}{x+4} \qquad \text{Invert rational expression after the division sign.}$$

$$= \frac{(3-x)}{(x+2)} \cdot \frac{(x+4)}{(x-3)(x^2+3x+9)} \cdot \frac{(x+2)(x^2-2x+4)}{x+4} \qquad \text{Factor.}$$

$$= \frac{\overset{-1}{\cancel{(3-x)}}}{\cancel{(x+2)}} \cdot \frac{\cancel{(x+4)}}{\cancel{(x-3)}(x^2+3x+9)} \cdot \frac{\cancel{(x+2)}(x^2-2x+4)}{\cancel{(x+4)}} \qquad \text{Simplify. Recall that } \tfrac{3-x}{x-3} = -1.$$

$$= \frac{-(x^2-2x+4)}{x^2+3x+9} \quad \text{or} \quad \frac{-x^2+2x-4}{x^2+3x+9} \qquad \text{Multiply } (-1 \cdot a = -a).$$

PROBLEM 4

Perform the indicated operations:

$$\frac{2-x}{x+3} \div \frac{x^3-8}{x-5} \cdot \frac{x^3+27}{x-5}$$

EXAMPLE 5 Operations involving multiplication and division of rational expressions

Perform the indicated operations:

$$\frac{x^3+2x^2-x-2}{x+3} \div \frac{x^3+8}{x^2-9} \cdot \frac{1}{x^2-1}$$

SOLUTION First rewrite the division as multiplication by inverting the rational expression following the division sign.

$$\frac{x^3+2x^2-x-2}{x+3} \div \frac{x^3+8}{x^2-9} \cdot \frac{1}{x^2-1}$$

$$= \frac{x^3+2x^2-x-2}{x+3} \cdot \frac{x^2-9}{x^3+8} \cdot \frac{1}{x^2-1}$$

Now we have the polynomial x^3+2x^2-x-2 with four terms in the first numerator. Factor it by grouping.

$$x^3+2x^2-x-2 = x^2(x+2)-1\cdot(x+2)$$
$$= (x+2)(x^2-1)$$
$$= (x+2)(x+1)(x-1)$$

Hence, $(x+2)(x+1)(x-1)$ replaces the first numerator. Factor the remaining numerators and denominators and simplify.

$$\frac{(x+2)(x+1)(x-1)}{(x+3)} \cdot \frac{(x+3)(x-3)}{(x+2)(x^2-2x+4)} \cdot \frac{1}{(x+1)(x-1)} \qquad \text{Factor.}$$

$$= \frac{\cancel{(x+2)}\cancel{(x+1)}\cancel{(x-1)}}{\cancel{(x+3)}} \cdot \frac{\cancel{(x+3)}(x-3)}{\cancel{(x+2)}(x^2-2x+4)} \cdot \frac{1}{\cancel{(x+1)}\cancel{(x-1)}} \qquad \text{Simplify.}$$

$$= \frac{x-3}{x^2-2x+4} \qquad \text{Multiply.}$$

PROBLEM 5

Perform the indicated operations:

$$\frac{x^3+x^2-x-1}{x+4} \div \frac{x^3+1}{x^2-16} \cdot \frac{1}{x^2-1}$$

Answers to PROBLEMS

4. $\dfrac{-(x^2-3x+9)}{x^2+2x+4}$ or $\dfrac{-x^2+3x-9}{x^2+2x+4}$ 5. $\dfrac{x-4}{x^2-x+1}$

Exercises 6.2

⟨A⟩ Multiplying Rational Expressions   In Problems 1–20, perform the indicated multiplication.

1. $\dfrac{3}{4} \cdot \dfrac{2}{5}$

2. $\dfrac{-9}{10} \cdot \dfrac{2}{3}$

3. $\dfrac{14x^2}{15} \cdot \dfrac{5}{7x}$

4. $\dfrac{-5x^3}{7y} \cdot \dfrac{4y^3}{9x^6}$

5. $\dfrac{-2xy^4}{9z^5} \cdot \dfrac{-3z}{7x^3y^3}$

6. $\dfrac{-35x^5z}{24x^3y^9} \cdot \dfrac{84x^3y^8}{15x^4y^7z}$

7. $\dfrac{10x + 50}{6x + 6} \cdot \dfrac{12}{5x + 25}$

8. $\dfrac{x + y}{xy - y^2} \cdot \dfrac{y^2}{x^2 - y^2}$

9. $\dfrac{6y + 3}{2y^2 - 3y - 2} \cdot \dfrac{y^2 - 4}{3y + 6}$

10. $\dfrac{y^2 + 9y + 18}{y - 2} \cdot \dfrac{2y - 4}{5y + 15}$

11. $\dfrac{y - x}{x^2 + 2xy} \cdot \dfrac{5x + 10y}{x^2 - y^2}$

12. $\dfrac{2 - 2x}{9x^2 - 25} \cdot \dfrac{6x - 10}{x^2 - 1}$

13. $\dfrac{3y^2 - 10y - 8}{3y + 2} \cdot \dfrac{3y + 12}{6(y^2 - 16)}$

14. $\dfrac{y^2 - 49}{7y} \cdot \dfrac{7y + 7}{y^2 + 8y + 7}$

15. $\dfrac{y^2 + 2y - 8}{y^2 + 7y + 12} \cdot \dfrac{y^2 + 2y - 3}{y^2 - 3y + 2}$

16. $\dfrac{y^2 - 3y + 2}{y^2 - 4y - 5} \cdot \dfrac{y^2 + 3y + 2}{y^2 + y - 2}$

17. $\dfrac{x^3 - 8}{4 - x^2} \cdot \dfrac{x^2 + x - 2}{x^2 + 2x + 4}$

18. $\dfrac{x^3 + y^3}{y^2 - x^2} \cdot \dfrac{x - y}{x^2 - xy + y^2}$

19. $\dfrac{a^3 + b^3}{a^3 - b^3} \cdot \dfrac{a^2 + ab + b^2}{a^2 - ab + b^2}$

20. $\dfrac{a^3 - 8}{a^2 + 2a + 4} \cdot \dfrac{a^2 + 3a + 9}{a^3 - 27}$

⟨B⟩ Dividing Rational Expressions   In Problems 21–40, perform the indicated division.

21. $\dfrac{3}{5} \div \dfrac{10}{9}$

22. $\dfrac{3}{7} \div \dfrac{-9}{14}$

23. $\dfrac{4}{5x^2} \div \dfrac{12}{25x^3}$

24. $\dfrac{6x^2}{7} \div \dfrac{30x}{28}$

25. $\dfrac{24a^2b}{7c^2d} \div \dfrac{8ab}{21cd^2}$

26. $\dfrac{16a^3b}{15a} \div \dfrac{12ab^2}{20b^4}$

27. $\dfrac{3x - 3}{x} \div \dfrac{x^2 - 1}{x^2}$

28. $\dfrac{5x^2 - 45}{x^3} \div \dfrac{x + 3}{x}$

29. $\dfrac{y^2 - 25}{y^2 - 4} \div \dfrac{3y - 15}{4y - 8}$

30. $\dfrac{y^2 + y - 12}{y^2 - 1} \div \dfrac{3y + 12}{4y^2 + 4y}$

31. $\dfrac{a^3 + b^3}{a^3 - b^3} \div \dfrac{a^2 - ab + b^2}{a^2 + ab + b^2}$

32. $\dfrac{a^2 + ab + b^2}{a^3 + b^3} \div \dfrac{a^2 + ab + b^2}{a^2 - ab + b^2}$

33. $\dfrac{8a^3 - 1}{6u^4w^3} \div \dfrac{1 - 2a}{3u^2w}$

34. $\dfrac{-b^2c}{27a^3 - 1} \div \dfrac{b^3c^2}{3a - 1}$

35. $\dfrac{x - x^3}{2x^2 + 6x} \div \dfrac{5x^2 - 5x}{2x + 6}$

36. $\dfrac{121y - y^3}{y^2 - 49} \div \dfrac{y^2 - 11y}{y + 7}$

37. $\dfrac{y^2 + y - 12}{y^2 - 8y + 15} \div \dfrac{3y^2 + 7y - 20}{2y^2 - 7y - 15}$

38. $\dfrac{3y^2 + 11y + 6}{4y^2 + 16y + 7} \div \dfrac{3y^2 - y - 2}{2y^2 - y - 28}$

39. $\dfrac{4x^2 - 12x + 9}{25 - 4x^2} \div \dfrac{6x^2 - 5x - 6}{6x^2 + 19x + 10}$

40. $\dfrac{4x^2 + 12x + 9}{9 - 4x^2} \div \dfrac{10x^2 + 27x + 18}{8x^2 - 2x - 15}$

C Using Multiplication and Division to Simplify Rational Expressions
In Problems 41–54, perform the indicated operations.

41. $\dfrac{x^2 + 2x - 3}{x - 5} \div \dfrac{x^2 - 9}{x^2} \cdot \dfrac{3x - 15}{9x}$

42. $\dfrac{8x^3}{x^2 - 25} \div \dfrac{16x^2}{2x - 10} \cdot \dfrac{x^2 + x - 12}{3 - x}$

43. $\dfrac{x^2 - 1}{x^2 + 3x - 10} \cdot \dfrac{x - 2}{x - 5} \div \dfrac{x^2 - 3x - 4}{x^2 - 25}$

44. $\dfrac{5 - x}{x + 7} \cdot \dfrac{x^2 - 49}{x^2 - 10x + 25} \div \dfrac{x + 7}{4x - 20}$

45. $\dfrac{x - 3}{x - 1} \cdot \dfrac{x^2 + 3x - 4}{6 - 2x} \div \dfrac{x^2 - 16}{10}$

46. $\dfrac{x^2 + 2x + 4}{x^3 - 8} \cdot \dfrac{x^2 + x - 2}{x^2 - 4} \div \dfrac{1 - x}{x^2 + 5x - 14}$

47. $\dfrac{x^2 - y^2}{x^2 - 2xy} \div \dfrac{x^2 + xy - 2y^2}{x^2 - 4y^2} \cdot \dfrac{x^2}{(x + y)^2}$

48. $\dfrac{x^2 + xy - 2y^2}{x^2 - 4y^2} \div \dfrac{x^2 - y^2}{x^2 - 2xy} \cdot \dfrac{(x + y)^2}{x^2}$

49. $\dfrac{x^2 + 2xy + y^2}{2y - 12} \div \dfrac{x + 3}{y - 6} \cdot \dfrac{4x + 12}{x^2 - y^2}$

50. $\dfrac{x^2 + 2xy - 8y^2}{x^2 + 7xy + 12y^2} \div \dfrac{x^2 - 3xy + 2y^2}{x^2 + 2xy - 3y^2} \cdot \dfrac{x^2 - 9}{9 - x^2}$

51. $\dfrac{x^3 + 2x^2 + 6x + 12}{x - 2} \div \dfrac{x^2 + 6}{x - 3} \cdot \dfrac{2 - x}{3 - x}$

52. $\dfrac{x^3 + 3x^2 + 9x + 27}{x - 4} \cdot \dfrac{6x - 2x^2}{2x} \div \dfrac{x^2 - 9}{4 - x}$

53. $\dfrac{x - 2}{x^2 - 9} \div \dfrac{x^3 - 8}{x + 3} \cdot \dfrac{x - 3}{x}$

54. $\dfrac{x - 1}{x^2 - 25} \div \dfrac{x - 3}{x^3 + 125} \cdot \dfrac{(x - 5)^2}{x - 1}$

❯❯❯ Applications

55. Prices If the price for the demand of x units of a product is given by
$$\dfrac{3x + 9}{4}$$
and the demand for these is given by
$$\dfrac{600}{x^2 + 3x}$$
find the product of the price and the demand.

56. Prices If the price for x units of a product is given by
$$\dfrac{4x + 8}{5}$$
and the demand is given by
$$\dfrac{200}{x^2 + 2x}$$
what is the product of the price and the demand?

57. Current In a simple electrical circuit, the current I is the quotient of the voltage E and the resistance R. If the resistance changes with the time t according to
$$R = \dfrac{t^2 + 9}{t^2 + 6t + 9}$$
and the voltage changes according to the formula
$$E = \dfrac{4t}{t + 3}$$
find the current I.

58. Current If in Problem 57 the resistance is given by
$$R = \dfrac{t^2 + 5}{t^2 + 4t + 4}$$
and the voltage is given by
$$E = \dfrac{5t}{t + 2}$$
find the current I.

Using Your Knowledge

Applications of Rational Expressions

59. When studying parallel resistors, the expression

$$R \cdot \frac{R_T}{R - R_T}$$

occurs, where R is a known resistance and R_T is a required resistance. Perform the indicated multiplication.

60. The molecular model predicts that the pressure of a gas is given by

$$\frac{2}{3} \cdot \frac{mv^2}{2} \cdot \frac{N}{v}$$

where m is the mass, N is a constant, and v is the velocity. Perform the indicated multiplication.

61. Suppose a store orders 3000 items each year. If it orders x units at a time, the number N of reorders is given by the equation

$$N = \frac{3000}{x}$$

If there is a fixed $20 reorder fee and a $3 charge per item, the cost of each order is

$$C = 20 + 3x$$

The yearly reorder cost R is then given by the equation

$$R = N \cdot C$$

Find R.

The formula for the area A of a rectangle is $A = LW$, where L is the length and W is the width of the rectangle. In Problems 62 and 63, find the area of the shaded rectangle.

62.

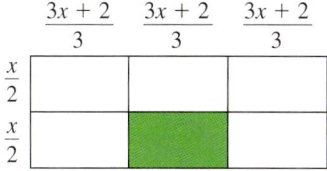

63.

Write On

64. Write in your own words the procedure you use to multiply two rational expressions.

65. Write in your own words the procedure you use to divide two rational expressions.

66. When multiplying $(x + 2)(x + 3)$, you get $x^2 + 5x + 6$. When multiplying

$$\frac{x + 2}{2} \cdot \frac{x + 3}{3}$$

most textbooks write the answer as

$$\frac{(x + 2)(x + 3)}{6}$$

Do you agree? Why or why not?

Concept Checker

Fill in the blank(s) with the correct word(s), phrase, or mathematical statement.

67. When simplifying the rational expression, $\frac{x^2 - 9}{3 - x}$, be sure to _____ the numerator and denominator first, if possible.

68. When dividing two rational expressions, you must _____ the second expression and then proceed with multiplication.

69. If multiplying two rational expressions whose numerators and denominators are all monomials, then it is not necessary to _____.

rational expression

factor

invert

reduce

undefined

››› Mastery Test

Perform the indicated operations:

70. $\dfrac{2-x}{x+3} \div \dfrac{x^3-8}{x-5} \cdot \dfrac{x^3+27}{x-5}$

71. $\dfrac{x^5 y}{z} \div \dfrac{x^3 y}{z^6}$

72. $\dfrac{3x^2-3}{5x+10} \div \dfrac{x+1}{5}$

73. $\dfrac{x+2}{x-2} \div (x^2+4x+4)$

74. $\dfrac{3-x}{x+2} \cdot \dfrac{x^2+5x+6}{x^2-9}$

75. $\dfrac{x^2+4x+3}{x^2+5x+6} \cdot \dfrac{x^2-2x-8}{x^2-2x-3}$

76. $\dfrac{x^3+x^2-x-1}{x+4} \div \dfrac{x^3+1}{x^2-16} \cdot \dfrac{1}{x^2-1}$

77. $\dfrac{x-2}{3x-2y} \cdot \dfrac{9x^2-4y^2}{2x^2-3x-2}$

78. $\dfrac{x^2-9}{9x^2-4y^2} \cdot \dfrac{3x^2-2xy}{3x+9}$

››› Skill Checker

Factor the polynomial:

79. $3xz + 3xy$

80. $25x^2 - 9$

81. $x^2 - 10x + 25$

82. $x^2 + 2x - 63$

Write the rational expression with the indicated denominator:

83. $\dfrac{3x^2}{5y^3}$ with denominator $20y^5$

84. $\dfrac{3x}{4(y+2)}$ with denominator $24(y+2)$

85. $\dfrac{7}{x-3}$ with denominator $(x-3)(x+2)$

86. $\dfrac{x+3}{x+4}$ with denominator x^2+6x+8

6.3 Addition and Subtraction of Rational Expressions

▶ Objectives

A ▶ Add and subtract rational expressions with the same denominator.

B ▶ Add and subtract rational expressions with different denominators.

▶ To Succeed, Review How To . . .

1. Add and subtract real numbers (pp. 18–20).
2. Use the distributive property (pp. 55–58).
3. Write a rational expression with a given denominator (pp. 431–432).
4. Factor polynomials (pp. 398–401).

▶ Getting Started

What Types of Materials Are More Likely to Be Recycled?

Using information from the Environmental Protection Agency (EPA), the fraction of waste generated in the United States that is recycled can be approximated by

$$P(t) = \dfrac{R(t)}{G(t)} = \dfrac{0.055t^2 - 0.7t + 6.86}{0.004t^2 + 3.52t + 86.5}$$

Paper—don't throw it away, recycle it!

where t is the number of years after 1960. Much of this waste is paper and paperboard. The approximate fraction of waste generated represented by paper and paperboard that will be recycled is given by

$$Q(t) = \frac{0.004t^2 - 1.25t + 30.28}{0.004t^2 + 3.52t + 86.5}$$

Can you find the fraction of the waste recycled that is not paper and paperboard? To do this, you need to find $P(t) - Q(t)$. Fortunately, $P(t)$ and $Q(t)$ have the same denominator, so you only need to subtract the numerators and keep the denominator. Make sure you understand that the subtraction sign must be distributed to every term in the numerator of the fraction that follows it, like this:

$$P(t) - Q(t) = \frac{0.055t^2 - 0.7t + 6.86}{0.004t^2 + 3.52t + 86.5} - \frac{0.004t^2 - 1.25t + 30.28}{0.004t^2 + 3.52t + 86.5}$$

$$= \frac{0.055t^2 - 0.7t + 6.86 - (0.004t^2 - 1.25t + 30.28)}{0.004t^2 + 3.52t + 86.5}$$

$$= \frac{0.055t^2 - 0.7t + 6.86 - 0.004t^2 + 1.25t - 30.28}{0.004t^2 + 3.52t + 86.5}$$

$$= \frac{0.051t^2 + 0.55t - 23.42}{0.004t^2 + 3.52t + 86.5}$$

If you let $t = 50$ in this expression, you can find, approximately, what fraction of the waste will not be paper or paperboard in the year 2010. You should come out with 48%. More of the material recycled (52%) will be paper or paperboard.

A › Adding and Subtracting Rational Expressions with the Same Denominator

> **Addition and Subtraction of Polynomials with the Same Denominator**
>
> In general, if P, Q, and T are polynomials ($T \neq 0$),
>
> $$\frac{P}{T} + \frac{Q}{T} = \frac{P + Q}{T} \quad \longleftarrow \text{Add numerators.}$$
> $$\quad\quad\quad\quad\quad\quad\quad \longleftarrow \text{Keep the denominator.}$$
> $$\frac{P}{T} - \frac{Q}{T} = \frac{P - Q}{T} \quad \longleftarrow \text{Subtract numerators.}$$
> $$\quad\quad\quad\quad\quad\quad\quad \longleftarrow \text{Keep the denominator.}$$

Adding and subtracting rational expressions in algebra follow the same rules we use to add and subtract rational numbers in arithmetic. If they have the *same* denominators, then add or subtract the numerators and *keep* the denominator. We will start with two arithmetic examples followed by some examples of adding and subtracting rational expressions.

Arithmetic:

1. $\frac{2}{7} + \frac{3}{7} = \frac{5}{7}$ Add numerators; keep denominator.

2. $\frac{5}{8} - \frac{3}{8} = \frac{2}{8} = \frac{1}{4}$ Subtract numerators; keep denominator; simplify.

450 Chapter 6 Rational Expressions

We can extend this concept in order to add and subtract rational expressions.

Rational Expressions:

1. $\dfrac{2}{x} + \dfrac{6}{x} = \dfrac{2+6}{x} = \dfrac{8}{x}$ ← Add numerators.
← Keep denominator.

2. $\dfrac{5x}{x^2+1} + \dfrac{2x}{x^2+1} = \dfrac{5x+2x}{x^2+1} = \dfrac{7x}{x^2+1}$

3. $\dfrac{3x}{7(x-1)^2} + \dfrac{4x}{7(x-1)^2} = \dfrac{3x+4x}{7(x-1)^2} = \dfrac{\cancel{7}x}{\cancel{7}(x-1)^2} = \dfrac{x}{(x-1)^2}$

4. $\dfrac{8x}{x^2+5} - \dfrac{2x}{x^2+5} = \dfrac{8x-2x}{x^2+5} = \dfrac{6x}{x^2+5}$

5. $\dfrac{10x}{9(x-3)^2} - \dfrac{x}{9(x-3)^2} = \dfrac{10x-x}{9(x-3)^2} = \dfrac{9x}{9(x-3)^2} = \dfrac{x}{(x-3)^2}$

As in arithmetic, the answer is written in simplified form.

PROCEDURE

Adding and Subtracting Rational Expressions with the Same Denominator
1. Combine like terms in the numerators.
2. Keep the like denominator.
3. Simplify, if possible.

EXAMPLE 1   Adding and subtracting rational expressions with the same denominator

Perform the indicated operations:

a. $\dfrac{3}{11x} + \dfrac{1}{11x}$

b. $\dfrac{5}{6x^2} - \dfrac{1}{6x^2}$

c. $\dfrac{8x}{3(x-2)} + \dfrac{x}{3(x-2)}$

d. $\dfrac{7x}{5(x+4)} + \dfrac{3x}{5x+20}$

e. $\dfrac{x}{x^2-1} - \dfrac{1}{x^2-1}$

f. $\dfrac{4x}{x+2} - \dfrac{3x-2}{x+2}$

SOLUTION   Since all these examples have the same denominators, we will add or subtract like terms in the numerators and keep the denominators. When possible we will simplify the final answer.

a. $\dfrac{3}{11x} + \dfrac{1}{11x} = \dfrac{4}{11x}$

b. $\dfrac{5}{6x^2} - \dfrac{1}{6x^2} = \dfrac{4}{6x^2} = \dfrac{2}{3x^2}$

c. $\dfrac{8x}{3(x-2)} + \dfrac{x}{3(x-2)} = \dfrac{8x+x}{3(x-2)} = \dfrac{\overset{3}{\cancel{9}}x}{\underset{1}{\cancel{3}}(x-2)} = \dfrac{3x}{x-2}$

d. The second denominator, when factored, is the same as the first.

$\dfrac{7x}{5(x+4)} + \dfrac{3x}{5x+20} = \dfrac{7x}{5(x+4)} + \dfrac{3x}{5(x+4)} = \dfrac{7x+3x}{5(x+4)} = \dfrac{\overset{2}{\cancel{10}}x}{\underset{1}{\cancel{5}}(x+4)} = \dfrac{2x}{(x+4)}$

PROBLEM 1
Perform the indicated operations:

a. $\dfrac{2}{7a} + \dfrac{4}{7a}$

b. $\dfrac{8}{5y^2} - \dfrac{3}{5y^2}$

c. $\dfrac{4x}{5(x+3)} + \dfrac{6x}{5(x+3)}$

d. $\dfrac{11x}{24(7+x)} + \dfrac{x}{24(7+x)}$

e. $\dfrac{3x}{9x^2-16} - \dfrac{4}{9x^2-16}$

f. $\dfrac{5x}{x+8} - \dfrac{4x-8}{x+8}$

Answers to PROBLEMS

1. a. $\dfrac{6}{7a}$ b. $\dfrac{1}{y^2}$ c. $\dfrac{2x}{x+3}$

d. $\dfrac{x}{2(7+x)}$ or $\dfrac{x}{14+2x}$

e. $\dfrac{1}{3x+4}$ f. 1

e. Both expressions have the same denominator, so we subtract numerators and use the same denominator. However, the answer can be simplified which you can see only when the denominator is completely factored. Make sure your denominator is in completely factored form so you can determine all the possible ways to simplify the answer.

$$\frac{x}{x^2 - 1} - \frac{1}{x^2 - 1} = \frac{x - 1}{x^2 - 1} \quad \leftarrow \text{Subtract numerators.}$$
$$\leftarrow \text{Keep denominator.}$$

$$= \frac{\cancel{x - 1}}{(x + 1)\cancel{(x - 1)}} \quad \text{Factor the denominator and divide out the common factor } (x - 1).$$

$$= \frac{1}{x + 1}$$

f. Since there are two terms in the second numerator, we indicate the subtraction of the *numerators* by using parentheses. Be careful with the signs when removing the parentheses.

$$\frac{4x}{x + 2} - \frac{3x - 2}{x + 2} = \frac{4x - (3x - 2)}{x + 2}$$

$$= \frac{4x - 3x + 2}{x + 2} \quad \begin{array}{l}\text{Recall that } -(3x - 2) = -1(3x - 2)\\ = -3x + 2\end{array}$$

$$= \frac{\cancel{x + 2}}{\cancel{x + 2}} = 1 \quad \text{Combine like terms and simplify.}$$

Note: Parentheses are necessary in subtracting numerators when the second numerator has more than one term.

B ▶ Adding and Subtracting Rational Expressions with Different Denominators

To add or subtract fractions with different denominators, first find a common denominator. It is most convenient to use the smallest one available, called the **least common denominator (LCD)**—the smallest multiple of the denominators. To add

$$\frac{5}{12} + \frac{7}{18}$$

we start by writing 12 and 18 as products of *primes*. We have

$$12 = 2 \cdot 2 \cdot 3 \quad = 2^2 \cdot 3 \quad \begin{array}{l}\text{The 2's are in one column and the}\\ \text{3's are written in another column.}\end{array}$$
$$18 = \quad 2 \cdot 3 \cdot 3 = 2 \cdot 3^2$$

Since we need the *smallest* number that is a multiple of 12 and 18, we select the factors raised to the *greater* power in each column. The product of these factors is the LCD. The LCD is $2^2 \cdot 3^2 = 4 \cdot 9 = 36$. We then write each fraction with a denominator of 36 and add:

$$\frac{5}{12} = \frac{5 \cdot 3}{12 \cdot 3} = \frac{15}{36} \quad \text{Multiply the denominator of } \frac{5}{12} \text{ by 3 (to get 36), and do the same to the numerator.}$$

$$\frac{7}{18} = \frac{7 \cdot 2}{18 \cdot 2} = \frac{14}{36} \quad \text{Multiply the numerator and denominator of } \frac{7}{18} \text{ by 2 to get 36 as the denominator.}$$

$$\frac{5}{12} + \frac{7}{18} = \frac{15}{36} + \frac{14}{36} = \frac{29}{36}$$

452   Chapter 6   Rational Expressions

The procedure used to find the LCD of two or more rational expressions is similar to that used to find the LCD of two fractions.

> **PROCEDURE**
>
> **Finding the LCD of Two or More Rational Expressions**
> 1. Factor each denominator. Place identical factors in columns. (Not necessary to factor monomials.)
> 2. From each column, select the factor with the greatest exponent.
> 3. The product of all the factors obtained in step 2 is the LCD.

If the denominators involved have no common factors, the LCD is the product of the denominators. For example, the denominators in $\frac{3}{5}$ and $\frac{1}{7}$ have no common factors. The LCD is $5 \cdot 7$. To subtract $\frac{1}{7}$ from $\frac{3}{5}$, we first write each fraction with a denominator of 35 and then subtract. Here are the steps.

1. The LCD is $5 \cdot 7 = 35$.
2. Write each fraction with 35 as the denominator. In Section 6.1, we called this "building up" the fraction.

$$\frac{3}{5} = \frac{3 \cdot 7}{5 \cdot 7} = \frac{21}{35} \quad \text{and} \quad \frac{1}{7} = \frac{1 \cdot 5}{7 \cdot 5} = \frac{5}{35}$$

3. Subtract: $\frac{3}{5} - \frac{1}{7} = \frac{21}{35} - \frac{5}{35} = \frac{16}{35}$

In general, to add or subtract fractions with different denominators, use the following procedure.

> **PROCEDURE**
>
> **Adding and Subtracting Rational Expressions with Different Denominators**
> 1. Find the LCD of the rational expressions.
> 2. "Build up" each rational expression to an equivalent one with the LCD as the denominator.
> 3. Combine like terms in the numerators; keep the LCD as the denominator.
> 4. Simplify, if possible.

EXAMPLE 2   Adding and subtracting rational expressions with different denominators

Perform the indicated operations:

a. $\dfrac{7}{12x} + \dfrac{3}{4x}$

b. $\dfrac{5}{2x^2} - \dfrac{2}{3x^4}$

c. $\dfrac{9x}{2x + 4} + \dfrac{3x}{x + 2}$

d. $\dfrac{2x}{x + 1} - \dfrac{x}{x + 2}$

SOLUTION

a. These denominators are not the same. By inspection their LCD is $12x$, the denominator of the first rational expression. We must "build up" the second rational expression so that it also has a denominator of $12x$.

PROBLEM 2

Perform the indicated operations:

a. $\dfrac{5}{6x} + \dfrac{11}{18x}$

b. $\dfrac{4}{5x^6} - \dfrac{1}{2x^2}$

c. $\dfrac{8x}{3x + 12} + \dfrac{6x}{2x + 8}$

d. $\dfrac{2x}{x + 3} - \dfrac{x}{x + 5}$

Answers to PROBLEMS

2. a. $\dfrac{13}{9x}$ b. $\dfrac{8 - 5x^4}{10x^6}$

   c. $\dfrac{17x}{3(x + 4)}$ d. $\dfrac{x^2 + 7x}{(x + 3)(x + 5)}$

$$\frac{7}{12x} + \frac{3}{4x} = \frac{7}{12x} + \frac{3 \cdot 3}{4x \cdot 3} \qquad \text{1. LCD} = 12x$$

$$= \frac{7}{12x} + \frac{9}{12x} \qquad \text{2. Build up.}$$

$$= \frac{16}{12x} \qquad \text{3. Combine terms.}$$

$$= \frac{\overset{4}{\cancel{16}}}{\underset{3}{\cancel{12x}}} = \frac{4}{3x} \qquad \text{4. Simplify.}$$

b. Since both of these denominators are monomials, it is not necessary to factor them. The LCD for 2 and 3 is 6 and the LCD for x^2 and x^4 is x^4. Putting the factors together, the LCD is $6x^4$.

$$\frac{5}{2x^2} - \frac{2}{3x^4} = \frac{5 \cdot 3x^2}{2x^2 \cdot 3x^2} - \frac{2 \cdot 2}{3x^4 \cdot 2} \qquad \begin{array}{l}\text{1. LCD} = 6x^4 \\ \text{2. Build up.}\end{array}$$

$$= \frac{15x^2}{6x^4} - \frac{4}{6x^4} \qquad \text{3. Combine terms.}$$

$$= \frac{15x^2 - 4}{6x^4} \qquad \text{4. It cannot be simplified.}$$

c. Since these denominators are not monomials, we must try to factor them. In the first denominator, we can factor out the GCF of 2. The second denominator is a prime polynomial. The LCD is the product of the factors of the denominator, so the LCD is $2(x + 2)$.

$$\frac{9x}{2x + 4} + \frac{3x}{x + 2} = \frac{9x}{2(x + 2)} + \frac{3x}{(x + 2)} \qquad \text{1. LCD} = 2(x + 2)$$

$$= \frac{9x \cdot 1}{2(x + 2) \cdot 1} + \frac{3x \cdot 2}{(x + 2) \cdot 2} \qquad \text{2. Build up.}$$

$$= \frac{9x}{2(x + 2)} + \frac{6x}{2(x + 2)} \qquad \text{3. Combine terms.}$$

$$= \frac{15x}{2(x + 2)} \qquad \text{4. It cannot be simplified.}$$

d. Since these denominators are not monomials and do not factor, their product is the LCD, which is $(x + 1)(x + 2)$.

$$\frac{2x}{x + 1} - \frac{x}{x + 2} = \frac{2x \cdot (x + 2)}{(x + 1) \cdot (x + 2)} - \frac{x \cdot (x + 1)}{(x + 2) \cdot (x + 1)} \qquad \begin{array}{l}\text{1. LCD} = (x + 1)(x + 2) \\ \text{2. Build up.}\end{array}$$

$$= \frac{2x^2 + 4x}{(x + 1)(x + 2)} - \frac{(x^2 + x)}{(x + 1)(x + 2)} \qquad \text{(Subtraction changes signs in second numerator.)}$$

$$= \frac{2x^2 + 4x - x^2 - x}{(x + 1)(x + 2)} \qquad \text{3. Combine terms.}$$

$$= \frac{x^2 + 3x}{(x + 1)(x + 2)} \qquad \text{4. It cannot be simplified.}$$

EXAMPLE 3 Adding and subtracting rational expressions with different denominators

Perform the indicated operations:

a. $\dfrac{2}{x^2 - 4} - \dfrac{3}{2 - x}$

b. $\dfrac{x + 2}{x + 3} + \dfrac{x - 4}{x - 3}$

c. $\dfrac{4x}{x^2 + x - 2} + \dfrac{5}{x^2 - 1}$

d. $\dfrac{x - 1}{x^2 - x - 6} - \dfrac{x}{x^2 - 9}$

PROBLEM 3

Perform the indicated operations:

a. $\dfrac{6}{x^2 - 25} - \dfrac{8}{5 - x}$

b. $\dfrac{x - 5}{x + 6} + \dfrac{x + 2}{x - 1}$

c. $\dfrac{7}{x^2 + 4x - 5} + \dfrac{6x}{x^2 - 25}$

d. $\dfrac{x + 3}{x + 6} - \dfrac{2x}{x^2 + 5x - 6}$

(continued)

Answers to PROBLEMS

3. a. $\dfrac{8x + 46}{(x + 5)(x - 5)}$ b. $\dfrac{2x^2 + 2x + 17}{(x + 6)(x - 1)}$ c. $\dfrac{6x^2 + x - 35}{(x + 5)(x - 1)(x - 5)}$ d. $\dfrac{x^2 - 3}{(x + 6)(x - 1)}$

...the rational expressions. Write the denominators in ... common factors in columns.

$$x^2 - 4 = \quad (x-2) \quad (x+2)$$
$$2 - x = -1 \quad (x-2)$$

...change the subtraction to +. Select the factor with the greater ... column, $(x-2)$ in the first column and $(x+2)$ in the second ... D is the product of these factors: $(x-2)(x+2)$.

...second rational expression to an equivalent one with the LCD ...inator.

$$-\frac{3}{2-x} = \frac{2}{(x-2)(x+2)} \cdot \frac{\cdot 1}{\cdot 1} - (-1)\frac{3}{(x-2)} \cdot \frac{\cdot (x+2)}{\cdot (x+2)}$$

$$= \frac{2}{(x-2)(x+2)} + \frac{3x+6}{(x-2)(x+2)}$$

...numerators and keep the denominator.

$$= \frac{3x+8}{(x-2)(x+2)}$$

... answer cannot be simplified.

...the four-step procedure.

...Since both denominators are prime polynomials, the LCD is their product: $(x+3)(x-3)$.

. Build up each rational expression to an equivalent one with the LCD as the denominator.

$$\frac{x+2}{x+3} + \frac{x-4}{x-3} = \frac{(x+2) \cdot (x-3)}{(x+3) \cdot (x-3)} + \frac{(x-4) \cdot (x+3)}{(x-3) \cdot (x+3)}$$

$$= \frac{x^2-x-6}{(x+3)(x-3)} + \frac{x^2-x-12}{(x+3)(x-3)}$$

3. Add the numerators and keep the denominator.

$$= \frac{2x^2 - 2x - 18}{(x+3)(x-3)}$$

4. The answer cannot be simplified.

c. Use the four-step procedure.

1. First find the LCD of the rational expressions. Write the denominators in factored form with common factors in a column.

$$x^2 + x - 2 = (x+2) \quad (x-1)$$
$$x^2 - 1 = \quad\quad\quad (x-1) \quad (x+1)$$

Select the factor with the greater exponent in each column, $(x+2)$ in column 1, $(x-1)$ in column 2, and $(x+1)$ in column 3. The LCD is the product of these factors.

$$(x+2)(x-1)(x+1)$$

2. Then build up each rational expression as an equivalent one with the LCD as the denominator.

$$\frac{4x}{x^2+x-2} = \frac{4x}{(x+2)(x-1)} = \frac{4x(x+1)}{(x+2)(x-1)(x+1)} = \frac{4x^2+4x}{(x+2)(x-1)(x+1)}$$

$$\frac{5}{x^2-1} = \frac{5}{(x+1)(x-1)} = \frac{5(x+2)}{(x+1)(x-1)(x+2)} = \frac{5x+10}{(x+1)(x-1)(x+2)}$$

3. $$\frac{4x}{x^2+x-2} + \frac{5}{x^2-1} = \frac{4x^2+4x}{(x+2)(x-1)(x+1)} + \frac{5x+10}{(x+2)(x-1)(x+1)}$$

$$= \frac{(4x^2+4x)+(5x+10)}{(x+2)(x-1)(x+1)} \quad \text{Add the numerators; keep the denominator.}$$

$$= \frac{4x^2+9x+10}{(x+2)(x-1)(x+1)} \quad \text{Combine like terms in the numerator.}$$

4. The answer cannot be simplified; $4x^2+9x+10$ is *not* factorable.

d. Again use the four-step procedure.

1. To find the LCD, factor the denominators keeping common factors in a column.

$$\begin{aligned} x^2 - x - 6 &= \quad\quad\quad (x-3) \quad (x+2) \\ x^2 - 9 &= (x+3) \quad (x-3) \end{aligned}$$

The LCD is $(x+3)(x-3)(x+2)$.

2. Build up each rational expression as an equivalent one with the LCD as the denominator.

$$\frac{x-1}{x^2-x-6} = \frac{(x-1)(x+3)}{(x-3)(x+2)(x+3)} = \frac{x^2+2x-3}{(x+3)(x-3)(x+2)}$$

$$\frac{x}{x^2-9} = \frac{x(x+2)}{(x+3)(x-3)(x+2)} = \frac{x^2+2x}{(x+3)(x-3)(x+2)}$$

3. $$\frac{x-1}{x^2-x-6} - \frac{x}{x^2-9} = \frac{x^2+2x-3}{(x+3)(x-3)(x+2)} - \frac{x^2+2x}{(x+3)(x-3)(x+2)}$$

$$= \frac{(x^2+2x-3)-(x^2+2x)}{(x+3)(x-3)(x+2)} \quad \text{Subtract the numerators; keep the denominator.}$$

$$= \frac{x^2+2x-3-x^2-2x}{(x+3)(x-3)(x+2)} \quad \text{Remember that } -(x^2+2x) = -x^2-2x.$$

$$= \frac{-3}{(x+3)(x-3)(x+2)} \quad \text{Combine like terms in the numerator.}$$

4. The answer is not reducible.

How would you add $\frac{6}{12} + \frac{1}{8}$? You can start by finding the LCD, 24. However, it is easier to reduce $\frac{6}{12}$ to $\frac{1}{2}$ first. Then add $\frac{1}{2} + \frac{1}{8} = \frac{4}{8} + \frac{1}{8} = \frac{5}{8}$. We illustrate a similar problem in Example 4.

EXAMPLE 4 Simplifying before adding or subtracting rational expressions

Perform the indicated operations:

a. $\dfrac{x+y}{x^2+2xy+y^2} + \dfrac{x-y}{x^2-2xy+y^2}$

b. $\dfrac{x}{(x+2)(x-2)} - \dfrac{2}{(2-x)(x+2)}$

PROBLEM 4

Perform the indicated operations:

a. $\dfrac{2x-2y}{x^2-2xy+y^2} + \dfrac{x-y}{x^2-y^2}$

b. $\dfrac{x}{(x+3)(x-3)} + \dfrac{3}{(3-x)(x+3)}$

(continued)

Answers to PROBLEMS

4. **a.** $\dfrac{3x+y}{(x-y)(x+y)}$ or $\dfrac{3x+y}{x^2-y^2}$

b. $\dfrac{1}{x+3}$

SOLUTION

a. $\dfrac{x+y}{x^2+2xy+y^2} = \dfrac{x+y}{(x+y)^2} = \dfrac{1}{x+y}$ Factor and simplify the first fraction.

$\dfrac{x-y}{x^2-2xy+y^2} = \dfrac{x-y}{(x-y)^2} = \dfrac{1}{x-y}$ Factor and simplify the second fraction.

$(x+y)$ and $(x-y)$ have no common factors, so the LCD is $(x+y)(x-y)$. We have

$\dfrac{x+y}{x^2+2xy+y^2} + \dfrac{x-y}{x^2-2xy+y^2}$

$= \dfrac{1}{x+y} + \dfrac{1}{x-y}$ Result of simplifying the fractions.

$= \dfrac{(x-y)}{(x+y)(x-y)} + \dfrac{(x+y)}{(x+y)(x-y)}$ Write each fraction with $(x+y)(x-y)$ as the denominator.

$= \dfrac{x-y+x+y}{(x+y)(x-y)}$ Combine like terms in the numerator.

$= \dfrac{2x}{(x+y)(x-y)}$

b. $2 - x = -1(x - 2)$, so rewrite the second rational expression.

$\dfrac{x}{(x+2)(x-2)} - (-1)\dfrac{2}{(x-2)(x+2)}$

$= \dfrac{x}{(x+2)(x-2)} + \dfrac{2}{(x-2)(x+2)}$ Simplify.

$= \dfrac{x+2}{(x+2)(x-2)}$ Add numerators; keep denominator.

$= \dfrac{\cancel{(x+2)}}{\cancel{(x+2)}(x-2)}$ Divide out $x+2$.

$= \dfrac{1}{x-2}$

🖩 Calculator Corner

Using Graphs to Verify Adding and Subtracting of Rational Expressions

The results of adding or subtracting rational expressions can be verified by graphing the original problem and the answer. Thus, to verify Example 4(b), use a decimal window to graph

$$Y_1 = \left(\dfrac{x}{(x+2)(x-2)}\right) - \left(\dfrac{2}{(2-x)(x+2)}\right)$$

and

$$Y_2 = \dfrac{1}{(x-2)}$$

and make sure that both graphs are identical.

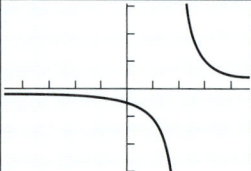

6.3 Addition and Subtraction of Rational Expressions

Boost your grade at mathzone.com!
> Practice Problems
> NetTutor
> Self-Tests
> e-Professors
> Videos

Exercises 6.3

< A > **Adding and Subtracting Rational Expressions with the Same Denominator** In Problems 1–10, perform the indicated operations.

1. $\dfrac{x}{5} + \dfrac{2x}{5}$

2. $\dfrac{x+1}{3x^2} + \dfrac{2x+7}{3x^2}$

3. $\dfrac{7x}{3} - \dfrac{2x}{3}$

4. $\dfrac{2x-1}{5x} - \dfrac{x+1}{5x}$

5. $\dfrac{3}{5x+10} + \dfrac{2x}{5(x+2)}$

6. $\dfrac{2x+1}{3(x+2)} + \dfrac{3x+1}{3x+6}$

7. $\dfrac{2x+1}{2(x+1)} - \dfrac{x-1}{2x+2}$

8. $\dfrac{3x-1}{4(x-1)} - \dfrac{4x-1}{4x-4}$

9. $\dfrac{2x+1}{3(x-1)} + \dfrac{x+3}{3x-3} - \dfrac{x-1}{3(x-1)}$

10. $\dfrac{3x-1}{5(x+1)} - \dfrac{x+1}{5x+5} + \dfrac{2x-5}{5(x+1)}$

< B > **Adding and Subtracting Rational Expressions with Different Denominators** In Problems 11–50, perform the indicated operations.

11. $\dfrac{5}{3x} + \dfrac{7}{15x}$

12. $\dfrac{11}{14x} - \dfrac{5}{7x}$

13. $\dfrac{5}{3m^2} - \dfrac{2}{4m^2}$

14. $\dfrac{8}{9y^3} - \dfrac{1}{2y^3}$

15. $\dfrac{3}{5a} - \dfrac{2}{3b}$

16. $\dfrac{4}{7x} + \dfrac{3}{8y}$

17. $\dfrac{11}{6y^3} + \dfrac{4}{5y^2}$

18. $\dfrac{6}{13x^4} - \dfrac{3}{2x^7}$

19. $\dfrac{2x}{x-1} + \dfrac{3x}{5x-5}$

20. $\dfrac{3}{2x+8} - \dfrac{5}{8x+32}$

21. $\dfrac{4a}{a-5} + \dfrac{2a}{a+2}$

22. $\dfrac{4}{b+6} - \dfrac{1}{b+3}$

23. $\dfrac{x}{x^2+3x-4} + \dfrac{x}{x^2-16}$

24. $\dfrac{x-2}{x^2-9} + \dfrac{x+1}{x^2-x-12}$

25. $\dfrac{1}{x^2-y^2} + \dfrac{5}{(x+y)^2}$

26. $\dfrac{3}{(x+y)^2} + \dfrac{5}{x+y}$

27. $\dfrac{x}{x^2+3x+2} - \dfrac{7}{x^2+5x+6}$

28. $\dfrac{2}{x^2+3xy+2y^2} - \dfrac{1}{x^2-xy-2y^2}$

29. $\dfrac{y}{y^2-1} + \dfrac{y}{y-1}$

30. $\dfrac{3y}{y^2-4} - \dfrac{y}{y+2}$

31. $\dfrac{5x+2y}{5x-2y} + \dfrac{5x-2y}{5x+2y}$

32. $\dfrac{x+3y}{x-5y} - \dfrac{x+5y}{x-3y}$

33. $\dfrac{x+1}{x^2-x-2} - \dfrac{x}{x^2-5x+4}$

34. $\dfrac{3}{x^2-4} + \dfrac{1}{2-x} - \dfrac{1}{2+x}$

35. $\dfrac{2}{5+x} + \dfrac{5x}{x^2-25} + \dfrac{7}{5-x}$

36. $\dfrac{1}{x^2+x-12} + \dfrac{2}{x^2+2x-15} + \dfrac{3}{x^2+9x+20}$

37. $\dfrac{x}{(x-y)(2-x)} - \dfrac{y}{(y-x)(2-x)} + \dfrac{y}{(x-y)(x-2)}$

38. $\dfrac{a}{(b-a)(c-a)} - \dfrac{b}{(b-c)(a-b)} + \dfrac{c}{(a-c)(b-c)}$

Hint: For Problems 39–49, first simplify the fractions.

39. $\dfrac{x+5}{x^3+125} + \dfrac{x-5}{x^2-25} - \dfrac{1}{x+5}$

40. $\dfrac{a+3}{2a+6} + \dfrac{a-3}{a^3-27}$

41. $\dfrac{x+2}{x^2-4} + \dfrac{x+3}{x^2-9}$

42. $\dfrac{x-4}{x^2-16} + \dfrac{x+3}{x^2-9}$

43. $\dfrac{x-3}{x^2-9} + \dfrac{x+3}{x^2+6x+9}$

44. $\dfrac{a-4}{a^2-16} + \dfrac{a+3}{a^2+5a+6}$

45. $\dfrac{a+3}{a^2+5a+6} + \dfrac{a+2}{a^2+6a+8}$

46. $\dfrac{a+3}{a^2+5a+6} - \dfrac{a-4}{a^2-16}$

47. $\dfrac{3a+3}{a^2+5a+4} - \dfrac{a-3}{a^2+a-12}$

48. $\dfrac{2a}{5a-7b} - \dfrac{5a+7b}{25a^2-49b^2}$

49. $\dfrac{5a-15}{a^2+2a-15} - \dfrac{a^2+5a}{a^2+8a+15}$

50. $\dfrac{3}{y^2-9} + \dfrac{2y}{y-3}$

>>> Applications

51. *Cantilever* The moment M of a cantilever beam of length L, x units from the end is given by the expression

$$-\dfrac{w_0 x^3}{6L} + \dfrac{w_0 Lx}{2} - \dfrac{w_0 L^2}{3}$$

Write this expression as a single rational expression in reduced form.

52. *Deflection of a beam* The deflection d of the beam of Problem 51 involves the expression

$$\dfrac{-x^4}{24L} + \dfrac{Lx^2}{4} - \dfrac{L^2 x}{3}$$

Write this expression as a single rational expression in reduced form.

53. *Planetary motion* In astronomy, planetary motion is given by the equation

$$\dfrac{p^2}{2mr^2} - \dfrac{gmM}{r}$$

Write this expression as a single rational expression in reduced form.

54. *Pendulum* The motion of a pendulum is given by the expression

$$\dfrac{P_1^2 + P_2^2}{2(h_1 + h_2)} + \dfrac{P_1^2 - P_2^2}{2(h_1 - h_2)}$$

Write this expression as a single rational expression in reduced form.

>>> Using Your Knowledge

Looking Ahead to Calculus In calculus, the derivative of a polynomial P is defined as the limiting value of

$$\dfrac{P(x+h) - P(x)}{h}$$

as h approaches zero. Let $P(x) = x^2$.

55. Find $P(x+h)$ and write it in expanded form.

56. Find $P(x+h) - P(x)$ and simplify it.

57. Find $\dfrac{P(x+h) - P(x)}{h}$ in simplified form.

58. Find $\dfrac{P(x+h) - P(x)}{h}$ for $P(x) = x^2 + x$.

›› Write On

59. Write in your own words the procedure you use to find the LCD of two or more rational expressions.

60. Write in your own words the procedure you use to add or subtract two rational expressions.

61. When adding two or more rational expressions, do you *always* have to find the LCD or can you use *any* common denominator? What is the advantage of using the LCD?

›› Concept Checker

Fill in the blank(s) with the correct word(s), phrase, or mathematical statement.

62. When adding rational expressions with the same denominator, keep the _____ and add the _____.

63. When adding or subtracting rational expressions, the final answer must be _____ to lowest terms.

64. To _____ $\frac{3x}{2}$ with the denominator of $8y$, both numerator and denominator must be multiplied by $4y$.

rational expression
factored
invert
reduced
rewrite

least common denominator
numerators
denominator

›› Mastery Test

Perform the indicated operation.

65. $\dfrac{x+1}{(x+3)(x-1)} + \dfrac{x+4}{x^2-1}$

66. $\dfrac{3}{x^2-x-2} - \dfrac{3}{x^2-4}$

67. $\dfrac{12}{x^2-2x+3} + \dfrac{4}{x+1}$

68. $\dfrac{x}{(x+3)(x-3)} + \dfrac{3}{(3-x)(x+3)}$

69. $\dfrac{5x}{x+1} - \dfrac{3x}{x+3}$

70. $\dfrac{4x}{5(x-2)} + \dfrac{6x}{5(x-2)}$

71. $\dfrac{11x}{3(x+2)^2} + \dfrac{4x}{3(x+2)^2}$

72. $\dfrac{x}{x^2-9} - \dfrac{3}{x^2-9}$

73. $\dfrac{5x}{x+3} - \dfrac{4x-3}{x+3}$

›› Skill Checker

Multiply.

74. $9\left(2 + \dfrac{2}{9}\right)$

75. $4\left(60 - \dfrac{15}{2}\right)$

76. $12xy\left(\dfrac{2}{y} + \dfrac{3}{2x}\right)$

77. $6ab\left(\dfrac{3}{a} - \dfrac{4}{b}\right)$

78. $x^2\left(1 - \dfrac{1}{x^2}\right)$

79. $x^3\left(1 - \dfrac{1}{x^3}\right)$

Simplify.

80. $\dfrac{14x^2}{28x^4}$

81. $\dfrac{a^2-b^2}{2a-2b}$

82. $\dfrac{12x+3y}{24x+6y}$

83. $\dfrac{x^2-y^2}{2x^2-xy-y^2}$

6.4 Complex Fractions

Objective

A ▸ Write a complex fraction as a simple fraction in reduced form.

To Succeed, Review How To . . .

1. Find the LCD of two or more rational expressions (pp. 451–456).
2. Remove parentheses using the distributive property (pp. 55–58).
3. Add, subtract, multiply, and divide rational expressions (pp. 440–444, 449–456).

▸ Getting Started

Rocking Along with Fractions

Suppose a disc jockey devotes $7\frac{1}{2}$ minutes each hour to commercials, leaving $60 - 7\frac{1}{2}$ minutes for music. If the songs she plays last an average of $3\frac{1}{4}$ minutes and it takes her about $\frac{1}{2}$ minute to get a song going, how many songs can she play each hour? The answer is

$$\frac{60 - 7\frac{1}{2}}{3\frac{1}{4} + \frac{1}{2}}$$

← Time allowed for music
← Time it takes to play each song

but you have to know how to simplify the expression to find out how many songs she plays each hour.

A ▸ Complex Fractions

The fraction

$$\frac{60 - 7\frac{1}{2}}{3\frac{1}{4} + \frac{1}{2}}$$

← Numerator of fraction
← Main fraction bar
← Denominator of fraction

contains other fractions in its numerator and denominator. A fraction whose numerator or denominator (or both) contains other fractions is called a **complex fraction.**

A fraction that is not complex is called a **simple fraction.** Thus,

$$\frac{\frac{1}{2}}{\frac{3}{4} + \frac{1}{5}}, \quad \frac{\frac{3x}{5} - \frac{1}{8}}{\frac{x}{7}}, \quad \frac{-\frac{1}{3}}{\frac{1}{9}}, \quad \text{and} \quad \frac{x}{\frac{7}{8}}$$

are all complex fractions, but $\frac{1}{7}$, $\frac{3}{5}$, and $\frac{x}{9}$ are simple fractions.

To simplify a complex fraction, it's necessary to recall that the main fraction bar indicates that the numerator of the fraction is to be divided by the denominator of the fraction. Thus,

$$\frac{60 - 7\frac{1}{2}}{3\frac{1}{4} + \frac{1}{2}} \quad \text{means} \quad \left(60 - 7\frac{1}{2}\right) \div \left(3\frac{1}{4} + \frac{1}{2}\right)$$

Use the ÷ sign instead of the bar.

Here are the procedures we use to simplify complex fractions.

> **PROCEDURES**
>
> **Procedures for Simplifying Complex Fractions**
> **Method 1.** Multiply the numerator and denominator of the complex fraction by the LCD of all the simple fractions; or
> **Method 2.** Perform the operations indicated in the numerator and denominator of the given complex fraction, and then divide the numerator by the denominator.

Now simplify

$$\frac{60 - 7\frac{1}{2}}{3\frac{1}{4} + \frac{1}{2}} = \frac{60 - \frac{15}{2}}{\frac{13}{4} + \frac{1}{2}}$$

using each of these methods.

Method 1. The LCD of $\frac{15}{2}$, $\frac{13}{4}$, and $\frac{1}{2}$ is 4, so we multiply by 1 in the form of $\frac{4}{4}$ to obtain

$$\frac{60 - \frac{15}{2}}{\frac{13}{4} + \frac{1}{2}} = \frac{4 \cdot \left(60 - \frac{15}{2}\right)}{4 \cdot \left(\frac{13}{4} + \frac{1}{2}\right)} \quad \text{Multiply numerator and denominator by 4, the LCD of } \frac{15}{2}, \frac{13}{4}, \text{ and } \frac{1}{2}.$$

$$= \frac{240 - 30}{13 + 2} \quad \leftarrow 4\left(60 - \frac{15}{2}\right) = 4 \cdot 60 - \overset{2}{\cancel{4}} \cdot \frac{15}{\cancel{2}} = 240 - 30$$
$$\phantom{= \frac{240 - 30}{13 + 2}} \leftarrow 4\left(\frac{13}{4} + \frac{1}{2}\right) = \cancel{4} \cdot \frac{13}{\cancel{4}} + \overset{2}{\cancel{4}} \cdot \frac{1}{\cancel{2}} = 13 + 2$$

$$= \frac{210}{15} \quad \leftarrow 240 - 30 = 210$$
$$\phantom{= \frac{210}{15}} \leftarrow 13 + 2 = 15$$

$$= 14 \quad \text{Divide.}$$

The disc jockey plays 14 songs each hour.

Method 2.

$$\frac{60 - \frac{15}{2}}{\frac{13}{4} + \frac{1}{2}} = \frac{\frac{120}{2} - \frac{15}{2}}{\frac{13}{4} + \frac{2}{4}} \quad \leftarrow \text{Write 60 as } \frac{120}{2}, \text{ so we can subtract } \frac{15}{2}.$$
$$\phantom{\frac{60 - \frac{15}{2}}{\frac{13}{4} + \frac{1}{2}} = \frac{\frac{120}{2} - \frac{15}{2}}{\frac{13}{4} + \frac{2}{4}}} \leftarrow \text{Write } \frac{1}{2} \text{ as } \frac{2}{4}, \text{ so we can add it to } \frac{13}{4}.$$

$$= \frac{\frac{105}{2}}{\frac{15}{4}} \quad \leftarrow \frac{120}{2} - \frac{15}{2} = \frac{105}{2}$$
$$\phantom{= \frac{\frac{105}{2}}{\frac{15}{4}}} \leftarrow \frac{13}{4} + \frac{2}{4} = \frac{15}{4}$$

$$= \frac{105}{2} \div \frac{15}{4} \quad \text{Replace the bar by the division sign, } \div.$$

$$= \overset{7}{\cancel{\frac{105}{2}}} \cdot \overset{2}{\cancel{\frac{4}{15}}} \quad \text{Multiply by the reciprocal of } \frac{15}{4}, \text{ which is } \frac{4}{15}, \text{ and simplify.}$$

$$= 14 \quad \text{As before.}$$

EXAMPLE 1 Simplifying a complex fraction using Method 1

Use Method 1 to write the following as a simple fraction in simplified form:

$$\frac{\frac{3}{a} - \frac{4}{b}}{\frac{1}{2a} + \frac{2}{3b}}$$

SOLUTION The LCD of

$$\frac{3}{a}, \frac{4}{b}, \frac{1}{2a}, \text{ and } \frac{2}{3b}$$

is $6ab$. Therefore, we multiply the numerator and denominator of the given fraction by $6ab$ to obtain

$$\frac{6ab \cdot \left(\frac{3}{a} - \frac{4}{b}\right)}{6ab \cdot \left(\frac{1}{2a} + \frac{2}{3b}\right)} = \frac{6ab \cdot \frac{3}{a} - 6ab \cdot \frac{4}{b}}{6ab \cdot \frac{1}{2a} + 6ab \cdot \frac{2}{3b}} \quad \text{Use the distributive property and simplify.}$$

$$= \frac{18b - 24a}{3b + 4a} \quad \begin{array}{l} \leftarrow 6b \cdot 3 = 18b \text{ and } 6a \cdot 4 = 24a \\ \leftarrow 3b \cdot 1 = 3b \text{ and } 2a \cdot 2 = 4a \end{array}$$

or $\quad \dfrac{6(3b - 4a)}{3b + 4a}$

PROBLEM 1

Simplify the complex rational expression using Method 1:

$$\frac{\frac{2}{b} - \frac{3}{a}}{\frac{1}{2b} - \frac{3}{4a}}$$

EXAMPLE 2 Simplifying a complex fraction using Method 1

Use Method 1 to simplify:

$$\frac{x - \frac{1}{x^3}}{x + \frac{1}{x^2}}$$

SOLUTION Here the LCD is x^3. Thus,

$$\frac{x - \frac{1}{x^3}}{x + \frac{1}{x^2}} = \frac{x^3 \cdot \left(x - \frac{1}{x^3}\right)}{x^3 \cdot \left(x + \frac{1}{x^2}\right)}$$

$$= \frac{x^3 \cdot x - \cancel{x^3} \cdot \frac{1}{\cancel{x^3}}}{x^3 \cdot x + \cancel{x^3} \cdot \frac{1}{\cancel{x^2}}}$$

$$= \frac{x^4 - 1}{x^4 + x} = \frac{(x^2 + 1)(x^2 - 1)}{x(x^3 + 1)} \quad \text{To simplify, completely factor the numerator and denominator.}$$

$$= \frac{(x^2 + 1)(\cancel{x + 1})(x - 1)}{x(\cancel{x + 1})(x^2 - x + 1)} \quad \text{Divide out } x + 1.$$

$$= \frac{(x^2 + 1)(x - 1)}{x(x^2 - x + 1)}$$

or $\quad \dfrac{x^3 - x^2 + x - 1}{x^3 - x^2 + x}$

PROBLEM 2

Use Method 1 to simplify:

$$\frac{2x^2 + \frac{1}{4x}}{4 - \frac{1}{x^2}}$$

Answers to PROBLEMS

1. 4

2. $\dfrac{x(4x^2 - 2x + 1)}{4(2x - 1)}$ or $\dfrac{4x^3 - 2x^2 + x}{8x - 4}$

EXAMPLE 3 Comparing Method 1 with Method 2

Simplify:

$$\frac{\frac{x}{x-2}+x}{1+\frac{3}{x^2-4}}$$

a. Use Method 1.
b. Use Method 2.

SOLUTION

a. First write $x^2 - 4$ as $(x + 2)(x - 2)$ to obtain

$$\frac{\frac{x}{x-2}+x}{1+\frac{3}{x^2-4}} = \frac{\frac{x}{x-2}+x}{1+\frac{3}{(x+2)(x-2)}}$$

The LCD of the fractions is $(x + 2)(x - 2)$. Multiply numerator and denominator by this LCD.

$$\frac{\frac{(x+2)(x-2)}{1}\cdot\left(\frac{x}{x-2}+x\right)}{\frac{(x+2)(x-2)}{1}\cdot\left[1+\frac{3}{(x+2)(x-2)}\right]}$$

$$= \frac{x(x+2) + x(x+2)(x-2)}{(x+2)(x-2) + 3} \quad \text{Distribute the LCD.}$$

$$= \frac{x^2 + 2x + x^3 - 4x}{x^2 - 4 + 3} \quad \text{Remove parentheses.}$$

$$= \frac{x^3 + x^2 - 2x}{x^2 - 1} \quad \text{Collect like terms.}$$

$$= \frac{x(x^2 + x - 2)}{(x+1)(x-1)} \quad \text{Factor out } x \text{ in numerator. Factor } x^2 - 1 = (x+1)(x-1) \text{ in denominator.}$$

$$= \frac{x(x-1)(x+2)}{(x+1)(x-1)} \quad \text{Factor } x^2 + x - 2 = (x-1)(x+2). \text{ Divide out } (x-1).$$

$$= \frac{x(x+2)}{x+1} \quad \text{or} \quad \frac{x^2 + 2x}{x+1}$$

b. Using Method 2, first perform the operations in the numerator and denominator, and then divide.

$$\frac{\frac{x}{x-2}+x}{1+\frac{3}{x^2-4}} = \frac{\frac{x}{x-2}+\frac{x(x-2)}{x-2}}{\frac{1(x^2-4)}{x^2-4}+\frac{3}{x^2-4}} \quad \begin{array}{l}\leftarrow \text{Rewrite } x \text{ as } \frac{x(x-2)}{x-2}.\\ \leftarrow \text{Rewrite } 1 \text{ as } \frac{1(x^2-4)}{x^2-4}.\end{array}$$

$$= \frac{\frac{x + x(x-2)}{x-2}}{\frac{1(x^2-4) + 3}{x^2-4}} \quad \text{Add.}$$

$$= \frac{\frac{x + x^2 - 2x}{x-2}}{\frac{x^2 - 4 + 3}{x^2 - 4}} \quad \text{Remove parentheses.}$$

$$= \frac{\frac{x^2 - x}{x-2}}{\frac{x^2 - 1}{x^2 - 4}} \quad \text{Simplify.}$$

PROBLEM 3

Simplify:

$$\frac{\frac{x}{x+3}+x}{1-\frac{7}{x^2-9}}$$

(continued)

Answers to PROBLEMS

3. $\frac{x(x-3)}{x-4}$ or $\frac{x^2 - 3x}{x-4}$

464 Chapter 6 Rational Expressions

$$= \frac{x^2 - x}{x - 2} \div \frac{x^2 - 1}{x^2 - 4}$$ Use the division sign, ÷, instead of the bar.

$$= \frac{x(x - 1)}{x - 2} \cdot \frac{(x + 2)(x - 2)}{(x + 1)(x - 1)}$$ Multiply by the reciprocal of $\frac{x^2 - 1}{x^2 - 4}$ and factor $x^2 - x$, $x^2 - 4$, and $x^2 - 1$.

$$= \frac{x(x - 1)(x + 2) \cdot (x - 2)}{(x - 2)(x + 1)(x - 1)}$$ Multiply. Divide out $x - 2$ and $x - 1$.

$$= \frac{x(x + 2)}{x + 1}$$

or $\quad \dfrac{x^2 + 2x}{x + 1}$

In comparing Method 1 with Method 2, the results are the same. Which is the easiest method? That will depend on the problem and the knowledge you bring to the problem.

EXAMPLE 4   Simplifying a complex fraction using Method 2

Simplify:

$$1 + \frac{a}{1 + \dfrac{1}{1 + a}}$$

SOLUTION   Start by working on the denominator of the fraction

$$\frac{a}{1 + \dfrac{1}{1 + a}}$$

Since we have to add 1 to

$$\frac{1}{1 + a}$$

we rewrite 1 as

$$\frac{1 + a}{1 + a}$$

$$1 + \frac{a}{1 + \dfrac{1}{1 + a}} = 1 + \frac{a}{\dfrac{1 + a}{1 + a} + \dfrac{1}{1 + a}}$$ Work on the denominator.

$$= 1 + \frac{a}{\dfrac{2 + a}{1 + a}}$$ $\dfrac{1+a}{1+a} + \dfrac{1}{1+a} = \dfrac{1+a+1}{1+a} = \dfrac{2+a}{1+a}$

$$= 1 + a \div \frac{2 + a}{1 + a}$$ Use the ÷ sign instead of the bar.

$$= 1 + a \cdot \frac{1 + a}{2 + a}$$ To divide a by $\frac{2+a}{1+a}$, multiply by the reciprocal $\frac{1+a}{2+a}$.

$$= 1 + \frac{a(1 + a)}{2 + a}$$

$$= \frac{2 + a}{2 + a} + \frac{a(1 + a)}{2 + a}$$ Write 1 as $\frac{2+a}{2+a}$.

$$= \frac{2 + a + a + a^2}{2 + a}$$ Add the numerator and keep the denominator.

$$= \frac{2 + 2a + a^2}{2 + a}$$

$$= \frac{a^2 + 2a + 2}{a + 2}$$ Write the numerator and denominator in descending order.

PROBLEM 4

Simplify:

$$2 + \frac{a}{2 + \dfrac{2}{2 + a}}$$

Answers to PROBLEMS

4. $\dfrac{a^2 + 6a + 12}{2a + 6}$ or $\dfrac{a^2 + 6a + 12}{2(a + 3)}$

Sometimes complex fractions are written using negative exponents. To simplify such expressions, begin by rewriting them using the definition of a negative exponent, that is,

$$a^{-n} = \frac{1}{a^n}$$

Using this definition,

$$(x-1)^{-1} = \frac{1}{x-1}, \ (x+2)^{-3} = \frac{1}{(x+2)^3}, \text{ and } x(x+y)^{-5} = \frac{x}{(x+y)^5}$$

Notice the x in $x(x+y)^{-5}$ does not have a negative exponent. It remains as x in the numerator.

We shall use these ideas in Example 5. Pay close attention to what we do in each step of the solution.

EXAMPLE 5 Simplifying a complex fraction involving negative exponents

Simplify: $\dfrac{x(x-3)^{-1} + x}{x(3-x)^{-1} - x}$

SOLUTION

$\dfrac{x(x-3)^{-1} + x}{x(3-x)^{-1} - x} = \dfrac{\dfrac{x}{x-3} + \dfrac{x}{1}}{\dfrac{x}{3-x} - \dfrac{x}{1}}$

Rewrite $x(x-3)^{-1}$ as $\frac{x}{x-3}$, $x(3-x)^{-1}$ as $\frac{x}{3-x}$, and x as $\frac{x}{1}$.

$= \dfrac{\dfrac{x}{x-3} + \dfrac{x}{1}}{\dfrac{x}{-(x-3)} - \dfrac{x}{1}}$

$3 - x$ and $x - 3$ are opposites, that is, $3 - x = -(x - 3)$, so we substitute $-(x-3)$ for $3 - x$.

$= \dfrac{(x-3)\left[\dfrac{x}{x-3} + \dfrac{x}{1}\right]}{(x-3)\left[\dfrac{x}{-(x-3)} - \dfrac{x}{1}\right]}$

The LCD of all the denominators is $x - 3$. Multiply the numerator and denominator by $(x - 3)$.

$= \dfrac{\dfrac{(x-3)x}{(x-3)} + \dfrac{(x-3)x}{1}}{\dfrac{(x-3)x}{-(x-3)} - \dfrac{(x-3)x}{1}}$

Use the distributive property to distribute the $(x - 3)$.

$= \dfrac{x + (x-3)x}{-x - (x-3)x}$

Divide out the $(x - 3)$ terms and note that $\frac{(x-3)x}{1} = (x-3)x$.

$= \dfrac{x + x^2 - 3x}{-x - (x^2 - 3x)}$

Multiply $(x - 3)x$ in the numerator and denominator.

$= \dfrac{x + x^2 - 3x}{-x - x^2 + 3x}$

Use the distributive property, $-(x^2 - 3x) = -x^2 + 3x$.

$= \dfrac{x^2 - 2x}{-x^2 + 2x}$

Collect like terms in the numerator and denominator.

$= \dfrac{x^2 - 2x}{-(x^2 - 2x)}$

Rewrite the denominator $-x^2 + 2x$ as $-(x^2 - 2x)$.

$= -1$

Since $\frac{a}{-a} = -1$, the answer is -1.

PROBLEM 5

Simplify:

$\dfrac{x + (x-2)^{-1}}{1 - x(2-x)^{-1}}$

Answers to PROBLEMS

5. $\dfrac{x-1}{2}$

NOTE

In Step 2 of Example 5, the numerator is

$$\frac{x}{x-3} + \frac{x}{1}$$

and the denominator is

$$\frac{x}{-(x-3)} - \frac{x}{1} = -\left(\frac{x}{x-3} + \frac{x}{1}\right)$$

which is the additive inverse of the numerator. Thus,

$$\frac{\frac{x}{x-3} + \frac{x}{1}}{\frac{x}{-(x-3)} - \frac{x}{1}} = \frac{\frac{x}{x-3} + \frac{x}{1}}{-\left(\frac{x}{x-3} + \frac{x}{1}\right)} = -1 \quad \longleftarrow \text{These two expressions are additive inverses of each other.}$$

Calculator Corner

Verifying Results of Simplifying Complex Fractions

In the last few sections we have been verifying the results of multiplying, dividing, adding, and subtracting rational expressions by making sure that the graph of the original problem and the graph of the final answer are identical. This can also be done with complex fractions. However, we want to explore the numerical capabilities of your calculator. In Example 3, we found that

```
                                    4
(X/(X-2)+X)/(1+3
/(X^2-4))
                                  4.8
X(X+2)/(X+1)
                                  4.8
```

$$\frac{\left(\frac{x}{(x-2)} + x\right)}{\left(1 + \frac{3}{(x^2-4)}\right)} = \frac{x(x+2)}{(x+1)}$$

To check the answer, substitute any convenient number for x and see if both sides are equal. You cannot choose numbers that will give you a zero denominator (-1, 2 and -2 will yield zero denominators). A simple number to use is $x = 4$. Now, go to the home screen and store the value 4 as x. With a TI-83 Plus press 4 STO▸ X,T,θ,n ENTER . The calculator confirms the entry by showing 4→X and the answer 4. Now enter the original complex fraction as

$$\left(\frac{x}{(x-2)} + x\right) \div \left(1 + \frac{3}{(x^2-4)}\right)$$

Be extremely careful with the parentheses. Press ENTER . The answer is 4.8, as shown in the window. Next, enter the simplified complex fraction as $x(x + 2)/(x + 1)$. Press ENTER again. You should get the same answer as before, 4.8. The final confirmation is shown on the screen.

You can check the rest of the problems involving one variable using this method. If you are checking Example 4, you can enter an x instead of an a when writing the complex fraction involved, but there are so many parentheses involved that it's probably easier to do the problem without the calculator. Try it if you don't believe this!

1. Verify the results of Example 2 numerically using the techniques shown.
2. Verify the results of Example 5 numerically using the techniques shown.
3. Can you verify the results of Example 1
   a. graphically? Explain.
   b. numerically? Explain.

Exercises 6.4

A Complex Fractions In Problems 1–42, perform the indicated operation and give the answer in simplified form.

1. $\dfrac{50 - 5\frac{1}{2}}{7\frac{3}{4} + \frac{1}{2}}$

2. $\dfrac{70 - 17\frac{1}{2}}{2\frac{1}{4} + 1\frac{1}{2}}$

3. $\dfrac{\frac{a}{b}}{\frac{c}{b}}$

4. $\dfrac{\frac{-a^2}{c}}{\frac{-b^2}{c}}$

5. $\dfrac{\frac{x}{y}}{\frac{x^2}{z}}$

6. $\dfrac{\frac{x^2}{y^2}}{\frac{x}{z}}$

7. $\dfrac{\frac{3x}{5y}}{\frac{3x}{2z}}$

8. $\dfrac{\frac{7x}{3y}}{\frac{14x}{5y}}$

9. $\dfrac{\frac{1}{2}}{2 - \frac{1}{2}}$

10. $\dfrac{\frac{1}{4}}{3 - \frac{1}{4}}$

11. $\dfrac{a - \frac{a}{b}}{1 + \frac{a}{b}}$

12. $\dfrac{1 - \frac{1}{a}}{1 + \frac{1}{a}}$

13. $\dfrac{\frac{1}{x} + 2}{2 - \frac{1}{x}}$

14. $\dfrac{3 - \frac{2}{y}}{\frac{1}{y} + 4}$

15. $\dfrac{\frac{2}{3} + x}{x - \frac{1}{2}}$

16. $\dfrac{x + \frac{1}{3}}{\frac{3}{4} - x}$

17. $\dfrac{y + \frac{2}{x}}{y^2 - \frac{4}{x^2}}$

18. $\dfrac{y - \frac{3}{x}}{y^2 - \frac{9}{x^2}}$

19. $\dfrac{\frac{x}{y^2} - \frac{y}{x^2}}{x^2 + xy + y^2}$

20. $\dfrac{\frac{x}{y^2} + \frac{y}{x^2}}{x^2 - xy + y^2}$

21. $3 - \dfrac{3}{3 - \frac{1}{2}}$

22. $2 - \dfrac{2}{2 - \frac{1}{2}}$

23. $a - \dfrac{a}{a + \frac{1}{2}}$

24. $a + \dfrac{a}{a + \frac{1}{2}}$

25. $x - \dfrac{x}{1 - \frac{x}{1 - x}}$

26. $2x - \dfrac{x}{2 - \frac{x}{2 - x}}$

27. $\dfrac{\frac{x-1}{x+1} + \frac{x+1}{x-1}}{\frac{x-1}{x+1} - \frac{x+1}{x-1}}$

28. $\dfrac{\frac{x-1}{x+1} - \frac{x+1}{x-1}}{\frac{x-1}{x+1} + \frac{x+1}{x-1}}$

29. $\dfrac{(x-y)^{-1} + (x+y)^{-1}}{(x-y)^{-1} - (x+y)^{-1}}$

30. $\dfrac{a(a+b)^{-1} + 1}{1 - b(a+b)^{-1}}$

31. $\dfrac{x(x-2)^{-1} - x}{x(2-x)^{-1} + x}$

32. $\dfrac{x + x(4-x)^{-1}}{x(x-4)^{-1}}$

33. $\dfrac{\frac{1}{x^2} + \frac{3}{x} - 4}{\frac{1}{x^2} + \frac{5}{x} + 4}$

34. $\dfrac{\frac{6}{v^2} - \frac{11}{v} - 10}{\frac{2}{v^2} + \frac{1}{v} - 15}$

35. $\dfrac{y + 3 - \frac{16}{y+3}}{y - 6 + \frac{20}{y+3}}$

36. $\dfrac{w + 2 - \frac{18}{w-5}}{w - 1 - \frac{12}{w-5}}$

37. $\dfrac{\frac{8x}{3x+1} - \frac{3x-1}{x}}{\frac{x}{3x+1} - \frac{2x-2}{x}}$

38. $\dfrac{\frac{3}{m-4} - \frac{16}{m-3}}{\frac{2}{m-3} - \frac{15}{m+5}}$

39. $\dfrac{\frac{c}{d} - \frac{d}{c}}{\frac{c}{d} - 2 + \frac{d}{c}}$

40. $\dfrac{\frac{a^2}{b^2} + 4 + \frac{4b^2}{a^2}}{\frac{a}{b} + \frac{2b}{a}}$

41. $\dfrac{\frac{a^2 - b^2}{a^2 + b^2} - \frac{a^2 + b^2}{a^2 - b^2}}{\frac{a-b}{a+b} - \frac{a+b}{a-b}}$

42. $\dfrac{1 + \frac{4uv}{(u-v)^2}}{1 + \frac{uv - 3v^2}{(u-v)^2}}$

››› Applications

43. **Electrical resistance** When connected in parallel, the combined resistance R of two electrical resistances R_1 and R_2 is given by

$$R = \dfrac{1}{\frac{1}{R_1} + \frac{1}{R_2}}$$

Simplify the expression for R.

44. **Electrical resistance** When connected in parallel, the combined resistance R of three electrical resistances R_1, R_2, and R_3 is given by

$$R = \dfrac{1}{\frac{1}{R_1} + \frac{1}{R_2} + \frac{1}{R_3}}$$

Simplify the expression for R.

45. *Doppler effect* The formula for the Doppler effect in light is

$$f = f_{\text{static}} \sqrt{\dfrac{1 + \dfrac{v}{c}}{1 - \dfrac{v}{c}}}$$

where f and f_{static} are frequencies, v is the velocity of the moving body, and c is the speed of light.
Simplify the expression for f.

46. *Hydrogen spectrum light* Balmer's formula for the wavelength λ (lambda) of the hydrogen spectrum light is given by

$$\lambda = \dfrac{1}{\dfrac{1}{m^2} - \dfrac{1}{n^2}}$$

Simplify the expression for λ.

》》》 Using Your Knowledge

Interest Rates and Planetary Orbits Do you have monthly payments on any type of loan? Do you know what your **A**nnual **P**ercentage **R**ate (**APR**) is? If you financed P dollars to be paid in N monthly payments of M dollars, your APR is

$$\dfrac{\dfrac{24(NM - P)}{N}}{P + \dfrac{NM}{12}}$$

47. Simplify the APR formula.

48. Use the simplified version of the APR formula to determine the APR on a $20,000 loan with payments of:

    a. $500 a month for 4 years.

    b. $400 a month for 5 years.

In the seventeenth century, the Dutch mathematician and astronomer Christian Huygens made a model of the solar system and found out that Saturn takes

$$29 + \dfrac{1}{2 + \dfrac{2}{9}}$$

years to go around the Sun. Now,

$$\dfrac{1}{2 + \dfrac{2}{9}} = \dfrac{9 \cdot 1}{9 \cdot \left(2 + \dfrac{2}{9}\right)}$$

$$= \dfrac{9}{18 + 2}$$

$$= \dfrac{9}{20}$$

Thus, it takes Saturn $29 + \dfrac{9}{20} = 29\dfrac{9}{20}$ years to go around the Sun. Use your knowledge to find the number of years it takes each of the following planets to go around the Sun by simplifying the fraction. (Write your answers as mixed numbers when necessary.)

49. Mercury: $\dfrac{1}{4 + \dfrac{1}{6}}$ yr

50. Venus: $\dfrac{1}{1 + \dfrac{2}{3}}$ yr

51. Jupiter: $11 + \dfrac{1}{1 + \dfrac{7}{43}}$ yr

52. Mars: $1 + \dfrac{1}{1 + \dfrac{3}{22}}$ yr

》》》 Write On

53. Write in your own words the definition of a complex fraction.

54. We have given two methods to simplify a complex fraction. Which method do you think is simpler and why?

55. Can you explain when one should use Method 1 to simplify a complex fraction and what types of fractions are easier to simplify using this method?

56. Can you explain when one should use Method 2 to simplify a complex fraction and what types of fractions are easier to simplify using this method?

⟩⟩⟩ Concept Checker

Fill in the blank(s) with the correct word(s), phrase, or mathematical statement.

57. One method of simplifying a complex fraction is to multiply the numerator and denominator by the _____ of the simple fractions within the complex fraction.

58. A fraction whose numerator or denominator (or both) contains other fractions is called a _____.

59. In one method of simplifying $\frac{\frac{3}{x}}{\frac{2}{y}}$, you must multiply $\frac{3}{x}$ by the _____ of $\frac{2}{y}$.

rational expression

complex fraction

invert

reciprocal

rewrite

least common denominator

numerators

denominator

⟩⟩⟩ Mastery Test

Simplify:

60. $\dfrac{\frac{12x}{5y}}{\frac{3x}{25y}}$

61. $\dfrac{\frac{6b^2}{7x}}{\frac{9b^3}{14x^2}}$

62. $\dfrac{\frac{x}{x+2} + x}{1 - \frac{5}{x^2 - 4}}$

63. $\dfrac{x + \frac{x}{x+3}}{\frac{1}{x^2 - 9} + 1}$

64. $\dfrac{\frac{2}{b} - \frac{3}{a}}{\frac{1}{2b} + \frac{3}{4a}}$

65. $\dfrac{\frac{3}{b} + \frac{2}{a}}{\frac{1}{2a} - \frac{3}{4b}}$

66. $\dfrac{x - \frac{1}{x^3}}{x - \frac{1}{x^2}}$

67. $\dfrac{\frac{1}{x^2} - x}{\frac{1}{x^3} - x}$

68. $\dfrac{x(x-4)^{-1} + x}{x(4-x)^{-1} - x}$

69. $\dfrac{y - y(y-4)^{-1}}{y + y(4-y)^{-1}}$

⟩⟩⟩ Skill Checker

Simplify:

70. $\dfrac{8x^4}{2x^3}$

71. $\dfrac{28x^7}{7x^5}$

72. $\dfrac{-30x^6}{6x^4}$

73. $\dfrac{-10x^2}{20x^4}$

74. $\dfrac{-30x}{10x^3}$

75. $\dfrac{-50}{10x^3}$

Multiply:

76. $x^2(5x + 5)$

77. $6x^2(x - 2)$

78. $3x^4(3x - 5)$

Factor completely:

79. $6x^2 + x - 2$

80. $6x^2 + 7x - 3$

81. $20x^2 - 7x - 6$

82. Subtract $6x^3 + 24x^2$ from $6x^3 + 25x^2 + 2x - 8$.

83. Subtract $5x^2 + 15x$ from $5x^2 + 9x - 18$.

Evaluate:

84. $P(2)$ if $P(x) = x^2 - 3x + 10$

85. $P(-5)$ if $P(x) = 3x^3 - 2x^2 + x - 12$

6.5 Division of Polynomials and Synthetic Division

Objectives

A Divide a polynomial by a monomial.

B Use long division to divide one polynomial by another.

C Completely factor a polynomial when one of the factors is known.

D Use synthetic division to divide a polynomial by a binomial.

E Use the remainder theorem to verify that a number is a solution of a given equation.

To Succeed, Review How To . . .

1. Simplify quotients using the properties of exponents (pp. 42–43).
2. Multiply polynomials (pp. 360–365).
3. Evaluate expressions using the order of operations (pp. 53–54).

Getting Started
Efficiency Quotients

How efficient is your car engine? The efficiency E of an engine is given by

$$E = \frac{Q_1 - Q_2}{Q_1}$$

where Q_1 is the horsepower rating of the engine and Q_2 is the horsepower delivered to the transmission. Can you do the indicated division in the rational expression

$$\frac{Q_1 - Q_2}{Q_1}$$

Follow the steps in the procedure.

$$\frac{Q_1 - Q_2}{Q_1} = (Q_1 - Q_2)\left(\frac{1}{Q_1}\right) \quad \text{Dividing by } Q_1 \text{ is the same as multiplying by the reciprocal of } Q_1, \text{ that is, } \frac{1}{Q_1}.$$

$$= Q_1\left(\frac{1}{Q_1}\right) - Q_2\left(\frac{1}{Q_1}\right) \quad \text{Use the distributive property.}$$

$$= \frac{Q_1}{Q_1} - \frac{Q_2}{Q_1} \quad \text{Multiply.}$$

$$= 1 - \frac{Q_2}{Q_1} \quad \frac{Q_1}{Q_1} = 1$$

Next, we will study how to divide many types of polynomials.

A > Dividing a Polynomial by a Monomial

To divide the trinomial $4x^4 - 8x^3 + 12x^2$ by $2x^2$, we proceed as in the *Getting Started*.

$$\frac{4x^4 - 8x^3 + 12x^2}{2x^2} = (4x^4 - 8x^3 + 12x^2)\left(\frac{1}{2x^2}\right)$$

$$= 4x^4\left(\frac{1}{2x^2}\right) - 8x^3\left(\frac{1}{2x^2}\right) + 12x^2\left(\frac{1}{2x^2}\right)$$

$$= \frac{4x^4}{2x^2} - \frac{8x^3}{2x^2} + \frac{12x^2}{2x^2}$$

$$= 2x^2 - 4x + 6$$

Or, more simply $\dfrac{4x^4 - 8x^3 + 12x^2}{2x^2} = \dfrac{4x^4}{2x^2} - \dfrac{8x^3}{2x^2} + \dfrac{12x^2}{2x^2} = 2x^2 - 4x + 6$

To avoid doing all these steps, we can state the following rule.

> **PROCEDURE**
>
> **Rule for Dividing a Polynomial by a Monomial**
>
> To divide a polynomial by a monomial, divide each term in the polynomial by the monomial.

EXAMPLE 1 Dividing a polynomial by a monomial

Divide:

a. $\dfrac{28x^5 - 14x^4 + 7x^3}{7x^2}$ **b.** $\dfrac{20x^4 - 15x^3 + 10x^2 - 30x + 50}{10x^3}$

SOLUTION

a. $\dfrac{28x^5 - 14x^4 + 7x^3}{7x^2} = \dfrac{28x^5}{7x^2} - \dfrac{14x^4}{7x^2} + \dfrac{7x^3}{7x^2}$

$\phantom{\dfrac{28x^5 - 14x^4 + 7x^3}{7x^2}} = 4x^3 - 2x^2 + x$

b. $\dfrac{20x^4 - 15x^3 + 10x^2 - 30x + 50}{10x^3} = \dfrac{20x^4}{10x^3} - \dfrac{15x^3}{10x^3} + \dfrac{10x^2}{10x^3} - \dfrac{30x}{10x^3} + \dfrac{50}{10x^3}$

$\phantom{\dfrac{20x^4 - 15x^3 + 10x^2 - 30x + 50}{10x^3}} = 2x - \dfrac{3}{2} + \dfrac{1}{x} - \dfrac{3}{x^2} + \dfrac{5}{x^3}$

In this case, the answer is *not* a polynomial.

PROBLEM 1

Divide:

a. $\dfrac{24x^7 - 18x^5 - 12x^3}{-6x^3}$

b. $\dfrac{16x^4 - 4x^3 + 8x^2 - 16x + 40}{8x^2}$

B › Dividing One Polynomial by Another Polynomial

If we wish to divide a polynomial (called the **dividend**) by another polynomial (called the **divisor**), we proceed very much as we did in long division in arithmetic. To show you that this is so, we shall perform the division of 337 by 16 and $(x^3 + 3x^2 + 3x + 1)$ by $(x^2 + x + 1)$ side by side.

1. $16\overline{)337}^{\,2}$ Divide 33 by 16. It goes twice. Write 2 over the 33.

$(x^2) + x + 1\overline{)x^3 + 3x^2 + 3x + 1}^{\,x}$ Write x over the $3x$.

— Think $\dfrac{x^3}{x^2} = x$. — 1st term inside / 1st term outside

2. $\begin{array}{r} 2 \\ 16\overline{)\,337\,} \\ -32 \\ \hline 1 \end{array}$ Multiply 16 by 2 and subtract the product 32 from 33, obtaining 1.

$\begin{array}{r} x \\ x^2 + x + 1\overline{)\,x^3 + 3x^2 + 3x + 1\,} \\ -(x^3 + x^2 + x) \\ \hline 0 + 2x^2 + 2x \end{array}$ Multiply $x(x^2 + x + 1)$. Subtract $(x^3 + x^2 + x)$ from $x^3 + 3x^2 + 3x + 1$. You can omit the zero.

3. $\begin{array}{r} 21 \\ 16\overline{)\,337\,} \\ -32 \\ \hline 17 \end{array}$ "Bring down" the 7. Now, divide 17 by 16. It goes once. Write 1 after the 2.

$\begin{array}{r} x + 2 \\ (x^2) + x + 1\overline{)\,x^3 + 3x^2 + 3x + 1\,} \\ -(x^3 + x^2 + x) \\ \hline 0 + (2x^2) + 2x + 1 \end{array}$ "Bring down" the 1. Write $+2$ after the x.

— Think $\dfrac{2x^2}{x^2} = 2$. — 1st term inside / 1st term outside

Answers to PROBLEMS

1. a. $-4x^4 + 3x^2 + 2$

b. $2x^2 - \dfrac{x}{2} + 1 - \dfrac{2}{x} + \dfrac{5}{x^2}$

472 Chapter 6 Rational Expressions

4. $16\overline{)337}$ Multiply 16 by 1 and sub-
   $\underline{-32}$ tract the result from 17.
   $\;\,17$ The remainder is 1.
   $\underline{-16}$
   $\;\;\,1$

5. The answer **(quotient)** can be written as
   21 R 1 (read "21 remainder 1") or as
   $21 + \frac{1}{16}$, which is $21\frac{1}{16}$.

6. You can check this answer by multiplying 21 by 16 (336) and adding the remainder 1 to obtain 337, the dividend.

$$\begin{array}{r} x + 2 \\ x^2 + x + 1 \overline{) x^3 + 3x^2 + 3x + 1} \\ \underline{-(x^3 + x^2 + x)} \\ 0 + 2x^2 + 2x + 1 \\ \underline{-(2x^2 + 2x + 2)} \\ -1 \end{array}$$

Multiply $2(x^2 + x + 1)$. Subtract $(2x^2 + 2x + 2)$ from $2x^2 + 2x + 1$. The remainder is -1.

The answer **(quotient)** can be written as $(x + 2)$ R -1 (read "$x + 2$ remainder -1"), or you can write the result more completely as

$$\underbrace{\frac{\overbrace{x^3 + 3x^2 + 3x + 1}^{\text{Dividend}}}{\underbrace{x^2 + x + 1}_{\text{Divisor}}}}_{} = \overbrace{x + 2}^{\text{Quotient}} - \frac{\overbrace{1}^{\text{Remainder}}}{\underbrace{x^2 + x + 1}_{\text{Divisor}}}$$

You can check the answer by multiplying $(x + 2)(x^2 + x + 1) = x^3 + 3x^2 + 3x + 2$ and adding the remainder -1 to get the dividend $x^3 + 3x^2 + 3x + 1$.

EXAMPLE 2 Dividing polynomials using long division
Divide: $(x^3 + 2x^2 - 17x) \div (x^2 + x - 3)$

SOLUTION

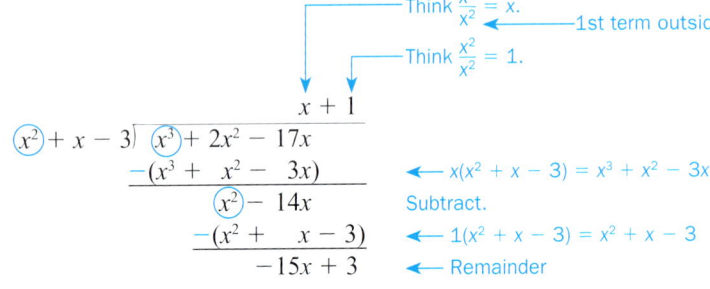

Think $\frac{x^3}{x^2} = x$. ← 1st term inside
← 1st term outside
Think $\frac{x^2}{x^2} = 1$.

$x(x^2 + x - 3) = x^3 + x^2 - 3x$
Subtract.
$1(x^2 + x - 3) = x^2 + x - 3$
Remainder

Thus,

$$\frac{x^3 + 2x^2 - 17x}{x^2 + x - 3} = x + 1 + \frac{-15x + 3}{x^2 + x - 3}$$

PROBLEM 2
Divide:
$(x^3 + 4x^2 - 15x) \div (x^2 - 2x + 3)$

If there are missing terms in the dividend, we insert zero coefficients, as shown in the next example.

Calculator Corner

Using Graphs to Verify Long Division with Polynomials

Division problems can be checked with your calculator. To check the result of Example 2, graph the original problem

$$Y_1 = \frac{(x^3 + 2x^2 - 17x)}{(x^2 + x - 3)}$$

and the answer

$$Y_2 = x + 1 + \frac{(-15x + 3)}{(x^2 + x - 3)}$$

using a standard window. (Note the parentheses when entering the expressions in your calculator.) The same graph is obtained in both cases, as shown in the window.

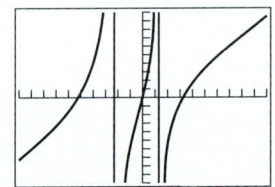

Answers to PROBLEMS

2. $x + 6 + \dfrac{-6x - 18}{x^2 - 2x + 3}$ or $(x + 6)$R$(-6x - 18)$

EXAMPLE 3 Using long division when there are missing terms

Divide: $(4x^3 - 4 - 8x) \div (4 + 4x)$

SOLUTION We write the polynomials in **descending** order, inserting $0x^2$ in the dividend.

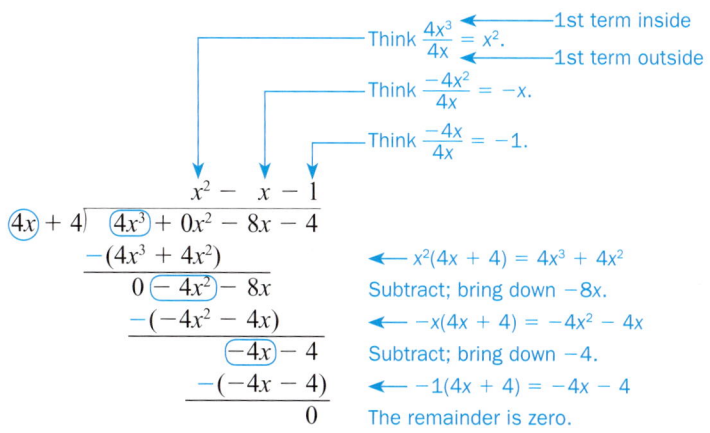

Thus,
$$\frac{4x^3 - 4 - 8x}{4 + 4x} = x^2 - x - 1$$

You can check this by multiplying $(4 + 4x)(x^2 - x - 1)$, obtaining $4x^3 - 4 - 8x$.

NOTE

When the remainder is zero, the denominator divides *exactly* into the numerator. We say $(4x + 4)$ is a factor of $4x^3 - 8x - 4$.

PROBLEM 3
Divide:

$(6x^3 + 3x - 9) \div (3x - 3)$

C › Factoring When One of the Factors Is Known

Suppose we wish to factor the polynomial

$$6x^3 + 23x^2 + 9x - 18$$

None of the methods we have studied so far will work here (try them!), so we need some more information. If we know that $(x + 3)$ is one of the factors, we can write $6x^3 + 23x^2 + 9x - 18 = (x + 3)P$, where P is a polynomial. Dividing both sides by $x + 3$ gives

$$\frac{6x^3 + 23x^2 + 9x - 18}{x + 3} = P$$

$(x + 3)$ is a factor, so we should anticipate zero as the remainder when we divide (as in Example 3).

Answers to PROBLEMS

3. $2x^2 + 2x + 3$

Let's do the division:

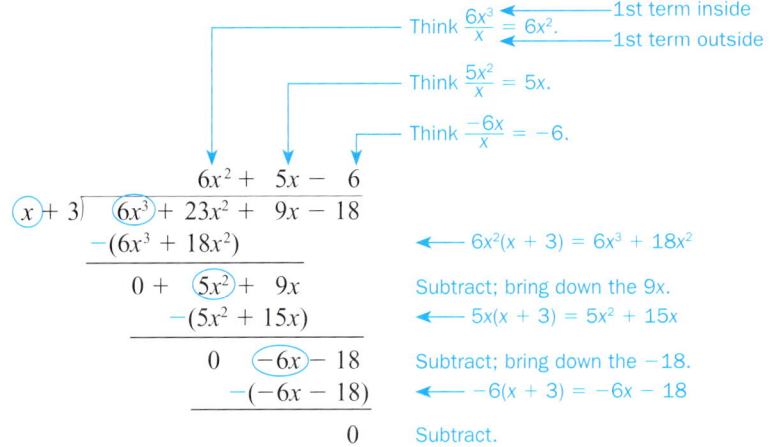

The polynomial P is $6x^2 + 5x - 6$, and we have

$$\frac{6x^3 + 23x^2 + 9x - 18}{x + 3} = 6x^2 + 5x - 6$$

Now $6x^2 + 5x - 6 = (3x - 2)(2x + 3)$, which gives

$$\frac{6x^3 + 23x^2 + 9x - 18}{x + 3} = (3x - 2)(2x + 3)$$

Multiplying both sides by $x + 3$ yields

$$6x^3 + 23x^2 + 9x - 18 = (x + 3)(3x - 2)(2x + 3)$$

If we wish to factor a polynomial when one of the factors is given, we can divide by the given factor. The product of the quotient obtained (factored, if possible) and the given factor gives the complete factorization of the polynomial.

EXAMPLE 4   Factoring when one factor is known
Factor $6x^3 + 25x^2 + 2x - 8$ if $(x + 4)$ is one of its factors.

SOLUTION   We start by dividing $6x^3 + 25x^2 + 2x - 8$ by $x + 4$.

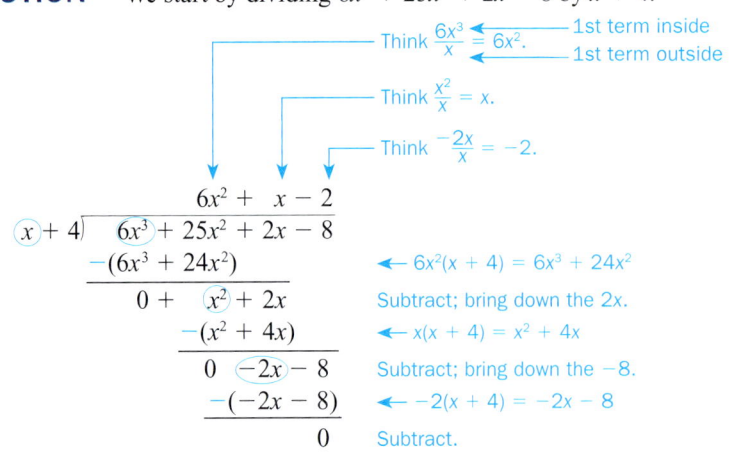

Now we know that the polynomial we are trying to factor is the product of $(x + 4)$ and the quotient we obtained, $6x^2 + x - 2$. To factor completely, we have to factor $6x^2 + x - 2$ as $(3x + 2)(2x - 1)$.

$$6x^3 + 25x^2 + 2x - 8 = (x + 4)(6x^2 + x - 2)$$
$$= (x + 4)(3x + 2)(2x - 1)$$

PROBLEM 4
Factor $2x^3 - x^2 - 18x + 9$ if $(x + 3)$ is one of its factors.

Answers to PROBLEMS

4. $(x + 3)(2x - 1)(x - 3)$

Calculator Corner

Using Graphs to Verify Factorization of Polynomials

Example 4 can be checked by graphing the original expression $Y_1 = 6x^3 + 25x^2 + 2x - 8$ and the factored $Y_2 = (x + 4)(3x + 2)(2x - 1)$.

D › Using Synthetic Division to Divide a Polynomial by a Binomial

The factorization process we have discussed can be made faster and more efficient if we can find a simpler method for doing the division. Look at the long division (left) and its simplified version (right):

$$\begin{array}{r} 3x^2 + 11x + 15 \\ x - 2 \overline{)3x^3 + 5x^2 - 7x + 10} \\ \underline{3x^3 - 6x^2} \\ 11x^2 \\ \underline{11x^2 - 22x} \\ 15x \\ \underline{15x - 30} \\ + 40 \end{array} \qquad \begin{array}{r} 3 + 11 + 15 \\ 1 - 2 \overline{)3 + 5 - 7 + 10} \\ 3 \;\; -6 \\ +11 \\ +11 \;\; -22 \\ +15 \\ +15 \;\; -30 \\ + 40 \end{array}$$

(left) Quotient / Divisor / Dividend / Remainder
(right) Quotient / Divisor / Dividend / Remainder

First, the diagram on the right omits the variables. Moreover, all the numbers in the boxes are repeated. Finally, the only numbers that we have to use to do the division are the circled ones and the ones in the answer (the quotient). We use these facts to write the division in an even shorter version using only three lines:

$$\begin{array}{c} 2 \overline{)3 \quad 5 \quad -7 \quad 10} \\ \downarrow \text{ Add} \\ \hline 3 \end{array}$$

1st line:
To make it easier, we replace the subtraction with addition by changing the sign of the constant in the indicated divisor (in this case -2 to 2) so that we can add at each step. Thus, the first line has the additive inverse of the constant of the divisor, followed by the coefficients of the dividend.

$$\begin{array}{c} 2 \overline{)3 \quad 5 \quad -7 \quad 10} \\ \text{Multiply} \searrow \; 6 \\ \hline 3 \end{array}$$

$$\begin{array}{c} 2 \overline{)3 \quad 5 \quad -7 \quad 10} \\ \quad 6 \; \downarrow \text{Add} \\ \hline 3 \quad 11 \end{array}$$

$$\begin{array}{c} 2 \overline{)3 \quad 5 \quad -7 \quad 10} \\ \text{Multiply} \searrow 6 \searrow 22 \\ \hline 3 \quad 11 \end{array}$$

$$\begin{array}{c} 2 \overline{)3 \quad 5 \quad -7 \quad 10} \\ \quad 6 \quad 22 \; \downarrow \text{Add} \\ \hline 3 \quad 11 \quad 15 \end{array}$$

$$\begin{array}{c} 2 \overline{)3 \quad 5 \quad -7 \quad 10} \\ \text{Multiply} \searrow 6 \searrow 22 \searrow 30 \\ \hline 3 \quad 11 \quad 15 \end{array}$$

2nd line:
The second line shows the results of multiplying each of the sums by 2, the additive inverse of the constant of the divisor. These numbers are the additive inverses of the numbers we circled above.

$$\begin{array}{c} 2 \overline{)3 \quad 5 \quad -7 \quad 10} \\ \downarrow \quad 6 \quad 22 \quad 30 \; \downarrow \text{Add} \\ \hline 3 \quad 11 \quad 15 \quad 40 \end{array}$$

3rd line:
Results of adding line 1 with line 2.

Quotient: $3x^2 + 11x + 15$ remainder 40. The degree (2) of the quotient is one less than the degree (3) of the dividend.

SYNTHETIC DIVISION

Synthetic division is a procedure used when dividing a polynomial by a binomial of the form $x - k$. The degree of the quotient is one less than the degree of the dividend.

EXAMPLE 5 Using synthetic division to divide a polynomial by a binomial

Use synthetic division to divide: $(2x^4 - 3x^2 + 5x - 7) \div (x + 3)$

SOLUTION Since synthetic division works only when dividing by binomials of the form $x - k$, write $x + 3$ as $x - (-3)$ to obtain the indicated divisor. The zero in the first line of the division is in place of the missing x^3 term in the dividend.

$$\begin{array}{r|rrrrr} -3 & 2 & 0 & -3 & +5 & -7 \\ & & -6 & +18 & -45 & +120 \\ \hline & 2 & -6 & +15 & -40 & +113 \end{array}$$

Quotient: $(2x^3 - 6x^2 + 15x - 40)$ R 113

The answer is read from the bottom row, $2x^3 - 6x^2 + 15x - 40$ with 113 as the remainder,

$$\frac{2x^4 - 3x^2 + 5x - 7}{x + 3} = 2x^3 - 6x^2 + 15x - 40 + \frac{113}{x + 3}$$

PROBLEM 5

Use synthetic division to divide:

$(3x^4 - 2x^3 - 9x + 1) \div (x - 2)$

E › Using the Remainder and Factor Theorems

Is there a quick way of checking at least part of the division process in Example 5? Amazingly enough, we can find the remainder in the division by using the *remainder theorem*, whose proof is given in more advanced courses.

THE REMAINDER THEOREM

If the polynomial $P(x)$ is divided by $x - k$, then the remainder is $P(k)$.

This theorem says that when $P(x)$ is divided by $x - k$, the remainder can be found by evaluating P at k, that is, by finding $P(k)$.

Thus, we can find the remainder in Example 5 by finding

$$P(-3) = 2(-3)^4 - 3(-3)^2 + 5(-3) - 7$$
$$= 2(81) - 3(9) - 15 - 7$$
$$= 162 - 27 - 15 - 7$$
$$= 113 \quad \text{Remainder}$$

But there is a more important application of the remainder theorem. If you divide a polynomial $P(x)$ by $x - k$ and the remainder is zero, then $P(k) = 0$, which means that k is a solution of the equation $P(k) = 0$. Thus, one way to show that $k = -3$ is a solution of $x^4 - 4x^3 - 5x^2 + 36x - 36 = 0$ is to substitute -3 for x in the equation. Another way is to use synthetic division and show that the remainder is zero, that is, $P(k) = 0$. We do this in Example 6.

Answers to PROBLEMS

5. $3x^3 + 4x^2 + 8x + 7 + \dfrac{15}{x - 2}$

EXAMPLE 6 Using the remainder theorem with synthetic division

Use synthetic division to show that -3 is a solution of

$$P(x) = x^4 - 4x^3 - 5x^2 + 36x - 36 = 0$$

SOLUTION We divide $x^4 - 4x^3 - 5x^2 + 36x - 36$ by $x - (-3)$.

$$\begin{array}{r|rrrrr} -3 & 1 & -4 & -5 & +36 & -36 \\ & & -3 & +21 & -48 & +36 \\ \hline & 1 & -7 & +16 & -12 & 0 \end{array}$$

The remainder is zero, $P(-3) = 0$, so -3 is a solution of the given equation.

PROBLEM 6

Use synthetic division to show that 1 is a solution of

$$P(x) = 2x^4 + x^3 - 35x^2 - 16x + 48 = 0$$

The remainder is zero when the polynomial given in Example 6 is divided by $x + 3$, so $x + 3$ must be a factor of the polynomial. Thus,

$$x^4 - 4x^3 - 5x^2 + 36x - 36 = (x + 3)(x^3 - 7x^2 + 16x - 12)$$

where the second factor is the quotient polynomial found using the last row of the synthetic division. Thus, the remainder theorem, used together with synthetic division, can help us evaluate a polynomial (by finding the remainder) and, if the remainder is zero, can even help us factor the polynomial. This last important result, which can be derived from the remainder theorem, is called the *factor theorem*.

THE FACTOR THEOREM

When a polynomial $P(x)$ has a factor $(x - k)$, it means that $P(k) = 0$.

Calculator Corner

Exploring the Remainder Theorem and Factor Theorem

We can do some exploring with the remainder theorem. In Example 6, we found out that -3 is a solution of $x^4 - 4x^3 - 5x^2 + 36x - 36 = 0$. How can we verify this graphically? We graph $Y_1 = x^4 - 4x^3 - 5x^2 + 36x - 36$. If we use a standard window, it seems that the curve cuts or touches the horizontal axis at $x = -3$, $x = 2$, and $x = 3$, but we cannot see the complete graph. To see more of the vertical axis, we let Ymin $= -100$ and Ymax $= 100$. We can clearly see that $x = -3$ is a solution (see Window 1).

If you have a TI-83 Plus, you can check the two other possible solutions, $x = 2$ and $x = 3$. Start with $x = 2$. Press [2nd] [TRACE] 2 and move the cursor to the left of 2. Press [ENTER]. Now move the cursor to the right of 2 (but less than 3) and press [ENTER]. When the calculator asks "Guess?" press [ENTER]. The root (solution) is $x = 2.0000012$. Do the same for 3. You can also verify that -3, 2, and 3 are the solutions of the equation by using the remainder theorem and checking that $P(-3) = 0$, $P(2) = 0$, and $P(3) = 0$. Here the calculator gives you the hint and the algebra confirms it.

Can we use all this information to factor the polynomial? Since -3, 2, and 3 are solutions of the equation, $x - (-3)$, $x - 2$, and $x - 3$ are factors of the polynomial; thus, $x^4 - 4x^3 - 5x^2 + 36x - 36 = (x + 3)(x - 2)(x - 3)Q(x)$. To find $Q(x)$, divide $x^4 - 4x^3 - 5x^2 + 36x - 36$ by $(x + 3)(x - 2)(x - 3)$, that is, by $x^3 - 2x^2 - 9x + 18$. Using long division, we get the answer, $x - 2$. Substituting $x - 2$ for $Q(x)$, we have $x^4 - 4x^3 - 5x^2 + 36x - 36 = (x + 3)(x - 2)(x - 3)(x - 2)$, which means that $x - 2$ is a factor twice. The number 2 is called a **double root** of the equation.

$Y_1 = x^4 - 4x^3 - 5x^2 + 36x - 36$

Window 1

Answers to PROBLEMS

6. $\begin{array}{r|rrrrr} 1 & 2 & +1 & -35 & -16 & +48 \\ & & 2 & +3 & -32 & -48 \\ \hline & 2 & +3 & -32 & -48 & 0 \end{array}$

Exercises 6.5

〈A〉 Dividing a Polynomial by a Monomial In Problems 1–10, perform the indicated divisions.

1. $\dfrac{3x^3 + 9x^2 - 6x}{3x}$

2. $\dfrac{6x^3 + 8x^2 - 4x}{2x}$

3. $\dfrac{10x^3 - 5x^2 + 15x}{-5x}$

4. $\dfrac{24x^3 - 12x^2 + 6x}{-6x}$

5. $\dfrac{8y^4 - 32y^3 + 12y^2}{-4y^2}$

6. $\dfrac{9y^4 - 45y^3 + 18y^2}{-3y^2}$

7. $\dfrac{10x^5 + 8x^4 - 16x^3 + 6x^2}{2x^3}$

8. $\dfrac{12x^4 + 18x^3 + 16x^2}{4x^3}$

9. $\dfrac{15x^3y^2 - 10x^2y + 15x}{5x^2y}$

10. $\dfrac{18x^4y^4 - 24x^2y^3 + 6xy^2}{3x^2y^2}$

〈B〉 Dividing One Polynomial by Another Polynomial In Problems 11–36, divide using long division.

11. $x^2 + 5x + 6$ by $x + 2$

12. $x^2 + 9x + 20$ by $x + 5$

13. $y^2 + 3y - 10$ by $y - 2$

14. $y^2 + 2y - 15$ by $y - 3$

15. $2x^3 - 4x - 2$ by $2x + 2$

16. $2x^3 + 5x^2 - x - 2$ by $2x - 1$

17. $3x^3 + 14x^2 + 13x - 6$ by $3x - 1$

18. $2x^3 - 5x^2 - 14x + 3$ by $2x - 1$

19. $2x^3 - 10x - 7x^2 + 24$ by $2x - 3$

20. $3x^3 + 8x + 13x^2 - 12$ by $3x - 2$

21. $2x^3 + 2x + 7x^2 - 2$ by $-1 + 2x$

22. $3x^3 + 3x + 8x^2 - 2$ by $-1 + 3x$

23. $y^4 - y^2 - 2y - 1$ by $y^2 + y + 1$

24. $y^4 - y^2 - 4y - 4$ by $y^2 + y + 2$

25. $8x^3 - 6x^2 + 5x - 9$ by $2x - 3$

26. $2x^4 - x^3 + 7x - 2$ by $2x + 3$

27. $x^3 + 8$ by $x + 2$

28. $x^3 + 64$ by $x + 4$

29. $8y^3 - 64$ by $2y - 4$

30. $27x^3 - 8$ by $3x - 2$

31. $a^4 - 4a^2 - 4a - 1$ by $a^2 + 2a + 1$

32. $b^4 - b^2 - 2b - 1$ by $b^2 - b - 1$

33. $x^5 - 5x + 12x^2$ by $x^2 + 5 - 2x$

34. $y^5 - y^4 + 10 - 27y + 7y^2$ by $y^2 + 5 - y$

35. $4x^4 - 13x^2 + 4x^3 - 3x - 21$ by $2x + 5$

36. $8y^4 - 75y^2 - 18y^3 + 46y + 121$ by $4y + 5$

〈C〉 Factoring When One of the Factors Is Known In Problems 37–42, factor completely.

37. $x^3 - 4x^2 + x + 6$ if $x + 1$ is one of the factors

38. $x^3 - 4x^2 + x + 6$ if $x - 3$ is one of the factors

39. $x^4 - 4x^3 + 3x^2 + 4x - 4$ if $x^2 - 4x + 4$ is one of the factors

40. $x^4 - 2x^3 - 13x^2 + 14x + 24$ if $x^2 - 6x + 8$ is one of the factors

41. $x^4 + 6x^3 + 3x + 140$ if $x^2 - 3x + 7$ is one of the factors

42. $x^4 - 22x^2 - 75$ if $x^2 + 3$ is one of the factors

⟨D⟩ Using Synthetic Division to Divide a Polynomial by a Binomial In Problems 43–52, use synthetic division to find the quotient and the remainder.

43. $(v^3 - 8v - 3) \div (v - 3)$

44. $(y^3 - 4y^2 - 25) \div (y - 5)$

45. $(x^3 + 4x^2 - 7x + 5) \div (x - 2)$

46. $(4w^3 - w^2 + 92) \div (w + 3)$

47. $(z^3 - 32z + 24) \div (z + 6)$

48. $(2y^4 - 3y^3 + y^2 - 3y) \div (y - 2)$

49. $(3y^4 - 41y^2 - 13y - 8) \div (y - 4)$

50. $(v^5 - 4v^3 + 5v^2 - 5) \div (v + 1)$

51. $(2y^4 - 13y^3 + 6y^2 + 5y - 30) \div (y - 6)$

52. $(4w^4 + 20w^3 - w^2 - 2w + 15) \div (w + 5)$

⟨E⟩ Using the Remainder and Factor Theorems In Problems 53–60, use synthetic division to show that the given number is a solution of the equation.

53. 4; $z^3 + 6z^2 - 6z - 136 = 0$

54. -7; $3y^3 + 13y^2 - 57y - 7 = 0$

55. -4; $5y^3 + 18y^2 - y + 28 = 0$

56. 6; $7x^3 - 39x^2 - 26x + 48 = 0$

57. 5; $3v^4 - 14v^3 - 7v^2 + 21v - 55 = 0$

58. -8; $8w^4 + 62w^3 - 15w^2 + 10w + 16 = 0$

59. -1; $y^5 + y^4 + 2y^3 + 5y^2 - 2y - 5 = 0$

60. 1; $3z^6 - z^5 + 5z^4 - 3z - 4 = 0$

⟩⟩⟩ Applications

Average cost In business, the average cost per unit, $\overline{C}$, is given by the equation

$$\overline{C} = \frac{C}{x}$$

where C is the total cost and x is the number of units.

61. Find the average cost when $C = 500 + 4x$.

62. Find the average cost when $C = 200 + 2x^2$.

⟩⟩⟩ Using Your Knowledge

Finding Possible Factors In this section we factored some polynomials by using a given first-degree factor. How did we find that one factor? Here is one way to do it.

If a polynomial of the form $c_n x^n + c_{n-1} x^{n-1} + \cdots + c_0$ with integer coefficients has a factor $ax + b$, where a and b are integers and $a > 0$, then a must divide c_n and b must divide c_0. For example, to find possible values of a and b for the polynomial $2x^3 + 3x^2 - 8x + 3$, we consider the positive divisors of 2 and all the divisors of 3. Thus, the possibilities for a are 1 and 2 and for b are 1, -1, 3, and -3. This means that only the binomials $x + 1$, $x - 1$, $x + 3$, $x - 3$, $2x + 1$, $2x - 1$, $2x + 3$, and $2x - 3$ have to be checked. It turns out that $x - 1$ is a factor and that division gives

$$2x^3 + 3x^2 - 8x + 3 = (x - 1)(2x^2 + 5x - 3)$$

Since $2x^2 + 5x - 3 = (2x - 1)(x + 3)$, we have

$$2x^3 + 3x^2 - 8x + 3 = (x - 1)(2x - 1)(x + 3)$$

63. What binomials should be checked as possible factors for the polynomial $x^3 + 3x^2 + 5x + 6$ if a and b are both positive?

64. What binomials should be checked as possible factors for the polynomial $3x^3 + 5x^2 - 3x - 2$ if a and b are both positive?

》》》 Write On

65. Describe the procedure you use to divide a polynomial by a binomial.

66. How can you check the result when dividing one polynomial by another? Use your answer to check that $(2x^3 + 5x^2 - 8x + 6) \div (x + 2) = (2x^2 + x - 10)$ R 26.

67. Describe two different procedures that can be used to determine that there is a zero remainder when dividing a polynomial by a binomial of the form $x - k$.

68. When dividing a polynomial by a monomial will the result always be a polynomial? Give an example and explain.

》》》 Concept Checker

Fill in the blank(s) with the correct word(s), phrase, or mathematical statement.

69. To divide a polynomial by a _____, divide each term in the polynomial by the _____.

70. If the polynomial, $x^2 + 5x - 10$ is divided by $x - 2$, then the _____ is $P(2)$.

71. _____ is a procedure used when dividing a polynomial by a binomial of the form $x - k$.

rational expression   polynomial
synthetic division   remainder
long division   denominator
reciprocal   numerator
monomial

》》》 Mastery Test

Use synthetic division to find the quotient and remainder when dividing the polynomials.

72. $x^3 - 3x^2 + 3$ by $x - 3$

73. $2x^4 - 13x^3 + 16x^2 - 9x + 20$ by $x - 5$

Use synthetic division to show that the given number is a solution of the equation.

74. 2; $\quad 2x^3 + 5x^2 - 8x - 20 = 0$

75. -2; $\quad 3x^4 + 5x^3 - x^2 + x - 2 = 0$

Factor completely:

76. $2x^3 - x^2 - 18x + 9$ if $x + 3$ is one of the factors

77. $z^3 - 3z^2 - 4z + 12$ if $z + 2$ is one of the factors

Use long division to divide the polynomials.

78. $6x^3 + 3x - 9$ by $x - 2$

79. $3x^3 - 2x^2 - x - 6$ by $x + 2$

Divide:

80. $\dfrac{24x^5 - 18x^4 + 12x^3}{6x^2}$

81. $\dfrac{16x^4 - 4x^3 + 8x^2 - 16x + 40}{8x^3}$

》》》 Skill Checker

Factor:

82. $5x + 15$

83. $2x^2 - 6x$

84. $x^2 - 25$

85. $x^2 - 64$

Solve:

86. $4x + 8 = 6x$

87. $5x + 10 = 7x$

88. $x(x + 2) - (x - 3)(x - 4) = 4x + 3$

89. $x(x + 1) - (x - 1)(x - 2) = 2x + 2$

6.6 Equations Involving Rational Expressions

Objectives

A Solve equations involving rational expressions.

B Solve proportions.

C Solve applications involving proportions.

To Succeed, Review How To . . .

1. Find the LCD of two or more fractions (pp. 451–456).
2. Factor polynomials (pp. 398–401).
3. Solve linear and quadratic equations (pp. 76–83, 405–409).

Getting Started
Play Ball with Equations Involving Rational Expressions

The baseball player in this picture wants to improve his batting average. His batting average near the beginning of the season (to three decimal places) is found by using the formula:

$$\text{Average} = \frac{\text{number of hits}}{\text{number of times at bat}} = \frac{6}{25} = 0.240$$

How many consecutive hits h would he need to bring his average to 0.320? Here is the information we have:

	Actual	New
Number of hits	6	$6 + h$
Times at bat	25	$25 + h$
New average = $0.320 = \frac{32}{100}$		$= \frac{6 + h}{25 + h}$

To find the answer, we must solve the equation. The first step in solving equations containing rational expressions is to multiply both sides of the equation by the LCD to clear the rational expressions; then we use the distributive property to clear parentheses and solve as usual. Here is the solution.

$$\frac{32}{100} = \frac{6 + h}{25 + h}$$

1. Multiply by the LCD, $100(25 + h)$.

$$\cancel{100}(25 + h)\frac{32}{\cancel{100}} = 100\cancel{(25 + h)}\frac{6 + h}{\cancel{(25 + h)}}$$

2. Reduce and use the distributive property.

$$800 + 32h = 600 + 100h$$

3. Subtract 600 from both sides.

$$200 + 32h = 100h$$

4. Subtract $32h$ from both sides.

$$200 = 68h$$

5. Divide by 68.

$$h = \frac{200}{68} \approx 2.941$$

He would need three consecutive hits to obtain a very respectable 0.320 average.

A › Solving Equations Involving Rational Expressions

So far in this chapter we have studied about rational expressions such as $\frac{1}{x+2}$ and $\frac{x}{x^2+x-7}$. We have simplified, evaluated, and performed operations with them. Next we will learn how to solve equations with one or more of these rational expressions.

What is the difference between the equations we have solved up to now and an equation such as the following?

$$\frac{32}{100} = \frac{6+h}{25+h}$$

This equation has a variable in the *denominator,* so we have to avoid values of the variable that make the denominator zero. Such equations are solved with a procedure similar to the one given on pages 79–80, except that when multiplying by the LCD of the fractions involved, we may end up with a quadratic equation, an equation that can be written in the form $ax^2 + bx + c = 0$. Here is the procedure we need.

PROCEDURE

Procedure for Solving Equations Containing Rational Expressions

1. Factor all denominators and multiply both sides of the equation by the LCD of all rational expressions in the equation.
2. Write the result in reduced form and use the distributive property to remove parentheses.
3. Determine whether the equation is *linear* (can be written in the form $ax + b = 0$) or *quadratic* ($ax^2 + bx + c = 0$) and solve accordingly. (For linear equations, see pages 76–83; for quadratic equations, see pages 405–409.)
4. Check that each *proposed* or *trial* solution satisfies the original equation. If it does not, discard it as an **extraneous solution,** a trial solution that does not satisfy the original equation.

Note: The guide to solving linear equations is CRAM. The guide to solving quadratic equations is OFF.

EXAMPLE 1 Solving equations containing rational expressions
Solve:

a. $\frac{3}{4x} = 9 - \frac{3}{2x}$

b. $\frac{7}{12y} + \frac{3}{4y} = \frac{1}{3}$

c. $\frac{4}{x} = \frac{6}{x+2}$

d. $\frac{1}{x+1} = \frac{2}{x+2}$

SOLUTION

a. Use the four-step procedure.
The LCD of $4x$ and $2x$ is $4x$. Multiply both sides of the equation by $4x$.

$$4x\left(\frac{3}{4x}\right) = 4x\left(9 - \frac{3}{2x}\right) \quad \text{1. Multiply by } 4x.$$

$$\frac{\overset{1}{\cancel{4x}}}{1} \cdot \frac{3}{\cancel{4x}} = 4x \cdot 9 - \frac{\overset{2}{\cancel{4x}}}{1} \cdot \frac{3}{\cancel{2x}} \quad \text{2. Simplify.}$$

$$3 = 36x - 6 \quad \text{3. Solve the linear equation.}$$

$$9 = 36x$$

PROBLEM 1
Solve:

a. $8 + \frac{8}{5x} = \frac{14}{2x}$

b. $\frac{1}{8} = \frac{5}{24y} - \frac{1}{12y}$

c. $\frac{-10}{2x+1} = \frac{5}{x}$

d. $\frac{6}{x+4} = \frac{5}{x-3}$

Answers to PROBLEMS

1. a. $\frac{27}{40}$ b. 1 c. $-\frac{1}{4}$ d. 38

$$\frac{9}{36} = x$$

$$\frac{1}{4} = x$$

$$\frac{3}{4(\frac{1}{4})} = 9 - \frac{3}{2(\frac{1}{4})} \qquad \text{4. Check the denominators.}$$

$$3 = 3$$

In the check we see that no denominator is equal to zero and $3 = 3$ so we conclude that $\frac{1}{4}$ is the solution and the solution set is $\{\frac{1}{4}\}$.

b. Use the four-step procedure.

The LCD of $12y$, $4y$, and 3 is $12y$. Multiply both sides of the equation by $12y$.

$$12y\left(\frac{7}{12y} + \frac{3}{4y}\right) = 12y\left(\frac{1}{3}\right) \qquad \text{1. Multiply by } 12y.$$

$$\frac{\cancel{12y}^{1}}{1} \cdot \frac{7}{\cancel{12y}} + \frac{\cancel{12y}^{3}}{1} \cdot \frac{3}{\cancel{4y}} = \frac{\cancel{12y}^{4}}{1}\left(\frac{1}{\cancel{3}}\right) \qquad \text{2. Simplify.}$$

$$7 + 9 = 4y \qquad \text{3. Solve the linear equation.}$$
$$16 = 4y$$
$$4 = y$$

$$\frac{7}{12(4)} + \frac{3}{4(4)} = \frac{1}{3} \qquad \text{4. Check the denominators.}$$

$$\frac{1}{3} = \frac{1}{3}$$

In the check we see that no denominator is equal to zero and $\frac{1}{3} = \frac{1}{3}$, so we conclude that 4 is the solution and the solution set is $\{4\}$.

c. We use the four-step procedure.

1. The LCD of

$$\frac{4}{x} \text{ and } \frac{6}{x+2}$$

is $x(x + 2)$. Multiplying both sides of the equation by the LCD,

$$x(x+2) = \frac{x(x+2)}{1}$$

gives

$$\frac{x(x+2)}{1} \cdot \frac{4}{x} = \frac{x(\cancel{x+2})}{1} \cdot \frac{6}{\cancel{x+2}}$$

2. Reduce and remove parentheses.

$$(x+2) \cdot 4 = x \cdot 6 \qquad \text{Divide out } x \text{ and } x+2.$$
$$4x + 8 = 6x \qquad \text{Remove parentheses.}$$

3. Solve.

$$8 = 2x \qquad \text{Subtract } 4x \text{ from both sides.}$$
$$4 = x \qquad \text{Divide by 2.}$$

(continued)

4. The proposed solution is 4. To check the answer, we substitute 4 for x in the original equation to obtain

$$\frac{4}{4} = \frac{6}{4+2}$$

or $1 = 1$. Therefore, the solution 4 is correct, and the solution set is $\{4\}$.

d. Again, we use the four-step procedure.

1. The LCD of

$$\frac{1}{x+1} \quad \text{and} \quad \frac{2}{x+2}$$

is $(x+1)(x+2)$. Multiplying both sides of the equation by

$$\frac{(x+1)(x+2)}{1}$$

we obtain

$$\frac{(x+1)(x+2)}{1} \cdot \frac{1}{x+1} = \frac{(x+1)(x+2)}{1} \cdot \frac{2}{x+2}$$

2.
$x + 2 = (x + 1) \cdot 2$ Simplify.
$x + 2 = 2x + 2$ Use the distributive property.

3.
$x = 2x$ Subtract 2.
$0 = x$ Subtract x.

4. The proposed solution is zero. The verification that zero is the actual solution is left to you.

As we mentioned, when variables occur in the denominator, it's possible to multiply both sides of the equation by the LCD of the fractions involved and obtain a solution of the resulting equation that does *not* satisfy the original equation. This points out the necessity of *checking,* by direct substitution in the original equation, any proposed solutions obtained after multiplying both sides of an equation by factors containing the unknown. If the proposed solution does not satisfy the original equation, it is called an *extraneous solution.*

> **CAUTION**
>
> By definition an extraneous solution does *not* satisfy the given equation and must *not* be listed as a solution.

EXAMPLE 2   Solving equations containing rational expressions

Solve:

a. $\dfrac{1}{x-4} - \dfrac{1}{x-2} = \dfrac{2x}{x^2 - 6x + 8}$ **b.** $\dfrac{x}{x-3} - \dfrac{x-4}{x+2} = \dfrac{4x+3}{x^2 - x - 6}$

PROBLEM 2

Solve:

a. $\dfrac{1}{x-6} - \dfrac{1}{x-4} = \dfrac{6}{(x-6)(x-2)}$

b. $\dfrac{x}{x-4} - \dfrac{x-5}{x+3} = \dfrac{12x+16}{x^2 - x - 12}$

SOLUTION

a. We use our four-step procedure.

1. First factor the denominator of the right-hand side of the equation to obtain

$$\frac{1}{x-4} - \frac{1}{x-2} = \frac{2x}{(x-4)(x-2)}$$

Answers to PROBLEMS

2. a. 5   **b.** No solution

Since the LCD of the fractions involved is

$$\frac{(x-4)(x-2)}{1}$$

we multiply each side of the equation by this LCD to get

$$\frac{(x-4)(x-2)}{1} \cdot \left[\frac{1}{x-4} - \frac{1}{x-2}\right] = \frac{(x-4)(x-2)}{1} \cdot \left[\frac{2x}{(x-4)(x-2)}\right]$$

2. Reduce and remove parentheses.

$$\frac{(x-4)(x-2)}{1} \cdot \frac{1}{x-4} - \frac{(x-4)(x-2)}{1} \cdot \frac{1}{x-2}$$
$$= \frac{(x-4)(x-2)}{1} \cdot \frac{2x}{(x-4)(x-2)}$$

3. Solve. $\quad (x-2) - (x-4) = 2x$

$\qquad x - 2 - x + 4 = 2x$

$\qquad \qquad \qquad 2 = 2x \qquad$ Combine like terms.

$\qquad \qquad \qquad x = 1 \qquad$ Divide by 2.

4. The proposed solution is 1. Substituting 1 for x in the original equation gives

$$\frac{1}{1-4} - \frac{1}{1-2} \stackrel{?}{=} \frac{2 \cdot 1}{1^2 - 6 \cdot 1 + 8}$$

$$\frac{1}{-3} - \frac{1}{-1} \stackrel{?}{=} \frac{2}{3}$$

$$-\frac{1}{3} + 1 = \frac{2}{3}$$

which is a true statement. Thus, our result is correct. The actual solution is 1 and the solution set is $\{1\}$.

b. We use our four-step procedure.

1. We first write the right-hand side of the equation with the denominator factored.

$$\frac{x}{x-3} - \frac{x-4}{x+2} = \frac{4x+3}{(x-3)(x+2)}$$

The LCD is

$$\frac{(x-3)(x+2)}{1}$$

We multiply both sides by the LCD.

$$\frac{(x-3)(x+2)}{1}\left[\frac{x}{x-3} - \frac{x-4}{x+2}\right] = \frac{(x-3)(x+2)}{1}\left[\frac{4x+3}{(x-3)(x+2)}\right]$$

2. Reduce and remove parentheses.

$$\frac{(x-3)(x+2)}{1} \cdot \frac{x}{x-3} - \frac{(x-3)(x+2)}{1} \cdot \frac{x-4}{x+2}$$
$$= \frac{(x-3)(x+2)}{1} \cdot \frac{4x+3}{(x-3)(x+2)}$$

$\qquad (x+2) \cdot x - (x-3)(x-4) = 4x + 3$

$\qquad x^2 + 2x - (x^2 - 7x + 12) = 4x + 3$

3. Solve. $\qquad \qquad 9x - 12 = 4x + 3 \qquad$ Combine like terms.

$\qquad \qquad \qquad \qquad 5x = 15 \qquad$ Add 12 and subtract 4x.

$\qquad \qquad \qquad \qquad x = 3 \qquad$ Divide by 5.

(continued)

4. The proposed solution is 3. However, if x is replaced by 3 in the original equation, the term

$$\frac{x}{x - 3}$$

yields $\frac{3}{0}$, which is meaningless. Consequently, the equation

$$\frac{x}{x - 3} - \frac{x - 4}{x + 2} = \frac{4x + 3}{x^2 - x - 6}$$

has no solution. Its solution set is $\varnothing$. 3 is an extraneous solution.

Finally, the equations that result from clearing denominators are not *always* linear equations. For example, to solve the equation

$$\frac{x^2}{x + 3} = \frac{9}{x + 3}$$

we first multiply by the LCD, $(x + 3)$, to obtain

$$(x+3)\frac{x^2}{(x+3)} = (x+3)\frac{9}{(x+3)} \quad \text{or} \quad x^2 = 9$$

In this equation, the variable x has 2 as an exponent; thus, it is a quadratic equation and can be solved when written in standard form—by writing the equation as

$$x^2 - 9 = 0 \quad \text{O} \quad \text{Subtract 9 to set } = 0.$$
$$(x + 3)(x - 3) = 0 \quad \text{F} \quad \text{Factor.}$$
$$x + 3 = 0 \quad \text{or} \quad x - 3 = 0 \quad \text{F} \quad \text{Factors } = 0.\ \text{Use the zero-product property.}$$
$$x = -3 \quad \text{or} \quad x = 3 \quad \text{Solve each equation.}$$

3 is a solution, since

$$\frac{3^2}{3 + 3} = \frac{9}{3 + 3}$$

However, for -3, the denominator $x + 3$ becomes zero. Thus, -3 is an extraneous solution. The only actual solution is 3, and the solution set is $\{3\}$.

EXAMPLE 3 Solving an equation having an extraneous solution
Solve:

$$1 + \frac{3}{x - 2} = \frac{12}{x^2 - 4}$$

SOLUTION Since $x^2 - 4 = (x + 2)(x - 2)$, the LCD is $(x + 2)(x - 2)$. We then write the equation with the denominator $x^2 - 4$ in factored form and multiply each term by the LCD, as before. Here are the steps.

1. Multiply each term by the LCD.

$$(x + 2)(x - 2) \cdot 1 + (x + 2)(x-2) \cdot \frac{3}{(x-2)} = (x+2)(x-2) \cdot \frac{12}{(x+2)(x-2)}$$

2. Reduce and remove parentheses.

$$(x^2 - 4) + 3(x + 2) = 12$$
$$x^2 - 4 + 3x + 6 = 12$$

PROBLEM 3
Solve:

$$3 - \frac{4}{x^2 - 1} = \frac{-2}{x - 1}$$

Answers to PROBLEMS

3. $-\frac{5}{3}$

3. Solve the resulting quadratic equation.

$$x^2 + 3x + 2 = 12$$
$$x^2 + 3x - 10 = 0 \quad \text{O Subtract 12 to set} = 0.$$
$$(x + 5)(x - 2) = 0 \quad \text{F Factor.}$$
$$x + 5 = 0 \quad \text{or} \quad x - 2 = 0 \quad \text{F Factors} = 0. \text{ Use the zero-product property.}$$
$$x = -5 \quad \text{or} \quad x = 2 \quad \text{Solve each equation.}$$

4. 2 makes the denominator $x - 2$ equal to zero, so it is an extraneous solution. The only actual solution is -5, and the solution set is $\{-5\}$. This solution can be checked in the original equation.

EXAMPLE 4 Solving an equation having an extraneous solution
Solve:

$$\frac{x - 3}{x^2 - 4x} = \frac{2}{x^2 - 16}$$

PROBLEM 4
Solve:

$$\frac{x - 7}{x^2 - 8x} = \frac{2}{x^2 - 64}$$

SOLUTION As usual, we solve by steps.

1. Write all denominators in factored form and then multiply by the LCD,

$$\frac{x(x - 4)(x + 4)}{1}$$

2. Reduce and remove parentheses.

$$\frac{x(x-4)(x+4)}{1} \cdot \frac{x - 3}{x(x-4)} = \frac{x(x-4)(x+4)}{1} \cdot \frac{2}{(x+4)(x-4)}$$
$$(x + 4)(x - 3) = 2x$$

3. Solve the resulting quadratic equation.

$$x^2 + x - 12 = 2x$$
$$x^2 - x - 12 = 0 \quad \text{O Subtract 2x to set} = 0.$$
$$(x + 3)(x - 4) = 0 \quad \text{F Factor.}$$
$$x + 3 = 0 \quad \text{or} \quad x - 4 = 0 \quad \text{F Factors} = 0. \text{ Use the zero-product property.}$$
$$x = -3 \quad \text{or} \quad x = 4 \quad \text{Solve.}$$

4. The proposed solutions are -3 and 4. If we substitute -3 for x in the original equation, we get a true statement. On the other hand, if we substitute 4 for x in the original equation, the denominators on both sides are zero, so the fractions are not defined. The only actual solution is -3, and 4 is an extraneous solution. The solution set is $\{-3\}$.

EXAMPLE 5 Solving an equation involving negative exponents
Solve: $3x(x + 2)^{-1} + 8(x - 3)^{-1} = 4$

PROBLEM 5
Solve:

$$2(x + 1)^{-1} + 4(x - 4)^{-1} = 1$$

SOLUTION At first, it seems that there are *no* rational expressions involved. However, since $a^{-1} = \frac{1}{a}$,

$$3x(x + 2)^{-1} = 3x \cdot \frac{1}{x + 2} = \frac{3x}{x + 2}$$

and

$$8(x - 3)^{-1} = 8 \cdot \frac{1}{x - 3} = \frac{8}{x - 3}$$

Thus, $3x(x + 2)^{-1} + 8(x - 3)^{-1} = 4$ becomes

$$\frac{3x}{x + 2} + \frac{8}{x - 3} = 4$$

(continued)

Answers to PROBLEMS

4. -7 5. $0, 9$

Using our four-step procedure, we have the following:

1. The only denominators are $x + 2$ and $x - 3$, so the LCD is $(x + 2)(x - 3)$. Multiply both sides of the equation by the LCD.

$$(x + 2)(x - 3)\left(\frac{3x}{x + 2} + \frac{8}{x - 3}\right) = (x + 2)(x - 3)4$$

2. Reduce and remove parentheses.

$$\cancel{(x + 2)}(x - 3) \cdot \frac{3x}{\cancel{(x + 2)}} + (x + 2)\cancel{(x - 3)} \cdot \frac{8}{\cancel{(x - 3)}} = (x + 2)(x - 3)4$$

$$(x - 3)3x + (x + 2)8 = (x^2 - x - 6)4$$
$$3x^2 - 9x + 8x + 16 = 4x^2 - 4x - 24$$

3. Solve the resulting quadratic equation.

$$3x^2 - x + 16 = 4x^2 - 4x - 24$$

Subtract $3x^2 - x + 16$ from both sides. $0 = x^2 - 3x - 40$

Factor. $0 = (x + 5)(x - 8)$

Factors = 0.

$x + 5 = 0$ or $x - 8 = 0$

$x = -5$ or $x = 8$ Solve each equation.

4. Neither of these proposed solutions makes a denominator zero. The solutions are -5 and 8 and the solution set is $\{-5, 8\}$. Check this.

B › Solving Proportions

Some equations containing rational expressions are special and can be solved with a different method. The equation used to solve the *Getting Started* problem, $\frac{32}{100} = \frac{6 + h}{25 + h}$, is an example of one. To be able to recognize this special case you must know a little about ratios and proportions. A **ratio** is the comparison of two numbers or quantities and is usually expressed as a fraction such as $\frac{2}{3}$ or $\frac{x}{x + 4}$. When two ratios are equal, the statement is called a **proportion.** Two examples of proportions are

$$\frac{5}{10} = \frac{1}{2} \quad \text{and} \quad \frac{2}{3} = \frac{x}{x + 4}$$

PROPERTY OF PROPORTIONS	If $\frac{a}{b} = \frac{c}{d}$ (where $b, d \neq 0$), then $a \cdot d = b \cdot c$. A proportion is true if the cross products are equal.

In the first example, we use this property to verify that the proportion is true.

$\frac{5}{10} = \frac{1}{2}$ Given

$5 \cdot 2 = 10 \cdot 1$ Cross products are equal.

$10 = 10$ Simplify.

10 does equal 10, so the proportion is true.

In the second example, we use this property to find the value of x that will make it a true proportion.

$$\frac{2}{3} = \frac{x}{x+4} \quad \text{Given}$$
$$3 \cdot x = 2 \cdot (x+4) \quad \text{Cross products are equal.}$$
$$3x = 2x + 8 \quad \text{Simplify.}$$
$$x = 8$$

As long as the variable, x, is replaced with 8, the proportion is true.
The procedure for solving a proportion follows.

> **PROCEDURE**
>
> **Procedure for solving proportions**
> 1. Set the cross products equal to each other.
> 2. Solve the resulting equation.
> 3. Check the proposed solutions for extraneous solutions.

Let's use this procedure to solve the proportions in Example 6.

EXAMPLE 6 Solving proportions

Solve:

a. $\frac{3}{2} = \frac{6}{x+5}$ b. $\frac{-2}{x-1} = \frac{4}{8x+7}$ c. $\frac{2}{3x+1} = \frac{5}{6x+2}$

SOLUTION

a. Use the three steps for solving proportions.

$$\frac{3}{2} = \frac{6}{x+5}$$
$$2(6) = 3(x+5) \quad \text{1. Set cross products equal.}$$
$$12 = 3x + 15$$
$$-3 = 3x \quad \text{2. Solve resulting equation.}$$
$$-1 = x$$
$$\frac{3}{2} = \frac{6}{(-1)+5} \quad \text{3. Check.}$$

Since $\frac{3}{2} = \frac{3}{2}$ the solution checks and the solution set is $\{-1\}$.

b. Use the three steps for solving proportions.

$$\frac{-2}{x-1} = \frac{4}{8x+7}$$
$$4(x-1) = -2(8x+7) \quad \text{1. Set cross products equal.}$$
$$4x - 4 = -16x - 14$$
$$20x = -10 \quad \text{2. Solve resulting equation.}$$
$$x = \frac{-1}{2}$$
$$\frac{-2}{\left(\frac{-1}{2}\right)-1} = \frac{4}{8\left(\frac{-1}{2}\right)+7} \quad \text{3. Check.}$$

Since $\frac{4}{3} = \frac{4}{3}$, the solution checks and the solution set is $\left\{\frac{-1}{2}\right\}$.

PROBLEM 6

Solve for the variable:

a. $\frac{-5}{2x} = \frac{10}{3x+4}$

b. $\frac{3}{2x-1} = \frac{6}{4x-5}$

c. $\frac{4}{2-x} = \frac{8}{-x+9}$

(continued)

Answers to PROBLEMS

6. a. $-\frac{4}{7}$ b. No solution   c. -5

c. Use the three steps for solving proportions.

$$\frac{2}{3x+1} = \frac{5}{6x+2}$$

$2(6x+2) = 5(3x+1)$ 1. Set cross products equal.

$12x + 4 = 15x + 5$

$-1 = 3x$ 2. Solve resulting equation.

$\frac{-1}{3} = x$

$\dfrac{2}{3\left(\frac{-1}{3}\right)+1} = \dfrac{5}{6\left(\frac{-1}{3}\right)+2}$ 3. Check.

Since the first denominator equals $-1 + 1$, which is 0, we know that this is an extraneous solution because in the real numbers it is impossible to divide by 0. Our conclusion is that this proportion has no solution, or the solution set is { }.

C ▶ Solving Applications Involving Proportions

A **rate** is a ratio of two different quantities such as

$$\frac{5 \text{ ft}}{2 \text{ min}} \quad \text{or} \quad \frac{30 \text{ mi}}{1 \text{ hr}}$$

When writing a proportion with rates, be sure the units are written in the same order in each of the ratios.

There are many real-life applications for proportions such as proportionally decreasing the size of a large object in a scale drawing, increasing the dosage of a medicine proportional to the weight of a person, or decreasing the proportion of the ingredients in a recipe.

Now that we have reviewed proportions and how to solve them, in Example 7 we will solve an application using a proportion.

EXAMPLE 7   Solving an application using a proportion

One of the ingredients in a recipe for braised steak with vegetables is 14.5 ounces of stewed tomatoes. The recipe will feed four people but you will be having six guests for dinner. Find the amount of stewed tomatoes needed to increase the recipe to feed six people.

SOLUTION   To solve this problem, we write a proportion in which the ratios compare ounces of stewed tomatoes with number of serving sizes.

$\dfrac{14.5 \text{ oz}}{4 \text{ people}}$ First ratio

$\dfrac{x \text{ oz}}{6 \text{ people}}$ Second ratio

$\dfrac{14.5 \text{ oz}}{4 \text{ people}} = \dfrac{x \text{ oz}}{6 \text{ people}}$ Proportion

To solve the proportion, we use the three-step procedure.

$14.5(6) = 4(x)$ 1. Set cross products equal.

$87 = 4x$ 2. Solve resulting equation.

$21.75 = x$ 3. The check is left for you.

To increase the recipe to feed 6 people, you will need 21.75 ounces of stewed tomatoes.

PROBLEM 7

If a map uses a scale of 1 in. = 25 mi, find the number of miles between two cities if the map indicates they are 3.5 in. apart.

Answers to PROBLEMS

7. 87.5 mi

EXAMPLE 8 Solving an application using a proportion

If a blueprint uses a scale of $\frac{1}{4}$ inch $= 2$ feet, find the scaled-down dimensions of a room that is 12 feet by 16 feet.

SOLUTION To solve this problem, we will write two proportions in which the ratios compare the scaled measurement with the actual measurement for the width (w) and the length (l) of the room.

For scaled width (w)

$\dfrac{\frac{1}{4} \text{ in. scaled}}{2 \text{ ft actual}}$ First ratio

$\dfrac{w \text{ in. scaled}}{12 \text{ ft actual}}$ Second ratio

$\dfrac{\frac{1}{4} \text{ in.}}{2 \text{ ft}} = \dfrac{w \text{ in.}}{12 \text{ ft}}$ Proportion

$2(w) = \left(\dfrac{1}{4}\right) 12$ Set cross products equal.

$2w = 3$

$w = \dfrac{3}{2} = 1\dfrac{1}{2}$ Solve equations.

For scaled length (l)

$\dfrac{\frac{1}{4} \text{ in. scaled}}{2 \text{ ft actual}}$

$\dfrac{l \text{ in. scaled}}{16 \text{ ft actual}}$

$\dfrac{\frac{1}{4} \text{ in.}}{2 \text{ ft}} = \dfrac{l \text{ in.}}{16 \text{ ft}}$

$2(l) = \left(\dfrac{1}{4}\right) 16$

$2l = 4$

$l = 2$

The dimensions of the scaled blueprint drawing of the room will be $1\frac{1}{2}$ inches by 2 inches.

PROBLEM 8

If 2 pounds of ground meat cost $3.98, how much would 5 pounds cost?

Answers to PROBLEMS

8. $9.95

Calculator Corner

Techniques for Solving Rational Equations

Now we shall discuss some techniques for solving rational equations: **graphing, finding roots (solutions), finding intersections,** and **using solve**. Let's select a typical equation, the one from Example 3:

$$1 + \dfrac{3}{x - 2} = \dfrac{12}{x^2 - 4}$$

To solve this equation by graphing, first subtract $\frac{12}{x^2-4}$ from both sides to obtain the equivalent rational equation $R(x) = 0$. To find the values of x for which $R(x) = 0$, that is, the values at which $R(x)$ crosses the horizontal axis, we graph $Y_1 = 1 + \frac{3}{(x-2)} - \frac{12}{(x^2-4)}$ (Window 1). The graph seems to cross the horizontal axis at $x = -5$. You can confirm this by using the TRACE and ZOOM keys as shown in Window 2. At $x = -5, y = -3E - 10 = -3 \times 10^{-10}$, which is nearly zero.

If your calculator has a **solve** feature, press MATH 0 and enter $1 + \frac{3}{(x-2)} - \frac{12}{(x^2-4)}$ (Window 3). Again, note the parentheses in the denominators. Press ENTER, then put in a guess for the answer. To have the calculator find the answer, press ALPHA ENTER. If you enter a number greater than 2 for your guess (say, 3), you will get an error message. If you enter a guess between -2 and 2 (say, 0), you will also get an error message. Finally, try a number less than -2 for your guess, say, -3. After you press ALPHA ENTER, the calculator shows the answer -5. To get to the final answer, it is essential that you understand what the graph looks like and where it is likely to have a solution. (This is your "guess.")

Now you can practice by verifying the rest of the examples using any one of these techniques available on your calculator.

Window 1

$y_1 = 1 + \dfrac{3}{(x-2)} - \dfrac{12}{(x^2-4)}$

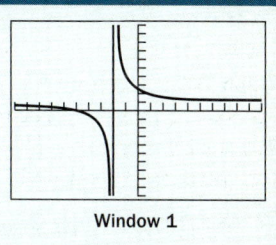

X=-5   Y=-3E-10

Window 2

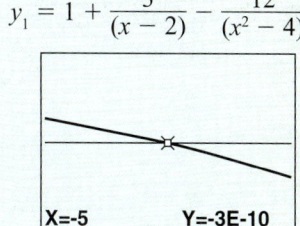

Window 3

Note the parentheses when entering the expression.

Exercises 6.6

⟨A⟩ Solving Equations Involving Rational Expressions In Problems 1–36, solve the given equation.

1. $\dfrac{x}{3} + \dfrac{x}{6} = 3$
2. $\dfrac{x}{2} + \dfrac{x}{4} = \dfrac{3}{8}$
3. $\dfrac{x}{5} - \dfrac{3x}{10} = \dfrac{1}{2}$

4. $\dfrac{x}{6} - \dfrac{x}{5} = \dfrac{1}{15}$
5. $\dfrac{1}{y} + \dfrac{4}{3y} = 7$
6. $\dfrac{10}{3y} - \dfrac{9}{2y} = \dfrac{7}{30}$

7. $\dfrac{2}{y-8} = \dfrac{1}{y-2}$
8. $\dfrac{2}{y-4} = \dfrac{3}{y-2}$
9. $\dfrac{3}{3z+4} = \dfrac{2}{5z-6}$

10. $\dfrac{2}{4z-1} = \dfrac{3}{2z+1}$
11. $\dfrac{-2}{2x+1} = \dfrac{3}{3x-1}$
12. $\dfrac{-5}{2x+3} = \dfrac{2}{3x-1}$

13. $\dfrac{-1}{x+1} = \dfrac{-2}{2x-1}$
14. $\dfrac{-5}{5x-2} = \dfrac{-3}{3x+1}$
15. $\dfrac{24}{3x+1} = \dfrac{4}{x}$

16. $\dfrac{3}{2x-1} = \dfrac{5}{4x}$
17. $\dfrac{2}{x^2-4} + \dfrac{5}{x+2} = \dfrac{7}{x-2}$
18. $\dfrac{3}{x^2-9} + \dfrac{5}{x+3} = \dfrac{8}{x-3}$

19. $\dfrac{t+2}{t^2-3t+2} = \dfrac{3}{t-1} - \dfrac{1}{t-2}$
20. $\dfrac{t+3}{t^2+4t+3} = \dfrac{4}{t+3} - \dfrac{1}{t+1}$
21. $\dfrac{x^2}{x^2-1} = 1 + \dfrac{1}{x+1}$

22. $\dfrac{x^2}{x^2-9} = 1 + \dfrac{1}{x-3}$
23. $\dfrac{1}{x^2-4x+3} + \dfrac{1}{x^2-2x-3} = \dfrac{1}{x^2-1}$

24. $\dfrac{1}{x^2+3x+2} + \dfrac{1}{x^2+x-2} = \dfrac{1}{x^2-1}$
25. $\dfrac{x+2}{3x^2+4x+1} = \dfrac{x+1}{3x^2+7x+2}$

26. $\dfrac{x+2}{2x^2+x-1} = \dfrac{x-2}{2x^2+x-1}$
27. $\dfrac{2z+13}{2z^2+5z-3} + \dfrac{3}{z+3} = \dfrac{4}{2z-1}$

28. $\dfrac{z-14}{2z^2-3z-2} + \dfrac{3}{z-2} = \dfrac{4}{2z+1}$
29. $\dfrac{3-x}{5x^2-4x-1} + \dfrac{2}{5x+1} = \dfrac{1}{x-1}$

30. $\dfrac{16-x}{4x^2-11x-3} + \dfrac{5}{4x+1} = \dfrac{2}{x-3}$
31. $4x^{-1} + 2 = 7$ (Recall that $x^{-1} = \dfrac{1}{x}$.)

32. $3 + 6x^{-1} = 5$
33. $4x^{-1} + 6x^{-1} = 15(x+1)^{-1}$
34. $6x^{-1} + 9x^{-1} = 25(x+2)^{-1}$

35. $2(x-8)^{-1} = (x-2)^{-1}$
36. $3(3y+4)^{-1} = 2(5y-6)^{-1}$

⟨B⟩ Solving Proportions In Problems 37–46, solve the given equation by using the cross products property of proportions.

37. $\dfrac{8}{x-3} = \dfrac{2}{5}$
38. $\dfrac{5x-3}{4} = \dfrac{x}{2}$
39. $\dfrac{2}{y-8} = \dfrac{1}{y-2}$

40. $\dfrac{2}{y-4} = \dfrac{3}{y-2}$
41. $\dfrac{3}{3z+4} = \dfrac{2}{5z-6}$
42. $\dfrac{2}{4z-1} = \dfrac{3}{2z+1}$

43. $\dfrac{-2}{2x+1} = \dfrac{3}{3x-1}$
44. $\dfrac{-5}{2x+3} = \dfrac{2}{3x-1}$
45. $\dfrac{-1}{x+1} = \dfrac{-2}{2x-1}$

46. $\dfrac{-5}{5x-2} = \dfrac{-3}{3x+1}$

⟨C⟩ Solving Applications Involving Proportions

47. *Film processing* A film processing department can process nine rolls of film in 2 hours. At that rate, how long will it take them to process 20 rolls of film?

48. *Blueprints* If a blueprint uses a scale of $\dfrac{1}{2}$ inch = 3 ft, find the scaled-down dimensions of a room that is 9 feet × 12 feet.

49. *Driving rate* Latrice and Bob want to drive from Miramar, Florida, to Pittsburgh, Pennsylvania, a distance of approximately 1000 miles. Each day they plan to drive 7 hours and cover 425 miles. At that rate,

a. How many driving hours should the trip take?

b. How many days should the trip take?

50. *Cycling* Marquel is an avid bicyclist and averages 16 miles per hour when he rides. He likes to ride 80 miles per day. He plans to take a 510-mile, round trip, between New Orleans and Panama City. At that rate,

a. How many hours of riding will the trip take?

b. How many days should the trip take?

51. *Batting average* Early in the season a softball player has a 0.250 batting average (5 hits for 20 times at bat). How many consecutive hits would she need to bring her average to 0.350? (*Hint:* See the *Getting Started.*)

52. *Medicine dosage* Norma takes a medicine that requires a dosage of 20 milliliters as a baseline amount plus 5 milliliters for every 30 pounds of body weight. Norma weighs 150 pounds.

a. How many milliliters should each dosage be?

b. If a teaspoon is 5 milliliters, how many teaspoons would the dosage be?

››› Using Your Knowledge

Solving for Variables   There are many instances in which a given formula must be changed to an equivalent form. For example, the formula

$$\frac{P}{R} = \frac{T}{V}$$

is frequently discussed in chemistry. Suppose you know P, R, and T. Can you find V? As before, we proceed by steps to solve for V.

Step 1. Since the LCD is RV, we multiply each term by RV to obtain

$$RV \cdot \frac{P}{R} = \frac{T}{V} \cdot RV$$

Step 2. Simplify.

$$VP = TR$$

Step 3. Divide by P.

$$V = \frac{TR}{P}$$

Thus,

$$V = \frac{TR}{P}$$

In Problems 53–57, use your knowledge of rational equations to solve for the indicated variable.

53. The area A of a trapezoid is

$$A = \frac{h(b_1 + b_2)}{2}$$

Solve for h.

54. In an electrical circuit, we have

$$\frac{1}{R} = \frac{1}{R_1} + \frac{1}{R_2}$$

Solve for R.

55. In refrigeration, we find the formula

$$\frac{Q_1}{Q_2 - Q_1} = P$$

Solve for Q_1.

56. When studying the expansion of metals, we use the formula

$$\frac{L}{1 + at} = L_0$$

Solve for t.

57. Students of photography use the formula

$$\frac{1}{f} = \frac{1}{a} + \frac{1}{b}$$

Solve for f.

⟩⟩⟩ Write On

58. Consider the expression

$$\frac{x}{2} + \frac{x}{3}$$

and the equation

$$\frac{x}{2} + \frac{x}{3} = 5$$

  a. What is the first step in simplifying the expression?
  b. What is the first step in solving the equation?
  c. What is the difference in the use of the LCD in adding two rational expressions as contrasted with solving an equation containing rational expressions?

59. Write your definition of an extraneous solution.

60. Do you have to check the proposed solutions to

$$\frac{x}{2} + \frac{x}{3} = 5$$

for extraneous proposed solutions? Why or why not?

61. Do you have to check the proposed solutions of

$$\frac{6}{x} + \frac{3}{x} = 4$$

for extraneous solutions? Why or why not?

62. In general, when would you check the solutions of an equation for extraneous solutions?

⟩⟩⟩ Concept Checker

Fill in the blank(s) with the correct word(s), phrase, or mathematical statement.

63. Two ratios set equal to one another are called a _____.

64. $\frac{\$1.98}{x \text{ oz}}$ is an example of a _____.

65. When solving a _____ start by multiplying by the LCD.

rational expression   **rate**
rational equation   **proportion**
ratio

⟩⟩⟩ Mastery Test

Solve:

66. $\frac{5}{2x} + 2 = \frac{10}{3x}$ **67.** $\frac{x-7}{x^2-8x} = \frac{2}{x^2-64}$ **68.** $1 - \frac{4}{x^2-1} = \frac{-2}{x-1}$ **69.** $\frac{3}{x} = \frac{5}{x+2}$ **70.** $\frac{2}{x+1} = \frac{3}{x+2}$

71. $4(x+3)^{-1} + 3x(x-2)^{-1} = -2$ **72.** $2x(x-3)^{-1} + 6(x+1)^{-1} = 6$

73. $\frac{1}{x-6} - \frac{1}{x-4} = \frac{6}{(x-6)(x-2)}$ **74.** $\frac{x}{x-4} - \frac{x-5}{x+3} = \frac{3x+16}{x^2-x-12}$

⟩⟩⟩ Skill Checker

Simplify:

75. $12x\left(\frac{1}{4} + \frac{1}{6}\right)$ **76.** $10y\left(\frac{3}{5} - 7\right)$ **77.** $18x\left(\frac{5}{6x} - \frac{7}{3}\right)$ **78.** $4y\left(\frac{5}{4} + 2\right)$

Solve using the RSTUV method:

79. The sum of three consecutive odd integers is 69. What are the integers?

80. An investor bought some municipal bonds yielding 5% annually and some certificates of deposit yielding 8%. If the total investment amounts to $10,000 and the annual interest is $680, how much is invested in bonds and how much in certificates of deposit?

81. How many gallons of a 20% salt solution must be mixed with 40 gallons of a 15% solution to obtain an 18% solution?

82. A car leaves a town traveling at 50 miles per hour. Two hours later, another car traveling at 60 miles per hour leaves on the same road in the same direction. How far from the town does the second car overtake the first?

6.7 Applications: Problem Solving

Objectives

A Solve integer problems.

B Solve work problems.

C Solve distance problems.

D Solve for a specified variable.

To Succeed, Review How To . . .

1. Solve equations involving rational expressions (pp. 482–488).
2. Use the RSTUV procedure to solve word problems (pp. 102–103).

Getting Started
Getting the Golden Rectangle

The unfinished canvas pictured here was painted by Leonardo da Vinci and is entitled *St. Jerome*. A Golden Rectangle fits so neatly around St. Jerome that experts conjecture that da Vinci painted the figure to conform to those proportions. For many years, it has been said that the Golden Rectangle is one of the most visually satisfying of all geometric forms. Do you know how to construct a Golden Rectangle? Such a rectangle has a special ratio of length to width of about 8 to 5 and can be described by writing

$$\frac{\text{Length of rectangle}}{\text{Width of rectangle}} = \frac{8}{5}$$

Now suppose you want to make a Golden Rectangle of your own, but you want the length to be 6 inches longer than the width. What should be the dimensions of your rectangle?

To solve this problem, you need to review the RSTUV procedure we used in Section 2.3.

The Golden Rectangle

1. **Read the problem.** You have to find the dimensions of the rectangle.
2. **Select the unknown.** Let w be the width.
3. **Think of a plan.** In Section 6.6, we learned how to write proportions that would solve a problem in which two ratios are the same.

Since you want the new dimensions,

$$\frac{\text{Length}}{\text{Width}} = \frac{w+6}{w} \quad \begin{array}{l}\leftarrow \text{Length is 6 inches more than the width.} \\ \leftarrow \text{Width is } w.\end{array}$$

to be in the same proportion as the Golden Rectangle,

$$\frac{\text{Length}}{\text{Width}} = \frac{8}{5}$$

we set the ratios equal.

$$\frac{w+6}{w} = \frac{8}{5}$$

4. Use the property for proportions and solve the resulting linear equation. To solve the proportion, set the cross products equal and simplify.

$$\frac{w+6}{w} = \frac{8}{5}$$

$$5(w+6) = 8(w)$$
$$5w + 30 = 8w \quad \text{Simplify.}$$
$$30 = 3w \quad \text{Subtract } 5w.$$
$$10 = w \quad \text{Divide by 3.}$$

Thus, the width is 10 inches and the length is 6 inches more than that, or 16 inches.

5. Verify the answer. The rectangle is 10 inches by 16 inches, so the ratio of length (16) to width (10) is $\frac{16}{10}$, or $\frac{8}{5}$, as desired.

A › Solving Integer Problems

We now discuss problems involving consecutive integers and other numerical properties. You can solve these problems using the RSTUV procedure.

EXAMPLE 1 Consecutive integers

There are two consecutive even integers such that the reciprocal of the first added to the reciprocal of the second is $\frac{3}{4}$. What are the integers?

SOLUTION We use the RSTUV method.

1. Read the problem. Make sure you understand the meaning of "consecutive even integers." For example, 6, 8 are consecutive even integers, as are 78, 80.

2. Select the unknown. Let n be the first integer. The next even integer is $n + 2$.

3. Think of a plan. First, translate the problem:

The reciprocal of the first	added to	the reciprocal of the second	is $\frac{3}{4}$.
$\frac{1}{n}$	$+$	$\frac{1}{n+2}$	$= \frac{3}{4}$

4. Use the procedure for solving equations with rational expressions. Start by finding the LCD, $4n(n + 2)$. Multiply both sides by this LCD.

$$4n(n+2)\left(\frac{1}{n} + \frac{1}{n+2}\right) = 4n(n+2) \cdot \frac{3}{4}$$

$$4n(n+2) \cdot \frac{1}{n} + 4n(n+2) \cdot \frac{1}{n+2} = 4n(n+2) \cdot \frac{3}{4}$$

$$4n + 8 + 4n = 3n^2 + 6n \quad \text{Remove parentheses.}$$
$$0 = 3n^2 - 2n - 8 \quad \text{Subtract } 8n \text{ and } 8 \text{ from both sides.}$$
$$0 = (3n + 4)(n - 2) \quad \text{Factor.}$$

By the zero-product property,

$$3n + 4 = 0 \quad \text{or} \quad n - 2 = 0$$
$$n = \frac{-4}{3} \quad \text{or} \quad n = 2 \quad \text{Solve } 3n + 4 = 0, n - 2 = 0.$$

Since n was assumed to be an integer, we discard the answer $\frac{-4}{3}$. Thus, the first even integer is 2 and the next one is 4.

5. Verify the answer. Verify that the sum of the reciprocals is $\frac{3}{4}$. Since $\frac{1}{2} + \frac{1}{4} = \frac{3}{4}$, our result is correct.

PROBLEM 1
The sum of the reciprocals of two consecutive odd integers is $\frac{-8}{15}$. What are the integers?

Answers to PROBLEMS

1. -5 and -3

B › Solving Work Problems

Have you ever wished for help with your taxes? The Internal Revenue Service estimates that it takes about 7 hours to file your 1040A form. (This includes record-keeping, familiarizing yourself with the form, and preparing and sending it.)

EXAMPLE 2 Work problems

A couple is about to file their form 1040A. One of them can complete it in 8 hours, and the other can do it in 6 hours. How long would it take if they work on it together?

PROBLEM 2

An accountant can finish form 1040A in 5 hours. An assistant can do it in 10 hours. How long would it take if they work together?

SOLUTION Again, we use the RSTUV method.

1. Read the problem. We need to find the total time it takes them when they work together.

2. Select the unknown. Let t be the time it takes the couple to complete the form working together.

3. Think of a plan. We concentrate on what happens each hour. Since one person can complete the form in 8 hours and the second can do it in 6 hours, the first person will complete $\frac{1}{8}$ of the form and the second will complete $\frac{1}{6}$ of the form each hour. They are working together and it takes t hours to do the whole thing, so they complete $\frac{1}{t}$ of the form each hour. Here is what we have in one hour:

$$\underbrace{\frac{1}{8}}_{\text{Work done by first person in 1 hr}} + \underbrace{\frac{1}{6}}_{\text{Work done by second person in 1 hr}} = \underbrace{\frac{1}{t}}_{\text{Work done together in 1 hr}}$$

4. Use the procedure for solving equations with rational expressions. First, find the LCD of $\frac{1}{8}, \frac{1}{6},$ and $\frac{1}{t}$. The LCD of these fractions is $24t$.

$$24t\left(\frac{1}{8} + \frac{1}{6}\right) = 24t \cdot \frac{1}{t} \quad \text{Multiply by } 24t.$$

$$\overset{3}{24t} \cdot \frac{1}{8} + \overset{4}{24t} \cdot \frac{1}{6} = 24t \cdot \frac{1}{t}$$

$$3t + 4t = 24 \quad \text{Simplify.}$$

$$7t = 24$$

$$t = \frac{24}{7} = 3\frac{3}{7} \text{ hr}$$

It takes $3\frac{3}{7}$ hours (about 3 hours 26 minutes) to complete the job.

5. Verify the answer. The verification is left to you.

Another type of problem can also be thought of as a work problem; this is the tank or pool problem. The idea is that pipes filling or emptying a tank or pool are doing the *work* to fill or empty the pool. Here is how we solve these problems.

EXAMPLE 3 Pool problems

A pool is filled by an intake pipe in 4 hours and is emptied by a drain pipe in 5 hours. How long will it take to fill the pool with both pipes open?

PROBLEM 3

Repeat Example 3 if the intake pipe can fill the pool in 6 hours and the drain pipe can empty it in 7 hours.

(continued)

Answers to PROBLEMS

2. $3\frac{1}{3}$ hr or 3 hr 20 min   **3.** 42 hr

SOLUTION As before, we use the RSTUV method.

1. **Read the problem.** We are asked for the time it takes to fill the pool.
2. **Select the unknown.** Let this time be T hours.
3. **Think of a plan.** In 1 hr, the intake pipe fills $\frac{1}{4}$ of the pool, the drain pipe empties $\frac{1}{5}$, and together they fill $\frac{1}{T}$ of the pool. Thus, in 1 hour

$$\underbrace{\frac{1}{4}}_{\text{Amount filled by intake pipe in 1 hr}} - \underbrace{\frac{1}{5}}_{\text{Amount emptied by drain pipe in 1 hr}} = \underbrace{\frac{1}{T}}_{\text{Amount filled by both in 1 hr}}$$

4. **Use algebra to solve the equation.** The LCD is $20T$.

$$\overset{5}{20T} \cdot \frac{1}{4} - \overset{4}{20T} \cdot \frac{1}{5} = 20T \cdot \frac{1}{T}$$
$$5T - 4T = 20$$
$$T = 20$$

It takes 20 hours to fill the pool if the intake and drain pipes are both open.

5. **Verify the answer.** The intake pipe can fill the pool in 4 hours. It can then fill the pool five times in 20 hours. The drain pipe can empty the pool in 5 hours, so it can empty the pool four times in 20 hours. Since the intake can fill the pool five times and the drain can empty it four times in 20 hours, the pool would be filled once at the end of 20 hours.

C › Solving Distance Problems

The ideas we have studied can be used to solve uniform motion problems like the ones discussed in Section 2.4. As you recall, when traveling at a constant rate R, the distance D traveled in time T is given by $D = RT$. We use this information in Example 4.

EXAMPLE 4 Distance problems

Why do people ride motorcycles? The most common answer in the United States is "the feeling of freedom." However, if there is a strong wind the cyclist will not be as "free" to travel as he would like because the wind could slow his speed or increase it. A cyclist decides to take a road trip when the wind current is 10 miles per hour. If he travels 108 miles with the wind in the same time it takes him to go 78 miles against the wind, what is the speed of the motorcycle in still air?

PROBLEM 4
Repeat Example 4 by finding the speed of the cyclist if the wind current is 8 miles per hour and the cyclist travels 32 miles with the wind in the same time it takes him to go 24 miles against the wind.

SOLUTION Once again, we use the RSTUV method.

1. **Read the problem.** We want to find the speed (rate) of the motorcycle in still air.
2. **Select the unknown.** Let R be the rate of the motorcycle in still air.
3. **Think of a plan.** Make a chart with D, R, and T as headings.

Answers to PROBLEMS
4. Speed in still air is 56 mph.

Category	D (mi)	R (mi/hr)	T (hr)
With the wind			
Against the wind			

Speed with the wind: $R + 10$ Air current helps, add 10 to R.
Speed against the wind: $R - 10$ Air current hinders, subtract 10 from R.
The time T is given by

$$T = \frac{D}{R}$$

(solving for T in $D = RT$). We then have

Time with the wind: $\dfrac{108}{R + 10}$

Time against the wind: $\dfrac{78}{R - 10}$

Enter this information in the chart.

Category	D (mi)	R (mi/hr)	$\left(T = \frac{D}{R}\right)$ (hr)
With the wind	108	$R + 10$	$\dfrac{108}{R + 10}$
Against the wind	78	$R - 10$	$\dfrac{78}{R - 10}$

Since it takes the same time, we have

$$T_{\text{with}} = T_{\text{against}}$$

$$\frac{108}{R + 10} = \frac{78}{R - 10}$$

4. Use algebra to solve the resulting equation. The LCD is $\dfrac{(R + 10)(R - 10)}{1}$, so we multiply both sides of the equation by this LCD.

$$\frac{\cancel{(R + 10)}(R - 10)}{1} \cdot \frac{108}{\cancel{R + 10}} = \frac{(R + 10)\cancel{(R - 10)}}{1} \cdot \frac{78}{\cancel{R - 10}}$$

$108(R - 10) = 78(R + 10)$
$18(R - 10) = 13(R + 10)$ Divide by 6.
$18R - 180 = 13R + 130$ Remove parentheses.
$5R - 180 = 130$ Subtract 13R.
$5R = 310$ Add 180.
$R = 62$ Divide by 5.

The speed of the motorcycle in still air is 62 miles per hour.

5. Verify the answer. The verification is left to you.

D > Solving for Specified Variables

Often when solving applications with a formula, it is more convenient to rewrite the formula so that it is solved for a specified variable. For example, $d = rt$ is the formula for finding the distance given the rate of speed and time traveled. This formula could be solved for t by dividing both sides by r.

$$d = rt \quad \text{Given formula}$$
$$\frac{d}{r} = \frac{rt}{r} \quad \text{Divide by } r.$$
$$\frac{d}{r} = t \quad \text{Solved for } t.$$

In Example 5 we will practice the technique of solving for a specified variable in a formula, and in Example 6 we will apply that to solving for a specified variable in an application involving a formula.

EXAMPLE 5 Solving for a specified variable

a. Given the formula, $P = a + b + c$, solve for b.
b. Given the formula, $V = \frac{1}{3}Bh$, solve for h.

SOLUTION

a. When solving for a specified variable, isolate the specified variable, b, on the right side by subtracting a and c from both sides.

Given: $\quad P = a + b + c \quad$ Solve for b.
$\quad\quad\quad\underline{-a - c \quad\quad -a \quad -c} \quad$ Subtract a and c.
$\quad\quad\quad P - a - c = \quad\quad b \quad\quad$ Solved for b.

b. This formula has a fraction, so we clear the fraction by multiplying both sides by 3. Then to isolate the variable h, divide by B on both sides.

Given: $\quad V = \frac{1}{3}Bh \quad$ Solve for h.
$\quad\quad\quad 3(V) = 3\left(\frac{1}{3}\right)Bh \quad$ Multiply by 3.
$\quad\quad\quad 3V = Bh \quad 3\left(\frac{1}{3}\right) = 1$
$\quad\quad\quad \frac{3V}{B} = \frac{Bh}{B} \quad$ Divide by B.
$\quad\quad\quad \frac{3V}{B} = h \quad$ Solved for h.

PROBLEM 5

a. Given $2l + 2w = P$, solve for w.
b. Given $A = \frac{1}{2}bh$, solve for b.

EXAMPLE 6 Golf Handicap

A golfer's handicap is intended to show a player's potential. Handicap systems are generally based on calculating an individual player's playing ability from his recent history of rounds.

This player can find his handicap differential H by the formula

$$H = \frac{(G - R)113}{S}$$

where G is the gross score; R is the course rating, which measures the average "good" score by a scratch golfer; and S is the course slope, which describes the difficulty of the course. Solve this formula for G and find out what his gross score

PROBLEM 6

Using the golfer's handicap formula

$$H = \frac{(G - R)113}{S}$$

solve for R and find the course rating R when the gross score is 85, the course slope is 124, and the handicap differential is 11.9.

Answers to PROBLEMS

5. a. $w = \frac{P - 2l}{2}$ b. $b = \frac{2A}{h}$

6. $R = \frac{113G - H(S)}{113}$; $R = 71.9$

G was when the course rating is 69.3, the course slope is 117, and his handicap differential is 15.2. (Round answer to a whole number.)

Source: http://en.wikipedia.org.

SOLUTION To solve for G in $H = \dfrac{(G-R)113}{S}$, we will clear the fraction and then isolate the variable G.

$$S \cdot H = \dfrac{(G-R)113}{S} \cdot S \quad \text{Clear fraction.}$$

$$S(H) = (G-R)113$$

$$S(H) = 113G - 113R \quad \text{Remove parentheses.}$$

$$S(H) + 113R = 113G \quad \text{Add } 113R.$$

$$\dfrac{S(H) + 113R}{113} = G \quad \text{Divide by 113.}$$

To find the gross score G substitute the other values into the formula and simplify.

$$G = \dfrac{S(H) + 113R}{113}$$

$$G = \dfrac{117(15.2) + 113(69.3)}{113}$$

$$G = 85 \text{ (Rounded to nearest whole number.)}$$

This means that before any adjusting the golfer had an 85 for his golf score.

There are more problems in the *Using Your Knowledge* section of the exercises in which you are asked to solve for a variable.

Calculator Corner

Using Tables to Solve Rational Equations

By this time you should be convinced that graphing is a powerful tool in solving algebra problems. But is that all a calculator does? No, we haven't even touched on an important feature of some calculators—the construction of tables listing values that may be solutions. This would be useful in the problem in the *Getting Started,* which required us to construct a Golden Rectangle whose length is 6 in. more than its width. As before, we solve the equation

$$\dfrac{w+6}{w} = \dfrac{8}{5} = 1.6$$

Now we will show you how to set up a table using a TI-83 Plus that will solve the problem for us. First, let's agree to use x instead of w in the solution. Start by pressing [2nd] [WINDOW]. The calculator wants to know the minimum at which you wish to start x. Let TblStart = 1. It then asks for ΔTbl=. This sets up the increments for x in your table. Let ΔTbl = 1. This means that the first x is 1, the next one is increased by 1 to 2, the next one is 3, and so on. For the "Indpnt" and "Depend," leave the setting in auto mode (Window 1). Now, press 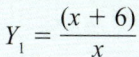 and enter

$$Y_1 = \dfrac{(x+6)}{x}$$

TABLE SETUP
TblStart=1
ΔTbl=1
Indpnt: **AUTO** Ask
Depend: **AUTO** Ask

Window 1

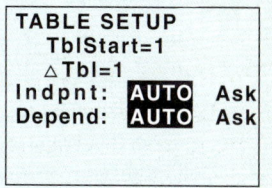

Window 2

(continued)

Then press [2nd] [GRAPH]. The first column shows successive values for x, while the second shows the corresponding values of

$$Y_1 = \frac{(x+6)}{x}$$

(see Window 2). Press the [▼] key while you are in the first column. It will show successive values of x and corresponding Y_1 values. You only need to move down until

$$Y_1 = \frac{8}{5} = 1.6$$

(see Window 3). This occurs when $x = 10$, the same answer as before, but it is more fun letting the calculator do the work for you.

X	Y₁
4	2.5
5	2.2
6	2
7	1.8571
8	1.75
9	1.6667
10	1.6

X=10

Window 3

> Exercises 6.7

Boost your grade at mathzone.com!
> Practice Problems
> NetTutor
> Self-Tests
> e-Professors
> Videos

< A > **Solving Integer Problems** In Problems 1–10, solve the integer problems.

1. The sum of an integer and its reciprocal is $\frac{65}{8}$. Find the integer.

2. The sum of an integer and its reciprocal is $\frac{50}{7}$. What is the integer?

3. One number is twice another. The sum of their reciprocals is $\frac{3}{10}$. Find the numbers.

4. One number is three times another. The sum of their reciprocals is $\frac{1}{3}$. Find the numbers.

5. Find two consecutive even integers the sum of whose reciprocals is $\frac{7}{24}$.

6. Find two consecutive odd integers such that the sum of their reciprocals is $\frac{16}{63}$.

7. The denominator of a fraction is 5 more than the numerator. If 3 is added to both numerator and denominator, the resulting fraction is $\frac{1}{2}$. Find the fraction.

8. The numerator of a certain fraction is 4 less than the denominator. If the numerator is increased by 8 and the denominator by 35, the resulting fraction is $\frac{1}{2}$. Find the fraction.

9. The current ratio of your business is defined by

$$\text{Current ratio} = \frac{\text{current assets}}{\text{current liabilities}}$$

By how much should you increase your current liabilities if they are $40,000 right now, your current assets amount to $90,000, and you wish the current ratio to be $\frac{3}{2}$?

10. Repeat Problem 9 where you want the current ratio to be 2.

< B > **Solving Work Problems** In Problems 11–23, solve the work problems.

11. If one word processor can finish a job in 3 hours while another word processor can finish in 5 hours, how long will it take both of them working together to finish the job?

12. A carpenter can finish a job in 8 hours, and another one can do it in 10 hours. How long will it take them to finish the job working together?

13. The world record for riveting is 11,209 rivets in 9 hours, by J. Mair of Ireland. If another person can rivet 11,209 rivets in 10 hours, how long will it take both of them working together to rivet the 11,209 rivets?

14. Mr. Gerry Harley of England shaved 130 men in 60 minutes. If another barber can shave all these men in 5 hours, how long will it take both of them working together to shave the 130 men?

15. A printing press can print the evening paper in half the time another press takes to print it. Together, the presses can print the paper in 2 hours. How long will it take each of them to print the paper?

16. A computer can do a job in 4 hours. With the help of a newer computer, the job can be completed in 1 hour. How long will it take the newer computer to do the job alone?

17. A tank can be filled by an intake pipe in 9 hours and drained by another pipe in 21 hours. If both pipes are open, how long will it take to fill the tank?

18. A faucet fills a tank in 12 hours, and the drain pipe empties it in 18 hours. If the faucet and the drain pipe are both open, how long does it take to fill the tank?

19. A pipe fills a pool in 7 hours, and another one fills it in 21 hours. How long will it take to fill the pool using both pipes?

20. One pipe fills a tank in 6 hours, and another fills it in 4 hours. How long will it take both pipes together to fill the tank?

21. The main engine of a rocket burns for 60 seconds on the fuel in the rocket's tank, while the auxiliary engine burns for 90 seconds on the same amount of fuel. How long do both engines burn if they are operated together on the single tank of fuel?

22. An in-flow pipe fills a pool in 12 hours, and another pipe drains it in 4 hours. How long does it take to empty the pool if both pipes are open simultaneously? (Assume that the pool is full at the start.)

23. A pipe fills a tank in 9 hours, but the drain empties it in 6 hours. How long does it take to empty the tank if both pipes are open simultaneously? (Assume that the tank is full at the start.)

〈 C 〉 Solving Distance Problems In Problems 24–30, solve the distance problems.

24. A boat travels 30 miles downstream in the same time it takes to go 20 miles upstream. If the river current flows at 5 miles per hour, what is the boat's speed in still water?

25. A small plane flies 240 miles against the wind in the same time it takes it to fly 360 miles with a tail wind. If the wind velocity is 30 miles per hour, find the plane's speed in still air.

26. A jet plane flies 700 miles against the wind in the same time it takes it to fly 900 miles with a tail wind. If the wind velocity is 50 miles per hour, what is the plane's speed in still air?

27. A small plane cruises at 120 miles per hour in still air. It takes this plane the same time to travel 270 miles against the wind as it does to travel 450 miles with a tail wind. What is the wind velocity?

28. A small plane can travel 200 miles against the wind in the same time it takes it to travel 260 miles with a tail wind. If the plane's speed in still air is 115 miles per hour, find the wind velocity.

29. An automobile travels 200 miles in the same time in which a small plane travels 1000 miles. Find their rates of speed if the airplane is 100 miles per hour faster than the automobile.

30. Janice ran 1000 meters in the same time that Paula ran 950 meters. If Paula's speed was $\frac{1}{4}$ meters per second less than Janice's, what was Janice's speed?

〈 D 〉 Solving for Specified Variables In Problems 31–34, solve the formula for the specified variable.

31. $V = \frac{4}{3}\pi r^3 h$, for h

32. $z = \frac{x - \bar{x}}{s}$, for x

33. $\frac{x^2}{a^2} + \frac{y^2}{b^2} = 1$, for x^2

34. $a = \frac{1}{2}\pi r^2 h$, for h

〉〉〉 Using Your Knowledge

Formulas, Formulas, and More Formulas In Section 2.2, we learned how to solve a formula for a specified variable. You will find that many formulas involve rational expressions. Use your knowledge of solving equations involving rational expressions to solve for the specified variables in the following equations.

35. To find the focal length of lenses, lens makers use the formula

$$\frac{1}{F} = \frac{1}{f_1} + \frac{1}{f_2}$$

Solve for F.

36. To find the radius of curvature R of a sphere, we use the formula

$$R = \frac{2AS}{L - 2S}$$

Solve for A.

37. The electric current i in a simple series circuit is given by the equation

$$i = \frac{2E}{R + 2r}$$

Solve for R.

38. Cowling's rule states that a child's dose c for a child A years old, where A is between 2 and 13, is given by the equation

$$c = \frac{A + 1}{24}d$$

where d is the adult dose. Solve for d.

39. In Problem 38, what would the adult dose be if a 5-year-old child's dose for aspirin is 3 tablets a day?

40. Is there an integer A for which the dosages are the same for Cowling's rule (see Exercise 38) and Young's rule?

$$\text{Young's rule: } c = \frac{A}{A + 12}d$$

41. Given this trigonometry identity, solve for $\cos(2u)$.

$$\cos^2 u = \frac{1 + \cos(2u)}{2}$$

42. Given this trigonometry identity, solve for the sum, $\tan u + \tan v$

$$\tan(u + v) = \frac{\tan u + \tan v}{1 - \tan u \tan v}$$

〉〉〉 Write On

43. There's another way to solve Example 2. If you assume that t is the time it takes for the couple to complete the form working together, then in t hr the first person will do $\frac{t}{6}$ of the job and the second person will do $\frac{t}{8}$ of the job. Working together, they will do $\frac{t}{6} + \frac{t}{8}$ of the job. To what should this sum be equal? Explain.

44. Using the method in Exercise 43, what equation would you use to solve Example 3?

〉〉〉 Mastery Test

45. A speed boat can travel 36 miles downstream in the same time it takes it to go 12 miles upstream. If the current is moving at 18 miles per hour, what is the speed of the boat in still water?

46. A pool is filled by an intake pipe in 5 hours and emptied by a drain pipe in 6 hours. How long would it take to fill the pool with both pipes open?

47. The sum of the reciprocals of two consecutive even integers is $\frac{5}{12}$. What are the integers?

48. According to Clark's rule, the dose c for a child weighing W pounds is

$$c = \frac{W}{150}d$$

where d is the adult dose. Solve for d.

〉〉〉 Skill Checker

Evaluate the expression for the specified values of the variables.

49. ksv^2; if $k = -2$, $s = 5$, and $v = 3$

50. $\frac{kl}{d^2}$; if $k = 6$, $l = 12$, and $d = 4$

51. $\frac{kx^3}{y^2}$; if $k = -1$, $x = -4$, and $y = 8$

Write an algebraic expression for the following statements.

52. k times the square of d

53. k times the cube root of m

54. k times the square root of l

55. k times the cube of h

56. k times the square of x times t

6.8 Variation

Objectives

Write an equation expressing:

A Direct variation.

B Inverse variation.

C Joint variation.

D Solve applications involving direct, inverse, and joint variation.

To Succeed, Review How To . . .

1. Evaluate an expression (pp. 51–55).
2. Solve linear equations (pp. 76–83).

Getting Started
Pendulums and Gas Mileage

What could pendulums and gas mileage possibly have in common? Each have measures associated with them that can be expressed in a formula referred to as a **variation**.

As the length L of the string of a pendulum increases, the time T it takes the pendulum to make a full back-and-forth swing increases. What is the formula relating the length L and the time T? Galileo Galilei discovered that the time T (in seconds) it takes for one swing of the pendulum varies directly as the square root of the length L of the pendulum.

$$k = \frac{T}{\sqrt{L}}$$

In the same manner, the number m of miles you drive a car is *proportional to,* or *varies directly as,* the number g of gallons of gas used. This means that the ratio

$\frac{m}{g}$ is a constant: $\frac{m}{g} = k$ or $m = kg$

In 2002 inventor Doug Malewicki, with a team of fellow enthusiasts, began construction of a "C2C" car that could go from California to New York (about 3500 miles) on one 25-gallon tank of standard gasoline. Using $m/g = k$ as before, what is k? Can you explain what k means?

A ▶ Direct Variation

Do you get higher grades when you study more hours? If this is the case, your grades vary directly or are directly proportional to the number of hours you study. Here is the definition for direct variation.

> **DIRECT VARIATION**
>
> y **varies directly as** x if there is a constant k such that
>
> $$y = kx$$
>
> (k is usually called the constant of variation.)

Other words can be used to indicate direct variation. Here's a list of some of these words and how they translate into an equation.

English Phrase	Translation
y varies with x	$y = kx$
y varies directly as t	$y = kt$
y is proportional to v	$y = kv$
v varies as the square of t	$v = kt^2$
p varies as the cube of r	$p = kr^3$
T varies as the square root of L	$T = k\sqrt{L}$

EXAMPLE 1 Solving a direct variation problem

The length L of a mustache varies directly as the time t that it takes to grow.

a. Write an equation of variation.

b. One of the longest mustaches on record was grown by Masuriya Din. His mustache grew 56 inches (on each side) over a 14-year period. Find k, the constant of variation, and explain what it represents.

SOLUTION

a. Since the length L varies directly as the time t,

$$L = kt$$

b. We know that when $L = 56$, $t = 14$. Thus,

$$56 = k \cdot 14$$
$$4 = k$$

The constant of variation is 4. This means that Mr. Din's mustache grew 4 inches each year.

PROBLEM 1

Hair length L is proportional to time t.

a. Write an equation of variation.

b. If your hair grew 6 inches in 2 months find k, the constant of variation.

B › Inverse Variation

Sometimes, as one quantity increases, a related quantity decreases proportionally. For example, the *more* time you spend practicing a task, the *less* time it will take you to do the task. In such cases, we say that the quantities *vary inversely* to one another.

INVERSE VARIATION

y varies inversely as x if there is a constant k such that

$$y = \frac{k}{x}$$

Here are some expressions that also mean "vary inversely."

English Phrase	Translation
y varies inversely with x	$y = \frac{k}{x}$
y is inversely proportional to x	$y = \frac{k}{x}$
v varies inversely as the square of t	$v = \frac{k}{t^2}$
p varies inversely as the cube of r	$p = \frac{k}{r^3}$
T varies inversely as the square root of L	$T = \frac{k}{\sqrt{L}}$

Answers to PROBLEMS

1. a. $L = kt$ b. 3 (in./mo)

EXAMPLE 2 Solving an inverse variation problem

The speed s at which a car travels is inversely proportional to the time t it takes to travel a given distance.

a. Write the equation of variation.
b. If a car travels at 60 miles per hour for 3 hours, what is k, the constant of variation, and what does it represent?

SOLUTION

a. The equation is
$$s = \frac{k}{t}$$

b. We know that $s = 60$ when $t = 3$. Substituting 60 for s and 3 for t,
$$60 = \frac{k}{3}$$
$$k = 180$$

In this case k, the constant of variation, represents the total distance traveled, 180 miles.

PROBLEM 2

The principal P invested is inversely proportional to the annual rate of interest r.

a. Write an equation of variation.
b. Find k, the constant of variation, if $r = 8\%$ and $P = \$100$.

EXAMPLE 3 Solving an inverse variation problem

Have you ever heard one of those loud boom boxes or a car sound system that makes your bones vibrate? The loudness L of sound is inversely proportional to the square of your distance d from the source.

a. Write an equation of variation.
b. The loudness of rock music coming from a boom box 5 feet away is 100 decibels. Find k, the constant of variation.
c. If you move to 10 ft away from the boom box, how loud is the sound?

SOLUTION

a. The equation is
$$L = \frac{k}{d^2}$$

b. We know that $L = 100$ for $d = 5$ so that
$$100 = \frac{k}{5^2} = \frac{k}{25}$$

Multiplying both sides by 25, we find that $k = 2500$. The constant of variation is 2500.

c. Since $k = 2500$,
$$L = \frac{2500}{d^2} \quad \text{Substitute 2500 for } k.$$

When $d = 10$,
$$L = \frac{2500}{10^2} = 25 \text{ dB}$$

100 decibels is only 20 decibels from the threshold of pain, which causes immediate and permanent hearing loss.

PROBLEM 3

The f-number on a camera varies inversely as the diameter a of the aperture when the distance is set at infinity.

a. Write an equation of variation.
b. Find k, the constant of variation, when the f-number is 6 and $a = \frac{1}{2}$.
c. Find a if the f-number is 18.

Answers to PROBLEMS

2. a. $P = \frac{k}{r}$ b. $8
3. a. $f = \frac{k}{a}$ b. 3   c. $\frac{1}{6}$

C › Joint Variation

Besides the direct and inverse variations we have discussed, there can be variation involving a third variable. A variable z can vary *jointly* with the variables x and y. For example, labor costs c vary jointly with the number of workers w used and the number of hours h that they work. The formal expression of joint variation is given here.

JOINT VARIATION

z varies jointly with x and y if there is a constant k such that

$$z = kxy$$

The statement *z is proportional to x and y* is sometimes used to mean *z varies jointly with variables x and y*.

Thus, the fact that labor costs c vary jointly with the number w of workers used and the number h of hours worked can be expressed as $c = kwh$, where k is a constant.

EXAMPLE 4   Solving a joint variation problem

The lifting force P exerted by the atmosphere on the wings of an airplane varies jointly with the wing area A in square feet and the square of the plane's speed V in miles per hour. Suppose the lift is 1200 pounds for a wing area of 100 square feet and a speed of 75 miles per hour.

a. Find an equation of variation.
b. Find k, the constant of variation.
c. Find the lifting force on a wing area of 60 square feet when $V = 125$.

SOLUTION

a. Since P varies jointly with the area A and the square of the velocity V, we have $P = kAV^2$.

b. When the lift $P = 1200$, we know that $A = 100$ and $V = 75$. Substituting these values in the equation $P = kAV^2$, we obtain

$$1200 = k \cdot 100 \cdot (75)^2$$

Dividing both sides by $100 \cdot 75^2$, we find

$$k = \frac{1200}{100 \cdot 75^2} = \frac{12}{75^2} = \frac{4}{1875}$$

The constant of variation is $\frac{4}{1875}$. Thus, $P = kAV^2$ becomes

$$P = \frac{4}{1875} AV^2$$

c. $P = \frac{4}{1875} AV^2$, where $A = 60$ and $V = 125$.

$$P = \frac{4}{1875}(60)(125^2)$$
$$= 2000 \text{ lb}$$

The lifting force is 2000 pounds.

PROBLEM 4

The wind force F on a vertical surface varies jointly with the area A of the surface and the square of the wind velocity V. Suppose the wind force on 1 square foot of surface is 2.2 pounds when $V = 20$ mi/hr.

a. Find an equation of variation.
b. Find k, the constant of variation.
c. Find the force on a 2 square-foot vertical surface when $V = 60$ miles per hour.

Answers to PROBLEMS

4. **a.** $F = kAV^2$ **b.** 0.0055
   c. 39.6 lb

D › Solving Applications Involving Variation

EXAMPLE 5 Lots of snow

Figure 6.1 shows the number of gallons of water, g (in millions), produced by an inch of snow in different cities. The larger the area of the city, the more gallons of water are produced, so g is directly proportional to A, the area of the city (in square miles).

a. Write an equation of variation.
b. If the area of St. Louis is about 62 square miles, what is k, the constant of variation?
c. Find the amount of water produced by 1 inch of snow falling in Anchorage, Alaska, with an area of 1700 square miles.

PROBLEM 5

Using the graph in Example 5, the formula found in part **a**, and the constant found in part **b**, find the approximate area of Chicago. (Round your answer to the nearest square mile.)

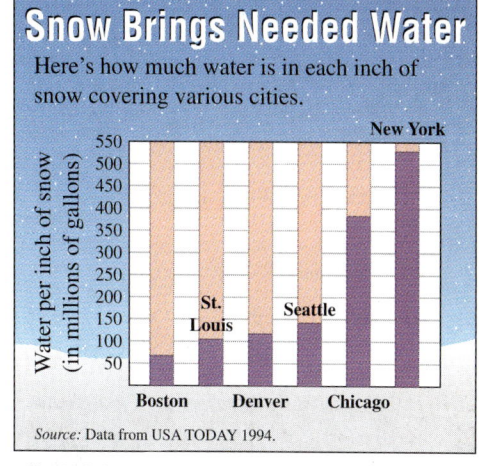

Snow Brings Needed Water
Here's how much water is in each inch of snow covering various cities.

Source: Data from USA TODAY 1994.

› Figure 6.1

SOLUTION

a. Since g is directly proportional to A, $g = kA$.
b. From Figure 6.1, we can see that $g = 100$ (million) is the number of gallons of water produced by 1 inch of snow in St. Louis. Since it is given that $A = 62$, $g = kA$ becomes

$$100 = k \cdot 62 \quad \text{or} \quad k = \frac{100}{62} = \frac{50}{31}$$

The constant of variation is $\frac{50}{31}$.

c. For Anchorage, $A = 1700$, thus, $g = \frac{50}{31} \cdot 1700 \approx 2742$ million gallons of water (rounded to the nearest million).

Calculator Corner

Using StatPlot and Lists to Verify Direct and Inverse Variation Equations

How do you recognize different types of variation with your calculator? If you have several points, you can graph them and examine the result. In Example 1, we concluded that Mr. Din's mustache grew 4 inches each year. This means that at the end of the first year, his mustache was 4 inches long; at the end of the second year, it was 8 inches long; and at the end of 14 years, it was 56 inches long. You now have three ordered pairs: (1, 4), (2, 8), and (14, 56). Before you input the list of ordered pairs, clear any equations on the Y= screen. You can make a list (clear lists by pressing 2nd   +   4   ENTER) by pressing STAT   1 and then entering 1, 2, and 14 under L_1 and 4, 8, and 56 under L_2. Press 2nd   Y=   ENTER   ENTER to turn plot 1 on if necessary; select the first type of graph, L_1, L_2, and ■. So that we don't have to contend with deciding what type of window to use, press ZOOM   9 . As you can see in Window 1, the points are on a line. (Remember that the length varies directly as the time.)

If you want to find the equation of the line passing through the three points in the graph, press STAT   ▶   4   ENTER . The calculator will tell you that $y = ax + b$, where $a = 4$ and $b = 0$, or $y = 4x$ as before.

To get a feel for inverse variation, in Example 2, select a $[-1, 200]$ by $[-1, 5]$ window with Xscl = 10 and Yscl = 1. Using Y_1 instead of s and x instead of t, graph

$$Y_1 = \frac{180}{x}$$

(Be sure the Star Plot is off.)
The result is shown in Window 2. Can you see that as x increases, Y_1 decreases? Try looking at the graph

$$L = \frac{2500}{d^2}$$

What window do you need? What does the graph look like?

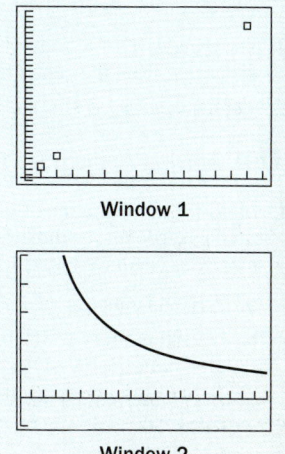

Window 1

Window 2

Answers to PROBLEMS

5. 233 mi²

Exercises 6.8

> Boost your grade at mathzone.com!
> - Practice Problems
> - NetTutor
> - Self-Tests
> - e-Professors
> - Videos

A Direct Variation   In Problems 1–5, write an equation of variation using k as the constant.

1. The tension T on a spring varies directly with the distance s it is stretched. (This is usually called Hooke's law.)

2. The distance s a body falls in t seconds is directly proportional to the square of t.

3. The weight W of a dam varies directly with the cube of its height h.

4. The kinetic energy KE of a moving body is proportional to the square of its velocity v.

5. The weight W of a human brain is directly proportional to the body weight B.

B Inverse Variation   In Problems 6–8, write an equation of variation using k as the constant.

6. In a circuit with constant voltage, the current I varies inversely with the resistance R of the circuit.

7. For a wire of fixed length, the resistance R varies inversely with the square of its diameter D.

8. The intensity of illumination I from a source of light varies inversely with the square of the distance d from the source.

C Joint Variation   In Problems 9–20, write an equation of variation using k as the constant.

9. The annual interest I received on a savings account varies jointly with the principal P (the amount in the account) and the interest rate r paid by the bank.

10. The cost C of a building varies jointly as the number w of workers used to build it and the cost of materials m.

11. The amount of oil A used by a ship traveling at a uniform speed varies jointly with the distance s and the square of the speed v.

12. The power P in an electric circuit varies jointly with the resistance R and the square of the current I.

13. The volume V of a rectangular container of fixed length varies jointly with its depth d and width w.

14. The force of attraction F between two spheres of mass m_1 and m_2 varies directly as the product of the masses and inversely as the square of the distance d between their centers.

15. The illumination I in foot-candles upon a wall varies directly with the intensity i in candlepower of the source of light and inversely with the square of the distance d from the light.

16. The strength S of a horizontal beam of rectangular cross section varies jointly as the breadth b and the square of the depth d and inversely as the length L.

17. The electrical resistance R of a wire of uniform cross section varies directly as its length L and inversely as its cross-sectional area A.

18. The electrical resistance R of a wire varies directly as the length L and inversely as the square of its diameter d.

19. The weight W of a body varies inversely as the square of its distance d from the center of the earth.

20. z varies directly as the cube of x and inversely as the square of y.

D Solving Applications Involving Variation

21. *Interest* The amount of annual interest I you receive on a savings account is directly proportional to the amount of money m you have in the account.
    a. Write an equation of variation.
    b. If $480 produces $26.40 in interest, what is k, the constant of variation?
    c. How much annual interest would you receive if the account had $750?

22. *Record revolutions* The number of revolutions, R (rev), a record makes as it is being played varies directly as the time t that it is on the turntable.
    a. Write an equation of variation.
    b. A record that lasted $2\frac{1}{2}$ minutes made 112.5 revolutions. What is k, the constant of variation?
    c. If a record makes 108 revolutions, how long does it take to play the entire record?

23. *Braking distance* The distance d an automobile travels after the brakes have been applied varies directly as the square of its speed s.

   a. Write an equation of variation.
   b. If the stopping distance for a car going 30 miles per hour is 54 feet, what is k, the constant of variation?
   c. What is the stopping distance for a car going 60 miles per hour?

24. *Weight* The weight of a person varies directly as the cube of the person's height h (in inches). The **threshold weight** T (in pounds) for a person is defined as "the crucial weight, above which the mortality (risk) for the patient rises astronomically."

   a. Write an equation of variation relating T and h.
   b. If $T = 196$ when $h = 70$, find k, the constant of variation written as a fraction.
   c. To the nearest pound, what is the threshold weight T for a person 75 inches tall?

25. *Music* The number S of new songs a rock band needs to stay on top each year is inversely proportional to the number of years y the band has been in the business.

   a. Write an equation of variation.
   b. If, after 3 years in the business, the band needs 50 new songs, how many songs will it need after 5 years?

26. *Camera lens* When the distance is set at infinity, the f-number on a camera lens varies inversely as the diameter d of the aperture (opening).

   a. Write an equation of variation.
   b. If the f-number on a camera is 8 when the aperture is $\frac{1}{2}$ in., what is k, the constant of variation?
   c. Find the f-number when the aperture is $\frac{1}{4}$ in.

27. *Weight* The weight W of an object varies inversely as the square of its distance d from the center of the Earth.

   a. Write an equation of variation.
   b. An astronaut weighs 121 pounds on the surface of the Earth. If the radius of the Earth is 3960 miles, find the value of k, the constant of variation for this astronaut. (Do not multiply out your answer.)
   c. What will this astronaut weigh when she is 880 miles above the surface of the Earth?

28. *Fuel* The number of miles m you can drive in your car is directly proportional to the amount of fuel g in your gas tank.

   a. Write an equation of variation.
   b. The greatest distance yet driven without refueling on a single fill in a standard vehicle is 1691.6 miles. If the twin tanks used to do this carried a total of 38.2 gallons of fuel, what is k, the constant of variation? (Round the answer to the nearest tenth.)
   c. How many miles per gallon is this?

29. *Distance traveled* The distance d (in miles) traveled by a car is directly proportional to the average speed s (in miles per hour) of the car, even when driving in reverse.

   a. Write an equation of variation.
   b. The highest average speed attained in any nonstop reverse drive of more than 500 mi is 28.41 miles per hour. If the distance traveled was 501 mi, find k, the constant of variation. (Round the answer to the nearest hundredth.)
   c. What does k represent?

30. *Radio contest* Have you called in on a radio contest lately? According to Don Burley, a radio talk-show host in Kansas City, the listener response to a radio call-in contest is directly proportional to the size of the prize.

   a. If 40 listeners call when the prize is $100, write an equation of variation using N for the number of listeners and P for the prize in dollars.
   b. How many calls would you expect for a $5000 prize?

31. *Crickets* The number of chirps C a cricket makes each minute is directly proportional to 37 less than the temperature F in degrees Fahrenheit.

   a. If a cricket chirps 80 times when the temperature is 57°F, what is the equation of variation?
   b. How many chirps per minute would the cricket make when the temperature is 90°F?

32. *Smoking* According to George Flick, the ship's surgeon of the SS Constitution, the number of hours H your life is shortened by smoking cigarettes varies jointly as N and $t + 10$, where N is the number of cigarettes you smoke and t is the time in minutes it takes you to smoke each cigarette. If it takes 5 minutes to smoke a cigarette and smoking 100 of them shortens your life span by 25 hours, how long would smoking 2 packs a day for a year (360 days) shorten your life span? (*Note:* There are 20 cigarettes in a pack.)

33. *Atmosphere* The concentration of carbon dioxide (CO_2) in the atmosphere has been increasing due to human activities such as automobile emissions, electricity generation, and deforestation. In 1965, CO_2 concentration was 319.9 parts per million (ppm), and 23 years later, it increased to 351.3 ppm. The *increase* of carbon dioxide concentration, I, in the atmosphere is directly proportional to number of years n elapsed since 1965.

   a. Write an equation of variation for I.
   b. Find k, the constant of variation. (Round to the nearest thousandth.)
   c. What would you estimate the CO_2 concentration to have been in the year 2000?

34. *College graduates* The increase, I, in the percent of college graduates in the United States among persons 25 years and older between 1930 and 1990 is proportional to the square of the number of years after 1930. In 1940, the increase was about 5%.

   a. Write an equation of variation for I if n is the number of years elapsed since 1930.
   b. Find k, the constant of variation.
   c. What would you estimate the percent increase to have been in the year 2000?

35. *Simple interest* The simple interest I on an account varies jointly as the time t and the principal P. After one quarter (3 months), an $8000 principal earned $100 in interest. How much would a $10,000 principal earn in 5 months?

36. *Water depth* At depths of more than 1000 meters (a kilometer), water temperature T (in degrees Celsius) in the Pacific Ocean varies inversely as the water depth d (in meters). If the water temperature at 4000 meters is 1°C, what would it be at 8000 meters?

37. *Anthropology* Anthropologists use the cephalic index C in the study of human races and groupings. This index is directly proportional to the width w and inversely proportional to the length L of the head. The width of the head in a skull found in 1921 and named Rhodesian man was 15 centimeters, and its length was 21 centimeters. If the cephalic index of Rhodesian man was 98, what would the cephalic index of Cro-Magnon man be, whose head was 20 centimeters long and 15 centimeters wide?

>>> Using Your Knowledge

The Pressure of Diving The equation for direct variation between x and y ($y = kx$) and the equation of a line of slope m passing through the origin ($y = mx$) are very similar. Look at the following table, which gives the water pressure in pounds per square inch exerted on a diver.

Depth of diver (ft)	10	25	40	55
Pressure on diver (lb/in.²)	4.2	10.5	16.8	23.1

38. Graph these points using x as the depth and y as the pressure.

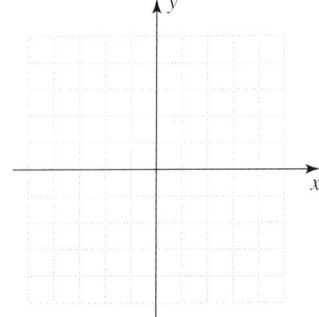

39. What is the slope of the resulting line?

40. As it turns out, the pressure p on a diver is directly proportional to the depth d. Write an equation of variation.

41. Use one of the points in the table to find k, the constant of variation.

42. What is the relationship between k, the constant of variation, and the slope found in Exercise 39?

43. Predict the pressure on a diver at a depth of 125 feet.

>>> Write On

44. Explain the difference between direct variation and inverse variation.

45. Find two different ways of expressing the idea that "y is directly proportional to x."

46. Find two different ways of expressing the idea that "y is inversely proportional to x."

47. Find two different ways of expressing the idea that "y and z are directly proportional to x."

>>> Concept Checker

Fill in the blank(s) with the correct word(s), phrase, or mathematical statement.

48. In the variation formula $C = kwm$, k is _____.

49. The variation formula $T = kh^3$ is an example of _____.

50. The variation formula $I = \frac{ki}{d^2}$ is an example of _____.

direct variation

inverse variation

joint variation

the constant of variation

>>> Mastery Test

51. The wind force F on a vertical surface varies jointly with the area A of the surface and the square of the wind velocity V. Suppose the wind force on 1 square foot of surface is 1.8 pounds when $V = 20$ miles per hour.
   a. Find an equation of variation.
   b. Find k, the constant of variation.
   c. Find the force on a 2-square-foot vertical surface when $V = 60$ miles per hour.

52. The f-number on a camera varies inversely as the diameter a of the aperture when the distance is set at infinity.
   a. Write an equation of variation.
   b. Find k, the constant of variation, when the f-number is 8 and $a = \frac{1}{2}$.
   c. Find a if the f-number is 16.

53. The principal P invested is inversely proportional to the annual rate of interest r.
   a. Write an equation of variation.
   b. Find k, the constant of variation, when $r = 10\%$ and $P = \$100$.

54. Hair length L is proportional to time t.
   a. Write an equation of variation.
   b. Find k, the constant of variation, if your hair grew 0.6 in. in 2 mo.

>>> Skill Checker

Simplify:

55. $\dfrac{x^{-5}}{x^3}$

56. $\dfrac{x^5}{x^{-3}}$

57. $(x^{-4})^{-5}$

58. $(x^{-4})^5$

59. $(-2xy^2)^3$

60. $(-2x^2y)^{-3}$

61. $x^{-9} \cdot x^7$

62. $x^9 \cdot x^{-11}$

63. $\left(\dfrac{a^{-4}}{b^3}\right)^2$

64. $\left(\dfrac{a^4}{b^{-3}}\right)^{-2}$

Perform the indicated operation:

65. $\dfrac{-6}{5} + \dfrac{3}{2}$

66. $\dfrac{-2}{3} + \left(\dfrac{-1}{4}\right)$

67. $\dfrac{2}{3} \cdot \left(\dfrac{-5}{8}\right)$

68. $\dfrac{-7}{4} \cdot (-8)$

> Collaborative Learning 6A

A bar graph is a chart that uses either horizontal or vertical bars to show comparisons among categories. One axis shows the categories being compared and the other axis represents the values. Some bar graphs show the bars being divided into subparts so that cumulative effects can be shown and are called **stacked bar graphs** as shown in Figure 6.2.

When making a stacked bar graph, examine the data to find:

1. the bar with the largest value that will help determine the range and the increments on the vertical axis.
2. how many bars will be needed for the horizontal axis.
3. whether the bars will be arranged from largest to smallest or chronologically.
4. the sequence of order to be used on the stacking of the subparts to each bar.

Make a stacked bar graph using the following information.

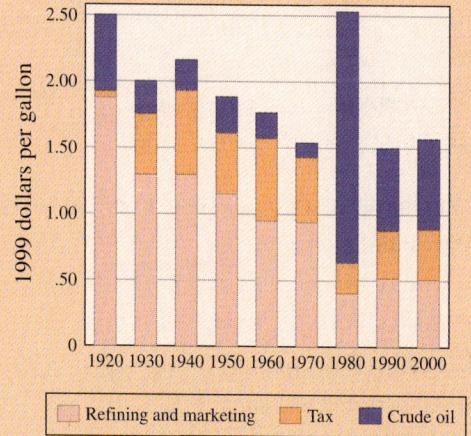

Components of U.S. Retail Gasoline Prices

> Figure 6.2

Source: Data from Cambridge Energy Research Associates.

<CONTINUED>

The University of Guelph in Ontario, Canada has published some statistics indicating the amount of tons of various recyclable waste materials during the years of 1999 to 2002. They are listed in the following table.

Waste Material	1999 Percentage of Total Tons	Tons	2000 Percentage of Total Tons	Tons	2001 Percentage of Total Tons	Tons	2002 Percentage of Total Tons	Tons
Landfill	66		57		58		72	
Mixed	12		12.5		11.5		0.5	
Manure	21		30		30		27.25	
Fine paper	0.5		0.25		0.2		0.15	
Misc.	0.5		0.25		0.3		0.1	
Total		3346		3935		3891		3608

1. Complete the table by finding the number of tons recycled for each type of waste material for each year.
2. Which year will have the longest bar in the graph, and what will be the increments used for the vertical axis?
3. How will the horizontal axis be arranged?
4. What order will be used to "stack" the sequence of different types of waste materials on each bar?
5. Make a stacked bar graph for these statistics using a poster board.
6. Discuss the trend in the waste recycling data at the University of Guelph over the 4-year period. What are some possible reasons for the changes you observed in the recycling of the waste?
7. Divide into groups, research waste recycling statistics in your area, and present the information to the class using a stacked bar graph.

› Collaborative Learning 6B

The Family Catering Company advertises they can cater events for 30 to 50 people. Recently they were asked to bid on a dinner for 45 people. Some of the company data for previous events is indicated in the following table. You will use it to help figure out what price to quote. Divide into groups and answer the questions.

Banquet Size	Cost of Food	Rate Per Person	Cost of Beverage	Rate Per Person	Cost of Labor	Rate Per Person	Total Cost	Rate Per Person	Amount Charged	Rate Per Person
30	$189		$33		$159		$381		$660	
35	$210		$35		$164.50		$409.50		$728	
40	$228		$36		$170		$434		$784	

1. Complete the table by computing the following rates:
   a. Cost of food per person
   b. Cost of beverage per person
   c. Cost of labor per person
   d. Total cost per person
   e. Amount charged per person
2. If the rates remain the same, what is the expected cost of the dinner for 45 people?
3. How much should the company bid for the dinner?
4. Sometimes the customer wants a quote that indicates the cost per person. What cost per person would you quote?
5. Discuss other applications for rates, research some data, and then present the results in a similar table.

> **Research Questions**

1. As we mentioned in *The Human Side of Algebra,* Tartaglia discovered new methods for solving cubic equations. But there is more to the story. Unfortunately, "an unprincipled genius who taught mathematics and practiced medicine in Milan, upon giving a solemn pledge of secrecy, wheedled the key to the cubic from Tartaglia." What was the name of this dastardly man, in what publication did Tartaglia's work appear, and what was the subject of the publication?

2. This "dastardly man" had a pupil who argued that his teacher received his information from del Ferro through a third party; the pupil then accused Tartaglia of plagiarism. What was the name of the pupil? Write a short paragraph about this pupil's contributions to mathematics.

3. Quadratic, cubic, and quartic equations were solved by formulas formed from the coefficients of the equation by using the four fundamental operations and taking radicals (which you shall study in the next chapter) of various sorts. Write a short paper detailing the struggles to solve quintic equations by these methods, which have culminated in the proof of the impossibility of obtaining such solutions by algebraic methods.

4. In the book *La Géométrie,* the author obtains the factor theorem as a major result. Who is the author of *La Géométrie,* and according to historians, what are the mathematical implications of this theorem?

> **Summary Chapter 6**

Section	Item	Meaning	Example
6.1	Fraction Rational expression	An expression denoting a division An expression of the form $\frac{P}{Q}$, where P and Q are polynomials, $Q \neq 0$	$\frac{3}{4}, \frac{-8}{7}$, and $\frac{1}{2}$ are fractions. $\frac{1}{x}, \frac{x}{x+y}, \frac{x+y}{z}$, and $\frac{x^2+21x-1}{x^3+3}$ are rational expressions.
6.1B	Fundamental property of rational expressions	If P, Q, and K are polynomials, $\frac{P}{Q} = \frac{P \cdot K}{Q \cdot K}$ for all values for which the denominators are not zero.	$\frac{3}{4} = \frac{3 \cdot 8}{4 \cdot 8}, \frac{x}{2} = \frac{x \cdot 5}{2 \cdot 5}$, and $\frac{3}{x+2} = \frac{3(x+5)}{(x+2)(x+5)}$
6.1C	Standard forms of a fraction	The forms $\frac{-a}{b}$ and $\frac{a}{b}$ are the standard forms of a fraction where a and b are positive.	$-\frac{5}{4}$ is written as $\frac{-5}{4}$, $-\frac{-8}{-x}$ is written as $\frac{-8}{x}$, and $\frac{5}{-4}$ is written as $\frac{-5}{4}$.
6.1D	Simplified (reduced) fraction	A fraction is simplified (reduced) if the numerator and denominator have no common factor.	$\frac{3}{4}, \frac{9}{8}$, and $\frac{x}{7}$ are reduced, but $\frac{3}{6}$ and $\frac{x+y}{x^2-y^2}$ are not.
6.2A	Multiplication of rational expressions	If P, Q, S, and T are polynomials ($Q \neq 0$, $T \neq 0$) $\frac{P}{Q} \cdot \frac{S}{T} = \frac{P \cdot S}{Q \cdot T}$	$\frac{3}{4} \cdot \frac{5}{7} = \frac{3 \cdot 5}{4 \cdot 7} = \frac{15}{28}$ $\frac{x}{x+y} \cdot \frac{3}{x-y} = \frac{3x}{(x+y)(x-y)}$ $= \frac{3x}{x^2-y^2}$

(continued)

Section	Item	Meaning	Example
6.2B	Division of rational expressions	If P, Q, S, and T are polynomials ($Q \neq 0$, $S \neq 0$, $T \neq 0$), $$\frac{P}{Q} \div \frac{S}{T} = \frac{P}{Q} \cdot \frac{T}{S}$$	$\frac{3}{4} \div \frac{7}{5} = \frac{3}{4} \cdot \frac{5}{7}$ $\frac{x}{x+y} \div \frac{x-y}{3} = \frac{x}{x+y} \cdot \frac{3}{x-y}$
6.3A	Addition and subtraction of rational expressions with the same denominator	If P, Q, and T are polynomials and $T \neq 0$, $$\frac{P}{T} + \frac{Q}{T} = \frac{P+Q}{T} \text{ and } \frac{P}{T} - \frac{Q}{T} = \frac{P-Q}{T}$$	$\frac{3}{5} + \frac{1}{5} = \frac{4}{5}$ and $\frac{2}{x} + \frac{1}{x} = \frac{3}{x}$ $\frac{3}{5} - \frac{1}{5} = \frac{2}{5}$ and $\frac{2}{x} - \frac{1}{x} = \frac{1}{x}$
6.3B	Addition and subtraction of rational expressions with different denominators	1. Find the LCD of the denominators. 2. "Build up" each rational expression to an equivalent one with the LCD as the denominator. 3. Combine like terms in the numerators; keep the denominator. 4. Simplify, if possible.	$\frac{3}{4} + \frac{1}{7} = \frac{21+4}{28} = \frac{25}{28}$ $\frac{4}{5} - \frac{1}{3} = \frac{12-5}{15} = \frac{7}{15}$
6.4A	Complex fraction	A fraction whose numerator or denominator (or both) contains other fractions	$\frac{\frac{1}{2}}{x+1}$, $\frac{3}{\frac{1}{5}+x}$, and $\frac{\frac{x}{2}}{x+\frac{1}{2}}$ are complex fractions.
	Simple fraction	A fraction that is not complex	$\frac{1}{2}$, $\frac{2x}{4}$, and $\frac{x+y}{x-y}$ are simple fractions.
6.5E	Remainder theorem	If the polynomial $P(x)$ is divided by $x-k$, then the remainder is $P(k)$.	If $P(x) = x^2 + 2x + 5$ is divided by $x - 3$, the remainder is $P(3) = 3^2 + 2(3) + 5 = 20$.
	Factor theorem	When $P(x)$ has a factor $(x-k)$, $P(k) = 0$.	$P(x) = x^2 + x - 6$ has $x - 2$ as a factor, thus, $P(2) = 0$.
6.6A	Extraneous solution	A trial solution that does not satisfy the equation	3 is an extraneous solution of $3 + \frac{1}{x-3} = \frac{1}{x-3}$.
6.6B	Proportion	Two equal ratios; cross products are equal	$\frac{6}{8} = \frac{9}{12}$ is a proportion and $6 \cdot 12 = 8 \cdot 9$.
6.7	**RSTUV** method for solving word problems	**R**ead the problem. **S**elect a variable for the unknown. **T**hink of a plan. **U**se algebra to solve. **V**erify the answer.	
6.8A	Direct variation	y varies directly as x if there is a constant k, such that $y = k \cdot x$.	"b varies directly as c" means $b = k \cdot c$.
6.8B	Inverse variation	y varies inversely as x if there is a constant k, such that $y = \frac{k}{x}$.	"y varies inversely as d" means $y = \frac{k}{d}$.
6.8C	Joint variation	z varies jointly with x and y if there is a constant k such that $z = kxy$.	"m varies jointly as p and q" means $m = kpq$.

Review Exercises Chapter 6

(If you need help with these exercises, look in the section indicated in brackets.)

1. ⟨**6.1A**⟩ *For what value(s) are the following rational expressions undefined?*

   a. $\dfrac{-3x}{x-4}$

   b. $\dfrac{x^2+9}{x^2+10x+9}$

   c. $\dfrac{8+x}{x^2+25}$

2. ⟨**6.1B**⟩ *Write the fraction with the indicated denominator.*

   a. $\dfrac{2x^2}{9y^4}$; denominator $36y^7$

   b. $\dfrac{2x^2}{9y^4}$; denominator $45y^8$

   c. $\dfrac{2x+1}{x+1}$; denominator x^2+6x+5

   d. $\dfrac{2x+1}{x+1}$; denominator x^2+7x+6

3. ⟨**6.1C**⟩ *Write in standard form.*

   a. $-\dfrac{-6}{y}$

   b. $\dfrac{-7}{-y}$

   c. $-\dfrac{-8}{-y}$

4. ⟨**6.1C**⟩ *Write in standard form.*

   a. $-\dfrac{x-y}{6}$

   b. $-\dfrac{7}{x-y}$

   c. $-\dfrac{x-y}{8}$

5. ⟨**6.1D**⟩ *Reduce to lowest terms.*

   a. $\dfrac{x^4 y^7}{xy^2}$

   b. $\dfrac{x^4 y^8}{xy^2}$

   c. $\dfrac{x^4 y^9}{xy^2}$

6. ⟨**6.1D**⟩ *Simplify.*

   a. $\dfrac{xy^2 + y^3}{x^2 - y^2}$

   b. $\dfrac{xy^3 + y^4}{x^2 - y^2}$

   c. $\dfrac{xy^4 + y^5}{x^2 - y^2}$

7. ⟨**6.1D**⟩ *Simplify.*

   a. $\dfrac{4y^2 - x^2}{x^3 + 8y^3}$

   b. $\dfrac{4y^2 - x^2}{x^3 - 8y^3}$

   c. $\dfrac{9y^2 - x^2}{x^3 - 27y^3}$

8. ⟨**6.2A**⟩ *Multiply.*

   a. $\dfrac{x-2}{3x-2y} \cdot \dfrac{9x^2 - 4y^2}{3x^2 - 5x - 2}$

   b. $\dfrac{x-2}{3x-2y} \cdot \dfrac{9x^2 - 4y^2}{4x^2 - 11x + 6}$

   c. $\dfrac{x-2}{3x-2y} \cdot \dfrac{9x^2 - 4y^2}{5x^2 - 8x - 4}$

9. ⟨**6.2B**⟩ *Divide.*

   a. $\dfrac{x+4}{x-2} \div (x^2 + 8x + 16)$

   b. $\dfrac{x+5}{x-2} \div (x^2 + 10x + 25)$

   c. $\dfrac{x+6}{x-2} \div (x^2 + 12x + 36)$

10. ⟨**6.2C**⟩ *Perform the indicated operations.*

    a. $\dfrac{2-x}{x+3} \div \dfrac{x^3 - 8}{x+6} \cdot \dfrac{x^3 + 27}{x+6}$

    b. $\dfrac{4-x}{x+3} \div \dfrac{x^3 - 64}{x+7} \cdot \dfrac{x^3 + 27}{x+7}$

    c. $\dfrac{2-x}{x+5} \div \dfrac{x^3 - 8}{x+8} \cdot \dfrac{x^3 + 125}{x+8}$

11. ⟨**6.3A**⟩ *Perform the indicated operations.*

    a. $\dfrac{x}{x^2 - 4} + \dfrac{2}{x^2 - 4}$

    b. $\dfrac{x}{x^2 - 9} + \dfrac{3}{x^2 - 9}$

    c. $\dfrac{x}{x^2 - 16} + \dfrac{4}{x^2 - 16}$

12. ⟨**6.3A**⟩ *Perform the indicated operations.*

    a. $\dfrac{x}{x^2 - 9} - \dfrac{3}{x^2 - 9}$

    b. $\dfrac{x}{x^2 - 16} - \dfrac{4}{x^2 - 16}$

    c. $\dfrac{x}{x^2 - 25} - \dfrac{5}{x^2 - 25}$

13. ⟨**6.3B**⟩ *Perform the indicated operations.*

  a. $\dfrac{x+1}{x^2+x-2} + \dfrac{x+5}{x^2-1}$

  b. $\dfrac{x+1}{x^2+x-2} + \dfrac{x+6}{x^2-1}$

  c. $\dfrac{x+1}{x^2+x-2} + \dfrac{x+7}{x^2-1}$

14. ⟨**6.3B**⟩ *Perform the indicated operations.*

  a. $\dfrac{x-4}{x^2-x-6} - \dfrac{x+1}{x^2-9}$

  b. $\dfrac{x-3}{x^2-x-6} - \dfrac{x+1}{x^2-9}$

  c. $\dfrac{x-1}{x^2-x-6} - \dfrac{x+1}{x^2-9}$

15. ⟨**6.4A**⟩ *Simplify.*

  a. $\dfrac{\dfrac{1}{x}+\dfrac{1}{x^4}}{\dfrac{1}{x}-\dfrac{1}{x^5}}$

  b. $\dfrac{\dfrac{1}{x^2}+\dfrac{1}{x^5}}{\dfrac{1}{x^2}-\dfrac{1}{x^6}}$

  c. $\dfrac{\dfrac{1}{x^3}+\dfrac{1}{x^6}}{\dfrac{1}{x^3}-\dfrac{1}{x^7}}$

16. ⟨**6.4A**⟩ *Simplify.*

  a. $4 + \dfrac{a}{4 + \dfrac{4}{4+a}}$

  b. $5 + \dfrac{a}{5 + \dfrac{5}{5+a}}$

  c. $6 + \dfrac{a}{6 + \dfrac{6}{6+a}}$

17. ⟨**6.5A**⟩ *Divide.*

  a. $\dfrac{18x^5 - 12x^3 + 6x^2}{6x^2}$

  b. $\dfrac{18x^5 - 12x^3 + 6x^2}{6x^3}$

  c. $\dfrac{18x^5 - 12x^3 + 6x^2}{6x^4}$

18. ⟨**6.5B**⟩ *Divide.*

  a. $2x^3 - 8 - 4x$ by $2 + 2x$

  b. $2x^3 - 9 - 4x$ by $2 + 2x$

  c. $2x^3 - 10 - 4x$ by $2 + 2x$

19. ⟨**6.5C**⟩ *Factor* $x^3 - 6x^2 + 11x - 6$ *if*

  a. $x - 1$ is one of its factors

  b. $x - 2$ is one of its factors

  c. $x - 3$ is one of its factors

20. ⟨**6.5D**⟩ *Use synthetic division to divide*

  $x^4 + 10x^3 + 35x^2 + 50x + 28$ by

  a. $x + 1$

  b. $x + 2$

  c. $x + 3$

21. ⟨**6.5E**⟩ *If* $P(x) = x^4 + 10x^3 + 35x^2 + 50x + 24$, *use synthetic division to show that*

  a. -1 is a solution of $P(x) = 0$

  b. -2 is a solution of $P(x) = 0$

  c. -3 is a solution of $P(x) = 0$

22. ⟨**6.6A**⟩ *Solve.*

  a. $\dfrac{x}{x+4} - \dfrac{x}{x-4} = \dfrac{x^2+16}{x^2-16}$

  b. $1 + \dfrac{4}{x-5} = \dfrac{40}{x^2-25}$

  c. $1 + \dfrac{5}{x-6} = \dfrac{60}{x^2-36}$

23. ⟨**6.6C**⟩ *Solve using a proportion.*

  a. The scale $\tfrac{1}{4}$ inch = 2 feet is used to draw a scaled-down drawing of a backyard. If the measurements on the drawing are 4 inches by 6 inches, what are the dimensions of the backyard?

  b. The average NBA player makes 75% of his free throws. At that rate, if an NBA player takes 10 free throws during a game, how many should he be expected to make?

24. ⟨**6.7A**⟩ *Find two consecutive even integers such that the sum of their reciprocals is:*

  a. $\dfrac{11}{60}$

  b. $\dfrac{13}{84}$

  c. $\dfrac{15}{112}$

25. **‹6.7B›** Jack can paint a room in 4 hr. Find how long it would take to paint the room if he is helped by Jill, who can paint the same room in:

   a. 5 hr
   b. 6 hr
   c. 7 hr

26. **‹6.7C›** A plane traveled 1200 miles with a 25-mile-per-hour tail wind. What is the plane's speed in still air if it took the plane the same time to travel the given mileage against the wind?

   a. 960 mi
   b. 1000 mi
   c. 1040 mi

27. **‹6.7D›** If $A = \dfrac{a + 2b + 3c}{2}$,

   a. Solve for a.
   b. Solve for b.
   c. Solve for c.

28. **‹6.8A›** The gas in a closed container exerts a pressure P on the walls of the container. This pressure varies directly as the temperature T of the gas.

   a. Write an equation of variation using k for the constant.
   b. If the pressure is 3 pounds per square inch when the temperature is 360°F, find k.

29. **‹6.8B›** If the temperature of a gas is held constant, the pressure P varies inversely as the volume V.

   a. Write an equation of variation using k for the constant.
   b. A pressure of 1600 pounds per square inch is exerted by 2 cubic feet of air in a cylinder fitted with a piston. Find k.

30. **‹6.8C›** If g varies jointly as x and the square of t, write an equation of variation using k for the constant.

Practice Test Chapter 6

(Answers on page 521)

Visit www.mhhe.com/bello to view helpful videos that provide step-by-step solutions to several of the problems below.

1. a. For what value(s) is $\frac{x-2}{3x+4}$ undefined?
    b. Write the fraction with the indicated denominator.
       $$\frac{2x^2}{9y^4}; \text{ denominator } 36y^7$$

2. Write in standard form.
    a. $-\frac{-5}{y}$
    b. $-\frac{x-y}{5}$

3. Reduce to lowest terms.
    a. $\frac{x^4 y^6}{xy^2}$
    b. $\frac{xy + y^2}{x^2 - y^2}$

4. Reduce $\frac{y^2 - x^2}{x^3 - y^3}$ to lowest terms.

5. Multiply: $\frac{x-2}{3x-2y} \cdot \frac{9x^2 - 4y^2}{2x^2 - x - 6}$

6. Divide: $\frac{x+3}{x-2}$ by $(x^2 + 6x + 9)$

7. Perform the indicated operations.
   $$\frac{2-x}{x+3} \div \frac{x^3 - 8}{x+5} \cdot \frac{x^3 + 27}{x+5}$$

8. Perform the indicated operations.
    a. $\frac{x}{x^2 - 1} + \frac{1}{x^2 - 1}$
    b. $\frac{x}{x^2 - 4} - \frac{2}{x^2 - 4}$

9. Perform the indicated operations.
    a. $\frac{5}{6x^2} + \frac{7}{12x^4}$
    b. $\frac{x-5}{x^2 - x - 6} - \frac{x+1}{x^2 - 9}$

10. Simplify: $\dfrac{x + \frac{1}{x^2}}{x - \frac{1}{x^3}}$

11. Simplify: $\dfrac{\frac{x}{x+3} + x}{1 - \frac{7}{x^2 - 9}}$

12. Divide: $\frac{28x^5 - 14x^3 + 7x^2}{7x^3}$

13. Divide: $2x^3 - 6 - 4x$ by $2 + 2x$

14. Factor $2x^3 + 3x^2 - 23x - 12$ if $x - 3$ is one of its factors.

15. Use synthetic division to divide
    $x^4 - 4x^3 - 7x^2 + 22x + 25$ by $x + 1$.

16. Use synthetic division to show that -1 is a solution of
    $x^4 - 4x^3 - 7x^2 + 22x + 24 = 0$.

17. Solve: $\frac{x}{x+3} - \frac{x}{x-3} = \frac{x^2 + 9}{x^2 - 9}$

18. Solve: $18x^{-2} - 3x^{-1} = 1$

19. A recipe for curried shrimp that normally serves four was once served to 200 guests at a wedding reception. One of the ingredients in the recipe is $1\frac{1}{2}$ cups of chicken broth.
    a. How much chicken broth was required to make the recipe for 200 people?
    b. If a medium-sized can of chicken broth contains 2 cups of broth, how many cans are necessary?

20. Find two consecutive even integers such that the sum of their reciprocals is $\frac{7}{24}$.

21. Jack can mow the lawn in 4 hours and Jill can mow it in 3 hours. How long would it take them to mow the lawn if they work together?

22. A plane traveled 990 miles with a 30-mile-per-hour tail wind in the same time it took to travel 810 miles against the wind. What is the plane's speed in still air?

23. The area A of a trapezoid is given by
    $$A = \frac{(B + b)h}{2}$$
    where B is the length of one base, b is the length of the other base, and h is the height of the trapezoid. Solve for h.

24. C is directly proportional to m.
    a. Write an equation of variation with k as the constant.
    b. Find k when $C = 12$ and $m = \frac{1}{3}$.

25. The force F with which the Earth attracts an object above the Earth's surface varies inversely as the square of the distance d from the center of the Earth.
    a. Write an equation of variation with k as the constant.
    b. A meteorite weighs 40 pounds on the Earth's surface. If the radius of the earth is about 4000 miles, find k.

Answers to Practice Test Chapter 6

Answer		If You Missed Question	Review Section	Examples	Page
1. a. $\frac{-4}{3}$	b. $\frac{8x^2y^3}{36y^7}$	1	6.1A, B	1, 2	430, 431–432
2. a. $\frac{5}{y}$	b. $\frac{y-x}{5}$	2	6.1C	3	433
3. a. x^3y^4	b. $\frac{y}{x-y}$	3	6.1D	4	435
4. $\frac{-(x+y)}{x^2+xy+y^2}$		4	6.1D	5	436
5. $\frac{3x+2y}{2x+3}$		5	6.2A	1, 2	441, 442
6. $\frac{1}{(x-2)(x+3)}$ or $\frac{1}{x^2+x-6}$		6	6.2B	3	443
7. $\frac{-(x^2-3x+9)}{x^2+2x+4}$		7	6.2C	4, 5	444
8. a. $\frac{1}{x-1}$	b. $\frac{1}{x+2}$	8	6.3A	1	450–451
9. a. $\frac{10x^2+7}{12x^4}$	b. $\frac{-5x-17}{(x+2)(x+3)(x-3)}$	9	6.3B	2, 3	452–455
10. $\frac{x(x^2-x+1)}{(x^2+1)(x-1)}$		10	6.4A	1, 2	462
11. $\frac{x(x-3)}{x-4}$ or $\frac{x^2-3x}{x-4}$		11	6.4A	3	463–464
12. $4x^2 - 2 + \frac{1}{x}$		12	6.5A	1	471
13. $(x^2 - x - 1)$ R -4		13	6.5B	2, 3	472, 473
14. $(x-3)(x+4)(2x+1)$		14	6.5C	4	474
15. $(x^3 - 5x^2 - 2x + 24)$ R 1		15	6.5D	5	476
16. When the division is done, the remainder is zero.		16	6.5E	6	477
17. No solution		17	6.6A	3, 4	486–487
18. $-6, 3$		18	6.6A	5	487–488
19. a. 75 cups	b. 38 cans	19	6.6B	7	490
20. 6 and 8		20	6.7A	1	496
21. $1\frac{5}{7}$ hr		21	6.7B	2	497
22. 300 mi/hr		22	6.7C	4	498–499
23. $h = \frac{2A}{B+b}$		23	6.7D	5	500
24. a. $C = km$	b. 36	24	6.8A	1	506
25. a. $F = \frac{k}{d^2}$	b. 640,000,000	25	6.8B	2, 3	507

Cumulative Review Chapters 1–6

1. Subtract: $-8 - 6$

2. Which property is illustrated by the following statement? $5 \cdot (8 \cdot 7) = (5 \cdot 8) \cdot 7$

3. Simplify:
$[(5x^2 - 1) + (x + 2)] - [(x - 6) + (8x^2 - 8)]$

4. Solve: $\dfrac{x+6}{5} - \dfrac{x-6}{7} = 2$

5. Graph: $x \geq -4$

6. Graph: $x + 6 \leq 7$ and $-4x < 16$

7. If $H = 2.75h + 73.49$, find h when $H = 136.74$.

8. The sum of three consecutive odd integers is 93. What are the three integers?

9. Find the slope of the line passing through the points $A(4, 2)$ and $B(-2, 4)$.

10. Find an equation of the line with slope -5 and passing through the point $(1, 1)$.

11. Find the slope and the y-intercept of the line $6x - 3y = 36$.

12. Graph: $2x - 5y \leq -10$

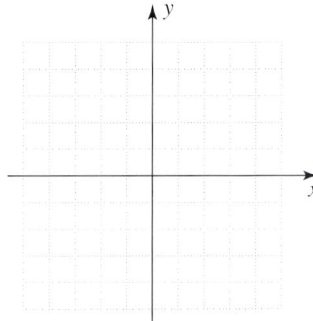

13. Graph: $|x + 4| > 2$

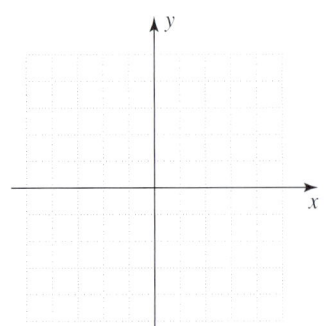

14. Use the graphical method to solve the system:
$y + x = -1$
$2y = -2x - 4$

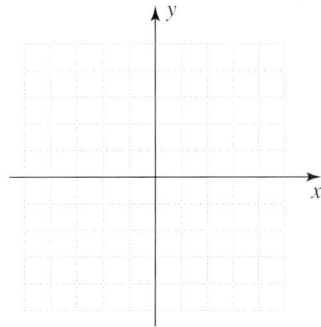

15. Use the substitution method to solve the system:
$x - 4y = 10$
$2x = 8y + 18$

16. Solve by the elimination method:
$3x + 4y = 18$
$x + y = 5$

17. The sum of two numbers is 52. If their difference is 18, find the numbers.

18. Solve:
$$x + 2y + z = 5$$
$$x + y - z = 6$$
$$3x + 5y + z = 7$$

19. The total height of a building and the flagpole on its roof is 210 feet. The building is 9 times as high as the flagpole. How high is the flagpole?

20. Find the value(s) for which the function $f(x) = \frac{5x}{2x + 3}$ is undefined.

21. Find the domain and range of $\{(x, y) \mid y = 4 + x\}$.

22. Add $2x^3 - 7x^2 - 2x - 3$ and $3 - 7x + 6x^2 - 7x^3$.

23. Multiply: $(2n - 1)(3n - 5)$

24. Factor: $12t^2 - 7t - 10$

25. Solve for x: $x^3 + 3x^2 - 16x - 48 = 0$

26. Perform the indicated operations:
$$\frac{1 - x}{x + 3} \div \frac{x^3 - 1}{x + 2} \cdot \frac{x^3 + 27}{x + 2}$$

27. Perform the indicated operations:
$$\frac{x - 1}{x^2 - 5x + 6} + \frac{x + 3}{x^2 - 4}$$

28. Solve for x: $1 + \frac{3}{x + 2} = \frac{-12}{x^2 - 4}$

29. Sandra can paint a kitchen in 3 hr and Roger can paint the same kitchen in 5 hr. How long would it take to paint the kitchen if they both worked together?

30. An enclosed gas exerts a pressure P on the walls of its container. This pressure is directly proportional to the temperature T of the gas. If the pressure is 8 lb/in.² when the temperature is 480°F, find k.

Chapter 7: Rational Exponents and Radicals

Section
- 7.1 Rational Exponents and Radicals
- 7.2 Simplifying Radicals
- 7.3 Operations with Radicals
- 7.4 Solving Equations Containing Radicals
- 7.5 Complex Numbers

The Human Side of Algebra

Imagine a universe in which only whole numbers and their ratios, some classified as "perfect" and "amicable," existed. Even numbers were "feminine" and odd numbers "masculine," while 1 was the generator of all other numbers. This was the universe of the Pythagoreans, an ancient Greek secret society (ca. 540–500 B.C.). And then, disaster. In the midst of these charming fantasies, a new type of number was discovered, a type so unexpected that the brotherhood tried to suppress its discovery—the set of irrational numbers!

Pythagoras was born between 580 and 569 B.C., but accounts of his eventual demise differ. One claims that he died in about 501 B.C. when popular revolt erupted and the meeting house of the Pythagoreans was set afire, which resulted in his death. A more dramatic ending claims that his disciples made a bridge over the fire with their bodies enabling the master to escape to Metapontum. In the ensuing fight, Pythagoras was caught between freedom and a field of sacred beans. Rather than trampling the plants, he valiantly chose to die at the hands of his enemies.

7.1 Rational Exponents and Radicals

Objectives

A Find the *n*th root of a number, if it exists.

B Evaluate expressions containing rational exponents.

C Simplify expressions involving rational exponents.

To Succeed, Review How To . . .

1. Use the properties of exponents (pp. 38–45).
2. Perform operations involving signed numbers (pp. 18–27).

Getting Started
Speeding Along to Exponents

Have you noticed that the speed limit on a curve is lower than on a straight road? The velocity v (in mi/hr) that a car can travel on a curved concrete highway of radius r feet without skidding is $v = \sqrt{9r}$. If the radius r of the curve is 100 feet, the velocity is $\sqrt{900}$ (read "the square root of 900"). A *square root* of 900 is a number whose square is 900. Since $30^2 = 900$, one square root of 900 is 30. Similarly, one square root of 25 is 5 (since $5^2 = 25$), and one square root of 36 is 6 (since $6^2 = 36$).

A Radicals

There are many numbers that are not square roots of perfect squares—for example, $\sqrt{2}$, $\sqrt{3}$, $\sqrt{10}$, and $\sqrt{24}$. These real numbers are *irrational*; they cannot be written as the quotient of two integers. The first irrational number was probably discovered by the Pythagoreans, an ancient Greek society of mathematicians who believed that everything was based on the whole numbers and that harmony consisted of numerical ratios. In this section we shall study rational exponents and radical expressions. Many of the important applications of mathematics use these two concepts. (See Problems 71–74.) We start by giving the definition of an *n*th root.

*n*th ROOT	If a and x are real numbers and n is a positive integer, then x is an *n*th root of a if $x^n = a$.

For example,

1. A square (second) root of 4 is 2 because $2^2 = 4$.
2. Another square root of 4 is -2 because $(-2)^2 = 4$.
3. A cube (third) root of 27 is 3 because $3^3 = 27$.
4. A cube (third) root of -64 is -4 because $(-4)^3 = -64$.
5. A fourth root of $\frac{16}{81}$ is $\frac{2}{3}$ because $\left(\frac{2}{3}\right)^4 = \frac{16}{81}$.
6. Another fourth root of $\frac{16}{81}$ is $-\frac{2}{3}$ because $\left(-\frac{2}{3}\right)^4 = \frac{16}{81}$.

Note that 4 has two square roots, 2 and -2, and $\frac{16}{81}$ has two real fourth roots, $-\frac{2}{3}$ and $\frac{2}{3}$. Sometimes it is necessary to name both square roots of a number but more often we only want the positive square root. So next we introduce the idea of the principal *n*th root.

PRINCIPAL nth ROOT

If n is a positive integer, then $\sqrt[n]{a}$ denotes the **principal nth root** of a, and

1. If a is positive ($a > 0$), $\sqrt[n]{a}$ is the *positive* nth root of a.
2. If a is negative ($a < 0$) and n is odd, $\sqrt[n]{a}$ is the *negative* nth root of a.
3. If a is negative and n is even, there is no real nth root.
4. If $a = 0$, then $\sqrt[n]{0} = 0$.
5. If $a = 1$, then $\sqrt[n]{1} = 1$.

In this definition, $\sqrt[n]{a}$ is called a **radical expression**, $\sqrt{}$ is called the **radical sign**, a is the **radicand**, and n is the **index** that tells you what root is being considered. By convention, the index 2 for square root is understood and usually not written, $\sqrt{x}$ means $\sqrt[2]{x}$. Thus, $\sqrt{9}$ means the principal square root of 9, and $\sqrt{25}$ means the principal square root of 25.

Now let's look at part 1 of the definition. It tells you that whenever the number under the radical sign is positive, the resulting root is also positive; "If $a > 0$, $\sqrt[n]{a} > 0$." Thus,

$$\sqrt{64} = 8 \quad \text{because} \quad 8^2 = 64$$
$$\sqrt[3]{8} = 2 \quad \text{because} \quad 2^3 = 8$$
$$\sqrt[4]{81} = 3 \quad \text{because} \quad 3^4 = 81$$

NOTE

A common mistake is to assume that $\sqrt{64}$ has two values. This is not correct. By our definition of principal root, $\sqrt{64} = 8$. If we wish to refer to the negative nth root, we write $-\sqrt[n]{a}$. Thus, $-\sqrt{64} = -8$, $-\sqrt{16} = -4$, and $-\sqrt{25} = -5$.

Part 2 of the definition tells you that if a is *negative* and n is *odd*, $\sqrt[n]{a}$ is negative, that is, "If $a < 0$ and n is odd, $\sqrt[n]{a} < 0$." Thus,

odd → $\sqrt[3]{\boxed{-27}} = -3$ because $(-3)^3 = -27$ Negative answer
odd → $\sqrt[5]{\boxed{-32}} = -2$ because $(-2)^5 = -32$ Negative answer
(negative radicands)

To illustrate part 3 of the definition consider $\sqrt{-16}$. There is no real number whose square is -16 because the square of a nonzero number is positive. Thus, $\sqrt{-16}$ is not a real number. Similarly, $\sqrt[4]{-81}$ is not a real number because there is no real number whose fourth power is -81. In both cases, the index n is even and the radicand is negative.

even → $\sqrt{\boxed{-16}} = $ no real number because $(n)^2 \neq -16$
even → $\sqrt[4]{\boxed{-81}} = $ no real number because $(n)^4 \neq -81$
(negative radicands)

"Perfect" numbers were mentioned in the *Human Side* of this chapter. When finding the roots of real numbers, it is important to be familiar with some of the "perfect" numbers by recognition. Perfect square numbers are found by multiplying each counting number by itself, perfect cube numbers by multiplying each counting number by itself twice, and so on. Table 7.1 is a partial list of some of these perfect numbers.

Table 7.1

Perfect Squares	Perfect Cubes	Perfect Fourths
1	1	1
4	8	16
9	27	81
16	64	256
25	125	
36	216	

This table will help when evaluating roots, as we will do in Example 1.

EXAMPLE 1 Finding the roots of numbers

Evaluate, if possible, in the real numbers:

a. $\sqrt[3]{-64}$ **b.** $\sqrt{-64}$ **c.** $\sqrt[3]{-\frac{1}{8}}$

SOLUTION

a. $\sqrt[3]{-64} = -4$, because $(-4)^3 = -64$.

b. $\sqrt{-64}$ is not a real number. $\sqrt{-64} \neq -8$, because $(-8)(-8) = 64$ and not -64.

c. $\sqrt[3]{-\frac{1}{8}} = -\frac{1}{2}$, because $\left(-\frac{1}{2}\right)^3 = -\frac{1}{8}$.

PROBLEM 1

Evaluate, if possible, in the real numbers:

a. $\sqrt{-25}$

b. $\sqrt[3]{-125}$

c. $\sqrt[3]{-\frac{1}{27}}$

B > Converting Rational Exponents to Radicals

We have used radicals to define the *n*th root of a number. We can also define *n*th roots using rational exponents. For example, what do you think $a^{1/3}$ means? To find out,

let $\quad\quad\quad\quad\quad\quad x = a^{1/3}$

then $\quad\quad\quad\quad\quad x^3 = (a^{1/3})^3 \quad\quad$ Cube both sides.

$\quad\quad\quad\quad\quad\quad\quad x^3 = a^{(1/3) \cdot (3)} \quad\quad$ Assume $(a^{1/n})^n = a^{(1/n) \cdot (n)}$.

$\quad\quad\quad\quad\quad\quad\quad x^3 = a^1$

$\quad\quad\quad\quad\quad\quad\quad x^3 = a$

Since $x^3 = a$, x must be the cube root of a; $x = \sqrt[3]{a}$. But $x = a^{1/3}$, so

$$a^{1/3} = \sqrt[3]{a}$$

Similarly, if $\sqrt[4]{a}$ is defined,

$$a^{1/4} = \sqrt[4]{a}$$

In general, the following definition establishes the relationship between rational exponents and roots.

Answers to PROBLEMS

1. a. Not a real number

b. -5 **c.** $-\frac{1}{3}$

7.1 Rational Exponents and Radicals

RATIONAL EXPONENTS AND THEIR ROOTS

If n is a positive integer and $\sqrt[n]{a}$ is a real number, then

$$a^{1/n} = \sqrt[n]{a}$$

When we write

$$a^{1/n} = \sqrt[n]{a}$$
(same)

the denominator of the rational exponent is the index of the radical:

$16^{1/2} = \sqrt{16} = 4$ $\quad 4^2 = 16.$ Remember $\sqrt{16}$ means $\sqrt[2]{16}$.

$(-8)^{1/3} = \sqrt[3]{-8} = -2$ $\quad (-2)^3 = -8$

$\left(\dfrac{1}{81}\right)^{1/4} = \sqrt[4]{\dfrac{1}{81}} = \dfrac{1}{3}$ $\quad \left(\dfrac{1}{3}\right)^4 = \dfrac{1}{81}$

EXAMPLE 2 Evaluating expressions containing rational exponents
Evaluate, if possible, in the real numbers:

a. $9^{1/2}$ **b.** $(-125)^{1/3}$ **c.** $\left(\dfrac{1}{16}\right)^{1/4}$

SOLUTION

a. $9^{1/2} = \sqrt{9} = 3$ $\quad 3^2 = 9$

b. $(-125)^{1/3} = \sqrt[3]{-125} = -5$ $\quad (-5)^3 = -125$

c. $\left(\dfrac{1}{16}\right)^{1/4} = \sqrt[4]{\dfrac{1}{16}} = \dfrac{1}{2}$ $\quad \left(\dfrac{1}{2}\right)^4 = \dfrac{1}{16}$

PROBLEM 2
Evaluate, if possible, in the real numbers:

a. $49^{1/2}$

b. $(-216)^{1/3}$

c. $\left(\dfrac{1}{10{,}000}\right)^{1/4}$

🖩 Calculator Corner

Two Ways to Find Roots

You can approximate the roots in Examples 1 and 2 in different ways, but be aware that different calculators have different procedures for this. To find $\sqrt[3]{-64}$ in Example 1(a) with a TI-83 Plus, press MATH 4 (-) 64) ENTER and you get -4. You can also use the $\sqrt[x]{}$ feature to find roots. Thus, to find $\sqrt[4]{16}$, press 4 MATH 5 16 ENTER and you get 2. The first entry is the index of the root you want (square, cube, fourth, fifth, and so on).

Table 7.2 shows a summary of our work so far.

Table 7.2 Rational Exponents and Radicals

	If a is a real number as specified and n is a positive integer	
	n even	**n odd**
For a positive number a ($a > 0$)	$\sqrt[n]{a} = a^{1/n}$ is positive.	$\sqrt[n]{a} = a^{1/n}$ is positive.
For a negative number a ($a < 0$)	$\sqrt[n]{a} = a^{1/n}$ is not a real number.	$\sqrt[n]{a} = a^{1/n}$ is negative.
For $a = 0$	$\sqrt[n]{a} = a^{1/n} = 0.$	$\sqrt[n]{a} = a^{1/n} = 0.$

Answers to PROBLEMS

2. **a.** 7   **b.** -6 **c.** $\dfrac{1}{10}$

We have already defined rational exponents of the form $1/n$. How do we define $a^{m/n}$, where m and n are positive integers, $n > 1$, and $\sqrt[n]{a}$ is a real number? If we assume that the power rule of exponents applies, then

$$a^{m/n} = (a^{1/n})^m = (\sqrt[n]{a})^m \text{ and } a^{m/n} = (a^m)^{1/n} = \sqrt[n]{a^m}$$

The following table demonstrates how $4^{3/2}$ can be evaluated either way to obtain the same result.

Evaluate $4^{3/2}$:

$(a^{1/n})^m$	$(a^m)^{1/n}$
$(4^{1/2})^3$	$(4^3)^{1/2}$
$(\sqrt{4})^3$	$\sqrt{4^3}$
$(2)^3$	$\sqrt{64}$
8	8

From this we arrive at the following definition.

RADICAL EXPRESSION WITH AN m/n EXPONENT

If m and n are positive integers, $n > 1$, and $\sqrt[n]{a}$ is a real number, then

$$a^{m/n} = (\sqrt[n]{a})^m = \sqrt[n]{a^m}$$

provided m/n is in its most simplified form.

The numerator of the exponent m/n is the exponent of the radical expression, and the denominator is the index of the radical

$$a^{m/n} = (\sqrt[n]{a})^m$$
(exponent → m; index → n)

For example,

$$a^{1/5} = \sqrt[5]{a} \quad \text{and} \quad a^{2/5} = (\sqrt[5]{a})^2 \quad \text{or} \quad \sqrt[5]{a^2}$$

EXAMPLE 3 Evaluating expressions containing rational exponents

Evaluate, if possible, in the real numbers:

a. $8^{2/3}$ **b.** $(-27)^{2/3}$ **c.** $(-25)^{3/2}$

SOLUTION

a. $8^{2/3}$ can be evaluated in two ways.
  1. $8^{2/3} = (\sqrt[3]{8})^2 = (2)^2 = 4$ $\sqrt[3]{8} = 2$
  2. $8^{2/3} = \sqrt[3]{8^2} = \sqrt[3]{64} = 4$

b. We can evaluate $(-27)^{2/3}$ in a similar manner.
  1. $(-27)^{2/3} = (\sqrt[3]{-27})^2$
   $= (-3)^2$ $\sqrt[3]{-27} = -3$
   $= 9$
  2. $(-27)^{2/3} = \sqrt[3]{(-27)^2} = \sqrt[3]{729} = 9$

c. $(-25)^{3/2} = \sqrt[2]{(-25)^3} = \sqrt{-15{,}625}$, which is not a real number.

PROBLEM 3

Evaluate, if possible, in the real numbers:

a. $27^{2/3}$ **b.** $(-64)^{2/3}$
c. $(-36)^{3/2}$

Answers to PROBLEMS

3. **a.** 9 **b.** 16
   c. Not a real number

> **Calculator Corner**
>
> **Two Ways to Find $a^{m/n}$**
> To do Example 3(b) with a TI-83 Plus, enter $((-27)^2)^{\wedge}(1/3)$ or $((-27)^{\wedge}(1/3))^2$ and you get 9. You can also enter $(-27)^{\wedge}(2/3)$ and get 9. These are all shown in the window.
> Now try Example 3(c). Your screen should indicate "ERR." It will say "NONREAL ANS."
>
> ```
> ((-27)^2)^(1/3)
>                9
> ((-27)^(1/3))^2
>                9
> (-27)^(2/3)
>                9
> ```

Finally, to define negative rational exponents, if m and n are positive integers with no common factors,

$$-\frac{m}{n} = \frac{-m}{n}$$

Thus, if the power rule of exponents applies,

$$a^{-m/n} = (a^{1/n})^{-m}$$
$$= \frac{1}{(a^{1/n})^m} \quad \text{By definition, } a^{-m} = \frac{1}{a^m}$$

Let's use this to consider evaluating $9^{-3/2}$

$$9^{-3/2} = (9^{1/2})^{-3}$$
$$= \frac{1}{(9^{1/2})^3} \quad \text{By definition } b^{-3} = \frac{1}{b^3}.$$
$$= \frac{1}{(\sqrt{9})^3}$$
$$= \frac{1}{3^3}$$
$$= \frac{1}{27}$$

We make the following definition for negative rational exponents.

DEFINITION OF $a^{-m/n}$

$$a^{-m/n} = \frac{1}{a^{m/n}}$$

where m and n are positive integers, $n > 1$, and $a^{1/n}$ a real number, $a \neq 0$.

Using this definition, we have the following.

1. $a^{-1/2} = \dfrac{1}{a^{1/2}} = \dfrac{1}{\sqrt{a}}$

2. $32^{-3/5} = \dfrac{1}{32^{3/5}} = \dfrac{1}{(\sqrt[5]{32})^3} = \dfrac{1}{2^3} = \dfrac{1}{8}$

3. $1000^{-2/3} = \dfrac{1}{1000^{2/3}} = \dfrac{1}{(\sqrt[3]{1000})^2} = \dfrac{1}{10^2} = \dfrac{1}{100}$

EXAMPLE 4 Evaluating expressions containing negative rational exponents

Evaluate, if possible, in the real numbers:

a. $16^{-3/4}$ **b.** $(-8)^{-4/3}$ **c.** $125^{-2/3}$

SOLUTION

a. $16^{-3/4} = \dfrac{1}{16^{3/4}} = \dfrac{1}{(\sqrt[4]{16})^3} = \dfrac{1}{2^3} = \dfrac{1}{8}$

b. $(-8)^{-4/3} = \dfrac{1}{(-8)^{4/3}} = \dfrac{1}{(\sqrt[3]{-8})^4} = \dfrac{1}{(-2)^4} = \dfrac{1}{16}$

c. $125^{-2/3} = \dfrac{1}{125^{2/3}} = \dfrac{1}{(\sqrt[3]{125})^2} = \dfrac{1}{5^2} = \dfrac{1}{25}$

PROBLEM 4

Evaluate, if possible, in the real numbers:

a. $81^{-3/4}$ **b.** $(-27)^{-4/3}$

c. $216^{-2/3}$

C ▶ Operations with Rational Exponents

The properties of exponents that we studied in Chapter 1 can be extended to rational exponents. If this is done, we have the following results.

> **PROPERTIES OF EXPONENTS**
>
> Let r, s, and t be rational numbers. If a and b are real numbers and the indicated expressions exist, then the following are true.
>
> **I.** $a^r \cdot a^s = a^{r+s}$ **II.** $\dfrac{a^r}{a^s} = a^{r-s}$ **III.** $(a^r)^s = a^{r \cdot s}$
>
> **IV.** $(a^r b^s)^t = a^{rt} b^{st}$ **V.** $\left(\dfrac{a^r}{b^s}\right)^t = \dfrac{a^{rt}}{b^{st}}$

EXAMPLE 5 Using the properties of exponents

If x and y are positive, simplify:

a. $x^{1/3} \cdot x^{1/4}$ **b.** $\dfrac{x^{-2/3}}{x^{1/5}}$ **c.** $(y^{3/5})^{-1/6}$

SOLUTION

a. $x^{1/3} \cdot x^{1/4} = x^{1/3 + 1/4}$ Property I

$\qquad\qquad\quad = x^{4/12 + 3/12}$ The LCD is 12.

$\qquad\qquad\quad = x^{7/12}$ Add exponents.

b. $\dfrac{x^{-2/3}}{x^{1/5}} = x^{-2/3 - 1/5}$ Property II

$\qquad\quad = x^{-10/15 - 3/15}$ The LCD is 15.

$\qquad\quad = x^{-13/15}$

$\qquad\quad = \dfrac{1}{x^{13/15}}$ $a^{-n} = \dfrac{1}{a^n}$

c. $(y^{3/5})^{-1/6} = y^{(3/5) \cdot (-1/6)}$ Property III

$\qquad\qquad\; = y^{-1/10}$ $\dfrac{3}{5} \cdot \left(-\dfrac{1}{6}\right) = -\dfrac{1}{10}$

$\qquad\qquad\; = \dfrac{1}{y^{1/10}}$

PROBLEM 5

If x and y are positive, simplify:

a. $x^{1/3} \cdot x^{2/5}$ **b.** $\dfrac{x^{-3/4}}{x^{-1/6}}$

c. $(y^{-3/2})^{-1/9}$

Answers to PROBLEMS

4. a. $\dfrac{1}{27}$ **b.** $\dfrac{1}{81}$ **c.** $\dfrac{1}{36}$ **5. a.** $x^{11/15}$ **b.** $x^{-7/12} = \dfrac{1}{x^{7/12}}$ **c.** $y^{1/6}$

EXAMPLE 6 Using the properties of exponents

If x and y are positive, simplify:

a. $(x^{1/5}y^{3/4})^{-20}$ **b.** $\dfrac{x^{1/3}y^{3/4}}{x^{2/3}y^{1/4}}$ **c.** $x^{2/3}(x^{-1/3} + y^{1/5})$

SOLUTION

a.
$$\begin{aligned}(x^{1/5}y^{3/4})^{-20} &= (x^{1/5})^{-20} \cdot (y^{3/4})^{-20} &&\text{Property IV}\\ &= x^{(1/5)(-20)} \cdot y^{(3/4)(-20)} &&\text{Property III}\\ &= x^{-4} \cdot y^{-15}\\ &= \dfrac{1}{x^4 y^{15}} &&\text{Definition of negative exponent}\end{aligned}$$

b.
$$\begin{aligned}\dfrac{x^{1/3}y^{3/4}}{x^{2/3}y^{1/4}} &= x^{1/3-2/3} \cdot y^{3/4-1/4} &&\text{Property II}\\ &= x^{-1/3} \cdot y^{1/2}\\ &= \dfrac{y^{1/2}}{x^{1/3}}\end{aligned}$$

c. We first use the distributive property and then simplify.
$$\begin{aligned}x^{2/3}(x^{-1/3} + y^{1/5}) &= x^{2/3} \cdot x^{-1/3} + x^{2/3} \cdot y^{1/5}\\ &= x^{2/3+(-1/3)} + x^{2/3}y^{1/5}\\ &= x^{1/3} + x^{2/3}y^{1/5}\end{aligned}$$

PROBLEM 6

If x and y are positive, simplify:

a. $(x^{-1/4}y^{3/5})^{-20}$ **b.** $\dfrac{x^{1/4}y^{2/3}}{x^{3/4}y^{1/3}}$

c. $x^{2/5}(x^{-1/5} + y^{1/4})$

The properties of exponents for rational numbers provide a good way of simplifying some of the expressions we have studied. In Example 3, we evaluated $8^{2/3}$ and $(-27)^{2/3}$. If we write 8 as 2^3, we can write

$$\begin{aligned}8^{2/3} &= (2^3)^{2/3}\\ &= 2^{(3)\cdot(2/3)} &&\text{Property III}\\ &= 2^2\\ &= 4\end{aligned}$$

Similarly, because $-27 = (-3)^3$,

$$\begin{aligned}(-27)^{2/3} &= [(-3)^3]^{2/3} &&\text{Property III}\\ &= (-3)^2\\ &= 9\end{aligned}$$

Use these ideas when working the exercises. One word of caution, however. We cannot evaluate $[(-2)^2]^{1/2}$ using Property III. If we do, we obtain

$$\begin{aligned}[(-2)^2]^{1/2} &= (-2)^{(2)(1/2)}\\ &= (-2)^{2/2}\\ &= (-2)^1\\ &= -2\end{aligned}$$

If we use the order of operations we have studied and square -2 first, we have

$$\begin{aligned}[(-2)^2]^{1/2} &= [4]^{1/2}\\ &= 2\end{aligned}$$

The problem is that $a^{m/n}$ must be defined differently when m and n have common factors. In this case, $m = 2$ and $n = 2$, so they have a common factor. To remedy this situation, we make the following definition.

Answers to PROBLEMS

6. **a.** $\dfrac{x^5}{y^{12}}$ **b.** $\dfrac{y^{1/3}}{x^{1/2}}$

   c. $x^{1/5} + x^{2/5}y^{1/4}$

DEFINITION OF $(a^m)^{1/n}$

If m and n are positive even integers, then

$$(a^m)^{1/n} = |a|^{m/n}$$

Thus, $[(-2)^2]^{1/2} = |2|^{2/2} = |2|^1 = 2$. $[(-2)^{1/2}]^2$ is not defined since $(-2)^{1/2} = \sqrt{-2}$, not a real number. We shall discuss this further in Section 7.2.

Calculator Corner

Making Tables of Roots

Before the time of calculators and computers you had to find square and cube roots using a long and boring table. No more! You can make your own table with a TI-83 Plus. For example, to construct a table of square roots, press [2nd] [WINDOW] and set Tblstart = 1 and ΔTbl = 1. Now to tell the calculator you want a square root table, enter $Y_1 = x^{(1\div 2)}$. Remember that to enter the exponent 1/2, you press [^] [(] 1 [÷] 2 [)] or the square root feature ([2nd] [x^2]). Now, press [2nd] [GRAPH]. You see two columns. The X column shows the number whose square root you want and the Y1 column shows the corresponding square root. In the window the second line shows a 2 for X and 1.4142 for Y1, meaning that the square root of 2 is approximately 1.4142. You can make tables of cube, fourth, or nth roots by entering $Y_1 = x^{1/3}$ or $x^{1/4}$ or $x^{1/n}$.

X	Y1
1	1
2	1.4142
3	1.7321
4	2
5	2.2361
6	2.4495
7	2.6458

X=2

There is at least one more way to approximate roots with a calculator. If you want the root of a specific number such as $\sqrt[5]{243}$, go to the home screen and enter 243^(1/5), then press [ENTER]. (Don't forget the parentheses when entering the exponent.) The answer is 3.

Many of the numerical evaluations in this section can be done with the [y^x] key of a nongraphing calculator. This key will raise the number y to the x power. Thus, to find 2^3, enter 2 [y^x] 3 [=]. The answer is 8. To find $\sqrt[3]{-64}$, first recall that $\sqrt[3]{-64} = (-64)^{1/3}$. Thus, we enter

6 4 [+/-] [y^x] [(] 1 [÷] 3 [)] [=]

Note that $\frac{1}{3}$ has to be entered in parentheses.

If your instructor permits, use the [y^x] key on your calculator to find the following values.

1. $\sqrt[3]{-\dfrac{1}{8}}$ 2. $(-125)^{1/3}$ 3. $\sqrt[4]{\dfrac{1}{16}}$

Boost your grade at mathzone.com!
> Practice Problems
> NetTutor
> Self-Tests
> e-Professors
> Videos

⟩ Exercises 7.1

⟨A⟩ **Radicals** In Problems 1–12, evaluate if possible.

1. $\sqrt{4}$ 2. $\sqrt{25}$ 3. $\sqrt[3]{8}$ 4. $\sqrt[3]{125}$

5. $\sqrt[3]{-8}$ 6. $\sqrt[3]{-125}$ 7. $\sqrt[3]{\dfrac{-1}{64}}$ 8. $\sqrt[3]{\dfrac{-1}{27}}$

9. $\sqrt[4]{16}$ 10. $\sqrt[4]{625}$ 11. $\sqrt[5]{32}$ 12. $\sqrt[5]{\dfrac{-1}{243}}$

‹ B › Converting Rational Exponents to Radicals In Problems 13–40, evaluate if possible.

13. $9^{1/2}$
14. $16^{1/2}$
15. $(-4)^{1/2}$
16. $-4^{1/2}$

17. $27^{1/3}$
18. $125^{1/3}$
19. $81^{1/4}$
20. $16^{1/4}$

21. $\left(\dfrac{-1}{8}\right)^{1/3}$
22. $\left(\dfrac{-1}{27}\right)^{1/3}$
23. $\left(-\dfrac{1}{256}\right)^{1/4}$
24. $\left(\dfrac{1}{256}\right)^{1/4}$

25. $27^{2/3}$
26. $(-27)^{2/3}$
27. $125^{2/3}$
28. $216^{2/3}$

29. $\left(\dfrac{1}{8}\right)^{2/3}$
30. $\left(\dfrac{1}{81}\right)^{3/4}$
31. $(-8)^{4/3}$
32. $(-27)^{4/3}$

33. $(32)^{4/5}$
34. $(-32)^{4/5}$
35. $-32^{4/5}$
36. $(-64)^{5/3}$

37. $64^{-2/3}$
38. $27^{-2/3}$
39. $[(-7)^4]^{1/4}$
40. $[(-11)^6]^{1/6}$

‹ C › Operations with Rational Exponents In Problems 41–70, simplify and write the expression with positive exponents. All variables represent positive real numbers.

41. $x^{1/7} \cdot x^{2/7}$
42. $y^{1/6} \cdot y^{1/6}$
43. $x^{-1/9} \cdot x^{-4/9}$
44. $y^{-5/2} \cdot y^{-3/2}$

45. $\dfrac{x^{4/5}}{x^{2/5}}$
46. $\dfrac{y^{5/7}}{y^{2/7}}$
47. $\dfrac{z^{2/3}}{z^{-1/3}}$
48. $\dfrac{a^{4/5}}{a^{-3/5}}$

49. $(x^{1/5})^{10}$
50. $(y^{1/3})^{12}$
51. $(z^{1/3})^{-6}$
52. $(a^{1/4})^{-8}$

53. $(b^{2/3})^{-6/5}$
54. $(c^{2/7})^{-7/8}$
55. $(a^{2/3}b^{3/4})^{-12}$
56. $(x^{1/8}y^{2/3})^{-24}$

57. $\left(\dfrac{a^{2/3}}{b^{3/5}}\right)^{-15}$
58. $\left(\dfrac{x^{1/2}}{y^{3/5}}\right)^{-20}$
59. $\left(\dfrac{x^{-2/5}}{y^{3/4}}\right)^{-40}$
60. $\left(\dfrac{x^{-1/3}}{y^{3/8}}\right)^{-48}$

61. $x^{1/3}(x^{2/3} + y^{1/2})$
62. $x^{-4/5}(y^{1/3} + x^{-1/5})$
63. $y^{3/4}(x^{1/2} - y^{1/2})$
64. $y^{2/3}(y^{1/2} - x^{2/3})$

65. $\dfrac{x^{1/6} \cdot x^{-5/6}}{x^{1/3}}$
66. $\dfrac{(x^{1/3} \cdot x^{1/2})^2}{x^{1/2}}$
67. $\dfrac{(x^{1/3} \cdot y^{-1/2})^6}{(y^{1/2})^{-4}}$
68. $\left(\dfrac{x^{4/3} \cdot y^{1/2}}{x^{1/3}}\right)^{-1/2}$

69. $\dfrac{(x^{1/4} \cdot y^2)^4}{(x^{2/3} \cdot y)^{-3}}$
70. $\left(\dfrac{-8a^{-3}b^{12}}{c^{15}}\right)^{-1/3}$

››› Applications

A falling body If air resistance is ignored, the terminal velocity v of a falling body in meters per second is given by the equation below where the height h is measured in meters and the initial velocity v_0 is measured in meters per second.

$$v = (20h + v_0)^{1/2}$$

71. Find v if $h = 10$ m and $v_0 = 25$ m/sec.

72. Find v if a body is dropped ($v_0 = 0$) from a height of 45 meters.

73. If the velocity as measured in feet per second is
$$v = (64h + v_0)^{1/2}$$
find v if $h = 12$ ft and $v_0 = 16$ ft/sec.

74. Find v if a body is dropped ($v_0 = 0$) from a height of 25 feet (see Problem 73).

Using Your Knowledge

Using a Calculator to Find Rational Roots and Radicals

75. You already know that $\sqrt{x} = x^{1/2}$. Use your knowledge to write

$$\sqrt{\sqrt{x}}$$

using an exponent

77. a. Write $\sqrt[3]{x}$ using an exponent
   b. Write $\sqrt[3]{\sqrt{x}}$ using a single exponent.
   c. If you have a calculator with a $\boxed{\sqrt{}}$ and a $\boxed{\sqrt[3]{}}$ key, find $\sqrt[6]{729}$.

76. If you have a calculator with a $\boxed{\sqrt{}}$ key, you can find $\sqrt{9}$ by pressing $\boxed{9}$ $\boxed{\sqrt{}}$. You can also find $\sqrt[4]{16}$ using the $\boxed{\sqrt{}}$ key and the result of Problem 75.
   a. Find $\sqrt[4]{16}$.
   b. Find $\sqrt[4]{4096}$.

Write On

78. Explain why -2 is a square root of 4 but $\sqrt{4} = 2$.

79. How many square roots does every positive real number a have? Name them when $a = 36$.

80. Explain why $\sqrt[n]{a} = a^{1/n}$ is not a real number when a is negative and n is even.

81. Explain what is meant by "the nth root of a real number a."

82. Explain why the even root of a negative number is not a real number. (For example, $\sqrt{-4}$ is not a real number.)

Concept Checker

Fill in the blank(s) with the correct word(s), phrase, or mathematical statement.

83. In the radical expression, $\sqrt[5]{32}$, "5" is the _____.

84. When evaluating $\sqrt[4]{16}$, "2" is the _____.

85. In the radical expression $\sqrt[3]{-27}$, "-27" is the _____.

86. When taking the square root with a negative radicand the result is _____.

radical | radicand
rational exponent | principal root
index | not a real number

Mastery Test

Simplify (x and y positive):

87. $(x^{1/4} y^{3/5})^{-20}$

88. $(x^{1/7} y^{3/14})^{-21}$

89. $\dfrac{x^{1/4} y^{2/3}}{x^{3/4} y^{1/3}}$

90. $\dfrac{x^{1/5} y^{3/7}}{x^{2/5} y^{2/7}}$

91. $x^{2/5}(x^{-1/5} + y^{1/4})$

92. $x^{3/7}(x^{-3/7} + y^{3/4})$

93. $x^{1/3} \cdot x^{1/5}$

94. $y^{1/5} \cdot y^{1/4}$

95. $\dfrac{x^{-2/3}}{x^{1/4}}$

96. $\dfrac{y^{-3/4}}{y^{1/5}}$

97. $(y^{3/4})^{-1/5}$

98. $(x^{2/3})^{-2/5}$

Evaluate if possible:

99. $81^{-3/4}$

100. $(-27)^{-4/3}$

101. $216^{-2/3}$

102. $27^{2/3}$

103. $(-64)^{-2/3}$

104. $(-36)^{3/2}$

105. $49^{1/2}$

106. $(-216)^{1/3}$

107. $\left(\dfrac{1}{81}\right)^{1/4}$

108. $\sqrt{-25}$

109. $\sqrt[3]{-125}$

110. $\sqrt[3]{\dfrac{-1}{27}}$

>>> Skill Checker

In Problems 111–116, evaluate $\sqrt{b^2 - 4ac}$ with the specified values.

111. $a = 1, b = 5, c = 4$ **112.** $a = 1, b = 3, c = 2$ **113.** $a = 2, b = -3, c = -20$

114. $a = \frac{1}{2}, b = -\frac{1}{12}, c = -1$ **115.** $a = \frac{1}{12}, b = \frac{1}{3}, c = -1$ **116.** $a = \frac{1}{12}, b = \frac{1}{2}, c = \frac{2}{3}$

Write the fraction with the specified denominator.

117. $\frac{2x}{xy^2}$ with a denominator of $8x^3y^3$ **118.** $\frac{1}{5xy}$ with a denominator of $125x^3y^2$ **119.** $\frac{3xy}{x^2y^3}$ with a denominator of $16x^4y^4$

120. $\frac{2}{27x^5}$ with a denominator of $81x^8$ **121.** $\frac{5}{8x^4}$ with a denominator of $32x^5$

Factor.

122. $x^2 - 6x + 9$ **123.** $x^2 + 16x + 64$ **124.** $y^2 + 20y + 100$ **125.** $y^2 - 18y + 81$

7.2 Simplifying Radicals

Objectives

A Simplify radical expressions.

B Rationalize the denominator of a fraction.

C Reduce the index of a radical expression.

To Succeed, Review How To . . .

1. Use the properties of exponents (pp. 38–45).
2. Factor perfect square trinomials (pp. 388–390).

Getting Started
Radical Speeding

Have you seen your local police measuring skid marks at the scene of an accident? The speed s (in mi/hr) a car was traveling if it skidded d feet after the brakes were applied on a dry concrete road is given by $s = \sqrt{24d}$. A car leaving a 50-foot skid mark was traveling at a speed $s = \sqrt{24 \cdot 50} = \sqrt{1200}$. How can we simplify this? By factoring 1200 into factors that are perfect squares. Thus, $\sqrt{1200} = \sqrt{100 \cdot 4 \cdot 3} = 20\sqrt{3}$ or about 35 miles per hour. We assumed $\sqrt{100 \cdot 4 \cdot 3} = \sqrt{100} \cdot \sqrt{4} \cdot \sqrt{3}$. Is this true?

In this section, we shall study how to simplify radical expressions by using properties analogous to the properties of rational exponents we studied in Section 7.1. First, let's recall the relationship between rational exponents and radicals. In general, the nth root of a number a has the following definition.

nth ROOT

If n is a positive integer, then
$$a^{1/n} = \sqrt[n]{a} \quad (a \geq 0)$$

From this definition, we can derive three important relationships involving radicals that can be proved using the properties of exponents discussed previously.

A › Properties of Radicals

PROPERTIES FOR SIMPLIFYING RADICAL EXPRESSIONS

I. $\sqrt[n]{a^n} = a \quad (a \geq 0)$

II. $\sqrt[n]{ab} = \sqrt[n]{a}\sqrt[n]{b} \quad (a, b \geq 0)$ — Product rule

III. $\sqrt[n]{\dfrac{a}{b}} = \dfrac{\sqrt[n]{a}}{\sqrt[n]{b}} \quad (a \geq 0, b > 0)$ — Quotient rule

The first property is equivalent to the definition of the principal nth root of a. Thus, $\sqrt[n]{a^n} = (a^n)^{1/n} = a^{n \cdot 1/n} = a$. The other two properties are obtained as follows:

II. $\sqrt[n]{ab} = (ab)^{1/n} = a^{1/n} \cdot b^{1/n} = \sqrt[n]{a} \cdot \sqrt[n]{b}$

III. $\sqrt[n]{\dfrac{a}{b}} = \left(\dfrac{a}{b}\right)^{1/n} = \dfrac{a^{1/n}}{b^{1/n}} = \dfrac{\sqrt[n]{a}}{\sqrt[n]{b}}$

We have already mentioned that when m and n are both even, $\sqrt[n]{a^m} = (a^m)^{1/n} = |a|^{m/n}$. For $m = n$, we have the following definition.

DEFINITION OF $\sqrt[n]{a^n}$

$\sqrt[n]{a^n} = |a|$ n is an **even** positive integer.

$\sqrt[n]{a^n} = a$ n is an **odd** positive integer.

Thus, for even indices, $\sqrt{3^2} = |3| = 3$, $\sqrt{(-3)^2} = |-3| = 3$, and $\sqrt[4]{(-x)^4} = |-x|$. When the index is odd, we do not use absolute value. Thus, $\sqrt[3]{-8} = -2$ and $\sqrt[5]{-x^5} = -x$.

EXAMPLE 1   Simplifying expressions containing radicals

Simplify:

a. $\sqrt[4]{(-2)^4}$ b. $\sqrt[8]{(-x)^8}$ c. $\sqrt[9]{(-x)^9}$ d. $\sqrt{x^2 + 8x + 16}$

SOLUTION

a. Since the index 4 is even, $\sqrt[4]{(-2)^4} = |-2| = 2$.

b. The index is even. Thus, $\sqrt[8]{(-x)^8} = |-x| = |x|$.

c. $\sqrt[9]{(-x)^9} = -x$. (Absolute value is not used because the index 9 is not even.)

d. We start by factoring: $x^2 + 8x + 16 = (x + 4)^2$. Thus,
$$\sqrt{x^2 + 8x + 16} = \sqrt{(x + 4)^2}$$
$$= |x + 4|$$

(Absolute value is needed because the index 2 is even.)

PROBLEM 1

Simplify:

a. $\sqrt[4]{(-5)^4}$ b. $\sqrt[6]{(-x)^6}$

c. $\sqrt[7]{(-x)^7}$ d. $\sqrt{x^2 - 10x + 25}$

Answers to PROBLEMS

1. a. $|-5| = 5$ b. $|-x| = |x|$
    c. $-x$ d. $|x - 5|$

Calculator Corner

Finding the Numerical Root

Most of the numerical problems in this section can be verified with a calculator. Verifying Example 1(a) can be done two ways. The first way is to enter the information using a rational exponent as shown in Window 1; press ENTER, and the answer "2" appears. Note how parentheses were used to enter the expression. Moreover, you have to know that taking the fourth root is the same as raising to the $\frac{1}{4}$ power.

The second way is to press 4 (for the index) MATH 5, the expression $((-2)^4)$ ENTER; the answer "2" appears (Window 2).

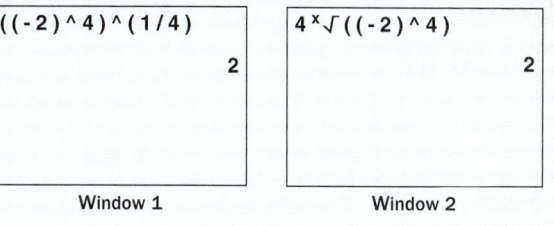

If the radicand does not have a perfect root, we can sometimes simplify it by factoring out any perfect roots. To help you, we list the first few square, cube, and fourth roots in Table 7.3.

Table 7.3 Partial List of Square, Cube, and Fourth Roots (Assume variables are positive real numbers.)

Square Roots		Cube Roots		Fourth Roots	
$\sqrt{0} = 0$	$\sqrt{x^2} = x$	$\sqrt[3]{0} = 0$		$\sqrt[4]{0} = 0$	
$\sqrt{1} = 1$	$\sqrt{x^4} = x^2$	$\sqrt[3]{1} = 1$	$\sqrt[3]{x^3} = x$	$\sqrt[4]{1} = 1$	$\sqrt[4]{x^4} = x$
$\sqrt{4} = 2$	$\sqrt{x^6} = x^3$	$\sqrt[3]{8} = 2$	$\sqrt[3]{x^6} = x^2$	$\sqrt[4]{16} = 2$	$\sqrt[4]{x^8} = x^2$
$\sqrt{9} = 3$	$\sqrt{x^8} = x^4$	$\sqrt[3]{27} = 3$	$\sqrt[3]{x^9} = x^3$	$\sqrt[4]{81} = 3$	$\sqrt[4]{x^{12}} = x^3$
$\sqrt{16} = 4$	$\sqrt{x^{10}} = x^5$	$\sqrt[3]{64} = 4$	$\sqrt[3]{x^{12}} = x^4$	$\sqrt[4]{256} = 4$	$\sqrt[4]{x^{16}} = x^4$
$\sqrt{25} = 5$	$\sqrt{x^{12}} = x^6$	$\sqrt[3]{125} = 5$	$\sqrt[3]{x^{15}} = x^5$	$\sqrt[4]{625} = 5$	$\sqrt[4]{x^{20}} = x^5$

To simplify a radical use the product rule to factor as two radicals so that the first radical contains a product of perfect powers as its radicand and the second has the nonperfect factors. Then, simplify the first radical. Variables that have perfect roots have exponents divisible by the index. This process is illustrated in Table 7.4.

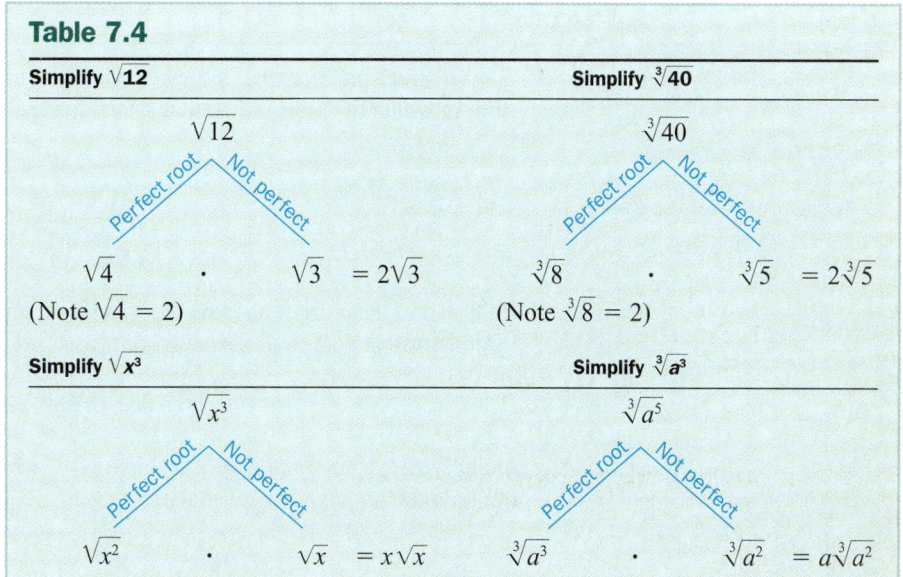

EXAMPLE 2 Simplifying square and cube roots

Simplify:

a. $\sqrt{48}$ b. $\sqrt{20x^5}$ c. $\sqrt[3]{54}$ d. $\sqrt[3]{128a^4b^6}$

PROBLEM 2
Simplify:

a. $\sqrt{32}$ b. $\sqrt{18y^2z^3}$ c. $\sqrt[3]{32}$ d. $\sqrt[3]{81a^6b^4}$

SOLUTION

a. Since $48 = 16 \cdot 3$

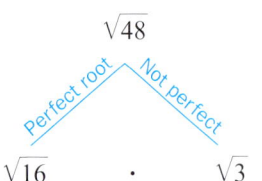

(Note $\sqrt{16} = 4$) $\sqrt{16} \cdot \sqrt{3} = 4\sqrt{3}$

b. Since $20 = 4 \cdot 5$ and $x^5 = x^4 \cdot x$

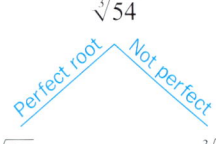

(Note $\sqrt{4x^4} = 2x^2$) $\sqrt{4x^4} \cdot \sqrt{5x} = 2x^2\sqrt{5x}$

c. Since $54 = 27 \cdot 2$

$\sqrt[3]{54}$

(Note $\sqrt[3]{27} = 3$) $\sqrt[3]{27} \cdot \sqrt[3]{2} = 3\sqrt[3]{2}$

d. We factor the radicand into factors that have perfect cube roots. 64 is a perfect cube and $128 = 64 \cdot 2$, a^3 is a perfect cube and $a^4 = a^3 \cdot a$, and b^6 is a perfect cube. Thus,

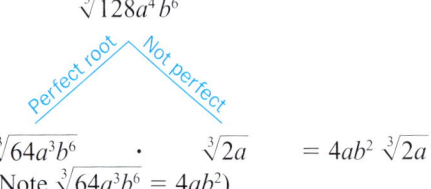

$\sqrt[3]{64a^3b^6} \cdot \sqrt[3]{2a} = 4ab^2\sqrt[3]{2a}$
(Note $\sqrt[3]{64a^3b^6} = 4ab^2$)

The third property mentioned in this section can be used to change a radical into a form in which the radicand contains no fractions. For example,

$$\sqrt{\frac{3}{16}} = \frac{\sqrt{3}}{\sqrt{16}} = \frac{\sqrt{3}}{4}$$

$$\sqrt[3]{\frac{7}{8}} = \frac{\sqrt[3]{7}}{\sqrt[3]{8}} = \frac{\sqrt[3]{7}}{2}$$

If the denominator does not have a perfect root, multiply the numerator and denominator by a factor that will yield a perfect root. To simplify

$$\sqrt{\frac{3}{8}}$$

multiply the denominator by 2 to obtain 16, which has a perfect root. You must also multiply the numerator by 2 so that you get an equivalent fraction.

$$\sqrt{\frac{3}{8}} = \sqrt{\frac{3 \cdot 2}{8 \cdot 2}} = \sqrt{\frac{6}{16}} = \frac{\sqrt{6}}{\sqrt{16}} = \frac{\sqrt{6}}{4}$$

Answers to PROBLEMS

2. a. $4\sqrt{2}$ b. $3|yz|\sqrt{2z}$ c. $2\sqrt[3]{4}$ d. $3a^2b\sqrt[3]{3b}$

EXAMPLE 3 Simplifying radical expressions containing fractions

Simplify:

a. $\sqrt{\dfrac{7}{32}}$ b. $\sqrt[3]{\dfrac{9}{x^3}}$ c. $\sqrt[4]{\dfrac{2}{27x^5}}$

SOLUTION

a. We multiply the denominator and numerator by 2 because $32 \cdot 2 = 64$ is a perfect square:

$$\sqrt{\dfrac{7}{32}} = \sqrt{\dfrac{7 \cdot 2}{32 \cdot 2}} = \sqrt{\dfrac{14}{64}} = \dfrac{\sqrt{14}}{\sqrt{64}} = \dfrac{\sqrt{14}}{8}$$

b. Since $\sqrt[3]{x^3} = x$, we simplify the denominator.

$$\sqrt[3]{\dfrac{9}{x^3}} = \dfrac{\sqrt[3]{9}}{\sqrt[3]{x^3}} = \dfrac{\sqrt[3]{9}}{x}$$

c. To obtain a perfect fourth power, multiply 27 by 3 to obtain $81 = 3^4$ and x^5 by x^3 to obtain x^8.

$$\sqrt[4]{\dfrac{2}{27x^5}} = \sqrt[4]{\dfrac{2 \cdot 3x^3}{27x^5 \cdot 3x^3}} = \sqrt[4]{\dfrac{6x^3}{81x^8}} = \dfrac{\sqrt[4]{6x^3}}{\sqrt[4]{81x^8}} = \dfrac{\sqrt[4]{6x^3}}{3x^2}$$

PROBLEM 3

Simplify:

a. $\sqrt{\dfrac{11}{12}}$ b. $\sqrt[3]{\dfrac{6}{x^3}}$ c. $\sqrt[4]{\dfrac{3}{8x^6}}$

Calculator Corner

Verifying Simplified Radical Expressions

To verify Example 3(a), enter the original problem,

$$\sqrt{\dfrac{7}{32}}$$

and the simplified version,

$$\dfrac{\sqrt{14}}{8}$$

```
√(7/32)
         .4677071733
√(14)/8
         .4677071733
```
Window 1

You should get the approximation 0.4677071733 in both cases, as shown in Window 1.

If an example to be verified contains a variable, you can graph the original problem and the answer to make sure the graphs are the same. For example, to verify the result of Example 3(c), let

$$Y_1 = (2/(27x^\wedge 5))^\wedge(1/4)$$

and

$$Y_2 = (\sqrt[4]{(6x^\wedge 3)})/(3x^2)$$

Graph both of these using a standard window first; then regraph using a $[-1, 10]$ by $[-1, 10]$ window for a better view. The graph is shown in Window 2.

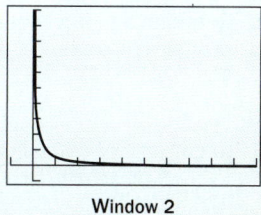
Window 2

B › Rationalizing the Denominator

As we saw in Example 3, in some cases the denominators of fractions contain radical expressions. For example,

$$\dfrac{\sqrt{3}}{\sqrt{5}}$$

has $\sqrt{5}$ in the denominator. To simplify

$$\dfrac{\sqrt{3}}{\sqrt{5}}$$

Answers to PROBLEMS

3. a. $\dfrac{\sqrt{33}}{6}$ b. $\dfrac{\sqrt[3]{6}}{x}$ c. $\dfrac{\sqrt[4]{6x^2}}{2x^2}$

we use the fundamental property of fractions to multiply numerator and denominator by $\sqrt{5}$, which causes the denominator to be rational. Thus,

$$\frac{\sqrt{3}}{\sqrt{5}} = \frac{\sqrt{3} \cdot \sqrt{5}}{\sqrt{5} \cdot \sqrt{5}} \quad \text{Fundamental property of fractions}$$

$$= \frac{\sqrt{15}}{\sqrt{5^2}} \quad \text{Product rule}$$

$$= \frac{\sqrt{15}}{5} \quad \sqrt[n]{a^n} = a, a > 0$$

This process is called **rationalizing the denominator**. To rationalize the denominator in the expression

$$\frac{\sqrt{7}}{\sqrt{3x}} \quad (x > 0)$$

we have to change the fraction to an equivalent one without a radical in the denominator. As before, we multiply the denominator and numerator by $\sqrt{3x}$ to obtain

$$\frac{\sqrt{7}}{\sqrt{3x}} = \frac{\sqrt{7} \cdot \sqrt{3x}}{\sqrt{3x} \cdot \sqrt{3x}} \quad \text{Fundamental property of fractions}$$

$$= \frac{\sqrt{21x}}{\sqrt{3^2 x^2}} \quad \text{Product rule}$$

$$= \frac{\sqrt{21x}}{3x} \quad \sqrt[n]{a^n} = a, a > 0$$

EXAMPLE 4 Rationalizing the denominator
Rationalize the denominator:

a. $\dfrac{\sqrt{11}}{\sqrt{6}}$ b. $\dfrac{\sqrt{3}}{\sqrt{5x}} \ (x > 0)$ c. $\dfrac{\sqrt{5}}{\sqrt{18x^2}} \ (x \neq 0)$

SOLUTION
a. To obtain a perfect square in the denominator, multiply by $\sqrt{6}$.

$$\frac{\sqrt{11}}{\sqrt{6}} = \frac{\sqrt{11} \cdot \sqrt{6}}{\sqrt{6} \cdot \sqrt{6}} = \frac{\sqrt{66}}{6}$$

b. To obtain a perfect square in the denominator, multiply by $\sqrt{5x}$.

$$\frac{\sqrt{3}}{\sqrt{5x}} = \frac{\sqrt{3} \cdot \sqrt{5x}}{\sqrt{5x} \cdot \sqrt{5x}} = \frac{\sqrt{15x}}{5x}$$

c. To convert $\sqrt{18}$ to a perfect square, multiply by $\sqrt{2}$.

$$\frac{\sqrt{5}}{\sqrt{18x^2}} = \frac{\sqrt{5} \cdot \sqrt{2}}{\sqrt{18x^2} \cdot \sqrt{2}} = \frac{\sqrt{10}}{\sqrt{36x^2}} = \frac{\sqrt{10}}{6x^2}$$

x^2 is *not* under the radical, so x^2 does not change.

PROBLEM 4
Rationalize the denominator:

a. $\dfrac{\sqrt{7}}{\sqrt{3}}$ b. $\dfrac{\sqrt{5}}{\sqrt{6x}}, x > 0$

c. $\dfrac{\sqrt{11}}{\sqrt{32x^3}}, x > 0$

When the radical in the denominator is of index n, we must make the radicand an exact nth power. For example, to rationalize

$$\frac{\sqrt[3]{5}}{\sqrt[3]{3x}}$$

we convert $\sqrt[3]{3}$ to a perfect cube root by multiplying by $\sqrt[3]{3^2}$, and we convert $\sqrt[3]{x}$ to a perfect cube root by multiplying by $\sqrt[3]{x^2}$. We can combine these two steps and multiply by $\sqrt[3]{3^2 x^2}$ to obtain

Answers to PROBLEMS

4. a. $\dfrac{\sqrt{21}}{3}$ b. $\dfrac{\sqrt{30x}}{6x}$ c. $\dfrac{\sqrt{22x}}{8x^2}$

$$\frac{\sqrt[3]{5}}{\sqrt[3]{3x}} = \frac{\sqrt[3]{5}\cdot\sqrt[3]{3^2\cdot x^2}}{\sqrt[3]{3x}\cdot\sqrt[3]{3^2\cdot x^2}}$$ Multiply the numerator and denominator by $\sqrt[3]{3^2\cdot x^2}$ to make the denominator a perfect cube root.

$$= \frac{\sqrt[3]{5\cdot 9\cdot x^2}}{\sqrt[3]{3^3 x^3}}$$ Property II

$$= \frac{\sqrt[3]{45x^2}}{3x}$$ Property I

EXAMPLE 5 Making the radicand an exact nth power

Rationalize the denominator:

a. $\dfrac{1}{\sqrt[3]{5x}}\;(x\neq 0)$ **b.** $\dfrac{\sqrt[4]{5}}{\sqrt[4]{8x}}\;(x>0)$

SOLUTION

a. To convert $\sqrt[3]{5x}$ to a perfect cube root, multiply by $\sqrt[3]{5^2 x^2}$.

$$\frac{1}{\sqrt[3]{5x}} = \frac{1\cdot\sqrt[3]{5^2 x^2}}{\sqrt[3]{5x}\cdot\sqrt[3]{5^2 x^2}}$$

$$= \frac{\sqrt[3]{25x^2}}{\sqrt[3]{5^3 x^3}}$$

$$= \frac{\sqrt[3]{25x^2}}{5x}$$

b. $\dfrac{\sqrt[4]{5}}{\sqrt[4]{8x}} = \dfrac{\sqrt[4]{5}}{\sqrt[4]{2^3 x}}$ Write $8x$ as $2^3 x$.

$$= \frac{\sqrt[4]{5}\cdot\sqrt[4]{2\cdot x^3}}{\sqrt[4]{2^3\cdot x}\cdot\sqrt[4]{2\cdot x^3}}$$ Use the fundamental property of fractions to make the denominator a perfect 4th root.

$$= \frac{\sqrt[4]{10x^3}}{\sqrt[4]{2^4 x^4}}$$ Property II

$$= \frac{\sqrt[4]{10x^3}}{2x}$$ Property I

PROBLEM 5

Rationalize the denominator:

a. $\dfrac{1}{\sqrt[3]{54x}}\;(x\neq 0)$

b. $\dfrac{\sqrt[5]{5}}{\sqrt[5]{27x^4}}\;(x\neq 0)$

C > Reducing the Index of a Radical Expression

The index of a radical expression can sometimes be reduced by writing the radical as a power with a rational exponent and then reducing the exponent. For example, if $x\geq 0$,

$$\sqrt[6]{x^3} = (x^3)^{1/6}$$
$$= x^{3\cdot 1/6}$$
$$= x^{1/2}$$
$$= \sqrt{x}$$

Similarly, for $x\geq 0$ and $y\geq 0$,

$$\sqrt[4]{64x^2y^2} = \sqrt[4]{(8xy)^2}$$
$$= [(8xy)^2]^{1/4}$$
$$= [8xy]^{2\cdot 1/4}$$
$$= (8xy)^{1/2} = \sqrt{8xy}$$

Perfect root   Not perfect

$\sqrt{4}$ · $\sqrt{2xy}$ = $2\sqrt{2xy}$

Answers to PROBLEMS

5. **a.** $\dfrac{\sqrt[3]{4x^2}}{6x}$ **b.** $\dfrac{\sqrt[5]{45x}}{3x}$

EXAMPLE 6   Reducing the index
Reduce the index of each radical.

a. $\sqrt[4]{\dfrac{4}{9}}$

b. $\sqrt[6]{27c^3 d^3}$ $(c \geq 0, d \geq 0)$

SOLUTION

a. $\sqrt[4]{\dfrac{4}{9}} = \left[\left(\dfrac{2}{3}\right)^2\right]^{1/4}$ $4 = 2^2$ and $9 = 3^2$

$= \left(\dfrac{2}{3}\right)^{1/2}$ Simplify exponents.

$= \sqrt{\dfrac{2}{3}}$ Write as a radical.

$= \dfrac{\sqrt{6}}{3}$ Rationalize the denominator. $\left(\dfrac{\sqrt{2} \cdot \sqrt{3}}{\sqrt{3} \cdot \sqrt{3}} = \dfrac{\sqrt{6}}{3}\right)$

b. $\sqrt[6]{27c^3 d^3} = \sqrt[6]{(3cd)^3}$ $27 = 3^3$

$= [(3cd)^3]^{1/6}$ Write with rational exponent.

$= [3cd]^{1/2}$ Simplify exponents.

$= \sqrt{3cd}$ Write as a radical.

PROBLEM 6
Reduce the index of each radical.

a. $\sqrt[4]{\dfrac{25}{x^2}}$ $(x > 0)$

b. $\sqrt[6]{4c^2 d^2}$ $(c \geq 0, d \geq 0)$

We have used different techniques to *simplify* radical expressions. To make sure that the resulting radicals are simplified, use these rules.

RULES FOR SIMPLIFYING RADICAL EXPRESSIONS

A radical expression is in **simplified** form if

1. All exponents in the radicand (the expression under the radical) are less than the index.
2. There are no fractions under the radical sign.
3. There are no radicals in the denominator.
4. The index is as low as possible.

EXAMPLE 7   Simplifying radical expressions
Simplify:

$\sqrt[6]{\dfrac{a^2}{16c^{10}}}$ where $a > 0, c > 0$

SOLUTION
To make the denominator a perfect sixth root, note that $16 = 2^4$, and then multiply the numerator and denominator of the fraction under the radical by $2^2 c^2$. We have

$\sqrt[6]{\dfrac{a^2}{16c^{10}}} = \sqrt[6]{\dfrac{a^2 \cdot 2^2 c^2}{2^4 c^{10} \cdot 2^2 c^2}} = \sqrt[6]{\dfrac{2^2 a^2 c^2}{2^6 c^{12}}}$

$= \dfrac{\sqrt[6]{2^2 a^2 c^2}}{2c^2}$ $\sqrt[6]{2^6 c^{12}} = 2c^2$

$= \dfrac{(2^2 a^2 c^2)^{1/6}}{2c^2}$ Rewrite numerator.

$= \dfrac{[(2ac)^2]^{1/6}}{2c^2}$

$= \dfrac{(2ac)^{1/3}}{2c^2}$ Reduce index.

$= \dfrac{\sqrt[3]{2ac}}{2c^2}$ Write in radical form.

PROBLEM 7
Simplify:

$\sqrt[6]{\dfrac{a^3}{8x^3}}$ $(a > 0, x > 0)$

Answers to PROBLEMS

6. a. $\sqrt{\dfrac{5}{x}} = \dfrac{\sqrt{5x}}{x}, x > 0$ b. $\sqrt[3]{2cd}$

7. $\dfrac{\sqrt{2ax}}{2x}$

Calculator Corner

Some of the simplifications we have made can be checked using a calculator with a $\sqrt[x]{y}$ key. To access this feature, you usually have to press the [2nd] or [2ndF] key first and then the $\sqrt[x]{y}$ key. The calculator will then find the xth root of y. In Example 2, we learned that $\sqrt[3]{54} = 3\sqrt[3]{2}$. To check this, enter [5] [4] [2nd] [$\sqrt[x]{y}$] [3] [ENTER]. The display shows 3.77976315. Now enter [3] [×] [2] [$\sqrt[x]{y}$] [3] [ENTER]. The same result appears, so our answer is correct. If your instructor agrees, check the numerical problems in this section (Problems 25–28, for example) with a calculator.

> Exercises 7.2

Boost your grade at mathzone.com!
> Practice Problems
> NetTutor
> Self-Tests
> e-Professors
> Videos

< A > Properties of Radicals
In Problems 1–24, simplify the expression given. (*Hint:* Some answers require absolute values.)

1. $\sqrt{(-5)^2}$
2. $\sqrt{5^2}$
3. $\sqrt[3]{-64}$
4. $\sqrt[3]{-125}$
5. $\sqrt[6]{(-x)^6}$
6. $\sqrt[5]{(-x)^5}$
7. $\sqrt{x^2 + 12x + 36}$
8. $\sqrt{4x^2 + 12x + 9}$
9. $\sqrt{9x^2 - 12x + 4}$
10. $\sqrt{16x^2 + 8x + 1}$
11. $\sqrt{16x^3y^3}$
12. $\sqrt{81x^3y^4}$
13. $\sqrt[3]{40x^4y}$
14. $\sqrt[3]{81x^3y^6}$
15. $\sqrt[4]{x^5y^7}$
16. $\sqrt[4]{162x^4y^7}$
17. $\sqrt[5]{-243a^{10}b^{17}}$
18. $\sqrt[5]{-32a^{15}b^{20}}$
19. $\sqrt{\dfrac{13}{49}}$
20. $\sqrt{\dfrac{17}{64}}$
21. $\sqrt{\dfrac{17}{4x^2}}$
22. $\sqrt{\dfrac{19}{64x^4}}$
23. $\sqrt[3]{\dfrac{3}{64x^3}}$
24. $\sqrt[3]{\dfrac{-7}{27x^6}}$

< B > Rationalizing the Denominator
In Problems 25–44, rationalize the denominator. (Assume all variables represent positive real numbers.)

25. $\sqrt{\dfrac{2}{3}}$
26. $\sqrt{\dfrac{4}{5}}$
27. $\dfrac{-\sqrt{2}}{\sqrt{7}}$
28. $\dfrac{-\sqrt{3}}{\sqrt{11}}$
29. $\sqrt{\dfrac{5}{2a}}$
30. $\sqrt{\dfrac{7}{36}}$
31. $\sqrt{\dfrac{5}{32ab}}$
32. $\sqrt{\dfrac{5}{8ab}}$
33. $-\sqrt{\dfrac{3}{2a^3b^3}}$
34. $-\sqrt{\dfrac{3}{8ab^3}}$
35. $\dfrac{\sqrt{x}\sqrt{xy^3}}{\sqrt{y}}$
36. $\dfrac{\sqrt{xy}\sqrt{xy^4}}{\sqrt{y}}$
37. $-\sqrt[3]{\dfrac{7}{9}}$
38. $-\sqrt[3]{\dfrac{3}{32}}$
39. $\sqrt[3]{\dfrac{3}{16x^2}}$
40. $\sqrt[3]{\dfrac{5}{16x}}$
41. $\sqrt[3]{\dfrac{1}{8x^2}}$
42. $\sqrt[3]{\dfrac{2}{4x^6}}$
43. $\sqrt[4]{\dfrac{3}{2}}$
44. $\sqrt[4]{\dfrac{1}{2x^3}}$

< C > Reducing the Index of a Radical Expression
In Problems 45–54, reduce the index (order) of the given radical and simplify. (Assume all variables represent positive real numbers.)

45. $\sqrt[6]{9}$
46. $\sqrt[4]{4}$
47. $\sqrt[4]{4a^2}$
48. $\sqrt[4]{9a^2}$
49. $\sqrt[4]{25x^6y^2}$
50. $\sqrt[4]{36x^2y^6}$
51. $\sqrt[4]{49x^{10}y^6}$
52. $\sqrt[4]{100x^{10}y^{10}}$
53. $\sqrt[6]{8a^3b^3}$
54. $\sqrt[6]{27a^3b^9}$

546 Chapter 7 Rational Exponents and Radicals

In Problems 55–59, simplify the radical expression. (Assume all variables represent positive real numbers.)

55. $\sqrt[6]{\dfrac{a^4}{b^8}}$ **56.** $\sqrt[4]{\dfrac{c^6}{4b^2}}$ **57.** $\sqrt[4]{\dfrac{64a^2}{9b^6}}$ **58.** $\sqrt[4]{\dfrac{4c^2 y^6}{9b^4}}$ **59.** $\sqrt[6]{\dfrac{b^3 a^3}{8x^3}}$

〉〉〉 Applications

60. *Measurements* A body starting at rest takes t seconds to fall a distance of d feet, where

$$t = \sqrt{\dfrac{d}{16}}$$

   a. Simplify this expression.

   b. How long would it take an object starting at rest to fall 100 feet?

61. *Measurements* The radius of a sphere is given by

$$r = \sqrt[3]{\dfrac{3V}{4\pi}}$$

where V is the volume of the sphere and π is about $\dfrac{22}{7}$.

   a. Simplify $\sqrt[3]{\dfrac{3V}{4\pi}}$.

   b. If the volume of a sphere is 36π cubic feet, what is its radius?

62. *Velocity of a gas particle* The root-mean-square velocity, $\bar{v}$, of a gas particle is given by the formula

$$\bar{v} = \dfrac{\sqrt{3kT}}{\sqrt{m}}$$

where k is a constant, T is the temperature (in degrees Kelvin), and m is the mass of the particle. Rationalize the denominator of the expression on the right-hand side.

63. *Mass of an object* The mass m of an object depends on its speed v and the speed of light c. The relationship is given by the formula

$$m = \dfrac{m_0}{\sqrt{1 - \dfrac{v^2}{c^2}}}$$

where m_0 is the *rest mass*, the mass when $v = 0$. Simplify the expression on the right-hand side and rationalize the denominator.

〉〉〉 Using Your Knowledge

From Radicals to Rational and Back   We've studied rational exponents and radical expressions. We will now use this knowledge to translate one notation to the other.

64. In Problem 63,

$$m = \dfrac{m_0}{\sqrt{1 - \dfrac{v^2}{c^2}}}$$

Rationalize the denominator and write the result using a rational exponent.

65. Simplify the expression in Problem 62 defining $\bar{v}$ and write the result using a rational exponent.

66. The period T of a pendulum is

$$T = \sqrt{\dfrac{2\pi L}{g}}$$

where L is the length of the pendulum and g is the gravity constant. Simplify this expression and write the result using a rational exponent.

67. The pressure P of a gas is related to its volume V by the formula $P = kV^{-7/5}$, where k is the constant of variation. Write this formula using radical notation.

68. The average speed $\bar{v}$ of oxygen molecules is given by the formula $\bar{v} = (3kT)^{1/2} m^{-1/2}$. Write this formula in simplified form using radicals.

〉〉〉 Write On

69. Property I for simplifying radical expressions states that $\sqrt[n]{a^n} = a$ for $a \geq 0$. What happens if $a < 0$? Explain and give examples.

70. Write the procedure you use to rationalize a radical denominator in a quotient.

71. State the conditions under which $\sqrt[n]{a^n} = (\sqrt[n]{a})^n = a$.

72. Use some counterexamples to show that $(a^2 + b^2)^{1/2} \neq a + b$.

73. Use some counterexamples to show that $(a^{1/2} + b^{1/2})^2 \neq a + b$.

〉〉〉 Concept Checker

Fill in the blank(s) with the correct word(s), phrase, or mathematical statement.

*n*th root	index
rationalizing	product rule
reducing	quotient rule
radicals	

74. Changing a fraction to an equivalent one without a radical in the denominator is a process called _____ the denominator.

75. $\sqrt{32} = \sqrt{16} \cdot \sqrt{2}$ is an example of the _____ used for simplifying radicals.

76. $27^{1/3} = \sqrt[3]{27}$ is an example of the _____ definition.

〉〉〉 Mastery Test

Simplify (assume variables are positive):

77. $\sqrt{\dfrac{11}{12}}$

78. $\sqrt[3]{\dfrac{6}{x^3}}$

79. $\sqrt[4]{\dfrac{3}{8x^6}}$

80. $\sqrt{32}$

81. $\sqrt[3]{32}$

82. $\sqrt[3]{81a^6b^4}$

83. $\sqrt[6]{\dfrac{a^3}{8x^3}}$

84. $\sqrt[8]{\dfrac{x^4}{16a^4}}$

85. $\sqrt[4]{(-10)^4}$

86. $\sqrt[6]{(-x)^6}$

87. $\sqrt[7]{(-x)^7}$

88. $\sqrt{x^2 + 10x + 25}$

Reduce the index (assume variables are positive):

89. $\sqrt[8]{\dfrac{81}{256}}$

90. $\sqrt[6]{4c^2d^2}$

Rationalize the denominator (assume variables are positive):

91. $\dfrac{\sqrt{7}}{\sqrt{5}}$

92. $\dfrac{\sqrt[3]{5}}{\sqrt[3]{4}}$

93. $\dfrac{\sqrt{5}}{\sqrt{6x}}$

94. $\dfrac{\sqrt{11}}{\sqrt{32x^3}}$

〉〉〉 Skill Checker

95. Multiply $(a + b)(a - b)$.

96. Use the result of Problem 95 to multiply $(\sqrt{x} + \sqrt{y})(\sqrt{x} - \sqrt{y})$.

97. Use the result of Problem 95 to multiply $(x^{3/2} + y^{3/2})(x^{3/2} - y^{3/2})$.

Simplify:

98. $\dfrac{4(x + 3)}{2}$

99. $\dfrac{3(y - 5)}{-3}$

100. $\dfrac{6 + 3y}{3}$

101. $\dfrac{3xy + 6x^2y}{3xy}$

102. $\dfrac{12x^2y^3 + 18xy^3}{6xy}$

Evaluate and simplify $\sqrt{b^2 - 4ac}$ for the specified values of each variable:

103. $a = 2, b = -1$, and $c = -6$

104. $a = 2, b = -5$, and $c = -12$

105. $a = 6, b = -4$, and $c = -2$

106. $a = -1, b = 1$, and $c = 12$

7.3 Operations with Radicals

Objectives

A ▸ Add and subtract similar radical expressions.

B ▸ Multiply and divide radical expressions.

C ▸ Rationalize the denominators of radical expressions involving sums or differences.

▸ To Succeed, Review How To . . .

1. Combine like terms (pp. 58–59).
2. Remove parentheses using the distributive property (pp. 55–58).
3. Write a fraction with a specified denominator (pp. 430–432).
4. Reduce fractions (pp. 433–436).

▸ Getting Started
Radical Flight

Have you heard of supersonic airplanes with a speed of Mach 2? The Mach number is named for the Austrian physicist **Ernst Mach** (1838–1916) and is used to indicate how fast one is going when compared to the speed of sound. Mach 2 means the speed of the plane is *twice* the speed of sound (747 miles per hour). How fast can the plane in this photo travel? The answer is classified information, but it is said that the plane's speed is more than Mach 2. The Mach number M can be found from the formula

$$M = \sqrt{\frac{2}{\gamma}} \sqrt{\frac{P_2 - P_1}{P_1}}$$

where P_1 and P_2 are air pressures and γ is the ratio of the specific heat at constant pressure to the specific heat at constant volume. This expression can be simplified by first multiplying the radical expressions and then simplifying by rationalizing the denominator. In this section we will add, subtract, multiply, and divide radical expressions.

A ▸ Adding and Subtracting Similar Radical Expressions

In Section 1.4 we combined like terms using the distributive property. Thus,

$$3x + 5x = (3 + 5)x = 8x$$
$$7x - 4x = (7 - 4)x = 3x$$

Similarly,

$$3\sqrt{2} + 5\sqrt{2} = (3 + 5)\sqrt{2} = 8\sqrt{2}$$
$$7\sqrt[3]{7} - 4\sqrt[3]{7} = (7 - 4)\sqrt[3]{7} = 3\sqrt[3]{7}$$

Thus, we can combine *like* (*similar*) radical expressions. Here is the definition.

LIKE (SIMILAR) RADICAL EXPRESSIONS

Radical expressions with the same index and the same radicand are **like (similar) expressions**.

If two or more expressions don't appear to be similar or like, we must try to simplify them first. Thus, to add $\sqrt{75} + \sqrt{27}$, we proceed as follows:

$$\sqrt{75} + \sqrt{27} = \sqrt{25 \cdot 3} + \sqrt{9 \cdot 3}$$
$$= \sqrt{25} \cdot \sqrt{3} + \sqrt{9} \cdot \sqrt{3} \quad \sqrt{ab} = \sqrt{a}\sqrt{b}$$
$$= 5\sqrt{3} + 3\sqrt{3}$$
$$= (5 + 3)\sqrt{3} = 8\sqrt{3} \quad \text{Add like radicals.}$$

The subtraction of similar (like) radicals is done in the same way. Thus,

$$\sqrt{80} - \sqrt{20} = \sqrt{16 \cdot 5} - \sqrt{4 \cdot 5}$$
$$= \sqrt{16} \cdot \sqrt{5} - \sqrt{4} \cdot \sqrt{5}$$
$$= 4\sqrt{5} - 2\sqrt{5}$$
$$= (4 - 2) \cdot \sqrt{5} = 2\sqrt{5} \quad \text{Subtract like radicals.}$$

EXAMPLE 1 Adding and subtracting radical expressions

Perform the indicated operations:

a. $\sqrt{175} + \sqrt{28}$ **b.** $\sqrt{98} - \sqrt{32}$

c. $3\sqrt{18x} - 5\sqrt{8x}$ **d.** $5\sqrt[3]{80x} - 3\sqrt[3]{270x}$

SOLUTION

a. $\sqrt{175} + \sqrt{28} = \sqrt{25 \cdot 7} + \sqrt{4 \cdot 7}$
$$= \sqrt{25} \cdot \sqrt{7} + \sqrt{4} \cdot \sqrt{7}$$
$$= 5\sqrt{7} + 2\sqrt{7} = 7\sqrt{7}$$

b. $\sqrt{98} - \sqrt{32} = \sqrt{49 \cdot 2} - \sqrt{16 \cdot 2}$
$$= \sqrt{49} \cdot \sqrt{2} - \sqrt{16} \cdot \sqrt{2}$$
$$= 7\sqrt{2} - 4\sqrt{2} = 3\sqrt{2}$$

c. $3\sqrt{18x} - 5\sqrt{8x} = 3 \cdot \sqrt{9 \cdot 2x} - 5 \cdot \sqrt{4 \cdot 2x}$
$$= 3 \cdot \sqrt{9} \cdot \sqrt{2x} - 5 \cdot \sqrt{4} \cdot \sqrt{2x}$$
$$= 3 \cdot 3 \cdot \sqrt{2x} - 5 \cdot 2 \cdot \sqrt{2x}$$
$$= 9\sqrt{2x} - 10\sqrt{2x}$$
$$= -\sqrt{2x}$$

d. This time we must find factors of 80 and 270 that are perfect cubes: $80 = 8 \cdot 10 = 2^3 \cdot 10$ and $270 = 27 \cdot 10 = 3^3 \cdot 10$. Thus,

$$5\sqrt[3]{80x} - 3\sqrt[3]{270x} = 5 \cdot \sqrt[3]{2^3 \cdot 10x} - 3 \cdot \sqrt[3]{3^3 \cdot 10x}$$
$$= 5 \cdot \sqrt[3]{2^3} \cdot \sqrt[3]{10x} - 3 \cdot \sqrt[3]{3^3} \cdot \sqrt[3]{10x}$$
$$= 5 \cdot 2 \cdot \sqrt[3]{10x} - 3 \cdot 3 \cdot \sqrt[3]{10x}$$
$$= 10\sqrt[3]{10x} - 9\sqrt[3]{10x}$$
$$= \sqrt[3]{10x}$$

PROBLEM 1

Perform the indicated operations:

a. $\sqrt{44} + \sqrt{99}$

b. $\sqrt{98} - \sqrt{50}$

c. $3\sqrt{20x} - 5\sqrt{45x}$

d. $2\sqrt[3]{250x} - 4\sqrt[3]{16x}$

We now show you how to combine radicals that contain fractions.

Answers to PROBLEMS

1. **a.** $5\sqrt{11}$ **b.** $2\sqrt{2}$ **c.** $-9\sqrt{5x}$ **d.** $2\sqrt[3]{2x}$

EXAMPLE 2 Subtracting radical expressions containing fractions

Perform the indicated operations:

a. $3\sqrt{\dfrac{1}{2}} - 5\sqrt{\dfrac{1}{8}}$
b. $\sqrt[3]{\dfrac{3}{16}} - \sqrt[3]{\dfrac{3}{2}}$

SOLUTION

a. We first simplify each of the radicals by making the denominator a perfect square.

$$\sqrt{\dfrac{1}{2}} = \sqrt{\dfrac{1\cdot 2}{2\cdot 2}} = \dfrac{\sqrt{2}}{2} \quad \text{and} \quad \sqrt{\dfrac{1}{8}} = \sqrt{\dfrac{1\cdot 2}{8\cdot 2}} = \dfrac{\sqrt{2}}{\sqrt{16}} = \dfrac{\sqrt{2}}{4}$$

Thus,

$$\begin{aligned}
3\sqrt{\dfrac{1}{2}} - 5\sqrt{\dfrac{1}{8}} &= 3\cdot\dfrac{\sqrt{2}}{2} - 5\cdot\dfrac{\sqrt{2}}{4} && \text{Substitute } \dfrac{\sqrt{2}}{2} \text{ for } \sqrt{\dfrac{1}{2}} \text{ and } \dfrac{\sqrt{2}}{4} \text{ for } \sqrt{\dfrac{1}{8}}. \\
&= \dfrac{3\sqrt{2}}{2} - \dfrac{5\sqrt{2}}{4} && \text{Multiply.} \\
&= \dfrac{2\cdot 3\sqrt{2}}{2\cdot 2} - \dfrac{5\sqrt{2}}{4} && \text{Since the LCD is 4, write the first fraction with 4 as the denominator.} \\
&= \dfrac{6\sqrt{2} - 5\sqrt{2}}{4} && \text{Use 4 as the denominator.} \\
&= \dfrac{\sqrt{2}}{4} && \text{Subtract.}
\end{aligned}$$

b. We first (1) make the denominators perfect cubes and find the cube root, (2) find the LCD of the resulting fractions, and then (3) subtract.

$$\begin{aligned}
\sqrt[3]{\dfrac{3}{16}} - \sqrt[3]{\dfrac{3}{2}} &= \sqrt[3]{\dfrac{3\cdot 4}{16\cdot 4}} - \sqrt[3]{\dfrac{3\cdot 4}{2\cdot 4}} \\
&= \sqrt[3]{\dfrac{12}{64}} - \sqrt[3]{\dfrac{12}{8}} && \text{(1) Make the denominators perfect cubes and find the cube root.}\\
&= \dfrac{\sqrt[3]{12}}{4} - \dfrac{\sqrt[3]{12}}{2} \\
&= \dfrac{\sqrt[3]{12}}{4} - \dfrac{\sqrt[3]{12}\cdot 2}{2\cdot 2} && \text{(2) Since the LCD is 4, multiply the numerator and denominator of the second fraction by 2.} \\
&= \dfrac{\sqrt[3]{12} - 2\sqrt[3]{12}}{4} \\
&= \dfrac{-\sqrt[3]{12}}{4} && \text{(3) Subtract.}
\end{aligned}$$

PROBLEM 2

Perform the indicated operations:

a. $5\sqrt{\dfrac{1}{2}} - 7\sqrt{\dfrac{1}{18}}$

b. $4\sqrt[3]{\dfrac{5}{4}} - \sqrt[3]{\dfrac{5}{32}}$

Calculator Corner

Verifying Radical Addition and Subtraction

Example 2(b) can be verified by evaluating the original problem, evaluating the answer, and comparing the results to be sure they yield the same value.

Original problem:

$\sqrt[3]{(3/16)} - \sqrt[3]{(3/2)}$

-0.5723571213

Answer:

$-\sqrt[3]{(12)}/4$

-0.5723571213

Answers to PROBLEMS

2. a. $\dfrac{4\sqrt{2}}{3}$ b. $\dfrac{7\sqrt[3]{10}}{4}$

B ❯ Multiplying and Dividing Radical Expressions

The distributive property, in conjunction with $\sqrt{a} \cdot \sqrt{b} = \sqrt{ab}$, can be used to simplify radical expressions that contain parentheses. For example,

$$\sqrt{2} \cdot (\sqrt{3} + \sqrt{5}) = \sqrt{2} \cdot \sqrt{3} + \sqrt{2} \cdot \sqrt{5} \quad \text{Distributive property}$$
$$= \sqrt{6} + \sqrt{10}$$

Similarly, if $x \geq 0$, then

$$\sqrt{2x} \cdot (\sqrt{x} + \sqrt{3}) = \sqrt{2x} \cdot \sqrt{x} + \sqrt{2x} \cdot \sqrt{3}$$
$$= \sqrt{2x^2} + \sqrt{6x}$$
$$= \sqrt{2}\sqrt{x^2} + \sqrt{6x} \quad \sqrt{ab} = \sqrt{a} \cdot \sqrt{b}, \text{ so } \sqrt{2x^2} = \sqrt{2} \cdot \sqrt{x^2}.$$
$$= x\sqrt{2} + \sqrt{6x} \quad \sqrt{x^2} = x \text{ since } x \geq 0.$$

Notice in the example that when $x \geq 0$, $(\sqrt{x})^2 = \sqrt{x} \cdot \sqrt{x} = \sqrt{x^2} = x$. This means that $(\sqrt{x})^2 = x$, and in general, $(\sqrt[n]{x})^n = x$ **when $x \geq 0$.**

EXAMPLE 3   Multiplying radical expressions
Perform the indicated operations. (Assume all variables are greater than or equal to 0.)

a. $\sqrt{3}(\sqrt{5} + \sqrt{12})$

b. $2\sqrt{3x}(\sqrt{x} - \sqrt{5})$

c. $\sqrt[3]{3x}(\sqrt[3]{9x^2} - \sqrt[3]{18x})$

SOLUTION

a. $\sqrt{3}(\sqrt{5} + \sqrt{12}) = \sqrt{3} \cdot \sqrt{5} + \sqrt{3} \cdot \sqrt{12}$
$= \sqrt{15} + \sqrt{36}$
$= \sqrt{15} + 6$

b. $2\sqrt{3x}(\sqrt{x} - \sqrt{5}) = 2\sqrt{3x}\sqrt{x} - 2\sqrt{3x}\sqrt{5}$
$= 2\sqrt{3x^2} - 2\sqrt{15x}$
$= 2x\sqrt{3} - 2\sqrt{15x}$

c. $\sqrt[3]{3x}(\sqrt[3]{9x^2} - \sqrt[3]{18x}) = \sqrt[3]{3x}\sqrt[3]{9x^2} - \sqrt[3]{3x}\sqrt[3]{18x}$
$= \sqrt[3]{27x^3} - \sqrt[3]{54x^2}$
$= \sqrt[3]{3^3 \cdot x^3} - \sqrt[3]{3^3 \cdot 2 \cdot x^2}$
$= 3x - 3\sqrt[3]{2x^2}$

PROBLEM 3
Perform the indicated operations. (Assume all variables are greater than or equal to 0.)

a. $\sqrt{2}(\sqrt{3} + \sqrt{10})$

b. $7\sqrt{5x}(\sqrt{x} - \sqrt{3})$

c. $\sqrt[3]{2x}(\sqrt[3]{4x^2} - \sqrt[3]{12x})$

If we wish to obtain the product of two binomials that contain radicals, we first simplify the radicals involved (if possible) and then use FOIL. For example, to find the product $(\sqrt{98} + \sqrt{27})(\sqrt{72} + \sqrt{75})$, we proceed as follows.

$$(\sqrt{98} + \sqrt{27})(\sqrt{72} + \sqrt{75})$$
$$= (\sqrt{49 \cdot 2} + \sqrt{9 \cdot 3})(\sqrt{36 \cdot 2} + \sqrt{25 \cdot 3}) \quad \text{Factor under each radical.}$$
$$= (7\sqrt{2} + 3\sqrt{3})(6\sqrt{2} + 5\sqrt{3}) \quad \text{Simplify.}$$

Answers to PROBLEMS

3. **a.** $\sqrt{6} + 2\sqrt{5}$ **b.** $7x\sqrt{5} - 7\sqrt{15x}$ **c.** $2x - 2\sqrt[3]{3x^2}$

$$(7\sqrt{2} + 3\sqrt{3})(6\sqrt{2} + 5\sqrt{3})$$

$$= \overbrace{7 \cdot 6 \cdot \sqrt{2} \cdot \sqrt{2}}^{F} + \overbrace{7 \cdot 5 \cdot \sqrt{2} \cdot \sqrt{3}}^{O} + \overbrace{3 \cdot 6 \cdot \sqrt{3} \cdot \sqrt{2}}^{I} + \overbrace{3 \cdot 5 \cdot \sqrt{3} \cdot \sqrt{3}}^{L}$$

Use FOIL.

$$= 42\sqrt{2^2} + 35\sqrt{2} \cdot \sqrt{3} + 18\sqrt{3} \cdot \sqrt{2} + 15\sqrt{3^2} \quad \text{Simplify.}$$
$$= 42 \cdot 2 + 35\sqrt{6} + 18\sqrt{6} + 15 \cdot 3 \quad \sqrt{2^2} = 2, \sqrt{3}\sqrt{2} = \sqrt{6}, \text{ and } \sqrt{3^2} = 3$$
$$= 84 + 53\sqrt{6} + 45 \quad \text{Multiply.}$$
$$= 129 + 53\sqrt{6} \quad \text{Combine like terms.}$$

EXAMPLE 4 Using FOIL to multiply binomials containing radicals

Find the product: $(\sqrt{63} + \sqrt{75})(\sqrt{28} - \sqrt{27})$

SOLUTION We first simplify the radicals and then use FOIL.

$$(\sqrt{63} + \sqrt{75})(\sqrt{28} - \sqrt{27}) = (\sqrt{9 \cdot 7} + \sqrt{25 \cdot 3})(\sqrt{4 \cdot 7} - \sqrt{9 \cdot 3})$$
$$= (3\sqrt{7} + 5\sqrt{3})(2\sqrt{7} - 3\sqrt{3})$$
$$\quad\ \ \ F \qquad\ \ O \qquad\ \ \ I \qquad\ \ L$$
$$= 6\sqrt{7^2} - 9\sqrt{21} + 10\sqrt{21} - 15\sqrt{3^2}$$
$$= 6 \cdot 7 + \sqrt{21} - 15 \cdot 3$$
$$= 42 + \sqrt{21} - 45$$
$$= -3 + \sqrt{21}$$

PROBLEM 4
Find the product:

$$(\sqrt{27} - \sqrt{28})(\sqrt{75} + \sqrt{63})$$

Calculator Corner

Verifying Radical Multiplication

Let's verify the results of Example 4. Enter $(\sqrt{(63)} + \sqrt{(75)})(\sqrt{(28)} - \sqrt{(27)})$ [ENTER]. It helps to know that $\sqrt{n} = n^{1/2}$, so you can enter $\sqrt{63}$ as $63^{1/2}$, (or [2nd] [x^2] 63); $\sqrt{75}$ as $75^{1/2}$ (or [2nd] [x^2] 75); and so on as you wish. Now enter the answer $-3 + \sqrt{(21)}$ [ENTER]. In both cases, the result is 1.582575695, the decimal approximation, as shown in the window.

```
(√(63)+√(75))(√(28)
-√(27))
            1.582575695
-3+√(21)
            1.582575695
```

EXAMPLE 5 Multiplying radical expressions
Multiply:

a. $(\sqrt{3} + 2)^2$ **b.** $(3 - 5\sqrt{2})^2$ **c.** $(\sqrt{3} + \sqrt{2})(\sqrt{3} - \sqrt{2})$

SOLUTION

a. Since $(x + a)^2 = x^2 + 2ax + a^2$,
$$(\sqrt{3} + 2)^2 = (\sqrt{3})^2 + 2 \cdot 2 \cdot \sqrt{3} + 2^2$$
$$= 3 + 4\sqrt{3} + 4$$
$$= 7 + 4\sqrt{3}$$

PROBLEM 5
Multiply:

a. $(\sqrt{2} + 5)^2$
b. $(2 - 3\sqrt{7})^2$
c. $(\sqrt{5} + \sqrt{3})(\sqrt{5} - \sqrt{3})$

Answers to PROBLEMS

4. $3 - \sqrt{21}$
5. **a.** $27 + 10\sqrt{2}$ **b.** $67 - 12\sqrt{7}$
   c. 2

b. Since $(x - a)^2 = x^2 - 2ax + a^2$,
$$(3 - 5\sqrt{2})^2 = 3^2 - 2 \cdot 5\sqrt{2} \cdot 3 + (5\sqrt{2})^2$$
$$= 9 - 30\sqrt{2} + 25 \cdot 2$$
$$= 59 - 30\sqrt{2}$$

c. Since $(x + a)(x - a) = x^2 - a^2$,
$$(\sqrt{3} + \sqrt{2})(\sqrt{3} - \sqrt{2}) = (\sqrt{3})^2 - (\sqrt{2})^2$$
$$= 3 - 2$$
$$= 1$$

In Chapter 8 some of the answers will be of the form

$$\frac{6 + \sqrt{8}}{2}$$

We can simplify this expression involving division in two ways.

Method 1. Write

$$\frac{6 + \sqrt{8}}{2}$$

in lowest terms by ① writing $\sqrt{8}$ as $\sqrt{4 \cdot 2} = 2\sqrt{2}$, ② factoring, and then ③ reducing or dividing the common factor. The procedure looks like this:

$$\frac{6 + \sqrt{8}}{2} \stackrel{①}{=} \frac{6 + 2\sqrt{2}}{2} \stackrel{②}{=} \frac{2(3 + \sqrt{2})}{2} \stackrel{③}{=} 3 + \sqrt{2}$$

Method 2. If we view

$$\frac{6 + \sqrt{8}}{2}$$

as a division of a binomial by a monomial, we can solve the problem by ① writing $\sqrt{8}$ as $2\sqrt{2}$; then we ② write each term in the numerator over the common denominator and ③ reduce or divide each term. The procedure looks like this:

$$\frac{6 + \sqrt{8}}{2} \stackrel{①}{=} \frac{6 + 2\sqrt{2}}{2} \stackrel{②}{=} \frac{6}{2} + \frac{2\sqrt{2}}{2} \stackrel{③}{=} 3 + \sqrt{2} \quad \text{As before.}$$

EXAMPLE 6 Simplifying radical expressions involving division
Simplify:

$$\frac{6 + \sqrt{18}}{3}$$

SOLUTION
Method 1. Since $\sqrt{18} = \sqrt{9 \cdot 2} = 3\sqrt{2}$, we have

$$\frac{6 + \sqrt{18}}{3} \stackrel{①}{=} \frac{6 + 3\sqrt{2}}{3} \stackrel{②}{=} \frac{3(2 + \sqrt{2})}{3} \stackrel{③}{=} 2 + \sqrt{2}$$

Method 2. We can also do this problem by dividing individual terms. Thus,

$$\frac{6 + \sqrt{18}}{3} \stackrel{①}{=} \frac{6 + 3\sqrt{2}}{3} \stackrel{②}{=} \frac{6}{3} + \frac{3\sqrt{2}}{3} \stackrel{③}{=} 2 + \sqrt{2}$$

PROBLEM 6
Simplify:

$$\frac{10 + \sqrt{75}}{5}$$

Answers to PROBLEMS

6. $2 + \sqrt{3}$

C ⟩ Rationalizing Denominators

We know how to rationalize the denominator in expressions of the form

$$\frac{a}{\sqrt{b}}$$

We now show you how to rationalize radical expressions that contain sums or differences involving square root radicals in the denominator. The procedure involves the concept of *conjugate expressions*.

> **CONJUGATE**   The expressions $a + b$ and $a - b$ are **conjugates** of each other.

Here are some numbers and their conjugates.

Number	Conjugate
$3 + \sqrt{2}$	$3 - \sqrt{2}$
$-4 + \sqrt{5}$	$-4 - \sqrt{5}$
$7 - \sqrt{3}$	$7 + \sqrt{3}$
$-8 - \sqrt{6}$	$-8 + \sqrt{6}$

Since $(a + b)(a - b) = a^2 - b^2$, the product of a number and its conjugate is $a^2 - b^2$. Now suppose we want to rationalize the denominator in

$$\frac{2}{5 + \sqrt{3}}$$

We use the fundamental property of fractions and multiply the numerator and denominator of

$$\frac{2}{5 + \sqrt{3}}$$

by the conjugate of $(5 + \sqrt{3})$, that is, by $(5 - \sqrt{3})$, to obtain

$$\frac{2}{5 + \sqrt{3}} = \frac{2 \cdot (5 - \sqrt{3})}{(5 + \sqrt{3})(5 - \sqrt{3})}$$

$$= \frac{2 \cdot (5 - \sqrt{3})}{5^2 - (\sqrt{3})^2} \quad \begin{array}{l} (a + b)(a - b) = a^2 - b^2 \text{, so} \\ (5 + \sqrt{3})(5 - \sqrt{3}) = 5^2 - (\sqrt{3})^2 \end{array}$$

$$= \frac{2 \cdot (5 - \sqrt{3})}{25 - 3} \quad \begin{array}{l} (\sqrt{3})^2 = \sqrt{3} \cdot \sqrt{3} \\ = \sqrt{9} \\ = 3 \end{array}$$

$$= \frac{\overset{1}{2} \cdot (5 - \sqrt{3})}{\underset{11}{22}}$$

$$= \frac{5 - \sqrt{3}}{11}$$

> **NOTE**
>
> To rationalize fractions that contain sums or differences involving square root radicals in the denominator, multiply the numerator and denominator by the conjugate of the denominator.

EXAMPLE 7 Rationalizing denominators

Rationalize the denominator:

a. $\dfrac{13}{4 + \sqrt{3}}$

b. $\dfrac{\sqrt{x}}{\sqrt{x} - 2}, x \geq 0, x \neq 4$

PROBLEM 7

Rationalize the denominator:

$\dfrac{\sqrt{y}}{\sqrt{y} + 4}, y \geq 0$

SOLUTION

a. We first multiply numerator and denominator by $(4 - \sqrt{3})$, the conjugate of $(4 + \sqrt{3})$.

$\dfrac{13}{4 + \sqrt{3}} = \dfrac{13}{(4 + \sqrt{3})} \cdot \dfrac{(4 - \sqrt{3})}{(4 - \sqrt{3})}$ Fundamental property of fractions

$= \dfrac{13(4 - \sqrt{3})}{(4)^2 - (\sqrt{3})^2}$ $(4 + \sqrt{3})(4 - \sqrt{3}) = (4)^2 - (\sqrt{3})^2$

$= \dfrac{13(4 - \sqrt{3})}{16 - 3}$ $(4)^2 = 16$ and $(\sqrt{3})^2 = 3$

$= \dfrac{\cancel{13}(4 - \sqrt{3})}{\cancel{13}}$ Simplify.

$= 4 - \sqrt{3}$ Reduce.

b. We first multiply numerator and denominator by $(\sqrt{x} + 2)$, the conjugate of $(\sqrt{x} - 2)$.

$\dfrac{\sqrt{x}}{\sqrt{x} - 2} = \dfrac{\sqrt{x}(\sqrt{x} + 2)}{(\sqrt{x} - 2)(\sqrt{x} + 2)}$ Fundamental property of fractions

$= \dfrac{\sqrt{x}(\sqrt{x} + 2)}{(\sqrt{x})^2 - (2)^2}$ $(\sqrt{x} - 2)(\sqrt{x} + 2) = (\sqrt{x})^2 - (2)^2$

$= \dfrac{\sqrt{x}(\sqrt{x} + 2)}{x - 4}$ $(\sqrt{x})^2 = x$ and $(2)^2 = 4$

$= \dfrac{\sqrt{x^2} + 2\sqrt{x}}{x - 4}$ Use the distributive property.

$= \dfrac{x + 2\sqrt{x}}{x - 4}$ $\sqrt{x^2} = x$ since $x \geq 0$

You can use a similar procedure to rationalize the *numerator* of a radical expression. We discuss how to do this in the *Using Your Knowledge* section of the Exercises.

Calculator Corner

1. Verify the result in Example 5(a).
2. Verify the result in Example 6.
3. Can you graphically verify the result in Example 7(b)? Explain.
4. The age A of a human fetus (in weeks) can be approximated by $A = 25W^{1/3}$, where W is the weight of the fetus in kilograms. (A kilogram, kg, is about 2.2 pounds.) If you want to make a table for $A = 25W^{1/3}$ using Y1 for the age A and X for the weight W, press `Y=` and enter $25x^{1/3}$, then press `2nd` `WINDOW` on a TI-83 Plus. Now answer these questions.
   a. What would you use for the minimum weight?
   b. What kind of increment (ΔTbl) would you use? (*Hint:* Look at the table shown in the window.)
   c. If a fetus weighs about 2.3 kilograms, how old is it?
   d. Assuming a 36-week pregnancy, what birth weight would you predict based on the table?

X	Y1
0	0
.1	11.604
.2	14.62
.3	16.736
.4	18.42
.5	19.843
.6	21.086
X=0	

Answers to PROBLEMS

7. $\dfrac{y - 4\sqrt{y}}{y - 16}$

> Exercises 7.3

‹A› Adding and Subtracting Similar Radical Expressions   In Problems 1–28, perform the indicated operations. (Where the index is even, assume all variables are positive.)

1. $12\sqrt{2} + 3\sqrt{2}$
2. $15\sqrt{3} + 2\sqrt{3}$
3. $\sqrt{80a} + \sqrt{125a}$
4. $\sqrt{98a} + \sqrt{32a}$
5. $\sqrt{50} - 4\sqrt{32}$
6. $\sqrt{75} + 7\sqrt{12}$
7. $\sqrt{50a^2} - \sqrt{200a^2}$
8. $\sqrt{48a^2} - \sqrt{363a^2}$
9. $2\sqrt{300} - 9\sqrt{12} - 7\sqrt{48}$
10. $\sqrt{175} + \sqrt{567} - \sqrt{63}$
11. $3x\sqrt{20x} - \sqrt{24x} + \sqrt{45x^3}$
12. $a\sqrt{18} + 4a\sqrt{8} - 5a\sqrt{3}$
13. $\sqrt[3]{40} + 3\sqrt[3]{625}$
14. $\sqrt[3]{54} + 2\sqrt[3]{16}$
15. $\sqrt[3]{81} - 3\sqrt[3]{375}$
16. $\sqrt[3]{24} - \sqrt[3]{81}$
17. $2\sqrt[3]{-24} - 4\sqrt[3]{-81} - \sqrt[3]{375}$
18. $10\sqrt[3]{-40} - 2\sqrt[3]{-135} + 4\sqrt[3]{-320}$
19. $\sqrt[3]{3a} - \sqrt[3]{24a} + \sqrt[3]{375a}$
20. $\sqrt[3]{r^5} - \sqrt[3]{8r^5} - r\sqrt[3]{64r^2}$
21. $\dfrac{3\sqrt[3]{3}}{2} - \dfrac{\sqrt[3]{3}}{3}$
22. $\dfrac{4}{5} - \dfrac{\sqrt[3]{2}}{2}$
23. $\sqrt{\dfrac{1}{2}} + \sqrt{\dfrac{1}{3}} + \sqrt{\dfrac{1}{6}}$
24. $\sqrt{\dfrac{25}{3}} - 2\sqrt{\dfrac{16}{3}} + 2\sqrt{\dfrac{4}{3}}$
25. $\sqrt{\dfrac{2}{3}} - \sqrt{\dfrac{1}{6}} + \sqrt{\dfrac{1}{2}}$
26. $\sqrt[3]{\dfrac{1}{5}} + \sqrt[3]{\dfrac{1}{40}}$
27. $6\sqrt[3]{\dfrac{3}{5}} + 6\sqrt[3]{\dfrac{81}{40}}$
28. $2a\sqrt[3]{\dfrac{a}{5}} + 6\sqrt[3]{\dfrac{a^4}{40}}$

‹B› Multiplying and Dividing Radical Expressions   In Problems 29–62, perform the indicated operations. (Where the index is even, assume all variables are positive.)

29. $3(5 - \sqrt{2})$
30. $-2(\sqrt{2} - 3)$
31. $\sqrt[3]{2}(\sqrt[3]{4} + 3)$
32. $\sqrt[3]{3}(\sqrt[3]{9} + 2)$
33. $2\sqrt{3}(7\sqrt{5} + 5\sqrt{3})$
34. $2\sqrt{5}(5\sqrt{2} + 3\sqrt{5})$
35. $3\sqrt[3]{5}(2\sqrt[3]{3} - \sqrt[3]{25})$
36. $4\sqrt[3]{2}(3\sqrt[3]{4} - 3\sqrt[3]{2})$
37. $-4\sqrt{7}(2\sqrt{3} - 5\sqrt{2})$
38. $-3\sqrt{2}(5\sqrt{7} - 2\sqrt{3})$
39. $(5\sqrt{3} + \sqrt{5})(3\sqrt{3} + 2\sqrt{5})$
40. $(2\sqrt{2} + 5\sqrt{3})(3\sqrt{2} + \sqrt{3})$
41. $(3\sqrt{6} - 2\sqrt{3})(4\sqrt{6} + 5\sqrt{3})$
42. $(3\sqrt{5} - 2\sqrt{3})(2\sqrt{5} + 3\sqrt{3})$
43. $(7\sqrt{5} - 11\sqrt{7})(5\sqrt{5} + 8\sqrt{7})$
44. $(2\sqrt{3} - 5\sqrt{2})(3\sqrt{3} + 2\sqrt{2})$
45. $(1 + \sqrt{2})(1 - \sqrt{2})$
46. $(2 + \sqrt{3})(2 - \sqrt{3})$
47. $(2 + 3\sqrt{3})(2 - 3\sqrt{3})$
48. $(5 + 5\sqrt{2})(5 - 5\sqrt{2})$
49. $(\sqrt{3} + \sqrt{2})^2$
50. $(\sqrt{2} + \sqrt{5})^2$
51. $(a + \sqrt{b})^2$
52. $(\sqrt{a} + b)^2$
53. $(\sqrt{3} - \sqrt{2})^2$
54. $(\sqrt{2} - \sqrt{5})^2$
55. $(a - \sqrt{b})^2$
56. $(\sqrt{b} - a)^2$
57. $(\sqrt{a} - \sqrt{b})^2$
58. $(\sqrt{b} - \sqrt{a})^2$
59. $\dfrac{3 + \sqrt{18}}{3}$
60. $\dfrac{5 + \sqrt{50}}{5}$
61. $\dfrac{6 - \sqrt{27}}{12}$
62. $\dfrac{8 - \sqrt{32}}{4}$

⟨ C ⟩ Rationalizing Denominators In Problems 63–73, rationalize the denominator. (Assume all variables represent positive numbers.)

63. $\dfrac{3 + \sqrt{3}}{\sqrt{2}}$

64. $\dfrac{2 + \sqrt{5}}{\sqrt{3}}$

65. $\dfrac{2}{3 - \sqrt{2}}$

66. $\dfrac{6}{2 - \sqrt{2}}$

67. $\dfrac{4a}{3 - \sqrt{5}}$

68. $\dfrac{3a}{4 - \sqrt{3}}$

69. $\dfrac{3a + 2b}{3 + \sqrt{2}}$

70. $\dfrac{5a + b}{2 + \sqrt{3}}$

71. $\dfrac{\sqrt{a} + b}{\sqrt{a} - b}$

72. $\dfrac{a + \sqrt{b}}{a - \sqrt{b}}$

73. $\dfrac{\sqrt{a} + \sqrt{2b}}{\sqrt{a} - \sqrt{2b}}$

⟩⟩⟩ Using Your Knowledge

Rationalizing Numerators in Calculus In this section we learned how to rationalize the denominator of a fraction involving the sum or difference of radical expressions. In calculus, we sometimes have to rationalize the *numerator* of a fraction that involves the sum or difference of radical expressions. The idea is the same: Multiply numerator and denominator by the *conjugate* of the numerator. To rationalize the numerator in

$$\dfrac{\sqrt{3} + \sqrt{2}}{5}$$

we proceed as follows.

$$\dfrac{\sqrt{3} + \sqrt{2}}{5} = \dfrac{(\sqrt{3} + \sqrt{2})(\sqrt{3} - \sqrt{2})}{5(\sqrt{3} - \sqrt{2})}$$ Multiply numerator and denominator by the conjugate of the numerator.

$$= \dfrac{3 - 2}{5(\sqrt{3} - \sqrt{2})}$$ $(\sqrt{3} + \sqrt{2})(\sqrt{3} - \sqrt{2}) = (\sqrt{3})^2 - (\sqrt{2})^2$

$$= \dfrac{1}{5(\sqrt{3} - \sqrt{2})}$$

In Problems 74–83, rationalize the numerator.

74. $\dfrac{\sqrt{5} + \sqrt{2}}{3}$

75. $\dfrac{\sqrt{5} + \sqrt{3}}{4}$

76. $\dfrac{\sqrt{x} - \sqrt{2}}{5}$

77. $\dfrac{\sqrt{5} - \sqrt{x}}{5}$

78. $\dfrac{\sqrt{x} + \sqrt{y}}{x}$

79. $\dfrac{\sqrt{x} - \sqrt{y}}{x}$

80. $\dfrac{\sqrt{x} + \sqrt{y}}{\sqrt{x}}$

81. $\dfrac{\sqrt{x} + \sqrt{y}}{\sqrt{y}}$

82. $\dfrac{\sqrt{x} - \sqrt{y}}{\sqrt{x}}$

83. $\dfrac{\sqrt{x} - \sqrt{y}}{\sqrt{y}}$

⟩⟩⟩ Write On

84. Why is it impossible to combine $\sqrt{3x} + \sqrt[3]{3x}$ into a single term?

85. Explain why $\sqrt{a + b} \neq \sqrt{a} + \sqrt{b}$ and give examples.

86. Explain why $\sqrt[3]{x} \cdot \sqrt[3]{x} \neq x$. What factor do you need in the box, $\sqrt[3]{x} \cdot \sqrt[3]{x} \cdot \square = x$, to make the statement true?

87. What does it mean when we say "rationalize the denominator"?

88. State what conditions have to be met for a radical expression to be simplified.

>>> Concept Checker

Fill in the blank(s) with the correct word(s), phrase, or mathematical statement.

89. To multiply $(6 + \sqrt{5})(2 - \sqrt{5})$ use the _____ process.
90. $-2 + 3\sqrt{5}$ and $-2 - 3\sqrt{5}$ are _____.
91. If two radical expressions have the same index and the same radicand, then they are called _____.

like radical expressions   radicals

equivalent   FOIL

conjugate expressions

>>> Mastery Test

Rationalize the denominator (assume variables are positive):

92. $\dfrac{\sqrt{y}}{\sqrt{y} + \sqrt{x}}$

93. $\dfrac{\sqrt{y}}{\sqrt{y} - \sqrt{x}}$

Reduce to lowest terms:

94. $\dfrac{10 + \sqrt{50}}{5}$

95. $\dfrac{20 + \sqrt{32}}{4}$

Perform the indicated operations (assume variables are positive):

96. $(\sqrt{27} + \sqrt{28})(\sqrt{75} - \sqrt{112})$

97. $(\sqrt{28} - \sqrt{27})(\sqrt{112} + \sqrt{75})$

98. $\sqrt{2}(\sqrt{3} + \sqrt{10})$

99. $\sqrt{5x}(\sqrt{x} - \sqrt{3})$

100. $\sqrt[3]{2x}(\sqrt[3]{4x^2} - \sqrt[3]{16x})$

101. $5\sqrt{\dfrac{1}{2}} - 7\sqrt{\dfrac{1}{8}}$

102. $4\sqrt[3]{\dfrac{5x}{4x^2}} - \sqrt[3]{\dfrac{5}{32x}}$

103. $2\sqrt[3]{250x} - 4\sqrt[3]{16x}$

104. $3\sqrt{20x} - 5\sqrt{45x}$

105. $\sqrt{98} - \sqrt{50}$

106. $\sqrt{44} + \sqrt{99}$

>>> Skill Checker

In Problems 107–111, solve the equation.

107. $x + 5 = 9$

108. $2x + 3 = 25$

109. $x^2 - 15x + 50 = 0$

110. $x^2 - 3x + 2 = 0$

111. $x^2 + 6x + 5 = 0$

Raise the following radicals to the power indicated. (Assume all radicands are positive.)

112. $(\sqrt{x + 3})^2$

113. $(-\sqrt{2x})^2$

114. $(\sqrt[3]{5x - 4})^3$

115. $(\sqrt[3]{6 - x})^3$

7.4 Solving Equations Containing Radicals

Objectives

A Solve equations involving radicals.

B Solve applications involving radical equations.

To Succeed, Review How To . . .

1. Solve linear and quadratic equations (pp. 76–83, 405–410).
2. Square a radical expression (pp. 551–553).

Getting Started
Radical Curves

If a traffic engineer wants the speed limit v on this curve to be 45 miles per hour, what radius should the curve have? The speed v (in miles per hour) a car can travel on a concrete highway curve without skidding is $v = \sqrt{9r}$, where r is the radius of the curve in feet. Since $v = 45$, to find the answer we must find r in the equation

$$45 = \sqrt{9r}$$
$$(45)^2 = (\sqrt{9r})^2 \quad \text{Square both sides.}$$
$$2025 = 9r \quad (\sqrt{9r})^2 = 9r$$
$$\frac{2025}{9} = r \quad \text{Divide by 9.}$$
$$225 = r$$

The curve must have a radius r of 225 feet or more. We can check this by substituting 225 for r in the equation

$$45 = \sqrt{9r}$$
to obtain $\quad 45 = \sqrt{9 \cdot 225} = \sqrt{9} \cdot \sqrt{225} = 3 \cdot 15$

which is a true statement. Thus, the curve must have at least a 225-foot radius.

This problem gives rise to the notion that sometimes it is necessary to know how to solve equations with radical terms, and that is what we will study in this section.

A Solving Equations Involving Radicals

In algebra, the equation $45 = \sqrt{9r}$ is called a **radical equation** and can be solved by squaring both sides of the equation. Sometimes, however, squaring both sides introduces **extraneous solutions**—solutions that do not satisfy the original equation.

For example, the equation $x = 2$ has one solution, 2. Squaring both sides gives $x^2 = 4$. The equation $x^2 = 4$ has two solutions, 2 and -2. We introduced the extraneous solution -2 when we squared both sides, so we got more solutions than we needed. However, all solutions of the equation $x = 2$ are solutions of $x^2 = 4$, so we didn't lose any of the solutions that we needed.

> **POWER RULE FOR EQUATIONS**
>
> All solutions of the equation $P = Q$ are solutions of the equation $P^n = Q^n$, where n is a natural number.

This rule tells us that when we raise both sides of an equation to a power, the solutions of the *original* equation are *always* solutions of the new equation. However, the new equation may have extraneous solutions that have to be discarded. Because of this, *the solutions of the new equations must be checked in the original equation and extraneous solutions discarded*. Here is the procedure we use to solve $\sqrt{x + 3} + 2 = 6$.

$\sqrt{x + 3} + 2 = 6$	Given
$\sqrt{x + 3} = 4$	Isolate the radical.
$(\sqrt{x + 3})^2 = (4)^2$	Square both sides.
$x + 3 = 16$	Solve resulting equation.
$x = 13$	
$\sqrt{13 + 3} + 2 \stackrel{?}{=} 6$	Check.
$\sqrt{16} + 2 = 6$	

Now we summarize the procedure for solving equations with radicals.

> **PROCEDURE**
>
> **To Solve Equations Containing Radicals**
> 1. **Isolate** one radical that contains variables on one side of the equation.
> 2. **Raise** each side of the equation to a power that is the same as the index of the radical isolated in Step 1.
> 3. **Simplify**.
> 4. **Repeat** Steps 1–3 *if* the equation still contains a radical term.
> 5. **Solve** the resulting linear or quadratic equation using an appropriate method.
> 6. **Check** all proposed (trial) solutions in the original equation.

EXAMPLE 1 Solving equations containing one radical
Solve:

a. $\sqrt{4x} = 8$ b. $\sqrt{x + 1} = x - 1$ c. $\sqrt{x - 1} - x = -1$

PROBLEM 1
Solve:

a. $\sqrt{5x} = 10$
b. $\sqrt{2x - 1} = x - 2$
c. $\sqrt{3 - x} + 1 = x$

SOLUTION

a. We use the six-step procedure. In this case, the radical containing the variable is already isolated.

1. $\sqrt{4x} = 8$ — Given
2. $(\sqrt{4x})^2 = (8)^2$ — Square each side.
3. $4x = 64$ — $(\sqrt{4x})^2 = 4x$
4. There are no radical terms left.
5. $x = 16$ — Linear equation—solve with CRAM.
6. Substituting the proposed solution 16 for x in the original equation gives

$$\sqrt{4(16)} \stackrel{?}{=} 8$$
$$\sqrt{64} = 8$$

a true statement. The solution of $\sqrt{4x} = 8$ is 16.

Answers to PROBLEMS

1. a. 20   b. 5   c. 2

b. Using the six-step procedure, we note that the radical is already isolated, so we square both sides to eliminate the radical.

1. $\sqrt{x+1} = x - 1$ Given
2. $(\sqrt{x+1})^2 = (x-1)^2$ Square each side.
3. $x + 1 = x^2 - 2x + 1$ Expand $(x-1)^2$, which equals $x^2 - 2x + 1$.
4. There are no radical terms left.
5. $0 = x^2 - 3x$ O   Quadratic equation—solve with OFF. Subtract x and 1 to write the resulting quadratic equation in standard form.

     $0 = x(x-3)$ F   Factor.
     $x = 0$ or $x - 3 = 0$ F   Factor = 0. Use the zero-factor property.
     $x = 0$ or $x = 3$ Solve $x - 3 = 0$.

Thus, the proposed solutions are 0 and 3.

Note: In Step 5 do not divide by x. It would eliminate one of the possible solutions to the equation, $x = 0$.

6. Substituting 0 for x in the original equation, we have

$$\sqrt{0+1} \stackrel{?}{=} 0 - 1$$
$$1 \stackrel{?}{=} -1 \quad \text{A false statement.}$$

Thus, zero is not a solution. Substituting 3 for x in $\sqrt{x+1} = x - 1$, we have

$$\sqrt{3+1} \stackrel{?}{=} 3 - 1$$
$$\sqrt{4} = 2 \quad \text{A true statement.}$$

The solution of $\sqrt{x+1} = x - 1$ is 3. (Discard the extraneous solution, 0.)

c. We first have to isolate the radical on one side (step 1), so we start by adding x on both sides. Then we proceed as before.

$\sqrt{x-1} - x = -1$ Given

1. $\sqrt{x-1} = x - 1$ Add x on both sides. (Now $\sqrt{x-1}$ is isolated.)
2. $(\sqrt{x-1})^2 = (x-1)^2$ Square each side.
3. $x - 1 = x^2 - 2x + 1$ Expand on the right.
4. There are no radicals left.
5. $0 = x^2 - 3x + 2$ O   Quadratic equation—solve with OFF. Subtract x and add 1.

     $0 = (x-2)(x-1)$ F   Factor.
     $x - 2 = 0$ or $x - 1 = 0$ F   Factors = 0. Use the zero-factor property.
     $x = 2$ or $x = 1$ Solve.

6. The proposed solutions are 2 and 1. Substituting 2 for x, we have

$$\sqrt{x-1} - x = -1$$
$$\sqrt{2-1} - 2 \stackrel{?}{=} -1$$
$$\sqrt{1} - 2 = -1 \quad \text{A true statement.}$$

2 is a solution. Now we check the proposed solution 1 by substitution.

$$\sqrt{x-1} - x = -1$$
$$\sqrt{1-1} - 1 \stackrel{?}{=} -1$$
$$\sqrt{0} - 1 = -1 \quad \text{A true statement.}$$

1 is also a solution. The solutions are 1 and 2.

Sometimes we have more than one radical in an equation. In such cases, we must isolate one of the radicals and raise both sides of the equation to the appropriate power. If we still have radicals on one side of the equation, we isolate one of them and square again (Step 4 in the procedure). To solve $\sqrt{x-11} = \sqrt{x} - 1$, we first square both sides of the equation. The right-hand side then contains an expression of the form $(x-a)^2 = x^2 - 2ax + a^2$. Using our six-step procedure, we have

1. $\sqrt{x-11} = \sqrt{x} - 1$ $\sqrt{x-11}$ is isolated.
2. $(\sqrt{x-11})^2 = (\sqrt{x} - 1)^2$ Square each side.
3. $x - 11 = x - 2 \cdot 1 \cdot \sqrt{x} + 1$ Simplify.
   $x - 11 = x - 2\sqrt{x} + 1$
4. $-11 = -2\sqrt{x} + 1$ Subtract x. — Since we still have a radical term ($\sqrt{x}$), we
   $-12 = -2\sqrt{x}$ Subtract 1.   isolate $\sqrt{x}$. This is step 4
   $6 = \sqrt{x}$ Divide by -2. — in the procedure.
5. $36 = x$ Square each side.
6. The solution is 36. You can verify this by substituting in the original equation.

EXAMPLE 2   Solving an equation containing two radicals

Solve: $\sqrt{x-5} - \sqrt{x} = -1$

SOLUTION In step 1, we add $\sqrt{x}$ to both sides of the equation so that $\sqrt{x-5}$ is isolated.

$\sqrt{x-5} - \sqrt{x} = -1$ Given
1. $\sqrt{x-5} = \sqrt{x} - 1$ Add $\sqrt{x}$ to isolate $\sqrt{x-5}$.
2. $(\sqrt{x-5})^2 = (\sqrt{x}-1)^2$ Square each side.
3. $x - 5 = x - 2\sqrt{x} + 1$ $(\sqrt{x} - 1)^2 = x - 2\sqrt{x} + 1$.
4. $-5 = -2\sqrt{x} + 1$ Subtract x.
   $-6 = -2\sqrt{x}$ Subtract 1.   We have to isolate $\sqrt{x}$.
   $3 = \sqrt{x}$ Divide both sides by -2.
5. $9 = x$ Square each side.
6. The proposed solution is 9. Since $\sqrt{9-5} - \sqrt{9} = \sqrt{4} - \sqrt{9} = 2 - 3 = -1$, 9 is the correct solution.

PROBLEM 2
Solve:

$\sqrt{x-3} - \sqrt{x} = -1$

The sides of an equation can be raised to powers greater than 2. For example, to solve the equation $\sqrt[4]{x} = 2$, we raise both sides of the equation to the fourth power. (Step 2 says that we have to *raise* both sides of the equation to a power that is the same as the index of the radical, 4.) Thus,

$$(\sqrt[4]{x})^4 = 2^4$$
$$x = 16$$

Here is another example.

EXAMPLE 3   Solving an equation containing a cube root

Solve: $\sqrt[3]{x-2} = 3$

SOLUTION The radical is isolated, so we go to step 2.

2. $(\sqrt[3]{x-2})^3 = 3^3$ Cube each side because 3 is the index.
3. $x - 2 = 27$ Simplify.

PROBLEM 3
Solve:

$\sqrt[3]{x-5} = -2$

Answers to PROBLEMS

2. 4   3. -3

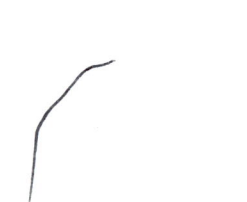

4. There are no radicals left.
5. $x = 29$ Linear equation—solve with CRAM. Add 2.
6. Substituting 29 for x in the original equation gives

$$\sqrt[3]{29 - 2} \stackrel{?}{=} 3$$
$$\sqrt[3]{27} = 3$$

Since this statement is true, 29 is the solution of $\sqrt[3]{x - 2} = 3$.

EXAMPLE 4 Solving an equation containing a fourth root

Solve: $\sqrt[4]{x - 1} + 3 = 0$

SOLUTION We use our six-step procedure.

$\sqrt[4]{x - 1} + 3 = 0$ Given
1. $\sqrt[4]{x - 1} = -3$ Subtract 3 to isolate $\sqrt[4]{x - 1}$.
2. $(\sqrt[4]{x - 1})^4 = (-3)^4$ Raise to the fourth power.
3. $x - 1 = 81$ Simplify.
4. There are no radicals left.
5. $x = 82$ Linear equation—solve with CRAM. Add 1 to solve $x - 1 = 81$.
6. Substitute 82 for x in $\sqrt[4]{x - 1} + 3 = 0$ to obtain

$$\sqrt[4]{82 - 1} + 3 \stackrel{?}{=} 0$$
$$\sqrt[4]{81} + 3 = 0$$
$$3 + 3 = 0 \quad \text{False}$$

Thus, 82 is an extraneous solution. There is no real-number solution for this equation. The solution set is $\varnothing$.

Now look at Step 1. Can you tell that there is no solution to the equation? How can you tell?

PROBLEM 4

Solve:

$\sqrt[4]{x + 3} + 4 = 6$

B › Solving Applications Involving Radical Equations

A common use of radicals occurs in problems involving free-falling objects. (See Problems 43 and 44.) Here is an example.

EXAMPLE 5 Free-falling object

If an object is dropped from a height of h feet, the relationship between its velocity v (in feet per second) when it hits the ground and the height h is $v^2 = 2gh$, where $g = 32$ ft/sec^2.

a. Solve for v.
b. If an object is dropped from a height of 81 feet, what is its velocity when it hits the ground?

PROBLEM 5

The distance a free-falling object has fallen from a position of rest is dependent upon the time of fall. The distance can be computed by this formula; the distance fallen after a time of t seconds is given by the formula

$$d = \tfrac{1}{2}gt^2$$

(continued)

Answers to PROBLEMS

4. 13
5. a. $t = \sqrt{\dfrac{2d}{g}}$ or $t = \dfrac{\sqrt{2dg}}{g}$ b. 4 sec

SOLUTION

a. We have to solve the equation $v^2 = 2gh$ for v. Taking the square root of both sides of $v^2 = 2gh$, we have $v = \sqrt{2gh}$.

b. Here we have to find v when $h = 81$ and $g = 32$. Substituting 81 for h and 32 for g in $v = \sqrt{2gh}$, we obtain

$$v = \sqrt{2gh} = \sqrt{2 \cdot 32 \cdot 81} = \sqrt{64 \cdot 81} = 8 \cdot 9 = 72$$

Thus, the velocity v of the object when it hits the ground is 72 feet per second. (Remember that v is in feet per second.)

where g is the acceleration of gravity (approximately 10 meters per square second on Earth).

a. Solve for t.

b. How long does it take a free-falling object to fall 80 meters?

EXAMPLE 6 Call lengths for cell phones

Do you have a cell phone? How long are your calls? According to the Cellular Telecommunications Industry Association, the average length of a call for 1999, 2000, and 2001, was 2.38, 2.56, and 2.74 minutes, respectively. The average length can be approximated by $L(t) = \sqrt{t + 5.5}$, where $L(t)$ represents the length of the call, in minutes, t years after 1999.

a. Use the formula to approximate the average length of a call for 2000. (The actual length was 2.56 minutes.)

b. In what year (to the nearest year) would you expect the average length of a call to be 4 minutes?

PROBLEM 6

Use Example 6 to do the following:

a. Use the formula to approximate the average length of a call for 2007.

b. In what year (to the nearest year) would you expect the average length of a call to be 5 minutes?

SOLUTION

a. Since t is the number of years after 1999, in 2000 $t = 1$ and
$L(1) = \sqrt{1 + 5.5} = \sqrt{6.5} \approx 2.55$ min.

b. To predict when the call length will be 4 min, we have to find t when $L(t) = 4$. Thus, we have to solve the equation

$$\sqrt{t + 5.5} = 4$$
$$t + 5.5 = 16 \quad \text{Square both sides.}$$
$$t = 10.5 \quad \text{Subtract 5.5 from both sides.}$$
$$t \approx 11 \quad \text{To the nearest year}$$

Approximately 11 years after 1999, in 2010, the average length of a cellular call is expected to be 4 minutes.

Answers to PROBLEMS

6. a. ≈ 3.67 min   b. 2019

> Exercises 7.4

Boost your grade at mathzone.com!
> Practice Problems
> NetTutor
> Self-Tests
> e-Professors
> Videos

< A > Solving Equations Involving Radicals In Problems 1–20, solve the given equation.

1. $\sqrt{x} = 4$
2. $\sqrt{3x} = 6$
3. $\sqrt{x + 6} = 7$
4. $\sqrt{x - 3} = 10$
5. $\sqrt{\dfrac{x}{2}} = 3$
6. $\sqrt{\dfrac{3x}{2}} = 3$
7. $\sqrt[4]{x + 1} + 2 = 0$
8. $\sqrt[4]{x + 3} + 5 = 0$
9. $\sqrt[3]{3x - 1} = \sqrt[3]{5x - 7}$
10. $\sqrt[3]{5x - 3} = \sqrt[3]{7x - 5}$
11. $\sqrt{x + 4} = x + 2$
12. $\sqrt{x + 3} = x + 1$
13. $\sqrt{x + 3} = x - 3$
14. $\sqrt{x + 9} = x - 3$
15. $\sqrt[3]{y + 8} = -2$

16. $\sqrt[3]{y+4} = -1$

17. $\sqrt{x+5} - x = -7$

18. $\sqrt{x+5} - x = -1$

19. $\sqrt{x-5} - x = -7$

20. $\sqrt{x+3} - x = 3$

In Problems 21–30, solve the given equation.

21. $\sqrt{y+1} = \sqrt{y} + 1$

22. $\sqrt{y-4} = 2 + \sqrt{y}$

23. $\sqrt{y+8} - \sqrt{y} = 2$

24. $\sqrt{y+5} - \sqrt{y} = 1$

25. $\sqrt{x+3} = \sqrt{x} + \sqrt{3}$

26. $\sqrt{x+5} = \sqrt{x} + \sqrt{5}$

27. $\sqrt{5x-1} + \sqrt{x+3} = 4$

28. $\sqrt{2x-1} + \sqrt{x+3} = 3$

29. $\sqrt{x-3} + \sqrt{2x+1} = 2\sqrt{x}$

30. $\sqrt{x+4} + \sqrt{3x+9} = \sqrt{x+25}$

In Problems 31–38, solve for x or y.

31. $\sqrt{x-a} = b$

32. $\sqrt{x+a} = b$

33. $\sqrt[3]{a-by} = c$

34. $\sqrt[3]{a^3+y} = b$

35. $\sqrt{\dfrac{x}{a}} = b$

36. $\sqrt{\dfrac{x}{b}} = \dfrac{a}{b}$

37. $\sqrt[3]{3x-a} = \sqrt[3]{b-a}$

38. $\sqrt[3]{2x-b} = \sqrt[3]{b-2a}$

⟨B⟩ Solving Applications Involving Radical Equations

39. *Volume* The formula for finding the length of the base, b, of a pyramid with a square base is $b = \sqrt{\dfrac{3V}{H}}$, where V is the volume and H is the height. If one of the Egyptian pyramids has a base that approximates a square with an approximate volume of 2,000,000 cubic meters and a height of 114 meters, what is the measure of one side of the square base of the pyramid? (Approximate to the nearest meter.)

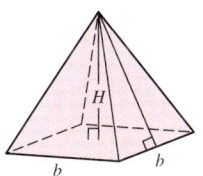

40. *Surface area* The formula for finding the length of the edge of a cube is $e = \sqrt{\dfrac{S}{6}}$, where S is the surface area of the cube. Find the length of the edge of the cube if the surface area is 108 square feet. (Approximate to the nearest tenth of a foot.)

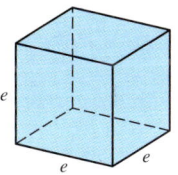

41. *Surface area* The radius r of a sphere is given by

$$r = \sqrt{\dfrac{S}{4\pi}}$$

where S is the surface area. If the surface area of a sphere is 942 square feet, find its radius. (Use 3.14 for π and round to the nearest hundredth.)

42. *Volume* The radius r of a cone is given by

$$r = \sqrt{\dfrac{3V}{\pi h}}$$

where V is the volume and h is the height. If a 10-centimeter-high cone contains 94.26 cubic centimeters of ice cream, what is its radius? (Use 3.142 for π and round to the nearest whole number.)

43. *Falling object* The time t (in seconds) it takes a body to fall d feet is given by the equation

$$t = \sqrt{\dfrac{2d}{g}}$$

where g is the gravitational acceleration.

a. Solve for d.

b. How far would a body fall in 3 seconds? (Use 32.2 feet per square second for g.)

44. *Falling object* After traveling d feet, the velocity v (in feet per second) of a falling body starting from rest is given by the equation

$$v = \sqrt{2gd}$$

a. Solve for d.

b. If a body that started from rest is traveling at 44 feet per second, how far has it fallen? (Use 32 feet per square second for g.)

45. *Pendulum* A pendulum of length L (in feet) takes

$$t = 2\pi\sqrt{\dfrac{L}{g}}$$

seconds to go through a complete cycle.

a. Solve for L.

b. If a pendulum takes 2 seconds to go through one complete cycle, how long is the pendulum? (Use 32 feet per square second for g and $\dfrac{22}{7}$ for π.)

⟩⟩⟩ Using Your Knowledge

Engineering and Radicals Suppose you are the engineer designing several roads. We mentioned at the beginning of this section that the velocity v (miles per hour) that a car can travel on a concrete highway curve without skidding is $v = \sqrt{9r}$, where r (in feet) is the radius of the curve.

Use your knowledge to determine the radius of the curve on a highway exit in which you want the speed to be as follows.

46. 25 miles per hour **47.** 30 miles per hour **48.** 35 miles per hour **49.** 40 miles per hour

⟩⟩⟩ Write On

50. Consider the equation $\sqrt{x+1} + 2 = 0$.

    a. What should be the first step in solving this equation?

    b. List reasons that show that this equation has no real-number solutions.

51. Consider the equation $\sqrt{x+3} = -\sqrt{2x-3}$. List reasons that show that this equation has no real-number solutions.

52. What is your definition of a "proposed" or "trial" solution when you solve equations involving radicals?

53. Why is it necessary to check proposed solutions in the original equation when you solve equations involving radicals?

⟩⟩⟩ Concept Checker

Fill in the blank(s) with the correct word(s), phrase, or mathematical statement.

54. When solving radical equations you must check for _____.

55. $5 - \sqrt{x+2} = 10$ is an example of a(n) _____.

56. If all the proposed solutions to a radical equation do not check then the equation has _____.

radical equation

extraneous solutions

power rule for equations

no solution

⟩⟩⟩ Mastery Test

Solve, if possible:

57. $\sqrt[3]{x-5} = 2$

58. $\sqrt[4]{x+3} + 16 = 0$

59. $\sqrt{x-3} - \sqrt{x} = -1$

60. $\sqrt{x+1} - x = 1$

61. $\sqrt{x+1} = x - 5$

62. $\sqrt{x+2} + 3 = 0$

63. $\sqrt{x-2} - 2 = 0$

64. $\sqrt{x} + \sqrt{2x+1} = 1$

65. The power used by an appliance is given by

$$I = \sqrt{\frac{P}{R}}$$

where I is the current (in amps), R is the resistance (in ohms), and P is the power (in watts).

    a. Solve for R.

    b. Find the resistance R for an electric oven rated at 1500 watts and drawing $I = 10$ amps of current.

⟩⟩⟩ Skill Checker

Simplify by removing parentheses and combining like terms:

66. $(5 + 4x) + (7 - 2x)$

67. $(3 + 4x) + (8 + 2x)$

68. $(9 + 2x) - (2 + 4x)$

69. $(6 + 5x) - (7 - 3x)$

70. $(8 + 3x) - (5 - 4x)$

Rationalize the denominator:

71. $\dfrac{2 + 3\sqrt{2}}{4 + \sqrt{2}}$

72. $\dfrac{4 + 4\sqrt{3}}{3 + 2\sqrt{3}}$

73. $\dfrac{2 - \sqrt{2}}{5 - 3\sqrt{2}}$

74. $\dfrac{3 - \sqrt{3}}{5 - \sqrt{3}}$

75. $\dfrac{\sqrt{x} - \sqrt{y}}{\sqrt{x} + \sqrt{y}}$

7.5 Complex Numbers

▶ Objectives

A ▶ Write the square root of a negative integer in terms of i.

B ▶ Add and subtract complex numbers.

C ▶ Multiply and divide complex numbers.

D ▶ Find powers of i.

▶ To Succeed, Review How To . . .

1. Remove parentheses and combine like terms in an expression (pp. 55–58).
2. Rationalize the denominator of an expression (pp. 541–543, 554–555).

▶ Getting Started

Sums and Products

Can you find two numbers whose sum is 10 and whose product is 40? Girolamo Cardan (1501–1576), an Italian mathematician (at right), claimed the answer is

$$5 + \sqrt{-15} \quad \text{and} \quad 5 - \sqrt{-15}$$

Obviously, the sum is 10, but what about the product? If we let $a = 5$ and $b = \sqrt{-15}$, we can use the product rule

$$(a + b)(a - b) = a^2 - b^2$$

and obtain $\quad 5^2 - (\sqrt{-15})^2 = 25 - (-15) = 40$

Too Complex?

What is the problem? Well, $\sqrt{-15}$ is not a real number! To solve this, Carl Friedrich Gauss (1777–1855) (at right) developed a new set of numbers containing elements that are square roots of negative numbers. One of these numbers is i, and it is defined as follows.

i is a number such that $i^2 = -1$; that is, $i = \sqrt{-1}$

We will learn how to operate with complex numbers in this section.

A ▸ Writing Square Roots of Negative Numbers in Terms of *i*

The complex numbers we will study in this section took several centuries to evolve, as indicated in the *Getting Started*. These complex numbers have applications in areas such as fluid flow and complex circuits. They belong to the area of mathematics known as complex analysis. We will begin our study of complex numbers by seeing how these numbers may occur when solving quadratic equations.

When trying to solve quadratic equations such as $x^2 = 25$, by inspection, we see that $+5$ or -5 are solutions to this equation. We could even consider writing the solutions as $x = \pm\sqrt{25} = \pm 5$. We will study more about solving these quadratic equations in Chapter 8, but for now we want to consider the kind of number that would occur if our quadratic equation was $x^2 = -25$ and $x = \pm\sqrt{-25}$. There is no real number that would solve this equation, but we could extend the real number system to include such numbers. Using the product rule for radicals, $\sqrt{-25} = \sqrt{25} \cdot \sqrt{-1}$. In fact, the square root of any negative number can be written with a factor of $\sqrt{-1}$. We define the *imaginary unit i* as follows.

THE IMAGINARY UNIT *i*

The imaginary unit *i* is equal to the square root of -1. In symbols,
$$i = \sqrt{-1}$$

With the preceding definition of *i*, the square root of any negative real number can be written as the product of a real number and *i*. Thus,

$$\sqrt{-4} = \sqrt{-1} \cdot \sqrt{4} = i2 \quad \text{or} \quad 2i$$
$$\sqrt{-3} = \sqrt{-1} \cdot \sqrt{3} = i\sqrt{3} \quad \text{or} \quad \sqrt{3}i$$

Note: Radical sign is not over *i*.

It is easy to confuse $\sqrt{3}i$ and $\sqrt{3i}$. When possible, we write products involving radicals and *i* as factors with the *i* in front; that is, we may write $i\sqrt{5}$ instead of $\sqrt{5}i$. Both notations are acceptable.

EXAMPLE 1 Writing expressions in terms of *i*
Write the given expression in terms of *i*.

a. $\sqrt{-9}$ **b.** $\sqrt{-18}$

SOLUTION

a. $\sqrt{-9} = \sqrt{-1 \cdot 9} = \sqrt{-1}\sqrt{9} = 3i$
b. $\sqrt{-18} = \sqrt{-1 \cdot 18} = \sqrt{-1}\sqrt{18} = i\sqrt{18} = i\sqrt{9 \cdot 2} = 3i\sqrt{2}$ or $3\sqrt{2}i$.

PROBLEM 1
Write the given expression in terms of *i*.

a. $\sqrt{-25}$ **b.** $\sqrt{-28}$

The numbers $3i$ and $3i\sqrt{3} = 3\sqrt{3}i$ are called **pure imaginary numbers.** We can form a new set of numbers by adding these imaginary numbers to real numbers as follows.

COMPLEX NUMBERS

If *a* and *b* are real numbers, then any number of the form
$$a + bi$$
(Real part) (Imaginary part)

is called a **complex number.**

Answers to PROBLEMS

1. a. $5i$ **b.** $2\sqrt{7}i$ or $2i\sqrt{7}$

In the complex number $a + bi$, a is called the *real* part and bi the *imaginary* part. The number $-3 + 4i$ is a complex number whose real part is -3 and whose imaginary part is $4i$. Similarly, $2 - 3i$ is a complex number with 2 as its real part and $-3i$ as its imaginary part.

> **NOTE**
>
> Real numbers and pure imaginary numbers are also complex numbers. For example, the real numbers $\sqrt{2}$, 0, $-0.\overline{3}$, and $\frac{1}{5}$ are complex numbers. The pure imaginary numbers $\sqrt{2}i$, $-3i$, and $-\frac{4}{5}i$ are also complex numbers.

B › Adding and Subtracting Complex Numbers

To add (or subtract) complex numbers, we add (or subtract) the real parts and the imaginary parts separately. The rules for these operations are as follows.

> **RULES FOR ADDING AND SUBTRACTING COMPLEX NUMBERS**
>
> For a, b, c, and d real numbers,
>
> $$(a + bi) + (c + di) = (a + c) + (b + d)i$$
> $$(a + bi) - (c + di) = (a - c) + (b - d)i$$

You will find that these operations are similar to combining like terms in a polynomial. For example, $(3 + 4i) + (8 + 2i) = (3 + 8) + (4 + 2)i = 11 + 6i$, and $(9 + 2i) - (5 + 4i) = (9 - 5) + (2 - 4)i = 4 - 2i$.

> **NOTE**
>
> The sum or difference of two complex numbers is always a complex number and should be written in the form $a + bi$.

EXAMPLE 2 Adding and subtracting complex numbers

Add or subtract:

a. $(5 + 4i) + (7 - 2i)$
b. $(6 + 5i) - (7 - 3i)$
c. $(7 + 6i) + (8 - 6i)$

SOLUTION

a. $(5 + 4i) + (7 - 2i) = (5 + 7) + [4 + (-2)]i$
$\qquad\qquad\qquad\qquad = 12 + 2i$

b. $(6 + 5i) - (7 - 3i) = (6 - 7) + [5 - (-3)]i$
$\qquad\qquad\qquad\qquad = -1 + 8i$

c. $(7 + 6i) + (8 - 6i) = (7 + 8) + (6 - 6)i = 15 + 0i$

(Even though the result is 15, we must express the sum in $a + bi$ form.)

PROBLEM 2

Add or subtract:

a. $(7 + 3i) + (2 + 4i)$
b. $(2 + 3i) - (4 + 5i)$
c. $(-9 + i) + (9 - 5i)$

Answers to PROBLEMS

2. a. $9 + 7i$ b. $-2 - 2i$
c. $0 - 4i$

C › Multiplying and Dividing Complex Numbers

To consider multiplication of complex numbers, we must consider the value of i^2. We begin with the definition

$$i = \sqrt{-1}$$

Next we square both sides as we did in solving equations with radicals,

$$(i)^2 = (\sqrt{-1})^2 \quad \text{Remember} \quad (\sqrt{a})^2 = \sqrt{a} \cdot \sqrt{a} = a$$

The result is

$$i^2 = -1$$

THE SQUARE OF i

The square of the imaginary unit i is -1.

$$i^2 = -1$$

The commutative, associative, and distributive properties of real numbers also apply to complex numbers. These properties can be used to find the product and quotient of complex numbers. We multiply complex numbers using the rule to multiply binomials (FOIL) and replacing i^2 with -1. Thus,

$$\begin{aligned} \overset{\text{F O I L}}{(3 + 4i)(2 + 3i)} &= 6 + 9i + 8i + 12i^2 \\ &= 6 + 9i + 8i - 12 \quad i^2 = -1, 12i^2 = -12 \\ &= -6 + 17i \end{aligned}$$

The answer is written in the form $a + bi$ because the product of two complex numbers is always a complex number.

EXAMPLE 3 Multiplying complex numbers
Find the product:

a. $(2 - 5i)(3 + 7i)$ **b.** $-3i(4 - 7i)$ **c.** $(5 + 2i)(5 - 2i)$

SOLUTION

a. $\overset{\text{F O I L}}{(2 - 5i)(3 + 7i)} = 6 + 14i - 15i - 35i^2$
$ = 6 - i + 35 \quad i^2 = -1, \text{ so } -35i^2 = 35$
$ = 41 - i$

b. $-3i(4 - 7i) = -12i + 21i^2 \quad$ Use the distributive property.
$ = -12i - 21 \quad i^2 = -1, \text{ so } 21i^2 = -21$
$ = -21 - 12i \quad a + bi \text{ form}$

c. $\overset{\text{F O I L}}{(5 + 2i)(5 - 2i)} = 25 - 10i + 10i - 4i^2$
$ = 25 - 4(-1)$
$ = 25 + 4$
$ = 29 \text{ or } 29 + 0i$

PROBLEM 3
Find the product:

a. $(2 - 4i)(2 + 6i)$
b. $-4i(5 - 8i)$
c. $(4 - 7i)(4 + 7i)$

Answers to PROBLEMS

3. a. $28 + 4i$ **b.** $-32 - 20i$
   c. $65 + 0i$

Expressions such as $\sqrt{-9}$ and $\sqrt{-4}$ should be written in the form bi before any other operations are carried out. Let's see why. For example, to multiply $\sqrt{-9} \cdot \sqrt{-4}$, we write in bi form first,

$$\sqrt{-9} \cdot \sqrt{-4} = 3i \cdot 2i = 6i^2 = -6$$

However, if we use the product rule for radicals before changing to bi form, we get $\sqrt{-9} \cdot \sqrt{-4} = \sqrt{36} = 6$, which is *not correct*.

It will be important to remember this next procedure.

bi FORM

If $a < 0$, then $\sqrt{a}$ must be written in bi form before any operations are performed.

EXAMPLE 4 Multiplying square roots of negative numbers

Multiply:

a. $\sqrt{-16}(3 + \sqrt{-8})$ b. $\sqrt{-36}(2 - \sqrt{-18})$

SOLUTION

a. We first write the square roots of negative numbers in terms of i and then proceed as usual. Since $\sqrt{-16} = \sqrt{-1 \cdot 16} = 4i$ and $\sqrt{-8} = \sqrt{-1 \cdot 4 \cdot 2} = 2i\sqrt{2}$, we write

$\sqrt{-16}(3 + \sqrt{-8}) = 4i(3 + 2i\sqrt{2})$ $\sqrt{-16} = 4i$ and $\sqrt{-8} = 2i\sqrt{2}$
$= 12i + 8i^2\sqrt{2}$ Use the distributive property.
$= 12i - 8\sqrt{2}$ $i^2 = -1$, so $8i^2\sqrt{2} = -8\sqrt{2}$
$= -8\sqrt{2} + 12i$ Write in the form $a + bi$.

b. Since $\sqrt{-36} = \sqrt{-1 \cdot 36} = 6i$, and $\sqrt{-18} = \sqrt{-1 \cdot 9 \cdot 2} = 3i\sqrt{2}$, we write

$\sqrt{-36}(2 - \sqrt{-18}) = 6i(2 - 3i\sqrt{2})$
$= 12i - 18i^2\sqrt{2}$
$= 12i + 18\sqrt{2}$ or $18\sqrt{2} + 12i$

PROBLEM 4

Multiply:

a. $\sqrt{-25}(6 + \sqrt{-8})$

b. $\sqrt{-49}(4 - \sqrt{-27})$

To find the *quotient* of two complex numbers, we use the rationalizing process developed in Section 7.3 and the assumption that

$$\frac{a + bi}{c} = \frac{a}{c} + \frac{b}{c}i$$

For example, to find

$$\frac{2 + 3i}{4 - i}$$

we proceed as follows.

$\frac{2 + 3i}{4 - i} = \frac{(2 + 3i)(4 + i)}{(4 - i)(4 + i)}$ Multiply the numerator and denominator by the conjugate of $(4 - i)$, that is, $(4 + i)$.

$= \frac{8 + 2i + 12i + 3i^2}{4^2 - i^2}$ $(4 - i)(4 + i) = 4^2 - i^2$

$= \frac{8 + 2i + 12i - 3}{16 - (-1)}$ $i^2 = -1$

$= \frac{5 + 14i}{17}$ Simplify.

$= \frac{5}{17} + \frac{14}{17}i$ Write in $(a + bi)$ form.

Answers to PROBLEMS

4. a. $-10\sqrt{2} + 30i$

b. $21\sqrt{3} + 28i$

PROCEDURE

Dividing One Complex Number by Another

To divide one complex number by another with two terms, as in $\frac{2+i}{3-i}$, multiply the numerator and denominator by the conjugate of the denominator. The conjugate of $a + bi$ is $a - bi$ and $(a + bi)(a - bi) = a^2 - (bi)^2 = a^2 + b^2$.

EXAMPLE 5 Dividing complex numbers

Divide:

a. $\frac{5 + 4i}{3 + 2i}$ b. $\frac{2 - 4i}{5 - 3i}$ c. $\frac{3 - 2i}{i}$

SOLUTION

a. $\frac{5 + 4i}{3 + 2i} = \frac{(5 + 4i)(3 - 2i)}{(3 + 2i)(3 - 2i)}$ Multiply by the conjugate of $(3 + 2i)$.

$= \frac{15 - 10i + 12i - 8i^2}{3^2 + 2^2}$ $(3 + 2i)(3 - 2i) = 3^2 + 2^2$

$= \frac{15 - 10i + 12i + 8}{13}$ $i^2 = -1$

$= \frac{23 + 2i}{13}$ Simplify.

$= \frac{23}{13} + \frac{2}{13}i$ Write in $(a + bi)$ form.

b. $\frac{2 - 4i}{5 - 3i} = \frac{(2 - 4i)(5 + 3i)}{(5 - 3i)(5 + 3i)}$ Multiply by the conjugate of $(5 - 3i)$.

$= \frac{10 + 6i - 20i - 12i^2}{5^2 + 3^2}$ $(5 - 3i)(5 + 3i) = 5^2 + 3^2$

$= \frac{10 + 6i - 20i + 12}{34}$ $i^2 = -1$

$= \frac{22 - 14i}{34}$ Simplify.

$= \frac{22}{34} - \frac{14}{34}i$ Write in $(a + bi)$ form.

$= \frac{11}{17} - \frac{7}{17}i$ Reduce.

c. The conjugate of $a + bi$ is $a - bi$, so the conjugate of $0 + 1i$ is $0 - 1i = -i$. Multiplying numerator and denominator of the fraction by $-i$, we have

$\frac{3 - 2i}{i} = \frac{(3 - 2i)(-i)}{i \cdot (-i)}$ Multiply by $(-i)$.

$= \frac{-3i + 2i^2}{-i^2}$ Distributive Property

$= \frac{-3i - 2}{1}$ $i^2 = -1$

$= -2 - 3i$ Write in $(a + bi)$ form.

PROBLEM 5

Divide:

a. $\frac{3 + 5i}{2 + 3i}$ b. $\frac{3 - 5i}{4 - 3i}$

c. $\frac{2 - 3i}{i}$

D › Finding Powers of i

We already know that, by definition, $i^2 = -1$. If we assume that the properties of exponents hold, we can write any power of i as one of the numbers $1, -1, i,$ or $-i$. Thus,

---

Answers to PROBLEMS

5. a. $\frac{21}{13} + \frac{1}{13}i$ b. $\frac{27}{25} - \frac{11}{25}i$

c. $-3 - 2i$

$$i^1 = i \qquad\qquad i^5 = i \cdot i^4 = i \cdot (1) = i$$
$$i^2 = -1 \qquad\qquad i^6 = i \cdot i^5 = i \cdot i = -1$$
$$i^3 = i \cdot i^2 = i(-1) = -i \qquad\qquad i^7 = i \cdot i^6 = i(-1) = -i$$
$$i^4 = i^2 \cdot i^2 = (-1)(-1) = 1 \qquad\qquad i^8 = i \cdot i^7 = i \cdot (-i) = 1$$

Since $i^4 = 1$, the easiest way to simplify higher powers of i is to write them in terms of i^4. To find i^{20}, we write

$$i^{20} = (i^4)^5 = (1)^5 = 1$$

Similarly,

$$i^{21} = (i^4)^5 \cdot i = 1 \cdot i = i$$
$$i^{22} = (i^4)^5 \cdot i^2 = 1 \cdot (-1) = -1$$
$$i^{23} = (i^4)^5 \cdot i^3 = 1 \cdot i^3 = -i$$

Dividing the exponent by 4 will give you the answer as follows:

If the remainder is 0, the answer is 1 (as in i^{20}).
If the remainder is 1, the answer is i (as in i^{21}).
If the remainder is 2, the answer is -1 (as in i^{22}).
If the remainder is 3, the answer is $-i$ (as in i^{23}).

After that, the answers repeat.

EXAMPLE 6 Finding powers of i
Find:

a. i^{53} **b.** i^{47} **c.** i^{-3} **d.** i^{-1} **e.** i^{-34}

SOLUTION

a. Dividing 53 by 4, we obtain 13 with a remainder of 1. Thus, the answer is i. To show this, write

$$i^{53} = (i^4)^{13} \cdot i$$
$$= 1 \cdot i = i$$

b. If we divide 47 by 4, the remainder is 3. Thus, the answer is $-i$.

$$i^{47} = (i^4)^{11} \cdot i^3$$
$$= 1 \cdot i^3$$
$$= -i$$

c. By definition of negative exponents,

$$i^{-3} = \frac{1}{i^3}$$

$$i^{-3} = \frac{1 \cdot i}{i^3 \cdot i} \qquad \text{Multiply by } i.$$

$$= \frac{i}{i^4} \qquad \text{Write with denominator of } i^4 = 1.$$

$$= \frac{i}{1} = i \qquad \text{Simplify.}$$

d. $i^{-1} = \dfrac{1}{i} = \dfrac{1 \cdot i^3}{i \cdot i^3} = \dfrac{i^3}{i^4} = \dfrac{i^3}{1} = -i \qquad$ Multiply by i^3 so denominator is $i^4 = 1$.

e. $i^{-34} = \dfrac{1}{i^{34}} = \dfrac{1 \cdot i^2}{i^{34} \cdot i^2} = \dfrac{i^2}{i^{36}} = \dfrac{i^2}{1} = -1 \qquad$ Multiply by i^2 so denominator is $i^{36} = 1$.

PROBLEM 6
Find:

a. i^{42} **b.** i^{27} **c.** i^{-8} **d.** i^{-2}

Answers to PROBLEMS

6. **a.** -1 **b.** $-i$
c. $\dfrac{1}{1} = 1$ **d.** $\dfrac{1}{-1} = -1$

The introduction of the complex numbers completes the development of our number system. We started with the natural numbers (N) and whole numbers (W), studied the integers (I) and the rational numbers (Q), and then concentrated on the real numbers (R). Now we have developed the complex numbers (C), which include all the other numbers we have discussed. In set language, the relationship is

$$N \subseteq W \subseteq I \subseteq Q \subseteq R \subseteq C$$

as shown in Figure 7.1. A diagram for the complex numbers is shown in Figure 7.2.

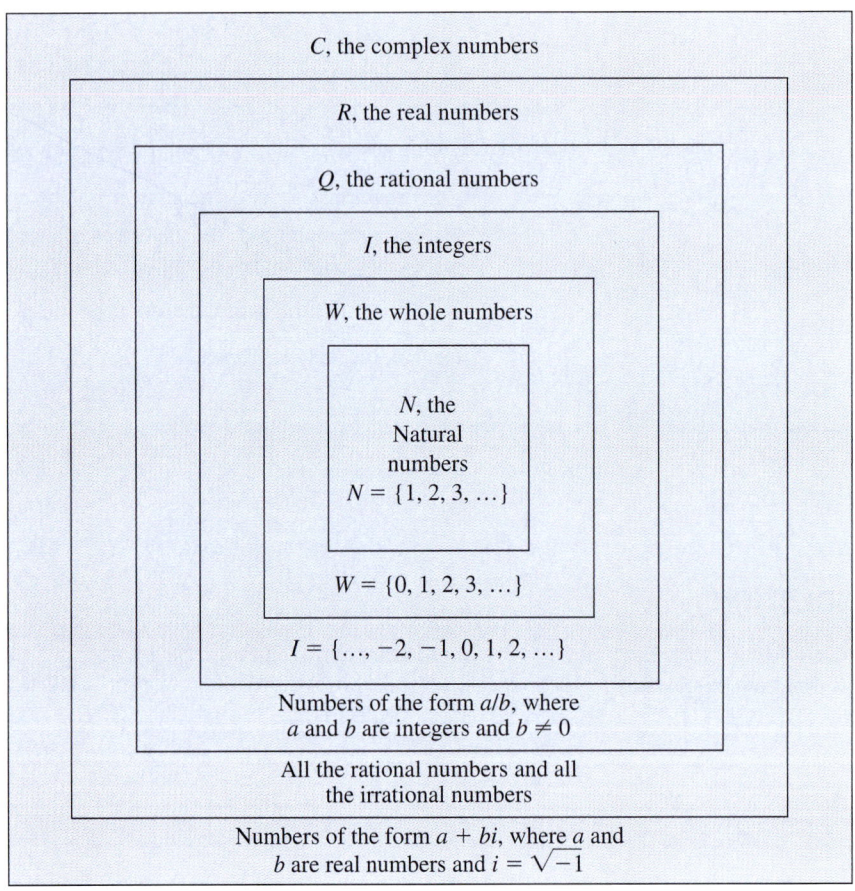

>**Figure 7.1** The Number System

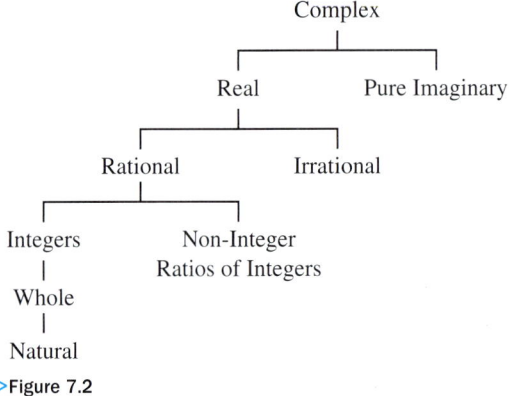

>**Figure 7.2**

Exercises 7.5

A Writing Square Roots of Negative Numbers in Terms of i In Problems 1–10, write the given expression in terms of i.

1. $\sqrt{-25}$
2. $\sqrt{-81}$
3. $\sqrt{-50}$
4. $\sqrt{-98}$
5. $4\sqrt{-72}$
6. $3\sqrt{-200}$
7. $-3\sqrt{-32}$
8. $-5\sqrt{-64}$
9. $4\sqrt{-28} + 3$
10. $7\sqrt{-18} + 5$

B Adding and Subtracting Complex Numbers In Problems 11–30, perform the indicated operations. (Write the answer in the form $a + bi$.)

11. $(4 + i) + (2 + 3i)$
12. $(7 + 3i) + (2 + i)$
13. $(3 - 2i) - (5 + 4i)$
14. $(4 - 5i) - (2 + 3i)$
15. $(-3 - 5i) + (-2 - i)$
16. $(-7 - 3i) + (-2 - i)$
17. $(3 + \sqrt{-4}) + (5 + \sqrt{-9})$
18. $(-2 - \sqrt{-16}) - (3 - \sqrt{-25})$
19. $(-5 + \sqrt{-1}) + (-2 + 3\sqrt{-1})$
20. $(-3 + 2\sqrt{-1}) + (-4 + 5\sqrt{-1})$
21. $(3 - 4i) + (5 + 3i)$
22. $(3 - 7i) + (3 + 4i)$
23. $(4 + \sqrt{-9}) + (6 + \sqrt{-4})$
24. $(-3 - \sqrt{-25}) + (5 - \sqrt{-16})$
25. $(2 - \sqrt{-2}) - (5 + \sqrt{-2})$
26. $(3 + \sqrt{-50}) - (7 + \sqrt{-2})$
27. $(-5 - \sqrt{-2}) - (-4 - \sqrt{-18})$
28. $(-8 - \sqrt{-125}) - (-2 - \sqrt{-5})$
29. $(-4 + \sqrt{-20}) + (-3 + \sqrt{-5})$
30. $(-7 + \sqrt{-24}) + (-3 + \sqrt{-6})$

C Multiplying and Dividing Complex Numbers In Problems 31–70, perform the indicated operations. (Write the answer in the form $a + bi$.)

31. $3(4 + 2i)$
32. $5(4 + 3i)$
33. $-4(3 - 5i)$
34. $-3(7 - 4i)$
35. $\sqrt{-4}(3 + 2i)$
36. $\sqrt{-9}(2 + 5i)$
37. $\sqrt{-3}(3 + \sqrt{-3})$
38. $\sqrt{-5}(2 - \sqrt{-5})$
39. $3i(3 + 2i)$
40. $7i(4 + 3i)$
41. $4i(3 - 7i)$
42. $-5i(2 - 3i)$
43. $-\sqrt{-16}(-5 - \sqrt{-25})$
44. $-\sqrt{-25}(-3 - \sqrt{-9})$
45. $(3 + i)(2 + 3i)$
46. $(2 + 3i)(4 + 5i)$
47. $(3 - 2i)(3 + 2i)$
48. $(4 - 3i)(4 + 3i)$
49. $(3 + 2\sqrt{-4})(4 - \sqrt{-9})$
50. $(-3 + 3\sqrt{-9})(-2 + 5\sqrt{-4})$
51. $(2 + 3\sqrt{-3})(2 - 3\sqrt{-3})$
52. $(4 + 2\sqrt{-5})(4 - 2\sqrt{-5})$
53. $\dfrac{3}{i}$
54. $\dfrac{5}{i}$
55. $\dfrac{6}{-i}$
56. $\dfrac{3}{-2i}$
57. $\dfrac{i}{1 + 2i}$
58. $\dfrac{2i}{1 + 3i}$
59. $\dfrac{3i}{1 - 2i}$
60. $\dfrac{4i}{2 - 3i}$
61. $\dfrac{3 + 4i}{1 - 2i}$
62. $\dfrac{3 + 5i}{1 - 3i}$
63. $\dfrac{4 + 3i}{2 + 3i}$
64. $\dfrac{5 + 4i}{3 + 2i}$
65. $\dfrac{3}{\sqrt{-4}}$
66. $\dfrac{-4}{\sqrt{-9}}$

67. $\dfrac{3 + \sqrt{-5}}{4 + \sqrt{-2}}$

68. $\dfrac{2 + \sqrt{-2}}{1 + \sqrt{-3}}$

69. $\dfrac{-1 - \sqrt{-2}}{-3 - \sqrt{-3}}$

70. $\dfrac{-1 - \sqrt{-3}}{-2 - \sqrt{-2}}$

⟨ D ⟩ Finding Powers of *i* In Problems 71–82, write the answer as $1, -1, i,$ or $-i$.

71. i^{40}
72. i^{28}
73. i^{19}
74. i^{38}
75. i^{21}
76. i^{-44}
77. i^{-32}
78. i^{53}
79. i^{65}
80. i^{16}
81. i^{-10}
82. i^{-27}

⟩⟩⟩ Applications

83. *Impedance* The impedance in a circuit is the measure of how much a circuit impedes (hinders) the flow of current through it. If the impedance of a resistor is $Z_1 = 5 + 3i$ ohms and the impedance of another resistor is $Z_2 = 3 - 2i$ ohms, what is the total impedance (sum) of the two resistors when placed in series?

84. *Impedance* Repeat Problem 83 where the impedance of the first resistor is $3 + 7i$ and the impedance of the second is $4 - 5i$.

85. *Impedance* If two resistors Z_1 and Z_2 are connected in parallel, their total impedance is given by

$$Z_T = \dfrac{Z_1 \cdot Z_2}{Z_1 + Z_2}$$

Find the total impedance of the resistors of Problem 83.

86. *Impedance* Use the formula in Problem 85 and find the total impedance of the resistors of Problem 84 if they are connected in parallel.

⟩⟩⟩ Using Your Knowledge

Absolute Values of Complex Numbers If x is a real number, the absolute value of x is defined as follows.

$$|x| = \begin{cases} x, & \text{if } x > 0 \\ 0, & \text{if } x = 0 \\ -x, & \text{if } x < 0 \end{cases}$$

Thus,

$|5| = 5$ $\quad 5 > 0$

$|0| = 0$

$|-8| = -(-8) = 8 \quad -8 < 0$

How can we define the absolute value of a complex number? The definition is

$$|a + bi| = \sqrt{a^2 + b^2}$$

In Problems 87–90, use this definition to find each value.

87. $|3 + 4i|$
88. $|12 + 5i|$
89. $|2 - 3i|$
90. $|5 - 7i|$

⟩⟩⟩ Write On

91. Explain why it is incorrect to use the product rule for radicals to multiply $\sqrt{-4} \cdot \sqrt{-9}$. What answer do you get if you follow the rule? What should the answer be? Under what conditions is $\sqrt{a} \cdot \sqrt{b} = \sqrt{a \cdot b}$?

92. Write in your own words the procedure you use to:
    a. find the conjugate of a complex number.
    b. add (or subtract) two complex numbers.
    c. find the quotient of two complex numbers.

93. Write a short paragraph explaining the relationship among the real numbers, the imaginary numbers, and the complex numbers.

⟩⟩⟩ Concept Checker

Fill in the blank(s) with the correct word(s), phrase, or mathematical statement.

94. The complex numbers are made up of the _____ and the _____.

95. $-\sqrt{2}$ is an example of a complex number that is real and _____.

96. The complex, real, rational number "0" is also a _____.

97. In the complex, real numbers the set $\{1, 2, 3, \ldots\}$ is called the _____.

complex numbers	irrational
real numbers	integer
pure imaginary numbers	whole number
	natural numbers
rational	

⟩⟩⟩ Mastery Test

Find:

98. i^{42}

99. i^{23}

100. i^{-8}

101. i^{-2}

102. $\dfrac{3 + 5i}{2 + 3i}$

103. $\dfrac{3 - 5i}{4 - 3i}$

104. $\dfrac{2 - 3i}{i}$

105. $\sqrt{-25}(6 + \sqrt{-8})$

106. $\sqrt{-49}(\sqrt{-3} - \sqrt{-27})$

107. $(2 - 4i)(2 + 6i)$

108. $-4i(5 - 8i)$

109. $(7 + 3i) - (4 + 5i)$

110. $(9 - 2i) - (12 - 3i)$

111. $(2 + 3i) + (-3 + 4i)$

112. $(-3 - 4i) + (-5 - 7i)$

Write in terms of i:

113. $\sqrt{-50}$

114. $\sqrt{-25}$

⟩⟩⟩ Skill Checker

Solve:

115. $x^2 - 4 = 0$

116. $x^2 - 16 = 0$

117. $x^2 = 25$

118. $x^2 = 36$

Rationalize the denominator:

119. $\sqrt{\dfrac{4}{5}}$

120. $\sqrt{\dfrac{25}{3}}$

121. $\sqrt{\dfrac{7}{9}}$

122. $\sqrt{\dfrac{5}{16}}$

Multiply:

123. $(x + 5)^2$

124. $(x - 4)^2$

Factor:

125. $x^2 - 16x + 64$

126. $x^2 + 20x + 100$

Collaborative Learning 7A

1. Complete the table to find the missing lengths.

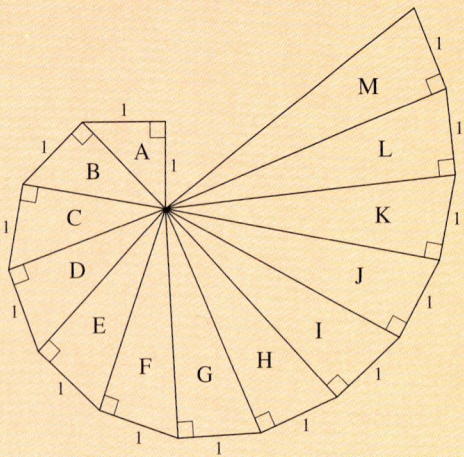

Source: Annenberg/CPB Learner.org.

Right Triangle	A	B	C	D	E	F	G	H	I	J	K	L	M
Leg lengths	1, 1	1, $\sqrt{2}$	1, ?	1, ?	1, ?	1, ?	1, ?	1, ?	1, ?	1, ?	1, ?	1, ?	1, ?
Hypotenuse length	$\sqrt{2}$	$\sqrt{3}$	?	?	?	?	?	?	?	?	?	?	?

2. Find the pattern in the lengths for the hypotenuses.
3. Make a table to find the length for the diagonal of each of the following five squares.

  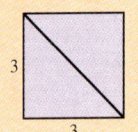 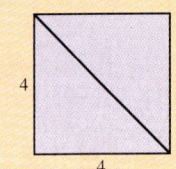 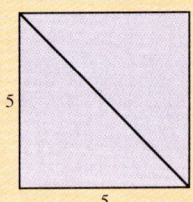

4. Find the pattern in the lengths of the diagonals of the squares.
5. Baseball "diamonds" are squares in which one of the diagonals of the square is the distance from home plate to second base. If the major league fields have 90 ft between bases, use the pattern found in number 4 to find how far the catcher has to throw from home plate to attempt to get a player out who is trying to steal second base.
6. If the adult fast pitch softball "diamond" has 60 ft between its bases, how far is it from home plate to second base?
7. Research and discuss other areas where patterns can help facilitate problem solving.

Collaborative Learning 7B

Divide into groups. You are going to investigate a car accident. We have the formula to find the rate of speed R of the vehicle based on the length L of the skid marks before the car stops. The formula is:

$$R = \sqrt{20L - 600}$$

1. Using the formula, complete the table to find the rate of speed R for the various skid mark lengths L.

2. Use the table of values from number 1 to make a two-dimensional graph where the horizontal axis represents the length of the skid marks L and the vertical axis represents the rate of speed R of the vehicle.

3. Your graph should indicate a linear model. Using your skills from Section 3.6, write the equation for the line of best fit for this graph.

4. Complete the following table by copying the rate results from column 2 of the previous table into column 2 in the following table. Use the equation from number 3 to complete column 3.

Length of Skid Marks	Rate of Speed in mi/hr
60 ft	
75 ft	
90 ft	
110 ft	
130 ft	

Length of Skid Marks	Rate Using $R = \sqrt{20L - 600}$	Rate Using Linear Equation
60 ft		
75 ft		
90 ft		
110 ft		
130 ft		

5. Comparing the results of columns 2 and 3 in the table in number 4, discuss which formula would be best to use to find the rate of speed R given the length of the skid marks L and why?

6. Research other formulas for finding the speed of a vehicle given the length of the skid marks including the one in the *Getting Started* in Section 7.2. How do they compare to your results and why?

Research Questions

1. Who invented the square root sign $\sqrt{}$ (called a radix)?

2. What mathematician coined the word *imaginary*, and in what book did the term first appear?

3. Who was the first mathematician to use the notation i for $\sqrt{-1}$, and in what book did this symbol first appear?

4. The author's *Algebra* (1770) remarks that "such numbers, which by their nature are impossible, are ordinarily called imaginary or fanciful numbers because they exist only in the imagination." Write a short paragraph about this author and his other contributions to mathematics.

5. Who was the first person to attempt (not successfully) to graphically represent the complex numbers, and what was the name of the book in which he tried to show his work?

6. The graphical representation of complex numbers occurred at nearly the same time to three mathematicians with no connection or knowledge of each other. Who were these three mathematicians?

7. Write a short paragraph about the mathematical contributions of each of the three people named in Question 6.

Summary Chapter 7

Section	Item	Meaning	Example
7.1A	nth root	If a and x are real numbers and n is a positive integer, x is an nth root of a if $x^n = a$.	One square root of 25 is 5, the cube root of -8 is -2, and the fourth root of -16 is not a real number.
7.1A	$\sqrt{}$ Radicand Index (order) $\sqrt[n]{a}$, (a radical expression)	A radical sign In $\sqrt[n]{a}$, a is the radicand. In $\sqrt[n]{a}$, the index is n. The nth root of a	In $\sqrt[3]{22}$, 22 is the radicand. In $\sqrt[3]{22}$, the index is 3. $\sqrt[3]{64} = 4$, $\sqrt[5]{-32} = -2$, and $\sqrt{-25}$ is not a real number.
7.1B	$a^{1/n}$ $a^{m/n}$ $a^{-m/n}$	$a^{1/n} = \sqrt[n]{a}$, if it exists. $a^{m/n} = (\sqrt[n]{a})^m = \sqrt[n]{a^m}$ $a^{-m/n} = \dfrac{1}{a^{m/n}}$	$8^{1/3} = \sqrt[3]{8} = 2$ and $32^{1/5} = \sqrt[5]{32} = 2$ $16^{3/2} = (\sqrt{16})^3 = 4^3 = 64$ $4^{-3/2} = \dfrac{1}{4^{3/2}} = \dfrac{1}{(\sqrt{4})^3} = \dfrac{1}{8}$
7.1C	$(a^m)^{1/n}$ (m and n positive, even integers)	$\lvert a \rvert^{m/n}$	$[(-4)^2]^{1/2} = \lvert -4 \rvert = 4$
		If r, s, and t are rational numbers and a is a real number, the following properties apply: I. $a^r \cdot a^s = a^{r+s}$ II. $\dfrac{a^r}{a^s} = a^{r-s}$ $(a \neq 0)$ III. $(a^r)^s = a^{rs}$ IV. $(a^r b^s)^t = a^{rt} b^{st}$ V. $\left(\dfrac{a^r}{b^s}\right)^t = \dfrac{a^{rt}}{b^{st}}$ $(a \neq 0, b \neq 0)$	I. $x^{3/5} \cdot x^{2/5} = x^{3/5 + 2/5} = x^{1/5}$ II. $\dfrac{x^{3/4}}{x^{1/4}} = x^{3/4 - 1/4} = x^{2/4} = x^{1/2}$ III. $(x^{2/3})^9 = x^{2/3 \cdot 9} = x^6$ IV. $(x^{1/5} y^{3/5})^{5/2} = x^{1/5 \cdot 5/2} \cdot y^{3/5 \cdot 5/2} = x^{1/2} y^{3/2}$ V. $\left(\dfrac{x^{1/8}}{y^{3/8}}\right)^8 = \dfrac{x^{1/8 \cdot 8}}{y^{3/8 \cdot 8}} = \dfrac{x}{y^3}$
7.2A	$\sqrt[n]{a^n}$ Product Rule Quotient Rule	$\sqrt[n]{a^n} = a$, for $a \geq 0$ $\sqrt[n]{ab} = \sqrt[n]{a} \cdot \sqrt[n]{b}$ $(a, b \geq 0)$ $\sqrt[n]{\dfrac{a}{b}} = \dfrac{\sqrt[n]{a}}{\sqrt[n]{b}}$ $(a \geq 0, b > 0)$	$\sqrt[4]{2^4} = 2$ $\sqrt{18} = \sqrt{9} \cdot \sqrt{2} = 3\sqrt{2}$ $\sqrt[3]{\dfrac{8}{27}} = \dfrac{\sqrt[3]{8}}{\sqrt[3]{27}} = \dfrac{2}{3}$
7.2C	Simplified form of a radical	A radical expression is in simplified form if the radicand has no factors with exponents greater than or equal to the index, there are no fractions under the radical sign or radicals in the denominator of a fraction, and the index is as low as possible.	$\dfrac{\sqrt[3]{2ac}}{2c^2}$ is in simplified form.
7.3A	Like radicals	Radicals with the same index and the same radicand	$\sqrt[3]{3x^2}$ and $-7\sqrt[3]{3x^2}$ are like radicals.

Section	Item	Meaning	Example
7.3C	Conjugates	$a + b$ and $a - b$ are conjugates.	$3 + \sqrt{2}$ and $3 - \sqrt{2}$ are conjugates.
7.4A	Extraneous solution	A proposed solution that does not satisfy the original equation	0 is an extraneous solution of $\sqrt{x + 1} = x - 1$.
7.5A	i	$i = \sqrt{-1}$; $i^2 = -1$	
	Complex number	A number that can be written in the form $a + bi$, where a and b are real numbers	$3 + 7i$ and $-4 - 8i$ are complex numbers.
7.5B	Addition and subtraction of complex numbers	$(a + bi) \pm (c + di) = (a \pm c) + (b \pm d)i$	$(3 + 2i) + (5 + 3i) = 8 + 5i$ $(4 - 2i) - (5 + 3i) = -1 - 5i$
7.5C	Multiplication of complex numbers	$(a + bi)(c + di)$ Use the rule to multiply binomials, FOIL, and replace i^2 by -1.	$(2 + 3i)(4 - 5i)$ $= 8 - 10i + 12i - 15i^2$ $= 8 + 2i + 15$ $= 23 + 2i$
7.5C	Division of complex numbers	$\dfrac{a + bi}{c + di}$, multiply numerator and denominator by the conjugate of $c + di$ which is $c - di$.	$\dfrac{2 + i}{5 + i} \cdot \dfrac{5 - i}{5 - i} = \dfrac{11}{26} + \dfrac{3}{26}i$

Review Exercises Chapter 7

(If you need help with these exercises, look in the section indicated in brackets.)

In Problems 1–8, evaluate if possible.

1. ⟨7.1A⟩
   a. $\sqrt{-8}$
   b. $\sqrt[3]{-64}$

2. ⟨7.1A⟩
   a. $\sqrt{-9}$
   b. $\sqrt[3]{-125}$

3. ⟨7.1B⟩
   a. $(-27)^{1/3}$
   b. $(-64)^{1/3}$

4. ⟨7.1B⟩
   a. $\left(\dfrac{1}{16}\right)^{1/4}$
   b. $\left(\dfrac{1}{256}\right)^{1/4}$

5. ⟨7.1B⟩
   a. $125^{2/3}$
   b. $64^{2/3}$

6. ⟨7.1B⟩
   a. $(-25)^{3/2}$
   b. $(-36)^{3/2}$

7. ⟨7.1B⟩
   a. $(-8)^{-2/3}$
   b. $(-64)^{-2/3}$

8. ⟨7.1B⟩
   a. $27^{-2/3}$
   b. $64^{-2/3}$

In Problems 9–18, simplify (x and y are positive).

9. ⟨7.1C⟩
   a. $x^{1/5} \cdot x^{1/3}$
   b. $x^{1/5} \cdot x^{1/4}$

10. ⟨7.1C⟩
    a. $\dfrac{x^{-1/4}}{x^{1/5}}$
    b. $\dfrac{x^{-1/3}}{x^{1/5}}$

11. ⟨7.1C⟩
    a. $(x^{1/3} y^{2/5})^{-15}$
    b. $(x^{1/6} y^{2/5})^{-30}$

12. ⟨7.1C⟩
    a. $x^{3/5}(x^{-2/5} + y^{3/5})$
    b. $x^{4/5}(x^{-1/5} + y^{3/5})$

13. ⟨7.2A⟩
    a. $\sqrt[4]{(-7)^4}$
    b. $\sqrt[4]{(-6)^4}$

14. ⟨7.2A⟩
    a. $\sqrt[8]{(-x)^8}$
    b. $\sqrt[4]{(-x)^4}$

15. ⟨7.2A⟩
    a. $\sqrt[3]{48}$
    b. $\sqrt[3]{56}$

16. ⟨7.2A⟩
    a. $\sqrt[3]{16x^4y^6}$
    b. $\sqrt[3]{16x^8y^{15}}$

17. ⟨7.2A⟩
    a. $\sqrt{\dfrac{5}{243}}$
    b. $\sqrt{\dfrac{5}{1024}}$

18. ⟨7.2A⟩
    a. $\sqrt[3]{\dfrac{1}{x^3}}$
    b. $\sqrt[3]{\dfrac{5}{x^3}}$

In Problems 19–22, rationalize the denominator.

19. ⟨7.2B⟩
    a. $\dfrac{\sqrt{5}}{\sqrt{11}}$
    b. $\dfrac{\sqrt{5}}{\sqrt{13}}$

20. ⟨7.2B⟩
    a. $\dfrac{\sqrt{2}}{\sqrt{5x}}, x > 0$
    b. $\dfrac{\sqrt{3}}{\sqrt{5x}}, x > 0$

21. ⟨7.2B⟩
    a. $\dfrac{1}{\sqrt[3]{5x}}, x \neq 0$
    b. $\dfrac{1}{\sqrt[3]{7x}}, x \neq 0$

22. ⟨7.2B⟩
    a. $\dfrac{\sqrt[4]{1}}{\sqrt[4]{16x}}, x > 0$
    b. $\dfrac{\sqrt[4]{5}}{\sqrt[4]{8x^3}}, x > 0$

In Problems 23–24, reduce the index of the radical.

23. ⟨7.2C⟩
    a. $\sqrt[4]{\dfrac{256}{81}}$
    b. $\sqrt[4]{\dfrac{625}{81}}$

24. ⟨7.2C⟩
    a. $\sqrt[6]{81c^4d^4}$
    b. $\sqrt[6]{625c^4d^4}$

25. ⟨7.2C⟩ Simplify.
    a. $\sqrt[6]{\dfrac{3a^2}{243c^{16}}}$
    b. $\sqrt[6]{\dfrac{9a^2}{81c^{16}}}$

In Problems 26–35, perform the indicated operations.

26. ⟨7.3A⟩
    a. $\sqrt{8} + \sqrt{32}$
    b. $\sqrt{18} + \sqrt{32}$

27. ⟨7.3A⟩
    a. $\sqrt{63} - \sqrt{28}$
    b. $\sqrt{112} - \sqrt{63}$

28. ⟨7.3A⟩
    a. $3\sqrt{\dfrac{1}{2}} - 3\sqrt{\dfrac{1}{8}}$
    b. $6\sqrt{\dfrac{1}{2}} - 3\sqrt{\dfrac{1}{8}}$

29. ⟨7.3A⟩
    a. $7\sqrt[3]{\dfrac{3}{4x}} - \sqrt[3]{\dfrac{3}{32x}}$
    b. $6\sqrt[3]{\dfrac{3}{4x}} - \sqrt[3]{\dfrac{3}{32x}}$

30. ⟨7.3B⟩
    a. $\sqrt{2}(\sqrt{18} + \sqrt{3})$
    b. $\sqrt{2}(\sqrt{32} + \sqrt{3})$

31. ⟨7.3B⟩
    a. $\sqrt[3]{2x}(\sqrt[3]{24x^2} - \sqrt[3]{81x})$
    b. $\sqrt[3]{3x}(\sqrt[3]{16x^2} - \sqrt[3]{54x})$

32. ⟨7.3B⟩
    a. $(\sqrt{27} + \sqrt{18})(\sqrt{12} + \sqrt{8})$
    b. $(\sqrt{12} + \sqrt{18})(\sqrt{12} + \sqrt{8})$

33. ⟨7.3B⟩
    a. $(\sqrt{3} + 4)(\sqrt{3} + 4)$
    b. $(\sqrt{3} + 5)(\sqrt{3} + 5)$

34. ⟨7.3B⟩
    a. $(7 - \sqrt{3})^2$
    b. $(4 - \sqrt{3})^2$

35. ⟨7.3B⟩
    a. $(\sqrt{8} + \sqrt{3})(\sqrt{8} - \sqrt{3})$
    b. $(\sqrt{7} + \sqrt{3})(\sqrt{7} - \sqrt{3})$

36. ⟨7.3B⟩ Reduce.
    a. $\dfrac{20 - \sqrt{50}}{5}$
    b. $\dfrac{30 - \sqrt{50}}{5}$

37. ⟨7.3C⟩ Rationalize the denominator.
    a. $\dfrac{5}{\sqrt{2} - 1}$
    b. $\dfrac{\sqrt{x}}{\sqrt{x} + 4}$

38. ⟨**7.4A**⟩ Solve.
  a. $\sqrt{x-2} = -2$
  b. $\sqrt{x-2} = -3$

39. ⟨**7.4A**⟩ Solve.
  a. $\sqrt{x+5} = x-1$
  b. $\sqrt{x+6} = x-6$

40. ⟨**7.4A**⟩ Solve.
  a. $\sqrt{x-7} - x = -7$
  b. $\sqrt{x-4} - x = -4$

41. ⟨**7.4A**⟩ Solve.
  a. $\sqrt{x-3} - \sqrt{x} = -1$
  b. $\sqrt{x-5} - \sqrt{x} = -1$

42. ⟨**7.4A**⟩ Solve.
  a. $\sqrt[3]{x-5} = 3$
  b. $\sqrt[3]{x-3} = 4$

43. ⟨**7.4A**⟩ Solve.
  a. $\sqrt[4]{x+1} + 1 = 0$
  b. $\sqrt[4]{x-2} + 16 = 0$

44. ⟨**7.4B**⟩ The distance d (in feet) from a light source of intensity I (in foot-candles) is
$$d = \sqrt{\frac{k}{I}}$$
where k is a constant.
  a. Solve for I.
  b. Solve for k.

45. ⟨**7.5A**⟩ Write in terms of i.
  a. $\sqrt{-100}$
  b. $\sqrt{-121}$

46. ⟨**7.5A**⟩ Write in terms of i.
  a. $\sqrt{-72}$
  b. $\sqrt{-50}$

47. ⟨**7.5B**⟩ Add.
  a. $(3 + 5i) + (7 - 2i)$
  b. $(4 + 7i) + (2 - 4i)$

48. ⟨**7.5B**⟩ Subtract.
  a. $(3 + 5i) - (7 - 2i)$
  b. $(4 + 7i) - (2 - 4i)$

49. ⟨**7.5C**⟩ Multiply.
  a. $(3 + 2i)(5 - 3i)$
  b. $(4 + 5i)(2 - 3i)$

50. ⟨**7.5C**⟩ Multiply.
  a. $\sqrt{-16}(4 - \sqrt{-72})$
  b. $\sqrt{-36}(4 - \sqrt{-72})$

51. ⟨**7.5C**⟩ Divide.
  a. $\dfrac{2 + 3i}{4 + 3i}$
  b. $\dfrac{3 - 5i}{4 - 3i}$

52. ⟨**7.5D**⟩ Write the answer as 1, -1, i, or $-i$.
  a. i^{38}
  b. i^{75}

53. ⟨**7.5D**⟩ Write the answer as 1, -1, i, or $-i$.
  a. i^{-14} b. i^{-27}

Practice Test Chapter 7

(Answers on pages 586–587)

Visit www.mhhe.com/bello to view helpful videos that provide step-by-step solutions to several of the problems below.

1. Find, if possible.
   a. $\sqrt[3]{-64}$
   b. $\sqrt{-36}$

2. Find, if possible.
   a. $(-27)^{1/3}$
   b. $\left(\dfrac{1}{16}\right)^{1/4}$

3. Evaluate, if possible.
   a. $8^{2/3}$
   b. $(-25)^{3/2}$

4. Evaluate, if possible.
   a. $(-27)^{-2/3}$
   b. $8^{-2/3}$

5. If x is positive, simplify.
   a. $x^{1/2} \cdot x^{2/3}$
   b. $\dfrac{x^{-1/3}}{x^{2/5}}$

6. If x and y are positive, simplify.
   a. $(x^{1/5}y^{3/4})^{-20}$
   b. $x^{3/5}(x^{-1/5} + y^{3/5})$

7. Simplify.
   a. $\sqrt[4]{(-5)^4}$
   b. $\sqrt[4]{(-x)^4}$

8. Simplify.
   a. $\sqrt[3]{40}$
   b. $\sqrt[3]{54a^4b^{12}}$

9. Simplify.
   a. $\sqrt{\dfrac{5}{48}}$
   b. $\sqrt[3]{\dfrac{5}{x^6}}$

10. Rationalize the denominator.
    a. $\dfrac{\sqrt{5}}{\sqrt{6}}$
    b. $\dfrac{\sqrt{2}}{\sqrt{5x}}, x > 0$

11. Rationalize the denominator.
    a. $\dfrac{1}{\sqrt[3]{3x}}$
    b. $\dfrac{\sqrt[5]{7}}{\sqrt[5]{16x^3}}$

12. Reduce the index.
    a. $\sqrt[4]{\dfrac{16}{81}}$
    b. $\sqrt[6]{81c^4d^4}$

13. Simplify. $\sqrt[6]{\dfrac{a^2}{16c^{16}}}, a \geq 0$

14. Perform the indicated operations.
    a. $\sqrt{32} + \sqrt{98}$
    b. $\sqrt{112} - \sqrt{28}$

15. Perform the indicated operations.
    a. $4\sqrt{\dfrac{1}{2}} - 3\sqrt{\dfrac{1}{8}}$
    b. $5\sqrt[3]{\dfrac{3}{4x}} - \sqrt[3]{\dfrac{3}{32x}}$

16. Perform the indicated operations.
    a. $\sqrt{2}(\sqrt{8} + \sqrt{5})$
    b. $\sqrt[3]{3x}(\sqrt[3]{9x^2} - \sqrt[3]{16x})$

17. Find the product.
    a. $(\sqrt{27} + \sqrt{50})(\sqrt{12} + \sqrt{18})$
    b. $(\sqrt{3} + 3)(\sqrt{3} + 3)$

18. Find the product.
    a. $(3 - \sqrt{3})^2$
    b. $(\sqrt{6} + \sqrt{3})(\sqrt{6} - \sqrt{3})$

19. Reduce. $\dfrac{10 - \sqrt{75}}{5}$

20. Rationalize the denominator in $\dfrac{2}{\sqrt{x} - 5}$.

21. a. Solve $\sqrt{x + 2} = -2$.
    b. Solve $\sqrt{x + 2} = x - 10$.

22. Solve $\sqrt{x - 3} - x = -3$.

23. Solve $\sqrt{x - 9} - \sqrt{x} = -1$.

24. Solve $\sqrt[3]{x - 3} - 3 = 0$.

25. The time t (in seconds) it takes a free-falling object to fall d feet is given by

$$t = \frac{\sqrt{d}}{4}$$

  a. Solve for d.

  b. An object dropped from the top of a building hits the ground after 4 seconds. How tall is the building?

26. Write in terms of i.

  a. $\sqrt{-64}$

  b. $\sqrt{-98}$

27. Add or subtract.

  a. $(4 + 5i) + (6 - 14i)$

  b. $(5 + 2i) - (7 - 8i)$

28. Multiply.

  a. $(2 + 2i)(4 - 3i)$

  b. $\sqrt{-9}(3 - \sqrt{-8})$

29. Divide.

  a. $\dfrac{4 + 3i}{2 + 3i}$

  b. $\dfrac{4 - 3i}{3 - 5i}$

30. Write the answer as 1, -1, i, or $-i$.

  a. i^{56}

  b. i^{-9}

Answers to Practice Test Chapter 7

Answer		If You Missed Question	Section	Review Examples	Page
1. a. -4	b. Not a real number	1	7.1A	1	528
2. a. -3	b. $\frac{1}{2}$	2	7.1B	2	529
3. a. 4	b. Not a real number	3	7.1B	3	530
4. a. $\frac{1}{9}$	b. $\frac{1}{4}$	4	7.1B	4	532
5. a. $x^{7/6}$	b. $\frac{1}{x^{11/15}}$	5	7.1C	5	532
6. a. $\frac{1}{x^4 y^{15}}$	b. $x^{2/5} + x^{3/5}y^{3/5}$	6	7.1C	6	533
7. a. 5	b. $|-x| = |x|$	7	7.2A	1	538
8. a. $2\sqrt[3]{5}$	b. $3ab^4\sqrt[3]{2a}$	8	7.2A	2	540
9. a. $\frac{\sqrt{15}}{12}$	b. $\frac{\sqrt[3]{5}}{x^2}$	9	7.2A	3	541
10. a. $\frac{\sqrt{30}}{6}$	b. $\frac{\sqrt{10x}}{5x}$	10	7.2B	4	542
11. a. $\frac{\sqrt[3]{9x^2}}{3x}$	b. $\frac{\sqrt[5]{14x^2}}{2x}$	11	7.2B	5	543
12. a. $\frac{2}{3}$	b. $\sqrt[3]{9c^2d^2}$	12	7.2C	6	544
13. $\frac{\sqrt[3]{2ac}}{2c^3}$, $a \geq 0$		13	7.2C	7	544
14. a. $11\sqrt{2}$	b. $2\sqrt{7}$	14	7.3A	1	549
15. a. $\frac{5\sqrt{2}}{4}$	b. $\frac{9\sqrt[3]{6x^2}}{4x}$	15	7.3A	2	550
16. a. $4 + \sqrt{10}$	b. $3x - 2\sqrt[3]{6x^2}$	16	7.3B	3	551
17. a. $48 + 19\sqrt{6}$	b. $12 + 6\sqrt{3}$	17	7.3B	4, 5	552–553
18. a. $12 - 6\sqrt{3}$	b. 3	18	7.3B	5	552–553
19. $2 - \sqrt{3}$		19	7.3B	6	553
20. $\frac{2\sqrt{x} + 10}{x - 25}$		20	7.3C	7	555
21. a. No real-number solution	b. 14	21	7.4A	1	560–561
22. 3 or 4		22	7.4A	1	560–561
23. 25		23	7.4A	2	562

Answer		If You Missed	Review		
		Question	Section	Examples	Page
24. 30		24	7.4A	3, 4	562–563
25. a. $d = 16t^2$	b. 256 ft	25	7.4B	5	563–564
26. a. $8i$	b. $7i\sqrt{2}$	26	7.5A	1	568
27. a. $10 - 9i$	b. $-2 + 10i$	27	7.5B	2	569
28. a. $14 + 2i$	b. $6\sqrt{2} + 9i$	28	7.5C	3, 4	570–571
29. a. $\frac{17}{13} - \frac{6}{13}i$	b. $\frac{27}{34} + \frac{11}{34}i$	29	7.5C	5	572
30. a. 1	b. $-i$	30	7.5D	6	573

Cumulative Review Chapters 1–7

1. The number $\sqrt{19}$ belongs to which of these sets: natural numbers, whole numbers, integers, rational numbers, irrational numbers, real numbers Name all that apply?

2. Simplify: $(2x^2y^{-4})^{-4}$

3. Evaluate: $[-9(9+6)] + 7$

4. Solve: $\dfrac{7}{6} - \dfrac{x}{18} = \dfrac{3(x+9)}{54}$

5. Graph: $-7 \leq -7x - 14 < 7$

6. The perimeter of a rectangle is $P = 2L + 2W$, where L is the length and W is the width. If the perimeter is 100 feet and the length is 30 feet more than the width, what are the dimensions of the rectangle?

7. Find the x- and y-intercepts of $y = 8x + 6$.

8. The line passing through $A(5, -2)$ and $B(-3, y)$ is perpendicular to a line with slope -4. Find y.

9. Find the standard form of the equation of the line that passes through the points $(-6, 4)$ and $(8, 7)$.

10. Graph: $y \leq x - 2$

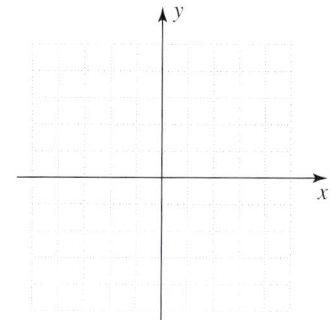

11. Graph: $|y| \leq 2$

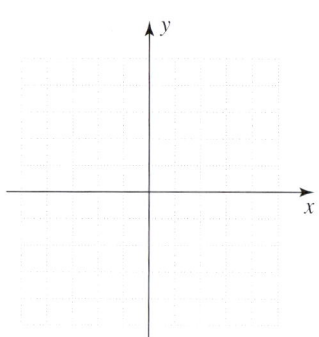

12. Solve the system by substitution:
$x - 2y = 3$
$x = -3 - y$

13. Solve the system:
$\dfrac{x}{4} + \dfrac{y}{3} = -6$
$\dfrac{x}{12} + \dfrac{y}{9} = -2$

14. Solve the system:
$2x + y + z = -5$
$-4x + y + z = -3$
$3y + 3z = -2$

15. Solve the system:
$$3x + 2y + 3z = 29$$
$$2x + y + z = 15$$
$$x + 2y - 2z = -1$$

16. Bart has three investments totaling $90,000. These investments earn interest at 7%, 9%, and 11%. Bart's total income from these investments is $8500. The income from the 11% investment exceeds the total income from the other two investments by $300. Find how much Bart has invested at each rate.

17. Find the range of the relation $\{(-3, -5), (3, -3), (-5, 0)\}$.

18. Find $f(2)$ given $f(x) = 4x - 3$.

19. Multiply: $(x - 4)(x^2 + 4x + 2)$

20. Factor: $9x^6 - 12x^5 + 15x^4 + 12x^3$

21. Factor: $9x^3y - 15x^2y^2 - 6xy^3$

22. Solve for x: $6x^2 - 7x = -2$

23. Reduce to lowest terms: $\dfrac{x^2 - 9}{27 - x^3}$

24. Factor $3x^3 + 16x^2 + 17x + 4$ if $(x + 4)$ is one of its factors.

25. Solve for x: $\dfrac{x}{x + 5} - \dfrac{x}{x - 5} = \dfrac{x^2 + 25}{x^2 - 25}$

26. Find two consecutive even integers such that the sum of their reciprocals is $\dfrac{11}{60}$.

27. If the temperature of a gas is held constant, the pressure P varies inversely as the volume V. A pressure of 1720 pounds per square inch is exerted by 4 cubic feet of air in a cylinder fitted with a piston. Find k constant of variation.

28. Evaluate: $9^{3/2}$

29. Assume x and y are positive and simplify:
$x^{2/5}(x^{-1/5} + y^{4/5})$

30. Add: $\sqrt{175} + \sqrt{112}$

31. Find the product: $(7 - \sqrt{7})^2$

32. Rationalize the denominator: $\dfrac{\sqrt{x}}{\sqrt{x} - \sqrt{5}}$

33. Solve: $\sqrt{x + 34} = x + 14$

34. Divide: $\dfrac{7 + 6i}{1 + 9i}$

35. Find: i^{49}

Section

8.1 Solving Quadratics by Completing the Square

8.2 The Quadratic Formula: Applications

8.3 The Discriminant and Its Applications

8.4 Solving Equations in Quadratic Form

8.5 Nonlinear Inequalities

Chapter

Quadratic Equations and Inequalities

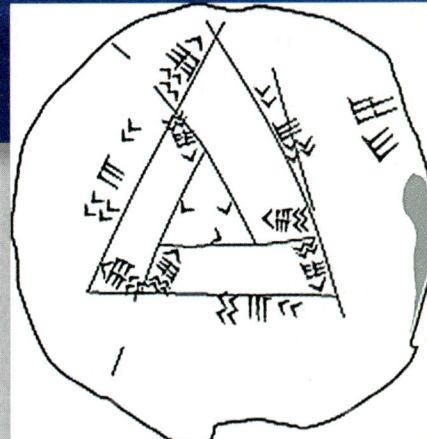

Problem with two concentric and parallel equilateral triangles. Babylonia, 19th c. B.C.

The Human Side of Algebra

Many civilizations, starting with the Babylonians in 2000 B.C., were familiar with quadratic equations, whose solutions were found by following verbal instructions given in a text. We believe that the Arabian mathematician al-Khowarizmi (ca. 820) was the first to classify quadratic equations into three types: $x^2 + ax = b$, $x^2 + b = ax$, and $x^2 = ax + b$, where a and b are positive. These three types of equations were solved by using a few general rules, and their correctness was demonstrated geometrically. In general, the quadratic equation $x^2 + px = q$ was solved by the method of "completion of squares."

Although the Arabs recognized the existence of two solutions for quadratic equations, they listed only the positive ones. In the twelfth century A.D., the Hindu mathematician Bhaskara affirmed the existence and validity of negative as well as positive solutions. What types of quadratic equations were being solved? A Babylonian clay tablet from 2000 B.C. states, "I have added the area and two-thirds of my square and it is $0;36$ $[\frac{36}{60}]$. What is the side of my square?" Can you translate this into an equation? The problem was solved by using a quadratic formula equivalent to the one we discuss in this chapter.

8.1 Solving Quadratics by Completing the Square

Objectives

A ▸ Solve quadratic equations of the form $ax^2 + c = 0$.

B ▸ Solve quadratic equations of the form $a(x + b)^2 = c$.

C ▸ Solve quadratic equations by completing the square.

D ▸ Applications Involving Quadratic Equations

▸ To Succeed, Review How To . . .

1. Find the square root of a number (pp. 526–528).
2. Rationalize the denominator of a fraction (pp. 541–543).
3. Add, subtract, multiply, and divide complex numbers (pp. 526–528).
4. Multiply $(x \pm a)^2$ (pp. 364–365).

▸ Getting Started

Baseball Diamonds and Quadratic Equations

Look at the picture of the baseball "diamond." If you pay close attention to the infield you will notice that it is a square with the four corners representing home plate, first base, second base, and third base. Since it is a square, the distance between consecutive bases is the same and we can refer to the length of one side as x ft. If the area of a square is x^2 and we know the area of the infield is 8100 ft^2, then we have $x^2 = 8100$. We solve this equation by taking (extracting) the square roots of both sides. Because a nonzero number has two square roots, we have

$$x = \sqrt{8100} = 90 \quad \text{or} \quad x = -\sqrt{8100} = -90$$

The solutions (or roots) are 90 and −90. This procedure is usually shortened to

$$x^2 = 8100$$
$$x = \pm\sqrt{8100} = \pm 90$$

where the notation ±90 (read "plus or minus 90") means that $x = 90$ or $x = -90$. If x represents the length of a side, the answer −90 must be discarded because the length of an object cannot be negative.

A ▸ Solving Quadratic Equations of the Form $ax^2 + c = 0$

In Chapter 2, we studied methods of solving *linear equations*—equations in which the variable involved has an exponent of 1 (the first power). We are now ready to discuss equations containing the *second* (but no higher) power of the unknown.

Such equations are called *second-degree,* or *quadratic,* equations; these can be written in a *standard form.* Here is the definition.

8.1 Solving Quadratics by Completing the Square

QUADRATIC EQUATION IN STANDARD FORM

$ax^2 + bx + c = 0 \quad$ (a, b, and c are real numbers, $a \neq 0$)

is a **quadratic equation in standard form.**

One of the procedures used to solve these equations is similar to that employed in Chapter 2 to solve linear equations and consists of applying certain transformations to obtain equivalent equations until you arrive at an equation whose solution set is evident.

In Section 5.7, we learned how to solve quadratic equations with factoring (using the OFF method). For example,

$$x^2 - 25 = 0 \quad \text{O} \quad \text{Given (equation is already set equal to 0).}$$
$$(x + 5)(x - 5) = 0 \quad \text{F} \quad \text{Factor (difference of squares).}$$
$$x + 5 = 0 \quad \text{or} \quad x - 5 = 0 \quad \text{F} \quad \text{Factors} = 0 \text{ (zero-product property).}$$
$$x = -5 \quad \text{or} \quad x = 5$$

This equation is a special kind of quadratic equation because it is missing the bx term, the term with x to the first power. In this case, we can use a different approach for solving. It will require that the x^2 term be isolated and then take the square roots of both sides.

$$x^2 - 25 = 0 \quad \text{Given}$$
$$x^2 = 25 \quad \text{Isolate } x^2 \text{ by adding 25 to both sides.}$$
$$\sqrt{x^2} = \pm\sqrt{25} \quad \text{Take the square roots of both sides.}$$
$$x = \pm 5 \quad \text{Simplify.}$$

The equation has two solutions, 5 and -5, as before.

If we compare the equation $x^2 = k$ to the standard form equation, $ax^2 + bx + c = 0$, we see that the bx term is missing from $x^2 = k$. The following is a procedure to solve such equations.

PROCEDURE

Solving a quadratic equation of the form $ax^2 + c = 0$

1. Isolate the x^2 on one side of the equation with the constant on the other side.
2. Take the square root of both sides, remembering it is $\pm$ when taking the square root of the constant.
3. Simplify the radical, if possible.
4. Check the solutions in the original equation.

For the special case of the quadratic equation $x^2 = k$, we have

SOLUTIONS OF $x^2 = k$

The solutions of $x^2 = k$, where k is a real number, are $\sqrt{k}$ and $-\sqrt{k}$, abbreviated as $\pm\sqrt{k}$.

To solve the equation $9x^2 - 5 = 0$, we notice that the bx term is missing. We need to isolate the x^2 and take the square roots of both sides. Here's how we do it:

$$9x^2 - 5 = 0 \quad \text{Given}$$
$$9x^2 = 5 \quad \text{Add 5 to both sides.}$$
$$x^2 = \frac{5}{9} \quad \text{Divide both sides by 9 (now } x^2 \text{ is by itself).}$$
$$\sqrt{x^2} = \pm\sqrt{\frac{5}{9}} \quad \text{Take square roots of both sides.}$$
$$x = \pm\frac{\sqrt{5}}{3} \quad \text{Simplify the radical } \left(\sqrt{\frac{5}{9}} = \frac{\sqrt{5}}{\sqrt{9}} = \frac{\sqrt{5}}{3}\right).$$

The solutions are

$$\frac{\sqrt{5}}{3} \quad \text{and} \quad -\frac{\sqrt{5}}{3}$$

and the solution set is

$$\left\{\frac{\sqrt{5}}{3}, -\frac{\sqrt{5}}{3}\right\}$$

EXAMPLE 1   Solving equations of the form $ax^2 + c = 0$

Solve:

a. $4x^2 - 7 = 0$ **b.** $3x^2 + 24 = 0$ **c.** $5x^2 - 4 = 0$

SOLUTION

a. The idea is to isolate x^2 and then take the square roots of both sides.

$4x^2 - 7 = 0$	Given
$4x^2 = 7$	Add 7 to both sides.
$x^2 = \frac{7}{4}$	Divide both sides by 4 (now x^2 is isolated).
$\sqrt{x^2} = \pm\sqrt{\frac{7}{4}}$	Take square roots of both sides $\left(\sqrt{\frac{7}{4}} = \frac{\sqrt{7}}{\sqrt{4}} = \frac{\sqrt{7}}{2}\right)$.
$x = \pm\frac{\sqrt{7}}{2}$	

The solutions are

$$\frac{\sqrt{7}}{2} \quad \text{and} \quad -\frac{\sqrt{7}}{2}$$

and the solution set is

$$\left\{\frac{\sqrt{7}}{2}, -\frac{\sqrt{7}}{2}\right\}$$

b.

$3x^2 + 24 = 0$	Given
$3x^2 = -24$	Add -24 to both sides.
$x^2 = -8$	Divide both sides by 3.
$\sqrt{x^2} = \pm\sqrt{-8}$	Take square roots of both sides.
$x = \pm 2i\sqrt{2}$	Simplify: $\sqrt{-8} = \sqrt{4}\sqrt{-2} = 2i\sqrt{2}$.

The solutions are $2i\sqrt{2}$ and $-2i\sqrt{2}$. The answers are non-real **complex numbers** given in the *simplified* form, $\pm 2i\sqrt{2}$ instead of $\pm i\sqrt{8}$.

c.

$5x^2 - 4 = 0$	Given
$5x^2 = 4$	Add 4 to both sides.
$x^2 = \frac{4}{5}$	Divide both sides by 5.
$\sqrt{x^2} = \pm\sqrt{\frac{4}{5}} = \pm\frac{2}{\sqrt{5}}$	Take square roots of both sides $\left(\sqrt{\frac{4}{5}} = \frac{\sqrt{4}}{\sqrt{5}} = \frac{2}{\sqrt{5}}\right)$.
$x = \pm\frac{2 \cdot \sqrt{5}}{\sqrt{5} \cdot \sqrt{5}}$	Rationalize the denominator.
$x = \pm\frac{2\sqrt{5}}{5}$	

PROBLEM 1

Solve:

a. $9x^2 - 4 = 0$

b. $3x^2 + 54 = 0$

c. $3x^2 - 16 = 0$

Answers to PROBLEMS

1. a. $\frac{2}{3}$ and $-\frac{2}{3}$ **b.** $3i\sqrt{2}$ and $-3i\sqrt{2}$ **c.** $\frac{4\sqrt{3}}{3}$ and $-\frac{4\sqrt{3}}{3}$

The solutions are

$$\frac{2\sqrt{5}}{5} \quad \text{and} \quad -\frac{2\sqrt{5}}{5}$$

and the solution set is

$$\left\{\frac{2\sqrt{5}}{5}, -\frac{2\sqrt{5}}{5}\right\}$$

The answer should be written with a *rationalized* denominator.

B › Solving Quadratic Equations of the Form $a(x + b)^2 = c$

The method used to solve equations of the form $ax^2 + c = 0$ can also be used to solve equations of the form $(x + b)^2 = c$. Thus, to solve the equation $(x + 3)^2 = 16$, we proceed as follows:

$(x + 3)^2 = 16$	Given
$\sqrt{(x + 3)^2} = \pm\sqrt{16} = \pm 4$	Take square roots of both sides.
$x + 3 = 4 \quad \text{or} \quad x + 3 = -4$	
$x = 1 \quad \text{or} \quad x = -7$	Solve $x + 3 = 4$ and $x + 3 = -4$.

The solutions are 1 and -7, and the solution set is $\{1, -7\}$.

To solve the equation $(x - 2)^2 - 9 = 0$, we add 9 to both sides and then take the square roots to obtain

$$(x - 2)^2 - 9 = 0$$
$$(x - 2)^2 = 9$$
$$\sqrt{(x - 2)^2} = \pm\sqrt{9} = \pm 3$$
$$x - 2 = 3 \quad \text{or} \quad x - 2 = -3$$
$$x = 5 \quad \text{or} \quad x = -1$$

The solutions for $(x - 2)^2 - 9 = 0$ are 5 and -1, and the solution set is $\{5, -1\}$.

EXAMPLE 2 Solving equations of the form $a(x + b)^2 = c$
Solve:
a. $(x - 1)^2 = 27$ **b.** $9(x + 5)^2 = 11$ **c.** $3(x - 2)^2 + 25 = 0$

SOLUTION

a.
$(x - 1)^2 = 27$	Given
$\sqrt{(x - 1)^2} = \pm\sqrt{27}$	Take the square root of both sides.
$x - 1 = \pm 3\sqrt{3}$	Simplify the radical ($\sqrt{27} = 3\sqrt{3}$).
$x - 1 = 3\sqrt{3} \quad \text{or} \quad x - 1 = -3\sqrt{3}$	
$x = 1 + 3\sqrt{3} \quad \text{or} \quad x = 1 - 3\sqrt{3}$	Solve for x by adding 1 to both sides.

The solutions are $1 + 3\sqrt{3}$ and $1 - 3\sqrt{3}$, and the solution set is $\{1 + 3\sqrt{3}, 1 - 3\sqrt{3}\}$.

PROBLEM 2
Solve:
a. $(x - 2)^2 = 24$
b. $4(x + 7)^2 = 3$
c. $5(x - 3)^2 + 36 = 0$

(continued)

Answers to PROBLEMS
2. **a.** $2 \pm 2\sqrt{6}$ **b.** $\frac{-14 \pm \sqrt{3}}{2}$
c. $3 \pm \frac{6\sqrt{5}}{5}i$

This can be verified by substituting the solutions into the original equation.

$(x-1)^2 \stackrel{?}{=} 27$				$(x-1)^2 \stackrel{?}{=} 27$	
$[(1+3\sqrt{3})-1]^2$	27	Original equation	Substitute $(1 \pm 3\sqrt{3})$ for x.	$[(1-3\sqrt{3})-1]^2$	27
$[1+3\sqrt{3}-1]^2$	27	Remove parentheses.		$[1-3\sqrt{3}-1]^2$	27
$[3\sqrt{3}]^2$	27	Combine like terms.		$[-3\sqrt{3}]^2$	27
$9 \cdot 3$	27	Square $[3\sqrt{3}]$.		$9 \cdot 3$	27
27	27			27	27

Both solutions resulted in a true statement when placed in the original equation, so the solutions are verified.

b. $9(x+5)^2 = 11$ Given

$(x+5)^2 = \dfrac{11}{9}$ Divide both sides by 9 [now $(x+5)^2$ is isolated].

$\sqrt{(x+5)^2} = \pm\sqrt{\dfrac{11}{9}}$ Take the square root of both sides.

$x+5 = \pm\dfrac{\sqrt{11}}{3}$ Simplify the radical.

$x = -5 \pm \dfrac{\sqrt{11}}{3}$ Subtract 5 from both sides.

$x = \dfrac{-15 \pm \sqrt{11}}{3}$ Write as one fraction with denominator 3 (multiply $-\dfrac{5}{1} \cdot \dfrac{3}{3}$).

c. $3(x-2)^2 + 25 = 0$ Given

$3(x-2)^2 = -25$ Subtract 25 from both sides.

$(x-2)^2 = \dfrac{-25}{3}$ Divide both sides by 3 [now the $(x-2)^2$ is isolated].

$\sqrt{(x-2)^2} = \pm\sqrt{\dfrac{-25}{3}}$ Take the square root of both sides.

$x-2 = \pm\dfrac{5i}{\sqrt{3}}$ Simplify the radical.

$x-2 = \pm\dfrac{5i \cdot \sqrt{3}}{\sqrt{3} \cdot \sqrt{3}}$ Rationalize the denominator.

$x = 2 \pm \dfrac{5\sqrt{3}}{3}i$ Solve for x by adding 2.

The solutions are

$$2 + \dfrac{5\sqrt{3}}{3}i \quad \text{and} \quad 2 - \dfrac{5\sqrt{3}}{3}i$$

and the solution set is

$$\left\{2 + \dfrac{5\sqrt{3}}{3}i, \ 2 - \dfrac{5\sqrt{3}}{3}i\right\}$$

We leave the verification to you.

EXAMPLE 3   Writing as $a(x+b)^2 = c$ and solving

Solve: $x^2 + 4x + 4 = 20$

SOLUTION   If we subtract 20 from both sides, we obtain

$$x^2 + 4x - 16 = 0$$

PROBLEM 3

Solve:

$$x^2 + 6x + 9 = 24$$

Answers to PROBLEMS

3. $-3 \pm 2\sqrt{6}$

This is *not* factorable using integer coefficients, so we try to factor the left side first.

$$x^2 + 4x + 4 = 20 \quad \text{Given}$$
$$(x + 2)^2 = 20 \quad \text{Factor } x^2 + 4x + 4 \text{ as } (x+2)^2 \text{ (perfect square trinomial).}$$
$$\sqrt{(x+2)^2} = \pm\sqrt{20} \quad \text{Take the square root of both sides.}$$
$$x + 2 = \pm 2\sqrt{5} \quad \text{Simplify the radical } (\sqrt{20} = \sqrt{4 \cdot 5} = 2\sqrt{5}).$$
$$x = -2 \pm 2\sqrt{5} \quad \text{Solve for } x \text{ by subtracting 2 from both sides.}$$

The solutions are

$$x = -2 + 2\sqrt{5} \quad \text{and} \quad x = -2 - 2\sqrt{5}$$

and the solution set is

$$\{-2 + 2\sqrt{5}, -2 - 2\sqrt{5}\}$$

We leave the verification to you.

C ▸ Solving Quadratic Equations by Completing the Square

The solutions of Examples 2 and 3 were obtained by writing the equation in the form $a(x + b)^2 = c$. Suppose we have a quadratic equation that is not in this form. We can put it in this form if we learn a technique called **completing the square.** As you recall,

$$(x + a)^2 = x^2 + 2ax + a^2 \quad \text{and} \quad (x - a)^2 = x^2 - 2ax + a^2$$

In both cases, the last term is the square of one-half the coefficient of x. How can we make $x^2 + 10x$ a perfect square trinomial? Since the coefficient of x is 10, and in a perfect square trinomial, the coefficient of x is $2a$, $10 = 2a$ and a must be 5. Thus, we make $x^2 + 10x$ a perfect square trinomial by adding 5^2. We then have

$$x^2 + 10x + 5^2 = (x + 5)^2$$

> **PROCEDURE**
>
> **Completing the square on the quadratic expression $x^2 + bx$**
> 1. Take $\frac{1}{2}$ of the b value and square it
> 2. The result becomes the constant of the expression or the "c" value.
>
> $x^2 + bx + \left(\frac{b}{2}\right)^2$ **is a perfect square trinomial and factors into** $\left(x + \frac{b}{2}\right)^2$.

Now consider the equation $x^2 + 8x = 12$. To make $x^2 + 8x$ a perfect square trinomial, we note that the coefficient of x is 8, so the number to be added is the square of one-half of 8, or 4^2. Adding 4^2 to both sides, we have

$$x^2 + 8x + 4^2 = 12 + 4^2$$
$$(x + 4)^2 = 12 + 4^2 \quad \text{Factor } x^2 + 8x + 4^2 \text{ as } (x+4)^2.$$
$$(x + 4)^2 = 12 + 16$$

And, $(x + 4)^2 = 28$, the form we want. Now take the square roots of both sides:

$$\sqrt{(x+4)^2} = \pm\sqrt{28} \quad \text{Take the square root of both sides.}$$
$$x + 4 = \pm 2\sqrt{7} \quad \text{Simplify the radical.}$$
$$x = -4 \pm 2\sqrt{7} \quad \text{Subtract 4 from both sides.}$$

The solutions are $-4 + 2\sqrt{7}$ and $-4 - 2\sqrt{7}$, and the solution set is $\{-4 + 2\sqrt{7}, -4 - 2\sqrt{7}\}$. Note that in the equation $x^2 + 8x = 12$, the variables x and x^2 are *isolated* on one side of the equation.

EXAMPLE 4 Solving by completing the square

Solve by completing the square: $x^2 + 10x - 2 = 0$

SOLUTION

$x^2 + 10x - 2 = 0$	Given
$x^2 + 10x = 2$	Add 2 to both sides so that only the variable terms are on one side.
$x^2 + 10x + 5^2 = 2 + 5^2$	Add 5^2, the square of one-half of the coefficient of x—add $[\frac{1}{2} \cdot 10]^2$ to both sides.
$(x + 5)^2 = 27$	Factor on the left; simplify on the right.
$\sqrt{(x + 5)^2} = \pm\sqrt{27}$	Take the square root of both sides.
$x + 5 = \pm 3\sqrt{3}$	Simplify the radical.
$x = -5 \pm 3\sqrt{3}$	Subtract 5.

The solutions are $-5 + 3\sqrt{3}$ and $-5 - 3\sqrt{3}$, and the solution set is $\{-5 + 3\sqrt{3}, -5 - 3\sqrt{3}\}$.

PROBLEM 4

Solve by completing the square:

$x^2 + 12x - 8 = 0$

When the coefficient of x^2 is not 1 and the left-hand side of the equation in standard form is not factorable using integer coefficients, we first isolate the terms containing the variable on one side of the equation and then divide each term by the coefficient of x^2. For example, to solve $3x^2 + 12x - 6 = 0$, we add 6 to both sides (so all the terms containing the variable are isolated on one side), and then divide by 3, the coefficient of x^2. Here are the steps.

$$3x^2 + 12x - 6 = 0 \quad \text{Given}$$

Step 1. Add 6 to isolate the variables on the left-hand side.

$$3x^2 + 12x = 6 \quad \text{Add 6 to both sides.}$$

Step 2. Divide each term by 3, the coefficient of x^2.

$$x^2 + 4x = 2 \quad \text{Divide each term by 3.}$$

Step 3. Add the square of one-half the x term's coefficient to both sides—add

$$\left[\tfrac{1}{2}(4)\right]^2 = (2)^2$$

to both sides

$$x^2 + 4x + (2)^2 = 2 + (2)^2 \quad \text{The first-degree term's coefficient is 4, so add } (\tfrac{1}{2} \cdot 4)^2 = (2)^2 \text{ to both sides.}$$

Step 4. Factor the left-hand side, and simplify the right-hand side.

$$(x + 2)^2 = 6$$

Step 5. Take the square root of both sides.

$$\sqrt{(x + 2)^2} = \pm\sqrt{6}$$

Step 6. Solve the resulting equation by subtracting 2 from both sides and simplifying.

$$x = -2 \pm \sqrt{6} \quad \text{Subtract 2.}$$

The solution set is

$$\{-2 + \sqrt{6}, -2 - \sqrt{6}\}$$

Answers to PROBLEMS

4. $-6 \pm 2\sqrt{11}$

Here is a summary of this procedure.

> **PROCEDURE**
>
> **Solving a Quadratic Equation by Completing the Square ($ax^2 + bx + c = 0$)**
>
> 1. Write the equation with the variables in descending order on the left and the constant on the right. ($ax^2 + bx = -c$)
> 2. If the coefficient of the x^2 term is not 1, divide each term by this coefficient. ($x^2 + \frac{b}{a}x = -\frac{c}{a}$)
> 3. Add the square of one-half of the coefficient of the x term to both sides. $\left(\frac{1}{2} \cdot \frac{b}{a}\right)^2$
> 4. Rewrite the left-hand side as a perfect square in factored form and simplify the right-hand side. $\left(x + \frac{b}{2a}\right)^2$
> 5. Take the square root of both sides and rationalize the denominator if necessary.
> 6. Solve the resulting equation.

EXAMPLE 5 Completing the square when the coefficient of x^2 is not 1

Solve by completing the square: $3x^2 - 3x - 1 = 0$

SOLUTION We use the six-step procedure.

$3x^2 - 3x - 1 = 0$ Given

Step 1. $3x^2 - 3x = 1$ Add 1 so the variable terms are isolated.

Step 2. $x^2 - x = \frac{1}{3}$ Divide each term by 3.

Step 3. $x^2 - x + \left(\frac{1}{2}\right)^2 = \frac{1}{3} + \left(\frac{1}{2}\right)^2$ The first-degree term's coefficient is -1, so add $\left[\frac{1}{2}(-1)\right]^2 = \left(-\frac{1}{2}\right)^2 = \left(\frac{1}{2}\right)^2$ to each side.

Step 4. $\left(x - \frac{1}{2}\right)^2 = \frac{7}{12}$ Factor the left-hand side, Simplify the right-hand side.

Step 5. $\sqrt{\left(x - \frac{1}{2}\right)^2} = \pm\sqrt{\frac{7}{12}} = \pm\frac{\sqrt{7} \cdot \sqrt{3}}{\sqrt{12} \cdot \sqrt{3}}$ Take the square root of both sides and rationalize the denominator.

$x - \frac{1}{2} = \pm\frac{\sqrt{21}}{6}$ Simplify.

Step 6. $x - \frac{1}{2} = \frac{\sqrt{21}}{6}$ or $x - \frac{1}{2} = -\frac{\sqrt{21}}{6}$

$x = \frac{1}{2} + \frac{\sqrt{21}}{6}$ or $x = \frac{1}{2} - \frac{\sqrt{21}}{6}$ Solve by adding $\frac{1}{2}$ to both sides.

The solutions are

$\frac{1}{2} + \frac{\sqrt{21}}{6}$ and $\frac{1}{2} - \frac{\sqrt{21}}{6}$, or $\frac{3 + \sqrt{21}}{6}$ and $\frac{3 - \sqrt{21}}{6}$

The solution set is thus,

$$\left\{\frac{3 + \sqrt{21}}{6}, \frac{3 - \sqrt{21}}{6}\right\}$$

The verification is left to you.

PROBLEM 5

Solve by completing the square:

$5x^2 - 5x - 1 = 0$

Answers to PROBLEMS

5. $\frac{5 \pm 3\sqrt{5}}{10}$ or $\frac{1}{2} \pm \frac{3\sqrt{5}}{10}$

D › Applications Involving Quadratic Equations

In Section 5.7 we used the Pythagorean theorem ($a^2 + b^2 = c^2$) to help solve applications involving a right triangle with integer solutions. Now we will practice with an example that does not have integer solutions.

Given the right triangle in Figure 8.1 with sides 3 inches and 6 inches, find the length of the hypotenuse c.

> Figure 8.1

$$a^2 + b^2 = c^2$$
$$3^2 + 6^2 = c^2 \quad \text{Substitute the values.}$$
$$9 + 36 = c^2 \quad \text{Simplify.}$$
$$45 = c^2$$
$$\pm\sqrt{45} = \sqrt{c^2} \quad \text{Take the square root of both sides.}$$
$$\pm\sqrt{9} \cdot \sqrt{5} = c \quad \text{Simplify the radical.}$$
$$\pm 3\sqrt{5} = c$$

Since measurements must be positive numbers, the length of the hypotenuse is $3\sqrt{5}$, which is approximately 6.7 inches.

In Example 6, we will solve an application involving the Pythagorean theorem.

EXAMPLE 6 Application involving the Pythagorean theorem

Cindy was at the furniture store and saw a sale on a home entertainment center. She wasn't sure if her 32-inch flat screen TV would fit in the opening, which was 30 by 25 inches. The 32-inch measurement of her flat screen TV is the diagonal measurement. If Cindy buys the home entertainment center will her TV fit?

SOLUTION To find out if Cindy's 32-inch, diagonally measured, TV will fit in this opening, we find the measurement of the diagonal of the opening. Then we will compare that measurement to the size of her 32-inch TV to make the final decision. Since the diagonal of a rectangle forms two right triangles, we can use the Pythagorean theorem to help solve this problem. We draw a diagram of the rectangular opening and label the parts.

$$a^2 + b^2 = c^2 \quad \text{Pythagorean theorem}$$
$$30^2 + 25^2 = x^2 \quad \text{Substitute the values.}$$
$$900 + 625 = x^2 \quad \text{Simplify.}$$
$$1525 = x^2$$
$$\sqrt{1525} = \sqrt{x^2} \quad \text{Take the square root of both sides.}$$
$$\sqrt{25} \cdot \sqrt{61} = x \quad \text{Simplify.}$$
$$5\sqrt{61} = x$$

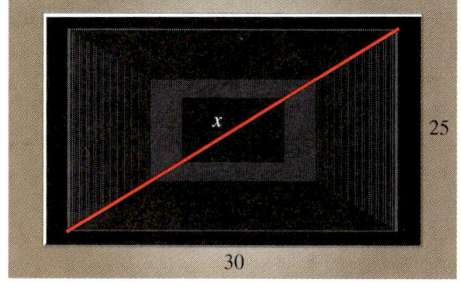

The length of the diagonal of the opening in the home entertainment center is exactly $5\sqrt{61}$, which is approximately 39 inches. Since the diagonal measurement of Cindy's TV is 32 inches, which is less than the 39-inch diagonal opening, she can buy the home entertainment center.

PROBLEM 6

Rework Example 6 to see if a 42-inch TV will fit in the 34- by 22-inch opening of a home entertainment center.

Answers to PROBLEMS

6. The diagonal of the opening measures approximately 40 inches and the diagonal of the TV is 42 inches, so it will **not** fit.

Calculator Corner

Using Graphs to Verify Solutions to Quadratic Equations

To graphically solve Example 5, place all variables on the left (Y_1) and the constant on the right (Y_2) of the equation. You have to do this algebraically. We then find the intersection of Y_1 and Y_2. The values of x thus obtained are the solutions of the equation. Start by rewriting the equation as $3x^2 - 3x = 1$; then graph $Y_1 = 3x^2 - 3x$ and $Y_2 = 1$ using a $[-1, 3]$ by $[-2, 3]$ window. To find the intersection with a TI-83 Plus, press [2nd] [TRACE] [5] : Move the cursor to a point on Y_1 and near the left intersection point. The calculator prompt is "First curve?" Press [ENTER]. Press [ENTER] again when the calculator asks "Second Curve?" and when it asks for a "Guess?" The intersection is given as $X = -.2637626$ and $Y = 1$. One solution is $x = -0.2637626$. Check to see that this decimal approximation corresponds to the solution

$$x = \frac{3 - \sqrt{21}}{6}$$

Now press [2nd] [TRACE] [5] again and move the cursor to a point near the right intersection point. Press [ENTER] three times. The intersection, as shown in the window, occurs when $X = 1.2637626$, which corresponds to

$$\frac{3 + \sqrt{21}}{6}$$

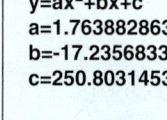

1. To solve Example 3 using the intersect feature of your calculator,
   a. What would you use for Y_1?
   b. What would you use for Y_2?
2. How can you tell if a quadratic equation has no real-number solutions by looking at its graph?
3. In some instances, we want to find the maximum or minimum of a quadratic function. For example, the numbers of robberies per 100,000 inhabitants for 1980, 1985, 1990, and 1991 were 251, 208, 257, and 273, respectively. Enter this information in two tables by pressing [STAT] [1] and then using L_1 for the year (1980 = 0, 1985 = 5, etc.) and L_2 for the number of robberies per 100,000 inhabitants. Now to find our quadratic function, press [STAT] [▶] [5] [ENTER] (see Window 1). You don't want to copy all this to graph it. So we tell the calculator to copy the equation produced from our statistical data by pressing [Y=] [VARS] [5] [▶] [▶] [1]. Finally, since we are covering a period of 11 years and the number of robberies varies from 208 to 273, use a $[-1, 11]$ by $[200, 300]$ window and press [GRAPH]. The graph is shown in Window 2.

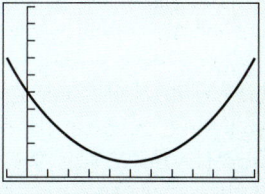

Window 1

   a. In what time interval is the number of robberies decreasing?
   b. In what time interval is the number of robberies increasing?
   c. What was the minimum number of robberies per year? (Answer to the closest whole number.)
   d. In what year did the minimum number of robberies occur? (Answer to the closest whole number.)
4. If you have a TI-83 Plus, the minimum of a function can be found by pressing [2nd] [TRACE] [3] and answering the calculator queries for "Left bound?", "Right bound?", and "Guess?". Use this method to find X and Y in Exercise 3 and explain what they mean.

Window 2

Exercises 8.1

A Solving Quadratic Equations of the Form $ax^2 + c = 0$ In Problems 1–20, solve the equation.

1. $x^2 = 64$
2. $x^2 = 81$
3. $x^2 = -121$
4. $x^2 = -144$
5. $x^2 - 169 = 0$
6. $x^2 - 100 = 0$
7. $x^2 + 4 = 0$
8. $x^2 + 25 = 0$

9. $36x^2 - 49 = 0$
10. $36x^2 - 81 = 0$
11. $4x^2 + 81 = 0$
12. $9x^2 + 64 = 0$
13. $3x^2 - 25 = 0$
14. $5x^2 - 16 = 0$
15. $5x^2 + 36 = 0$
16. $11x^2 + 49 = 0$
17. $3x^2 - 100 = 0$
18. $4x^2 - 13 = 0$
19. $13x^2 + 81 = 0$
20. $11x^2 + 4 = 0$

B. Solving Quadratic Equations of the Form $a(x + b)^2 = c$
In Problems 21–40, solve the equation.

21. $(x + 5)^2 = 4$
22. $(x + 3)^2 = 9$
23. $x^2 + 4x + 4 = -25$
24. $x^2 + 2x + 1 = -16$
25. $(x - 6)^2 = 18$
26. $(x - 2)^2 = 50$
27. $x^2 - 2x + 1 = -28$
28. $x^2 - 6x + 9 = -32$
29. $(x - 1)^2 - 50 = 0$
30. $(x - 2)^2 - 18 = 0$
31. $(x - 5)^2 - 32 = 0$
32. $(x - 2)^2 - 4 = 0$
33. $(x - 9)^2 + 64 = 0$
34. $(x - 2)^2 + 25 = 0$
35. $3x^2 + 6x + 3 = 96$
36. $3x^2 + 30x + 75 = 72$
37. $7(x - 2)^2 - 350 = 0$
38. $3(x - 1)^2 - 54 = 0$
39. $7(x - 5)^2 + 189 = 0$
40. $5(x - 3)^2 + 250 = 0$

C. Solving Quadratic Equations by Completing the Square
In Problems 41–70, solve by completing the square.

41. $x^2 + 6x + 5 = 0$
42. $x^2 + 4x + 3 = 0$
43. $x^2 + 8x + 15 = 0$
44. $x^2 + 8x + 7 = 0$
45. $x^2 + 6x + 10 = 0$
46. $x^2 + 12x + 37 = 0$
47. $x^2 - 10x + 24 = 0$
48. $x^2 + 12x - 28 = 0$
49. $x^2 - 10x + 21 = 0$
50. $x^2 - 2x - 143 = 0$
51. $x^2 - 8x + 17 = 0$
52. $x^2 - 14x + 58 = 0$
53. $2x^2 + 4x + 3 = 0$
54. $2x^2 + 7x + 6 = 0$
55. $3x^2 + 6x + 78 = 0$
56. $9x^2 + 6x + 2 = 0$
57. $25y^2 - 25y + 6 = 0$
58. $4y^2 - 16y + 15 = 0$
59. $4y^2 - 4y + 5 = 0$
60. $9x^2 - 12x + 13 = 0$
61. $4x^2 - 7 = 4x$
62. $2x^2 - 18 = -9x$
63. $2x^2 + 1 = 4x$
64. $2x^2 + 3 = 6x$
65. $(x + 3)(x - 2) = -4$
66. $(x + 4)(x - 1) = -6$
67. $2x(x + 5) - 1 = 0$
68. $2x(x - 4) = 2(9 - 8x) - x$
69. $2x(x + 3) - 10 = 0$
70. $4x(x + 1) - 5 = 0$

D. Applications Involving Quadratic Equations

71. *Pythagorean theorem* A 25-foot ladder is leaning against a building. If the foot of the ladder is 10 feet from the base of the building, how far up the side of the building does the ladder reach? (Approximate to the nearest tenth.)

72. *Pythagorean theorem* Beau wants to stretch a rope ladder from his tree house which is vertically 20 feet above the ground to a spot on the ground which is 24 feet from the base of the tree. How long should the rope ladder be? (Approximate to the nearest tenth.)

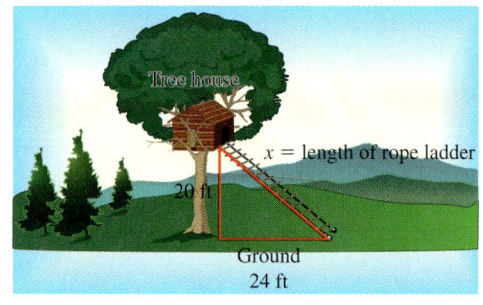

73. *Distance traveled* The distance traveled in t seconds by an object dropped from a distance d is given by $d = 16t^2$. How long would it take an object dropped from a distance of 64 feet to hit the ground?

74. *Time* Use the formula given in Problem 73 to find the time it takes an object dropped from a height of 32 feet to hit the ground.

75. *Interest rate* The amount of money A received at the end of 2 years when P dollars are invested at a compound rate r is $A = P(1 + r)^2$. Find the rate of interest r (written as a percent) if a person invested $100 and received $121 at the end of 2 years.

76. *Interest rate* Use the formula of Problem 75 to find the rate of interest r (written as a percent) if a person invested $100 and received $144 at the end of 2 years.

▶▶▶ Using Your Knowledge

Average, Demand, and Bacteria Many applications of mathematics require finding the maximum or the minimum of certain functions. A certain business may wish to find the price at which a product will bring *maximum* profits, while a team of engineers may be interested in *minimizing* the amount of carbon monoxide produced by automobiles. Now suppose you are the manufacturer of a product whose average manufacturing cost $\overline{C}$ (in dollars), based on producing x (thousand) units, is given by the expression

$$\overline{C} = x^2 - 8x + 18$$

How many units should be produced to minimize the cost per unit? If we consider the right-hand side of the equation, we can complete the square and leave the equation unchanged by adding and subtracting the appropriate number. Thus,

$$\overline{C} = x^2 - 8x + 18$$
$$= (x^2 - 8x +) + 18$$
$$= (x^2 - 8x + 4^2) + 18 - 4^2$$

Then

$$\overline{C} = (x - 4)^2 + 2$$

Now for $\overline{C}$ to be as small as possible (minimizing the cost), we make $(x - 4)^2$ zero by letting $x = 4$; then $\overline{C} = 2$. This tells us that when 4 (thousand) units are produced, the minimum cost is 2. That is, the minimum average cost per unit is $2.

Use your knowledge about completing the square to solve the following problems.

77. A manufacturer's average cost $\overline{C}$ (in dollars), based on manufacturing x (thousand) items, is given by

$$\overline{C} = x^2 - 4x + 6$$

  a. How many units should be produced to minimize the cost per unit?

  b. What is the minimum average cost per unit?

78. The demand D for a certain product depends on x, the number of units produced (in thousands), and is given by

$$D = x^2 - 2x + 3$$

For what number of units is the demand at its lowest?

79. Have you seen people adding chlorine to their swimming pools? This is done to reduce the number of bacteria present in the water. Suppose that after t days, the number of bacteria per cubic centimeter is given by the expression

$$B = 20t^2 - 120t + 200$$

In how many days will the number of bacteria be at its lowest?

▶▶▶ Write On

80. Explain why $\sqrt{49}$ has only one value, 7, but $x^2 = 49$ has two solutions, 7 and -7.

81. We have solved $x^2 - 49 = 0$ by adding 49 to both sides and then extracting roots. Describe another procedure you can use to solve $x^2 - 49 = 0$. Does your method work when solving $x^2 - 2 = 0$?

82. What is the first step in solving a quadratic equation by completing the square?

⟩⟩⟩ Concept Checker

Fill in the blank(s) with the correct word(s), phrase, or mathematical statement.

83. One way to solve the quadratic equation, $x^2 = 10$, is to take the __square root__ of both sides.

84. When completing the square to solve a quadratic equation, the coefficient of the x^2 term must eventually become __one__ if it isn't already.

85. When __completing the square__ on the expression $x^2 - 16x$, the result is the trinomial $x^2 - 16x + 64$.

quadratic
standard form
completing the square
simplifying

square root
one
product
quotient

⟩⟩⟩ Mastery Test

Solve by completing the square:

86. $5x^2 - 5x = 1$
87. $2x^2 - 2x - 1 = 0$
88. $x^2 + 12x = 8$
89. $x^2 + 6x - 1 = 0$

Solve:

90. $2x^2 + 12x + 18 = 27$
91. $3x^2 + 6x + 3 = 20$
92. $5(x - 3)^2 + 36 = 0$

93. $2(x - 1)^2 + 49 = 0$
94. $(x - 2)^2 = 24$
95. $(x + 1)^2 = 18$

96. $3x^2 - 16 = 0$
97. $3x^2 + 54 = 0$
98. $5x^2 + 60 = 0$

99. $9x^2 - 4 = 0$

⟩⟩⟩ Skill Checker

Evaluate

$$\frac{-b \pm \sqrt{b^2 - 4ac}}{2a}$$

for the given values of a, b, and c:

100. $a = 1, b = -9, c = 0$
101. $a = 1, b = -6, c = 0$
102. $a = 1, b = -2, c = -2$
103. $a = 1, b = -4, c = -4$
104. $a = 8, b = 7, c = -1$
105. $a = 3, b = -2, c = -5$
106. $a = 3, b = -8, c = 7$
107. $a = 4, b = -3, c = 5$
108. $a = 1, b = 2, c = 6$
109. $a = 1, b = 2, c = 5$

8.2 The Quadratic Formula: Applications

Objectives

A ▶ Solve equations using the quadratic formula.

B ▶ Solve factorable cubic equations.

C ▶ Solve applications involving quadratic equations.

▶ To Succeed, Review How To . . .

1. Find the square root of a number (pp. 526–528).
2. Simplify square roots (pp. 538–541).
3. Write fractions in lowest terms (pp. 433–436).
4. Solve a quadratic equation by completing the square (pp. 597–598).
5. Factor the sum or difference of cubes (pp. 392–394).

▶ Getting Started
Rockets and Quadratics

The man standing by the first liquid-fueled rocket is Dr. Robert Goddard. His rocket went up 41 feet. Do you know how long it took it to go up that high? The height (in feet) of the rocket is given by

$$h = -16t^2 + v_0 t$$

where v_0 is the initial velocity (51.225 feet per second). We can substitute 41 for h, the height, and 51.225 for v_0, the initial velocity, and solve for t in the equation

$$-16t^2 + 51.225t = 41$$

Unfortunately, this equation is not factorable, and to complete the square, we would have to divide by -16 and add

$$\left(-\frac{51.225}{32}\right)^2$$

to both sides. There must be a better way to solve this equation. In this section, we derive a formula that will give us a more efficient method for solving this type of problem.

To derive a formula for solving quadratic equations we start with the general equation $ax^2 + bx + c = 0$ $(a \neq 0)$ and use the six-step procedure we learned for solving by completing the square in Section 8.1. To facilitate the understanding of each step, we will parallel the solving of the specific equation $2x^2 + 5x + 1 = 0$ along side the solving of the general quadratic equation.

	Specific Equation	*General Equation*	
	$2x^2 + 5x + 1 = 0$	$ax^2 + bx + c = 0 \quad (a \neq 0)$	Given
Add -1 to both sides.	$2x^2 + 5x = -1$	$ax^2 + bx = -c$	Add $-c$ to both sides.
Divide each term by 2.	$x^2 + \frac{5}{2}x = -\frac{1}{2}$	$x^2 + \frac{b}{a}x = -\frac{c}{a}$	Divide each term by a.

Add $(\frac{1}{2}\cdot\frac{5}{2})^2$ to both sides.	$x^2 + \frac{5}{2}x + (\frac{5}{4})^2 = (\frac{5}{4})^2 - \frac{1}{2}$	$x^2 + \frac{b}{a}x + (\frac{b}{2a})^2 = (\frac{b}{2a})^2 - \frac{c}{a}$	Add the square of one-half the coefficient of x, that is, $(\frac{1}{2}\cdot\frac{b}{a})^2$, to both sides;
Factor the left-hand side; simplify the right-hand side.	$(x + \frac{5}{4})^2 = \frac{25}{16} - \frac{1}{2}$	$(x + \frac{b}{2a})^2 = \frac{b^2}{4a^2} - \frac{c}{a}$	Factor the left-hand side; simplify the right-hand side. Since the LCD on the right is $4a^2$, write
Write $\frac{1}{2}$ as $\frac{8}{16}$.	$(x + \frac{5}{4})^2 = \frac{25}{16} - \frac{8}{16}$	$(x + \frac{b}{2a})^2 = \frac{b^2}{4a^2} - \frac{4ac}{4a^2}$	$-\frac{c}{a}$ as $-\frac{4ac}{4a^2}$.
Write as one fraction.	$(x + \frac{5}{4})^2 = \frac{25 - 8}{16}$	$(x + \frac{b}{2a})^2 = \frac{b^2 - 4ac}{4a^2}$	Then combine $\frac{b^2}{4a^2}$ and $-\frac{4ac}{4a^2}$.
Take the square root of both sides.	$x + \frac{5}{4} = \frac{\pm\sqrt{25-8}}{4}$	$x + \frac{b}{2a} = \frac{\pm\sqrt{b^2-4ac}}{2a}$	Take the square root of both sides.
Add $-\frac{5}{4}$ to both sides.	$x = -\frac{5}{4} \pm \frac{\sqrt{17}}{4}$	$x = -\frac{b}{2a} \pm \frac{\sqrt{b^2-4ac}}{2a}$	Add $-\frac{b}{2a}$.
Write as one fraction.	$x = \frac{-5 \pm \sqrt{17}}{4}$	$x = \frac{-b \pm \sqrt{b^2-4ac}}{2a}$	Write as one fraction.

Remember, we only have to do this once. Now we have the **quadratic formula** to use!

A › Solving Equations Using the Quadratic Formula

If an equation in one variable has 2 as the highest exponent for any variable term in the equation, it is considered a quadratic or second-degree equation. The standard form for the quadratic equation is $ax^2 + bx + c = 0$. If it is in standard form, we can find the solutions by substituting the values for a, b, and c in the quadratic formula.

SOLUTIONS OF A QUADRATIC EQUATION IN STANDARD FORM

The solutions of $ax^2 + bx + c = 0$ are

$$x = \frac{-b \pm \sqrt{b^2 - 4ac}}{2a} \qquad a \neq 0$$

It is important to note that the values of a, b, and c that are used in the quadratic formula must come from the equation in **standard form,** $ax^2 + bx + c = 0$, where a, b, and c are integers and a is positive. In Example 1, we will practice putting the quadratic equations in standard form and identifying the a-, b-, and c-values.

EXAMPLE 1 Writing quadratic equations in standard form and identifying a, b, and c

Write the equation in standard form and identify the values of a, b, and c.

a. $12 = x^2 - 5x$ **b.** $\frac{x^2}{3} = 1 + \frac{x}{4}$

PROBLEM 1

Write the equation in standard form and identify the values of a, b, and c.

a. $2 = x^2 + 5x - 8$
b. $6 - 4x^2 = -5x$

SOLUTION

a. Remember in standard form the a-value is positive. In this equation, we start by subtracting 12 from both sides.

$$12 = x^2 - 5x$$
$$0 = x^2 - 5x - 12 \qquad \text{Subtract 12 from both sides.}$$

This is standard form. The coefficients of the three terms indicate $a = 1$, $b = -5$, and $c = -12$.

Answers to PROBLEMS

1. a. $x^2 + 5x - 10 = 0$; $a = 1$, $b = 5$, and $c = -10$ **b.** $4x^2 - 5x - 6 = 0$; $a = 4$, $b = -5$, and $c = -6$

b. In order to have standard form we must multiply both sides by the LCD to clear the fractions. Then we set the equation to 0.

$$\frac{x^2}{3} = 1 + \frac{x}{4}$$

$$12 \cdot \frac{x^2}{3} = 12 \cdot 1 + 12 \cdot \frac{x}{4} \quad \text{Multiply both sides by the LCD, 12.}$$

$$4x^2 = 12 + 3x \quad \text{Simplify.}$$

$$4x^2 = 12 + 3x \quad \text{Subtract 3x and 12 from both sides.}$$

$$4x^2 - 3x - 12 = 0$$

This is standard form and $a = 4$, $b = -3$, and $c = -12$.

EXAMPLE 2 Using the quadratic formula to solve equations

Solve $8x^2 + 7x + 1 = 0$ by using the quadratic formula.

SOLUTION The equation is second degree and is written in standard form:

$$\underbrace{8}_{a=8}x^2 + \underbrace{7}_{b=7}x + \underbrace{1}_{c=1} = 0$$

It is clear that $a = 8$, $b = 7$, and $c = 1$. Substituting the values of a, b, and c in the formula, we obtain

$$x = \frac{-7 \pm \sqrt{(7)^2 - 4(8)(1)}}{2(8)} \quad \text{Let } a = 8, b = 7, \text{ and } c = 1.$$

$$= \frac{-7 \pm \sqrt{49 - 32}}{16} \quad \text{Since } (7)^2 = 49 \text{ and } -4(8)(1) = -32$$

$$= \frac{-7 \pm \sqrt{17}}{16} \quad \sqrt{49 - 32} = \sqrt{17}$$

Thus,

$$x = \frac{-7 + \sqrt{17}}{16} \quad \text{or} \quad x = \frac{-7 - \sqrt{17}}{16}$$

The solution set is

$$\left\{ \frac{-7 + \sqrt{17}}{16}, \frac{-7 - \sqrt{17}}{16} \right\}$$

PROBLEM 2

Solve $3x^2 + 2x - 5 = 0$ by using the quadratic formula.

By the way, we left Dr. Goddard's rocket up in the air! How long *did* it fly to reach the height of 41 feet? The equation was

$$-16t^2 + 51.225t = 41$$

or, in standard form,

$$16t^2 - 51.225t + 41 = 0$$

Here $a = 16$, $b = -51.225$, and $c = 41$, so the quadratic formula gives

$$t = \frac{51.225 \pm \sqrt{(-51.225)^2 - 4(16)(41)}}{2 \cdot (16)}$$

With a calculator, we get $t \approx 1.6$. We don't mention the negative solution since time in this problem is expressed in positive numbers. Thus, it took the rocket about 1.6 seconds to reach 41 feet.

Answers to PROBLEMS

2. $-\frac{5}{3}, 1$

Calculator Corner

Solving a Quadratic Equation by Finding the x-Intercepts

To solve Example 2, we graph $Y_1 = 8x^2 + 7x + 1$ using a $[-3, 3]$ by $[-3, 3]$ window.

To find where Y_1 intersects the horizontal axis, we can enter $Y_2 = 0$ (which is the horizontal axis) and use [2nd] [TRACE] 5, the intersect feature of your calculator, and [ENTER] [ENTER] [ENTER]. You will get one solution -0.1798059. This will confirm that

$$-0.1798059 \approx \frac{-7 + \sqrt{17}}{16}$$

is indeed one of the solutions. The other solution is given as

$$-0.6951941 \approx \frac{-7 - \sqrt{17}}{16}$$

and can be found by pressing [TRACE] and using the left arrow key to move the cursor to the 2nd point of intersection. Once there, again use [2nd] [TRACE] 5 [ENTER] [ENTER] [ENTER].

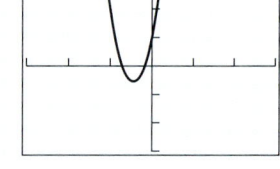

EXAMPLE 3 Using the quadratic formula to solve a second-degree equation

Solve $2x^2 = 2x + 1$ by using the quadratic formula.

SOLUTION Since it is a second-degree equation, we proceed by steps as before. We write the equation in standard form by subtracting $2x$ and 1 from both sides of $2x^2 = 2x + 1$ to obtain

$$\underbrace{2}_{a=2}x^2 - \underbrace{2}_{b=-2}x - \underbrace{1}_{c=-1} = 0 \quad \text{In standard form}$$

where $a = 2$, $b = -2$, and $c = -1$. Substituting these values in the quadratic formula, we have

$$x = \frac{-(-2) \pm \sqrt{(-2)^2 - 4(2)(-1)}}{2(2)} \quad \text{Let } a = 2, b = -2, \text{ and } c = -1.$$

$$= \frac{2 \pm \sqrt{4 + 8}}{4} \quad \text{Since } (-2)^2 = 4 \text{ and } -4(2)(-1) = 8$$

$$= \frac{2 \pm \sqrt{12}}{4} \quad \sqrt{4 + 8} = \sqrt{12}$$

$$= \frac{2 \pm \sqrt{4 \cdot 3}}{4} \quad \sqrt{12} = \sqrt{4 \cdot 3}$$

$$= \frac{2 \pm 2\sqrt{3}}{4} \quad \sqrt{4 \cdot 3} = \sqrt{4} \cdot \sqrt{3} = 2\sqrt{3}$$

$$= \frac{\overset{1}{\cancel{2}}(1 \pm \sqrt{3})}{\underset{2}{\cancel{4}}} \quad \text{Reduce.}$$

Thus,

$$x = \frac{1 + \sqrt{3}}{2} \quad \text{or} \quad x = \frac{1 - \sqrt{3}}{2}$$

The solution set is

$$\left\{\frac{1 + \sqrt{3}}{2}, \frac{1 - \sqrt{3}}{2}\right\}$$

PROBLEM 3

Solve $x^2 = 4x + 4$ by using the quadratic formula.

Answers to PROBLEMS

3. $2 + 2\sqrt{2}, 2 - 2\sqrt{2}$

EXAMPLE 4 Using the quadratic formula when there is no constant term

Solve $9x = x^2$ by using the quadratic formula.

SOLUTION Subtracting $9x$ from both sides of $9x = x^2$, we have

$$0 = x^2 - 9x$$

or

$$\underset{a=1}{x^2} - \underset{b=-9}{9x} + \underset{c=0}{0} = 0 \quad \text{In standard form}$$

where $a = 1$, $b = -9$, and $c = 0$ (because the c term is missing). Substituting these values in the quadratic formula, we obtain

$$x = \frac{-(-9) \pm \sqrt{(-9)^2 - 4(1)(0)}}{2(1)} \quad \text{Let } a = 1, b = -9, \text{ and } c = 0.$$

$$= \frac{9 \pm \sqrt{81 - 0}}{2} \quad \text{Since } (-9)^2 = 81 \text{ and } -4(1)(0) = 0$$

$$= \frac{9 \pm \sqrt{81}}{2} \quad \sqrt{81 - 0} = \sqrt{81}$$

$$= \frac{9 \pm 9}{2} \quad \pm\sqrt{81} = \pm 9$$

Thus,

$$x = \frac{9 + 9}{2} = \frac{18}{2} = 9 \quad \text{or} \quad x = \frac{9 - 9}{2} = \frac{0}{2} = 0$$

The solutions are 9 and 0, and the solution set is $\{9, 0\}$.

PROBLEM 4
Solve $x^2 = 15x$ by using the quadratic formula.

> **NOTE**
> The expression $x^2 - 9x = x(x - 9) = 0$ could have been solved by factoring. Try factoring the equation *before* you use the quadratic formula.

EXAMPLE 5 Clearing fractions before using the quadratic formula

Solve $\frac{x^2}{4} + \frac{2}{3}x = -\frac{1}{3}$ by using the quadratic formula.

SOLUTION We have to write the equation in standard form, but first we clear fractions by multiplying each term by the LCM of 4 and 3, namely 12:

$$\overset{3}{\cancel{12}} \cdot \frac{x^2}{\cancel{4}} + \overset{4}{\cancel{12}} \cdot \frac{2}{\cancel{3}}x = -\frac{1}{\cancel{3}} \cdot \overset{4}{\cancel{12}}$$

$$3x^2 + 8x = -4$$

We then add 4 to obtain the equivalent equation

$$\underset{a=3}{3x^2} + \underset{b=8}{8x} + \underset{c=4}{4} = 0 \quad \text{In standard form}$$

PROBLEM 5
Solve $\frac{x^2}{4} - \frac{3}{8}x = \frac{1}{4}$ by using the quadratic formula.

(continued)

Answers to PROBLEMS

4. 0, 15   **5.** $-\frac{1}{2}$, 2

where $a = 3$, $b = 8$, and $c = 4$. Substituting in the quadratic formula

$$x = \frac{-8 \pm \sqrt{(8)^2 - 4(3)(4)}}{2(3)} \quad \text{Let } a = 3, b = 8, \text{ and } c = 4.$$

$$= \frac{-8 \pm \sqrt{64 - 48}}{6} \quad \text{Since } (8)^2 = 64 \text{ and } -4(3)(4) = -48$$

$$= \frac{-8 \pm \sqrt{16}}{6} \quad \sqrt{64 - 48} = \sqrt{16}$$

$$= \frac{-8 \pm 4}{6} \quad \pm\sqrt{16} = \pm 4$$

Thus,

$$x = \frac{-8 + 4}{6} = \frac{-4}{6} = -\frac{2}{3} \quad \text{or} \quad x = \frac{-8 - 4}{6} = \frac{-12}{6} = -2$$

The solutions are $-\frac{2}{3}$ and -2, and the solution set is $\left\{-\frac{2}{3}, -2\right\}$.

Some quadratic equations have non-real complex-number solutions. Such solutions can be obtained by using the quadratic formula, as shown in Example 6.

EXAMPLE 6 Solving quadratic equations with non-real complex number solutions

Solve: $3x^2 + 3x = -2$

SOLUTION We add 2 to both sides of $3x^2 + 3x = -2$ so the equation is in standard form. We then have

$$\underbrace{3x^2}_{a=3} + \underbrace{3x}_{b=3} + \underbrace{2}_{c=2} = 0$$

where $a = 3$, $b = 3$, and $c = 2$. Now we have

$$x = \frac{-3 \pm \sqrt{(3)^2 - 4(3)(2)}}{2(3)}$$

$$= \frac{-3 \pm \sqrt{9 - 24}}{6}$$

$$= \frac{-3 \pm \sqrt{-15}}{6}$$

$$= \frac{-3 \pm i\sqrt{15}}{6} \quad \sqrt{-15} = i\sqrt{15}$$

The solutions are thus,

$$\frac{-3 + i\sqrt{15}}{6} \quad \text{and} \quad \frac{-3 - i\sqrt{15}}{6}$$

or

$$-\frac{1}{2} + \frac{\sqrt{15}}{6}i \quad \text{and} \quad -\frac{1}{2} - \frac{\sqrt{15}}{6}i$$

and the solution set is

$$\left\{-\frac{1}{2} + \frac{\sqrt{15}}{6}i, -\frac{1}{2} - \frac{\sqrt{15}}{6}i\right\}$$

PROBLEM 6
Solve:

$$3x^2 + 2x = -1$$

Answers to PROBLEMS

6. $-\frac{1}{3} + \frac{\sqrt{2}}{3}i, -\frac{1}{3} - \frac{\sqrt{2}}{3}i$

Calculator Corner

Recognizing Complex Number Solutions from the Graph of a Quadratic Equation

What happens if you try to solve $3x^2 + 3x = -2$ in Example 5 using your graphing calculator? First, write the equation as $3x^2 + 3x + 2 = 0$ and graph $Y_1 = 3x^2 + 3x + 2$ using a $[-3, 3]$ by $[-3, 3]$ window. The curve does not cut or touch the horizontal axis, as shown in the window.

This means that there are no real-number solutions to this equation. The solutions are *complex* numbers, and you have to find them algebraically. (Remember, even with the best calculator available, you have to know your algebra!)

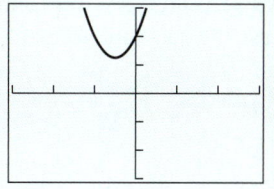

B > Solving Factorable Cubic Equations

Now let's consider $x^3 - 27 = 0$, which is *not* a quadratic equation. We will solve it by factoring (using the OFF method). First, recall the procedure for factoring sums and differences of cubes.

$$x^3 + a^3 = (x + a)(x^2 - ax + a^2)$$
$$x^3 - a^3 = (x - a)(x^2 + ax + a^2)$$

Here's how we do it.

$x^3 - 27 = 0$	O	Given (equation is set equal to 0).
$(x - 3)(x^2 + 3x + 9) = 0$	F	Factor.
$x - 3 = 0$ or $x^2 + 3x + 9 = 0$	F	Factors = 0 (zero-product property).
$x = 3$ or $x^2 + 3x + 9 = 0$		

Since the second equation does not factor, we use the quadratic formula to solve it. For this equation, $a = 1$, $b = 3$, and $c = 9$. Substituting in the quadratic formula, we have

$$x = \frac{-3 \pm \sqrt{3^2 - 4(1)(9)}}{2 \cdot 1}$$
$$= \frac{-3 \pm \sqrt{9 - 36}}{2}$$
$$= \frac{-3 \pm \sqrt{-27}}{2}$$
$$= \frac{-3 \pm \sqrt{9 \cdot 3 \cdot (-1)}}{2}$$
$$= \frac{-3 \pm 3i\sqrt{3}}{2}$$

The solutions of $x^3 - 27 = 0$ are thus,

$$3, \quad \frac{-3 + 3i\sqrt{3}}{2}, \quad \text{and} \quad \frac{-3 - 3i\sqrt{3}}{2}$$

and the solution set is

$$\left\{ 3, \frac{-3}{2} + \frac{3\sqrt{3}}{2}i, \frac{-3}{2} - \frac{3\sqrt{3}}{2}i \right\}$$

The equation $x^3 - 27 = 0$ *cannot* be completely solved by taking cube roots—that is, by writing

$x^3 - 27 = 0$	
$x^3 = 27$	Add 27.
$\sqrt[3]{x^3} = \sqrt[3]{27}$	Take cube roots of both sides.
$x = 3$	

As you can see, this method yields only one real-number solution when there are actually three solutions: one real-number and two non-real complex-number solutions.

EXAMPLE 7 Solving a factorable cubic equation

Solve: $8x^3 - 27 = 0$

SOLUTION We factor the equation, use the zero-product property, and then use the quadratic formula.

$$8x^3 - 27 = 0 \quad \text{Given. Equation is set} = 0.$$
$$(2x - 3)(4x^2 + 6x + 9) = 0 \quad \text{Factor.}$$
$$2x - 3 = 0 \quad \text{or} \quad 4x^2 + 6x + 9 = 0 \quad \text{Factors} = \text{zero (zero-product property).}$$
$$x = \frac{3}{2} \quad \text{or} \quad 4x^2 + 6x + 9 = 0$$

The second equation is a quadratic equation with $a = 4$, $b = 6$, and $c = 9$. We solve it with the quadratic formula:

$$x = \frac{-6 \pm \sqrt{6^2 - 4(4)(9)}}{2 \cdot 4}$$
$$= \frac{-6 \pm \sqrt{36 - 144}}{8}$$
$$= \frac{-6 \pm \sqrt{-108}}{8}$$
$$= \frac{-6 \pm \sqrt{36 \cdot 3 \cdot (-1)}}{8}$$
$$= \frac{-6 \pm 6i\sqrt{3}}{8}$$
$$= \frac{2(-3 \pm 3i\sqrt{3})}{2 \cdot 4}$$
$$= \frac{-3 \pm 3i\sqrt{3}}{4}$$

The solutions of $8x^3 - 27 = 0$ are

$$\frac{3}{2}, \quad \frac{-3 + 3i\sqrt{3}}{4}, \quad \text{and} \quad \frac{-3 - 3i\sqrt{3}}{4}$$

and the solution set is

$$\left\{ \frac{3}{2}, -\frac{3}{4} + \frac{3\sqrt{3}}{4}i, -\frac{3}{4} - \frac{3\sqrt{3}}{4}i \right\}$$

PROBLEM 7

Solve:

$$64x^3 + 1 = 0$$

Calculator Corner

Finding the Real Solutions for a Cubic Equation

Try to find the solutions of $8x^3 - 27 = 0$. The graph is shown in the window and, as before, only one solution, $x = 1.5 = \frac{3}{2}$, is a real number.

You have to use the quadratic equation to get the other two solutions! (By the way, if you want to see a more complete graph, try a $[-2, 2]$ by $[-50, 50]$ window.)

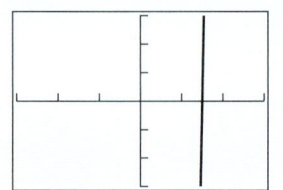

C › Solving Applications Involving Quadratic Equations

Quadratic equations are used in many fields: engineering, economics, and marketing, to name a few. Here are some examples.

Answers to PROBLEMS

7. $\frac{1}{8} + \frac{\sqrt{3}}{8}i, \frac{1}{8} - \frac{\sqrt{3}}{8}i, -\frac{1}{4}$

EXAMPLE 8 Equilibrium point: What's the price?

If in a certain business, when the price p (in dollars) of a product increases, then the demand d decreases and is given by $d = 300/p$. On the other hand, when the price p increases, the supply s producers are willing to sell increases and is given by $s = 100p - 50$. In economic theory, the point at which the supply equals the demand, $s = d$, is called the **equilibrium point.** Find the price p at the equilibrium point.

SOLUTION Since $s = d$ at equilibrium, we have

$$100p - 50 = \frac{300}{p}$$

$$p(100p - 50) = p \cdot \frac{300}{p} \qquad \text{Multiply by } p.$$

$$100p^2 - 50p = 300$$

$$2p^2 - p = 6 \qquad \text{Divide by 50.}$$

$$2p^2 - p - 6 = 0 \qquad \text{Subtract 6.}$$

Since $a = 2$, $b = -1$, and $c = 6$, we have

$$p = \frac{-(-1) \pm \sqrt{(-1)^2 - 4(2)(-6)}}{2 \cdot 2}$$

$$= \frac{1 \pm \sqrt{1 + 48}}{4}$$

$$= \frac{1 \pm \sqrt{49}}{4}$$

$$= \frac{1 \pm 7}{4}$$

Thus,

$$p = \frac{1 + 7}{4} = 2 \quad \text{or} \quad p = \frac{1 - 7}{4} = -\frac{3}{2}$$

Since the price must be positive, we use $p = \$2$. In this case, we obtain two rational roots, which means that the original equation was factorable. Before you use the quadratic formula, always try to factor. You will often save time!

PROBLEM 8

Find the price p (in dollars) at the equilibrium point if the demand and supply are given by

$$d = \frac{50}{p} \quad \text{and} \quad s = 100p - 50$$

EXAMPLE 9 Beef and poultry consumption: Where's the beef?

According to the U.S. Department of Agriculture, per capita consumption of beef increased from 2002 to 2006, while poultry consumption fluctuated. If $B(t)$ and $P(t)$ represent per capita consumption of beef and poultry (in pounds), respectively, and t represents the number of years after 2002, beef and poultry consumption can be described by these equations.

Beef: $B(t) = 3t + 55$
Poultry: $P(t) = 0.2t^2 - 1.5t + 9$

Will the consumption of beef and poultry ever be the same?

PROBLEM 9

If the supply equation for manufacturing a certain item is $s = p^2 - 8$ and the demand equation is $d = -p + 4$, find the equilibrium point.

(continued)

Answers to PROBLEMS

8. $1   9. $p = 3$

SOLUTION To find out whether the consumption of beef and poultry will ever be the same, we have to determine whether $B(t) = P(t)$ has a real-number solution. So we assume $B(t) = P(t)$, or

$$3t + 55 = 0.2t^2 - 1.5t + 9$$

$$0.2t^2 - 4.5t - 46 = 0 \qquad \text{Subtract } 3t \text{ and } 55.$$

Here $a = 0.2$, $b = -4.5$, and $c = -46$. Substituting in the quadratic formula, we have

$$t = \frac{-(-4.5) \pm \sqrt{(-4.5)^2 - 4(0.2)(-46)}}{2(0.2)}$$

$$= \frac{4.5 \pm \sqrt{20.25 + 36.8}}{0.4}$$

$$= \frac{4.5 \pm \sqrt{57.05}}{0.4}$$

$$= \frac{4.5 \pm 7.55}{0.4}$$

Thus,

$$t = \frac{12.05}{0.4} = 30.13 \quad \text{or} \quad t = \frac{-3.05}{0.4} = -7.63$$

which means that the consumption of beef and poultry will be the same for the specified positive value of t. That is, in approximately 30 years from 2002 or in 2032, the consumption will be the same.

Calculator Corner

Using a Calculator with the Quadratic Formula

Any non-graphing calculator that has store (STO) and recall (RCL) keys can be extremely helpful in finding the solutions of a quadratic equation with the quadratic formula. The solutions you obtain are being approximated by decimals. It's especially convenient to start with the radical part in the solution of the quadratic equation and then store this value so you can evaluate both solutions without having to backtrack or write down any intermediate steps. Let's look at the equation of Example 2:

$$8x^2 + 7x + 1 = 0$$

Using the quadratic formula, one solution is obtained by entering

[7] [x^2] [−] [4] [×] [8] [×] [1] [=] [√x] [STO] [1] [7] [+/−] [+] [RCL] [1] [=] [÷] [2] [=] [÷] [8] [=]

The display shows -0.1798059 (given as

$$\frac{-7 + \sqrt{17}}{16}$$

in the example). To obtain the other solution, enter

[7] [+/−] [−] [RCL] [1] [=] [÷] [2] [=] [÷] [8] [=]

which yields -0.6951194.

If $b^2 - 4ac < 0$, the calculator will give you an error message when you press [√x]. In such cases, you will have to change the sign before pressing [√x] and calculate the real and imaginary parts separately. (Try this in Example 6.)

Exercises 8.2

‹ A › **Solving Equations Using the Quadratic Formula** In Problems 1–4, write the equation in standard form and identify the values of a, b, and c.

1. $12 + 5x^2 = 4x$
2. $3x^2 + 12x = 8$
3. $3 + \dfrac{x^2}{2} = \dfrac{5x}{4}$
4. $\dfrac{x^2}{3} = \dfrac{5x}{2}$

In Problems 5–30, use the quadratic formula to solve the equation.

5. $x^2 + x - 2 = 0$
6. $x^2 + 4x - 1 = 0$
7. $x^2 + 4x = -1$
8. $x^2 + 6x = -5$
9. $x^2 - 3x = 2$
10. $x^2 - 4x = 12$
11. $7y^2 = 12y - 5$
12. $7x^2 = 6x - 1$
13. $5y^2 + 8y = -5$
14. $5y^2 + 6y = -5$
15. $7y + 6 = -2y^2$
16. $7y + 3 = -2y^2$
17. $\dfrac{x^2}{5} - \dfrac{x}{2} = -\dfrac{3}{10}$
18. $\dfrac{x^2}{4} - \dfrac{x}{2} = -\dfrac{1}{8}$
19. $\dfrac{x^2}{7} + \dfrac{x}{2} = -\dfrac{3}{14}$
20. $\dfrac{x^2}{10} - \dfrac{x}{5} = \dfrac{3}{2}$
21. $\dfrac{x^2}{8} = -\dfrac{x}{4} - \dfrac{1}{8}$
22. $\dfrac{x^2}{12} = -\dfrac{x}{4} - \dfrac{1}{3}$
23. $6x = 4x^2 + 1$
24. $6x = 9x^2 - 4$
25. $3x = 1 - 3x^2$
26. $3x = 2x^2 - 5$
27. $x(x + 2) = 2x(x + 1) - 4$
28. $x(4x - 7) - 10 = 6x^2 - 7x$
29. $6x(x + 5) = (x + 15)^2$
30. $6x(x + 1) = (x + 3)^2$

‹ B › **Solving Factorable Cubic Equations** In Problems 31–36, solve the equation.

31. $x^3 - 8 = 0$
32. $x^3 - 1 = 0$
33. $8x^3 - 1 = 0$
34. $27x^3 - 1 = 0$
35. $x^3 + 27 = 0$
36. $x^3 + 64 = 0$

‹ C › **Solving Applications Involving Quadratic Equations**

37. *New home sales* The trend in new home sales has fluctuated over the last 45 years. In a certain geographical region over a period of time t, the new home sales can be approximated by $N(t) = 6t^2 - 45t + 375$. During that same time period, the volume of mortgage loan applications for new purchases has declined and can be approximated by $M(t) = -85t + 495$. Will the number of new home sales equal the number of mortgages for new homes?
Source: Census Bureau; http://bigpicture.typepad.com.

38. *Equilibrium point* Find the price p (in dollars) at the equilibrium point if the supply is $s = 30p - 50$ and the demand is $d = \dfrac{20}{p}$.

39. *Equilibrium point* Find the price p (in dollars) at the equilibrium point if the supply is $s = 30p - 50$ and the demand is $d = \dfrac{10}{p}$.

40. *Equilibrium point* Find the price p (in dollars) at the equilibrium point if the supply is $s = 20p - 60$ and the demand is $d = \dfrac{30}{p}$.

41. *Bending moment* The bending moment M of a simple beam is given by $M = 20x - x^2$. For what values of x is $M = 40$?

42. *Bending moment* Use the formula given in Problem 41 to find the values of x for which $M = 60$.

Support beams The maximum safe length L for which a beam will support a load d is given by $aL^2 + bL + c = d$, where a, b, c, and d depend on the materials and structures used. In Problems 43 and 44, find:

43. L when $a = 400$, $b = 200$, $c = 200$, and $d = 800$

44. L when $a = 5$, $b = 0$, $c = 100$, and $d = 180$

⟩⟩⟩ Using Your Knowledge

A Different Way of Finding the Quadratic Formula In this section, we derived the quadratic formula by completing the square. The procedure depends on making the x^2 coefficient 1. But there is another way to derive the quadratic formula. See if you can state the reasons for each of the steps, given consecutively in Problems 45–51. ($ax^2 + bx + c = 0$ is given.)

45. $4a^2x^2 + 4abx + 4ac = 0$

46. $4a^2x^2 + 4abx = -4ac$

47. $4a^2x^2 + 4abx + b^2 = b^2 - 4ac$

48. $(2ax + b)^2 = b^2 - 4ac$

49. $2ax + b = \pm\sqrt{b^2 - 4ac}$

50. $2ax = -b \pm\sqrt{b^2 - 4ac}$

51. $x = \dfrac{-b \pm \sqrt{b^2 - 4ac}}{2a}$

⟩⟩⟩ Write On

52. Why do we have the restriction $a \ne 0$ when solving the equation $ax^2 + bx + c = 0$?

53. The term $\sqrt{a}$ is a real number only when a is nonnegative. Use this to write a procedure that enables you to determine whether a quadratic equation with real coefficients has real-number solutions or non-real complex number solutions.

54. Explain the difference between linear, quadratic, and cubic equations.

⟩⟩⟩ Concept Checker

Fill in the blank(s) with the correct word(s), phrase, or mathematical statement.

55. In the quadratic equation $3x^2 - 16x + 2 = 0$, -16 is the value of _____.

56. To solve the quadratic equation, $5x + 3x^2 - 4 = 0$ using the quadratic formula, it is more convenient if the left side is in _____.

57. It is often best to use the _____ when solving a quadratic equation that doesn't factor.

quadratic formula	a
standard form	b
completing the square	c
descending order	

⟩⟩⟩ Mastery Test

Solve:

58. $27x^3 - 8 = 0$

59. $64x^3 - 1 = 0$

60. $3x^2 + 2x = -1$

61. $2x^2 + 3x = -2$

62. $\dfrac{x^2}{4} - \dfrac{3}{8}x = \dfrac{1}{4}$

63. $\dfrac{x^2}{2} - \dfrac{1}{4}x = \dfrac{3}{2}$

64. $6x = x^2$

65. $7x - x^2 = 0$

66. $x^2 = 4x + 4$

67. $x^2 = 2 + 2x$

68. $3x^2 + 2x - 5 = 0$

69. $2x^2 - 3x = 4$

70. The supply s of a certain product that producers are willing to sell is given by $s = 100p - 50$, where p is the price in dollars. If the demand d for this product is given by $d = \dfrac{50}{p}$, find the price p when $s = d$, the equilibrium point.

Skill Checker

Simplify $\sqrt{b^2 - 4ac}$ using:

71. $a = 3, b = -2, c = -1$

72. $a = 2, b = 3, c = -1$

73. $a = 3, b = -5, c = 4$

74. $a = 3, b = -1, c = 1$

Find the product:

75. $(2x + 1)(3x - 4)$

76. $(3x + 1)(2x - 5)$

77. $(3x - 7)(4x + 3)$

78. $(4x - 8)(3x + 5)$

8.3 The Discriminant and Its Applications

Objectives

A Use the discriminant to determine the number and type of solutions of a quadratic equation.

B Use the discriminant to determine whether a quadratic expression is factorable and then factor it.

C Find a quadratic equation with specified solutions.

D Verify the solutions of a quadratic equation.

To Succeed, Review How To . . .

1. Evaluate and simplify expressions that contain radicals (pp. 538–540).
2. Multiply two binomials (pp. 362–363).

Getting Started
Breaking Even Through Discriminants

A merchant in this mall wants to know if she is going to "break even." First, her daily costs, C, are represented by

$$C = 0.001x^2 + 10x + 100$$

where x is the number of items sold, and the corresponding revenue is $R = 20x - 0.01x^2$. The break-even point occurs when the cost C equals the revenue R—that is, when

$$C = R$$
$$0.001x^2 + 10x + 100 = 20x - 0.01x^2$$
$$0.011x^2 - 10x + 100 = 0 \quad \text{In standard form}$$

She can break even only *if* this equation has real-number solutions. How do we ascertain that? By using the $b^2 - 4ac$ under the radical in the quadratic formula. Keep reading to see how.

The quadratic formula

$$x = \frac{-b \pm \sqrt{b^2 - 4ac}}{2a}$$

gives the solutions to any quadratic equation in standard form, so we can find out what type of solutions the equation has by looking at the expression under the radical, $b^2 - 4ac$. To do this, we need the following definition.

DISCRIMINANT

The expression $b^2 - 4ac$ under the radical in the quadratic formula is called the **discriminant** D; that is, $D = b^2 - 4ac$.

To find the discriminant of $0.011x^2 - 10x + 100 = 0$, substitute $a = 0.011$, $b = -10$, and $c = 100$ into the expression for D. Thus,

$$b^2 - 4ac = (-10)^2 - 4(0.011)(100)$$
$$= 100 - 4.4$$
$$= 95.6$$

This means that the solutions of the equation are

$$\frac{-(-10) + \sqrt{95.6}}{2(0.011)} \approx 898.98 \quad \text{and} \quad \frac{-(-10) - \sqrt{95.6}}{2(0.011)} \approx 10.11$$

There are two positive real-number solutions to the equation, and our merchant can break even.

A > Using the Discriminant to Classify Solutions

If we calculate the discriminant of a quadratic equation with rational-number coefficients, we can predict whether solutions will be rational, irrational, or non-real complex numbers as well as the number of solutions we will have. The first case to consider would be a discriminant that is positive but not a perfect square, like 18. In the radical part of the quadratic formula, that means we would have $\pm\sqrt{18}$, which can be simplified to $\pm 3\sqrt{2}$ but would still result in two irrational number solutions. The second case would be a discriminant that is positive and a perfect square, like 36. In the radical part of the quadratic formula, we would have $\pm\sqrt{36}$, which simplifies to ± 6 and results in two rational number solutions. The third case would be a discriminant that is a negative number like -25. Again examining the radical part of the quadratic formula, we would have $\pm\sqrt{-25}$, which simplifies to $\pm 5i$ and results in two non-real complex number solutions. The fourth and final case is a discriminant that is 0. The radical part of the formula would be $\pm\sqrt{0}$, which simplifies to ± 0. This results in one rational solution since adding or subtracting 0 from any number leads to the same answer.

Being able to predict the solutions to a quadratic equation can be important in applied problems where irrational or non-real number solutions are not always acceptable. A summary of the different cases are listed in Table 8.1.

Table 8.1 Solutions to $ax^2 + bx + c = 0$ Based on the Discriminant

The discriminant of $ax^2 + bx + c = 0$ is $b^2 - 4ac$. If a, b, and c are real numbers ($a \neq 0$), the type and number of solutions are as follows:

Discriminant, $b^2 - 4ac$	Solutions, $\frac{-b \pm \sqrt{b^2 - 4ac}}{2a}$
Positive, not perfect square	Two different real, *irrational* solutions
Positive, and perfect square	Two different real, *rational* solutions
Negative	Two different *non-real complex* solutions
Zero	One real, *rational* solution

Here are some examples. Notice each equation is in standard form. It must be to get the correct value of the determinant.

Equation	$b^2 - 4ac$	D	Solutions (Roots)
$4x^2 - 3x - 5 = 0$	$(-3)^2 - 4(4)(-5) = 89$	Positive, not perfect square	Two real, irrational numbers
$x^2 - 2x - 3 = 0$	$(-2)^2 - 4(1)(-3) = 16$	Positive, perfect square	Two real, rational numbers
$4x^2 - 3x + 5 = 0$	$(-3)^2 - 4(4)(5) = -71$	Negative	Two non-real complex numbers
$4x^2 - 4x + 1 = 0$	$(-4)^2 - 4(4)(1) = 0$	Zero	One real, rational number

EXAMPLE 1 Using the discriminant to classify the solutions

a. Given $2x^2 + x - 4 = 0$, find the value of the discriminant and classify the solutions.

b. Given $3x^2 - 5x = 2$, find the value of the discriminant and classify the solutions.

c. Given $4x^2 - kx = -1$, find the discriminant and then find k so that the equation has exactly one real, rational solution.

SOLUTION

a. The equation is in standard form so $a = 2$, $b = 1$, and $c = -4$.
$$b^2 - 4ac = (1)^2 - 4(2)(-4)$$
$$= 1 + 32$$
$$= 33$$

Since 33 is positive but not a perfect square, the equation will have two real, irrational solutions.

b. First write the equation in standard form by adding -2 to both sides to obtain $3x^2 - 5x - 2 = 0$. Thus, $a = 3$, $b = -5$, $c = -2$, and
$$b^2 - 4ac = (-5)^2 - 4(3)(-2)$$
$$= 25 + 24$$
$$= 49$$

Since 49 is a positive perfect square, the equation will have two real, rational solutions.

c. We first write the equation in standard form by adding 1 to both sides to obtain $4x^2 - kx + 1 = 0$. Thus, $a = 4$, $b = -k$, $c = 1$, and
$$b^2 - 4ac = (-k)^2 - 4(4)(1)$$
$$= k^2 - 16$$

For this equation to have one rational solution, the discriminant $k^2 - 16$ must be zero:
$$0 = k^2 - 16$$
$$16 = k^2$$
$$\pm\sqrt{16} = k$$
$$\pm 4 = k$$

When k is 4 or -4, the discriminant is zero, and the equation has one real, rational number as its solution.

PROBLEM 1

a. Given $x^2 + 4x + 4 = 0$, find the value of the discriminant and classify the solutions.

b. Given $2x^2 = 6x - 7$, find the value of the discriminant and classify the solutions.

c. Given $x^2 + kx = -9$, find the value of the discriminant and then find k so that the equation has exactly one real rational solution.

Answers to PROBLEMS

1. a. 0; one real, rational solution
   b. -20; 2 non-real complex solutions   c. $k^2 - 36$; $-6, 6$

B › Determining Whether a Quadratic Expression Is Factorable

We have now learned that if the discriminant D is a positive, perfect square, $ax^2 + bx + c = 0$ has two real, rational solutions, say r and s. Thus, $a(x - r)(x - s) = 0$ has the same solution set as $ax^2 + bx + c = 0$, which means that $ax^2 + bx + c$ is factorable into factors with *integer* coefficients. Here is our result.

FACTORABLE QUADRATIC EXPRESSIONS	If the quadratic expression $ax^2 + bx + c$ has a, b, and c as rational numbers and $b^2 - 4ac$ is a positive perfect square, then $ax^2 + bx + c$ is factorable using integer coefficients.

To find out whether $12x^2 + 20x - 25$ is factorable, we must find $b^2 - 4ac = (20)^2 - 4(12)(-25) = 1600$. Since 1600 is a positive perfect square $[(40)^2 = 1600]$, $12x^2 + 20x - 25$ is factorable. (Contrast this technique with the ac test we learned in Section 5.4 or with trial and error.)

EXAMPLE 2 Using the discriminant to determine whether an expression is factorable

Use the discriminant to determine whether $20x^2 + 10x - 32$ is factorable.

SOLUTION Here $a = 20$, $b = 10$, and $c = -32$.

$$b^2 - 4ac = (10)^2 - 4(20)(-32)$$
$$= 100 + 2560$$
$$= 2660$$

Since 2660 is not a perfect square, $20x^2 + 10x - 32$ is not factorable.

PROBLEM 2

Use the discriminant to determine whether the following quadratic expression is factorable.

$$20x^2 + 10x - 30$$

Now that we know how to use the discriminant to determine whether a quadratic expression is factorable, we shall learn how to factor it. To factor $12x^2 + 20x - 25$, we first solve the corresponding quadratic equation $12x^2 + 20x - 25 = 0$. Recall that $b^2 - 4ac = 1600$; thus,

$$x = \frac{-20 \pm \sqrt{1600}}{2(12)} = \frac{-20 \pm 40}{24}$$

The solutions are

$$\frac{-20 + 40}{24} = \frac{5}{6} \quad \text{and} \quad \frac{-20 - 40}{24} = -\frac{5}{2}$$

We now *reverse* the steps for solving a quadratic equation by factoring.

	Solution 1 = $\frac{5}{6}$	Solution 2 = $-\frac{5}{2}$	
	$x = \frac{5}{6}$	$x = -\frac{5}{2}$	
Multiply by 6.	$6x = 5$	$2x = -5$	Multiply by 2.
Subtract 5.	$6x - 5 = 0$	$2x + 5 = 0$	Add 5.

That is, $(6x - 5)(2x + 5) = 0$. Thus, the expression, $12x^2 + 20x - 25 = (6x - 5)(2x + 5)$. (You can check this by multiplying.)

Answers to PROBLEMS

2. Yes; $D = 2500 = (50)^2$

EXAMPLE 3 Using the discriminant to determine whether an expression is factorable

Factor if possible: $12x^2 + x - 35$

SOLUTION

$$b^2 - 4ac = (1)^2 - 4(12)(-35) = 1 + 1680 = 1681$$

Since $1681 = 41^2$, the expression is factorable. We use the quadratic formula to solve the related equation $12x^2 + x - 35 = 0$.

$$x = \frac{-1 \pm \sqrt{1681}}{2(12)} = \frac{-1 \pm 41}{24}$$

$$x = \frac{40}{24} = \frac{5}{3} \quad \text{or} \quad x = -\frac{42}{24} = -\frac{7}{4}$$

Now we reverse the steps for solving a quadratic equation by factoring.

$$x = \frac{5}{3} \quad \text{or} \quad x = -\frac{7}{4}$$

Multiply by 3.	$3x = 5$	or	$4x = -7$	Multiply by 4.
Subtract 5.	$3x - 5 = 0$	or	$4x + 7 = 0$	Add 7.

Multiplying, $(3x - 5)(4x + 7) = 0$. Thus, the expression, $12x^2 + x - 35 = (3x - 5)(4x + 7)$. You can check that the factorization is correct by multiplying the factors to obtain $12x^2 + x - 35$.

PROBLEM 3

Factor if possible:

$$21x^2 + x - 10$$

C > Finding Quadratic Equations with Specified Solutions

The process just discussed can be used to find a quadratic equation with specified solutions as shown in Example 4.

EXAMPLE 4 Finding an equation with specified solutions

A professor wants to create a test question involving a quadratic equation whose solution set is $\{\frac{4}{5}, -\frac{2}{3}\}$. What should the quadratic equation be?

SOLUTION In essence, you have to work backward. Since the solutions are $\frac{4}{5}$ and $-\frac{2}{3}$, the professor wants:

$$x = \frac{4}{5} \quad \text{or} \quad x = -\frac{2}{3}$$

Multiply by 5.	$5x = 4$	or	$3x = -2$	Multiply by 3.
Subtract 4.	$5x - 4 = 0$	or	$3x + 2 = 0$	Add 2.

$(5x - 4)(3x + 2) = 0$ By the zero-product property.

$15x^2 + 10x - 12x - 8 = 0$ Multiply.

$15x^2 - 2x - 8 = 0$ Simplify.

Thus, $15x^2 - 2x - 8 = 0$ is a quadratic equation whose solution set is $\{\frac{4}{5}, -\frac{2}{3}\}$.

PROBLEM 4

Write the quadratic equation whose solution set is $\{-2, \frac{3}{4}\}$.

D > Verifying the Solutions of a Quadratic Equation

In the process of solving Example 3, we found out that the solutions of the equation $12x^2 + x - 35 = 0$ are $\frac{5}{3}$ and $-\frac{7}{4}$. To verify this, we can substitute these values in the original equation. But there is another way. If the equation is

$$ax^2 + bx + c = 0$$

Answers to PROBLEMS

3. $(3x - 2)(7x + 5)$
4. $4x^2 + 5x - 6 = 0$

then dividing by a, we can rewrite it as

$$x^2 + \frac{b}{a}x + \frac{c}{a} = 0 \qquad (1)$$

If the solutions are r_1 and r_2, we can also write

$$(x - r_1)(x - r_2) = 0$$
$$x^2 - r_1 x - r_2 x + r_1 r_2 = 0 \quad \text{Multiply.}$$
$$x^2 - (r_1 + r_2)x + r_1 r_2 = 0 \quad \text{Factor } [-r_1 x - r_2 x = -(r_1 + r_2)x]. \qquad (2)$$

Comparing equations (1) and (2), we see that

$$\frac{b}{a} = -(r_1 + r_2) \quad \text{and} \quad \frac{c}{a} = r_1 r_2$$

This discussion can be summarized as follows.

SUM AND PRODUCT OF THE SOLUTIONS OF A QUADRATIC EQUATION

If r_1 and r_2 are the solutions of the equation $ax^2 + bx + c = 0$, then

$$r_1 + r_2 = -\frac{b}{a} \quad \text{and} \quad r_1 r_2 = \frac{c}{a}$$

That is, the sum of the solutions of a quadratic equation is $-\frac{b}{a}$, and the product of the solutions is $\frac{c}{a}$.

We can now verify that $\frac{5}{3}$ and $-\frac{7}{4}$ are solutions of $12x^2 + x - 35 = 0$. The sum of the numbers is

$$\frac{5}{3} + \left(-\frac{7}{4}\right) = \frac{20}{12} + \left(-\frac{21}{12}\right) = -\frac{1}{12} = -\frac{b}{a}$$

The product is

$$\frac{5}{3} \cdot \left(-\frac{7}{4}\right) = -\frac{35}{12} = \frac{c}{a}$$

So our results are correct.

EXAMPLE 5 Using the sum and product properties to verify solutions

Use the sum and product properties to see whether the solutions of $3x^2 + 5x - 2 = 0$ are:

a. $-\frac{1}{3}$ and 2 　　　　 b. $\frac{1}{3}$ and -2

SOLUTION

a. In the equation $3x^2 + 5x - 2 = 0$, $a = 3$, $b = 5$, and $c = -2$. Using $-\frac{b}{a}$ for the sum of the solutions we get

$$-\frac{b}{a} = -\frac{5}{3}$$

Finding the sum of the proposed solutions, we get

$$-\frac{1}{3} + 2 = -\frac{1}{3} + \frac{6}{3} = \frac{5}{3}$$

Since $-\frac{b}{a} = -\frac{5}{3}$ and $r_1 + r_2 = \frac{5}{3}$, $-\frac{1}{3}$ and 2 cannot be the solutions.

PROBLEM 5

Use the sum and product properties to see whether the solutions of $4x^2 - 12x + 5 = 0$ are:

a. $\frac{1}{2}$ and $-\frac{5}{2}$ 　　 b. $\frac{1}{2}$ and $\frac{5}{2}$

Answers to PROBLEMS

5. a. No　b. Yes

b. The sum of the proposed solutions is $\frac{1}{3} + (-\frac{6}{3}) = -\frac{5}{3}$, which also equals $-\frac{b}{a}$. The product of the solutions of $3x^2 + 5x - 2 = 0$ must be

$$\frac{c}{a} = -\frac{2}{3}$$

The product of the proposed solutions is $\frac{1}{3} \cdot (-2) = -\frac{2}{3}$. Since $\frac{c}{a}$ is the same as the product, $-\frac{2}{3}$, the numbers $\frac{1}{3}$ and -2 are the correct solutions.

Calculator Corner

Using the Graph to Determine the Number of Solutions to a Quadratic Equation

Here's how you use a calculator to determine the number of solutions of a quadratic equation:

Case 1. If the graph of a quadratic cuts the horizontal axis in *two* places, the corresponding quadratic equation has *two* solutions, but you *still* have to use the discriminant to determine if the solutions are rational or irrational.

Case 2. If the graph cuts the horizontal axis in *one* place, there is *one* solution, but we don't know if it is rational or irrational. [Try $x^2 - 4x + 4 = 0$ whose solution is $x = 2$, which is rational, and $(x - \sqrt{2})^2 = 0$ whose solution is $x = \sqrt{2}$, which is irrational.]

Case 3. If the graph does not touch the horizontal axis, there are no real-number solutions. The *two* resulting solutions are non-real complex numbers. (Try $x^2 + 4 = 0$.)

In Example 9 of Section 8.2, we were asked to determine whether the consumption of beef and poultry could ever be the same. The answer was obtained by solving the equation $B(t) = P(t)$. We could answer the question more quickly by looking at the discriminant of $B(t) - P(t) = 0$. If the discriminant $b^2 - 4ac > 0$, the equation has two solutions. Use this idea to solve the following problem. According to the U.S. Department of Agriculture, the consumption of citrus and noncitrus fruits (in pounds) t years after 1979 can be approximated by

Citrus: $C(x) = -0.013x^2 - 0.24x + 29$

Noncitrus: $N(x) = x + 50$

1. Use the discriminant of $C(x) - N(x)$ to find out whether citrus and noncitrus consumption can ever be the same.
2. Verify your result by graphing $C(x)$ and $N(x)$.

Boost your grade at mathzone.com!
> Practice Problems
> NetTutor
> Self-Tests
> e-Professors
> Videos

› Exercises 8.3

‹A› Using the Discriminant to Classify Solutions In Problems 1–10, find the discriminant and determine the number and type of solutions.

1. $3x^2 + 5x - 2 = 0$
2. $3x^2 - 2x + 5 = 0$
3. $4x^2 = 4x - 1$

4. $x^2 - 10x = -25$
5. $2x^2 = 2x + 5$
6. $x^2 - 5x = 5$

7. $4x^2 - 5x + 3 = 0$
8. $5x^2 - 7x + 8 = 0$
9. $x^2 - 2 = \frac{5}{2}x$

10. $x^2 + \frac{1}{5} = \frac{2}{5}x$

In Problems 11–20, determine the value of k that will make the given equation have exactly one rational solution.

11. $x^2 - 4kx + 64 = 0$
12. $3x^2 + kx + 3 = 0$
13. $kx^2 - 10x = 5$
14. $2kx^2 - 12x = -9$
15. $2x^2 = kx - 8$
16. $3x^2 = kx - 3$
17. $25x^2 - kx = -4$
18. $4x^2 + 9kx = -1$
19. $x^2 + 8x = k$
20. $2x^2 - 4x = k$

⟨ B ⟩ Determining Whether a Quadratic Expression Is Factorable In Problems 21–30, use the discriminant to determine whether the given polynomial is factorable into factors with integer coefficients. If it is, use the technique of Example 3 to factor it.

21. $10x^2 - 7x + 8$
22. $10x^2 - 7x + 1$
23. $12x^2 - 17x + 6$
24. $27x^2 + 51x - 56$
25. $12x^2 - 17x + 2$
26. $15x^2 + 52x - 83$
27. $15x^2 + 52x - 84$
28. $27x^2 - 57x - 40$
29. $12x^2 - 61x + 60$
30. $30x^2 - 19x - 140$

⟨ C ⟩ Finding Quadratic Equations with Specified Solutions In Problems 31–40, find a quadratic equation with integer coefficients and the given solution set.

31. $\{3, 4\}$
32. $\{-1, 3\}$
33. $\{-5, -7\}$
34. $\{-3, -4\}$
35. $\left\{3, -\frac{2}{3}\right\}$
36. $\left\{-5, -\frac{2}{7}\right\}$
37. $\left\{\frac{1}{2}, -\frac{1}{2}\right\}$
38. $\left\{\frac{1}{3}, -\frac{1}{3}\right\}$
39. $\left\{0, -\frac{1}{5}\right\}$
40. $\left\{-\frac{3}{4}, 0\right\}$

⟨ D ⟩ Verifying the Solutions of a Quadratic Equation In Problems 41–45, use the sum and product properties to (a) find the sum of the solutions, (b) find the product of the solutions, and (c) determine if the two given values are the solutions of the given equation.

41. $4x^2 - 6x + 5 = 0$; the proposed solutions are $\frac{1}{2}$ and $\frac{5}{2}$.

42. $2x^2 + 9x = 35$; the proposed solutions are $-\frac{7}{2}$ and -1.

43. $5x^2 + 13x = 6$; the proposed solutions are $\frac{2}{5}$ and -3.

44. $4 - 3x = 7x^2$; the proposed solutions are $\frac{7}{4}$ and 1.

45. $-2 - 5x = 2x^2$; the proposed solutions are $\frac{1}{2}$ and 2.

46. If d is a constant and 3 is one solution of the equation $2x^2 - dx + 5 = 0$, use the product property to find the other solution.

47. If k is a constant and -5 is one solution of $3x^2 + kx = 40$, use the product property to find the other solution.

48. If the sum of the solutions of the equation $2x^2 - kx = 4$ is 3, find the value of k.

49. If the sum of the solutions of the equation $10x^2 + (k - 2)x = 3$ is $-\frac{13}{10}$, find k.

50. If the sum of the solutions of $3x^2 + (2k - 5)x + 8 = 0$ is 4, find k.

〉〉〉 Using Your Knowledge

Take a Dive The highest regularly performed head-first dives are made at La Quebrada in Acapulco, Mexico. The height h (in meters) above the water of the diver after t seconds is given by $h = -t^2 + 2t + 27$. Use the discriminant to find out whether:

51. The diver will ever be 27.5 meters above the water. How many times will this occur?

52. The diver will ever be 28 meters above the water. How many times will this occur?

53. The diver will ever be 29 meters above the water.

Write On

54. Why do you think the expression $b^2 - 4ac$ is called the discriminant?

55. Can a quadratic equation with integer coefficients have exactly one complex solution? Explain why or why not.

56. Can a quadratic equation with integer coefficients have exactly one irrational solution? Explain why or why not.

57. Can a quadratic equation with integer coefficients have exactly one rational and one irrational solution? Explain why or why not.

Concept Checker

Fill in the blank(s) with the correct word(s), phrase, or mathematical statement.

58. If the value of the discriminant is 20, then the solution set of the quadratic equation will be two _____.

59. If the value of the discriminant is -20, then the solution set of the quadratic equation will be two _____.

60. If the value of the discriminant is 100, then the solution set of the quadratic equation will be two _____.

61. When the _____ has a value of 0, then there will be one rational solution to the quadratic equation.

discriminant   rational numbers

positive perfect square   irrational numbers

non-real, complex numbers   not a real number

Mastery Test

Use the sum and product properties to see whether the solutions of $4x^2 - 12x + 5 = 0$ are:

62. $\frac{1}{2}$ and $-\frac{5}{2}$

63. $\frac{1}{2}$ and $\frac{5}{2}$

Use the discriminant to determine whether the expression is factorable. If it is, factor it.

64. $12x^2 + 23x + 10$

65. $12x^2 + x - 35$

Find a quadratic equation with integer coefficients whose solution set is:

66. $\{-2, 4\}$

67. $\left\{-1, \frac{2}{3}\right\}$

68. Consider $2x^2 + 3x = 7$.

    a. Find the discriminant.

    b. Classify the solutions without solving.

Skill Checker

Solve:

69. $x^2 + 6x + 5 = 0$

70. $x^2 - 14x - 15 = 0$

71. $x^2 - x - 4 = 0$

72. $x^2 + 3x - 1 = 0$

Find the LCM.

73. $(x - 2), (x^2 - 4)$

74. $(x^2 - 25), (x + 5)$

75. $(x + 1), (x - 3), 4$

76. $5, (x + 6)$

8.4 Solving Equations in Quadratic Form

Objectives

A Solve equations involving rational expressions by converting them to quadratic equations.

B Solve equations that are quadratic in form by substitution.

To Succeed, Review How To . . .

1. Find the LCD of two or more rational expressions (pp. 451–456).
2. Solve quadratic equations by factoring or by using the quadratic formula (pp. 405–409, 606–610).

Getting Started

Work Project

This man and his son can split a supply of firewood in two days. If each works alone, the son takes three days more than the father. How long would it take each of them working alone to finish the job? If we assume that the man can finish in d days, he will do $1/d$ of the work each day. The son will finish in $d + 3$ days and do

$$\frac{1}{d+3}$$

of the work each day. The work done each day is

work done by father	+	work done by son	=	work done together
$\frac{1}{d}$	+	$\frac{1}{d+3}$	=	$\frac{1}{2}$

To solve this equation, we multiply each term by the LCD, $2d(d + 3)$, to obtain

$$2d(d+3) \cdot \frac{1}{d} + 2d(d+3) \cdot \frac{1}{d+3} = 2d(d+3) \cdot \frac{1}{2}$$

$2(d + 3) + 2d = d(d + 3)$	Simplify.
$2d + 6 + 2d = d^2 + 3d$	Remove parentheses.
$0 = d^2 - d - 6$	Set equation equal to 0 (standard form).
$0 = (d - 3)(d + 2)$	Factor.
$d - 3 = 0$ or $d + 2 = 0$	Factors = 0 by the zero-product property.
$d = 3$ or $d = -2$	Solve.

Since d is the number of days, $d = -2$ has to be discarded. Thus, it takes the father working alone three days to finish and the son working alone three days more—6 days—to finish.

A › Solving Equations That Contain Rational Expressions

As we have seen, equations involving rational expressions can lead to quadratic equations that can be solved by factoring or by using the quadratic formula. When solving such equations, make sure that the proposed solutions are checked in the original equations to avoid zero denominators. If a zero denominator occurs, discard the proposed solution as an *extraneous* solution.

EXAMPLE 1 Solving equations that contain rational expressions

Solve:

a. $\dfrac{4}{x^2 - 4} - \dfrac{1}{x - 2} = 1$

b. $\dfrac{-12}{x^2 - 9} + \dfrac{1}{x - 3} = 1$

SOLUTION

a. Since $x^2 - 4 = (x + 2)(x - 2)$, the LCD is $(x + 2)(x - 2)$. Multiplying each term by the LCD of the fractions, we have

$$(x+2)(x-2) \cdot \dfrac{4}{(x^2-4)} - (x+2)(x-2) \cdot \dfrac{1}{x-2} = (x+2)(x-2) \cdot 1$$

Simplify.	$4 - (x + 2) = x^2 - 4$
Remove parentheses.	$4 - x - 2 = x^2 - 4$
Combine like terms.	$2 - x = x^2 - 4$
Subtract 2 and add x.	$0 = x^2 + x - 6$
0 Set equation = 0 (standard form).	$x^2 + x - 6 = 0$
F Factor.	$(x + 3)(x - 2) = 0$
F Factors = 0 by zero-product property.	$x + 3 = 0$ or $x - 2 = 0$
Solve.	$x = -3$ or $x = 2$

The proposed solutions are -3 and 2. However, $x = 2$ is not a solution, since

$$\dfrac{1}{x - 2}$$

is not defined for $x = 2$. Discard $x = 2$ as an extraneous solution. The only solution is -3, which you can check in the original equation.

b. Since $x^2 - 9 = (x + 3)(x - 3)$, the LCD is $(x + 3)(x - 3)$. Multiplying each term by the LCD, we have

$$(x+3)(x-3) \cdot \dfrac{-12}{x^2-9} + (x+3)(x-3) \cdot \dfrac{1}{x-3} = (x+3)(x-3) \cdot 1$$

Simplify.	$-12 + x + 3 = x^2 - 9$
0 Set equation = 0 (standard form).	$x^2 - x = 0$
F Factor.	$x(x - 1) = 0$
F Factors = 0.	$x = 0$ or $x - 1 = 0$
Solve.	$x = 0$ or $x = 1$

This time both solutions satisfy the original equation. Thus, the solutions are 0 and 1. Check this.

PROBLEM 1

Solve:

a. $\dfrac{10}{x^2 - 25} - \dfrac{1}{x - 5} = 1$

b. $\dfrac{10}{x^2 - 16} + \dfrac{1}{x - 4} = 1$

Answers to PROBLEMS

1. a. -6 b. $-5, 6$

B › Solving Equations by Substitution

Some equations that are not quadratic equations can be written in quadratic form and solved by appropriate substitutions. The way to recognize if an equation can be written in quadratic form is to see if the exponents on the variable are in a 2 to 1 ratio like the exponents in a quadratic equation. That is, one exponent is twice the other. Examples of variables with such exponents would be,

$$x^4 \text{ and } x^2$$
$$x^{1/2} \text{ and } x^{1/4}$$
$$x \text{ and } \sqrt{x}, \text{ which mean } x^1 \text{ and } x^{1/2}$$
$$x^{-4} \text{ and } x^{-2}$$

In each of these cases if we substitute u^2 and u for the two variable terms, the resulting quadratic equation in u can be solved. Then, using substitution again, the original equation can be solved.

Let's consider solving the equation $x^4 + 4x^2 - 12 = 0$. We know it is not a quadratic equation but since the ratio of the exponents on the variable are 4 to 2 which reduces to a 2 to 1 ratio, then we can use the "u" substitution method as follows.

$x^4 + 4x^2 - 12 = 0$	Given
$u^2 + 4u - 12 = 0$	Let $u = x^2$ and $u^2 = x^4$.
$(u - 2)(u + 6) = 0$	This is a quadratic equation and we will factor to solve.
$u = 2$ or $u = -6$	The solutions for u
$x^2 = 2$ or $x^2 = -6$	Since $u = x^2$, we replace x^2 for each "u."
$x = \pm\sqrt{2}$ or $x = \pm\sqrt{-6} = \pm i\sqrt{6}$	To solve, take the square root of both sides and simplify.

Remember in solving a quadratic equation we anticipated two solutions, but this equation has four solutions in its solution set, $\{-\sqrt{2}, \sqrt{2}, -i\sqrt{6}, i\sqrt{6}\}$.

The procedure for the "u" substitution method for solving quadratic equations is summarized next.

PROCEDURE

Solving a Quadratic-Type Equation by "u" Substitution

1. Substitute u^2 and u for the appropriate variables.
2. Solve the resulting quadratic equation for u, either by the factoring (OFF) method or the quadratic formula.
3. Substitute the variable that represented u into the equations with the solutions for u.
4. Solve the resulting equations.
5. Verify the solutions in the original equation.

We illustrate this procedure of using substitution in Examples 2–6.

EXAMPLE 2 Solving an equation by substitution

Solve: $x^4 - 10x^2 = -9$

SOLUTION Recognizing that the exponents on the variables are in a 2 to 1 ratio, we can solve this as a quadratic-type equation by substituting u^2 and u for x^4 and x^2, respectively.

PROBLEM 2

Solve:

$$x^4 - 17x^2 + 16 = 0$$

Answers to PROBLEMS

2. $\pm 1, \pm 4$

8.4 Solving Equations in Quadratic Form

$$x^4 - 10x^2 = -9 \quad \text{Given}$$
$$(x^2)^2 - 10(x^2) = -9 \quad \text{Substitute } u = x^2.$$
$$u^2 - 10u + 9 = 0 \quad \text{Set equation} = 0 \text{ (standard form).}$$
$$(u - 9)(u - 1) = 0 \quad \text{Factor.}$$
$$u - 9 = 0 \quad \text{or} \quad u - 1 = 0 \quad \text{Factors} = 0.$$
$$u = 9 \quad \text{or} \quad u = 1 \quad \text{Solve.}$$
$$x^2 = 9 \quad \text{or} \quad x^2 = 1 \quad \text{Substitute } x^2 = u.$$
$$x = \pm 3 \quad \text{or} \quad x = \pm 1 \quad \text{Take the square root of both sides.}$$

The solutions of $x^4 - 10x^2 = -9$ are 3, 1, -3, and -1. The verification is left to you.

We can also solve the equation in Example 2, $x^4 - 10x^2 = -9$, by the factoring method. We start by setting the equation equal to 0.

$$x^4 - 10x^2 = -9 \quad \text{Given}$$
$$x^4 - 10x^2 + 9 = 0 \quad \text{Set equation} = 0.$$
$$(x^2 - 1)(x^2 - 9) = 0 \quad \text{Factor.}$$
$$x^2 - 1 = 0 \quad \text{or} \quad x^2 - 9 = 0 \quad \text{Factors} = 0$$
$$x^2 = 1 \quad \text{or} \quad x^2 = 9 \quad \text{Isolate } x^2.$$
$$x = \pm\sqrt{1} = \pm 1 \text{ or } x = \pm\sqrt{9} = \pm 3 \quad \text{To solve, take the square root of both sides and simplify.}$$

The solutions are 1, -1, 3, and -3.

EXAMPLE 3 Solving equations that are quadratic in form

Solve: $(x^2 - x)^2 - (x^2 - x) - 30 = 0$

SOLUTION Notice that the binomials in parentheses are the same but one is raised to the second power and the other to the first power. This equation is quadratic in form. If we let $u = (x^2 - x)$, we can write

$$(x^2 - x)^2 - (x^2 - x) - 30 = 0 \quad \text{Given}$$
$$u^2 - u - 30 = 0$$
$$(u - 6)(u + 5) = 0 \quad \text{Factor.}$$
$$u = 6 \quad \text{or} \quad u = -5 \quad \text{Solve } u - 6 = 0 \text{ and } u + 5 = 0.$$
$$x^2 - x = 6 \quad \text{or} \quad x^2 - x = -5 \quad \text{Substitute } u = x^2 - x.$$
$$x^2 - x - 6 = 0 \quad \text{or} \quad x^2 - x + 5 = 0 \quad \text{In standard form}$$

The first equation can be solved by factoring:

$$x^2 - x - 6 = (x - 3)(x + 2) = 0$$

Thus, $x = 3$ or $x = -2$. We use the quadratic formula for $x^2 - x + 5 = 0$. Here $a = 1$, $b = -1$, and $c = 5$. Thus,

$$x = \frac{-(-1) \pm \sqrt{(-1)^2 - 4(1)(5)}}{2(1)} = \frac{1 \pm \sqrt{-19}}{2}$$
$$= \frac{1 \pm i\sqrt{19}}{2}$$

The solutions of $(x^2 - x)^2 - (x^2 - x) - 30 = 0$ are

$$3, -2, \frac{1}{2} + \frac{\sqrt{19}}{2}i, \text{ and } \frac{1}{2} - \frac{\sqrt{19}}{2}i$$

PROBLEM 3

Solve:

$$(x^2 + x)^2 - 11(x^2 + x) - 12 = 0$$

Answers to PROBLEMS

3. $-4, 3, -\frac{1}{2} \pm \frac{\sqrt{3}}{2}i$

EXAMPLE 4 Solving an equation containing rational exponents

Solve: $x^{1/2} - 8x^{1/4} + 15 = 0$

SOLUTION Recognizing that the exponents on the variables are in a 2 to 1 ratio, we can solve this as a quadratic-type equation by substituting u^2 and u for $x^{1/2}$ and $x^{1/4}$, respectively.

$x^{1/2} - 8x^{1/4} + 15 = 0$		Given
$(x^{1/4})^2 - 8(x^{1/4}) + 15 = 0$		Substitute $u = x^{1/4}$.
$u^2 - 8u + 15 = 0$		Set equation = 0 (standard form).
$(u - 3)(u - 5) = 0$		Factor.
$u - 3 = 0$ or	$u - 5 = 0$	Factors = 0.
$u = 3$ or	$u = 5$	Solve.
$x^{1/4} = 3$ or	$x^{1/4} = 5$	Substitute $x^{1/4} = u$.
$(x^{1/4})^4 = (3)^4$ or	$(x^{1/4})^4 = (5)^4$	Raise each side to the fourth power.
$x = 81$ or	$x = 625$	Simplify.

The solutions are 81 and 625. We verify these solutions by substituting them into the original equation to see if they make true statements.

Check: $x = 81$	$x = 625$
$x^{1/2} - 8x^{1/4} + 15 = 0$	$x^{1/2} - 8x^{1/4} + 15 = 0$
$\sqrt{81} - 8\sqrt[4]{81} + 15 \stackrel{?}{=} 0$	$\sqrt{625} - 8\sqrt[4]{625} + 15 \stackrel{?}{=} 0$
$9 - 8(3) + 15 \stackrel{?}{=} 0$	$25 - 8(5) + 15 \stackrel{?}{=} 0$
$9 - 24 + 15 \stackrel{?}{=} 0$	$25 - 40 + 15 \stackrel{?}{=} 0$
$24 - 24 \stackrel{?}{=} 0$	$40 - 40 \stackrel{?}{=} 0$
$0 = 0$	$0 = 0$

Both solutions check.

PROBLEM 4

Solve:

$$x^{1/2} - 7x^{1/4} + 10 = 0$$

Next we solve some equations involving radicals. For example, the equation $x + 2\sqrt{x} - 3 = 0$ can be written as a quadratic equation if we let $u = \sqrt{x}$. This makes $u^2 = x$ so that we can write

$x + 2\sqrt{x} - 3 = 0$		
$u^2 + 2u - 3 = 0$		Substitute u for $\sqrt{x}$ and u^2 for x.
$(u + 3)(u - 1) = 0$		Factor.
$u = -3$ or	$u = 1$	Solve $u + 3 = 0$ and $u - 1 = 0$.
$\sqrt{x} = -3$ or	$\sqrt{x} = 1$	Substitute $u = \sqrt{x}$.

But the principal square root of x is positive, so the equation $\sqrt{x} = -3$ has no solution. The solution of $\sqrt{x} = 1$ is 1. Thus, the equation $x + 2\sqrt{x} - 3 = 0$ has only one solution, 1. You can verify that this solution is correct by direct substitution into the equation.

EXAMPLE 5 Using substitution to solve an equation containing a radical

Solve: $x - 4\sqrt{x} + 3 = 0$

SOLUTION We let $u = \sqrt{x}$, which makes $u^2 = x$.

$$x - 4\sqrt{x} + 3 = 0$$

PROBLEM 5

Solve:

$$x - 3\sqrt{x} - 10 = 0$$

Answers to PROBLEMS

4. 16, 625   **5.** 25

8.4 Solving Equations in Quadratic Form

$$u^2 - 4u + 3 = 0 \quad \text{Substitute.}$$
$$(u - 3)(u - 1) = 0 \quad \text{Factor.}$$
$$u = 3 \quad \text{or} \quad u = 1 \quad \text{Solve } u - 3 = 0 \text{ and } u - 1 = 0.$$
$$\sqrt{x} = 3 \quad \text{or} \quad \sqrt{x} = 1 \quad \text{Substitute } u = \sqrt{x}.$$
$$x = 9 \quad \text{or} \quad x = 1 \quad \text{Square both sides.}$$

We leave it to you to verify that both solutions satisfy the original equation.

EXAMPLE 6 Solving an equation containing negative exponents

Solve: $5x^{-4} - 4x^{-2} - 1 = 0$

SOLUTION We could solve this equation by making the substitution $u = x^{-2}$, but this time we first convert the equation to one with positive exponents. Since

$$x^{-4} = \frac{1}{x^4} \quad \text{and} \quad x^{-2} = \frac{1}{x^2}$$

we have

$$5x^{-4} - 4x^{-2} - 1 = \frac{5}{x^4} - \frac{4}{x^2} - 1 = 0$$

Multiplying each term by the LCD, x^4, we obtain

$$x^4 \cdot \frac{5}{x^4} - x^4 \cdot \frac{4}{x^2} - x^4 \cdot 1 = 0$$

$$5 - 4x^2 - x^4 = 0 \quad \text{Simplify.}$$
$$-x^4 - 4x^2 + 5 = 0 \quad \text{Write in descending order.}$$
$$x^4 + 4x^2 - 5 = 0 \quad \text{Multiply by } -1.$$
$$u^2 + 4u - 5 = 0 \quad \text{Let } u = x^2.$$
$$(u + 5)(u - 1) = 0 \quad \text{Factor.}$$
$$u = -5 \quad \text{or} \quad u = 1 \quad \text{Solve } u + 5 = 0 \text{ and } u - 1 = 0.$$
$$x^2 = -5 \quad \text{or} \quad x^2 = 1 \quad \text{Substitute } x^2 \text{ for } u.$$
$$x = \pm\sqrt{5}i \quad \text{or} \quad x = \pm 1 \quad \text{Take the square root of both sides.}$$

We leave it to you to verify that the four solutions satisfy the original equation.

PROBLEM 6

Solve:

$$2x^{-2} - 5x^{-1} - 3 = 0$$

Calculator Corner

Solve Equations by Graphing

Your calculator can be used in several ways to solve the equations we have studied. One way is to use the solve feature. When doing so, it's usually easy to obtain one solution, but in many cases, the second solution is hard to obtain. Try Example 1(b) if you don't believe this. Moreover, there may be cases in which the graph does not cross the horizontal axis. This means that there are no real-number solutions.

Now let's try to solve Example 2 by graphing. It will be easier to graph $Y_1 = x^4 - 10x^2 + 9$ and $Y_2 = 0$ using a standard window and then pressing [2nd] [TRACE] 5 [ENTER] [ENTER] [ENTER] to find one intersection point. To find the other intersection points, press [TRACE] and use the arrow keys to move the cursor near the intersection point. Then press [2nd] [TRACE] 5 and [ENTER] three times. The solutions are -3, -1, 1, and 3, as shown in Window 1.

Window 1

(continued)

Answers to PROBLEMS

6. $\frac{1}{3}$, -2

What about Example 6? Since some of the solutions are real and some imaginary, what does this graph look like? Enter $Y_1 = 5x^{-4} - 4x^{-2} - 1$ and graph. You can see only the real solutions, -1 and 1, as shown in Window 2 using a window of $[-10, 10]$ by $[-2, 12]$. How do you know there are more solutions? You need more information! The original equation is equivalent to $x^4 + 4x^2 - 5 = 0$, which has degree 4, so the equation must have four solutions (some may repeat). Since we found only two solutions, we suspect that the other two solutions may be imaginary. (An nth-degree equation has n solutions but some of them may repeat. At this time, the only way to check for repeated solutions is to solve the problem algebraically.)

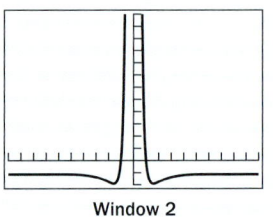

Window 2

1. Verify the results of Example 4 using your calculator. If you have a root feature, use it.
2. Verify the results of Example 5 using your calculator.

Boost your grade at mathzone.com!
> Practice Problems
> NetTutor
> Self-Tests
> e-Professors
> Videos

> Exercises 8.4

⟨ A ⟩ **Solving Equations That Contain Rational Expressions** In Problems 1–10, solve the equation.

1. $\dfrac{x}{x+4} + \dfrac{x}{x+1} = 0$
2. $\dfrac{x}{x+2} + \dfrac{x}{x+3} = 0$
3. $\dfrac{x-1}{x+11} - \dfrac{2}{x-1} = 0$
4. $\dfrac{x+1}{x-2} - \dfrac{8}{x-1} = 0$
5. $\dfrac{x}{x-1} - \dfrac{x}{x+1} = 0$
6. $\dfrac{x}{x+4} + \dfrac{x}{x-2} = -\dfrac{1}{2}$
7. $\dfrac{x}{x+2} - \dfrac{x}{x+1} = -\dfrac{1}{6}$
8. $\dfrac{x}{x+1} - \dfrac{x}{x-1} = -\dfrac{3}{4}$
9. $\dfrac{x}{x+4} + \dfrac{x}{x+2} = -\dfrac{4}{3}$
10. $\dfrac{2x}{x-2} + \dfrac{x}{x-1} = \dfrac{7}{6}$

⟨ B ⟩ **Solving Equations by Substitution** In Problems 11–39, solve by substitution.

11. $x^4 - 13x^2 + 36 = 0$
12. $x^4 - 5x^2 + 4 = 0$
13. $4x^4 + 35x^2 = 9$
14. $3x^4 + 2x^2 = 8$
15. $3y^4 = 5y^2 + 2$
16. $6y^4 = 7y^2 - 2$
17. $x^6 + 7x^3 - 8 = 0$
18. $x^6 - 26x^3 - 27 = 0$
19. $(x+1)^2 - 3(x+1) = 40$
20. $(x+2)^2 - 2(x+2) = 8$
21. $(y^2 - y)^2 - 8(y^2 - y) = 9$
22. $(y^2 - y)^2 - 4(y^2 - y) = 12$
23. $x^{1/2} + 3x^{1/4} - 10 = 0$
24. $x^{1/2} + 4x^{1/4} - 12 = 0$
25. $y^{2/3} - 5y^{1/3} = -6$
26. $y^{2/3} + 5y^{1/3} = -6$
27. $x + \sqrt{x} - 6 = 0$
28. $x - \sqrt{x} - 30 = 0$
29. $(x^2 - 4x) - 8\sqrt{x^2 - 4x} + 15 = 0$
30. $(x^2 + 3x) + 5\sqrt{x^2 + 3x} - 14 = 0$
31. $z + 3 - \sqrt{z+3} - 6 = 0$
32. $z + 4 - \sqrt{z+4} - 12 = 0$
33. $3\sqrt{x} - 5\sqrt[4]{x} + 2 = 0$
34. $x^{-2} + 2x^{-1} - 3 = 0$
35. $x^{-2} + 2x^{-1} - 8 = 0$
36. $8x^{-4} - 9x^{-2} + 1 = 0$
37. $3x^{-4} - 5x^{-2} - 2 = 0$
38. $6x^{-4} + x^{-2} - 1 = 0$
39. $6x^{-4} + 5x^{-2} - 4 = 0$

>>> Applications

40. Work rate Two workers can complete a job in 6 hours if they work together. If they work alone, one of them takes 9 hours more than the other to finish. How long does it take for each of them working alone to finish the job?

41. Work rate Jack and Jill can shovel the snow in the driveway in 6 hours if they work together. It takes Jack 5 hours more than Jill to do the job by himself. How long does it take each of them working alone to finish the job?

42. x-intercepts Given the equation $x^4 - 6x^2 + 9 = 0$ of the graph, find the exact value of the x-intercepts.

43. x-intercepts Given the equation $x^4 - 10x^2 + 25 = 0$ of the graph, find the exact value of the x-intercepts.

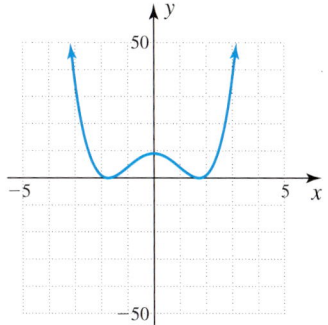

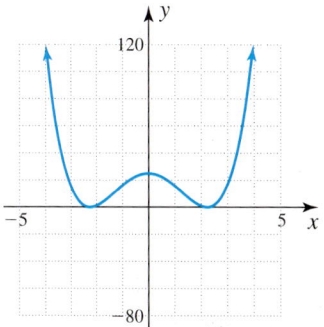

>>> Using Your Knowledge

A Rental Problem

44. A group of students rented a cabin for $1600. When two of the group failed to pay their shares, the cost to each of the remaining students was $40 more. Use your knowledge of the RSTUV procedure to find how many students were in the group.

n = number of students in the group

$n - 2$ = number of students who paid

$\dfrac{1600}{n}$ = amount each student should have paid in dollars

45. A group of students rented a bus for $720. If there had been six more students, the price per student would have been $6 less. How many students were in the group?

>>> Write On

46. Find the number of solutions for $x - 1 = 0$, $x^2 - 1 = 0$, and $x^3 - 1 = 0$. Make a conjecture regarding the number of solutions for $x^4 - 1 = 0$. (If you want to solve the equation $x^4 - 1 = 0$ first, do it by factoring.)

47. Which method would you use to solve $x + 2\sqrt{x} - 3 = 0$, the substitution method or isolating the radical and squaring both sides? Explain. Which method has fewer steps?

48. In Example 6, we solved the problem by transforming the given equation into an equivalent one with positive exponents. Now solve Example 6 using the substitution $u = x^{-2}$. Which method do you prefer? Explain why.

49. The equation $x^4 - 1 = 0$ has four solutions, two real and two non-real complex. (Solve $x^4 - 1 = 0$ by factoring if you don't believe this.) How many solutions do you think the equation $x^6 - 1 = 0$ has? How many do you think are real? How many do you think are non-real complex? Make a conjecture about the number and nature of the solutions (real or non-real complex) of $x^{2n} - 1 = 0$.

>>> Concept Checker

Fill in the blank(s) with the correct word(s), phrase, or mathematical statement.

50. A method for solving the equation $x^6 - 5x^3 + 6 = 0$ is called _____.

51. After clearing the rational expressions in the equation, $\frac{10}{x^2 - 16} + \frac{1}{x - 4} = 1$, the result is $x^2 - x - 30 = 0$, which is a _____ equation.

52. If a zero denominator occurs when checking a proposed solution in the original equation, then discard the proposed solution as a(n) _____.

extraneous
solution
quadratic
linear

rational
u-substitution
simplifying

>>> Mastery Test

Solve:

53. $x - 5\sqrt{x} + 4 = 0$

54. $2x - 3\sqrt{x} = -1$

55. $x^{1/2} - 5x^{1/4} + 6 = 0$

56. $x^{1/4} - 5x^{1/2} = -4$

57. $x^{-4} - 9x^{-2} + 14 = 0$

58. $x^{-4} - 7x^{-2} = -12$

59. $(x^2 - x)^2 - (x^2 - x) - 2 = 0$

60. $4(x^2 + 1)^2 - 7(x^2 + 1) = 2$

61. $\frac{12}{x^2 - 36} - \frac{1}{x - 6} = 1$

62. $\frac{6}{x^2 - 4} + \frac{1}{x - 2} = 1$

63. $x^4 - 5x^2 + 4 = 0$

64. $x^4 - 6x^2 + 5 = 0$

>>> Skill Checker

Solve:

65. $\frac{x}{5} - \frac{x}{3} \leq \frac{x - 5}{5}$

66. $\frac{7x + 2}{6} \geq \frac{3x - 2}{4}$

67. $\frac{8x - 23}{6} + \frac{1}{3} \geq \frac{5x}{2}$

True or False:

68. $x^2 - 3x \geq 5$, if $x = -1$

69. $x^2 - 2x \leq 0$, if $x = 2$

70. $(x + 2)(x - 3) < 0$, if $x = 1$

71. $(x - 4)(x - 1) \geq 0$, if $x = 3$

8.5 Nonlinear Inequalities

Objectives

A Solve quadratic inequalities.

B Solve polynomial inequalities of degree 3 or higher.

C Solve rational inequalities.

D Solve an application involving inequalities.

To Succeed, Review How To . . .

1. Solve a quadratic equation (pp. 405–409, 606–610).
2. Solve and graph linear inequalities (pp. 123–129).

Getting Started

Skidding to Quadratics

Have you seen an officer measuring skid marks at the scene of an accident? The distance d (in ft) in which a car traveling v mi/hr can be stopped is given by

$$d = 0.05v^2 + v$$

The skid marks of one accident were more than 40 feet long. The accident occurred in a 20-mi/hr zone. Was the driver going over the speed limit? To answer this question, we must solve the quadratic inequality $0.05v^2 + v > 40$. Let's start by solving the related quadratic equation $0.05v^2 + v = 40$.

$0.05v^2 + v = 40$	Given
$5v^2 + 100v = 4000$	Multiply by 100.
$v^2 + 20v = 800$	Divide by 5.
$v^2 + 20v - 800 = 0$	Standard form
$(v + 40)(v - 20) = 0$	Factor.
$v = -40$ or $v = 20$	Solve $v + 40 = 0$ and $v - 20 = 0$.

Now divide a number line into three intervals, A, B, and C, using the solutions (*critical values*) -40 and 20 as boundaries (see the following table). To find the regions where $0.05v^2 + v > 40$, choose a test point in each of the regions, and test whether $0.05v^2 + v > 40$ for each of the points. Let's select the points -50, 0, and 30 from intervals A, B, and C, respectively.

Test	For $v = -50$	For $v = 0$	For $v = 30$
In original:	$0.05v^2 + v > 40$	$0.05v^2 + v > 40$	$0.05v^2 + v > 40$
	$0.05(-50)^2 + (-50) > 40$	$0.05(0)^2 + 0 > 40$	$0.05(30)^2 + 30 > 40$
	$125 - 50 > 40$	$0 > 40$ (FALSE)	$45 + 30 > 40$
	$75 > 40$ (TRUE)		$75 > 40$ (TRUE)
	Thus, $0.05v^2 + v > 40$. Use this interval: $(-\infty, -40)$.	Thus, $0.05v^2 + v < 40$. Discard this interval.	Thus, $0.05v^2 + v > 40$. Use this interval: $(20, \infty)$.

The solution set is the union of the two intervals where the inequality $0.05v^2 + v > 40$ is true, that is, $(-\infty, -40) \cup (20, \infty)$. The end points -40 and 20 are *not* included in the solution set. Also, since the velocity must be positive, we discard the interval $(-\infty, -40)$ and conclude that the car was going more than 20 miles per hour.

A › Solving Quadratic Inequalities

The procedure used in *Getting Started* is based on the fact that, if you select a test point in one of the chosen intervals and the inequality is satisfied by the test point, then *every* point in the interval satisfies the inequality. The procedure we just used to solve the speed-limit problem can be generalized to solve a quadratic inequality. Here are the steps.

PROCEDURE
Procedure to Solve Quadratic Inequalities
1. Write the quadratic inequality in standard form—the quadratic expression is on the left side in descending powers of the variable and zero is on the right side.
2. Find the *critical values* by solving the related equation.
3. Separate the number line into intervals using the critical values found in Step 2 as boundaries.
4. Choose a test point in each interval and determine whether the point satisfies the **original quadratic inequality.**
5. State the solution set, which consists of all the intervals in which the test point satisfies the original quadratic inequality.

EXAMPLE 1 Using the five-step procedure to solve a quadratic inequality
Solve and write the solution in interval notation: $(x - 1)(x + 3) < 0$

PROBLEM 1
Solve and write the solution in interval notation:
$$(x - 4)(x + 2) \leq 0$$

SOLUTION We use the five-step procedure.
1. The inequality $(x - 1)(x + 3) < 0$ has zero on the right side.
2. The related equation is $(x - 1)(x + 3) = 0$, and the critical values are $x = 1$ and $x = -3$.
3. Separate the number line into three intervals using -3 and 1 as boundaries.
4. Select the test points: -4 from A, 0 from B, and 2 from C.

Test	For $x = -4$	For $x = 0$	For $x = 2$
In original:	$(x - 1)(x + 3) < 0$	$(x - 1)(x + 3) < 0$	$(x - 1)(x + 3) < 0$
	$(-4 - 1)(-4 + 3) < 0$	$(0 - 1)(0 + 3) < 0$	$(2 - 1)(2 + 3) < 0$
	$(-5)(-1) < 0$	$(-1)(3) < 0$	$(1)(5) < 0$
	$5 < 0$ (FALSE)	$-3 < 0$ (TRUE)	$5 < 0$ (FALSE)
	Thus, $(x - 1)(x + 3) > 0$. Discard this interval.	Thus, $(x - 1)(x + 3) < 0$. Use this interval: $(-3, 1)$.	Thus, $(x - 1)(x + 3) > 0$. Discard this interval.

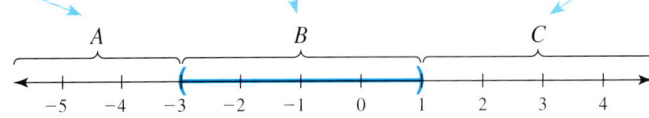

5. The solution set is the interval $(-3, 1)$.

Answers to PROBLEMS
1. $[-2, 4]$

EXAMPLE 2 Using the five-step procedure to solve a quadratic inequality

Solve, graph, and write the solution in interval notation: $x^2 - 6x \geq -7$

SOLUTION Again, we use the five-step procedure.

1. We first add 7 to both sides so that the inequality is in the standard form, $x^2 - 6x + 7 \geq 0$.

2. The related equation is $x^2 - 6x + 7 = 0$. Since $x^2 - 6x + 7$ is *not* factorable, we use the quadratic formula with $a = 1$, $b = -6$, and $c = 7$, and we obtain

$$x = \frac{-(-6) \pm \sqrt{(-6)^2 - 4(1)(7)}}{2(1)} = \frac{6 \pm \sqrt{8}}{2} = \frac{6 \pm 2\sqrt{2}}{2} = 3 \pm \sqrt{2}$$

The critical values are $x = 3 + \sqrt{2}$ and $x = 3 - \sqrt{2}$.

3. Separate the number line into three regions using $3 + \sqrt{2}$ and $3 - \sqrt{2}$ as boundaries. Approximating $\sqrt{2}$, as 1.4, the boundaries are $3 + 1.4 = 4.4$ and $3 - 1.4 = 1.6$.

4. Select the test points: 0 from A, 3 from B, and 6 from C.

Test	For $x = 0$	For $x = 3$	For $x = 6$
In original:	$x^2 - 6x \geq -7$	$x^2 - 6x \geq -7$	$x^2 - 6x \geq -7$
	$(0)^2 - 6(0) \geq -7$	$(3)^2 - 6(3) \geq -7$	$(6)^2 - 6(6) \geq -7$
	$0 - 0 \geq -7$	$9 - 18 \geq -7$	$36 - 36 \geq -7$
	$0 \geq -7$ (TRUE)	$-9 \geq -7$ (FALSE)	$0 \geq -7$ (TRUE)
	Thus, $x^2 - 6x \geq -7$.	Thus, $x^2 - 6x < -7$.	Thus, $x^2 - 6x \geq -7$.
	Use this interval: $(-\infty, 3 - \sqrt{2})$.	Discard this interval.	Use this interval: $(3 + \sqrt{2}, \infty)$.

5. The solution set is $(-\infty, 3 - \sqrt{2}] \cup [3 + \sqrt{2}, \infty)$. The critical values $3 - \sqrt{2}$ and $3 + \sqrt{2}$ are part of the solution set because they satisfy the inequality $x^2 - 6x \geq -7$. (Check this, but first read the caution note for a good rule of thumb.)

PROBLEM 2

Solve, graph, and write the solution in interval notation:

$x^2 - 5 > -2$

CAUTION

The end points of an interval are included in the solution set when the original inequality sign is $\leq$ or $\geq$; they are omitted when the original inequality sign is $>$ or $<$.

Example:

$x > 2$ $x \geq 2$

Answers to PROBLEMS

2. $(-\infty, -\sqrt{3}) \cup (\sqrt{3}, \infty)$

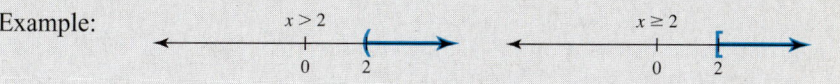

B › Solving Polynomial Inequalities

Suppose we wish to solve the inequality $(x + 3)(x - 2)(x - 4) \geq 0$, which is not a quadratic inequality. This would be called a *polynomial inequality* and the steps to solving it are the same as the steps for solving a quadratic inequality. The procedure follows.

> **PROCEDURE**
>
> **Procedure to Solve Polynomial Inequalities**
> 1. Write the polynomial inequality in standard form—the polynomial expression is on the left side and zero is on the right side.
> 2. Find the *critical values* by solving the related equation.
> 3. Separate the number line into intervals using the critical values found in Step 2 as boundaries.
> 4. Choose a test point in each interval and determine whether the point satisfies the **original polynomial inequality.**
> 5. State the solution set, which consists of all the intervals in which the test point satisfies the original polynomial inequality.

The original inequality, $(x + 3)(x - 2)(x - 4) \geq 0$, is in standard form and the polynomial on the left is in factored form. This makes it easy to set each factor equal to zero and solve for the three critical numbers, $x = -3$, $x = 2$, and $x = 4$. We separate the number line into *four* intervals using -3, 2, and 4 as boundaries. Now select the points -4 from A, 0 from B, 3 from C, and 5 from D and determine which of these points satisfies the original polynomial inequality

$$(x + 3)(x - 2)(x - 4) \geq 0.$$

Test	For $x = -4$	For $x = 0$	For $x = 3$	For $x = 5$
In original:	$(x + 3)(x - 2)(x - 4) \geq 0$	$(x + 3)(x - 2)(x - 4) \geq 0$	$(x + 3)(x - 2)(x - 4) \geq 0$	$(x + 3)(x - 2)(x - 4) \geq 0$
	$(-4 + 3)(-4 - 2)(-4 - 4) \geq 0$	$(0 + 3)(0 - 2)(0 - 4) \geq 0$	$(3 + 3)(3 - 2)(3 - 4) \geq 0$	$(5 + 3)(5 - 2)(5 - 4) \geq 0$
	$(-1)(-6)(-8) \geq 0$	$(3)(-2)(-4) \geq 0$	$(6)(1)(-1) \geq 0$	$(8)(3)(1) \geq 0$
	$-48 \geq 0$ (FALSE)	$24 \geq 0$ (TRUE)	$-6 \geq 0$ (FALSE)	$24 \geq 0$ (TRUE)
	Thus,	Thus,	Thus,	Thus,
	$(x + 3)(x - 2)(x - 4) < 0$.	$(x + 3)(x - 2)(x - 4) \geq 0$.	$(x + 3)(x - 2)(x - 4) < 0$.	$(x + 3)(x - 2)(x - 4) \geq 0$.
	Discard this interval.	Use this interval: $[-3, 2]$.	Discard this interval.	Use this interval: $[4, \infty)$.

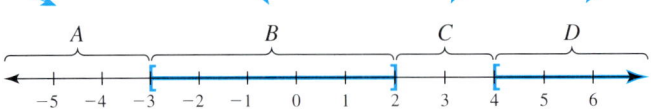

The solution set is the union of the two intervals selected, $[-3, 2] \cup [4, \infty)$. -3, 2, and 4 are part of the solution set because they satisfy the inequality $(x + 3)(x - 2)(x - 4) \geq 0$.

EXAMPLE 3 Using the five-step procedure to solve polynomial inequalities

Solve, graph, and write the solution in interval notation:

$$(x + 1)(x - 3)(x + 4) \leq 0$$

SOLUTION We follow our five-step procedure.

1. The inequality is in standard form.
2. The related equation is $(x + 1)(x - 3)(x + 4) = 0$, and the critical values are -1, 3, and -4.
3. Separate the number line into four intervals using -1, 3, and -4 as boundaries.
4. Select the test points: -5 from A, -3 from B, 0 from C, and 4 from D.

Test	For $x = -5$	For $x = -3$	For $x = 0$	For $x = 4$
In original:	$(x + 1)(x - 3)(x + 4) \leq 0$	$(x + 1)(x - 3)(x + 4) \leq 0$	$(x + 1)(x - 3)(x + 4) \leq 0$	$(x + 1)(x - 3)(x + 4) \leq 0$
	$(-5 + 1)(-5 - 3)(-5 + 4) \leq 0$	$(-3 + 1)(-3 - 3)(-3 + 4) \leq 0$	$(0 + 1)(0 - 3)(0 + 4) \leq 0$	$(4 + 1)(4 - 3)(4 + 4) \leq 0$
	$(-4)(-8)(-1) \leq 0$	$(-2)(-6)(1) \leq 0$	$(1)(-3)(4) \leq 0$	$(5)(1)(8) \leq 0$
	$-32 \leq 0$ (TRUE)	$12 \leq 0$ (FALSE)	$-12 \leq 0$ (TRUE)	$40 \leq 0$ (FALSE)
	Thus,	Thus,	Thus,	Thus,
	$(x + 1)(x - 3)(x + 4) \leq 0$. Use this interval: $(-\infty, -4]$.	$(x + 1)(x - 3)(x + 4) > 0$. Discard this interval.	$(x + 1)(x - 3)(x + 4) \leq 0$. Use this interval: $[-1, 3]$.	$(x + 1)(x - 3)(x + 4) > 0$. Discard this interval.

5. The solution set is the union of the two intervals selected, $(-\infty, -4] \cup [-1, 3]$.

PROBLEM 3

Solve, graph, and write the solution in interval notation:

$$(x + 3)(x - 2)(x + 1) \leq 0$$

C > Solving Rational Inequalities

In solving rational equations, we were able to clear the fractions by multiplying both sides of the equation by the LCD. It will not be possible to clear the fractions in a *rational inequality* if any denominator contains a variable. Remember if you divide or multiply an inequality by a negative number, then the inequality symbol must be reversed. Thus, if the denominator has a variable expression, we can't be certain that its value will always be positive when multiplying both sides of the inequality by it. So we can't solve by clearing the fractions.

What method can we use to solve these rational inequalities? We will use a method similar to what was used in solving quadratic inequalities. We are now ready to consider solving the *rational inequality*

$$\frac{x - 3}{x + 2} \leq 0$$

This time, the **critical values** are the numbers that make the *numerator* $(x - 3)$ and *denominator* $(x + 2)$ zero, that is, 3 and -2. The critical values that make the denominator zero are excluded from the solution set. As before, we separate the number line into three intervals, A, B, and C, using -2 and 3 as boundaries, and then select test points from each of these regions to determine whether they satisfy our inequality. Let's select -3 from A, 0 from B, and 4 from C.

Answers to PROBLEMS

3. $(-\infty, -3] \cup [-1, 2]$

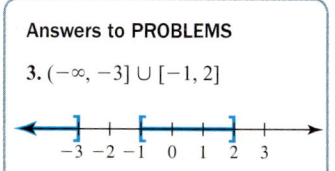

640　Chapter 8　Quadratic Equations and Inequalities

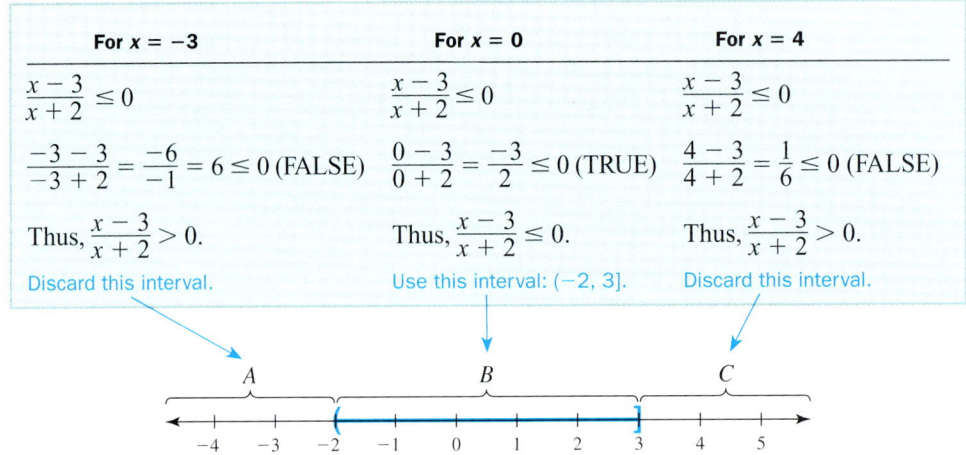

The solution set is the interval $(-2, 3]$.

Notice we enclosed -2 with a parenthesis and not a bracket like we used to enclose 3. This means we will not be able to include -2 as part of the solution set. The fraction would be undefined if we allowed x to equal -2 because it would give the denominator the value of 0.

The procedure for solving rational inequalities is summarized next.

PROCEDURE

Procedure to Solve Rational Inequalities

1. Write the rational inequality so that there is one rational expression on the left side and 0 on the right side.
2. Find the *critical values* by setting the numerator $= 0$ and solving and setting the denominator $= 0$ and solving.
3. Separate the number line into intervals using the critical values found in Step 2 as boundaries.
4. Choose a test point in each interval and determine whether the point satisfies the **original rational inequality.**
5. State the solution set, which consists of all the intervals in which the test point satisfies the original rational inequality.

EXAMPLE 4　Using the five-step procedure to solve rational inequalities

Solve, graph, and write the solution in interval notation:

a. $\dfrac{3x}{x+4} > 0$　　b. $\dfrac{x}{x-2} \geq 2$

SOLUTION　We still follow our five-step procedure.

a. 1. The inequality has zero on the right side.
   2. The critical points of 0 and -4 are obtained by setting the numerator and denominator equal to 0; then solve each.
   3. Separate the number line into three intervals using 0 and -4 as the boundaries.

PROBLEM 4

Solve, graph, and write the solution in interval notation:

a. $\dfrac{x+3}{2x-1} \leq 0$　b. $\dfrac{x}{x-1} > 1$

Answers to PROBLEMS

4. a. $[-3, \frac{1}{2})$　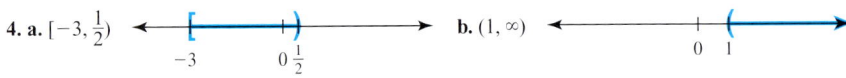　b. $(1, \infty)$

4. Select the test points: -5 from A, -1 from B, and 2 from C.

Test	For $x = -5$	For $x = -1$	For $x = 2$
In original:	$\dfrac{3x}{x+4} > 0$	$\dfrac{3x}{x+4} > 0$	$\dfrac{3x}{x+4} > 0$
	$\dfrac{3(-5)}{-5+4} = \dfrac{-15}{-1} = 15 > 0$	$\dfrac{3(-1)}{-1+4} = \dfrac{-3}{3} = -1 > 0$	$\dfrac{3(2)}{2+4} = \dfrac{6}{6} = 1 > 0$
	(TRUE)	(FALSE)	(TRUE)
	Use this interval: $(-\infty, -4)$	Discard this interval.	Use this interval: $(0, \infty)$.

A   B   C

(number line from -5 to 3)

5. The solution set is $(-\infty, -4) \cup (0, \infty)$.

b. 1. To write the inequality so that zero is on the right side, we first subtract 2 from both sides and get a common denominator to simplify.

$\dfrac{x}{x-2} \geq 2$ Given

$\dfrac{x}{x-2} - 2 \geq 0$ Subtract 2.

$\dfrac{x}{x-2} - \dfrac{2(x-2)}{x-2} \geq 0$ $2 = \dfrac{2(x-2)}{x-2}$

$\dfrac{x - 2x + 4}{x-2} \geq 0$ Remove parentheses in the numerator and write as one fraction.

$\dfrac{4-x}{x-2} \geq 0$ Simplify.

2. The critical points are 4 and 2. The 2 will cause the denominator to be zero, so it is not included in the solution set. The 4 is included because it will make the rational expression zero, thus yielding a true statement.

3. Separate the number line into three intervals using 2 and 4 as boundaries.

4. Select the test points: 0 from A, 3 from B, and 5 from C.

Test	For $x = 0$	For $x = 3$	For $x = 5$
In original:	$\dfrac{x}{x-2} \geq 2$	$\dfrac{x}{x-2} \geq 2$	$\dfrac{x}{x-2} \geq 2$
	$\dfrac{0}{0-2} = 0 \geq 2$ (FALSE)	$\dfrac{3}{3-2} = 3 \geq 2$ (TRUE)	$\dfrac{5}{5-2} = \dfrac{5}{3} \geq 2$ (FALSE)
	Thus, $\dfrac{x}{x-2} < 2$.	Thus, $\dfrac{x}{x-2} \geq 2$.	Thus, $\dfrac{x}{x-2} < 2$.
	Discard this interval.	Use this interval: $(2, 4]$.	Discard this interval.

A   B   C

(number line from -1 to 7)

5. The solution set is $(2, 4]$, which does not include 2 but does include 4, because 4 satisfies the inequality, $\dfrac{x}{x-2} \geq 2$, and 2 does not.

You have probably noticed that all the solution sets we've obtained are either intervals or unions of intervals. There are other possibilities, and they are based on the fact that for any real number a, $a^2 \geq 0$. Table 8.2 gives you some of these unusual solution sets, where a and x represent real numbers.

Table 8.2 Unusual Solution Sets

Explanation	Example	Solution Set
$a^2 \geq 0$ for every real number a.	$(x - 1)^2 \geq 0$ for every real number x.	The solution set of $(x - 1)^2 \geq 0$ is the set of all real numbers.
$a^2 \leq 0$ only when $a = 0$.	$(x - 1)^2 \leq 0$ only when $x - 1 = 0$, that is, when $x = 1$.	The solution set of $(x - 1)^2 \leq 0$ is $\{1\}$.
a^2 is never negative, so $a^2 < 0$ is false.	$(x - 1)^2$ is never negative, so $(x - 1)^2 < 0$ is false.	The solution set of $(x - 1)^2 < 0$ is the empty set, $\emptyset$.

You can check all these examples by using the same five-step procedure we've been using.

D > Solving Applications Involving Inequalities

Example 5 shows how to apply the material we've been studying to a projectile motion problem. Remember our RSTUV method? We will use it again here.

EXAMPLE 5 Problem for a rocket scientist

The height h (in ft) of a rocket is given by $h = 64t - 16t^2$, where t is the time in seconds. During what time interval will the rocket be more than 48 feet above the ground?

SOLUTION We'll use the RSTUV method.

1. **Read the problem.** The problem asks for a certain time interval in which $h > 48$.
2. **Select the unknown.** The unknown is the time t.
3. **Think of a plan.** We have to find values of t for which $h = 64t - 16t^2 > 48$.
4. **Use the five-step procedure to solve the inequality.**
   1. Subtract 48 from both sides and rewrite in descending order. $-16t^2 + 64t - 48 > 0$
   2. The related equation is $-16t^2 + 64t - 48 = 0$

   $t^2 - 4t + 3 = 0$ Divide by -16.
   $(t - 1)(t - 3) = 0$ Factor.

   The critical values are 1 and 3.
   3. Separate the number line into three intervals, A, B, and C with 1 and 3 as boundaries.
   4. Select the test points: 0 from A, 2 from B, and 4 from C. Test each in the original inequality, $64t - 16t^2 > 48$.

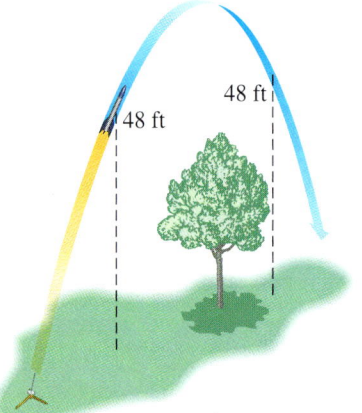

PROBLEM 5

The bending moment of a beam is $M = 30 - x^2$, where x is the distance, in feet, from one end of the beam. At what distances will the bending moment be more than 5?

Answers to PROBLEMS

5. Between 0 and 5 ft

For $t = 0$, $64(0) - 16(0)^2 = 0 > 48$. (FALSE)   Discard interval A.
For $t = 2$, $64(2) - 16(2)^2 = 128 - 64 = 64 > 48$. (TRUE)   Use interval B: (1, 3).
For $t = 4$, $64(4) - 16(4)^2 = 256 - 256 = 0 > 48$. (FALSE)   Discard interval C.

```
         A         B         C
    ←——+——+——(——+——)——+——+——→
      -1  0  1  2  3  4  5
```

5. The solution set is (1, 3), which means that the rocket is more than 48 ft high when the time t is between 1 and 3 sec.

5. Verify the solution. When $t = 1$, the height of the rocket is $h = 64(1) - 16(1)^2 = 48$, exactly 48 feet. If t is more than 1 and less than 3, the height is more than 48. Now on the way down, when $t = 3$, the height is $h = 64(3) - 16(3)^2 = 48$ ft again.

Calculator Corner

Solve a Polynomial Inequality

Your calculator simplifies the work in this section more than in any other topic we've encountered. To solve a polynomial inequality with a calculator,

1. Write the inequality with zero on the right side.
2. Graph the related equation.
3. The solution set consists of the intervals on the **horizontal** axis, the **x-axis,** for which the corresponding portions of the graph satisfy the inequality (above the x-axis if the inequality is > or ≥ and below the x-axis if the inequality is < or ≤).

Thus, to do Example 1, graph the related equation $Y_1 = (x - 1)(x + 3)$ in a standard window. The portion of the graph *below* the horizontal axis (less than zero) has corresponding x-values between -3 and 1. Hence, the solution set is as before, the interval $(-3, 1)$ (see Window 1).

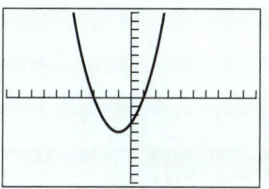

Window 1
$Y_1 = (x - 1)(x + 3)$

In Example 2, enter $Y_1 = x^2 - 6x + 7$. Enter $Y_2 = 0$, then press **2nd** **TRACE** **5** with **ENTER** **ENTER** **ENTER** to get the points at which the graph cuts the x-axis, which can be approximated by 1.59 and 4.41. Since we want the points for which the graph is ≥ 0, the solution set consists of the points in the union of intervals $(-\infty, 1.59) \cup (4.41, \infty)$ (see Window 2). Now you try Example 3.

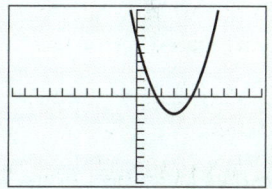

Window 2
$Y_1 = x^2 - 6x + 7$

For Example 4(b), graph

$$Y_1 = \frac{(4 - x)}{(x - 2)}$$

using a $[-5, 5]$ by $[-5, 5]$ window or, if you want to avoid some of the algebra, graph

$$Y_1 = \left(\frac{x}{(x - 2)}\right) - 2$$

an equivalent form of

$$\frac{(4 - x)}{(x - 2)}$$

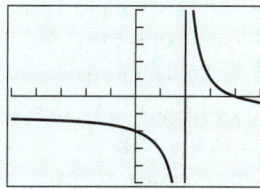

Window 3
$Y_1 = \frac{4 - x}{x - 2}$

In either case, be very careful with the placement of parentheses! Clearly, the solution set consists of the x-values whose corresponding y-values (vertical) are above the x-axis—the values between 2 and 4. The 2 is excluded because it will yield a zero denominator, so the solution set is as before, (2, 4] (see Window 3).

You can also graph

$$Y_1 = \frac{x}{(x - 2)}$$

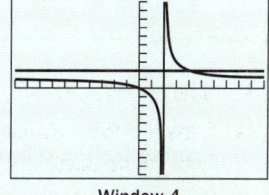

Window 4

and $Y_2 = 2$ and find the interval where Y_1 is above Y_2 as shown in Window 4.

Exercises 8.5

A **Solving Quadratic Inequalities**   In Problems 1–16, solve and graph the given inequality.

1. $(x + 1)(x - 3) > 0$
2. $(x - 1)(x + 2) < 0$
3. $x(x + 4) \leq 0$

4. $(x - 1)x \geq 0$
5. $x^2 - x - 2 \leq 0$
6. $x^2 - x - 6 \leq 0$

7. $x^2 - 3x \geq 0$
8. $x^2 + 2x \leq 0$
9. $x^2 - 3x + 2 < 0$

10. $x^2 - 2x - 3 > 0$
11. $x^2 + 2x - 3 < 0$
12. $x^2 + x - 2 < 0$

13. $x^2 + 10x \leq -25$
14. $x^2 + 8x \leq -16$
15. $x^2 - 8x \geq -16$

16. $x^2 - 6x \geq -9$

In Problems 17–20, use the quadratic formula to approximate the critical values.

17. $x^2 - x \geq 1$ (use 2.2 for $\sqrt{5}$)
18. $x^2 - x \geq 3$ (use 3.6 for $\sqrt{13}$)
19. $x^2 - x \leq 4$ (use 4.1 for $\sqrt{17}$)
20. $x^2 - x \leq 5$ (use 4.6 for $\sqrt{21}$)

B **Solving Polynomial Inequalities**   In Problems 21–24, solve and graph the inequality.

21. $(x + 1)(x - 2)(x + 3) \geq 0$
22. $(x - 1)(x + 2)(x - 3) \geq 0$

23. $(x - 1)(x - 2)(x - 3) \leq 0$
24. $(x - 2)(x - 3)(x - 4) \leq 0$

C **Solving Rational Inequalities**   In Problems 25–34, solve and graph the inequality.

25. $\dfrac{2}{x - 2} \geq 0$
26. $\dfrac{3}{x - 1} \leq 0$

27. $\dfrac{x+5}{x-1} > 0$

28. $\dfrac{2x-3}{x+3} < 0$

29. $\dfrac{3x-4}{2x-1} < 0$

30. $\dfrac{x-1}{x+5} < 1$

31. $\dfrac{x+5}{x-1} > 2$

32. $\dfrac{3x-1}{x+4} > 1$

33. $\dfrac{1}{x-1} < \dfrac{1}{x-2}$

34. $\dfrac{3}{x} + 1 < \dfrac{1}{x} - 2$

In Problems 35–38, find all values of x for which the given expression is a real number. (Hint: $\sqrt{a}$ is a real number if $a \geq 0$.)

35. $\sqrt{x^2 - 9}$

36. $\sqrt{x^2 - 4x + 4}$

37. $\sqrt{x^2 - 6x + 5}$

38. $\sqrt{3x - 8}$

⟨D⟩ Solving Applications Involving Inequalities

39. *Resistance* The equivalent resistance of two electric circuits is given by $R^2 - 3R + 1$. When is this resistance more than 5 ohms?

40. *Moment of a beam* The bending moment of a beam is given by $M = 20 - x^2$, where x is the distance, in feet, from one end of the beam. At what distances will the bending moment be more than 4?

41. *Water mites* The number N of water mites in a water sample depends on the temperature T in degrees Fahrenheit and is given by $N = 110T - T^2$. At what temperatures will the number of mites exceed 1000?

42. *Profit* The profit P in a business varies in an 8-hour day according to the formula $P = 15t - 5t^2$, where t is the time in hours. During what hours is there a profit; that is, during what hours is $P > 0$?

43. *Projectile* The height h (in ft) of a projectile is $h = 48t - 16t^2$, where t is the time in seconds. During what time interval will the projectile be more than 32 feet above the ground?

44. *Stopping distance* The distance d (in feet) in which a car traveling v miles per hour can be stopped is given by

$$d = 0.05v^2 + v$$

At what speeds will it take more than 120 feet to stop the car?

⟩⟩⟩ Using Your Knowledge

Stopping Jaguars In Problems 45–47, give your answers to the nearest tenth of a unit. We already know that the distance d (in feet) in which a car traveling v mi/hr can be stopped is given by

$$d = 0.05v^2 + v$$

45. For what values of v is $50 \leq d \leq 60$? (Hint: $\sqrt{11} \approx 3.32$ and $\sqrt{13} \approx 3.61$.)

46. For what values of v is $80 \leq d \leq 90$? (Hint: $\sqrt{17} \approx 4.12$ and $\sqrt{19} \approx 4.36$.)

47. According to the *Guinness Book of World Records*, the longest skid mark on a public road was made by a Jaguar automobile involved in an accident in England. The skid mark was 950 ft long. If you assume that the car actually stopped after the brakes were applied and that it traveled for 950 ft, how fast was the car traveling? (Use a calculator to solve this problem.)

>>> Write On

Explain in your own words:

48. Why the solution set of $(x - 1)^2 \geq 0$ is the set of all real numbers.

49. Why the solution set of $(x - 1)^2 \leq 0$ is the single point $\{1\}$.

50. Why the solution set of $(x - 1)^2 < 0$ is the empty set $\emptyset$.

51. Why the solution set of the rational inequality

$$\frac{ax + b}{cx + d} \leq 0$$

cannot include more than one end point (where a, b, and c are integers).

>>> Concept Checker

Fill in the blank(s) with the correct word(s), phrase, or mathematical statement.

52. In the inequality $(x + 3)(x - 2) < 0$, -3 and $+2$ are called the _____.

53. $x^2 + 3x \geq 28$ is an example of a _____.

54. When solving a rational inequality be sure a single rational expression is on the left side and _____ is on the right before finding the critical values.

quadratic inequality

linear inequality

critical values

rational inequality

zero

>>> Mastery Test

Solve and graph the solution set:

55. $\dfrac{3}{x - 2} \leq 0$

56. $\dfrac{x + 2}{x} > 0$

57. $(x + 3)(x - 2)(x + 1) \geq 0$

58. $(x - 4)(x + 3)(x - 1) \leq 0$

59. $x^2 + x \geq 2$

60. $x^2 + x \leq 6$

61. $(x + 3)(x - 2) < 0$

62. $(x + 1)(x - 3) \geq 0$

63. The profit P for a restaurant varies in an 8-hour day according to the formula $P = 3t - t^2$, where t is the time in hours. During what hours is there a profit; that is, during what hours is $P > 0$?

>>> **Skill Checker**

Find y for the given value of x:

64. $y = 3x + 6$, $x = 2$

65. $y = 3x + 6$, $x = -2$

66. $y = -\frac{2}{3}x + 4$, $x = 3$

67. $y = -\frac{2}{3}x + 4$, $x = -3$

Find x for the given value of y:

68. $y = 3x + 6$, $y = 0$

69. $y = 2x + 8$, $y = 0$

Find the x- and y-intercepts:

70. $5x - 3y = 15$

71. $-4y = x + 8$

Complete the square on the quadratic expression and then write the perfect square trinomial in factored form:

72. $x^2 - 6x$

73. $x^2 + 10x$

>Collaborative Learning 8A

Are the data in Graph 1 and Graph 2 related to a quadratic or linear function? Use Graph 1 and Graph 2 to answer the following questions.

1. Which graph (1 or 2) would best be modeled as a quadratic function?
2. Write the quadratic equation that would approximate this data.
3. Which graph (1 or 2) would best be modeled as a linear function?
4. Write the linear equation of the line of best fit.
5. Discuss how both graphs can indicate rainfall measures and yet have different shaped graphs and equation models.

Divide into groups and research the monthly rainfall in your area over the period of a year and then over the period of one month.

6. Make a table of your data, draw a line graph, and discuss your findings.
7. Discuss how your findings compare to Graph 1 and Graph 2.

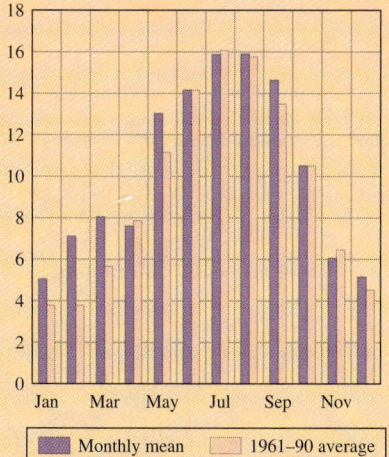

Graph 1

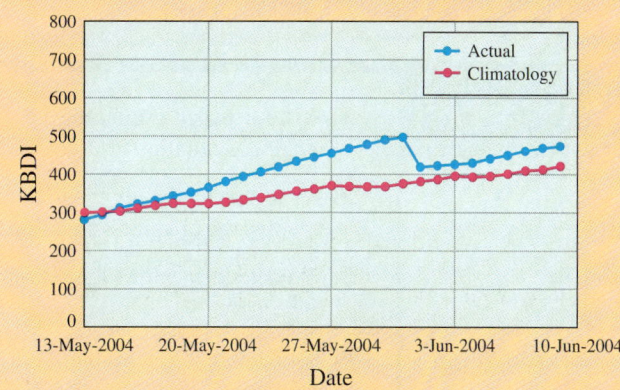

Graph 2

Collaborative Learning 8B

How does a few minutes of exercise affect your pulse rate? That depends on whether you are Mabel, a 36-year-old office worker, or Albert, a 26-year-old former college athlete. The graph indicates the pulse rate in beats per minute of the two, Mabel and Albert, over a 3-minute exercise period.

Divide into groups, answer the questions, and then conduct your own research regarding pulse rate and exercise with respect to age.

Exercise and Pulse Rate

1. Comparing pulse rates for Mabel and Albert,

   a. Whose pulse rate was the highest in beats per minute? What was the rate?

   b. Whose pulse rate was the lowest? What was the rate?

2. The graphs resemble those of a quadratic function. Write the equations for the two graphs based on the data in the following table. Using your calculator you can enter the list and calculate a quadratic regression. If you do not have a calculator, you can solve the system of three equations created by using $y = ax^2 + bx + c$ and entering each ordered pair as the respective x and y values each time.

Albert		Mabel	
Time in Minutes	Pulse Rate in Beats per Minute	Time in Minutes	Pulse Rate in Beats per Minute
1	80	1	62
3	120	3	150
5	62	5	80

3. Using the equations found in Question 2 for Albert and Mabel's pulse rates, what is the maximum of each parabola? How does that compare with the actual maximum point for each in the graph?

4. Measure the pulse rate in beats per minute of two people with significant age differences before, during, and after an exercise workout. Put the data in a table, graph it, and then discuss how it compares to Albert and Mabel's pulse rates during their workout. How do you account for any discrepancies?

Research Questions

1. Write a paper explaining how Euclid used theorems on areas to solve the quadratic equation $x^2 + b^2 = ax$.

2. Write a paper detailing Al-Khowarizmi's geometric demonstration of his rules to solve the quadratic equation $x^2 + 10x = 39$.

3. Why did they call Al-Khowarizmi's method of solving quadratics "completing the square"?

4. We have mentioned that Al-Khowarizmi was probably the foremost Arabic algebraist. The other one was dubbed "The Reckoner from Egypt." Who was this person, what did he write, and on what subjects?

> Summary Chapter 8

Section	Item	Meaning	Example
8.1	Quadratic equation in standard form $\pm\sqrt{a}$	An equation of the form $ax^2 + bx + c = 0 \ (a \neq 0)$ The solutions of $x^2 = a$	$2x^2 - 3x + 7 = 0$ is a quadratic equation in standard form. $\pm\sqrt{3}$ are the solutions of $x^2 = 3$.
8.1C	Completing the square	1. Write the equation with the variables in descending order on the left and the constants on the right. 2. Divide each term by the coefficient of $x^2 \ (\neq 1)$. 3. Add the square of one-half the coefficient of x to both sides. 4. Factor the perfect square trinomial on the left; simplify on the right. 5. Solve the resulting equation.	$5x^2 - 5x - 1 = 0$ Given. 1. $5x^2 - 5x = 1$ 2. $x^2 - x = \frac{1}{5}$ 3. $x^2 - x + \left(-\frac{1}{2}\right)^2 = \frac{1}{5} + \left(-\frac{1}{2}\right)^2$ 4. $\left(x - \frac{1}{2}\right)^2 = \frac{9}{20}$ 5. $x - \frac{1}{2} = \pm\sqrt{\frac{9}{20}}$ $x - \frac{1}{2} = \pm\frac{3\sqrt{5}}{10}$ $x = \frac{1}{2} \pm \frac{3\sqrt{5}}{10} = \frac{5 \pm 3\sqrt{5}}{10}$
8.2A	Quadratic formula	The solutions of $ax^2 + bx + c = 0$ are $x = \frac{-b \pm \sqrt{b^2 - 4ac}}{2a}$	The solutions of $3x^2 + 2x - 5 = 0$ are $\frac{-2 \pm \sqrt{4 + 60}}{6}$ which equal 1 and $-\frac{5}{3}$.
8.3	Discriminant	The discriminant of $ax^2 + bx + c = 0$ is $D = b^2 - 4ac$.	The discriminant of $3x^2 + 2x - 5 = 0$ is $b^2 - 4ac = 64$.
8.3A	Types of solutions for $ax^2 + bx + c = 0$ where a, b, and c are rational numbers	1. If $D > 0$ and D is not a perfect square, two different real, irrational solutions 2. If $D > 0$ and $D = N^2$ (a perfect square), two different real, rational solutions 3. If $D < 0$, two different non-real complex solutions 4. If $D = 0$, one rational solution	1. $4x^2 - 3x - 5 = 0 \ (D = 89)$ has two different irrational solutions. 2. $x^2 - 2x - 3 = 0 \ (D = 16)$ has two different rational solutions. 3. $4x^2 - 3x + 5 = 0 \ (D = -71)$ has two different non-real solutions. 4. $4x^2 - 4x + 1 = 0 \ (D = 0)$ has one rational solution.

(continued)

Section	Item	Meaning	Example
8.3B	Factorable quadratic expressions	If D is a perfect square, $ax^2 + bx + c$ is factorable.	$20x^2 + 10x - 30$ is factorable ($D = 2500$). $20x^2 + 10x - 30 = 10(2x + 3)(x - 1)$
8.3D	Sum and product of the solutions	The sum and product of the solutions of $ax^2 + bx + c = 0$ are $-\frac{b}{a}$ and $\frac{c}{a}$, respectively.	The sum and product of the solutions of $4x^2 - 12x + 5 = 0$ are 3 and $\frac{5}{4}$, respectively.
8.4B	Equations that are quadratic in form	Equations that can be written as quadratics by use of appropriate substitutions	$x^4 - 5x^2 + 4 = 0$ $\Rightarrow u^2 - 5u + 4 = 0$ $(x^2 - x)^2 - (x^2 - x) - 2 = 0$ $\Rightarrow u^2 - u - 2 = 0$
8.5B	Polynomial inequality	An inequality that can be written with a polynomial on the left side and zero on the right side. The symbol can be $<, >, \leq,$ or $\geq$.	$x^2 - x - 2 < 0$ $(x + 1)(x - 2) < 0$ Critical numbers: $-1, 2$ Solution: $(-1, 2)$
8.5C	Rational inequality	An inequality that can be written with a rational expression on the left side and zero on the right side. The symbol can be $<, >, \leq,$ or $\geq$.	$\frac{2x}{x - 5} > 0$ Critical numbers: $0, 5$ Solution: $(-\infty, 0) \cup (5, \infty)$

❯ Review Exercises Chapter 8

(If you need help with these exercises, look in the section indicated in brackets.)

1. ❮ **8.1A** ❯ Solve.
   a. $16x^2 - 49 = 0$
   b. $25x^2 - 16 = 0$

2. ❮ **8.1A** ❯ Solve.
   a. $5x^2 + 30 = 0$
   b. $6x^2 + 42 = 0$

3. ❮ **8.1B** ❯ Solve.
   a. $(x - 3)^2 = 32$
   b. $(x - 5)^2 = 50$

4. ❮ **8.1B** ❯ Solve.
   a. $2(x - 2)^2 + 25 = 0$
   b. $3(x - 3)^2 + 64 = 0$

5. ❮ **8.1B** ❯ Solve.
   a. $5x^2 - 10x + 5 = 12$
   b. $12x^2 + 12x + 3 = 16$

6. ❮ **8.1C** ❯ Solve by completing the square.
   a. $x^2 - 8x - 9 = 0$
   b. $x^2 + 12x + 32 = 0$

7. ❮ **8.1C** ❯ Solve by completing the square.
   a. $4x^2 + 4x - 3 = 0$
   b. $16x^2 - 24x + 7 = 0$

8. ❮ **8.2A** ❯ Write the equation in standard form and name the values of a, b, and c.
   a. $x = 7x^2 - 4$
   b. $\frac{x^2}{5} = \frac{1}{10} - 9x$

9. ❮ **8.2A** ❯ Solve by the quadratic formula.
   a. $3x^2 = 2x + 4$
   b. $4x^2 = 6x + 3$

10. ⟨**8.2A**⟩ *Solve by the quadratic formula.*
   a. $16x = x^2$
   b. $12x = x^2$

11. ⟨**8.2A**⟩ *Solve by the quadratic formula.*
   a. $x^2 + \dfrac{x}{15} = \dfrac{1}{3}$
   b. $\dfrac{x^2}{2} + \dfrac{9x}{10} = \dfrac{1}{5}$

12. ⟨**8.2A**⟩ *Solve by the quadratic formula.*
   a. $3x^2 - 2x = -1$
   b. $5x^2 - 2x = -4$

13. ⟨**8.2B**⟩ *Solve.*
   a. $8x^3 - 125 = 0$
   b. $125x^3 - 8 = 0$

14. ⟨**8.2C**⟩ *The demand d is given by*
$$d = \dfrac{450}{p}$$
   and the supply s is given by $s = 100p - 150$.
   a. Find the equilibrium point (at which $d = s$).
   b. Find the equilibrium point when
$$d = \dfrac{50}{p}$$
   and $s = 150p - 100$.

15. ⟨**8.3A**⟩
   a. Given $16x^2 - 3x = -1$, find the value of the discriminant and classify the solutions.
   b. Given $8x^2 - kx + 2 = 0$, find the discriminant and then find k so that the equation has exactly one rational solution.

16. ⟨**8.3B**⟩ *Use the discriminant to determine whether the given quadratic is factorable into factors with integer coefficients. If so, factor it.*
   a. $3x^2 - 11x - 6$
   b. $18x^2 + 13x + 2$

17. ⟨**8.3B**⟩ *Factor into factors with integer coefficients if possible.*
   a. $18x^2 - 9x - 5$
   b. $18x^2 + 13x + 1$

18. ⟨**8.3C**⟩ *Find a quadratic equation with integer coefficients whose solution set is:*
   a. $\{-2, 3\}$
   b. $\left\{\dfrac{1}{4}, -\dfrac{2}{3}\right\}$

19. ⟨**8.3D**⟩ *Without solving the equation, find the sum, using $-\dfrac{b}{a}$, and the product, using $\dfrac{c}{a}$, of the solutions of:*
   a. $15x^2 + 4x - 3 = 0$
   b. $9x^2 - 12x - 5 = 0$

20. ⟨**8.3D**⟩ *Use the sum and product properties to check whether:*
   a. $\dfrac{1}{3}$ and $-\dfrac{3}{5}$ are the solutions of the equation in Problem 19, part **a**.
   b. $\dfrac{1}{3}$ and $-\dfrac{5}{3}$ are the solutions of the equation in Problem 19, part **b**.

21. ⟨**8.4A**⟩ *Solve.*
   a. $\dfrac{8}{x^2 - 16} - \dfrac{1}{x - 4} = 1$
   b. $\dfrac{-24}{x^2 - 36} + \dfrac{2}{x + 6} = 1$

22. ⟨**8.4B**⟩ *Solve.*
   a. $(x^2 + x)^2 + 2(x^2 + x) - 8 = 0$
   b. $(x^2 - 3x)^2 - 4(x^2 - 3x) - 12 = 0$

23. ⟨**8.4B**⟩ *Solve.*
   a. $x^{1/2} - 4x^{1/4} = -3$
   b. $x^{1/2} + x^{1/4} = 6$

24. ⟨**8.4B**⟩ *Solve.*
   a. $x^{2/3} + x^{1/3} = 12$
   b. $x^{2/3} - 5x^{1/3} = 6$

25. ⟨**8.4B**⟩ *Solve.*
   a. $x - 2\sqrt{x} = 3$
   b. $x - 4\sqrt{x} = 5$

26. ⟨**8.4B**⟩ *Solve.*
   a. $3x^{-4} - 4x^{-2} + 1 = 0$
   b. $3x^{-4} - 2x^{-2} - 1 = 0$

For Exercises 27–32, write the solutions in interval notation.

27. ⟨**8.5A**⟩ *Solve and graph.*
  a. $(x-2)(x+3) < 0$

  b. $(x+2)(x-3) < 0$

28. ⟨**8.5A**⟩ *Solve and graph.*
  a. $x^2 + 4x \geq 0$

  b. $x^2 - 3x \geq 0$

29. ⟨**8.5A**⟩ *Solve and graph.*
  a. $x^2 + 4x < 45$

  b. $x^2 - 2x < 2$

30. ⟨**8.5B**⟩ *Solve and graph.*
  a. $(x-1)(x-2)(x-3) \leq 0$

  b. $(x+1)(x+2)(x-3) \leq 0$

31. ⟨**8.5B**⟩ *Solve and graph.*
  a. $(x+1)(x+2)(x-3) > 0$

  b. $(x-1)(x+2)(x+3) > 0$

32. ⟨**8.5C**⟩ *Solve and graph.*
  a. $\dfrac{x+2}{x-2} \leq 0$

  b. $\dfrac{-12}{x-6} < 3$

Practice Test Chapter 8

(Answers on page 654)

Visit www.mhhe.com/bello to view helpful videos that provide step-by-step solutions to several of the problems below.

1. Solve.
   a. $25x^2 - 4 = 0$
   b. $18x^2 + 3 = 0$

2. Solve.
   a. $(x - 1)^2 = 45$
   b. $2(x - 2)^2 = 49$

3. Solve $3x^2 + 6x + 3 = 5$.

4. Solve by completing the square: $x^2 - 6x + 5 = 0$.

5. Solve by completing the square: $9x^2 - 6x - 1 = 0$.

6. Solve by the quadratic formula: $7x^2 + 5x - 2 = 0$.

7. Solve by the quadratic formula: $3x^2 = 3x + 2$.

8. Solve by the quadratic formula: $32x = x^2$.

9. Solve by the quadratic formula:
$$\frac{x^2}{16} + \frac{5x}{4} = \frac{11}{4}$$

10. Solve by the quadratic formula: $4x^2 + 3x = -2$.

11. Solve $64x^3 - 27 = 0$.

12. If the demand d is given by
$$d = \frac{500}{p}$$
and the supply s is given by $s = 200p - 150$, find the equilibrium point (at which $s = d$).

13. a. Given $4x^2 - 5x = -6$, find the value of the discriminant and classify the solutions.
    b. Given $4x^2 - kx + 9 = 0$, find k so that the equation has exactly one rational solution.

14. Use the discriminant to determine whether $12x^2 - 4x - 21$ is factorable into factors with integer coefficients.

15. Factor $35x^2 - 32x - 12$ into factors with integer coefficients if possible.

16. Find a quadratic equation with integer coefficients whose solution set is $\{\frac{2}{3}, -\frac{1}{4}\}$.

17. a. Find the sum and the product of the solutions of the equation $6x^2 + 7x - 5 = 0$.
    b. Use the sum and product properties to check whether the solutions are $-\frac{5}{3}$ and $\frac{1}{2}$.

18. Solve.
    a. $\dfrac{6}{x^2 - 9} - \dfrac{1}{x - 3} = 1$
    b. $\dfrac{2}{x^2 - 1} - \dfrac{3}{x - 1} = 1$

19. Solve.
    a. $x^4 - 7x^2 + 6 = 0$
    b. $(x^2 - 2x)^2 - (x^2 - 2x) - 6 = 0$

20. Solve.
    a. $x^{1/2} - 3x^{1/4} + 2 = 0$
    b. $x - 5\sqrt{x} + 6 = 0$

21. Solve $3x^{-4} - 2x^{-2} - 1 = 0$.

For Problems 22–25, write the solutions in interval notation.

22. Solve and graph $(x + 1)(x - 2) < 0$.

23. Solve and graph $x^2 + 2x \geq 15$.

24. Solve and graph $(x + 1)(x - 2)(x - 3) \leq 0$.

25. Solve and graph $\dfrac{x + 4}{x - 7} \leq 0$.

Answers to Practice Test Chapter 8

Answer	If You Missed Question	Review Section	Examples	Page
1. a. $\frac{2}{5}, -\frac{2}{5}$ b. $\frac{\sqrt{6}}{6}i, -\frac{\sqrt{6}i}{6}$	1	8.1A	1	594–595
2. a. $1 \pm 3\sqrt{5}$ b. $2 \pm \frac{7\sqrt{2}}{2}$ or $\frac{4 \pm 7\sqrt{2}}{2}$	2	8.1B	2	595–596
3. $-1 \pm \frac{\sqrt{15}}{3}$ or $\frac{-3 \pm \sqrt{15}}{3}$	3	8.1B	3	596–597
4. $1, 5$	4	8.1C	4	598
5. $\frac{1 \pm \sqrt{2}}{3}$	5	8.1C	5	599
6. $\frac{2}{7}, -1$	6	8.2A	2	607
7. $\frac{3 \pm \sqrt{33}}{6}$	7	8.2A	3	608
8. $0, 32$	8	8.2A	4	609
9. $2, -22$	9	8.2A	5	609–610
10. $-\frac{3}{8} \pm \frac{\sqrt{23}}{8}i$	10	8.2A	6	610
11. $\frac{3}{4}, -\frac{3}{8} \pm \frac{3\sqrt{3}}{8}i$	11	8.2B	7	612
12. When price is $2	12	8.2C	8	613
13. a. -71; two non-real complex solutions b. $k = \pm 12$	13	8.3A	1	619
14. Yes, $D = 32^2$	14	8.3B	2	620
15. $(5x - 6)(7x + 2)$	15	8.3B	3	621
16. $12x^2 - 5x - 2 = 0$	16	8.3C	4	621
17. a. Sum: $-\frac{7}{6}$; product: $-\frac{5}{6}$ b. Yes	17	8.3D	5	622–623
18. a. -4 b. $0, -3$	18	8.4A	1	627
19. a. $\pm 1, \pm \sqrt{6}$ b. $3, -1, 1 \pm i$	19	8.4B	2, 3	628–629
20. a. $1, 16$ b. $4, 9$	20	8.4B	4, 5	630–631
21. $\pm 1, \pm \sqrt{3}i$	21	8.4B	6	631
22. $(-1, 2)$	22	8.5A	1	636
23. $(-\infty, -5] \cup [3, \infty)$	23	8.5A	2	637
24. $(-\infty, -1] \cup [2, 3]$	24	8.5B	3	639
25. $[-4, 7)$	25	8.5C	4	640–641

Cumulative Review Chapters 1–8

1. Use braces to list the elements of the set of even natural numbers less than 14.

2. Write $\frac{11}{15}$ as a decimal.

3. Find: $|-11|$

4. Evaluate: $[-3(7 + 6)] + 7$

5. Solve: $x + 4 = 2(5x - 3)$

6. Graph $\{x \mid x > -1 \text{ and } x < 1\}$

7. Graph: $|3x + 1| > 2$

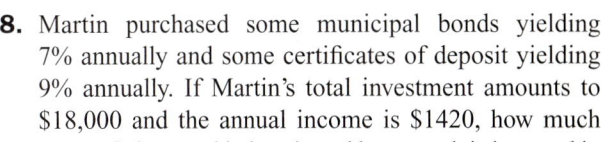

8. Martin purchased some municipal bonds yielding 7% annually and some certificates of deposit yielding 9% annually. If Martin's total investment amounts to $18,000 and the annual income is $1420, how much money is invested in bonds and how much is invested in certificates of deposit?

9. Graph: $2x - y = 2$

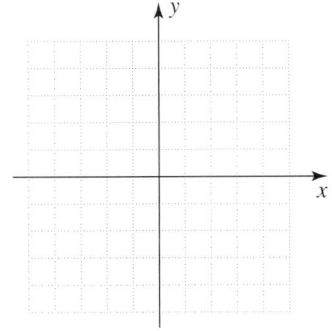

10. A line L_1 has slope $\frac{3}{11}$. Find whether the line through $(-7, 5)$ and $(-4, -6)$ is parallel or perpendicular to line L_1.

11. Find an equation of the line with slope -1 and passing through the point $(-3, 2)$.

12. Find an equation of the line that passes through the point $(-6, 4)$ and is parallel to the line $8x - y = 3$.

13. Find the domain and range of $\{(x, y) \mid y = 5 + x\}$.

14. Use the graphical method to solve the system:
$x + y \leq -1$
$y \leq 3x - 5$

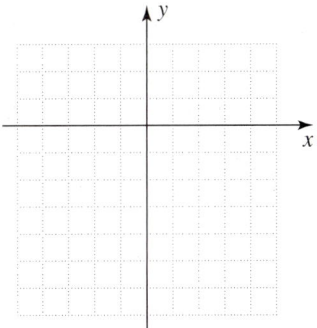

15. Solve by the elimination method:
$3x + 4y = -1$
$x + y = -1$

16. Solve:
$x + 2y = 6$
$4x + 2y = 7$

17. Jose has $3.20 in nickels and dimes. He has 10 more nickels than dimes. How many of each coin does he have?

18. Subtract $6x^3 - 6x^2 + 4x - 9$ from $4x^3 + 3x^2 + 9$.

19. Factor: $36x^2 + 60xy + 25y^2$

20. Factor: $2x^4 + 8x^2$

21. Solve $x^3 + x^2 - 9x - 9 = 0$

22. Divide: $\dfrac{-12x^7 + 12x^3 + 6x^2}{6x^5}$

23. Perform the indicated operations: $\dfrac{x}{x^2 - 81} - \dfrac{9}{x^2 - 81}$

24. Simplify: $\dfrac{x - \dfrac{1}{x^2}}{x - \dfrac{1}{x^3}}$

25. Solve $1 - \dfrac{1}{x + 1} = \dfrac{2}{x^2 - 1}$

26. Find two consecutive even integers such that the sum of their reciprocals is $\dfrac{9}{40}$.

27. Find, if possible in the real numbers: $(-16)^{1/2}$

28. Simplify: $\sqrt[3]{192a^7b^9}$

29. Rationalize the denominator: $\sqrt{\dfrac{5}{6k}}$

30. Perform the indicated operations:
$\sqrt[3]{3x}\left(\sqrt[3]{9x^2} + \sqrt[3]{16x}\right)$

31. Solve: $\sqrt{x + 8} - x = 8$

32. Find: $(1 - 4i) - (4 + i)$

33. Multiply: $(-1 - 8i)(-4 - 3i)$

34. Solve $16x^2 - 9 = 0$

35. Solve by completing the square: $x^2 + 2x - 3 = 0$

36. Solve by the quadratic formula: $x^2 = -4x + 2$

37. If the demand (d) is given by $d = \dfrac{480}{p}$ and the supply is given by $s = 200p - 680$, find the equilibrium point (at which $s = d$).

38. Determine whether $12x^2 + 7x - 10$ is factorable into factors with integer coefficients.

39. Solve $x^4 - 8x^2 + 7 = 0$

40. Solve $(x + 6)(x - 4) < 0$

Chapter 9: Quadratic Functions and the Conic Sections

Sections

- 9.1 Quadratic Functions (Parabolas) and Their Graphs
- 9.2 Circles and Ellipses
- 9.3 Hyperbolas and Identification of Conics
- 9.4 Nonlinear Systems of Equations
- 9.5 Nonlinear Systems of Inequalities

The Human Side of Algebra

Why do we study conic sections? The answer is a tangled tale based in antiquity. The first 300 years of Greek mathematics yielded three famous problems: the squaring of the circle, the trisection of an angle, and the duplication of the cube. The third problem was first mentioned by a Greek poet describing mythical king Minos as dissatisfied with the size of the tomb erected for his son Glaucus. "You have embraced too little space; quickly double it without spoiling its beautiful cubical form." When the construction was done incorrectly, legend says, local geometers were hastily summoned to solve the problem. A different version has the Athenians appealing to the oracle at Delos to get rid of pestilence. "Apollo's cubical altar must be doubled in size" was the reply. When workmen did the work, again incorrectly, "the indignant god made the pestilence even worse than before."

Hippocrates of Chios (ca. 440–380 B.C.) finally showed that the duplication of the cube could be done by using curves with certain properties, but it was the Greek mathematician Menaechmus (350 B.C.), the tutor of Alexander the Great, reputed to have discovered the curves later known as the ellipse, the parabola, and the hyperbola (the conics) who solved the problem.

9.1 Quadratic Functions (Parabolas) and Their Graphs

Objectives

A Graph a parabola of the form $y = f(x) = ax^2 + k$.

B Graph a parabola of the form $y = f(x) = a(x - h)^2 + k$.

C Graph a parabola of the form $y = f(x) = ax^2 + bx + c$.

D Graph parabolas that are not functions with form $x = a(y - k)^2 + h$ and $x = ay^2 + by + c$.

E Solve applications involving parabolas.

To Succeed, Review How To . . .

1. Graph points in the Cartesian coordinate system (pp. 166–168).
2. Find x- and y-intercepts (pp. 171–173).
3. Use the quadratic formula (pp. 606–610).
4. Complete the square in a quadratic equation (pp. 597–599).
5. Find the discriminant of a quadratic equation (pp. 618–619).

Getting Started

The Fountain of Parabolas

Have you seen any parabolas lately? The streams of water in fountains follow the path of a quadratic function called a parabola. Parabolas, ellipses, circles, and hyperbolas are called **conic sections** because they can be obtained by intersecting (slicing) a cone with a plane, as shown in Figure 9.1. As you will see later, these conic sections occur in many practical applications. For example, your satellite dish, your flashlight lens, and your telescope lens have a parabolic shape, while comets travel in orbits that are either elliptical (those may be seen from Earth more than once) or hyperbolic (once in a lifetime viewing!). We shall begin our study of conic sections by examining the parabola.

The streams of water are in the shapes of a parabola.

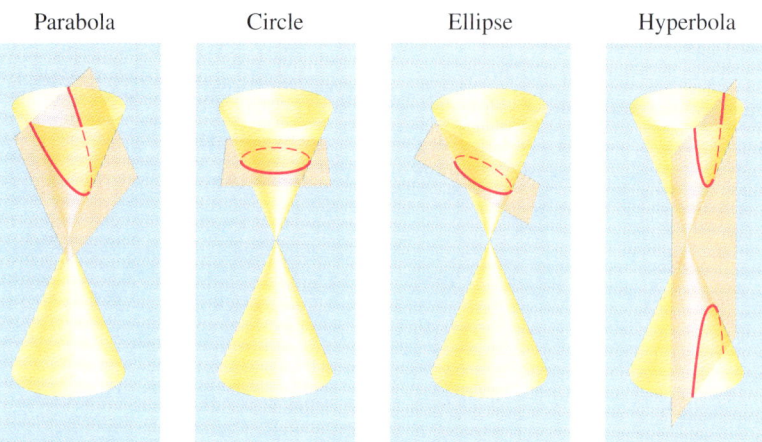

> Figure 9.1

A › Graphing a Parabola of the Form $y = f(x) = ax^2 + k$

In Chapter 3, we studied *linear functions* like $f(x) = 2x + 5$ and $g(x) = -3x - 4$ whose graphs were straight lines. In this chapter, we will study equations (functions) defined by quadratic (second-degree) polynomials of the form

$$f(x) = ax^2 + bx + c$$

These functions are **quadratic functions** and their graphs are **parabolas**.

Just as the "simplest" *line* to graph is $y = f(x) = x$, the "simplest" *parabola* to graph is $y = f(x) = x^2$. To draw this graph, we select values for x and find the corresponding values of y:

x-Value	f(x) = y-Value	Ordered Pair (x, y)
$x = -2$	$f(-2) = (-2)^2 = 4$	$(-2, 4)$
$x = -1$	$f(-1) = (-1)^2 = 1$	$(-1, 1)$
$x = 0$	$f(0) = (0)^2 = 0$	$(0, 0)$
$x = 1$	$f(1) = (1)^2 = 1$	$(1, 1)$
$x = 2$	$f(2) = (2)^2 = 4$	$(2, 4)$

Then we make a table of ordered pairs, plot the ordered pairs on a coordinate system, and draw a smooth curve through the plotted points as in Figure 9.2.

An important feature of this parabola is its **symmetry** with respect to the y-axis. If you folded the graph of $y = x^2$ along the y-axis, the two halves of the graph would coincide because the same value of y is obtained for any value of x and its opposite $-x$. The points on one side of the axis of symmetry are mirror images of the points on the other side. For instance, $x = 2$ and $x = -2$ both give $y = 4$. (See the preceding table.) Because of this symmetry, the y-axis is called the **axis of symmetry** or the **axis** of the parabola. The point $(0, 0)$, where the parabola crosses its axis, is called the **vertex** of the curve. The arrows on the curve in Figure 9.2 mean that the parabola goes on without end.

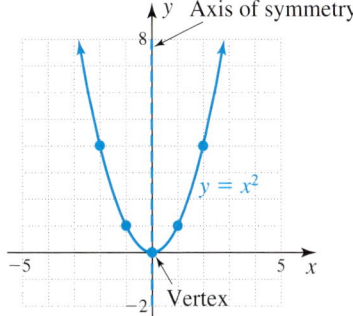
›Figure 9.2

EXAMPLE 1 Graphing a parabola that opens downward
Graph the function and name the coordinates of the vertex: $y = f(x) = -x^2$

SOLUTION   We could make a table of x- and y-values as before. However, for any x-value, the y-value will be the *negative* of the y-value on the parabola $y = x^2$. (If you don't believe this, go ahead and make the table and check it, but it's easier to copy the table for $y = x^2$ with the negatives of the y-values entered as shown.) Thus, the parabola $y = -x^2$ has the same shape as $y = x^2$, but it is turned in the *opposite* direction (opens *downward*). The graph of $y = -x^2$ is shown in Figure 9.3, and the vertex $(0, 0)$ is the highest point on the parabola.

x	y	(x, y)
-2	-4	$(-2, -4)$
-1	-1	$(-1, -1)$
0	0	$(0, 0)$
1	-1	$(1, -1)$
2	-4	$(2, -4)$

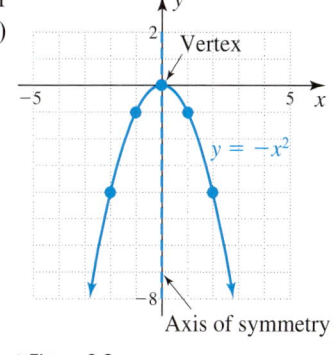
›Figure 9.3

PROBLEM 1
Graph the function and name the coordinates of the vertex: $y = f(x) = -2x^2$

Answers to PROBLEMS

1.
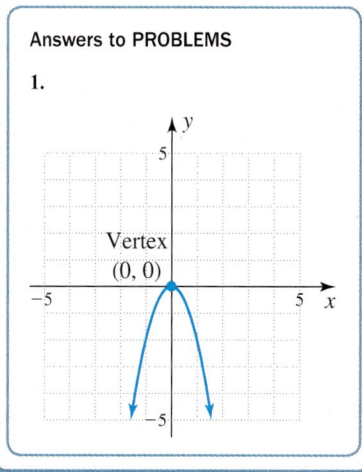

As you can see from the preceding examples, when the coefficient of x^2 is positive (as in $y = x^2 = 1x^2$), the parabola opens upward (is **concave up**), but when the coefficient of x^2 is negative (as in $y = -x^2 = -1x^2$), the parabola opens downward (is **concave down**). In either case, the vertex is at (0, 0). To see the effect of a in $f(x) = ax^2$ in general, let's plot some points and see how the graphs of $g(x) = 2x^2$ and $h(x) = \frac{1}{2}x^2$ compare to the graph of $f(x) = x^2$. All three graphs are shown in Figure 9.4.

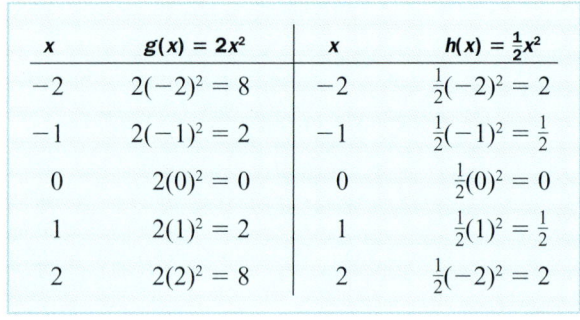

x	$g(x) = 2x^2$	x	$h(x) = \frac{1}{2}x^2$
-2	$2(-2)^2 = 8$	-2	$\frac{1}{2}(-2)^2 = 2$
-1	$2(-1)^2 = 2$	-1	$\frac{1}{2}(-1)^2 = \frac{1}{2}$
0	$2(0)^2 = 0$	0	$\frac{1}{2}(0)^2 = 0$
1	$2(1)^2 = 2$	1	$\frac{1}{2}(1)^2 = \frac{1}{2}$
2	$2(2)^2 = 8$	2	$\frac{1}{2}(-2)^2 = 2$

>Figure 9.4

The graph of $g(x) = 2x^2$ is narrower than that of $f(x) = x^2$, while the graph of $h(x) = \frac{1}{2}x^2$ is wider. The vertex and line of symmetry is the same for the three curves. In general, we have the following.

PROPERTIES OF THE PARABOLA $g(x) = ax^2$

The graph of $g(x) = ax^2$ is a parabola with the vertex at the origin and the y-axis as its line of symmetry.

If a is *positive*, the parabola opens *upward*, if a is *negative*, the parabola opens *downward*.

If $|a|$ is greater than 1 ($|a| > 1$), the parabola is narrower than the parabola $f(x) = x^2$.

If $|a|$ is between 0 and 1 ($0 < |a| < 1$), the parabola is wider than the parabola $f(x) = x^2$.

Using this information, you can draw the graph of any parabola of the form $g(x) = ax^2$, as we illustrate in Example 2.

EXAMPLE 2 Graphing a parabola of the form $y = ax^2$

Graph the functions and name the coordinates of the vertices:

a. $f(x) = 3x^2$ **b.** $g(x) = -3x^2$ **c.** $h(x) = \frac{1}{3}x^2$

SOLUTION

a. By looking at the properties of the parabola $g(x) = ax^2$, we know that the vertex of the parabola $f(x) = 3x^2$ is at the origin (0, 0) and that the y-axis is its line of symmetry. Since $3 > 0$, we also know that the parabola $f(x) = 3x^2$ opens upward and because $3 > 1$ the graph is narrower than the parabola $y = x^2$. We pick three easy points to complete our graph, as shown in Figure 9.5.

x	$y = 3x^2$	(x, y)
-1	$y = 3(-1)^2 = 3$	$(-1, 3)$
0	$y = 3(0)^2 = 0$	$(0, 0)$ ← Vertex
1	$y = 3(1)^2 = 3$	$(1, 3)$

We picked three points, but we could have picked the first two points and obtained the third by using the principles of symmetry. There must be a point on the other side of $x = 0$ that is 1 unit away and on the same horizontal line as $(-1, 3)$. Thus, the mirror-image point of $(-1, 3)$ is **(1, 3)**.

PROBLEM 2

Graph the functions and name the coordinates of the vertices:

a. $f(x) = -4x^2$

b. $g(x) = 4x^2$

c. $h(x) = \frac{1}{4}x^2$

Answer on page 661

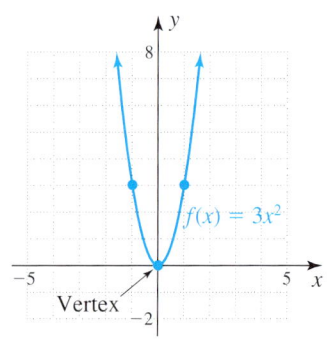

>Figure 9.5

b. The parabola $g(x) = -3x^2$ opens downward but is again narrower than the parabola $y = x^2$. The parabola $g(x) = -3x^2$ is the reflection of the parabola $f(x) = 3x^2$ across the x-axis. Again, we pick three points to complete the graph, shown in Figure 9.6.

x	$y = -3x^2$	(x, y)
-1	$y = -3(-1)^2 = -3$	$(-1, -3)$
0	$y = -3(0)^2 = 0$	$(0, 0)$ ← Vertex
1	$y = -3(1)^2 = -3$	$(1, -3)$

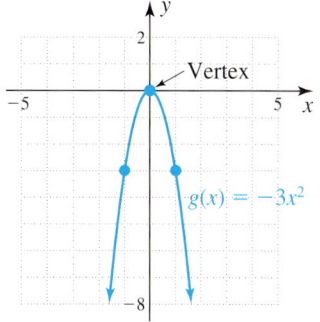
>Figure 9.6

We picked three points, but we could have picked the first two points and obtained the third by using the principles of symmetry. There must be a point on the other side of $x = 0$ that is 1 unit away and on the same horizontal line as $(-1, -3)$. Thus, the mirror-image point of $(-1, -3)$ is **$(1, -3)$**.

c. The parabola $h(x) = \frac{1}{3}x^2$ opens upward, since $\frac{1}{3} > 0$, but is wider than the parabola $y = x^2$. This time, instead of selecting $x = -1$, 0, and 1, we select $x = -3$, 0, and 3 for ease of computation; the completed graph is shown in Figure 9.7.

x	$y = \frac{1}{3}x^2$	(x, y)
-3	$y = \frac{1}{3}(-3)^2 = 3$	$(-3, 3)$
0	$y = \frac{1}{3}(0)^2 = 0$	$(0, 0)$ ← Vertex
3	$y = \frac{1}{3}(3)^2 = 3$	$(3, 3)$

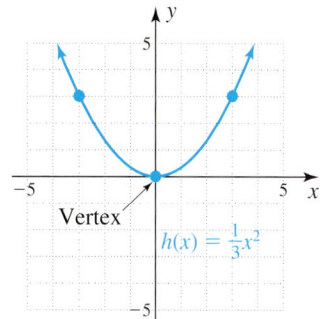
>Figure 9.7

We picked three points, but we could have picked the first two points and obtained the third by using the principles of symmetry. There must be a point on the other side of $x = 0$ that is 3 units away and on the same horizontal line as $(-3, 3)$. Thus, the mirror-image point of $(-3, 3)$ is **$(3, 3)$**.

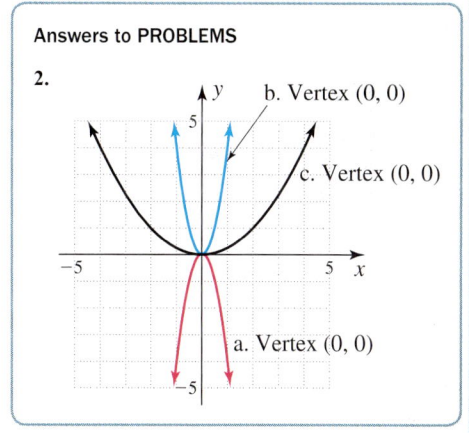

Answers to PROBLEMS

2. a. Vertex $(0, 0)$ b. Vertex $(0, 0)$ c. Vertex $(0, 0)$

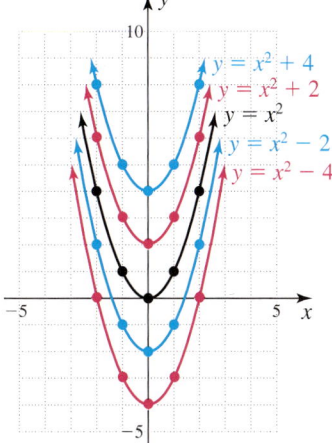
>Figure 9.8

What do you think will happen if we graph the parabola $y = x^2 + 2$? Two things: first, the parabola opens upward because the coefficient of x^2 is understood to be $+1$. Second, all of the points will be 2 units higher than those for the same value of x on the parabola $y = x^2$. Thus, we can make the graph of $y = x^2 + 2$ by following the pattern of $y = x^2$. The graphs of $y = x^2 + 2$, $y = x^2 + 4$, $y = x^2 - 2$, and $y = x^2 - 4$ are shown in Figure 9.8. The points used to make the graphs are listed in the following table.

For $y = x^2 + 2$		For $y = x^2 + 4$		For $y = x^2 - 2$		For $y = x^2 - 4$	
x	y	x	y	x	y	x	y
Vertex → 0	2	0	4	0	-2	0	-4
± 1	3	± 1	5	± 1	-1	± 1	-3
± 2	6	± 2	8	± 2	2	± 2	0

Adding or subtracting a positive number k on the right-hand side of the equation $y = x^2$ raises or lowers the graph (and the vertex) by k units.

EXAMPLE 3 Graphing a parabola opening downward

Graph the function and name the coordinates of the vertex: $y = -x^2 - 2$

SOLUTION Since the coefficient of x^2 (understood to be -1) is negative, the parabola opens downward. It is also 2 units lower than the graph of $y = -x^2$. Thus, the graph of $y = -x^2 - 2$ is a parabola opening downward with its vertex at $(0, -2)$. Letting $x = 1$, we get $y = -3$ and for $x = 2$, $y = -6$. Graph the two points $(1, -3)$ and $(2, -6)$ and, by symmetry, the points $(-1, -3)$ and $(-2, -6)$. The parabola passing through all these points is shown in Figure 9.9.

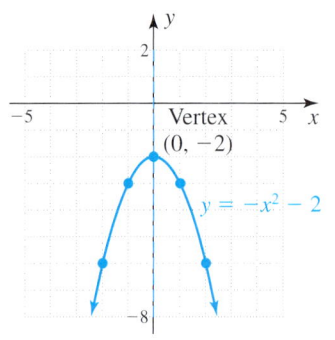

>Figure 9.9

PROBLEM 3

Graph the function and name the coordinates of the vertex:

$y = -x^2 - 1$

B > Graphing a Parabola of the Form $y = f(x) = a(x - h)^2 + k$

So far, we have graphed only parabolas of the form $y = ax^2 + k$. What do you think the graph of $y = (x - 1)^2$ looks like? As before, we make a table of values.

x	$y = (x - 1)^2$
$x = -1$,	$y = (-1 - 1)^2 = (-2)^2 = 4$
$x = 0$,	$y = (0 - 1)^2 = (-1)^2 = 1$ ←y-intercept
$x = 1$,	$y = (1 - 1)^2 = (0)^2 = 0$ ←Vertex
$x = 2$,	$y = (2 - 1)^2 = 1^2 = 1$
$x = 3$,	$y = (3 - 1)^2 = 2^2 = 4$

or

x	y
-1	4
0	1
1	0
2	1
3	4

The graph appears in Figure 9.10.

The shape of the graph is identical to that of $y = x^2$ but it is shifted 1 unit to the *right*. Thus, the vertex is at $(1, 0)$ and the axis of symmetry is as shown in Figure 9.10.

Similarly, the graph of $y = -(x + 1)^2$ is identical to that of $y = -x^2$ but shifted 1 unit to the *left*. Thus, the vertex is at $(-1, 0)$ and the axis of symmetry is as shown in Figure 9.11. Some easy points to plot are $x = 0$, $y = -(1)^2 = -1$ and $x = 1$, $y = -(1 + 1)^2 = -2^2 = -4$. When we plot the points $(0, -1)$ and $(1, -4)$, then by symmetry the points $(-2, -1)$ and $(-3, -4)$ are also on the graph.

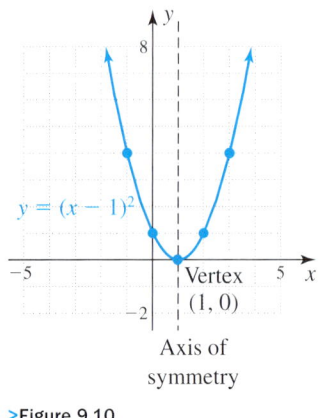

>Figure 9.10

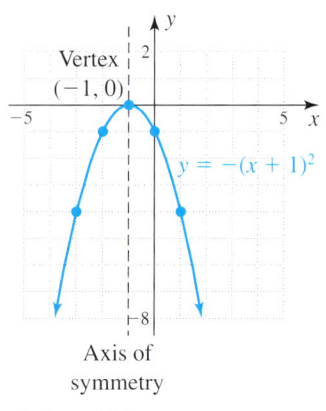

>Figure 9.11

Answers to PROBLEMS

3.

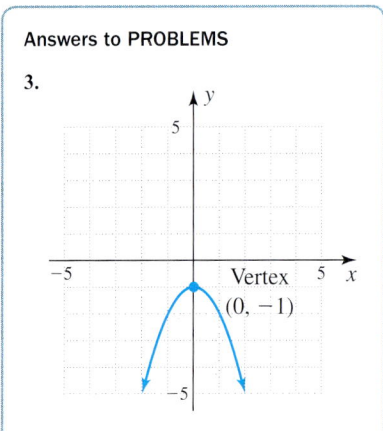

EXAMPLE 4 Graphing a parabola of the form $y = a(x - h)^2 + k$

Graph the function and name the coordinates of the vertex: $y = (x - 1)^2 - 2$

SOLUTION The graph of this equation is identical to the graph of $y = x^2$ except for its position. The new parabola is shifted 1 unit to the right (because of the -1) and 2 units down (because of the -2). Thus, the vertex is $(1, -2)$. Note the axis of symmetry. Figure 9.12 indicates these two facts and shows the completed graph of $y = (x - 1)^2 - 2$.

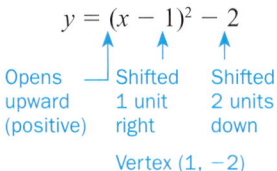

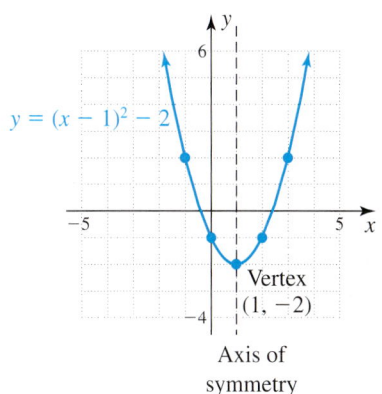

> Figure 9.12

PROBLEM 4

Graph the function and name the coordinates of the vertex: $y = (x - 2)^2 - 1$

From the examples presented we can see that

1. The graph of $y = -x^2 - 2$ (Example 3) is exactly the same as the graph of $y = -x^2$ (Example 1) but moved 2 units *down*. In general, the graph of $y = ax^2 + k$ is the same as the graph of $y = ax^2$ but moved vertically k units. The vertex is at $(0, k)$.
2. The graph of $y = (x - 1)^2$ is the same as that of $y = x^2$ but moved 1 unit *right*. The vertex is at $(1, 0)$.
3. The graph of $y = (x - 1)^2 - 2$ (Example 4) is exactly the same as the graph of $y = (x - 1)^2$ but moved 2 units *down*. The vertex is at $(1, -2)$.

Here is the summary of this discussion.

PROPERTIES OF THE PARABOLA $y = a(x - h)^2 + k$

The graph of the parabola $y = a(x - h)^2 + k$ is the same as that of $y = ax^2$ but moved h units horizontally and k units vertically. The *vertex* is at the point (h, k), and the axis of symmetry is $x = h$.

In conclusion, follow the given directions to graph an equation of the form

$$y = a(x - h)^2 + k \quad \text{Vertex } (h, k)$$

Opens upward for $a > 0$, downward for $a < 0$

Shifts the graph right or left

Moves the graph up or down

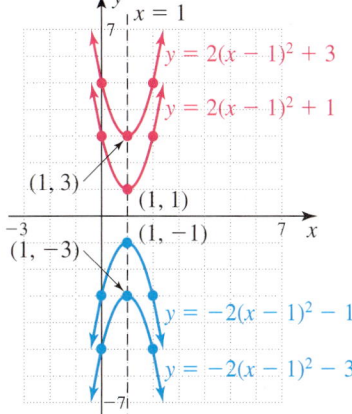

The graphs of $y = 2(x - 1)^2 + 1$, $y = 2(x - 1)^2 + 3$, $y = -2(x - 1)^2 - 1$, and $y = -2(x - 1)^2 - 3$ are shown in Figure 9.13. For each of these parabolas, the axis of symmetry is $x = 1$.

> Figure 9.13

Answers to PROBLEMS

4.

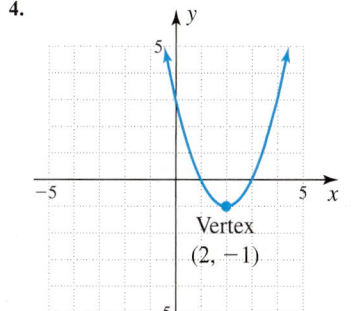

Vertex $(2, -1)$

Calculator Corner

Selecting Window for $y = ax^2 + k$

The graph of $f(x) = ax^2 + k$ can be obtained with a calculator. The only difficulty here is in selecting a window that will show the vertex $(0, k)$. Thus, to graph $f(x) = 2x^2 + 6$, whose vertex is at $(0, 6)$, make sure that $(0, 6)$ is part of the graph by selecting a standard window. If you were graphing $f(x) = 2x^2 + 10$, the vertex would be at $(0, 10)$, so you would have to adjust the window to see more of the graph. Try a $[-15, 15]$ by $[-15, 15]$ window. The graph of $y = 2x^2 + 6$ in a standard window is shown.

Use this idea and a standard window to check the graphs in Examples 1, 2, 3, and 4.

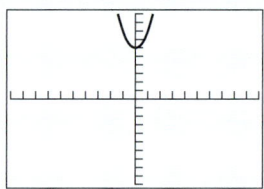

C Graphing a Parabola of the Form $y = f(x) = ax^2 + bx + c$

How can we graph $f(x) = ax^2 + bx + c$? If we learn to write $f(x) = ax^2 + bx + c$ as $f(x) = a(x - h)^2 + k$, we can do it by using the techniques we just learned.

We do this by completing the square. This was previously done in Section 8.2, pages 605 and 606, paralleling a numerical example beside the standard form of the quadratic equation, $ax^2 + bx + c = 0$. The difference here is the equation we start with is $f(x) = ax^2 + bx + c$.

$f(x) = ax^2 + bx + c$ Given

$= (ax^2 + bx) + c$ Group.

$= a\left[x^2 + \dfrac{b}{a}x + \right] + c$ Factor out a.

$= a\left[x^2 + \dfrac{b}{a}x + \left(\dfrac{b}{2a}\right)^2 - \left(\dfrac{b}{2a}\right)^2\right] + c$ To complete the square, add and subtract $\left(\dfrac{1}{2}\cdot\dfrac{b}{a}\right)^2 = \left(\dfrac{b}{2a}\right)^2$ inside the brackets.

$= a\left[\left(x + \dfrac{b}{2a}\right)^2 - \left(\dfrac{b}{2a}\right)^2\right] + c$ Factor inside the brackets.

$= a\left(x + \dfrac{b}{2a}\right)^2 - a\left(\dfrac{b}{2a}\right)^2 + c$ Use the distributive property.

$= a\left(x + \dfrac{b}{2a}\right)^2 - a\cdot\dfrac{b^2}{4a^2} + c$ Square $\dfrac{b}{2a}$.

$= a\left(x + \dfrac{b}{2a}\right)^2 - \dfrac{b^2}{4a} + c$ Multiply $-a \cdot \dfrac{b^2}{4a^2} = -\dfrac{b^2}{4a}$.

$= a\left(x + \dfrac{b}{2a}\right)^2 + \dfrac{4ac - b^2}{4a}$ Find the LCD of $\dfrac{-b^2}{4a}$ and c and write as one fraction.

Thus, to write

$$f(x) = a\underbrace{\left(x + \dfrac{b}{2a}\right)^2} + \underbrace{\dfrac{4ac - b^2}{4a}}$$

as

$$f(x) = a(x - h)^2 + k$$

we must have

$$h = -\dfrac{b}{2a} \quad \text{and} \quad k = \dfrac{4ac - b^2}{4a}$$

which are the coordinates of the vertex. You *do not* have to memorize the *y*-coordinate of the vertex. After you use the *vertex formula*, $\dfrac{-b}{2a}$, to find the *x*-coordinate, substitute in the equation and find *y*.

Here is a summary of our discussion.

9.1 Quadratic Functions (Parabolas) and Their Graphs

PROCEDURE

Graphing the Parabola $y = f(x) = ax^2 + bx + c$

1. To find the vertex use one of the following methods:
   Method 1. Let $x = -\frac{b}{2a}$ in the equation and solve for y,
   or
   Method 2. Complete the square and compare with $y = a(x - h)^2 + k$.
2. Let $x = 0$. The result, c, is the y-intercept.
3. Since the parabola is symmetric with respect to its axis, use this symmetry to find additional points.
4. Let $y = 0$. Find x by solving $ax^2 + bx + c = 0$. If the solutions are real numbers, they are the x-intercepts. If not, the parabola does not intersect the x-axis.
5. Draw a smooth curve through the points found in Steps 1–4. Remember that if $a > 0$, the parabola opens *upward;* if $a < 0$, the parabola opens *downward.*

We demonstrate this procedure in Example 5.

EXAMPLE 5 Graphing a parabola of the form $y = ax^2 + bx + c$

Graph the function and name the coordinates of the vertex: $y = x^2 + 3x + 2$

SOLUTION

1. We first find the vertex using either of the two methods.

Method 1

Use the vertex formula for the x-coordinate. Since $a = 1$, $b = 3$, and $c = 2$,

$$x = -\frac{b}{2a} = -\frac{3}{2}$$

Substituting for x in the equation gives

$$y = x^2 + 3x + 2$$
$$= \left(-\frac{3}{2}\right)^2 + 3\left(-\frac{3}{2}\right) + 2$$
$$= \frac{9}{4} - \frac{9}{2} + 2$$
$$= \frac{9}{4} - \frac{18}{4} + \frac{8}{4} = -\frac{1}{4}$$

The vertex is at $\left(-\frac{3}{2}, -\frac{1}{4}\right)$.

Method 2

Complete the square.

$$y = [x^2 + 3x +] + 2$$
$$= \left[x^2 + 3x + \left(\frac{3}{2}\right)^2\right] + 2 - \left(\frac{3}{2}\right)^2$$
$$= \left(x + \frac{3}{2}\right)^2 + 2 - \frac{9}{4}$$
$$= \left(x + \frac{3}{2}\right)^2 - \frac{1}{4}$$

The vertex is at $\left(-\frac{3}{2}, -\frac{1}{4}\right)$.

PROBLEM 5

Graph the function and name the coordinates of the vertex:

$y = x^2 + 2x - 3$

(continued)

Answers to PROBLEMS

5.

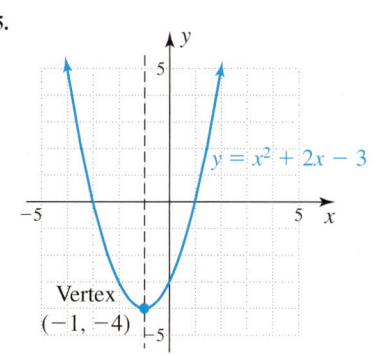

2. Let $x = 0$; then $y = x^2 + 3x + 2 = (0)^2 + 3(0) + 2 = 2$. The y-intercept is at $(0, 2)$.
3. By symmetry, the point $(-3, 2)$ is also on the graph.
4. Let $y = 0$; $y = x^2 + 3x + 2$ becomes
$$0 = x^2 + 3x + 2$$
$$0 = (x + 2)(x + 1)$$

Thus, $x = -2$ or $x = -1$. The graph intersects the x-axis at $(-2, 0)$ and $(-1, 0)$.
5. Since the coefficient of x^2 is 1, $a > 0$ and the parabola opens upward.

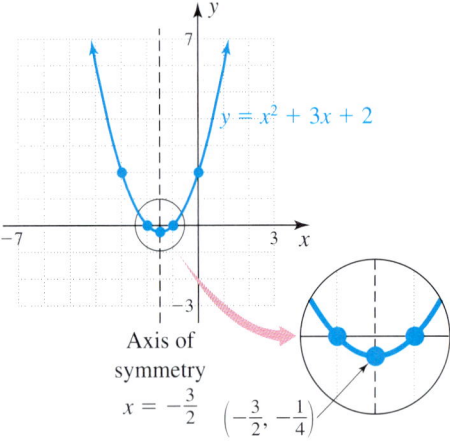

> Figure 9.14

We draw a smooth curve through these points to obtain the graph of the parabola as shown in Figure 9.14.

In Example 5, $x^2 + 3x + 2$ can be factored, and we can find the points at which the parabola crosses the x-axis. If the equation of the parabola cannot be factored, look at the discriminant $D = b^2 - 4ac$ and determine what kind of solutions the equation has.

PROCEDURE

Using the Discriminant, D, to Graph Quadratics, $y = f(x) = ax^2 + bx + c$
Compute $D = b^2 - 4ac$
1. If $D < 0$, there are no real solutions and the graph will *not* cross the x-axis.
2. If $D = 0$, there is one real solution and the graph will have one x-intercept, the vertex.
3. If $D \geq 0$, either factor or use the quadratic formula to find x and approximate the answers so you can graph them as the two x-intercepts.

The possibilities are shown in Figures 9.15–9.17.

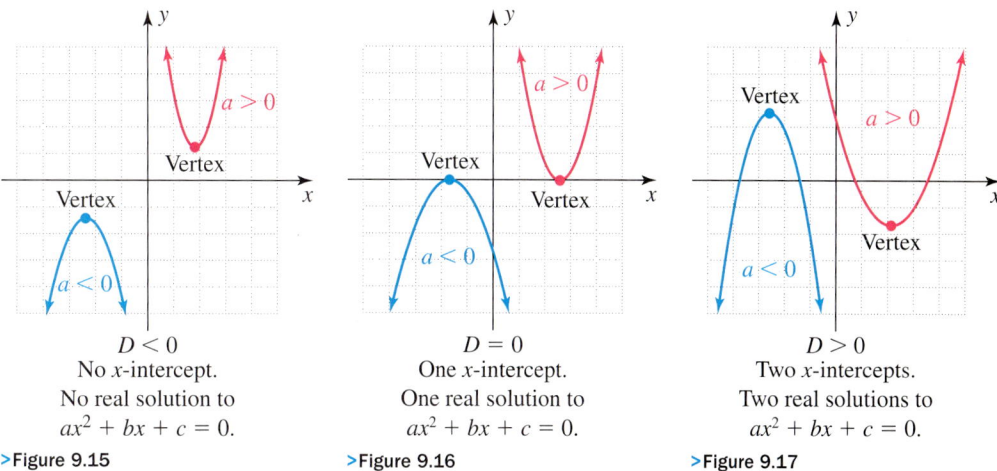

Figure 9.15
$D < 0$
No x-intercept.
No real solution to $ax^2 + bx + c = 0$.

Figure 9.16
$D = 0$
One x-intercept.
One real solution to $ax^2 + bx + c = 0$.

Figure 9.17
$D > 0$
Two x-intercepts.
Two real solutions to $ax^2 + bx + c = 0$.

EXAMPLE 6 Graphing a parabola whose equation cannot be factored

Graph the function and name the coordinates of the vertex: $y = -2x^2 + 4x - 3$.

SOLUTION

1. To find the vertex, we can use either of these methods:

Method 1

Use the vertex formula.
Here $a = -2$ and $b = 4$, so

$$x = -\frac{b}{2a}$$
$$= \frac{-4}{2(-2)}$$
$$= 1$$

If we substitute $x = 1$ in $y = -2x^2 + 4x - 3$,

$$y = -2(1)^2 + 4(1) - 3$$
$$= -2 + 4 - 3$$
$$= -1$$

The vertex is at $(1, -1)$.

Method 2

Complete the square.

$$y = -2x^2 + 4x - 3$$
$$= -2(x^2 - 2x +) - 3$$
$$= -2(x^2 - 2x + 1 - 1) - 3$$
$$= -2[(x - 1)^2 - 1] - 3$$
$$= -2(x - 1)^2 + 2 - 3$$
$$= -2(x - 1)^2 - 1$$

The vertex is at $(1, -1)$.

2. If $x = 0$, $y = -2x^2 + 4x - 3 = -2(0)^2 + 4(0) - 3 = -3$. The y-intercept is at $(0, -3)$.

3. We graph the vertex $(1, -1)$ and the y-intercept -3. To make a more accurate graph, we need some more points.

PROBLEM 6

Graph the function and name the coordinates of the vertex:
$y = -x^2 - 4x - 3$.

(continued)

Answers to PROBLEMS

6.

Vertex $(-2, 1)$

$y = -x^2 - 4x - 3$

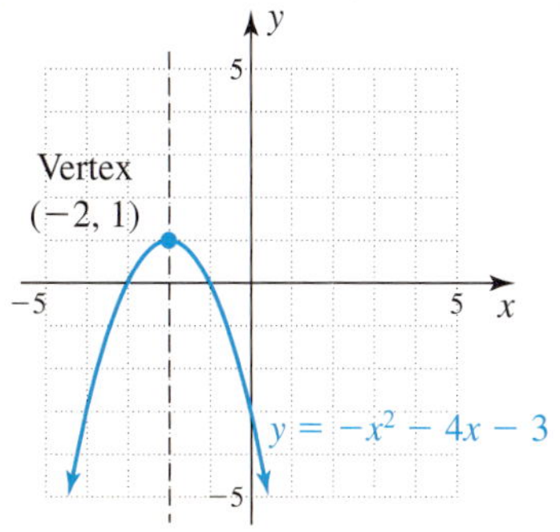

The parabola is symmetric, so we know there must be a point on the other side of the axis of symmetry from the y-intercept that is the same distance away, 1 unit. The axis of symmetry is at $x = 1$ and one unit away from that would be $x = 2$. The mirror image point of the y-intercept, $(0, -3)$, is $(2, -3)$.

4. For $y = 0$, $0 = -2x^2 + 4x - 3$. However, the right-hand side is not factorable. The discriminant of the equation is $4^2 - 4(-2)(-3) = 16 - 24 = -8$. This means that this equation has no real number solutions and there are no x-intercepts. The graph does not cross the x-axis.
5. Since $a = -2 < 0$, the parabola opens downward. The completed graph is shown in Figure 9.18.

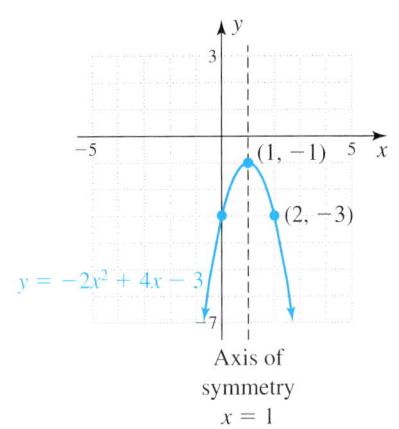

> Figure 9.18

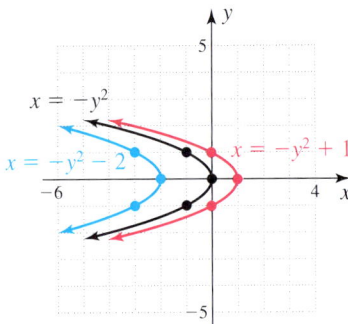

> Figure 9.19

D ⟩ Graphing Parabolas of the Form $x = a(y - k)^2 + h$ or $x = ay^2 + by + c$

Do you remember how the graph of $y = x^2$ looks? The graph of $x = y^2$ has a similar shape but opens *horizontally to the right*. Similarly, the graph of $x = y^2 + 1$ looks like that of $x = y^2$ but is shifted *right* 1 unit, as shown in Figure 9.19. When $y = 0$, $x = 1$ so the vertex is at $(1, 0)$. The graph of $x = y^2 - 2$ is similar to that of $x = y^2$ but is shifted *left* 2 units. Here, when $y = 0$, $x = -2$ so the vertex is at $(-2, 0)$.

Similarly, the graphs of $x = -y^2$, $x = -y^2 + 1$, and $x = -y^2 - 2$ look like the graph of $x = -y^2$, opening *horizontally to the left* and then shifted right or left the correct number of units. For $x = -y^2 + 1$, when $y = 0$, $x = 1$ and the vertex is at $(1, 0)$. For $x = -y^2 - 2$, when $y = 0$, $x = -2$ so the vertex is at $(-2, 0)$. The graphs are shown in Figure 9.20. **None of these are graphs of functions.** (Remember the vertical line test? You can find several vertical lines that will intersect all of these graphs at more than one point.)

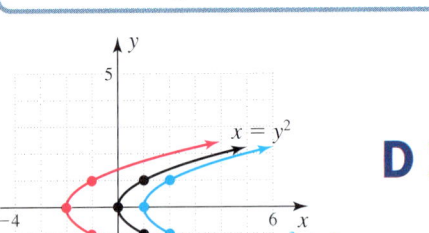

> Figure 9.20

PROPERTIES OF THE PARABOLA $x = a(y - k)^2 + h$

The graph of the parabola $x = a(y - k)^2 + h$ is the same as that of $x = ay^2$ but moved h units horizontally and k units vertically. The *vertex* is at the point (h, k), and the axis of symmetry is $y = k$.

In conclusion, follow the directions given in blue below to graph an equation of the form

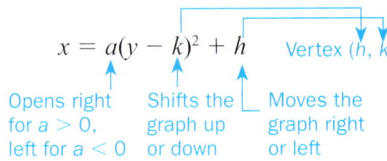

$x = a(y - k)^2 + h$ Vertex (h, k)

Opens right for $a > 0$, left for $a < 0$ Shifts the graph up or down   Moves the graph right or left

EXAMPLE 7 Graphing a parabola of the form $x = a(y - k)^2 + h$

Graph the equation and name the coordinates of the vertex: $x = 2(y - 1)^2 - 3$

SOLUTION In this problem, the roles of x and y are reversed, so the graph will look like that of $y = 2(x - 1)^2 - 3$ but opening horizontally.

The vertex of $x = 2(y - 1)^2 - 3$ is at $(-3, 1)$. The curve opens to the right, the positive x-direction. The graph is shown in Figure 9.21. You can verify that the graph is correct by letting $x = -1$, which gives $y = 0$ or $y = 2$.

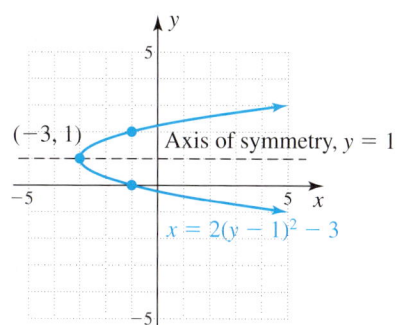

> Figure 9.21

PROBLEM 7

Graph the equation and name the coordinates of the vertex:
$x = (y + 1)^2 - 2$

EXAMPLE 8 Graphing a parabola of the form $x = ay^2 + by + c$

Graph the equation and name the coordinates of the vertex: $x = y^2 + 3y + 2$

SOLUTION The graph is similar to that of $y = x^2 + 3x + 2$, but it opens horizontally (see Example 5). The vertex occurs where

$$y = -\frac{b}{2a} = -\frac{3}{2}$$

Substituting for y in the equation gives $x = (-\frac{3}{2})^2 + 3(-\frac{3}{2}) + 2 = -\frac{1}{4}$. The vertex is at $(-\frac{1}{4}, -\frac{3}{2})$. The x-intercept is 2 and the y-intercepts are where $x = 0$. Thus,

$$0 = y^2 + 3y + 2$$
$$= (y + 2)(y + 1)$$

That is, $y = -2$ or $y = -1$. The parabola opens to the right because the coefficient of y^2 is $1 > 0$, and the completed graph is shown in Figure 9.22.

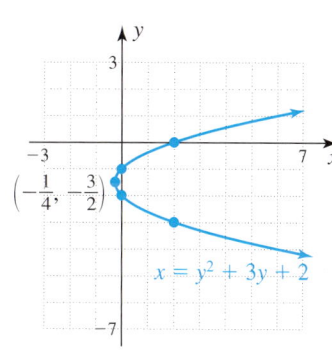

> Figure 9.22

PROBLEM 8

Graph the equation and name the coordinates of the vertex:
$x = y^2 + 2y - 3$

🖩 Calculator Corner

Graphing Parabolas That Are Not Functions

To graph $x = 2(y - 1)^2 - 3$ in Example 7, first solve for y to obtain

$$y = 1 \pm \sqrt{\frac{x + 3}{2}}$$

(continued)

Answers to PROBLEMS

7.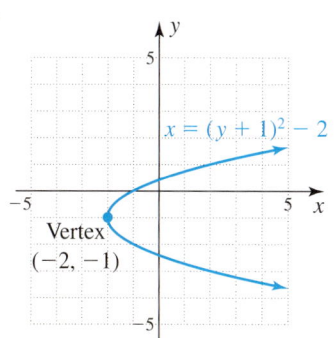

8.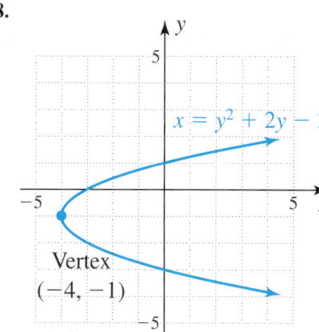

Since the graph consists of two branches, graph

$$Y_1 = 1 + \sqrt{\frac{x+3}{2}}$$

and $\quad Y_2 = 1 - \sqrt{\frac{x+3}{2}}$

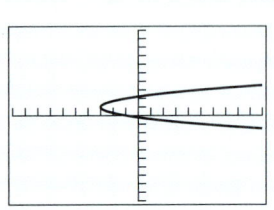

The two graphs together form the graph of $x = 2(y-1)^2 - 3$ as shown in the window.

How would you graph $x = y^2 + 3y + 2$ in Example 8? You have to complete the square and then solve for y or use the quadratic formula. Try it and see if you agree that it's easier to do it algebraically.

E › Applications Involving Parabolas

Every parabola of the form $y = ax^2 + bx + c$ that we have graphed has its vertex at either its maximum (highest) or minimum (lowest) point on the graph. If the graph opens upward, the vertex is the minimum (Figure 9.23) and if the graph opens downward, the vertex is the maximum (Figure 9.24).

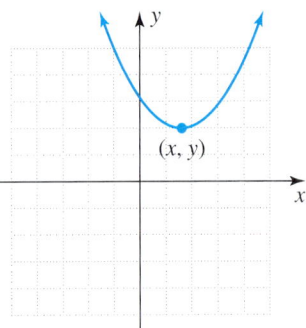

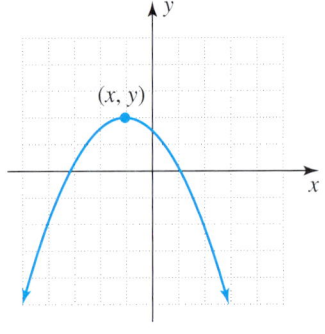

›**Figure 9.23** $f(x)$ has a minimum at the vertex (x, y).

›**Figure 9.24** $f(x)$ has a maximum at the vertex (x, y).

If we are dealing with a quadratic function, we can find its maximum or minimum by finding the vertex of the corresponding parabola. This idea can be used to solve many real-world applications. For example, suppose that a CD company manufactures and sells x CDs per week. If the revenue is given by $R = 10x - 0.01x^2$, we can use the techniques we've just studied to maximize the revenue. We do that next.

EXAMPLE 9   Recording a maximum revenue

If $R = 10x - 0.01x^2$, how many CDs does the company have to sell to obtain maximum revenue? (R is revenue and x is the number of CDs made and sold per week.)

SOLUTION   We first write the equation as $R = -0.01x^2 + 10x$. Since the coefficient of x^2 is negative, the parabola opens downward (is concave down), and the vertex is its **highest point,** a maximum. Letting $a = -0.01$ and $b = 10$, we have

$$x = -\frac{b}{2a} = -\frac{10}{-0.02} = 500$$

$R = 10(500) - 0.01(500)^2 = 5000 - 2500 = 2500$. The vertex is at $(500, 2500)$, thus, when the company sells $x = 500$ CDs a week, the revenue is a maximum: $2500.

PROBLEM 9

A farmer has 200 feet of fencing he wants to use to enclose a rectangular plot of land for his wife's vegetable garden. What dimensions will give her the maximum area?

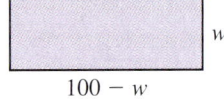

Answers to PROBLEMS

9. 50 ft × 50 ft

We end this section by presenting in Table 9.1 a summary of the material we have studied.

Table 9.1 Summary of Graphing Parabolas

Description	Graph when $a > 0$	Graph when $a < 0$
$f(x) = ax^2 + k$ A parabola with vertex at $(0, k)$. When $|a| > 1$, the graph is narrower than the graph of $y = x^2$. When $0 < |a| < 1$, it is wider.	Vertex at $(0, k)$ For $a > 0$, the parabola opens upward.	Vertex at $(0, k)$ For $a < 0$, the parabola opens downward.
$f(x) = a(x - h)^2 + k$ A parabola with vertex at (h, k). When $|a| > 1$, the graph is narrower than the graph of $y = x^2$. When $0 < |a| < 1$, it is wider.	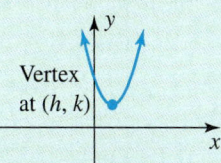 Vertex at (h, k) For $a > 0$, the parabola opens upward.	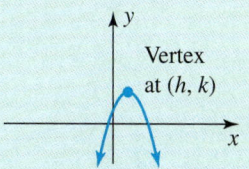 Vertex at (h, k) For $a < 0$, the parabola opens downward.
$f(x) = ax^2 + bx + c$ A parabola with vertex at $\left(-\frac{b}{2a}, f\left(-\frac{b}{2a}\right)\right)$. When $|a| > 1$, the graph is narrower than the graph of $y = x^2$. When $0 < |a| < 1$, it is wider.	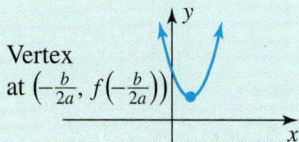 Vertex at $\left(-\frac{b}{2a}, f\left(-\frac{b}{2a}\right)\right)$ For $a > 0$, the parabola opens upward.	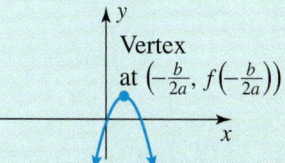 Vertex at $\left(-\frac{b}{2a}, f\left(-\frac{b}{2a}\right)\right)$ For $a < 0$, the parabola opens downward.
$x = a(y - k)^2 + h$ A parabola with vertex at (h, k) and is not the graph of a function. When $|a| > 1$, the graph is narrower than the graph of $x = y^2$. When $0 < |a| < 1$, it is wider.	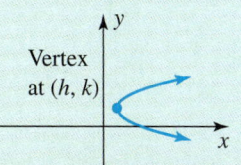 Vertex at (h, k) For $a > 0$, the parabola opens to the right.	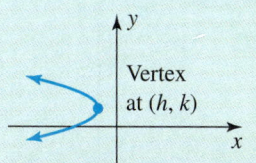 Vertex at (h, k) For $a < 0$, the parabola opens to the left.

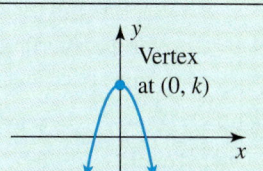

Calculator Corner

1. Graph the horizontal parabola of Example 8, $x = y^2 + 3y + 2$.
2. Graph the parabolas $g(x) = ax^2$ for $a = \frac{1}{4}, \frac{1}{2}, 2,$ and 4. *Note:* Some calculators can graph "families" of curves. If your calculator is one of these, enter $Y_1 = \{\frac{1}{4}, \frac{1}{2}, 2, 4\}x^2$
   a. What happens to the graph as a increases?
   b. What are the domain and range of $g(x) = ax^2$ for $a > 0$?
3. Graph the parabolas $g(x) = ax^2$ for $a = -\frac{1}{4}, -\frac{1}{2}, -2,$ and -4.
   a. What happens to the graph as a decreases?
   b. What are the domain and range of $g(x) = ax^2$ for $a < 0$?
   c. How does the sign of a in $g(x) = ax^2$ affect the graph?
4. Graph the parabolas $u(x) = x^2 + k$ for $k = -2, -1, 1,$ and 2.
   a. What effect does k have on the graph?
   b. What are the domain and range of $u(x)$?
5. Graph the parabolas $v(x) = (x - h)^2$ for $h = -2, -1, 1,$ and 2.
   a. What effect does h have on the graph?
   b. What are the domain and range of $v(x)$?

Exercises 9.1

A **Graphing a Parabola of the Form $y = f(x) = ax^2 + k$** In Problems 1–8, graph the given equations on the same coordinate axes and name the coordinates of the vertices.

1. a. $y = 2x^2$
   b. $y = 2x^2 + 2$
   c. $y = 2x^2 - 2$

2. a. $y = 3x^2 + 1$
   b. $y = 3x^2 + 3$
   c. $y = 3x^2 - 2$

3. a. $y = -2x^2$
   b. $y = -2x^2 + 1$
   c. $y = -2x^2 - 1$

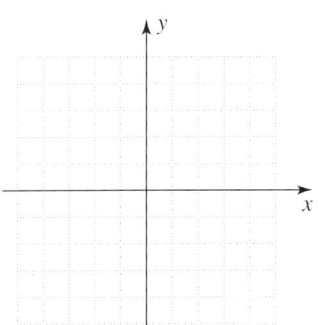
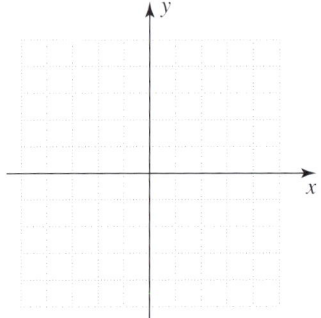
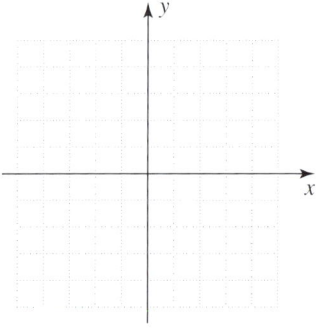

4. a. $y = -4x^2$
   b. $y = -4x^2 + 1$
   c. $y = -4x^2 - 1$

5. a. $y = \frac{1}{4}x^2$
   b. $y = -\frac{1}{4}x^2$

6. a. $y = \frac{1}{5}x^2$
   b. $y = -\frac{1}{5}x^2$

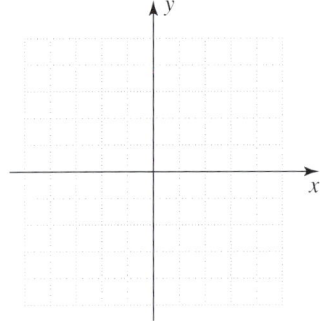
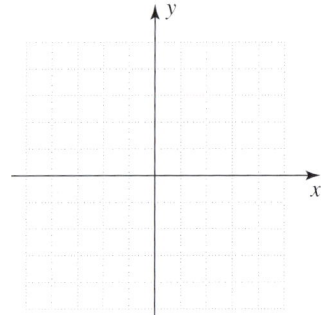
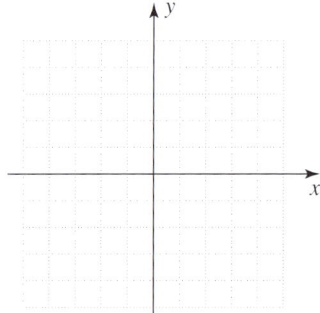

7. a. $y = \frac{1}{3}x^2 + 1$
   b. $y = -\frac{1}{3}x^2 + 1$

8. a. $y = \frac{1}{4}x^2 + 1$
   b. $y = -\frac{1}{4}x^2 + 1$

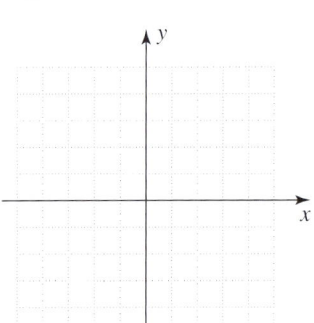
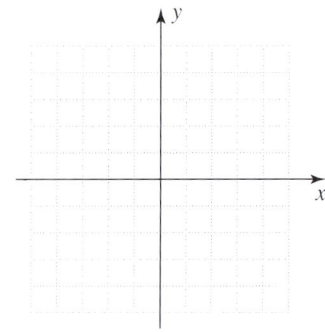

‹ B › Graphing a Parabola of the Form $y = f(x) = a(x - h)^2 + k$ In Problems 9–16, graph the given equations on the same coordinate axes and name the coordinates of the vertices.

9. a. $y = (x + 2)^2 + 3$
   b. $y = (x + 2)^2$
   c. $y = (x + 2)^2 - 2$

10. a. $y = (x - 2)^2 + 2$
    b. $y = (x - 2)^2$
    c. $y = (x - 2)^2 - 2$

11. a. $y = -(x + 2)^2 - 2$
    b. $y = -(x + 2)^2$
    c. $y = -(x + 2)^2 - 4$

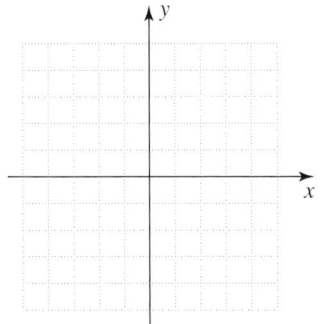 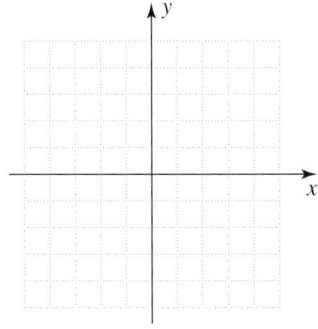 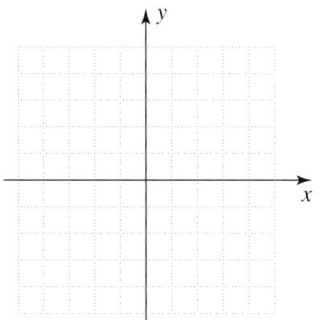

12. a. $y = -(x - 1)^2 + 1$
    b. $y = -(x - 1)^2$
    c. $y = -(x - 1)^2 + 2$

13. a. $y = -2(x + 2)^2 - 2$
    b. $y = -2(x + 2)^2$
    c. $y = -2(x + 2)^2 - 4$

14. a. $y = -2(x - 1)^2 + 1$
    b. $y = -2(x - 1)^2$
    c. $y = -2(x - 1)^2 + 2$

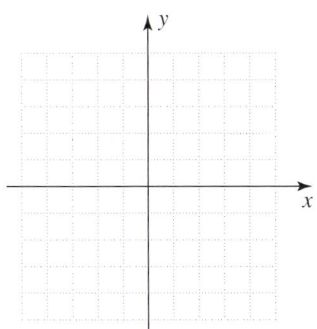 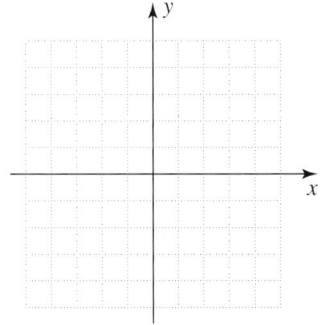 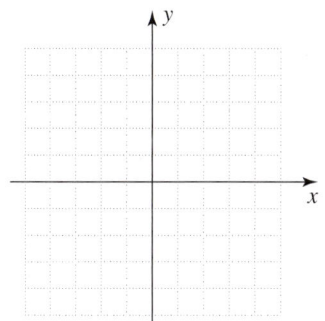

15. a. $y = 2(x + 1)^2 + \dfrac{1}{2}$
    b. $y = 2(x + 1)^2$

16. a. $y = 2(x + 1)^2 - \dfrac{1}{2}$
    b. $y = 2(x + 1)^2$

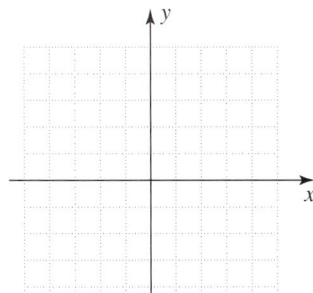 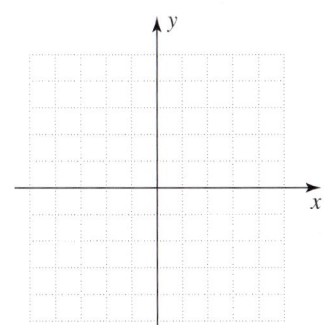

⟨ C ⟩ **Graphing a Parabola of the Form** $y = f(x) = ax^2 + bx + c$ In Problems 17–28, use the five-step procedure in the text to sketch the graph. Label the vertex and the intercept(s). For irrational intercepts, approximate the values to one decimal place.

17. $y = x^2 + 2x + 1$

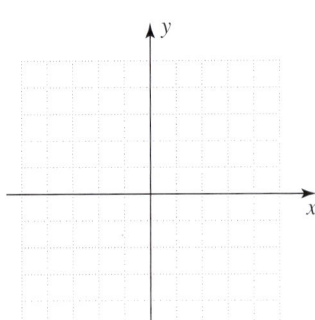

18. $y = x^2 + 4x + 4$

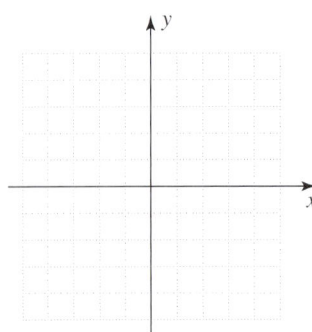

19. $y = -x^2 + 2x + 1$

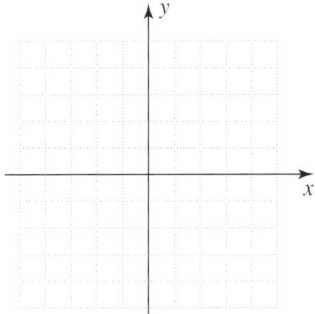

20. $y = -x^2 + 4x - 2$

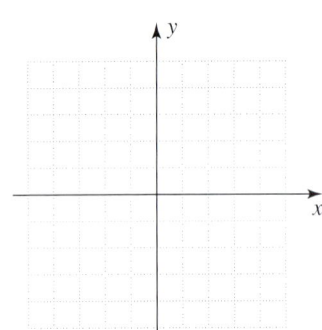

21. $y = -x^2 + 4x - 5$

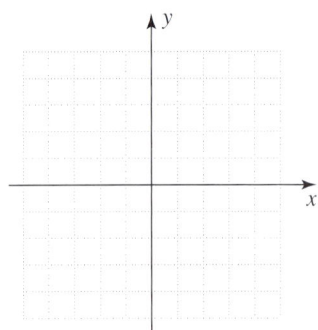

22. $y = -x^2 + 4x - 3$

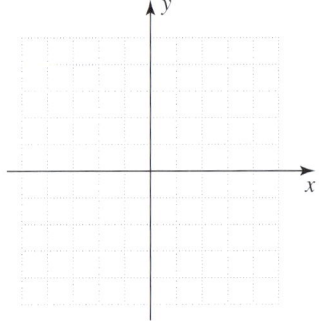

23. $y = 3 - 5x + 2x^2$

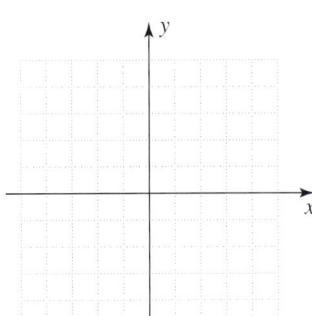

24. $y = 3 + 5x + 2x^2$

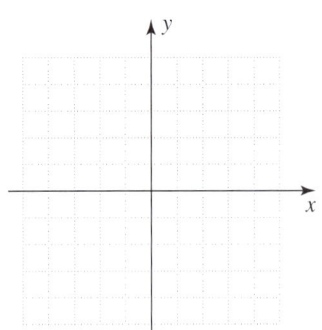

25. $y = 5 - 4x - 2x^2$
(Hint: $\sqrt{56} \approx 7.5$)

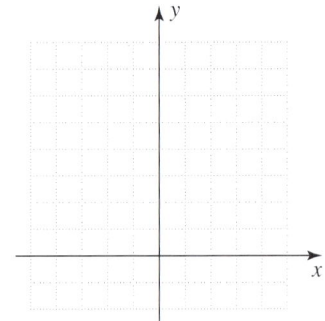

26. $y = 3 - 4x - 2x^2$
(Hint: $\sqrt{40} \approx 6.3$)

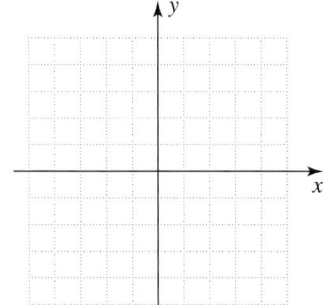

27. $y = -3x^2 + 3x + 2$
(Hint: $\sqrt{33} \approx 5.7$)

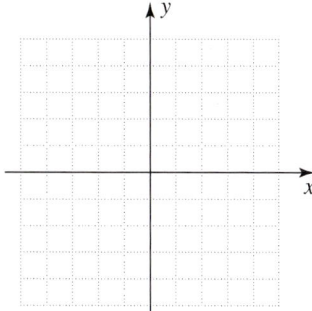

28. $y = -3x^2 + 3x + 1$
(Hint: $\sqrt{21} \approx 4.6$)

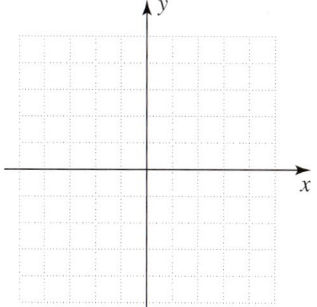

⟨ **D** ⟩ **Graphing Parabolas of the Form** $x = a(y - k)^2 + h$ **or** $x = ay^2 + by + c$ In Problems 29–34, graph the given equations on the same coordinate axes and name the coordinates of the vertices.

29. **a.** $x = (y + 2)^2 + 3$
   b. $x = (y + 2)^2$

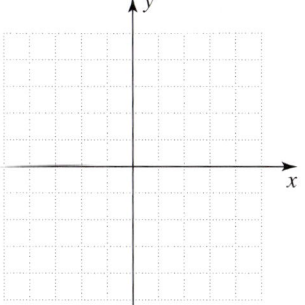

30. **a.** $x = (y - 2)^2 + 2$
   b. $x = (y - 2)^2$

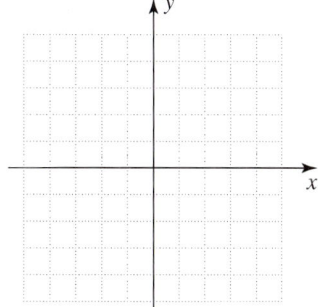

31. **a.** $x = -(y + 2)^2 - 2$
   b. $x = -(y + 2)^2$

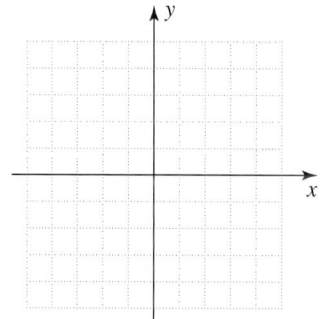

32. **a.** $x = -(y - 1)^2 + 1$
   b. $x = -(y - 1)^2$

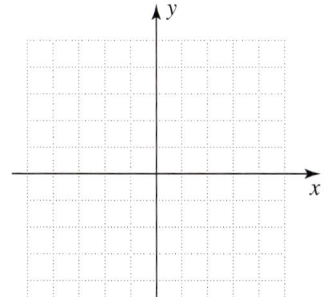

33. **a.** $x = -y^2 + 2y + 1$
   b. $x = -y^2 + 2y + 4$

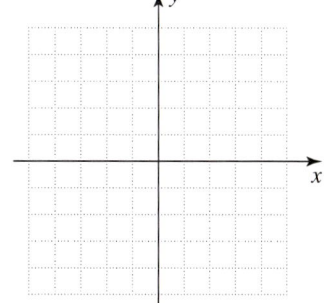

34. **a.** $x = -y^2 + 4y - 5$
   b. $x = -y^2 + 4y - 3$

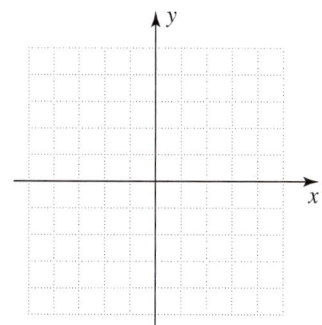

⟨ **E** ⟩ **Applications Involving Parabolas**

35. *Profit* The profit P (in dollars) for a company is modeled by the equation $P = -5000 + 8x - 0.001x^2$, where x is the number of items sold each month. How many items does the company have to sell to obtain maximum profit? What is this profit?

36. *Revenue* The revenue R for Shady Glasses is given by the equation $R = 1500p - 75p^2$, where p is the price of each pair of sunglasses (R and p in dollars). What should the price be to maximize revenue?

37. Sales After spending x thousand dollars on an advertising campaign, the number of units N sold is given by the equation $N = 50x - x^2$. How much should be spent in the campaign to obtain maximum sales?

38. Sales The number N of units of a product sold after a television commercial blitz is given by the equation $N = 40x - x^2$, where x is the amount spent in thousands of dollars. How much should be spent on television commercials to obtain maximum sales?

39. Height of a ball If a ball is batted upward at 160 feet per second, its height h feet after t seconds is given by $h = -16t^2 + 160t$. Find the maximum height reached by the ball.

40. Height of a ball If a ball is thrown upward at 20 feet per second, its height h feet after t seconds is given by $h = -16t^2 + 20t$. How many seconds does it take for the ball to reach its maximum height, and what is this height?

41. Farming If a farmer digs potatoes today, she will have 600 bushels worth \$1 per bushel. Every week she waits, the crop increases by 100 bushels, but the price decreases 10¢ a bushel. Show that she should dig and sell her potatoes at the end of 2 weeks.

42. Area A man has a large piece of property along Washington Street. He wants to fence the sides and back of a rectangular plot. If he has 400 feet of fencing, what dimensions will give him the maximum area?

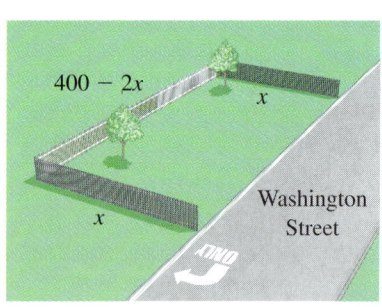

43. Frog leaps Have you read the story "The Jumping Frog of Calaveras County"? According to the *Guinness Book of World Records*, the second greatest distance covered by a frog in a triple jump is 21 feet, $5\frac{3}{4}$ inches at the annual Calaveras Jumping Jubilee; this occurred on May 18, 1986.

  a. If Rosie the Ribiter's (the winner) path in her first jump is approximated by the equation $R = -\frac{1}{98}x^2 + \frac{6}{7}x$ (where x is the horizontal distance covered in inches), what are the coordinates of the vertex of Rosie's path?

  b. Find the maximum height attained by Rosie in her first jump.

  c. Use symmetry to find the horizontal length of Rosie's first jump.

  d. Make a sketch of R showing the initial position (0, 0), the vertex, and Rosie's ending position after her first jump.

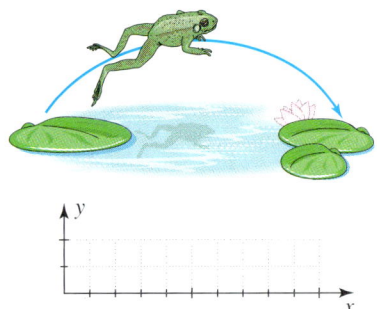

44. Frog leaps Amazingly, Rosie's is not the best triple jump on record. (See Problem 43.) That distinction belongs to Santjie, a South African frog who jumped 33 feet, $5\frac{1}{2}$ inches on May 21, 1977.

  a. If Santjie's path in his first jump is approximated by the equation $S = -\frac{1}{200}x^2 + \frac{7}{10}x$ (where x is the horizontal distance covered in inches), what are the coordinates of the vertex of Santjie's path?

  b. Find Santjie's maximum height in his first jump.

  c. Use symmetry to find the horizontal length of Santjie's first jump.

  d. Make a sketch of S showing the initial position (0, 0), the vertex, and Santjie's ending position after his first jump.

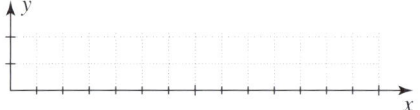

45. Baseball A baseball is hit at an angle of 35° with a velocity of 130 miles per hour. Its trajectory can be approximated by the equation $d = -\frac{1}{400}x^2 + x$, where x is the distance the ball travels in feet.

  a. What are the coordinates of the vertex of the trajectory?

  b. Find the maximum height attained by the ball.

  c. Use symmetry to find how far the ball travels horizontally.

  d. Make a sketch of d showing the initial position (0, 0), the vertex, and the ending position of the baseball.

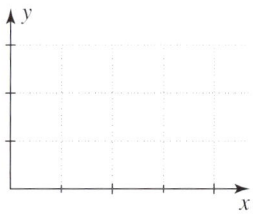

46. *SAT Scores* Examing the data at a university for the 7th through 12th years, the graph of the average men's SAT verbal scores is nearly a parabola.

   a. What is the maximum average verbal score for men in this period?

   b. What is the minimum average verbal score for men in this period?

   c. If the function approximating the average men's verbal score is

   $$f(x) = ax^2 + bx + c$$

   what can you say about *a*?

Source: Data from USA TODAY, 1993.

47. *Poverty* According to the U.S. Census Bureau the number of people in poverty (in millions) *t* years after 1998 can be approximated by the equation $P(t) = t^2 - 4t + 35$.

   a. How many millions were in poverty in 1998?

   b. What is the vertex of $P(t)$?

   c. Is the vertex a minimum or a maximum?

   d. What is the minimum number of people in poverty, and in what year does the minimum occur?

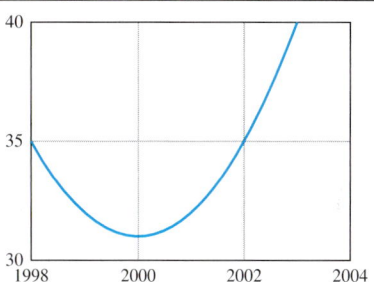

48. *Poverty* Referring to the graph in Problem 47:

   a. In what interval of time is poverty decreasing?

   b In that interval of time is poverty increasing?

   c. If you use $P(t)$ as your model, what would you predict the number of people in poverty be in 2010?

49. *Gas mileage* How many miles per gallon do you get in your car? You probably know that the best mileage happens when you don't drive too quickly. Suppose the mileage $M(s)$ in miles per gallon when driving at *s* miles per hour is given by the equation $M(s) = -0.04s^2 + 3s - 30$.

   a. What is the vertex of $M(s)$?

   b. At what speed *s* do you have to drive to get the best mileage?

   c. How many miles per gallon do you get when you drive at 45 miles per hour?

50. *Sensitivity to a drug* The sensitivity $S(m)$ to a drug is dependent on the dosage *m* in milligrams and given by the equation $S(m) = 1000m - m^2$. What dosage *m* gives the maximum sensitivity to the drug?

〉〉〉 Using Your Knowledge

Parabolas Revisited   Here is another way of defining a parabola.

PARABOLA

A **parabola** is the set of all points equidistant from a fixed point $F(0, p)$ (called the **focus**) and a fixed line $y = -p$ (called the **directrix**).

If $P(x, y)$ is a point on the parabola, this definition says that $FP = DP$; that is, the distance from *F* to *P* is the same as the distance from *D* to *P*:

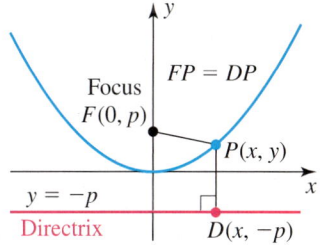

51. Find *FP*. The distance between two points can be found by using: $\sqrt{(x_2 - x_1)^2 + (y_2 - y_1)^2}$.

52. Find *DP*.

53. Set $FP = DP$ and solve for x^2.

54. For the parabola $x^2 = 4y$,

   a. Locate the focus.

   b. Write the equation of the directrix.

Many applications of the parabola depend on an important focal property of the curve. If the parabola is a mirror, a ray of light parallel to the axis reflects to the focus, and a ray originating at the focus reflects parallel to the axis. (This can be proved by methods of calculus.)

If the parabola is revolved about its axis, a surface called a *paraboloid of revolution* is formed. This is the shape used for automobile headlights and searchlights that throw a parallel beam of light when the light source is placed at the focus; it's also the shape of a radar dish or a reflecting telescope mirror that collects parallel rays of energy (light) and reflects them to the focus.

We can find the equation of the parabola needed to generate a paraboloid of revolution by using the equation $x^2 = 4py$ as follows: Suppose a parabolic mirror has a diameter of 6 feet and a depth of 1 foot. Then, we find the value of p that makes the parabola pass through the point (3, 1). This means that we substitute into the equation and solve for p. Thus, we have

$$3^2 = 4p(1)$$

so that $4p = 9$ and $p = 2.25$. The equation of the parabola is $x^2 = 9y$ and the focus is at (0, 2.25).

55. A radar dish has a diameter of 10 feet and a depth of 2 feet. The dish is in the shape of a paraboloid of revolution. Find an equation for a parabola that would generate this dish and locate the focus.

56. The cables of a suspension bridge hang very nearly in the shape of a parabola. A cable on such a bridge spans a distance of 1000 feet and sags 50 feet in the middle. Find an equation for this parabola.

❯❯❯ Write On

57. Explain how you determine whether the graph of a quadratic function opens up or down.

58. Explain what causes the graph of the function $f(x) = ax^2$ to be wider or narrower than the graph of $f(x) = x^2$.

59. Explain the effect of the constant k on the graph of the function $f(x) = ax^2 + k$.

60. Explain how a parabola that has two x-intercepts and vertex at (1, 1) opens—that is, does it open up or down? Why?

61. Explain why the graph of a function never has two y-intercepts.

62. Explain how you can tell if the vertex of a parabola is the maximum or the minimum point on the graph of a parabola with a vertical axis.

❯❯❯ Concept Checker

Fill in the blank(s) with the correct word(s), phrase, or mathematical statement.

63. When a is positive, the parabola $g(x) = ax^2$ opens _____.

64. When a is negative, the parabola $g(x) = ax^2$ opens _____.

65. The graph of the parabola $y = a(x - h)^2 + k$ has the same shape as the graph of _____.

66. The vertex of $y = a(x - h)^2 + k$ is the point _____.

67. The graph of the parabola $x = a(y - k)^2 + h$ has the same shape as the graph of _____.

68. The vertex of $x = a(y - k)^2 + h$ is _____.

y = ax²

x = ay²

downward

upward

(k, h)

(h, k)

❯❯❯ Mastery Test

Graph the equation and name the coordinates of the vertex.

69. $x = y^2 + 2y - 3$

70. $x = -y^2 + 2y + 3$

71. $x = 2(y - 1)^2 + 3$

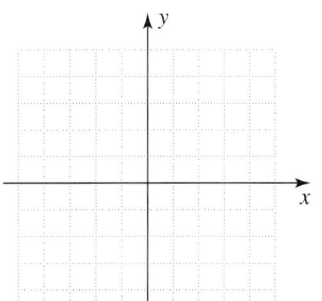

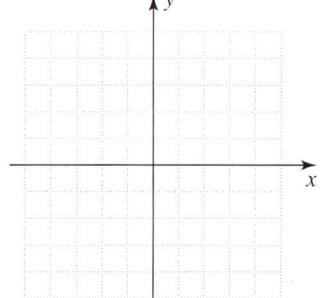

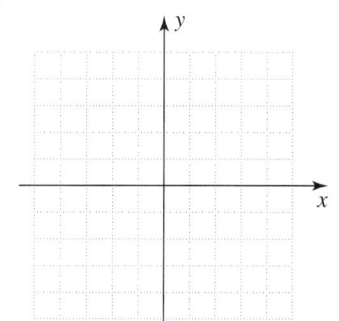

72. $x = -2(y-1)^2 + 3$

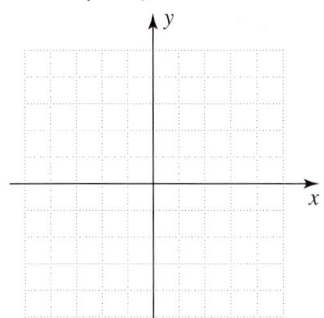

73. $y = -2x^2 - 4x - 3$

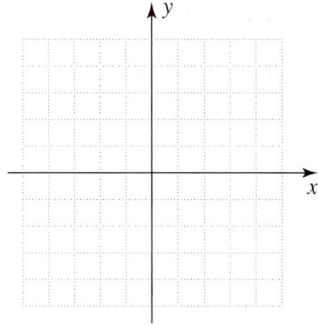

74. $y = x^2 + 2x - 3$

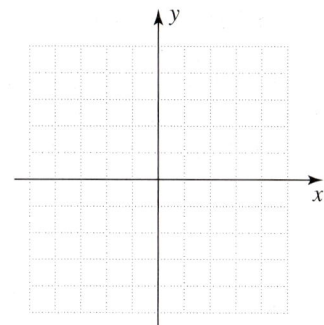

75. $y = (x-2)^2 + 3$

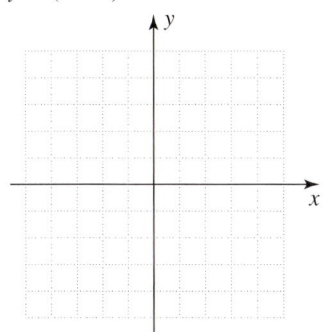

76. $y = -(x-3)^2 + 2$

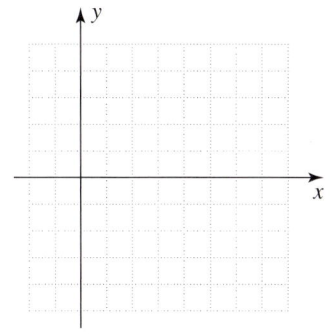

77. $f(x) = -x^2 + 4$

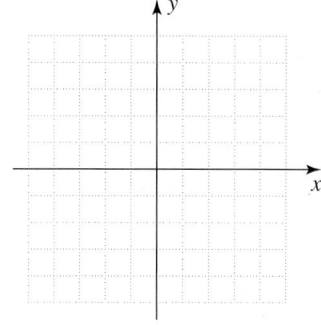

78. $f(x) = x^2 - 4$

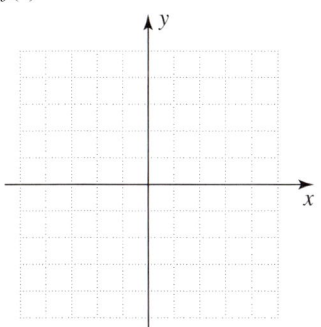

79. $f(x) = 2x^2$

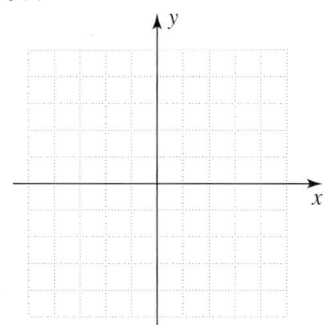

80. $g(x) = \frac{1}{2}x^2$

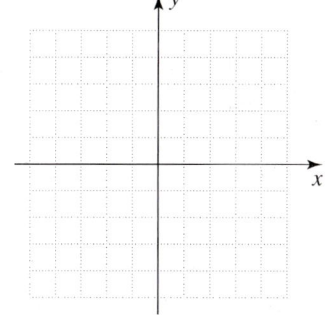

81. $h(x) = -2x^2$

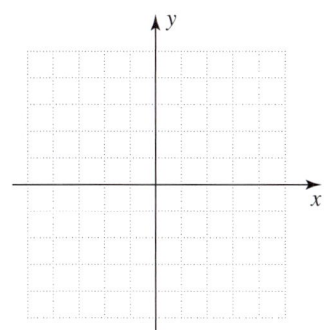

82. $f(x) = -\frac{1}{2}x^2$

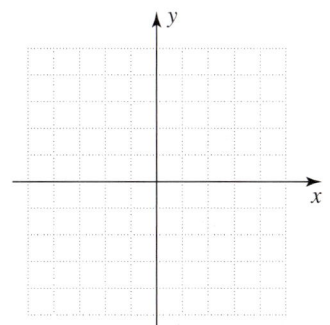

83. The revenue R for a company is modeled by the equation $R = 300p - 15p^2$, where p is the price of each unit (R and p are in dollars). What should the price p be so that the revenue R is maximized?

〉〉〉 Skill Checker

Simplify.

84. $\sqrt{27}$

85. $\sqrt{32}$

Complete the square for the quadratic expression.

86. $x^2 - 6x$

87. $y^2 + 8y$

88. Find the hypotenuse of a right triangle whose legs are 4 cm and 6 cm.

9.2 Circles and Ellipses

Objectives

A > Find the distance between two points.

B > Find an equation of a circle with a given center and radius.

C > Find the center and radius and sketch the graph of a circle when its equation is given.

D > Graph an ellipse when its equation is given.

E > Solve applications involving circles and ellipses.

To Succeed, Review How To . . .

1. Simplify radicals (pp. 538–541).
2. Find the hypotenuse of a right triangle using the Pythagorean theorem (pp. 411–412).
3. Complete the square for a quadratic equation (pp. 597–599).

Getting Started
Comets and Conics

This diagram shows the orbit of the comet Kohoutek with respect to the orbit of the Earth. The comet's orbit is an *ellipse*, whereas the Earth's orbit is nearly a perfect *circle*. In this section we continue our study of the conic sections by discussing circles and ellipses.

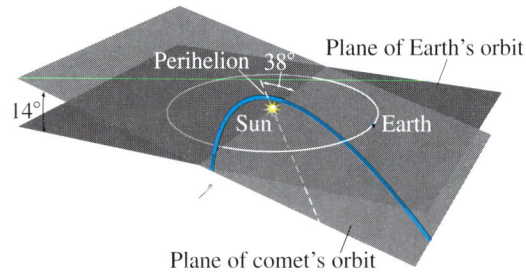

Architecture and Conics

Usually we think of the conic sections as the results of intersecting a cone with a plane. However, any cylinder sliced on an angle will reveal an ellipse in the cross section, as seen in the Tycho Brahe Planetarium in Copenhagen.

A > The Distance Formula

In the triangle of Figure 9.25, the distance a between $(-2, -1)$ and $(6, -1)$ is $6 - (-2) = 8$, and the distance b between $(6, 5)$ and $(6, -1)$ is $5 - (-1) = 6$. To find the distance c, we need to use the Pythagorean theorem. According to that theorem, if the legs of a right triangle are a and b and the hypotenuse is c, then

$$c^2 = a^2 + b^2$$

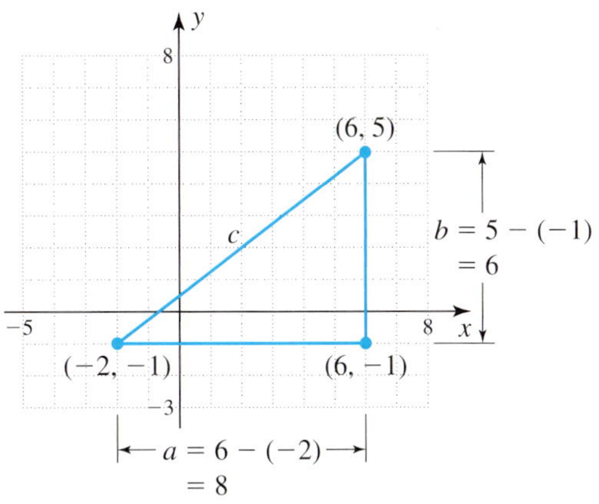

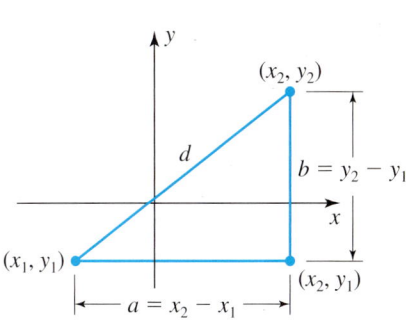

> Figure 9.25

> Figure 9.26

Thus,
$$c^2 = 8^2 + 6^2 = 64 + 36$$
$$c^2 = 100$$
$$c = \pm\sqrt{100} = \pm 10$$

Since c represents a distance, we discard the negative answer and conclude that $c = 10$. We can repeat a similar argument to find the distance d between any two points (x_1, y_1) and (x_2, y_2), as shown in Figure 9.26. As before, the distance a is $|x_2 - x_1|$ and the distance b is $|y_2 - y_1|$. By the Pythagorean theorem, we have

$$d^2 = |x_2 - x_1|^2 + |y_2 - y_1|^2$$
$$d = \sqrt{(x_2 - x_1)^2 + (y_2 - y_1)^2}$$

The square of a number is the same as the square of its opposite, so we don't need the absolute-value signs when we square a quantity. Also, the order of x_1 and x_2 does not matter. That is, $(x_2 - x_1)^2 = (x_1 - x_2)^2$. Here is a summary of what we have done.

THE DISTANCE FORMULA

The distance between the points (x_1, y_1) and (x_2, y_2) is

$$d = \sqrt{(x_2 - x_1)^2 + (y_2 - y_1)^2}$$

EXAMPLE 1 Using the distance formula
Find the distance between the given points:

a. $A(1, 1)$ and $B(5, 4)$ **b.** $C(2, -2)$ and $D(-2, 5)$
c. $E(-2, 1)$ and $F(-2, 3)$

SOLUTION
a. If we let $x_1 = 1$, $y_1 = 1$, $x_2 = 5$, and $y_2 = 4$, then
$$d = \sqrt{(5 - 1)^2 + (4 - 1)^2} = \sqrt{(4)^2 + (3)^2} = \sqrt{25} = 5$$

b. Here $x_1 = 2$, $y_1 = -2$, $x_2 = -2$, and $y_2 = 5$. Thus,
$$d = \sqrt{(-2 - 2)^2 + (5 + 2)^2} = \sqrt{(-4)^2 + (7)^2} = \sqrt{65}$$

c. Now $x_1 = -2$, $y_1 = 1$, $x_2 = -2$, and $y_2 = 3$. Hence,
$$d = \sqrt{[-2 - (-2)]^2 + (3 - 1)^2} = \sqrt{(0)^2 + (2)^2} = \sqrt{4} = 2$$

Note that EF is a vertical line, so that $d = |3 - 1| = 2$. You do not need the distance formula when finding the distance between two points that are on a horizontal or vertical line.

PROBLEM 1
Find the distance between the given points:

a. $A(2, 2)$ and $B(8, 10)$
b. $C(-3, 2)$ and $D(5, 4)$
c. $E(-4, 3)$ and $F(-4, 7)$

Answers to PROBLEMS

1. **a.** 10 **b.** $\sqrt{68} = 2\sqrt{17}$ **c.** 4

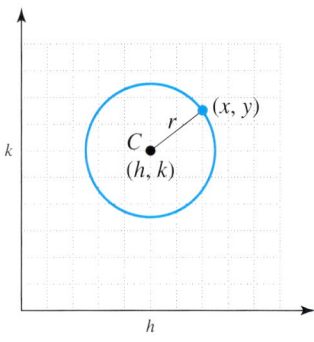

> Figure 9.27 The distance from (h, k) to (x, y) is r.

B ⟩ Finding an Equation of a Circle

Can you define a circle? A **circle** is defined as a set of points in a plane equidistant from a fixed point. The fixed point is called the c*enter* and the given distance is the *radius*. To find the equation of a circle of radius r, suppose the center is at a point $C(h, k)$; see Figure 9.27. The distance from C to any point $P(x, y)$ on the circle is found by the distance formula. Since r is the radius, this distance must be r. Thus,

$\sqrt{(x_2 - x_1)^2 + (y_2 - y_1)^2} = d$ Distance formula

$\sqrt{(x - h)^2 + (y - k)^2} = r$ Substitute (h, k) for (x_2, y_2).

$(x - h)^2 + (y - k)^2 = r^2$ Square both sides.

We then have the following definition.

> **GRAPHING FORM OF THE EQUATION OF A CIRCLE CENTERED AT (h, k)**
>
> An equation of a circle with **radius** r and with **center** at $C(h, k)$ is
>
> $$(x - h)^2 + (y - k)^2 = r^2$$

EXAMPLE 2   Finding an equation of a circle
Find the equation of the circle with center at $(3, -5)$ and radius 2.

SOLUTION   Here, the center $(h, k) = (3, -5)$ and $r = 2$. This means $h = 3$, $k = -5$, and $r = 2$. Using the formula, we have

$(x - h)^2 + (y - k)^2 = r^2$
$(x - 3)^2 + [(y - (-5)]^2 = 2^2$ Substitute $h = 3$, $k = -5$, $r = 2$.
$(x - 3)^2 + (y + 5)^2 = 4$ Simplify.

PROBLEM 2
Find an equation of the circle with center at $(-1, 2)$ and radius 3.

EXAMPLE 3   Finding an equation of a circle centered at the origin $(0, 0)$
Find an equation of a circle of radius 3 and with the center at the origin $(0, 0)$.

SOLUTION   The center is at $(h, k) = (0, 0)$. Thus, $h = 0$, $k = 0$, and $r = 3$. Substituting $h = 0$, $k = 0$, $r = 3$ in $(x - h)^2 + (y - k)^2 = r^2$ gives

$(x - 0)^2 + (y - 0)^2 = 3^2$
$x^2 + y^2 = 9$

PROBLEM 3
Find an equation of a circle of radius $\sqrt{3}$ and with center at the origin.

In general, we have the following definition.

> **GRAPHING FORM OF A CIRCLE CENTERED AT THE ORIGIN $(0, 0)$**
>
> An equation of a circle of radius r with center at the origin $(0, 0)$ is
>
> $$x^2 + y^2 = r^2$$

C ⟩ Finding the Center and Radius of a Circle

If we have an equation of a circle, we can write it in graphing form $(x - h)^2 + (y - k)^2 = r^2$ and find the center and radius. For example, if a circle has equation $(x - 3)^2 + (y - 4)^2 = 5^2$, then $h = 3$, $k = 4$, and $r = 5$. Thus, the equation $(x - 3)^2 + (y - 4)^2 = 5^2$ is the equation of a circle of radius 5 with center at $(3, 4)$.

Answers to PROBLEMS
2. $(x + 1)^2 + (y - 2)^2 = 9$
3. $x^2 + y^2 = 3$

EXAMPLE 4 Finding the center and radius of a circle

Find the center and radius and sketch the graph of the circle whose equation is

$$(x + 2)^2 + (y - 1)^2 = 9.$$

SOLUTION We first write the equation in graphing form

$$(x - h)^2 + (y - k)^2 = r^2$$

Thus,

$$[x - (-2)]^2 + (y - 1)^2 = 3^2$$

We have $h = -2$, $k = 1$, and $r = 3$. The center is at $(h, k) = (-2, 1)$ and the radius is 3. The sketch is shown in Figure 9.28.

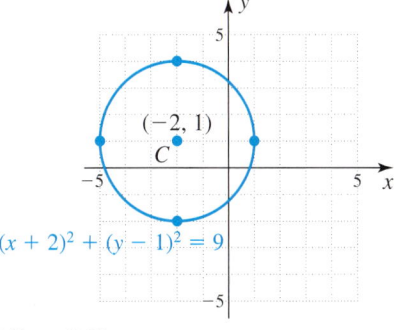

>Figure 9.28

PROBLEM 4

Find the center and radius and sketch the graph of the circle whose equation is

$$(x - 3)^2 + (y + 5)^2 = 1$$

Answer on page 684

EXAMPLE 5 Sketching the graph of a circle centered at the origin

Find the center and radius and sketch the graph of

$$x^2 + y^2 = 16$$

SOLUTION This equation can be written as $x^2 + y^2 = 4^2$, the equation of a circle of radius 4 centered at the origin $(0, 0)$. The graph is shown in Figure 9.29.

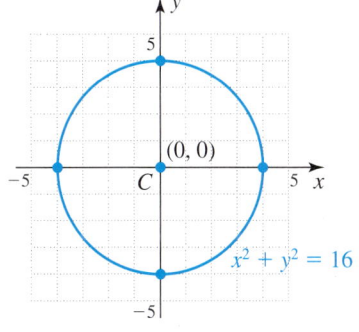

>Figure 9.29

PROBLEM 5

Find the center and radius and sketch the graph of $x^2 + y^2 = 10$.

Answer on page 684

Calculator Corner

Graphing Circles

Most calculators graph only certain types of algebraic expressions. Do you know what types? Functions! Would your calculator graph $x^2 + y^2 = 25$ from Example 5? The vertical line test should convince you that $x^2 + y^2 = 25$ doesn't represent a function. How can we graph it? We do it by first solving for y to obtain $y = \pm\sqrt{25 - x^2}$. Then we graph each of the two halves of the circle, $Y_1 = \sqrt{25 - x^2}$ (the top half) and $Y_2 = -\sqrt{25 - x^2}$ (the bottom half). Unfortunately, in a standard window the result looks like an ellipse (Window 1).

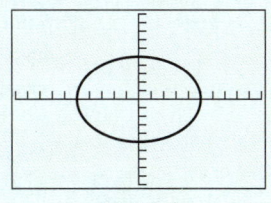
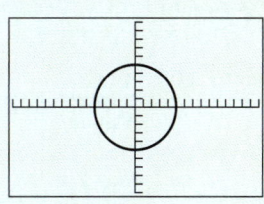

Window 1   Window 2

Do you know why? Because the units on the x-axis are longer than the units on the y-axis; the graph is wider than it is high. To fix it, use a "square window" (ZOOM 5 on a TI-83 Plus) to obtain the graph shown in Window 2.

(You can save time when you graph $y = \pm\sqrt{25 - x^2}$ if you enter $Y_1 = \sqrt{25 - x^2}$ and then $Y_2 = -Y_1$. With a TI-83 Plus, place the cursor by $Y_2 =$ and press (−)   VARS ▶ 1 1 .)

In Section 8.1 we used a process called completing the square to solve quadratic equations. In Example 6, we will use completing the square to find the center of the circle.

EXAMPLE 6 Sketching the graph of a circle not centered at the origin

Find the center and radius and sketch the graph of

$$x^2 - 4x + y^2 + 6y + 8 = 0.$$

SOLUTION We must find the center and the radius by writing the equation in graphing form $(x - h)^2 + (y - k)^2 = r^2$. We can do this by subtracting 8 from both sides and then completing the squares on x and y.

$x^2 - 4x + y^2 + 6y + 8 = 0$	Given
$x^2 - 4x + \underline{} + y^2 + 6y + \underline{} = -8$	
$x^2 - 4x + 4 + y^2 + 6y + 9 = -8 + 4 + 9$	To complete the squares add $(\frac{1}{2} \cdot 4)^2$ and $(\frac{1}{2} \cdot 6)^2$.
$(x - 2)^2 + [y - (-3)]^2 = 5$	
$(x - 2)^2 + [y - (-3)]^2 = (\sqrt{5})^2$	Remember that $(\sqrt{5})^2 = 5$.

Thus, the center is at $(2, -3)$, and the radius is $\sqrt{5} \approx 2.2$. See Figure 9.30.

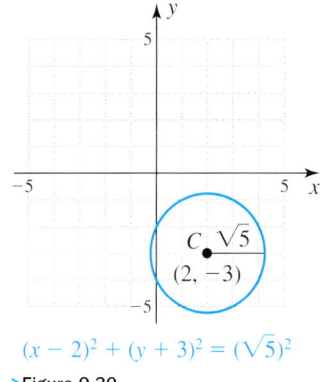

$(x - 2)^2 + (y + 3)^2 = (\sqrt{5})^2$

>Figure 9.30

PROBLEM 6

Find the center and radius and sketch the graph of

$$x^2 - 4x + y^2 - 6y + 9 = 0.$$

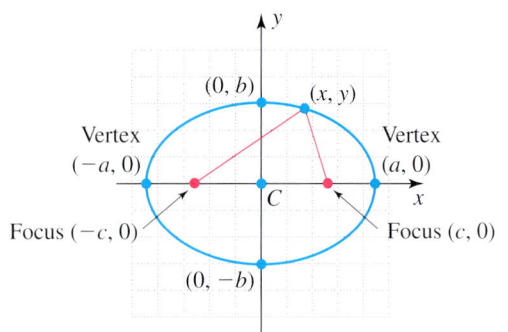

>Figure 9.31

D > Graphing Ellipses

If you look at the diagram in the *Getting Started* at the beginning of this section, you will see the drawing of part of an ellipse. An *ellipse* is the set of points in a plane such that the sum of the distances of each point from two fixed points (called the **foci**; singular, *focus*) is a constant. If the coordinates of the foci are $(c, 0)$ and $(-c, 0)$, then the center of the ellipse is at the origin, and the x- and y-intercepts are given by $x = \pm a$ and $y = \pm b$. The points $(a, 0)$ and $(-a, 0)$ are called the vertices (singular, vertex). The graph is shown in Figure 9.31.

Answers to PROBLEMS

4. Center $(3, -5)$; $r = 1$

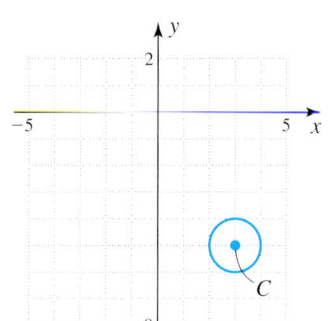

5. Center $(0, 0)$; $r = \sqrt{10} \approx 3.2$

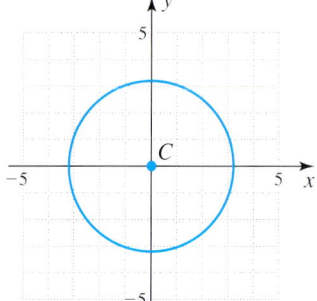

6. Center $(2, 3)$; $r = 2$

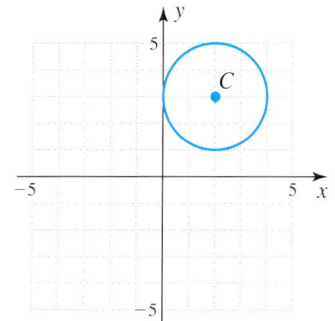

We can use the distance formula to find the equation of an ellipse (see *Using Your Knowledge* in the Exercises) but for the time being, we assume the following.

> **GRAPHING FORM OF THE EQUATION OF AN ELLIPSE WITH CENTER AT (0, 0)**
>
> The equation of the **ellipse** with center at (0, 0) whose *x*-intercepts are $(a, 0)$ and $(-a, 0)$ and whose *y*-intercepts are $(0, b)$ and $(0, -b)$ is
>
> $$\frac{x^2}{a^2} + \frac{y^2}{b^2} = 1, \quad \text{where } a^2 > b^2, a \neq 0, b \neq 0$$
>
> The vertices are at $(a, 0)$ and $(-a, 0)$.
>
> The equation of the **ellipse** with center at (0, 0) whose *x*-intercepts are $(b, 0)$ and $(-b, 0)$ and whose *y*-intercepts are $(0, a)$ and $(0, -a)$ is
>
> $$\frac{x^2}{b^2} + \frac{y^2}{a^2} = 1, \quad \text{where } a^2 > b^2, a \neq 0, b \neq 0$$
>
> The vertices are at $(0, a)$ and $(0, -a)$.
>
> If *a* and *b* are equal, the **ellipse** is a **circle**.

EXAMPLE 7   Graphing an ellipse with center at (0, 0)

Find the coordinates of the center, the *x*- and *y*-intercepts, and the vertices, and sketch the graph of

$$4x^2 + 25y^2 = 100$$

SOLUTION   To make sure we have an ellipse with center at (0, 0), we write the equation in graphing form

$$\frac{x^2}{a^2} + \frac{y^2}{b^2} = 1$$

with 1 on the right side.

If we divide each term by 100 (to make the right side 1), we have

$$\frac{x^2}{25} + \frac{y^2}{4} = 1$$

where $a^2 = 25$ with $a = 5$ and $b^2 = 4$ with $b = 2$.

The *x*-intercepts are $(5, 0)$ and $(-5, 0)$, the *y*-intercepts are $(0, 2)$ and $(0, -2)$, and the center is at $(0, 0)$. The vertices are at $V_1 = (5, 0)$ and $V_2 = (-5, 0)$. We then pass the ellipse through the four intercepts, as shown in Figure 9.32.

We can graph the original equation $4x^2 + 25y^2 = 100$ by letting $x = 0$ to obtain

$$25y^2 = 100$$
$$y^2 = 4$$
$$y = \pm\sqrt{4} = \pm 2$$

Letting $y = 0$ will yield $x = \pm 5$, as before.

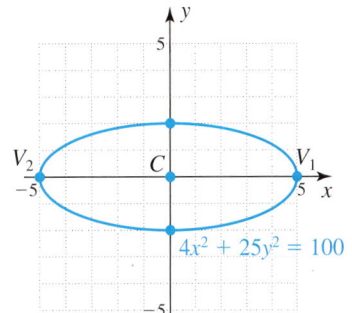
> Figure 9.32

PROBLEM 7

Find the center, *x*- and *y*-intercepts, and the vertices, and sketch the graph of

$$9x^2 + 4y^2 = 36$$

Answers to PROBLEMS

7. Center (0, 0); *x*-int (± 2, 0), *y*-int (0, ± 3); $V_1 = (0, 3), V_2 = (0, -3)$

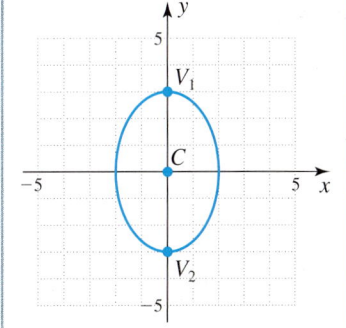

GRAPHING FORM OF THE EQUATION OF AN ELLIPSE WITH CENTER AT (h, k)

The equation of the ellipse with center at (h, k) is

$$\frac{(x-h)^2}{a^2} + \frac{(y-k)^2}{b^2} = 1, \quad \text{where } a^2 > b^2, a \neq 0, b \neq 0$$

The vertices are horizontally $\pm a$ units from (h, k).

The equation of the ellipse with center at (h, k) is

$$\frac{(x-h)^2}{b^2} + \frac{(y-k)^2}{a^2} = 1, \quad \text{where } a^2 > b^2, a \neq 0, b \neq 0$$

The vertices are vertically $\pm a$ units from (h, k).

EXAMPLE 8 Graphing an ellipse not centered at the origin

Find the coordinates of the center, the vertices, and sketch the graph of

$$\frac{(x-3)^2}{4} + \frac{(y+1)^2}{9} = 1$$

SOLUTION The center of this ellipse is at $(3, -1)$. Use the values of a and b to determine the dimensions of the ellipse, as shown in Figure 9.33. The vertices are vertically ± 3 units from $(3, -1)$. The coordinates of the vertices are $V_1 = (3, 2)$ and $V_2 = (3, -4)$.

The graph of the ellipse is shown in Figure 9.34.

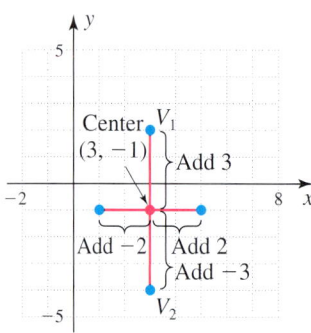
>Figure 9.33

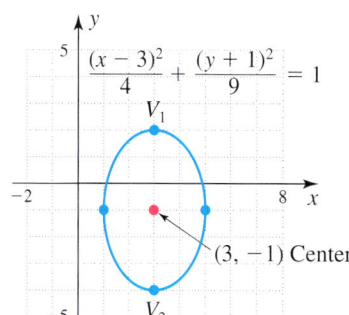
>Figure 9.34

PROBLEM 8

Find the coordinates of the center, the vertices, and sketch the graph of

$$\frac{(x+2)^2}{9} + \frac{(y-1)^2}{4} = 1$$

Answers to PROBLEMS

8. Center $(-2, 1)$; $V_1 = (1, 1)$, $V_2 = (-5, 1)$

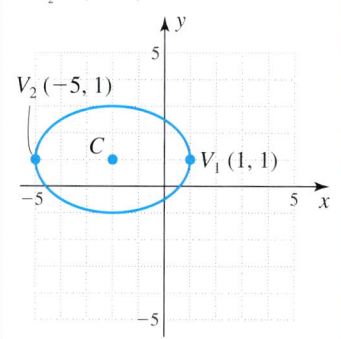

Calculator Corner

Graphing Ellipses

The ellipses in Examples 7 and 8 can be graphed using the same technique we used to graph the circle of Example 5. Thus, if we solve for y in Example 7, we have

$$y = \pm\sqrt{\frac{100 - 4x^2}{25}}$$

Graphing the top and bottom halves of the ellipse in a square window produces the graph of $4x^2 + 25y^2 = 100$, as shown in the window. Now try Example 8.

We can find the equation of an ellipse with foci on one of the axes and with center at the origin if we know its *x*- and *y*-intercepts. The ellipse of Example 7 passes through (± 5, 0) and (0, ± 2) and has equation

$$\frac{x^2}{5^2} + \frac{y^2}{2^2} = 1$$

Similarly, if the ellipse passes through (± 3, 0) and (0, ± 4), its equation would be

$$\frac{x^2}{3^2} + \frac{y^2}{4^2} = 1$$

We will use this idea in Problems 62–65. Finally, make sure you know how to tell the difference between the graph of a circle and that of an ellipse. The equation

$$Ax^2 + By^2 = C \quad (A, B, \text{ and } C \text{ positive})$$

has a **circle** as its graph when $A = B$ and an **ellipse** when $A \neq B$.

E > Applications Involving Circles and Ellipses

EXAMPLE 9   Solving an application with a conic equation

To protect the baseball mound in rainy weather it is covered with a tarp. The tarp shown here has a diameter of 20 feet. Assuming the center of the tarp is placed at the center of an *x-y* coordinate system, write the equation that describes the circular edge of the circular tarp.

SOLUTION   The outer edge of the mound is a circle. The assumption that its center coincides with (0, 0) allows us to use the graphing form of the equation of a circle, $x^2 + y^2 = r^2$. To complete the equation of the circle we need to know the radius. The radius can be obtained by taking half of the diameter or $r = \frac{1}{2}$ of $20 = 10$. Thus, the equation of the outer edge of the circular tarp is $x^2 + y^2 = 100$.

PROBLEM 9

The shapes of leaf types vary. The one shown in Figure 9.35 is roughly the shape of an ellipse. Assuming the center of the leaf is placed at the center of an *x-y* coordinate system with the longest measurement coinciding with the *x*-axis, write the equation of the ellipse. The longer measurement or axis of the leaf is 4 inches and the shorter axis is 2 inches.

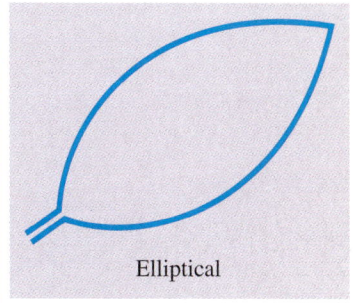

Elliptical

> Figure 9.35

Answers to PROBLEMS

9. $\frac{x^2}{4} + \frac{y^2}{1} = 1$

Exercises 9.2

Boost your grade at mathzone.com!
> Practice Problems
> NetTutor
> Self-Tests
> e-Professors
> Videos

< A > The Distance Formula   In Problems 1–10, find the distance between the points.

1. $A(2, 4), B(-1, 0)$
2. $A(3, -2), B(8, 10)$
3. $C(-4, -5), D(-1, 3)$
4. $C(5, 7), D(-2, 3)$
5. $E(4, 8), G(1, -1)$
6. $H(-2, -2), I(6, -4)$
7. $A(3, -1), B(-2, -1)$
8. $C(-2, 3), D(4, 3)$
9. $E(-1, 2), F(-1, -4)$
10. $G(-3, 2), H(-3, 5)$

< B > Finding the Equation of a Circle   In Problems 11–20, find an equation of a circle with the given center and radius.

11. Center $(3, 8)$, radius 2
12. Center $(2, 5)$, radius 3
13. Center $(-3, 4)$, radius 5
14. Center $(-5, 2)$, radius 5
15. Center $(-3, -2)$, radius 4
16. Center $(-1, -7)$, radius 9
17. Center $(2, -4)$, radius $\sqrt{5}$
18. Center $(3, -5)$, radius $\sqrt{7}$
19. Radius 3, center at the origin
20. Radius 2, center at the origin

< C > Finding the Center and Radius of a Circle   In Problems 21–42, find the center and radius of the circle and sketch the graph. (For irrational radii, approximate to the nearest tenth.)

21. $(x - 1)^2 + (y - 2)^2 = 9$

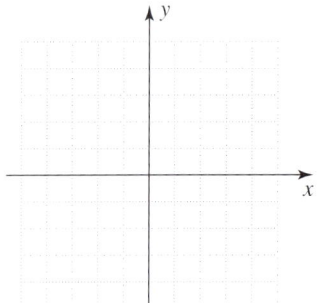

22. $(x - 2)^2 + (y - 1)^2 = 4$

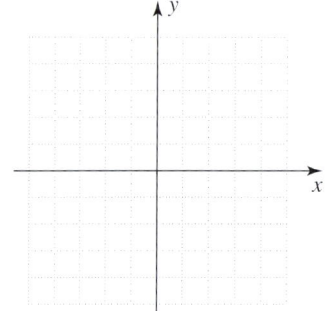

23. $(x + 1)^2 + (y - 2)^2 = 4$

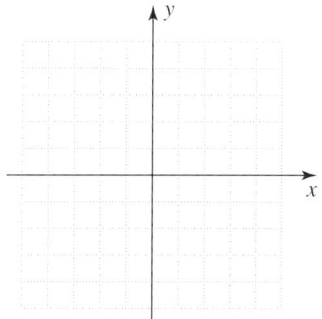

24. $(x + 2)^2 + (y - 1)^2 = 9$

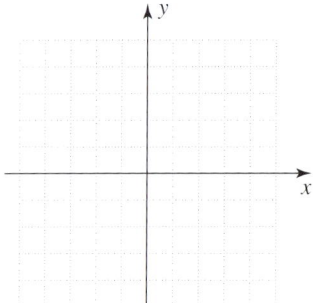

25. $(x - 1)^2 + (y + 2)^2 = 1$

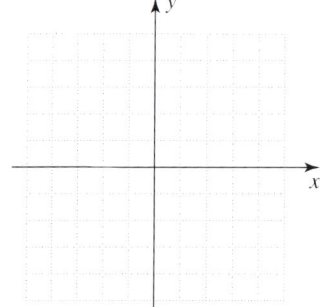

26. $(x - 2)^2 + (y + 1)^2 = 4$

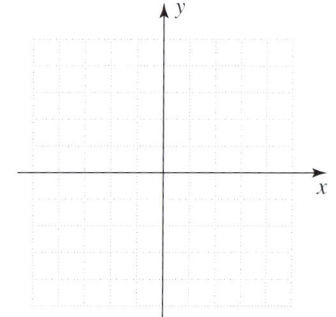

27. $(x + 2)^2 + (y + 1)^2 = 9$

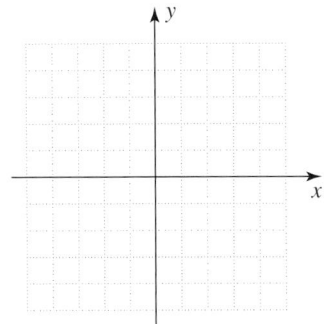

28. $(x + 3)^2 + (y + 1)^2 = 4$

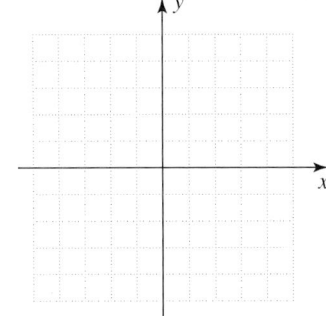

29. $(x - 1)^2 + (y - 1)^2 = 7$

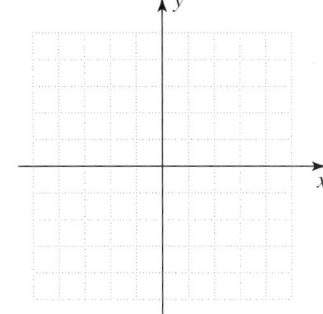

30. $(x - 1)^2 + (y - 1)^2 = 3$

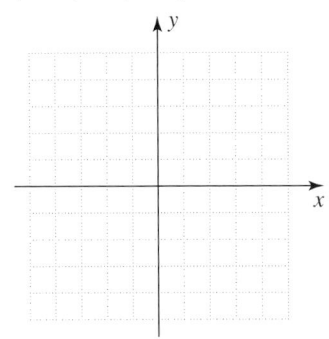

31. $x^2 - 6x + y^2 - 4y + 9 = 0$

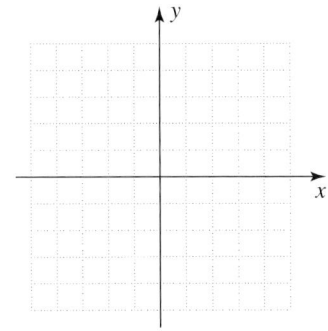

32. $x^2 - 6x + y^2 - 2y + 9 = 0$

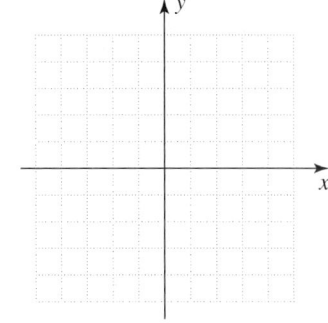

33. $x^2 + y^2 - 4x + 2y - 4 = 0$

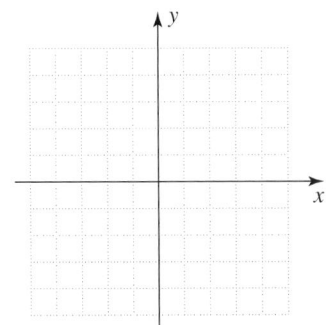

34. $x^2 + y^2 + 2x - 4y - 4 = 0$

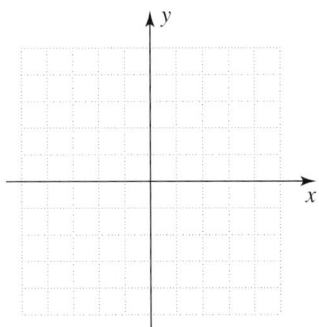

35. $x^2 + y^2 - 25 = 0$

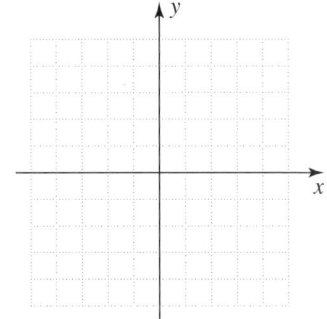

36. $x^2 + y^2 - 9 = 0$

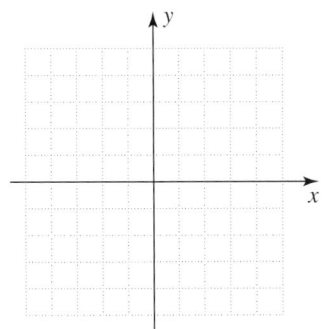

37. $x^2 + y^2 - 7 = 0$

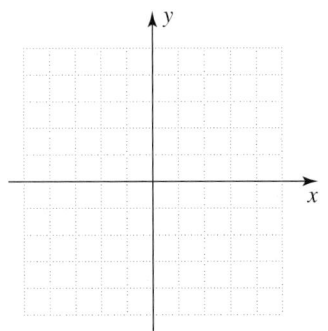

38. $x^2 + y^2 - 3 = 0$

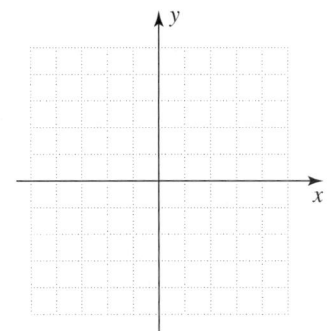

39. $x^2 + y^2 + 6x - 2y = -6$

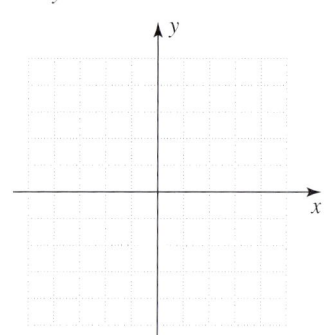

40. $x^2 + y^2 + 4x - 2y = -4$

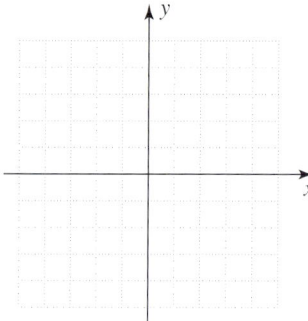

41. $x^2 + y^2 - 6x - 2y + 6 = 0$

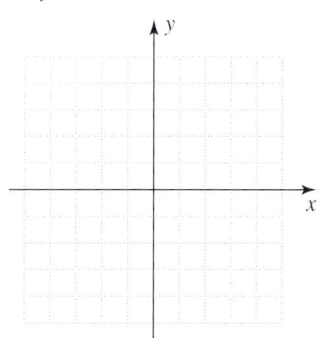

42. $x^2 + y^2 - 4x - 6y + 12 = 0$

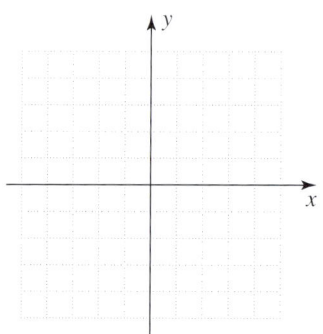

〈D〉 Graphing Ellipses In Problems 43–56, graph the ellipse. Give the coordinates of the center and the values of a and b.

43. $25x^2 + 4y^2 = 100$

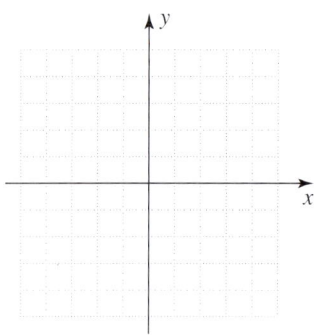

44. $9x^2 + 4y^2 = 36$

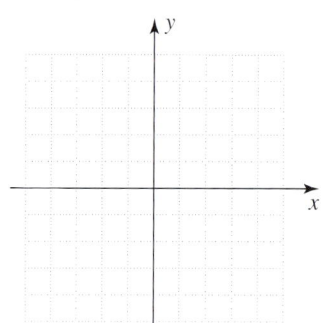

45. $x^2 + 4y^2 = 4$

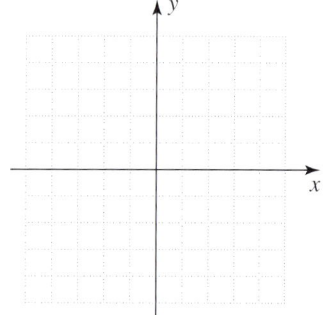

46. $x^2 + 9y^2 = 9$

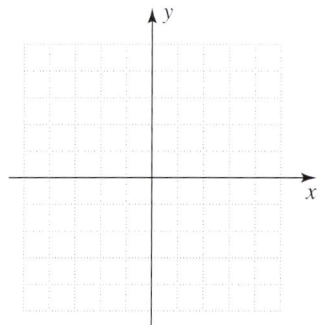

47. $x^2 + 4y^2 = 16$

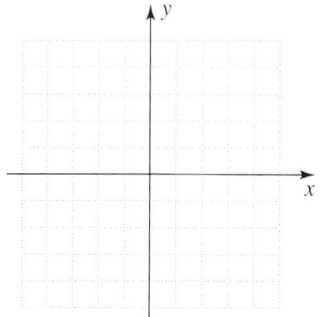

48. $x^2 + 9y^2 = 25$

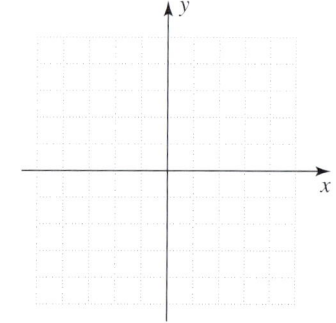

49. $\dfrac{x^2}{9} + \dfrac{y^2}{16} = 1$

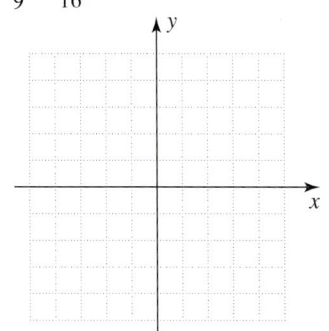

50. $\dfrac{x^2}{4} + \dfrac{y^2}{1} = 1$

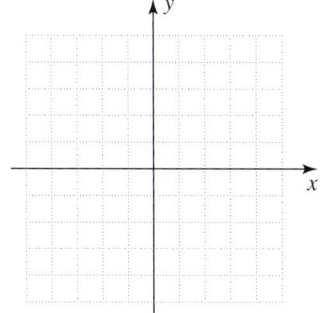

51. $\dfrac{(x-1)^2}{4} + \dfrac{(y-2)^2}{9} = 1$

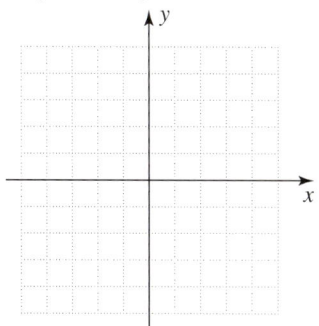

52. $\dfrac{(x-2)^2}{9} + \dfrac{(y-1)^2}{4} = 1$

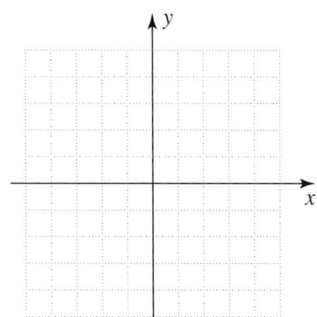

53. $\dfrac{(x-2)^2}{9} + \dfrac{(y+3)^2}{4} = 1$

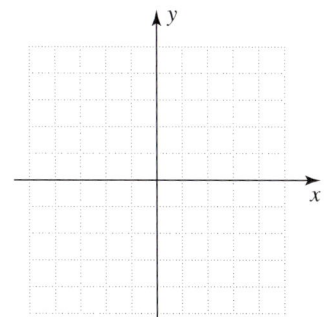

54. $\dfrac{(x-1)^2}{4} + \dfrac{(y+2)^2}{9} = 1$

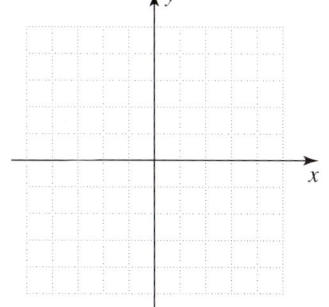

55. $\dfrac{(x-1)^2}{16} + \dfrac{(y-1)^2}{9} = 1$

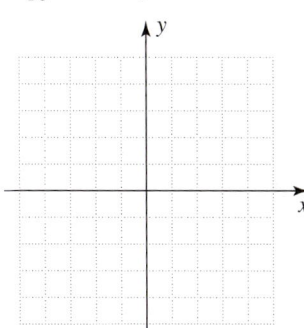

56. $\dfrac{(x-2)^2}{9} + \dfrac{(y-1)^2}{16} = 1$

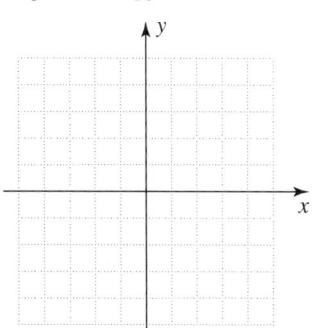

In Problems 57–61, find an equation of the circle centered at the origin and

57. Passing through the point (4, 3)

58. Passing through the point (3, 4)

59. Passing through the point (−5, −12)

60. x-intercepts ± 5

61. y-intercepts ± 3

In Problems 62–65, find an equation of the ellipse centered at the origin and passing through the given points.

62. Points (± 7, 0) and (0, ± 2)

63. Points (± 2, 0) and (0, ± 6)

64. Points (± 3, 0) and (0, ± 7)

65. Points (± 6, 0) and (0, ± 4)

E Applications Involving Circles and Ellipses

66. *Bridges* A circular arch for a bridge has a 100-foot span. If the height of the arch above the water is 25 feet, find an equation of the circle containing the arch if the center is at the origin as shown. [*Hint:* If the radius is r, $(50, r - 25)$ must satisfy the equation $x^2 + y^2 = r^2$.]

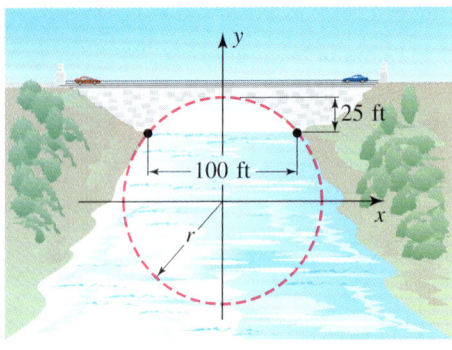

67. *Cylinders* A cylindrical drum is cut to make a barbecue grill. The end of the resulting grill is 5 inches high and 20 inches wide at the top. What was the radius r of the original drum?

68. *Semicircle* The larger semicircle has a radius of 15 feet.
   a. If the x-axis is placed at ground level, what is the equation of the circle?
   b. If the 19 vertical bars inside the smaller semicircle are 1 foot apart, what is the length of the longest vertical bar?
   c. What is the length of the bar 1 foot to the right of the longest bar?

69. *Semicircle* A portion of a circle with a 15-foot radius (blue) is shown. If the vertical bar (red) is 5 feet, how long is the horizontal bar (green)? (*Hint:* Look at the diagram.)

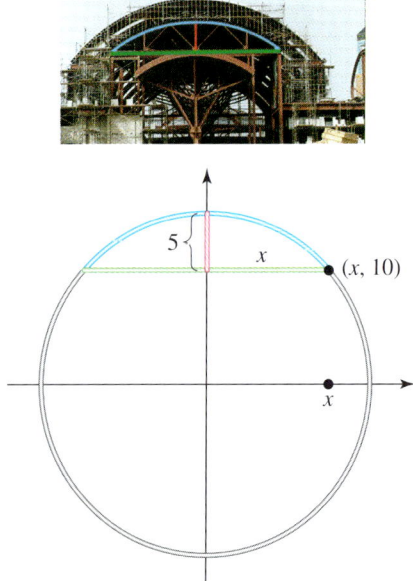

70. *Drain pipe* The elliptical drain pipe is 150 centimeters wide and 100 centimeters high. If the origin is placed at the center of the pipe, what is the equation in graphing form of the elliptical opening?

71. *Signs* The elliptical portion of the sign is 8 feet wide and 5 feet high. What is the graphing form of the equation of this ellipse if the origin is at the center of the sign?

72. *Signs* The top half of a sign is half of an ellipse 7 feet high and 13 feet wide. If the origin is at the center of the sign, what is the equation of the complete ellipse in graphing form?

73. *Trucks* The elliptical portion of the tanker truck is 8 feet wide and 6 feet high. What is the graphing form of equation of the ellipse whose origin is at the center of the elliptical portion of the truck?

74. *Running tracks* An elliptical running track is 100 yards long and 50 yards wide.

  a. If the origin is at its center and the longer side is along the x-axis, what is the equation of the ellipse?

  b. A running strip for pole vaulting is parallel to the y-axis 20 yards from the right-hand end of the track. Both ends of this strip are 5 yards from the running track. How long is the strip?

75. *Earth's orbit* The orbit of the Earth around the sun (one of the foci) is an ellipse as shown. The equation of the ellipse is written as

$$\frac{x^2}{a^2} + \frac{y^2}{b^2} = 1$$

where x and y are in millions of miles.

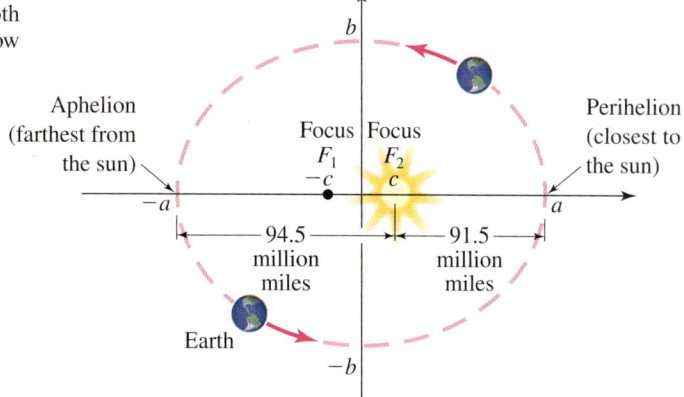

  a. Find a.
  b. Find c.
  c. If $b^2 = a^2 - c^2$, find b. Round the answer to two decimal places.

76. Arch A semielliptic arch is supporting a bridge spanning a river 50 feet wide. The center of the arch is 20 feet above the center of the river. Write an equation of the ellipse in which the x-axis coincides with the water level and the y-axis passes through the center of the arch.

77. Plastic plates Have you eaten from plastic food plates lately? These plates are made using an elliptical mold 12 inches long and 9 inches wide.

  a. Write in standard form the equation of the outside ellipse of the mold.

  b. Find the width of the dish at a distance of 4 inches from the center. (This is the width in the direction perpendicular to the x-axis.)

78. Arch A semielliptic arch spanning a river has the dimensions shown. What is the height h of the arch at a distance of 10 feet from the center?

79. Arch An 8-foot-wide boat with a mast whose top is 15 feet above the water is about to go under the bridge of Problem 76. How close can it get to the bank on the right side of the river and still fit under the bridge?

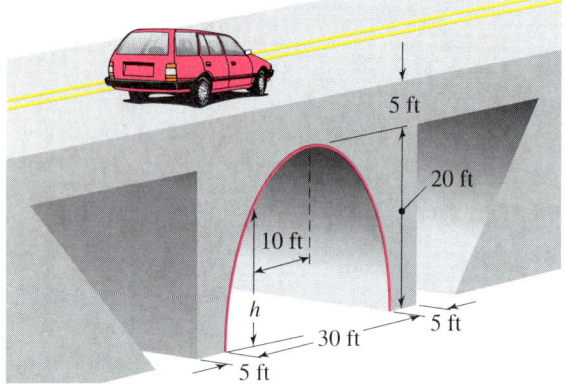

Eccentricity Use the following information for Problems 80–82. The **eccentricity** e of a circle or an ellipse is defined by $e = c/a$, where $c^2 = a^2 - b^2$. The closer to a circle, the smaller the eccentricity. Pluto has the most **eccentric** orbit, measured at 0.248, but the eccentricity of Venus and Earth, which have almost circular orbits, is very close to 0. The eccentricity of a circle is indeed 0.

Source: http://esep10.phys.utk.edu.

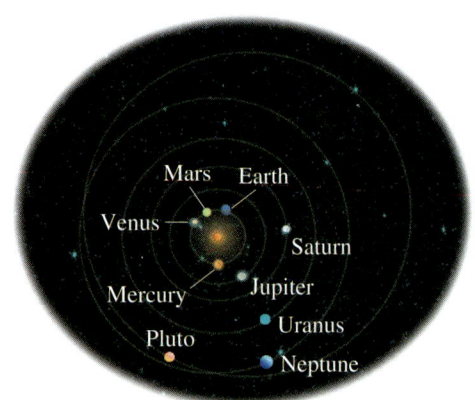

80. Rewrite the equation of the circle in Example 5 as $\frac{x^2}{25} + \frac{y^2}{25} = 1$. Find:

  a. a^2 **b.** b^2 **c.** c^2
  d. c **e.** e

81. Rewrite the equation of the ellipse in Example 7 as $\frac{x^2}{25} + \frac{y^2}{4} = 1$. Find:

  a. a^2 **b.** b^2 **c.** c^2
  d. c **e.** e

82. Rewrite the equation of the ellipse in Problem 7 as $\frac{x^2}{4} + \frac{y^2}{9} = 1$. Find:

  a. a^2 **b.** b^2 **c.** c^2
  d. c **e.** e

❯❯❯ Using Your Knowledge

Ellipses Revisited   The definition of an ellipse is as follows:

ELLIPSE

An **ellipse** is the set of all points, the sum of whose distances from two fixed points $(c, 0)$ and $(-c, 0)$ is a constant $2a$ ($a > c$). Each fixed point is called a **focus**.

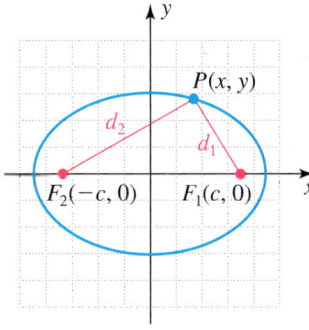

In Problems 83–88, we will prove that this definition leads to the equation of the ellipse that we gave in the text.

83. Suppose $P(x, y)$ is a point on the ellipse. Find the distance d_1 from P to F_1.

84. Find the distance d_2 from P to F_2.

85. The sum of the distances found in Problems 83 and 84 must be $2a$. Thus,

$$\sqrt{(x-c)^2 + y^2} + \sqrt{(x+c)^2 + y^2} = 2a$$

Rewrite this equation with one radical on each side. Then square both sides and simplify. What is your answer?

86. Rewrite your answer for Problem 85 with the radical on one side and then square both sides again and simplify. What is your answer?

87. In your answer for Problem 86, since $a > c$, let $a^2 - c^2 = b^2$. Isolate all the variables on the left side. What is your answer?

88. Divide all terms of the answer you obtained in Problem 87 by a^2b^2. What is your answer? If everything went well, you should have

$$\frac{x^2}{a^2} + \frac{y^2}{b^2} = 1$$

the equation of an ellipse.

❯❯❯ Write On

If we remove the restriction that $a^2 > b^2$, then the general equation for an ellipse centered at the origin and with foci on one of the axes is

$$\frac{x^2}{a^2} + \frac{y^2}{b^2} = 1$$

Discuss the resulting graph:

89. When $a > b$.

90. When $a < b$.

91. When $a = b$.

92. Can you explain why a circle is a special case of an ellipse?

Concept Checker

Fill in the blank(s) with the correct word(s), phrase, or mathematical statement.

93. The distance d between (x_1, y_1) and (x_2, y_2) is _____.

94. The equation of a circle of radius r with center at (h, k) is _____.

95. The equation of a circle of radius r with center at the origin is _____.

96. The equation of the ellipse with center at $(0, 0)$ whose x-intercepts are $(a, 0)$ and $(-a, 0)$ and whose y-intercepts are $(0, b)$ and $(0, -b)$ is _____.

$$\frac{x^2}{b^2} + \frac{y^2}{a^2} = 1 \qquad d = \sqrt{(x_2 - x_1)^2 + (y_2 - y_1)^2}$$

$$\frac{x^2}{a^2} + \frac{y^2}{b^2} = 1 \qquad d = \sqrt{(x_2 - x_1)^2 + (y_2 - y_1)^2}$$

$$r^2 + y^2 = x^2 \qquad (x - h)^2 + (y - k)^2 = r^2$$

$$x^2 + y^2 = r^2$$

Mastery Test

Graph:

97. $\dfrac{(x + 3)^2}{4} + \dfrac{(y + 1)^2}{9} = 1$

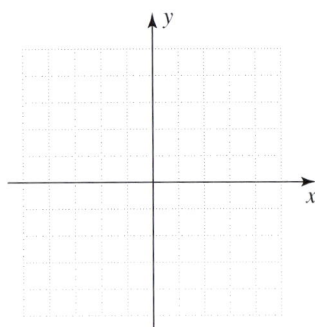

98. $\dfrac{(x - 3)^2}{4} + \dfrac{(y - 1)^2}{9} = 1$

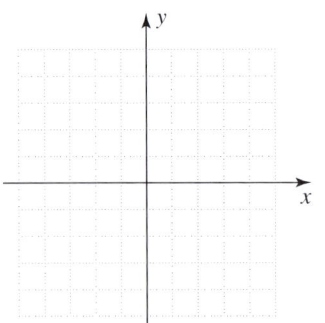

99. $4x^2 + 9y^2 = 36$

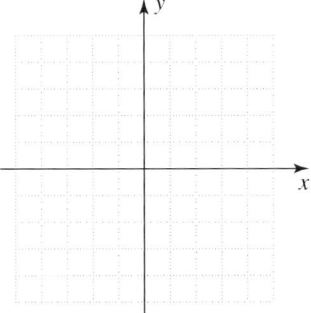

100. $9x^2 + 4y^2 = 36$

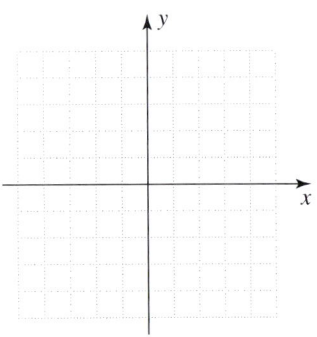

101. $x^2 - 6x + y^2 - 4y + 9 = 0$

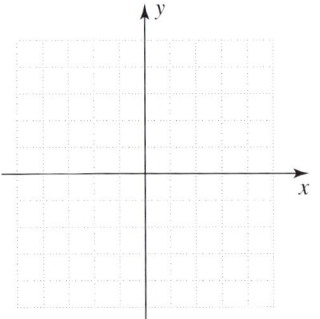

102. $x^2 + 6x + y^2 + 2y + 7 = 0$

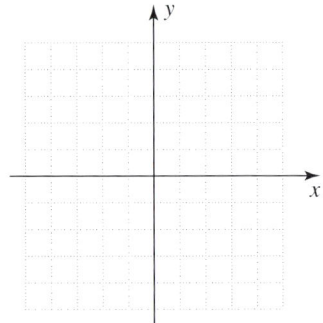

103. $x^2 + y^2 = 4$

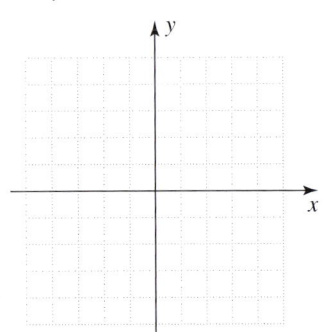

104. $x^2 + y^2 - 6 = 0$

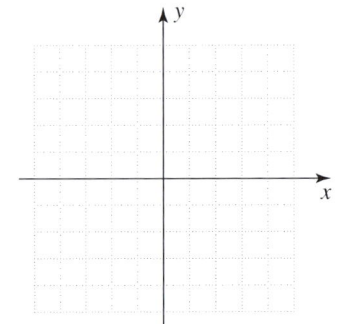

Find the center and radius and sketch the graph of the circle with the given characteristics.

105. A circle whose equation is $(x - 3)^2 + (y - 1)^2 = 4$.

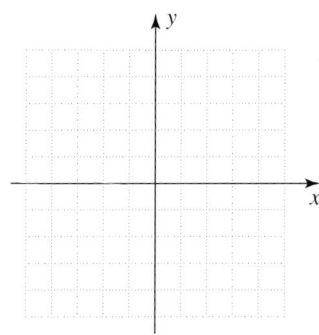

106. A circle whose equation is $(x + 3)^2 + (y + 1)^2 - 5 = 0$.

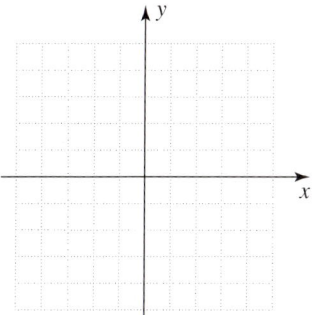

Find an equation of the circle with the given characteristics.

107. A circle of radius 5 and with center at the origin.

108. A circle of radius 4 and with center at the origin.

109. A circle with center at $(-3, 6)$ and radius 3.

110. A circle with center at $(-2, -3)$ and radius 2.

〉〉〉 Skill Checker

111. If you solve the equation $x^2 + y^2 = 25$ for x, you obtain two answers. What are these answers?

112. One of the answers in Problem 111 is always nonnegative. Look at the graph of $x^2 + y^2 = 25$ in Example 5. To what part of the graph does the positive answer correspond?

113. One of the answers in Problem 111 is always nonpositive. Look at the graph of $x^2 + y^2 = 25$ in Example 5. To what part of the graph does the negative answer correspond?

9.3 Hyperbolas and Identification of Conics

Objectives

A Graph hyperbolas.

B Identify conic sections by examining their equations.

To Succeed, Review How To . . .

1. Graph points on the Cartesian plane (pp. 166–168).
2. Find the *x*- and *y*-intercepts (pp. 171–173).

Getting Started

Hyperbolas in the Night

Have you seen any hyperbolas lately? Next time you are outside at night, look at building lights. Many of the beams of light you see are hyperbolas. Hyperbolas may even be in your living room, as in the wall in the photo. If you've studied chemistry or physics, you might also know that alpha particles (one of three types of radiation resulting from natural radioactivity) have trajectories (paths) that are hyperbolas. In this section, we shall study hyperbolas by examining their equations and then graphing them.

Hyperbolas in the Day

When traveling during the day, look for buildings or towers that might be hyperbolic. At the right is a hyperboloid tower in Kobe, Japan. Below is the St. Louis Science Center's James S. McDonnell Planetarium. The roof is a hyperboloid.

A › Graphing Hyperbolas

A **hyperbola** is the set of points in a plane such that the difference of the distances of each point from two fixed points (called the **foci**) is a constant. (See *Using Your Knowledge* in the Exercises.) Consider the equation

$$\frac{x^2}{4} - \frac{y^2}{9} = 1$$

When $x = 0$, $y^2 = -9$, so there are **no** y-intercepts because $y^2 = -9$ has **no** real-number solution. When $y = 0$, $x^2 = 4$, and $x = \pm 2$ are the x-intercepts, called the **vertices** of the hyperbola. The graph is shown in Figure 9.36.

Similarly, the graph of

$$\frac{y^2}{9} - \frac{x^2}{4} = 1$$

has no x-intercept because $y = 0$ yields $x^2 = -4$, which has no real-number solution. The y-intercepts are ± 3, the vertices. When the equation of the hyperbola is in this form, the vertices will be a units from the center and a can be found by taking the square root of the denominator of the positive variable term. Thus, in this equation, $a = 3$. The graph is shown in Figure 9.37.

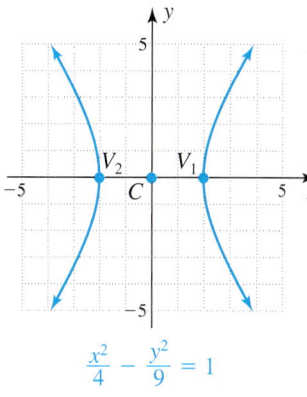

> Figure 9.36 $\quad\frac{x^2}{4} - \frac{y^2}{9} = 1$

> Figure 9.37 $\quad\frac{y^2}{9} - \frac{x^2}{4} = 1$

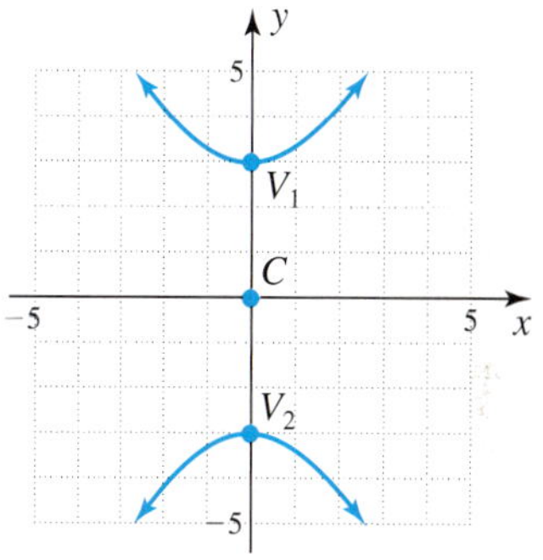

The hyperbola

$$\frac{x^2}{4} - \frac{y^2}{9} = 1$$

has $a^2 = 4$ and $a = 2$. The vertices are $(\pm 2, 0)$.

We can use the denominator of y^2 to help us with the graph. If we draw an auxiliary rectangle with sides parallel to the x- and y-axes and passing through the x-intercepts and the points on the y-axis corresponding to the square root of the denominator of

$$\frac{y^2}{9}$$

in this case ± 3, and then connect opposite corners of the rectangle with lines, the graph of the hyperbola will approach these lines, called *asymptotes*. The asymptotes are *not* part of the hyperbola, but are used to help graph it. The hyperbola never touches the asymptotes but gets closer and closer to them as x and y get larger and larger in absolute value. The graphs of the hyperbolas

$$\frac{x^2}{4} - \frac{y^2}{9} = 1 \quad \text{and} \quad \frac{y^2}{9} - \frac{x^2}{4} = 1$$

are shown in Figures 9.38 and 9.39.

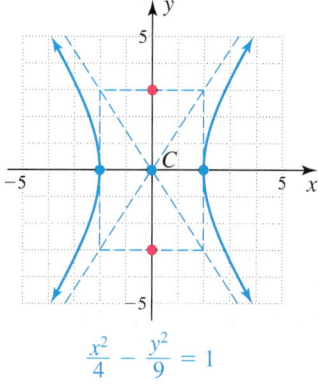

> Figure 9.38 $\quad\frac{x^2}{4} - \frac{y^2}{9} = 1$

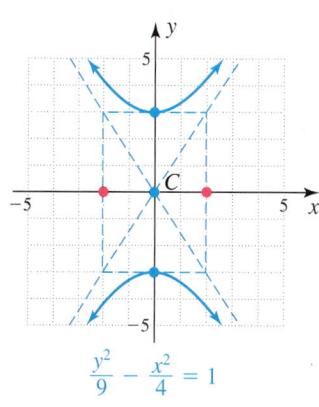

> Figure 9.39 $\quad\frac{y^2}{9} - \frac{x^2}{4} = 1$

Here is a summary of this discussion.

GRAPHING FORMS OF EQUATIONS OF HYPERBOLAS WITH CENTER AT (0, 0)

The graph of the equation

$$\frac{x^2}{a^2} - \frac{y^2}{b^2} = 1 \quad (1)$$

is a **hyperbola** centered at (0, 0) with vertices $x = (\pm a, 0)$

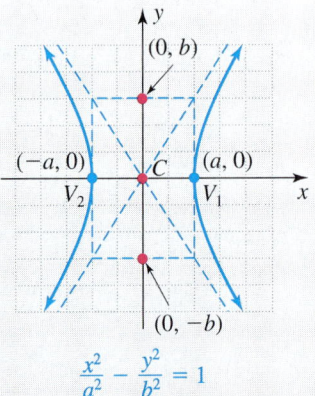

$$\frac{x^2}{a^2} - \frac{y^2}{b^2} = 1$$

The graph of the equation

$$\frac{y^2}{a^2} - \frac{x^2}{b^2} = 1 \quad (2)$$

is a **hyperbola** centered at (0, 0) with vertices $y = (0, \pm a)$

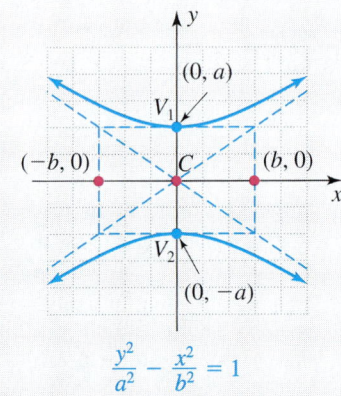

$$\frac{y^2}{a^2} - \frac{x^2}{b^2} = 1$$

Note: a^2 is always the denominator of the positive variable, and either $a^2 = b^2$, $a^2 > b^2$, or $a^2 < b^2$.

The **asymptotes** of either hyperbola are the lines through opposite corners of the auxiliary rectangle whose sides pass through $(\pm a, 0)$ and $(0, \pm b)$ for hyperbola (1), and $(0, \pm a)$ and $(\pm b, 0)$ for hyperbola (2).

PROCEDURE

Graphing a Hyperbola with Center at (0, 0)

1. Find and graph the points

   $(\pm a, 0)$ and $(0, \pm b)$ for $\dfrac{x^2}{a^2} - \dfrac{y^2}{b^2} = 1$

   or $(0, \pm a)$ and $(\pm b, 0)$ for $\dfrac{y^2}{a^2} - \dfrac{x^2}{b^2} = 1$

   (*Note:* a^2 is always the denominator of the positive variable.)

2. Connect the opposite corners of the auxiliary rectangle whose sides pass through the points in step 1 with lines called asymptotes.

3. Start the graph from the vertices $(\pm a, 0)$ or $(0, \pm a)$ and draw the hyperbola so that it approaches (but does not touch) the asymptotes.

EXAMPLE 1 Graphing hyperbolas with center at (0, 0)

Name the coordinates of the center and the vertices and graph:

a. $\dfrac{y^2}{4} - \dfrac{x^2}{25} = 1$

b. $25x^2 - 4y^2 = 100$

SOLUTION

a. The hyperbola is centered at (0, 0), the origin. Since the y^2 term is positive, $a^2 = 4$ and $a = 2$. It has vertices at $V_1 = (0, 2)$ and $V_2 = (0, -2)$. (There are no x-intercepts.) Our auxiliary rectangle will pass through $(0, \pm 2)$ and through $(\pm 5, 0)$, the square root of the denominator of x^2. We then connect opposite corners to complete our asymptotes, as shown in Figure 9.40. Since our hyperbola has vertices $(0, \pm 2)$, we start our graph from the vertex $(0, 2)$ and approach the asymptotes, obtaining the top half of the hyperbola. The bottom half is obtained similarly by starting at the vertex $(0, -2)$. (See Figure 9.41.)

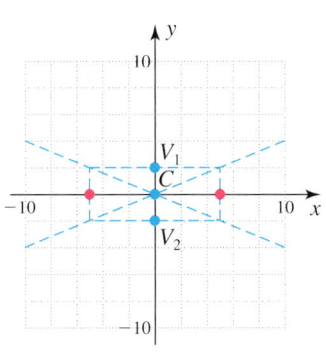

>Figure 9.40   >Figure 9.41

b. Divide each term by 100 to obtain a 1 on the right-hand side. We then have

$$\dfrac{x^2}{4} - \dfrac{y^2}{25} = 1,$$

which indicates the center at (0, 0), $a^2 = 4$ with $a = 2$ and $b^2 = 25$ with $b = 5$.

This time, we will show our auxiliary rectangle and the hyperbola on the same graph. Since the x^2 term is positive, the hyperbola has vertices at $V_1 = (2, 0)$ and $V_2 = (-2, 0)$. Our auxiliary rectangle will pass through $(\pm 2, 0)$ and through $(\pm 5, 0)$. We then complete the auxiliary rectangle, the asymptotes, and the graph of the hyperbola with vertices $(\pm 2, 0)$, as shown in Figure 9.42.

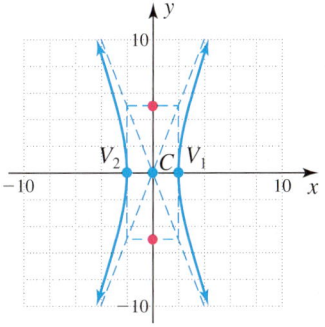

>Figure 9.42

PROBLEM 1

Find the coordinates of the vertices and graph:

a. $\dfrac{y^2}{16} - \dfrac{x^2}{9} = 1$

b. $\dfrac{x^2}{16} - \dfrac{y^2}{9} = 1$

Answers to PROBLEMS

1. a. vertices $(0, \pm 4)$

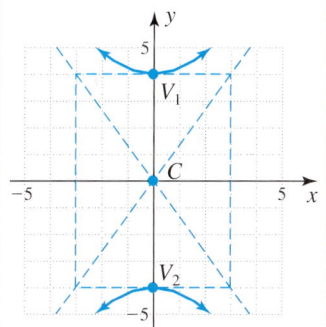

b. vertices $(\pm 4, 0)$

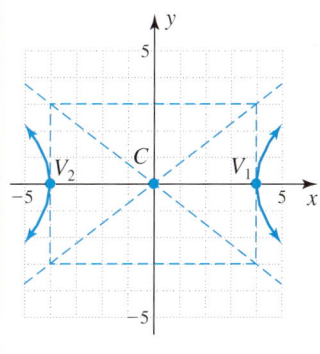

Calculator Corner

Graphing a Hyperbola

As we've mentioned, calculators only graph functions. Is a hyperbola the graph of a function? If you apply the vertical line test to the hyperbolas shown in Example 1, you will see that the graphs are *not* graphs of functions. To graph

$$\frac{y^2}{4} - \frac{x^2}{25} = 1$$

we first solve for y by adding

$$\frac{x^2}{25}$$

to both sides of the equation, multiplying both sides of the equation by 4, and taking the square root of both sides of the equation to obtain

$$y = \pm\sqrt{4 + \frac{4x^2}{25}}$$

We then use a square window and graph

$$Y_1 = \sqrt{4 + \frac{4x^2}{25}}$$

and

$$Y_2 = -\sqrt{4 + \frac{4x^2}{25}}$$

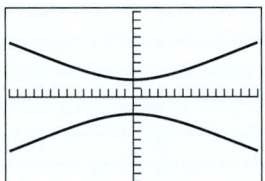

The graphs for Y_1 (top part) and Y_2 (bottom part) are shown in the window. Remember, you can save time when you enter the expressions to be graphed by entering

$$Y_1 = \sqrt{4 + \frac{4x^2}{25}} \quad \text{and} \quad Y_2 = -Y_1$$

Now try to graph Example 1(b).

GRAPHING FORMS OF EQUATIONS OF HYPERBOLAS WITH CENTER AT (h, k)

The equation of the hyperbola with center at (h, k) is

$$\frac{(x-h)^2}{a^2} - \frac{(y-k)^2}{b^2} = 1$$

The vertices are horizontally $\pm a$ units from (h, k).

The equation of the hyperbola with center at (h, k) is

$$\frac{(y-k)^2}{a^2} - \frac{(x-h)^2}{b^2} = 1$$

The vertices are vertically $\pm a$ units from (h, k).

Note: a^2 is always the denominator of the positive variable and either $a^2 = b^2$, $a^2 > b^2$, or $a^2 < b^2$.

EXAMPLE 2   Graphing a hyperbola not centered at the origin

Name the coordinates of the center and the vertices and graph:

$$\frac{(x-2)^2}{16} - \frac{(y+1)^2}{4} = 1$$

SOLUTION   The center of this hyperbola is at $(2, -1)$. Since the x^2 term is positive, $a^2 = 16$ with $a = 4$, and $b^2 = 4$ with $b = 2$. Now construct the auxiliary rectangle with sides parallel to $x = 2$ that are ± 4 units away from $x = 2$, and sides parallel to $y = -1$ that are ± 2 units away from $y = -1$, as shown in Figure 9.43. The vertices of the hyperbola are horizontally ± 4 units from $(2, -1)$, thus, the hyperbola has vertices at $V_1 = (6, -1)$ and $V_2 = (-2, -1)$. Use the auxiliary rectangle to draw the asymptotes. Use the vertices to sketch the graph of the hyperbola, as shown in Figure 9.44.

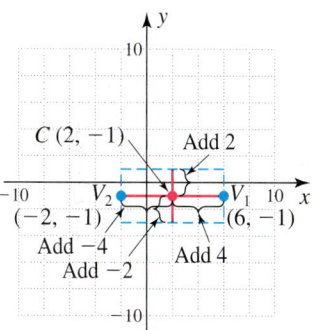
> Figure 9.43

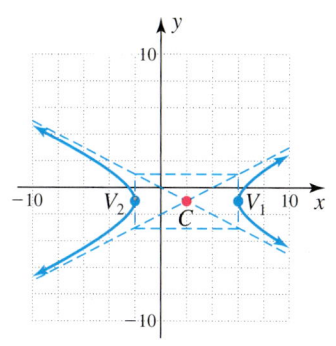
> Figure 9.44

PROBLEM 2

Name the coordinates of the center and the vertices and graph:

$$\frac{(x+1)^2}{9} - \frac{(y+3)^2}{9} = 1$$

Answers to PROBLEMS

2. Center at $(-1, -3)$;
$V_1 = (2, -3)$ and $V_2 = (-4, -3)$

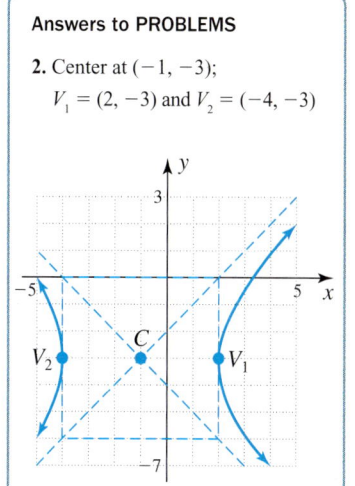

You have probably noticed that the equations whose graphs are hyperbolas are similar to the equations whose graphs are ellipses (when written in standard form, hyperbolas are written as differences). We examine this idea in more detail next.

B ▸ Identifying Conic Sections by Their Equations

How would you know the shape of the graph of a conic by studying its equation? Table 9.2 will help you with this.

Table 9.2   Identifying Conic Sections

Equation	Graph	Description	Identification
$y = a(x-h)^2 + k$	(parabola opening upward, $a > 0$, vertex (h, k))	Parabola with vertex at (h, k). Opens upward for $a > 0$, downward for $a < 0$.	y is not squared.
$y = ax^2 + bx + c$	(parabola opening downward, $a < 0$, axis $x = -\frac{b}{2a}$)	Parabola with vertex at $x = -\frac{b}{2a}$. Opens upward for $a > 0$, downward for $a < 0$.	y is not squared.

(continued)

Table 9.2 (continued)

Equation	Graph	Description	Identification
$x = a(y - k)^2 + h$		Parabola with vertex at (h, k). Opens right if $a > 0$, left if $a < 0$.	x is not squared.
$x = ay^2 + by + c$		Parabola with vertex at $y = -\frac{b}{2a}$. Opens right if $a > 0$, left if $a < 0$.	x is not squared.
$(x - h)^2 + (y - k)^2 = r^2$		Circle of radius r, centered at (h, k).	The coefficients of $(x - h)^2$ and $(y - k)^2$ are positive and equal when the variables are on the same side of the equation.
$\dfrac{x^2}{a^2} + \dfrac{y^2}{b^2} = 1$		Ellipse with x-intercepts $\pm a$, y-intercepts $\pm b$.	The coefficients of x^2 and y^2 are positive and not equal when variables are on the same side.
$\dfrac{x^2}{b^2} + \dfrac{y^2}{a^2} = 1$		Ellipse with x-intercepts $\pm b$, y-intercepts $\pm a$.	The coefficients of x^2 and y^2 are positive and not equal when variables are on the same side.
$\dfrac{x^2}{a^2} - \dfrac{y^2}{b^2} = 1$		Hyperbola with vertices $(\pm a, 0)$. Auxiliary rectangle passing through $(\pm a, 0)$ and $(0, \pm b)$. Asymptotes drawn through the corners of the auxiliary rectangle.	x^2 has positive coefficient, y^2 has negative coefficient when variables are on the same side.
$\dfrac{y^2}{a^2} - \dfrac{x^2}{b^2} = 1$		Hyperbola with vertices $(0, \pm a)$. Auxiliary rectangle passing through $(0, \pm a)$ and $(\pm b, 0)$. Asymptotes drawn through the corners of the auxiliary rectangle.	y^2 has positive coefficient, x^2 has negative coefficient when variables are on the same side.

EXAMPLE 3 Identifying conic sections

Identify:

a. $x^2 = 9 - y^2$

b. $y = x^2 - 4$

c. $4x^2 = 36 - 9y^2$

d. $9x^2 = 36 + 4y^2$

SOLUTION If both variables appear to the second power, we write all variables on the left to make the identification easier.

a. In $x^2 = 9 - y^2$, both variables appear to the second power. Thus, $x^2 = 9 - y^2$ is written as $x^2 + y^2 = 9$. The square terms have the same coefficient (1) and are both positive. The equation $x^2 = 9 - y^2$ represents a circle centered at the origin and with radius 3.

b. In this case, only one variable is squared, the x. Thus, the conic is a parabola with the vertex at $(0, -4)$ and opening upward.

c. Both variables are squared. We then write $4x^2 = 36 - 9y^2$ as $4x^2 + 9y^2 = 36$. Here the square terms have different coefficients and are both positive. The equation corresponds to an ellipse centered at the origin. The x-intercepts are found by letting $y = 0$ to obtain $x = \pm 3$. Similarly, the y-intercepts are $y = \pm 2$. To confirm that $4x^2 = 36 - 9y^2$ is an ellipse, we write the equation in standard form by dividing each term in the equation by 36 so we obtain a 1 on the right-hand side of the equation $4x^2 + 9y^2 = 36$. We then have

$$\frac{4x^2}{36} + \frac{9y^2}{36} = \frac{36}{36}$$

$$\frac{x^2}{9} + \frac{y^2}{4} = 1$$

$$\frac{x^2}{3^2} + \frac{y^2}{2^2} = 1$$

confirming that the conic is an ellipse with x-intercepts at $(\pm 3, 0)$ and y-intercepts at $(0, \pm 2)$.

d. Again, both variables are squared. Thus, we write $9x^2 = 36 + 4y^2$ as $9x^2 - 4y^2 = 36$. The minus sign indicates that the conic is a hyperbola. To confirm that $9x^2 - 4y^2 = 36$ corresponds to a hyperbola, divide each term by 36 to obtain

$$\frac{9x^2}{36} - \frac{4y^2}{36} = \frac{36}{36}$$

$$\frac{x^2}{2^2} - \frac{y^2}{3^2} = 1$$

which confirms that the conic is a hyperbola. The vertices are at $(\pm 2, 0)$

PROBLEM 3

Identify:

a. $4x^2 = 36 + 9y^2$

b. $y = x^2 + 3$

c. $y^2 = 9 - x^2$

d. $9x^2 = 36 - 4y^2$

Answers to PROBLEMS

3. **a.** Hyperbola; center $(0, 0)$, vertices $(\pm 3, 0)$
   b. Parabola; vertex $(0, 3)$, opens upward
   c. Circle; center $(0, 0)$, radius 3
   d. Ellipse; center $(0, 0)$, x-intercepts $(\pm 2, 0)$ and y-intercepts $(0, \pm 3)$

Exercises 9.3

A **Graphing Hyperbolas** In Problems 1–12, draw the auxiliary rectangle and asymptotes, name the coordinates of the vertices, and graph.

1. $\dfrac{x^2}{25} - \dfrac{y^2}{9} = 1$

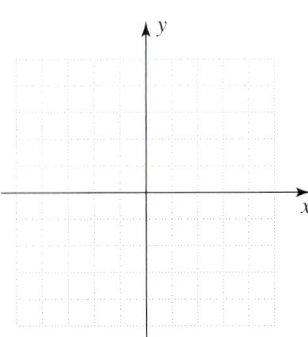

2. $\dfrac{y^2}{9} - \dfrac{x^2}{25} = 1$

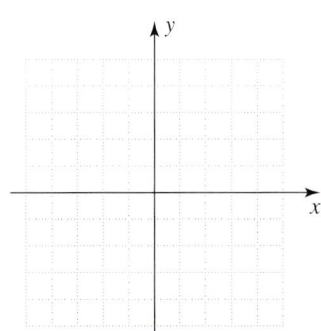

3. $\dfrac{y^2}{9} - \dfrac{x^2}{9} = 1$

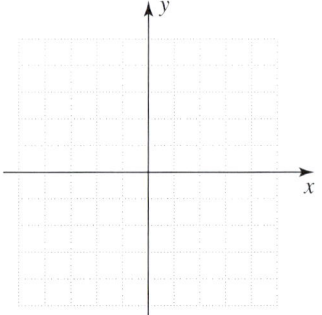

4. $\dfrac{x^2}{9} - \dfrac{y^2}{9} = 1$

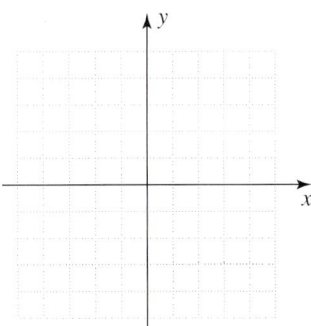

5. $\dfrac{x^2}{9} - \dfrac{y^2}{1} = 1$

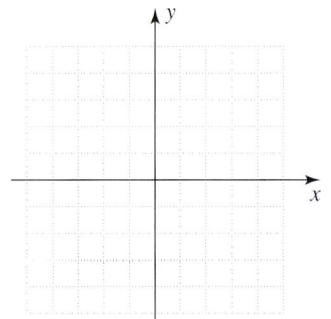

6. $\dfrac{y^2}{16} - \dfrac{x^2}{1} = 1$

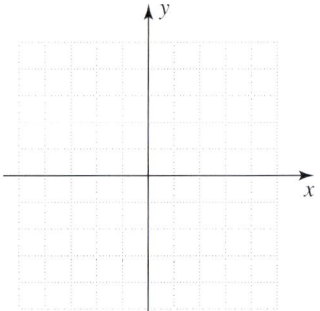

7. $\dfrac{x^2}{64} - \dfrac{y^2}{49} = 1$

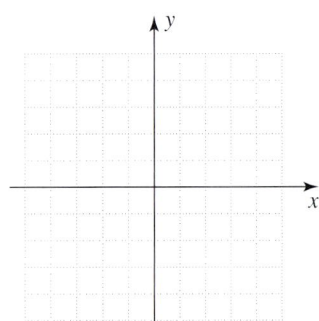

8. $\dfrac{y^2}{49} - \dfrac{x^2}{64} = 1$

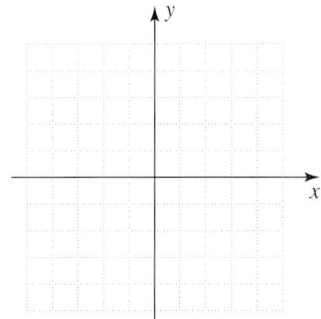

9. $\dfrac{y^2}{\frac{16}{9}} - \dfrac{x^2}{\frac{9}{16}} = 1$

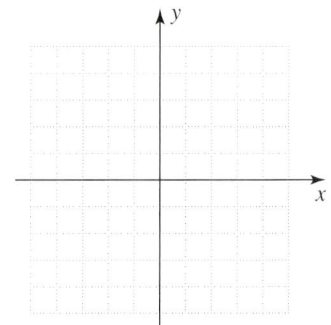

10. $\dfrac{x^2}{\frac{9}{4}} - \dfrac{y^2}{\frac{4}{9}} = 1$

11. $y^2 - 9x^2 = 9$

12. $x^2 - 16y^2 = 16$

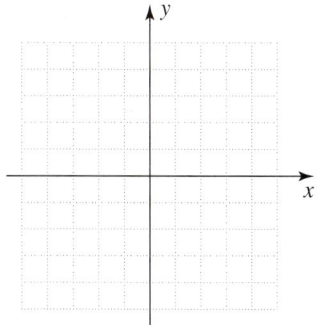
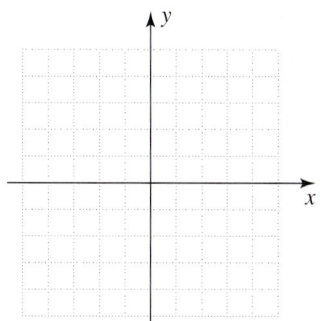
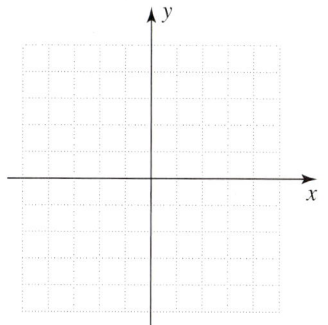

In Problems 13–16, sketch the hyperbola. Name the coordinates of the center and give the values of a and b. (*Hint:* They are not centered at the origin.)

13. $\dfrac{(x-1)^2}{4} - \dfrac{(y+1)^2}{9} = 1$

14. $\dfrac{(x-2)^2}{9} - \dfrac{(y+1)^2}{4} = 1$

15. $\dfrac{(y-1)^2}{9} - \dfrac{(x-2)^2}{4} = 1$

16. $\dfrac{(y-2)^2}{4} - \dfrac{(x-1)^2}{9} = 1$

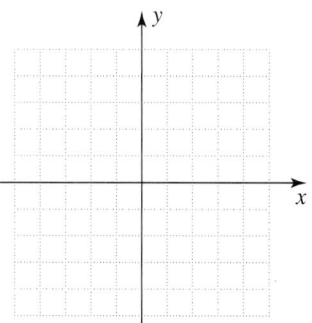
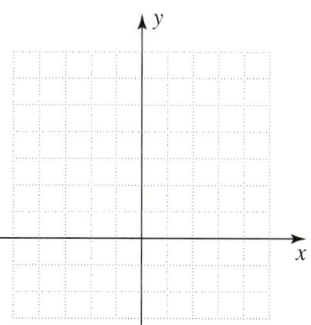

‹ B › Identifying Conic Sections by Their Equations In Problems 17–30, identify the conic and name the intercepts. If the conic is a parabola, name the vertex only.

17. $x^2 + y^2 = 25$

18. $x^2 - y^2 = 25$

19. $x^2 - y^2 = 36$

20. $x^2 + y^2 = 36$

21. $x^2 - y = 9$

22. $x^2 + y = 9$

23. $y^2 - x = 4$

24. $y^2 + x = 4$

25. $9x^2 = 36 - 9y^2$

26. $4x^2 = 16 - 4y^2$ **27.** $9x^2 = 36 + 9y^2$ **28.** $4x^2 = 36 - 9y^2$

29. $x^2 = 9 - 9y^2$ **30.** $y^2 = 4 - 4x^2$

〉〉〉 Applications

31. *Semicircular plate* A semicircular plate of diameter D with a circular opening of diameter d is to be constructed. If the area of the plate is π square inches, the relationship between D and d is given by the equation

$$\frac{D^2}{8} - \frac{d^2}{4} = 1$$

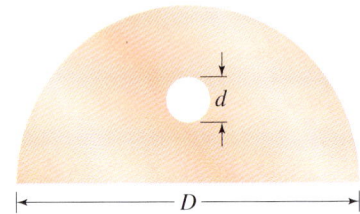

a. What type of conic corresponds to this equation?

b. Sketch the graph of

$$\frac{D^2}{8} - \frac{d^2}{4} = 1$$

(Use 2.8 for $\sqrt{8}$.)

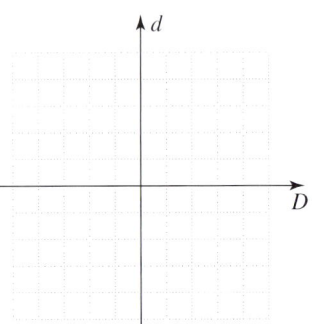

32. *Semicircular plate* If three holes of diameter d were drilled in a semicircular plate of diameter D and the remaining area was π square inches, the relationship between D and d would be

$$\frac{D^2}{8} - \frac{3d^2}{4} = 1$$

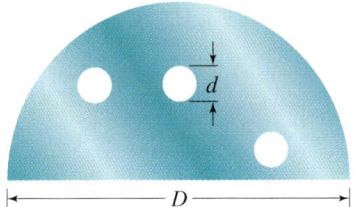

a. What type of conic corresponds to this equation?

b. Show that you can write the equation of the conic as

$$\frac{D^2}{8} - \frac{d^2}{\frac{4}{3}} = 1$$

c. Sketch the graph of

$$\frac{D^2}{8} - \frac{3d^2}{4} = 1$$

(Use 2.8 for $\sqrt{8}$ and 1.15 for $\sqrt{\frac{4}{3}}$.)

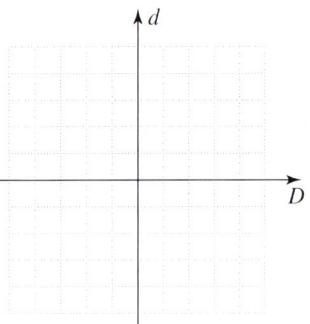

33. *Kinetic energy* The total kinetic energy of a spinning body moving through the air is 144 foot-pound (ft-lb). The velocity v through the air and the spinning velocity ω are related by the equation $4v^2 + 9\omega^2 = 144$. Graph this equation.

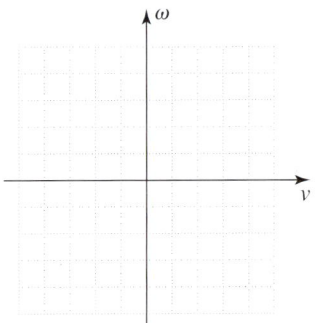

34. *Graphing* If the equation in Problem 33 is $16v^2 + 4\omega^2 = 256$, graph the equation.

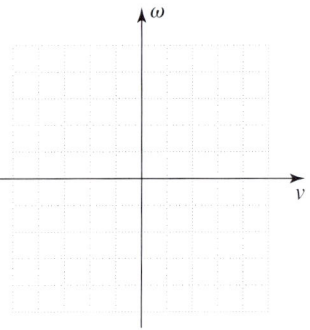

Eccentricity Use the following information for Problems 35–40. The **eccentricity** e of a hyperbola is defined by $e = c/a$, where $c^2 = a^2 + b^2$. The eccentricity is a number that describe the "flatness" of the hyperbola. The larger the eccentricity, the more the hyperbola resembles two parallel lines. The photo shows a hyperbola with eccentricity 2.5, where the light rays from one focus are refracted to the other focus.

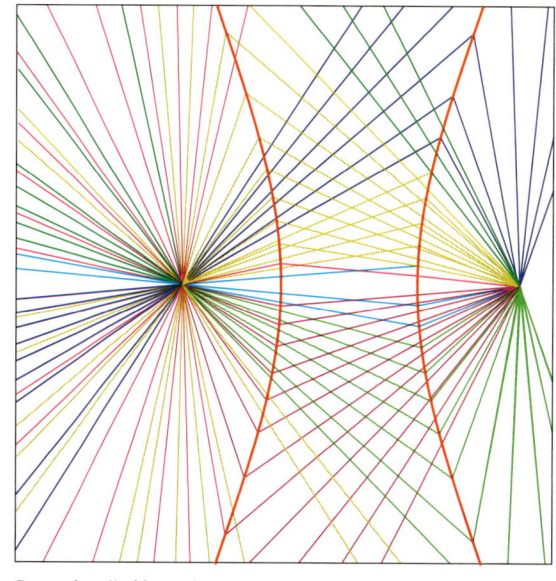

Source: http://xahlee.org/.

35. *Eccentricity* Find the eccentricity of the hyperbola.

$$\frac{x^2}{4} - \frac{y^2}{9} = 1$$

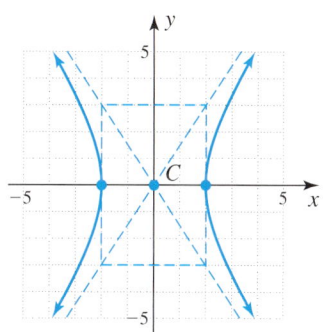

36. *Eccentricity* Find the eccentricity of the hyperbola.

$$\frac{x^2}{4} - \frac{y^2}{25} = 1$$

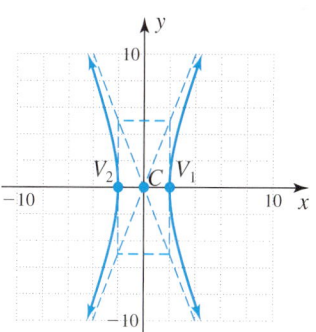

37. *Eccentricity* Find the eccentricity of the hyperbola.

$$\frac{y^2}{4} - \frac{x^2}{25} = 1$$

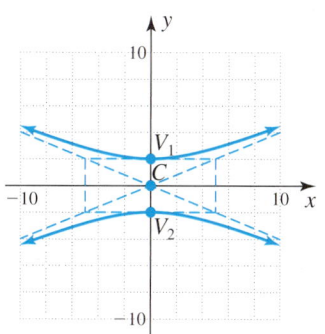

38. *Eccentricity* Find the eccentricity of the hyperbola.

$$\frac{y^2}{9} - \frac{x^2}{4} = 1$$

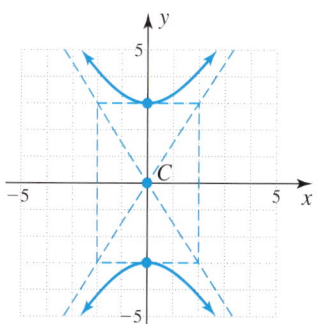

39. *Eccentricity* Which hyperbola is "flatter," Problem 35 or Problem 36? Which has the larger eccentricity?

40. *Eccentricity* Which hyperbola is flatter, Problem 37 or Problem 38? Which has the larger eccentricity?

>>> Using Your Knowledge

Hyperbolas Revisited The definition for a hyperbola is as follows.

HYPERBOLA

A **hyperbola** is the set of points in a plane such that the difference between the distances of each point from two fixed points (called the **foci**) is a constant.

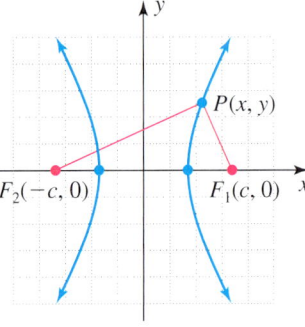

In Problems 41–46, we will prove that this definition leads to the equation of the hyperbola we gave in the text.

41. Suppose (x, y) is a point on the hyperbola. Find the distance from P to F_1.

42. Find the distance from P to F_2.

43. Let the difference between the distances found in Problems 41 and 42 be $2a$. Thus,

$$\sqrt{(x-c)^2 + y^2} - \sqrt{(x+c)^2 + y^2} = 2a$$

Rewrite this equation with one radical on each side. Then square both sides and simplify. What is your answer?

44. Rewrite your answer for Problem 43 with the radical on one side. Then square both sides again and simplify. What is your answer?

45. In your answer for Problem 44, let $c^2 - a^2 = b^2$, where $a < c$. Isolate all the variables on the left side. What is your answer?

46. Divide all terms of the answer you obtained in Problem 45 by a^2b^2. What is your answer? If everything went well, you should have

$$\frac{x^2}{a^2} - \frac{y^2}{b^2} = 1$$

the equation of a hyperbola.

We have shown you how to graph the asymptotes of a hyperbola. We are now ready to use this knowledge to find the equation of these asymptotes. Consider the hyperbola

$$\frac{x^2}{a^2} - \frac{y^2}{b^2} = 1$$

47. The expression on the left is the difference of two squares. Factor it.

48. Isolate $\frac{x}{a} - \frac{y}{b}$ on the left. The expression on the right is a complex fraction with 1 as the numerator. What is it?

49. Look at the denominator of the complex fraction in Problem 48. If x and y are positive and very large, what happens to the denominator? What happens to the complex fraction?

50. If you answered that the complex fraction is very small, you are correct. In mathematics we say that $\frac{x}{a} - \frac{y}{b} \to 0$ (the expression approaches zero). Thus, for very large positive x and y, $\frac{x}{a} - \frac{y}{b} \approx 0$. This means that $\frac{x}{a} - \frac{y}{b} = 0$ is an asymptote. Solve for y and find its equation.

51. We can show in the same way that $\frac{x}{a} + \frac{y}{b} = 0$ is an asymptote. Solve for y and find its equation.

52. a. In summary, what are the equations of the asymptotes for the hyperbola

$$\frac{x^2}{a^2} - \frac{y^2}{b^2} = 1?$$

b. What about the asymptotes for the hyperbola

$$\frac{y^2}{a^2} - \frac{x^2}{b^2} = 1?$$

››› Write On

53. Write an explanation of the procedure you would use to determine whether the graph of an equation is an ellipse or a hyperbola.

54. Write your own definition of the asymptote of a hyperbola.

55. Consider the equation $Ax^2 + By^2 = C$, where A, B, and C are real numbers. Under what conditions will the graph of this equation be one of the following conics?

  a. A circle   **b.** An ellipse
  c. A hyperbola

››› Concept Checker

Fill in the blank(s) with the correct word(s), phrase, or mathematical statement.

56. The graph of the equation $\frac{x^2}{a^2} - \frac{y^2}{b^2} = 1$ is a _____ centered at (0, 0).

57. An equation of a hyperbola centered at (h, k) is _____.

ellipse

$\frac{(x - h)^2}{a^2} + \frac{(y - k)^2}{b^2}$

$\frac{(x - h)^2}{a^2} - \frac{(y - k)^2}{b^2}$

hyperbola

››› Mastery Test

Identify each equation as a parabola, circle, ellipse, or hyperbola.

58. $9x^2 = 36 - 4y^2$

59. $y^2 = 9 - x^2$

60. $y^2 = 9 - 4x^2$

61. $y = x^2 + 3$

62. $4x^2 = 36 + 9y^2$

Graph the equation.

63. $\frac{y^2}{16} - \frac{x^2}{9} = 1$

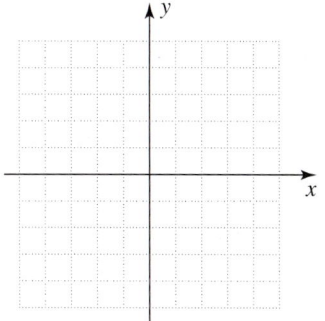

64. $\frac{x^2}{16} - \frac{y^2}{9} = 1$

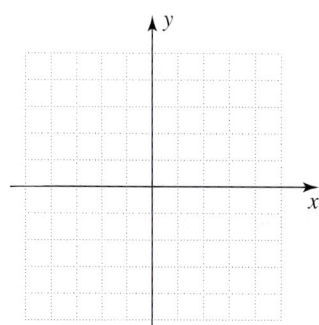

65. $9x^2 - 25y^2 = 225$

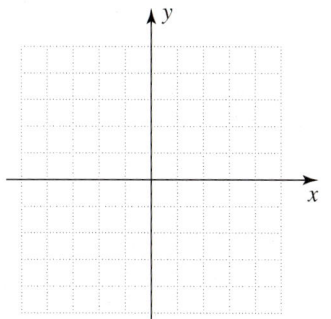

66. $25y^2 - 9x^2 = 225$

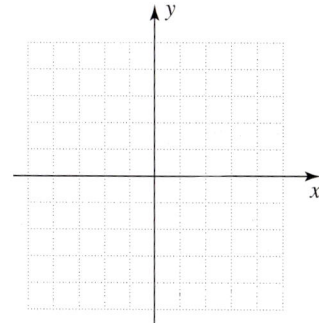

Skill Checker

Graph the equation:

67. $y = x - 4$

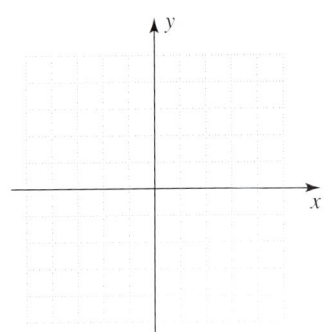

68. $x^2 + y^2 = 4$

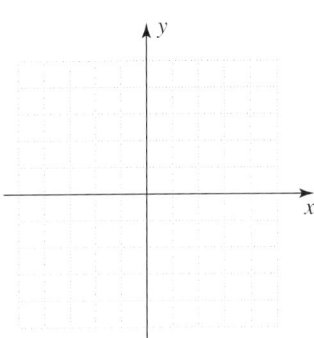

69. $y = x^2 + 1$

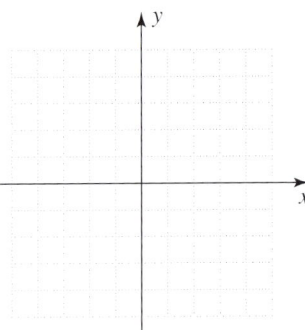

70. $y = x^2 - 1$

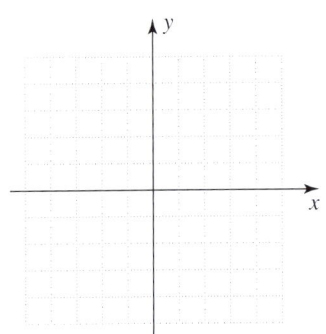

71. $4x^2 + 9y^2 = 36$

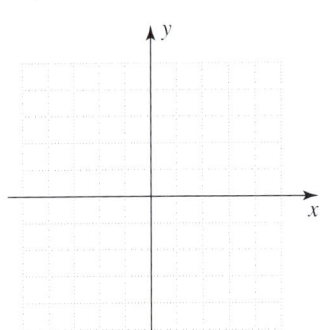

72. $9x^2 - 4y^2 = 36$

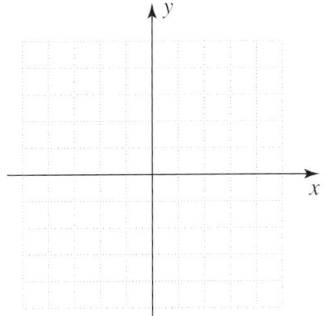

9.4 Nonlinear Systems of Equations

▶ Objectives

A ▶ Solve nonlinear systems by substitution.

B ▶ Solve systems with two second-degree equations by elimination.

C ▶ Solve applications involving nonlinear systems.

▶ To Succeed, Review How To . . .

1. Graph lines and conics (pp. 168–174, 659–669, 682–687, 698–704).
2. Use the graphical and substitution methods to solve systems of equations (pp. 285–292).

▶ Getting Started

Consumer's Demand and Supply

How do supply and demand determine the market price and quantity of wheat available for sale? As the price of wheat *decreases*, the quantity demanded by consumers *increases*. If the price *increases*, the demand *decreases*. On the other hand, as the price *increases*, the amount the suppliers are willing to sell also *increases*.

The point C of intersection of the two curves is called the **equilibrium point**. At this point, the price of a bushel of wheat is $3, and the amount demanded by the consumers (12 million bushels per month) exactly equals the amount supplied by producers. The equations that describe this system are an example of a nonlinear system of equations. Graphical results may be difficult to confirm so we use the substitution method instead. We show you how to do this next.

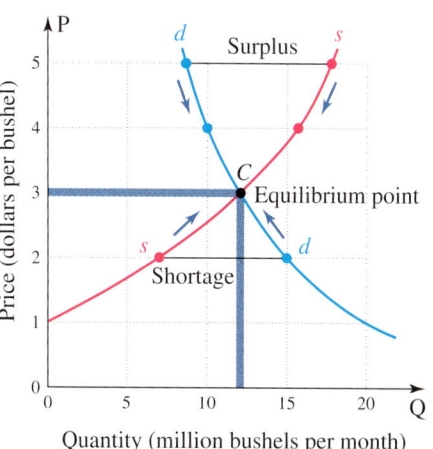

How Supply and Demand Determine Market Price and Quantity

A > Solving Nonlinear Systems by Substitution

The systems of equations we studied in Chapter 4 consisted of linear equations. An equation in which some terms have more than one variable or have a variable of degree 2 or higher is a **nonlinear equation**. A system of equations including at least one nonlinear equation is called a **nonlinear system of equations**. Such systems may have **one, more than one,** or **no** real solutions. Let's start with a system consisting of two parabolas, solve them graphically, and confirm the results by substitution.

Suppose the demand curve for a product is given by $y = (x - 5)^2$ (a parabola), where x is the number of units produced and y is the price, and the supply curve is given by the equation $y = x^2 + 2x + 13$ (another parabola), where y is the price and x the number of units available. To find the equilibrium point, we sketch both curves to find the intersection, as shown in Figure 9.45.

The equilibrium point seems to be the point (1, 16). How can we be sure? We use the substitution method to solve the system and confirm our result.

$$y = (x - 5)^2 \quad \text{A parabola} \tag{1}$$
$$y = x^2 + 2x + 13 \quad \text{A parabola} \tag{2}$$

We wish to find an ordered pair (x, y) that is a solution of *both* equations (1) and (2). Using the substitution method, we substitute $(x - 5)^2$ for y on the left-hand side of equation (2) to obtain

$$(x - 5)^2 = x^2 + 2x + 13$$
$$x^2 - 10x + 25 = x^2 + 2x + 13 \quad \text{Multiply.}$$
$$-10x + 25 = 2x + 13 \quad \text{Subtract } x^2.$$
$$-12x = -12 \quad \text{Subtract 2x and 25.}$$
$$x = 1 \quad \text{Divide by } -12.$$

If $x = 1$ in equation (2), then

$$y = (1)^2 + 2(1) + 13 = 16$$

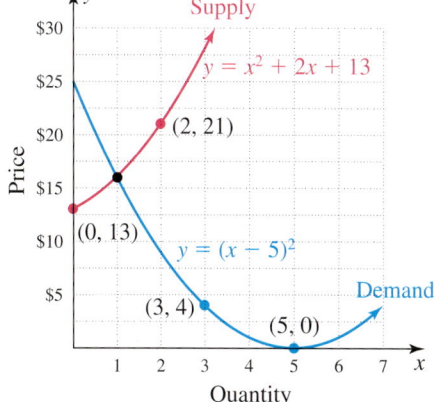

> Figure 9.45

Hence, the solution of the given system is (1, 16), as you can verify by substituting $x = 1$ and $y = 16$ in the two equations. This means when 1 unit of the product is produced, the amount demanded by the consumers exactly equals the amount supplied by the producers.

EXAMPLE 1 Solving a nonlinear system by substitution

Find the solution of the given system by the substitution method. Check the solution by sketching the graphs of the equations.

$$x^2 + y^2 = 25 \quad \text{A circle} \quad (1)$$
$$x + y = 5 \quad \text{A line} \quad (2)$$

SOLUTION We first rewrite equation (2) in the equivalent form $y = 5 - x$ to obtain

$$x^2 + y^2 = 25 \quad (1)$$
$$y = 5 - x \quad (3)$$

Replacing y in equation (1) by $(5 - x)$, we get

$$\begin{aligned} x^2 + (5 - x)^2 &= 25 && \text{Substitute } (5 - x) \text{ in equation (1).} \\ x^2 + 25 - 10x + x^2 &= 25 && \text{Multiply.} \\ 2x^2 - 10x &= 0 && \text{Simplify; subtract 25.} \\ x^2 - 5x &= 0 && \text{Divide by 2.} \\ x(x - 5) &= 0 && \text{Factor.} \end{aligned}$$

$x = 0$ or $x - 5 = 0$
$x = 0$ or $x = 5$ Solve for x.

We now let $x = 0$ and $x = 5$ in equation (3) to obtain the corresponding y-values:

$$y = 5 - 0 = 5 \quad \text{and} \quad y = 5 - 5 = 0$$

Thus, when $x = 0$, $y = 5$, and when $x = 5$, $y = 0$. Therefore, the solutions of the system are (0, 5) and (5, 0).

If we had substituted $x = 0$ and $x = 5$ in equation (1) rather than in equation (3), we would have obtained

$$0^2 + y^2 = 25 \quad \text{and} \quad 5^2 + y^2 = 25$$

That is, $y = \pm 5$ and $y = 0$. In this case, the solutions obtained would have been (0, 5), (0, −5), and (5, 0). However, (0, −5) is *not* a solution of equation (3), since $-5 \ne 5 - 0$. Therefore, the only solutions are (0, 5) and (5, 0), as before.

For this reason, we double-check our work by graphing the given system (see Figure 9.46) and verify that our solutions are correct.

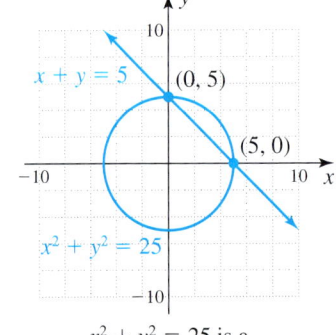

$x^2 + y^2 = 25$ is a circle of radius 5.

> Figure 9.46

PROBLEM 1

Find the solutions of the system.

$$x + y = 4$$
$$x^2 + y^2 = 16$$

CAUTION

If the degrees of the equations are different, one component of a solution should be substituted in the *lower-degree* equation to find the ordered pairs satisfying *both* equations.

Answers to PROBLEMS

1. (0, 4) and (4, 0)

Calculator Corner

Graphing to Check Solutions of a Nonlinear System

You can use your calculator to check the solutions of systems of nonlinear equations. For Example 1, start by graphing $x^2 + y^2 = 25$. First, solve for y to obtain $y = \pm\sqrt{25 - x^2}$ and then graph $y = 5 - x$ using a square window. As you can see, the graphs intersect at two points, so there are two solutions.

To find the solutions with a TI-83 Plus, press [2nd] [TRACE] [5], and three [ENTER]'s. (To make things easier, "turn off" $Y_2 = -Y_1 = -\sqrt{25 - x^2}$.) The calculator will show the points of intersection $(0, 5)$ and $(5, 0)$. Now try Example 2.

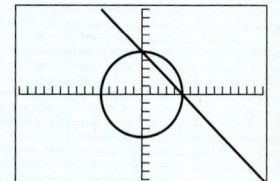

EXAMPLE 2 Solving nonlinear systems by substitution and checking with a graph

Find the solution of the given system by the substitution method. Check the solution by sketching the graphs of the equations.

$$x^2 + y^2 = 9 \quad (1)$$
$$x + y = 5 \quad (2)$$

SOLUTION Rewriting equation (2) in the form $y = 5 - x$, we obtain the equivalent system

$$x^2 + y^2 = 9 \quad (1)$$
$$y = 5 - x \quad (3)$$

Substituting $y = 5 - x$ in equation (1), we get

$$x^2 + (5 - x)^2 = 9$$
$$x^2 + 25 - 10x + x^2 = 9 \quad \text{Multiply.}$$
$$2x^2 - 10x + 25 = 9 \quad \text{Simplify.}$$
$$2x^2 - 10x + 16 = 0 \quad \text{Subtract 9.}$$
$$x^2 - 5x + 8 = 0 \quad \text{Divide by 2.}$$

Using the quadratic formula with $a = 1$, $b = -5$, $c = 8$, we get

$$x = \frac{5 \pm \sqrt{25 - 4 \cdot 8}}{2} = \frac{5 \pm \sqrt{-7}}{2} = \frac{5 \pm i\sqrt{7}}{2}$$

Substituting these values in equation (3), we obtain

$$y = 5 - \frac{5 + i\sqrt{7}}{2} \quad \text{and} \quad y = 5 - \frac{5 - i\sqrt{7}}{2}$$

That is,

$$y = \frac{5}{2} - \frac{\sqrt{7}}{2}i \quad \text{and} \quad y = \frac{5}{2} + \frac{\sqrt{7}}{2}i$$

Hence, the solutions of the system are

$$\left(\frac{5}{2} + \frac{\sqrt{7}}{2}i, \frac{5}{2} - \frac{\sqrt{7}}{2}i\right) \quad \text{and} \quad \left(\frac{5}{2} - \frac{\sqrt{7}}{2}i, \frac{5}{2} + \frac{\sqrt{7}}{2}i\right)$$

as can be checked in the original equations. The graphs of the two equations are shown in Figure 9.47. As you can see, the graphs *do not intersect*. When the solutions of a system of equations are non-real complex numbers, there are no points of intersection for the graphs. This is because the coordinates of points in the real plane are *real* numbers.

PROBLEM 2

Find the solutions of the system.

$$x + y = 3$$
$$x^2 + y^2 = 3$$

> Figure 9.47

Answers to PROBLEMS

2. $\left(\frac{3}{2} + \frac{\sqrt{3}}{2}i, \frac{3}{2} - \frac{\sqrt{3}}{2}i\right)$ and $\left(\frac{3}{2} - \frac{\sqrt{3}}{2}i, \frac{3}{2} + \frac{\sqrt{3}}{2}i\right)$

B › Solving Systems with Two Second-Degree Equations

When both equations in a system are of degree 2, it's easier to use the elimination method, as in Example 3.

EXAMPLE 3 Solving a system with two second-degree equations

Find the solutions of the system and verify the solution by graphing.

$$x^2 - 2y^2 = 1 \quad (1)$$
$$x^2 + 4y^2 = 25 \quad (2)$$

SOLUTION To eliminate y^2, we multiply the first equation by 2 and add the result to the second equation:

$$2x^2 - 4y^2 = 2$$
$$\underline{x^2 + 4y^2 = 25}$$
$$3x^2 = 27$$
$$x^2 = 9$$
$$x = \pm 3$$

The x-coordinates of the point of intersection are 3 and -3. Substituting in the second equation,

$$(\pm 3)^2 + 4y^2 = 25$$
$$4y^2 = 16$$
$$y^2 = 4$$
$$y = \pm 2$$

Thus, the four points of intersection are $(3, 2)$, $(3, -2)$, $(-3, 2)$, and $(-3, -2)$, as you can check in the original equations. The graph for the two equations is shown in Figure 9.48.

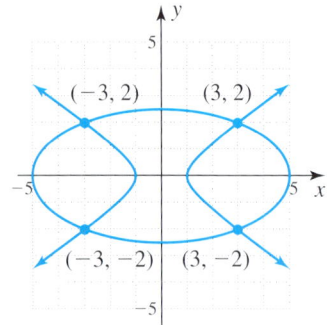

> Figure 9.48

PROBLEM 3 Find the solutions of the system.

$$x^2 + y^2 = 9$$
$$x^2 - 9y^2 = 9$$

Calculator Corner

Solving a Nonlinear System

To do Example 3, graph

$$y = \pm\sqrt{\frac{x^2 - 1}{2}} \quad \text{and} \quad y = \pm\sqrt{\frac{25 - x^2}{4}}$$

Press **2nd** **TRACE** **5** . Then you will need to use the ▼ key to select the part of each graph that the calculator is looking at. For instance, the top half of the ellipse (Y_3) intersects the top half of the hyperbola (Y_1) twice. Press **ENTER** to select the part of each graph, then once more to find a point of intersection. The graphs and one of the points of intersection, point $(-3, 2)$, are shown in the window.

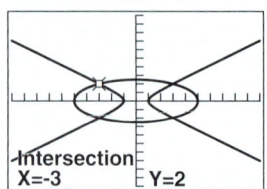

NOTE

An independent system of equations with at least one being quadratic can have four, three, two, one, or no real-number solutions.

Answers to PROBLEMS

3. $(3, 0)$ and $(-3, 0)$

C > Applications of Nonlinear Systems

The break-even point is the point at which enough units have been sold so that the cost C and the revenue R are equal. Example 4 shows how to use the methods we've studied to find the break-even point.

EXAMPLE 4 Breaking even

The total cost C for manufacturing and selling x units of a product each week is given by the equation $C = 30x + 100$, whereas the revenue R is given by $R = 81x - 0.5x^2$. How many items must be manufactured and sold for the company to break even—that is, for C to equal R?

SOLUTION We are asked to find the value of x for which $C = R$, that is,

$$\underbrace{30x + 100}_{C} = \underbrace{81x - 0.5x^2}_{R}$$

or, in standard form,

$$0.5x^2 - 51x + 100 = 0$$

where $a = 0.5$, $b = -51$, and $c = 100$. Using the quadratic formula, we get

$$x = \frac{51 \pm \sqrt{(-51)^2 - 4(0.5)(100)}}{2(0.5)}$$

$$= \frac{51 \pm \sqrt{2601 - 200}}{1}$$

$$= \frac{51 \pm \sqrt{2401}}{1}$$

$$= 51 \pm 49$$

Thus, x is $51 - 49 = 2$ or $x = 51 + 49 = 100$; the company will break even when 2 or 100 items are sold (see Figure 9.49). The company makes a profit when they sell between 2 and 100 items.

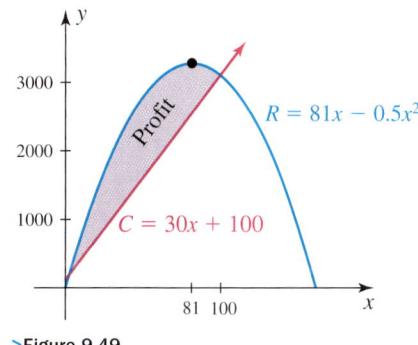

>Figure 9.49

PROBLEM 4

If the total cost for manufacturing and selling x units of a product each year is represented by the equation $C = 40x - 500$ and the revenue is represented by the equation $R = 135x - x^2$, find how many units must be manufactured and sold for the company to break even.

Calculator Corner

Using a Graph to Find Break-Even Points

In Example 4, we want to find the points at which $C = 30x + 100$ is the same as $R = 81x - 0.5x^2$. To do this, let

$$C = Y_1 = 30x + 100,$$
$$R = Y_2 = 81x - 0.5x^2$$

and find the points at which $C = R$, the points at which the graphs intersect.

To obtain the complete graph, we must be careful with the window we use. Start with a $[-10, 100]$ by $[-10, 3500]$ window. When $x = 100$.

$$R = 81(100) - 0.5 \cdot 100^2 = 3100$$

Since we didn't get a complete graph (see Window 1), change the domain for x to $[-10, 200]$.

We now have a complete graph and are able to find the points of intersection. With a TI-83 Plus, use the intersect feature to find these points. The graph with one of the points of intersection, point (100, 3100), is shown in Window 2.

You should find the second point to convince yourself that the answers are as before.

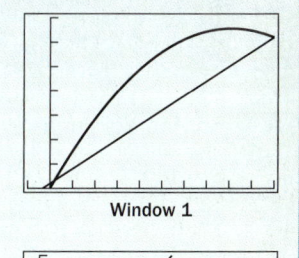

Window 1

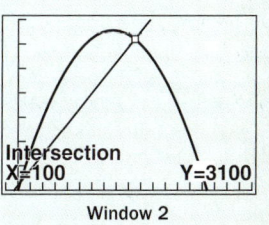

Window 2

Answers to PROBLEMS

4. 100 units

EXAMPLE 5 A new dimension in desserts

Can you guess the dimensions of one of the largest shortcakes ever made? It covered 360 square feet and had a 106-foot perimeter. Instead of guessing, we will use the RSTUV method to find the dimensions of this rectangular shortcake.

SOLUTION

1. Read the problem. We are asked to find the dimensions, and we know that the area is 360 square feet and the perimeter is 106 feet.

2. Select the unknowns. Let L be the length and W the width.

3. Think of a plan. Translate the problem.

Area A of the rectangle: $\quad A = LW$
Perimeter P of the rectangle: $\quad P = 2L + 2W$
In our case $A = 360$ and $P = 106$, thus, $\quad 360 = LW$
$\quad 106 = 2L + 2W$

4. Use the techniques we have studied to solve this system.

First, solve $360 = LW$ for W to obtain $\quad W = \dfrac{360}{L}$

Substitute $\dfrac{360}{L}$ for W in $\quad 106 = 2L + 2W$

We then have $\quad 106 = 2L + 2\left(\dfrac{360}{L}\right)$

$\quad 106L = 2L^2 + 720 \quad$ Multiply each term by L.
$\quad L^2 - 53L + 360 = 0 \quad$ Divide by 2 and write in standard form.
$\quad (L - 45)(L - 8) = 0 \quad$ Factor.
$\quad L = 45 \quad \text{or} \quad L = 8$

If $L = 45$, $W = \dfrac{360}{L} = \dfrac{360}{45} = 8$. If $L = 8$, $W = \dfrac{360}{8} = 45$.

Since the length is usually longer than the width, we select the first case in which $L = 45$ and $W = 8$. Thus, the shortcake is 45 feet long and 8 feet wide.

5. Verify the answer. Since $L = 45$ and $W = 8$, the area is $A = 45(8) = 360$ square feet, and the perimeter is $P = 2(45) + 2(8) = 106$ feet, as required.

PROBLEM 5

If the area of a picture frame is 88 square inches and its perimeter is 38 inches, find the dimensions of the picture frame.

Answers to PROBLEMS

5. 8 in. × 11 in.

Boost your grade at mathzone.com!
> Practice Problems
> NetTutor
> Self-Tests
> e-Professors
> Videos

› Exercises 9.4

‹ A › Solving Nonlinear Systems by Substitution In Problems 1–16, solve the system and check by graphing.

1. $x^2 + y^2 = 16$
$\quad x + y = 4$

2. $x^2 + y^2 = 9$
$\quad x + y = 3$

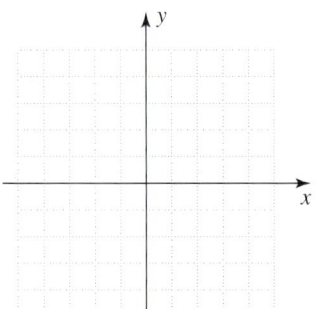

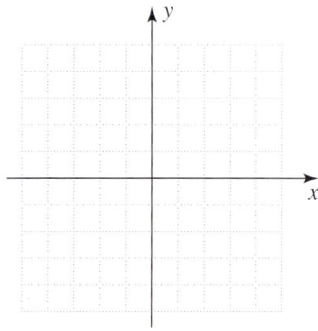

3. $x^2 + y^2 = 25$
$y - x = 5$

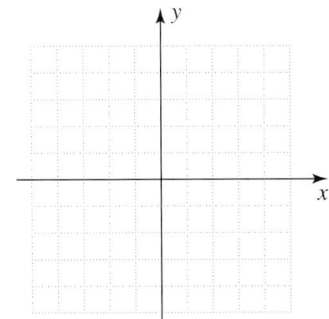

4. $x^2 + y^2 = 9$
$y - x = 3$

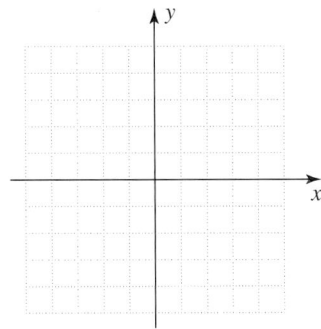

5. $x^2 + y^2 = 25$
$y - x = 1$

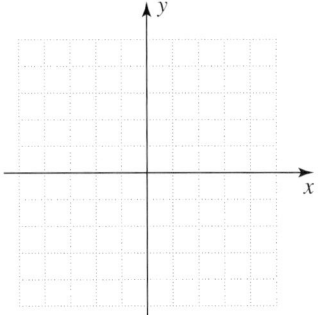

6. $x^2 + y^2 = 5$
$y - x = 1$

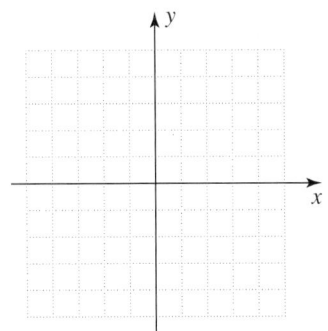

7. $y = x^2 - 5x + 4$
$x - y = 1$

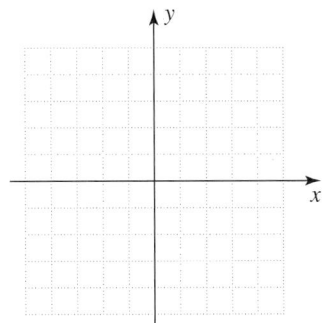

8. $y = x^2 - 2x + 1$
$x - y = 1$

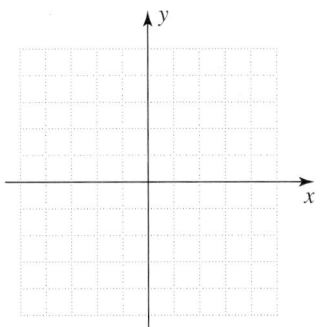

9. $y = (x - 1)^2$
$y - x = 1$

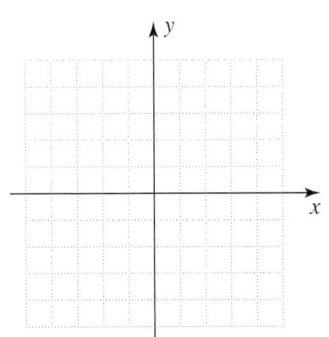

10. $y = (x + 3)^2$
$x + y = -1$

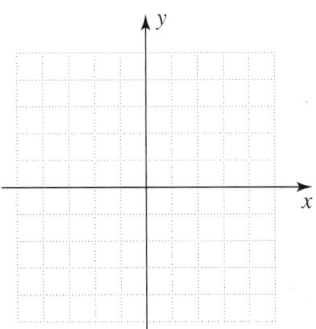

11. $4x^2 + 9y^2 = 36$
$3y - 2x = 6$

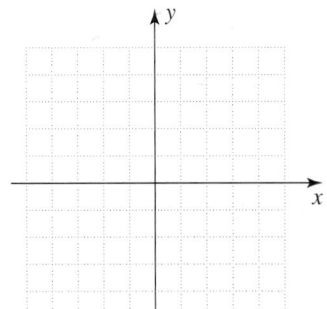

12. $4x^2 + 9y^2 = 36$
$3y + 2x = 6$

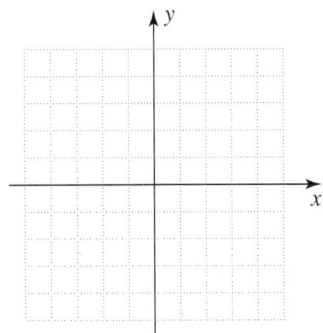

13. $x^2 - y^2 = 16$
$x + 4y = 4$

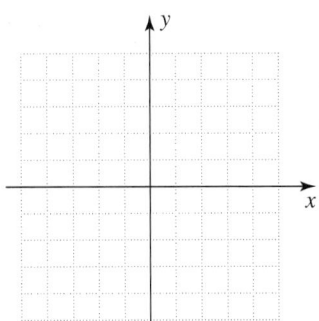

14. $x^2 - y^2 = 9$
$x + 3y = 3$

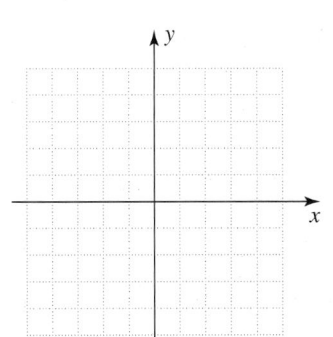

15. $x^2 + y^2 = 4$
$y - x = 5$

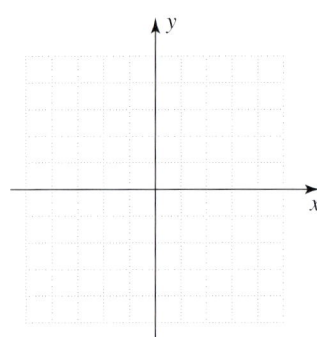

16. $x^2 + y^2 = 4$
$y - x = 3$

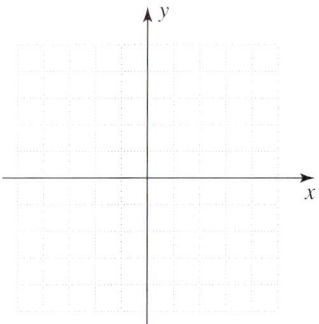

⟨B⟩ Solving Systems with Two Second-Degree Equations In Problems 17–30, solve the system and check by graphing.

17. $y = 4 - x^2$
$y = x^2 - 4$

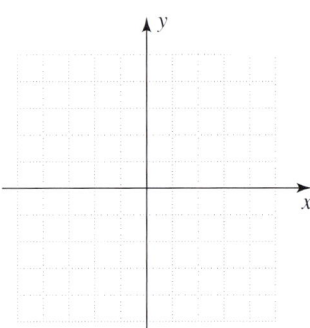

18. $y = 2 - x^2$
$y = x^2 - 2$

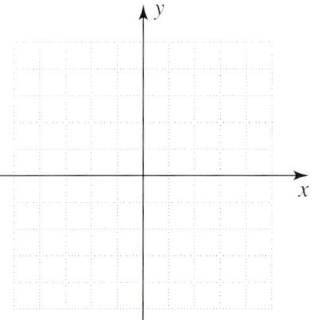

19. $x^2 + y^2 = 25$
$x^2 - y^2 = 7$

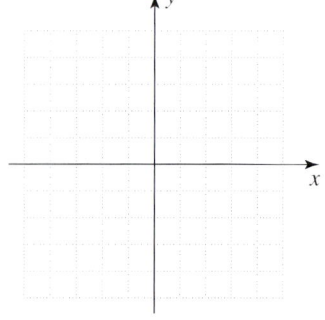

20. $x^2 + y^2 = 20$
$x^2 - y^2 = 2$

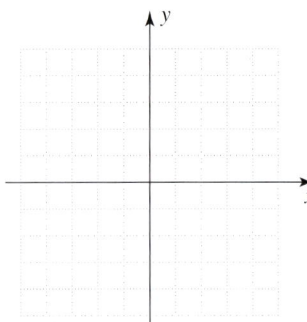

21. $x^2 + y^2 = 16$
$x^2 + 16y^2 = 16$

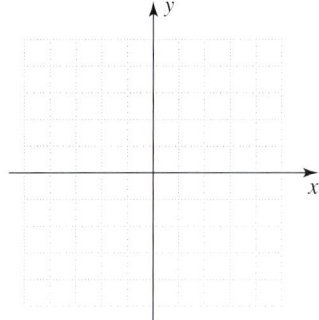

22. $x^2 + y^2 = 9$
$x^2 + 9y^2 = 9$

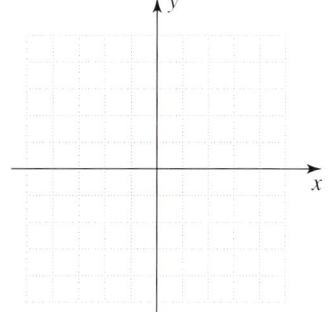

23. $3x^2 - y^2 = 2$
$x^2 + 2y^2 = 3$

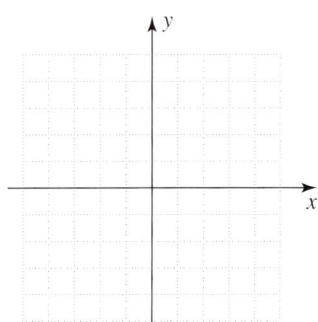

24. $x^2 - 4y^2 = 4$
$9x^2 + 4y^2 = 36$

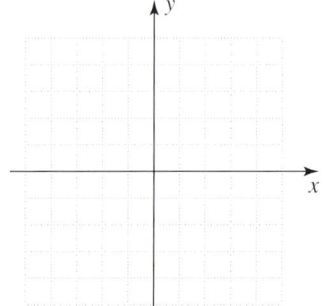

25. $x^2 + 2y^2 = 11$
$2x^2 + y^2 = 19$

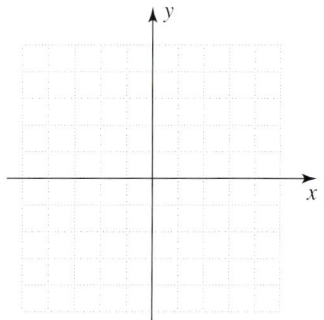

26. $4x^2 + 9y^2 = 52$
$9x^2 + 4y^2 = 52$

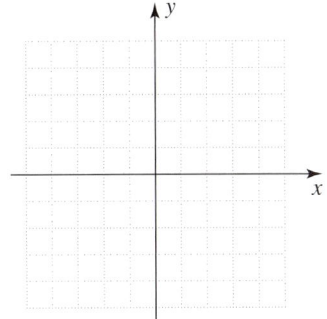

27. $x^2 + y^2 = 4$
$x^2 - y^2 = 9$

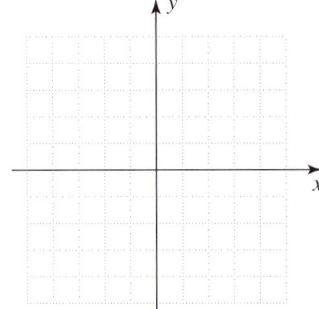

28. $x^2 + y^2 = 9$
$x^2 - y^2 = 16$

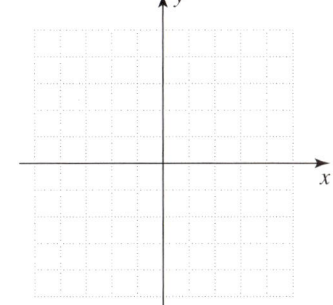

29. $x^2 + y^2 = 1$
$4x^2 + 9y^2 = 36$

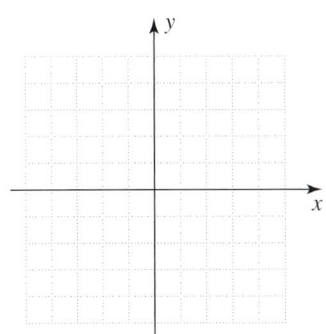

30. $x^2 + y^2 = 25$
$4x^2 + 9y^2 = 36$

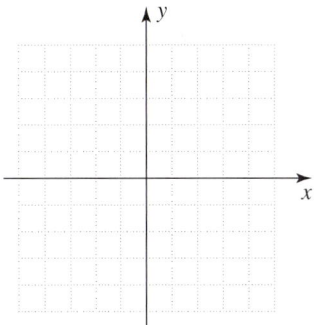

⟨ C ⟩ Applications of Nonlinear Systems

31. *Break-even point* The total cost C (in thousands of dollars) for manufacturing and selling x (thousand) items of a product each month is given by the equation $C = x + 4$, and the revenue is $R = 6x - x^2$. How many items must be manufactured and sold for the company to break even?

32. *Break-even point* The total cost C (in thousands of dollars) for manufacturing and selling x (thousand) items of a product each month is given by the equation $C = 2x + 6$, and the revenue is given by the equation $R = 150 + 20x - x^2$. How many items must be manufactured and sold for the company to break even?

33. *Numbers* There are two very interesting numbers. Their sum is 15 and the difference of their squares is also 15. Find the numbers.

34. *Dimensions* Two squares differ in area by 108 square feet and their sides differ by 6 feet. Find the lengths of the sides of these squares. (*Hint:* Make a picture!)

35. *Numbers* Find two numbers such that their product is 176 and the sum of their squares is 377.

36. *Dimensions* Find the dimensions of a rectangle whose area is 96 square inches and whose perimeter is 44 inches.

37. *Dimensions* Have you written any big checks lately? The world's *physically* largest check had an area of 2170 square feet and a perimeter of 202 feet. If the check was rectangular, what were its dimensions?

38. *Dimensions* The largest ancient carpet (A.D. 743) was a gold-enriched silk carpet that covered 54,000 square feet. If the carpet was rectangular and its perimeter was 960 feet, what were the dimensions of the carpet?

39. *Simple interest* A person receives $340 interest from an amount loaned at simple interest for 1 year. If the interest rate had been 1% higher, she would have received $476. What was the amount loaned and what was the interest rate? (*Hint:* $I = Pr$, where I is the interest, P is the principal, and r is the rate.)

40. *Numbers* There are two interesting positive numbers such that their product, their difference, and the difference of their squares are all equal. Find the numbers. (*Hint:* The numbers are irrational.)

41. *Small apartments* Do you feel cramped in your apartment? Not as cramped as the man in the picture. He lives in possibly the smallest apartment in the world. But we have a smaller one! Our rectangular apartment has an area of 60 square feet.

   a. If the area A of a rectangle is LW, where L is the length and W is the width, write an equation representing A.

   b. The perimeter P of a rectangle is $P = 2L + 2W$, and our apartment has a perimeter of 32 feet. Write an equation for the perimeter P.

   c. Solve for L in $LW = 60$ and substitute the result in $2L + 2W = 32$. Then solve for W.

   d. What are the dimensions of the apartment?

42. *Small apartments* Follow the steps in Problem 41 and find the dimensions of the apartment in the photo, with 62 square feet of area and 32.4 feet of perimeter.

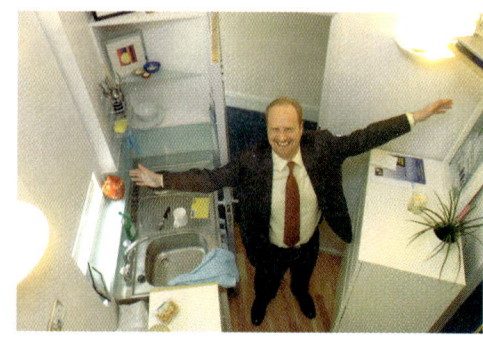

43. *Small apartments* New York has a contest for the smallest, coolest apartment under 650 square feet. Suppose your rectangular apartment is exactly 650 square feet and has a perimeter of 102 feet. What are the dimensions of the apartment?

44. *Aspect ratio* The 50-inch HDTV screen has a 16:9 aspect ratio (width to height). This means that $\frac{W}{H} = \frac{16}{9}$.

   a. Write an equation using the Pythegorean theorem relating H, W, and 50.

   b. Solve the system of equations using the equation from part **a**, and the equation $\frac{W}{H} = \frac{16}{9}$ to find H and W. Round your answers to the nearest whole number.

45. *Aspect ratio* Use a system of equations similar to the one in Problem 44 to find the height and width (round your answers to the nearest whole number) of the screen on a 60-inch TV with a 16:9 aspect ratio.

46. *Aspect ratio for computers* The aspect ratio of a computer monitor is 4:3. Use a system of equations to find the height and the width (round your answer to the nearest whole number) of a 36-centimeter computer monitor.

⟩⟩⟩ Using Your Knowledge

100-year-old Problems! Do you think the algebra we have been studying is relatively new? What does it have to do with the price of eggs? Here are a couple of problems from *Elementary Algebra*, published by Macmillan in 1894. See if you have the knowledge to solve these problems.

47. "The number of eggs which can be bought for 25 cents is equal to twice the number of cents which 8 eggs cost. How many eggs can be bought for 25 cents?" (*Hint:* If you let p be the price of eggs, then $\frac{25}{p}$ is the number of eggs you can buy for 25¢.)

48. "One half of the number of cents which a dozen apples cost is greater by 2 than twice the number of apples which can be bought for 30 cents. How many can be bought for $2.50?" (*Hint:* If you let d be the cost of a dozen apples, then $\frac{d}{12}$ is the cost of one apple and

$$\frac{30}{\frac{d}{12}}$$

is the number of apples that can be bought for 30¢.)

Write On

49. Consider the system

$$y = ax + b$$
$$y = cx + d$$

a. If $b \neq d$, what is the maximum number of solutions for this system? Write an explanation to support your answer. How many solutions are possible if b is equal to d?

b. Write a set of conditions under which this system will have no solution. What will the relationship between a and c be then?

50. Consider the system

$$y = ax + b$$
$$x^2 + y^2 = r^2$$

a. What is the maximum number of real solutions for this system? Write an explanation to support your answer.

b. Make sketches showing a system similar to the given one and that has no real solution, one solution, and two solutions. What is the geometric interpretation for the number of solutions in a system?

51. What is the maximum number of solutions you can have for a system of equations with the given characteristics. Explain.

a. Two hyperbolas

b. Two circles

c. Two ellipses

d. A hyperbola and a circle

Concept Checker

Fill in the blank(s) with the correct word(s), phrase, or mathematical statement.

52. An equation in which some terms have more than one variable, or have a variable of degree 2 or higher, is a _____ equation.

53. A nonlinear system of equations may have _____, _____ or _____ solution.

quadratic   no

one   nonlinear

more than one

Mastery Test

54. The total cost C for manufacturing and selling x units of a product is given by $C = 30x + 100$, whereas the revenue is given by $R = 57x - 0.5x^2$. How many items must be manufactured and sold for the company to break even?

55. The perimeter of a rectangle is 44 centimeters and its area is 120 square centimeters. What are the dimensions of this rectangle?

Solve each system.

56. $x^2 - 9y^2 = 9$
$x_2 + y_2 = 9$

57. $4y^2 - x^2 = 4$
$x_2 + y_2 = 1$

58. $x^2 + y^2 = 3$
$x + y = 3$

59. $x^2 + y^2 = 4$
$x + y = 4$

60. $x^2 + 4y^2 = 16$
$x + 2y = 4$

61. $x^2 + y^2 = 1$
$x + y = 1$

Skill Checker

Graph the inequalities.

62. $x + y < 2$

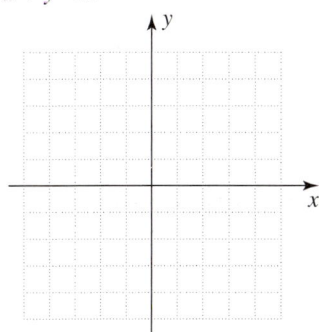

63. $x - y < 4$

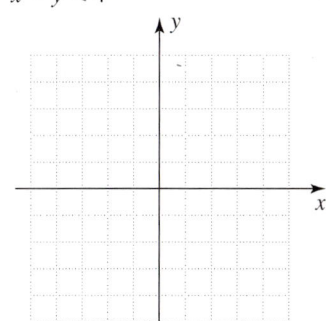

64. $y - x < 4$

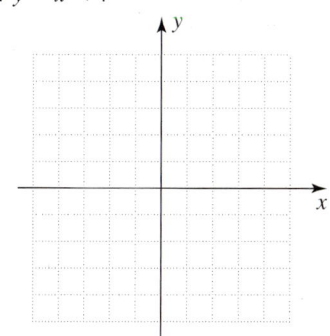

65. $2x - 3y \geq 6$

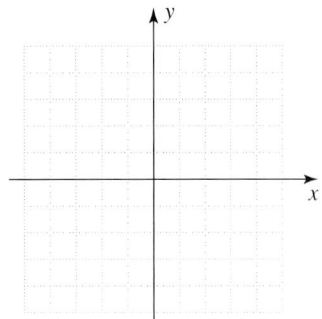

66. $3x - 2y \geq 6$

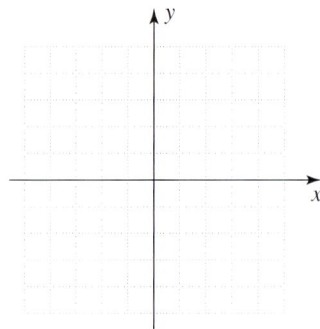

67. $y \geq 2x + 4$

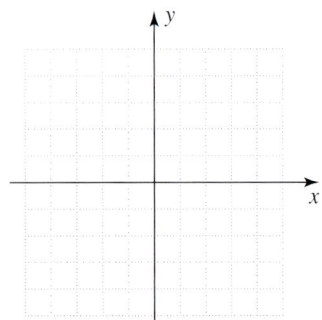

68. $y \geq 3x + 6$

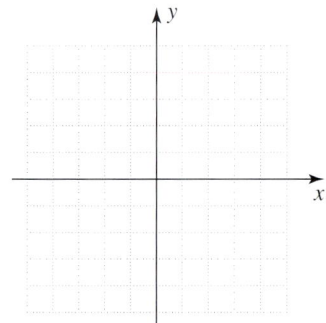

9.5 Nonlinear Systems of Inequalities

Objectives

A Graph second-degree inequalities.

B Graph the solution set of a system of nonlinear inequalities.

To Succeed, Review How To . . .

1. Graph linear inequalities in two variables (pp. 211–214).
2. Graph conic sections (pp. 659–669, 682–687, 698–704).

Getting Started
Profits and Losses

This graph offers a wealth of information if you know how to read it. The line represents the cost C of a product, whereas the parabola represents the revenue R obtained when the product is sold. Here are some facts about the graph:

1. Where the graph representing the cost C lies above the graph of the revenue R, there is a loss.
2. The revenue equals the cost at a point (or points) called a break-even point.

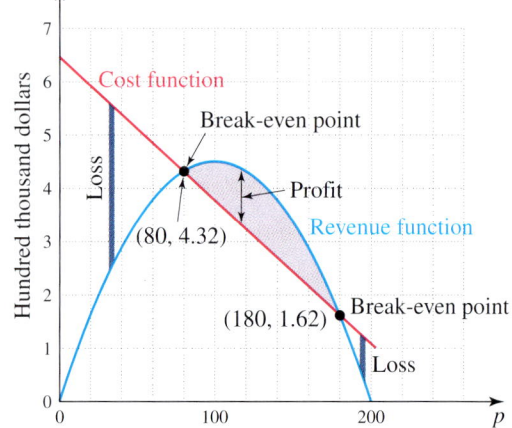

3. Where the graph representing the revenue R lies above the graph of the cost C, there is a profit.

This is an example of a nonlinear system of inequalities, a concept we shall investigate in detail in this section.

A › Graphing Second-Degree Inequalities

In Section 3.4, we graphed linear inequalities by graphing the boundary, selecting a test point not on the boundary, and determining the regions corresponding to the solution set. The region under the parabola in the graph on the previous page is obtained by graphing a **second-degree inequality** using a similar procedure.

EXAMPLE 1 Solving a second-degree inequality involving a parabola
Graph: $y \leq -x^2 + 3$

SOLUTION The boundary of the required region is the parabola $y = -x^2 + 3$ with its vertex at (0, 3), as shown in Figure 9.50. Since the inequality sign is $\leq$, the boundary *is included* in the solution set. To determine which region represents the solution set, select a test point not on the boundary and test the original inequality. A convenient point is (0, 0). When $x = 0$ and $y = 0$, $y \leq -x^2 + 3$ becomes $0 \leq 0 + 3$, a true statement. Thus, all points on the same side as (0, 0) will be in the solution set. The graph of the solution set is shown shaded in Figure 9.51.

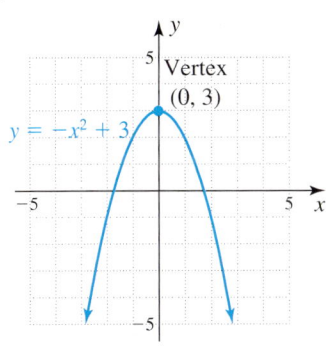
›Figure 9.50

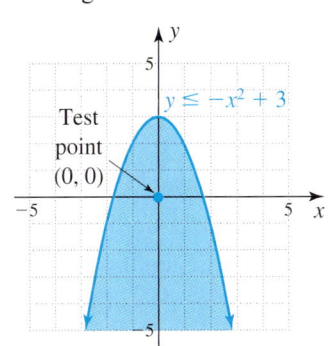
›Figure 9.51

PROBLEM 1
Graph: $y \geq x^2 - 2$

Answers to PROBLEMS

1.
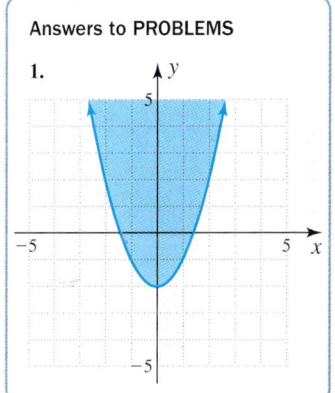

EXAMPLE 2 Solving a second-degree inequality involving a circle
Graph: $x^2 + y^2 > 4$

SOLUTION The boundary is the circle $x^2 + y^2 = 4$ with radius 2 and centered at the origin. The inequality sign is $>$, so that the boundary is *not* included in the solution set. It is shown dashed in Figure 9.52. For a test point, select the point (0, 0). For $x = 0$ and $y = 0$, $x^2 + y^2 > 4$ becomes $0 + 0 > 4$, a false statement. This means that the region containing (0, 0) is *not* in the solution set. We then shade the region *outside* the circle to represent the solution set, as shown in Figure 9.52.

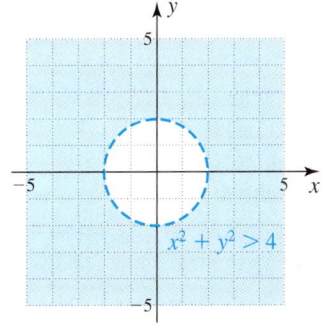
›Figure 9.52

PROBLEM 2
Graph: $x^2 + y^2 < 9$

Answers to PROBLEMS

2.
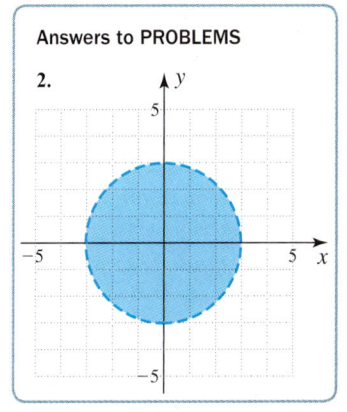

EXAMPLE 3 Solving a second-degree inequality involving a hyperbola
Graph: $9y^2 - 4x^2 \leq 36$

SOLUTION Since all the variables are on the left and there is a minus sign between the terms, the boundary is a hyperbola, with equation $\frac{y^2}{4} - \frac{x^2}{9} = 1$. The center is at $(0, 0)$ and the vertices are $(0, \pm 2)$. Since the inequality is $\leq$, the boundary *is* part of the solution set. If we use $(0, 0)$ for our test point, $9y^2 - 4x^2 \leq 36$ becomes $0 - 0 \leq 36$, which is true. Thus, the point $(0, 0)$ is part of the complete solution set, shown shaded in Figure 9.53.

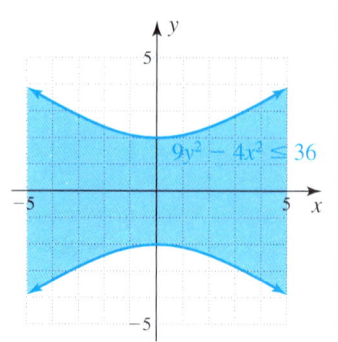

> Figure 9.53

PROBLEM 3
Graph: $4x^2 - 4y^2 \geq 16$

Answers to PROBLEMS

3.

B › Graphing Systems of Nonlinear Inequalities

We have already solved a linear system of inequalities in two variables using the graphical method. To solve a *nonlinear system of inequalities* graphically, we use the following procedure.

> **PROCEDURE**
>
> **Graphing a Nonlinear System of Inequalities**
> 1. Graph each of the inequalities on the same set of axes.
> 2. Find the region common to both graphs, the intersection of both graphs. The result is the solution set of the system.

We illustrate this procedure next.

EXAMPLE 4 Graphing a nonlinear system of inequalities
Graph: The system $x^2 + y^2 \leq 4$ and $y \geq x^2 - 3$

SOLUTION The boundary for the first inequality is a circle of radius 2 centered at the origin. The solution set of this inequality includes the boundary and the points inside the circle and is shown in red in Figure 9.54. The boundary for the second inequality is a parabola with vertex at $(0, -3)$, and the solution set includes the boundary and all the points above the parabola, shown in blue.

The solution set for the system is the intersection of the two solution sets, the region inside *both* the circle and the parabola and the points on the circle *above* or *on* the parabola where the red and blue regions overlap.

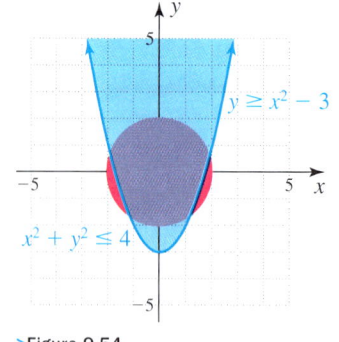

> Figure 9.54

PROBLEM 4
Graph the system:
$$x^2 + y^2 \leq 9$$
$$y \geq x^2$$

Answers to PROBLEMS

4.

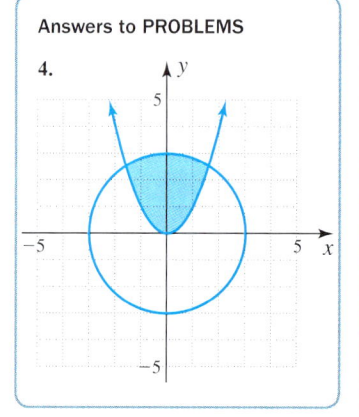

Calculator Corner

Graphing Nonlinear System of Inequalities

To do Example 4 with your calculator, solve for y in $x^2 + y^2 = 4$ to obtain $y = \pm\sqrt{4 - x^2}$. The corresponding inequality is $y \leq \pm\sqrt{4 - x^2}$—that is, the points *on* the boundary and *inside* the circle $x^2 + y^2 = 4$. Remember to use a square window (ZOOM 5 on a TI-83 Plus) to make the circle appear round. The solution set of $y \geq x^2 - 3$ consists of the points *on* the boundary and *above* the parabola $y = x^2 - 3$. The intersection of these two sets consists of the points as shaded in Figure 9.54.

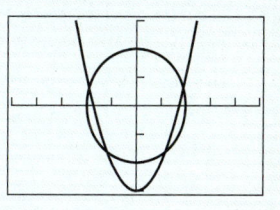

Exercises 9.5

‹ A › Graphing Second-Degree Inequalities In Problems 1–12, graph the solution set of the inequality.

1. $x^2 + y^2 > 16$ **2.** $x^2 + y^2 < 16$ **3.** $x^2 + y^2 \leq 1$

4. $x^2 + y^2 \geq 1$

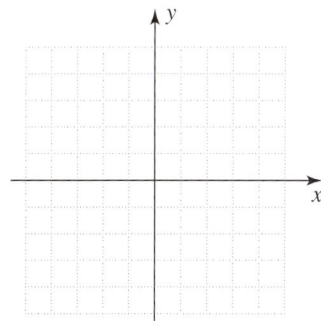

5. $y < x^2 - 2$

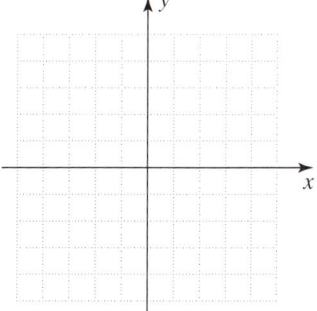

6. $y > x^2 - 2$

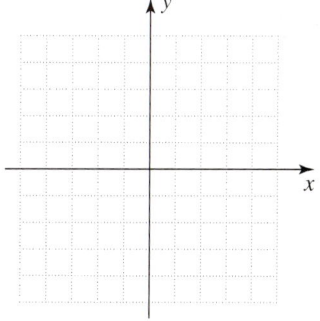

7. $y \leq x^2 + 3$

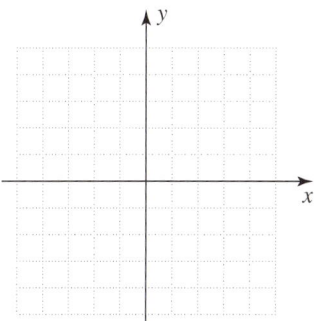

8. $y \geq -x^2 + 3$

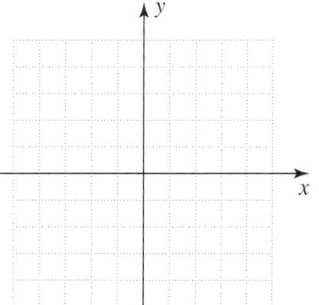

9. $4x^2 - 9y^2 > 36$

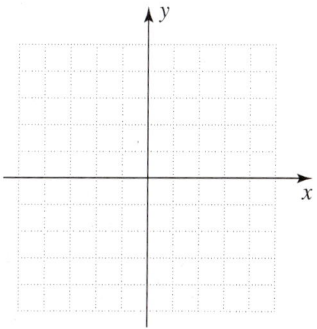

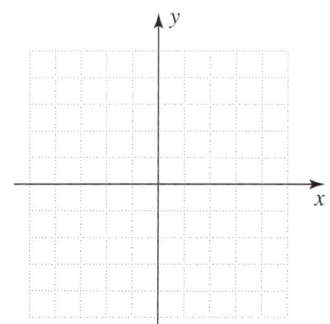
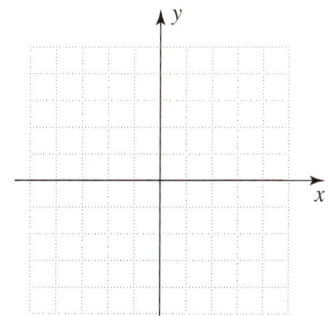
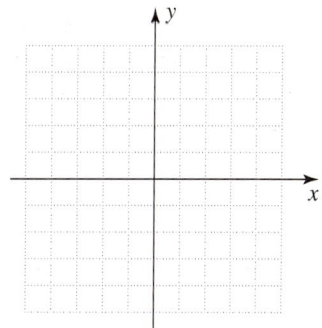

10. $4x^2 - 9y^2 < 36$

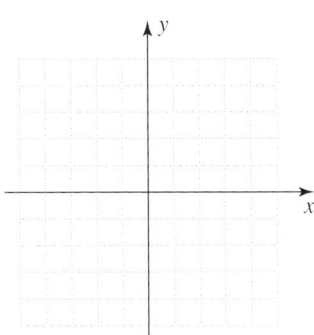

11. $x^2 - y^2 \geq 1$

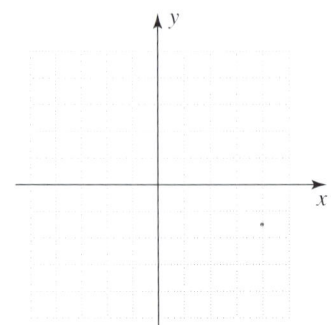

12. $x^2 - y^2 \leq 1$

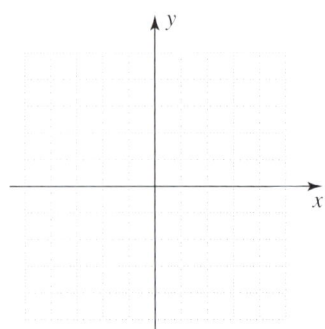

⟨ B ⟩ Graphing Systems of Nonlinear Inequalities In Problems 13–24, graph the solution set of the system.

13. $x^2 + y^2 \leq 25$
  $y \geq x^2$

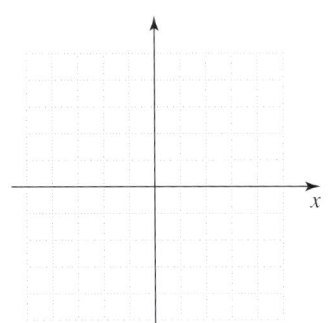

14. $x^2 + y^2 \leq 25$
  $y \leq x^2$

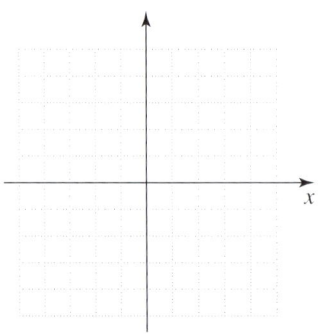

15. $x^2 + y^2 \geq 25$
  $y \leq x^2$

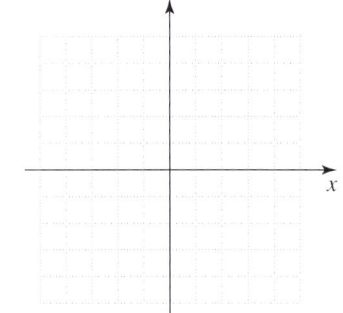

16. $x^2 + y^2 \geq 25$
  $y \geq x^2$

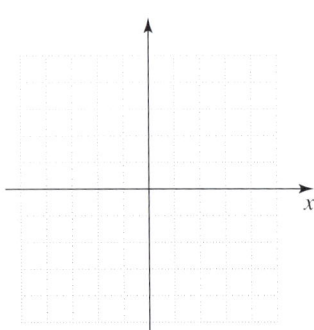

17. $y < x^2 + 2$
  $y > x^2 - 2$

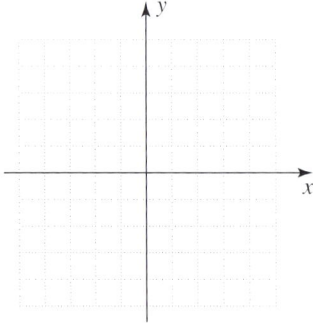

18. $y < x^2 + 2$
  $y < x^2 - 2$

19. $y \geq x^2 + 2$
  $y \geq x^2 - 2$

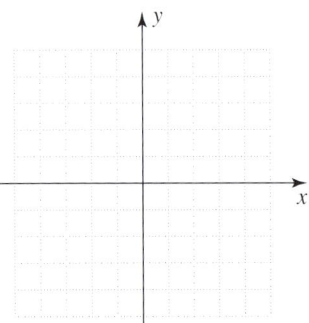

20. $y \geq x^2 + 2$
  $y \leq x^2 - 2$

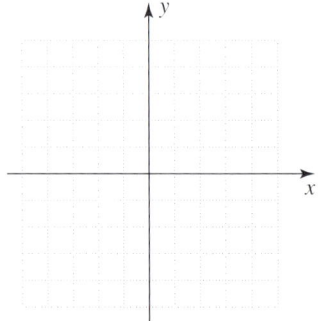

21. $\dfrac{x^2}{4} - \dfrac{y^2}{4} \geq 1$

  $\dfrac{x^2}{25} + \dfrac{y^2}{4} \leq 1$

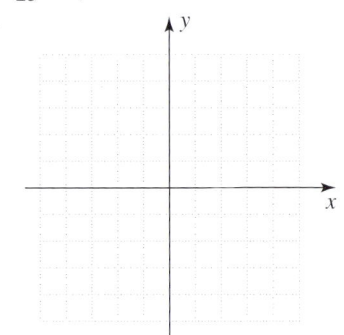

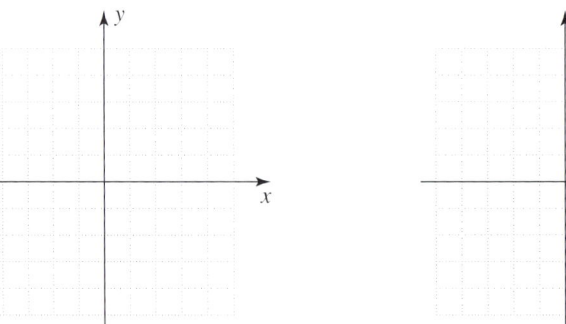

22. $\dfrac{x^2}{4} - \dfrac{y^2}{4} \geq 1$

$\dfrac{x^2}{25} + \dfrac{y^2}{4} \geq 1$

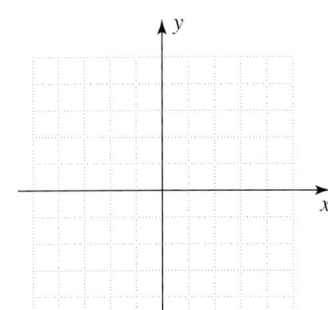

23. $\dfrac{x^2}{36} + \dfrac{y^2}{16} < 1$

$\dfrac{x^2}{16} + \dfrac{y^2}{36} < 1$

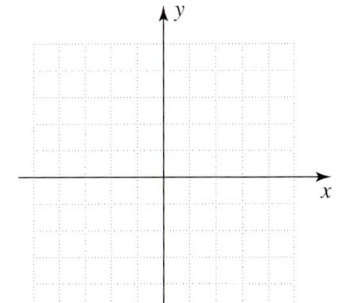

24. $\dfrac{x^2}{36} + \dfrac{y^2}{16} > 1$

$\dfrac{x^2}{16} + \dfrac{y^2}{36} < 1$

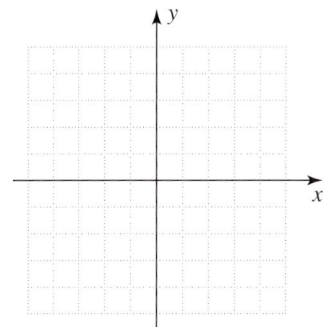

25. Fill in each blank with an inequality symbol so that the system will have the indicated number of solutions.

$\dfrac{x^2}{36} + \dfrac{y^2}{16} \;\underline{\quad}\; 1$ (ellipse)

$\dfrac{x^2}{16} + \dfrac{y^2}{16} \;\underline{\quad}\; 1$ (circle)

a. No solution.

b. Two solutions.

26. Can you fill in each blank with an inequality symbol so that the system will have no solution?

$\dfrac{x^2}{25} - \dfrac{y^2}{4} \;\underline{\quad}\; 1$ (hyperbola)

$\dfrac{x^2}{25} + \dfrac{y^2}{4} \;\underline{\quad}\; 1$ (ellipse)

〉〉〉 Using Your Knowledge

Give Me a Break-Even Point! You may have wondered about the equations representing the cost C and the revenue R in the graph in *Getting Started* at the beginning of this section. Both the cost C and the revenue R are given in terms of p, the price per unit. Can we find the break-even point—the point at which the revenue R equals the cost C? Let's try.

27. The demand equation—the number x of units retailers are likely to buy at p dollars per unit—is given by the equation $x = 9000 - 45p$, and the cost C is given by the equation $C = 108{,}000 + 60x$. Substitute $x = 9000 - 45p$ in $C = 108{,}000 + 60x$ and find C in terms of the price p.

28. The revenue is $R = xp$—the revenue is the product of the number of units retailers are likely to buy and the price p per unit. To find R in terms of p, substitute $x = 9000 - 45p$ into $R = xp$. What is R in terms of p? What is the shape of the graph of R?

29. The graph in *Getting Started* on page 724 shows two points at which the revenue R equals the cost—$R = C$. Substitute the expressions for R and C from Problems 27 and 28 in the equation $R = C$ and solve for p.

〉〉〉 Write On

30. Write your own definition of a second-degree inequality.

31. The system of inequalities we solved in Example 4 has a solution set with infinitely many points. Can you find a system of nonlinear inequalities with the given characteristics.

  a. A solution set consisting of only one point. Make a sketch to show the system.

  b. A solution set consisting of exactly two points. Make a sketch to show the system.

32. How do you decide whether the boundary of a region should be included in the solution set?

33. Write the procedure you use to solve a system of nonlinear inequalities graphically.

››› Concept Checker

Fill in the blank(s) with the correct word(s), phrase, or mathematical statement.

34. When graphing a nonlinear system of two inequalities, the solution is the region _____ to both graphs.

35. The solution of a nonlinear system of two inequalities is the _____ of both graphs.

containing   intersection
common   union

››› Mastery Test

Graph the solution set of the system:

36. $x^2 + y^2 \leq 9$ and $y \geq x^2 + 2$

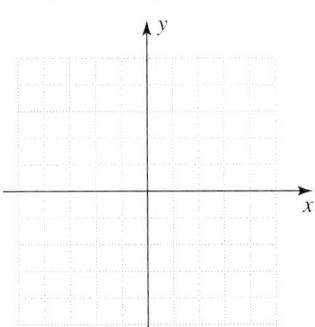

37. $x^2 + 4y^2 \geq 4$ and $y \geq x^2 + 1$

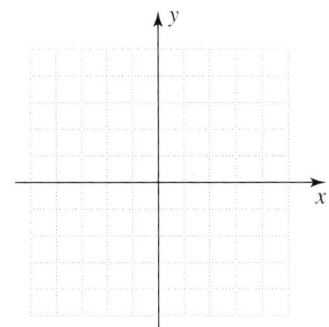

38. $4y^2 - 9x^2 \leq 36$

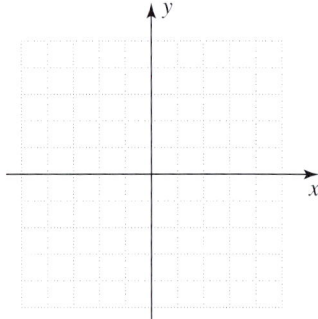

39. $x^2 - 9y^2 \geq 9$

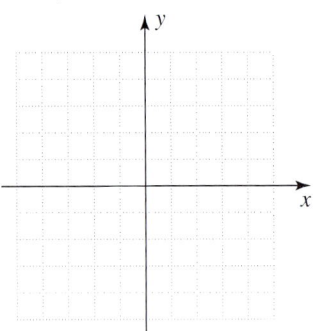

40. $x^2 + y^2 > 9$

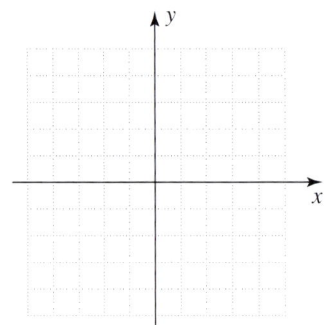

41. $x^2 > 4 - y^2$

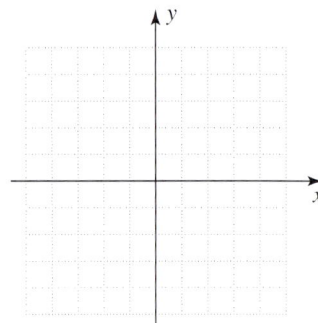

42. $y \leq -x^2 - 3$

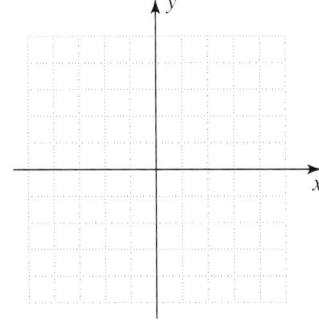

43. $y \geq -x^2 + 2$

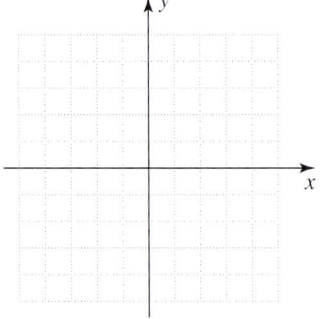

››› Skill Checker

Find the value of $y = 3x + 5$ for the specified value of x.

44. $x = -2$

45. $x = -5$

If $P(x) = x^2 - 9$ and $Q(x) = x + 3$, find:

46. The sum and the difference of $P(x)$ and $Q(x)$.

47. The product and the quotient of $P(x)$ and $Q(x)$.

Collaborative Learning 9A

A resort rents 40 bicycles each day (on the average) and charges an $8 rental fee for each. They are considering raising their rental fee. After investigating studies done on bicycle rentals, there is an indication that for every $1 increase in the rental fee, they will lose two rentals. Divide into groups and answer the questions.

Assuming the data from these studies is accurate, how much should the resort charge to maximize their revenue (total income)?

1. Complete the table.

Price Increase	Rental Fee Per Bicycle	Number of Rentals	Revenue
Present Fee	$8	40	$8(40) = $320
1 rental increase	$8 + 1($1)	40 − 1(2)	$9(38) = $342
2 rental increases	$8 + 2($1)	40 − 2(2)	$10(36) = $360
3 rental increases			
4 rental increases			
x rental increases			(?)(?) = R

2. Use the information from the table to write the equation that will find the revenue, R, given x rental increases.
3. Draw the graph of the equation and find the vertex of the curve.
4. What is the maximum point on the curve?
5. How much should the resort charge for each bicycle rental to maximize the revenue?
6. Each group should find an application of maximizing or minimizing and present the results to the class for discussion. Include in your discussion why the equation of a parabola is appropriate for this type of problem.

Collaborative Learning 9B

The National Statuary Hall in the Capitol building in Washington, D.C., is a semicircular room with a ceiling in the shape of an ellipse. It was the meeting place for the House of Representatives from 1819 to 1857. Now it is an exhibition hall commemorating outstanding U.S. citizens chosen by each state.

When John Quincy Adams was a member of the House of Representatives he took advantage of the Hall's acoustics to eavesdrop on other members conversing on the opposite side of the room. He placed his desk at a focal point of the elliptical ceiling and was able to eavesdrop on the private conversations of other House members sitting near the other focal point. Can you calculate where he placed his desk? Divide into groups and answer the questions.

<CONTINUED>

The equation of an ellipse with center at the origin is $\frac{x^2}{a^2} + \frac{y^2}{b^2} = 1$, where $a > b$. One-half the length of the major axis is a, one-half the length of the minor axis is b, and one-half the distance between the two foci is c. To find c, use the formula, $c = \sqrt{a^2 - b^2}$.

1. Find the distance a, b, and c for the given ellipses.

   a. $\frac{x^2}{9} + \frac{y^2}{16} = 1$ b. $\frac{x^2}{4} + \frac{y^2}{1} = 1$

   c. $\frac{x^2}{16} + \frac{y^2}{4} = 1$ d. $\frac{x^2}{1} + \frac{y^2}{9} = 1$

2. Sketch the graphs for the ellipses in Question 1 and name the coordinates for the foci.
3. If the estimated dimensions for Statuary Hall indicate a major axis of approximately 240 feet and a minor axis of approximately 100 feet, find c, one-half the length between the foci.
4. Using the origin of an x-y coordinate system as the center of the ellipse that approximates Statuary Hall, sketch its graph. Be sure to label and name the coordinates of the foci.
5. How far from the center of Statuary Hall did John Quincy Adams place his desk so he could hear the Congressmen talking at the other focal point?
6. Each group should research other acoustical phenomena to see if they have any connection to the ellipse or other conic sections. Be sure to include information about the acoustics at the Mayan ruins found at http://www.tomzap.com/sounds.html.
7. Present your research findings to the class and have a discussion about which may be related to the conic sections. If they are not, then is there a scientific explanation for the acoustical phenomena?

Source: Architect of the Capitol; National Statuary Hall (gives most of the information about John Quincy Adams).

> Research Questions

1. Write a paper detailing the origin and meaning of the words *ellipse*, *hyperbola*, and *parabola*.
2. Who was the first person that used the terminology "ellipse, hyperbola, and parabola" and from what book or paper were these terms adapted?
3. Who was "The Great Geometer"?
4. Write a short paper about Hippocrates of Chios, his discoveries, his works, and his connection to the conics.
5. Write a short paper about Menaechmus, his discoveries, his works, and his connection to the conics.
6. Write a short paper about Apollonius of Perga, his discoveries, his works, and his connection to the conics.
7. One of the three famous problems in the first 300 years of Greek mathematics was "the duplication of the cube." What were the other two famous problems?
8. How are conics used in mathematics, science, art, and other areas?
9. Write a paper detailing Girard Desargues' contribution to the study of the conic sections.
10. A famous French mathematician wrote *A Complete Work on Conics*. Who was this mathematician and what did the book contain?

Summary Chapter 9

Section	Item	Meaning	Example
9.1	Conic sections	Curves obtained by slicing a cone with a plane	Parabolas, circles, ellipses, and hyperbolas
9.1A	Vertex of a parabola	The highest or lowest point of a parabola opening vertically	The vertex of $y = x^2 + 2$ is at $(0, 2)$.
9.1B, C	Parabola (vertical axis)	A curve with equation $y = a(x - h)^2 + k$ or $y = ax^2 + bx + c$	$y = x^2$, $y = 3(x - 1)^2 + 2$, and $y = -2x^2 + 3x - 7$
9.1D	Parabola (horizontal axis)	A curve with equation $x = a(y - k)^2 + h$ or $x = ay^2 + by + c$	$x = y^2$, $x = 3(y - 1)^2 + 2$, and $x = -4(y + 1)^2 - 3$
9.2A	Distance formula	The distance between (x_1, y_1) and (x_2, y_2) is $d = \sqrt{(x_2 - x_1)^2 + (y_2 - y_1)^2}$	For $(x_1, y_1) = (3, 4)$ and $(x_2, y_2) = (-3, 12)$ $d = \sqrt{(-3 - 3)^2 + (12 - 4)^2}$ $= \sqrt{36 + 64}$ $= \sqrt{100}$ $= 10$
9.2B	Circle centered at (h, k) with radius r	A curve with the equation $(x - h)^2 + (y - k)^2 = r^2$	$(x - 3)^2 + (y + 4)^2 = 9$ is a circle with center at $(3, -4)$ and radius 3.
9.2D	Ellipse centered at (h, k)	A curve with equation $\frac{(x - h)^2}{a^2} + \frac{(y - k)^2}{b^2} = 1$ or $\frac{(x - h)^2}{b^2} + \frac{(y - k)^2}{a^2} = 1$, where $a^2 > b^2$, $a \neq 0$ and $b \neq 0$	$\frac{(x - 4)^2}{9} + \frac{(y + 2)^2}{4} = 1$ is an ellipse.
9.3A	Hyperbola centered at (h, k)	A curve with equation $\frac{(x - h)^2}{a^2} - \frac{(y - k)^2}{b^2} = 1$ or $\frac{(y - k)^2}{a^2} - \frac{(x - h)^2}{b^2} = 1$ where a^2 is always the denominator of the positive variable.	$\frac{(y + 3)^2}{16} - \frac{(x - 2)^2}{4} = 1$ is a hyperbola.
	Asymptotes	Lines associated with a hyperbola, going through the opposite corners of the rectangle whose sides pass through $\pm a$ and $\pm b$	The asymptotes for $\frac{x^2}{a^2} - \frac{y^2}{b^2} = 1$ are shown

(continued)

Section	Item	Meaning	Example
9.4A	Nonlinear systems	A system of equations containing at least one second-degree equation	$x^2 + y^2 = 9$ $x - y = 3$ is a nonlinear system.
	Substitution method	A method of solving nonlinear systems in which substitution is made from one of the equations into the other	To solve $x^2 + y^2 = 9$ $x = y + 3$ by substitution, replace x by $y + 3$ in $x^2 + y^2 = 9$.
9.5	Second-degree inequality	An inequality containing at least one second-degree term	$y \leq x^2 + 2$ and $x^2 + y^2 > 9$ are second-degree inequalities.

❯ Review Exercises Chapter 9

(If you need help with these exercises, look in the section indicated in brackets.)

1. ❮ **9.1A** ❯ *Graph.*

a. $y = 9x^2$

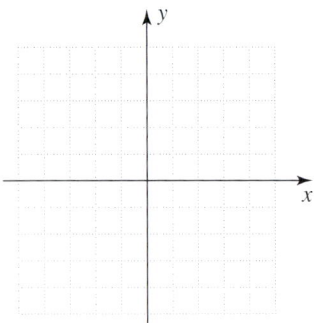

b. $y = -9x^2$

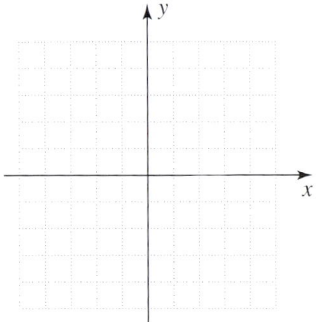

2. ❮ **9.1B** ❯ *Find the vertex and graph.*

a. $y = (x - 1)^2 - 2$

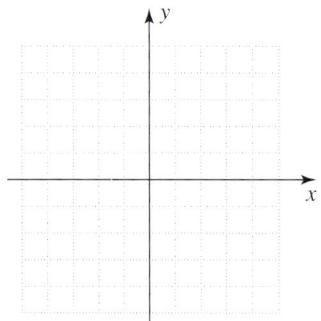

b. $y = -(x - 1)^2 + 2$

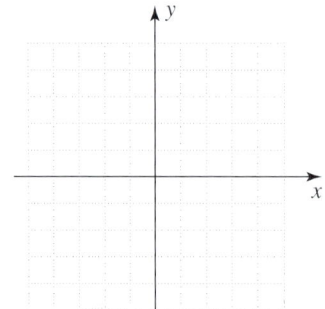

3. ❮ **9.1C** ❯ *Find the vertex and graph.*

a. $y = x^2 - 4x + 2$

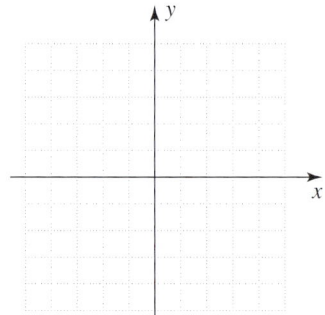

b. $y = -x^2 + 6x - 5$

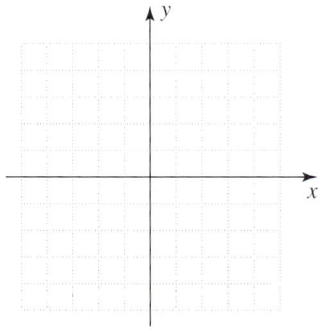

4. ⟨**9.1C**⟩ *Find the vertex and graph.*
   a. $y = 2x^2 - 4x + 3$

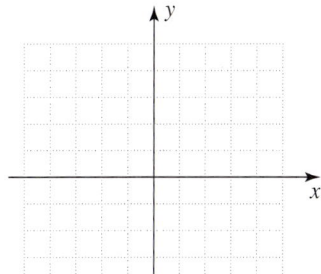

   b. $y = -2x^2 + 4x - 5$

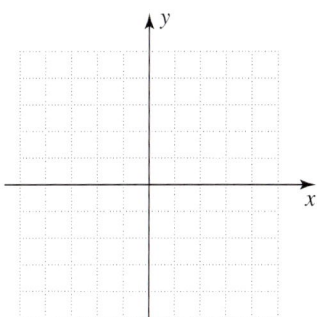

5. ⟨**9.1D**⟩ *Find the vertex and graph.*
   a. $x = 2(y - 2)^2 - 2$

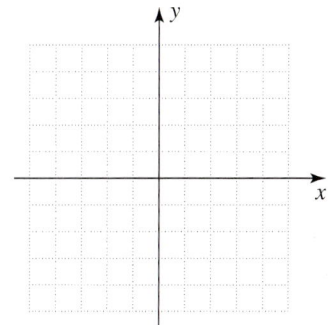

   b. $x = -2(y - 3)^2 + 1$

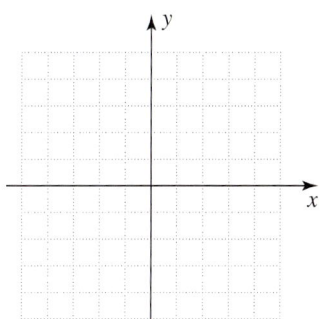

6. ⟨**9.1D**⟩ *Find the vertex and graph.*
   a. $x = y^2 - 4y + 1$

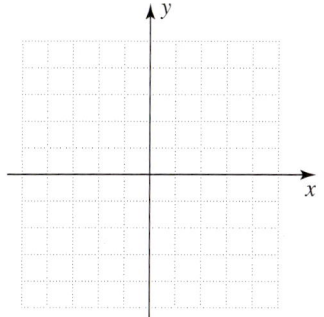

   b. $x = y^2 - 2y + 3$

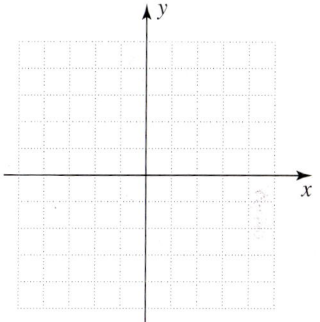

7. ⟨**9.1E**⟩ *Find the value of x that gives the maximum revenue R if:*
   a. $R = 20x - 0.01x^2$
   b. $R = 10x - 0.02x^2$

8. ⟨**9.1E**⟩ *If a ball is thrown upward at 24 feet per second, its height h, in feet, after t seconds is given by the equation $h = -16t^2 + 24t$. How many seconds does it take for the ball to reach its maximum height? What is this height?*

9. ⟨**9.2A**⟩ *Find the distance between each pair of points.*
   a. $(5, -3)$ and $(2, 8)$
   b. $(10, 10)$ and $(2, -4)$
   c. $(-3, 3)$ and $(-3, 8)$

10. ⟨**9.2B**⟩ *Find an equation of the circle of radius 3 and with the given center.*
    a. $(-2, 2)$
    b. $(3, 2)$

11. ⟨**9.2C**⟩ *Find the center, the radius, and sketch the graph.*
    a. $(x + 2)^2 + (y - 1)^2 = 4$

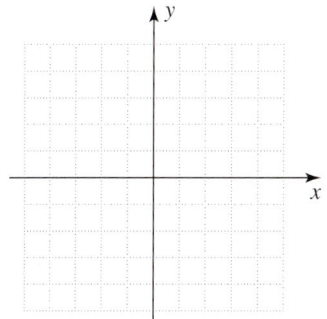

    b. $(x - 1)^2 + (y + 2)^2 = 9$

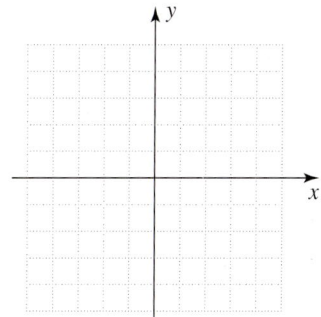

12. ‹9.2C› Sketch the graph.
    a. $x^2 + y^2 = 4$

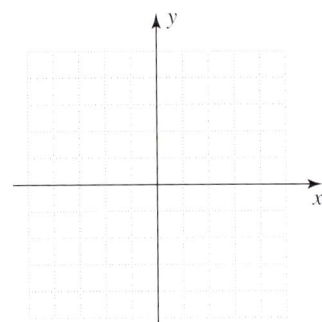

13. ‹9.2C› Find the center, radius, and sketch the graph.
    a. $x^2 + y^2 + 2x + 2y - 2 = 0$

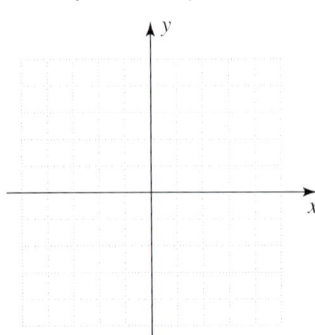

14. ‹9.2D› Graph.
    a. $4x^2 + 9y^2 = 36$

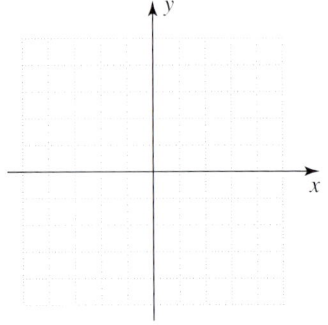

b. $x^2 + y^2 = 25$

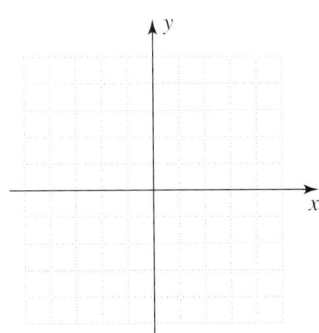

b. $x^2 + y^2 - 4x + 6y + 9 = 0$

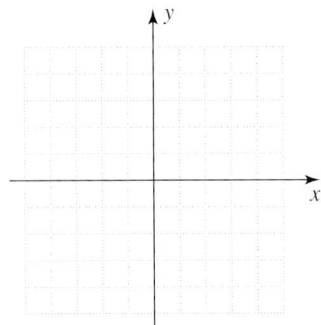

b. $9x^2 + y^2 = 9$

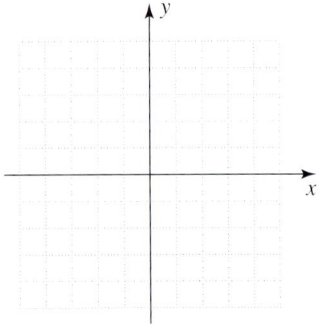

15. ‹9.2D› Graph.
    a. $\dfrac{(x-1)^2}{4} + \dfrac{(y-2)^2}{9} = 1$

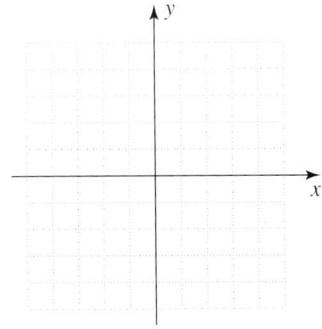

16. ‹9.3A› Graph.
    a. $\dfrac{x^2}{9} - \dfrac{y^2}{16} = 1$

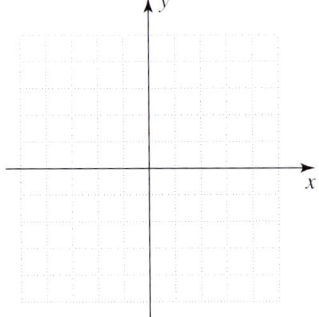

17. ‹9.3A› Graph.
    a. $\dfrac{y^2}{9} - \dfrac{x^2}{16} = 1$

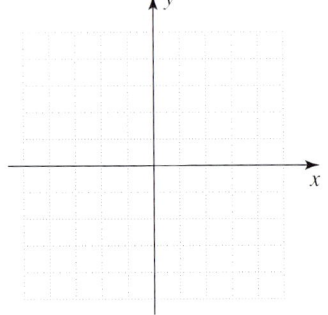

b. $\dfrac{(x+2)^2}{9} + \dfrac{(y-2)^2}{4} = 1$

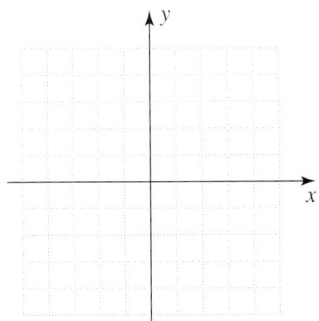

b. $\dfrac{x^2}{16} - \dfrac{y^2}{9} = 1$

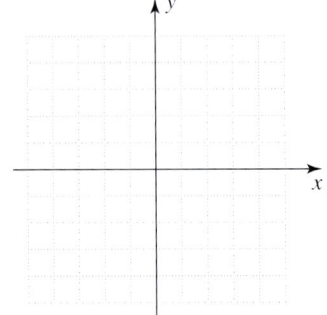

b. $\dfrac{y^2}{16} - \dfrac{x^2}{9} = 1$

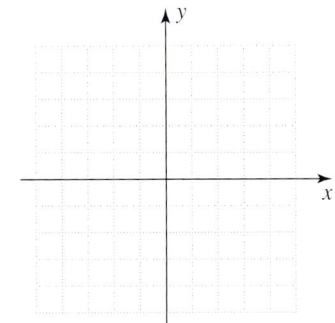

18. ⟨**9.3B**⟩ *Identify each of the curves.*
   a. $x = 1 - y^2$
   b. $x^2 = 4y^2 - 4$
   c. $y^2 = 9 - x^2$
   d. $4y^2 = 36 - 9x^2$

19. ⟨**9.4A**⟩ *Solve the system by the substitution method.*
   a. $x^2 + y^2 = 1$
      $x + y = 1$
   b. $x^2 + y^2 = 10$
      $x + y = 4$

20. ⟨**9.4A**⟩ *Solve the system by the substitution method.*
   a. $x^2 - y^2 = 16$
      $2x = y$
   b. $x^2 + y^2 = 4$
      $x + y = 3$

21. ⟨**9.4B**⟩ *Solve the system.*
   a. $x^2 - y^2 = 5$
      $x^2 + 2y^2 = 17$
   b. $x^2 - y^2 = 3$
      $2x^2 + y^2 = 9$

22. ⟨**9.4C**⟩
   a. The cost C of manufacturing and selling x units of a product is modeled by the equation $C = 10x + 400$, and the corresponding revenue R is $R = x^2 - 200$. Find the break-even value of x.
   b. Repeat part a if $C = 6x + 80$ and $R = x^2 - 200$.

23. ⟨**9.5A**⟩ *Graph the inequality.*
   a. $y \leq 1 - x^2$

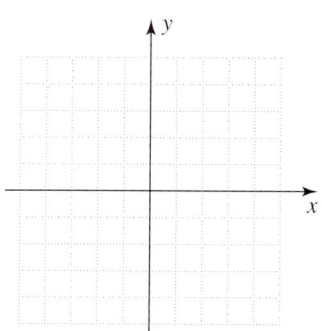

   b. $x \leq 4 - y^2$

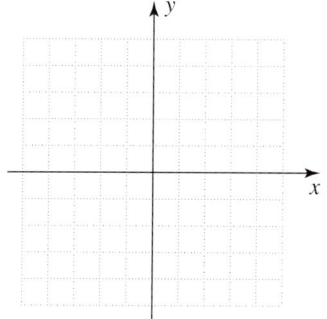

24. ⟨**9.5A**⟩ *Graph the inequality.*
   a. $x^2 + y^2 \leq 4$

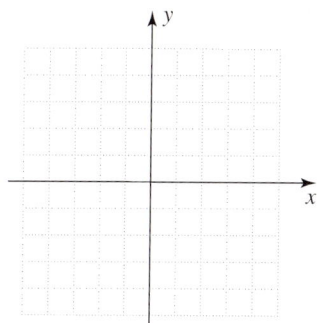

   b. $x^2 + y^2 > 9$

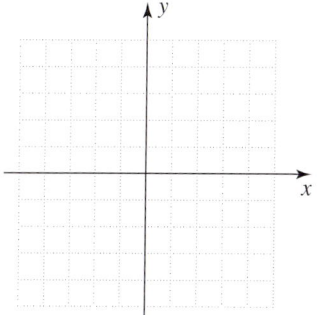

25. ⟨ **9.5A** ⟩ *Graph the inequality.*

a. $4x^2 - y^2 \leq 4$

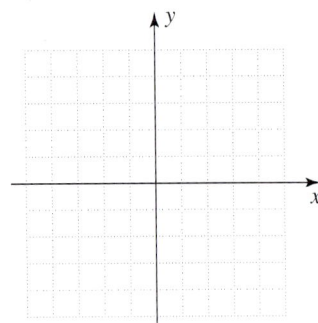

b. $x^2 - 4y^2 \leq 4$

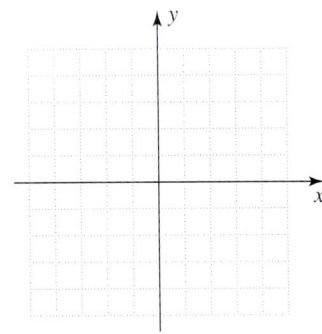

26. ⟨ **9.5B** ⟩ *Graph the system.*

a. $x^2 + y^2 \leq 4$ and $y \leq 2 - x^2$

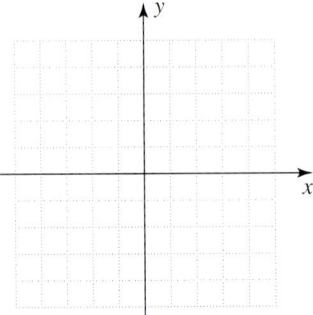

b. $x^2 + y^2 \leq 4$ and $y \geq 4x^2$

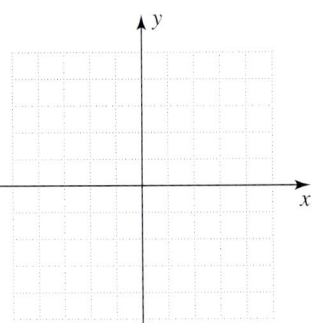

Practice Test Chapter 9

(Answers on pages 740–744)

Visit www.mhhe.com/bello to view helpful videos that provide step-by-step solutions to several of the problems below.

1. Graph the parabola $y = -x^2 - 4$.

2. Name the coordinates of the vertex and graph the parabola $y = (x - 2)^2 + 2$.

3. Name the coordinates of the vertex and graph the parabola $y = -x^2 - 4x - 1$.

4. Name the coordinates of the vertex and graph the parabola $y = -2x^2 + 4x + 1$.

5. Name the coordinates of the vertex and graph the parabola $x = 2(y + 1)^2 - 1$.

6. Name the coordinates of the vertex and graph the parabola $x = y^2 + 2y - 1$.

7. If the revenue is given by the equation $R = 60x - 0.03x^2$, find the value of x that yields the maximum revenue.

8. Find the equation of the circle of radius 2 with its center at $(1, -2)$.

9. Find an equation of the circle of radius $\sqrt{3}$ with its center at the origin, $(0, 0)$.

10. Find the center and the radius and sketch the graph of $(x + 1)^2 + (y - 2)^2 = 9$.

11. Sketch the graph of $x^2 + y^2 = 16$.

12. Find the center and the radius and sketch the graph of $x^2 + y^2 + 4x - 2y - 4 = 0$.

13. Graph $x^2 + 9y^2 = 9$.

14. Graph $\dfrac{(x - 2)^2}{4} + \dfrac{(y + 1)^2}{9} = 1$.

15. Graph $\dfrac{y^2}{9} - \dfrac{x^2}{25} = 1$.

16. Graph $\dfrac{x^2}{9} - \dfrac{y^2}{25} = 1$.

17. Identify each of the following curves.
    a. $x = y^2 - 4$
    b. $16y^2 = 144 - 9x^2$
    c. $9y^2 = 144 + 16x^2$
    d. $x^2 = 16 - y^2$

18. Use the substitution method to solve the system.
$$x^2 + y^2 = 4$$
$$x + y = 2$$

19. Use the substitution method to solve the system.
$$x^2 + y^2 = 1$$
$$x + y = 2$$

20. Solve the system.
$$x^2 + y^2 = 20$$
$$x^2 - 2y^2 = 8$$

21. The cost C of manufacturing and selling x units of a product is modeled by the equation $C = 20x + 50$, and the corresponding revenue R is $R = x^2 - 75$. Find the break-even value of x.

22. Graph the inequality $y \leq -x^2 - 1$.

23. Graph the inequality $x^2 + y^2 < 9$.

24. Graph the inequality $y^2 - 4x^2 \leq 4$.

25. Graph the system $x^2 + y^2 \leq 4$ and $y \leq 2 - x^2$.

Answers to Practice Test Chapter 9

Answer	If You Missed Question	Review Section	Examples	Page
1. [graph: downward parabola, vertex near (0, −3), passing through y = −7]	1	9.1A	1, 2, 3	659–662
2. Vertex (2, 2) [graph: upward parabola]	2	9.1B	4	663
3. Vertex (−2, 3) [graph: downward parabola]	3	9.1C	5, 6	665–668
4. Vertex (1, 3) [graph: downward parabola]	4	9.1C	5, 6	665–668

Answer	If You Missed	Review		
	Question	Section	Examples	Page
5. Vertex $(-1, -1)$	5	9.1D	7	669
6. Vertex $(-2, -1)$	6	9.1D	8	669
7. $x = 1000$	7	9.1E	9	670
8. $(x - 1)^2 + (y + 2)^2 = 4$	8	9.2B	2	682
9. $x^2 + y^2 = 3$	9	9.2B	3	682
10. Center $(-1, 2)$; $r = 3$	10	9.2C	4	683
11.	11	9.2C	5	683

Answer	If You Missed	Review		
	Question	Section	Examples	Page
12. Center $(-2, 1)$; $r = 3$	12	9.2C	6	684
13.	13	9.2D	7	685
14.	14	9.2D	8	686
15.	15	9.3A	1a	701

Answer	If You Missed		Review	
	Question	Section	Examples	Page
16. [graph of hyperbola with vertices V_2, V_1 and center C]	16	9.3A	1b	701
17. a. Parabola **b.** Ellipse **c.** Hyperbola **d.** Circle	17	9.3B	3	705
18. $(2, 0)$ and $(0, 2)$	18	9.4A	1	714
19. $\left(1 + \frac{\sqrt{2}}{2}i, 1 - \frac{\sqrt{2}}{2}i\right)$, $\left(1 - \frac{\sqrt{2}}{2}i, 1 + \frac{\sqrt{2}}{2}i\right)$	19	9.4A	2	715
20. $(4, 2), (4, -2), (-4, 2), (-4, -2)$	20	9.4B	3	716
21. $x = 25$	21	9.4C	4	717
22. [graph of downward parabola shaded region]	22	9.5A	1	725
23. [graph of shaded disk with dashed circle]	23	9.5A	2	725

Answer	If You Missed		Review	
	Question	Section	Examples	Page
24.	24	9.5A	3	726
25.	25	9.5B	4	726

Cumulative Review Chapters 1–9

1. Graph the additive inverse of $\frac{4}{3}$ on a number line.

2. Perform the indicated operation and simplify: $\frac{45x^8}{15x^{-7}}$

3. Evaluate: $-4^3 + \frac{12-8}{2} + 15 \div 3$

4. Solve: $\left|\frac{3}{2}x + 7\right| + 3 = 8$

5. Graph: $x + 1 \leq 4$ and $-2x < 8$

6. Solve for A in $B = \frac{7}{8}(A - 10)$.

7. A woman's salary was increased by 10% to $31,900. What was her salary before the increase?

8. Find the x- and y-intercepts of $y = -2x - 5$.

9. A line passes through the point $(-5, -2)$ and has slope $-\frac{1}{2}$. Graph this line.

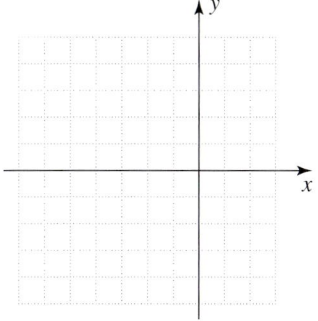

10. Graph: $2x - 5y \leq -10$

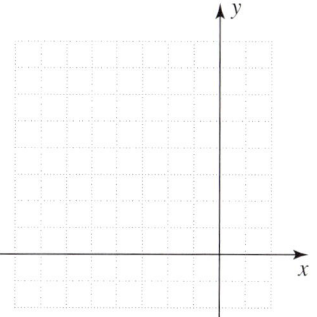

11. Find the slope of the line passing through the points $A(3, 6)$ and $B(7, 3)$.

12. Find the domain of the relation $\{(-2, -3), (5, -2), (-3, 1)\}$.

13. Find $f(3)$ given $f(x) = -2x + 2$.

14. Solve the system by the elimination method.
$$5x + 2y = 14$$
$$10x + 4y = 4$$

15. Solve the system:
$$x + y + z = 4$$
$$-x + 2y + z = 3$$
$$2x + y + z = 6$$

16. Graph: $x < 2$
$y \geq -1$

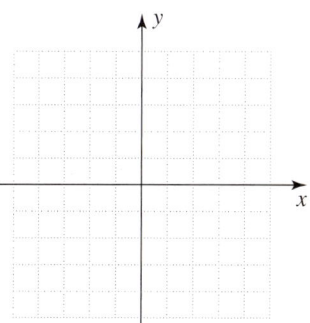

17. A motorboat can go 12 miles downstream on a river in 20 minutes. It takes 30 minutes for this boat to go back upstream the same 12 miles. Find the speed of the boat.

18. Multiply: $(3j + 1)(3j - 1)$

19. Factor completely: $12x^6 - 3x^4 + 8x^3 - 2x$

20. Factor completely: $32x^2y^2 - 48xy^3 + 18y^4$

21. Solve for x: $2x^2 + 9x = -9$

22. Multiply: $\dfrac{x+1}{3x+y} \cdot \dfrac{9x^2 - y^2}{2x^2 - x - 3}$

23. Solve for x: $\dfrac{x}{x+7} - \dfrac{x}{x-7} = \dfrac{x^2 + 49}{x^2 - 49}$

24. Janet can paint a kitchen in 5 hours and Charles can paint the same kitchen in 4 hours. How long would it take for both working together to paint the kitchen?

25. An enclosed gas exerts a pressure P on the walls of the container. This pressure is directly proportional to the temperature T of the gas. If the pressure is 6 pounds per square inch when the temperature is 360°F, find k.

26. Assume x is positive and simplify: $x^{2/3} \cdot x^{5/2}$

27. Reduce the order (index) of: $\sqrt[6]{16c^4d^4}$

28. Find the product: $(\sqrt{125} + \sqrt{18})(\sqrt{20} + \sqrt{50})$

29. Solve over the real numbers: $\sqrt{x+3} = -2$

30. Divide and simplify: $\dfrac{1+2i}{9+4i}$

31. Solve for x: $7x^2 + 42x + 63 = 11$

32. Solve for x: $27x^3 - 64 = 0$

33. Find the sum and the product of the roots of the equation $4x^2 - 5x - 4 = 0$.

34. Solve for x: $x^2 + x \geq 56$

35. Graph the parabola and name the vertex: $y = -3x^2 - 4x + 3$

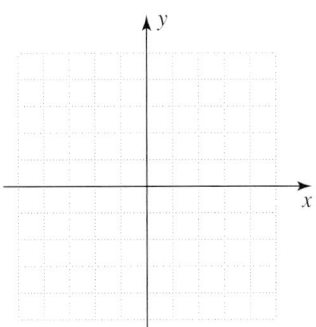

36. If the revenue is given by $R = 20x - 0.02x^2$, find the value of x that yields the maximum revenue.

37. Find an equation of the circle of radius 9 with center at the origin.

38. Graph and name the center: $\dfrac{(x-2)^2}{25} + \dfrac{(y+4)^2}{16} = 1$

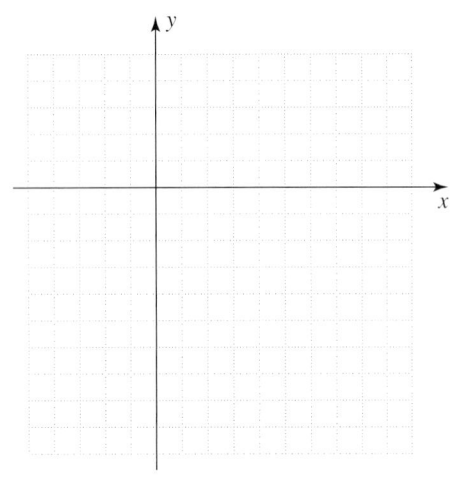

39. Use the substitution method to solve the system:
$x^2 + y^2 = 25$
$x + y = 1$

40. Graph the inequality: $x^2 + y^2 < 4$

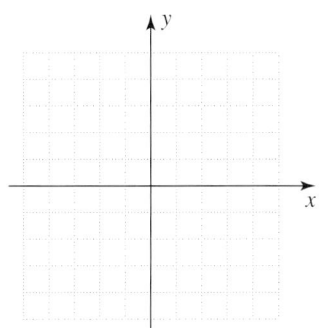

Chapter 10
ten
Functions—Inverse, Exponential, and Logarithmic

Section

- 10.1 The Algebra of Functions
- 10.2 Inverse Functions
- 10.3 Exponential Functions
- 10.4 Logarithmic Functions and Their Properties
- 10.5 Common and Natural Logarithms
- 10.6 Exponential and Logarithmic Equations and Applications

The Human Side of Algebra

Who invented logarithms? It's sometimes implied that the invention of logarithms was the work of one man: John Napier. Interestingly enough, Napier was not a professional mathematician but a Scottish laird (landowner) and Baron of Murchiston who dabbled in many controversial topics. (In a commentary on the Book of Revelations, he argued that the pope at Rome was the Antichrist!) Napier claimed that he worked on the invention of logarithms for 20 years before he published his *Mirifici logarithmorum canonis descriptio (A description of the Marvelous Rule of Logarithms)* in 1614. While working on his invention, Napier lived near an old mill. *It is reported that the noise of the cascade, being constant, never gave him uneasiness, but that the clack of the mill, which was only occasional, greatly disturbed his thoughts.* What did Napier do? He ordered the mill stopped so that *"the train of his ideas might not be interrupted."*

In his book, Napier showed that all numbers can be expressed in exponential form, so when multiplying $100 = 10^2$ by $100,000 = 10^5$ the 100 and 100,000 can be represented by their exponents (logarithms) 2 and 5 and the product by the sum of the exponents, 7. This is easier than writing 100 times 100,000. His book not only described and explained his invention but included his first set of logarithmic tables. These tables were a stroke of genius and a big hit with astronomers and scientists.

John Napier

In 1617, the year of Napier's death, Henry Briggs constructed the first table of Briggsian or common logarithms. But there is one more participant in the invention of logarithms. Jobst Bürgi of Switzerland independently developed the idea of logarithms as early as 1588. Unfortunately, his results were not published until 1620. In his scheme, Bürgi used "red" and "black" numbers, with the "red" numbers appearing on the side of the page and "black" numbers in the body of the table, thus creating what could now be described as an antilogarithm table.

10.1 The Algebra of Functions

Objectives

A ▸ Find the sum, difference, product, and quotient of two functions.

B ▸ Find the composite of two functions.

C ▸ Find the domain of $(f + g)(x)$, $(f - g)(x)$, $(fg)(x)$, and $\left(\frac{f}{g}\right)(x)$.

D ▸ Applications involving operations with functions.

▸ To Succeed, Review How To . . .

1. Evaluate expressions (pp. 51–55).
2. Add, subtract, multiply, and divide polynomials (pp. 352–354, 360–365, 470–473).

▸ Getting Started

A Lot of Garbage!

How many pounds of garbage does each person generate each day? The average is more than 4 pounds and growing! Some of this waste is recovered and recycled into other products. The amount of solid waste recovered (in millions of tons) is a **function** of time and can be approximated by the equation

$$R(t) = 0.04t^2 - 0.59t + 7.42$$

where t is the number of years after 1960.

On the other hand, the amount of solids *not* recovered can be approximated by the equation

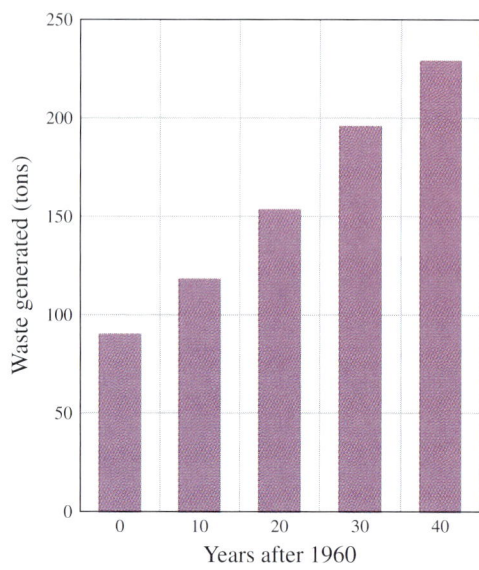

$$N(t) = 2.93t + 82.58$$

Can we find the total amount of solid waste generated? This amount is the **sum** of $R(t)$ and $N(t)$,

$$R(t) + N(t) = (0.04t^2 - 0.59t + 7.42) + (2.93t + 82.58)$$
$$= 0.04t^2 + 2.34t + 90$$

The items recovered most often are paper and paperboard. The amount of paper and paperboard is also a function of time and can be approximated by $P(t) = 0.02t^2 - 0.25t + 6$. You may be surprised to learn what fraction of the total recovered waste $R(t)$ is actually paper and paperboard. That fraction is the **quotient:**

$$\frac{P(t)}{R(t)}$$

Now suppose you want to know what this fraction was in 1990. Since 1990 is 30 (1990 − 1960) years after 1960, this fraction for 1990 is

$$\frac{P(30)}{R(30)} = \frac{0.02(30)^2 - 0.25(30) + 6}{0.04(30)^2 - 0.59(30) + 7.42} = \frac{16.5}{25.72} \approx 0.64$$

Thus, 64% of the total amount of recovered waste in 1990 was paper and paperboard.

Is the situation getting better? That is, are we recycling more paper and paperboard? In 2003, 40 million tons of paper and paperboard were recovered out of the 72 tons of recovered material, which means that $\frac{32}{56} \approx 57\%$ of the paper and paperboard was recovered in 2003.

We can perform the fundamental operations of addition, subtraction, multiplication, and division using functions. We shall study such operations in this section.

A › Using Operations with Functions

Here are the definitions we need to add, subtract, multiply, and divide functions.

OPERATIONS WITH FUNCTIONS

If f and g are functions and x is in the domain of both functions, then

$(f + g)(x) = f(x) + g(x)$ The sum of f and g
$(f - g)(x) = f(x) - g(x)$ The difference of f and g
$(fg)(x) = f(x) \cdot g(x)$ The product of f and g
$\left(\frac{f}{g}\right)(x) = \frac{f(x)}{g(x)}, \; g(x) \neq 0$ The quotient of f and g

EXAMPLE 1   Performing operations with functions

If $f(x) = x^2 + 4$ and $g(x) = x + 2$, find:

a. $(f + g)(x)$ **b.** $(f - g)(x)$ **c.** $(fg)(x)$ **d.** $\left(\frac{f}{g}\right)(x)$

SOLUTION

a. $(f + g)(x) = f(x) + g(x)$
$= (x^2 + 4) + (x + 2)$
$= x^2 + x + 6$ Simplify.

b. $(f - g)(x) = f(x) - g(x)$
$= (x^2 + 4) - (x + 2)$ Recall that $-(a + b) = -a - b$.
$= x^2 + 4 - x - 2$ Simplify.
$= x^2 - x + 2$

c. $(fg)(x) = f(x) \cdot g(x)$
$= (x^2 + 4)(x + 2)$ Multiply (use FOIL method).
$= x^3 + 2x^2 + 4x + 8$

d. $\left(\frac{f}{g}\right)(x) = \frac{f(x)}{g(x)}$
$= \frac{x^2 + 4}{x + 2}$

The denominator can't be zero, so $x + 2 \neq 0$, that is, $x \neq -2$.

PROBLEM 1

If $f(x) = x^2 - 9$ and $g(x) = x - 3$, find:

a. $(f + g)(x)$ **b.** $(f - g)(x)$
c. $(fg)(x)$ **d.** $\left(\frac{f}{g}\right)(x)$

Answers to PROBLEMS

1. **a.** $x^2 + x - 12$
b. $x^2 - x - 6$
c. $x^3 - 3x^2 - 9x + 27$
d. $x + 3, x \neq 3$

We shall talk about domains of quotients later in this section.

EXAMPLE 2 Performing operations with functions

If $f(x) = 3x + 1$, find:

$$\frac{f(x) - f(a)}{x - a}, \quad x \neq a$$

SOLUTION Since $f(x) = 3x + 1$, and $f(a) = 3a + 1$, we have

$$\frac{f(x) - f(a)}{x - a} = \frac{(3x + 1) - (3a + 1)}{x - a}$$

$$= \frac{3x + 1 - 3a - 1}{x - a} \quad \text{Simplify.}$$

$$= \frac{3x - 3a}{x - a}$$

$$= \frac{3(x - a)}{x - a} \quad \text{Factor.}$$

$$= 3, \quad x \neq a$$

PROBLEM 2

If $f(x) = 2x - 3$, find:

$$\frac{f(x) - f(a)}{x - a}, \quad x \neq a$$

B › Finding Composite Functions

There are many instances in which some quantity depends on a variable that, in turn, depends on another variable. For instance, the tax you pay on your house depends on the assessed value and the millage rate. (A 1-mill rate means that you pay $1 for each $1000 of assessed value.) Many states offer a Homestead Exemption, exempting a certain amount of the house value from taxes. In Florida, this exemption is $25,000. Thus, the function g giving the correspondence between the assessed value x of a home in Florida and its taxable value is $g(x) = x - 25,000$.

A house assessed at $150,000 will have a taxable value given by $g(150,000) = 150,000 - 25,000 = \$125,000$. When the tax rate is 5 mills, the function f that computes the tax on the house is

$$f(V) = \frac{5}{1000} \cdot V = 0.005V$$

where V is the taxable value of the house. Thus, a house valued at $150,000 with a $25,000 Homestead Exemption will pay $0.005 \cdot (150,000 - 25,000) = 0.005 \cdot 125,000 = \$625$ in taxes. If you look at the table, you will see that a house assessed at $150,000 should indeed pay $625 in taxes. Can you find a function h that will find the tax? If you guessed $h(x) = 0.005(x - 25,000)$, you guessed correctly.

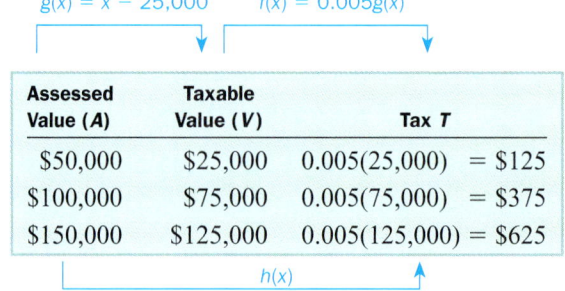

The taxable value V of a house with a $25,000 exemption is $V = x - 25,000$. To find the actual tax, we find

$$f(V) = f(x - 25,000) = f(g(x)) = 0.005(x - 25,000)$$

This gives a formula for h: $h(x) = 0.005(x - 25,000)$. The function h is called the composite of f and g and is denoted by $f \circ g$ (read "f of g"). Here is the definition.

Answers to PROBLEMS

2. $2, x \neq a$

COMPOSITE FUNCTION

If f and g are functions, then

$$(f \circ g)(x) = f(g(x))$$

is the **composite of f with g**.

NOTE

The domain of $f \circ g$ is the set of all x in the domain of g such that $g(x)$ is in the domain of f.

Thus, if $f(x) = x^2$ and $g(x) = x + 2$,
$$(f \circ g)(x) = f(g(x)) = f(x + 2)$$
$$= (x + 2)^2$$

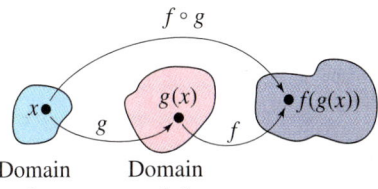

EXAMPLE 3 Finding composite functions
If $f(x) = x^3$ and $g(x) = x - 1$, find:

a. $(f \circ g)(3)$ **b.** $(f \circ g)(x)$ **c.** $(g \circ f)(x)$

SOLUTION

a. Substitute 3 for x into "g." The result is substituted for x into "f."
$$(f \circ g)(3) = f(g(3)) = f(3 - 1) = f(2) = 2^3 = 8$$

b. Substituting $x - 1$ for $g(x)$ in $f(g(x))$, we have
$$(f \circ g)(x) = f(g(x)) = f(x - 1)$$
$$= (x - 1)^3$$

c. Substituting x^3 for $f(x)$ in $g(f(x))$, we have
$$(g \circ f)(x) = g(f(x)) = g(x^3) = x^3 - 1$$

PROBLEM 3
If $f(x) = x^2 - 1$ and $g(x) = 3x + 1$, find:

a. $(f \circ g)(-1)$ **b.** $(f \circ g)(x)$

c. $(g \circ f)(x)$

NOTE
In Example 3, $(f \circ g)(x) = (x - 1)^3$ and $(g \circ f)(x) = x^3 - 1$. Thus, $(f \circ g)(x) \neq (g \circ f)(x)$.

C ▸ Finding the Domains of Combinations of Functions

To find

$$(f + g)(a), \quad (f - g)(a), \quad (fg)(a), \quad \text{and} \quad \left(\frac{f}{g}\right)(a)$$

we first have to find $f(a)$ and $g(a)$. To do this, we must determine whether a is in the domain of f and g.

Answers to PROBLEMS

3. a. 3 **b.** $9x^2 + 6x$ **c.** $3x^2 - 2$

EXAMPLE 4   Determining whether a is in the domain of f and g

Let $f(x) = \frac{3}{x}$ and $g(x) = \frac{x+2}{x-1}$. Find the domain of $f + g$, $f - g$, and fg.

SOLUTION   Since division by zero is not defined,

$$f(x) = \frac{3}{x}$$

is not defined when $x = 0$, and

$$g(x) = \frac{x+2}{x-1}$$

is not defined when $x - 1 = 0$ (when $x = 1$). Thus,

$$\text{Domain of } f = \{x \mid x \text{ is a real number and } x \neq 0\}$$

and

$$\text{Domain of } g = \{x \mid x \text{ is a real number and } x \neq 1\}$$

To find $f(a) + g(a)$, $f(a) - g(a)$, and $fg(a)$, we have to be sure that a is in *both* the domain of f and the domain of g. Hence,

$$\text{Domain of } f + g = \text{domain of } f - g = \text{domain of } fg$$
$$= \{x \mid x \text{ is a real number and } x \neq 0 \text{ and } x \neq 1\}$$

PROBLEM 4

Let $f(x) = \frac{1}{2x}$ and $g(x) = \frac{3x-1}{x+4}$. Find the domain of $f + g$, $f - g$, and fg.

Now let's talk about quotients of functions. Suppose that in Example 1(d), we wanted to find

$$\left(\frac{f}{g}\right)(-2)$$

We can easily find $f(-2)$ and $g(-2)$. Since $f(x) = x^2 + 4$, $f(-2) = (-2)^2 + 4 = 8$, and $g(x) = x + 2$, so $g(-2) = -2 + 2 = 0$. But then,

$$\left(\frac{f}{g}\right)(-2) = \frac{f(-2)}{g(-2)} = \frac{8}{0}$$

which is not defined. Even though -2 is in the domain of *both* f and g, -2 is *not* in the domain of $\frac{f}{g}$.

EXAMPLE 5   Finding the domain of $\frac{f}{g}$

Find the domain of $\frac{f}{g}$, if

$$f(x) = \frac{2}{x-3} \quad \text{and} \quad g(x) = \frac{4}{x+1}$$

SOLUTION   Since the domain of $f = \{x \mid x \text{ is a real number and } x \neq 3\}$ and the domain of $g = \{x \mid x \text{ is a real number and } x \neq -1\}$, we conclude that the domain of $\frac{f}{g}$ is the set of all real numbers *except* 3, -1, and any other values of x for which $g(x) = 0$. Since

$$g(x) = \frac{4}{x+1}$$

is never zero, there are no other values of x such that $g(x) = 0$. Hence,

$$\text{Domain of } \frac{f}{g} = \{x \mid x \text{ is a real number and } x \neq 3 \text{ and } x \neq -1\}$$

PROBLEM 5

Find the domain of $\frac{f}{g}$, if

$$f(x) = \frac{5}{x+6} \quad \text{and} \quad g(x) = \frac{10}{x-1}$$

Answers to PROBLEMS

4. Domain of $f + g$ = domain of $f - g$ = domain of $fg = \{x \mid x \text{ is a real number and } x \neq 0 \text{ and } x \neq -4\}$
5. $\{x \mid x \text{ is a real number and } x \neq -6 \text{ and } x \neq 1\}$

EXAMPLE 6 Finding the domain of $\frac{f}{g}$

Find: The domain of $\frac{f}{g}$, if

$$f(x) = \frac{2}{x-3} \quad \text{and} \quad g(x) = \frac{4x}{x+1}$$

SOLUTION As before, the domain of $f = \{x \mid x \text{ is a real number and } x \neq 3\}$ and the domain of $g = \{x \mid x \text{ is a real number and } x \neq -1\}$. Thus, the domain of $\frac{f}{g}$ is the set of all real numbers *except* 3, -1, and any other values of x for which $g(x) = 0$. Since

$$g(x) = \frac{4x}{x+1} = 0$$

when $x = 0$, we conclude that

Domain of $\frac{f}{g} = \{x \mid x \text{ is a real number and } x \neq 3, x \neq -1, \text{ and } x \neq 0\}$

PROBLEM 6

Find the domain of $\frac{f}{g}$, if

$$f(x) = \frac{1}{x+7} \quad \text{and} \quad g(x) = \frac{x+4}{x-1}$$

In general, we have the following procedure.

> **PROCEDURE**
>
> **Finding the domain of a sum, difference, product, or quotient of two functions**
> 1. Find the domain of each function.
> 2. The domain of the **sum, difference,** or **product** is the set of all values common to *both* domains.
> 3. The domain of the **quotient** is the set of all values common to both domains excluding any value that would result in division by zero.

D › Applications Involving Operations with Functions

Functions and their operations are used frequently in business. For example, suppose the cost C of making x items and the resulting revenue R are given as functions of x.

When does one make a profit? Since the profit (or loss) is the difference between the revenue and the cost, the profit P is modeled by the equation.

$$P(x) = R(x) - C(x)$$

EXAMPLE 7 The profit function

The revenue (in dollars) obtained from selling x units of a product is given by the equation

$$R(x) = 200x - \frac{x^2}{30}$$

and the cost is given by the equation $C(x) = 72{,}000 + 60x$.

a. Find the profit function, $P(x)$.
b. How many units must be made and sold to yield the maximum profit? Find this profit.

PROBLEM 7

Do Example 7 if
$C(x) = 50{,}000 + 50x$.

(continued)

Answers to PROBLEMS

6. $\{x \mid x \text{ is a real number and } x \neq -7, x \neq 1, \text{ and } x \neq -4\}$

7. a. $P(x) = -\frac{x^2}{30} + 150x - 50{,}000$ **b.** 2250 units must be made and sold to give a maximum profit of \$118,750.

SOLUTION

a. $P(x) = R(x) - C(x)$

$= \left(200x - \dfrac{x^2}{30}\right) - (72{,}000 + 60x)$

$= -\dfrac{x^2}{30} + 140x - 72{,}000$

b. To find the value of x that gives the maximum profit, note that the graph of the profit function is a parabola opening downward. Thus, the maximum value of P is at the vertex. The vertex is at the point where

$$x = -\dfrac{b}{2a} = -\dfrac{140}{-\dfrac{2}{30}} = (15)(140) = 2100$$

2100 units must be made and sold to give the maximum profit. This profit is P dollars, where P is given by

$$P(2100) = -\dfrac{(2100)^2}{30} + 140(2100) - 72{,}000$$

$$= 75{,}000$$

So the maximum profit is $75,000.

Calculator Corner

Finding the Maximum or Minimum Value of a Function

You can use a TI-83 Plus to find the maximum or minimum values of the functions discussed in this section. In Example 7, the profit function can be graphed using an appropriate window (try $[-1000, 4000]$ by $[-1000, 75{,}000]$ with Xscl = 1000, Yscl = 1000). Using the **Y=**, enter

$$Y_1 = -\dfrac{x^2}{30} + 140x - 72{,}000$$

To find the maximum, press **2nd** **TRACE** **4**. Use the arrow key to move the blinking cursor to a point on the curve on the left side of the maximum and press **ENTER**. Now use the arrow key to move the blinking cursor to a point on the curve that is on the right side of the maximum and press **ENTER**. When prompted with "GUESS," press **ENTER**. The calculator gives the coordinates of the maximum for the function, $X = 2100.0003$ and $Y = 75000$, which you can approximate to 2100 and 75,000, as before.

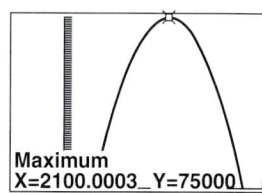

Boost your grade at mathzone.com!
> Practice Problems
> NetTutor
> Self-Tests
> e-Professors
> Videos

> Exercises 10.1

< A > **Using Operations with Functions** In Problems 1–6, use $f(x) = x + 4$, $g(x) = x^2 - 5x + 4$, and $h(x) = x^2 + 16$, to perform the following operations:

1. $(f + g)(x)$

2. $(f - g)(x)$

3. $(hf)(x)$

4. $\left(\dfrac{h}{f}\right)(x)$

5. $\left(\dfrac{f}{h}\right)(x)$

6. $(f + g - h)(x)$

In Problems 7–12, find

$$\frac{f(x) - f(a)}{x - a}, \quad x \neq a$$

7. $f(x) = 3x - 2$ **8.** $f(x) = 5x - 1$ **9.** $f(x) = x^2$

10. $f(x) = x^3$ **11.** $f(x) = x^2 + 3x$ **12.** $f(x) = x^2 - 2x$

⟨B⟩ Finding Composite Functions In Problems 13–20, find the following compositions:

  a. $(f \circ g)(1)$ **b.** $(f \circ g)(x)$ **c.** $(g \circ f)(x)$

13. $f(x) = x^2, \quad g(x) = \sqrt{x}, x > 0$ **14.** $f(x) = x - 1, \quad g(x) = x^2$

15. $f(x) = 3x - 2, \quad g(x) = x + 1$ **16.** $f(x) = x^2, \quad g(x) = x - 1$

17. $f(x) = \sqrt{x + 1}, x > -1, \quad g(x) = x^2 - 1$ **18.** $f(x) = \sqrt{x^2 + 1}, \quad g(x) = 2x + 1$

19. $f(x) = 3, \quad g(x) = -1$ **20.** $f(x) = ax, \quad g(x) = bx$

In Problems 21–30, evaluate the indicated combination of f and g for the given values of the independent variable. (If this is not possible, state the reason.)

21. $f(x) = \sqrt{x}, x > 0, \quad g(x) = x^2 - 1$
  a. $(f + g)(4)$
  b. $(f - g)(4)$

22. $f(x) = \sqrt{x - 2}, x > 2, \quad g(x) = x^2 + 1$
  a. $(f + g)(1)$
  b. $(f - g)(6)$

23. $f(x) = |x|, \quad g(x) = 3$
  a. $\left(\frac{f}{g}\right)(3)$ **b.** $\left(\frac{f}{g}\right)(0)$

24. $f(x) = x^2 - 4, \quad g(x) = x + 2$
  a. $\left(\frac{f}{g}\right)(2)$ **b.** $\left(\frac{f}{g}\right)(-2)$

25. $f(x) = x - 3, \quad g(x) = (x + 3)(x - 3)$
  a. $\left(\frac{f}{g}\right)(3)$
  b. $\left(\frac{g}{f}\right)(3)$

26. $f(x) = x + 5, \quad g(x) = (x + 5)(x - 5)$
  a. $\left(\frac{f}{g}\right)(5)$
  b. $\left(\frac{f}{g}\right)(-5)$

27. $f(x) = \sqrt{x}, x > 0, \quad g(x) = x^2 + 1$
  a. $(f \circ g)(1)$
  b. $(g \circ f)(-1)$
  c. $(f \circ g)(x)$
  d. $(g \circ f)(x)$

28. $f(x) = \sqrt{x + 1}, x > -1, \quad g(x) = x^2$
  a. $(f \circ g)(-1)$
  b. $(g \circ f)(-1)$
  c. $(f \circ g)(x)$
  d. $(g \circ f)(x)$

29. $f(x) = \frac{1}{x^2 - 2}, \quad g(x) = \sqrt{x}, x > 0$
  a. $(f \circ g)(2)$
  b. $(g \circ f)(2)$
  c. $(f \circ g)(x)$
  d. $(g \circ f)(x)$

30. $f(x) = \frac{1}{x^2}, \quad g(x) = \sqrt{x}, x > 0$
  a. $(f \circ g)(-1)$
  b. $(g \circ f)(-1)$
  c. $(f \circ g)(x)$
  d. $(g \circ f)(x)$

⟨C⟩ Finding the Domains of Combinations of Functions

In Problems 31–40, determine the domain of the sum, difference, product, and quotient of f and g.

31. $f(x) = x^2$

$g(x) = x - 1$

32. $f(x) = -2x^2$

$g(x) = x - 2$

33. $f(x) = 2x + 1$

$g(x) = -2x + 2$

34. $f(x) = -3x + 1$

$g(x) = 3x - 2$

35. $f(x) = \dfrac{1}{x - 1}$

$g(x) = -\dfrac{3}{x + 2}$

36. $f(x) = -\dfrac{3}{x - 2}$

$g(x) = -\dfrac{4}{x + 5}$

37. $f(x) = \dfrac{2x}{x + 4}$

$g(x) = \dfrac{x}{x - 1}$

38. $f(x) = \dfrac{5x}{x + 2}$

$g(x) = \dfrac{x}{x + 3}$

39. $f(x) = \dfrac{x - 2}{x + 3}$

$g(x) = \dfrac{x - 1}{x - 2}$

40. $f(x) = \dfrac{x + 3}{x - 4}$

$g(x) = \dfrac{x + 3}{x + 2}$

⟨D⟩ Applications Involving Operations with Functions

41. *Profit* The dollar revenue obtained from selling x copies of a textbook is modeled by the equation $R(x) = 40x - 0.0005x^2$. The production cost C is modeled by the equation $C(x) = 120{,}000 + 6x$. Find the profit function $P(x)$.

42. *Profit* Repeat Problem 41 if the production cost increases by $20,000.

43. *Clams and crabs* The clam and crab catch (in millions of pounds) in New England in a recent decade can be approximated by the equation $L(x) = -0.28x^2 + 2.8x + 15$ and $R(x) = 0.2x + 7$, respectively, where x is the number of years after the beginning of the decade.

  a. What is the total catch of clams and crabs?

  b. How many pounds of clams and crabs were caught at the beginning of the decade?

  c. How many pounds of clams and crabs were caught by the end of the decade?

  d. How many more pounds of clams than of crabs were caught by the end of the decade?

44. *Fish* The total fish catch (in millions of pounds) in Hawaii in a recent decade can be approximated by the equation $C(x) = 1.5x + 10.5$ where x is the number of years after the beginning of the decade. The tuna catch during the same period was $T(x) = 0.7x + 7$.

  a. How many pounds of fish other than tuna were caught?

  b. How many pounds of tuna did they catch in Hawaii by the end of the decade ($x = 10$)?

  c. How many pounds of fish other than tuna were caught in Hawaii by the end of the decade ($x = 10$)?

  d. By the end of the decade, what fraction of the total catch in Hawaii was tuna?

  e. What percent of the total catch in Hawaii by the end of the decade was tuna? Round the answer to one decimal digit.

45. Medicare costs The total Medicare costs (in billions of dollars) in a 5-year period can be approximated by the function $C(t) = 2.5t^2 + 8.5t + 111$, where t is the number of years (0 to 5). The total U.S. population age 65 and older in the same period is

$$P(t) = -0.46t^2 + 1.14t + 31.08 \text{ (in millions)}$$

a. Find a function that represents the average cost of Medicare for persons 65 and older during the 5-year period.

b. Find the average cost of Medicare for persons 65 and older for $t = 0$. Round to the nearest dollar.

c. Repeat part **b** for $t = 5$.

46. Reaction and braking distance The reaction distance modeled by the equation $R(v) = 0.75v$ (in feet) is the distance a car moving at v miles per hour travels while a driver with a reaction time of 0.5 sec is reacting to apply the brakes. If the braking distance for the car is modeled by the equation $B(v) = 0.06v^2$, find a function that gives the total distance (in feet) a car moving at v miles per hour travels during a panic stop. What is this distance if the car is moving at 30 mi/hr?

47. Temperature The function $C(F) = \frac{5}{9}(F - 32)$ converts the temperature F in degrees Fahrenheit to C in degrees Celsius. The function $K(C) = C + 273$ converts degrees Celsius to kelvins.

a. Find a composite function that converts degrees Fahrenheit to kelvins.

b. If the temperature is 41°F, what is the Celsius temperature?

c. If water boils at 212°F, at what kelvin temperature does water boil?

48. Toys A parent realizes that her son's demand for birthday toys (in dollars) depends on the average number of hours of TV that he watches each week. If h is the average number of hours of TV that he watches each week, the cost of the birthday toys is modeled by the equation $D(h) = 5h + 8$.

a. Find $D(7)$.

b. If the number of hours of TV he is allowed to watch each week is $h(A) = A + 2$, where A is his age, find a composite function that gives the son's demand for birthday toys as a function of his age.

c. What is the cost of the toys that the son demands when he is 11?

49. Dress sizes The function F giving the correspondence between dress sizes in the United States and France is $F(x) = x + 32$, where x is the U.S. size and $F(x)$ the French size. The function $E(F) = F - 30$ gives the correspondence between dress sizes in France and those in England.

a. What French size corresponds to a U.S. size 6?

b. Find a function that will give a correspondence between U.S. and English dress sizes.

c. What English size corresponds to a U.S. size 8?

50. Dress sizes The correspondence between dress sizes in the United States and Italy is $I(x) = 2x + 22$ where x is the U.S. dress size and the correspondence between Italian and English dress sizes is $f(I) = \frac{1}{2}I - 9$.

a. What Italian size corresponds to a U.S. size 8?

b. What English size corresponds to a U.S. size 10?

c. Find a composite function that will give the correspondence between U.S. and English dress sizes.

51. Cricket chirps There are many interesting functions that can be defined using the ideas of this section. For example, we have mentioned that the frequency with which a cricket chirps is a function of the temperature. The table shows the number of chirps per minute and the temperature in degrees Fahrenheit. If f is the function that relates the number of chirps per minute, c, and the temperature x, find

a. $f(40)$ b. $f(42)$ c. $f(44)$

Temperature (°F)	40	41	42	43	44
Chirps per minute	0	4	8	12	16

52. Cricket chirps The function relating the number of chirps per minute of the cricket and the temperature x (in degrees Fahrenheit) is given by $f(x) = 4(x - 40)$. If the temperature is 80°F, how many chirps per minute will you hear from your friendly house cricket?

53. Temperature conversion The function $F(x) = \frac{9}{5}x + 32$ converts the temperature from degrees Celsius to degrees Fahrenheit:

a. Find a composite function that would relate the number of chirps per minute a cricket makes (Problem 52) to the temperature in degrees Celsius.

b. How many chirps per minute would a cricket make when the temperature is 10°C?

54. Ants The distance (in centimeters per second) traveled by a certain type of ant when the temperature is C (in degrees Celsius) is $d(C) = \frac{1}{6}(C - 4)$. The function $C(F) = \frac{5}{9}(F - 32)$ converts the temperature from degrees Fahrenheit to Celsius.

a. Find a composite function that would relate the distance the ant travels to the temperature in degrees Fahrenheit.

b. How fast is the ant traveling when the temperature is 50°F?

Using Your Knowledge

Projecting Domains Look at the graphs of the functions f and g shown in Figure 10.1. As you can see,

$$\text{Domain of } f = \{x \mid -1 \leq x \leq 4\}$$
$$\text{Domain of } g = \{x \mid 1 \leq x \leq 5\}$$

These domains can be regarded as the "projections" of f and g on the x-axis. Thus, the domains of $f + g$, $f - g$, and fg are $\{x \mid 1 \leq x \leq 4\}$—the values common to the domains of f and g. To find the domain of $\frac{f}{g}$, $g(2) = 0$, thus,

$$\text{Domain of } \frac{f}{g} = \{x \mid 1 \leq x \leq 4 \text{ and } x \neq 2\}$$

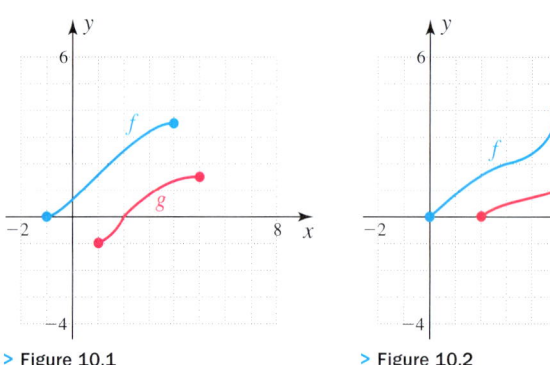

> Figure 10.1   > Figure 10.2

Now refer to Figure 10.2 and find:

55. The domain of f

56. The domain of g

57. The domain of $f + g$, $f - g$, and fg

58. The domain of $\frac{f}{g}$

Write On

59. Explain why $(f + g)(x) = (g + f)(x)$ for every value of x but $(f - g)(x) \neq (g - f)(x)$.

60. Explain why
$$\left(\frac{f}{g}\right)(x) \neq \left(\frac{g}{f}\right)(x)$$

61. Is $(f \circ g)(x) = (g \circ f)(x)$ for every value of x? Explain why or why not.

62. If $f(x) = \frac{1}{2}x$ and $g(x) = 2x$,
  a. Find $(f \circ g)(x)$ and $(g \circ f)(x)$. Are they the same?
  b. Can you find other functions f and g so that $(f \circ g)(x) = (g \circ f)(x)$?

Concept Checker

Fill in the blank(s) with the correct word(s), phrase, or mathematical statement.

63. $(f + g)(x) =$ _____.

64. $(f - g)(x) =$ _____.

65. $(fg)(x) =$ _____.

66. $\left(\frac{f}{g}\right)(x) =$ _____, where $g(x) \neq 0$.

67. The composite of f and g, $(f \circ g)(x) =$ _____.

68. The domain of $f \circ g$ is the set of all x in the domain of g and such that $g(x)$ is in the _____ of f.

range $f(x) \cdot g(x)$
domain $f(x) - g(x)$
$\frac{f(x)}{g(x)}$ $g(x) - f(x)$
             $f(x) + g(x)$
$f(g(x))$
$g(f(x))$

⟩⟩⟩ Mastery Test

69. The revenue (in dollars) obtained by selling x units of a product is given by the equation

$$R(x) = 300x - \frac{x^2}{30}$$

and the cost $C(x)$ of making x units is given by $C(x) = 72{,}000 + 60x$. Find the profit function $P(x)$.

If $f(x) = x^3$ and $g(x) = x + 1$, find each composition.

70. $(f \circ g)(x)$

71. $(g \circ f)(x)$

72. $(f \circ g)(3)$

73. $(g \circ f)(-3)$

74. Find $\dfrac{f(x) - f(a)}{x - a}$, $x \neq a$ if $f(x) = 2x + 1$.

75. Find $\dfrac{f(x) - f(a)}{x - a}$, $x \neq a$ if $f(x) = x^2 + 1$.

If $f(x) = x^2 + 4$ and $g(x) = x - 2$, perform the operation, if possible.

76. $(f + g)(x)$

77. $(f - g)(x)$

78. $(g - f)(x)$

79. $\left(\dfrac{f}{g}\right)(-2)$

80. $\left(\dfrac{g}{f}\right)(x)$

81. $\left(\dfrac{g}{f}\right)(-2)$

82. $\left(\dfrac{f}{g}\right)(2)$

Find the domains of each of the following.

83. $f + g$, $f - g$, fg, and $\dfrac{f}{g}$ if $f(x) = \dfrac{1}{x}$ and $g(x) = \dfrac{3}{x+1}$

84. $\dfrac{f}{g}$ if $f(x) = \dfrac{3}{x}$ and $g(x) = \dfrac{x-1}{x+2}$

⟩⟩⟩ Skill Checker

Graph the equation

85. $x + y = 3$

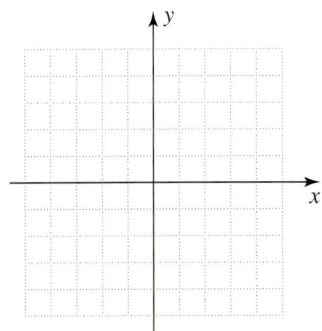

86. $2x - y = 2$

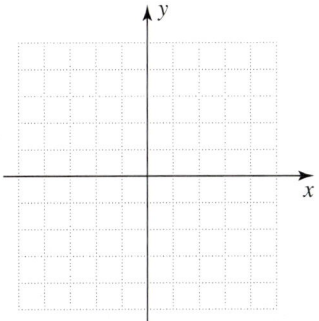

87. $2x + \dfrac{1}{2}y = 2$

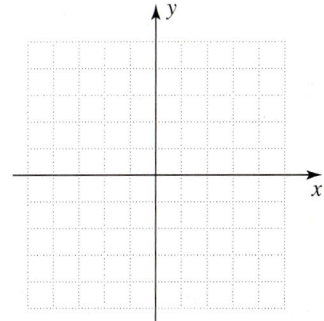

88. $y = -x - 3$

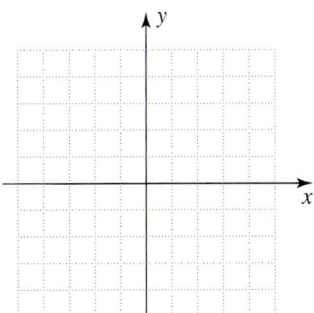

89. $y = -2x + 4$

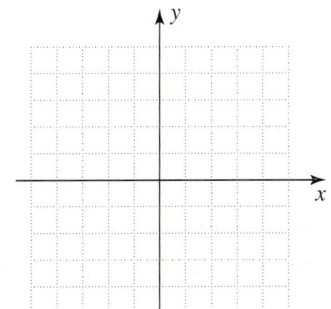

90. $y = -3x + 6$

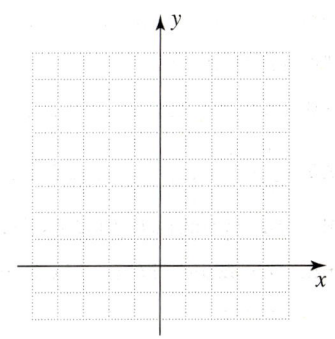

10.2 Inverse Functions

Objectives

A ▶ Find the inverse of a relation when the relation is given as a set of ordered pairs.

B ▶ Find the equation of the inverse of a relation.

C ▶ Graph a function and its inverse and determine whether the inverse is a function.

D ▶ Solve applications involving inverse functions.

▶ To Succeed, Review How To . . .

1. Find the range and domain of a relation (pp. 224–228).
2. Solve an equation for a specified variable (pp. 87–89).
3. Graph linear and quadratic equations (pp. 168–174, 659–669).

▶ Getting Started

Money, Money, Money

If you're planning on traveling somewhere, it's a good idea to know how much your dollars are worth in the country you plan to visit. The converted amount column in the table shows the value of $1 (U.S.) in different currencies in a recent year. If you look at the exchange rate column, the value of a Canadian dollar in U.S. dollars is $0.8760. This means that the number of dollars D you get for C Canadian dollars is given by the equation

$$D = 0.8760\,C$$

To find how many Canadian dollars you get for one U.S. dollar, solve for C in $D = 0.8760\,C$ to obtain

$$C = \frac{1D}{0.8760} \approx 1.1416\,D$$

You get 1.1415 Canadian dollars for every U.S. dollar. You can check this in the table.

In this section we learn how to find the inverse of a function, which uses a similar procedure.

	COUNTRY	ISO 4217	CURRENCY		EXCHANGE RATE (USD/Unit)	CONVERTED AMOUNT (Units/USD)
	AUSTRALIA	AUD	Australian Dollars		0.7884	1.2684
	BRAZIL	BRL	Brazilian Real		0.4621	2.1640
	CANADA	CAD	Canadian Dollars		0.8760	1.1415
	CHINA	CNY	Chinese Renminbi		0.1277	7.8309
	DENMARK	DKK	Danish Kroner		0.1776	5.6306
	EURO	EUR	Euro		1.3259	0.7542
	GREAT BRITAIN	GBP	United Kingdom Pounds		1.9689	0.5079
	HONG KONG	HKD	Hong Kong Dollars		0.1286	7.7760
	INDIA	INR	Indian Rupees		0.0224	44.6429
	JAPAN	JPY	Japanese Yen		0.0087	115.559
	MEXICO	MXP	Mexican Pesos		0.0911	10.9769
	NORWAY	NOK	Norwegian Kroner		0.1623	6.1614
	RUSSIA	RUR	Russian Rubles		0.0381	26.2467
	SINGAPORE	SGD	Singapore Dollars		0.6498	1.5389
	SWEDEN	SEK	Swedish Krona		0.1460	6.8493
	SWITZERLAND	CHF	Swiss Francs		0.8343	1.1986
	UNITED STATES	USD	United States Dollars		1.0000	1.0000

A ▸ Finding the Inverse of a Relation

C Canadian Dollars	D U.S. Dollars
1	0.88
10	8.76
100	87.60
1000	876.00

Let's look again at how we found the exchange rate for Canadian dollars in the *Getting Started*. If we think of D as the definition for the function $f(C)$, we can make a table to two decimal places and write the function f as the ordered pairs (C, D).

$$f = \{(1, 0.88), (10, 8.76), (100, 87.60), (1000, 876.00)\}$$

On the other hand, if you exchange U.S. currency for Canadian, the number of Canadian dollars you get for $0.88, $8.76, $87.60, and $876 (U.S.), respectively, corresponds to the set of ordered pairs (D, C) and is given by

$$g = \{(0.88, 1), (8.76, 10), (87.60, 100), (876, 1000)\}$$

The relation g obtained by reversing the order of the coordinates in each ordered pair in f is called the *inverse of f*. As you can see, the domain of f is the range of g and the range of f is the domain of g. Here is the definition we need.

INVERSE OF A RELATION

If f is a relation, then the **inverse of f**, denoted by f^{-1} (read "f inverse," or "the inverse of f") is the relation obtained by reversing the order of x and y in each ordered pair (x, y) in f.

NOTE

The -1 in f^{-1} is *not* an exponent. Here it denotes the inverse of f.

EXAMPLE 1 Finding the inverse of a relation

Let $S = \{(1, 2), (3, 4), (5, 4)\}$ and find:

a. The domain and range of S
b. S^{-1}
c. The domain and range of S^{-1}
d. The graphs of S and S^{-1} on the same coordinate axes

SOLUTION

a. The domain of S is $\{1, 3, 5\}$. The range is $\{2, 4\}$.
b. $S^{-1} = \{(2, 1), (4, 3), (4, 5)\}$
c. The domain of S^{-1} is $\{2, 4\}$; the range is $\{1, 3, 5\}$.
d. The graphs of S (in blue) and S^{-1} (in red) are shown in Figure 10.3. As you can see, the two graphs are symmetric with respect to the line $y = x$ (shown dashed in Figure 10.3).

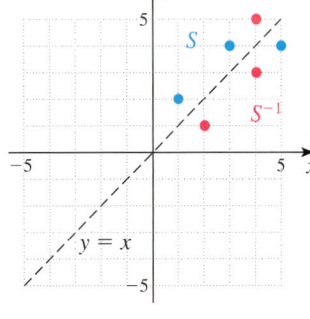
▸Figure 10.3

PROBLEM 1

Let $S = \{(4, 3), (3, 2), (2, 1)\}$ and find:

a. The domain and range of S
b. S^{-1}
c. The domain and range of S^{-1}
d. The graphs of S and S^{-1}

Answers to PROBLEMS

1. a. $D = \{2, 3, 4\}; R = \{1, 2, 3\}$
   b. $\{(3, 4), (2, 3), (1, 2)\}$
   c. $D = \{1, 2, 3\}; R = [2, 3, 4]$
   d.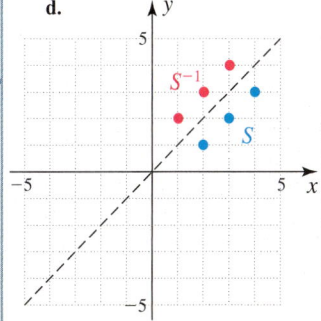

Calculator Corner

Graphing Ordered Pairs

With some calculators (the TI-83 Plus, for example), you can graph ordered pairs by entering the domain and range using the list feature. To do Example 1(a), clear any lists in the calculator, then press STAT 1 1, 3, and 5 (the domain) under L1 and 2, 4, and 4 (the range) under L2. The result is shown in Window 1.

To tell the calculator to plot these points, first press ZOOM 6 for a standard window, then press 2nd Y= 1 and select ON, [⣿] (the first type of graph sometimes called a scatter plot), L1, L2, and □. Finally, press GRAPH. The result using a standard window is shown in Window 2.

Follow this procedure to graph S^{-1} using lists L3 (2, 4, 4) and L4 (1, 3, 5) for the domain and range of S^{-1}. Press 2nd Y= 2 and select ON, [⣿], L3, L4, and [+]. Now press GRAPH to obtain the graph shown in Window 3. We used a square window so that S and S^{-1} appear as reflections of each other across the line $y = x$ which is also shown.

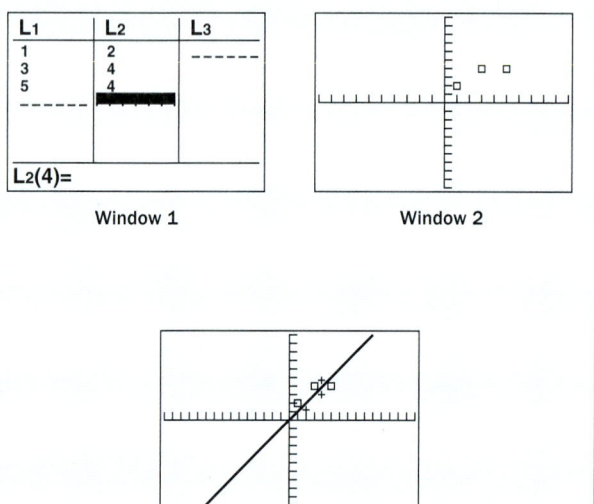

Window 1

Window 2

Window 3

B › Finding the Equation of the Inverse Function

Previously, we found the inverse of

$$f = \{(1, 0.88), (10, 8.76), (100, 87.60), (1000, 876.00)\}$$

by reversing the *order* of the coordinates in each ordered pair in f. A *function* is a relation in which no two different ordered pairs have the same first coordinate. To find the inverse of the function $y = f(x) = 0.880x$, we interchange the *variables* in $y = 0.880x$ to obtain

$$x = 0.880y$$

$$y = \frac{1}{0.880}x \quad \text{Solve for y.}$$

$$y \approx 1.1363x \quad \text{Divide 1 by 0.880.}$$

Thus, $f^{-1}(x) = 1.1363x$. The procedure is summarized below.

PROCEDURE

Finding the equation of an inverse function
1. Interchange the roles of x and y in the equation for f.
2. Solve for y.

For example, consider the function $f(x) = 4x - 4$ or

$$y = 4x - 4 \tag{1}$$

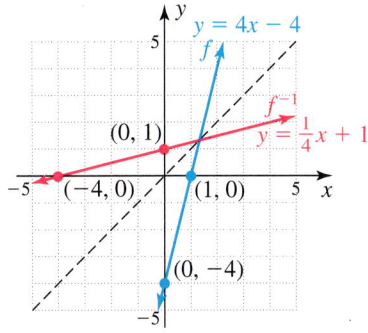

>Figure 10.4

The inverse of this function, f^{-1}, is obtained by first interchanging the x- and y-coordinates—by writing $x = 4y - 4$ and then solving for y:

$$x + 4 = 4y \quad \text{Add 4.}$$
$$\frac{x+4}{4} = y \quad \text{Divide by 4.}$$

We have

$$f^{-1}(x) = y = \frac{x+4}{4}, \text{ which can be written } y = \frac{1}{4}x + 1 \quad (2)$$

The graphs of equation (1) (in blue) and its inverse, equation (2) (in red), are shown in Figure 10.4. Clearly, the graphs are symmetric to each other with respect to the line $y = x$, shown dashed. This is to be expected, because one function was obtained from the other by interchanging x and y.

EXAMPLE 2 Finding and graphing an inverse function

Let $f(x) = y = 4x - 2$:

a. Find $f^{-1}(x)$. **b.** Graph f and its inverse.

SOLUTION

a. Since $y = 4x - 2$, we interchange the variables x and y and solve for y to obtain

$$x = 4y - 2$$
$$x + 2 = 4y \quad \text{Add 2.}$$
$$y = \frac{x+2}{4} \quad \text{Divide by 4.}$$

Thus, the inverse of f is

$$f^{-1}(x) = \frac{x+2}{4} = \frac{1}{4}x + \frac{1}{2}$$

b. We can graph $y = 4x - 2$ by putting the y-intercept at $(0, -2)$ and using the slope of $\frac{4}{1}$; going up 4 from $(0, -2)$ and to the right one to the point $(1, 2)$. Connect the points. The graph is shown in blue in Figure 10.5. We then graph

$$y = \frac{1}{4}x + \frac{1}{2}$$

in a similar manner. This graph is shown in red and is symmetric to the graph of $y = 4x - 2$ with respect to the line $y = x$, shown dashed. We could have obtained the graph of the inverse function by reflecting the graph of $y = 4x - 2$ about the line $y = x$.

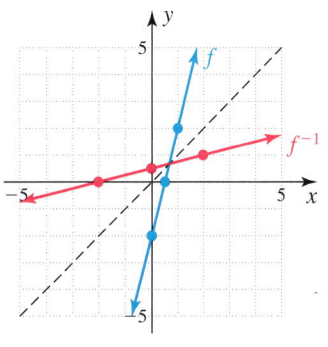

>Figure 10.5

PROBLEM 2

Let $f(x) = y = 2x - 3$

a. Find $f^{-1}(x)$

b. Graph f and its inverse

Answers to PROBLEMS

2. **a.** $f^{-1}(x) = \frac{x+3}{2} = \frac{1}{2}x + \frac{3}{2}$

b.

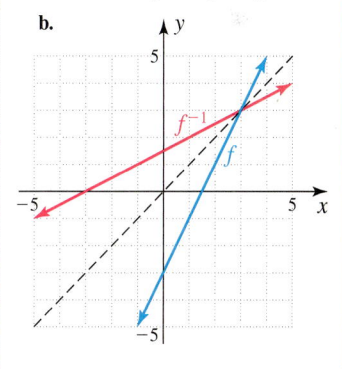

C > Graphing Functions and Their Inverses

Every function is a set of ordered pairs (a relation), so every function has an inverse. Is this inverse always a function? We can see that it is not. For example, if S is the function defined by $S = \{(1, 2), (3, 2)\}$, the inverse of S is $S^{-1} = \{(2, 1), (2, 3)\}$, which is *not* a function, because two distinct ordered pairs have the same first component, 2. On the other hand, if G is the function defined by $G = \{(3, 4), (5, 6)\}$, the inverse is $G^{-1} = \{(4, 3), (6, 5)\}$, which *is* a function. The reason that the inverse of S is not a function is that

S has two ordered pairs with the same *second* component. A function in which no two distinct ordered pairs have the same second component is called a **one-to-one function.** The inverse of such a function is always a function. We summarize this discussion as follows.

ONE-TO-ONE FUNCTION

If the function $y = f(x)$ is one-to-one, then the inverse of f is also a function and is denoted by $y = f^{-1}(x)$.

To determine whether the inverse of a function is a function, we must ascertain whether the original function is one-to-one. To do this, we must return to the definition of a one-to-one function—a one-to-one function *cannot* have two ordered pairs with the same second component. Since any two points with the same second coordinate will be on a *horizontal* line parallel to the x-axis, if any horizontal line intersects the graph of a function more than once, the function will not be one-to-one. Its inverse will not be a function. Thus, we have the following **horizontal line test.**

HORIZONTAL LINE TEST TO DETERMINE IF A FUNCTION IS ONE-TO-ONE

If a horizontal line intersects the graph of a function more than once, then it is not a one-to-one function and the inverse of the function is not a function.

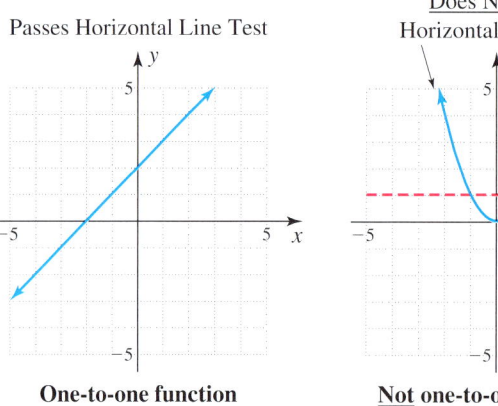

Passes Horizontal Line Test — One-to-one function

Does Not Pass Horizontal Line Test — Not one-to-one function

Nonconstant linear functions have inverses that are functions (horizontal lines will intersect the graph only once), but **quadratic functions** do not have inverses that are functions. This is because quadratic functions have graphs that can be intersected by a horizontal line more than once.

PROCEDURE

Finding the inverse of a function $y = f(x)$

1. Replace $f(x)$ by y, if necessary.
2. Interchange the roles of x and y.
3. Solve the resulting equation for y, if possible.
4. Replace y by $f^{-1}(x)$. (If the original function was one-to-one, f^{-1} is a function.)

EXAMPLE 3 Finding the inverse of a function and using the horizontal line test

Find the inverse of $f(x) = x^2$. Is the inverse a function?

SOLUTION We use the four-step procedure.

1. Replace $f(x)$ by y. $\quad y = x^2$
2. Interchange the roles of x and y. $\quad x = y^2$
3. Solve for y. $\quad y = \pm\sqrt{x}$
4. Replace y by $f^{-1}(x)$. $\quad f^{-1}(x) = \pm\sqrt{x}$

The inverse of $f(x)$ is *not* a function, because we can draw a horizontal line that intersects the graph of $y = x^2$ at more than one point.

The function $f(x) = x^2$ (in blue) and its inverse (in red) are shown in the graph in Figure 10.6.

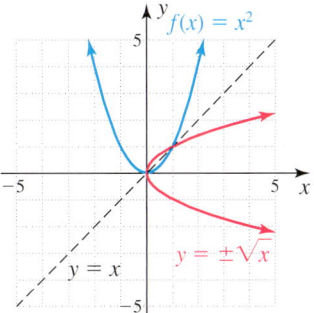

> Figure 10.6

PROBLEM 3

Find the inverse of $f(x) = y = x^3$. Is the inverse a function?

Answers to PROBLEMS

3. $f^{-1}(x) = \sqrt[3]{x}$; yes

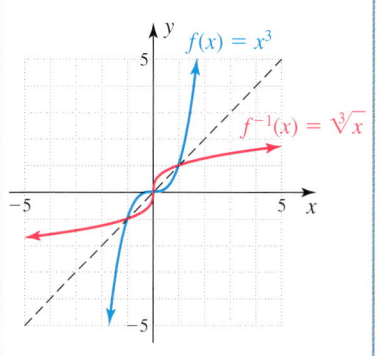

EXAMPLE 4 Finding the inverse of a function and using the horizontal line test

Graph the function $f(x) = y = 3^x$. Is the inverse a function?

SOLUTION Following is a table of values to graph $y = 3^x$. If we examine the graph of $y = 3^x$ (in blue in Figure 10.7), we can see that any horizontal line will intersect the graph only once. Thus, the inverse is a function. To try to find the inverse, we interchange the x and y in $y = 3^x$ to obtain $x = 3^y$.

Note that we are not able to solve for y at this time (we will have to wait until later in this chapter to do this). However, we can still graph the inverse $x = 3^y$ by reflecting the graph across the dotted line $y = x$ and giving values to y and finding the corresponding x-values, as shown in the following table on the right. The graph of the inverse relation $x = 3^y$ is in red.

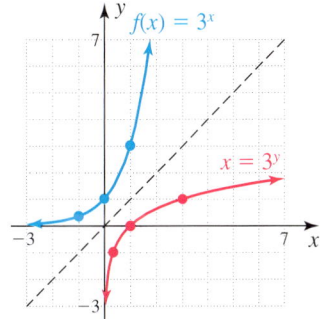

> Figure 10.7

x	$y = 3^x$
-1	$3^{-1} = \frac{1}{3}$
0	$3^0 = 1$
1	$3^1 = 3$

$x = 3^y$	y
$3^{-1} = \frac{1}{3}$	-1
$3^0 = 1$	0
$3^1 = 3$	1

PROBLEM 4

Graph $f(x) = y = 2^x$. Is the inverse a function?

Answers to PROBLEMS

4. Yes. The graph of the inverse is in red.

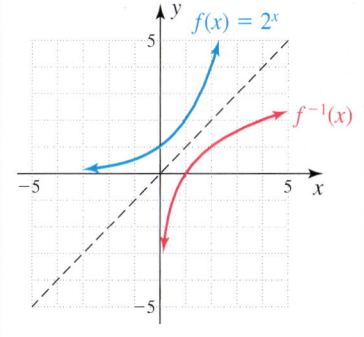

Calculator Corner

Graphing Inverses

To do Example 4, you need a calculator with a draw feature. Enter $Y_1 = 3^x$ and graph it using a square window. To graph the inverse using the draw feature on a TI-83 Plus, press `2nd` `PRGM` `8` to tell the calculator you want an inverse. Then press `VARS` `▶` `1` `1` `ENTER` to tell the calculator you specifically want the inverse of Y_1. The result is shown in the window. The calculator still doesn't tell you how to find the inverse or what the equation for this inverse is. To find out, you have to learn the techniques in Section 10.4.

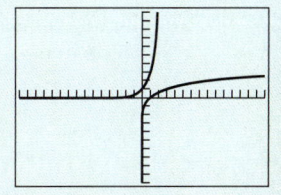

D › Applications Involving Inverse Functions

Often in applications the variables that are used have specific meanings. This could cause some confusion when interchanging the variables to find the inverse of a function as described in the procedure on page 764. For this reason when applying the inverse function procedure in an application, we will keep the letters the same and solve for the other variable.

Can you tell yet what the temperature is by listening to crickets? Let's return to our cricket problem one more time and solve it yet another way.

EXAMPLE 5 Cricket chirping

The relationship between the temperature F (in degrees Fahrenheit) and the number of chirps c a cricket makes in 1 minute is given by the function
$$c = n(F) = 4(F - 40)$$

a. Find the inverse of n.

b. If a cricket is chirping 120 times a minute, what is the temperature in degrees Fahrenheit?

SOLUTION

a. Since this is an application we solve for the other variable F.

1. Given.	$c = 4(F - 40)$
2. Divide by 4.	$\frac{c}{4} = F - 40$
3. Add 40.	$\frac{c}{4} + 40 = F$
4. The inverse of n.	$F = \frac{c}{4} + 40$

b. If the cricket is chirping 120 times a minute, the temperature is
$$n^{-1}(120) = \frac{120}{4} + 40 = 70°F$$

CHECK Does the cricket make 120 chirps per minute when the temperature is 70°F? Using the equation $n(F) = 4(F - 40)$, $n(70) = 4(70 - 40) = 120$ as expected.

PROBLEM 5

The speed S (in centimeters per second) of a certain type of ant is given by the equation $S = n(C) = \frac{1}{6}(C - 4)$, where C is the temperature in degrees Celsius.

a. Find the inverse of n.

b. If an ant is moving at 2 centimeters per second, what is the temperature in degrees Celsius?

Calculator Corner

1. Window 1 shows the coordinates of six different points. Let f be the set of ordered pairs whose coordinates are shown using the symbol □ and g be the set of ordered pairs whose coordinates are shown using the symbol ◇.
   a. Write f as a set of ordered pairs.
   b. Write g as a set of ordered pairs.
   c. Is g the inverse of f?
2. Window 2 shows the graphs of two linear functions and the graph of the function $f(x) = y = x$.
   a. Are the functions symmetric with respect to the line $y = x$?
   b. Are the functions inverses of each other?
   c. Can you find the equation of each of the functions?

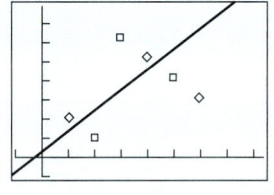

Window 1

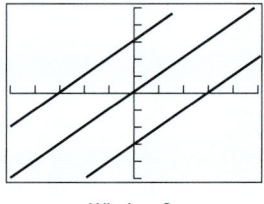

Window 2

Answers to PROBLEMS

5. a. $n^{-1}(C) = 6C + 4$ **b.** 16°C

> Exercises 10.2

⟨ A ⟩ Finding the Inverse of a Relation
⟨ C ⟩ Graphing Functions and Their Inverses

In Problems 1–4, find f^{-1}, draw the graphs of f and f^{-1} on the same axes, and determine whether f^{-1} is a function.

1. $f = \{(1, 3), (2, 4), (3, 5)\}$

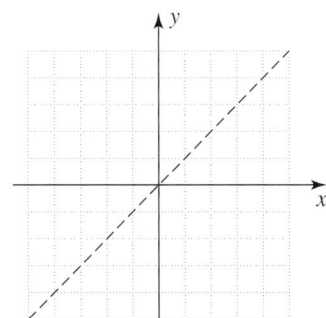

2. $f = \{(2, 3), (3, 4), (4, 5)\}$

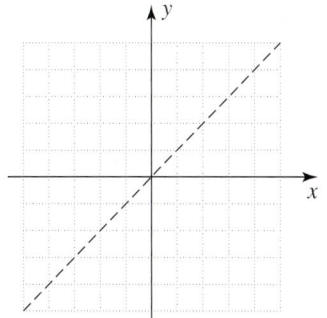

3. $f = \{(-1, 5), (-3, 4), (-4, 4)\}$

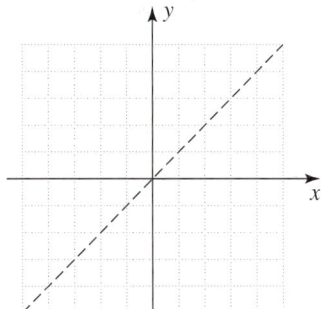

4. $f = \{(-2, 4), (-3, 3), (-5, 3)\}$

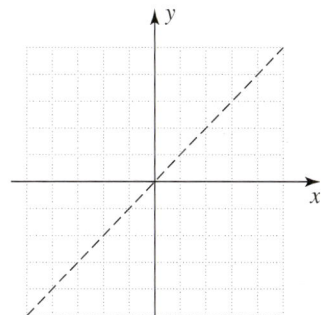

⟨ B ⟩ Finding the Equation of the Inverse Function
⟨ C ⟩ Graphing Functions and Their Inverses

In Problems 5–14, find the equation of the inverse, graph it, and state whether the inverse is a function.

5. $y = 3x + 3$

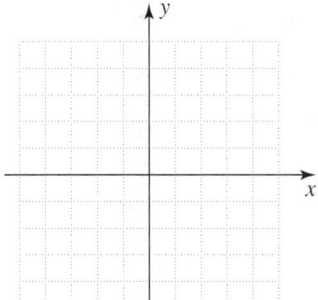

6. $y = 2x + 4$

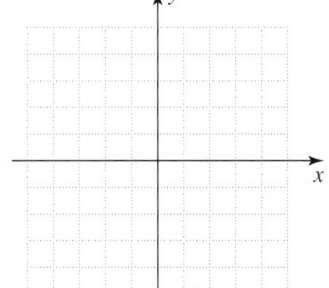

7. $y = 2x - 4$

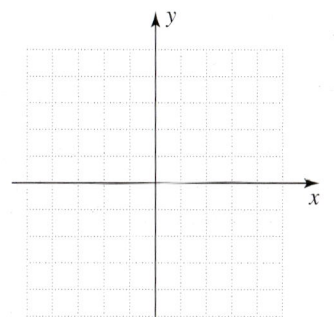

8. $y = 3x - 3$

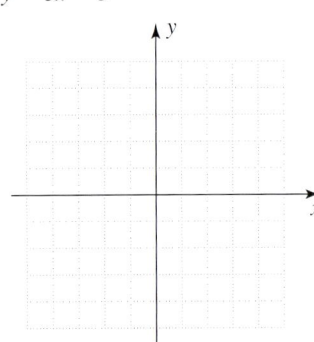

9. $y = 2x^2$

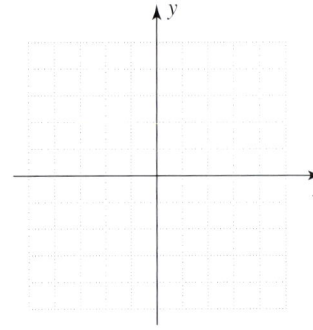

10. $y = x^2 + 1$

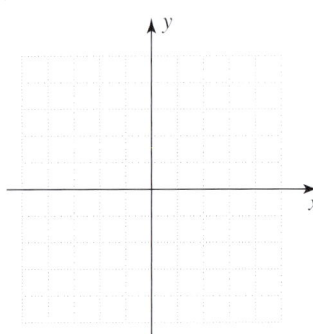

11. $y = x^2 - 1$

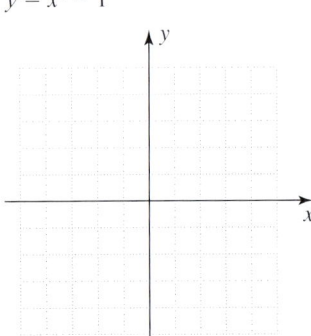

12. $y = x^3 - 1$

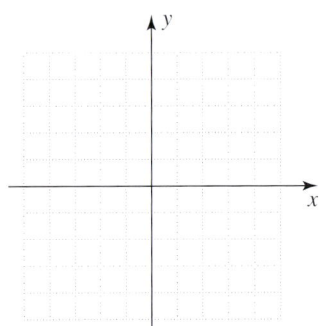

13. $y = -x^3$

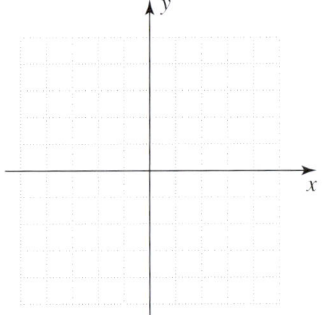

14. $y = -2x^3$

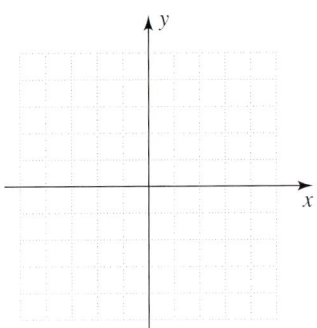

‹ C › **Graphing Functions and Their Inverses** In Problems 15–20, use a table of values (see Example 4) to graph the function and its inverse and determine whether the inverse is a function.

15. $y = f(x) = 2^{x+1}$

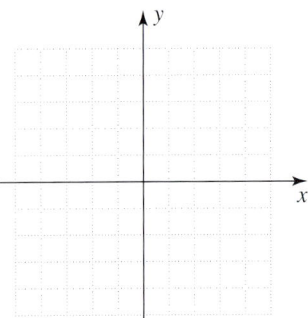

16. $y = f(x) = 3^{x+1}$

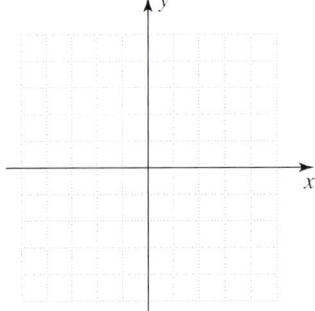

17. $y = f(x) = \left(\dfrac{1}{3}\right)^x$

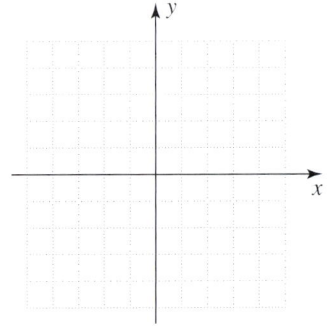

18. $y = f(x) = \left(\frac{1}{2}\right)^x$

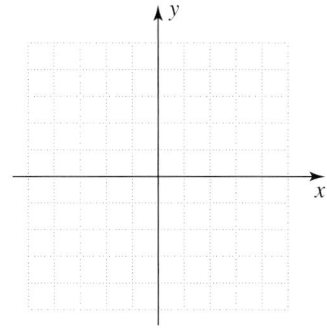

19. $y = f(x) = 2^{-x}$

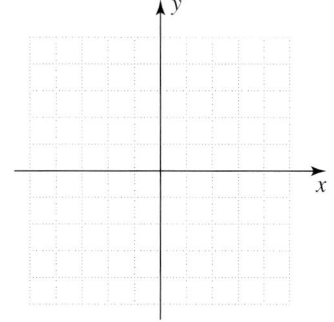

20. $y = f(x) = 3^{-x}$

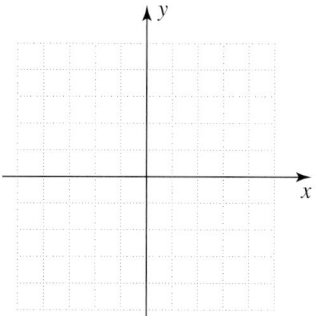

21. If $f(x) = 4x + 4$, $f^{-1}(x) = \frac{x-4}{4}$. Find:
   a. $f(f^{-1}(3))$ **b.** $f^{-1}(f(-1))$

22. If $f(x) = 2x - 2$, $f^{-1}(x) = \frac{x+2}{2}$. Find:
   a. $f(f^{-1}(-1))$ **b.** $f^{-1}(f(x))$

23. If $y = f(x) = \frac{1}{x}$, find $f^{-1}(x)$.

24. If $y = f(x) = \frac{2}{x}$, find $f^{-1}(x)$.

In Problems 25–35, graph the given function and determine the equation of its inverse. Graph the inverse by reflecting the given function along the line $y = x$ (shown dashed).

25. $y = 2x$

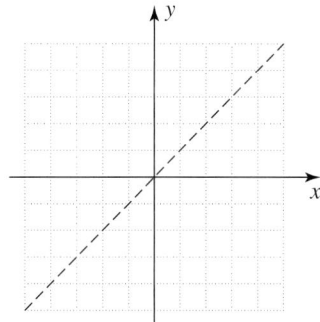

26. $y = -3x$

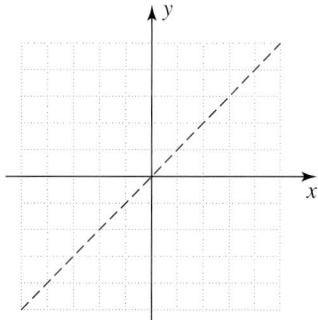

27. $y = -x^2 + 2$

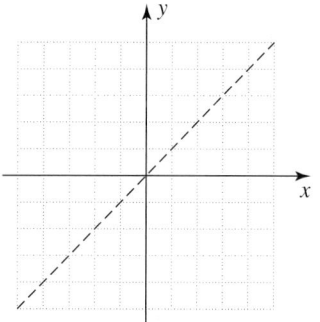

28. $y = x^2 + 1$

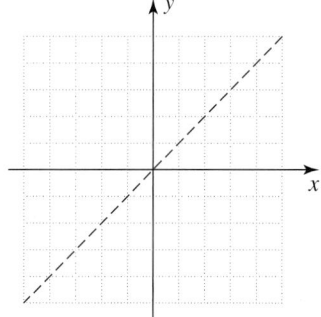

29. $y = \sqrt{4 - x^2}$

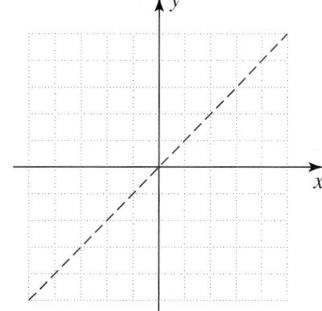

30. $y = -\sqrt{4 - x^2}$

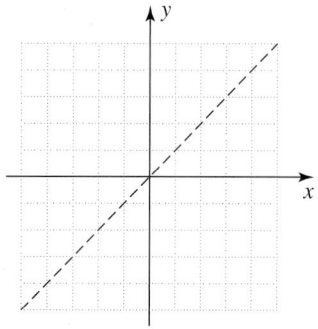

31. $y = \sqrt{9 - x^2}$

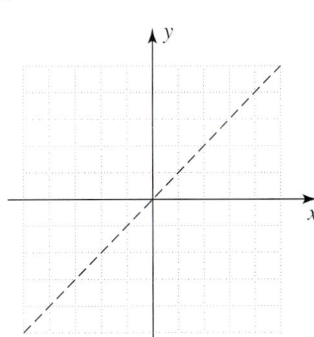

32. $y = -\sqrt{9 - x^2}$

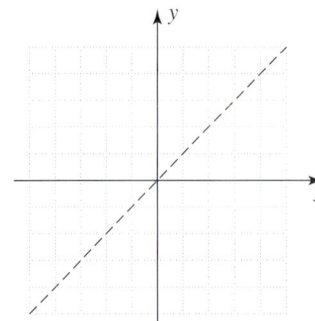

33. $y = x^3$

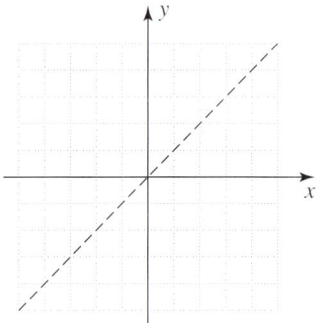

34. $y = -x^3$

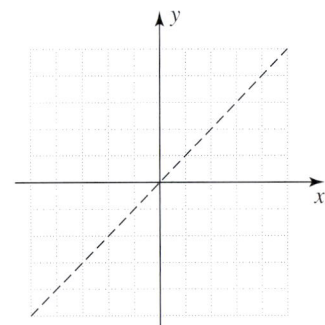

35. $y = |x|$

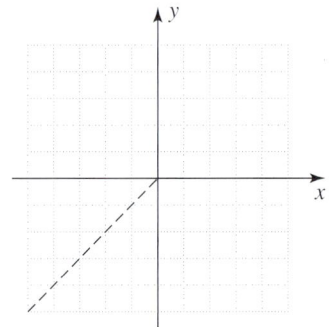

‹ D › Applications Involving Inverse Functions

36. *Shoe size for men* Your shoe size S is a function of the length of your foot L in inches. For men, the function giving this correspondence is modeled by the equation

$$S = f(L) = 3L - 22$$

a. Find $f^{-1}(S)$.
b. If a man's shoe size is 7, what is the length of his foot?

37. *Shoe size for women* Your shoe size S is a function of the length of your foot L in inches. For women, the function giving the correspondence between S and L is modeled by the equation

$$S = f(L) = 3L - 21$$

a. Find $f^{-1}(S)$.
b. If a woman's shoe size is 7, what is the length of her foot?

38. *Dress size* There is a correspondence between dress sizes d in the United States and France. If the U.S. dress size is x, the corresponding size in France is modeled by the equation

$$d = f(x) = x + 32$$

a. Find $f^{-1}(d)$.
b. What U.S. dress size corresponds to a French size 40?

39. *Dress size* In the United States, dress sizes d are a function of waist sizes w (in inches) and are given by the equation

$$d = f(w) = w - 16$$

a. What dress size corresponds to a 32-inch waist?
b. Find $f^{-1}(d)$.
c. If a woman wears a size 12, what is her waist size?

40. *Dress size* There is a relationship between bust size b (in inches) and dress size d. The correspondence is defined by the function

$$d = f(b) = b - 24$$

a. If $b = 38$, what is d?
b. Find $f^{-1}(d)$.
c. If $d = 16$, what is b?

41. *Olympic predictions* Can you predict Olympic outcomes? You can come close if you start with the right function. The winning time in the Women's Olympic 400-meter track relay w is a function of the year x in which the event was run and is approximately given by the equation

$$w = f(x) = -0.12x + 280 \quad \text{(seconds)}$$

a. Predict the winning time for the 1988 Olympics (the actual time was 41.98 seconds).
b. Find $f^{-1}(w)$.
c. Use $f^{-1}(w)$ to predict in what year the winning time was 40 seconds.

42. Olympic predictions The winning time in the Women's Olympic 200-meter dash w is a function of the year x in which the event was run, starting with 1948, and is given by the equation

$$w = f(x) = -0.0661x + 152.8 \quad \text{(seconds)}$$

a. Predict the winning time for the 1988 Olympics. (The actual time was 21.34 seconds.)
b. Find $f^{-1}(w)$.
c. Use $f^{-1}(w)$ to predict in approximately what year the winning time will be 20 seconds.

❯❯❯ Using Your Knowledge

The "Undoer" What does the inverse of a function do? You can think of the inverse as "undoing" the operations performed on the variable by the function. We will demonstrate this with the function from Example 2, $f(x) = 4x - 2$. The following table shows how.

To construct the inverse, we start with x at the bottom of column (4) and work our way up. Columns (2) and (3), respectively, show the operations performed on x and their "undoing."

Function (1)	Function (2)	Inverse Operation (3)	Inverse Function (4)
			Finish here.
$f(x) = 4x - 2$	Multiply by 4.	undo → Divide by 4.	$f^{-1}(x) = \dfrac{x+2}{4}$
	Subtract 2.	undo → Add 2.	$x + 2$
			x
			Start here.

Now let's examine the function from Example 5, $c = n(F) = 4(F - 40)$, which as you will remember became $F = 4(c - 40)$. So we shall "undo" the operations on c. We proceed in a similar manner.

Function (1)	Function (2)	Inverse Operation (3)	Inverse Function (4)
			Finish here.
$c = n(F) = 4(F - 40)$	Subtract 40.	undo → Add 40.	$n^{-1}(c) = \dfrac{c}{4} + 40$
	Multiply by 4.	undo → Divide by 4.	$\dfrac{c}{4}$
			c
			Start here.

Use this method to construct the inverse for the given functions.

43. $f(x) = 3x - 2$
44. $f(x) = 5x + 2$
45. $f(x) = \dfrac{x+1}{2}$
46. $f(x) = \dfrac{x-1}{2}$
47. $f(x) = x^3 + 1$
48. $f(x) = x^3 - 1$
49. $f(x) = \sqrt{x}$
50. $f(x) = \sqrt[3]{x}$

❯❯❯ Write On

51. Explain why the function $f(x) = ax + b$ ($a \neq 0$) always has an inverse that is a function.

52. Explain why the function $f(x) = ax^2 + bx + c$ ($a \neq 0$) never has an inverse that is a function.

53. If f and f^{-1} are functions that are inverses of each other, explain the result of the composition of f and f^{-1}; that is, what happens when you take $(f \circ f^{-1})(x)$?

54. Under the same conditions as in Problem 53, what happens when you take $(f^{-1} \circ f)(x)$?

55. In view of your answers for Problems 53 and 54, how could you verify that f and f^{-1} are inverses of each other?

⟩⟩⟩ Concept Checker

Fill in the blank(s) with the correct word(s), phrase, or mathematical statement.

56. If f is a relation, the inverse f^{-1} of f is the relation obtained by _____ the order of x and y in each ordered pair (x, y) of f.

57. A function is a _____ in which no two different ordered pairs have the _____ first coordinate.

deleting   same
reversing   relation

⟩⟩⟩ Mastery Test

58. Graph $y = f(x) = 2^x$. Is the inverse a function?

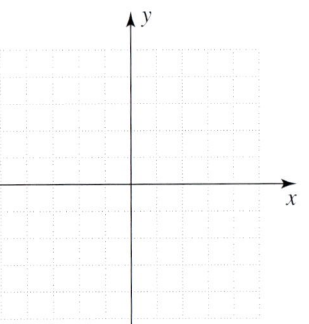

59. Find the inverse of $f(x) = x^3$. Is the inverse a function? Graph it.

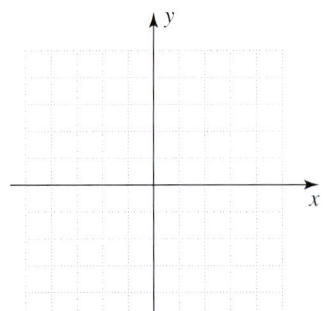

60. Let $y = f(x) = 2x + 4$.
  a. Find $f^{-1}(x)$.
  b. Graph f and its inverse.

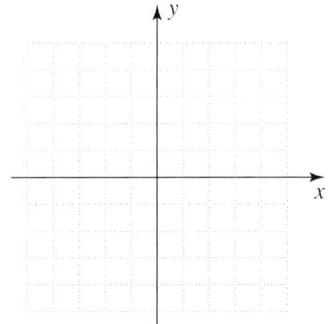

In Problems 61–64, let $S = \{(4, 3), (3, 2), (2, 1)\}$:

61. Find the domain and range of S.

62. Find S^{-1}.

63. Find the domain and range of S^{-1}.

64. Graph S and S^{-1} on the same coordinate axes.

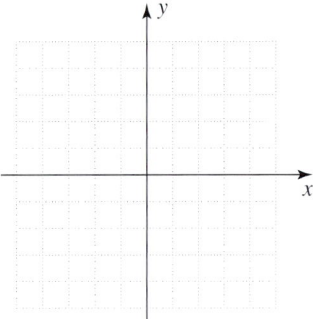

⟩⟩⟩ Skill Checker

Evaluate:

65. 3^2

66. 3^{-2}

67. $\left(\frac{1}{3}\right)^{-3}$

68. $\left(\frac{1}{2}\right)^{x/3}$ when $x = 6$

69. $\left(\frac{1}{2}\right)^{x/3}$ when $x = -6$

70. $\left(\frac{1}{3}\right)^{x/2}$ when $x = -4$

10.3 Exponential Functions

Objectives

A ▶ Graph exponential functions of the form a^x or a^{-x} ($a > 0$ and $a \neq 1$).

B ▶ Determine whether exponential functions are increasing or decreasing.

C ▶ Solve applications involving exponential functions.

▶ To Succeed, Review How To . . .

1. Understand the concept of base and exponent (pp. 38–45).
2. Interpret and evaluate expressions containing rational exponents (pp. 528–532).

▶ Getting Started

What Do Cells Know About Exponents?

Are you taking biology? Have you studied cell reproduction? The photographs show a cell reproducing by a process called *mitosis*. In mitosis, a single cell or bacterium divides and forms two identical daughter cells. Each daughter cell then doubles in size and divides. As you can see, the number of bacteria present is a function of time. If we start with 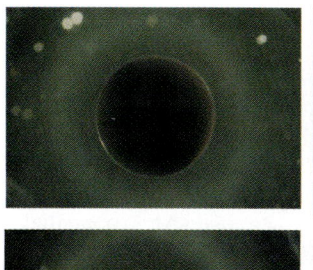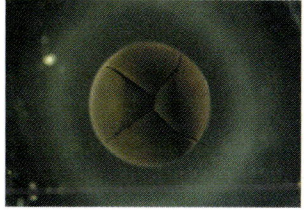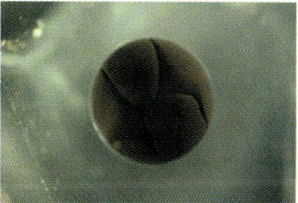 one cell and assume that each cell divides after 10 minutes, then the number of bacteria present at the end of the first 10-minute period ($t = 10$) is

$$2 = 2^1 = 2^{10/10}$$

At the end of the second 10-minute period ($t = 20$), the two cells divide, and the number of bacteria present is

$$4 = 2^2 = 2^{20/10}$$

Similarly, at the end of the third 10-minute period ($t = 30$), the number is

$$8 = 2^3 = 2^{30/10}$$

Thus, we can see that the number of bacteria present at the end of t minutes is given by the function

$$f(t) = 2^{t/10}$$

This also gives the correct result for $t = 0$, because $2^0 = 1$.

The function $f(t) = 2^{t/10}$ is called an *exponential function* because the variable t is in the exponent. In this section we shall learn more about graphing exponential functions, and we shall see how such functions can be used to solve real-world problems.

A › Graphing Exponential Functions

Exponential functions take many forms. For example, the following functions are all exponential functions:

$$f(x) = 3^x, \quad F(y) = \left(\frac{1}{2}\right)^y, \quad H(z) = (1.02)^{z/2}$$

In general, we have the following definition.

EXPONENTIAL FUNCTION

An **exponential function** is a function defined for all real values of x by

$$f(x) = b^x \quad (b > 0, b \neq 1)$$

NOTE

The variable b must not equal 1 because $f(x) = 1^x = 1$ is a constant function, *not* an exponential function.

In this definition, b is a constant called the **base**, and the **exponent** x is the variable. It is proved in more advanced courses that for $b > 0$, b^x has a unique real value for each real value of x. We assume this in all the following work.

An exponential function is frequently not in the form given in our definition, but it can be put in that form. For example, $f(t) = 2^{t/10}$ can be written as $(2^{1/10})^t$ so that the base is $2^{1/10}$ and the exponent is t.

The exponential function defined by $f(t) = 2^{t/10}$ can be graphed and used to predict the number of bacteria present after time t. To make this graph, we first construct a table giving the value of the function for certain convenient times:

t	0	10	20	30
$f(t) = 2^{t/10}$	1	2^1	2^2	2^3

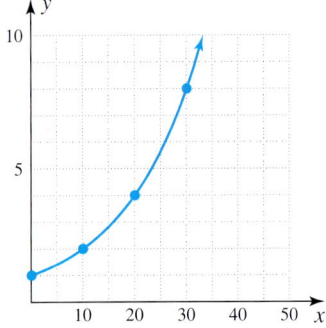
›Figure 10.8

The corresponding points can then be graphed and joined with a smooth curve, as shown in Figure 10.8. In general, we graph an exponential function by plotting several points calculated from the function and then drawing a smooth curve through these points.

EXAMPLE 1 Graphing exponential functions

Graph on the same coordinate system:

a. $f(x) = 2^x$ **b.** $g(x) = \left(\frac{1}{2}\right)^x$

SOLUTION

a. We first make a table with convenient values for x and find the corresponding values for $f(x)$. We then graph the points and connect them with a smooth curve, as shown in blue in Figure 10.9.

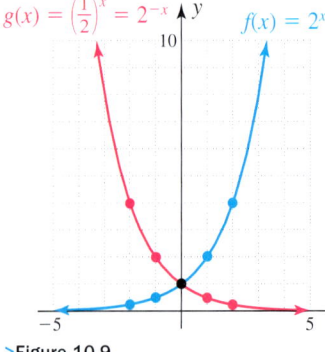
›Figure 10.9

x	-2	-1	0	1	2
$f(x) = 2^x$	$2^{-2} = \frac{1}{4}$	$2^{-1} = \frac{1}{2}$	$2^0 = 1$	$2^1 = 2$	$2^2 = 4$

PROBLEM 1

Graph $f(x) = 3^x$ and $g(x) = \left(\frac{1}{3}\right)^x$ on the same coordinate system.

Answers to PROBLEMS

1.
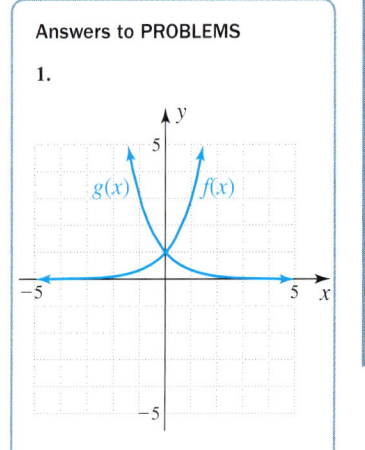

b. If we let $x = -2$,

$$g(-2) = \left(\frac{1}{2}\right)^{-2} = \frac{1}{\left(\frac{1}{2}\right)^2} = 4$$

Similarly, for $x = -1$,

$$g(-1) = \left(\frac{1}{2}\right)^{-1} = \frac{1}{\left(\frac{1}{2}\right)^1} = 2$$

For $x = 0, 1,$ and 2, the function values are $\left(\frac{1}{2}\right)^0 = 1$, $\left(\frac{1}{2}\right)^1 = \frac{1}{2}$, and $\left(\frac{1}{2}\right)^2 = \frac{1}{4}$, as shown in the table.

x	-2	-1	0	1	2
$g(x) = \left(\frac{1}{2}\right)^x$	4	2	1	$\frac{1}{2}$	$\frac{1}{4}$

We can save time if we realize that $\left(\frac{1}{2}\right)^x = (2^{-1})^x = (2^x)^{-1}$, so we can use the values of 2^x shown in the table for $f(x)$ in part **a**. The graph of $g(x) = \left(\frac{1}{2}\right)^x = 2^{-x}$ is shown in red in Figure 10.9.

The two graphs in Figure 10.9 are symmetric to each other with respect to the y-axis. In general, we have the following fact.

SYMMETRIC GRAPHS

The graphs of $y = b^x$ and $y = b^{-x}$ are **symmetric** to each other with respect to the y-axis.

CAUTION

The two functions graphed in Figure 10.9, $f(x) = 2^x$ and $g(x) = \left(\frac{1}{2}\right)^x$ are NOT inverse functions.

B › Determining Whether Exponential Functions Are Increasing or Decreasing

As you have seen, the graphs of some functions increase and some decrease. We'll make the idea more precise next.

INCREASING AND DECREASING FUNCTIONS

If the graph of a function rises from left to right, the function is an **increasing function.** If the graph falls from left to right, the function is a **decreasing function.**

Thus, we see that $f(x) = 2^x$ is an increasing function and $g(x) = \left(\frac{1}{2}\right)^x$ is a decreasing function. (See Figure 10.9.)

In our definition of the function $y = b^x$, it was required only that $b > 0$ and $b \neq 1$. For many practical applications, however, there is a particularly important base, the irrational number e. The value of e is approximately 2.7182818. When e is used as the base in an exponential function, the function is referred to as the natural exponential function.

THE NATURAL EXPONENTIAL FUNCTION, BASE e

The natural exponential function is defined by the equation

$$f(x) = e^x$$

The irrational real number e has the approximate value **2.7182818**.

The reasons for using this base are made clear in more advanced mathematics courses, but for our purposes we need only note that e is defined as the value that the quantity $(1 + \frac{1}{n})^n$ approaches as n increases indefinitely. In symbols,

$$\left(1 + \frac{1}{n}\right)^n \rightarrow e \approx 2.7182818 \quad \text{as} \quad n \rightarrow \infty$$

To show this, we use increasing values of n (1000, 10,000, 100,000, 1,000,000) and evaluate the expression $(1 + \frac{1}{n})^n$ using a calculator with a $\boxed{x^y}$ key.

For $n = 1000$, $\left(1 + \frac{1}{n}\right)^n = (1.001)^{1000} \approx 2.7169239$ Enter 1.001 $\boxed{x^y}$ 1000 $\boxed{=}$.

For $n = 10{,}000$, $\left(1 + \frac{1}{n}\right)^n = (1.0001)^{10{,}000} \approx 2.7181459$

For $n = 100{,}000$, $\left(1 + \frac{1}{n}\right)^n = (1.00001)^{100{,}000} \approx 2.7182682$

For $n = 1{,}000{,}000$, $\left(1 + \frac{1}{n}\right)^n = (1.000001)^{1{,}000{,}000} \approx 2.7182805$

As you can see, the value of $(1 + \frac{1}{n})^n$ is indeed getting closer to $e \approx 2.7182818$.

To graph the functions $f(x) = e^x$ and $g(x) = e^{-x}$, we make a table giving x different values (say $-2, -1, 0, 1, 2$) and finding the corresponding $y = e^x$ and $y = e^{-x}$ values. This can be done with a calculator with an $\boxed{e^x}$ key. On such calculators, you usually have to enter $\boxed{\text{INV}}$ or $\boxed{\text{2nd}}$ to find the value of e^x. [Enter 1 $\boxed{\text{2nd}}$ (or $\boxed{\text{INV}}$) $\boxed{e^x}$, and the calculator will give the value 2.7182818.] We use these ideas next.

EXAMPLE 2 Graphing increasing and decreasing functions

Use the values in the table to graph: $f(x) = e^x$ and $g(x) = e^{-x}$ and determine which of these functions is increasing and which is decreasing. (Use the same coordinate system.)

x	−2	−1	0	1	2
e^x	0.1353	0.3679	1	2.7183	7.3891

x	−2	−1	0	1	2
e^{-x}	7.3891	2.7183	1	0.3679	0.1353

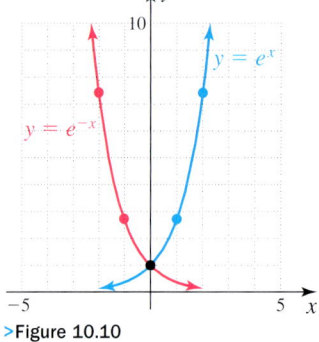

▶Figure 10.10

SOLUTION Plotting the given values, we obtain the graphs of $f(x) = e^x$ and $g(x) = e^{-x}$ shown in Figure 10.10. The functions e^x and e^{-x} are symmetric to each other with respect to the y-axis. Also, e^x is increasing and e^{-x} is decreasing.

PROBLEM 2

Use the table to graph $f(x) = -e^x$ and $g(x) = -e^{-x}$. Which function is increasing and which is decreasing?

Answers to PROBLEMS

2.

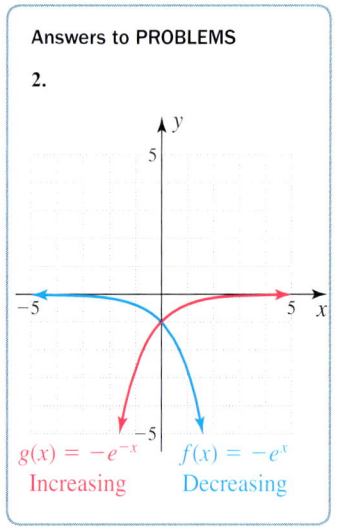

$g(x) = -e^{-x}$ $f(x) = -e^x$
Increasing   Decreasing

C ▶ Applications Involving Exponential Functions

Do you have some money invested? Is it earning interest compounded annually, quarterly, monthly, or daily? Does the frequency of compounding make a difference? Some banks have instituted *continuous interest compounding*. We can compare continuous compounding and n compoundings per year by examining their formulas:

Continuous compounding: $A = Pe^{rt}$

n compoundings per year: $A = P\left(1 + \frac{r}{n}\right)^{nt}$

where A = compound amount
P = principal
r = interest rate
t = time in years
and n = periods per year (second formula only)

As the number of times the interest is compounded increases, n increases, and it can be shown that

$$\left(1 + \frac{r}{n}\right)^{nt}$$

gets closer to e^{rt}. Let's see what the two formulas earn.

EXAMPLE 3 Maximizing your investment

Find the compound amount for each situation.

a. $100 compounded continuously for 18 months at 6%.
b. $100 compounded quarterly for 18 months at 6%.

SOLUTION

a. Use the formula, $A = Pe^{rt}$, for compounding continuously. Substitute $P = 100$, $r = 0.06$, and $t = 1.5$ (18 months), so

$$A = Pe^{rt}$$
$$= 100e^{(0.06)(1.5)}$$
$$= 100e^{0.09}$$

A calculator gives the value

$$e^{0.09} \approx 1.0942$$

Thus,

$$A \approx (100)(1.0942) = 109.42$$

and the compound amount is $109.42.

b. Use the formula, $A = P\left(1 + \frac{r}{n}\right)^{nt}$, for compounding quarterly. $P = 100$, $r = 0.06$, $t = 1.5$, and $n = 4$, so

$$A = P\left(1 + \frac{r}{n}\right)^{nt} = 100\left(1 + \frac{0.06}{4}\right)^{4(1.5)}$$
$$= 100(1 + 0.015)^6$$
$$= 100(1.015)^6$$
$$\approx 109.34$$

Thus, the compound amount for $100 at the same rate, compounded quarterly, is given by $A = 100(1.015)^6 \approx 109.34$. At 18 months, the difference between continuous and quarterly compounding is only 8¢. For more comparisons, see the *Using Your Knowledge* in the exercises.

PROBLEM 3

a. Find the compound amount for $100 compounded continuously for 30 months at 6%.

b. Find the compound amount for $100 compounded quarterly for 30 months at 6%.

Answers to PROBLEMS

3. **a.** $116.18   **b.** $116.05

EXAMPLE 4 Calculating radioactive decay

A radioactive substance decays so that G, the number of grams present, is given by the equation

$$G = 1000e^{-1.2t}$$

where t is the time in years. Find, to the nearest gram, the amount of the substance present at the given time.

a. At the start **b.** In 2 years

SOLUTION

a. Here, $t = 0$, so $G = 1000e^0 = 1000(1) = 1000$, and 1000 grams of the substance are present at the start.

b. Since $t = 2$, $G = 1000e^{-1.2(2)} = 1000e^{-2.4}$. To evaluate G, we use a calculator to obtain

$$e^{-2.4} \approx 0.090718$$

so that

$$G \approx (1000)(0.090718)$$
$$= 90.718$$

So, after starting with 1000 grams, there are about 91 grams present in 2 years.

PROBLEM 4

Repeat Example 4(b) if the time is 21 months.

Calculator Corner

Radioactive Decay

In Example 4, let $Y_1 = 1000e^{-1.2x}$ using a $[-1, 10]$ by $[-100, 1000]$ window and Yscl $= 100$. Do you now see that the substance is *decaying?* Now let's calculate the value of the function when $x = 2$ as required in the example. Press `2nd` `TRACE` `1`. Answer the calculator question by entering $X = 2$ and pressing `ENTER`. The result is $Y = 90.717953$ as shown in the window. Rounding the answer to the nearest gram, the answer is 91, as before.

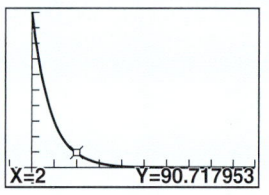

EXAMPLE 5 Predicting U.S. population

Thomas Robert Malthus invented a model for predicting population based on the idea that, when the birth rate (B) and the death rate (D) are constant and no other factors are considered, the population P is given by the equation

$$P = P_0 e^{kt}$$

where
- P = population at any time t
- P_0 = initial population
- k = annual growth rate $(B - D)$
- t = time in years after a given base year

According to the *Statistical Abstract of the United States,* the population in 1980 was 226,546,000, the birth rate was 0.016, and the death rate 0.0086. Use this information to predict the number of people in the United States in the specified year.

a. 2000 **b.** 2010

SOLUTION

a. To predict the population in the year 2000, we identify the values to use in the formula. The population for 1980 is given, so 1980 is the base year. The initial population is $P_0 = 226{,}546{,}000$, $k = 0.016 - 0.0086 = 0.0074$, and the number of years from 1980 to 2000 is $t = 20$.

PROBLEM 5

Using the population formula for Example 5, what would the prediction be for 2020?

Answers to PROBLEMS

4. About 122 g **5.** About 304,584,000

$$P = P_0 \cdot e^{kt}$$
$$= 226{,}546{,}000 e^{0.0074(20)}$$
$$= 226{,}546{,}000 e^{0.148}$$
$$\approx 262{,}683{,}009$$

The predicted population for 2000 is 262,683,009. The actual population in 2000 was about 281,000,000. Why do you think there is a discrepancy?

b. The number of years from 1980 to 2010 is $t = 30$.

$$P = P_0 \cdot e^{kt}$$
$$= 226{,}546{,}000 e^{0.0074(30)}$$
$$= 226{,}546{,}000 e^{0.222}$$
$$\approx 282{,}858{,}851$$

Thus, the predicted population is near 282,859,000. The Census Bureau predicts about 300 million by the year 2006. Why the discrepancy?

Exercises 10.3

⟨ A ⟩ Graphing Exponential Functions In Problems 1–6, find the value of the given exponential for the indicated values of the variable.

1. 5^x
  a. $x = -1$
  b. $x = 0$
  c. $x = 1$

2. 5^{-x}
  a. $x = -1$
  b. $x = 0$
  c. $x = 1$

3. 3^t
  a. $t = -2$
  b. $t = 0$
  c. $t = 2$

4. 3^{-t}
  a. $t = -2$
  b. $t = 0$
  c. $t = 2$

5. $10^{t/2}$
  a. $t = -2$
  b. $t = 0$
  c. $t = 2$

6. $10^{-t/2}$
  a. $t = -2$
  b. $t = 0$
  c. $t = 2$

⟨ A ⟩ Graphing Exponential Functions
⟨ B ⟩ Determining Whether Exponential Functions Are Increasing or Decreasing

In Problems 7–16, graph the functions given in parts **a** and **b** on the same coordinate system. State whether the function is increasing or decreasing.

7. a. $f(x) = 5^x$
  b. $g(x) = 5^{-x}$

8. a. $f(t) = 3^t$
  b. $g(t) = 3^{-t}$

9. a. $f(x) = 10^x$
  b. $g(x) = 10^{-x}$

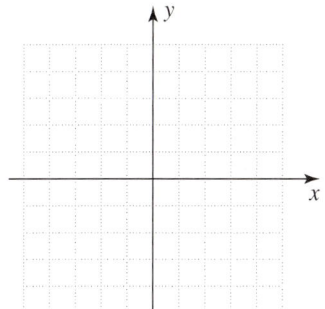
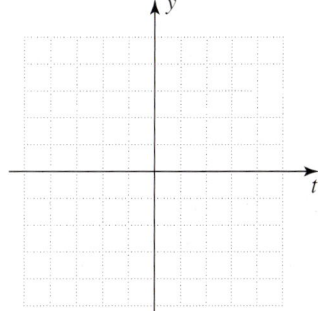
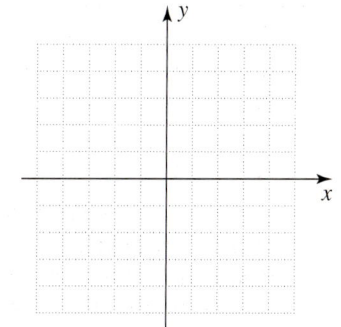

10. a. $f(t) = 10^{t/2}$
    b. $g(t) = 10^{-t/2}$

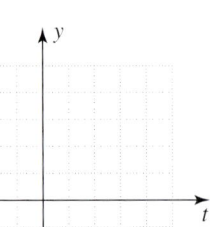

11. a. $f(x) = e^{2x}$
    b. $g(x) = e^{-2x}$

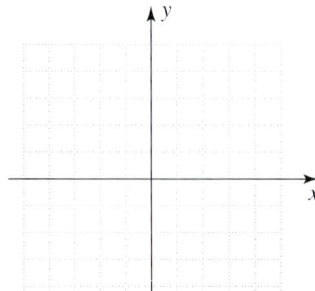

12. a. $f(t) = e^{t/4}$
    b. $g(t) = e^{-t/4}$

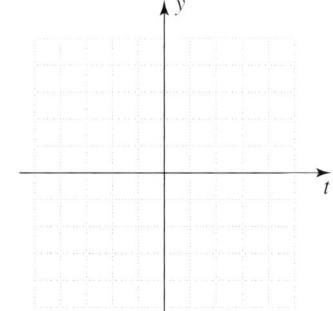

13. a. $f(x) = 3^x + 1$
    b. $g(x) = 3^{-x} + 1$

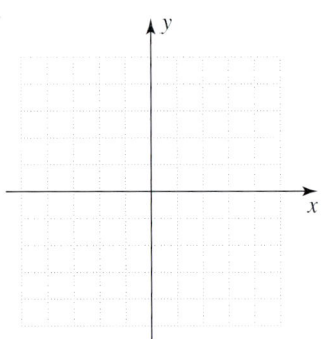

14. a. $f(x) = \left(\frac{1}{3}\right)^x + 1$
    b. $g(x) = \left(\frac{1}{3}\right)^{-x} + 1$

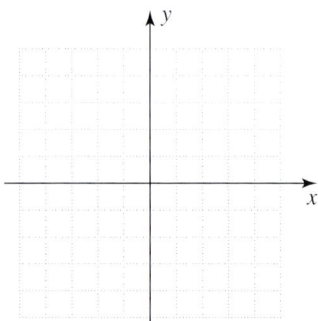

15. a. $f(x) = 2^{x+1}$
    b. $g(x) = 2^{-x+1}$

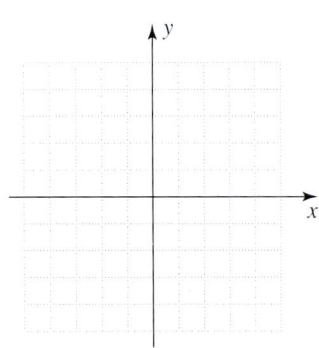

16. a. $f(x) = \left(\frac{1}{2}\right)^{x+1}$
    b. $g(x) = \left(\frac{1}{2}\right)^{-x+1}$

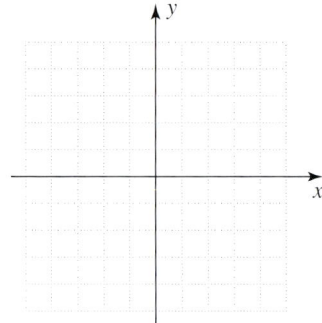

⟨ C ⟩ **Applications Involving Exponential Functions**

Compound Interest In Problems 17–20, find the compound amount if the compounding is **a** continuous or **b** quarterly.

17. $1000 at 9% for 10 years

18. $1000 at 9% for 20 years

19. $1000 at 6% for 10 years

20. $1000 at 6% for 20 years

21. *Population* The population of a town is given by the equation $P = 2000(2^{0.2t})$, where t is the time in years from 1985. Find the population in

   a. 1985   **b.** 1990   **c.** 1995

22. *Bacteria growth* A colony of bacteria grows so that their number, B, is given by the equation $B = 1200(2^t)$, where t is in days. Find the number of bacteria

   a. At the start.   **b.** In 5 days.   **c.** In 10 days.

23. *Radioactive decay* A radioactive substance decays so that G, the number of grams present, is given by the equation $G = 2000e^{-1.05t}$, where t is the time in years. To the nearest tenth of a gram, find the amount of the substance present

   a. At the start.   **b.** In 1 year.   **c.** In 2 years.

24. *Radioactive decay* Solve Problem 23 where the equation is $G = 2000e^{-1.1t}$.

25. *Hispanic population* In 1980, the number of persons of Hispanic origin living in the United States was 14,609,000. If their birth rate is 0.0232 and their death rate is 0.004, predict the number of persons of Hispanic origin living in the United States for the year 2000. Round to the nearest thousand. (*Hint:* See Example 5.)

26. *African-American population* In 1980, the number of African-Americans living in the United States was 26,683,000. If their birth rate is 0.0221 and their death rate is 0.0088, predict the number of African-Americans living in the United States in the year 2000. Round to the nearest thousand. (*Hint:* See Example 5.)

27. *Compact disk sales* The number of compact discs (CDs) sold in the United States (in millions) since 1985 can be approximated by the exponential function $S(t) = 32(10)^{0.19t}$, where t is the number of years after 1985. (Round to the nearest million)

   a. Predict the number of CDs sold in 1990.

   b. Predict the number that will be sold in the year 2010.

28. *Cellular phone sales* The number of cellular phones sold in the United States (in thousands) since 1989 can be approximated by the exponential function $S(t) = 900(10)^{0.27t}$, where t is the number of years after 1989. (Round to the nearest thousand)

   a. Predict the number of cellular phones sold in 1990.

   b. Predict the number that will be sold in the year 2010.

29. *Recycling* According to the *Statistical Abstract of the United States,* about $\frac{2}{3}$ of all aluminum cans distributed are recycled. If a company distributes 500,000 cans, the number of recycled cans in use after t years is given by the exponential function $N(t) = 500{,}000\left(\frac{2}{3}\right)^t$. How many recycled cans are in use after the specified amount of time?

   a. 1 year   **b.** 2 years   **c.** 10 years

30. *Depreciation* If the value of an item each year is about 60% of its value the year before, after t years the salvage value of an item originally costing C dollars is given by $S(t) = C(0.6)^t$. Find the salvage value of a computer costing \$10,000

   a. 1 year after it was purchased.

   b. 10 years after it was purchased.

31. *Atmospheric pressure* The atmospheric pressure A (in pounds per square inch) can be approximated by the exponential function $A(a) = 14.7(10)^{-0.000018a}$, where a is the altitude in feet. (Round to the nearest hundredth)

   a. The highest mountain in the world is Mount Everest, about 29,000 feet high. Find the atmospheric pressure at the top of Mount Everest.

   b. In the United States, the highest mountain is Mount McKinley in Alaska, whose highest point is about 20,000 feet. Find the atmospheric pressure at the top of Mount McKinley.

32. *Atmospheric pressure* The atmospheric pressure A (in pounds per square inch) can also be approximated by the equation $A(a) = 14.7e^{-0.21a}$, where a is the altitude in miles. (See Problem 31. Round to the nearest hundredth.)

   a. If we assume that the altitude of Mount Everest is about 6 miles, what is the atmospheric pressure at the top of Mount Everest?

   b. If we assume that the altitude of Mount McKinley is about 4 miles, what is the atmospheric pressure at the top of Mount McKinley?

33. *Plutonium decay* The number of kilograms $P(t)$ of radioactive plutonium left after t years is modeled by the equation:

$$P(t) = 2^{-0.0001754t}$$

Find, to four decimal places, the amount of plutonium left after the specified amount of years.

   a. $t = 0$ years

   b. $t = 1000$ years

   c. $t = 10{,}000$ years

34. *Bacteria present* The number $N(t)$ of bacteria present after t minutes is given by the formula

$$N(t) = 1000(3)^{0.1t}$$

Find, to the nearest whole number, the number of bacteria after the specified amount of time.

   a. 5 minutes

   b. 10 minutes

35. *AIDS cases* In 1987, the number of cases of AIDS in the United States was estimated at 50,000. This number was doubling every year. If this trend continued the number of cases of AIDS after 1987 will be given by the equation

$$f(t) = 50{,}000(2)^t$$

a. How many cases of AIDS were there in 1990?

b. How many cases does the model predict in the year 2007?

c. Is it possible that the number of AIDS cases doubled every year until 2007?

36. *E. coli* The bacterium *Escherichia coli*, E. coli for short, is found in your digestive tract. If a colony of E. coli started with a single cell and divided into two cells every 20 minutes ($\frac{1}{3}$ hour) after its birth, and assuming that no deaths occur, the number of E. coli cells after t hours is modeled by the equation $N(t) = 2^{3t}$.

a. How many E. coli will there be after 10 hours? (Leave answer in exponential form.)

b. If the mass of a single E. coli cell is approximately $5 \cdot 2^{-43}$ grams, what is the mass of the colony after the 10 hours?

⟩⟩⟩ Using Your Knowledge

Compounding Your Money In this section you learned that for continuous compounding, the compound amount is given by $A = Pe^{rt}$. For ordinary compound interest, the compound amount is given by $A = P(1 + \frac{r}{n})^{nt}$, where r is the annual interest rate and n is the number of periods per year. Suppose you have \$1000 to put into an account where the interest rate is 6%. How much more would you have at the end of 2 years for continuous compounding than for monthly compounding?

For continuous compounding, the amount is given by the equation

$$A = 1000e^{(0.06)(2)} = 1000e^{0.12}$$
$$\approx (1000)(1.1275) \quad \text{Use a calculator.}$$
$$= 1127.50$$

The amount is \$1127.50. For monthly compounding,

$$\frac{r}{n} = \frac{0.06}{12} = 0.005 \quad \text{and} \quad nt = 24$$

The amount is given by the equation $A = 1000(1 + 0.005)^{24} = 1000(1.005)^{24}$. Compound interest tables or your calculator will give the value

$$(1.005)^{24} \approx 1.1271598$$

so that $A = 1127.16$ (to the nearest hundredth). The amount is \$1127.16. Continuous compounding earns you only 34¢ more. But, see what happens when the period t is extended in Problems 37 and 38.

37. Make the same comparison where the time is 10 years.

38. Make the same comparison where the time is 20 years.

⟩⟩⟩ Write On

39. The definition of the exponential function $f(x) = b^x$ does not allow $b = 1$.

a. What type of graph will $f(x)$ have when $b = 1$?

b. Is $f(x) = b^x$ a function when $b = 1$? Explain.

c. Does $f(x) = b^x$ have an inverse when $b = 1$? Explain.

40. List some reasons to justify the condition $b > 0$ in the definition of the exponential function $f(x) = b^x$.

41. Discuss the relationship between the graphs of $f(x) = b^x$ and $g(x) = b^{-x}$.

42. In Example 5, we predicted the U.S. population for the year 2000 to be 262,683,009. The actual population in 2000 was near 281,000,000. Can you give some reasons for this discrepancy?

❯❯❯ Concept Checker

Fill in the blank(s) with the correct word(s), phrase, or mathematical statement.

43. An exponential function is a function defined by $f(x) =$ _____, $b > 0, b \neq 1$.

44. If the graph of a function rises from left to right, the function is _____.

45. If the graph of a function falls from left to right, the function is _____.

46. The natural exponential function is defined by $f(x) =$ _____.

constant
decreasing
e^x
b^x
increasing

❯❯❯ Mastery Test

47. A radioactive substance decays so that the number of grams present, G, is given by the equation $G = 1000e^{-1.2t}$, where t is the time in years. Find, to the nearest gram, the amount of substance present in 18 months.

48. The compound amount A for a principal P at rate r compounded continuously for t years is given by the equation $A = Pe^{rt}$. Find the compound amount for $100 compounded continuously for 36 months where the rate is 6%.

In Problems 49–52, graph the exponential function.

49. $f(x) = e^{x/2}$

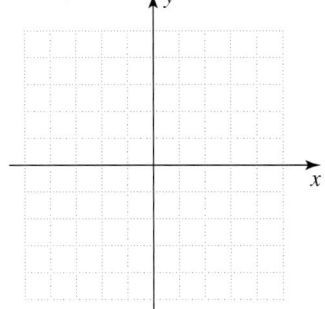

50. $f(x) = e^{-x/2}$

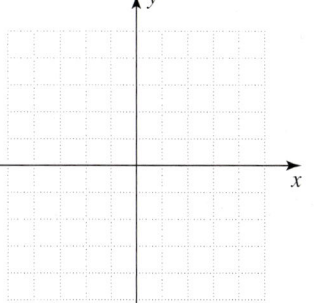

51. $f(x) = 6^x$

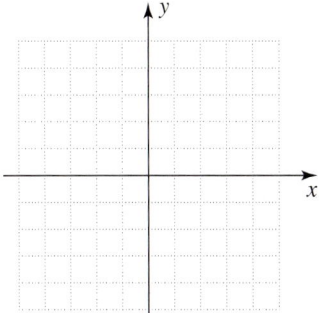

52. $f(x) = 6^{-x}$

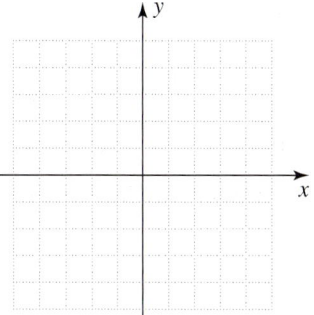

❯❯❯ Skill Checker

State the property of exponents being applied in each situation.

53. $(10^x)(10^y) = 10^{x+y}$

54. $\dfrac{10^x}{10^y} = 10^{x-y}$

55. $(10^x)^3 = 10^{3x}$

56. $[(2)(10^x)]^y = (2^y)(10^{xy})$

10.4 Logarithmic Functions and Their Properties

Objectives

A ▸ Graph logarithmic functions.

B ▸ Write an exponential equation in logarithmic form and a logarithmic equation in exponential form.

C ▸ Solve logarithmic equations.

D ▸ Use the properties of logarithms to simplify logarithms of products, quotients, and powers.

E ▸ Solve applications involving logarithmic functions.

▸ To Succeed, Review How To . . .

1. Use the rules of exponents in multiplication, division, and raising variables to a power (pp. 38–45).
2. Simplify algebraic expressions (pp. 55–58).

▸ Getting Started

Shake, Rattle, and Roll

The damage done by the 1989 earthquake in San Francisco was tremendous, as can be seen in the photo. If I_0 denotes the minimum intensity of an earthquake (used for comparison purposes), the intensity of this quake was $10^{7.1} \cdot I_0$ or $10^{7.1}$ times as intense as the minimum intensity.

The magnitude of an earthquake is usually measured on the Richter scale. The Richter scale is logarithmic. In this scale the magnitude R of an earthquake of intensity $I = 10^{7.1} \cdot I_0$ is reported as

$$R = \log_{10} \frac{I}{I_0} = \log_{10} \frac{10^{7.1} \cdot I_0}{I_0} = \log_{10} 10^{7.1} = 7.1$$

The magnitude was reported as 7.1 on the Richter scale. 7.1 is the *exponent* of 10. The exponent 7.1 is called the **logarithm**, base 10, of $10^{7.1}$ and is written as

$$\log_{10} 10^{7.1} = 7.1$$

In this section we shall study logarithmic functions and their properties.

A ▸ Graphing Logarithmic Functions

In Section 10.2, we graphed the function $f(x) = 3^x$ and its inverse $x = 3^y$ and promised to show how to find $f^{-1}(x)$. Here are the steps we shall use:

$$f(x) = 3^x \quad \text{Given}$$

1. Replace $f(x)$ by y. $y = 3^x$
2. Interchange x and y. $x = 3^y$
3. Solve for y. $y =$ the exponent to which we raise 3 to get x
4. Replace y by $f^{-1}(x)$. $f^{-1}(x) =$ the exponent to which we raise 3 to get x

Since a logarithm is an exponent, we define "the exponent to which we raise 3 to get x" as follows.

DEFINITION OF $\log_3 x$

$\log_3 x$ ("the logarithm base 3 of x") means "the **exponent** to which we raise 3 to get x."

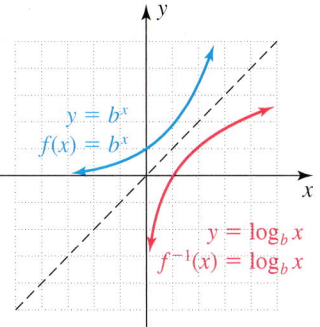

>Figure 10.11

If we use this definition in Step 4, $f^{-1}(x) = \log_3 x$. $f^{-1}(9) = \log_3 9 = 2$ because 2 is the exponent to which we raise the base 3 to get 9; $\log_3 9 = 2$ means $3^2 = 9$.

In general, for any exponential function $f(x) = b^x$, the inverse is $f^{-1}(x) = \log_b x$. The graph of $f^{-1}(x) = \log_b x$ can be drawn by reflecting the graph of $f(x) = b^x$ across the line $y = x$ (see Figure 10.11). (Recall that the definition of exponential function requires $b > 0$ and $b \neq 1$.) Using these ideas, we have the following definition.

DEFINITION OF LOGARITHMIC FUNCTION

$f(x) = y = \log_b x$ is equivalent to $b^y = x$ $(b > 0, b \neq 1,$ and $x > 0)$

This definition means that the logarithm of a number is an exponent. For example,

$4 = \log_2 16$	is equivalent to	$2^4 = 16$ — The logarithm is the exponent.
$2 = \log_5 25$	is equivalent to	$5^2 = 25$ — The logarithm is the exponent.
$-3 = \log_{10} 0.001$	is equivalent to	$10^{-3} = 0.001$ — The logarithm is the exponent.

Here are some properties to help you graph logarithmic functions.

PROPERTIES OF LOGARITHMIC GRAPHS

1. The point $(1, 0)$ is always on the graph because $\log_b 1 = 0$, since $b^0 = 1$.
2. For $b > 1$, the graph will rise from left to right (increase).
3. For $0 < b < 1$, the graph will fall from left to right (decrease).
4. The y axis is the vertical asymptote (the curve gets closer and closer to it but never reaches it) for $f(x)$.
5. The domain of $f(x) = \log_b x$ is all positive real numbers and the range is all real numbers.

EXAMPLE 1 Graphing a logarithmic function

Graph: $y = f(x) = \log_4 x$.

SOLUTION By the definition of logarithm, $y = \log_4 x$ is equivalent to $4^y = x$. We can graph $4^y = x$ by first assigning values to y and calculating the corresponding x-values.

y	$x = 4^y$	x	Ordered Pair (x, y)
0	$x = 4^0 =$	1	$(1, 0)$
1	$x = 4^1 =$	4	$(4, 1)$
2	$x = 4^2 =$	16	$(16, 2)$
-1	$x = 4^{-1} =$	$\frac{1}{4}$	$(\frac{1}{4}, -1)$
-2	$x = 4^{-2} =$	$\frac{1}{16}$	$(\frac{1}{16}, -2)$

We then graph the ordered pairs and connect them with a smooth curve, the graph of $y = \log_4 x$. To confirm that our graph is correct, graph $y = 4^x$ on the same coordinate axes (see Figure 10.12). As expected, the graphs are reflections of each other across the line $y = x$.

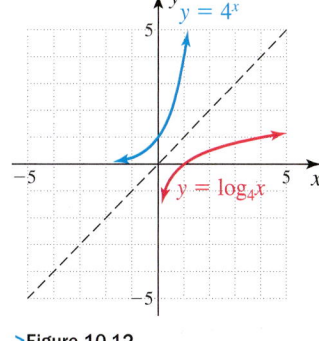

>Figure 10.12

PROBLEM 1

Graph $y = f(x) = \log_3 x$.

Answers to PROBLEMS

1.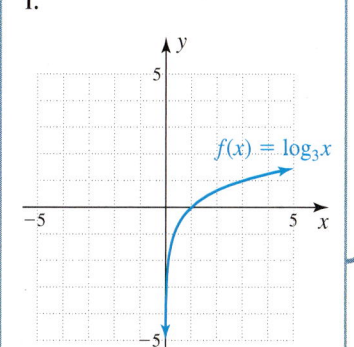

B › Converting Exponential to Logarithmic Equations and Vice Versa

The definition of logarithm states "$y = \log_b x$ is equivalent to $b^y = x$." We can use this definition to write logarithmic equations as exponential equations, and vice versa. *The logarithm is an exponent for a given base.* Thus,

$8 = 2^3$	is equivalent to	$\log_2 8 = 3$	The logarithm 3 is the exponent for the base 2.
$25 = 5^2$	is equivalent to	$\log_5 25 = 2$	The logarithm 2 is the exponent for the base 5.
$9 = b^a$	is equivalent to	$\log_b 9 = a$	The logarithm a is the exponent for the base b.

Similarly,

$y = \log_5 3$	is equivalent to	$5^y = 3$	The logarithm y is the exponent for the base 5.
$-2 = \log_b 4$	is equivalent to	$b^{-2} = 4$	The logarithm -2 is the exponent for the base b.
$a = \log_b c$	is equivalent to	$b^a = c$	The logarithm a is the exponent for the base b.

EXAMPLE 2 Converting an exponential equation to logarithmic form
Write in logarithmic form: $125 = 5^x$

SOLUTION By the definition of logarithm,

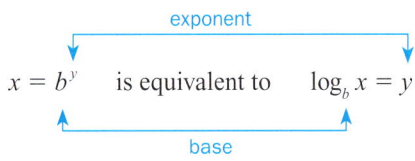

$x = b^y$ is equivalent to $\log_b x = y$ (exponent; base)

$125 = 5^x$ is equivalent to $\log_5 125 = x$.

PROBLEM 2
Write $2^x = 1024$ in logarithmic form.

EXAMPLE 3 Converting a logarithmic equation to exponential form
Write in exponential form and check its accuracy: $\log_{32} 64 = \frac{6}{5}$

SOLUTION By the definition of logarithm, $\log_b x = y$ means $x = b^y$. Hence, $\log_{32} 64 = \frac{6}{5}$ means $32^{6/5} = 64$. Because $32^{6/5} = (\sqrt[5]{32})^6 = 2^6 = 64$, the equation $\log_{32} 64 = \frac{6}{5}$ is correct.

PROBLEM 3
Write in exponential form and check: $\log_4 32 = \frac{5}{2}$.

C › Solving Logarithmic Equations

In Example 2, we pointed out that $125 = 5^x$ is equivalent to $\log_5 125 = x$. Can we find x? Yes, if we write both sides of the equation $125 = 5^x$ as exponentials with the same base, set the exponents equal, and then solve the resulting equation. This can be done because the exponential function $f(x) = b^x$ is one-to-one, so when $b^x = b^y$, $x = y$. Thus, if we rewrite the equation $125 = 5^x$ as $5^3 = 5^x$, it's easy to see that $3 = x$. In doing this, we have used the following property.

EQUIVALENCE PROPERTY For any $b > 0$, $b \neq 1$, $b^x = b^y$ is equivalent to $x = y$.

Answers to PROBLEMS

2. $\log_2 1024 = x$

3. $4^{5/2} = 32$ or $(\sqrt{4})^5 = 32$ (true)

10.4 Logarithmic Functions and Their Properties

EXAMPLE 4 Solving logarithmic equations
Solve:

a. $\log_3 x = -2$ **b.** $\log_x 9 = 2$

SOLUTION

a. The equation $\log_3 x = -2$ is equivalent to
$$3^{-2} = x$$
$$\frac{1}{9} = x$$

CHECK Substitute $x = \frac{1}{9}$ in the original equation to obtain $\log_3 \frac{1}{9} = -2$, equivalent to $3^{-2} = \frac{1}{9}$, a true statement.

b. The equation $\log_x 9 = 2$ is equivalent to
$$x^2 = 9$$
$$x = \pm 3 \quad \text{Take the square roots.}$$

But x represents the base, which must be positive, so $x = 3$ is the only answer. (We discard $x = -3$, which is negative and cannot be the base.)

CHECK Substitute $x = 3$ in the original equation to obtain $\log_3 9 = 2$, equivalent to $3^2 = 9$, a true statement.

PROBLEM 4
Solve:

a. $\log_3 x = -5$

b. $\log_x 81 = 2$

EXAMPLE 5 Finding logarithms
Find:

a. $\log_2 32$ **b.** $\log_4\left(\frac{1}{4}\right)$

c. $\log_{10} 1000$ **d.** $\log_{10} 0.01$

SOLUTION

a. We know that $\log_2 32$ means the exponent to which we must raise the base 2 to get 32. Since $2^5 = 32$, the exponent is 5 and $\log_2 32 = 5$. Alternatively, since we are looking for $\log_2 32$, we can write
$$\log_2 32 = x$$
which is equivalent to $\quad 2^x = 32$
or $\quad 2^x = 2^5$

Since the exponents must be equal, $x = 5$, as before.

b. Since $\log_4(\frac{1}{4})$ is the exponent to which we must raise the base 4 to get $\frac{1}{4}$ and since $4^{-1} = \frac{1}{4}$, the exponent is -1 and $\log_4(\frac{1}{4}) = -1$. Alternatively, since we are looking for $\log_4(\frac{1}{4})$, we can write
$$\log_4\left(\frac{1}{4}\right) = x$$
which is equivalent to $\quad 4^x = \frac{1}{4}$
or $\quad 4^x = 4^{-1}$

Since the exponents must be equal, $x = -1$, as before.

PROBLEM 5
Find:

a. $\log_2 64$ **b.** $\log_2\left(\frac{1}{8}\right)$

c. $\log_{10} 100$ **d.** $\log_{10} 0.1$

(continued)

Answers to PROBLEMS

4. a. $x = \frac{1}{243}$ **b.** $x = 9$

5. a. 6 **b.** -3 **c.** 2 **d.** -1

c. Since $10^3 = 1000$, $\log_{10} 1000 = 3$. Alternatively, let

$$\log_{10} 1000 = x$$

which is equivalent to $\quad\quad 10^x = 1000 = 10^3 \quad$ Since the exponents must be equal
and $\quad\quad\quad\quad\quad\quad\quad\quad x = 3$

d. Since $10^{-2} = \frac{1}{100} = 0.01$, $\log_{10} 0.01 = -2$. Alternatively, let

$$\log_{10} 0.01 = x$$

which is equivalent to $\quad\quad 10^x = 0.01 = 10^{-2} \quad$ Since the exponents must be equal
and $\quad\quad\quad\quad\quad\quad\quad\quad x = -2$

D › Using Properties of Logarithms

Logarithms have three important properties that are the counterparts of the corresponding properties of exponents. By the definition of logarithm, if $x = \log_b M$ and $y = \log_b N$, then $M = b^x$ and $N = b^y$. From the properties of exponents, it follows that

$$MN = b^x b^y = b^{x+y}$$

so that

$$\log_b MN = x + y = \log_b M + \log_b N$$

This means that the *logarithm of a product is the sum of the logarithms of its factors.* Similarly,

$$\frac{M}{N} = \frac{b^x}{b^y} = b^{x-y}$$

which shows that

$$\log_b \frac{M}{N} = x - y = \log_b M - \log_b N$$

Thus, the *logarithm of M divided by N is the logarithm of M minus the logarithm of N.* If we have a power of a number such as M^r, then

$$M^r = (b^x)^r = b^{rx}$$

means that $\log_b M^r = rx = r \log_b M$. In words, the *logarithm of M^r is r times the logarithm of M.* We summarize these results as follows.

PROPERTIES OF LOGARITHMS

$\log_b MN = \log_b M + \log_b N \quad$ Product property

$\log_b \dfrac{M}{N} = \log_b M - \log_b N \quad$ Quotient property

$\log_b M^r = r \log_b M \quad$ Power property

EXAMPLE 6 Using the power and quotient properties of logarithms

Use the properties of logarithms to show that

$$\log_b\left(\frac{x^4}{16}\right) = 4 \log_b x - \log_b 16$$

SOLUTION Use the properties to rewrite the right side

$4 \log_b x - \log_b 16 = \log_b x^4 - \log_b 16 \quad$ Power property

$\quad\quad\quad\quad\quad\quad\quad\quad = \log_b\left(\dfrac{x^4}{16}\right) \quad$ Quotient property

PROBLEM 6
Use the properties of logarithms to show that

$$\log_b\left(\frac{a^2}{c^3}\right) = 2 \log_b a - 3 \log_b c$$

Answers to PROBLEMS

6. $2 \log_b a - 3 \log_b c = \log_b a^2 - \log_b c^3$
$\quad\quad\quad\quad\quad\quad\quad\quad = \log_b\left(\dfrac{a^2}{c^3}\right)$

EXAMPLE 7 Using the product and quotient properties of logarithms

Assume the logarithms are all to base 10 and use the properties of logarithms to show that

$$\log\left(\frac{\sqrt[3]{pq}}{r}\right) = \frac{1}{3}\log p + \frac{1}{3}\log q - \log r$$

where base 10 is understood throughout.

SOLUTION Use the properties to rewrite the right side of the equation

$$\frac{1}{3}\log p + \frac{1}{3}\log q - \log r = \log p^{1/3} + \log q^{1/3} - \log r \quad \text{Power property}$$
$$= \log(pq)^{1/3} - \log r \quad \text{Product property}$$
$$= \log\left(\frac{\sqrt[3]{pq}}{r}\right) \quad \text{Quotient property}$$

PROBLEM 7

Assume the logarithms are all to base 10 and use the properties of logarithms to show that

$$\log\left(\frac{\sqrt{xy}}{z^3}\right) = \frac{1}{2}\log x + \frac{1}{2}\log y - 3\log z$$

Answer on page 790

There are three more properties of logarithms worth mentioning. We state why these properties are true by writing the given logarithmic equation as an equivalent exponential equation.

OTHER PROPERTIES OF LOGARITHMS

$\log_b 1 = 0$ Because $b^0 = 1$

$\log_b b = 1$ Because $b^1 = b$

$\log_b b^x = x$ Because $b^x = b^x$

E) Applications Involving Logarithmic Functions

We can use logarithmic functions to solve a variety of applications, from measuring earthquake intensity to compounding interest.

EXAMPLE 8 Earthquake intensity

In the *Getting Started* we mentioned the 1989 San Francisco earthquake, but earthquakes are commonplace in California. One California earthquake was $10^{6.4}$ times as intense as that of a quake of minimum intensity, I_0. What was the magnitude of that quake on the Richter scale? (Recall the formula in the *Getting Started*, $R = \log_{10} \frac{I}{I_0}$.)

SOLUTION We are given that $I = 10^{6.4} I_0$, so the magnitude on the Richter scale is

$$R = \log_{10} \frac{I}{I_0} = \log_{10} 10^{6.4} = 6.4$$

PROBLEM 8

Do Example 8 with $10^{6.4}$ replaced by $10^{7.2}$.

Answer on page 790

EXAMPLE 9 Compounding the purchase price of Manhattan

In 1626, Peter Minuit bought the island of Manhattan from the Native Americans for the equivalent of about $24. If this money had been invested at 5% compounded annually, in 2001 the money would have been worth $24(1.05)^{375}$. Before calculators were available, we could have used logs to calculate the amount of money, if it's known that $\log_{10} 24 \approx 1.3802$ and $\log_{10} 1.05 \approx 0.0212$, find $\log_{10} 24(1.05)^{375}$.

PROBLEM 9

Do Example 9 when the rate is changed to 6% if it is given that $\log_{10} 1.06 \approx 0.0253$.

Answer on page 790

(continued)

SOLUTION

$$\log_{10} 24(1.05)^{375} = \log_{10} 24 + \log_{10} (1.05)^{375}$$
$$= \log_{10} 24 + 375(\log_{10} 1.05)$$
$$\approx 1.3802 + 375(0.0212)$$
$$= 1.3802 + 7.9500$$
$$= 9.3302 \quad \text{Log of the accumulated amount}$$

Answers to PROBLEMS

7. $\frac{1}{2} \log x + \frac{1}{2} \log y - 3 \log z$
$= \log x^{1/2} + \log y^{1/2} - \log z^3$
$= \log(xy)^{1/2} - \log z^3$
$= \log\left(\dfrac{\sqrt{xy}}{z^3}\right)$

8. 7.2   9. 10.8677

Now suppose the accumulated amount is A. We have

$$\log_{10} A \approx 9.3302$$

or $10^{9.3302} = A \approx 2{,}139{,}000{,}000$ Rounded from 2,138,946,884

Over two billion dollars. Do you think you could buy Manhattan for two billion dollars today?

Calculator Corner

1. Assuming the logarithms are all to base 10 in Example 6, confirm graphically by making sure that both sides of the equation

$$\log\left(\frac{x^4}{16}\right) = 4 \log x - \log 16$$

produce the same graph. To do this, you need to make some decisions.
   a. What should Y_1 be?   b. What should Y_2 be?
2. Graph $f(x) = \log x$ using a $[-5, 5]$ by $[-5, 5]$ window and, based on the graph, answer the following questions regarding the logarithmic function $f(x) = \log x$.
   a. What is the domain of $f(x)$?   b. What is the range of $f(x)$?   c. What is the x-intercept of $f(x)$?
   d. Is there a y-intercept?   e. What happens to $f(x)$ as x increases?

Boost your grade at **mathzone.com**!
> Practice Problems
> NetTutor
> Self-Tests
> e-Professors
> Videos

> Exercises 10.4

〈A〉 Graphing Logarithmic Functions   In Problems 1–8, graph the equation.

1. $y = \log_2 x$

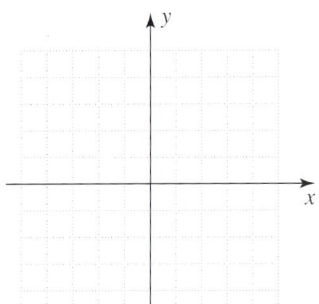

2. $y = \log_3 x$

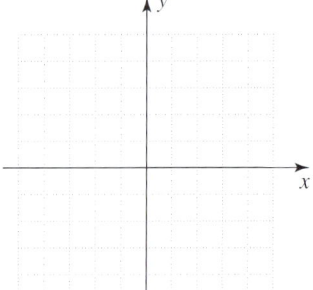

3. $y = \log_5 x$

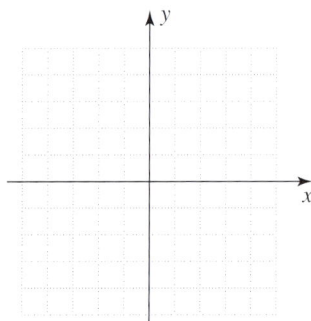

4. $f(x) = \log_6 x$

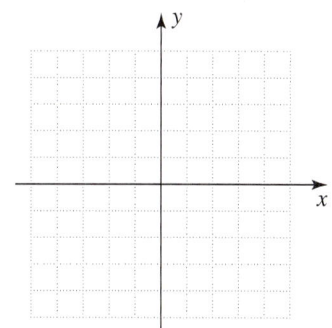

5. $f(x) = \log_{1/2} x$

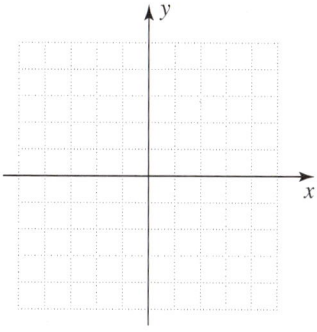

6. $f(x) = \log_{1/3} x$

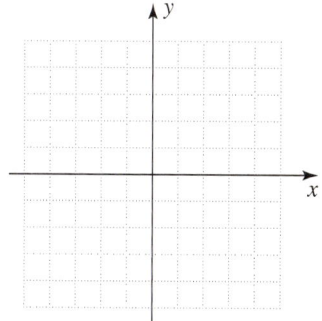

7. $y = \log_{1.5} x$

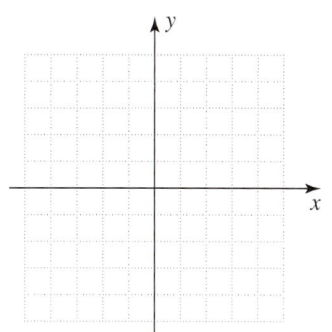

8. $y = \log_{2.5} x$

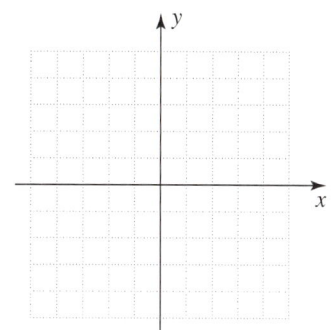

‹B› Converting Exponential to Logarithmic Equations and Vice Versa
In Problems 9–18, write the equation in logarithmic form.

9. $2^x = 128$

10. $3^x = 81$

11. $10^t = 1000$

12. $10^{-t} = 0.001$

13. $81^{1/2} = 9$

14. $16^{1/2} = 4$

15. $216^{1/3} = 6$

16. $64^{1/6} = 2$

17. $e^3 = t$

18. $e^2 = 7.389056$

In Problems 19–32, write the equation in exponential form and check its accuracy (if possible).

19. $\log_9 729 = 3$

20. $\log_7 343 = 3$

21. $\log_2 \frac{1}{256} = -8$

22. $\log_5 \frac{1}{125} = -3$

23. $\log_{81} 27 = \frac{3}{4}$

24. $\log_{64} 32 = \frac{5}{6}$

25. $x = \log_4 16$

26. $t = \log_5 10$

27. $-2 = \log_{10} 0.01$

28. $-3 = \log_{10} 0.001$

29. $\log_e 30 \approx 3.4012$

30. $\log_e 40 \approx 3.6889$

31. $\log_e 0.3166 \approx -1.15$

32. $\log_e 0.2592 \approx -1.35$

‹C› Solving Logarithmic Equations
In Problems 33–50, solve the equation.

33. $\log_3 x = 2$

34. $\log_2 x = 3$

35. $\log_3 x = -3$

36. $\log_2 x = -3$

37. $\log_4 x = \frac{1}{2}$

38. $\log_9 x = \frac{1}{2}$

39. $\log_8 x = \frac{1}{3}$

40. $\log_x 4 = 2$

41. $\log_x 16 = 4$

42. $\log_x 8 = 3$

43. $\log_x 27 = 3$

44. $\log_8 \frac{1}{8} = x$

45. $\log_2 \frac{1}{4} = x$

46. $\log_3 1 = x$

47. $\log_{16} \frac{1}{2} = x$

48. $\log_{32} \frac{1}{2} = x$

49. $\log_3 \frac{1}{9} = x$

50. $\log_2 \frac{1}{8} = x$

In Problems 51–74, find the value of the logarithm.

51. $\log_2 256$ **52.** $\log_2 128$ **53.** $\log_3 81$ **54.** $\log_3 243$

55. $\log_2 \frac{1}{8}$ **56.** $\log_3 \frac{1}{27}$ **57.** $\log_{10} 1{,}000{,}000$ **58.** $\log_{10} 0.001$

59. $\log_3 1$ **60.** $\log_{10} 1$ **61.** $\log_{10} 10$ **62.** $\log_2 2$

63. $\log_e e$ **64.** $\log_e \frac{1}{e}$ **65.** $\log_5 \frac{1}{5}$ **66.** $\log_{27} 3$

67. $\log_8 2$ **68.** $\log_e e^2$ **69.** $\log_e e^{-3}$ **70.** $\log_e e^{-5}$

71. $\log_{10} 10^t$ **72.** $\log_e e^x$ **73.** $\log_4 4^t$ **74.** $\log_5 5^t$

⟨ D ⟩ **Using Properties of Logarithms** In Problems 75–84, use the properties of logarithms to transform the left-hand side into the right-hand side of the stated equation. Assume all the logarithms are to base 10.

75. $\log \frac{26}{7} - \log \frac{15}{63} + \log \frac{5}{26} = \log 3$

76. $\log 9 - \log 8 - \log \sqrt{75} + \log \sqrt{\frac{25}{27}} = -3 \log 2$

77. $\log b^3 + \log 2 - \log \sqrt{b} + \log \frac{\sqrt{b^3}}{2} = 4 \log b$

78. $\log k^2 - \log k^{-2} - \log \sqrt{k} - \log k^{-1} = \frac{9}{2} \log k$

79. $\log k^{3/2} + \log r - \log k - \log r^{3/4} = \frac{1}{4}(\log k^2 r)$

80. $\log a - \frac{1}{6} \log b - \frac{1}{2} \log a + \frac{1}{3} \log b = \frac{1}{6} \log a^3 b$

81. $2 \log b + 6 \log a - 3 \log b = \log\left(\frac{a^6}{b}\right)$

82. $2 \log b + 2 \log a - 4 \log b = \log\left(\frac{a^2}{b^2}\right)$

83. $\frac{1}{3} \log x + \frac{1}{3} \log y^2 - \log z = \log\left(\frac{\sqrt[3]{xy^2}}{z}\right)$

84. $\frac{1}{4} \log x^3 + \frac{1}{4} \log y - 2 \log z = \log\left(\frac{\sqrt[4]{x^3 y}}{z^2}\right)$

⟨ E ⟩ Applications Involving Logarithmic Functions

85. *Earthquakes* The worst earthquake ever recorded occurred in the Pacific Ocean near Colombia in 1906. The intensity of this earthquake was $10^{8.9}$ times as great as that of an earthquake of minimum intensity I_0. What was the magnitude of this earthquake on the Richter scale? (Recall the formula in the *Getting Started*, $R = \log_{10} \frac{I}{I_0}$.)

86. *Earthquakes* The San Francisco earthquake of 1906 was $10^{8.3}$ times as intense as an earthquake of minimum intensity I_0. What was the magnitude of the 1906 San Francisco earthquake on the Richter scale? (Recall the formula in the *Getting Started*, $R = \log_{10} \frac{I}{I_0}$.)

87. *Compounding* When Johnny was born, his father deposited $5000 in an account that paid 6% compounded monthly. This was to help pay Johnny's college expenses starting on his eighteenth birthday. The compound amount in this account was $A = \$5000(1.005)^{216}$. If $\log_{10} 5000 \approx 3.69897$ and $\log_{10} 1.005 \approx 0.00217$, find $\log_{10} A$ to five decimal places using the properties of logarithms.

88. *Compounding* Suppose that the account in Problem 87 paid 10% compounded quarterly. The compound amount would then be given by $A = \$5000(1.025)^{72}$. If $\log_{10} 1.025 \approx 0.01072$, find $\log_{10} A$ to five decimal places using the properties of logarithms.

89. *Height of boys* By the age of 2, most children have reached about 50% of their mature height. Thus, if you measure the height of a 2-year-old child and double this height, the result should be close to the child's mature height. The percent P of adult height attained by a boy can be approximated by the function

$$P(A) = 29 + 50 \log_{10}(A + 1)$$

where A is the age in years ($0 < A < 17$).

Use a calculator to answer the following questions. (Approximate the answers to the nearest percent.)

a. What percent of his mature height is a 2-year-old boy?

b. What percent of his mature height is an 8-year-old boy?

c. What percent of his mature height is a 16-year-old boy?

d. Graph $P(A)$.

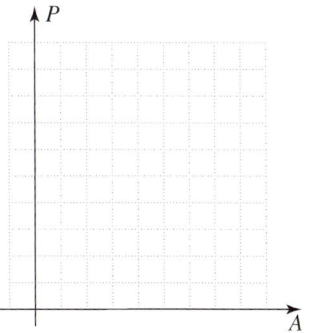

90. *Height of girls* The height H of a girl (in inches) can be approximated by the function $H(A) = 11 + 19.44 \log_e A$, where A is the age in years and $6 < A < 15$.

a. What should the height of a 7-year-old girl be?

b. What should the height of an 11-year-old girl be?

c. What should the height of a 13-year-old girl be?

d. Graph $H(A)$. Remember that the domain is $\{A \mid 6 < A < 15\}$.

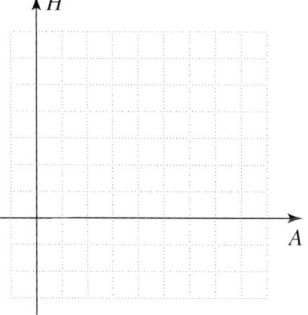

e. If a 12-year-old boy is 60 inches tall, how tall, to the nearest inch, would you expect him to be when he is an adult?

Threshold Limit Values Use the following information for Problems 91–94. The threshold limit values (TLVs) for noise are designed to protect the population against a noise induced hearing loss after a lifetime of occupational exposure. The formula for the TLV is

$$\text{TLV} = \log_{10} t, \text{ where } t \text{ is the duration of the noise in seconds per day}$$

If the value exceeds 1, the TLV is considered to be exceeded!

Source: American Conference of Governmental Industrial Hygienists. 1994–1995 Threshold Limit Values for Chemical and Physical Agents and Biological Exposure Indices, 1994, pages 104–105.

91. If the noise in your job lasts 300 seconds per day, use the properties of logarithms and the fact that $\log_{10} 3 \approx 0.4771$ to find the TLV. Is the TLV greater than 1?

92. The TLV can also be approximated by $\text{TLV} = -0.100868 \text{ dB} + 13.0521$, where dB is the sound level in decibels. Use this formula to approximate the TLV of a 105-decibel sound. Is the answer close to the one you obtained in Problem 91?

93. a. Use TLV = $\log_{10} t$ and $\log_{10} 9 \approx 0.9542$ to find the TLV of a sound lasting 900 seconds per day.
b. Use TLV = -0.100868 dB $+ 13.0521$ to find the TLV of a 100-decibel sound.

94. a. Use TLV = $\log_{10} t$ and $\log_{10} 18 \approx 1.2553$ to find the TLV of a sound lasting 1800 seconds per day.
b. Use TLV = -0.100868 dB $+ 13.0521$ to find the TLV of a 97-decibel sound.

Volcanic Eruptions Use the following information for Problems 95–96. The intensity of a volcanic eruption depends on the mass of the lava flow in kilograms per second (kg/sec) and is given by the formula: Intensity = $\log_{10} m + 3$, where m is the mass eruption rate (lava flow) in kilograms per second. The intensity scale is from 1 (lowest) to 12 (highest).
Source: Pyle DM. Sizes of volcanic eruptions, pages 263–269 (page 264). In: Sigurdsson H, Houghton B, et al. (editors). *Encyclopedia of Volcanoes*. Academic Press, 2000.

95. Find the intensity of a volcanic eruption in which the lava flow was 10,000 kilograms per second.

96. The Kilauea volcano in Hawaii had an eruption rate of 1 million kilograms per second in 1965. What was the intensity of this eruption?

>>> Using Your Knowledge

Which Is Better? One way of comparing the compound amounts in Problems 87 and 88 is to find their ratio. If we denote this ratio by R, then

$$R = \frac{5000(1.005)^{216}}{5000(1.025)^{72}} = \frac{(1.005)^{216}}{(1.025)^{72}}$$

so that

$$\log_{10} R = \log_{10}(1.005)^{216} - \log_{10}(1.025)^{72}$$
$$= 216 \log_{10} 1.005 - 72 \log_{10} 1.025$$

Using the values of the logarithms given in Problems 87 and 88, we get

$$\log_{10} R \approx (216)(0.00217) - (72)(0.01072) = -0.30312$$

The negative value for $\log_{10} R$ means that R is less than 1. Can you see why? This means that the compound amount in Problem 87 is less than that in Problem 88. (A calculator gives the value of R as about 0.49631 so that the first amount is less than half the second.)

97. Compare an investment of $1000 at 8% compounded quarterly with the same amount invested at 8.5% compounded annually for a period of 10 years. You will need the values $\log_{10} 1.02 \approx 0.00860$ and $\log_{10} 1.085 \approx 0.03543$.

98. Compare an investment of $1000 at 6% compounded quarterly with the same amount invested at 6.25% compounded annually for a period of 10 years. You will need the values $\log_{10} 1.015 \approx 0.00647$ and $\log_{10} 1.0625 \approx 0.02633$.

>>> Write On

99. Explain the relationship between the exponential function $f(x) = b^x$ and the logarithmic function $g(x) = \log_b x$.

100. Explain why the logarithm of a negative number is not defined.

101. Explain why the number 1 is not allowed as a base for the logarithmic function $f(x) = \log_b x$.

102. Explain why $\log_b 1 = 0$.

>>> Concept Checker

Fill in the blank(s) with the correct word(s), phrase, or mathematical statement.

103. $f(x) = y = \log_b x$ is equivalent to _____. ($b > 0$, $b \neq 1$, and $x > 0$)

104. $b^x = b^y$ is equivalent to _____ for $b > 0$, $b \neq 1$.

b

x

0

$\log_b M + \log_b N$

105. $\log_b MN = $ _____.
106. $\log_b \dfrac{M}{N} = $ _____.
107. $\log_b M^r = $ _____.
108. $\log_b 1 = $ _____.
109. $\log_b b = $ _____.
110. $\log_b b^x = $ _____.

$r \log_b M$ $b^x = y$

$M \log_b r$ $b^y = x$

$\log_b N - \log_b M$ 1

$\log_b M - \log_b N$ $x = y$

>>> Mastery Test

111. A recent California earthquake was $10^{6.4}$ times as intense as that of an earthquake of minimum intensity I_0. What was the magnitude of this earthquake on the Richter scale? Recall: $\left(R = \log_{10} \dfrac{I}{I_0}\right)$.

In Problems 112–113, use the properties of logarithms to justify the equation:

112. $\log 6 - \log \sqrt{2} + \log \sqrt{50} = \log 30$

113. $2 \log x + 4 \log y - 6 \log z = \log\left(\dfrac{xy^2}{z^3}\right)^2$

114. Write the equation $\log_4 32 = \tfrac{5}{2}$ in exponential form

115. Write the equation $2^{10} = 1024$ in logarithmic form

Find the logarithm.

116. $\log_2 64$ 117. $\log_2 \dfrac{1}{8}$ 118. $\log_{10} 100$ 119. $\log_{10} 0.1$

Solve the logarithmic equation.

120. $\log_3 x = 4$ 121. $\log_3 x = -2$ 122. $\log_x 16 = 2$ 123. $\log_3 \dfrac{1}{27} = x$ 124. $\log_{16} x = \dfrac{1}{4}$

Graph.

125. $y = \log_7 x$

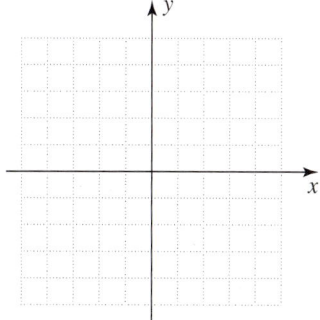

126. $f(x) = \log_8 x$

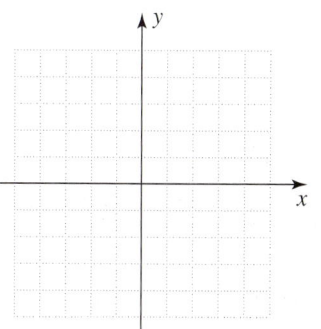

>>> Skill Checker

Write in scientific notation:

127. 32.68   128. 326.8   129. 0.002387

130. 0.0004392   131. 0.0000569   132. 0.000006731

10.5 Common and Natural Logarithms

Objectives

A Find common logarithms.

B Find natural logarithms.

C Change the base of a logarithm.

D Graph exponential and logarithmic functions base e.

E Solve applications involving common and natural logarithms.

To Succeed, Review How To . . .

1. Write numbers in scientific notation (pp. 45–46).
2. Write a radical expression in exponential form (pp. 528–531).

Getting Started
How Many Decibels Was That?

The scale for measuring the loudness of sound is called the *decibel* scale. When the loudness L of a sound of intensity I is measured in decibels (dB), it is expressed as

$$L = 10 \log_{10} \frac{I}{I_0}$$

where I_0 is the minimum intensity detectable by the human ear. For example, the sound of a riveting machine 30 feet away is 10^{10} times as intense as the minimum intensity I_0, and hence its loudness in decibels is expressed as

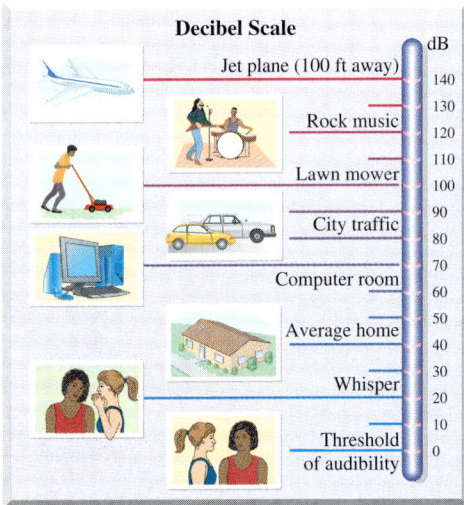

$$L = 10 \log_{10} \frac{10^{10} I_0}{I_0} = 10 \log_{10} 10^{10} = (10 \cdot 10) \log_{10} 10$$

$$= 100 \text{ dB}$$

The decibel scale uses logarithms to the base 10; these are called *common* logarithms and we shall study them and another type of logarithm called *natural* logarithms in this section.

A ⟩ Finding Common Logarithms

The decibel scale is just one example of the many applications of logarithms that use the base 10. Logarithms to base 10 are called **common logarithms** because they use the same base as our "common" decimal system; it's customary to omit the base when working with these logarithms. Be sure to keep in mind that when the base is omitted,

$$\log M = \log_{10} M$$

↑
no base (10 is understood)

10.5 Common and Natural Logarithms

If you have a scientific calculator, you can find the logarithm of a number by using the LOG key. To find log 396 using a scientific calculator, enter 396 and press LOG. The result is 2.5976952 or, to four decimal places, 2.5977. Before calculators were widely used, the logarithm of a number was found using a table.

We should be able to estimate the common log for a number like 396 even without a calculator or table if we compare it to the following list of powers of 10 and their related logarithmic fact.

$$10^0 = 1 \qquad \log 1 = 0$$
$$10^1 = 10 \qquad \log 10 = 1$$
$$10^2 = 100 \qquad \log 100 = 2$$
$$10^3 = 1000 \qquad \log 1000 = 3$$

Since 396 is a number between 100 and 1000 we anticipate its common logarithm to be a number between 2 and 3.

$$10^2 = 100 \qquad \log 100 = 2$$
$$10^? = 396 \qquad \log 396 \approx ? \quad 2.\_\_$$
$$10^3 = 1000 \qquad \log 1000 = 3$$

This is a good way to see if your answer is reasonable when using a calculator to find common logarithms.

EXAMPLE 1   Finding common logarithms
Find the value, rounded to four decimal places:

a. log 42,500   **b.** log 0.000425

SOLUTION

a. Enter 42,500 and press LOG to obtain

$$\log 42{,}500 \approx 4.6284 \quad \text{Rounded to four decimal places}$$

$10^4 = 10{,}000$ and $10^5 = 100{,}000$. Since 42,500 is between 10,000 and 100,000, its logarithm should be 4.____.

b. Enter 0.000425 and press LOG to obtain

$$\log 0.000425 \approx -3.3716$$

PROBLEM 1
Find the value, rounded to four decimal places:

a. log 734,000

b. log 0.0000734

Calculator Corner

Find Common Logarithms

To do Example 1(a), go to your home screen. Now to find log 42,500, press LOG; then enter 42,500 and press ENTER. You will obtain 4.62838893, which can be rounded to the desired number of decimal places.

As you recall, the inverse of a logarithmic function is an exponential function. The inverse of a logarithm is an **antilogarithm** or **inverse logarithm**. To find the inverse logarithm, we find the power of the base. Thus,

$$\text{if} \quad f(x) = \log x, \quad f^{-1}(x) = \text{inverse log } x = 10^x$$

Answers to PROBLEMS

1. **a.** 5.8657 **b.** −4.1343

For example, if log x = 2.5105, then inverse log 2.5105 = 10^x = $10^{2.5105}$. Most calculators don't have an antilog key, so you have to know that to find the inverse, you must use the 10^x key, if your calculator has one. If it doesn't you can use the x^y key. Sometimes, the LOG key is used as the 10^x key after using the 2nd or INV key. In either case, to find inverse log 2.5105, enter 2.5105 and press 2nd LOG. The result is 324 (rounded to the nearest whole number).

EXAMPLE 2 Finding inverse logarithms
Find the value (round to four decimal places):

a. inv log 0.8176
b. inv log(−2.1824)

SOLUTION

a. Using a scientific calculator, enter 0.8176 and press 2nd LOG to obtain

inverse log 0.8176 ≈ 6.5705   *Rounded to four decimal places*

b. Using a scientific calculator, enter −2.1824 2nd LOG to obtain

inverse log(−2.1824) ≈ 0.0066   *Rounded to four decimal places*

To enter −2.1824, you have to enter 2.1824 and press the $+/-$ key (if available) to change the sign.

PROBLEM 2
Find the value (round to four decimal places):

a. inv log 0.6243
b. inv log(−3.1963)

Calculator Corner

Find Inverse Logarithm

To find the inv $\log_{10}$ on a graphing calculator, we use the 10^x feature. Thus, to find inv $\log_{10}$ 0.8176, press 2nd LOG . 8 1 7 6 and ENTER to obtain 6.57052391.

B ▶ Finding Natural Logarithms

The irrational number e ≈ 2.718281828 was discussed in Section 10.3. This number is used as the base of an important system of logarithms called **natural logarithms** or **Napierian logarithms** in honor of John Napier (1550–1617), the inventor of logarithms (see *The Human Side of Algebra* at the beginning of this chapter). To distinguish natural from common logarithms, the abbreviation ln (pronounced *el-en*) is used instead of $\log_e$.

NATURAL LOGARITHMIC FUNCTION

The natural logarithmic function is defined by $f(x) = \ln x$, where ln x means $\log_e x$ and $x > 0$.

The calculator key LN or ln x is used for natural logarithms.

EXAMPLE 3 Finding a natural logarithm
Find the value of ln 3252 (round to four decimal places).

SOLUTION
Enter 3252 and press LN to obtain

ln 3252 ≈ 8.0870   *Rounded to four decimal places*

PROBLEM 3
Find the value of ln 4563 (round to four decimal places).

Answers to PROBLEMS

2. a. 4.2102 b. 0.0006
3. 8.4257

Inverse logs base *e* can be found using the e^x key, if your calculator has one. If it doesn't, use the x^y key and approximate *e*. Sometimes, the LN key is used as the e^x key after using the 2nd or INV key.

EXAMPLE 4 Finding the inverse log of a natural logarithm
Find the value of inv ln(−3.4865) (round to four decimal places).

SOLUTION Enter −3.4865 (you may have to enter 3.4865 and press +/−). Now, to find the inv ln, press 2nd (or INV) LN to obtain

$$\text{inv ln}(-3.4865) \approx 0.0306 \quad \text{Rounded to four decimal places}$$

PROBLEM 4
Find the value of inv ln(−4.3874) (round to four decimal places).

C › Changing Logarithmic Bases

Calculators are used to find log *x* and ln *x*, but can we find $\log_4 x$ or $\log_5 x$? To do this, we need the following conversion formula.

CHANGE-OF-BASE FORMULA

For any logarithms with base *a* and *b* and any number $M > 0$,

$$\log_b M = \frac{\log_a M}{\log_a b}$$

This can be proved as follows: Let $x = \log_b M$. By the definition of logarithm,

$$b^x = M$$
$$\log_a b^x = \log_a M \quad \text{Take the logarithm base } a \text{ of both sides.}$$
$$x \log_a b = \log_a M \quad \text{Use the power rule on the left side.}$$
$$x = \frac{\log_a M}{\log_a b} \quad \text{Solve for } x.$$
$$\log_b M = \frac{\log_a M}{\log_a b} \quad \text{Since we let } x = \log_b M$$

When using calculators to do Examples 5 and 6, you can avoid rounding errors by not rounding the intermediate values of $\log_{10} 20$ and $\log_{10} 4$ in Example 5 and ln 20 and ln 4 in Example 6. The calculator keystrokes are indicated in the solution to each example.

EXAMPLE 5 Using the change-of-base formula with common logarithms
Using common logarithms, find $\log_4 20$.

SOLUTION Using the change-of-base formula with $b = 4$, $a = 10$,

$$\log_4 20 = \frac{\log_{10} 20}{\log_{10} 4}$$

We then use a calculator to find $\log_{10} 20 \approx 1.3010$ and $\log_{10} 4 \approx 0.6021$ and substitute the values in the equation. Thus,

$$\log_4 20 = \frac{\log_{10} 20}{\log_{10} 4}$$
$$\approx 2.1608$$

To avoid rounding errors, **do not** round the intermediate values $\log_{10} 20$ and $\log_{10} 4$. For example, enter

2 0 log ÷ 4 log =

and then round the answer 2.160964047 to 2.1610.

PROBLEM 5
Use common logarithms to find $\log_5 20$.

Answers to PROBLEMS

4. 0.0124   5. 1.8614

EXAMPLE 6 Using the change-of-base formula with natural logarithms

Using natural logarithms, find $\log_4 20$.

SOLUTION Using the change-of-base formula with $b = 4$, $a = e$,

$$\log_4 = \frac{\ln 20}{\ln 4}$$

We then use a calculator to find $\ln 20 \approx 2.9957$ and $\ln 4 \approx 1.3863$ and substitute the values in the equation. Thus,

$$\log_4 20 = \frac{\ln 20}{\ln 4}$$
$$\approx 2.1609$$

To avoid rounding errors, **do not** round the intermediate values $\ln 20$ and $\ln 4$. For example, enter

2 0 [LN] ÷ 4 [LN] [=]

and *then* round the answer 2.160964047 to 2.1610. You will obtain the same answer as in Example 5.

PROBLEM 6
Use natural logarithms to find $\log_5 20$.

Answers to PROBLEMS
6. 1.8614

NOTE
It's best to wait until the final step to round the answer.

D › Graphing Exponential and Logarithmic Functions Base e

The logarithmic function $f(x) = \ln x$ is the **inverse** of the exponential function $f(x) = e^x$. Because of that, they are reflections across the line $y = x$ as shown in Figure 10.13. What will the graphs of $f(x) = e^{ax}$, $f(x) = -e^{ax}$, and $f(x) = e^x + b$ look like? The answers are given in Table 10.1.

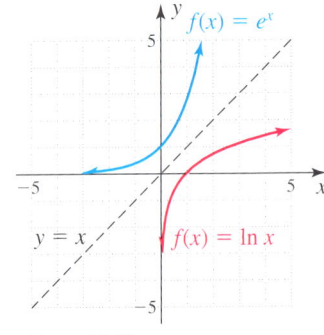

> Figure 10.13

Table 10.1 Graphs of $f(x) = e^{ax}$, $f(x) = -e^{ax}$, and $f(x) = e^x + b$

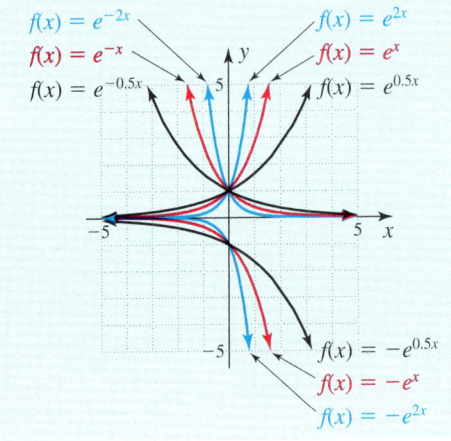

1. The graph of $f(x) = e^{ax}$, where a is positive, looks like that of $f(x) = e^x$. The larger the value of a, the "steeper" the graph. The graph of $f(x) = e^{2x}$ is steeper than the graph of $f(x) = e^{0.5x}$.

2. The graph of $f(x) = -e^{ax}$, where a is positive, is the reflection across the x-axis of the graph of $f(x) = e^{ax}$. The graph of $f(x) = -e^{2x}$ is the reflection across the x-axis of the graph of $f(x) = e^{2x}$.

3. The graph of $f(x) = e^{-ax}$, where a is positive, is the reflection across the y-axis of the graph of $f(x) = e^{ax}$. The graph of $f(x) = e^{-2x}$ is the reflection across the y-axis of the graph of $f(x) = e^{2x}$.

Table 10.1 (continued)

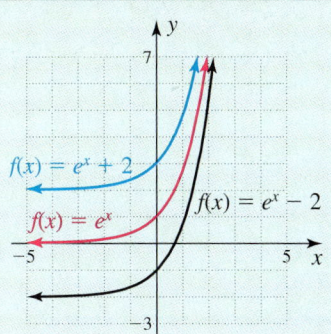

The graph of $f(x) = e^x + b$ is identical to that of $f(x) = e^x$ shifted b units (upward when b is positive, downward when b is negative).

EXAMPLE 7 Graphing exponential functions

Graph:

a. $f(x) = e^{(1/3)x}$ **b.** $f(x) = -e^{(1/3)x}$ **c.** $f(x) = e^x + 3$

SOLUTION

a. We use convenient values for x (say $x = -3, 0,$ and 3) and find the $f(x) = y$-values using a calculator; then we plot the resulting points and draw the graph (Figure 10.14). For example, when $x = 3$, $f(3) = e^1 \approx 2.72$.

x	$y = e^{(1/3)x}$
−3	0.37
0	1
3	2.72

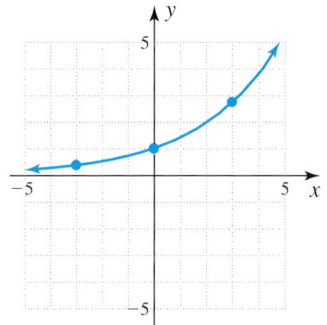

> Figure 10.14

b. Before you draw the graph, recall that $f(x) = e^{(1/3)x}$ and $f(x) = -e^{(1/3)x}$ are reflections of each other across the x-axis. As before, use convenient points such as $x = -3, 0,$ and 3 but note that the y-values are the *negatives* of the y-values obtained in part **a.** Plot the resulting points and draw the graph (Figure 10.15).

x	$y = -e^{(1/3)x}$
−3	−0.37
0	−1
3	−2.72

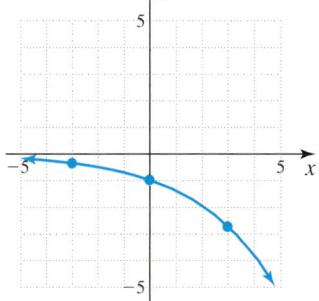

> Figure 10.15

PROBLEM 7

Graph:

a. $f(x) = e^{(1/4)x}$

b. $f(x) = -e^{(1/4)x}$

c. $f(x) = e^{(1/4)x} + 2$

(continued)

Answers to PROBLEMS

7.

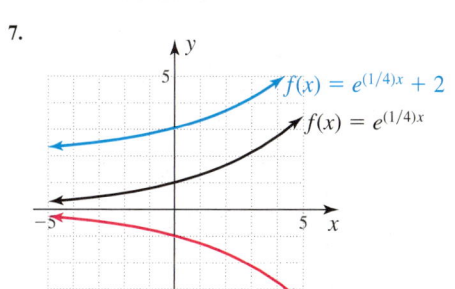

c. The graph of $f(x) = e^x + 3$ is identical to that of $f(x) = e^x$ but shifted 3 units up. Using $x = -1, 0$, and 1 and a calculator, we find the points shown in the table. We plot these points to obtain the graph shown in Figure 10.16.

x	y = e^x + 3
-1	3.37
0	4
1	5.72

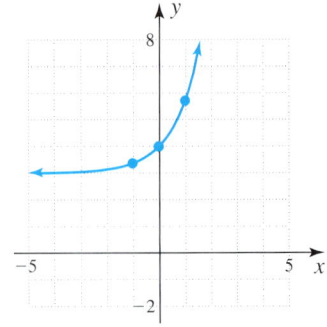

> Figure 10.16

Calculator Corner

Graphing Exponential Functions

To do Example 7(a) with a TI-83 Plus, press WINDOW and set Xmin = -5, Xmax = 5, Xscl = 1, Ymin = 5, Ymax = 5, and Yscl = 1. To select the function to be entered, press Y= and enter 2nd LN (1 ÷ 3) X,T,θ,n) . Now press GRAPH. The result is shown in the window.

You can graph Example 7(b), $f(x) = -e^{(1/3)x}$ by entering $Y_2 =$ (-) VARS ▶ 1 1 GRAPH, that is, by making $Y_2 = -Y_1 = -e^{(1/3)x}$. $f(x) = e^{(1/3)x}$ and $f(x) = -e^{(1/3)x}$ are reflections of each other across the x-axis, as shown in the second window.

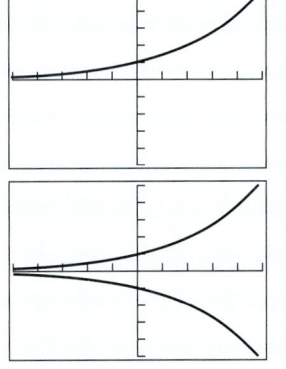

CAUTION

Remember that $\ln(x + 1)$ and $\ln x + 1$ are **different**. The 1 added inside the parentheses in $\ln(x + 1)$ shifts the curve $\ln x$ 1 unit *left*. The 1 added to $\ln x$ in $\ln x + 1$ shifts the curve 1 unit *up*. In general, the *a* in $\ln(x + a)$ shifts the curve *a* units *right* or *left* (right if *a* is *negative*, left if *a* is *positive*). The *a* in $\ln x + a$ shifts the curve *up* or *down* (up if *a* is *positive*, down if *a* is *negative*).

We use these ideas next.

EXAMPLE 8 Graphing logarithmic functions
Graph:

a. $f(x) = \ln(x + 2)$ **b.** $f(x) = \ln x + 2$

SOLUTION

a. The graph of $f(x) = \ln(x + 2)$ is the same as that of $f(x) = \ln x$ shifted 2 units left. Since $\ln x$ is only defined for positive values of *x*, we must select *x*'s that make $x + 2$ positive so that $\ln(x + 2)$ is defined. This means that $x + 2 > 0$, so $x > -2$. Using convenient values for *x*, say $x = 0, 1,$ and 2 and a calculator to find the corresponding *y*-values, we construct a table and then plot the resulting points to obtain the graph shown in Figure 10.17.

PROBLEM 8
Graph:

a. $f(x) = \ln(x + 1)$ **b.** $f(x) = \ln x + 1$

Answer on page 803

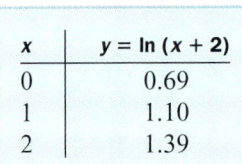

x	y = ln (x + 2)
0	0.69
1	1.10
2	1.39

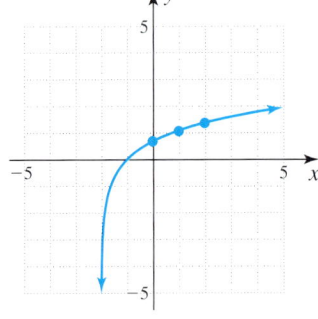

> Figure 10.17

b. The graph of $f(x) = \ln x + 2$ is the same as that of $f(x) = \ln x$ shifted 2 units up. Using the values $x = 1, 2,$ and 3 and a calculator, we obtain the numbers in the table, plot the resulting points, and draw the graph (Figure 10.18).

x	y = ln x + 2
1	2
2	2.69
3	3.10

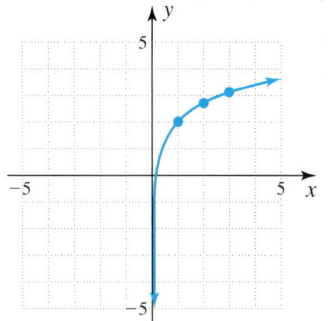

> Figure 10.18

Calculator Corner

Graphing Logarithmic Functions

To do Example 8(a), enter $Y_1 = \ln(x + 2)$, making sure you use parentheses. The result using a $[-5, 5]$ by $[-5, 5]$ window is shown in the window.

The domain consists of all real numbers x greater than -2. The line $x = -2$ is an *asymptote* for the function.

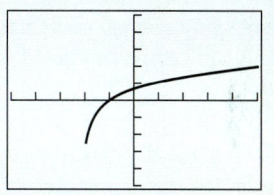

E › Applications Involving Common and Natural Logarithms

Here are some examples of the many problems that involve logarithms for their solution.

EXAMPLE 9 Finding the pH of a solution

In chemistry, the pH (a measure of acidity) of a solution is defined by the formula $pH = -\log[H^+]$, where $[H^+]$ is the hydrogen ion concentration of the solution in moles per liter. Find the pH of a solution for which $[H^+] = 8 \times 10^{-6}$. Round to four decimal places.

PROBLEM 9

Find the pH of a solution for which $[H^+] = 4.2 \times 10^{-7}$. Round to four decimal places.

(continued)

Answers to PROBLEMS

8.

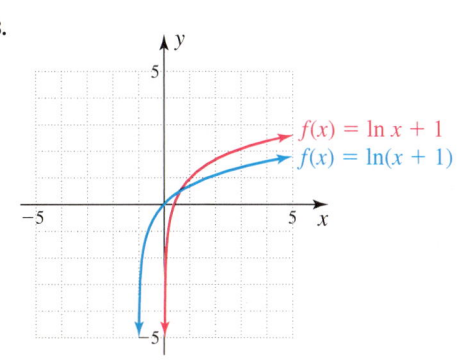

9. 6.3768

SOLUTION The pH is defined to be $-\log[H^+]$, so for this solution

$$\begin{aligned}
pH &= -\log(8 \times 10^{-6}) \\
&= -(\log 8 + \log 10^{-6}) \\
&= -(\log 8 - 6) \\
&= 6 - \log 8 \\
&\approx 6 - 0.9031 \\
&\approx 5.0969
\end{aligned}$$

The pH of the solution is 5.0969.

EXAMPLE 10 Diastolic pressure

Have you taken your blood pressure lately? Normal blood pressures are stated as 120/80, where 120 is the *systolic* pressure (when the heart is contracting) and 80 is the *diastolic* pressure (when the heart is relaxed). Over short periods, the diastolic pressure in the aorta of a normal adult is a function of time and can be approximated by the equation

$$P = 90e^{-0.5t}$$

a. What is the diastolic pressure when the aortic valve is closed ($t = 0$)?

b. What is the diastolic pressure when $t = 0.3$ sec? Round to two decimal places.

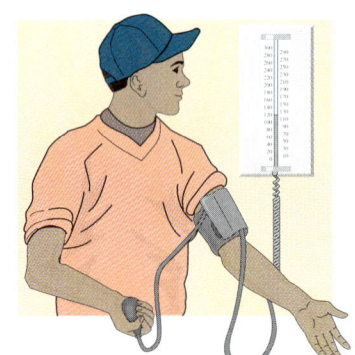

PROBLEM 10

What is the diastolic pressure when:

a. $t = 0.35$?

b. $t = 0.4$?

Round to two decimal places.

SOLUTION

a. At $t = 0$, $P = 90e^{-0.5(0)} = 90e^0 = 90(1) = 90$

b. When $t = 0.3$,

$$\begin{aligned}
P &= 90e^{-0.5(0.3)} \\
&= 90e^{-0.15} \\
&= 90(0.8607) \quad \text{Use a calculator.} \\
&\approx 77.46
\end{aligned}$$

Calculator Corner

1. Use your calculator to confirm the results of Example 9.
2. Graph the function $f(x) = 90e^{-0.5x}$ of Example 10.
   a. If you have a TI-83 Plus, find the value of $f(x)$ at $x = 0.3$ by pressing [2nd] [TRACE] 1 and entering 0.3 for the *x*-value.
   b. What happens to $f(x)$ as x increases?
   c. What does that mean for your diastolic blood pressure?
   d. What is the *y*-intercept for $f(x)$, and what does that mean for your diastolic pressure?
3. Graph the functions $f(x) = e^x$ and $f(x) = e^x + k$ for $k = 1, 2,$ and 3. What happens to the graph of $f(x)$ when a constant k is added to it? (*Hint:* Consider two cases, k positive and k negative.)
4. What happens to the graph of $f(x) = e^x$ when a positive constant k is multiplied by $f(x)$ to obtain $g(x) = ke^x$?

Answers to PROBLEMS

10. a. 75.55 b. 73.69

> Exercises 10.5

<A> **Finding Common Logarithms** In Problems 1–12, find the common logarithm of the given number. (Round to four decimal places.)

1. 74.5
2. 952
3. 1840
4. 3.05
5. 0.0437
6. 0.0673
7. 50.18
8. 94.44
9. 0.01238
10. 0.01004
11. 0.008606
12. 0.0004632

In Problems 13–20, find the inverse logarithm base 10 (inv log). (Round to two decimal places.)

13. inv log 1.2672
14. inv log 2.4409
15. inv log(-2.2328)
16. inv log(-1.4045)
17. inv log 1.4630
18. inv log 2.9408
19. inv log(-0.134)
20. inv log(-1.2275)

<B> **Finding Natural Logarithms** In Problems 21–30, find the natural logarithm. (Round to four decimal places.)

21. ln 3
22. ln 4
23. ln 52
24. ln 62
25. ln 2356
26. ln 208.3
27. ln 0.054
28. ln 0.049
29. ln 0.00062
30. ln 0.00132

In Problems 31–40, find the inverse logarithm (inv ln). (Round to two decimal places.)

31. inv ln 1.2528
32. inv ln 2.2925
33. inv ln 4.1744
34. inv ln 4.6052
35. inv ln 0.0392
36. inv ln 0.0198
37. inv ln(-2.3025)
38. inv ln(-0.3566)
39. inv ln(-4.6051)
40. inv ln(-4.8281)

<C> **Changing Logarithmic Bases** In Problems 41–50, use the change-of-base formula to find the logarithm using (a) common logarithms and (b) natural logarithms. (Round to four decimal places.)

41. $\log_3 20$
42. $\log_3 40$
43. $\log_{100} 40$
44. $\log_{200} 50$
45. $\log_{0.2} 3$
46. $\log_{0.1} 4$
47. $\log_4 0.4$
48. $\log_2 0.16$
49. $\log_{\sqrt{2}} 0.8$
50. $\log_{\sqrt{2}} 0.16$

<D> **Graphing Exponential and Logarithmic Functions Base e** In Problems 51–70, graph the function.

51. $f(x) = e^{3x}$
52. $f(x) = e^{4x}$
53. $f(x) = -e^{3x}$

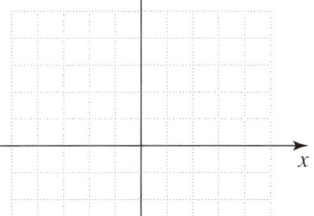
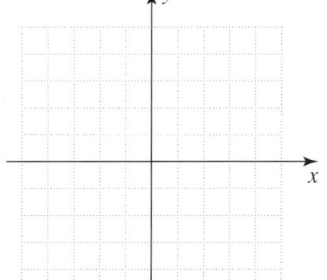
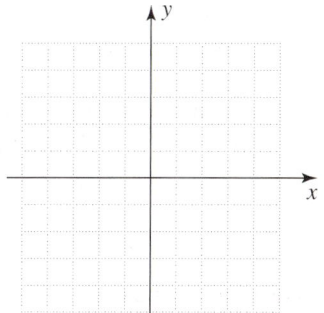

54. $f(x) = -e^{4x}$

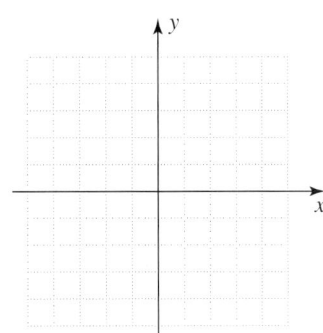

55. $f(x) = e^{-3x}$

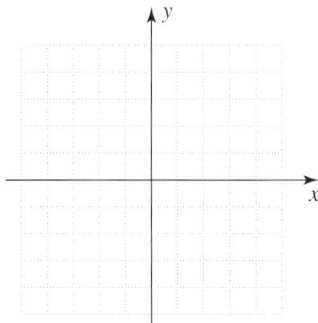

56. $f(x) = e^{-4x}$

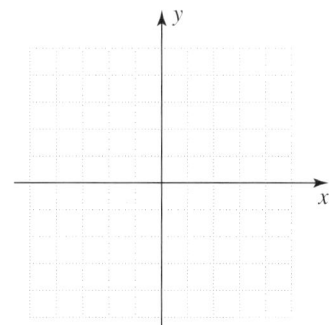

57. $f(x) = e^{(1/2)x}$

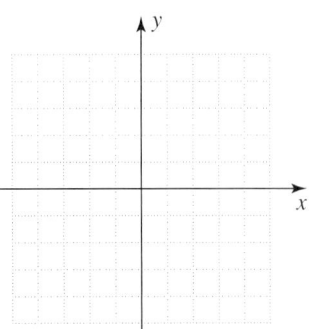

58. $f(x) = -e^{(1/2)x}$

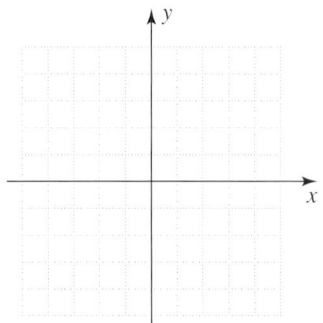

59. $f(x) = e^x + 1$

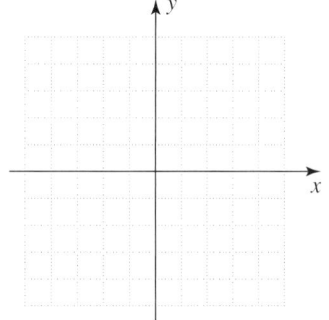

60. $f(x) = e^x - 1$

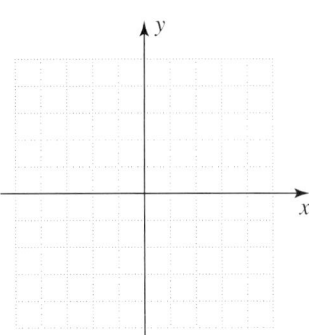

61. $f(x) = e^{0.5x} + 1$

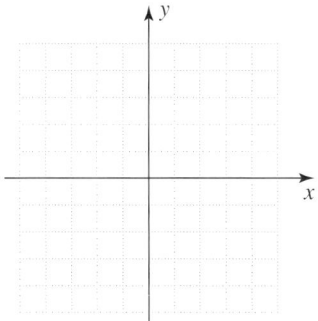

62. $f(x) = e^{0.5x} - 1$

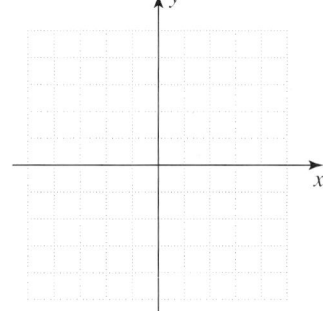

63. $f(x) = 2e^x$

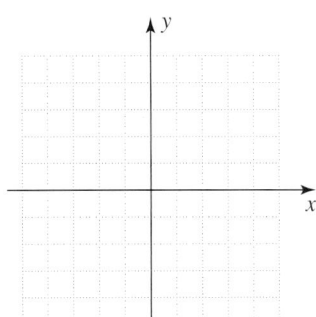

64. $f(x) = -2e^x$

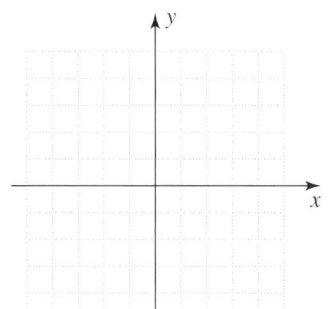

65. $f(x) = \ln(x + 1)$

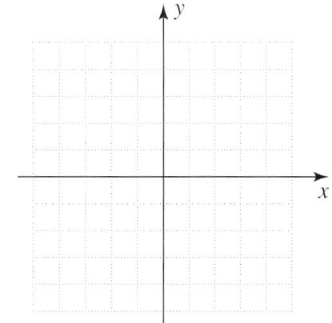

66. $f(x) = \ln x + 1$ **67.** $f(x) = \ln x + 4$ **68.** $f(x) = \ln(x + 4)$

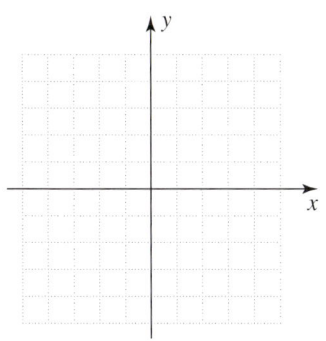 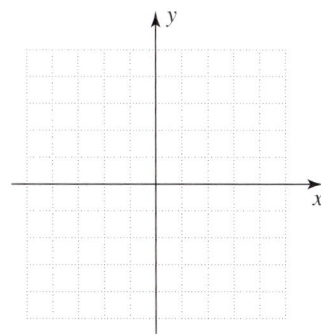 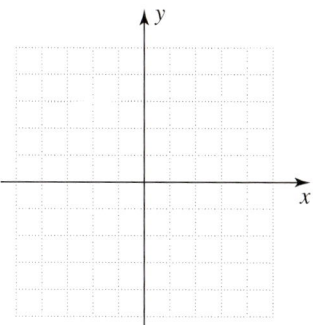

69. $f(x) = \ln(x - 1)$ **70.** $f(x) = \ln x - 1$

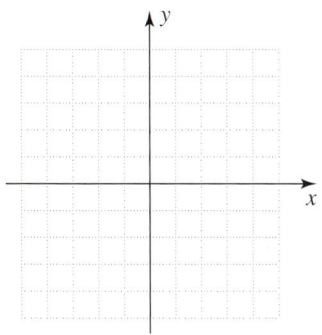 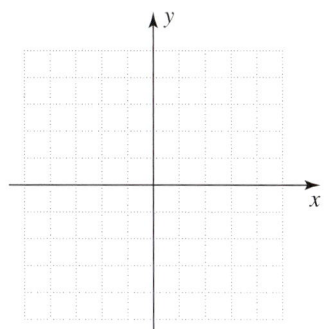

〈 E 〉 Applications Involving Common and Natural Logarithms

pH Levels In Problems 71–76, find the pH of a solution with the given [H⁺]. Round to one decimal place. (*Hint:* pH = $-\log[H^+]$.)

71. $[H^+] = 7 \times 10^{-7}$

72. $[H^+] = 1.5 \times 10^{-9}$

73. Eggs whose $[H^+]$ is 1.6×10^{-8}

74. Tomatoes whose $[H^+]$ is 6.3×10^{-5}

75. Milk whose $[H^+]$ is 4×10^{-7}

76. $[H^+] = 5 \times 10^{-8}$

77. *Hispanic population* The number of Hispanic persons in the United States can be approximated by the equation

$$H = 33{,}000{,}000 e^{0.03t}$$

where *t* is the number of years after 2000.

  a. What was the number of Hispanic persons in 2000?

  b. Using this approximation, what will be the number of Hispanic persons in the United States in the year 2010?

78. *African-American population* The number of African-Americans in the United States can be approximated by the equation

$$B = 36{,}000{,}000 e^{0.01t}$$

where *t* is the number of years after 2000.

  a. What was the number of African-Americans in 2000?

  b. Using this approximation, what will be the number of African-Americans in the United States in the year 2010?

79. *Bacteria* The number of bacteria present in a certain culture can be approximated by the equation

$$B = 50{,}000 e^{0.2t}$$

where *t* is measured in hours and $t = 0$ corresponds to 12 noon. Find the number of bacteria present at the specified times.

  a. noon
  b. 2 P.M.
  c. 6 P.M.

80. *Bacteria* If a bactericide (a bacteria killer) is introduced into a bacteria culture, the number of bacteria can be approximated by the equation

$$B = 50{,}000 e^{-0.1t}$$

where *t* is measured in hours. Find the number of bacteria present

  a. when $t = 0$.
  b. when $t = 1$.
  c. when $t = 10$.

81. **Sales** Sales begin to decline d days after the end of an advertising campaign and can be approximated by the equation

$$S = 1000e^{-0.1d}$$

   a. How many sales will be made on the last day of the campaign—when $d = 0$?

   b. How many sales will be made 10 days after the end of the campaign?

82. **Supply and demand** The demand function for a certain commodity is approximated by the equation

$$p = 100e^{-q/2}$$

where q is the number of units demanded at a price of p dollars per unit.

   a. If there is a 100-unit demand for the product, what will be its price?

   b. If there is no demand for the product, what will be its price?

83. **Medicine** The concentration C of a drug in the bloodstream at time t (in hours) can be approximated by the equation

$$C = 100(1 - e^{-0.5t})$$

   a. What will the concentration be when $t = 0$?

   b. What will the concentration be after 1 hour? Round to one decimal place.

84. **Rumors** The number of people $N(t)$ reached by a particular rumor at time t is approximated by the equation

$$N(t) = \frac{5050}{1 + 100e^{-0.06t}}$$

   a. Find $N(0)$.

   b. Find $N(10)$.

85. **Stellar magnitude** The stellar magnitude M of a star is defined by the equation

$$M = -2.5 \log\left(\frac{B}{B_0}\right)$$

where B is the brightness of the star and B_0 the minimum of brightness.

   a. Find the stellar magnitude of the North Star, 2.1 times as bright as B_0. Round to four decimal places.

   b. Find the stellar magnitude of Venus, 36.2 times as bright as B_0. Round to four decimal places.

86. **Height of adult male** The percent P of adult height a male has reached at age A ($13 \leq A \leq 18$) is modeled by the equation

$$P = 16.7 \log(A - 12) + 87$$

   a. What percent of adult height has a 13-year-old male reached? Round to the nearest percent.

   b. What percent of adult height has an 18-year-old male reached? Round to the nearest thousandth of a percent.

〉〉〉 Using Your Knowledge

Exponential and Logarithmic Functions Is there a relationship between the graph of an exponential function and the graph of the logarithm of the function? Let's use what we learned in this section to find out.

Body Fat Use the following information for Problems 87–90. Dr. Angel Ledesma, a physician for Nutripac in Mexico, developed the following formulas that measure the percent of body fat in men (BFM) and women (BFW)

$$BFM = 14.83 \ln S - 36.45$$
$$BFW = 14.71 \ln S - 29.04$$

where S is the sum of the measurements (in millimeters, mm) of four skin folds on the biceps, triceps, shoulder blade, and supra iliac region. (See photos)

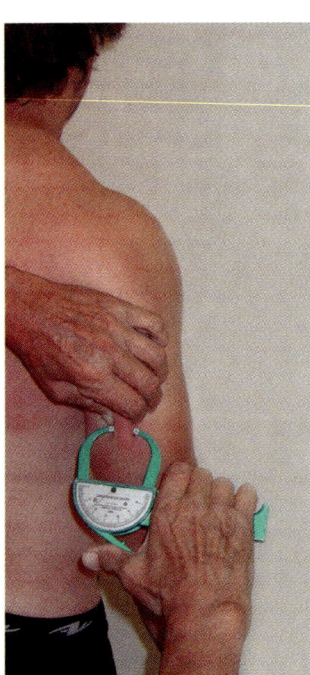
Triceps measurement

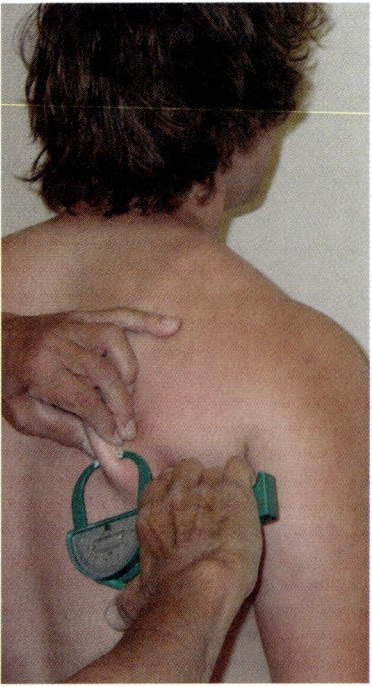
Sub scapular measurement

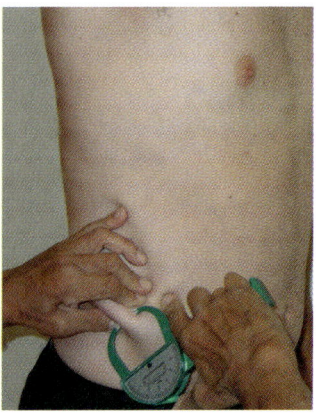
Supra iliac measurement

87. Find the percent of body fat on Tyrone whose four measurements S added to 40 millimeters. Answer to the nearest whole number.

88. Find the percent of body fat on Latasha whose four measurements S added to 50 millimeters. Answer to the nearest whole number.

89. Luigi has 20% body fat. Use the formula BFM = 14.83 ln S − 36.45 to find S, the sum of the measurements of the four skin folds in Luigi's body.
(*Hint:* Solve for ln S and use ln $S = N$ to mean $S = e^N$.)
Answer to the nearest whole number.

90. Narayana has 17% body fat. Use the formula BFW = 14.71 ln S − 29.04 to find S, the sum of the measurements of the four skin folds in Narayana's body.
(*Hint:* Solve for ln S and use ln $S = N$ to mean $S = e^N$.)
Answer to the nearest whole number.

91. a. Graph $f(x) = 2^x$.

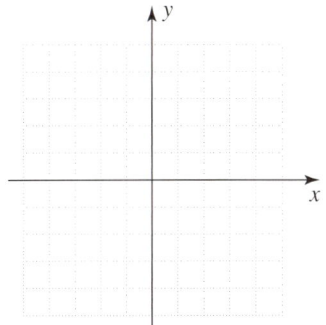

92. a. Graph $f(x) = 3^x$.

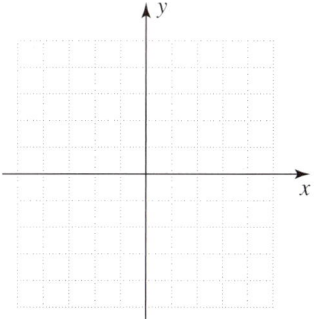

b. Graph $g(x) = \log 2^x$.

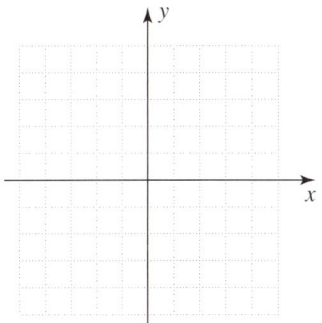

b. Graph $g(x) = \log 3^x$.

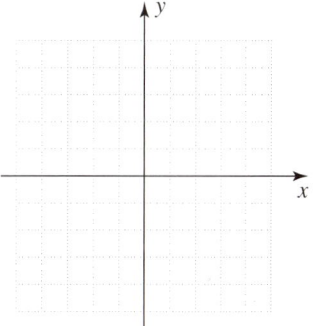

c. What is the slope of $g(x) = \log 2^x$? (*Hint:* Use the properties of logarithms to simplify $\log 2^x$.)

c. What is the slope of $g(x) = \log 3^x$?

93. Compare the base used for the function f in Problems 91 and 92 with the slope of the function g. What is the relationship between the two?

〉〉〉 Write On

94. Explain the difference between **common** logarithms and **natural** logarithms.

95. Explain what the antilog$_{10}$ of a number is.

96. Explain the usefulness of the change-of-base formula.

97. Explain the relationship between the graphs of $f(x) = 10^x$ and $g(x) = \log x$.

98. Explain the relationship between the graphs of $f(x) = e^x$ and $g(x) = \ln x$.

99. Explain the meaning of "the graph of $f(x) = e^{2x}$ is steeper than the graph of $g(x) = e^{0.5x}$."

〉〉〉 Concept Checker

Fill in the blank(s) with the correct word(s), phrase, or mathematical statement.

100. The natural logarithm function is defined by $f(x) = $ _____, $x > 0$.

101. For any logarithm with base a and b, and $M > 0$, $\log_b M = $ _____.

e^x $\dfrac{\log_a M}{\log_a b}$

$\dfrac{\log_a b}{\log_a M}$ $\ln x$

»» Mastery Test

Find each value to four decimal places.

102. $\ln 3120$

103. $\operatorname{inv} \ln(-1.5960)$

104. $\log_4 40$

105. $\log_6 20$

106. $\log 41{,}500$

107. $\log 0.000415$

108. $\operatorname{inv} \log 0.8432$

109. $\operatorname{inv} \log(-2.4683)$

Graph the equation.

110. $f(x) = e^{(1/4)x}$

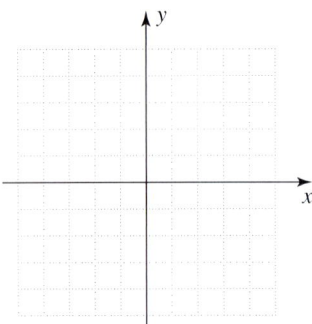

111. $f(x) = -e^{(1/4)x}$

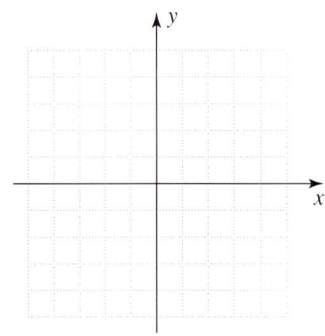

112. $f(x) = e^{(-1/4)x}$

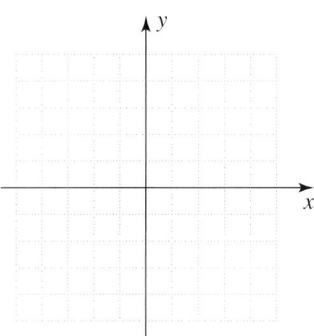

113. $f(x) = e^{(1/4)x} + 1$

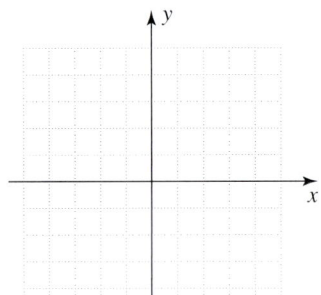

114. $f(x) = \ln(x + 3)$

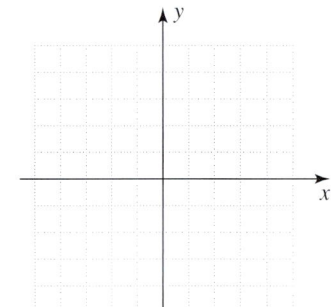

115. $f(x) = \ln x + 3$

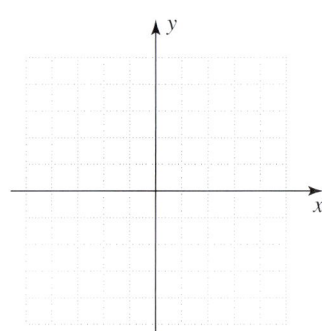

116. Over short periods, the systolic pressure of a normal adult can be approximated by the equation

$$P = 130e^{-0.5t}$$

  a. What is the systolic pressure when $t = 0$?

  b. What is the systolic pressure when $t = 0.5$ sec?

»» Skill Checker

Solve for k. (Round to four decimal places.)

117. $\log 3 = k \log 2$

118. $\log 25 = 0.15k$

119. $1000 = 10^k$

120. $\dfrac{1}{125} = 5^k$

10.6 Exponential and Logarithmic Equations and Applications

Objectives

A Solve exponential equations.

B Solve logarithmic equations.

C Solve applications involving exponential or logarithmic equations.

To Succeed, Review How To...

1. Use the properties of logarithms (pp. 788–789).
2. Use the laws of exponents (pp. 38–45).
3. Solve linear equations (pp. 76–83).
4. Evaluate logarithms (pp. 796–799).

Getting Started

Don't Drink and Drive!

Is there a relationship between blood alcohol level (BAC) and the probability of having an accident? Absolutely! As the chart shows, the probability $P(b)$ of having an accident, written as a percent, is a function of your blood alcohol level b. The formula is given by the equation

$$P(b) = e^{kb} \quad (1)$$

As you can see from the chart, this probability is 25% when the BAC is 0.15%. Can we find k using this information? To do this, let $P(b) = 25$ and $b = 0.15$ in equation (1). We then have

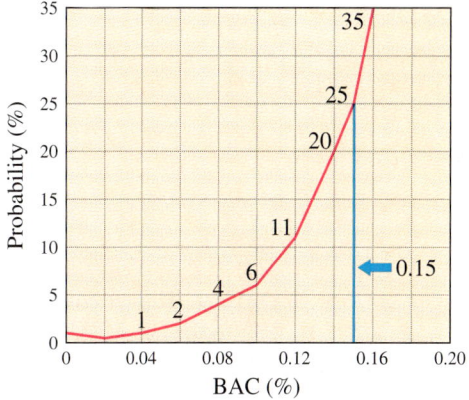

$$25 = e^{0.15k} \quad (2)$$

Equation (2) is an example of an *exponential* equation, because the variable k occurs in the exponent. We shall solve this equation in Example 3. We will even be able to predict what BAC theoretically leads to certain disaster—what alcohol level b corresponds to a 100% probability of an accident.

A ❯ Solving Exponential Equations

In all the equations we have solved, we have seldom used variables as exponents. When this happens, we have an exponential equation.

> **EXPONENTIAL EQUATION**
>
> An **exponential equation** is an equation in which the variable occurs in an exponent.

$6^x = 14$, $3^{5x} = 20$, and $2^{6x} = 32$ are exponential equations. The equation $2^{6x} = 32$ can be rewritten using powers of the base 2, as

$$2^{6x} = 2^5$$

Because we are using the same base, the exponents must be equal;

$$6x = 5 \quad \text{and} \quad x = \frac{5}{6}$$

Remember the equivalence property from Section 10.4? We restate it here for your convenience.

EQUIVALENCE PROPERTY

For $b > 0$, $b \neq 1$, $b^x = b^y$ is equivalent to $x = y$.

We can use this equivalence property to help us solve exponential equations in which the exponential terms on each side can be expressed in the same base. The procedure follows.

PROCEDURE

To solve exponential equations when it is possible to write each side as a power of the same base.

1. Be sure the exponential terms are on opposite sides. ($b^x = c$ or $b^x = c^t$)
2. Rewrite each side as a power of the same base. ($b^x = b^y$)
3. Equate the exponents. (Equivalence property) ($x = y$)
4. Solve the equation from Step 3.
5. Verify the solution.

EXAMPLE 1 Solving exponential equations

Solve:

a. $3^{2x-1} = 81$ **b.** $2^{x+1} = 8^{x-1}$

SOLUTION

a. Since $3^4 = 81$, we write each side of $3^{2x-1} = 81$ as a power of 3.

$$3^{2x-1} = 3^4$$

Since the base 3 is the same, the exponents must be equal. Thus,

$$2x - 1 = 4 \quad \text{Equate the exponents.}$$
$$2x = 5 \quad \text{Add 1.}$$
$$x = \frac{5}{2} \quad \text{Divide by 2.}$$

CHECK Letting $x = \frac{5}{2}$ in $3^{2x-1} = 81$, we obtain

$$3^{2 \cdot (5/2) - 1} = 81$$

or

$$3^4 = 81 \quad \text{A true statement}$$

The solution is $\frac{5}{2}$.

PROBLEM 1

Solve:

a. $3^{3x-4} = 9$ **b.** $2^{x+3} = 8^{x-1}$

Answers to PROBLEMS

1. **a.** $x = 2$ **b.** $x = 3$

b. The idea is to write both sides of the equation using the same base. Since $8 = 2^3$, the equation can be rewritten as

$$2^{x+1} = (2^3)^{x-1}$$
$$2^{x+1} = 2^{3x-3} \quad \text{Simplify.}$$

Since the base is the same, 2, the exponents must be equal. Thus,

$$x + 1 = 3x - 3 \quad \text{Equate the exponents.}$$
$$-2x + 1 = -3 \quad \text{Subtract } 3x.$$
$$-2x = -4 \quad \text{Subtract } 1.$$
$$x = 2 \quad \text{Divide by } -2.$$

CHECK If we let $x = 2$ in the original equation, we obtain

$$2^{2+1} = 8^{2-1}$$
$$2^3 = 8^1 \quad \text{A true statement}$$

The solution is 2.

How can we solve $6^x = 14$? Since it isn't possible to write each side of $6^x = 14$ as a power of the same base, we make use of a fundamental property of logarithms.

EQUIVALENCE PROPERTY FOR LOGARITHMS

If M, N, and b are all positive numbers and $b \neq 1$, then

$$\log_b M = \log_b N \quad \text{is equivalent to} \quad M = N$$

This means that

$$\text{if } \log_b M = \log_b N, \quad \text{then} \quad M = N$$

and conversely

$$\text{if } M = N, \quad \text{then} \quad \log_b M = \log_b N$$

To solve exponential equations when it is not possible to write each side as a power of the same base, we use the following procedure.

PROCEDURE

To solve exponential equations when it is not possible to write each side as a power of the same base.

1. Be sure the base and power are isolated on one side and the constant is on the other side. $(b^x = c)$
2. Take the log of both sides. (Equivalence property for logarithms)
3. Rewrite $\log_b M^r$ as $r \log_b M$. (Power property of logarithms)
4. Solve.
5. Verify the solution.

EXAMPLE 2 Solving exponential equations

Solve: $6^x = 14$

SOLUTION

$$6^x = 14 \quad \text{Given}$$
$$\log 6^x = \log 14 \quad \text{Take the logarithm of both sides.}$$
$$x \log 6 = \log 14 \quad \text{Since } \log 6^x = x \log 6 \text{ (Power property of logarithms)}$$
$$x = \frac{\log 14}{\log 6} \quad \text{Divide by log 6.}$$

This is the *exact* answer. If we wish to approximate it, use a calculator to obtain

$$x = \frac{\log 14}{\log 6} \approx 1.4729 \quad \text{(Rounded to four decimal places)}$$

Remember to wait until the final step to round your answer.

PROBLEM 2

Solve (round to four decimal places): $5^x = 12$

Now let's return to the *Getting Started* and solve equation (2), $25 = e^{0.15k}$. Keep in mind that the property "$\log_b M = \log_b N$ is equivalent to $M = N$" works when the base $b = e$, that is,

$$M = N \quad \text{is equivalent to} \quad \ln M = \ln N$$

EXAMPLE 3 Using an equivalency property to solve an exponential equation

Solve (round to the nearest tenth): $25 = e^{0.15k}$

SOLUTION

$$25 = e^{0.15k} \quad \text{Given}$$
$$\ln 25 = \ln e^{0.15k} \quad \text{Take the natural logarithm of both sides.}$$
$$\ln 25 = 0.15k \quad \text{Since } \log_b b^x = x, \ln e^{0.15k} = 0.15k \text{ (Power property of logarithms).}$$
$$\frac{\ln 25}{0.15} = k \quad \text{Divide both sides by 0.15.}$$

Thus,

$$k = \frac{\ln 25}{0.15} \approx \frac{3.2189}{0.15} \approx 21.5 \quad \text{(to the nearest tenth)}$$

PROBLEM 3

Solve (round to the nearest hundredth): $20 = e^{0.15k}$

EXAMPLE 4 Accident rate and blood alcohol level

If we substitute $k = 21.5$ into equation (2) in *Getting Started*, the formula for $P(b)$ becomes $P(b) = e^{21.5b}$. At what blood alcohol level b will the probability of having an accident be 100%? Round to the nearest hundredth.

SOLUTION We use the RSTUV method.

1. **Read the problem.**
2. **Select the unknown.** We are asked to find b so that $P(b) = 100$. To do this we must solve the equation

$$100 = e^{21.5b}$$

3. **Think of a plan.** First, we take natural logarithms of both sides of the equation, then solve for b.

PROBLEM 4

At what alcohol level b will the probability of having an accident be 50%? Round to the nearest hundredth.

Answers to PROBLEMS

2. $x = \frac{\log 12}{\log 5} \approx 1.5440$

3. $k = \frac{\ln 20}{0.15} \approx 19.97$

4. About 0.18%

4. **Use algebra to solve the problem.**

$$\ln 100 = \ln e^{21.5b} \quad \text{Take the natural logarithm of both sides.}$$
$$\ln 100 = 21.5b \quad \text{Power property of logarithms } (\log_b b^x = x)$$
$$\frac{\ln 100}{21.5} = b \quad \text{Divide by 21.5.}$$

or $\quad b = \dfrac{\ln 100}{21.5}$

This is the exact value. Using a calculator to approximate b to the nearest hundredth, $b \approx 0.21\%$.

5. **Verify the answer.** The verification that $100 = e^{21.5(0.214)}$ is left to you (use your calculator). When the blood alcohol level is about 0.21%, the probability of an accident is 100%. If your blood alcohol level is 0.21%, you are probably too drunk to even get in your car.

B › Solving Logarithmic Equations

We have already solved certain types of **logarithmic equations**—equations containing *logarithmic expressions*—in Section 10.4.

Let's recall the definition of a logarithmic function.

$$y = \log_b x \quad \text{is equivalent to} \quad b^y = x,$$

where $b > 0$, $b \neq 1$, and $x > 0$.

When we are solving logarithmic equations we must discard any values of the variable that do not satisfy the part of the definition that says $x > 0$. We will use this definition of a logarithm and convert the given equation into an exponential equation. This technique will be used in Example 5.

In general, we use the following procedure to solve logarithmic equations.

> **PROCEDURE**
>
> **Solving Logarithmic Equations**
> 1. Write the equation as an equivalent one with a single logarithmic expression on one side; write the equation in the form $\log_b M = N$.
> 2. Write the equivalent exponential equation $b^N = M$ and solve.
> 3. Always check your answer. Discard any values of the variable for which $M \leq 0$.

EXAMPLE 5   Solving logarithmic equations

Solve: $\log_5(2x - 3) = 2$

SOLUTION

$$\log_5(2x - 3) = 2 \quad \text{Given}$$
$$2x - 3 = 5^2 \quad \text{Since } \log_b x = y \text{ is equivalent to } b^y = x$$
$$2x - 3 = 25 \quad \text{Simplify.}$$
$$2x = 28 \quad \text{Add 3.}$$
$$x = 14 \quad \text{Divide by 2.}$$

The solution is 14. You can check this by letting $x = 14$ in $\log_5(2x - 3) = 2$ to obtain

$$\log_5(2 \cdot 14 - 3) = 2$$
$$\log_5(25) = 2$$
$$5^2 = 25 \quad \text{A true statement}$$

PROBLEM 5

Solve: $\log_5(2x - 1) = 2$

Answers to PROBLEMS

5. $x = 13$

EXAMPLE 6 Solving logarithmic equations

Solve:

a. $\log(x + 3) + \log x = 1$ **b.** $\log_3(x - 1) - \log_3(x - 3) = 1$

SOLUTION

a.
$\log(x + 3) + \log x = 1$	Given
$\log[(x + 3)x] = 1$	Use the product property of logarithms.
$(x + 3)x = 10^1$	Write as an equivalent exponential equation.
$x^2 + 3x = 10$	Simplify.
$x^2 + 3x - 10 = 0$	Subtract 10 to set equation = 0.
$(x + 5)(x - 2) = 0$	Factor.
$x + 5 = 0$ or $x - 2 = 0$	Factors = 0. By the zero-factor property
$x = -5$ or $x = 2$	

CHECK For $x = -5$, the expression $\log x$ becomes $\log(-5)$, which isn't defined because we cannot find the logarithm of a negative number. We discard that value and check $x = 2$.

$$\log(x + 3) + \log x = 1$$
becomes
$$\log(2 + 3) + \log 2 = 1$$
$$\log 5 + \log 2 = 1$$
or
$$\log(5 \cdot 2) = 1 \quad \text{Use the product property of logarithms.}$$
$$10^1 = 5 \cdot 2 \quad \text{Use the definition of logarithm.}$$

Since the result is a true statement, the solution is 2.

b.
$\log_3(x - 1) - \log_3(x - 3) = 1$	Given
$\log_3\left(\dfrac{x - 1}{x - 3}\right) = 1$	Use the quotient property of logarithms.
$\dfrac{x - 1}{x - 3} = 3^1$	Write as an equivalent exponential equation.
$x - 1 = 3(x - 3)$	Multiply both sides by $(x - 3)$.
$x - 1 = 3x - 9$	By the distributive property
$-1 = 2x - 9$	Subtract x from both sides.
$8 = 2x$	Add 9 to both sides.
$4 = x$	Divide both sides by 2.

CHECK For $x = 4$,

$$\log_3(x - 1) - \log_3(x - 3) = 1$$
becomes
$$\log_3(4 - 1) - \log_3(4 - 3) = 1$$
or
$$\log_3 3 - \log_3 1 = 1$$
$$1 - 0 = 1 \quad \text{Since } \log_3 3 = 1 \text{ and } \log_3 1 = 0$$

The resulting statement $1 - 0 = 1$ is true, so the solution is 4.

PROBLEM 6

Solve:

a. $\log(x - 3) + \log x = 1$

b. $\log_3(x + 1) - \log_3(x - 3) = 1$

C > Solving Applications Involving Exponential or Logarithmic Equations

Exponential and logarithmic equations have many applications in such areas as business, engineering, social science, psychology, and science. Examples 7–10 will give you an idea of the variety and range of their use.

Answers to PROBLEMS

6. a. $x = 5$ **b.** $x = 5$

EXAMPLE 7   Compound interest

With continuous compounding, a principal of P dollars accumulates to an amount A given by the equation

$$A = Pe^{rt}$$

where r is the interest rate and t is the time in years. If the interest rate is 6%, how long would it take for the money in your bank account to double?

SOLUTION   We use the RSTUV method.

1. **Read the problem.**
2. **Select the unknown.** We are looking for the time t.
3. **Think of a plan.** With $A = 2P$ and $r = 0.06$, the equation becomes

$$2P = Pe^{0.06t}$$

or

$$2 = e^{0.06t} \quad \text{Divide by } P.$$

4. **Use algebra to solve the problem.** We want to solve this equation for t, so we take natural logarithms of both sides:

$$\ln 2 = \ln e^{0.06t} \quad \text{Take natural logarithm of both sides.}$$
$$\ln 2 = 0.06t \quad \text{Power property of logarithms, } \log_b b^x = x$$
$$\frac{\ln 2}{0.06} = t \quad \text{Divide by 0.06.}$$
$$t = \frac{\ln 2}{0.06}$$

Using a calculator,

$$t \approx 11.6$$

This means that it would take about 11.6 years for your money to double.

5. **Verify the solution.** The verification is left to you.

PROBLEM 7

In Example 7, how long would it take if the rate is 8%?

EXAMPLE 8   World population

In 2000, the population of the world exceeded 6 billion for the first time reaching about 6.1 billion with a yearly growth rate of 1.2%. The equation giving the population P in terms of the time t is

$$P = 6.1e^{0.012t}$$

Estimate the world population P in the year 2010.

SOLUTION   We use the RSTUV method.

1. **Read the problem.**
2. **Select the unknown.** We are looking for the population P in the year 2010.
3. **Think of a plan.** Since $P = 6.1$ for $t = 0$, the equation shows that t is measured from the year 2000. To estimate the population in 2010, we use $t = 10$ in the equation:

$$P = 6.1e^{(0.012)(10)} = 6.1e^{0.12}$$

4. **Use arithmetic to solve the problem.** Using a calculator and approximating to the nearest tenth,

$$P = 6.1e^{0.12} \approx 6.9$$

Our estimate for the population in 2010 is about 6.9 billion.

5. **Verify the answer.** The verification is left to you.

PROBLEM 8

Estimate the world population in the year 2020.

Answers to PROBLEMS

7. About 8.7 yr   **8.** About 7.8 billion

EXAMPLE 9 Bacteria growth

If B is the number of bacteria present in a laboratory culture after t minutes, then, under ideal conditions,

$$B = Ke^{0.05t}$$

where K is a constant. If the initial number of bacteria is 1000, how long would it take for there to be 50,000 bacteria present?

SOLUTION   We use the RSTUV method.

1. **Read the problem.**
2. **Select the unknown.** Since $B = 1000$ for $t = 0$, we have

$$1000 = Ke^0 = K$$

The equation for B is then

$$B = 1000e^{0.05t}, \text{ where } t \text{ is the unknown number of minutes.}$$

3. **Think of a plan.** Now let $B = 50,000$:

$$50,000 = 1000e^{0.05t}$$
$$50 = e^{0.05t} \quad \text{Divide by 1000.}$$

4. **Use algebra to solve the problem.** To solve for t, take natural logarithms of both sides:

$$\ln 50 = \ln e^{0.05t} \quad \text{Take the natural logarithm of both sides.}$$
$$\ln 50 = 0.05t \quad \text{Power property of logarithms, } \log_b b^x = x.$$
$$\frac{\ln 50}{0.05} = t \quad \text{Divide by 0.05.}$$

Using a calculator,

$$t = \frac{\ln 50}{0.05} \approx 78.2 \text{ min}$$

It will take approximately 78.2 minutes for there to be 50,000 bacteria present.

5. **Verify the solution.** The verification is left to you.

PROBLEM 9

In Example 9, how long would it take to have 100,000 bacteria?

EXAMPLE 10   Half-life of cesium-137

The half-life of a substance is found by using the equation

$$A(t) = A_0 e^{-kt}$$

where $A(t)$ is the amount present at time t (years), k is the decay rate, and A_0 is the initial amount of the substance present. The element cesium-137 decays at the rate of 2.3% per year. Find the half-life (the time it takes for $\frac{1}{2}$ of the element to decay) of this element. Round to one decimal place.

SOLUTION   We use the RSTUV method.

1. **Read the problem.**
2. **Select the unknown.** In this problem,

$$k = 2.3\% = 0.023, \quad A(t) = \frac{1}{2}A_0$$

and we want to find t.

PROBLEM 10

Find the half-life of an element that decays at the rate of 4% per year. Round to one decimal place.

Answers to PROBLEMS

9. About 92.1 min
10. About 17.3 yr

3. **Think of a plan.** With this information, the basic equation becomes

$$\frac{1}{2}A_0 = A_0 e^{-0.023t}$$

$$\frac{1}{2} = e^{-0.023t} \quad \text{Divide by } A_0.$$

$$0.5 = e^{-0.023t} \quad \text{Rewrite } \tfrac{1}{2} \text{ as 0.5.}$$

4. **Use algebra to solve the problem.** Now take natural logarithms of both sides:

$$\ln 0.5 = \ln e^{-0.023t} \quad \text{Take the natural logarithm of both sides.}$$

$$\ln 0.5 = -0.023t \quad \text{Power property of logarithms, } \log_b b^x = x.$$

$$\frac{\ln 0.5}{-0.023} = t \quad \text{Divide by } -0.023.$$

Using a calculator,

$$t = \frac{\ln 0.5}{-0.023} \approx 30.1 \text{ yr}$$

The half-life of cesium-137 is about 30.1 years.

5. **Verify the solution.** The verification is left to you.

Calculator Corner

Solving Equations

Do you remember how to solve equations using your calculator? The idea is to graph both sides of the equation, call them Y_1 and Y_2 and use TRACE and ZOOM (or the intersect feature) to find the point of intersection (the solution). In Example 1(a), graph $Y_1 = 3^{(2x-1)}$ (note the parentheses around the $2x - 1$) and $Y_2 = 81$ using a $[-3, 3]$ by $[-10, 100]$ window with Yscl = 10. On the TI-83 Plus, press 2nd TRACE 5 and ENTER three times to answer the calculator questions "First Curve?" "Second Curve?" "Guess?". The intersection is $X = 2.5$, as shown in Window 1. Now try Example 1(b).

For Example 2, let $Y_1 = 6^x$ and $Y_2 = 14$ and follow the procedure of Example 1(a) using a $[-1, 10]$ by $[-1, 20]$ window. The intersection is $X = 1.4728859$, as shown in Window 2. To do Example 3, let $Y_1 = 25$ and $Y_2 = e^{0.15x}$ with a $[-5, 30]$ by $[-5, 30]$ window and Xscl = Yscl = 5. As usual, press 2nd TRACE 5 and ENTER three times to find the intersection $X = 21.459172$, as shown in Window 3.

What type of window do you need for Example 4? Note that $Y_1 = 100$. After you find the appropriate window, you can check the results of Example 4.

In Example 5, we have to graph $Y_1 = \log_5(2x - 3)$, but calculators don't have a $\log_5$ key. Here, *you* have to remember that the change-of-base formula states that

$$Y_1 = \log_5(2x - 3) = \frac{\log(2x - 3)}{\log 5}$$

Then let $Y_2 = 2$. If you use a standard window, the point of intersection **does not** show (try it!), so we suggest a $[0, 20]$ by $[-1, 3]$ window. Press 2nd TRACE 5 and ENTER three times to obtain the intersection $X = 14$, as shown in Window 4. To do Example 6(a), use the LOG function in your calculator. Let $Y_1 = \log(x + 3) + \log x$ (there is no need to simplify Y_1) and $Y_2 = 1$ and use a standard window. Press 2nd TRACE 5 and move the cursor so that it shows a positive value for x. Then press ENTER three times. The intersection $X = 2$ is shown in Window 5.

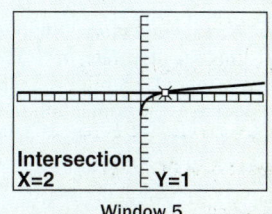

Window 5

Exercises 10.6

⟨A⟩ Solving Exponential Equations
In Problems 1–26, solve the equation. Round the final answer to four decimal places where necessary (do not round intermediate values).

1. $5^x = 25$
2. $3^x = 81$
3. $2^x = 32$
4. $3^{2x} = 81$
5. $5^{3x} = 625$
6. $6^{-2x} = 216$
7. $7^{-3x} = 343$
8. $5^x = 4$
9. $7^x = 512$
10. $3^{x+1} = 729$
11. $5^{x-2} = 625$
12. $2^{3x+1} = 128$
13. $3^x = 2$
14. $3^x = 20$
15. $2^{3x-2} = 32$
16. $3^{4x-3} = 27$
17. $5^{3x} \cdot 5^{x^2} = 25$
18. $3^{4x} \cdot 3^{x^2} = 243$
19. $e^x = 10$
20. $e^x = 100$
21. $e^{-x} = 0.1$
22. $e^{-x} = 0.01$
23. $30 = e^{2k}$
24. $40 = e^{3k}$
25. $10 = e^{-2k}$
26. $20 = e^{-3k}$

⟨B⟩ Solving Logarithmic Equations
In Problems 27–50, solve the equation. Round the final answer to four decimal places where necessary (do not round intermediate values).

27. $\log_2 x = 3$
28. $\log_3 x = 2$
29. $\log_2 x = -3$
30. $\log_3 x = -2$
31. $\ln x = 1$
32. $\ln x = -1$
33. $\ln x = 3$
34. $\ln x = -3$
35. $\log_2(3x - 5) = 1$
36. $\log_3(2x - 1) = 4$
37. $\log_4(3x - 1) = 2$
38. $\log_6(2x + 1) = 2$
39. $\log_5(3x + 1) = 2$
40. $\log_2(4x - 1) = 4$
41. $\log x + \log(x - 3) = 1$
42. $\log x + \log(x - 11) = 1$
43. $\log_2(x + 1) + \log_2(x + 3) = 3$
44. $\log_3(x + 4) + \log_3(x - 2) = 3$
45. $\log(x + 1) - \log x = 1$
46. $\log(x - 1) - \log x = 1$
47. $\log_2(3 + x) - \log_2(7 - x) = 2$
48. $\log_3(2 + x) - \log_3(8 - x) = 2$
49. $\log_2(x^2 + 4x + 7) = 2$
50. $\log_2(x^2 + 4x + 3) = 3$

⟨C⟩ Solving Applications Involving Exponential or Logarithmic Equations

Compound Interest   In Problems 51–54, assume continuous compounding and follow the procedure in Example 7 to find how long it takes a given amount to double at the given interest rate. Round to two decimal places.

51. $r = 5\%$
52. $r = 7\%$
53. $r = 6.5\%$
54. $r = 7.5\%$

55. *World population*   Suppose that the population of the world grows at the rate of 1.5% and that the population in 1984 was about 4.8 billion. Follow the procedure of Example 8 to estimate the population in the year 2000.

56. *World population*   Repeat Problem 55 for growth rate of 1.75%.

Bacteria Growth In Problems 57–60, assume that the number of bacteria present in a culture after t minutes is given by $B = 1000e^{0.04t}$. Find the time it takes (to the nearest tenth of a minute) for the specified number of bacteria to be present.

57. 2000

58. 5000

59. 25,000

60. 50,000

61. *Bacteria* When a bacteria-killing solution is introduced into a certain culture, the number of live bacteria is given by the equation $B = 100{,}000e^{-0.2t}$, where t is the time in hours. Find the number of live bacteria present at the following times.

  a. $t = 0$ **b.** $t = 2$ **c.** $t = 10$ **d.** $t = 20$

62. *Bees* The number of honey bees in a hive is growing according to the equation $N = N_0 e^{0.015t}$, where t is the time in days. If the bees swarm when their number is tripled, find how many days until this hive swarms.

Half-life In Problems 63–66, follow the procedure of Example 10 to find the half-life of the substance.

63. Plutonium, whose decay rate is 0.003% per year

64. Krypton, whose decay rate is 6.3% per year

65. A radioactive substance whose decay rate is 5.2% per year

66. A radioactive substance whose decay rate is 0.2% per year

67. *Atmospheric pressure* The atmospheric pressure P in pounds per square inch at an altitude of h feet above the Earth is given by the equation $P = 14.7e^{-0.00005h}$. Find the pressure at the specified altitude. Round to one decimal place.

  a. 0 feet   **b.** 5000 feet   **c.** 10,000 feet

68. *Atmospheric pressure* If the atmospheric pressure in Problem 67 is measured in inches of mercury, then $P = 30e^{-0.207h}$, where h is the altitude in miles. Find the pressure at the specified altitude. Round to one decimal place.

  a. At sea level.   **b.** At 5 mi above sea level.

69. *Football* According to the National Football League Players Association (NFLPA), average NFL salaries are as shown in the graph and can be approximated by the equation

$$S = 540{,}000(1.09)^t$$

where t is the number of years after 1992.

  a. In how many years will average salaries be $800,000? Answer to the nearest year.

  b. Based on this approximation formula, in how many years will salaries reach the 1 million dollar mark? Answer to the nearest year.

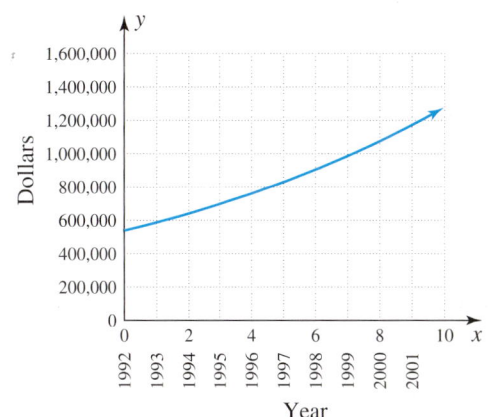

70. *Recycling* According to the *Statistical Abstract of the United States*, about $\frac{2}{3}$ of all aluminum cans distributed are recycled. If a company distributes 500,000 cans, the number in use after t years is approximated by the equation

$$N(t) = 500{,}000\left(\frac{2}{3}\right)^t$$

How many years will it take for the number of cans to reach 100,000? Answer to the nearest year.

71. *Flying* Do you have a fear of flying? The U.S. Department of Transportation has good news for nervous fliers: The number of general aviation accidents A has gone down significantly in the last 30 years. It can be approximated by the equation

$$A = 5000e^{-0.04t}$$

where t is the number of years after 1970.

  a. How many accidents were there in 1970?

  b. How many accidents were there in 1990?

  c. In what year do you predict the number of accidents to be 1000 (answer to the nearest year)?

72. Diastolic blood pressure After exercise the diastolic blood pressure of normal adults is a function of time and can be approximated by the equation

$$P = 90e^{-0.5t}$$

where t is the time in minutes. How long would it be before the diastolic pressure comes down to 80? Answer to two decimal places.

73. Credit cards How much do you spend on your credit cards? According to the Federal Reserve Board and the Bankcard Holders of America, credit card spending (in millions of dollars) is on the increase and can be approximated by the equation

$$S = 54e^{0.15t}$$

where t is the number of years after 1980.

a. How many millions of dollars were spent in 1980?

b. In what year did the amount spent on credit cards reach 500 million dollars? Answer to the nearest year.

74. Height The percentage of adult height P attained by a boy can be approximated by the equation

$$P = h(A) = 29 + 50 \log(A + 1)$$

where P is the percentage of adult height and A is the age in years ($0 < A < 17$). At what age will a boy reach 89% of his adult height?

75. Height The height H of a girl (in inches) can be approximated by the function

$$H = h(A) = 11 + 19.44 \ln A$$

where A is the age in years ($6 < A < 15$). To the nearest year, at what age would you expect a girl to be the specified height?

a. 60 inches tall

b. 50 inches tall

76. Attendance According to the National Football League, paid attendance (in millions) is as shown in the graph and can be approximated by the equation

$$A = 17.4 + 0.32 \ln t$$

where t is the number of years after 1988.

a. Based on the equation, in what year did attendance reach 18 million? Answer to the nearest year.

b. Based on the equation, in what year will attendance reach 18.5 million? Answer to the nearest year.

c. What will the estimated attendance be in 2010?

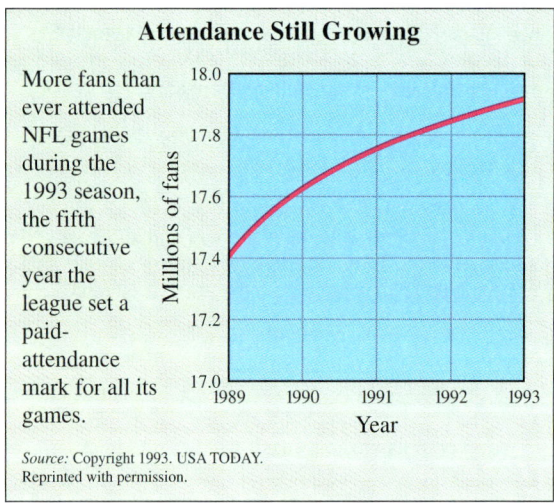

Attendance Still Growing

More fans than ever attended NFL games during the 1993 season, the fifth consecutive year the league set a paid-attendance mark for all its games.

Source: Copyright 1993. USA TODAY. Reprinted with permission.

77. Interest If $1000 is invested in an account earning 1% interest each month, the number of months t it takes the account to grow to an amount A can be approximated by the equation

$$t = -694.2 + 231.4 \log A$$

where $A \geq 1000$. How long would it take for this account to grow to 1 million dollars?

78. Ointment A research company has developed a new ointment to help heal skin wounds in diabetics. If originally the area of a wound is A_0 square millimeters (mm²), after n days of applying the ointment the area A of the wound is given by the equation $A = A_0 e^{-0.43n}$. If a diabetic wound originally measures 10 square millimeters, find the area of the wound after 10 days of applying the ointment. Answer to four decimal places.

79. Wound treatment The area A of a wound after n days of treatment is given by the equation $A = A_0 e^{-0.43n}$, where A_0 is the area of the original wound. If the area A of a wound after 20 days of applying an ointment is 0.1 square millimeters (mm²), what was the area of the original wound? Answer to the nearest square millimeter.

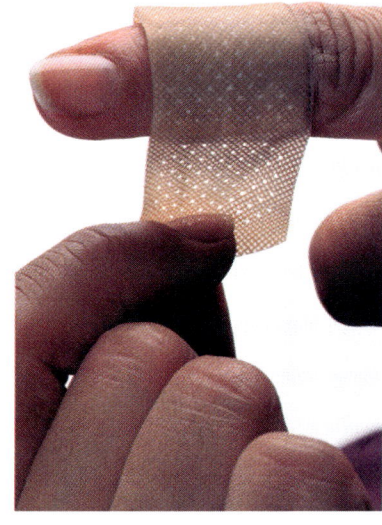

››› Using Your Knowledge

Fax Machines Have you ever wondered how the approximations in this exercise set are done? To construct a growth equation based on time t, we start with the basic equation

$$A = A_0 e^{kt}.$$

When $t = 0$, then $A = A_0 e^0 = A_0$. Now, if we know the value A for a certain time t, we can solve the resulting equation and find k. Use this idea and the fact that the sale of fax machines soared from 580,000 in 1987 to 11 million in 1992 to solve Problems 80–85.

80. If the number N of fax machines sold (in thousands) is given by the equation $N = N_0 e^{kt}$, where t is the number of years after 1987, what is N_0?

81. Using the N_0 obtained in Problem 80, we can write $N = 580 e^{kt}$. We know that $N = 11{,}000$ in 1992, when $t = 5$ ($1992 - 1987 = 5$). Thus, $11{,}000 = 580 e^{5k}$. Solve for k (to two decimal places) in this equation.

82. Use the values obtained in Problems 80 and 81 to find an exponential equation $N = N_0 e^{kt}$ for the number of fax machines sold from 1987 to 1992.

83. According to the equation obtained in Problem 82, how many fax machines were sold in 1990?

84. a. According to the equation obtained in Problem 82, how many fax machines were sold in 1992?

  b. What was the actual number of fax machines sold in 1992?

  c. What was the percent error between your approximation and actual number of fax machines sold?

85. How many fax machines would you predict were sold in the year 2000?

››› Write On

In Problems 80–85, you constructed your own mathematical model, $A = A_0 e^{kt}$, based on a given set of data. How do you know what type of model to use for a particular set of data? Two of the possibilities are logarithmic or exponential:

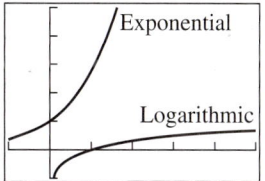

86. The number of cellular phones sold from 1985 to 1994 is shown in the graph. What type of mathematical model would you use to approximate the number of phones sold based on the information provided in the graph? Explain why.

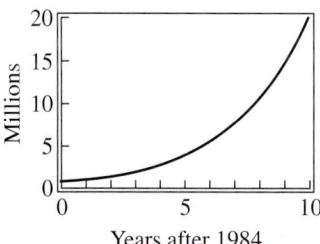

87. If you use the exponential model $S = S_0 e^{kt}$ to approximate the information in the graph for Problem 86, explain the procedure you would use to find S_0.

88. Explain how you would find k in $S = S_0 e^{kt}$.

89. The number of cigarettes produced per capita from 1969 to 1990 is shown in the graph. What type of model would you use to approximate this information? Explain.

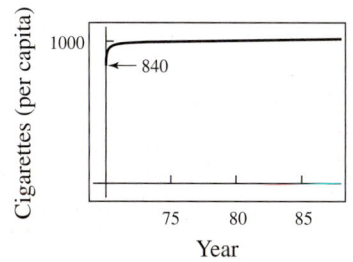

90. If you assume that the equation modeling the graph in Problem 89 is of the form $N = a \ln x + b$, what information do you need to find b? Explain.

91. For the equation in Problem 90, what other information do you need to find a?

92. Explain the procedure you would use to find a in the equation in Problem 90.

››› Concept Checker

Fill in the blank(s) with the correct word(s), phrase, or mathematical statement.

93. An exponential equation is an equation in which the variable occurs as an _____.
94. $b^x = b^y$ is equivalent to _____, for $b > 0$ and $b \neq 1$.
95. $\log_b M = \log_b N$ is equivalent to _____ for M, and N positive and $b > 0$ and $b \neq 1$.
96. If $M = N$ are positive, $\log_b M = $ _____.

$x = y$ $\log_b N$
$M = N$ exponent

››› Mastery Test

97. The number of bacteria B present in a laboratory culture after t minutes is given by the equation $B = Ke^{0.05t}$. If the initial number of bacteria is 1000, how long (to the nearest tenth of a minute) would it take for there to be 20,000 bacteria present?

98. After a bactericide is introduced, the number of bacteria present in a laboratory culture after t minutes is given by the equation $B = Ke^{-0.02t}$. If the initial number of bacteria is 50,000, how long (to the nearest tenth of a minute) would it be before this number is reduced to 10,000?

Solve. Round the final answer to four decimal places where necessary (do not round intermediate values).

99. $\log(x + 9) + \log x = 1$
100. $\log_2(x + 3) - \log_2 x = 1$
101. $\log_6(4x - 4) = 2$
102. $50 = e^{0.15k}$
103. $5^x = 10$
104. $2^{3x-1} = 4$
105. $2^{2x+1} = 8^{x-1}$

››› Skill Checker

106. Evaluate the expression $\frac{1}{2}n(n - 1)$ when:
    a. $n = 1$ **b.** $n = 3$ **c.** $n = 10$

107. Find the value of the function $f(n) = 2n + 1$ when:
    a. $n = 1$ **b.** $n = 5$ **c.** $n = 10$

› Collaborative Learning 10A

What are the trends in the population growth of the United States? The growth rate of the population takes into consideration births, deaths, and immigration. It seems the U.S. birth rate is on the decline since 1970. However, immigration has been rapidly increasing since 1930 as shown in the bar graph.

The population growth will have an impact on the environment and other issues in our American society. For the government to be able to plan for, keep up with, and possibly limit immigration, it would be important to be able to approximate the immigration population for a certain period.

Divide into groups, answer the questions, and then conduct your own research.

1. Make a table that contains as its first column the numbers 10–80 representing the 8 decades of immigration data. The second column should be the approximation for the amount of immigration population for each of the bars.
2. Use the data in the table to make a curved line graph.
3. The graph resembles an exponential function. If the exponential equation, $A = 800{,}000\, e^{0.35t}$, is used to approximate the U.S. immigration population since 1930 where t is the time in 10-year increments and A is the immigration population, find the immigration population for each of the 8 decades. Make a new column on your table for your results.

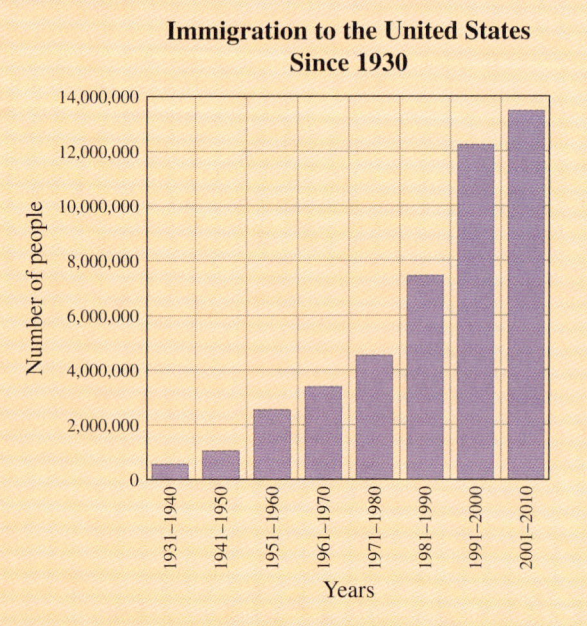

Immigration to the United States Since 1930

4. How do the answers, found using the exponential equation for the immigration population in the new column, compare to the values you approximated for the table from the bar graph? How do you account for any discrepancies?

5. Using the exponential equation, approximate what the immigration population will be in the decade 2011–2020.

6. Research the immigration population growth for other major countries in the world and see how it compares to the United States.

7. Each group should present their findings to the class and then discuss how this growth may affect the environment and other issues in society.

Collaborative Learning 10B

Do you know how quickly a baby chick embryo can grow? The way a baby chick embryo grows is by repeated cell division, beginning with one cell. Divide into groups, answer the questions, do the research, and then have a class discussion.

1. Make a two-column table that will describe the number of cells after successive divisions, the entries for the first line being one division and two cells. Continue the table for four more lines.

2. Graph the ordered pairs on an x-y coordinate plane.

3. Write the equation that would describe this cell division.

4. Soon after fertilization of the egg occurs, cell division starts. As long as the egg remains warmer than 67°, cell division will continue. If cell division were to occur every 20 minutes for 1 day, how many cells would be produced during that day?

The growth of cell populations can be modeled using the assumption that each cell divides into two. However, this could be misleading. The population growth rate may be reduced if the length of the cell cycle increases, and it could be reduced if only a fraction of the cells are dividing.

5. Do some research and find exponential equations that describe other cell divisions. How do they compare to the equation that you found in Question 3. Discuss the differences.

Fresh Egg
Note the small whitish spot on the germ from which the life of the chick begins.

1st Division $2 = 2^1$

2nd Division $4 = 2^2$

3rd Division $8 = 2^3$

And so on …

Research Questions

1. Who was the first person to publish a book describing the rules of logarithms, what was the name of the book he published, and what does "logarithm" mean?

2. Who wrote the book from which the words "mantissa" and "characteristic" are derived, and what is the meaning of each of these words?

3. Write a few paragraphs about Jobst Bürgi and describe how he used "red" and "black" numbers in his logarithmic table.

4. Name the author and the title of his book published in 1748 that "gave his approval for e to represent the base of natural logarithms."

5. The symbol e was first used in 1731. Name the circumstances under which the symbol e was used.

Summary Chapter 10

Section	Item	Meaning	Example
10.1A	$(f+g)(x)$	$f(x) + g(x)$	If $f(x) = x^2$ and $g(x) = x + 1$, then:
	$(f-g)(x)$	$f(x) - g(x)$	$(f+g)(x) = x^2 + x + 1$
	$(fg)(x)$	$f(x) \cdot g(x)$	$(f-g)(x) = x^2 - x - 1$
	$\left(\dfrac{f}{g}\right)(x)$	$\dfrac{f(x)}{g(x)}, g(x) \neq 0$	$(fg)(x) = x^3 + x^2$
			$\left(\dfrac{f}{g}\right)(x) = \dfrac{x^2}{x+1}, x \neq -1$
10.1B	$(f \circ g)(x)$	$f(g(x))$, the composite of f with g	If $f(x) = x^2$ and $g(x) = x + 1$, then $(f \circ g)(x) = f(g(x)) = (x+1)^2$
10.2A	f^{-1}	The inverse of a relation, obtained by reversing the order of the coordinates in each ordered pair in f	If $f = \{(1, 2), (4, 6), (6, 9)\}$, then $f^{-1} = \{(2, 1), (6, 4), (9, 6)\}$
10.2C	Horizontal line test	If any horizontal line intersects the graph of $f(x)$ more than once, then $f(x)$ is not a one-to-one function and f^{-1} is not a function.	$f(x) = x^2$. The graph is a parabola that opens upward. Thus, any horizontal line $y = b > 0$ cuts the graph in two points. $f(x)$ is not a one-to-one function and the inverse, $f^{-1}(x) = \pm\sqrt{x}$, is not a function.
10.3A	Exponential function	A function defined for all real values of x by $f(x) = b^x$, $b > 0$, $b \neq 1$	$f(x) = 2^x$
10.3B	Increasing function	A function whose graph rises from left to right	$f(x) = 2^x$
	Decreasing function	A function whose graph falls from left to right	$f(x) = 2^{-x}$ or $f(x) = \left(\dfrac{1}{2}\right)^x$
	Natural exponential function	$f(x) = e^x$, $e \approx 2.7182818$	$f(x) = e^x$ is an increasing function.
10.4A	Logarithm	$\log_b x = y$ means $x = b^y$, where $b > 0$, $b \neq 1$, and $x > 0$	$\log_2 8 = 3$ because $2^3 = 8$.
10.4D	Logarithm of a product	$\log_b MN = \log_b M + \log_b N$	$\log_2(4 \cdot 8) = \log_2 4 + \log_2 8$
	Logarithm of a quotient	$\log_b \dfrac{M}{N} = \log_b M - \log_b N$	$\log_2 \dfrac{4}{8} = \log_2 4 - \log_2 8$
	Logarithm of a power	$\log_b M^r = r \log_b M$	$\log_2 4^3 = 3 \log_2 4$
10.5A	Common logarithm	The logarithm to base 10	$\log 10 = 1$, $\log 100 = 2$
	Inverse logarithm	The number that corresponds to a given logarithm	inv log $2.3010 \approx 200$
10.5B	Natural logarithms	Logarithms to base e	ln $2 \approx 0.69315$
10.5C	Change-of-base formula	$\log_b M = \dfrac{\log_a M}{\log_a b}$	$\log_5 20 = \dfrac{\log 20}{\log 5} \approx 1.86135$
10.6A	Exponential equation	An equation in which the variable occurs in an exponent	$10^{2x} = 5$
10.6B	Logarithmic equation	Equations containing logarithmic expressions	$\log(2x - 1) = 3$ and $\log_2(3x - 1) = 2$ are logarithmic equations.

Review Exercises Chapter 10

(If you need help with these exercises, look in the section indicated in brackets.)

1. ⟨**10.1A**⟩ Let $f(x) = 2 - x^2$ and $g(x) = 2 + x$. Find the following.
   a. $(f + g)(x)$
   b. $(f - g)(x)$
   c. $(fg)(x)$
   d. $\left(\dfrac{f}{g}\right)(x)$

2. ⟨**10.1A**⟩ Let $f(x) = 3 - x^2$ and $g(x) = 3 + x$. Find the following.
   a. $(f + g)(x)$
   b. $(f - g)(x)$
   c. $(fg)(x)$
   d. $\left(\dfrac{f}{g}\right)(x)$

3. ⟨**10.1A**⟩ Find $\dfrac{f(x) - f(a)}{x - a}$, $(x \neq a)$ if:
   a. $f(x) = 6x + 1$
   b. $f(x) = 7x + 1$

4. ⟨**10.1B**⟩ If $f(x) = x^3$ and $g(x) = 2 - x$, find:
   a. $(g \circ f)(2)$
   b. $(f \circ g)(x)$
   c. $(g \circ f)(x)$

5. ⟨**10.1B**⟩ If $f(x) = x^3$ and $g(x) = 3 - x$, find
   a. $(f \circ g)(x)$
   b. $(g \circ f)(x)$
   c. $(g \circ f)(2)$

6. ⟨**10.1C**⟩ Let $f(x) = \dfrac{9}{x - 1}$ and $g(x) = \dfrac{x + 3}{x - 4}$. Find the domain of $f + g$, $f - g$, and fg.

7. ⟨**10.1C**⟩ Let $f(x) = \dfrac{x + 1}{x - 3}$ and $g(x) = \dfrac{x - 1}{x + 4}$. Find the domain of $\dfrac{f}{g}$.

8. ⟨**10.1D**⟩
   a. The revenue (in dollars) obtained from selling x units of a product is given by the equation $R(x) = 100x - 0.02x^2$, and the cost per unit is given by the equation $C(x) = 30{,}000 + 30x$. Find the profit function $P(x) = R(x) - C(x)$.
   b. If the revenue $R(x) = 100x - 0.02x^2$ and the cost per unit is $C(x) = 40{,}000 + 40x$, find the profit function $P(x)$.

9. ⟨**10.2A**⟩ Let $S = \{(4, 4), (6, 6), (8, 8)\}$ and find
   a. The domain and range of S
   b. S^{-1}
   c. The domain and range of S^{-1}
   d. The graphs of S and S^{-1}

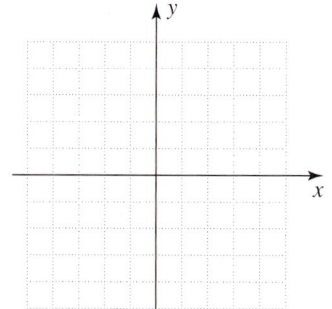

10. ⟨**10.2A**⟩ Let $S = \{(4, 5), (6, 7), (8, 9)\}$ and find
    a. The domain and range of S
    b. S^{-1}
    c. The domain and range of S^{-1}
    d. The graphs of S and S^{-1}

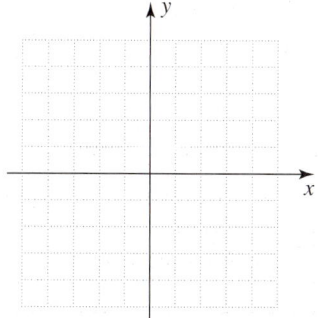

11. ⟨**10.2B**⟩ Let $f(x) = y = 3x - 3$.
   a. Find $f^{-1}(x)$.
   b. Graph f and its inverse.

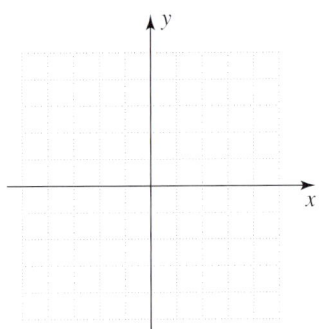

12. ⟨**10.2B**⟩ Let $f(x) = y = 4x - 4$.
   a. Find $f^{-1}(x)$.
   b. Graph f and its inverse.

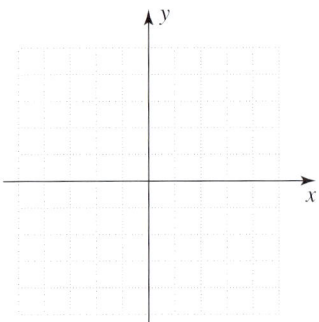

13. ⟨**10.2C**⟩ Find the inverse of $f(x) = y = 4x^2$. Is the inverse a function?

14. ⟨**10.2C**⟩ Find the inverse of $f(x) = y = 5x^2$. Is the inverse a function?

15. ⟨**10.2C**⟩ Graph $y = 3^x$. Is the inverse a function?

16. ⟨**10.2C**⟩ Graph $y = 4^x$. Is the inverse a function?

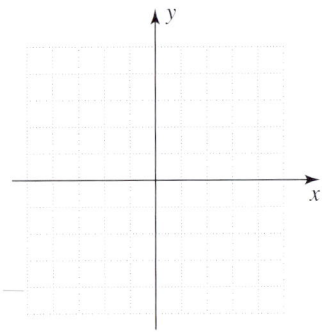

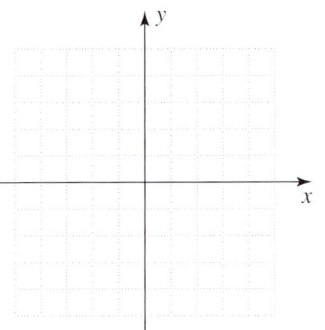

17. ⟨**10.2D**⟩ The relationship between dress size d and waist size w (in inches) is given by the equation $d = f(w) = w - 16$.
   a. What dress size corresponds to a 28-inch waist?
   b. Find $f^{-1}(d)$.
   c. If a woman wears a size 10 dress, what is her waist size?

18. ⟨**10.3A**⟩ Graph the function:
   a. $f(x) = 2^{x/2}$
   b. $f(x) = 2^{-x/2}$

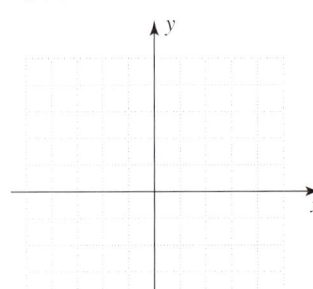

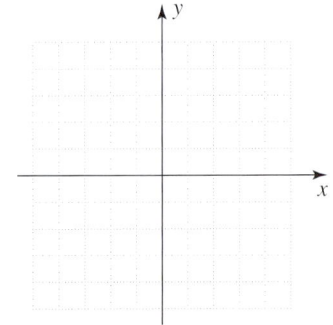

19. ⟨**10.3A**⟩ *Graph the function:*

   a. $g(x) = \left(\frac{1}{2}\right)^{x/2}$

   b. $g(x) = \left(\frac{1}{2}\right)^{-x/2}$

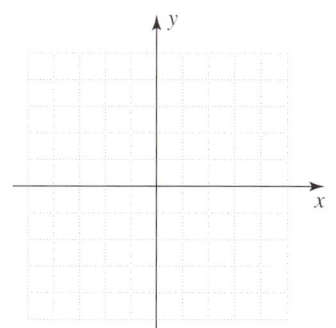
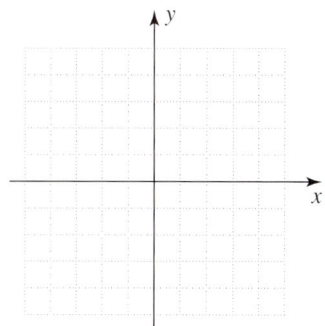

20. ⟨**10.3C**⟩ A radioactive substance decays so that G, the number of grams present, is given by the equation

$$G = 1000e^{-1.4t}$$

where *t* is the time in years. Find, to the nearest gram, the amount of the substance present at the specified time

   a. At the start

   b. In 2 years

21. ⟨**10.4A**⟩ Graph $f(x) = \log_5 x$.

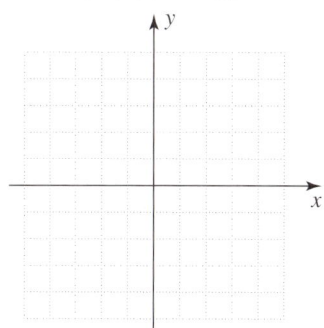

22. ⟨**10.4B**⟩ *Write in logarithmic form.*

   a. $243 = 3^5$

   b. $\frac{1}{8} = 2^{-3}$

23. ⟨**10.4B**⟩ *Write in exponential form.*

   a. $\log_2 32 = 5$

   b. $\log_3 \frac{1}{81} = -4$

24. ⟨**10.4C**⟩ *Solve.*

   a. $\log_4 x = -2$

   b. $\log_x 16 = 2$

25. ⟨**10.4C**⟩ *Find.*

   a. $\log_2 16$

   b. $\log_3 \frac{1}{27}$

26. ⟨**10.4D**⟩ *Fill in the blank with the correct expression.*

   a. $\log_b MN = $ _____

   b. $\log_b M - \log_b N = $ _____

   c. $\log_b M^r = $ _____

In Problems 27–31, use a calculator to find each value and round to four decimal places where necessary.

27. ⟨**10.5A**⟩

   a. log 975

   b. log 837

28. ⟨**10.5A**⟩

   a. log 0.00759

   b. log 0.000648

29. ⟨**10.5A**⟩

   a. inv log 2.8215

   b. inv log −3.3904

30. ⟨**10.5B**⟩

   a. ln 2850

   b. ln 0.345

31. ⟨**10.5B**⟩

   a. inv ln 2.0855

   b. inv ln 2.7183

32. ⟨**10.5C**⟩ Use the change of base formula to find each value. Round the final answer to four decimal places (do not round intermediate values).

   a. $\log_3 10$

   b. $\log_3 100$

33. ⟨**10.5D**⟩ Graph.

a. $f(x) = e^{(1/2)x}$

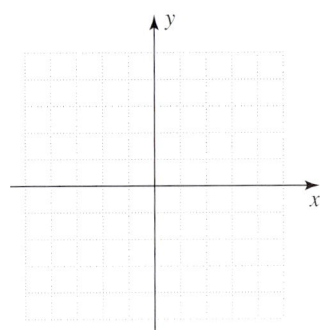

b. $g(x) = -e^{(1/2)x}$

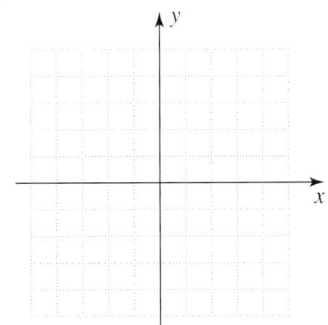

34. ⟨**10.5D**⟩ Graph.

a. $f(x) = \ln(x + 1)$

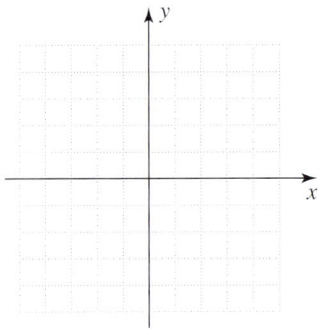

b. $g(x) = \ln x + 1$

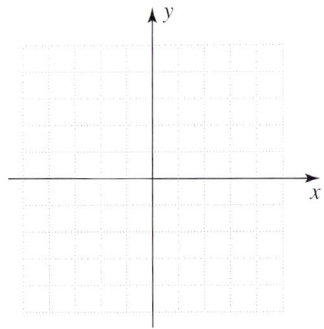

35. ⟨**10.5E**⟩ The pH of a solution is defined by pH $= -\log[\text{H}^+]$, where $[\text{H}^+]$ is the hydrogen ion concentration of the solution in moles per liter. Find the pH, to one decimal place, of a solution for which $[\text{H}^+] = 4 \times 10^{-6}$.

36. ⟨**10.6A**⟩ Solve.

a. $2^{2x-1} = 32$

b. $3^{x+1} = 9^{x-1}$

37. ⟨**10.6A**⟩ Solve. Round the final answer to four decimal places (do not round intermediate values).

a. $2^x = 3$

b. $5^{2x} = 2.5$

38. ⟨**10.6A**⟩ Solve. Round the final answer to four decimal places (do not round intermediate values).

a. $e^{5.6x} = 2$

b. $e^{-0.33x} = 2$

39. ⟨**10.6B**⟩ Solve.

a. $\log x + \log(x - 10) = 1$

b. $\log_3(x + 1) - \log_3(x - 1) = 1$

40. ⟨**10.6C**⟩ The compound amount with continuous compounding is given by the equation $A = Pe^{rt}$, where P is the principal, r the rate, and t the time in years. Using 0.69315 for ln 2, find how long it takes for the money to double—for A to equal $2P$—for the given rate. Round to one decimal place.

a. 5%

b. 8.5%

41. ⟨**10.6C**⟩ The number of bacteria in a culture after t minutes is given by $N = 1000e^{kt}$. If there are 1804 bacteria after the given time, find k to four decimal places.

a. 2 min

b. 4 min

42. ⟨**10.6C**⟩ A radioactive substance decays so that the amount present in t years is given by the equation $A = A_0 e^{-kt}$. Use -0.69315 for ln 0.5 and find the half-life for the given values of k.

a. $k = 0.5$

b. $k = 0.02$

> Practice Test Chapter 10

(Answers on pages 833–835)

Visit www.mhhe.com/bello to view helpful videos that provide step-by-step solutions to several of the problems below.

1. Let $f(x) = x^2 + 16$ and $g(x) = 4 - x$. Find the following.
   a. $(f + g)(x)$ b. $(f - g)(x)$ c. $(fg)(x)$ d. $\dfrac{f}{g}(x)$

2. $f(x) = 7x - 2$, find $\dfrac{f(x) - f(a)}{x - a}$, where $x \neq a$.

3. If $f(x) = x^2 + 2$ and $g(x) = x + 3$, find:
   a. $(g \circ f)(-2)$ b. $(f \circ g)(x)$ c. $(g \circ f)(x)$

4. If $f(x) = -\dfrac{1}{5x}$ and $g(x) = \dfrac{3x + 1}{2x - 2}$, find the domain of $f + g, f - g,$ and fg.

5. Find the domain of $\dfrac{f}{g}$ if $f(x) = \dfrac{3}{x + 3}$ and $g(x) = \dfrac{4x}{x - 6}$

6. Let $S = \{(3, 5), (5, 7), (7, 9)\}$ and find
   a. The domain and range of S
   b. S^{-1}
   c. The domain and range of S^{-1}
   d. The graph of S and S^{-1}

7. Let $f(x) = y = 4x - 4$.
   a. Find $f^{-1}(x)$.   b. Graph f and its inverse.

8. Find the inverse of $f(x) = y = 3x^2$. Is the inverse a function?

9. Given $y = 3^x$.
   a. Graph the equation.
   b. Is $y = 3^x$ increasing or decreasing?

10. A radioactive substance decays so that the number of grams present after t years is modeled by the equation
    $$G = 1000e^{-1.4t}$$
    Find, to the nearest gram, the amount of the substance present
    a. At the start.   b. In 2 years.

11. Graph on the same coordinate axes.
    a. $f(x) = 2^x$ b. $g(x) = \log_2 x$

12. Write the equation
    a. $27 = 3^x$ in logarithmic form.
    b. $\log_5 25 = x$ in exponential form.

13. Solve.
    a. $\log_4 x = -1$ b. $\log_x 16 = 2$

14. Use the properties of logarithms to show that $3 \log x - \log 12 = \log\left(\dfrac{x^3}{12}\right)$.

15. Use the properties of logarithms to show that $\dfrac{1}{4} \log r + \dfrac{1}{4} \log t = \log \sqrt[4]{rt}$.

16. Find. Round to four decimal places.
    a. log 325   b. inv log 3.5502

17. Find. Round to four decimal places.
    a. ln 325   b. inv ln 1.1618

18. a. Use the change-of-base formula to fill in the blank: $\log_3 10 = $ _____
    b. Use the result of part a to find a numerical approximation for $\log_3 10$.

19. Graph.
    a. $f(x) = e^{(1/2)x}$ b. $g(x) = -e^{(1/2)x}$

20. Graph.
    a. $f(x) = \ln(x + 1)$ b. $g(x) = \ln x + 1$

21. Solve.
  a. $5^{2x+1} = 25$
  b. $3^{x+1} = 9^{2x-1}$

22. Solve. Round to two decimal places.
  a. $3^x = 2$
  b. $50 = e^{0.20k}$

23. Solve.
  a. $\log(x+2) + \log(x-7) = 1$
  b. $\log_3(x+5) - \log_3(x-1) = 1$

24. The compound amount with continuous compounding is given by the equation $A = Pe^{rt}$, where P is the principal, r is the interest rate, and t is the time in years. If the rate is 8%, find how long it takes for the money to double—for A to equal $2P$ (use 0.69315 for $\ln 2$).

25. A radioactive substance decays so that the amount A present at time t (years) is given by the equation $A = A_0 e^{-0.5t}$. Find the half-life (time for half to decay) of this substance (use -0.69315 for $\ln 0.5$).

Answers to Practice Test Chapter 10

Answer	If You Missed Question	Section	Review Examples	Page
1. a. $x^2 - x + 20$ **b.** $x^2 + x + 12$ **c.** $-x^3 + 4x^2 - 16x + 64$ **d.** $\dfrac{x^2 + 16}{4 - x}, x \neq 4$	1	10.1A	1	749
2. $7, x \neq a$	2	10.1A	2	750
3. a. 9 **b.** $x^2 + 6x + 11$ **c.** $x^2 + 5$	3	10.1B	3	751
4. $\{x \mid x \text{ is a real number and } x \neq 0 \text{ and } x \neq 1\}$	4	10.1C	4	752
5. $\{x \mid x \text{ is a real number and } x \neq -3, x \neq 6, \text{ and } x \neq 0\}$	5	10.1C	5, 6	752–753
6. a. $D = \{3, 5, 7\}; R = \{5, 7, 9\}$ **b.** $\{(5, 3), (7, 5), (9, 7)\}$ **c.** $D = \{5, 7, 9\}; R = \{3, 5, 7\}$ **d.**	6	10.2A	1	761
7. a. $f^{-1}(x) = \dfrac{x + 4}{4}$ **b.**	7	10.2B	2	763
8. $y = \pm\sqrt{\dfrac{x}{3}}$ or $\pm\dfrac{\sqrt{3x}}{3}$; no	8	10.2C	3	765

Answer	If You Missed		Review	
	Question	Section	Examples	Page
9. a. [graph of $y = 3^x$]	9	10.3A, B	1, 2	774–776
b. Increasing				
10. a. 1000 g b. 61 g	10	10.3C	4	778
11. [graph of $f(x) = 2^x$ and $g(x) = \log_2 x$]	11	10.4A	1	785
12. a. $\log_3 27 = x$ b. $5^x = 25$	12	10.4B	2, 3	786
13. a. $\frac{1}{4}$ b. 4	13	10.4C	4	787
14. $3 \log x - \log 12 = \log x^3 - \log 12$ $= \log\left(\frac{x^3}{12}\right)$	14	10.4D	6	788
15. $\frac{1}{4} \log r + \frac{1}{4} \log t = \log r^{1/4} + \log t^{1/4}$ $= \log(rt)^{1/4}$ $= \log \sqrt[4]{rt}$	15	10.4D	7	789
16. a. 2.5119 b. 3550	16	10.5A	1, 2	797–798
17. a. 5.7838 b. 3.1957	17	10.5B	3, 4	798–799
18. a. $\frac{\log 10}{\log 3}$ or $\frac{\ln 10}{\ln 3}$ b. 2.0959	18	10.5C	5, 6	799–800

Answer	If You Missed		Review	
	Question	Section	Examples	Page
19.	19	10.5D	7	801–802

(Graph showing $f(x) = e^{(1/2)x}$ and $g(x) = -e^{(1/2)x}$)

20.	20	10.5D	8	802–803

(Graph showing $g(x) = \ln x + 1$ and $f(x) = \ln(x + 1)$)

21. a. $\frac{1}{2}$ b. 1	21	10.6A	1	812–813
22. a. $\frac{\log 2}{\log 3} \approx 0.63$ b. $\frac{\ln 50}{0.2} \approx 19.56$	22	10.6A	2, 3	814
23. a. 8 b. 4	23	10.6B	6	816
24. About 8.66 yr	24	10.6C	7	817
25. About 1.386 yr	25	10.6C	10	818–819

Cumulative Review Chapters 1–10

1. Simplify:
 $[(3x^2 - 2) + (8x + 3)] - [(x - 2) + (2x^2 - 6)]$

2. Simplify: $(4x^4y^{-2})^2$

3. Solve: $0.02P + 0.04(1700 - P) = 65$

4. Solve: $|x - 4| = |x - 8|$

5. Graph on a number line: $\{\,x \mid x < -4 \text{ or } x \geq 4\}$

6. Graph on a number line: $|6x - 9| \leq 3$

7. If $H = 2.85h + 72.69$, find h when $H = 135.39$.

8. The perimeter of a rectangle is $P = 2L + 2W$, where L is the length and W is the width. If the perimeter is 160 feet and the length is 20 feet more than the width, what are the dimensions?

9. Find the distance between the points $E(5, 2)$ and $F(-2, 1)$.

10. The line passing through $A(3, -4)$ and $B(-1, y)$ is perpendicular to a line with slope $\frac{4}{9}$. Find y.

11. Find an equation of the line that passes through the point $(-6, 6)$ and is parallel to the line $8x + 2y = 2$.

12. Graph on an x-y coordinate system: $|x + 3| > 2$

    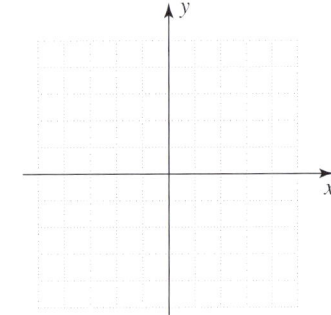

13. If the temperature of a gas is held constant, the pressure P varies inversely as the volume V. A pressure of 1850 pounds per square inch is exerted by 5 cubic feet of air in a cylinder fitted with a piston. Find k.

14. Use the substitution method to solve the system:
 $x - 4y = -19$
 $-3x = -12y + 61$

15. Solve the system:
 $2x = 5y - 28$
 $2y = 5x + 28$

16. Solve the system:
 $3x + y + z = -17$
 $x + 2y - z = -15$
 $3x + y - z = -21$

17. Graph the solution set of the system:
 $2x + y \leq 4$
 $y \geq 1$

    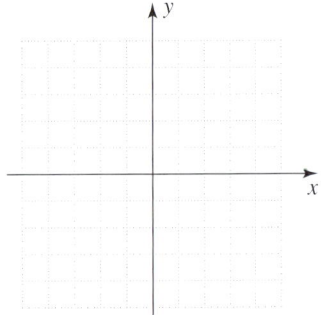

18. The total height of a building and the flagpole on the roof is 252 feet. The building is 8 times as tall as the flagpole. How tall is the building?

19. If $P(x) = x^2 + 5x + 1$, find $P(-4)$.

20. Multiply: $(3h + 5)(6h - 7)$

21. Factor completely: $48x^4y + 20x^3y^2 - 12x^2y^3$

22. Factor completely: $27n^3 + 8$

23. Solve for x: $x^3 + 4x^2 - x - 4 = 0$

24. Factor $3x^3 + 22x^2 + 37x + 10$ if $(x + 5)$ is one of its factors.

25. Divide: $\dfrac{x + 7}{x - 7} \div (x^2 + 14x + 49)$

26. Perform the indicated operations:
$$\frac{x-3}{x^2-5x+6} - \frac{x-2}{x^2-4}$$

27. Find two consecutive even integers such that the sum of their reciprocals is $\frac{7}{24}$.

28. Evaluate: $(27)^{-4/3}$

29. Rationalize the denominator: $\dfrac{\sqrt[5]{7}}{\sqrt[5]{16d^3}}$

30. Simplify: $\sqrt{18} + \sqrt{50}$

31. Reduce: $\dfrac{4 + \sqrt{8}}{2}$

32. Solve: $\sqrt{x+7} = x+5$

33. Multiply: $(-5 + 9i)(-3 - 5i)$

34. Solve by the quadratic formula: $\dfrac{x^2}{72} + \dfrac{x}{9} = \dfrac{1}{8}$

35. Solve for x: $x^{1/2} - 3x^{1/4} + 2 = 0$

36. Solve for x: $(x+5)(x-2)(x-7) \le 0$

37. Solve for x: $\dfrac{-2x+10}{x-1} \le 0$

38. Graph the parabola:
$y = -(x-3)^2 + 2$

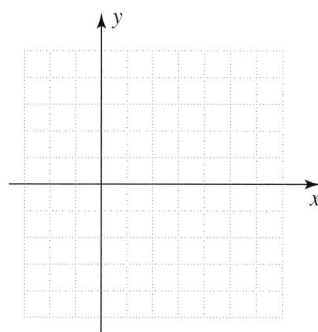

39. Find the center and the radius of $x^2 + y^2 - 6x - 10y + 30 = 0$.

40. Graph: $\dfrac{y^2}{16} - \dfrac{x^2}{4} = 1$

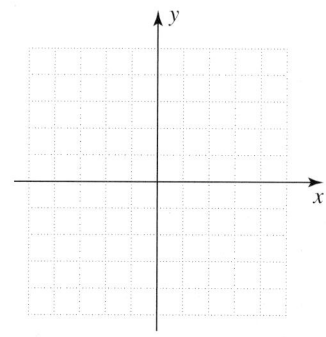

41. Identify the following curve: $x = y^2 - 4$

42. The cost C of manufacturing and selling x units of a product is given by the equation $C = 23x + 85$, and the corresponding revenue R is $R = x^2 - 55$. Find the break-even value of x.

43. Find the domain of $y = \sqrt{x+9}$.

44. If $f(x) = x^4$ and $g(x) = 4 - x$, find $(f \circ g)(x)$.

45. Let $f(x) = y = 4x - 6$. Graph f and its inverse.

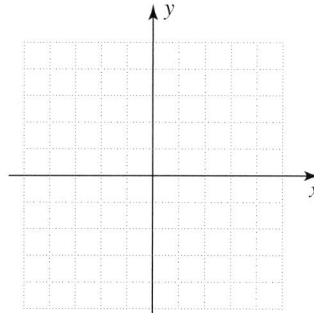

46. Graph $f(x) = 2^x$. Is the inverse a function?

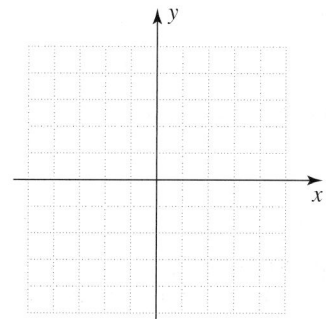

47. Show that $\frac{1}{2}\log_b 21 - \frac{1}{2}\log_b 83 = \log_b \sqrt{\frac{21}{83}}$

48. Find x to four decimal places if $e^{9.6x} = 7$, and you are given $\ln 7 \approx 1.94591$.

49. Use the change of base formula to find $\log_3 5$ to four decimal places.

50. The number of bacteria present in a culture after t minutes is given by the equation $B = 1000e^{kt}$. If there are 8392 bacteria present after 7 min, find k to four decimal places.

Appendix A

Section

- **A.1** Sequences and Series
- **A.2** Arithmetic Sequences and Series
- **A.3** Geometric Sequences and Series

(A.1–A.3 can be found online at www.mhhe.com/bello)

- **A.4** Matrices
- **A.5** Determinants and Cramer's Rule
- **A.6** The Binomial Expansion

A.4 Matrices

Objectives

A ▸ Perform elementary operations on systems of equations.

B ▸ Solve systems of linear equations using matrices.

C ▸ Solve applications using matrices.

▸ To Succeed, Review How To...

1. Solve a system of three equations in three unknowns (pp. 306–310).
2. Recognize whether a system is inconsistent or dependent (pp. 309–310).

▸ Getting Started

Tweedledee and Tweedledum

Do you remember Tweedledee and Tweedledum? Here's what they said:

Tweedledee: The sum of your weight and twice mine is 361 pounds.

Tweedledum: Contrariwise, the sum of your weight and twice mine is 360 pounds.

Alice meets Tweedledee and Tweedledum in *Through the Looking Glass.*

If Tweedledee weighs x pounds and Tweedledum weighs y pounds, the two sentences can be translated as

$$2x + y = 361$$
$$x + 2y = 360$$

We are going to solve this system again, but this time we shall also write the equivalent operations using *matrices*.

In this section we will solve systems of three equations using matrices.

MATRIX — A **matrix** (plural matrices) is a rectangular array of numbers enclosed in brackets.

For example,

$$\begin{bmatrix} 2 & 1 \\ 1 & 1 \end{bmatrix} \quad \text{and} \quad \begin{bmatrix} -5 & 3 & 2 \\ 4 & 0 & 1 \end{bmatrix}$$

are matrices. The first matrix has two rows and two columns (2×2), while the second one has two rows and three columns (2×3). A matrix derived from a system of linear equations (each written in standard form with the constant terms on the right) is called the **augmented matrix** of the system. For example, the augmented matrix of the system in the *Getting Started* is

$$\begin{matrix} 2x + y = 361 \\ x + 2y = 360 \end{matrix} \quad \text{is} \quad \begin{bmatrix} 2 & 1 & | & 361 \\ 1 & 2 & | & 360 \end{bmatrix}$$

Now compare the solution of the two-equation system with its solution using matrices:

$$2x + y = 361$$
$$x + 2y = 360$$
$$\begin{bmatrix} 2 & 1 & | & 361 \\ 1 & 2 & | & 360 \end{bmatrix}$$

$$2x + y = 361$$
$$-2x - 4y = -720$$
$$\begin{bmatrix} 2 & 1 & | & 361 \\ -2 & -4 & | & -720 \end{bmatrix}$$ Multiply the second equation by -2.

$$2x + y = 361$$
$$-3y = -359$$
$$\begin{bmatrix} 2 & 1 & | & 361 \\ 0 & -3 & | & -359 \end{bmatrix}$$ Copy first equation.
Result of adding the two equations

$$2x + y = 361$$
$$y = 119\tfrac{2}{3}$$
$$\begin{bmatrix} 2 & 1 & | & 361 \\ 0 & 1 & | & 119\tfrac{2}{3} \end{bmatrix}$$ Copy first equation.
Divide the second equation by -3.

Now you can substitute $119\tfrac{2}{3}$ for y in $2x + y = 361$ and solve for x. Did you note that the operations performed on the equations were identical to those performed on the matrices?

A ❯ Perform Elementary Operations on Systems of Equations

Suppose we wish to solve a system of three linear equations in three unknowns. This means that we wish to find all sets of values of (x, y, z) that make all three equations true. We first need to consider what changes can be made in the system to yield an **equivalent system,** a system that has exactly the same solutions as the original system. We need to consider only three simple operations.

PROCEDURE

Elementary Operations on Systems of Equations

1. **The order of the equations may be changed.** This clearly cannot affect the solutions.
2. **Any of the equations may be multiplied by any nonzero real number.** If (m, n, p) is a solution of $ax + by + cz = d$, then, for $k \neq 0$, it is also a solution of $kax + kby + kcz = kd$, and conversely.
3. **Any equation of the system may be replaced by the sum (term by term) of itself and any other equation of the system.** (You can show this by doing Problem 11 of the Exercises.)

These three **elementary operations** are used to simplify systems of equations and to find their solutions. Let's use them to solve the system

System I
(original system)
$$2x - y + z = 3$$
$$x + y = -1$$
$$3x - y - 2z = 7$$

Step 1. In the second equation of the system, x and y have unit coefficients, so we interchange the first two equations to get the following more convenient arrangement:

System II
$$x + y = -1 \quad (1)$$
$$2x - y + z = 3 \quad (2)$$
$$3x - y - 2z = 7 \quad (3)$$

Step 2. To make the coefficients of x in equations (1) and (2) the same in absolute value but opposite in sign, multiply both sides of equation (1) by -2:

$$-2x - 2y = 2$$

Step 3. To eliminate x between equations (1) and (2), add the equation obtained in step 2 to equation (2) to get

$$-3y + z = 5 \qquad \text{Replaces 2nd equation} \qquad (4)$$

and restore equation (1) by dividing out the -2. The system now reads

System III
$$\begin{aligned} x + y &= -1 & (1) \\ -3y + z &= 5 & (4) \\ 3x - y - 2z &= 7 & (3) \end{aligned}$$

Now we proceed in a similar way to eliminate x between equations (1) and (3).

Step 4. Multiply both sides of equation (1) by -3:

$$-3x - 3y = 3$$

Step 5. Add this last equation to equation (3) to get

$$-4y - 2z = 10 \qquad \text{Replaces 3rd equation} \qquad (5)$$

and restore equation (1) by dividing out the -3. The system now reads

System IV
$$\begin{aligned} x + y &= -1 & (1) \\ -3y + z &= 5 & (4) \\ -4y - 2z &= 10 & (5) \end{aligned}$$

We eliminate y between equations (4) and (5) in the following way:

Step 6. Multiply both sides of equation (4) by -4 and both sides of equation (5) by 3 to get

$$\begin{aligned} 12y - 4z &= -20 \\ -12y - 6z &= 30 \end{aligned}$$

Step 7. Add the last two equations to get

$$-10z = 10 \qquad \text{Replaces 3rd equation} \qquad (6)$$

and restore the second equation by dividing out the -4. The system is now

System V
$$\begin{aligned} x + y &= -1 & (1) \\ -3y + z &= 5 & (4) \\ -10z &= 10 & (6) \end{aligned}$$

Step 8. It's easy to solve system V: Equation (6) immediately gives $z = -1$. Then, by substitution into equation (4), we get

$$\begin{aligned} -3y - 1 &= 5 & (4) \\ -3y &= 6 \\ y &= -2 \end{aligned}$$

By substituting $y = -2$ into equation (1) we find

$$\begin{aligned} x - 2 &= -1 & (1) \\ x &= 1 \end{aligned}$$

The solution of the system is $x = 1, y = -2, z = -1$. This is easily checked in the given set of equations.

B › Solve Systems of Linear Equations Using Matrices

A general system of the same form as system V may be written

System VI
$$ax + by + cz = d$$
$$ey + fz = g$$
$$hz = k$$

A system such as system VI, in which the first unknown, x, is missing from the second and third equations and the second unknown, y, is missing from the third equation, is said to be in **echelon form.** A system in echelon form is easy to solve. The third equation immediately yields the value of z. Back-substitution of this value into the second equation yields the value of y. Finally, back-substitution of the values of y and z into the first equation yields the value of x. Briefly, we say that we solve the system by **back-substitution.**

Every system of linear equations can be brought into echelon form by the use of the elementary operations. It remains for us only to organize and make the procedure more efficient by employing the augmented matrix.

Let's compare the augmented matrices of system I and system V:

System I (original System) Main → diagonal $\begin{bmatrix} 2 & -1 & 1 & | & 3 \\ 1 & 1 & 0 & | & -1 \\ 3 & -1 & -2 & | & 7 \end{bmatrix}$

System V Main → diagonal $\begin{bmatrix} 1 & 1 & 0 & | & -1 \\ 0 & -3 & 1 & | & 5 \\ 0 & 0 & -10 & | & 10 \end{bmatrix}$

The augmented matrix for system V shows that the system is in echelon form because it has only 0's below the main diagonal (the diagonal left to right of the coefficient matrix). We should be able to obtain the second matrix from the first by performing operations corresponding to the elementary operations on the equations. These operations are called **elementary row operations,** and they always yield matrices of equivalent systems. Such matrices are called **row-equivalent.** If two matrices A and B are row-equivalent, we write $A \sim B$. The elementary row operations are as follows.

PROCEDURE

Elementary Row Operations on Matrices
1. Change the order of the rows.
2. Multiply all the elements of a row by any nonzero number.
3. Replace any row by the element-by-element sum of itself and any other row.

These operations are performed as necessary to write the equivalent system in echelon form.

NOTE

Compare the elementary row operations with the elementary operations on a system. They are analogous!

We illustrate the procedure by showing the transition from the matrix of the original system I to that of system V. To explain what is happening at each step, we use the notation R_1, R_2, and R_3 for the respective rows of the matrix, along with the following typical abbreviations:

Step	Notation	Meaning
1	$R_1 \leftrightarrow R_2$	Interchange R_1 and R_2.
2	$2 \times R_1$	Multiply each element of R_1 by 2.
3	$2 \times R_1 + R_2 \to R_2$	Replace R_2 by $2 \times R_1 + R_2$.

We write Step 1 like this:

$$\begin{bmatrix} 2 & -1 & 1 & | & 3 \\ 1 & 1 & 0 & | & -1 \\ 3 & -1 & -2 & | & 7 \end{bmatrix} \sim \begin{bmatrix} 1 & 1 & 0 & | & -1 \\ 2 & -1 & 1 & | & 3 \\ 3 & -1 & -2 & | & 7 \end{bmatrix}$$
$$R_1 \leftrightarrow R_2$$

Next, we proceed to get 0's in the second and third rows of the first column:

$$\begin{bmatrix} 1 & 1 & 0 & | & -1 \\ 2 & -1 & 1 & | & 3 \\ 3 & -1 & -2 & | & 7 \end{bmatrix} \sim \begin{bmatrix} 1 & 1 & 0 & | & -1 \\ 0 & -3 & 1 & | & 5 \\ 0 & -4 & -2 & | & 10 \end{bmatrix}$$
$$-2 \times R_1 + R_2 \to R_2$$
$$-3 \times R_1 + R_3 \to R_3$$

To complete the procedure, we get a 0 in the third row of the second column:

$$\begin{bmatrix} 1 & 1 & 0 & | & -1 \\ 0 & -3 & 1 & | & 5 \\ 0 & -4 & -2 & | & 10 \end{bmatrix} \sim \begin{bmatrix} 1 & 1 & 0 & | & -1 \\ 0 & -3 & 1 & | & 5 \\ 0 & 0 & -10 & | & 10 \end{bmatrix}$$
$$-4 \times R_2 + 3 \times R_3 \to R_3$$

Now we need to verify the result, the augmented matrix of system V. Once we obtain this last matrix, we can solve the system by back-substitution, as before.

We shall now provide additional illustrations of this procedure, but these will help you only if you take a pencil and paper and carry out the detailed row operations as they are indicated.

EXAMPLE 1 Using matrices to solve a system of equations with one solution

Use matrices to solve the system:

$$2x - y + 2z = 3$$
$$2x + 2y - z = 0$$
$$-x + 2y + 2z = -12$$

SOLUTION The augmented matrix is

$$\begin{bmatrix} 2 & -1 & 2 & | & 3 \\ 2 & 2 & -1 & | & 0 \\ -1 & 2 & 2 & | & -12 \end{bmatrix} \sim \begin{bmatrix} 2 & -1 & 2 & | & 3 \\ 0 & 3 & -3 & | & -3 \\ 0 & 3 & 6 & | & -21 \end{bmatrix}$$
$$-R_1 + R_2 \to R_2$$
$$R_1 + 2 \times R_3 \to R_3$$

$$\sim \begin{bmatrix} 2 & -1 & 2 & | & 3 \\ 0 & 1 & -1 & | & -1 \\ 0 & 0 & 9 & | & -18 \end{bmatrix}$$
$$-R_2 + R_3 \to R_3$$
$$\tfrac{1}{3} \times R_2 \to R_2$$

PROBLEM 1
Use matrices to solve the system:

$$2x + 3y + z = 14$$
$$x - y + 2z = 8$$
$$-x + 4y - z = 2$$

Answers to PROBLEMS

1. $(2, 2, 4)$

This last matrix is in echelon form and corresponds to the system
$$2x - y + 2z = 3 \quad (1)$$
$$y - z = -1 \quad (2)$$
$$9z = -18 \quad (3)$$

We solve this system by back-substitution. Equation (3) immediately yields $z = -2$. Equation (2) then becomes
$$y + 2 = -1$$
so that
$$y = -3$$
Equation (1) then becomes
$$2x + 3 - 4 = 3$$
so that
$$x = 2$$
The final answer, $x = 2$, $y = -3$, $z = -2$, can be checked in the given system.

Calculator Corner

Solve a System Using Matrices

Calculators use matrix operations (too advanced to discuss in detail here) to solve systems of equations. Let's illustrate the procedure to solve Example 1. Let A be the coefficient matrix and B the constant matrix defined by:

$$A = \begin{bmatrix} 2 & -1 & 2 \\ 2 & 2 & -1 \\ -1 & 2 & 2 \end{bmatrix}$$

$$B = \begin{bmatrix} 3 \\ 0 \\ -12 \end{bmatrix}$$

To enter A, press [2nd] [x^{-1}] [▶][▶] [1] [3] [ENTER] [3] [ENTER]. This tells the calculator you are about to enter a 3 × 3 matrix. Now, enter the values for A by pressing [2] [ENTER] [(−)] [1] [ENTER] [2] [ENTER] [2] [ENTER] [2] [ENTER] [(−)] [1] [ENTER] [(−)] [1] [ENTER] [2] [ENTER] [2] [ENTER].

To enter B press [2nd] [x^{-1}] [▶][▶] [2] [3] [ENTER] [1] [ENTER]. Next, enter the values for B by pressing [3] [ENTER] [0] [ENTER] [(−)] [1] [2] [ENTER]. Press [2nd] [MODE] to go to the home screen. Finally, press [2nd] [x^{-1}] [1] [x^{-1}] [2nd] [x^{-1}] [2] [ENTER]. The solution is as shown in the window.

This means that $x = 2$, $y = -3$, and $z = -2$, as obtained in Example 1. See if you can follow the same procedure to solve Example 2 but do not get alarmed if you get an error message. (There is no solution to the system in Example 2.)

```
[A]⁻¹[B]
         [[2 ]
          [-3]
          [-2]]
```

EXAMPLE 2 Using matrices to solve a system of equations with no solution

Use matrices to solve the system:

$$2x - y + 2z = 3$$
$$2x + 2y - z = 0$$
$$4x + y + z = 5$$

SOLUTION

$$\begin{bmatrix} 2 & -1 & 2 & | & 3 \\ 2 & 2 & -1 & | & 0 \\ 4 & 1 & 1 & | & 5 \end{bmatrix} \sim \begin{bmatrix} 2 & -1 & 2 & | & 3 \\ 0 & 3 & -3 & | & -3 \\ 0 & 3 & -3 & | & -1 \end{bmatrix}$$

$-R_1 + R_2 \longrightarrow R_2$
$-2 \times R_1 \times R_3 \longrightarrow R_3$

$$\sim \begin{bmatrix} 2 & -1 & 2 & | & 3 \\ 0 & 3 & -3 & | & -3 \\ 0 & 0 & 0 & | & 2 \end{bmatrix}$$

$-R_2 + R_3 \longrightarrow R_3$

The final matrix is in echelon form, and the last line corresponds to the equation

$$0x + 0y + 0z = 2$$

which is false for all values of x, y, z. Hence the given system has *no solution*.

PROBLEM 2

Use matrices to solve the system:

$$x + 2y - z = 4$$
$$5x - 3y + 2z = 1$$
$$6x - y + z = 3$$

NOTE

If reduction to echelon form introduces any row with all 0's to the left and a nonzero number to the right of the vertical line, then the system has no solution.

EXAMPLE 3 Using matrices to solve a system of equations with infinitely many solutions

Solve the system:

$$2x - y + 2z = 3$$
$$2x + 2y - z = 0$$
$$4x + y + z = 3$$

SOLUTION

$$\begin{bmatrix} 2 & -1 & 2 & | & 3 \\ 2 & 2 & -1 & | & 0 \\ 4 & 1 & 1 & | & 3 \end{bmatrix} \sim \begin{bmatrix} 2 & -1 & 2 & | & 3 \\ 0 & 3 & -3 & | & -3 \\ 0 & 3 & -3 & | & -3 \end{bmatrix}$$

$-R_1 + R_2 \longrightarrow R_2$
$-2 \times R_1 + R_3 \longrightarrow R_3$

$$\sim \begin{bmatrix} 2 & -1 & 2 & | & 3 \\ 0 & 3 & -3 & | & -3 \\ 0 & 0 & 0 & | & 0 \end{bmatrix}$$

$-R_2 + R_3 \longrightarrow R_3$

PROBLEM 3

Use matrices to solve the system of equations.

$$-4x + y + z = 4$$
$$2x - y + 3z = 5$$
$$6x - 2y + 2z = 1$$

Answers to PROBLEMS

2. No solution   **3.** Infinitely many solutions such that for any real number, k, solutions can be obtained from the ordered triple $(2k - 4\frac{1}{2}, 7k - 14, k)$.

The last matrix is in echelon form, and the last line corresponds to the equation

$$0x + 0y + 0z = 0$$

which is true for all values of x, y, and z. Any solution of the first two equations will be a solution of the system. The first two equations from the equivalent matrix in echelon form are

$$2x - y + 2z = 3 \quad \textbf{(1)}$$
$$3y - 3z = -3 \quad \textbf{(2)}$$

This system is equivalent to

$$2x - y = 3 - 2z \quad \text{Subtract 2z from both sides.} \quad \textbf{(3)}$$
$$y = -1 + z \quad \text{Solve equation (2) for y.} \quad \textbf{(4)}$$

Suppose we let $z = k$, where k is any real number. Then equations (3) and (4) become

$$2x - y = 3 - 2k \quad \textbf{(5)}$$
$$y = -1 + k \quad \textbf{(6)}$$

Substitution of $y = -1 + k$ into equation (5) results in

$$2x - (-1 + k) = 3 - 2k$$

and we solve for x.

$$2x + 1 - k = 3 - 2k \quad \text{Remove parentheses.}$$
$$2x = 2 - k \quad \text{Subtract 1 and add k.}$$
$$x = 1 - \tfrac{1}{2}k \quad \text{Divide by 2.}$$

Thus, if k is any real number, then $x = 1 - \tfrac{1}{2}k$, $y = k - 1$, $z = k$ is a solution of the system. You can verify this by substitution in the original system. We see that the system in this example has infinitely many solutions, because the value of k may be arbitrarily chosen. For instance, if $k = 2$, then the solution is $x = 0$, $y = 1$, $z = 2$; if $k = 5$, then $x = -\tfrac{3}{2}$, $y = 4$, $z = 5$; if $k = -4$, then $x = 3$, $y = -5$, $z = -4$; and so on. In conclusion, we can say this system has infinitely many solutions and for any real number, k, solutions can be obtained from the ordered triple $(1 - \tfrac{1}{2}k, k - 1, k)$.

In the final echelon form, the system has infinite solutions if there is no row with all 0's to the left and a nonzero number to the right of the vertical line but there *is* a row with all 0's to both the left and the right.

Examples 1–3 illustrate the three possibilities for three linear equations in three unknowns. The system may have **one unique solution** as in Example 1; the system may have **no solution** as in Example 2; or the system may have **infinitely many solutions** as in Example 3. The final echelon form of the matrix always shows which case is at hand.

C › Solving Applications Using Matrices

Here is a problem from *Using Your Knowledge,* Section 4.3. Now let's solve this problem using matrices.

EXAMPLE 4 Creating a diet

A dietitian wants to arrange a diet composed of three basic foods A, B, and C. The diet must include 170 units of calcium, 90 units of iron, and 110 units of vitamin B. The table gives the number of units per ounce of each of the needed ingredients contained in each of the basic foods.

	Units per Ounce		
Nutrient	Food A	Food B	Food C
Calcium	15	5	20
Iron	5	5	10
Vitamin B	10	15	10

SOLUTION

1. **Read the problem.** If a, b, and c are the number of ounces of basic foods A, B, and C taken by an individual, find the number of ounces of each of the basic foods needed to meet the diet requirements.

2. **Select the unknown.** We want to find the values of a, b, and c, the number of ounces of basic foods A, B, and C taken by an individual.

3. **Think of a plan.** We write a system of equations: What is the amount of calcium needed? The individual gets 15 units of calcium from A, 5 from B, and 20 from C, so the amount of calcium is

$$15a + 5b + 20c = 170$$

What is the amount of iron needed?

$$5a + 5b + 10c = 90$$

What is the amount of vitamin B needed?

$$10a + 15b + 10c = 110$$

4. **Use matrices or your calculator to solve the system.** Write the equations obtained using matrices. The simplified system of three equations and three unknowns, obtained by dividing each term in each of the equations by 5, is written as

$$3a + b + 4c = 34$$
$$a + b + 2c = 18$$
$$2a + 3b + 2c = 22$$

The corresponding augmented matrix is

$$\begin{bmatrix} 3 & 1 & 4 & | & 34 \\ 1 & 1 & 2 & | & 18 \\ 2 & 3 & 2 & | & 22 \end{bmatrix}$$

$$\begin{bmatrix} 3 & 1 & 4 & | & 34 \\ 1 & 1 & 2 & | & 18 \\ 2 & 3 & 2 & | & 22 \end{bmatrix} \sim \begin{bmatrix} 1 & 1 & 2 & | & 18 \\ 3 & 1 & 4 & | & 34 \\ 2 & 3 & 2 & | & 22 \end{bmatrix} \sim \begin{bmatrix} 1 & 1 & 2 & | & 18 \\ 0 & -2 & -2 & | & -20 \\ 0 & 1 & -2 & | & -14 \end{bmatrix}$$

$$R_1 \leftrightarrow R_2 \qquad R_2 - 3R_1 \rightarrow R_2$$
$$R_3 - 2R_1 \rightarrow R_3$$

$$\sim \begin{bmatrix} 1 & 1 & 2 & | & 18 \\ 0 & 1 & 1 & | & 10 \\ 0 & 1 & -2 & | & -14 \end{bmatrix} \sim \begin{bmatrix} 1 & 1 & 2 & | & 18 \\ 0 & 1 & 1 & | & 10 \\ 0 & 0 & -3 & | & -24 \end{bmatrix}$$

$$\frac{R_2}{-2} \rightarrow R_2 \qquad R_3 - R_2 \rightarrow R_3$$

PROBLEM 4

A landscaping company placed three orders with a nursery. The first order was for 24 bushes, 16 flowering plants, and 10 trees and totaled $374. The second order was for 36 bushes, 48 flowering plants, and 20 trees and totaled $828. The third order was for 50 bushes, 40 flowering plants, and 30 trees and totaled $970. The bill does not list the per item price. What is the cost of one bush, one flowering plant, and one tree?

Answers to PROBLEMS

4. One bush is $4, one flowering plant is $8, and one tree is $15.

From the third row, $-3c = -24$, or $c = 8$. Substituting in the second row, we have $b + 8 = 10$, or $b = 2$. Finally, substituting in the first row, we have $a + 2 + 2(8) = 18$, or $a = 0$.

5. Verify the solution. Substitute $a = 0$, $b = 2$, and $c = 8$ into the first equation to obtain $15 \cdot 0 + 5 \cdot 2 + 20 \cdot 8 = 10 + 160 = 170$. Then use the same procedure to check the second and third equations. The required answer is 0 ounces of food A, 2 ounces of food B, and 8 ounces of food C.

EXAMPLE 5 Matrices and Construction

Tom Jones, who was building a workshop, went to the hardware store and bought 1 pound each of three types of nails: small, medium, and large. After completing part of the work, Tom found that he had underestimated the number of small and large nails he needed. So he bought another pound of the small nails and 2 pounds more of the large nails. After some more work, he again ran short of nails and had to buy another pound of each of the small and the medium nails. Upon looking over his bills, he found that the hardware store charged him $2.10 for nails the first time, $2.30 the second time, and $1.20 the third time. The prices for the various sizes of nails were not listed. Find these prices.

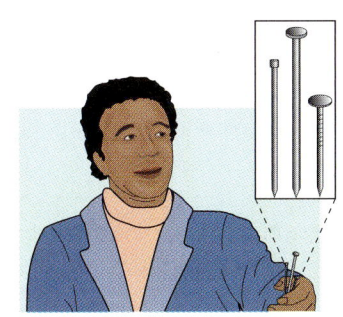

SOLUTION

1. Read the problem. There were three purchases of nails costing $2.10, $2.30, and $1.20.

2. Select the unknown. We let x, y, and z be the prices in cents per pound for the small, medium, and large nails, respectively.

3. Think of a plan. Then we know that

$$x + y + z = 210$$
$$x + 2z = 230$$
$$x + y = 120$$

4. Use matrices to solve the system. We solve this system as follows:

$$\begin{bmatrix} 1 & 1 & 1 & | & 210 \\ 1 & 0 & 2 & | & 230 \\ 1 & 1 & 0 & | & 120 \end{bmatrix} \sim \begin{bmatrix} 1 & 1 & 1 & | & 210 \\ 0 & 1 & -1 & | & -20 \\ 0 & 0 & 1 & | & 90 \end{bmatrix}$$

$$R_1 - R_2 \rightarrow R_2$$
$$R_1 - R_3 \rightarrow R_3$$

The second matrix is in echelon form, and the solution of the system is easily found by back-substitution to be $x = 50$, $y = 70$, $z = 90$, giving the schedule of prices shown.

Tom's Schedule of Nail Prices

Nail Size	Price per Pound
Small	50¢
Medium	70¢
Large	90¢

5. Verify the solution. Substitute $x = 50$, $y = 70$, and $z = 90$ in $x + y + z = 210$ to obtain $50 + 70 + 90 = 210$, which is true.

PROBLEM 5

After Tom built his workshop (see Example 5), he had another project requiring small, medium, and large nails. This time he went to a different hardware store. On the first trip he bought 1 pound of each size nail for $2.15. On the second trip, he bought 4 pounds of small nails and 1 pound of medium nails for $2.85. He needed one more trip, where he bought 1 pound of medium nails and 2 pounds of large nails and paid $2.55. How much did this hardware store charge for each size of nail?

Answers to PROBLEMS

5. Small: 55¢/lb; medium: $65¢/lb; large: 95¢/lb

Exercises A.4

‹A› Perform Elementary Operations on Systems of Equations
‹B› Solve Systems of Linear Equations Using Matrices

‹AB› In Problems 1–10, find all the solutions (if possible).

1. $\begin{aligned} x + y - z &= 3 \\ x - 2y + z &= -3 \\ 2x + y + z &= 4 \end{aligned}$

2. $\begin{aligned} x + 2y - z &= 5 \\ 2x + y + z &= 1 \\ x - y + z &= -1 \end{aligned}$

3. $\begin{aligned} 2x - y + 2z &= 5 \\ 2x + y - z &= -6 \\ 3x + 2z &= 3 \end{aligned}$

4. $\begin{aligned} x + 2y - z &= 0 \\ 2x + 3y &= 3 \\ 2y + z &= -1 \end{aligned}$

5. $\begin{aligned} 3x + 2y + z &= -5 \\ 2x - y - z &= -6 \\ 2x + y + 3z &= 4 \end{aligned}$

6. $\begin{aligned} 4x + 3y - z &= 12 \\ 2x - 3y - z &= -10 \\ x + y - 2z &= -5 \end{aligned}$

7. $\begin{aligned} x + y + z &= 3 \\ x - 2y + z &= -3 \\ 3x + 3z &= 5 \end{aligned}$

8. $\begin{aligned} x + y + z &= 3 \\ x - 2y + 3z &= 5 \\ 5x - 4y + 11z &= 20 \end{aligned}$

9. $\begin{aligned} x + y + z &= 3 \\ x - 2y + z &= -3 \\ x + z &= 1 \end{aligned}$

10. $\begin{aligned} x - y - 2z &= -1 \\ x + 2y + z &= 5 \\ 5x + 4y - z &= 13 \end{aligned}$

‹C› Solving Applications Using Matrices   In Problems 11–16, use matrices to solve the system.

11. *Money* Nancy's change drawer contains $58 in dimes, quarters, and one-dollar coins. If twice the number of quarters is 80 more than the number of one-dollar coins, and the number of quarters plus twice the number of dimes is three times the number of one-dollar coins, find the number of each type of coin in Nancy's drawer.

12. *Money* The sum of $8.50 is made up of nickels, dimes, and quarters. The number of dimes is equal to the number of quarters plus twice the number of nickels. The value of the dimes exceeds the combined value of the nickels and the quarters by $1.50. How many of each coin are there?

13. *Vending Machines* The Mechano Distributing Company has three types of vending machines, which dispense snacks as listed in the table. Mechano fills all the machines once a day and finds them all sold out before the next day. The total daily sales are candy, 760; peanuts, 380; sandwiches, 660. How many of each type of machine does Mechano have?

Mechano Distributing Company Data

Snack	Vending Machine Type		
	I	II	III
Candy	20	24	30
Peanuts	10	18	10
Sandwiches	0	30	30

14. *Vending Machines* Suppose the total daily income from the various types of vending machines in Problem 13 is as follows: type I, $32.00; type-II, $159.00; type III, $192.00. What is the selling price for each type of snack? (Use your answers from Problem 13.)

15. *Fertilizer* Gro-Kwik Garden Supply has three types of fertilizer, which contain chemicals A, B, and C in the percentages shown in the table. In what proportions must Gro-Kwik mix these three types to get an 8-8-8 fertilizer (one that has 8% of each of the three chemicals)?

Gro-Kwik Garden Supply Data

Chemical	Type of Fertilizer		
	I	II	III
A	6%	8%	12%
B	6%	12%	8%
C	8%	4%	12%

16. *Water Supply* Three water supply valves, A, B, and C, are connected to a tank. If all three valves are opened, the tank is filled in 8 hours. The tank can also be filled by opening A for 8 hours and B for 12 hours, while keeping C closed, or by opening B for 10 hours and C for 28 hours, while keeping A closed. Find the time needed by each valve to fill the tank by itself. (*Hint:* Let x, y, and z, respectively, be the fractions of the tank that valves A, B, and C can fill alone in 1 hour.)

17. *Matrices* A 2×2 matrix (2 rows and 2 columns)

$$\begin{bmatrix} a & b \\ c & d \end{bmatrix}$$

is said to be **singular** if $ad - bc = 0$. Otherwise, it is nonsingular. Determine whether the following matrices are singular or nonsingular.

a. $\begin{bmatrix} 1 & 2 \\ 2 & 4 \end{bmatrix}$ b. $\begin{bmatrix} 2 & -3 \\ 3 & 5 \end{bmatrix}$ c. $\begin{bmatrix} 0 & 2 \\ 2 & 4 \end{bmatrix}$

18. Matrices The product of two matrices,

$$\begin{bmatrix} a & b \\ c & d \end{bmatrix} \text{ and } \begin{bmatrix} x & y \\ z & w \end{bmatrix}$$

is found by a row-column multiplication defined as follows:

$$\begin{bmatrix} a & b \\ c & d \end{bmatrix} \begin{bmatrix} x & y \\ z & w \end{bmatrix} = \begin{bmatrix} ax + bz & ay + bw \\ cx + dz & cy + dw \end{bmatrix}$$

If this product equals

$$\begin{bmatrix} 1 & 0 \\ 0 & 1 \end{bmatrix}$$

each of the two matrices is said to be the **multiplicative inverse** of the other. To find the inverse of

$$\begin{bmatrix} a & b \\ c & d \end{bmatrix}$$

you have to find a matrix

$$\begin{bmatrix} x & y \\ z & w \end{bmatrix}$$

such that

$$\begin{bmatrix} a & b \\ c & d \end{bmatrix} \begin{bmatrix} x & y \\ z & w \end{bmatrix} = \begin{bmatrix} 1 & 0 \\ 0 & 1 \end{bmatrix}$$

This means that you must solve the systems

$$ax + bz = 1 \qquad ay + bw = 0$$
$$cx + dz = 0 \text{ and } cy + dw = 1$$

To solve the first system when $c \neq 0$, we can write

$$\begin{bmatrix} a & b & | & 1 \\ c & d & | & 0 \end{bmatrix} \sim \begin{bmatrix} ac & bc & | & c \\ ac & ad & | & 0 \end{bmatrix} \sim \begin{bmatrix} ac & bc & | & c \\ 0 & ad - bc & | & -c \end{bmatrix}$$

Now we see that the second equation has a unique solution if and only if $ad - bc \neq 0$.

Use these ideas to find the inverse of

a. $\begin{bmatrix} 3 & 2 \\ 2 & 1 \end{bmatrix}$ **b.** $\begin{bmatrix} 1 & -2 \\ 2 & 1 \end{bmatrix}$

>>> Using Your Knowledge

Finding Your Identity Instead of stopping with the echelon form of the matrix and then using back-substitution to solve a system of equations, many people prefer to transform the matrix of coefficients into the **identity matrix** (a square matrix with 1's along the diagonal and 0's everywhere else) and then read off the solution by inspection. If the final augmented matrix reads

$$\begin{bmatrix} 1 & 0 & 0 & | & a \\ 0 & 1 & 0 & | & b \\ 0 & 0 & 1 & | & c \end{bmatrix}$$

then the solution of the system is $x = a, y = b, z = c$. You need only read the column to the right of the vertical line.

Suppose, for example, that we have reduced the augmented matrix to the form

$$\begin{bmatrix} 2 & -1 & 2 & | & 3 \\ 0 & 3 & -3 & | & -3 \\ 0 & 0 & 2 & | & 5 \end{bmatrix}$$

We can now divide R_2 by 3 and R_3 by 2 to obtain

$$\begin{bmatrix} 2 & -1 & 2 & | & 3 \\ 0 & 1 & -1 & | & -1 \\ 0 & 0 & 1 & | & \frac{5}{2} \end{bmatrix}$$

To get 0's in the off-diagonal places of the matrix to the left of the vertical line, we first add R_2 to R_1 to get the new R_1:

$$\begin{bmatrix} 2 & 0 & 1 & | & 2 \\ 0 & 1 & -1 & | & -1 \\ 0 & 0 & 1 & | & \frac{5}{2} \end{bmatrix}$$

Next, we subtract R_3 from R_1 and add R_3 to R_2, with the result

$$\begin{bmatrix} 2 & 0 & 0 & | & -\frac{1}{2} \\ 0 & 1 & 0 & | & \frac{3}{2} \\ 0 & 0 & 1 & | & \frac{5}{2} \end{bmatrix}$$

Finally, we divide R_1 by 2 to obtain

$$\begin{bmatrix} 1 & 0 & 0 & | & -\frac{1}{4} \\ 0 & 1 & 0 & | & \frac{3}{2} \\ 0 & 0 & 1 & | & \frac{5}{2} \end{bmatrix}$$

from which we can see by inspection that the solution of the system is $x = -\frac{1}{4}, y = \frac{3}{2}, z = \frac{5}{2}$.

After you bring the augmented matrix into echelon form, you perform additional elementary row operations to get 0's in all the off-diagonal places of the coefficient matrix and 1's on the diagonal. We assume that the system has a unique solution. If this is not the case, you will see what the situation is when you obtain the echelon form, and you will also see that it's impossible to transform the coefficient matrix into the identity matrix if the system doesn't have a unique solution. Use these ideas to solve Problems 1–10 in this exercise set.

〉〉〉 Write On

When solving systems of equations using matrices, explain how you recognize each type of system.

19. A consistent system

20. An inconsistent system

21. A dependent system

22. Describe the relationship between elementary row operations on matrices and elementary operations on systems of equations.

〉〉〉 Mastery Test

Solve the systems using matrices.

23. $2x + 3y - z = -1$
$3x + 4y + 2z = 14$
$x - 6y - 5z = 4$

24. $5x + 2y + 4z = -5$
$7x + 8y - 2z = 13$
$2x - 5y + 3z = 4$

25. $5x + 6y - 30z = 13$
$2x + 4y - 12z = 6$
$x + 2y - 6z = 8$

26. $x + y + z = 4$
$x - 2y + z = 7$
$2x - y + 2z = 11$

27. $3x - y + z = 3$
$x - 2y + 2z = 1$
$2x + y - z = 2$

28. $x - y + z = 3$
$x + 2y - z = 3$
$2x - 2y + 2z = 6$

29. The sum of $14.10 is made up of nickels, dimes, and quarters. The number of quarters is equal to the number of dimes plus three times the number of nickels. The value of the quarters exceeds the combined value of the dimes and nickels by $8.90. How many of each coin are there?

〉〉〉 Skill Checker

Evaluate:

30. $(8)(-5) - (22)(5)$

31. $(13)(-4) - (21)(3)$

32. $(2)(-4) - (5)(93)$

33. $\dfrac{(5)(-1) - (3)(1)}{(1)(-1) - (3)(1)}$

34. $\dfrac{(1)(3) - (3)(5)}{(1)(-3) - (1)(93)}$

35. $\dfrac{(-3)(1) - (-2)(5)}{(2)(1) - (-1)(-3)}$

A.5 Determinants and Cramer's Rule

Objectives

A Evaluate a 2 × 2 determinant.

B Use Cramer's rule to solve a system of two equations in two unknowns.

C Use minors to evaluate 3 × 3 determinants.

D Use Cramer's rule to solve a system of three equations.

To Succeed, Review How To . . .

Evaluate expressions involving integers (pp. 18–27).

Getting Started
Leibniz and Matrices

Gottfried Wilhelm Leibniz developed the theory of determinants. In 1693, Leibniz studied and used *determinants* to solve systems of simultaneous equations. In this section we will use determinants to solve systems of equations.

Gottfried Wilhelm Leibniz (1646–1716)

A Evaluating 2 × 2 Determinants

A **determinant** is a square array of numbers of the form

$$\begin{vmatrix} a_1 & b_1 \\ a_2 & b_2 \end{vmatrix}$$

The numbers a_1, a_2, b_1, and b_2 are called the *elements* of the determinant. As you can see, this determinant has *two* rows and *two* columns. For this reason,

$$\det A = \begin{vmatrix} a_1 & b_1 \\ a_2 & b_2 \end{vmatrix}$$

is called a two-by-two (2 × 2) determinant. Every **square matrix** (a matrix with the same number of rows and columns) has a determinant associated with it.

Each 2 × 2 determinant has a real number associated with it. The value of

$$\det A = \begin{vmatrix} a_1 & b_1 \\ a_2 & b_2 \end{vmatrix}$$

is defined to be

$$a_1 b_2 - a_2 b_1$$

which can be obtained by multiplying along the diagonals, as indicated in the following definition.

DETERMINANT

The determinant of the matrix $\begin{bmatrix} a_1 & b_1 \\ a_2 & b_2 \end{bmatrix}$ is denoted by $\begin{vmatrix} a_1 & b_1 \\ a_2 & b_2 \end{vmatrix}$ and is defined as

$$\begin{vmatrix} a_1 & b_1 \\ a_2 & b_2 \end{vmatrix} = a_1 b_2 - a_2 b_1$$

For example,

$$\begin{vmatrix} 2 & 5 \\ -3 & -9 \end{vmatrix} = (2)(-9) - (-3)(5) = -18 + 15 = -3$$

EXAMPLE 1 Evaluating determinants

Evaluate:

a. $\begin{vmatrix} -3 & 7 \\ -5 & 4 \end{vmatrix}$

b. $\begin{vmatrix} -3 & 6 \\ -5 & 10 \end{vmatrix}$

SOLUTION

a. $\begin{vmatrix} -3 & 7 \\ -5 & 4 \end{vmatrix} = (-3)(4) - (-5)(7) = -12 - (-35) = 23$

b. $\begin{vmatrix} -3 & 6 \\ -5 & 10 \end{vmatrix} = (-3)(10) - (-5)(6) = -30 - (-30) = 0$

PROBLEM 1

Evaluate:

a. $\begin{vmatrix} -2 & 6 \\ -3 & -5 \end{vmatrix}$

b. $\begin{vmatrix} 2 & -8 \\ -3 & -12 \end{vmatrix}$

In Example 1(b) the second-column elements of the determinant are both the same multiple of the corresponding first-column elements:

$$6 = (-2)(-3) \quad \text{and} \quad 10 = (-2)(-5)$$

In general, if the elements of one row are just some constant k times the elements of the other row (or if this is true of the columns), then the value of the determinant is zero. This is easy to see:

$$\begin{vmatrix} a & b \\ ka & kb \end{vmatrix} = akb - bka = 0$$

This result will be of use to us in *Using Your Knowledge* in Exercise A.5.

B › Using Cramer's Rule to Solve a System of Two Equations

One of the important applications of determinants is in the solution of a system of linear equations. Let's look at the simplest case, two equations in two unknowns. Such a system can be written

$$a_1 x + b_1 y = d_1 \tag{1}$$
$$a_2 x + b_2 y = d_2 \tag{2}$$

We can eliminate y from this system by multiplying equation (1) by b_2 and equation (2) by b_1 and then subtracting, as follows.

$\quad\quad a_1 b_2 x + b_1 b_2 y = d_1 b_2$ This is b_2 times equation (1).
$(-)\, a_2 b_1 x + b_1 b_2 y = d_2 b_1$ This is b_1 times equation (2).
$\quad\quad a_1 b_2 x - a_2 b_1 x = d_1 b_2 - d_2 b_1$ Subtract the second equation from the first.
$\quad\quad (a_1 b_2 - a_2 b_1) x = d_1 b_2 - d_2 b_1$ Factor out x.

Answers to PROBLEMS

1. a. 28 b. -48

Now if the quantity $a_1b_2 - a_2b_1 \neq 0$, we can divide by this quantity to get

$$x = \frac{d_1b_2 - d_2b_1}{a_1b_2 - a_2b_1} \quad \text{Solve for } x.$$

The denominator, $a_1b_2 - a_2b_1$, which we shall denote by D, can be written as the determinant

$$D = \begin{vmatrix} a_1 & b_1 \\ a_2 & b_2 \end{vmatrix} = a_1b_2 - a_2b_1$$

This determinant is naturally called the **determinant of the coefficients.** The numerator, $d_1b_2 - d_2b_1$, can be obtained from the denominator by replacing the coefficients of x (the a's) by the corresponding constant terms, the d's. We denote the numerator of x by D_x. Thus,

Constant terms replace the a's. $\qquad D_x = \begin{vmatrix} d_1 & b_1 \\ d_2 & b_2 \end{vmatrix} \qquad$ y coefficients unchanged

We can now write the solution for x in the form

$$x = \frac{D_x}{D}$$

A similar procedure shows that the solution for y is

$$y = \frac{a_1d_2 - a_2d_1}{a_1b_2 - a_2b_1}$$

The denominator, $a_1b_2 - a_2b_1$, is again the determinant D. The numerator, $a_1d_2 - a_2d_1$, which we shall denote by D_y, can be formed from D by replacing the coefficients of y (the b's) by the corresponding d's. Hence,

x coefficients unchanged $\qquad D_y = \begin{vmatrix} a_1 & d_1 \\ a_2 & d_2 \end{vmatrix} \qquad$ Constant terms replace the y coefficients.

and we can write the solution for y in the form

$$y = \frac{D_y}{D}$$

We summarize these results as follows.

CRAMER'S RULE FOR SOLVING A SYSTEM OF TWO EQUATIONS IN TWO UNKNOWNS

The system

$$a_1x + b_1y = d_1$$
$$a_2x + b_2y = d_2$$

1. Has the unique solution

$$x = \frac{D_x}{D} \quad \text{and} \quad y = \frac{D_y}{D}$$

where $D \neq 0$ and

$$D_x = \begin{vmatrix} d_1 & b_1 \\ d_2 & b_2 \end{vmatrix}, \quad D_y = \begin{vmatrix} a_1 & d_1 \\ a_2 & d_2 \end{vmatrix}, \quad D = \begin{vmatrix} a_1 & b_1 \\ a_2 & b_2 \end{vmatrix}$$

(continued)

2. The system is inconsistent and has no solution if $D = 0$ and any one of D_x or D_y is different from 0.
3. The system has no unique solution if $D = 0$ and $D_x = D_y = 0$. (In this case, the system either has no solution or infinitely many solutions. This situation is studied more fully in advanced algebra.)

EXAMPLE 2   Using Cramer's rule to solve a system of two equations

Use Cramer's rule to solve the system:

$$2x + 3y = 7$$
$$5x + 9y = 11$$

SOLUTION

$$D = \begin{vmatrix} 2 & 3 \\ 5 & 9 \end{vmatrix} = 18 - 15 = 3 \quad \text{Use the coefficients of the variables.}$$

$$D_x = \begin{vmatrix} 7 & 3 \\ 11 & 9 \end{vmatrix} = 63 - 33 = 30 \quad \begin{array}{l}\text{These are the constant terms.}\\ \text{These are the coefficients of } y.\end{array}$$

$$D_y = \begin{vmatrix} 2 & 7 \\ 5 & 11 \end{vmatrix} = 22 - 35 = -13 \quad \begin{array}{l}\text{These are the coefficients of } x.\\ \text{These are the constant terms.}\end{array}$$

Therefore,

$$x = \frac{D_x}{D} = \frac{30}{3} = 10 \quad y = \frac{D_y}{D} = \frac{-13}{3} = -\frac{13}{3}$$

CHECK   Substituting $x = 10$, $y = -\frac{13}{3}$ in the equations, we obtain

$$(2)(10) + (3)\left(-\frac{13}{3}\right) = 20 - 13 = 7$$

$$(5)(10) + (9)\left(-\frac{13}{3}\right) = 50 - 39 = 11$$

The answers do satisfy the given equations, and the solution is $x = 10$, $y = -\frac{13}{3}$, or $\left(10, -\frac{13}{3}\right)$.

PROBLEM 2

Use Cramer's rule to solve the system:

$$2x + 3y = 16$$
$$5x - 4y = 17$$

C › Using Minors to Evaluate 3 × 3 Determinants

We can also solve a system of three linear equations in three unknowns by using determinants. A 3 × 3 determinant—a determinant with three rows and three columns—and is written as

$$\det A = \begin{vmatrix} a_1 & b_1 & c_1 \\ a_2 & b_2 & c_2 \\ a_3 & b_3 & c_3 \end{vmatrix}$$

The value of a 3 × 3 determinant can be defined by using the idea of a *minor* of an element in a determinant. Here is the definition of minor.

Answers to PROBLEMS

2. (5, 2)

MINOR

In the determinant

$$\begin{vmatrix} a_1 & b_1 & c_1 \\ a_2 & b_2 & c_2 \\ a_3 & b_3 & c_3 \end{vmatrix}$$

the **minor** of an element is the determinant that remains after deleting the row and column in which the element appears. Thus, the minor of a_1 is

$$\begin{vmatrix} b_2 & c_2 \\ b_3 & c_3 \end{vmatrix}$$

the minor of b_1 is

$$\begin{vmatrix} a_2 & c_2 \\ a_3 & c_3 \end{vmatrix}$$

and the minor of c_1 is

$$\begin{vmatrix} a_2 & b_2 \\ a_3 & b_3 \end{vmatrix}$$

The value of a 3×3 determinant is defined as follows:

$$\begin{vmatrix} a_1 & b_1 & c_1 \\ a_2 & b_2 & c_2 \\ a_3 & b_3 & c_3 \end{vmatrix} = a_1 \begin{vmatrix} b_2 & c_2 \\ b_3 & c_3 \end{vmatrix} - b_1 \begin{vmatrix} a_2 & c_2 \\ a_3 & c_3 \end{vmatrix} + c_1 \begin{vmatrix} a_2 & b_2 \\ a_3 & b_3 \end{vmatrix}$$

The 2×2 determinants in this equation are clearly minors of the elements in the first row of the 3×3 determinant on the left; the right-hand side is called the **expansion** of the 3×3 determinant by minors along the first row. An expansion with numbers is carried out in Example 3.

EXAMPLE 3 Using minors to evaluate a 3×3 determinant

Evaluate the determinant by expanding by minors along the first row:

$$\begin{vmatrix} 1 & 1 & 1 \\ 1 & 2 & 1 \\ 1 & 1 & 2 \end{vmatrix}$$

SOLUTION

$$\begin{vmatrix} 1 & 1 & 1 \\ 1 & 2 & 1 \\ 1 & 1 & 2 \end{vmatrix} = (1)\begin{vmatrix} 2 & 1 \\ 1 & 2 \end{vmatrix} - (1)\begin{vmatrix} 1 & 1 \\ 1 & 2 \end{vmatrix} + (1)\begin{vmatrix} 1 & 2 \\ 1 & 1 \end{vmatrix}$$

$$= (1)(3) - (1)(1) + (1)(-1)$$

$$= 3 - 1 - 1 = 1$$

PROBLEM 3

Evaluate the determinant by expanding by minors along the first row:

$$\begin{vmatrix} 2 & -3 & 4 \\ -1 & 2 & 1 \\ 1 & 1 & -2 \end{vmatrix}$$

It's also possible to expand a determinant by the minors of *any row* or *any column*. To do this, it's necessary to define the *sign array* of a determinant.

Answers to PROBLEMS

3. -19

SIGN ARRAY

For a 3 × 3 determinant, the **sign array** is the following arrangement of alternating signs:

$$\begin{vmatrix} + & - & + \\ - & + & - \\ + & - & + \end{vmatrix}$$

To obtain the expansion of

$$\begin{vmatrix} a_1 & b_1 & c_1 \\ a_2 & b_2 & c_2 \\ a_3 & b_3 & c_3 \end{vmatrix}$$

along a particular row or column, we write the corresponding sign from the array in front of each term in the expansion. For example, to expand

$$\begin{vmatrix} 1 & 1 & 1 \\ 1 & 2 & 1 \\ 1 & 1 & 2 \end{vmatrix}$$

along the *second row,* we write

$$\begin{vmatrix} 1 & 1 & 1 \\ 1 & 2 & 1 \\ 1 & 1 & 2 \end{vmatrix} = -(1)\begin{vmatrix} 1 & 1 \\ 1 & 2 \end{vmatrix} + (2)\begin{vmatrix} 1 & 1 \\ 1 & 2 \end{vmatrix} - (1)\begin{vmatrix} 1 & 1 \\ 1 & 1 \end{vmatrix}$$

$$= -1(1) + 2(1) - 1(0) = 1$$

We used the signs of the *second row* of the sign array, − + −, for the first, second, and third terms, respectively.

EXAMPLE 4 Expanding determinants by minors

Expand the given determinant along the third column:

a. $\begin{vmatrix} 0 & 1 & 1 \\ 1 & 2 & -1 \\ 1 & -1 & 3 \end{vmatrix}$
b. $\begin{vmatrix} 1 & 1 & 0 \\ 0 & -1 & 1 \\ 2 & -1 & -3 \end{vmatrix}$

SOLUTION

a. $\begin{vmatrix} 0 & 1 & 1 \\ 1 & 2 & -1 \\ 1 & -1 & 3 \end{vmatrix} = +(1)\begin{vmatrix} 1 & 2 \\ 1 & -1 \end{vmatrix} - (-1)\begin{vmatrix} 0 & 1 \\ 1 & -1 \end{vmatrix} + (3)\begin{vmatrix} 0 & 1 \\ 1 & 2 \end{vmatrix}$

$$= (1)(-3) + (1)(-1) + (3)(-1) = -3 - 1 - 3 = -7$$

b. $\begin{vmatrix} 1 & 1 & 0 \\ 0 & -1 & 1 \\ 2 & -1 & -3 \end{vmatrix} = +(0)\begin{vmatrix} 0 & -1 \\ 2 & -1 \end{vmatrix} - (1)\begin{vmatrix} 1 & 1 \\ 2 & -1 \end{vmatrix} + (-3)\begin{vmatrix} 1 & 1 \\ 0 & -1 \end{vmatrix}$

$$= 0 - (1)(-3) + (-3)(-1) = 6$$

PROBLEM 4

Expand the given determinant along the third column:

a. $\begin{vmatrix} -2 & -1 & 4 \\ 0 & 1 & -1 \\ -1 & 2 & 1 \end{vmatrix}$

b. $\begin{vmatrix} -1 & 3 & -2 \\ -4 & 2 & 0 \\ 0 & -1 & 1 \end{vmatrix}$

NOTE

When expanding a 3 × 3 determinant, it's easier to expand along the row or column containing the most zeros, since the coefficient of the resulting minors would then be zero. This will save you time.

Answers to PROBLEMS

4. a. −3 b. 2

D ▶ Cramer's Rule for Solving a System of Three Equations

We can solve the system

$$a_1 x + b_1 y + c_1 z = d_1$$
$$a_2 x + b_2 y + c_2 z = d_2$$
$$a_3 x + b_3 y + c_3 z = d_3$$

in exactly the same manner as we solved the system of equations with two unknowns. The details are similar to those for a system of two equations. We state these results here.

CRAMER'S RULE FOR SOLVING A SYSTEM OF THREE EQUATIONS IN THREE UNKNOWNS

The system

$$a_1 x + b_1 y + c_1 z = d_1$$
$$a_2 x + b_2 y + c_2 z = d_2$$
$$a_3 x + b_3 y + c_3 z = d_3$$

1. Has the unique solution

$$x = \frac{D_x}{D}, \quad y = \frac{D_y}{D}, \quad z = \frac{D_z}{D}$$

where $D \neq 0$ and

$$D = \begin{vmatrix} a_1 & b_1 & c_1 \\ a_2 & b_2 & c_2 \\ a_3 & b_3 & c_3 \end{vmatrix}, \quad D_x = \begin{vmatrix} d_1 & b_1 & c_1 \\ d_2 & b_2 & c_2 \\ d_3 & b_3 & c_3 \end{vmatrix},$$

$$D_y = \begin{vmatrix} a_1 & d_1 & c_1 \\ a_2 & d_2 & c_2 \\ a_3 & d_3 & c_3 \end{vmatrix}, \quad \text{and} \quad D_z = \begin{vmatrix} a_1 & b_1 & d_1 \\ a_2 & b_2 & d_2 \\ a_3 & b_3 & d_3 \end{vmatrix}.$$

2. Is *inconsistent* and has no solution if $D = 0$ and any one of D_x, D_y, D_z is different from zero.

3. Has *no unique* solution if $D = 0$ and $D_x = D_y = D_z = 0$. (In this case, the system either has no solution or infinitely many solutions. This situation is studied more fully in advanced algebra.)

D is the determinant of the coefficients; D_x is formed from D by replacing the coefficients of x (the a's) by the corresponding d's; D_y is formed from D by replacing the coefficients of y (the b's) by the corresponding d's; and D_z is formed similarly.

EXAMPLE 5 Using Cramer's rule to solve a system of three equations in three unknowns

Use Cramer's rule to solve the system:

$$x + y + 2z = 7$$
$$x - y - 3z = -6$$
$$2x + 3y + z = 4$$

SOLUTION To use Cramer's rule, we first have to evaluate D, the determinant of the coefficients. If D is not zero, then we calculate the other three determinants, D_x, D_y, and D_z. For the given system,

PROBLEM 5

Use Cramer's rule to solve the system:

$$x + y + z = 6$$
$$x - y + z = 2$$
$$x + y - 2z = -3$$

(continued)

Answers to PROBLEMS

5. $(1, 2, 3)$

$$D = \begin{vmatrix} 1 & 1 & 2 \\ 1 & -1 & -3 \\ 2 & 3 & 1 \end{vmatrix} = (1)\begin{vmatrix} -1 & -3 \\ 3 & 1 \end{vmatrix} - (1)\begin{vmatrix} 1 & -3 \\ 2 & 1 \end{vmatrix} + (2)\begin{vmatrix} 1 & -1 \\ 2 & 3 \end{vmatrix}$$

$$= (1)(-1 + 9) - (1)(1 + 6) + (2)(3 + 2)$$

$$= 8 - 7 + 10 = 11$$

Since $D \neq 0$, we calculate the other three determinants.

$$D_x = \begin{vmatrix} 7 & 1 & 2 \\ -6 & -1 & -3 \\ 4 & 3 & 1 \end{vmatrix} = (7)\begin{vmatrix} -1 & -3 \\ 3 & 1 \end{vmatrix} - (1)\begin{vmatrix} -6 & -3 \\ 4 & 1 \end{vmatrix} + (2)\begin{vmatrix} -6 & -1 \\ 4 & 3 \end{vmatrix}$$

$$= (7)(-1 + 9) - (1)(-6 + 12) + (2)(-18 + 4)$$

$$= 56 - 6 - 28 = 22$$

$$D_y = \begin{vmatrix} 1 & 7 & 2 \\ 1 & -6 & -3 \\ 2 & 4 & 1 \end{vmatrix} = (1)\begin{vmatrix} -6 & -3 \\ 4 & 1 \end{vmatrix} - (7)\begin{vmatrix} 1 & -3 \\ 2 & 1 \end{vmatrix} + (2)\begin{vmatrix} 1 & -6 \\ 2 & 4 \end{vmatrix}$$

$$= (1)(-6 + 12) - (7)(1 + 6) + (2)(4 + 12)$$

$$= 6 - 49 + 32 = -11$$

$$D_z = \begin{vmatrix} 1 & 1 & 7 \\ 1 & -1 & -6 \\ 2 & 3 & 4 \end{vmatrix} = (1)\begin{vmatrix} -1 & -6 \\ 3 & 4 \end{vmatrix} - (1)\begin{vmatrix} 1 & -6 \\ 2 & 4 \end{vmatrix} + (7)\begin{vmatrix} 1 & -1 \\ 2 & 3 \end{vmatrix}$$

$$= (1)(-4 + 18) - (1)(4 + 12) + (7)(3 + 2)$$

$$= 14 - 16 + 35 = 33$$

By Cramer's rule,

$$x = \frac{D_x}{D} = \frac{22}{11} = 2, \quad y = \frac{D_y}{D} = \frac{-11}{11} = -1, \quad z = \frac{D_z}{D} = \frac{33}{11} = 3$$

You can check the solution $(2, -1, 3)$ by substituting in the given equations.

EXAMPLE 6 Using Cramer's rule to solve a system of three equations in three unknowns

Use Cramer's rule to solve the system:

$$x + y - z = 2$$
$$2x + y + z = 4$$
$$-x - y + z = 3$$

SOLUTION By Cramer's rule, if there is a unique solution, it is given by

$$x = \frac{D_x}{D}, \quad y = \frac{D_y}{D}, \quad \text{and} \quad z = \frac{D_z}{D}, \text{ where } D \neq 0$$

However,

$$D = \begin{vmatrix} 1 & 1 & -1 \\ 2 & 1 & 1 \\ -1 & -1 & 1 \end{vmatrix} = (1)\begin{vmatrix} 1 & 1 \\ -1 & 1 \end{vmatrix} - (1)\begin{vmatrix} 2 & 1 \\ -1 & 1 \end{vmatrix} + (-1)\begin{vmatrix} 2 & 1 \\ -1 & -1 \end{vmatrix}$$

$$= 2 - 3 + 1 = 0$$

PROBLEM 6

Use Cramer's rule to solve the system:

$$x - y - z = 2$$
$$x + 2y + z = 6$$
$$-x + y + z = 4$$

Answers to PROBLEMS

6. Since $D = 0$ and $D_x \neq 0$, the system is inconsistent. There is no solution.

Since $D = 0$, we must try calculating each of the other three determinants to see if we get one of them to be different from zero.

$$D_x = \begin{vmatrix} 2 & 1 & -1 \\ 4 & 1 & 1 \\ 3 & -1 & 1 \end{vmatrix}$$

$$= (2)\begin{vmatrix} 1 & 1 \\ -1 & 1 \end{vmatrix} - (1)\begin{vmatrix} 4 & 1 \\ 3 & 1 \end{vmatrix} + (-1)\begin{vmatrix} 4 & 1 \\ 3 & -1 \end{vmatrix}$$

$$= 4 - 1 + 7 = 10 \ne 0$$

D_x is different from zero. The system is inconsistent and has no solution.

Calculator Corner

Solving a System by Matrices

Most calculators have a feature that calculates the determinant of a matrix. Let's start by erasing any matrices you may have in memory by pressing `2nd` `+` `2` `5` and `DEL` as many times as there are matrices listed. Let's enter the matrix A from Example 1(a) and find its determinant. As you recall,

$$A = \begin{bmatrix} -3 & 7 \\ -5 & 4 \end{bmatrix}$$

Press `2nd` `x⁻¹` ◁ ▷ `ENTER` to edit matrix A. Now enter `2` `ENTER` `2` `ENTER` for the number of rows and columns and then the values of the matrix A. Go to the home screen by pressing `2nd` `MODE`. To find the determinant on the TI-83 Plus, press `2nd` `x⁻¹` ▷ `1` `2nd` `x⁻¹` `1` `)` `ENTER` and you get the answer 23 (Window 1). Now let's do Example 2. Since the calculator uses the matrices A, B, C, D, and so on, let $A = D$, $B = D_x$, and $C = D_y$. Enter the values for A, B, and C as defined in Example 2. In your home screen, matrices A and B should look like those shown in Window 2. Since $x = \frac{B}{A}$, press `2nd` `x⁻¹` ▷ `1` `2nd` `x⁻¹` `2` `)` (this will enter det $[B]$ on the home screen). Then press `÷` `2nd` `x⁻¹` ▷ `1` `2nd` `x⁻¹` `1` `)` `ENTER` (this will divide by det $[A]$). The result is 10, as shown in Window 3. You can use the same procedure to find y. Now that you know how to find determinants, you can do the rest of the examples.

Window 1
Det $[A]$ from Example 1

Window 2
Matrices A and B from Example 2

```
det ([B])/det ([A])
                 10
```

Window 3
Finding the solution for x in Example 2

Exercises A.5

⟨ A ⟩ Evaluating 2 × 2 Matrix
In Problems 1–10, evaluate the determinant.

1. $\begin{vmatrix} 1 & 1 \\ 0 & 2 \end{vmatrix}$

2. $\begin{vmatrix} 2 & -1 \\ 4 & 3 \end{vmatrix}$

3. $\begin{vmatrix} -3 & -2 \\ 5 & 1 \end{vmatrix}$

4. $\begin{vmatrix} 2 & -1 \\ -3 & 1 \end{vmatrix}$

5. $\begin{vmatrix} -2 & 0 \\ 5 & -3 \end{vmatrix}$

6. $\begin{vmatrix} 5 & 2 \\ -10 & -4 \end{vmatrix}$

7. $\begin{vmatrix} \frac{1}{2} & \frac{-1}{4} \\ \frac{1}{2} & \frac{3}{4} \end{vmatrix}$

8. $\begin{vmatrix} \frac{1}{5} & \frac{1}{10} \\ \frac{1}{2} & \frac{1}{4} \end{vmatrix}$

9. $\begin{vmatrix} \frac{3}{5} & \frac{1}{2} \\ \frac{-1}{4} & \frac{-1}{2} \end{vmatrix}$

10. $\begin{vmatrix} \frac{4}{5} & \frac{-1}{3} \\ \frac{-1}{2} & \frac{1}{2} \end{vmatrix}$

⟨ B ⟩ Using Cramer's Rule to Solve a System of Two Equations
In Problems 11–30, solve the system using Cramer's rule. If the system is dependent or inconsistent, state that fact.

11. $x + y = 5$
    $3x - y = 3$

12. $x + y = 9$
    $x - y = 3$

13. $x + y = 9$
    $x - y = -1$

14. $2x + y = -1$
    $x - 2y = -13$

15. $4x + 9y = 3$
    $3x + 7y = 2$

16. $5x + 2y = 32$
    $3x + y = 18$

17. $x - y = -1$
    $x - 2y = -6$

18. $x - 2y = -13$
    $3x - 2y = -19$

19. $2x + 3y = -13$
    $6x + 9y = -39$

20. $4x + 5y = -2$
    $12x + 15y = -6$

21. $x - y = 1$
    $x - 2y = 4$

22. $x - 2y = 4$
    $4x - 5y = 7$

23. $x + 3y = 6$
    $2x + 6y = 5$

24. $x - 2y = 3$
    $-x + 2y = 6$

25. $x = 7y + 3$
    $2x + 3y = 23$

26. $x = 3y + 1$
    $2x + 3y = 20$

27. $y = -3x + 17$
    $2x - y = 8$

28. $y = -2x + 14$
    $3x - y = 11$

29. $\frac{x}{2} - \frac{y}{3} = \frac{-1}{6}$
    $\frac{x}{3} + \frac{y}{4} = \frac{-7}{12}$

30. $\frac{x}{3} - \frac{y}{5} = \frac{4}{3}$
    $\frac{x}{4} - \frac{y}{3} = \frac{1}{12}$

⟨ C ⟩ Using Minors to Evaluate 3 × 3 Determinants
In Problems 31–40, evaluate the determinant.

31. $\begin{vmatrix} 1 & 3 & 2 \\ 2 & 4 & 1 \\ 3 & 6 & 5 \end{vmatrix}$

32. $\begin{vmatrix} 1 & 3 & 5 \\ 2 & 0 & 10 \\ -3 & 1 & -15 \end{vmatrix}$

33. $\begin{vmatrix} 1 & 2 & 3 \\ 4 & 5 & 6 \\ 7 & 8 & 9 \end{vmatrix}$

34. $\begin{vmatrix} 1 & 1 & 1 \\ 2 & 3 & 1 \\ 2 & 4 & 1 \end{vmatrix}$

35. $\begin{vmatrix} 2 & 1 & 3 \\ 1 & 2 & -1 \\ 3 & 1 & 5 \end{vmatrix}$

36. $\begin{vmatrix} -1 & 1 & -1 \\ -2 & 2 & -6 \\ 3 & -3 & 4 \end{vmatrix}$

37. $\begin{vmatrix} 1 & 1 & 6 \\ 1 & 1 & 4 \\ 1 & -1 & 2 \end{vmatrix}$

38. $\begin{vmatrix} 1 & 4 & 0 \\ 1 & -3 & 1 \\ 0 & 8 & -1 \end{vmatrix}$

39. $\begin{vmatrix} 0 & -1 & 2 \\ 2 & 1 & -3 \\ 1 & -3 & 1 \end{vmatrix}$

40. $\begin{vmatrix} -3 & 2 & -4 \\ 1 & -1 & 3 \\ 1 & 2 & 10 \end{vmatrix}$

⟨ D ⟩ Cramer's Rule for Solving a System of Three Equations
In Problems 41–50, solve the system using Cramer's rule. Use minors to expand the resulting determinants.

41. $x + y + z = 6$
    $2x - 3y + 3z = 5$
    $3x - 2y - z = -4$

42. $x + y + z = 13$
    $3x + y - 3z = 5$
    $x - 2y + 4z = 10$

43. $6x + 5y + 4z = 5$
    $5x + 4y + 3z = 5$
    $4x + 3y + z = 7$

44. $3x + 2y + z = 4$
$4x + 3y + z = 5$
$5x + y + z = 9$

45. $x - 2y + 3z = 15$
$5x + 7y - 11z = -29$
$-13x + 17y + 19z = 37$

46. $2x - y + z = 3$
$x + 2y + z = 12$
$4x - 3y + z = 1$

47. $5x + 3y + 5z = 3$
$3x + 5y + z = -5$
$2x + 2y + 3z = 7$

48. $x + y = 5$
$y + z = 3$
$x + z = 7$

49. $2y + z = 9$
$-2y + z = 1$
$x + y + z = 1$

50. $x - y = 3$
$y - z = 3$
$x + z = 9$

In Problems 51–60, show that the statement is true.

51. $\begin{vmatrix} a & b & 0 \\ c & d & 0 \\ e & f & 0 \end{vmatrix} = 0$

52. $\begin{vmatrix} a & b & c \\ d & e & f \\ 0 & 0 & 0 \end{vmatrix} = 0$

53. $\begin{vmatrix} a & b & c \\ 1 & 2 & 3 \\ a & b & c \end{vmatrix} = 0$

54. $\begin{vmatrix} 1 & a & a \\ 2 & b & b \\ 3 & c & c \end{vmatrix} = 0$

55. $\begin{vmatrix} 1 & 2 & 3 \\ 3 & 1 & 2 \\ 3k & 2k & k \end{vmatrix} = k \begin{vmatrix} 1 & 2 & 3 \\ 3 & 1 & 2 \\ 3 & 2 & 1 \end{vmatrix}$

56. $\begin{vmatrix} 1 & 2 & 3k \\ 3 & 2 & k \\ 3 & 1 & 2k \end{vmatrix} = k \begin{vmatrix} 1 & 2 & 3 \\ 3 & 2 & 1 \\ 3 & 1 & 2 \end{vmatrix}$

57. $\begin{vmatrix} kb_1 & b_1 & 1 \\ kb_2 & b_2 & 2 \\ kb_3 & b_3 & 3 \end{vmatrix} = 0$

58. $\begin{vmatrix} b_1 & b_2 & b_3 \\ kb_1 & kb_2 & kb_3 \\ 1 & 2 & 3 \end{vmatrix} = 0$

59. $\begin{vmatrix} 1 & 1 & 1 \\ 2 & a & a \\ 3 & b & b \end{vmatrix} = 0$

60. $\begin{vmatrix} 0 & 0 & 0 \\ a & b & c \\ d & e & f \end{vmatrix} = 0$

〉〉〉 Using Your Knowledge

Determining the Equations of Lines Determinants provide a convenient and neat way of writing the equation of a line through two points. Suppose the line is to pass through the points (x_1, y_1) and (x_2, y_2); then an equation of the line can be written in the form

$$\begin{vmatrix} x & y & 1 \\ x_1 & y_1 & 1 \\ x_2 & y_2 & 1 \end{vmatrix} = 0$$

It's easy to verify. First, think of expanding the determinant by minors along the first row. The coefficients of x and y will be constants, so the equation is linear. Next, you can see that (x_1, y_1) and (x_2, y_2) both satisfy the equation, because if you substitute either of these pairs for (x, y) in the first row of the determinant, the result is a determinant with two identical rows and this you know is zero. (See Problem 53.) The equation is that of the desired line.

As an illustration, let's find an equation of the line through $(1, 3)$ and $(-5, -2)$. The determinant form of this equation is

$$\begin{vmatrix} x & y & 1 \\ 1 & 3 & 1 \\ -5 & -2 & 1 \end{vmatrix} = 0$$

or

$$[3 - (-2)]x - [1 - (-5)]y + [-2 - (-15)](1) = 0$$

where the quantities in brackets are the expanded minors of the first-row elements. The final equation is

$$5x - 6y + 13 = 0$$

In Problems 61–66, use the method described above to find the equation of the line passing through the two given points.

61. $(2, 7)$ and $(0, 3)$

62. $(10, 12)$ and $(-7, 1)$

63. $(-1, 4)$ and $(8, 2)$

64. $(5, 0)$ and $(0, -3)$

65. $(a, 0)$ and $(0, b)$, $ab \neq 0$

66. (a, b) and $(0, 0)$, $(a, b) \neq (0, 0)$

>>> Write On

67. How can you determine if a system of equations is consistent when using determinants to solve it?

68. How can you determine if a system of equations is inconsistent when using determinants to solve it?

69. How can you determine if a system of equations is dependent when using determinants to solve it?

70. How can you determine if a system of linear equations has a determinant $D = 0$, what do you know about the system?

71. Explain why a determinant that has a column or row of zeros has a value of zero. (See Problems 51 and 52.)

>>> Mastery Test

Expand the determinant by minors:

72. Along the third row

$$\begin{vmatrix} 0 & 1 & 1 \\ 1 & 2 & -1 \\ 1 & -1 & 3 \end{vmatrix}$$

73. Along the third row

$$\begin{vmatrix} 1 & 1 & 0 \\ 0 & -1 & 1 \\ 2 & -1 & -3 \end{vmatrix}$$

74. Along the first column

$$\begin{vmatrix} 1 & 1 & 1 \\ 1 & 2 & 1 \\ 1 & 1 & 2 \end{vmatrix}$$

75. Use Cramer's rule to solve the system:

$$x - y - z = 2$$
$$x + 2y + z = 6$$
$$-x + y + z = 4$$

76. Use Cramer's rule to solve the system:

$$x + y + z = 6$$
$$x - y + z = 2$$
$$x + y - 2z = 4$$

77. Evaluate the determinant

$$\begin{vmatrix} 2 & -3 & 4 \\ -1 & 2 & 1 \\ 1 & 1 & -2 \end{vmatrix}$$

78. Use Cramer's rule to solve the system:

$$2x + 3y = 13$$
$$5x - 4y = 21$$

79. Evaluate: $\begin{vmatrix} -2 & 6 \\ -3 & 5 \end{vmatrix}$

80. Evaluate: $\begin{vmatrix} -2 & 8 \\ -3 & 12 \end{vmatrix}$

>>> Skill Checker

Determine whether the point $(-3, 1)$ satisfies the inequality.

81. $3x + 2y > 6$

82. $2x + y \geq -2$

83. $-2y < 6$

84. $x + 3 \leq 5$

85. $y < x$

86. $y \geq -x$

A.6 The Binomial Expansion

Objectives

A Use the binomial expansion to expand and simplify a power of a binomial.

B Find the coefficient of a term in a binomial expansion.

C Solve applications involving binomial expansions.

To Succeed, Review How To . . .

1. Simplify an algebraic expression (pp. 56–59).
2. Expand binomials of the form $(x + y)^n$, where $n < 6$ (pp. 360–365).

Getting Started

It's Chinese to Me

This drawing made in A.D. 1303 by Chu Shi-kie turns out to be a nice depiction of Pascal's triangle (Pascal having lived in the 1600s). As you can see, every row starts and ends with the symbol ⊖. If we make ⊖ = 1, we can say the triangle starts with the number 1. The next row has the symbols ⊖ and ⊖, or in our translation, the numbers 1 and 1. The next row has the symbols ⊖, ⊜, and ⊖—that is, the numbers 1, 2, and 1. What are the "numbers" in the next row? Can you see a pattern emerging? It looks like this:

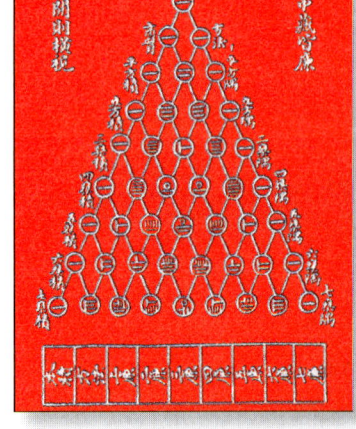

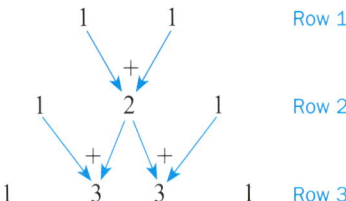

Row 1
Row 2
Row 3

The 2 in the second row is the sum of the two elements in the first row. The first 3 in the third row is the sum of the first two elements in the second row, and so on. If we continue to construct rows in this way, we get the following triangular array

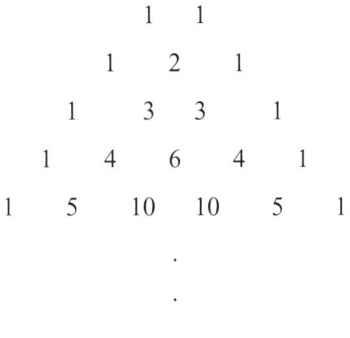

This array is known as Pascal's triangle, in honor of the French mathematician Blaise Pascal. We shall examine Pascal's triangle as our introduction to binomial expansions.

A ▶ Pascal's Triangle and Binomial Expansions

The seventeenth-century mathematician Blaise Pascal wrote a book about the triangle presented in *Getting Started*. In the book, he stated that if we have five coins, the relationship between the number of heads that can come up and the number of ways in which these heads can occur is as follows:

Number of Heads	5	4	3	2	1	0
Number of Occurrences	1	5	10	10	5	1

For example, if five coins are tossed, five heads can only occur one way (HHHHH), but four heads can occur in five ways (THHHH, HTHHH, HHTHH, HHHTH, HHHHT). What do these results have to do with algebra? As it turns out, there is a somewhat different way to arrive at the numbers in Pascal's triangle. From our previous work, you can verify that

$$(x + y)^1 = x + y$$
$$(x + y)^2 = x^2 + 2xy + y^2$$
$$(x + y)^3 = x^3 + 3x^2y + 3xy^2 + y^3$$
$$(x + y)^4 = x^4 + 4x^3y + 6x^2y^2 + 4xy^3 + y^4$$
$$(x + y)^5 = x^5 + 5x^4y + 10x^3y^2 + 10x^2y^3 + 5xy^4 + y^5$$

Look at the coefficients (the numbers that multiply x and y) of the terms in the expansion of $(x + y)^5$. They are 1, 5, 10, 10, 5, and 1. Now compare this with the fifth row in Pascal's triangle. The numbers are identical.

Can we predict the expansion of $(x + y)^6$ by studying these expansions? The pattern for the powers of x and y is easy. In going from left to right, the powers of x decrease from n to 0 and the powers of y increase from 0 to n. Ignoring the coefficients, the expansion of $(x + y)^n$ has the form

$$(x + y)^n = x^n y^0 + \Box x^{n-1}y^1 + \Box x^{n-2}y^2 + \cdots + \Box x^{n-k}y^k + \cdots + y^n x^0$$

where $\Box$ represents the missing coefficients. (The sum of the powers of x and y is always n.) To write the expansion of $(x + y)^n$, we need only know how to find these missing coefficients. In the case of $(x + y)^6$, the coefficients should be 1, 6, 15, 20, 15, 6, and 1, and they are obtained by constructing the sixth row of Pascal's triangle, starting and ending with a 1 and finding each coefficient by the addition pattern we have just discussed. In going from left to right, the powers of x decrease from $n = 6$ to 0, the powers of y increase from 0 to 6, and the sum of the exponents of x and y is always 6. Thus, we have

$$(x + y)^6 = 1x^6y^0 + 6x^5y^1 + 15x^4y^2 + 20x^3y^3 + 15x^2y^4 + 6x^1y^5 + 1x^0y^6$$
$$= x^6 + 6x^5y + 15x^4y^2 + 20x^3y^3 + 15x^2y^4 + 6xy^5 + y^6$$

Here is the generalization for the expression of $(x + y)^n$.

PROCEDURE

The Expansion of $(x + y)^n$

1. The first term is x^n and the last term is y^n.
2. The powers of x *decrease* by 1 from term to term, the powers of y *increase* by 1 from term to term, and the sum of the powers of x and y in any term is always n.
3. The coefficients for the expansion correspond to those in the nth row of Pascal's triangle and can be written as $\binom{n}{r}$ where r is the power of x.

Using this notation,

$$(x + y)^6 = \binom{6}{6}x^6 + \binom{6}{5}x^5y + \binom{6}{4}x^4y^2 + \binom{6}{3}x^3y^3 + \binom{6}{2}x^2y^4 + \binom{6}{1}xy^5 + \binom{6}{0}y^6$$
$$= x^6 + 6x^5y + 15x^4y^2 + 20x^3y^3 + 15x^2y^4 + 6xy^5 + y^6$$

where

$$\binom{6}{6} = 1, \binom{6}{5} = 6, \binom{6}{4} = 15, \binom{6}{3} = 20, \binom{6}{2} = 15, \binom{6}{1} = 6, \text{ and } \binom{6}{0} = 1$$

To facilitate defining $\binom{n}{r}$, we first introduce a special symbol, $n!$ (read "n factorial"), for which we have the following definition.

FACTORIAL

For any natural number n,

$$n! = n(n-1)(n-2) \cdots (1)$$

Thus,

$$5! = 5 \cdot 4 \cdot 3 \cdot 2 \cdot 1 = 120$$
$$7! = 7 \cdot 6 \cdot 5 \cdot 4 \cdot 3 \cdot 2 \cdot 1 = 5040$$

This notation can also be used to denote the product of consecutive integers beginning with integers different from 1. For example,

$$\frac{7!}{4!} = 7 \cdot 6 \cdot 5$$

since

$$\frac{7!}{4!} = \frac{7 \cdot 6 \cdot 5 \cdot \cancel{4} \cdot \cancel{3} \cdot \cancel{2} \cdot \cancel{1}}{\cancel{4} \cdot \cancel{3} \cdot \cancel{2} \cdot \cancel{1}}$$

Similarly,

$$\frac{9!}{5!} = 9 \cdot 8 \cdot 7 \cdot 6 \quad \text{and} \quad \frac{10!}{6!} = 10 \cdot 9 \cdot 8 \cdot 7$$

Also,

$$n! = n[(n-1)(n-2)(n-3)\ldots(1)]$$
$$= n(n-1)!$$

Thus, we can write

$$n! = n(n-1)!$$

With this notation,

$$9! = 9 \cdot 8!$$
$$13! = 13 \cdot 12!$$

If we let $n = 1$ in $n! = n(n-1)!$, we obtain

$$1! = 1 \cdot 0!$$

Hence, we define $0!$ as follows:

ZERO FACTORIAL

$$0! = 1$$

EXAMPLE 1 Calculations with factorials

Find:

a. $4!$ **b.** $\dfrac{8!}{5!}$

SOLUTION

a. Since $n! = n(n-1)(n-2)\cdots(1)$

$$4! = 4 \cdot 3 \cdot 2 \cdot 1 = 24$$

b.
$$8! = 8 \cdot 7 \cdot 6 \cdot 5 \cdot 4 \cdot 3 \cdot 2 \cdot 1$$
$$\text{and } 5! = 5 \cdot 4 \cdot 3 \cdot 2 \cdot 1$$

Thus,

$$\dfrac{8!}{5!} = \dfrac{8 \cdot 7 \cdot 6 \cdot \cancel{5} \cdot \cancel{4} \cdot \cancel{3} \cdot \cancel{2} \cdot \cancel{1}}{\cancel{5} \cdot \cancel{4} \cdot \cancel{3} \cdot \cancel{2} \cdot \cancel{1}} = 336$$

PROBLEM 1

a. $8!$ **b.** $\dfrac{9!}{7!}$

Calculator Corner

Factorials

You can use the numerical capabilities of your calculator in this section. Suppose you want to do Example 1. With a TI-83 Plus, go to the home screen by pressing [2nd] [MODE] and enter the number 4. To get 4!, press [MATH], move the cursor three places to the right so you are in the PRB column and then press 4. The home screen now shows 4!. Press [ENTER] and the answer 24 appears as shown in Window 1.

Now suppose you want to find $\binom{8}{4}$. Most calculators use the notation nCr instead of $\binom{n}{r}$. Thus, to find $\binom{8}{4}$, start at the home screen, enter 8, then press [MATH], move to the PRB column, and press [3] [4] [ENTER]. The answer 70 is obtained, as shown in Window 2.

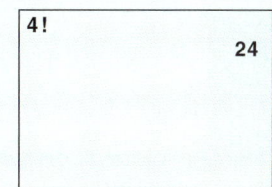

Window 1: 4! 24

Window 2: 8 nCr 4 70

We are now ready to use factorial notation to define the binomial coefficient.

BINOMIAL COEFFICIENT

The **binomial coefficient**, denoted $\binom{n}{r}$, is defined as

$$\binom{n}{r}! = \dfrac{n!}{r!(n-r)!}$$

Thus,

$$\binom{5}{2} = \dfrac{5!}{2!(5-2)!} = \dfrac{5!}{2!3!} = \dfrac{5 \cdot 4 \cdot \cancel{3!}}{2!\cancel{3!}} = 10$$

$$\binom{5}{5} = \dfrac{5!}{5!(5-5)!} = \dfrac{5!}{5!0!} = 1$$

EXAMPLE 2 Evaluating a binomial coefficient

Evaluate: $\binom{6}{4}$

SOLUTION By definition,

$$\binom{n}{r} = \dfrac{n!}{r!(n-r)!}$$

$$\binom{6}{4} = \dfrac{6!}{4!2!} = \dfrac{6 \cdot 5 \cdot \cancel{4!}}{\cancel{4!}2!} = \dfrac{6 \cdot 5}{2} = 15$$

PROBLEM 2

Evaluate: $\binom{7}{5}$

Answers to PROBLEMS

1. **a.** 40,320 **b.** 72   2. 21

Now that we know the meaning of $\binom{n}{r}$, we can use it to write the coefficients of a binomial expansion.

EXAMPLE 3 Writing the coefficients of a binomial expansion

Given that $(a + b)^4 = a^4 + 4a^3b + 6a^2b^2 + 4ab^3 + b^4$, write each coefficient in the expansion using the $\binom{n}{r}$ notation.

SOLUTION

$$(a + b)^4 = a^4 + 4a^3b + 6a^2b^2 + 4ab^3 + b^4$$

$$= \binom{4}{4}a^4 + \binom{4}{3}a^3b + \binom{4}{2}a^2b^2 + \binom{4}{1}ab^3 + \binom{4}{0}b^4$$

You can easily verify that

$$\binom{4}{4} = 1, \binom{4}{3} = 3, \binom{4}{2} = 6, \binom{4}{1} = 4, \text{ and } \binom{4}{0} = 1$$

PROBLEM 3

Given that $(a + b)^5 = a^5 + 5a^4b + 10a^3b^2 + 10a^2b^3 + 5ab^4 + b^5$ write each coefficient using $\binom{n}{r}$ notation.

Following the pattern in Example 3, the general formula for $(x + y)^n$ can be written using the $\binom{n}{r}$ notation, where n represents the exponent of the binomial being expanded and r is the power of x. The powers of x *decrease* by 1 from term to term, while those of y *increase* by 1 from term to term. Thus, we have

THE GENERAL BINOMIAL EXPANSION

$$(x + y)^n = \binom{n}{n}x^n + \binom{n}{n-1}x^{n-1}y^1 + \binom{n}{n-2}x^{n-2}y^2$$
$$+ \cdots + \binom{n}{n-k}x^{n-k}y^k + \cdots + \binom{n}{0}y^n$$

Using this formula, we have

$$(x + y)^5 = \binom{5}{5}x^5 + \binom{5}{4}x^4y^1 + \binom{5}{3}x^3y^2 + \binom{5}{2}x^2y^3 + \binom{5}{1}x^1y^4 + \binom{5}{0}y^5$$

$$= x^5 + 5x^4y + 10x^3y^2 + 10x^2y^3 + 5xy^4 + y^5$$

To avoid confusion, we first write the expansion completely, and *then* substitute the coefficients

$$\binom{5}{5} = 1, \binom{5}{4} = 5, \binom{5}{3} = 10, \binom{5}{2} = 10, \binom{5}{1} = 5, \text{ and } \binom{5}{0} = 1$$

in the expansion. Note that

$$\binom{5}{5} = \binom{5}{0}, \binom{5}{4} = \binom{5}{1}, \text{ and } \binom{5}{3} = \binom{5}{2}$$

In general,

$$\binom{n}{k} = \binom{n}{n-k}$$

This tells you that the coefficients of the first and last terms, second and next to last, and so on, are equal, thus saving you time and computation when you expand binomials.

Answers to PROBLEMS

3. $\binom{5}{5} = 1; \binom{5}{4} = 5; \binom{5}{3} = 10;$
$\binom{5}{2} = 10; \binom{5}{1} = 5; \binom{5}{0} = 1$

EXAMPLE 4 Using the binomial expansion

Expand: $(a - 2b)^5$

SOLUTION We use the binomial expansion with $x = a$, $y = -2b$, and $n = 5$. This gives

$$(a - 2b)^5 = \binom{5}{5}a^5 + \binom{5}{4}a^4(-2b)^1 + \binom{5}{3}a^3(-2b)^2 + \binom{5}{2}a^2(-2b)^3$$
$$+ \binom{5}{1}a^1(-2b)^4 + \binom{5}{0}(-2b)^5$$

We know that

$$\binom{5}{5} = \binom{5}{0} = 1, \binom{5}{4} = \binom{5}{1} = \frac{5!}{1!4!} = 5 \text{ and } \binom{5}{3} = \binom{5}{2} = \frac{5!}{2!3!} = 10$$

Thus,

$$(a - 2b)^5$$
$$= a^5 + 5a^4(-2b)^1 + 10a^3(-2b)^2 + 10a^2(-2b)^3 + 5a^1(-2b)^4 + (-2b)^5$$
$$= a^5 - 10a^4b + 40a^3b^2 - 80a^2b^3 + 80ab^4 - 32b^5$$

PROBLEM 4

Expand: $(2x - y)^4$

B › Finding the Coefficient of a Term in a Binomial Expansion

As you recall, when we expand $(x + y)^n$, in the coefficient $\binom{n}{r}$, n represents the exponent of the binomial being expanded and r is the power of x. We shall use this to find the coefficient of a particular term in an expansion without writing out the whole expansion.

EXAMPLE 5 Finding the coefficient of a term in a binomial expansion

Find the coefficient of a^4b^3 in the expansion of $(a - 3b)^7$.

SOLUTION In our case, $n = 7$, $x = a$, $y = -3b$, and the power of a is $r = 4$. Thus, the coefficient of a^4b^3 can be obtained by simplifying

$$\binom{7}{4}a^4(-3b)^3 = \frac{7!}{4!3!}a^4(-3b)^3 = \frac{7 \cdot 6 \cdot 5}{3 \cdot 2 \cdot 1}a^4(-27b^3) = -945a^4b^3$$

Hence, the coefficient of a^4b^3 is -945.

PROBLEM 5

Find the coefficient of x^3 in the expansion of $(x - 2)^7$.

C › Applications Involving Binomial Expansions

The binomial coefficients for $(x + y)^n$ furnish us with a row of Pascal's triangle. As you recall, the numbers in such a row also tell us how many ways a stated number of heads can come up if n coins are tossed. For example, the first coefficient $\binom{n}{n} = 1$ tells us that the number of ways that n heads can occur when n coins are tossed is 1. The second coefficient,

$$\binom{n}{n-1} = n$$

tells us that the number of ways that $n - 1$ heads can occur when n coins are tossed is n, the third coefficient

$$\binom{n}{n-2} = \frac{n(n-1)}{1 \cdot 2}$$

Answers to PROBLEMS

4. $16x^4 - 32x^3y + 24x^2y^2 - 8xy^3 + y^4$

5. 560

tells us that the number of ways that $n - 2$ heads can occur when n coins are tossed is

$$\frac{n(n-1)}{1 \cdot 2}$$

and so on. We use this idea in Example 6.

EXAMPLE 6 Using the binomial expansion
Six coins are tossed. In how many ways can exactly two heads come up?

SOLUTION In this case, $n = 6$ and $r = 2$. Thus, the number of ways in which exactly two heads can come up when six coins are tossed is

$$\binom{6}{2} = \frac{6!}{2!4!} = \frac{6 \cdot 5}{2 \cdot 1} = 15$$

PROBLEM 6
Seven coins are tossed. In how many ways can exactly three heads come up?

It is shown in probability theory that if p is the probability of an event occurring favorably in *one* trial, then the probability of its occurring favorably r times in n trials is

$$\binom{n}{r} p^r (1-p)^{n-r}$$

For example, if a fair coin is tossed, the probability of its coming up heads is $\frac{1}{2}$. So, if the coin is tossed six times, the probability of getting exactly four heads is

$$\binom{6}{4}\left(\frac{1}{2}\right)^4\left(1 - \frac{1}{2}\right)^2 = \left(\frac{6 \cdot 5}{2 \cdot 1}\right)\left(\frac{1}{16}\right)\left(\frac{1}{4}\right) = \frac{30}{128} = \frac{15}{64}$$

EXAMPLE 7 Using binomial expansion
A fair coin is tossed eight times. Find the probability of getting exactly four heads.

SOLUTION Here $n = 8$, $r = 4$, and $p = \frac{1}{2}$. Substituting in the probability formula, we get

$$\binom{8}{4}\left(\frac{1}{2}\right)^4\left(1-\frac{1}{2}\right)^4 = 70\left(\frac{1}{2}\right)^8 = \frac{70}{256} = \frac{35}{128}$$

PROBLEM 7
A fair coin is tossed 10 times. Find the probability of getting exactly five heads.

> Exercises A.6

Boost your grade at mathzone.com!
> Practice Problems
> NetTutor
> Self-Tests
> e-Professors
> Videos

< A > Pascal's Triangle and Binomial Expansions In Problems 1–14, evaluate the given expression.

1. $3!$
2. $9!$
3. $10!$
4. $2!$
5. $\frac{6!}{2!}$
6. $\frac{11!}{10!}$
7. $\frac{9!}{6!}$
8. $\frac{3!}{0!}$
9. $\binom{6}{2}$
10. $\binom{6}{5}$
11. $\binom{11}{1}$
12. $\binom{11}{0}$
13. $\binom{4}{0}$
14. $\binom{7}{4}$

Answers to PROBLEMS

6. 35   7. $\frac{252}{1024} = \frac{63}{256}$

In Problems 15–26, use the binomial expansion to expand the expressions.

15. $(a + 3b)^4$ **16.** $(a - 3b)^4$

17. $(x + 4)^4$ **18.** $(4x - 1)^4$

19. $(2x - y)^5$ **20.** $(x + 2y)^5$

21. $(2x + 3y)^5$ **22.** $(3x - 2y)^5$

23. $\left(\dfrac{1}{x} - \dfrac{y}{2}\right)^4$ **24.** $\left(\dfrac{x}{2} + \dfrac{3}{y}\right)^4$

25. $(x + 1)^6$ **26.** $(y - 1)^6$

〈 **B** 〉 **Finding the Coefficient of a Term in a Binomial Expansion** In Problems 27–34, find the coefficient of the indicated term in the expansion.

27. Given $(x - 3)^6$; find the coefficient of x^3 **28.** Given $(x + 2)^6$; find the coefficient of x^3

29. Given $(x + 2y)^7$; find the coefficient of x^2y^5 **30.** Given $(y - 2z)^7$; find the coefficient of y^2z^5

31. Given $(2x - 1)^8$; find the coefficient of x^4 **32.** Given $(y + 2z)^8$; find the coefficient of y^4z^4

33. Given $\left(\dfrac{a}{2} - 1\right)^5$; find the coefficient of a^2 **34.** Given $\left(y + \dfrac{z}{2}\right)^5$; find the coefficient of y^3z^2

〈 **C** 〉 **Applications Involving Binomial Expansions**

35. *Coin tosses* Six coins are tossed. In how many ways can exactly three heads come up?

36. *Coin tosses* Six coins are tossed. In how many ways can exactly two heads come up?

37. *Coin tosses* Nine coins are tossed. In how many ways can exactly three heads come up?

38. *Coin tosses* Nine coins are tossed. In how many ways can exactly four heads come up?

39. *Coin tosses* A fair coin is tossed six times. Find the probability of getting exactly two heads.

40. *Coin tosses* A fair coin is tossed six times. Find the probability of getting exactly three heads.

41. *Coin tosses* A fair coin is tossed nine times. Find the probability of getting exactly three heads.

42. *Coin tosses* A fair coin is tossed nine times. Find the probability of getting exactly four heads.

43. *Rolling a die* A single die (plural, dice) is rolled four times. Find the probability that a 4 comes up exactly twice. (*Note:* the probability that a specified number, 1, 2, 3, 4, 5, or 6, comes up in a single throw is $\dfrac{1}{6}$.)

〉〉〉 **Using Your Knowledge**

Follow the Pattern How do you relate the pattern used to construct Pascal's triangle with the $\binom{n}{r}$ notation? As you recall, the numbers in Pascal's triangle are obtained by adding the numbers above and to the left and right of the number in question. Using $\binom{n}{r}$ notation, this fact is written as

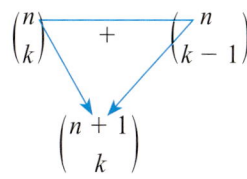

44. Prove that
$$\binom{n}{k} + \binom{n}{k-1} = \binom{n+1}{k}$$

45. In Example 4, we used $\binom{5}{5} = \binom{5}{0}$, $\binom{5}{4} = \binom{5}{1}$ and so on. Prove that
$$\binom{n}{k} = \binom{n}{n-k}$$
where $k \leq n$.

▶▶▶ Write On

46. Write an explanation of how Pascal's triangle is constructed.

47. Give your own definition of $n!$.

48. In the binomial expansion, which part is the "binomial"?

▶▶▶ Mastery Test

Evaluate:

49. $6!$

50. $\dfrac{7!}{3!}$

51. $\binom{6}{2}$

52. $\binom{7}{3}$

Expand:

53. $(2a - b)^4$

54. $(3a - 2b)^5$

55. Find the coefficient of $a^2 b^4$ in the expansion of $(2a - 3b)^6$.

56. Find the coefficient of ab^3 in the expansion of $(3a - b)^4$.

57. Five coins are tossed. In how many ways can exactly 4 heads come up?

58. A fair coin is tossed 7 times. Find the probability of getting exactly 5 tails.

Selected Answers

The brackets preceding answers for the Chapter Review Exercises indicate the chapter, section, and objective for you to review for further study. For example, [3.4C] appearing before an answer means that exercise corresponds to Chapter 3, Section 4, Objective C.

Chapter 1

Exercises 1.1

1. $\{1, 2\}$ 3. $\{5, 6, 7\}$ 5. $\{-1, -2, -3\}$ 7. $\{0, 1, 2, 3\}$
9. $\{1, 2, 3, \ldots\}$ 11. $\{x | x$ is an integer between 0 and 4$\}$
13. $\{x | x$ is an integer between -3 and 3$\}$
15. $\{x | x$ is an even number between 19 and 78$\}$ 17. False   19. True
21. True   23. True   25. True   27. $0.\overline{6}$ 29. 0.875 31. 2.5 33. $1.1\overline{6}$
35.   37.   39.   41.   43.

Set	$\frac{-3}{8}$	$\sqrt{8}$	$0.\overline{3}$	0.9	3.1416
Natural numbers					
Whole numbers					
Integers					
Rational numbers	✓		✓	✓	✓
Irrational numbers		✓			
Real numbers	✓	✓	✓	✓	✓

45. True   47. False   49. -8 51. 7 53. $-\frac{3}{4}$ 55. $\frac{1}{5}$ 57. -0.5
59. $-0.\overline{2}$ 61. $1.\overline{36}$ 63. $-\pi$ 65. 10 67. $\frac{3}{5}$ 69. $0.\overline{5}$ 71. $-\pi$
73. 22 75. 32 77. 3 79. $<$ 81. $>$ 83. $>$ 85. $<$
87. $<$ 89. (a) {Saudi Arabia, Russia, Norway, Iran};
(b) {Kazakhstan, Qatar}   91. (a) Japan with approximately $140 billion; (b) United States, approximately –$135 billion; Great Britain, approximately –$25 billion; (c) >   93. 0.11; Burger King   95. Going from a bachelor's degree to a doctoral degree will give a person a larger percent of increase in earnings since $0.50 > 0.18$.   97. Monjane $8'0.75''$ 99. False. 5 is not a set. To be correct, $\{5\} \subseteq N$ or $5 \in N$.   105. set; element   107. roster notation
109. constant   111. 0.125 113. -8 115. 2 117. $\{x|x$ is 3 or an integer obtained when 3, 4, 5 ... is added in succession$\}$
119. $>$ 121. 2 123. $\frac{1}{2}$ 125. $\frac{13}{8}$ 127. $\frac{7}{6}$ 129. $\frac{10}{63}$ 131. $\frac{2}{21}$
133. $\frac{25}{6}$ 135. 11.58 137. 8.68 139. 3.402 141. 17.44 143. 60

Exercises 1.2

1. $\frac{2}{5}$ 3. -0.1 5. -4 7. -0.8 9. $\frac{1}{5}$ 11. -7 13. -0.3
15. $-\frac{13}{56}$ 17. -14 19. -0.6 21. $-\frac{41}{63}$ 23. -1 25. -4 27. -0.1
29. $\frac{22}{21}$ 31. -40 33. -12 35. 0 37. 60 39. -30 41. 2.86
43. -120 45. 2 47. $-\frac{25}{42}$ 49. $\frac{1}{4}$ 51. $-\frac{15}{4}$ 53. -2 55. -4
57. 2 59. 0 61. Undefined   63. -2 65. 9 67. 3 69. 1
71. -4 73. -7 75. $-\frac{21}{20}$ 77. $\frac{4}{7}$ 79. $-\frac{5}{7}$ 81. $-\frac{1}{2}$ or -0.5
83. $\frac{1}{6}$ or $0.1\overline{6}$ 85. $\frac{14}{15}$ 87. Commutative property of addition
89. Associative property of multiplication   91. Multiplicative inverse
93. Multiplicative identity   95. $A = a(b + c) = ab + ac$
97. $3500°C$ 99. $46   101. $14°C$ 103. (a) 8.848 km; (b) 5.5 mi
105. (a) 52 m; (b) $378,500$ L   107. -1.35 109. -3.045
111. 3.875 115. reciprocal/multiplicative inverse   117. dividend; divisor; quotient   119. multiplicative identity   121. $\frac{16}{25}$ 123. 2
125. $-\frac{59}{56}$ 127. -12.8 129. 16 131. Additive inverse
133. Associative property of multiplication
135. Commutative property of multiplication
137. 16 139. -16 141. $-7; \frac{1}{7}$ 143. 0; Undefined   145. -1
147. 1 149. -24 151. 0

Exercises 1.3

1. -16 3. 25 5. -125 7. 1296 9. -32 11. $\frac{1}{16}$ 13. $\frac{1}{x^5}$
15. $\frac{2}{x^6}$ 17. $\frac{m^2}{n}$ 19. $\frac{6x^2}{yz^3}$ 21. $\frac{1}{64}$ 23. $12x^2$ 25. $-15y^2$ 27. $\frac{20}{a^5}$
29. $-\frac{30y^3}{x^2}$ 31. $-\frac{24y^6}{x^4}$ 33. $-\frac{40}{a^2b^3}$ 35. -30 37. $2x^4$ 39. $\frac{a^2}{2}$
41. $-2x^3y^2$ 43. $-\frac{x}{2}$ 45. $\frac{2}{3a^3}$ 47. $\frac{3}{4}$ 49. $\frac{3b^3}{2a^6}$ 51. $\frac{8x^9}{y^6}$ 53. $-\frac{4y^6}{x^4}$
55. $-\frac{1}{27x^9y^6}$ 57. $\frac{1}{x^{12}y^6}$ 59. $x^{12}y^{12}$ 61. $\frac{a^2}{b^6}$ 63. $-\frac{8b^6}{27a^3}$
65. a^8b^4 67. $\frac{1}{x^{15}y^6}$ 69. $x^{27}y^6$ 71. 2.68×10^8 73. 2.4×10^{-4}
75. $8,000,000$ 77. 0.23 79. 2×10^3 81. 3×10^8 83. 31 yr
85. Approximately 200 times greater   87. 3.34×10^5 89. (a) $\boxed{7.3 \quad 10}$;
(b) $\boxed{1.23 \quad -07}$ 97. scientific notation   99. base; exponent
101. $-\frac{9y^3}{x^3}$ 103. $-3y^{11}$ 105. $\frac{1}{625}$ 107. $-\frac{8x^{12}}{y^{15}}$ 109. $-\frac{1}{243}$
111. $\frac{9x^8y^4}{4}$ 113. 3.87×10^8 115. -48 117. -36
119. -11 121. 1 123. -21 125. 0

Exercises 1.4

1. (a) -26; (b) -70 3. (a) 27; (b) 3 5. -47 7. 20 9. -15
11. -13 13. 57 15. 3 17. -6 19. 1 21. -36 23. -10
25. -4 27. 0 29. 1 31. -20 33. 8 35. -24 37. -33
39. $11,800$ 41. 37 43. $1050   45. 13 47. 32 49. 60
51. $4x - 4y$ 53. $-9a + 9b$ 55. $1.2x - 0.6$ 57. $-\frac{3a}{2} + \frac{6}{7}$ or $\frac{-21a + 12}{14}$
59. $-2x + 6y$ 61. $-2.1 - 3y$ 63. $-4a^2b - 20ab$
65. $-6x - xy$ 67. $8xy - 8y^2$ 69. $2a^3 - 7a^2b$ 71. $0.5x + 0.5y - 1$
73. $-\frac{6}{5}a + \frac{6}{5}b - 6$ 75. $-2x^2 + 2xy - 6xz - 10x$
77. $-0.3x - 0.3y + 0.6z + 1.8$ 79. $-\frac{5}{2}a + 5b - \frac{5}{2}c - 5d + 5$
81. $9x - 6$ 83. $-9x - 8$ 85. $11L - 4W$ 87. $-3xy + 1$
89. $\frac{x}{9} + 2$ 91. $6a + 2b$ 93. $3x^2 - 4xy$ 95. $x - 5y + 36$
97. $-2x^3 + 9x^2 - 3x + 12$ 99. $\frac{2}{7}x^2 + \frac{4}{5}x - \frac{3}{4}$ 101. $4a - 11$
103. $-7a + 10b - 3$ 105. $-4.8x + 3.4y + 5$
107. (a) $82°F$; (b) $191°C$ 109. (a) $1380 interest; (b) $5980 total
111. $v_a = \frac{1}{2}v_1 + \frac{1}{2}v_2$ 113. KE $= \frac{1}{2}mv_1^2 + \frac{1}{2}mv_2^2$ 117. numerical coefficient   119. numerical expression or algebraic expression
121. Terms   123. -3 125. $-2x + 6y - 4z + 8$ 127. $\frac{1}{4}x + 5$
129. $-2a^3 + 2a^2 - a - 17$ 131. 6 133. -8 135. 0 137. -36
139. -3 141. 1 143. 0 145. $6x - 21$ 147. $5x - 8$

Review Exercises

1. [1.1A] (a) $\{4, 5, 6, 7, 8\}$; (b) $\{5, 6, 7\}$ 2. [1.1B] (a) 0.2; (b) 0.4
3. [1.1B] (a) $0.\overline{1}$; (b) $0.\overline{2}$ 4. [1.1C]

Set	0.3	0	$\frac{-3}{4}$	-5	$\sqrt{3}$
Natural numbers					
Whole numbers		✓			
Integers		✓		✓	
Rational numbers	✓	✓	✓	✓	
Irrational numbers					✓
Real numbers	✓	✓	✓	✓	✓

5. [1.1D] (a) 3.5; (b) $-\frac{3}{4}$ 6. [1.1E] (a) 9; (b) 4.2 7. [1.1E] (a) 21; (b) 3 8. [1.1F] (a) $<$; (b) $<$ 9. [1.1F] (a) $>$; (b) $=$

SA-1

10. [1.1F] (a) <; (b) = **11.** [1.2A] (a) −11; (b) −3
12. [1.2A] (a) $-\frac{2}{7}$; (b) −0.6 **13.** [1.2A] (a) 12; (b) 4
14. [1.2A] (a) $\frac{19}{20}$; (b) $\frac{13}{12}$ **15.** [1.2A] (a) −36; (b) −14.4
16. [1.2A] (a) $-\frac{21}{32}$; (b) $\frac{5}{21}$ **17.** [1.2A] (a) 0; (b) Undefined
18. [1.2A] (a) $-\frac{5}{3}$; (b) $\frac{1}{0.3}$ or $\frac{10}{3}$ **19.** [1.2A] (a) $-\frac{9}{4}$; (b) −3
20. [1.2B] (a) Commutative property of addition; (b) Associative property of addition **21.** [1.3A] (a) 81; (b) −81
22. [1.3B] (a) 1; (b) 1; (c) $-\frac{1}{512}$; (d) $\frac{1}{x^{10}}$
23. [1.3B] (a) $\frac{12}{x^2}$; (b) $\frac{-4x^3}{y}$
24. [1.3C] (a) $-\frac{15y^{10}}{x^4}$; (b) $-\frac{24}{x^{11}y^8}$ **25.** [1.3C] (a) $\frac{3}{x^5}$; (b) $-4x^{11}$
26. [1.3C] (a) $-\frac{x}{3}$; (b) $-\frac{2}{x^{11}}$ **27.** [1.3D] (a) $-\frac{8x^{21}}{y^{18}}$; (b) $\frac{16}{x^{24}y^{24}}$
28. [1.3D] (a) $\frac{1}{x^{24}y^{12}}$; (b) $x^{25}y^{15}$
29. [1.3E] (a) 3.4×10^5; (b) 4.7×10^{-5}
30. [1.3E] (a) 37,000; (b) 0.0078 **31.** [1.4A] (a) −75; (b) 50
32. [1.4B] (a) 9; (b) 20 **33.** [1.4B] (a) −7; (b) −73
34. [1.4C] (a) 352 sq in.; (b) −33
35. [1.4D] (a) $-3x + 21$; (b) $2x + 17$
36. [1.4E, F] (a) $-x + 10y - 16$; (b) $3x^2 + 3x + 8$

Chapter 2

Exercises 2.1

1. Yes **3.** Yes **5.** Yes **7.** No **9.** No **11.** 4 **13.** 1 **15.** 2 **17.** 2
19. 7 **21.** −1 **23.** −8 **25.** −4 **27.** 0 **29.** 0 **31.** $\frac{1}{13}$ **33.** 8
35. $-\frac{11}{80}$ **37.** 24 **39.** 1 **41.** 15 **43.** 10 **45.** 4 **47.** 10 **49.** −5
51. 0 **53.** −1 **55.** $-\frac{10}{11}$ **57.** 4 **59.** −6 **61.** 57 **63.** 0 **65.** 16
67. 21.5 **69.** 1500 **71.** Infinitely many solutions, R; identity
73. One solution, $x = 2$; conditional **75.** No solutions, $\emptyset$; contradiction
77. 11 in. **79.** 8 **81.** 12 in. **83.** (a) 47; (b) 48 **87.** contradiction
89. equivalent **91.** real numbers **93.** 10 **95.** 16 **97.** −1 **99.** No
101. $-3x - 7$ **103.** 2 **105.** 50 **107.** 5 **109.** −1 **111.** $\frac{8}{3}$

Exercises 2.2

1. $h = \frac{V}{\pi r^2}$ **3.** $w = \frac{V}{lh}$ **5.** $b = P - s_1 - s_2$ **7.** $s = \frac{A - \pi r^2}{\pi r}$ or $s = \frac{A}{\pi r} - r$ **9.** $V_2 = \frac{P_1 V_1}{P_2}$ **11.** $y = 4 - \frac{2}{3}x$ or $y = \frac{12 - 2x}{3}$
13. (a) $T = \frac{D}{R}$; (b) 4 hr **15.** (a) $A = 34 - 2H$; (b) 18 yr
17. (a) $c = (o)(n) - e$; (b) $29,500 **19.** (a) $B = 180 - A - C$;
(b) 14° **21.** (a) $L = \frac{H - 32}{1.88}$; (b) No; (c) Yes; (d) $L = \frac{H - 29}{1.95}$; (e) Yes
23. $4m + 18$ **25.** $\frac{t}{3}$ **27.** $\frac{1}{2}b - 7$ **29.** $(x + y)(x - y)$
31. (a) $C = 2000 + 309m$; (b) $14,051 **33.** (a) $C = 75 + 150h$
(b) 4hr **35.** (a) $C = 0.36 + 0.08(t - 1)$, $t \geq 1$; (b) 47 min
37. $x = -5$; angles = 35° each **39.** $x = 20$; angles = 140° each
41. $x = 3$; angles = 50° each **43.** $x = -5.5$; $(42 - 6x)° = 75°$; $(50 - 10x)° = 105°$ **45.** 2025 **47.** 3.04 ft **51.** formula
53. literal equations **55.** $x = -20$; angles = 170° each
57. (a) $C = 0.39 + 0.24(w - 1)$, $w \geq 1$; (b) $w = \frac{C - 0.15}{0.24}$; (c) $2.55
59. (a) $C = 6S + 4$; (b) 16°C **61.** $3n + 3$ **63.** 30 **65.** 90 **67.** 30

Exercises 2.3

1. $12m = m + 9$ **3.** $\frac{1}{2}x - 12 = \frac{2x}{3}$ **5.** $500,000 + M = 950,000$
7. $0.12(2500) = I$ **9.** $4x + 5 = 29$; $x = 6$ **11.** $3x + 8 = 35$; $x = 9$
13. $3x - 2 = 16$; $x = 6$ **15.** $5x = 12 + 2x$; $x = 4$ **17.** $\frac{1}{3}x - 2 = 10$;
$x = 36$ **19.** 75% **21.** 8 yr **23.** 130 mi **25.** 44; 46; 48 **27.** −11;
−9; −7 **29.** 87; 92 **31.** (a) 95 mi/hr; (b) 155 mi/hr **33.** 213 lb
35. 47 awards **37.** 12 ft by 5000 ft **39.** 518 ft by 716 ft **41.** 72°
43. 135° **45.** $x = 19$; angles: 67°, 23° **47.** $x = 39$; angles: 127°, 53°
49. 30°; 60°; 90° **51.** 5 **59.** complementary angles
61. straight angle **63.** 31; 33; 35 **65.** $x - 12 = 7x$ **67.** $x = 27$;
angles: 85°, 95° **69.** Approx. 16 million **71.** 50 **73.** $50x - 150$
75. 200 **77.** 45 **79.** 2000 **81.** 5

Exercises 2.4

Translate This

1. H **3.** F **5.** L

1. (a) 64.5; (b) 9.75 **3.** $10.81 **5.** 98.8 million **7.** $23.13
9. Markup: $24; 44.4% of selling price **11.** 40% **13.** 50
15. 20% **17.** 30 mi/gal **19.** 650 **21.** $1100
23. $6.146 billion, $31.169 billion, $6.585 billion

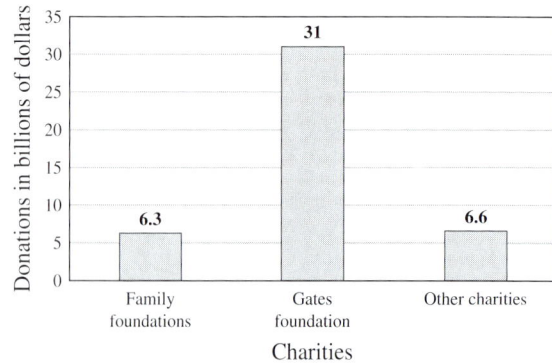

25. 81, 54.5, 53.8, 60.8

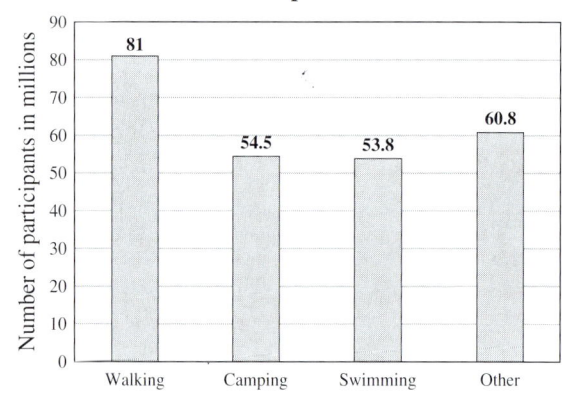

27. $5000 at 6%; $15,000 at 8% **29.** $8000 **31.** (a) 2 hr;
(b) 288 km **33.** 141 mi **35.** 10 mi **37.** 25.2 **39.** 10
41. -290.4 billion **43.** $381.675 billion **49.** $3000 at 5%; $5000
at 10% **51.** 24 million **53.** 20% **55.** −6 **57.** $4y$ **59.** 39
61. $-2x - 1$ **63.** > **65.** > **67.** −7

Exercises 2.5

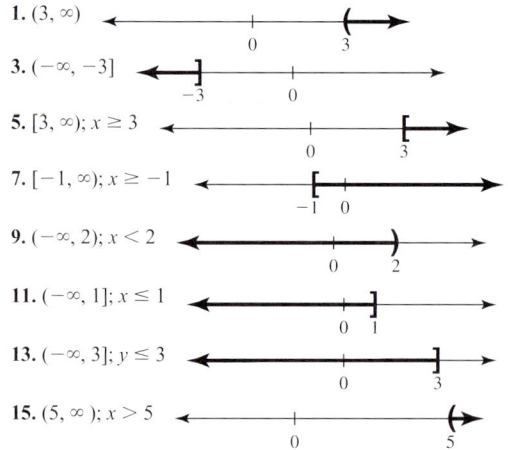

1. $(3, \infty)$

3. $(-\infty, -3]$

5. $[3, \infty)$; $x \geq 3$

7. $[-1, \infty)$; $x \geq -1$

9. $(-\infty, 2)$; $x < 2$

11. $(-\infty, 1]$; $x \leq 1$

13. $(-\infty, 3]$; $y \leq 3$

15. $(5, \infty)$; $x > 5$

17. $(-\infty, 2]; a \le 2$
19. $(-\infty, -4); z < -4$
21. $(-\infty, -1]; x \le -1$
23. $(-\infty, 0]; x \le 0$
25. $\left(-\infty, -\tfrac{11}{80}\right); x < -\tfrac{11}{80}$
27. $[-20, \infty); x \ge -20$
29. $[-2, \infty); x \ge -2$
31. $x < -4$ or $x > 3$; $(-\infty, -4) \cup (3, \infty)$
33. $x < 1$ or $x > 2$; $(-\infty, 1) \cup (2, \infty)$
35. $x \le 4$ or $x > 5$; $(-\infty, 4] \cup (5, \infty)$
37. $x < 2$ or $x > 2$; $(-\infty, 2) \cup (2, \infty)$
39. $x \le 2$ or $x > 3$; $(-\infty, 2] \cup (3, \infty)$
41. $-2 \le x \le 4; [-2, 4]$
43. $2 < x \le 6; (2, 6]$
45. $1 < x < 3; (1, 3)$
47. No solutions; ∅
49. $1 \le x \le 4; [1, 4]$
51. $2 < x < 3; (2, 3)$
53. $1 < x < 3; (1, 3)$
55. $-2 < x < 5; (-2, 5)$
57. $-5 < x < 1; (-5, 1)$
59. $3 < x < 4; (3, 4)$
61. $-2 < x < 4; (-2, 4)$
63. $-6 < y < 1; (-6, 1)$
65. $4 \le y \le 6; [4, 6]$
67. $-2 < x < 4; (-2, 4)$
69. $-3 < a < 3; (-3, 3)$

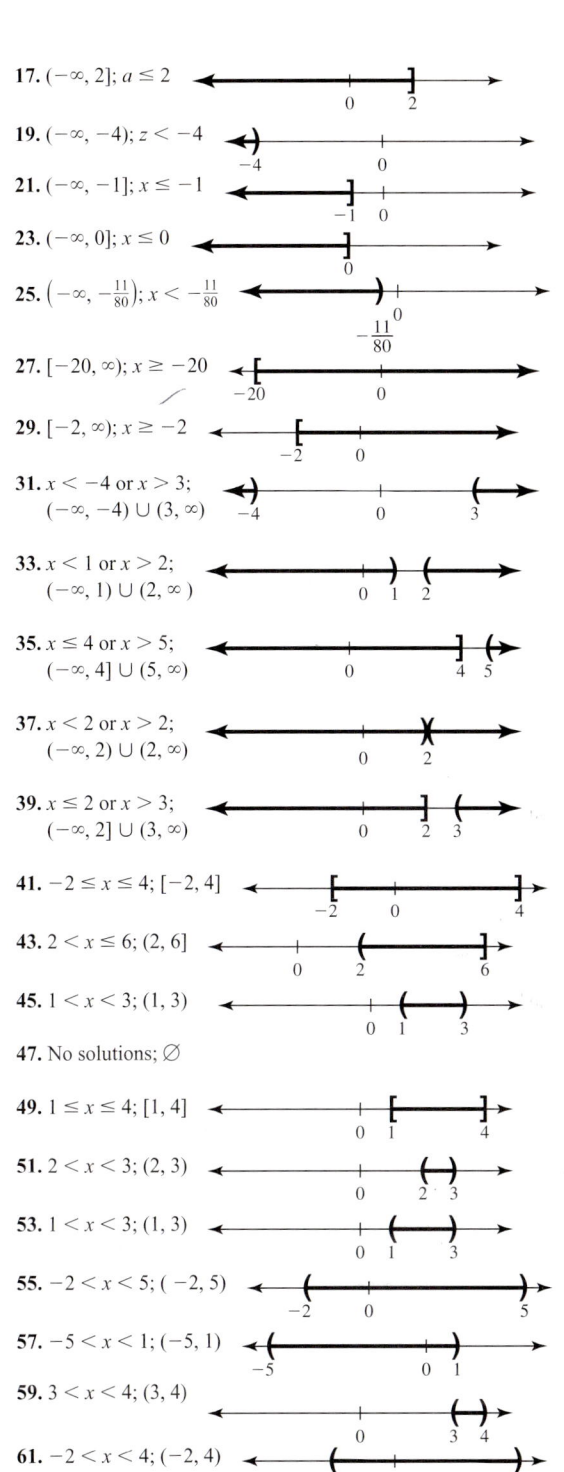

71. $h \le 29{,}028$ 73. $e \ge 2$ 75. $n \ge 4 \times 10^{25}$ 77. $20{,}001$
79. when $h < 13$ hr   81. (a) $4c \le \tfrac{1}{2}(2200)$ or carb grams ≤ 275; (b) Yes, $2(4)(30) + 2(4)(47) + 2(4)(59) = 1088 \le 1100$

83. (a) $0.15(1600) + 0.30(1600) + c \le 1600$ or carb calories ≤ 880; (b) No, $\tfrac{1}{2}(1600) \ne 880$ 85. $x > 46; 2006$ 87. all real numbers greater than 1.   89. all real numbers greater than 4.
91. since $a > b$, $b - a < 0$.   93. intersection
95. negative infinity   97. $p \ge 45$

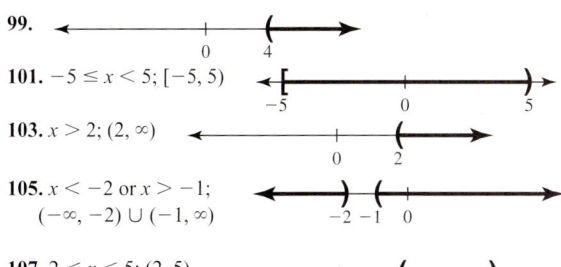

99.
101. $-5 \le x < 5; [-5, 5)$
103. $x > 2; (2, \infty)$
105. $x < -2$ or $x > -1$; $(-\infty, -2) \cup (-1, \infty)$
107. $2 < x < 5; (2, 5)$
109. >   111. <   113. 9   115. 0.34
117.
119.

Exercises 2.6

Translate This
1. G   3. D

1. $-13; 13$ 3. $-2.3; 2.3$ 5. 0   7. No solutions   9. $-9; -5$
11. $-2; 6$ 13. $-\tfrac{6}{5}; 2$ 15. $-20; 4$ 17. $-9; 18$ 19. -3
21. $-5; -\tfrac{1}{3}$ 23. $-7; 5$ 25. No solutions   27. All real numbers
29. All real numbers
31. $-4 < x < 4; (-4, 4)$
33. $-2 \le z \le 2; [-2, 2]$
35. $-4 \le a \le 4; [-4, 4]$
37. $-1 < x < 3; (-1, 3)$
39. No solutions
41. $-2 \le x \le -1; [-2, -1]$
43. $-2 < x < 1; (-2, 1)$
45. $x < -2$ or $x > 2$; $(-\infty, -2) \cup (2, \infty)$
47. $z \le -1$ or $z \ge 1$; $(-\infty, -1] \cup [1, \infty)$
49. $a \le -3$ or $a \ge 3$; $(-\infty, -3] \cup [3, \infty)$
51. $x < 0$ or $x > 2$; $(-\infty, 0) \cup (2, \infty)$

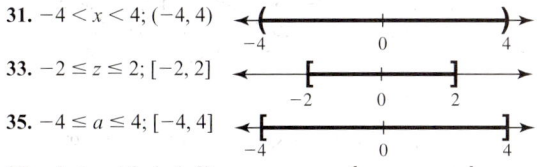

53. All real numbers; $(-\infty, \infty)$
55. $x \le -2$ or $x \ge -1$; $(-\infty, -2] \cup [-1, \infty)$
57. $x < -1$ or $x > \tfrac{5}{2}$; $(-\infty, -1) \cup (\tfrac{5}{2}, \infty)$

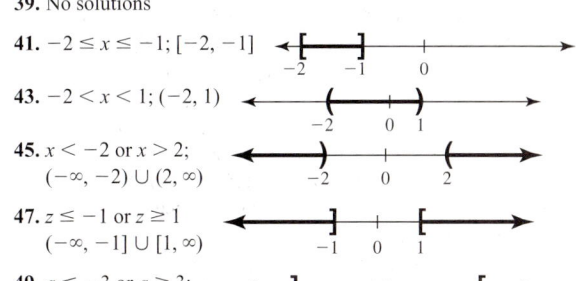

59. $x \le -2$ or $x \ge 1$; $(-\infty, -2] \cup [1, \infty)$
61. $a < 2$ or $a > 6$; $(-\infty, 2) \cup (6, \infty)$

63. All real numbers; $(-\infty, \infty)$

65. No solutions; $\emptyset$

67. $\frac{1}{3} \leq x \leq 1$; $[\frac{1}{3}, 1]$

69. $\$450 \leq a \leq \$550$ 71. Yes   77. equivalent   79. open
81. $\frac{9}{2}$ 83. $\frac{4}{3}; -2$ 85. $-\frac{21}{2}; \frac{15}{2}$
87. $x < -2$ or $x > 3$; $(-\infty, -2) \cup (3, \infty)$

89. $x \leq -1$ or $x \geq \frac{7}{3}$; $(-\infty, -1] \cup [\frac{7}{3}, \infty)$

91. -59 93. $x^2 + 11x + 7$ 95. $-10x^2 - 2x + 3$ 97. 3
99. $y = -2x + 8$

Review Exercises

1. [2.1A] (a) No; (b) No; (c) Yes   2. [2.1B] (a) 12; (b) 15; (c) 18
3. [2.1B] (a) 2; (b) 3; (c) 4   4. [2.1C] (a) 14; (b) 27; (c) 28
5. [2.1C] (a) 3; (b) 4; (c) 3   6. [2.1C] (a) $P = 500$; (b) $P = 2000$;
(c) $P = 4000$ 7. [2.2A] (a) $h = \frac{H - 72.48}{2.5}$; $h = 4$;
(b) $h = \frac{H - 77.48}{2.5}$; $h = 2$; (c) $h = \frac{H - 84.98}{2.5}$; $h = -1$
8. [2.2A] (a) $A = \frac{7B + 14}{2}$; (b) $A = \frac{7B + 15}{3}$; (c) $A = \frac{5B + 28}{4}$
9. [2.2A] (a) $L = \frac{P - 2W}{2}$; 40 ft by 50 ft;
(b) $L = \frac{P - 2W}{2}$; 50 ft by 60 ft; (c) $L = \frac{P - 2W}{2}$; 60 ft by 70 ft
10. [2.2D] (a) $x = 10$; angles: 50° each; (b) $x = 15$;
angles: 57° each; (c) $x = 20$; angles: 58° each   11. [2.3B] (a) 100 mi;
(b) 150 mi; (c) 200 mi   12. [2.3C] (a) 49; 51; 53; (b) 51; 53; 55;
(c) 67; 69; 71   13. [2.4A] (a) \$20,000; (b) \$30,000;
(c) \$15,000   14. [2.4A] 186, 80, 46, 30, 23, 15

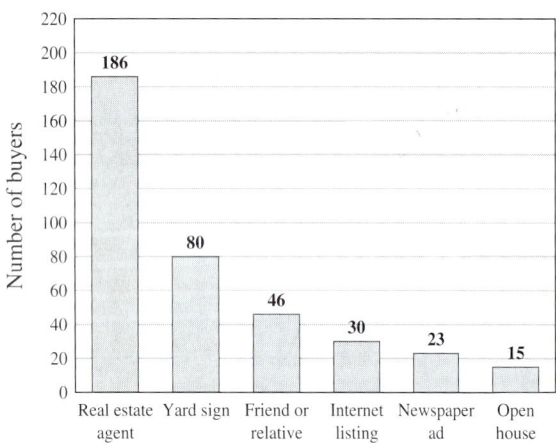

Home Buyer Survey

15. [2.4B] (a) \$5000 in bonds; \$15,000 in CDs;
(b) \$17,000 in bonds; \$3000 in CDs; (c) \$10,000 in bonds;
\$10,000 in CDs   16. [2.4C] (a) 200 mi; (b) 300 mi; (c) 120 mi
17. [2.4D] (a) 100 L; (b) 25 L; (c) 0 L

18. [2.5A] (a)
(b)
(c)

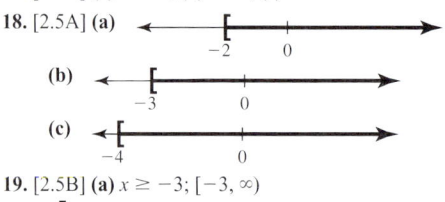

19. [2.5B] (a) $x \geq -3$; $[-3, \infty)$

(b) $x \geq -2$; $[-2, \infty)$

(c) $x \geq -2$; $[-2, \infty)$

20. [2.5B] (a) $x > 3$; $(3, \infty)$

(b) $x > 3$; $(3, \infty)$

(c) $x > 3$; $(3, \infty)$

21. [2.5C] (a) $(-\infty, -2) \cup [3, \infty)$

(b) $(-\infty, -3) \cup [2, \infty)$

(c) $(-\infty, -4) \cup [1, \infty)$

22. [2.5C] (a) $-2 < x \leq 2$; $(-2, 2]$

(b) $-3 < x \leq 3$; $(-3, 3]$

(c) $-2 < x \leq 4$; $(-2, 4]$

23. [2.5C] (a) $-5 < x \leq -1$; $(-5, -1]$

(b) $-4 < x \leq -1$; $(-4, -1]$

(c) $-3 < x \leq 1$; $(-3, 1]$

24. [2.5D] $4p \leq 0.15(2000)$ or protein grams ≤ 75
25. [2.6A] (a) 7; -21; (b) 14; -28; (c) 21; -35
26. [2.6A] (a) 2; (b) 4; (c) 6
27. [2.6B] (a) $-\frac{1}{3} \leq x \leq 1$; $[-\frac{1}{3}, 1]$

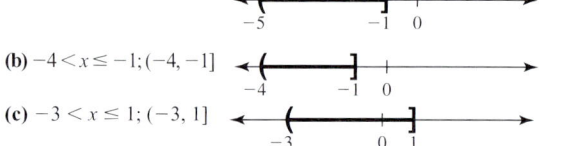

(b) $-\frac{1}{2} \leq x \leq 1$; $[-\frac{1}{2}, 1]$

(c) $-\frac{3}{5} \leq x \leq 1$; $[-\frac{3}{5}, 1]$

28. [2.6B] (a) $x \leq -\frac{1}{3}$ or $x \geq 1$; $(-\infty, -\frac{1}{3}] \cup [1, \infty)$

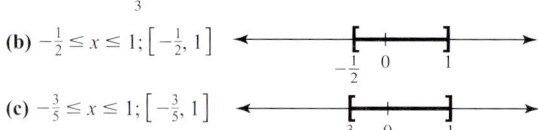

(b) $x \leq -\frac{1}{2}$ or $x \geq 1$; $(-\infty, -\frac{1}{2}] \cup [1, \infty)$

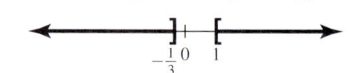

(c) $x \leq -\frac{3}{5}$ or $x \geq 1$; $(-\infty, -\frac{3}{5}] \cup [1, \infty)$

Cumulative Review Chapters 1–2

1. $\{2, 4, 6\}$ 2. $\frac{19}{100}$ 3. -11 4. -20 5. $-5x^2 + 6x + 5$
6. $5x^{14}$ 7. -9 8. No   9. $\frac{8}{5}$ 10. 58   11. 1850   12. $-\frac{14}{3}; -\frac{98}{3}$
13.
14.
15.
16.

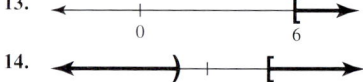

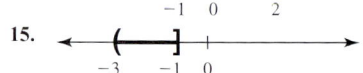

17. $A = \frac{7B + 66}{6}$ 18. 20 ft by 30 ft   19. \$23,000   20. 174

Chapter 3
Exercises 3.1

1, 3, 5, 7.

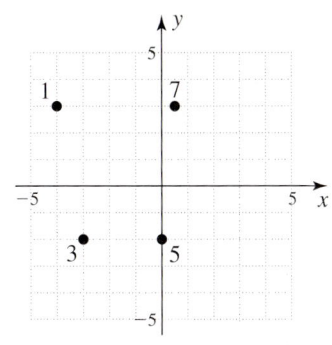

9. QI; QII **11.** QIII; QIV **13.** (3, 5) **15.** (−2, 3) **17.** (−4, 0)
19. (0, −3)
21. (−2, 1); (−1, 2); (0, 3); (1, 4); (2, 5)

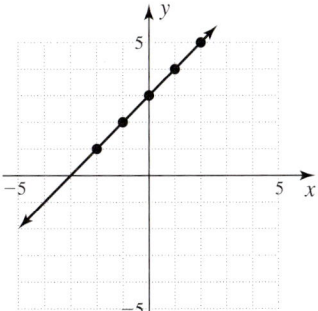

23. (−1, −5); (0, −4); (1, −3)

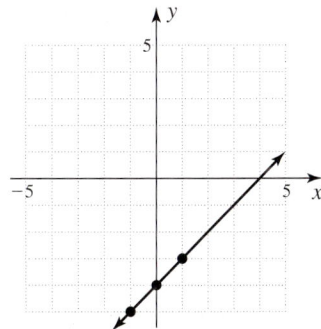

25. (−1, −5); (0, −3); (1, −1)

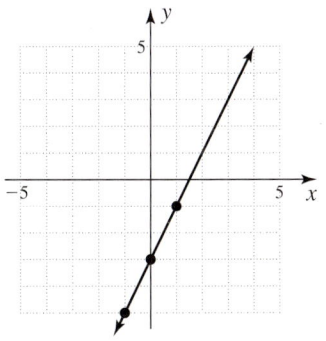

27. (5, 0); (0, −5)

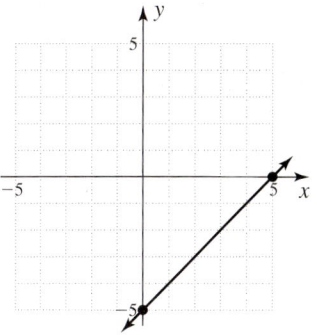

29. (3, 0); (0, 2)

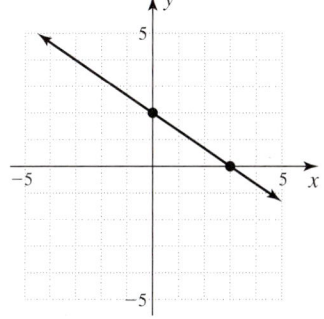

31. (2, 0); (0, −4)

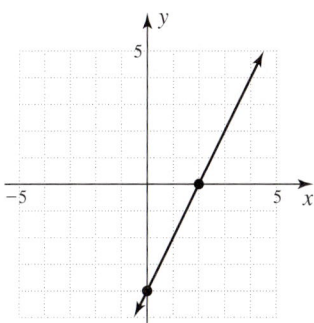

33. (2, 0); (0, 4)

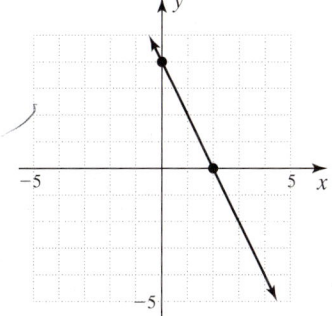

35. (0, 0)

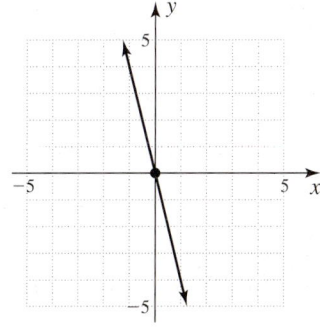

37. $(-5, 0)$; $(0, 2)$

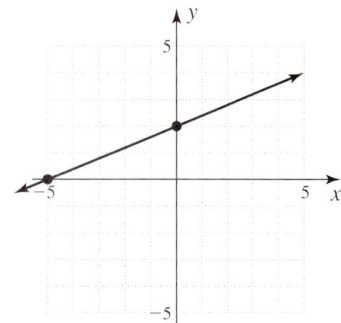

39. $(5, 0)$; $(0, -5)$

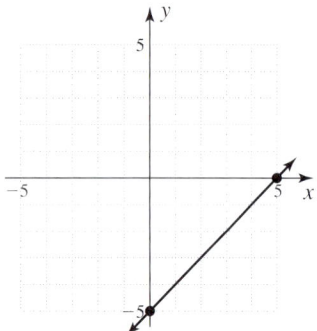

41. Vertical; $x = -4$

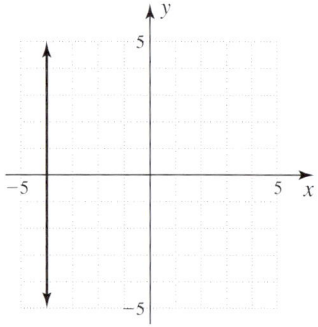

43. Vertical; $x = 4$

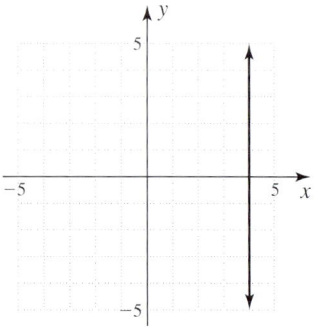

45. Vertical; $x = -4$

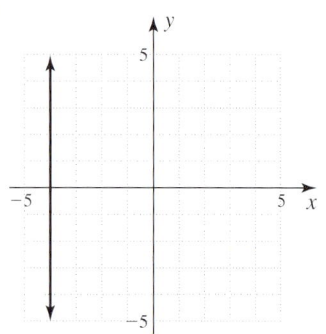

47. Horizontal; $y = 1$

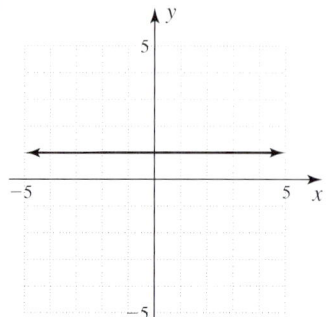

49. Vertical; $x = 2$

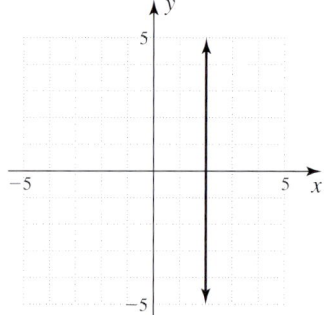

51.

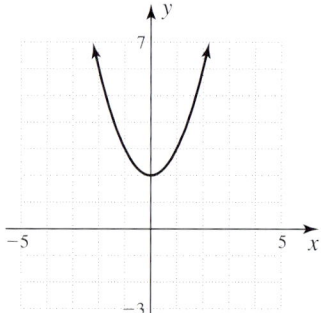

53.

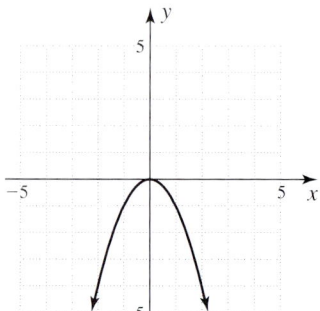

55.

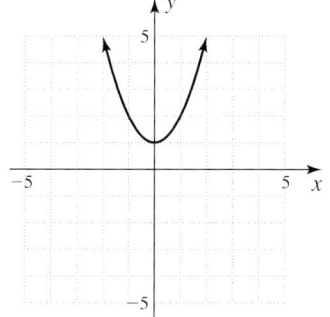

57.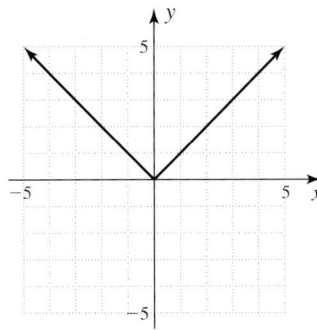

59.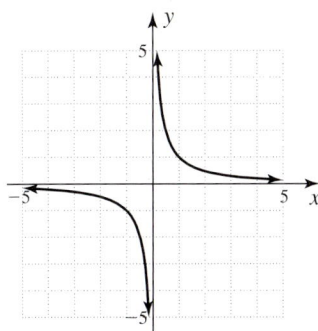

61. (a) Yes; (b) 80; (c) 160   **63.** (a) $23,000; (b) $20,000; (c) $8000; (d) 8 yr; (e) Answers will vary.   **65.** (20, 140)   **67.** (45, 148)
69. 86°F   **71.** Less than 10%   **73.** 11°F   **83.** quadrants
85. linear equation
87.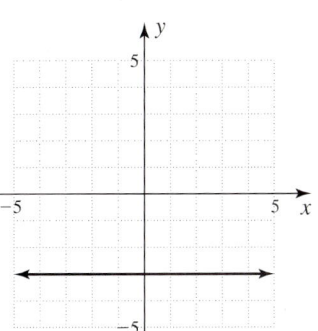

89.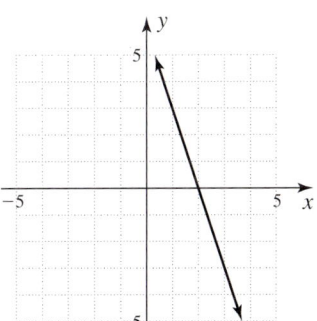

91.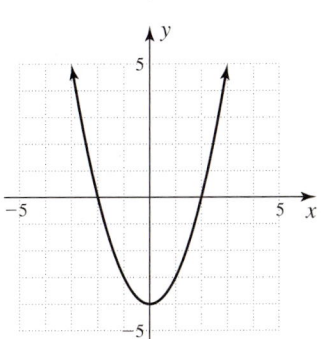

93. $A(-2, 3)$; $B(-3, 0)$; $C(4, -2)$ **95.** 0   **97.** 6   **99.** -14
101. undefined   **103.** -4 **105.** 18

Exercises 3.2

1. $m = \frac{4}{3}$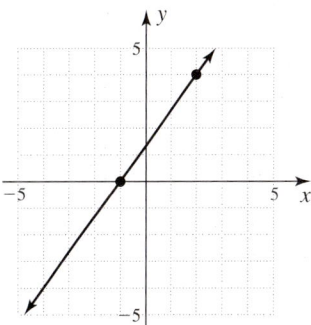

3. $m = \frac{8}{3}$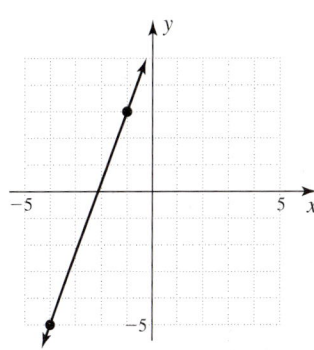

5. $m = 3$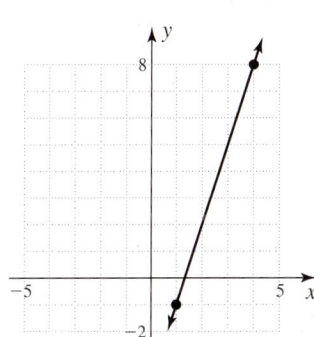

7. $m = 0$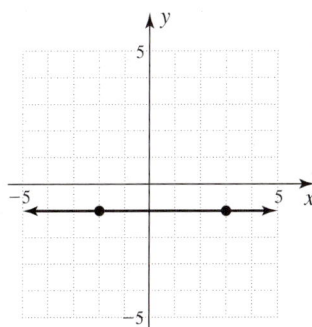

9. (a) $m = \frac{2000}{3}$; (b) The salary increased approximately $667 per year.
11. Parallel   **13.** Perpendicular   **15.** Neither   **17.** Perpendicular
19. Parallel   **21.** $x = 2$ **23.** $x = 4$ **25.** $x = -\frac{16}{3}$ **27.** $y = -8$
29. $x = 3$

31.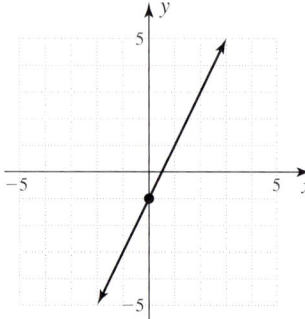

33.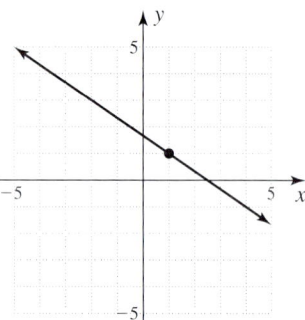

35.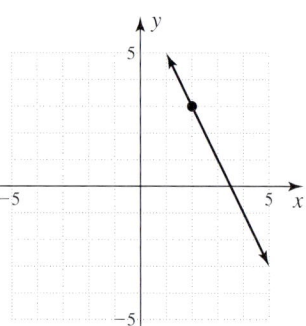

37.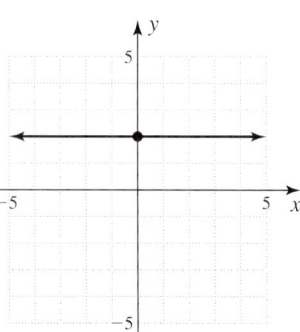

39.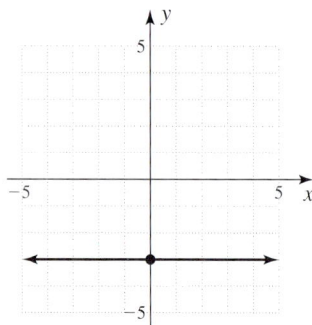

41. (5, 3) **43.** (−4, −4) **45.** $\left(-\frac{11}{2}, 0\right)$ **47.** $m = 1$; y-int: −4 **49.** $m = -\frac{1}{3}$; y-int: −2 **51.** $m = -2$; y-int: 4 **53.** (83, 35) **55.** No **57.** Yes **59.** Yes **61. (a)** 246 million; **(b)** 268 million; **(c)** 290 million; **(d)** year 2020; **(e)**

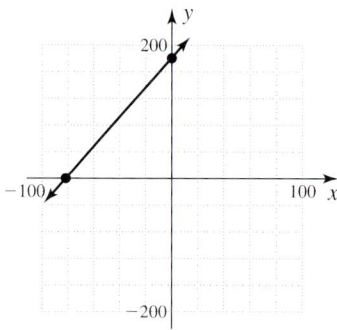

63. (a) $6.00; **(b)** $10.20; **(c)** 38.5 lb; **(d)**

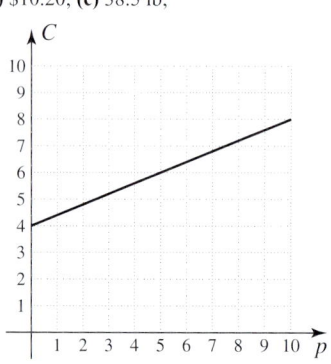

65. $\frac{3}{2}$ **67.** −2.3 **69.** −3 **71.** 1 **77.** The line falls from left to right. **79.** rate of change **81.** negative **83.**

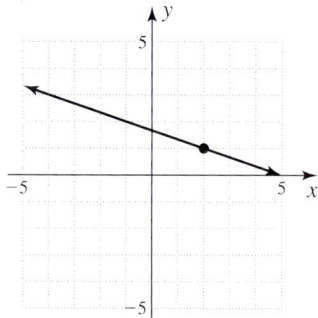

85. $y = 4$ **87. (a)** Perpendicular; **(b)** Parallel **89.** $y = 3x + 10$ **91.** $y = \frac{2}{3}x - 4$ **93.** $y = 2x - 5$ **95.** $2x - y = -1$ **97.** $x + 3y = 5$ **99.** $3x - 4y = 23$

Exercises 3.3

1. $3x - y = 4$ **3.** $x + y = 5$ **5.** $2x + y = 4$ **7.** $2x - y = -11$ **9.** $3x + y = -5$ **11.** $x - y = 0$ **13.** $4x - y = 6$ **15.** $y = 5x + 2$

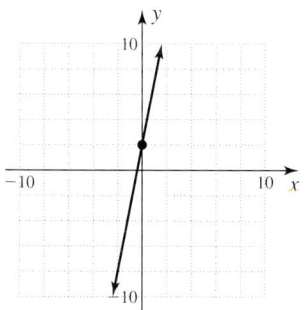

17. $y = -\frac{1}{5}x - \frac{1}{3}$

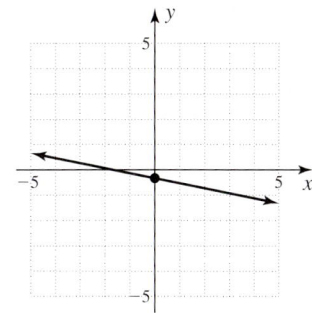

19. $y = 2x - 4$ **21.** $y = -3x - 12$ **23.** $y = -2x + 3$
25. $y = -x - 6$ **27.** $y = 4$ **29.** $x = -2$ **31.** $x = -2$ **33.** $y = 2$
35. (a) $d + 2p = 14$; **(b)** 10 pairs **37. (a)** $s - 2p = -20$; **(b)** $10;
(c) 60 **39.** $4
41. $y = 2.5x - 10$

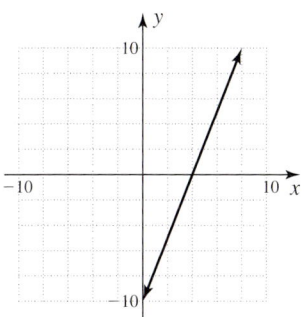

43. (a) $y = -0.6x + 152$; **(b)** 132 BPM
45. $y = 2x + 50$ **47. (a)** $2; **(b)** $75
53. y-intercept **55.** slope-intercept form
57. (a) $3x - y = 2$; **(b)** $x + 3y = 4$
59. $m = 0$; $b = 3$ **61.** $x = -2$
63. $2x - y = 5$

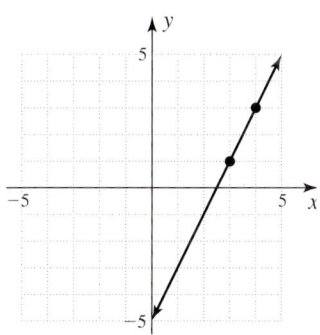

65. y-intercept; $(0, 2)$ x-intercept; $\left(-\frac{2}{3}, 0\right)$
67. y-intercept; $(0, -3)$ no x-intercept **69.** $(1, -3), (0, 1)$
71. $-3 < x < 3$ **73.** $y \leq -5$ or $y \geq 5$

Exercises 3.4
1.

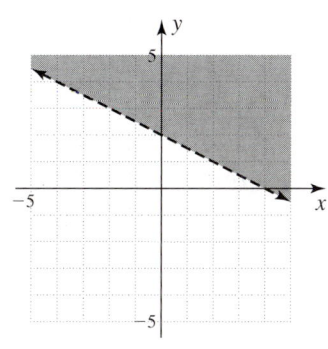

3.

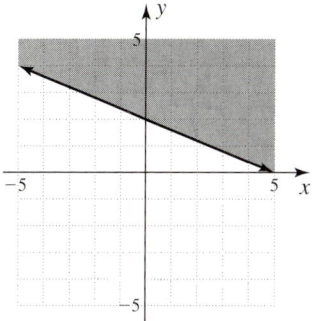

5.

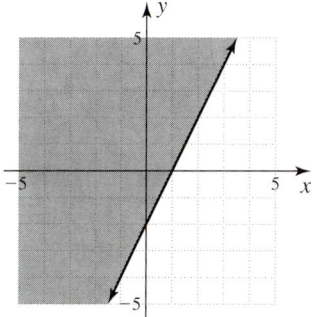

7.

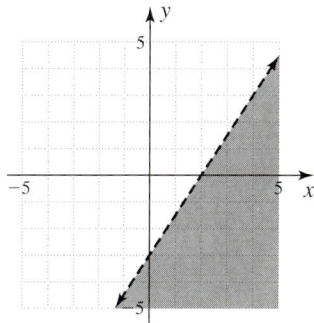

9.

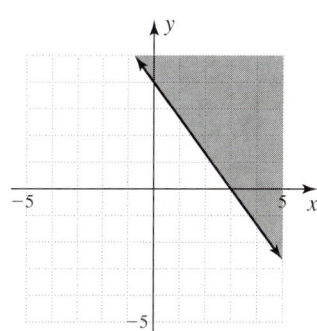

11.

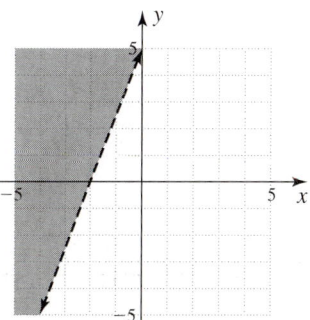

13.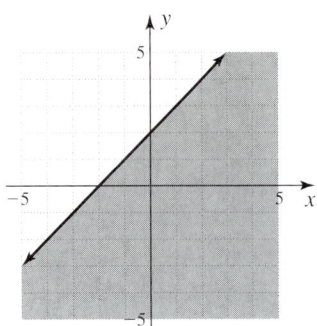

15.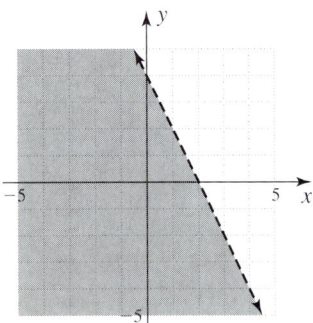

17.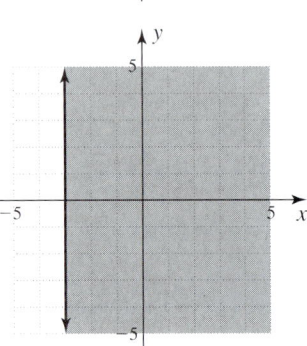

19.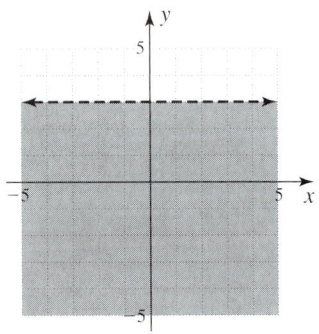

21.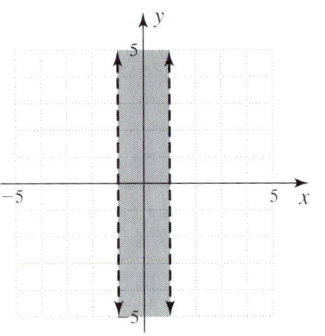

23.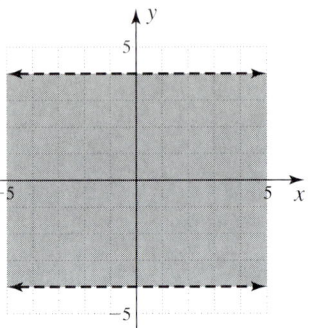

25.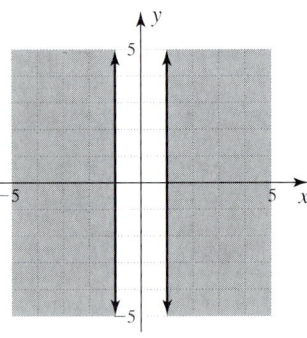

27.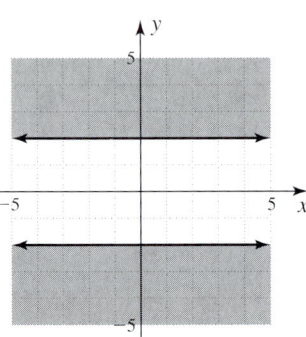

29.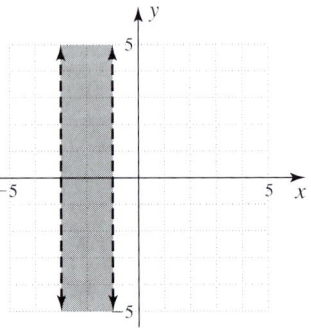

31.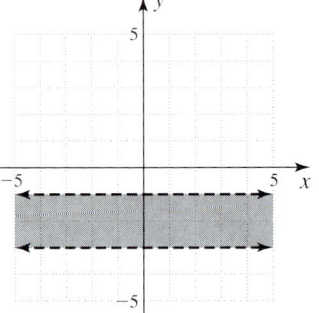

33.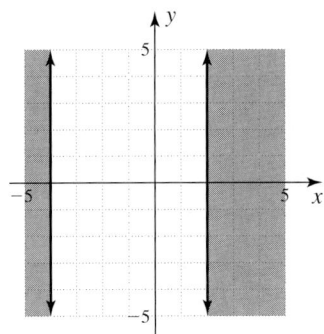

35.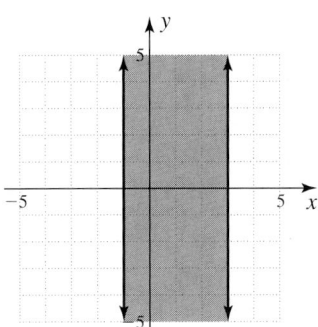

37.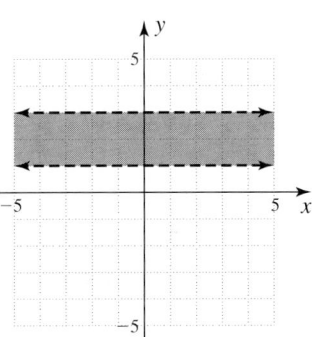

39. (a) $0.125m + 75d \leq 250$;
(b)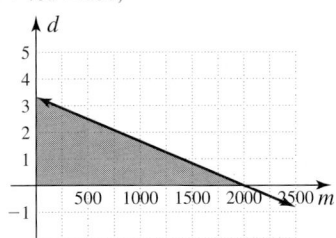
(c) No; **(d)** Yes

41. (a) $2M + 6H \leq 40$;
(b)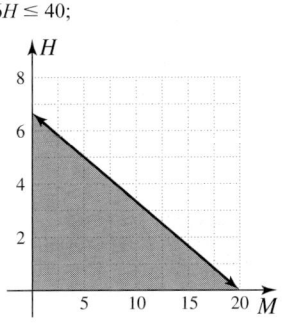

(c) Answers will vary. (10, 2) is in the shaded region and means a worker can make 10 parts by machinery, 2 parts by hand, and still work less than or equal to 40 hr. per week.
43. (a) 33 mi; **(b)** Rental A   **45.** If you plan to drive more than 33 mi, Rental A is cheaper.   **51.** above   **53.** dashed

55.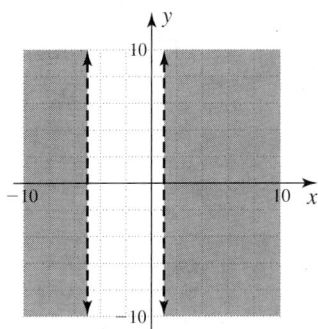

57.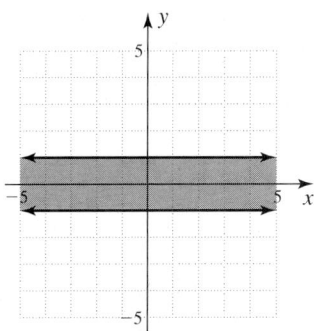

59.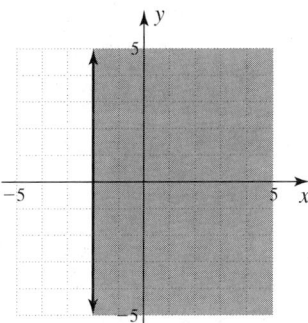

61.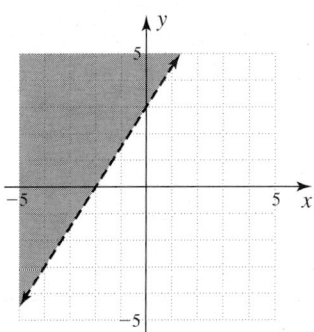

63. $48d + 0.32m < 160$

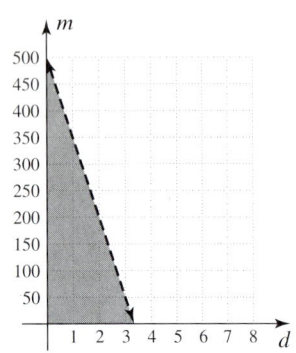

65. No, the vertical line has one point of intersection.
67. Yes, the vertical line has two points of intersection. The x-value of the ordered pair is repeated when naming the intersection points
69. -2 **71.** $-\frac{2}{3}$ **73.** 10 **75.** 4

Exercises 3.5

1. $D = \{-3, -2, -1\}$; $R = \{0, 1, 2\}$; a function **3.** $D = \{3, 4, 5\}$; $R = \{0\}$; a function **5.** $D = \{1, 2\}$; $R = \{2, 3\}$; not a function
7. $D = \{1, 3, 5, 7\}$; $R = \{-1\}$; a function **9.** $D = \{2\}$; $R = \{1, 0, -1, -2\}$; not a function **11.** $D = \{2, 6, 11\}$; $R = \{February, June, November\}$; a function
13. $D = \{-2, 0, 1\}$; $R = \{-3, 1, 2, 4\}$

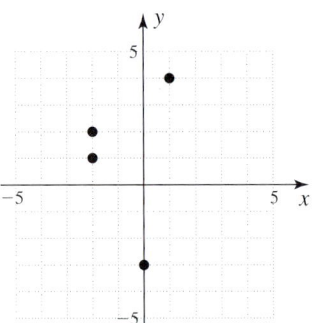

15. $D = \{Real\ numbers\}$; $R = \{Real\ numbers\}$

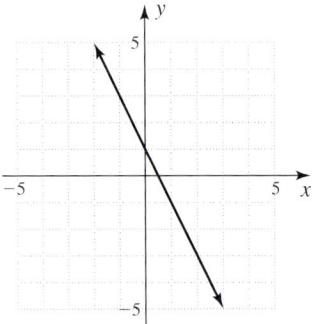

17. $D = \{x | -5 \le x \le 5\}$; $R = \{y | -5 \le y \le 5\}$; not a function
19. $D = \{x | -5 \le x \le 5\}$; $R = \{y | 0 \le y \le 5\}$; a function
21. $D = \{x | 0 \le x \le 5\}$; $R = \{y | -5 \le y \le 5\}$; not a function
23. $D = \{x | x \text{ is a real number}\}$; $R = \{y | y \ge 0\}$; a function
25. $D = \{x | x \text{ is a real number}\}$; $R = \{y | y \le 0\}$; a function
27. $D = \{x | x \ge 0\}$; $R = \{y | y \text{ is a real number}\}$; not a function
29. $D = \{x | x \text{ is a real number}\}$; $R = \{y | y \text{ is a real number}\}$; a function
31. $D = \{x | x \text{ is a real number}\}$; $R = \{y | y \text{ is a real number}\}$; a function
33. $D = \{x | -3 \le x \le 3\}$; $R = \{y | -2 \le y \le 2\}$; not a function
35. $D = \{x | x \ge 2 \text{ or } x \le -2\}$; $R = \{y | y \text{ is a real number}\}$; not a function **37.** A function **39.** A function **41.** A function
43. $D = \{x | x \ge 5\}$ **45.** $D = \{x | x \le 2\}$ **47.** $D = \{x | x \ge -1\}$

49. $D = \{x | x \text{ is a real number and } x \ne 5\}$ **51.** $D = \{x | x \text{ is a real number and } x \ne -5\}$ **53.** $D = \{x | x \text{ is a real number and } x \ne -2 \text{ or } x \ne -1\}$ **55.** (a) 1; (b) 7; (c) -5 **57.** (a) 0; (b) 2; (c) 5
59. (a) $\frac{1}{4}$; (b) $\frac{3}{28}$; (c) $\frac{1}{28}$ **61.** (a) 5; (b) 19; (c) 24 **63.** (a) -10; (b) 7; (c) -70 **65.** (a) 3; (b) 4; (c) 7 **67.** (a) 7; (b) 8; (c) 56
69. (a) $P(x) = -0.0005x^2 + 24x - 100,000$; (b) $90,000
71. (a) $U(50) = 140$; (b) $U(60) = 130$ **73.** (a) 160 lb; (b) 78 in.
75. (a) 639 lb/ft²; (b) 6390 lb/ft² **77.** (a) L; (b) S; (c) Size 11; (d) Size 12 **79.** (a) 619.2; (b) 1352.8 per 100,000 **81.** (a) Yes; (b) $D = \{x | 1 \le x \le 6\}$; $R = \{y | 0 \le y \le 0.11\}$;
(c) $D = \{x | 1 \le x \le 4\}$; $R = \{y | 0 \le y \le 0.10\}$ **87.** False **89.** range
91. independent variable **93.** Yes **95.** Yes **97.** Yes
99. (a) 3; (b) -1; (c) 4
101. $D = \{x | x \text{ is a real number and } x \ne 2\}$ **103.** Not a function
105. $D = \{x | -3 \le x \le 3\}$; $R = \{y | -3 \le y \le 3\}$
107. $D = \{x | -3 \le x \le 3\}$; $R = \{y | -3 \le y \le 0\}$
109. $D = \{7, 8, 9\}$; $R = \{8, 9, 10\}$ **111.** (a) $\frac{3}{2}$; (b) 5; (c) $-\frac{1}{8}$;
(d) $-\frac{4}{3}$ **113.** $\frac{1}{3}$ **115.** 0 **117.** 2

Exercises 3.6

1. Nonlinear function **3.** Not a function **5.** Linear decreasing function **7.** Linear decreasing function **9.** Nonlinear function
11. Nonlinear function **13.** Linear increasing function; $f(x) = \frac{1}{5}x - 2$
15. Linear constant function; $g(x) = -4$ **17.** Not a function
19. Linear decreasing function; $p(x) = -3x + 5$
21. Nonlinear function **23.** Linear increasing function; $f(x) = x - 8$
25. $f(x) = -4$ **27.** $f(x) = \frac{2}{3}x + 2$ **29.** (a) $N = 0.25x + 5.5$, $f(x) = 0.25x + 5.5$; (b) 0.25 million increase each year; (c) $x = 25$; (d) 11.75 million **31.** (a) $C = 85x, f(x) = 85x$; (b) $85 per ft²; (c) $125,375; (d) 1624 ft² **33.** Not linearly related
35. Linearly related
37.

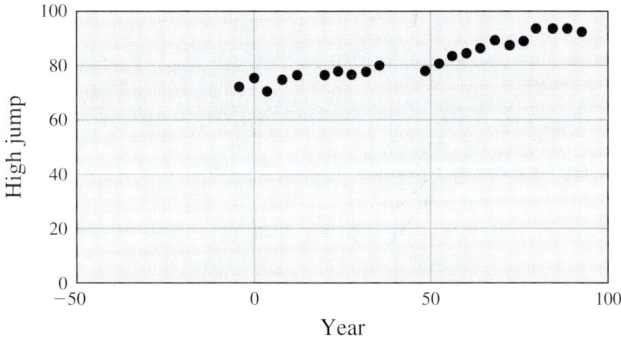

Linearly related

39. (a)

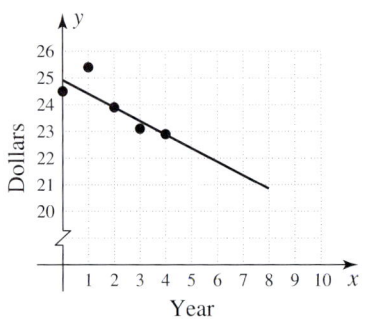

(b) (2, 23.9); (4, 22.9); (c) $y = -\frac{1}{2}x + 24.9$; (d) $20.9;
(e) Each year it decreases by $0.5.

41. (a)

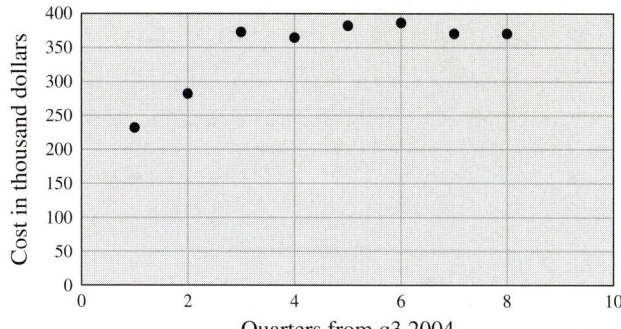

(b)

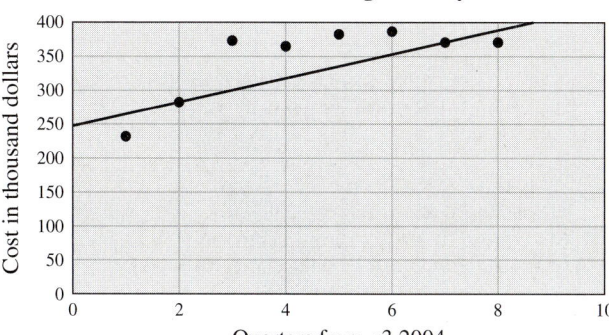

(c) $y = 18.12x + 250.16$; (d) $866,240   **43.** False
47. data set   **49.** increasing function   **51.** $y = 20x$; increasing function
53. Constant function   **55.** Increasing
57.

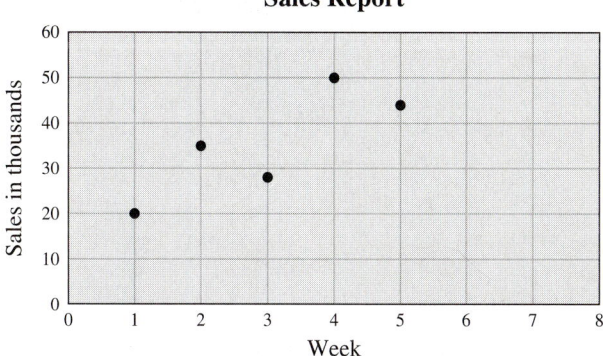

Linear function

59. x-int: $(3, 0)$; y-int: $(0, 4.5)$ **61.** no x-intercept; y-int: $(0, -8)$
63. $m = \frac{3}{5}$; y-int: $(0, 2)$ **65.** $m = 0$; y-int: $(0, -5)$
67.

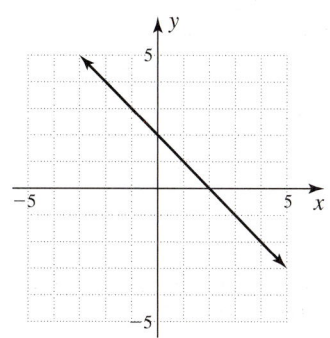

69.

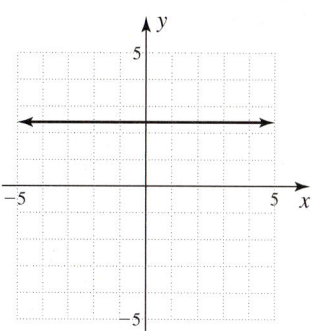

Review Exercises

1. [3.1A] **(a)**

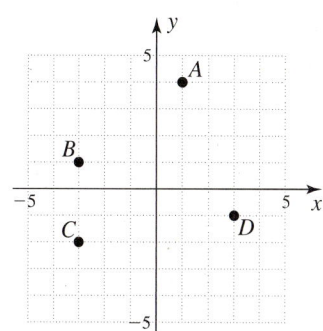

(b)

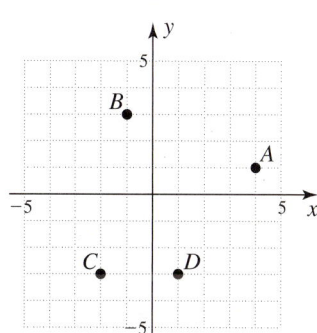

2. [3.1A] $A(2, 0)$; $B(-1, 1)$; $C(-3, -3)$; $D(4, -4)$
3. [3.1A] $A(2, 1)$; $B(0, 3)$; $C(-3, -1)$; $D(3, -3)$
4. [3.1B] **(a)**

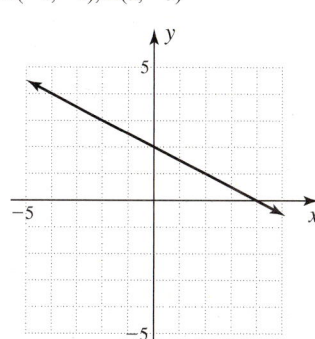

(b)

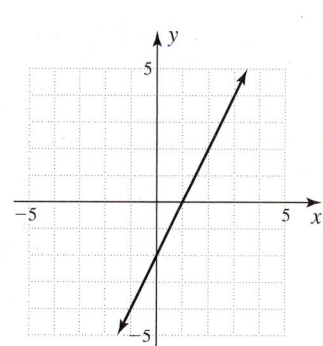

5. [3.1C] **(a)** $(-1, 0)$; $(0, 3)$

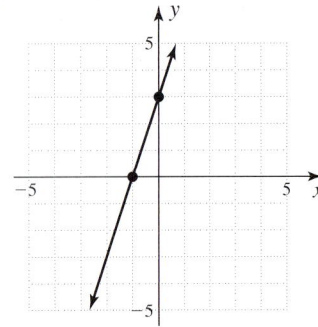

(b) $(2, 0)$; $(0, -4)$

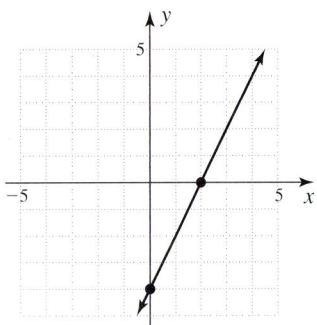

6. [3.1D] **(a–b)**

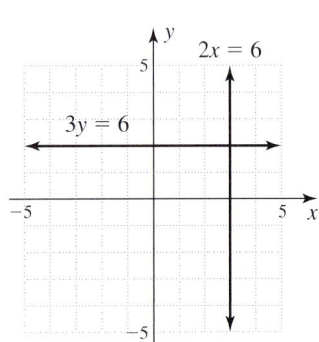

7. [3.1D] **(a–b)**

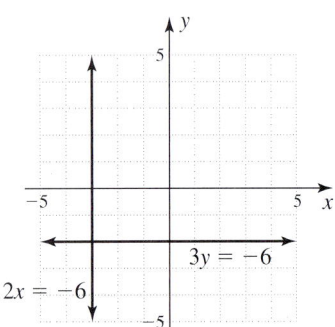

8. [3.2A] **(a)** $m = -\frac{1}{2}$; **(b)** Undefined
9. [3.2A] **(a)** $m = -1$; **(b)** $m = -\frac{3}{2}$ **10.** [3.2B] **(a)** Perpendicular; **(b)** Parallel **11.** [3.2B] **(a)** Parallel; **(b)** Neither
12. [3.2B] **(a)** $y = 2$; **(b)** $y = 5\frac{1}{2}$ or $\frac{11}{2}$
13. [3.2C] **(a)**

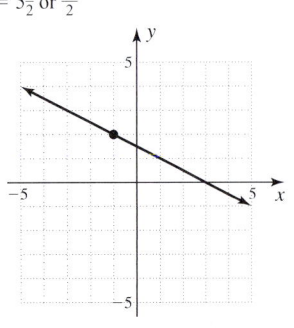

(b)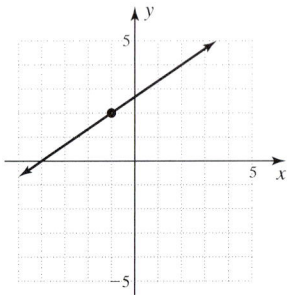

14. [3.2D] **(a)** $m = 6$; y-intercept $= -2$; **(b)** $m = \frac{1}{2}$; y-intercept $= -2$ **15.** [3.3A] **(a)** $x - y = -3$; **(b)** $5x + 2y = -14$ **16.** [3.3B] **(a)** $2x - y = 1$; **(b)** $2x - y = 10$
17. [3.3C] **(a)** $y = 3x + 2$; **(b)** $y = -3x + 4$
18. [3.3D] **(a)** $y = -2x + 5$; **(b)** $y = 3x - 5$
19. [3.3D] **(a)** $y = \frac{3}{2}x - 2$; **(b)** $y = -\frac{2}{3}x + \frac{7}{3}$
20. [3.3E] **(a)** $y = 7$; **(b)** $x = -3$
21. [3.4A] **(a)**

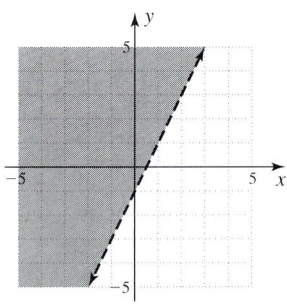

(b)

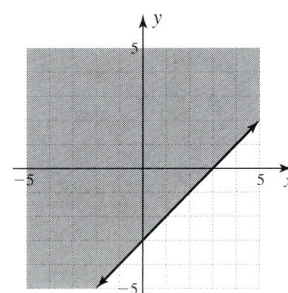

(c)

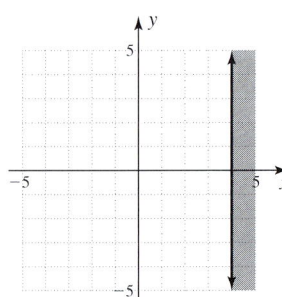

(d)

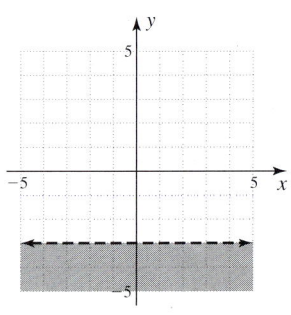

22. [3.4B] (a)

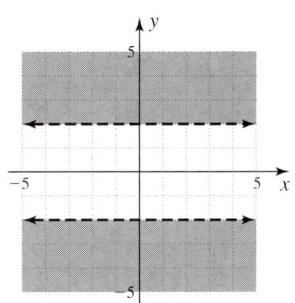

(b)

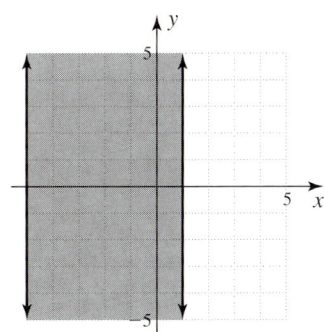

23. [3.5A] (a) $D = \{0, 2, 3, 5\}; R = \{5, 8, 9, 10\}$;
(b) $D = \{0, 2, 3, 5\}; R = \{6, 9, 10, 11\}$
24. [3.5A] (a) $D = \{x \mid x \text{ is a real number}\}$;
$R = \{y \mid y \text{ is a real number}\}$

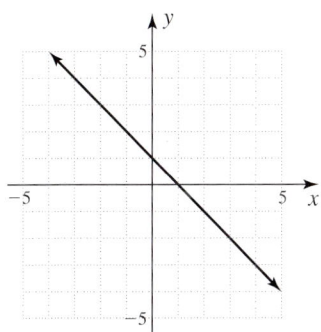

(b) $D = \{x \mid x \text{ is a real number}\}$;
$R = \{y \mid y \text{ is a real number}\}$

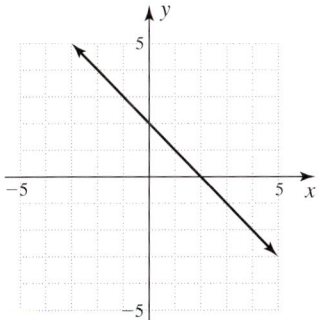

25. [3.5A] (a) $D = \{x \mid -2 \leq x \leq 2\}; R = \{y \mid -2 \leq y \leq 2\}$;
(b) $D = \{x \mid -3 \leq x \leq 3\}; R = \{y \mid -3 \leq y \leq 3\}$
26. [3.5A] (a) $D = \{x \mid -2 \leq x \leq 2\}; R = \{y \mid 0 \leq y \leq 2\}$;
(b) $D = \{x \mid -3 \leq x \leq 3\}; R = \{y \mid 0 \leq y \leq 3\}$ 27. [3.5B] (a) Yes;
(b) No 28. [3.5C] (a) $D = \{x \mid x \text{ is a real number and } x \neq 1\}$;
(b) $D = \{x \mid x \text{ is a real number and } x \neq 2\}$
29. [3.5C] (a) $D = \{x \mid x \geq 3\}$; (b) $D = \{x \mid x \geq 4\}$

30. [3.5D] (a) -2; (b) -3; (c) 1   31. [3.5D] (a) 1; (b) -2; (c) 3
32. [3.5D] (a) 0; (b) -1; (c) 1   33. [3.5D] (a) 1; (b) 0; (c) 1
34. [3.6A] (a) Nonlinear function; (b) linear function, $f(x) = -5x + 3$, decreasing function; (c) linear function, $f(x) = 1$, constant function
35. [3.6B] $y = f(x) = -\frac{1}{3}x + 1$ 36. [3.6C] $y = \frac{3}{5}x + \frac{6}{5}$ (Answers may vary.)

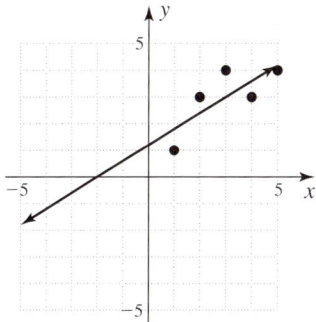

Cumulative Review Chapters 1–3

1. Irrational numbers; Real numbers
2.

$-\frac{5}{2}$

3. $\frac{2}{7}$ 4. $\frac{8x^{12}}{y^{12}}$ 5. -336 6. 14   7. 7

8.

9.

10. 26   11. 23; 25; 27   12. 240 mi   13. $\{-3, -2, -4\}$
14. $D = \{\text{all real numbers}\}; R = \{\text{all real numbers}\}$ 15. $\{x \mid x \geq -25\}$
16. 1   17. 134
18.

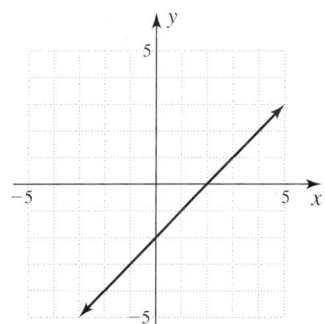

19. x-intercept: $\frac{2}{5}$; y-intercept: 2
20.

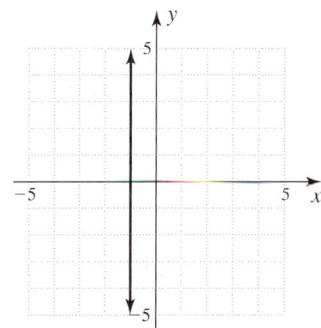

21. Perpendicular   22. $2x + y = -3$ 23. $m = 2$; y-intercept: -12

24. (a) $P(a) = 6.25a - 57$; (b) Approximately 61.75%; (c) Increases by approximately 6.25% each year
25.

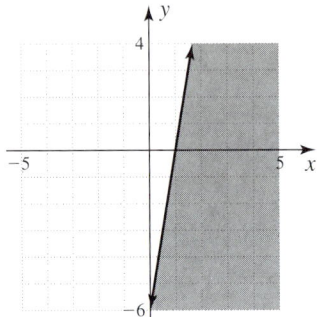

Chapter 4

Exercises 4.1

1. (3, 1)   **3.** Parallel lines; no solution
5. Intersecting lines; Solution: $(-2, -4)$; (1) $x - 2y = 6$; (2) $y = 2x$

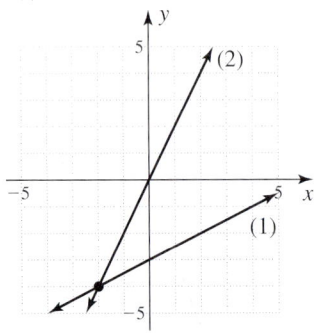

7. Intersecting lines; Solution: $(1, -2)$; (1) $y = x - 3$; (2) $y = 2x - 4$

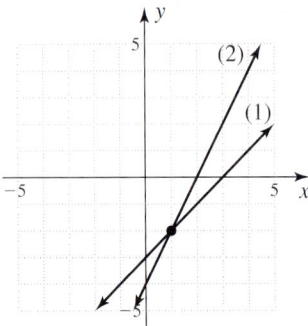

9. Intersecting lines; Solution: $(1\frac{1}{3}, 1\frac{1}{3})$; (1) $2y = -x + 4$; (2) $y = -2x + 4$

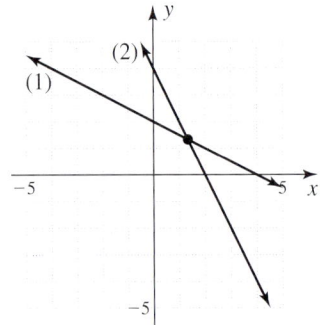

11. Parallel lines; No solution; (1) $2x - y = -2$; (2) $y = 2x + 4$

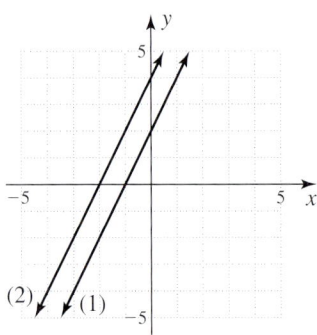

13. Coinciding lines; Infinitely many solutions; (1) $3x + 4y = 12$; (2) $8y = 24 - 6x$

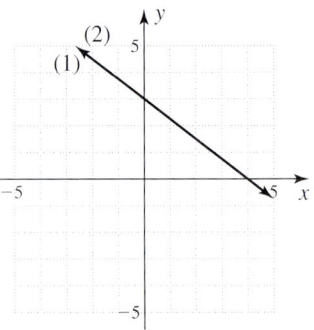

15. (2, 0); consistent   **17.** $(1, \frac{3}{2})$; consistent   **19.** No solution; inconsistent   **21.** $(3, \frac{5}{2})$; consistent   **23.** Infinitely many solutions; dependent   **25.** $(-1, -1)$; consistent   **27.** $(-\frac{4}{3}, \frac{4}{3})$; consistent
29. (4, 1); consistent   **31.** (5, 3)   **33.** $(0, \frac{1}{2})$ **35.** (0, 1)   **37.** (2, 3)
39. $(\frac{5}{2}, -\frac{1}{2})$ **41.** No solution   **43.** (3, 2)   **45.** Infinitely many solutions   **47.** $(\frac{1}{3}, 2)$ **49.** (5, -2)   **51.** $(-\frac{1}{2}, -\frac{2}{3})$ **53.** (8, -12)
55. (6, 8)   **57.** (4, -3)   **59.** $p = 14$ **61.** $p = 200$ **63.** $p = 40$
65. $p = 28$; $D(28) = 512$ **67.** $p = 15$; $D(15) = S(15) = 435$
69. (a) $x + y = 90$; $y = x + 15$; (b) $x = 37.5$; $y = 52.5$
71. (a) $x + y = 180$; $y = 4x$; (b) $x = 36$; $y = 144$
73. (a) $x + y = 465$; $x = y + 15$; (b) 240 lb; 225 lb
75. Japan: 2537; United States: 2100   **77.** 8.5 shekels   **79.** 36 and 34
81. The shorter building is 1483 ft; the taller one is 1667 ft
83. Tweedledee: $120\frac{2}{3}$ lb; Tweedledum: $119\frac{2}{3}$ lb   **89.** coinciding
91. (2, -1)   **93.** Infinitely many solutions   **95.** (2, 1)   **97.** $(-4, -\frac{8}{3})$
99. $(\frac{24}{5}, \frac{8}{5})$ **101.** No solution; inconsistent
(1) $2x + y = 4$; (2) $2y + 4x = 6$

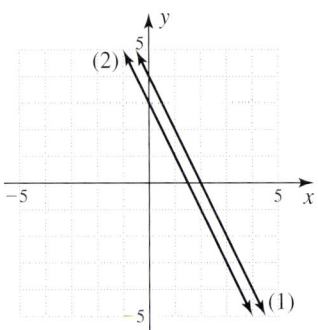

103. Infinitely many solutions; dependent
(1) $x + \frac{1}{2}y = -2$; (2) $y = -2x - 4$

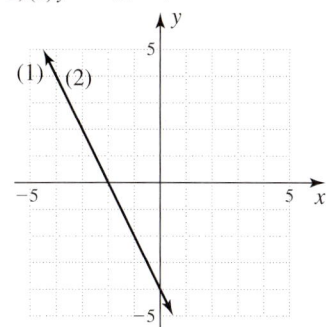

53.

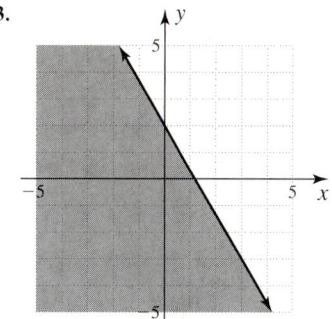

55. Solution: $(-5, -3)$

105. $(-1, 0)$; consistent
(1) $x - y = -1$; (2) $y = -x - 1$

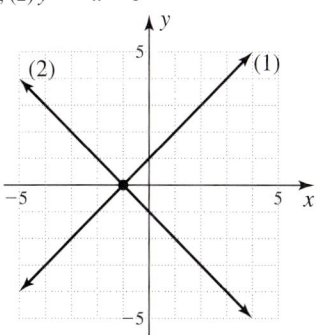

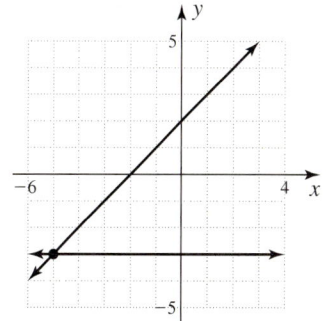

Exercises 4.4

1.

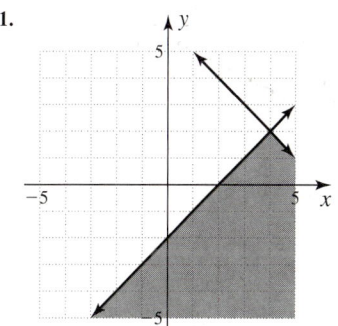

107. $-\frac{3}{4}$ **109.** 0 **111.** $y = 6x - 3$ **113.** $y = \frac{2}{3}x - 5$

Exercises 4.2

1. $(5, 3, 4)$; consistent **3.** $(-1, 1, 4)$; consistent **5.** $(3, 4, 1)$;
consistent **7.** No solution; inconsistent **9.** $(\frac{1}{2}, \frac{1}{4}, \frac{1}{3})$; consistent
11. No solution; inconsistent **13.** No solution; inconsistent
15. $(\frac{9}{2}, \frac{1}{2}, \frac{5}{2})$; consistent **17.** $(6, 3, -1)$; consistent **19.** $(-2, -3, -4)$;
consistent **21.** 8; 16; 24 **23.** 30°; 60°; 90° **25.** President: $400,000;
vice president: $202,900; chief justice: $202,900
27. B: $260 million; C: $350 million; E: $275 million **29.** Piano:
21 million; guitar: 19 million; organ: 6 million **31.** Israel: 216 days;
Japan: 243 days; United States: 180 days **33.** $d = \frac{5n + 16}{216}$
41. no **43.** Infinitely many solutions; dependent
45. No solution; inconsistent **47.** $(-1, 1, 4)$; consistent
49. $12,000 at 5%; $8000 at 7% **51.** 32 gal **53.** 250 mi/hr

Exercises 4.3

1. 50 dimes; 25 nickels **3.** 10 nickels; 20 quarters **5.** 5 pennies;
5 nickels **7.** 20 tens; 5 twenties **9.** 13 nickels; 9 dimes; 22 quarters
11. 43 and 59 **13.** 21 and 105 **15.** 4 and 20 **17.** Longs Peak:
14,255 ft; Pikes Peak: 14,110 ft **19.** Butterscotch: $2033\frac{1}{3}$ lb; caramel:
$2033\frac{1}{3}$ lb; chocolate: $2633\frac{1}{3}$ lb **21.** Plane: 210 mi/hr; wind: 30 mi/hr
23. Current: 3 mi/hr; boat: 12 mi/hr **25.** Wind: 50 mi/hr; plane:
450 mi/hr **27.** $6000 at 8%; $4000 at 6% **29.** $10,000 at 6%;
$5000 at 8%; $10,000 at 10% **31.** $L = 505$ ft; $W = 255$ ft
33. $L = 134$ ft; $W = 85$ ft **35.** Approximately 3.09 yr after 1995 or
about 1998; 397 Euros **37.** Approximately 1.7 yr after 1990 or about
1992; $1,000,000 **39.** $15a + 5b + 20c = 170$
41. $10a + 15b + 10c = 110$ **45.** $3000 at 6%; $7000 at 8%;
$10,000 at 10% **47.** 20 nickels; 10 dimes **49.** Lincoln: $6\frac{1}{3}$ ft tall;
Madison: $5\frac{1}{4}$ ft tall **51.** $x = 56$

3.

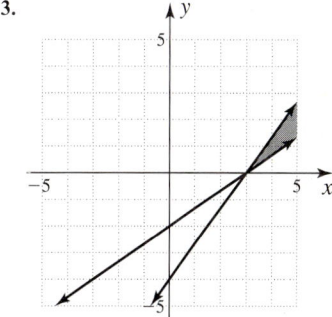

5.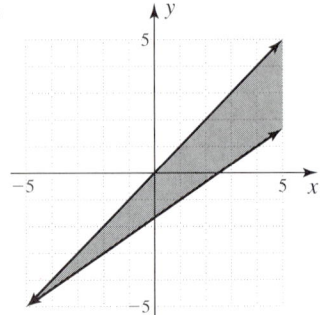

7.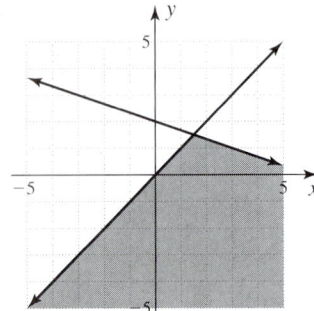

9.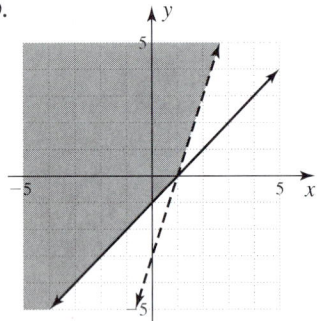

11.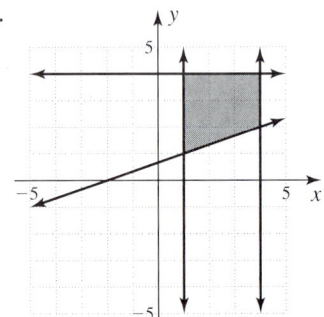

13.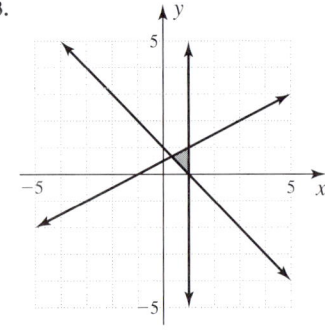

15.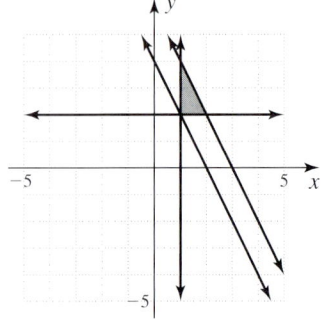

17. 80 cars; 20 trucks 21. solution set 23. vertical
25.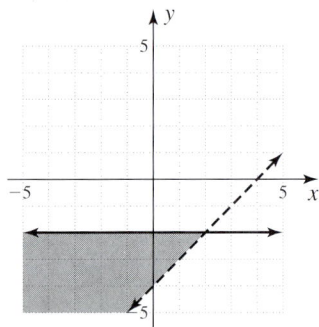

27. base is x; exponent 5 29. -96 31. $-5x - 17$ 33. $-5x + 12$

Review Exercises

1. [4.1A] (a) Solution: $(1, 0)$
$$5(1) + 0 = 5 \qquad -6(1) + 0 = -6$$
$$5 + 0 = 5 \qquad -6 + 0 = -6$$
$$5 = 5 \qquad -6 = -6$$
(b) No solution; parallel lines
2. [4.1A]
(a) (1) $2x - y = 2$; (2) $y = 3x - 4$

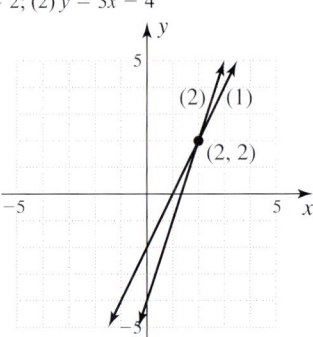

(b) (1) $x - 2y = 0$; (2) $y = x - 2$

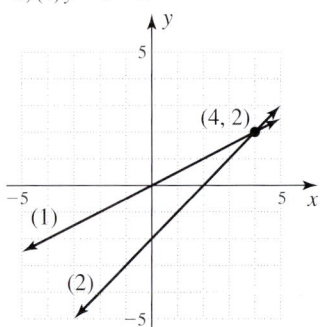

3. [4.1A]
(a) (1) $2y - x = 3$; (2) $4y = 2x + 8$

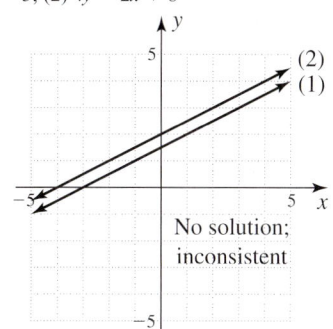
No solution; inconsistent

(b) (1) $3y + x = 5$; (2) $2x = 8 - 6y$

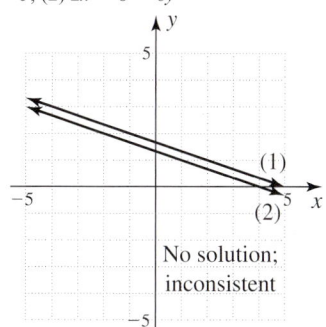
No solution; inconsistent

4. [4.1A]
(a) (1) $3x + 2y = 6$; (2) $y = 3 - \frac{3}{2}x$

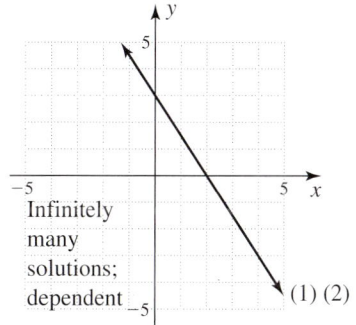
Infinitely many solutions; dependent

(b) (1) $x + 2y = 4$; (2) $2x = 8 - 4y$

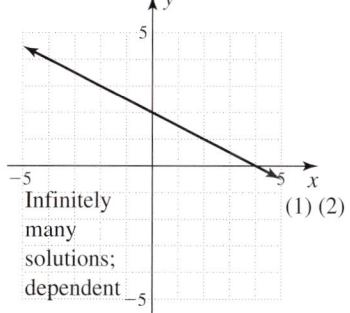
Infinitely many solutions; dependent

5. [4.1B] (a) $(3, 2)$; (b) $(\frac{7}{5}, \frac{12}{5})$ **6.** [4.1B] (a) No solution;
(b) No solution   **7.** [4.1B] (a) $\{(x, y) | y = \frac{1}{2}x + \frac{5}{2}\}$; infinitely many
solutions; (b) $\{(x, y) | y = -\frac{1}{5}x + 1\}$; infinitely many solutions
8. [4.1C] (a) $(4, -1)$; (b) $(-1, 2)$ **9.** [4.1C] (a) No solution;
(b) No solution   **10.** [4.1C] (a) $\{(x, y) | y = -\frac{2}{5}x + \frac{2}{5}\}$; infinitely
many solutions; (b) $\{(x, y) | y = \frac{4}{3}x - 8\}$; infinitely many solutions
11. [4.1D] $93.13   **12.** [4.2A] (a) $(4, 2, -1)$; (b) $(7, 2, -12)$
13. [4.2A] (a) No solution; (b) Infinitely many solutions
14. [4.3A] (a) 30 nickels; 25 dimes; (b) 10 nickels; 15 dimes

15. [4.3B] (a) 180 ft; (b) 160 ft   **16.** [4.3C] (a) 6 mi/hr; (b) 3 mi/hr
17. [4.3D] (a) $10,000 at 4%; $10,000 at 6%; $20,000 at 8%;
(b) $10,000 at 4%; $15,000 at 6%; $20,000 at 8%
18. [4.3E] (a) 10 in. by 40 in.; (b) 10 in. by 30 in.
19. [4.3F] Approximately 0.73 yr after 1995 or about 1996; 93 Euros
20. [4.4A, B]
(a)

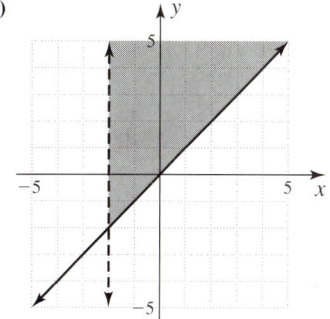

(b)

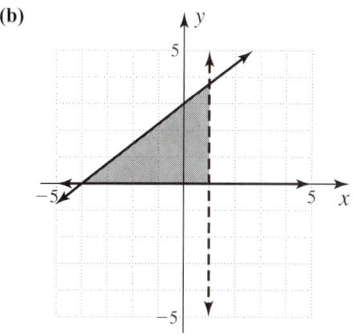

Cumulative Review Chapters 1–4

1. Irrational numbers; real numbers   **2.** $-2x^2 + 5x + 7$ **3.** $4x^{13}$
4. $\frac{9x^6}{y^8}$ **5.** -50 **6.** $\frac{6}{7}$ **7.** 580   **8.** 7
9.

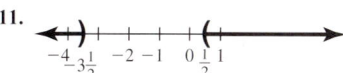

10.

11.

12. 23   **13.** $27,000   **14.** 42 gallons
15.

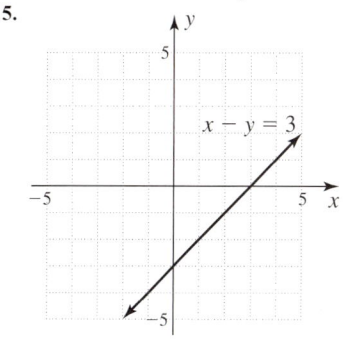

16. $\frac{15}{7}$

17.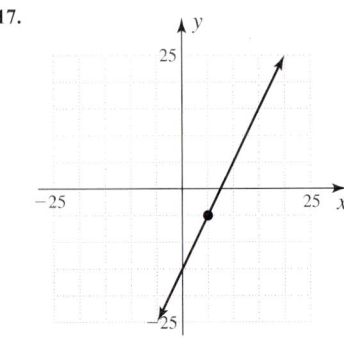

18. $m = -3$; y-intercept $= -18$ **19.** $4x + 3y = 30$
20.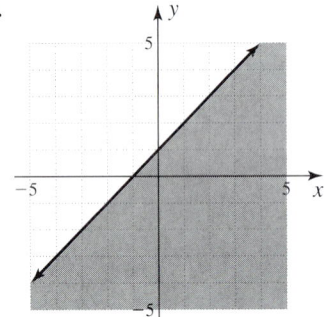

21.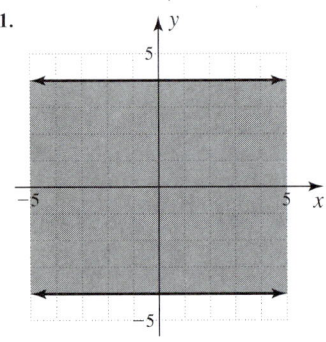

22. $D = \{\text{all real numbers}\}$; $R = \{\text{all real numbers}\}$ **23.** 4
24. 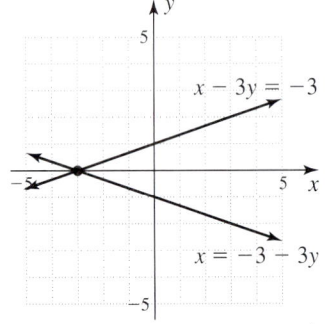   Solution: $(-3, 0)$

25. Infinitely many solutions: $\{(x, y) | y = \frac{1}{2}x - 1\frac{1}{2}\}$ **26.** $(6, 15)$
27. $(-1, 3, -2)$ **28.** 30 mi/hr

Chapter 5
Exercises 5.1

1. Monomial; 4   **3.** Binomial; 3   **5.** Trinomial; 3   **7.** Trinomial; 5
9. Zero polynomial; no degree   **11.** $-x^4 + 3x^2 + 5x + 7$
13. $-8x^2 + 9x + 20$ **15.** $-5x^2 + x + 2$ **17.** -8 **19.** 4   **21.** -1
23. 16   **25.** -136 **27.** (a) 0; (b) -9; (c) -9 **29.** (a) 8; (b) 2; (c) 6
31. $6x^2 - 5$ **33.** $x^2 - 2x - 4$ **35.** $3x^2 + 4x - 9$ **37.** $-11y^2 - y - 3$
39. $4x^3 - 12x^2 + 9x - 6$ **41.** $7y^2 + 12y - 11$ **43.** $3v^3 - 2v^2 + v - 7$
45. $-5u^3 - 5u^2 - 3u + 10$ **47.** $4x^3 + 2xy - 1$ **49.** $x^3 - x^2 - 4$
51. $-2a^2 + 5a$ **53.** $4y$ **55.** $3x^2 + 7y$ **57.** 2   **59.** 0
61. Commutative property of addition   **63.** Distributive property
65. Associative property of addition   **67.** Commutative property of addition   **69.** Distributive property   **71.** (a) 48 ft; (b) 64 ft
73. (a) $42,500; (b) $70,000   **75.** (a) $50,000; (b) $0 (no value)
77. $20   **79.** $1000   **81.** $30,000   **83.** $x^2 + 15x$ **85.** $4x^2 + 12x$
91. terms   **93.** (a) $P = -0.2x^2 + 47x - 100$; (b) $2600
95. 4   **97.** (a) Binomial; (b) Polynomial; (c) Monomial;
(d) Trinomial   **99.** -3 **101.** $-5x + 7$ **103.** $12x^3y$ **105.** $4x^3y^3$
107. $4x^2$ **109.** $-2x - 2$ **111.** $-3x + 1$

Exercises 5.2

1. $12x^2 - 6x$ **3.** $-3x^3 + 9x^2$ **5.** $-24x^3 + 16x^2 - 8x$
7. $-18x^3y^2 - 9xy^4 + 21xy^2$ **9.** $6x^3y^6 - 10x^2y^5 + 2x^2y^4$
11. $x^3 + 4x^2 + 8x + 15$ **13.** $x^3 + 3x^2 - x + 12$
15. $x^3 + 2x^2 - 5x - 6$ **17.** $x^3 - 8$ **19.** $x^4 - x^3 + x^2 + x - 2$
21. $9x^2 + 9x + 2$ **23.** $5x^2 + 11x - 12$ **25.** $3a^2 + 14a - 5$
27. $2y^2 + 7y - 15$ **29.** $x^2 - 8x + 15$ **31.** $6x^2 - 7x + 2$
33. $4x^2 + 4ax - 15a^2$ **35.** $x^2 + 15x + 56$ **37.** $4a^2 + 10ab + 4b^2$
39. $16u^2 + 8uv + v^2$ **41.** $4y^2 + 4yz + z^2$ **43.** $9a^2 - 6ab + b^2$
45. $a^2 - b^2$ **47.** $25x^2 - 4y^2$ **49.** $b^2 - 9a^2$ **51.** $3x^3 + 9x^2 + 6x$
53. $-3x^3 + 12x^2 - 9x$ **55.** $x^3 + 6x^2 + 9x$ **57.** $-2x^3 + 4x^2 - 2x$
59. $4x^2y^2 - y^4$ **61.** $x^2 + \frac{3}{2}x + \frac{9}{16}$ **63.** $4y^2 - \frac{4}{5}y + \frac{1}{25}$
65. $\frac{9}{16}p^2 + \frac{3}{10}pq + \frac{1}{25}q^2$ **67.** $9x^2 + 24xy + 16y^2 + 6x + 8y + 1$
69. $9x^2 - 24xy + 16y^2 - 6x + 8y + 1$
71. $9x^2 + 12xy + 4y^2 - 6x - 4y + 1$
73. $16p^2 - 24pq + 9q^2 + 8p - 6q + 1$ **75.** (a) $R = 1000p - 30p^2$;
(b) $8000   **77.** $A = 2x^2 + 13x + 6$; when $x = 2$, the area is 40
square units.   **79.** $T_1^4 - T_2^4$ **81.** (a) 9; (b) 5; (c) No
83. (a) x^2; (b) xy; (c) y^2; (d) xy **85.** $(x + y)^2 = x^2 + 2xy + y^2$
91. FOIL or vertical scheme   **93.** vertical scheme or distribute
95. $12x^2 + 17xy + 6y^2$ **97.** $25x^2 - 30xy + 9y^2$ **99.** $9x^2 - y^2$
101. $x^3 - 6x^2 + 5x + 6$ **103.** $20x^6 + 12x^5 - 8x^4 - 20x^3$
105. $9x^2 - 24xy + 16y^2 + 6x - 8y + 1$ **107.** $3(x + y)$
109. $2x(z + y)$ **111.** $3a(b - c)$ **113.** $3x^2y + 3x^7$ **115.** $3x^3y$

Exercises 5.3

1. $8(x + 2)$ **3.** $-5(y - 5)$ **5.** $-8(x + 3)$ **7.** $4x(x + 9)$
9. $-5x^2(1 + 7x^2)$ **11.** $3x(x^2 + 2x + 13)$ **13.** $9y(7y^2 - 2y + 3)$
15. $6x^2(6x^4 + 2x^3 - 3x^2 + 5)$ **17.** $8y^3(6y^5 + 2y^2 - 3y + 2)$
19. $\frac{1}{7}(4x^3 + 3x^2 - 9x + 3)$ **21.** $\frac{1}{8}y^2(7y^7 + 3y^4 - 5y^2 + 5)$
23. $(x + 5)(x^2 + 4)$ **25.** $(4x + 9)(z - 1)$ **27.** $(x + 2)(x^2 + 3)$
29. $(y - 3)(y^2 + 1)$ **31.** $(2x + 3)(2x^2 + 5)$ **33.** $(3x - 1)(2x^2 + 1)$
35. $(y + 2)(4y^2 + 3)$ **37.** $(2a^2 + 3)(a^4 + 1)$ **39.** $(x^2 + 4)(3x^3 + 5)$
41. $(2y^2 + 3)(3y^3 + 1)$ **43.** $y^2(y^2 + 3)(4y^3 + 1)$
45. $a^2(a^2 - 2)(3a^3 - 2)$ **47.** $a^2(2a - 3)(4a^2 - 5)$
49. $x^3(x - 2)(x^2 + 2)$ **51.** $(x - 4)(2x + 5)$ **53.** $\alpha L(t_2 - t_1)$
55. $(R - 1)(R - 1)$ or $(R - 1)^2$ **57.** $-w(l - z)$ **59.** $a(a + 2s)$
61. $-16(t^2 - 5t - 15)$ **67.** factor by grouping
69. $x(x - 2)(3x^2 - 1)$ **71.** $(x + 1)(2x^2 + 3)$ **73.** $7x^2(2x^2 + x - 7)$
75. $4x^2(1 - 8x)$ **77.** $-3y(y - 7)$ **79.** $x^2 + 7x + 12$ **81.** $x^2 + 3x - 10$
83. $25x^2 + 20xy + 4y^2$ **85.** $25x^2 - 20xy + 4y^2$ **87.** $u^2 - 36$

Exercises 5.4

1. $(x + 2)(x + 3)$ **3.** $(a + 2)(a + 5)$ **5.** $(x + 4)(x - 3)$
7. $(x + 2)(x - 1)$ **9.** $(x + 1)(x - 2)$ **11.** $(x + 2)(x - 5)$
13. $(a - 7)(a - 9)$ **15.** $(y - 2)(y - 11)$ **17.** $(9x + 1)(x + 4)$
19. $(3a + 1)(a - 2)$ **21.** $(2y + 5)(y - 4)$ **23.** $(4x - 3)(x - 2)$
25. $(2x + 3)(3x - 4)$ **27.** $(3a + 2)(7a - 1)$ **29.** $(2x + 3y)(3x - y)$
31. $x^2(7x - 3y)(x - y)$ **33.** $y^3(3x + y)(5x - 2y)$
35. $xy^2(15x^2 - 2xy - 2y^2)$ **37.** $-(2b - 5)(b - 4)$
39. $-(4y - 3)(3y + 4)$ **41.** $(2y + 7)(y + 1)$ **43.** $(2x - 3)(x - 3)$

45. $-(a+1)^4$ **47.** $(2g+7)(g-3)$ **49.** $(2R-1)(R-1)$
51. $(L-3)(2L-3)$ **53.** $(5t-7)(t-1)$ **57.** ac test **59.** prime
61. $(3x-1)(2x-5)$ **63.** $(x-6)(x+1)$ **65.** $(x-5y)(x-2y)$
67. $(2x+3y)(x-y)$ **69.** $(3x+2)(x-2)$ **71.** $-(7x-1)(3x+2)$
73. $-(9y-17)(y+2)$ **75.** $9a^2+12ab+4b^2$ **77.** $4a^2-12ab+9b^2$
79. $81a^2-4b^2$ **81.** $25x^2-49y^2$ **83.** x^3+64

Exercises 5.5

1. $(x+1)^2$ **3.** $(y+11)^2$ **5.** $(1+2x)^2$ **7.** $(3x+5y)^2$ **9.** $4(3a+2)^2$
11. $(y-1)^2$ **13.** $(7-x)^2$ **15.** $(7a-2x)^2$ **17.** $(4x-3y)^2$
19. $(3x^2+2)^2$ **21.** $(4x^2-3)^2$ **23.** $(1+x^2)^2$ **25.** $(y+8)(y-8)$
27. $(a+\frac{1}{3})(a-\frac{1}{3})$ **29.** $(8+b)(8-b)$ **31.** $(6a+7b)(6a-7b)$
33. $(\frac{x}{3}+\frac{y}{4})(\frac{x}{3}-\frac{y}{4})$ **35.** $(a+2b+c)(a+2b-c)$
37. $(2x-y+1)(2x-y-1)$ **39.** $(3y-2x+5)(3y-2x-5)$
41. $(4a+x+3y)(4a-x-3y)$ **43.** $(y+a-b)(y-a+b)$
45. $(x+5)(x^2-5x+25)$ **47.** $(1+a)(1-a+a^2)$
49. $(2x+y)(4x^2-2xy+y^2)$ **51.** $(x-1)(x^2+x+1)$
53. $(5a-2b)(25a^2+10ab+4b^2)$
55. $(x+2)(x-2)(x^2-2x+4)(x^2+2x+4)$
57. $(x+\frac{1}{2})(x-\frac{1}{2})(x^2-\frac{1}{2}x+\frac{1}{4})(x^2+\frac{1}{2}x+\frac{1}{4})$
59. $(\frac{x}{2}+1)(\frac{x}{2}-1)(\frac{x^2}{4}-\frac{x}{2}+1)(\frac{x^2}{4}+\frac{x}{2}+1)$
61. $(x-y+1)(x^2-2xy+y^2-x+y+1)$
63. $(1+x+2y)(1-x-2y+x^2+4xy+4y^2)$
65. $(y-2x-1)(y^2-4xy+4x^2+y-2x+1)$
67. $(3-x-2y)(9+3x+6y+x^2+4xy+4y^2)$
69. $(4+x^2-y^2)(16-4x^2+4y^2+x^4-2x^2y^2+y^4)$
71. $(10+x)(10-x)$ **73.** $(2x+1)(4x^2-2x+1)$ **77.** 9
79. $(a^2-4a+8)(a^2+4a+8)$ **81.** prime polynomial
83. $(4-\frac{1}{3}x)(16+\frac{4}{3}x+\frac{1}{9}x^2)$ **85.** $(x+1)(x-1)(x^2-x+1)$
(x^2+x+1) **87.** $(4x+1)(4x-1)$ **89.** Not factorable
91. Not factorable **93.** $(x-8)^2$ **95.** $(x-y)^2(x-y+1)(x-y-1)$
97. $(x-y+5)(x^2-2xy+y^2-5x+5y+25)$ **99.** $3x(xy+2x^3z-3)$
101. $(x+7)(3x^2+4)$ **103.** $(x-5y)(x+3y)$

Exercises 5.6

1. $3x^2(x+2)(x-3)$ **3.** $5x^2(x+4y)(x-2y)$ **5.** $-3x^4(x^2+2x+7)$
7. $2x^4y(x^2-2xy-5y^2)$ **9.** $-2x^4(2x^2+6xy+9y^2)$
11. $2y^2(x+2)(3x^2+1)$ **13.** $-3xy(x+1)(3x^2+2)$
15. $-2x(x+y)(2x^2-y)$ **17.** $3y^2(x+4y)^2$ **19.** $-2k(3x+2y)^2$
21. $4xy^2(2x-3y)^2$ **23.** Not factorable **25.** $3x^3(x+2y)^2$
27. $2x^4(3x+y)^2$ **29.** $3x^2y^2(2x-3y)^2$ **31.** $6(x+2)(x+1)(x-1)$
33. $7(x^2+y^2)(x+y)(x-y)$ **35.** $2x^2(x^2+4y^2)(x+2y)(x-2y)$
37. $-2(x+3)^2$ **39.** $-3(x+2)^2$ **41.** $-x^2(2x+y)^2$
43. $-y^2(3x+2y)^2$ **45.** $-2y^2(2x-3y)^2$ **47.** $-2x(3x+2y)^2$
49. $-2x(3x+5y)^2$ **51.** $-x(x+y)(x-y)$ **53.** $-x^2(x+2y)(x-2y)$
55. $-x^2(2x+3y)(2x-3y)$ **57.** $-2x(2x+3y)(2x-3y)$
59. $-2x^2(3x+2y)(3x-2y)$ **61.** $x^2(3-x)(9+3x+x^2)$
63. $x^4(x-2)(x^2+2x+4)$ **65.** $x^4(3+2x)(9-6x+4x^2)$
67. $x^4(3x+4y)(9x^2-12xy+16y^2)$ **69.** $(x+y+2)(x-y+2)$
71. Not factorable **73.** $(x+y+2)(x-y-2)$ **75.** $-(3x-5y)^2$
77. $2x(3x-5y)^2$ **79.** $\frac{2\pi A}{360}(R+Kt)$ **81.** $\frac{3S}{2bd^3}(d+2z)(d-2z)$
87. $(x-2)(3x+1)(3x-1)$ **89.** $x^2(2x^2+xy+2y^2)$
91. $(x+y-5)(x-y-5)$ **93.** $-(4y-7)(2y+3)$ **95.** $4x^3(3x+y)^2$
97. $8x^3(x^2+9)$ **99.** $\frac{7}{3}$ **101.** -37 **103.** $x^2-8x+16$

Exercises 5.7

1. $-1, -2$ **3.** $1, -4, -3$ **5.** $\frac{1}{2}, \frac{1}{3}$ **7.** $0, 3$ **9.** $8, -8$ **11.** $9, -9$
13. $0, -6$ **15.** $0, 3$ **17.** $3, 9$ **19.** $-1, -5$ **21.** $5, -3$
23. $-\frac{2}{3}, -1$ **25.** $1, \frac{1}{2}$ **27.** $1, -\frac{1}{2}$ **29.** $2, -6$ **31.** $1, \frac{1}{2}$ **33.** $-2, -4$
35. $4, 1$ **37.** $-2, -\frac{5}{2}$ **39.** 1 **41.** $2, -2, -4$ **43.** $5, 3, -3$
45. $2, -2, -1$ **47.** $6, 8, 10$ **49.** 10 in., 24 in., 26 in. **51.** 1 sec
53. 1 sec **55.** 40 **57.** 10 **59.** (a) $5.50-x$; (b) $550+450x-100x^2$;
(c) $0.50 or $4; (d) $0.50 **61.** $2\frac{1}{5}$ ft by 5 ft **63.** 4 sec $(3+1)$

65. 400 ft **71.** quadratic equation **73.** $-2, 3, -3$ **75.** $\frac{1}{4}, -\frac{4}{3}$
77. $-3, 2$ **79.** $0, -9$ **81.** $\frac{1}{5x^2}$ **83.** $2x^{10}$ **85.** $\frac{2}{x^{12}}$ **87.** $7y(x^2-y)$
89. $(5x+3)(5x-3)$ **91.** $(2x+3)(x-5)$

Review Exercises

1. [5.1A, B] (a) Binomial; degree 6; (b) Monomial; degree 8;
(c) Trinomial; degree 5 **2.** [5.1B] (a) $3x^4 - x^2 - 5x + 2$;
(b) $4x^3 - x^2 + 3x$; (c) $6x^2 + x - 2$ **3.** [5.1C, E] (a) $20,000; (b) $0
4. [5.1C] (a) 9; (b) 3; (c) 23 **5.** [5.1D] (a) $x^3 + 8x^2 - 10x + 7$;
(b) $5x^3 - 3x^2 - 3x + 8$; (c) $3x^3 + 5x^2 - 7x + 6$
6. [5.1D] (a) $-3x^3 + 7x^2 - 3x + 3$; (b) $10x^2 + x - 6$;
(c) $-7x^3 + 13x^2 - 12x + 4$ **7.** [5.2A] (a) $-2x^4y - 6x^3y^2 + 4x^2y^4$;
(b) $-3x^4y^2 - 9x^3y^3 + 6x^2y^5$; (c) $-4x^3y^2 - 12x^2y^3 + 8xy^5$
8. [5.2B] (a) $x^3 - 4x^2 + x + 2$; (b) $x^3 - 5x^2 + 4x + 4$;
(c) $x^3 - 11x - 6$ **9.** [5.2C] (a) $8x^2 + 2xy - 15y^2$;
(b) $6x^2 + 13xy + 6y^2$; (c) $10x^2 + 9xy - 9y^2$
10. [5.2D] (a) $4x^2 + 20xy + 25y^2$; (b) $9x^2 + 42xy + 49y^2$;
(c) $16x^2 + 72xy + 81y^2$ **11.** [5.2D] (a) $9x^2 - 12xy + 4y^2$;
(b) $16x^2 - 56xy + 49y^2$; (c) $25x^2 - 60xy + 36y^2$
12. [5.2E] (a) $9x^2 - 4y^2$; (b) $16x^2 - 9y^2$; (c) $25x^2 - 9y^2$
13. [5.3A] (a) $5x^2(3x^3 - 4x^2 + 2x + 5)$; (b) $3x^2(3x^3 - 4x^2 + 2x + 5)$;
(c) $2x^2(3x^3 - 4x^2 + 2x + 5)$ **14.** [5.3B] (a) $x(3x^2 - 1)(2x^3 + 5)$;
(b) $x(3x^2 - 4)(2x^3 + 5)$; (c) $x(3x^2 - 2)(2x^3 + 3)$
15. [5.4A] (a) $(x - 6y)(x + 3y)$; (b) $(x - 6y)(x + 2y)$;
(c) $(x - 6y)(x + y)$ **16.** [5.4B] (a) $(2x + 5y)(x - 6y)$; (b) $(2x + 5y)$
$(x - 4y)$; (c) $(2x + 5y)(x - 5y)$ **17.** [5.4C] (a) $-3x^2y(3x + 2y)$
$(2x - y)$; (b) $-5x^2y(3x + 2y)(2x + y)$; (c) $6x^2y(3x + 2y)(2x - 3y)$
18. [5.5A] (a) $(2x - 7y)^2$; (b) $(3x - 7y)^2$; (c) $(4x - 7y)^2$
19. [5.5A] (a) $(3x + 4y)^2$; (b) $(3x + 5y)^2$; (c) $9(x + 2y)^2$
20. [5.5B] (a) $(9x^2 + y^2)(3x + y)(3x - y)$; (b) $(x^2 + 4y^2)(x + 2y)$
$(x - 2y)$; (c) $(9x^2 + 4y^2)(3x + 2y)(3x - 2y)$
21. [5.5B] (a) $(x + y - 2)(x - y - 2)$; (b) $(x + y - 3)(x - y - 3)$;
(c) $(x + y + 4)(x - y + 4)$ **22.** [5.5C] (a) $(3x + 2y)(9x^2 - 6xy + 4y^2)$;
(b) $(3x + 4y)(9x^2 - 12xy + 16y^2)$; (c) $(4x + 3y)(16x^2 - 12xy + 9y^2)$
23. [5.5C] (a) $(3x - 2y)(9x^2 + 6xy + 4y^2)$;
(b) $(3x - 4y)(9x^2 + 12xy + 16y^2)$; (c) $(4x - 3y)(16x^2 + 12xy + 9y^2)$
24. [5.6A] (a) $x^3(3x - 2y)(9x^2 + 6xy + 4y^2)$;
(b) $x^4(3x - 4y)(9x^2 + 12xy + 16y^2)$;
(c) $x^5(4x - 3y)(16x^2 + 12xy + 9y^2)$
25. [5.6A] (a) $3x^4(9x^2 + 1)$; (b) $4x^4(x^2 + 16)$; (c) $2x^4(x + 3)(x - 3)$
26. [5.6A] (a) $3x^2(3x + 2y)^2$; (b) $4x^2(3x - 2y)^2$; (c) $5x^2(3x + y)^2$
27. [5.6A] (a) $3x^2(3x - y)^2$; (b) $4x^2(3x - 2y)^2$; (c) $5x^2(3x + 2y)^2$
28. [5.6A] (a) $4xy(3x + y)(x - 4y)$; (b) $5xy(3x + y)(x + 4y)$;
(c) $6xy(3x + 2y)(x - 4y)$ **29.** [5.6A] (a) $(2x - 1)(x + 1)(x - 1)$;
(b) $(2x - 1)(3x + 1)(3x - 1)$; (c) $(2x - 1)(4x + 1)(4x - 1)$
30. [5.7A] (a) $3, -4$; (b) $4, -5$; (c) $4, -6$ **31.** [5.7A] (a) $\frac{1}{3}, -\frac{1}{2}$;
(b) $\frac{1}{4}, -\frac{1}{2}$; (c) $\frac{1}{5}, -\frac{1}{2}$ **32.** [5.7A] (a) $1, -1, -2$; (b) $1, -1, -4$;
(c) $3, -2, -3$ **33.** [5.7B] (a) $5, 12, 13$; (b) $9, 12, 15$; (c) $12, 16, 20$
34. [5.7C] $\frac{2}{3}$ ft by 9 ft

Cumulative Review Chapters 1–5

1. $\{2, 4, 6\}$ **2.** $\frac{19}{100}$ **3.** 15 **4.** 6×10^{-3} **5.** -66 **6.** 14
7. $\frac{3}{4}, -\frac{39}{4}$
8.
9.
10.
11. $A = \frac{5B + 22}{2}$ **12.** 495 mi **13.** x-intercept $= -\frac{5}{3}$; y-intercept $= -5$

14.

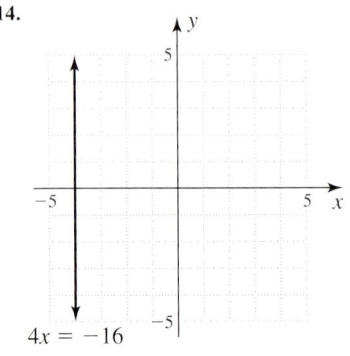

$4x = -16$

15. Parallel **16.** $5x - 4y = 19$ **17.** $y = -3x - 1$
18.

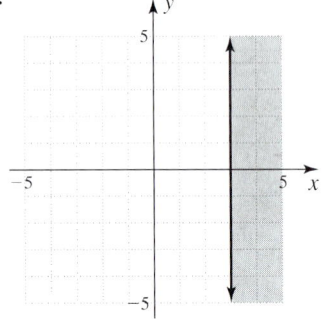

19. $(-3, -1)$ **20.** $(2, 1, 1)$ **21.** 33 nickels; 21 dimes
22. 31 **23.** $4x^2 - 28xy + 49y^2$ **24.** $(2c + 5)(4c^2 - 10c + 25)$
25. $-\frac{3}{2}; -3$

Chapter 6

Exercises 6.1

1. $x = 0$ **3.** $x = -3$ **5.** $m = -2$ **7.** $m = -1; 2$ **9.** $p = -3; 3$
11. None, defined for all values of v **13.** $\frac{4xy^2}{6y^3}$ **15.** $\frac{30}{18(x-7)}$ **17.** $\frac{6x+6y}{(x+y)^2}$
19. $\frac{x(x-y)}{x^2-y^2}$ or $\frac{x^2-xy}{x^2-y^2}$ **21.** $\frac{-x(2x+3y)}{4x^2-9y^2}$ or $\frac{-2x^2-3xy}{4x^2-9y^2}$ **23.** $\frac{4x(x-2)}{x^2-x-2}$ or
$\frac{4x^2-8x}{x^2-x-2}$ **25.** $\frac{-5(x-2)}{x^2+x-6}$ or $\frac{-5x+10x}{x^2+x-6}$ **27.** $\frac{3(x^2-xy+y^2)}{x^3+y^3}$ or $\frac{3x^2-3xy+3y^2}{x^3+y^3}$
29. $\frac{x(x^2+xy+y^2)}{x^3-y^3}$ or $\frac{x^3+x^2y+xy^2}{x^3-y^3}$ **31.** $\frac{y}{2}$ **33.** $\frac{x}{5-x}$ **35.** $\frac{-2x}{5y}$ **37.** $\frac{x+y}{x-y}$
39. $\frac{1}{x-2}$ **41.** $\frac{x^3}{y^3}$ **43.** 3 **45.** $\frac{1}{3x+2y}$ **47.** $\frac{(x-y)^2}{x+y}$ **49.** $y - 1$ **51.** $\frac{x+y}{x-y}$
53. $\frac{y-5}{y+6}$ **55.** -1 **57.** $-(x+3)$ **59.** $-(y^2+2y+4)$
61. -1 **63.** $-(x+5)$ **65.** $2-x$ **67.** $\frac{-1}{x+6}$ **69.** $\frac{1}{x-2}$
71. (a) \$525 million; (b) \$1400 million or \$1.4 billion; (c) \$3150 million or \$3.15 billion; (d) No; Denominator = 0 for $p = 100$. As p increases toward 100, price increases without bound. **73.** (a) $5(x-1)$: (b) \$45: (c) \$495 (when $x = 100$) **75.** y is a term in the numerator, not a factor
77. No **81.** rational equation **83.** undefined **85.** x^4y^4
87. $\frac{-(x+y)}{y^2+xy+x^2}$ **89.** $4 - x$ **91.** $x = -1; 1$ **93.** $\frac{4}{x}$ **95.** $\frac{14}{16}$
97. $\frac{4x^2-11x-3}{x^2-x-6}$ **99.** $4x^3$ **101.** $-4x^2$ **103.** $(x-5)(x+3)$
105. $(2x+3y)(2x-3y)$ **107.** $(x-2)(x^2+2x+4)$
109. $(3x+2)(9x^2-6x+4)$ **111.** $\frac{-2}{3}$ **113.** 2 **115.** -4

Exercises 6.2

1. $\frac{3}{10}$ **3.** $\frac{2x}{3}$ **5.** $\frac{2y}{21x^2z^4}$ **7.** $\frac{4}{x+1}$ **9.** 1 **11.** $\frac{-5}{x(x+y)}$ **13.** $\frac{1}{2}$ **15.** 1
17. $1-x$ **19.** $\frac{a+b}{a-b}$ **21.** $\frac{27}{50}$ **23.** $\frac{5x}{3}$ **25.** $\frac{9ad}{c}$ **27.** $\frac{3x}{x+1}$
29. $\frac{4(y+5)}{3(y+2)}$ or $\frac{4y+20}{3y+6}$ **31.** $\frac{a+b}{a-b}$ **33.** $\frac{-(4a^2+2a+1)}{2a^2w^2}$ **35.** $\frac{-(1+x)}{5x}$ or $\frac{-1-x}{5x}$
37. $\frac{2y+3}{3y-5}$ **39.** $\frac{2x-3}{5-2x}$ **41.** $\frac{x(x-1)}{3(x-3)}$ or $\frac{x^2-x}{3x-9}$ **43.** $\frac{x-1}{x-4}$ **45.** $\frac{-5}{x-4}$ or $\frac{5}{4-x}$

47. $\frac{x}{x+y}$ **49.** $\frac{2(x+y)}{x-y}$ or $\frac{2x+2y}{x-y}$ **51.** $x+2$ **53.** $\frac{1}{x(x^2+2x+4)}$ or
$\frac{1}{x^3+2x^2+4x}$ **55.** $\frac{450}{x}$ **57.** $\frac{4t(t+3)}{t^2+9}$ or $\frac{4t^2+12t}{t^2+9}$ **59.** $\frac{RR_T}{R-R_T}$
61. $R = \frac{60,000+9000x}{x}$ **63.** $\frac{2w^2-Lw}{6}$ **67.** factor **69.** factor **71.** x^2z^5
73. $\frac{1}{(x-2)(x+2)}$ or $\frac{1}{x^2-4}$ **75.** $\frac{x-4}{x-3}$ **77.** $\frac{3x+2y}{2x+1}$ **79.** $3x(z+y)$
81. $(x-5)(x-5)$ or $(x-5)^2$ **83.** $\frac{12x^2y^2}{20y^5}$ **85.** $\frac{7x+14}{(x-3)(x+2)}$

Exercises 6.3

1. $\frac{3x}{5}$ **3.** $\frac{5x}{3}$ **5.** $\frac{3+2x}{5(x+2)}$ **7.** $\frac{x+2}{2(x+1)}$ **9.** $\frac{2x+5}{3(x-1)}$ **11.** $\frac{32}{15x}$ **13.** $\frac{7}{6m^2}$
15. $\frac{9b-10a}{15ab}$ **17.** $\frac{55+24y}{30y^3}$ **19.** $\frac{13x}{5(x-1)}$ **21.** $\frac{6a^2-2a}{(a-5)(a+2)}$
23. $\frac{2x^2-5x}{(x-1)(x+4)(x-4)}$ **25.** $\frac{6x-4y}{(x+y)^2(x-y)}$ **27.** $\frac{x^2-4x-7}{(x+1)(x+2)(x+3)}$
29. $\frac{y^2+2y}{(y+1)(y-1)}$ **31.** $\frac{50x^2+8y^2}{(5x-2y)(5x+2y)}$ **33.** $\frac{4-3x}{(x-2)(x-4)(x-1)}$
35. $\frac{-45}{(x+5)(x-5)}$ **37.** $\frac{x}{(x-y)(2-x)}$ **39.** $\frac{1}{x^2-5x+25}$ **41.** $\frac{2x-5}{(x-2)(x-3)}$
43. $\frac{2}{x+3}$ **45.** $\frac{2a+6}{(a+2)(a+4)}$ **47.** $\frac{2}{a+4}$ **49.** $\frac{15-a^2}{(a+5)(a+3)}$
51. $\frac{-w_0x^3+3w_0L^2x-2w_0L^3}{6L}$ **53.** $\frac{p^2-2gm^2rM}{2mr^2}$ **55.** $P(x+h) = x^2 + 2xh + h^2$
57. $\frac{P(x+h)-P(x)}{h} = 2x+h$ **63.** reduced **65.** $\frac{2x^2+9x+13}{(x+3)(x-1)(x+1)}$
67. $\frac{4x}{(x-3)(x+1)}$ **69.** $\frac{2x^2+12x}{(x+1)(x+3)}$ **71.** $\frac{5x}{(x+2)^2}$ **73.** 1 **75.** 210
77. $18b - 24a$ **79.** $x^3 - 1$ **81.** $\frac{a+b}{2}$ **83.** $\frac{x+y}{2x+y}$

Exercises 6.4

1. $\frac{178}{33}$ **3.** $\frac{a}{c}$ **5.** $\frac{z}{xy}$ **7.** $\frac{2z}{5y}$ **9.** $\frac{1}{3}$ **11.** $\frac{ab-a}{b+a}$ **13.** $\frac{1+2x}{2x-1}$ **15.** $\frac{4+6x}{6x-3}$
17. $\frac{x}{xy-2}$ **19.** $\frac{x-y}{x^2y^2}$ **21.** $\frac{9}{5}$ **23.** $\frac{2a^2-a}{2a+1}$ **25.** $\frac{x^2}{2x-1}$ **27.** $\frac{-(x^2+1)}{2x}$ **29.** $\frac{x}{y}$
31. -1 **33.** $\frac{1-x}{1+x}$ **35.** $\frac{y+7}{y-2}$ **37.** $\frac{x^2-1}{5x^2-4x-2}$ **39.** $\frac{c+d}{c-d}$ **41.** $\frac{ab}{a^2+b^2}$
43. $R = \frac{R_1R_2}{R_2+R_1}$ **45.** $f = f_{static}\sqrt{\frac{c+v}{c-v}}$ **47.** $\frac{288(NM-P)}{N(12P+NM)}$ **49.** $\frac{6}{25}$ yr
51. $11\frac{43}{50}$ yr **57.** least common denominator **59.** reciprocal **61.** $\frac{4x}{3b}$
63. $\frac{x(x-3)(x+4)}{x^2-8}$ **65.** $\frac{4(2b+3a)}{2b-3a}$ **67.** $\frac{x(x^2+x+1)}{(x+1)(x^2+1)}$ **69.** 1 **71.** $4x^2$
73. $\frac{-1}{2x^2}$ **75.** $\frac{-5}{x^3}$ **77.** $6x^3 - 12x^2$ **79.** $(3x+2)(2x-1)$
81. $(5x+2)(4x-3)$ **83.** $-6x-18$ **85.** -442

Exercises 6.5

1. $x^2 + 3x - 2$ **3.** $-2x^2 + x - 3$ **5.** $-2y^2 + 8y - 3$
7. $5x^2 + 4x - 8 + \frac{3}{x}$ **9.** $3xy - 2 + \frac{3}{xy}$ **11.** $x + 3$ **13.** $y + 5$
15. $x^2 - x - 1$ **17.** $x^2 + 5x + 6$ **19.** $x^2 - 2x - 8$
21. $(x^2 + 4x + 3)$ R 1 **23.** $y^2 - y - 1$ **25.** $(4x^2 + 3x + 7)$ R 12
27. $x^2 - 2x + 4$ **29.** $4y^2 + 8y + 16$ **31.** $a^2 - 2a - 1$
33. $x^3 + 2x^2 - x$ **35.** $(2x^3 - 3x^2 + x - 4)$ R -1
37. $(x+1)(x-3)(x-2)$ **39.** $(x-2)(x-2)(x+1)(x-1)$
41. $(x^2-3x+7)(x+5)(x+4)$ **43.** v^2+3v+1
45. (x^2+6x+5) R 15 **47.** $z^2 - 6z + 4$
49. $(3y^3 + 12y^2 + 7y + 15)$ R 52 **51.** $2y^3 - y^2 + 5$
53. Rem. = 0; 4 is a solution **55.** Rem. = 0; -4 is a solution
57. Rem. = 0; 5 is a solution **59.** Rem. = 0; -1 is a solution
61. $\frac{500}{x} + 4$ **63.** $x+1, x+2, x+3, x+6$ **69.** monomial; monomial
71. Synthetic division **73.** $2x^3 - 3x^2 + x - 4$ **75.** Rem. = 0; -2 is a solution **77.** $(z+2)(z-2)(z-3)$ **79.** $(3x^2 - 8x + 15)$ R -36
81. $2x - \frac{1}{2} + \frac{1}{x} - \frac{2}{x^2} + \frac{5}{x^3}$ **83.** $2x(x-3)$ **85.** $(x-8)(x+8)$ **87.** 5
89. 2

Exercises 6.6

1. 6 **3.** -5 **5.** $\frac{1}{3}$ **7.** -4 **9.** $\frac{26}{9}$ **11.** $\frac{-1}{12}$ **13.** No solution **15.** $\frac{1}{3}$
17. -11 **19.** 7 **21.** 2 **23.** -3 **25.** $\frac{-3}{2}$ **27.** No solution **29.** 0
31. $\frac{4}{5}$ **33.** 2 **35.** -4 **37.** 23 **39.** -4 **41.** $\frac{26}{9}$ **43.** $\frac{-1}{12}$
45. No solution **47.** $4\frac{4}{9}$ hr **49.** (a) Approximately $16\frac{1}{2}$ hr;
(b) Approximately $2\frac{1}{2}$ days **51.** Approximately 3 consecutive hits

53. $h = \frac{2A}{b_1 + b_2}$ **55.** $Q_1 = \frac{PQ_2}{1+P}$ **57.** $f = \frac{ab}{a+b}$ **61.** Yes
63. proportion **65.** rational equation/proportion **67.** -7
69. 3 **71.** -4 and 1 **73.** 5 **75.** $3x + 2x = 5x$ **77.** $15 - 42x$
79. 21, 23, 25 **81.** 60 gal

Exercises 6.7

1. 8 **3.** 5 and 10 **5.** 6 and 8 **7.** $\frac{2}{7}$ **9.** $20,000 **11.** $1\frac{7}{8}$ hr **13.** $4\frac{14}{19}$ hr
15. 6 hr; 3 hr **17.** $15\frac{3}{4}$ hr **19.** $5\frac{1}{4}$ hr **21.** 36 sec **23.** 18 hr
25. 150 mi/hr **27.** 30 mi/hr **29.** Auto 25 mi/hr; plane 125 mi/hr
31. $h = \frac{3V}{4\pi r^3}$ **33.** $x^2 = \frac{a^2b^2 - a^2y^2}{b^2}$ **35.** $F = \frac{f_1 f_2}{f_1 + f_2}$ **37.** $R = \frac{2E - 2ri}{i}$
39. 12 tablets a day **41.** $\cos(2u) = 2\cos^2 u - 1$ **45.** 36 mi/hr
47. 4 and 6 **49.** -90 **51.** 1 **53.** $k\sqrt[3]{m}$ **55.** kh^3

Exercises 6.8

1. $T = ks$ **3.** $W = kh^3$ **5.** $W = kB$ **7.** $R = \frac{k}{D^2}$ **9.** $I = kPr$
11. $A = ksv^2$ **13.** $V = kdw$ **15.** $I = \frac{ki}{d^2}$ **17.** $R = \frac{kL}{A}$ **19.** $W = \frac{k}{d^2}$
21. (a) $I = km$; **(b)** 0.055 or 5.5%; **(c)** $41.25 **23. (a)** $d = ks^2$;
(b) 0.06; **(c)** 216 ft **25. (a)** $S = \frac{k}{y}$; **(b)** 30 new songs
27. (a) $W = \frac{k}{d^2}$; **(b)** $121(3960)^2$; **(c)** 81 lb **29. (a)** $d = ks$;
(b) 17.63; **(c)** The number of hours needed to travel d distance
at s speed **31. (a)** $C = 4(F - 37)$; **(b)** 212 chirps **33. (a)** $I = kn$;
(b) 1.365; **(c)** 367.675 ppm **35.** $208.33 **37.** 102.9
39. 0.42 **41.** 0.42 **43.** 52.5 lb/in.² **49.** direct variation
51. (a) $F = kAV^2$; **(b)** 0.0045; **(c)** 32.4 lb **53. (a)** $P = \frac{k}{r}$;
(b) 10 **55.** $\frac{1}{x^8}$ **57.** x^{20} **59.** $-8x^3 y^6$ **61.** $\frac{1}{x^2}$ **63.** $\frac{1}{a^8 b^6}$ **65.** $\frac{3}{10}$
67. $\frac{-5}{12}$

Review Exercises

1. [6.1A] **(a)** $x = 4$; **(b)** $x = -1; -9$; **(c)** None, defined for all values
2. [6.1B] **(a)** $\frac{8x^2 y^3}{36y^7}$; **(b)** $\frac{10x^2 y^4}{45y^8}$; **(c)** $\frac{2x^2 + 11x + 5}{x^2 + 6x + 5}$; **(d)** $\frac{2x^2 + 13x + 6}{x^2 + 7x + 6}$
3. [6.1C] **(a)** $\frac{6}{y}$; **(b)** $\frac{7}{y}$; **(c)** $\frac{-8}{y}$ **4.** [6.1C] **(a)** $\frac{y-x}{6}$; **(b)** $\frac{7}{y-x}$; **(c)** $\frac{y-x}{8}$
5. [6.1D] **(a)** $x^3 y^5$; **(b)** $x^3 y^6$; **(c)** $x^3 y^7$ **6.** [6.1D] **(a)** $\frac{y^2}{x-y}$; **(b)** $\frac{y^3}{x-y}$;
(c) $\frac{y^4}{x-y}$ **7.** [6.1D] **(a)** $\frac{2y-x}{x^2 - 2xy + 4y^2}$; **(b)** $\frac{-(x+2y)}{x^2 + 2xy + 4y^2}$; **(c)** $\frac{-(x+3y)}{x^2 + 3xy + 9y^2}$
8. [6.2A] **(a)** $\frac{3x+2y}{3x+1}$; **(b)** $\frac{3x+2y}{4x-3}$; **(c)** $\frac{3x+2y}{5x+2}$ **9.** [6.2B] **(a)** $\frac{1}{(x-2)(x+4)}$;
(b) $\frac{1}{(x-2)(x+5)}$; **(c)** $\frac{1}{(x-2)(x+6)}$ **10.** [6.2C] **(a)** $\frac{-(x^2 - 3x + 9)}{x^2 + 2x + 4}$;
(b) $\frac{-(x^2 - 3x + 9)}{x^2 + 4x + 16}$; **(c)** $\frac{-(x^2 - 5x + 25)}{x^2 + 2x + 4}$ **11.** [6.3A] **(a)** $\frac{1}{x-2}$; **(b)** $\frac{1}{x-3}$;
(c) $\frac{1}{x-4}$ **12.** [6.3A] **(a)** $\frac{1}{x+3}$; **(b)** $\frac{1}{x+4}$; **(c)** $\frac{1}{x+5}$ **13.** [6.3B]
(a) $\frac{2x^2 + 9x + 11}{(x-1)(x+2)(x+1)}$; **(b)** $\frac{2x^2 + 10x + 13}{(x-1)(x+2)(x+1)}$; **(c)** $\frac{2x^2 + 11x + 15}{(x-1)(x+2)(x+1)}$
14. [6.3B] **(a)** $\frac{-4x - 14}{(x-3)(x+2)(x+3)}$; **(b)** $\frac{-3x - 11}{(x-3)(x+2)(x+3)}$;
(c) $\frac{-x - 5}{(x-3)(x+2)(x+3)}$ **15.** [6.4A] **(a)** $\frac{x(x^2 - x + 1)}{(x^2 + 1)(x - 1)}$; **(b)** $\frac{x(x^2 - x + 1)}{(x^2 + 1)(x - 1)}$;
(c) $\frac{x(x^2 - x + 1)}{(x^2 + 1)(x - 1)}$ **16.** [6.4A] **(a)** $\frac{a^2 + 20a + 80}{4a + 20}$; **(b)** $\frac{a^2 + 30a + 150}{5a + 30}$;
(c) $\frac{a^2 + 42a + 252}{6a + 42}$ **17.** [6.5A] **(a)** $3x^3 - 2x + 1$; **(b)** $3x^2 - 2 + \frac{1}{x}$;
(c) $3x - \frac{2}{x} + \frac{1}{x^2}$ **18.** [6.5B] **(a)** $(x^2 - x - 1)$ R -6;
(b) $(x^2 - x - 1)$ R -7; **(c)** $(x^2 - x - 1)$ R -8
19. [6.5C] **(a)** $(x - 1)(x - 2)(x - 3)$; **(b)** $(x - 2)(x - 1)(x - 3)$;
(c) $(x - 3)(x - 1)(x - 2)$ **20.** [6.5D] **(a)** $(x^3 + 9x^2 + 26x + 24)$ R 4;
(b) $(x^3 + 8x^2 + 19x + 12)$ R 4; **(c)** $(x^3 + 7x^2 + 14x + 8)$ R 4
21. [6.5E] **(a)** Rem. $= 0$, so -1 is a solution; **(b)** Rem. $= 0$, so -2 is a
solution; **(c)** Rem. $= 0$, so -3 is a solution **22.** [6.6A] **(a)** No solution;
(b) -9; **(c)** -11 **23.** [6.6C] **(a)** 32 ft by 48 ft;
(b) Approximately 7 or 8 **24.** [6.7A] **(a)** 10 and 12; **(b)** 12 and 14;

(c) 14 and 16 **25.** [6.7B] **(a)** $2\frac{2}{9}$ hr; **(b)** $2\frac{2}{5}$ hr; **(c)** $2\frac{6}{11}$ hr
26. [6.7C] **(a)** 225 mi/hr; **(b)** 275 mi/hr; **(c)** 350 mi/hr
27. [6.7D] **(a)** $a = 2A - 2b - 3c$; **(b)** $b = \frac{2A - a - 3c}{2}$ or $b = A - \frac{a}{2} - \frac{3c}{2}$;
(c) $c = \frac{2A - a - 2b}{3}$ or $c = \frac{2A}{3} - \frac{a}{3} - \frac{2b}{3}$ **28.** [6.8A] **(a)** $P = kT$;
(b) $\frac{1}{120}$ **29.** [6.8B] **(a)** $P = \frac{k}{V}$; **(b)** 3200 **30.** [6.8C] $g = kxt^2$

Cumulative Review Chapters 1–6

1. -14 **2.** Associative property of multiplication **3.** $-3x^2 + 15$
4. -1 **5.**

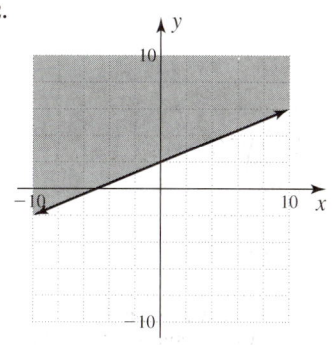

7. 23 **8.** 29, 31, 33 **9.** $\frac{-1}{3}$ **10.** $5x + y = 6$
11. $m = 2$; y-int. $= -12$
12.

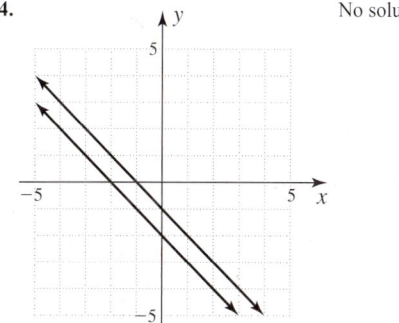

13.

14.   No solution

15. No solution **16.** (2, 3) **17.** 35 and 17 **18.** No solution
19. 21 ft **20.** $-\frac{3}{2}$ **21.** $D = \{\text{all real numbers}\}$;
$R = \{\text{all real numbers}\}$ **22.** $-5x^3 - x^2 - 9x$ **23.** $6n^2 - 13n + 5$
24. $(3t + 2)(4t - 5)$ **25.** $-3, -4, 4$ **26.** $\frac{-(x^2 - 3x + 9)}{x^2 + x + 1}$
27. $\frac{2x^2 + x - 11}{(x-3)(x-2)(x+2)}$ **28.** -1 **29.** $1\frac{7}{8}$ hr **30.** $\frac{1}{60}$

Chapter 7

Exercises 7.1

1. 2 3. 2 5. -2 7. $-\frac{1}{4}$ 9. 2 11. 2 13. 3 15. Not a real number
17. 3 19. 3 21. $-\frac{1}{2}$ 23. Not a real number 25. 9 27. 25 29. $\frac{1}{4}$
31. 16 33. 16 35. -16 37. $\frac{1}{16}$ 39. 7 41. $x^{3/7}$ 43. $\frac{1}{x^{5/9}}$ 45. $x^{2/5}$
47. z 49. x^2 51. $\frac{1}{z^2}$ 53. $\frac{1}{b^{4/5}}$ 55. $\frac{1}{a^8 b^5}$ 57. $\frac{b^9}{a^{10}}$ 59. $x^{16}y^{30}$
61. $x + x^{1/3}y^{1/2}$ 63. $x^{1/2}y^{3/4} - y^{5/4}$ 65. $\frac{1}{x}$ 67. $\frac{x^2}{y}$ 69. x^3y^{11}
71. $v = 15$ m/sec 73. $v = 28$ ft/sec 75. $x^{1/4}$ 77. (a) $x^{1/3}$; (b) $x^{1/6}$; (c) 3 83. index 85. radicand 87. $\frac{1}{x^5 y^{12}}$ 89. $\frac{y^{1/3}}{x^{1/2}}$ 91. $x^{1/5} + x^{2/5}y^{1/4}$
93. $x^{8/15}$ 95. $\frac{1}{x^{11/12}}$ 97. $\frac{1}{y^{3/20}}$ 99. $\frac{1}{27}$ 101. $\frac{1}{36}$ 103. $\frac{1}{16}$ 105. 7
107. $\frac{1}{3}$ 109. -5 111. 3 113. 13 115. $\frac{2}{3}$ 117. $\frac{16x^3y}{8x^3y^3}$ 119. $\frac{48x^3y^2}{16x^4y^4}$
121. $\frac{20x}{32x^5}$ 123. $(x+8)^2$ 125. $(y-9)^2$

Exercises 7.2

1. 5 3. -4 5. $|-x| = |x|$ 7. $|x+6|$ 9. $|3x-2|$ 11. $4|xy|\sqrt{xy}$
13. $2x\sqrt[3]{5xy}$ 15. $|xy|\sqrt[4]{xy^3}$ 17. $-3a^2b^3\sqrt[5]{b^2}$ 19. $\frac{\sqrt{13}}{7}$ 21. $\frac{\sqrt{17}}{2|x|}$
23. $\frac{\sqrt{3}}{4x}$ 25. $\frac{\sqrt{6}}{3}$ 27. $-\frac{\sqrt{14}}{7}$ 29. $\frac{\sqrt{10a}}{2a}$ 31. $\frac{\sqrt{10ab}}{8ab}$ 33. $-\frac{\sqrt{6ab}}{2a^2b^2}$ 35. xy
37. $-\frac{\sqrt[3]{21}}{3}$ 39. $\frac{\sqrt[3]{12x}}{4x}$ 41. $\frac{\sqrt[3]{x}}{2x}$ 43. $\frac{\sqrt[3]{24}}{2}$ 45. $\sqrt[3]{3}$ 47. $\sqrt{2a}$
49. $x\sqrt{5xy}$ 51. $x^2y\sqrt{7xy}$ 53. $\sqrt{2ab}$ 55. $\frac{\sqrt[3]{a^2b^2}}{b^2}$ 57. $\frac{2\sqrt{6ab}}{3b^2}$
59. $\frac{\sqrt{2abx}}{2x}$ 61. (a) $\sqrt[3]{\frac{6\pi^2 V}{2\pi}}$; (b) 3 ft 63. $m = \frac{m_0c\sqrt{c^2-v^2}}{c^2-v^2}$ 65. $\overline{v} = \left(\frac{3kTm}{m}\right)^{1/2}$
67. $P = \frac{k}{\sqrt[3]{V^3}}; P = \frac{k\sqrt[3]{V^3}}{V^2}$ 75. product rule 77. $\frac{\sqrt{33}}{6}$ 79. $\frac{\sqrt[3]{6x^2}}{2x^2}$
81. $2\sqrt[3]{4}$ 83. $\frac{\sqrt{2ax}}{2x}$ 85. 10 87. $-x$ 89. $\frac{\sqrt{3}}{2}$ 91. $\frac{\sqrt{35}}{5}$ 93. $\frac{\sqrt{30x}}{6x}$
95. $a^2 - b^2$ 97. $x^3 - y^3$ 99. $-y + 5$ or $5 - y$ 101. $1 + 2x$
103. 7 105. 8

Exercises 7.3

1. $15\sqrt{2}$ 3. $9\sqrt{5a}$ 5. $-11\sqrt{2}$ 7. $-5a\sqrt{2}$ 9. $-26\sqrt{3}$
11. $9x\sqrt{5x} - 2\sqrt{6x}$ 13. $17\sqrt[3]{5}$ 15. $-12\sqrt[3]{3}$ 17. $3\sqrt[3]{3}$ 19. $4\sqrt[3]{3a}$
21. $\frac{7\sqrt{3}}{6}$ 23. $\frac{3\sqrt{2}+2\sqrt{3}+\sqrt{6}}{6}$ 25. $\frac{\sqrt{6}+3\sqrt{2}}{6}$ 27. $3\sqrt[3]{75}$ 29. $15 - 3\sqrt{2}$
31. $2 + 3\sqrt[3]{2}$ 33. $14\sqrt{15} + 30$ 35. $6\sqrt[3]{15} - 15$
37. $-8\sqrt{21} + 20\sqrt{14}$ 39. $55 + 13\sqrt{15}$ 41. $42 + 21\sqrt{2}$
43. $-441 + \sqrt{35}$ 45. -1 47. -23 49. $5 + 2\sqrt{6}$
51. $a^2 + 2a\sqrt{b} + b$ 53. $5 - 2\sqrt{6}$ 55. $a^2 - 2a\sqrt{b} + b$
57. $a - 2\sqrt{ab} + b$ 59. $1 + \sqrt{2}$ 61. $\frac{2-\sqrt{3}}{4}$ 63. $\frac{3\sqrt{2}+\sqrt{6}}{2}$
65. $\frac{6+2\sqrt{2}}{7}$ 67. $3a + a\sqrt{5}$ 69. $\frac{9a - 3a\sqrt{2} + 6b - 2b\sqrt{2}}{7}$ 71. $\frac{a + 2b\sqrt{a} + b^2}{a - b^2}$
73. $\frac{a + 2\sqrt{2ab} + 2b}{a - 2b}$ 75. $\frac{1}{2\sqrt{5} - 2\sqrt{3}}$ 77. $\frac{5-x}{5\sqrt{5}+5\sqrt{x}}$ 79. $\frac{x-y}{x\sqrt{x}+x\sqrt{y}}$
81. $\frac{x-y}{\sqrt{xy}-y}$ 83. $\frac{x-y}{\sqrt{xy}+y}$ 89. FOIL 91. like radical expressions
93. $\frac{y+\sqrt{xy}}{y-x}$ 95. $5 + \sqrt{2}$ 97. $11 - 2\sqrt{21}$ 99. $x\sqrt{5} - \sqrt{15x}$
101. $\frac{3\sqrt{2}}{4}$ 103. $2\sqrt[3]{2x}$ 105. $2\sqrt{2}$ 107. 4 109. 5, 10
111. $-1, -5$ 113. $2x$ 115. $6 - x$

Exercises 7.4

1. 16 3. 43 5. 18 7. No real-number solution 9. 3 11. 0
13. 6 15. -16 17. 11 19. 9 21. 0 23. 1 25. 0 27. 1 29. 4
31. $x = a + b^2$ 33. $y = \frac{a-c^3}{b}$ 35. $x = ab^2$ 37. $x = \frac{b}{3}$
39. Approximately 229m 41. 8.66 ft 43. (a) $d = \frac{gt^2}{2}$; (b) 144.9 ft
45. (a) $L = \frac{gt^2}{4\pi^2}$; (b) $\frac{392}{121}$ ft $\approx$ 3.2 ft 47. 100 ft 49. $\frac{1600}{9} \approx 177.8$ ft
55. radical equation 57. 13 59. 4 61. 8 63. 6 65. (a) $R = \frac{P}{I^2}$;
(b) 15 ohms 67. $11 + 6x$ 69. $-1 + 8x$ 71. $\frac{1+5\sqrt{2}}{7}$ 73. $\frac{4+\sqrt{2}}{7}$
75. $\frac{x - 2\sqrt{xy} + y}{x - y}$

Exercises 7.5

1. $5i$ 3. $5i\sqrt{2}$ 5. $24i\sqrt{2}$ 7. $-12i\sqrt{2}$ 9. $3 + 8i\sqrt{7}$ 11. $6 + 4i$
13. $-2 - 6i$ 15. $-5 - 6i$ 17. $-2 + 5i$ 19. $-7 + 4i$ 21. $8 - i$
23. $10 + 5i$ 25. $-3 - 2i\sqrt{2}$ 27. $-1 + 2i\sqrt{2}$ 29. $-7 + 3i\sqrt{5}$
31. $12 + 6i$ 33. $-12 + 20i$ 35. $-4 + 6i$ 37. $-3 + 3i\sqrt{3}$
39. $-6 + 9i$ 41. $28 + 12i$ 43. $-20 + 20i$ 45. $3 + 11i$
47. $13 + 0i$ 49. $24 + 7i$ 51. $31 + 0i$ 53. $0 - 3i$ 55. $0 + 6i$
57. $\frac{2}{5} + \frac{1}{5}i$ 59. $-\frac{6}{5} + \frac{3}{5}i$ 61. $-1 + 2i$ 63. $\frac{17}{13} - \frac{6}{13}i$ 65. $0 - \frac{3}{2}i$
67. $\frac{12+\sqrt{10}}{18} + \frac{4\sqrt{5}-3\sqrt{2}}{18}i$ 69. $\frac{3+\sqrt{6}}{12} + \frac{3\sqrt{2}-\sqrt{3}}{12}i$ 71. 1 73. $-i$
75. i 77. 1 79. i 81. -1 83. $Z_1 + Z_2 = (8+i)$ ohms
85. $Z_T = \frac{167}{65} - \frac{29}{65}i$ 87. 5 89. $\sqrt{13}$ 95. irrational
97. natural numbers 99. $-i$ 101. -1 103. $\frac{27}{25} - \frac{11}{25}i$
105. $-10\sqrt{2} + 30i$ 107. $28 + 4i$ 109. $3 - 2i$ 111. $-1 + 7i$
113. $5i\sqrt{2}$ 115. $2, -2$ 117. $5, -5$ 119. $\frac{2\sqrt{5}}{5}$
121. $\frac{\sqrt{7}}{3}$ 123. $x^2 + 10x + 25$ 125. $(x-8)^2$

Review Exercises

1. [7.1A] (a) Not a real number; (b) -4 2. [7.1A] (a) Not a real number; (b) -5 3. [7.1B] (a) -3; (b) -4 4. [7.1B] (a) $\frac{1}{2}$; (b) $\frac{1}{4}$
5. [7.1B] (a) 25; (b) 16 6. [7.1B] (a) Not a real number; (b) Not a real number 7. [7.1B] (a) $\frac{1}{4}$; (b) $\frac{1}{16}$ 8. [7.1B] (a) $\frac{1}{9}$; (b) $\frac{1}{16}$
9. [7.1C] (a) $x^{8/15}$; (b) $x^{9/20}$ 10. [7.1C] (a) $\frac{1}{x^{9/20}}$; (b) $\frac{1}{x^{8/15}}$
11. [7.1C] (a) $\frac{1}{x^5y^6}$; (b) $\frac{1}{x^5y^{12}}$ 12. [7.1C] (a) $x^{1/5} + x^{3/5}y^{3/5}$; (b) $x^{3/5} + x^{4/5}y^{3/5}$
13. [7.2A] (a) 7; (b) 6 14. [7.2A] (a) $|-x| = |x|$; (b) $|-x| = |x|$
15. [7.2A] (a) $2\sqrt[3]{6}$; (b) $2\sqrt[3]{7}$ 16. [7.2A] (a) $2xy^2\sqrt[3]{2x}$; (b) $2x^2y^5\sqrt[3]{2x^2}$
17. [7.2A] (a) $\frac{\sqrt{15}}{27}$; (b) $\frac{\sqrt{5}}{32}$ 18. [7.2A] (a) $\frac{1}{x}$; (b) $\frac{\sqrt{5}}{x}$ 19. [7.2B] (a) $\frac{\sqrt{55}}{11}$;
(b) $\frac{\sqrt{65}}{13}$ 20. [7.2B] (a) $\frac{\sqrt{10x}}{5x}$; (b) $\frac{\sqrt{15x}}{5x}$ 21. [7.2B] (a) $\frac{\sqrt{25x^2}}{5x}$; (b) $\frac{\sqrt{49x^2}}{7x}$
22. [7.2B] (a) $\frac{\sqrt[4]{x^3}}{2x}$; (b) $\frac{\sqrt[4]{10x}}{2x}$ 23. [7.2C] (a) $\frac{4}{3}$; (b) $\frac{5}{3}$
24. [7.2C] (a) $\sqrt[3]{9c^2d^2}$; (b) $\sqrt[3]{25c^2d^2}$ 25. [7.2C] (a) $\frac{\sqrt[3]{3ac}}{3c^3}$; (b) $\frac{\sqrt[3]{9ac}}{3c^3}$
26. [7.3A] (a) $6\sqrt{2}$; (b) $7\sqrt{2}$ 27. [7.3A] (a) $\sqrt{7}$; (b) $\sqrt{7}$
28. [7.3A] (a) $\frac{3\sqrt{2}}{4}$; (b) $\frac{9\sqrt{2}}{4}$ 29. [7.3A] (a) $\frac{13\sqrt{6x^2}}{4x}$; (b) $\frac{11\sqrt{6x^2}}{4x}$
30. [7.3B] (a) $6 + \sqrt{6}$; (b) $8 + \sqrt{6}$ 31. [7.3B] (a) $2x\sqrt[3]{6} - 3\sqrt[3]{6x^2}$;
(b) $2x\sqrt[3]{6} - 3\sqrt[3]{6x^2}$ 32. [7.3B] (a) $30 + 12\sqrt{6}$; (b) $24 + 10\sqrt{6}$
33. [7.3B] (a) $19 + 8\sqrt{3}$; (b) $28 + 10\sqrt{3}$ 34. [7.3B] (a) $52 - 14\sqrt{3}$;
(b) $19 - 8\sqrt{3}$ 35. [7.3B] (a) 5; (b) 4 36. [7.3B] (a) $4 - \sqrt{2}$;
(b) $6 - \sqrt{2}$ 37. [7.3C] (a) $5\sqrt{2} + 5$; (b) $\frac{x - 4\sqrt{x}}{x - 16}$
38. [7.4A] (a) No real-number solution; (b) No real-number solution
39. [7.4A] (a) 4; (b) 10 40. [7.4A] (a) 7, 8; (b) 4, 5
41. [7.4A] (a) 4; (b) 9 42. [7.4A] (a) 32; (b) 67
43. [7.4A] (a) No real-number solution; (b) No real-number solution
44. [7.4B] (a) $I = \frac{k}{d^2}$; (b) $k = d^2I$ 45. [7.5A] (a) $10i$; (b) $11i$
46. [7.5A] (a) $6i\sqrt{2}$; (b) $5i\sqrt{2}$ 47. [7.5B] (a) $10 + 3i$; (b) $6 + 3i$
48. [7.5B] (a) $-4 + 7i$; (b) $2 + 11i$ 49. [7.5C] (a) $21 + i$; (b) $23 - 2i$
50. [7.5C] (a) $24\sqrt{2} + 16i$; (b) $36\sqrt{2} + 24i$ 51. [7.5C] (a) $\frac{17}{25} + \frac{6}{25}i$;
(b) $\frac{27}{25} - \frac{11}{25}i$ 52. [7.5D] (a) -1; (b) $-i$ 53. [7.5D] (a) -1; (b) i

Cumulative Review Chapters 1–7

1. Irrational numbers, real numbers 2. $\frac{y^{16}}{16x^8}$ 3. -128 4. 6
5. ←——(——)—|——→
   $\quad\quad -3\quad -1\ \ 0$

6. 10 ft by 40 ft 7. x-intercept; $-\frac{3}{4}$; y-intercept; 6 8. -4

9. $3x - 14y = -74$

10.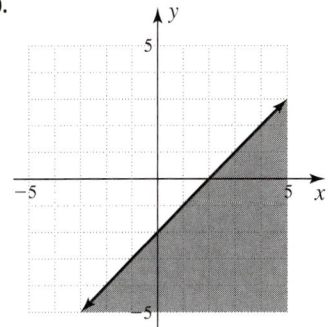

11.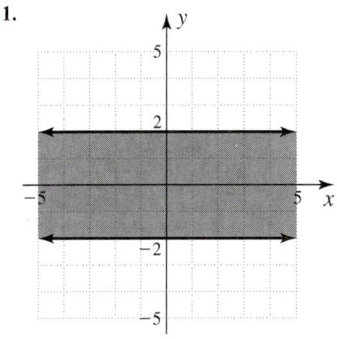

12. $(-1, -2)$ **13.** Infinitely many solutions such that $y = -\frac{3}{4}x - 18$
14. No solution **15.** $(5, 1, 4)$ **16.** \$20,000 at 7%; \$30,000 at 9%;
\$40,000 at 11% **17.** $R = \{-5, -3, 0\}$ **18.** 5 **19.** $x^3 - 14x - 8$
20. $3x^3(3x^3 - 4x^2 + 5x + 4)$ **21.** $3xy(3x + y)(x - 2y)$ **22.** $\frac{1}{2}; \frac{2}{3}$
23. $\frac{-(x+3)}{9+3x+x^2}$ **24.** $(x+4)(x+1)(3x+1)$ **25.** No solution
26. 10 and 12 **27.** 6880 **28.** 27 **29.** $x^{1/5} + x^{2/5}y^{4/5}$ **30.** $9\sqrt{7}$
31. $56 - 14\sqrt{7}$ **32.** $\frac{x+\sqrt{5}x}{x-5}$ **33.** -9 **34.** $\frac{61}{82} - \frac{57}{82}i$ **35.** i

Chapter 8
Exercises 8.1
1. ± 8 **3.** $\pm 11i$ **5.** ± 13 **7.** $\pm 2i$ **9.** $\pm \frac{7}{6}$ **11.** $\pm \frac{9}{2}i$ **13.** $\pm \frac{5\sqrt{3}}{3}$
15. $\pm \frac{6\sqrt{5}}{5}i$ **17.** $\pm \frac{10\sqrt{3}}{3}$ **19.** $\pm \frac{9\sqrt{13}}{13}i$ **21.** $-3, -7$ **23.** $-2 \pm 5i$
25. $6 \pm 3\sqrt{2}$ **27.** $1 \pm 2i\sqrt{7}$ **29.** $1 \pm 5\sqrt{2}$ **31.** $5 \pm 4\sqrt{2}$ **33.** $9 \pm 8i$
35. $-1 \pm 4\sqrt{2}$ **37.** $2 \pm 5\sqrt{2}$ **39.** $5 \pm 3i\sqrt{3}$ **41.** $-1, -5$
43. $-5, -3$ **45.** $-3 \pm i$ **47.** $4, 6$ **49.** $3, 7$ **51.** $4 \pm i$
53. $-1 \pm \frac{\sqrt{2}}{2}i$ **55.** $-1 \pm 5i$ **57.** $\frac{3}{5}, \frac{2}{5}$ **59.** $\frac{1}{2} \pm i$ **61.** $\frac{1 \pm 2\sqrt{2}}{2}$
63. $\frac{2 \pm \sqrt{2}}{2}$ **65.** $1, -2$ **67.** $\frac{-5 \pm 3\sqrt{3}}{2}$ **69.** $\frac{-3 \pm \sqrt{29}}{2}$
71. $5\sqrt{21}$; approximately 22.9 ft **73.** 2 sec **75.** 10% **77. (a)** 2000;
(b) \$2 **79.** 3 days **81.** Factoring; no **83.** square root
85. completing the square **87.** $\frac{1 \pm \sqrt{3}}{2}$ **89.** $-3 \pm \sqrt{10}$
91. $\frac{-3 \pm 2\sqrt{15}}{3}$ **93.** $1 \pm \frac{7\sqrt{2}}{2}i$ **95.** $-1 \pm 3\sqrt{2}$ **97.** $\pm 3i\sqrt{2}$
99. $\pm \frac{2}{3}$ **101.** $0, 6$ **103.** $2 \pm 2\sqrt{2}$ **105.** $\frac{5}{3}, -1$ **107.** $\frac{3}{8} \pm \frac{\sqrt{71}}{8}i$
109. $-1 \pm 2i$

Exercises 8.2
1. $5x^2 - 4x + 12 = 0; a = 5, b = -4, c = 12$
3. $2x^2 - 5x + 12 = 0; a = 2, b = -5, c = 12$ **5.** $1, -2$
7. $-2 \pm \sqrt{3}$ **9.** $\frac{3 \pm \sqrt{17}}{2}$ **11.** $\frac{5}{7}, 1$ **13.** $-\frac{4}{5} \pm \frac{3}{5}i$ **15.** $-\frac{3}{2}, -2$
17. $\frac{3}{2}, 1$ **19.** $-\frac{1}{2}, -3$ **21.** -1 **23.** $\frac{3 \pm \sqrt{5}}{4}$ **25.** $\frac{-3 \pm \sqrt{21}}{6}$ **27.** ± 2
29. $\pm 3\sqrt{5}$ **31.** $2, -1 \pm i\sqrt{3}$ **33.** $\frac{1}{2}, -\frac{1}{4} \pm \frac{\sqrt{3}}{4}i$ **35.** $-3, \frac{3}{2} \pm \frac{3\sqrt{3}}{2}i$
37. Yes, in approximately 2.2 yr **39.** \$1.85 **41.** $10 \pm 2\sqrt{15}$ **43.** 1

45. Multiply both sides by $4a$. **47.** Add b^2 to both sides.
49. Take square root of both sides. **51.** Divide both sides by $2a$.
55. b **57.** quadratic formula **59.** $\frac{1}{4}, -\frac{1}{8} \pm \frac{\sqrt{3}}{8}i$ **61.** $-\frac{3}{4} \pm \frac{\sqrt{7}}{4}i$
63. $-\frac{3}{2}, 2$ **65.** $0, 7$ **67.** $1 \pm \sqrt{3}$ **69.** $\frac{3 \pm \sqrt{41}}{4}$ **71.** 4 **73.** $i\sqrt{23}$
75. $6x^2 - 5x - 4$ **77.** $12x^2 - 19x - 21$.

Exercises 8.3
1. $D = 49$: two rational numbers **3.** $D = 0$: one rational number
5. $D = 44$: two irrational numbers **7.** $D = -23$: two non-real
complex numbers **9.** $D = \frac{57}{4}$: two irrational numbers **11.** ± 4
13. -5 **15.** ± 8 **17.** ± 20 **19.** -16 **21.** Not factorable
23. $(4x - 3)(3x - 2)$ **25.** Not factorable **27.** $(5x - 6)(3x + 14)$
29. $(4x - 15)(3x - 4)$ **31.** $x^2 - 7x + 12 = 0$
33. $x^2 + 12x + 35 = 0$ **35.** $3x^2 - 7x - 6 = 0$ **37.** $4x^2 - 1 = 0$
39. $5x^2 + x = 0$ **41. (a)** $\frac{6}{4} = \frac{3}{2}$; **(b)** $\frac{5}{4}$; **(c)** No **43. (a)** $-\frac{13}{5}$; **(b)** $-\frac{6}{5}$;
(c) Yes **45. (a)** $-\frac{5}{2}$; **(b)** 1; **(c)** No **47.** $\frac{8}{3}$ **49.** $k = 15$
51. Yes: occurs twice **53.** Does not occur ($b^2 - 4ac$ is negative)
55. No **57.** No **59.** non-real, complex numbers
61. discriminant **63.** Yes **65.** $(3x - 5)(4x - 7)$
67. $3x^2 + x - 2 = 0$ **69.** $-1, -5$ **71.** $\frac{1 \pm \sqrt{17}}{2}$
73. $(x + 2)(x - 2)$ **75.** $4(x + 1)(x - 3)$

Exercises 8.4
1. $0, -\frac{5}{2}$ **3.** $-3, 7$ **5.** 0 **7.** $2, 1$ **9.** $-\frac{16}{5}, -1$ **11.** $\pm 3, \pm 2$
13. $\pm \frac{1}{2}, \pm 3i$ **15.** $\pm \sqrt{2}, \pm \frac{\sqrt{3}}{3}i$ **17.** $1, -2, 1 \pm i\sqrt{3}, -\frac{1}{2} \pm \frac{\sqrt{3}}{2}i$
19. $7, -6$ **21.** $\frac{1 \pm \sqrt{37}}{2}, \frac{1}{2} \pm \frac{\sqrt{3}}{2}i$ **23.** 16 **25.** 8, 27 **27.** 4
29. $2 \pm \sqrt{29}, 2 \pm \sqrt{13}$ **31.** 6 **33.** $1, \frac{16}{81}$ **35.** $\frac{1}{2}, -\frac{1}{4}$
37. $\pm i\sqrt{3}, \pm \frac{\sqrt{2}}{2}$ **39.** $\pm \sqrt{2}, \pm \frac{\sqrt{3}}{2}i$ **41.** 10 hr, 15 hr
43. x-intercepts: $(\sqrt{5}, 0); (-\sqrt{5}, 0)$ **45.** 24 students **49.** 6
51. quadratic **53.** 1, 16 **55.** 16, 81 **57.** $\pm \frac{\sqrt{2}}{2}, \pm \frac{\sqrt{7}}{7}$
59. $2, -1, \frac{1}{2} \pm \frac{\sqrt{3}}{2}i$ **61.** -7 **63.** $\pm 1, \pm 2$ **65.** $x \geq 3$
67. $x \leq -3$ **69.** True **71.** False

Exercises 8.5

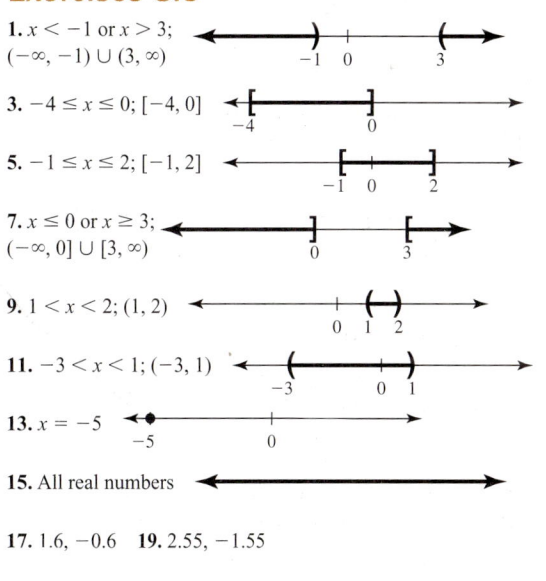

1. $x < -1$ or $x > 3$; $(-\infty, -1) \cup (3, \infty)$

3. $-4 \leq x \leq 0$; $[-4, 0]$

5. $-1 \leq x \leq 2$; $[-1, 2]$

7. $x \leq 0$ or $x \geq 3$; $(-\infty, 0] \cup [3, \infty)$

9. $1 < x < 2$; $(1, 2)$

11. $-3 < x < 1$; $(-3, 1)$

13. $x = -5$

15. All real numbers

17. $1.6, -0.6$ **19.** $2.55, -1.55$

21. $-3 \leq x \leq -1$ or $x \geq 2$; $[-3, -1] \cup [2, \infty)$

23. $x \leq 1$ or $2 \leq x \leq 3$; $(-\infty, 1] \cup [2, 3]$

25. $x > 2$; $(2, \infty)$

27. $x < -5$ or $x > 1$; $(-\infty, -5) \cup (1, \infty)$

29. $\frac{1}{2} < x < \frac{4}{3}$; $\left(\frac{1}{2}, \frac{4}{3}\right)$

31. $1 < x < 7$; $(1, 7)$

33. $x < 1$ or $x > 2$; $(-\infty, 1) \cup (2, \infty)$

35. $x \le -3$ or $x \ge 3$; $(-\infty, -3] \cup [3, \infty)$
37. $x \le 1$ or $x \ge 5$; $(-\infty, 1] \cup [5, \infty)$ **39.** $R > 4$ **41.** $10° < T < 100°$
43. $1 < t < 2$ **45.** $23.2 \le v \le 26.1$ **47.** 128.2 mi/hr
53. quadratic inequality

55. $x < 2$; $(-\infty, 2)$;

57. $-3 \le x \le -1$ or $x \ge 2$; $[-3, -1] \cup [2, \infty)$;

59. $x \le -2$ or $x \ge 1$; $(-\infty, -2] \cup [1, \infty)$;

61. $-3 < x < 2$; $(-3, 2)$;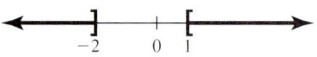

63. During the first 3 hr **65.** 0 **67.** 6 **69.** -4 **71.** x-int: $(-8, 0)$; y-int: $(0, -2)$ **73.** $x^2 + 10x + 25$; $(x + 5)^2$

Review Exercises

1. [8.1A] (a) $\pm\frac{7}{4}$; (b) $\pm\frac{4}{5}$ **2.** [8.1A] (a) $\pm i\sqrt{6}$; (b) $\pm i\sqrt{7}$
3. [8.1B] (a) $3 \pm 4\sqrt{2}$; (b) $5 \pm 5\sqrt{2}$ **4.** [8.1B] (a) $2 \pm \frac{5\sqrt{2}}{2}i$;
(b) $3 \pm \frac{8\sqrt{3}}{3}i$ **5.** [8.1B] (a) $\frac{5 \pm 2\sqrt{15}}{5}$; (b) $\frac{-3 \pm 4\sqrt{3}}{6}$ **6.** [8.1C] (a) 9, -1;
(b) -4, -8 **7.** [8.1C] (a) $\frac{1}{2}$, $-\frac{3}{2}$; (b) $\frac{3 \pm \sqrt{2}}{4}$
8. [8.2A] (a) $7x^2 - x - 4 = 0$; $a = 7, b = -1, c = -4$;
(b) $2x^2 + 90x - 1 = 0$; $a = 2, b = 90, c = -1$
9. [8.2A] (a) $\frac{1 \pm \sqrt{13}}{3}$; (b) $\frac{3 \pm \sqrt{21}}{4}$
10. [8.2A] (a) 0, 16; (b) 0, 12 **11.** [8.2A] (a) $\frac{-1 \pm \sqrt{301}}{30}$; (b) $\frac{1}{5}$, -2
12. [8.2A] (a) $\frac{1}{3} \pm \frac{\sqrt{2}}{3}i$; (b) $\frac{1}{5} \pm \frac{\sqrt{19}}{5}i$
13. [8.2B] (a) $\frac{5}{2}$, $-\frac{5}{4} \pm \frac{5\sqrt{3}}{4}i$; (b) $\frac{2}{5}$, $-\frac{1}{5} \pm \frac{\sqrt{3}}{5}i$
14. [8.2C] (a) $p = 3$; (b) $p = 1$ **15.** [8.3A] (a) $D = -55$; two non-real complex solutions; (b) $D = k^2 - 64$; $k = \pm 8$
16. [8.3B] (a) Not factorable; (b) $(2x + 1)(9x + 2)$
17. [8.3B] (a) $(6x - 5)(3x + 1)$; (b) Not factorable
18. [8.3C] (a) $x^2 - x - 6 = 0$; (b) $12x^2 + 5x - 2 = 0$
19. [8.3D] (a) Sum $= -\frac{4}{15}$; product $= -\frac{1}{5}$; (b) Sum $= \frac{4}{3}$; product $= -\frac{5}{9}$
20. [8.3D] (a) Yes; (b) No **21.** [8.4A] (a) -5; (b) 0, 2
22. [8.4B] (a) $1, -2, -\frac{1}{2} \pm \frac{\sqrt{15}}{2}i$; (b) $1, 2, \frac{3 \pm \sqrt{33}}{2}$
23. [8.4B] (a) 1, 81; (b) 16 **24.** [8.4B] (a) 27, -64; (b) 216, -1
25. [8.4B] (a) 9; (b) 25 **26.** [8.4B] (a) ± 1, $\pm\sqrt{3}$; (b) ± 1, $\pm i\sqrt{3}$

27. [8.5A] (a) $(-3, 2)$

(b) $(-2, 3)$

28. [8.5A] (a) $(-\infty, -4] \cup [0, \infty)$
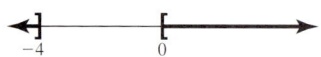

(b) $(-\infty, 0] \cup [3, \infty)$

29. [8.5A] (a) $(-9, 5)$

(b) $(1 - \sqrt{3}, 1 + \sqrt{3})$

30. [8.5B] (a) $(-\infty, 1] \cup [2, 3]$

(b) $(-\infty, -2] \cup [-1, 3]$

31. [8.5B] (a) $(-2, -1) \cup (3, \infty)$

(b) $(-3, -2) \cup (1, \infty)$

32. [8.5C] (a) $[-2, 2)$

(b) $(-\infty, 2) \cup (6, \infty)$

Cumulative Review Chapters 1–8

1. $\{2, 4, 6, 8, 10, 12\}$ **2.** $0.7\overline{3}$ **3.** 11 **4.** -32 **5.** $\frac{10}{9}$

6.

7.

8. $10,000 in bonds; $8000 in CDs **9.** $2x - y = 2$

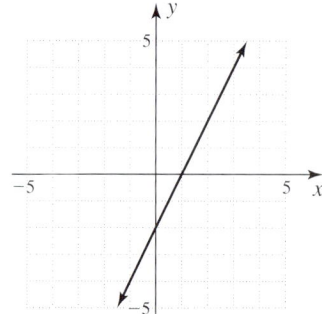

10. Perpendicular **11.** $x + y = -1$ **12.** $8x - y = -52$
13. $D = \{$all real numbers$\}$; $R = \{$all real numbers$\}$
14. $x + y \le -1$; $y \le 3x - 5$

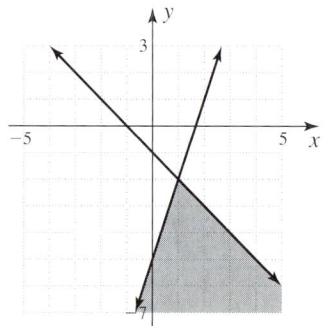

15. $(-3, 2)$ **16.** $\left(\frac{1}{3}, \frac{17}{6}\right)$ **17.** 28 nickels; 18 dimes
18. $-2x^3 + 9x^2 - 4x + 18$ **19.** $(6x + 5y)^2$ **20.** $2x^2(x^2 + 4)$
21. $-1, -3, 3$ **22.** $-2x^2 + \frac{2}{x^2} + \frac{1}{x^3}$ **23.** $\frac{1}{x+9}$ **24.** $\frac{x(x^2+x+1)}{(x+1)(x^2+1)}$
25. 2 **26.** 8 and 10 **27.** Not a real number **28.** $4a^2b^3\sqrt[3]{3a}$
29. $\frac{\sqrt{30k}}{6k}$ **30.** $3x + 2\sqrt[3]{6x^2}$ **31.** $-8, -7$ **32.** $-3 - 5i$
33. $-20 + 35i$ **34.** $\frac{3}{4}, -\frac{3}{4}$ **35.** $-3, 1$
36. $-2 \pm \sqrt{6}$ **37.** when price is $4 **38.** Yes **39.** $\pm 1, \pm\sqrt{7}$
40. $-6 < x < 4$

Chapter 9

Exercises 9.1

1. (a) $(0, 0)$; **(b)** $(0, 2)$; **(c)** $(0, -2)$

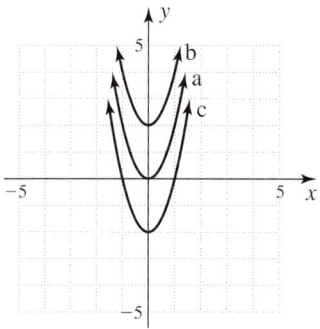

3. (a) $(0, 0)$; **(b)** $(0, 1)$; **(c)** $(0, -1)$

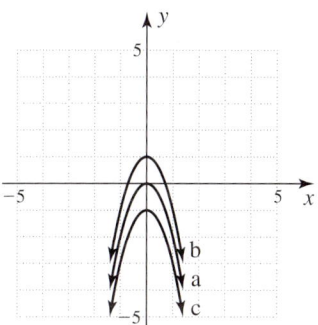

5. (a) $(0, 0)$; **(b)** $(0, 0)$

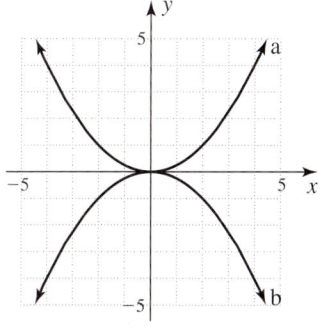

7. (a) $(0, 1)$; **(b)** $(0, 1)$

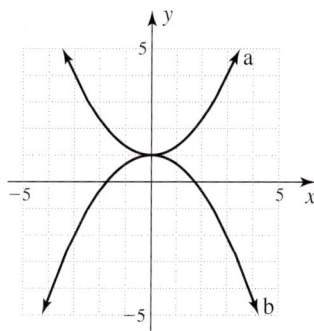

9. (a) $(-2, 3)$; **(b)** $(-2, 0)$; **(c)** $(-2, -2)$

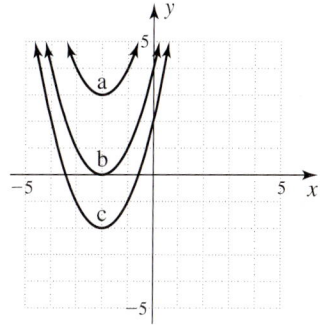

11. (a) $(-2, -2)$; **(b)** $(-2, 0)$; **(c)** $(-2, -4)$

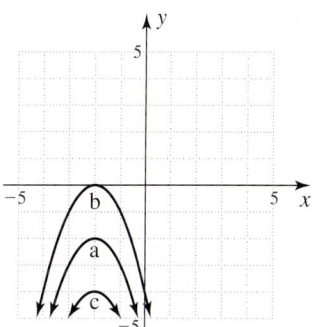

13. (a) $(-2, -2)$; **(b)** $(-2, 0)$; **(c)** $(-2, -4)$

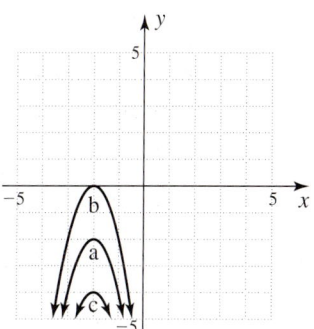

15. (a) $\left(-1, \frac{1}{2}\right)$; **(b)** $(-1, 0)$

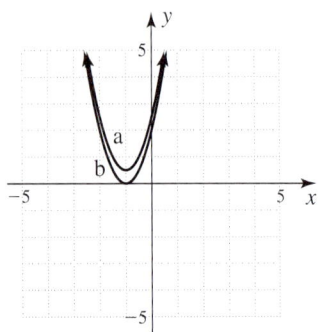

17. Int: $(-1, 0)$, $(0, 1)$

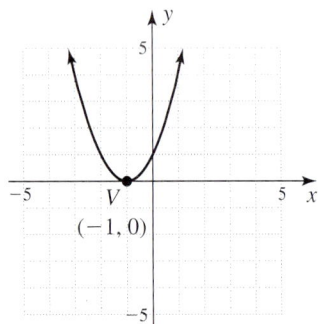

19. Int: $(2.4, 0)$, $(-0.4, 0)$, $(0, 1)$

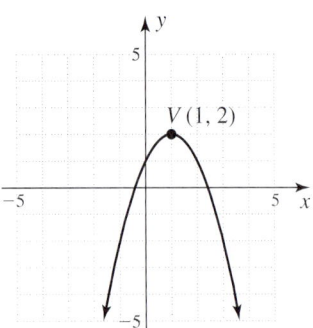

21. Int: $(0, -5)$

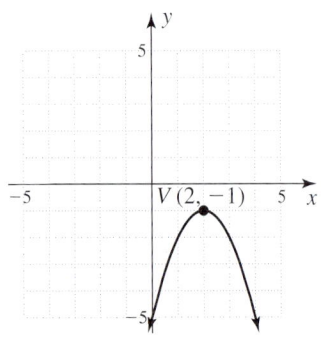

23. Int: $\left(\frac{3}{2}, 0\right)$, $(1, 0)$, $(0, 3)$

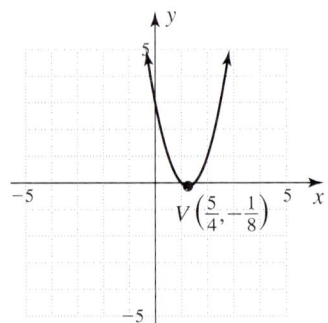

25. Int: $(-2.9, 0)$, $(0.9, 0)$, $(0, 5)$

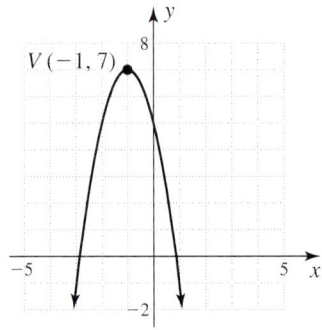

27. Int: $(1.5, 0)$, $(-0.5, 0)$, $(0, 2)$

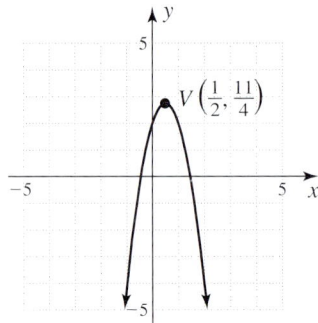

29. (a) $(3, -2)$; **(b)** $(0, -2)$

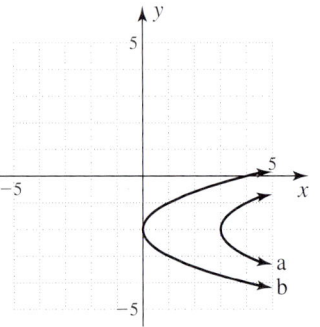

31. (a) $(-2, -2)$; (b) $(0, -2)$

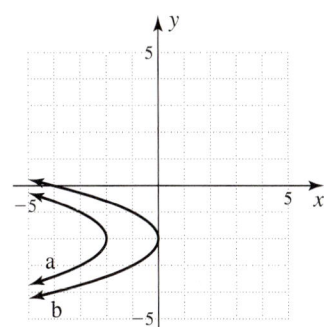

33. (a) $(2, 1)$; (b) $(5, 1)$

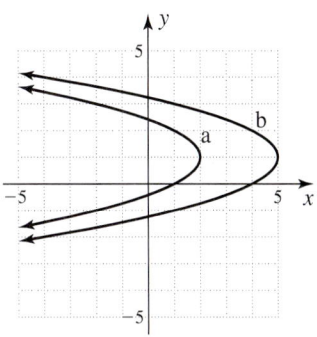

35. $x = 4000$; $P = \$11,000$ **37.** $25 thousand (\$25,000) **39.** 400 ft
41. $P = (600 + 100W)(1 - 0.10W)$; P = price, W = weeks elapsed. Max P occurs when $W = 2$ (at end of 2 weeks).
43. (a) $(42, 18)$; (b) 18 in.; (c) 84 in.;
(d)

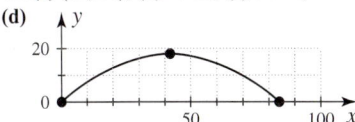

45. (a) $(200, 100)$; (b) 100 ft; (c) 400 ft;
(d)

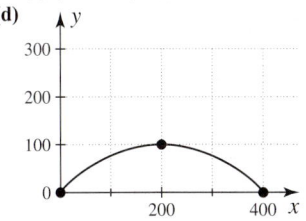

47. (a) 35; (b) $(2, 31)$; (c) minimum; (d) 31 million; 2000
49. (a) $(37.5, 26.25)$; (b) at 37.5 mph; (c) 24 mpg
51. $FP = \sqrt{x^2 + (y - p)^2}$ **53.** $x^2 = 4py$ **55.** $x^2 = 12.5y$; Focus: $(0, 3.125)$ **57.** $f(x) = ax^2 + bx + c$ opens up if $a > 0$, down if $a < 0$. **59.** Moves the graph up or down. **63.** upward
65. $y = ax^2$ **67.** $x = ay^2$
69.

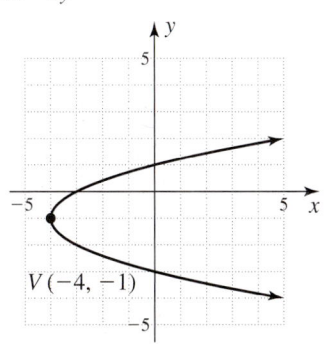

71.

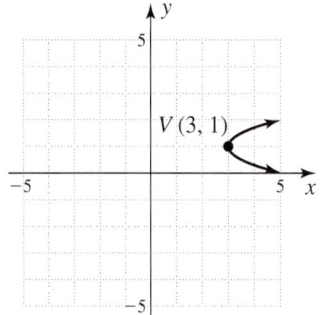

73.

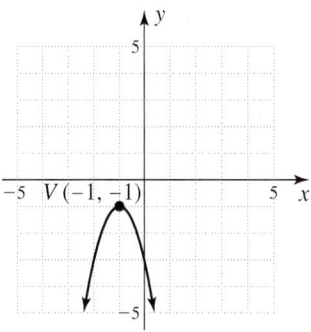

75.

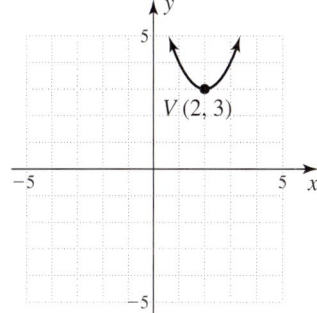

77.

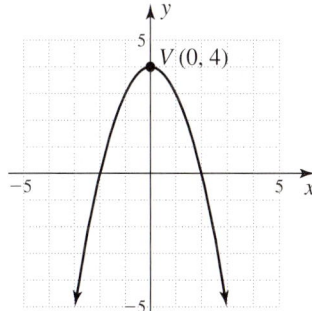

79.

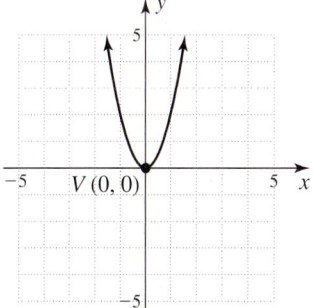

81.

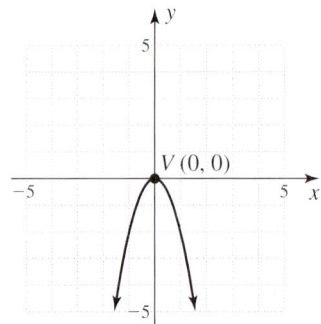

83. $10   **85.** $4\sqrt{2}$ **87.** $y^2 + 8y + 16$

Exercises 9.2
1. 5   **3.** $\sqrt{73}$ **5.** $\sqrt{90} = 3\sqrt{10}$ **7.** 5   **9.** 6
11. $(x - 3)^2 + (y - 8)^2 = 4$ **13.** $(x + 3)^2 + (y - 4)^2 = 25$
15. $(x + 3)^2 + (y + 2)^2 = 16$ **17.** $(x - 2)^2 + (y + 4)^2 = 5$
19. $x^2 + y^2 = 9$
21. $C(1, 2); r = 3$

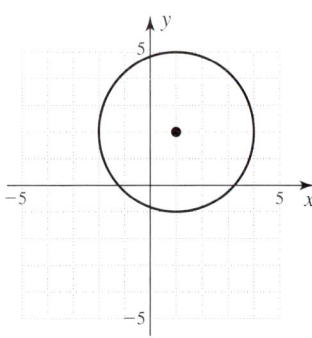

23. $C(-1, 2); r = 2$

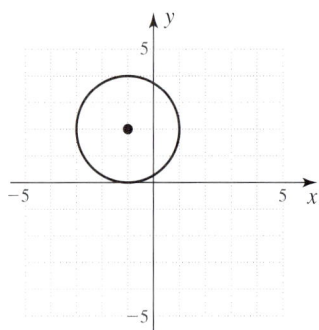

25. $C(1, -2); r = 1$

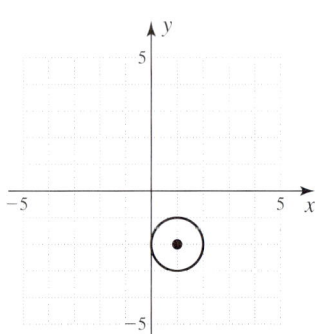

27. $C(-2, -1); r = 3$

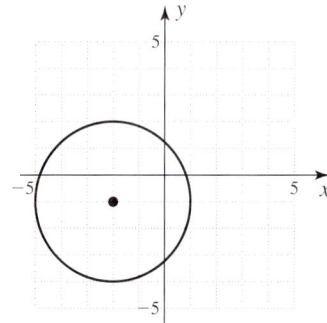

29. $C(1, 1); r = \sqrt{7} \approx 2.6$

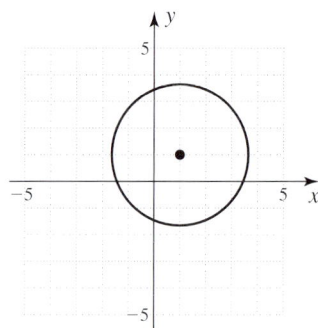

31. $C(3, 2); r = 2$

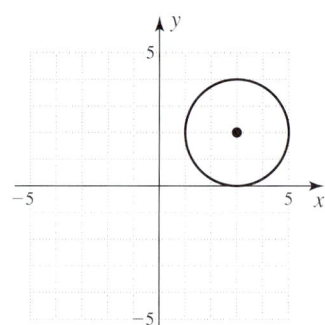

33. $C(2, -1); r = 3$

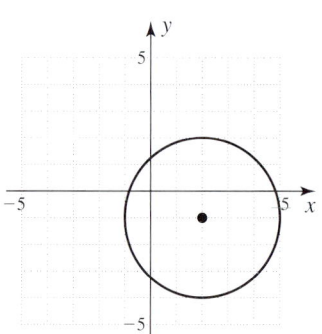

35. $C(0, 0); r = 5$

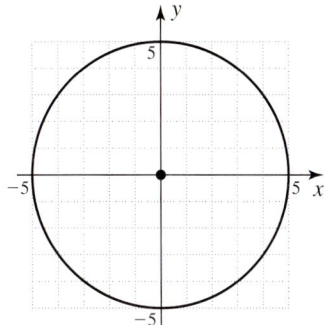

37. $C(0, 0); r = \sqrt{7} \approx 2.6$

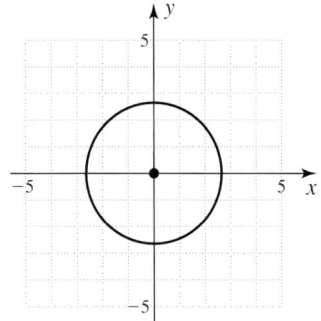

39. $C(-3, 1); r = 2$

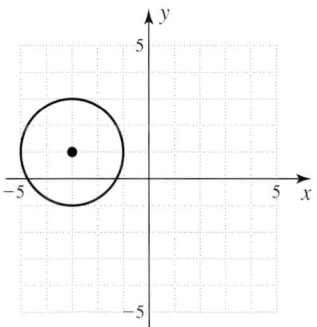

41. $C(3, 1); r = 2$

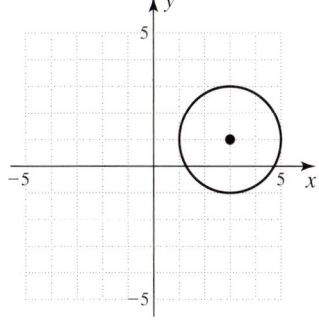

43. $C(0, 0); a = 5, b = 2$

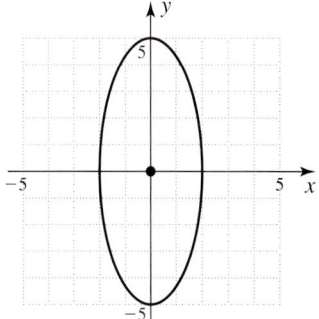

45. $C(0, 0); a = 2, b = 1$

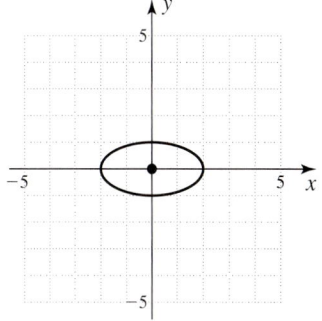

47. $C(0, 0); a = 4, b = 2$

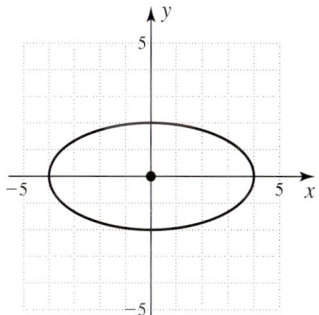

49. $C(0, 0); a = 4, b = 3$

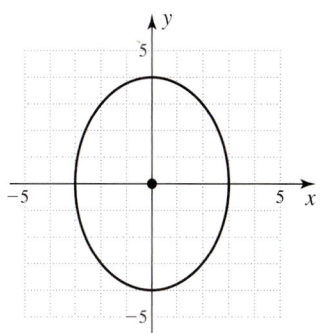

51. $C(1, 2); a = 3, b = 2$

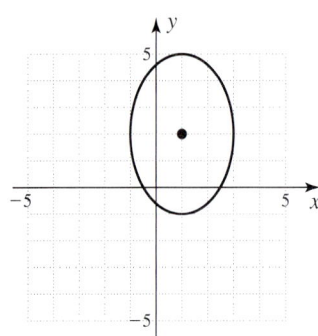

53. $C(2, -3); a = 3, b = 2$

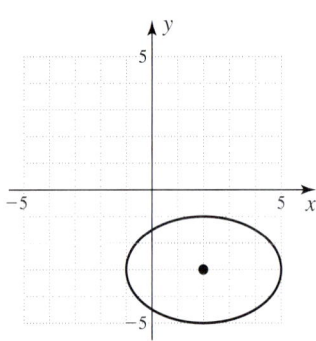

55. $C(1, 1); a = 4, b = 3$

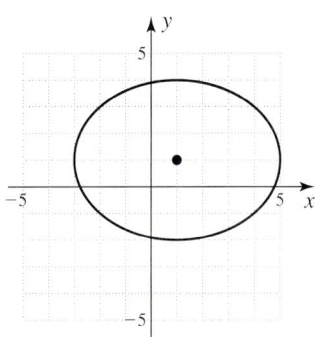

57. $x^2 + y^2 = 25$ **59.** $x^2 + y^2 = 169$ **61.** $x^2 + y^2 = 9$
63. $\frac{x^2}{4} + \frac{y^2}{36} = 1$ **65.** $\frac{x^2}{36} + \frac{y^2}{16} = 1$ **67.** 12.5 in. **69.** $10\sqrt{5} \approx 22.4$ ft
71. $\frac{x^2}{4^2} + \frac{y^2}{2.5^2} = 1$ or $\frac{x^2}{16} + \frac{y^2}{6.25} = 1$ **73.** $\frac{x^2}{16} + \frac{y^2}{9} = 1$
75. (a) 93 million mi; (b) 1.5 million mi; (c) 92.99 million mi
77. (a) $\frac{x^2}{6^2} + \frac{y^2}{4.5^2} = 1$ or $\frac{x^2}{36} + \frac{y^2}{20.25} = 1$; (b) $3\sqrt{5} \approx 6.71$ in.
79. $21 - \frac{25\sqrt{7}}{4} \approx 4.5$ ft between side of boat and riverbank
81. (a) 25; (b) 4; (c) 21; (d) $\sqrt{21}$; (e) $\frac{\sqrt{21}}{5}$ **83.** $PF_1 = \sqrt{(x-c)^2 + y^2}$
85. $a^2 + cx = a\sqrt{(x+c)^2 + y^2}$ or $a^2 - cx = a\sqrt{(x-c)^2 + y^2}$
87. $b^2x^2 + a^2y^2 = a^2b^2$
93. $d = \sqrt{(x_2 - x_1)^2 + (y_2 - y_1)^2}$ **95.** $x^2 + y^2 = r^2$

97.

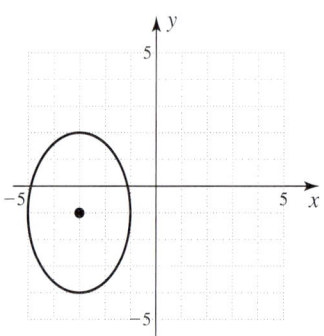

99.

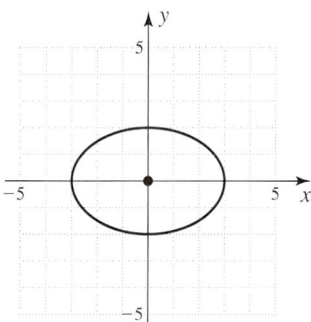

101.

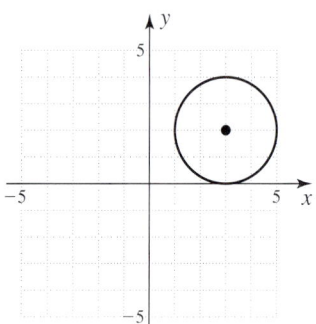

103.

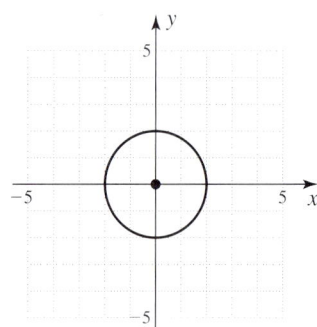

105. $C(3, 1); r = 2$

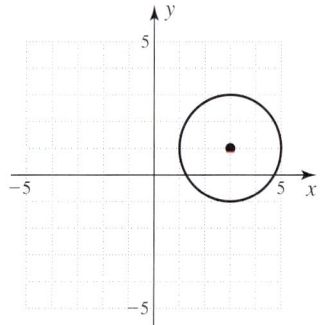

107. $x^2 + y^2 = 25$ **109.** $(x+3)^2 + (y-6)^2 = 9$
111. $x = \pm\sqrt{25 - y^2}$
113. Left half of the circle

Exercises 9.3
1. $V: (\pm 5, 0)$

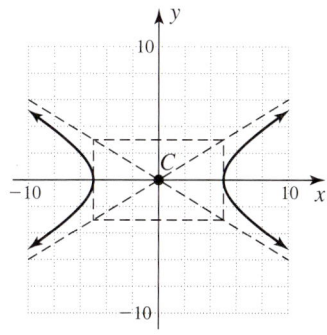

3. $V: (0, \pm 3)$

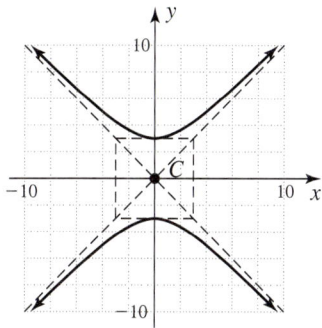

5. $V: (\pm 3, 0)$

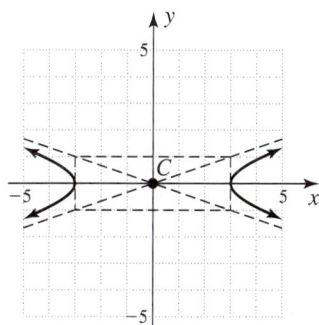

7. $V: (\pm 8, 0)$

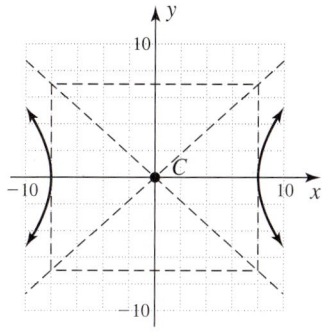

9. $V: \left(0, \pm\frac{4}{3}\right)$

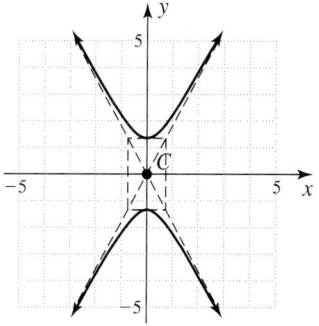

11. $V: (0, \pm 3)$

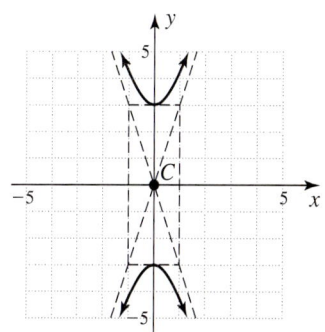

13. $C(1, -1); a = 2, b = 3$

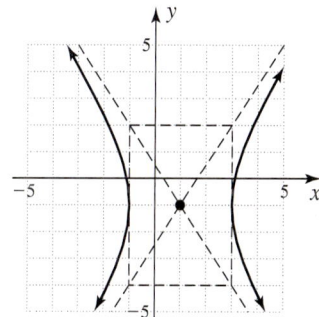

15. $C(2, 1); a = 3, b = 2$

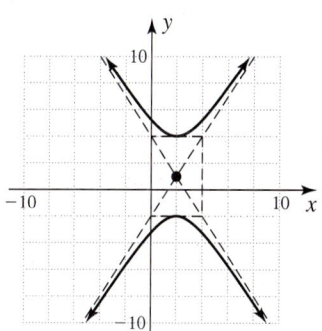

17. Circle; $(\pm 5, 0), (0, \pm 5)$
19. Hyperbola; $(\pm 6, 0)$
21. Parabola; $(0, -9)$ **23.** Parabola; $(-4, 0)$
25. Circle; $(\pm 2, 0), (0, \pm 2)$
27. Hyperbola; $(\pm 2, 0)$
29. Ellipse; $(\pm 3, 0), (0, \pm 1)$

31. (a) Hyperbola;
(b)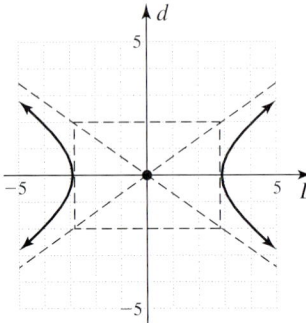

33.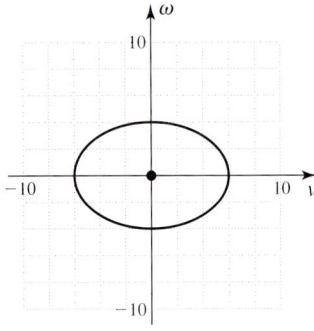

35. $\frac{\sqrt{13}}{2}$ **37.** $\frac{\sqrt{29}}{2}$ **39.** Problem 36; Problem 36
41. $PF_1 = \sqrt{(x-c)^2 + y^2}$ **43.** $a^2 + cx = -a\sqrt{(x+c)^2 + y^2}$
45. $b^2x^2 - a^2y^2 = a^2b^2$ **47.** $\left(\frac{x}{a} + \frac{y}{b}\right)\left(\frac{x}{a} - \frac{y}{b}\right) = 1$
49. The denominator becomes very large. The fraction approaches zero.
51. $y = -\frac{b}{a}x$ **57.** $\frac{(x-h)^2}{a^2} - \frac{(y-k)^2}{b^2} = 1$ **59.** Circle **61.** Parabola
63.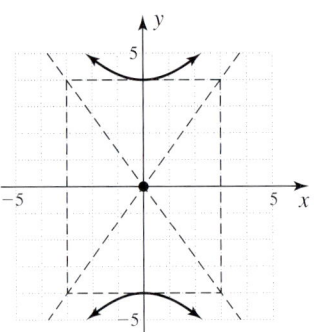

65.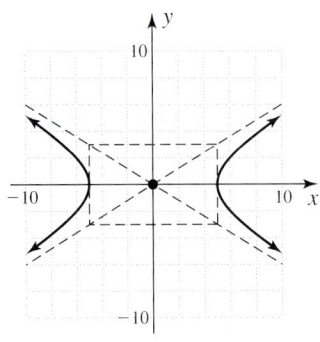

67.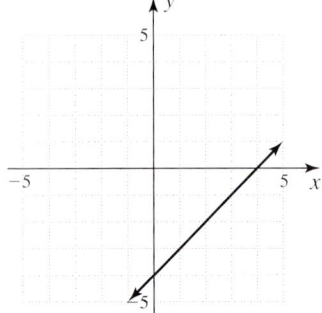

69.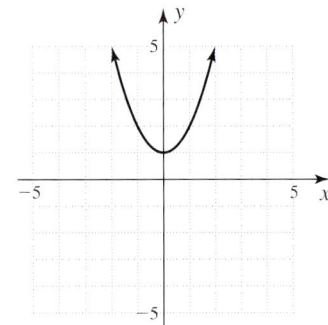

71.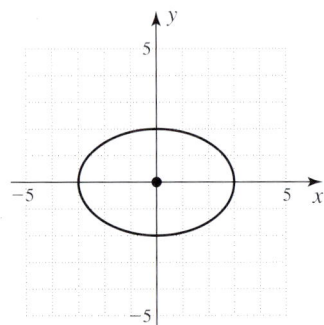

Exercises 9.4

1. (0, 4), (4, 0)

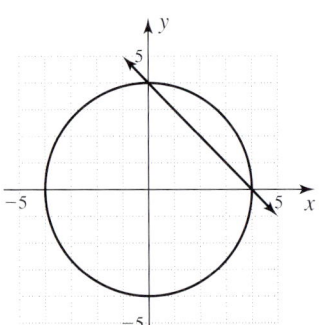

3. (−5, 0), (0, 5)

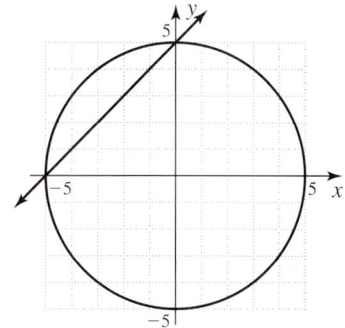

5. (−4, −3), (3, 4)

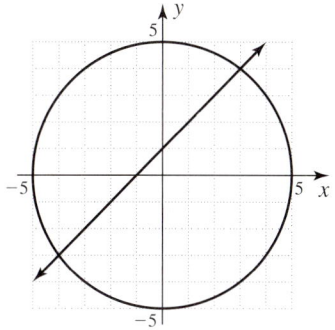

7. (1, 0), (5, 4)

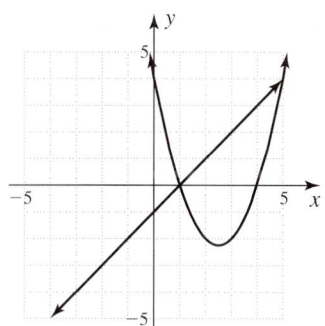

9. (0, 1), (3, 4)

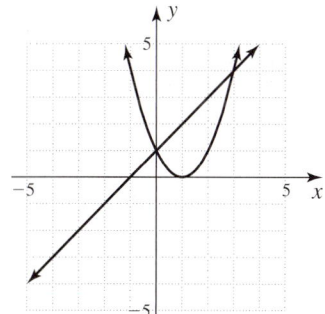

11. (−3, 0), (0, 2)

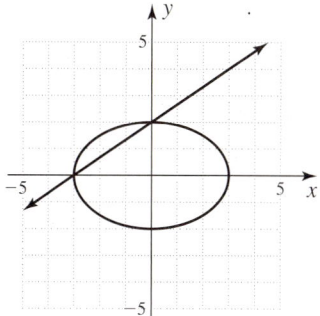

13. (4, 0), $\left(-\frac{68}{15}, \frac{32}{15}\right)$

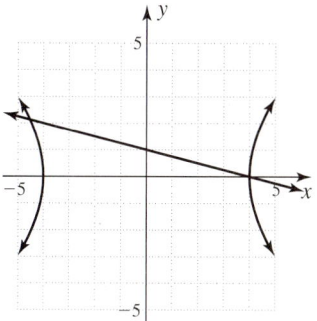

15. No solution

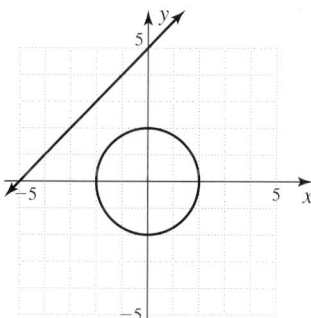

17. (−2, 0), (2, 0)

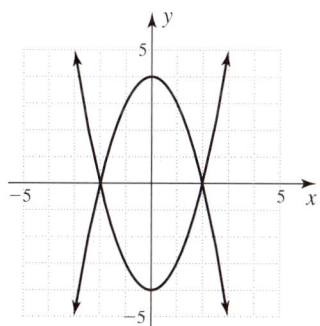

19. $(4, 3), (4, -3), (-4, 3), (-4, -3)$

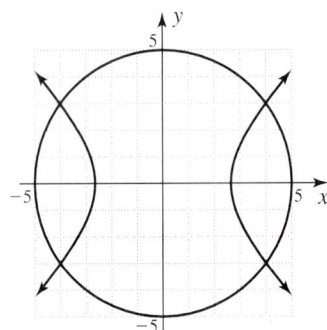

21. $(-4, 0), (4, 0)$

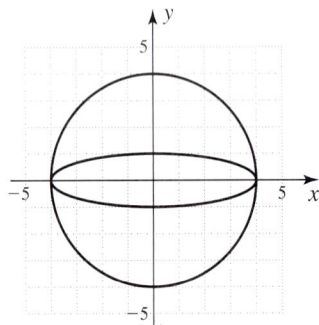

23. $(1, -1), (-1, -1), (1, 1), (-1, 1)$

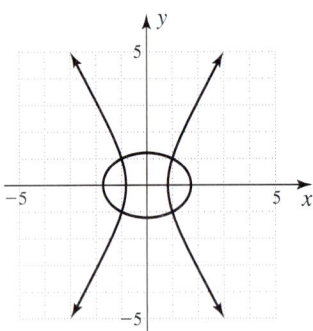

25. $(-3, 1), (-3, -1), (3, 1), (3, -1)$

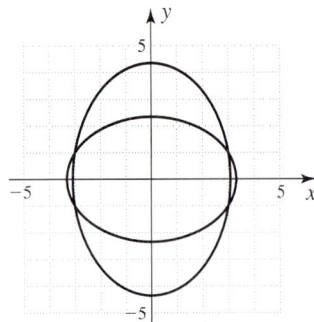

27. No solution

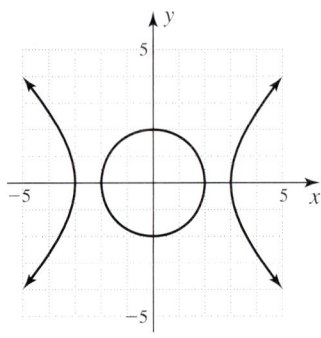

29. No solution

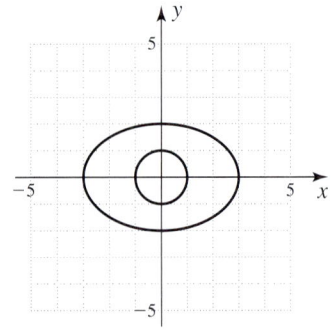

31. 4000 or 1000 **33.** 7 and 8 **35.** 11 and 16 or -11 and -16
37. 70 ft by 31 ft **39.** $P = \$13{,}600$; $r = 2.5\%$ **41. (a)** $LW = 60$;
(b) $2L + 2W = 32$; **(c)** $W = 6$ or $W = 10$; **(d)** 6 ft by 10 ft
43. 26 ft by 25 ft **45.** $W = 52$ in., $H = 29$ in. **47.** 20 eggs
53. one, more than one, no **55.** 12 cm by 10 cm **57.** $(0, 1), (0, -1)$
59. No solution **61.** $(1, 0), (0, 1)$
63.

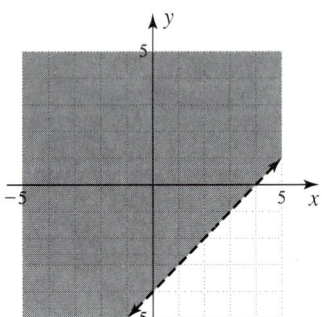

65.

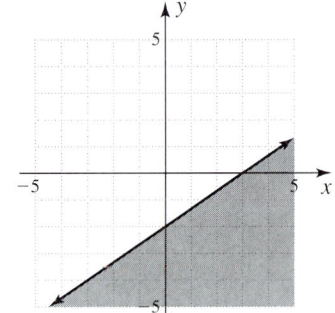

67.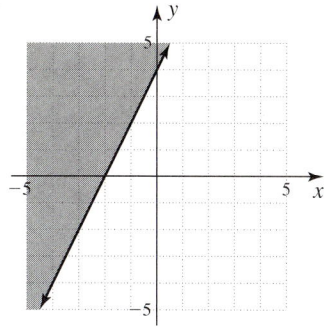

Exercises 9.5

1.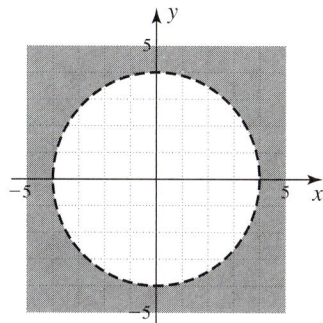

3.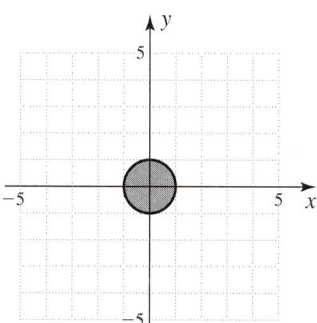

5.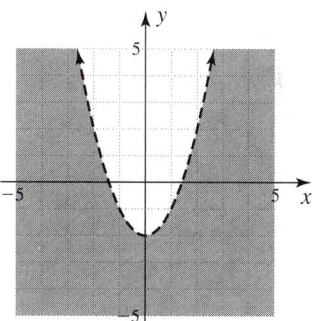

7.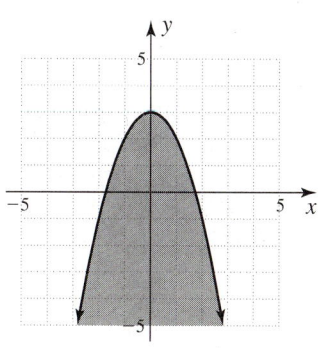

9.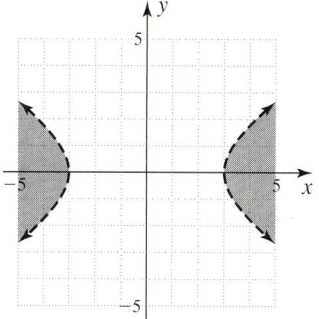

11.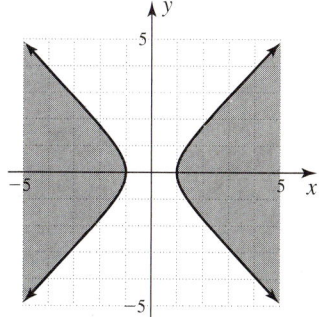

13.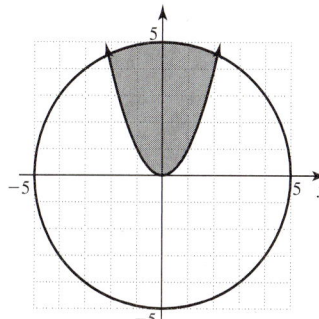

15.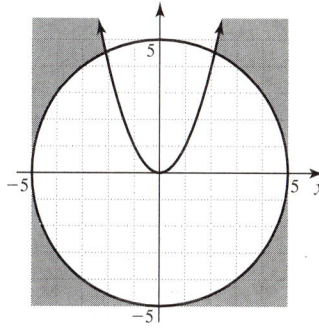

17.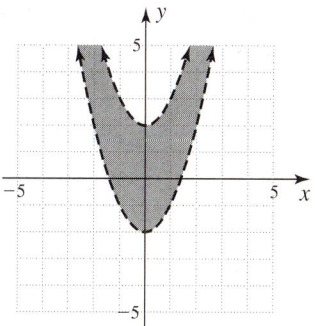

19.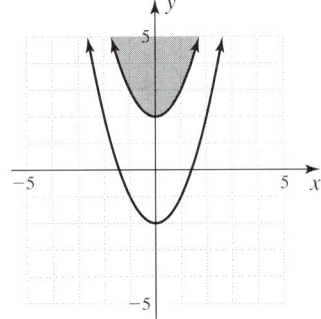

21.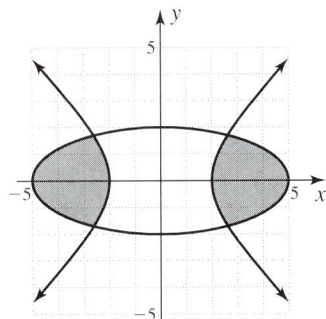

23.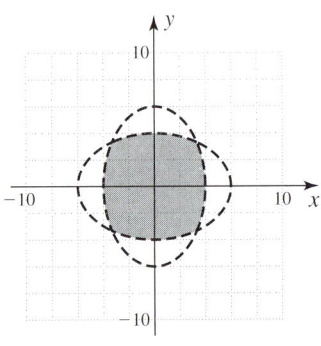

25. (a) $\frac{x^2}{36} + \frac{y^2}{16} > 1$; $\frac{x^2}{16} + \frac{y^2}{16} < 1$; (b) $\frac{x^2}{36} + \frac{y^2}{16} \geq 1$; $\frac{x^2}{16} + \frac{y^2}{16} \leq 1$
27. $C = 648{,}000 - 2700p$ **29.** $p = 180$ or $p = 80$ **35.** intersection

37.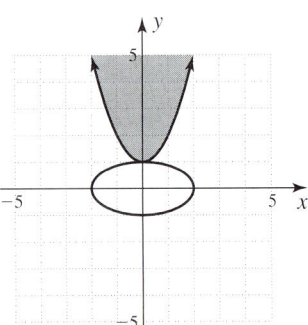

39.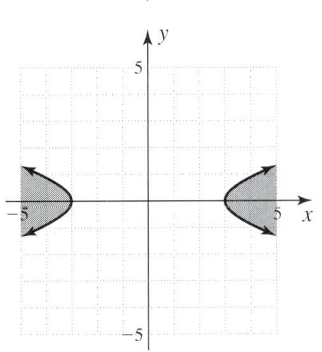

41.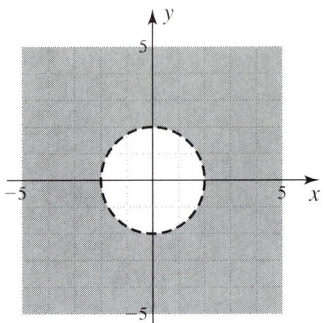

43.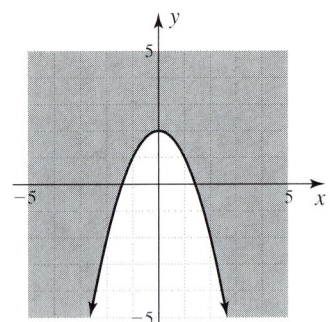

45. -10 **47.** $P(x) \cdot Q(x) = x^3 + 3x^2 - 9x - 27$; $\frac{P(x)}{Q(x)} = x - 3$

Review Exercises

1. [9.1A] (a)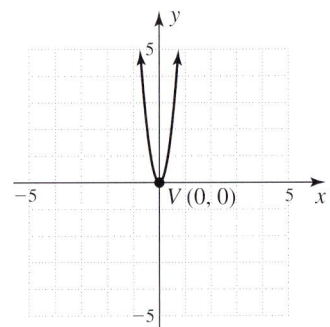

(b)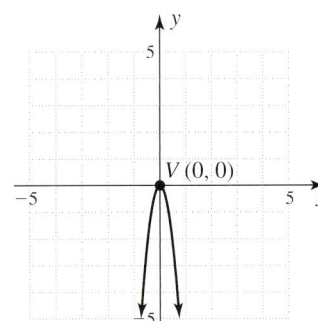

2. [9.1B] (a)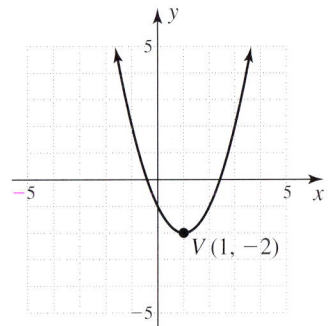

(b)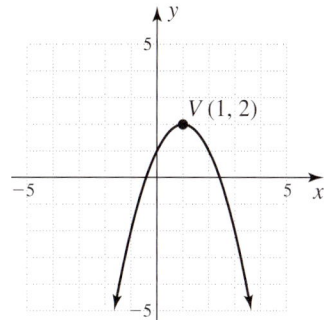

3. [9.1C] (a)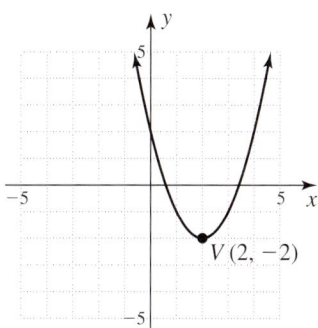

(b)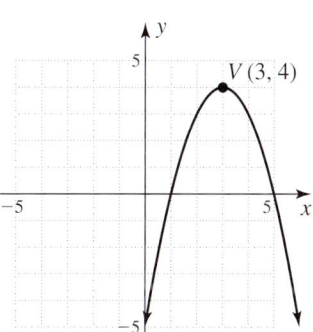

4. [9.1C] (a)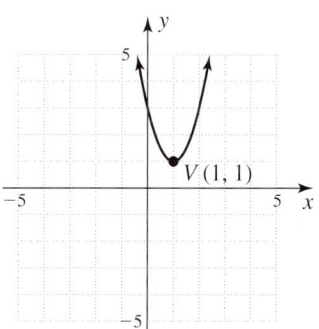

(b)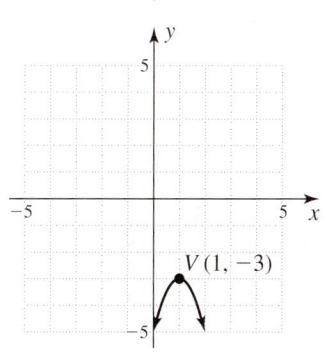

5. [9.1D] (a)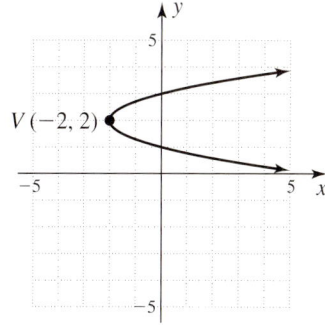

(b)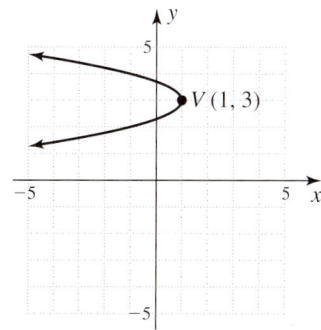

6. [9.1D] (a)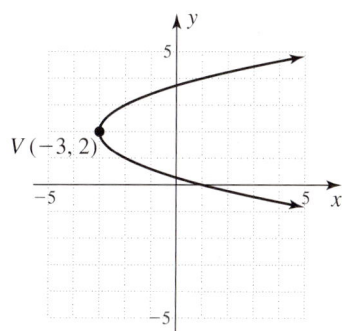

(b)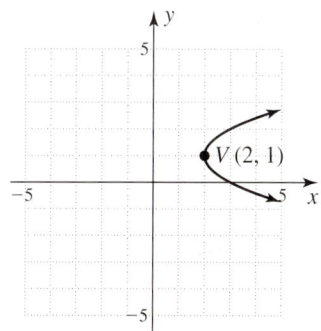

7. [9.1E] (a) 1000; (b) 250   8. [9.1E] $\frac{3}{4}$ sec; 9 ft
9. [9.2A] (a) $\sqrt{130}$; (b) $2\sqrt{65}$; (c) 5
10. [9.2B] (a) $(x + 2)^2 + (y - 2)^2 = 9$; (b) $(x - 3)^2 + (y + 2)^2 = 9$

11. [9.2C] **(a)** $C(-2, 1); r = 2$

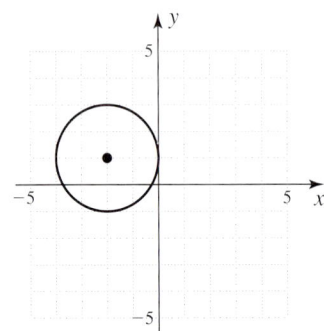

(b) $C(1, -2); r = 3$

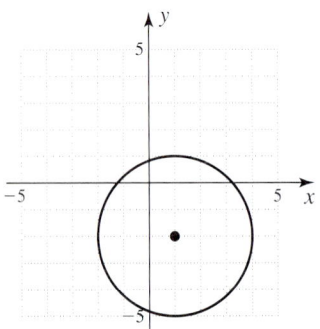

12. [9.2C] **(a)**

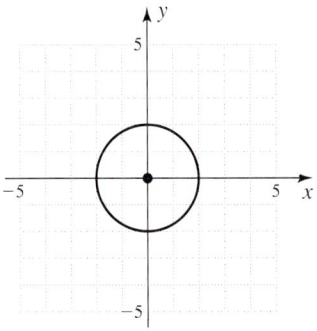

(b)

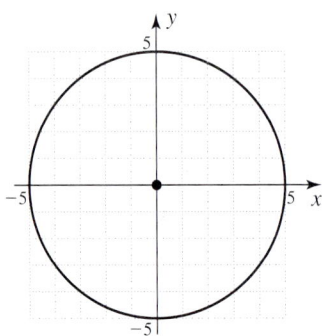

13. [9.2C] **(a)** $C(-1, -1); r = 2$

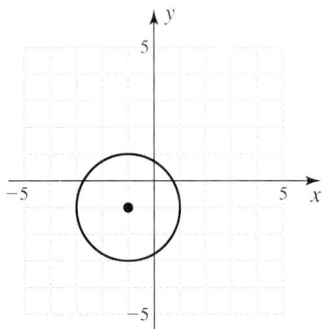

(b) $C(2, -3); r = 2$

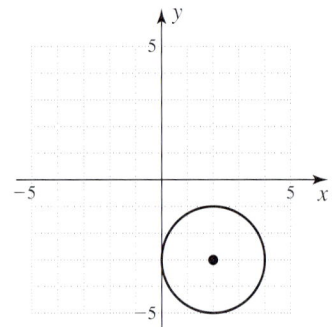

14. [9.2D] **(a)**

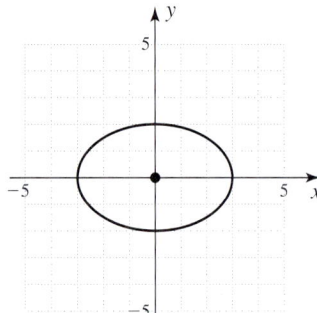

(b)

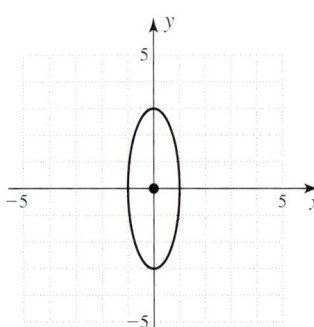

15. [9.2D] **(a)**

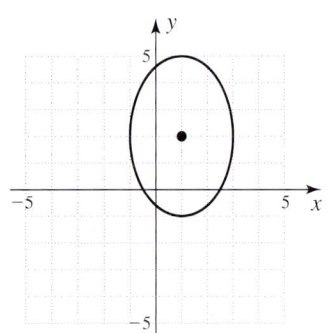

(b)
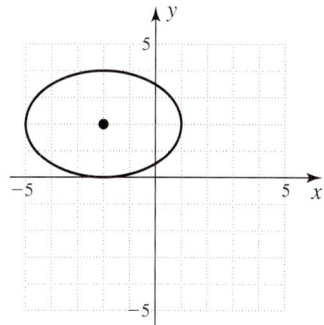

16. [9.3A] **(a)**
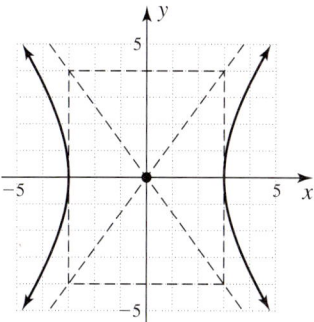

(b)
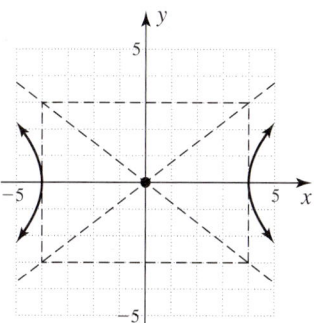

17. [9.3A] **(a)**
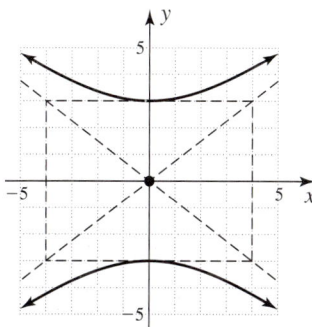

(b)
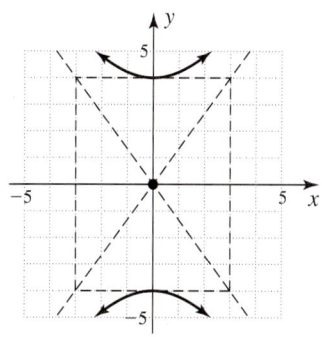

18. [9.3B] **(a)** Parabola; **(b)** Hyperbola; **(c)** Circle; **(d)** Ellipse
19. [9.4A] **(a)** (0, 1), (1, 0); **(b)** (1, 3), (3, 1)
20. [9.4A] **(a)** No solution
(b) No solution
21. [9.4B] **(a)** (3, 2), (3, −2), (−3, 2), (−3, −2);
(b) (2, 1), (2, −1), (−2, 1), (−2, −1) **22.** [9.4C] **(a)** 30; **(b)** 20
23. [9.5A] **(a)**
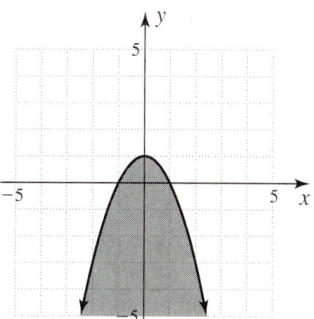

(b)
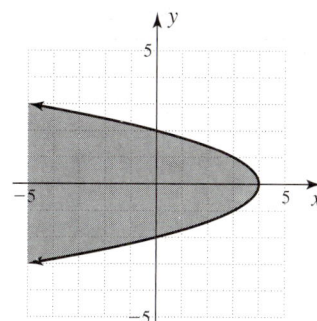

24. [9.5A] **(a)**
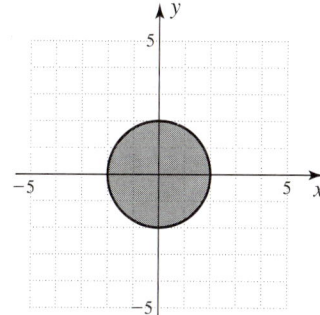

(b)
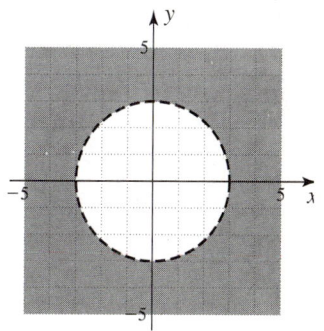

25. [9.5A] **(a)**

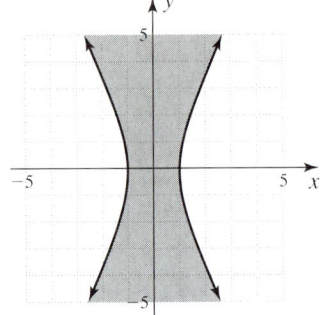

(b)

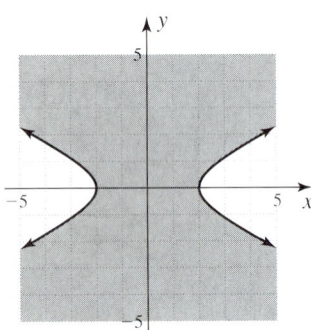

26. [9.5B] **(a)**

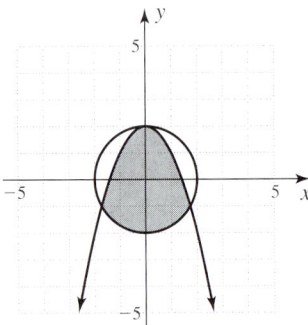

(b)

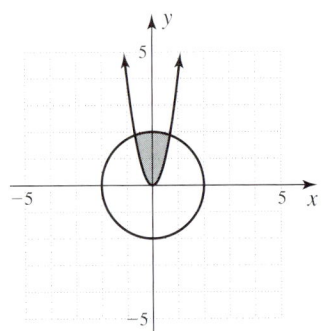

Cumulative Review Chapters 1–9

1.

2. $3x^{15}$ **3.** -57 **4.** $-\frac{4}{3}$; 8
5. $x + 1 \leq 4$; $-2x < 8$

6. $A = \frac{8B + 70}{7}$ or $A = \frac{8}{7}B + 10$ **7.** $29,000

8. x-intercept: $\left(-\frac{5}{2}, 0\right)$; y-intercept: $(0, -5)$
9.

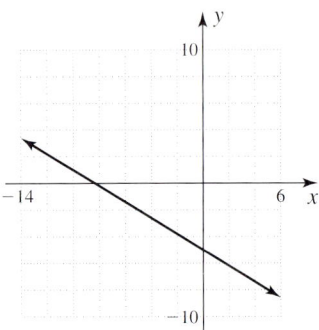

10.

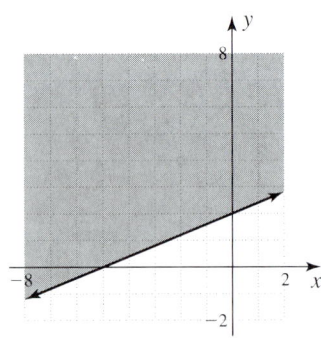

11. $-\frac{3}{4}$ **12.** $D = \{-2, 5, -3\}$ **13.** -4 **14.** No solution
15. $(2, 3, -1)$
16.

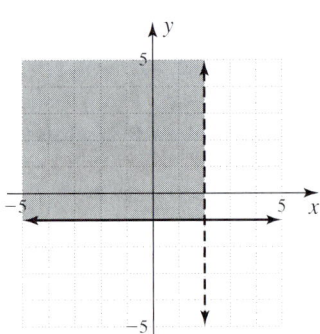

17. 30 mi/hr **18.** $9j^2 - 1$ **19.** $x(3x^3 + 2)(2x + 1)(2x - 1)$
20. $2y^2(4x - 3y)^2$ **21.** $-\frac{3}{2}, -3$
22. $\frac{3x - y}{2x - 3}$ **23.** No solution **24.** $2\frac{2}{9}$ hr **25.** $\frac{1}{60}$ **26.** $x^{19.6}$
27. $\sqrt[3]{4c^2d^2}$ **28.** $80 + 31\sqrt{10}$ **29.** No real-number solution
30. $\frac{17}{97} + \frac{14}{97}i$ **31.** $\frac{-21 \pm \sqrt{77}}{7}$ **32.** $\frac{4}{3}, -\frac{2}{3} \pm \frac{2\sqrt{3}}{3}i$
33. Sum: $\frac{5}{4}$; product: -1 **34.** $x \leq -8$ or $x \geq 7$
35.

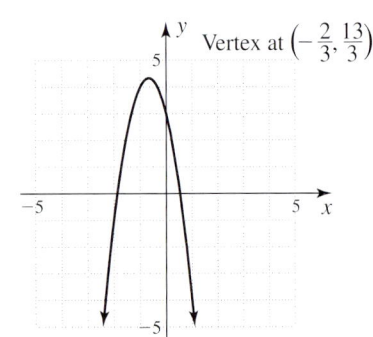

Vertex at $\left(-\frac{2}{3}, \frac{13}{3}\right)$

36. 500   **37.** $x^2 + y^2 = 81$
38.

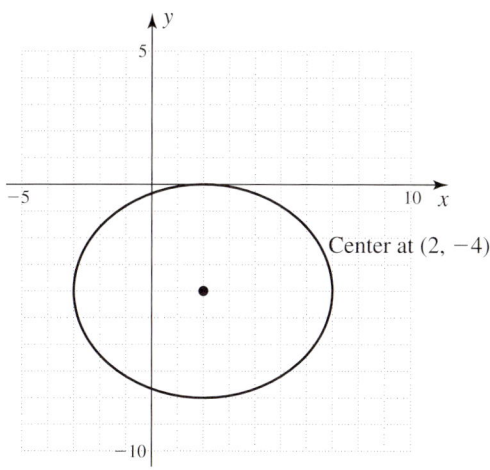
Center at $(2, -4)$

39. $(-3, 4), (4, -3)$
40.

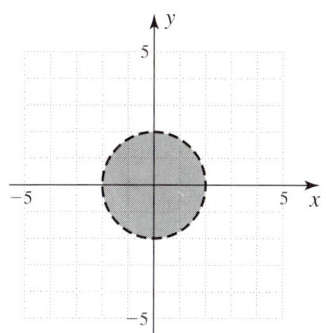

55. $\{x \mid 0 \le x \le 5\}$ **57.** $\{x \mid 2 \le x \le 5\}$ **63.** $f(x) + g(x)$
65. $f(x) \cdot g(x)$ **67.** $f(g(x))$ **69.** $P(x) = -\frac{x^2}{30} + 240x - 72{,}000$
71. $x^3 + 1$ **73.** -26 **75.** $x + a, x \ne a$ **77.** $x^2 - x + 6$ **79.** -2
81. $-\frac{1}{2}$ **83.** The domain of the sum, difference, product, and quotient is: $\{x \mid x \text{ is a real number and } x \ne 0 \text{ and } x \ne -1\}$
85.

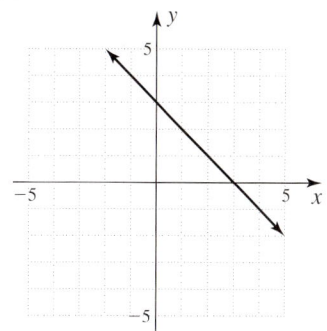

87.

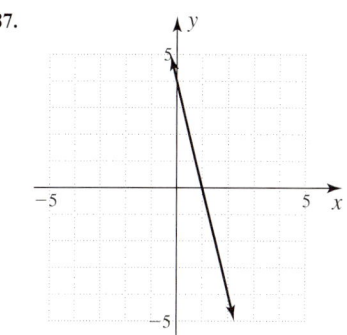

89.

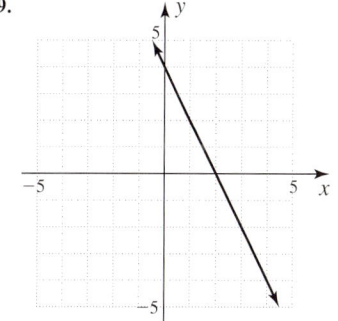

Chapter 10

Exercises 10.1

1. $x^2 - 4x + 8$ **3.** $x^3 + 4x^2 + 16x + 64$ **5.** $\frac{x+4}{x^2+16}$ **7.** 3   **9.** $x + a$
11. $x + a + 3$ **13.** (a) 1; (b) x; (c) $|x|$ **15.** (a) 4; (b) $3x + 1$; (c) $3x - 1$
17. (a) 1; (b) $|x|$; (c) x **19.** (a) 3; (b) 3; (c) -1 **21.** (a) 17; (b) -13
23. (a) 1; (b) 0   **25.** (a) Undefined for $x = 3$; (b) Undefined for $x = 3$
27. (a) $\sqrt{2}$; (b) Undefined for $x \le 0$; (c) $\sqrt{x^2 + 1}$; (d) $x + 1, x > 0$
29. (a) Undefined for $x = 2$; (b) $\sqrt{\frac{1}{2}} = \frac{\sqrt{2}}{2}$; (c) $\frac{1}{x-2}, x > 0$;
(d) $\sqrt{\frac{1}{x^2-2}}$ or $\frac{\sqrt{x^2-2}}{x^2-2}, |x| \sqrt{z}$ **31.** The domain of the sum, difference, and product is: $\{x \mid x \text{ is a real number}\}$; The domain of the quotient is: $\{x \mid x \text{ is a real number and } x \ne 1\}$ **33.** The domain of the sum, difference, and product is: $\{x \mid x \text{ is a real number}\}$; The domain of the quotient is: $\{x \mid x \text{ is a real number and } x \ne 1\}$ **35.** The domain of the sum, difference, and product is: $\{x \mid x \text{ is a real number and } x \ne 1 \text{ and } x \ne -2\}$; The domain of the quotient is: $\{x \mid x \text{ is a real number and } x \ne 1 \text{ and } x \ne -2\}$ **37.** The domain of the sum, difference, and product is: $\{x \mid x \text{ is a real number and } x \ne 1 \text{ and } x \ne -4\}$; The domain of the quotient is: $\{x \mid x \text{ is a real number and } x \ne 1 \text{ and } x \ne -4 \text{ and } x \ne 0\}$ **39.** The domain of the sum, difference, and product is: $\{x \mid x \text{ is a real number and } x \ne -3 \text{ and } x \ne 2\}$; The domain of the quotient is: $\{x \mid x \text{ is a real number and } x \ne -3 \text{ and } x \ne 2 \text{ and } x \ne 1\}$ **41.** $P(x) = -0.0005x^2 + 34x - 120{,}000$
43. (a) $L(x) + R(x) = -0.28x^2 + 3x + 22$; (b) $L(0) + R(0) = 22$ million; (c) $L(10) + R(10) = 24$ million; (d) $L(10) - R(10) = 6$ million
45. (a) $\frac{C(t)}{P(t)} = \frac{2.5t^2 + 8.5t + 111}{-0.46t^2 + 1.14t + 31.08}$ (in thousands); (b) \$3571;
(c) \$8544   **47.** (a) $(K \circ C)(F) = \frac{5}{9}(F - 32) + 273$; (b) $C(41) = 5°$;
(c) $(K \circ C)(212) = 373$ K   **49.** (a) 38; (b) $(E \circ F)(x) = x - 2$;
(c) 10   **51.** (a) 0; (b) 8; (c) 16   **53.** (a) $(f \circ F)(x) = 4\left(\frac{9}{5}x - 8\right)$; (b) 40

Exercises 10.2

1. Yes

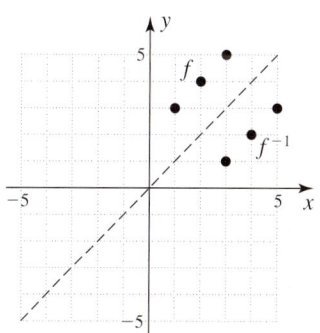

3. No

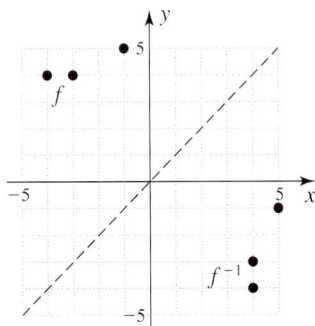

5. $y = \frac{1}{3}x - 1$; yes

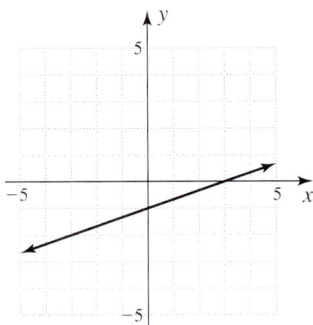

7. $y = \frac{1}{2}x + 2$; yes

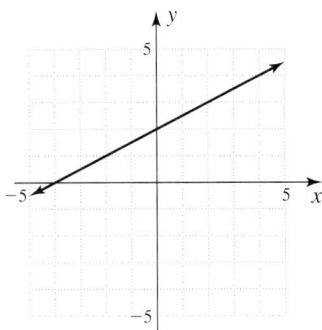

9. $y = \pm \frac{\sqrt{2x}}{2}$; no

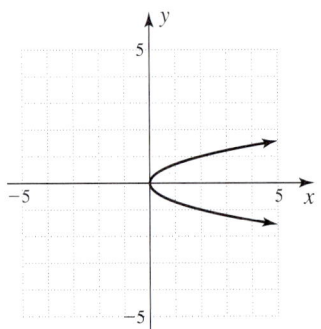

11. $y = \pm\sqrt{x+1}$; no

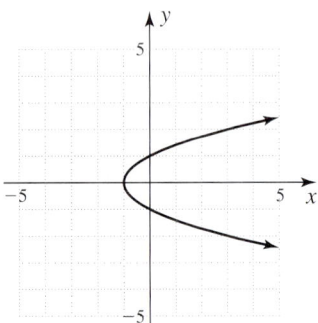

13. $y = -\sqrt[3]{x}$; yes

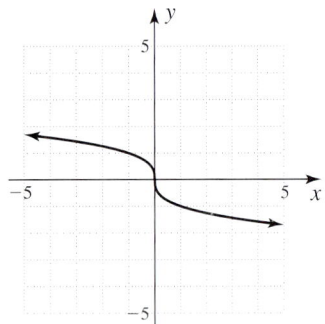

15. yes

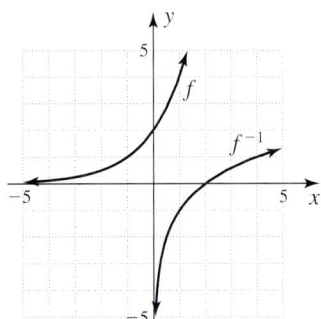

17. yes

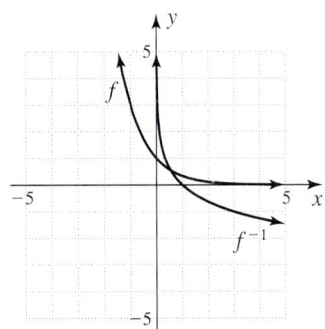

19. yes

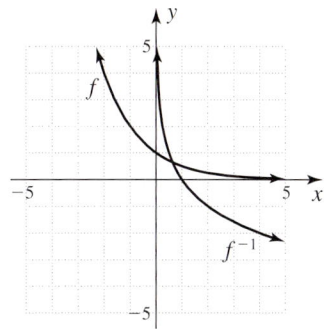

21. (a) 3; **(b)** -1 **23.** $\frac{1}{x}$
25. $f^{-1}(x) = \frac{1}{2}x$

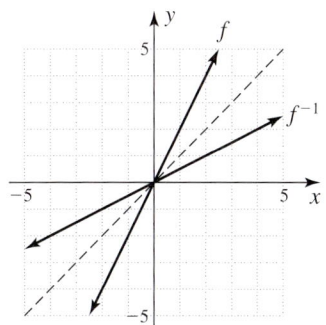

27. $f^{-1}(x) = \pm\sqrt{2-x}$

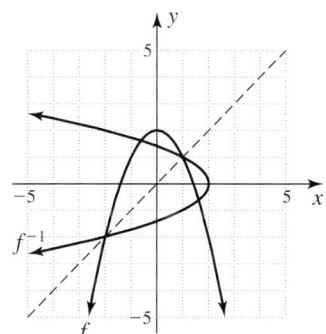

29. $f^{-1}: x = \sqrt{4-y^2}$

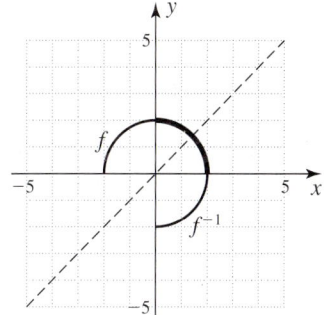

31. $f^{-1}: x = \sqrt{9-y^2}$

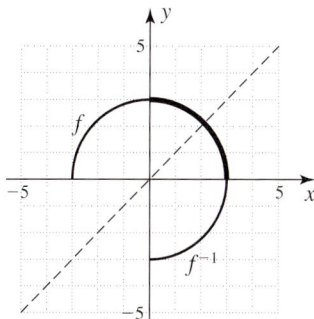

33. $f^{-1}(x) = \sqrt[3]{x}$

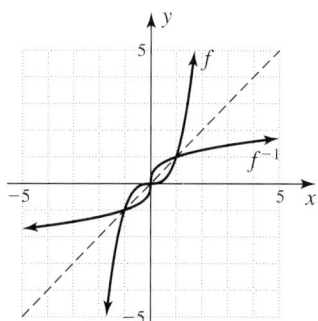

35. $f^{-1}: x = |y|$

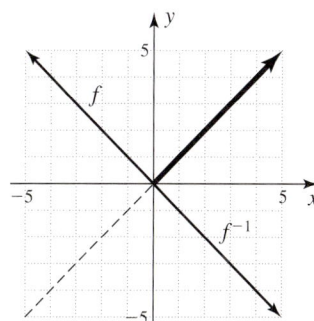

37. (a) $f^{-1}(S) = \frac{S+21}{3}$; **(b)** $9\frac{1}{3}$ in.   **39. (a)** 16; **(b)** $f^{-1}(d) = d + 16$; **(c)** 28 in.   **41. (a)** $f(1988) = 41.44$ sec; **(b)** $f^{-1}(w) = \frac{280-w}{0.12}$; **(c)** In the year 2000   **43.** $f^{-1}(x) = \frac{x+2}{3}$ **45.** $f^{-1}(x) = 2x - 1$
47. $f^{-1}(x) = \sqrt[3]{x-1}$ **49.** $f^{-1}(x) = x^2, x \geq 0$ **57.** relation, same
59. $f^{-1}(x) = x^{1/3} = \sqrt[3]{x}$; yes

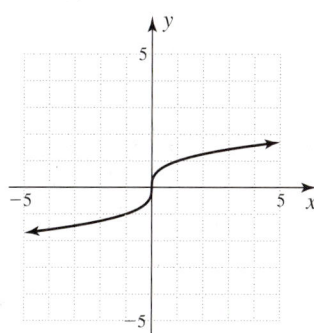

61. $D = \{2, 3, 4\}$; $R = \{1, 2, 3\}$ **63.** $D = \{1, 2, 3\}$; $R = \{2, 3, 4\}$
65. 9   **67.** 27   **69.** 4

Exercises 10.3

1. (a) $\frac{1}{5}$; **(b)** 1; **(c)** 5   **3. (a)** $\frac{1}{9}$; **(b)** 1; **(c)** 9   **5. (a)** $\frac{1}{10}$; **(b)** 1; **(c)** 10
7. (a) Increasing; **(b)** Decreasing

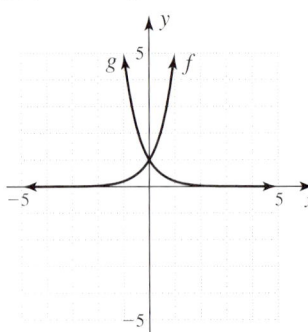

9. (a) Increasing; **(b)** Decreasing

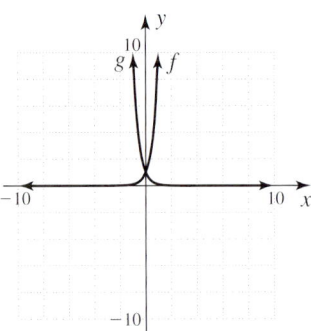

11. (a) Increasing; **(b)** Decreasing

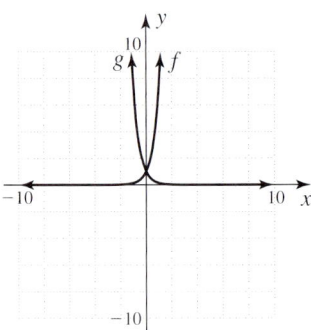

13. (a) Increasing; **(b)** Decreasing

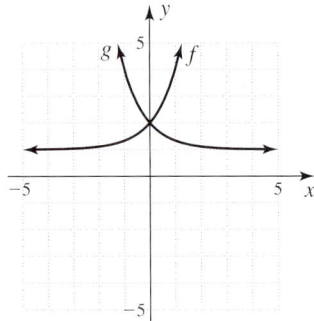

15. (a) Increasing; **(b)** Decreasing

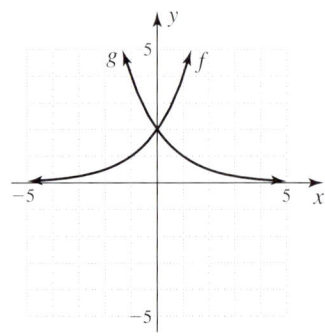

17. (a) $2459.60; **(b)** $2435.19   **19. (a)** $1822.12; **(b)** $1814.02
21. (a) 2000; **(b)** 4000; **(c)** 8000   **23. (a)** 2000 g; **(b)** 699.9 g;
(c) 244.9 g   **25.** 21,448,000   **27. (a)** 285 million; **(b)** 1,799,492 million
29. (a) 333,333; **(b)** 222,222; **(c)** 8671   **31. (a)** 4.42 lb/in.^{2};
(b) 6.42 lb/in.^{2} **33. (a)** 1.0000 kg; **(b)** 0.8855 kg; **(c)** 0.2965 kg
35. (a) 400,000; **(b)** 52,428,800,000; **(c)** No. The number in part **b** is larger than the entire U.S. population.   **37.** Continuous = $1822.12; Monthly = $1819.40; Continuous is $2.72 more.
39. (a) A horizontal line; **(b)** Yes; **(c)** No   **43.** b^x **45.** decreasing
47. 165 g
49.

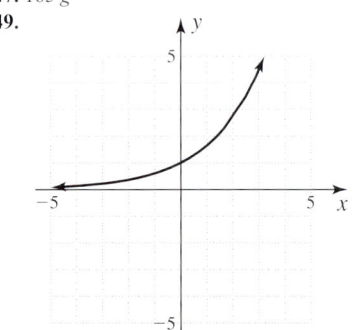

51.

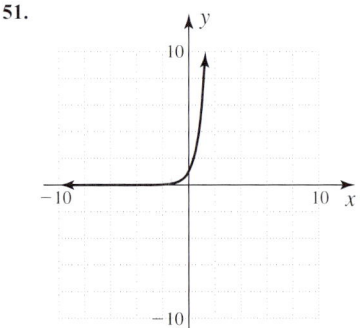

53. Product property   **55.** Power property

Exercises 10.4

1.

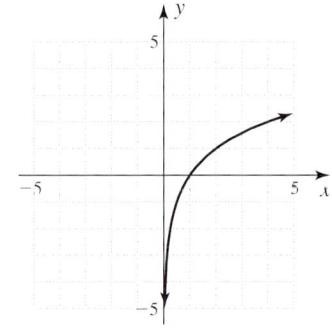

3.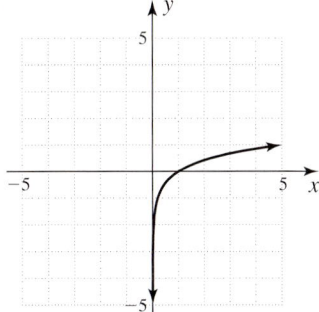

5.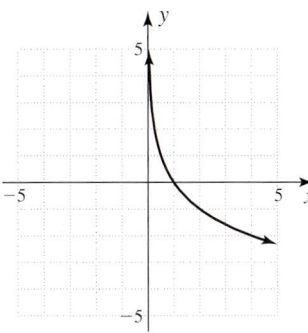

7.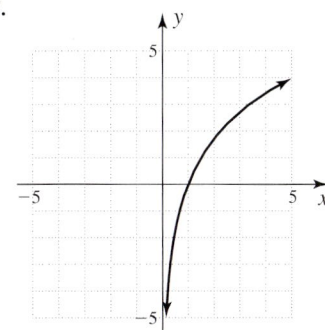

9. $\log_2 128 = x$ **11.** $\log_{10} 1000 = t$ **13.** $\log_{81} 9 = \frac{1}{2}$
15. $\log_{216} 6 = \frac{1}{3}$ **17.** $\log_e t = 3$ **19.** $9^3 = 729$; true
21. $2^{-8} = \frac{1}{256}$; true **23.** $81^{3/4} = 27$; true **25.** $4^x = 16$
27. $10^{-2} = 0.01$; true **29.** $e^{3.4012} \approx 30$; true **31.** $e^{-1.15} \approx 0.3166$; true
33. 9 **35.** $\frac{1}{27}$ **37.** 2 **39.** 2 **41.** 2 **43.** 3
45. -2 **47.** $-\frac{1}{4}$ **49.** -2 **51.** 8 **53.** 4 **55.** -3 **57.** 6
59. 0 **61.** 1 **63.** 1 **65.** -1 **67.** $\frac{1}{3}$ **69.** -3 **71.** t **73.** t
75. $\log \frac{26}{7} - \log \frac{15}{63} + \log \frac{5}{26} = \log\left(\frac{26}{7} \div \frac{15}{63} \cdot \frac{5}{26}\right)$

$= \log\left(\frac{\frac{26}{7}}{1} \cdot \frac{\frac{63}{15}}{3} \cdot \frac{\frac{5}{26}}{1}\right) = \log\left(\frac{9}{3}\right) = \log 3$

77. $\log b^3 + \log 2 - \log \sqrt{b} + \log \frac{\sqrt{b^3}}{2}$

$= \log\left(b^3 \cdot 2 \div \sqrt{b} \cdot \frac{\sqrt{b^3}}{2}\right) = \log\left(b^3 \cdot 2 \cdot \frac{1}{\sqrt{b}} \cdot \frac{b\sqrt{b}}{2}\right)$

$= \log b^4 = 4 \log b$

79. $\log k^{3/2} + \log r - \log k - \log r^{3/4}$

$= \log(k^{3/2} \cdot r \div k \div r^{3/4}) = \log(k^{3/2-1} \cdot r^{1-3/4})$
$= \log(k^{1/2} r^{1/4}) = \log(k^{2/4} r^{1/4}) = \log(k^2 r)^{1/4}$
$= \frac{1}{4} \log (k^2 r)$

81. $2 \log b + 6 \log a - 3 \log b$
$= \log b^2 + \log a^6 - \log b^3$
$= \log(b^2 \cdot a^6 \div b^3)$
$= \log\left(b^2 \cdot a^6 \cdot \frac{1}{b^3}\right)$
$= \log\left(\frac{a^6}{b}\right)$

83. $\frac{1}{3} \log x + \frac{1}{3} \log y^2 - \log z$
$= \log x^{1/3} + \log y^{2/3} - \log z$
$= \log(x^{1/3} \cdot y^{2/3} \div z)$
$= \log(\sqrt[3]{xy^2} \div z)$
$= \log\left(\frac{\sqrt[3]{xy^2}}{z}\right)$

85. $R = 8.9$ **87.** $\log_{10} A = 4.16769$
89. (a) 53%; (b) 77%; (c) 91%;
(d)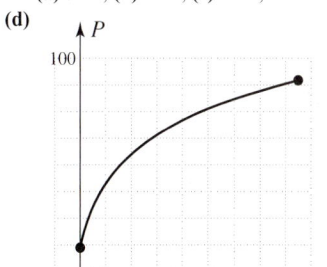

(e) 71 in. **91.** $\log_{10} 300 = \log_{10}(3 \times 100) = \log_{10} 3 + 2 \approx 2.4771$; yes
93. (a) 2.9542; (b) 2.9653 **95.** 7
97. $R = \frac{(1.02)^{40}}{(1.085)^{10}}$; $\log_{10} R = -0.01030$; 8.5% annually is greater.
103. $b^y = x$ **105.** $\log_b M + \log_b N$ **107.** $r \log_b M$ **109.** 1 **111.** 6.4
113. $2 \log x + 4 \log y - 6 \log z$
$= 2(\log x + 2 \log y - 3 \log z)$
$= 2(\log x + \log y^2 - \log z^3)$
$= 2[\log(x \cdot y^2 \div z^3)]$
$= 2\left[\log\left(\frac{xy^2}{z^3}\right)\right]$
$= \log\left(\frac{xy^2}{z^3}\right)^2$

115. $\log_2 1024 = 10$ **117.** -3 **119.** -1 **121.** $\frac{1}{9}$ **123.** -3
125.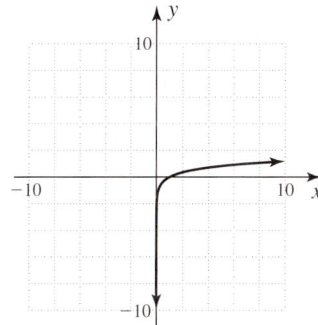

127. 3.268×10^1 **129.** 2.387×10^{-3} **131.** 5.69×10^{-5}

Exercises 10.5

1. 1.8722 **3.** 3.2648 **5.** -1.3595 **7.** 1.7005 **9.** -1.9073
11. -2.0652 **13.** 18.50 **15.** 0.01 **17.** 29.04 **19.** 0.73
21. 1.0986 **23.** 3.9512 **25.** 7.7647 **27.** -2.9188 **29.** -7.3858
31. 3.50 **33.** 65.00 **35.** 1.04 **37.** 0.10 **39.** 0.01 **41.** 2.7268
43. 0.8010 **45.** -0.6826 **47.** -0.6610 **49.** -0.6439
51.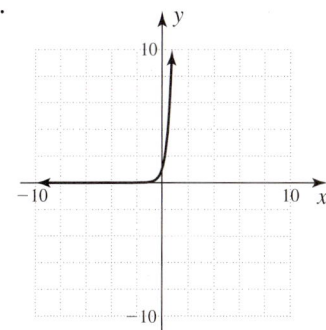

53.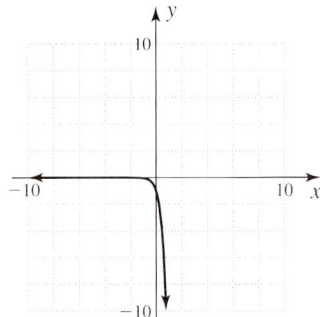

55.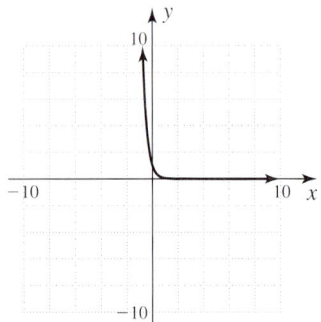

57.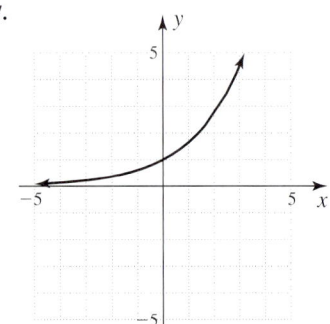

59.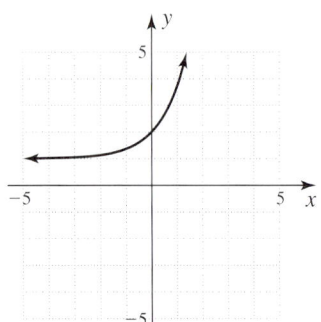

61.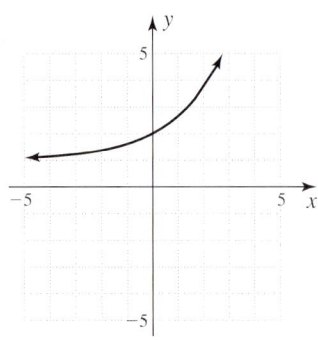

63.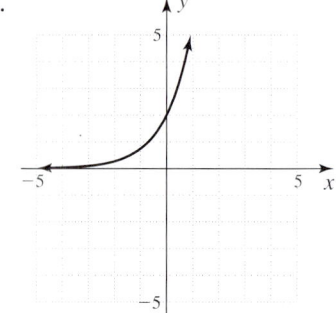

65.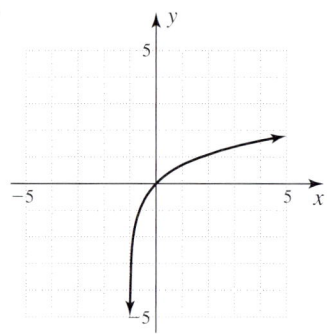

67.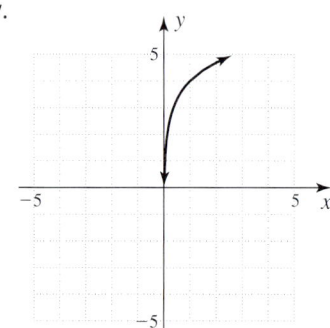

69.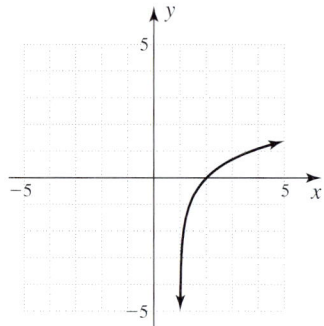

71. 6.2   **73.** 7.8   **75.** 6.4   **77. (a)** 33,000,000; **(b)** 44,545,341
79. (a) 50,000; **(b)** 74,591; **(c)** 166,006   **81. (a)** 1000; **(b)** 368
83. (a) 0; **(b)** 39.3   **85. (a)** -0.8055; **(b)** -3.8968 **87.** 18%
89. 45 mm

91. (a)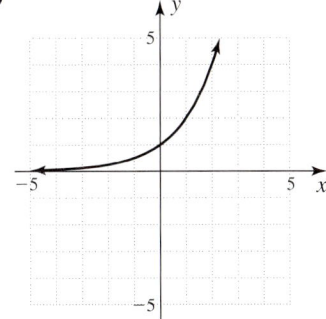

(b)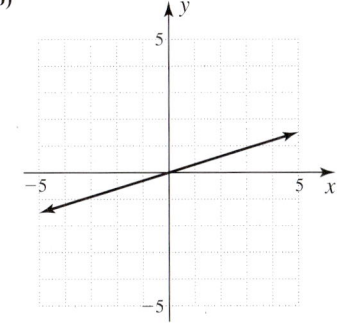

(c) Slope is log 2 ≈ 0.3.   **93.** The slope of $g(x)$ = log of base of $f(x)$.
101. $\dfrac{\log_a M}{\log_a b}$ **103.** 0.2027   **105.** 1.6720   **107.** −3.3820
109. 0.0034
111.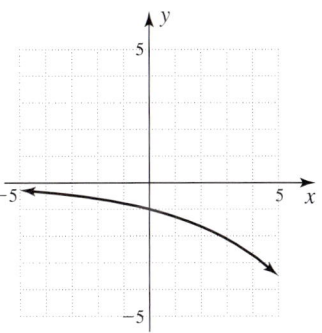

113.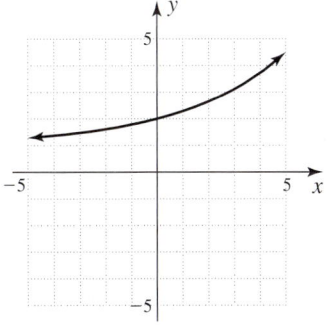

115.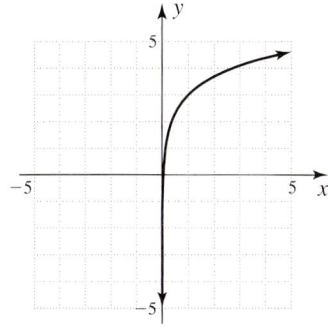

117. 1.5850   **119.** 1.3979

Exercises 10.6

1. 2   **3.** 5   **5.** $\frac{4}{3}$ **7.** −1   **9.** 3.2059   **11.** 6   **13.** 0.6309   **15.** $\frac{7}{3}$
17. $\dfrac{-3 \pm \sqrt{17}}{2}$ **19.** 2.3026   **21.** 2.3026   **23.** 1.7006   **25.** −1.1513
27. 8   **29.** $\frac{1}{8}$ **31.** $e \approx 2.7183$ **33.** $e^3 \approx 20.0855$ **35.** $\frac{7}{3}$ **37.** $\frac{17}{3}$
39. 8   **41.** 5   **43.** 1   **45.** $\frac{1}{9}$ **47.** 5   **49.** −3, −1   **51.** 13.86 yr
53. 10.66 yr   **55.** About 6.1 billion   **57.** 17.3 min   **59.** 80.5 min
61. (a) 100,000; **(b)** 67,032; **(c)** 13,534; **(d)** 1832   **63.** 23,104.9 yr
65. 13.3 yr   **67. (a)** 14.7 lb/in.²; **(b)** 11.4 lb/in.²; **(c)** 8.9 lb/in.²
69. (a) 5 yr after 1992 (1997); **(b)** 7 yr after 1992 (1999)
71. (a) 5000; **(b)** 2247; **(c)** In 2010   **73. (a)** 54 million; **(b)** In 1995
75. (a) 12 yr old; **(b)** 7 yr old   **77.** About 694 months   **79.** 543mm²
81. $k = 0.59$ **83.** 3,405,095   **85.** 1,242,987,238   **89.** Logarithmic
93. exponent   **95.** $M = N$ **97.** 59.9 min   **99.** $x = 1$ **101.** $x = 10$
103. $x = 1.4307$ **105.** $x = 4$ **107. (a)** 3; **(b)** 11; **(c)** 21

Review Exercises

1. [10.1A] **(a)** $4 + x - x^2$; **(b)** $-x - x^2$; **(c)** $4 + 2x - 2x^2 - x^3$;
(d) $\dfrac{2 - x^2}{2 + x}, x \neq -2$ **2.** [10.1A] **(a)** $6 + x - x^2$; **(b)** $-x - x^2$;
(c) $9 + 3x - 3x^2 - x^3$; **(d)** $\dfrac{3 - x^2}{3 + x}, x \neq -3$ **3.** [10.1A] **(a)** 6; **(b)** 7
4. [10.1B] **(a)** −6; **(b)** $(2 - x)^3$; **(c)** $2 - x^3$ **5.** [10.1B] **(a)** $(3 - x)^3$;
(b) $3 - x^3$; **(c)** −5   **6.** [10.1C] $\{x \mid x$ is a real number and $x \neq 1$ and
$x \neq 4\}$ **7.** [10.1C] $\{x \mid x$ is a real number and $x \neq 1$ and $x \neq 3$ and
$x \neq -4\}$ **8.** [10.1D] **(a)** $P(x) = -0.02x^2 + 70x - 30,000$;
(b) $P(x) = -0.02x^2 + 60x - 40,000$ **9.** [10.2A] **(a)** $D = \{4, 6, 8\}$;
$R = \{4, 6, 8\}$; **(b)** $S^{-1} = \{(4, 4), (6, 6), (8, 8)\}$; **(c)** $D = \{4, 6, 8\}$;
$R = \{4, 6, 8\}$
(d)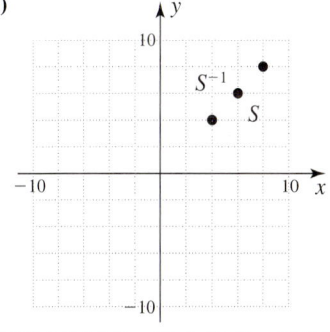

10. [10.2A] **(a)** $D = \{4, 6, 8\}$; $R = \{5, 7, 9\}$; **(b)** $S^{-1} = \{(5, 4), (7, 6), (9, 8)\}$; **(c)** $D = \{5, 7, 9\}$; $R = \{4, 6, 8\}$;
(d)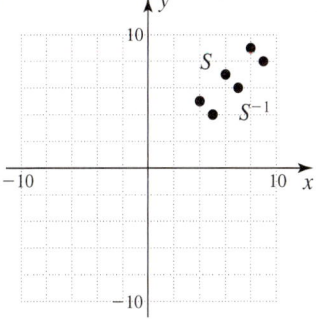

11. [10.2B] **(a)** $f^{-1}(x) = \dfrac{x + 3}{3}$;

(b)

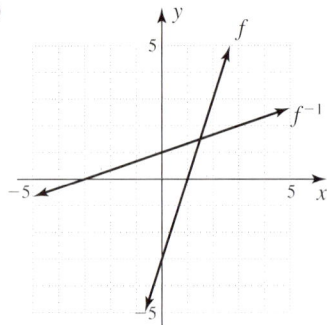

12. [10.2B] **(a)** $f^{-1}(x) = \frac{x+4}{4}$;
(b)

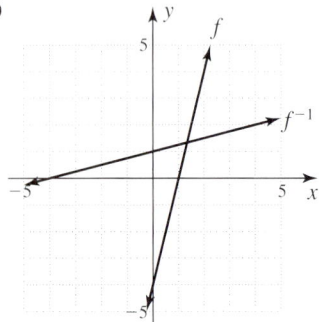

13. [10.2C] $f^{-1}(x) = \pm\frac{\sqrt{x}}{2}$; no
14. [10.2C] $f^{-1}(x) = \pm\frac{\sqrt{5x}}{5}$; no
15. [10.2C] Yes

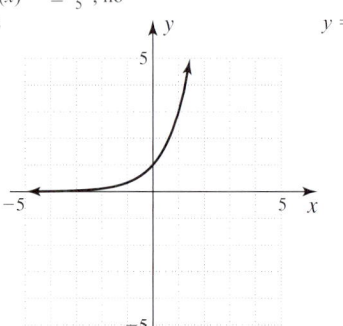

16. [10.2C] Yes

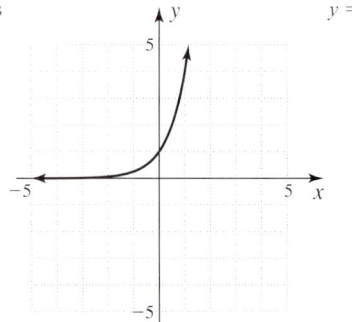

17. [10.2D] **(a)** 12; **(b)** $f^{-1}(d) = d + 16$; **(c)** 26 in.
18. [10.3A] **(a)** $f(x) = 2^{x/2}$

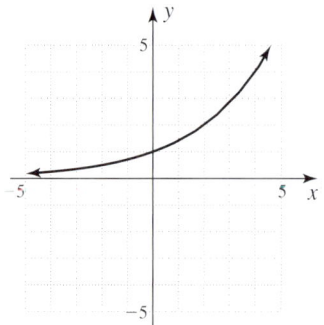

(b) $f(x) = 2^{-x/2}$

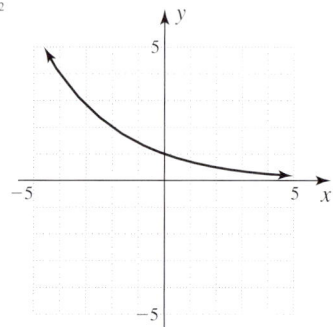

19. [10.3A] **(a)** $g(x) = \left(\frac{1}{2}\right)^{x/2}$

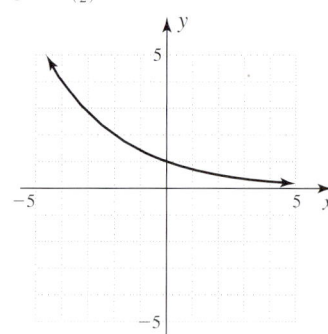

(b) $g(x) = \left(\frac{1}{2}\right)^{-x/2}$

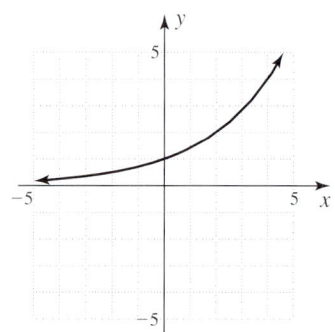

20. [10.3C] **(a)** 1000 g; **(b)** 61 g
21. [10.4A] $f(x) = \log_5 x$

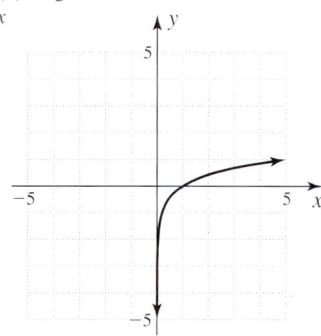

22. [10.4B] **(a)** $\log_3 243 = 5$; **(b)** $\log_2\left(\frac{1}{8}\right) = -3$
23. [10.4B] **(a)** $2^5 = 32$; **(b)** $3^{-4} = \frac{1}{81}$ **24.** [10.4C] **(a)** $\frac{1}{16}$;
(b) 4 **25.** [10.4C] **(a)** 4; **(b)** -3 **26.** [10.4D] **(a)** $\log_b M + \log_b N$;
(b) $\log_b \frac{M}{N}$; **(c)** $r \log_b M$ **27.** [10.5A] **(a)** 2.9890; **(b)** 2.9227
28. [10.5A] **(a)** -2.1198; **(b)** -3.1884
29. [10.5A] **(a)** 662.9793; **(b)** 0.0004
30. [10.5B] **(a)** 7.9551; **(b)** -1.0642
31. [10.5B] **(a)** 8.0486; **(b)** 15.1545
32. [10.5C] **(a)** 2.0959; **(b)** 4.1918

33. [10.5D] **(a)**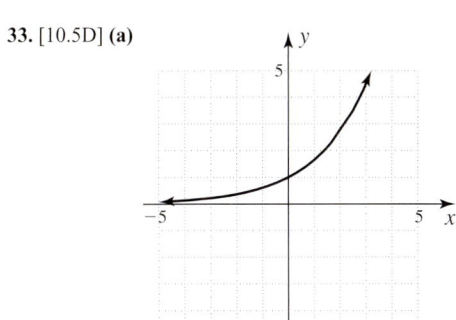

(b)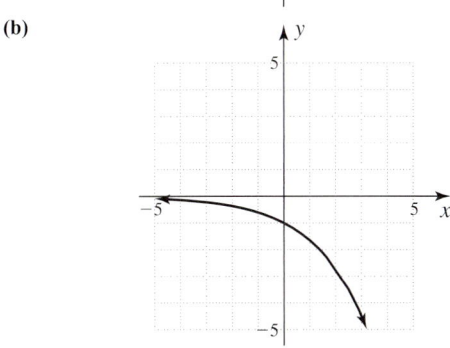

34. [10.5D] **(a)**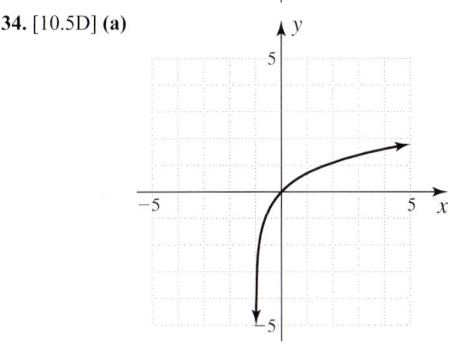

(b)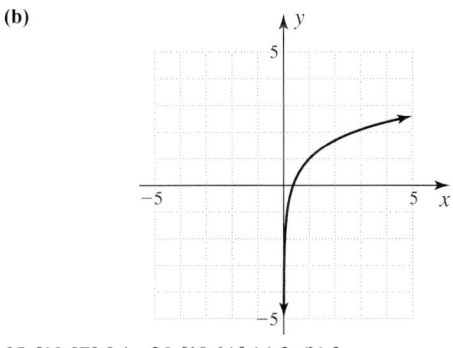

35. [10.5E] 5.4 **36.** [10.6A] **(a)** 3; **(b)** 3
37. [10.6A] **(a)** 1.5850; **(b)** 0.2847;
38. [10.6A] **(a)** 0.1238; **(b)** −2.1004
39. [10.6B] **(a)** $5 + \sqrt{35} \approx 10.9161$; **(b)** 2
40. [10.6C] **(a)** About 13.9 yr; **(b)** About 8.2 yr
41. [10.6C] **(a)** 0.2950; **(b)** 0.1475
42. [10.6C] **(a)** 1.3863 yr; **(b)** 34.6575 yr

Cumulative Review Chapters 1–10

1. $x^2 + 7x + 9$ **2.** $\frac{16x^8}{y^4}$ **3.** 150 **4.** 6
5.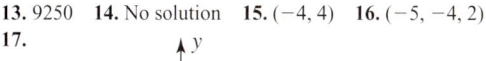
6.

7. 22 **8.** 30 ft × 50 ft **9.** $5\sqrt{2}$ **10.** 5 **11.** $4x + y = -18$
12.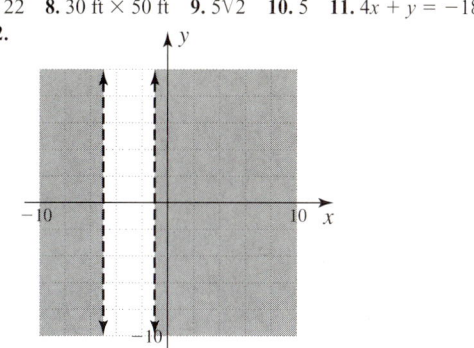

13. 9250 **14.** No solution **15.** (−4, 4) **16.** (−5, −4, 2)
17.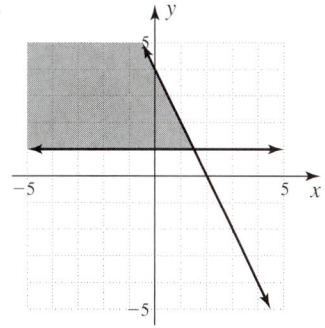

18. 224 ft **19.** −3 **20.** $18h^2 + 9h - 35$
21. $4x^2y(3x - y)(4x + 3y)$ **22.** $(3n + 2)(9n^2 - 6n + 4)$
23. −4, −1, 1 **24.** $(x + 5)(x + 2)(3x + 1)$ **25.** $\frac{1}{x^2 - 49}$
26. $\frac{4}{(x + 2)(x - 2)}$ **27.** 6 and 8 **28.** $\frac{1}{81}$ **29.** $\frac{\sqrt[5]{14d^2}}{2d}$
30. $8\sqrt{2}$ **31.** $2 + \sqrt{2}$ **32.** −3 **33.** $60 - 2i$ **34.** −9, 1
35. 16, 1 **36.** $x \leq -5$ or $2 \leq x \leq 7$; $(-\infty, -5] \cup [2, 7]$
37. $x < 1$ or $x \geq 5$; $(-\infty, 1) \cup [5, \infty)$
38.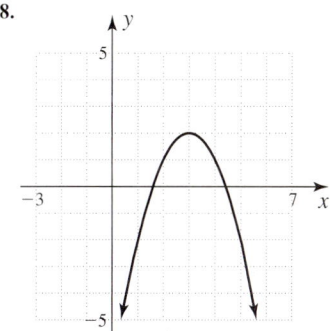

39. Center (3, 5); $r = 2$
40.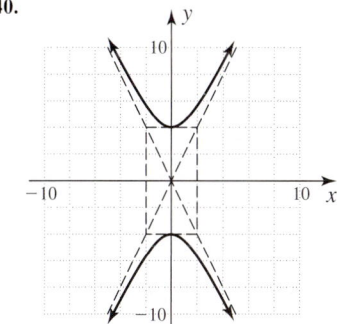

41. Parabola **42.** 28 **43.** $\{x \mid x \geq -9\}$ **44.** $(4 - x)^4$

45.

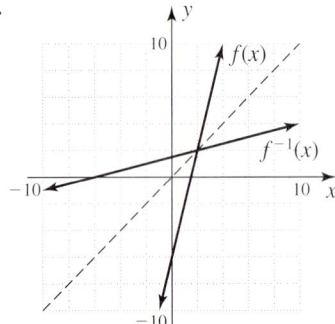

46.

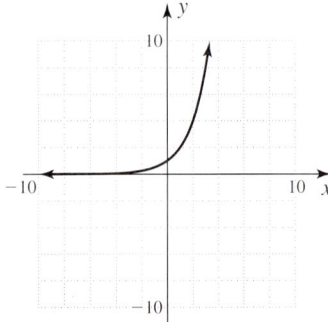

Yes, the inverse is a function.

47. $\frac{1}{2}\log_b 21 - \frac{1}{2}\log_b 83$
$= \log_b 21^{1/2} - \log_b 83^{1/2}$
$= \log_b \left(\frac{21}{83}\right)^{1/2} = \log_b \sqrt{\frac{21}{83}}$

48. 0.2027   **49.** 1.4650   **50.** 0.3039

Appendix A

Exercises A.4

1. (1, 2, 0)   **3.** (−1, −1, 3)   **5.** (−2, −1, 3)   **7.** No solution
9. (1 − k, 2, k), k is any real number   **11.** 30 dimes; 60 quarters;
40 one-dollar coins   **13.** Type I: 8; type II: 10; type III: 12
15. 50% type I; 25% type II; 25% type III   **17. (a)** Singular;
(b) Nonsingular; **(c)** Nonsingular   **23.** (6, −3, 4)
25. No solution   **27.** (1, k, k), k is any real number
29. Dimes: 22; nickels: 8; quarters: 46   **31.** −115   **33.** 2   **35.** −7

Exercises A.5

1. 2   **3.** 7   **5.** 6   **7.** $\frac{1}{2}$ **9.** $-\frac{7}{40}$ **11.** (2, 3)   **13.** (4, 5)   **15.** (3, −1)
17. (4, 5)   **19.** $\left(x, \frac{-2x-13}{3}\right)$; dependent   **21.** (−2, −3)
23. No solution; inconsistent   **25.** (10, 1)   **27.** (5, 2)   **29.** (−1, −1)
31. −7   **33.** 0   **35.** −1   **37.** −4   **39.** −9   **41.** (1, 2, 3)
43. (3, −1, −2)   **45.** (3, 0, 4)   **47.** (−5, 1, 5)   **49.** (−6, 2, 5)

51. $\begin{vmatrix} a & b & 0 \\ c & d & 0 \\ e & f & 0 \end{vmatrix} = a\begin{vmatrix} d & 0 \\ f & 0 \end{vmatrix} - b\begin{vmatrix} c & 0 \\ e & 0 \end{vmatrix} + 0\begin{vmatrix} c & d \\ e & f \end{vmatrix} = a(0) - b(0) + 0 = 0$

53. $\begin{vmatrix} a & b & c \\ 1 & 2 & 3 \\ a & b & c \end{vmatrix} = -1\begin{vmatrix} b & c \\ b & c \end{vmatrix} + 2\begin{vmatrix} a & c \\ a & c \end{vmatrix} - 3\begin{vmatrix} a & b \\ a & b \end{vmatrix}$
$= -1(bc - bc) + 2(ac - ac) - 3(ab - ab)$
$= -1(0) + 2(0) - 3(0) = -0 + 0 - 0 = 0$

55. $\begin{vmatrix} 1 & 2 & 3 \\ 3 & 1 & 2 \\ 3k & 2k & k \end{vmatrix} = 3k\begin{vmatrix} 2 & 3 \\ 1 & 2 \end{vmatrix} - 2k\begin{vmatrix} 1 & 3 \\ 3 & 2 \end{vmatrix} + k\begin{vmatrix} 1 & 2 \\ 3 & 1 \end{vmatrix}$
$= 3k(1) - 2k(-7) + k(-5) = 3k + 14k - 5k$

$k\begin{vmatrix} 1 & 2 & 3 \\ 3 & 1 & 2 \\ 3 & 2 & 1 \end{vmatrix} = k\left[3\begin{vmatrix} 2 & 3 \\ 1 & 2 \end{vmatrix} - 2\begin{vmatrix} 1 & 3 \\ 3 & 2 \end{vmatrix} + \begin{vmatrix} 1 & 2 \\ 3 & 1 \end{vmatrix}\right]$
$= k[3(1) - 2(-7) + 1(-5)]$
$= k(3 + 14 - 5)$
$= 3k + 14k - 5k$

$\therefore \begin{vmatrix} 1 & 2 & 3 \\ 3 & 1 & 2 \\ 3k & 2k & k \end{vmatrix} = k\begin{vmatrix} 1 & 2 & 3 \\ 3 & 1 & 2 \\ 3 & 2 & 1 \end{vmatrix}$

57. $\begin{vmatrix} kb_1 & b_1 & 1 \\ kb_2 & b_2 & 2 \\ kb_3 & b_3 & 3 \end{vmatrix} = kb_1\begin{vmatrix} b_2 & 2 \\ b_3 & 3 \end{vmatrix} - kb_2\begin{vmatrix} b_1 & 1 \\ b_3 & 3 \end{vmatrix} + kb_3\begin{vmatrix} b_1 & 1 \\ b_2 & 2 \end{vmatrix}$

$= kb_1(3b_2 - 2b_3) - kb_2(3b_1 - b_3) + kb_3(2b_1 - b_2)$
$= 3kb_1b_2 - 2kb_1b_3 - 3kb_1b_2 + kb_2b_3 + 2kb_1b_3 - kb_2b_3$
$= 3kb_1b_2 - 3kb_1b_2 - 2kb_1b_3 + 2kb_1b_3 + kb_2b_3 - kb_2b_3$
$= 0$

59. $\begin{vmatrix} 1 & 1 & 1 \\ 2 & a & a \\ 3 & b & b \end{vmatrix} = 1\begin{vmatrix} a & a \\ b & b \end{vmatrix} - 2\begin{vmatrix} 1 & 1 \\ b & b \end{vmatrix} + 3\begin{vmatrix} 1 & 1 \\ a & a \end{vmatrix}$

$= 1(ab - ab) - 2(b - b) + 3(a - a)$
$= 1(0) - 2(0) + 3(0) = 0$

61. $2x - y + 3 = 0$ **63.** $2x + 9y - 34 = 0$ **65.** $bx + ay - ab = 0$

73. $2\begin{vmatrix} 1 & 0 \\ -1 & 1 \end{vmatrix} - (-1)\begin{vmatrix} 1 & 0 \\ 0 & 1 \end{vmatrix} + (-3)\begin{vmatrix} 1 & 1 \\ 0 & -1 \end{vmatrix} = 2(1) + 1(1) - 3(-1)$
$= 2 + 1 + 3$
$= 6$

75. No solution   **77.** −19   **79.** 8
81. No   **83.** Yes   **85.** No

Exercises A.6

1. 6   **3.** 3,628,800   **5.** 360   **7.** 504   **9.** 15   **11.** 11   **13.** 1
15. $a^4 + 12a^3b + 54a^2b^2 + 108ab^3 + 81b^4$
17. $x^4 + 16x^3 + 96x^2 + 256x + 256$
19. $32x^5 - 80x^4y + 80x^3y^2 - 40x^2y^3 + 10xy^4 - y^5$
21. $32x^5 + 240x^4y + 720x^3y^2 + 1080x^2y^3 + 810xy^4 + 243y^5$
23. $\frac{1}{x^4} - \frac{2y}{x^3} + \frac{3y^2}{2x^2} - \frac{y^3}{2x} + \frac{y^4}{16}$
25. $x^6 + 6x^5 + 15x^4 + 20x^3 + 15x^2 + 6x + 1$ **27.** −540   **29.** 672
31. 1120   **33.** $-\frac{5}{2}$ **35.** 20   **37.** 84   **39.** $\frac{15}{64}$ **41.** $\frac{21}{128}$ **43.** $\frac{25}{216}$
45. $\binom{n}{k} = \frac{n!}{k!(n-k)!} = \frac{n!}{(n-k)!k!} = \binom{n}{n-k}$ **49.** 720   **51.** 15
53. $16a^4 - 32a^3b + 24a^2b^2 - 8ab^3 + b^4$ **55.** 4860   **57.** 5

Photo Credits

Chapter 1

Opener(left): © Scala/Art Resource, NY; **opener(right):** © Erich Lessing/Art Resource, NY; **30:** © Jeff Greenberg/PhotoEdit; **34(top left):** © Alan Kearney/Getty Royalty Free; **34(top right)** © Denis Poroy/AFP/Getty Images); **34(low left):** © Robyn Beck/AFP/Getty Images; **34(low right):** Courtesy Bold Data Technology, Inc; **37:** © BenjaminStein/All Too Flat; **49:** © Ryan McVay/PhotoDisc/Getty Images RF; **51:** © Image 100/Royalty Free Corbis; **55:** © John A. Rizzo/Getty Images RF; **62:** © Burke/Triolo/Brand X Pictures/Alamy RF.

Chapter 2

Opener(right): © Photri-Microstock; **opener(left):** © Art Resource, NY; **74:** © Dennis Johnson; Papilio/Corbis; **87:** Courtesy Ignacio Bello; **100:** © BananaStock/PunchStock RF; **102:** © Marianna Day Massey/ZUMA/Corbis; **103:** © A. Ramey/PictureQuest; **111(left):** © TRBfoto/Getty Images RF; **111(center):** © Tony Freeman/Photo Edit; **111(right):** © Royalty Free/Corbis; **123:** © Lawrence Manning/Corbis RF; **142:** © PhotoDisc/Getty Images.

Chapter 3

Opener: © Bettmann/Corbis; **166:** © John Bigelow Taylor/Art Resource, NY; **207(left):** © PM Images/Getty; **220:** © The McGraw-Hill Companies, Inc./Connie Mueller Photo; **223:** © Tim Laman/NG/Getty; **231:** © Tom Pantages Stock Photos; **242:** © C Squared Studios/PhotoDisc/Getty Images RF.

Chapter 4

Opener(left): © The Art Archive/Academie des Sciences Paris/Dagli Orti/The Picture Desk, Inc.; **opener(right):** © Michael Nicholson/Corbis; **305:** © Rob Meinychuk/Getty Images RF; **310:** © PhotoLibrary; **315:** © The McGraw-Hill Companies, Inc.; **317:** Courtesy F. Hopf; **318:** © Philippe Colombi/Getty Images RF; **320:** © Royalty Free/Corbis; **323:** © PIXTAL/age fotostock RF; **324:** © Tom Pantages; **325:** Courtesy F. Hopf; **326:** © Bettmann/Corbis.

Chapter 5

Opener: © 2004 age fotostock; **348:** © Dallas and John Heaton/PictureQuest; **359:** © Donovan Reese/Getty Images/RF; **379:** © Philip Rostron/Masterfile; **388:** © Morton Beebe/Corbis; **404:** © Gilles Mingasson/Liaison/Getty; **412:** © Master File RF; **418:** © C Squared Studios/Getty Images.

Chapter 6

Opener: © The Granger Collection; **428:** Courtesy F. Hopf; **440:** © The McGraw-Hill Companies, Inc.; **448:** Russell Illig/Getty Images RF; **460:** © Steve Finn/Getty Images; **470:** © Royalty Free/Corbis; **481:** Courtesy F. Hopf; **495:** © Scala/Art Resource, NY; **498:** Courtesy F. Hopf; **500:** Courtesy F. Hopf; **505(top):** © Tony Freeman/PhotoEdit; **505(lower):** Reprinted by permission of the Orange County Register © 2002.

Chapter 7

Opener: © Bettmann/Corbis; **526:** © Mike Dobel/Masterfile; **537:** Courtesy Ignacio Bello; **548:** NASA; **559:** © PhotoLink/Getty Images RF; **567(top):** © Mary Evans Picture Library; **567(lower):** © Bettmann/Corbis; **578:** © Dennis MacDonald/PhotoEdit.

Chapter 8

Opener: © Martin Schoyen/The Schoyen Collection MS2192; **592:** © Chase Swift/Corbis; **600:** © Omri Waisman/Beateworks/Corbis RF; **602:** Courtesy F. Hopf; **605:** © Bettmann/Corbis; **615:** © PhotoDisc/Getty Images RF; **617:** © Neil Beer/Getty Images RF; **626:** © Rob Crandal/The Image Works; **635:** © David Young-Wolff/PhotoEdit.

Chapter 9

Opener(left): © Bob Kristi/Corbis; **opener(right):** © Joseph Sohm/ChromoSohm Inc./Corbis; **658:** Courtesy Ignacio Bello; **680:** © Bob Kristi/Corbis; **687:** Courtesy Patent pending Wind Weighted™ Baseball Tarps by Aer-Flo, Inc., Brandenton, FL.; **692(both), 693(all):** Courtesy Ignacio Bello; **698(top):** © 2004 age fotostock; **698(right):** © Robert Essel NYC/Corbis; **698(left):** © Joseph Sohm/ChromoSohm Inc./Corbis; **713:** © PhotoLink/Getty Images RF; **722 (top):** © AP Photo/Michael Stephens; **722 (bottom):** © The McGraw-Hill Companies, Inc./Connie Mueller Photo, Balloons: © Royalty Free/Corbis; **731(top):** © Kelly-Mooney Photography/Corbis; **731(lower):** © Royalty Free/Corbis.

Chapter 10

Opener: © George Bernard/SPL/Photo Researchers, Inc.; **748:** © D. Falconer/PhotoLink/Getty Images; **773:** © Hans Plletschinger/Peter Arnold, Inc.; **784:** © John Swart/AP Photo; **808(all):** © The McGraw-Hill Companies, Inc./Connie Mueller Photo; **822:** © PhotoDisc V18/Getty Images RF.

Index

A

Absolute value
  on calculator, 10
  definition of, 8
  equations, 142–145
  inequalities, 145–149
    graphing, 214–215
  of real number, 9
Ac test, 382–385
Addition
  of complex numbers, 569
  of functions, 749–750
  of polynomials, 352–353
  of radical expressions, 548–550
  of rational expressions, 449–456
  of signed numbers, 18
Addition property of inequality, 126
Additive identity, 18
Additive inverses
  and addition, 18
  definition of, 7–8
Ahmes, 73
Al-Khowarizmi, Mohammed ibn Musa, 73, 591
Algebra, history of, 73, 347
Alternate interior angles, 94
Angle measures, 94–95
Angles
  alternate interior, 94
  corresponding, 94
  exterior, 94
  interior, 94
  right, 106
  straight, 106
  supplementary, 106
  vertical, 94
Antilogarithm, 797–798
Associative property, 28
Asymptotes, of hyperbola, 700
Axis of symmetry, 659

B

Bar graphs, 513–514
Base
  of exponential function, 774
  of logarithms, changing, 799–800
Binomial
  definition of, 349
  factoring of, 373
  multiplication of, 362–363
  squaring of, 364–365
Boundary lines, 211
Bounded interval, 131
Brescia, Nicolo de, 427
Briggs, Henry, 747
Bürgi, Jobst, 747

C

Calculator
  absolute value on, 10
  break-even points on, 717
  circles on, 683
  complex fractions on, 466
  division of fractions on, 27
  division of polynomials on, 472
  domain on, 230
  ellipses on, 686
  equations on, 819
  exponential functions on, 802
  exponents on, 39
  factor theorem on, 477
  finding equations given two points, 201
  fractions on, 19
  functions on, 232, 754
  graphing inverses on, 765
  graphing on, 631–632
  hyperbolas on, 702
  inequalities on, 216
  logarithms on, 790, 797, 798, 803
  maxima on, 754
  minima on, 754
  multiplication of polynomials on, 360–361
  negative sign *vs.* subtraction key, 8
  nonlinear systems on, 715, 716, 727
  numerical calculations on, 5
  order of operations on, 54
  ordered pairs on, 762, 765
  parabolas on, 664, 669–670
  perfect square trinomial factoring on, 389
  pi on, 10
  polynomial inequalities on, 643
  quadratic formula on, 614
  radical expressions on, 541
  range on, 230
  rational equations on, 491, 501–502
  rational expressions on, 456
  remainder theorem on, 477
  roots on, 530, 534, 539
  scatter plots on, 251
  scientific notation on, 47
  slope on, 192
  systems of equations on, 287, 291, 292
  systems of inequalities on, 727
  variation on, 509
Cardan, Girolamo, 567
Cartesian coordinate system, 166
Center, of circle, 682–683
Change-of-base formula, 799–800
Circle
  on calculator, 683
  as conic section, 658
  definition of, 682
  eccentricity of, 694
  equation of, 682, 704
  finding center of, 682–683
  finding radius of, 682–683
Closure property, 28
Coefficients, 349
Coin problems, 316–317
Common logarithms, 796–798
Commutative property, 28
Complete factoring, 398
Completing the square, 597–599
Complex fractions, 460–466
Complex numbers
  addition of, 569
  definition of, 568
  division of, 570–572
  history of, 567
  multiplication of, 570–572
  and quadratic formula, 610–611
  subtraction of, 569
Composite functions, 750–751
Compound inequalities, 129–135
Concave down, 660
Concave up, 660
Conditional equations, 83
Conic sections. *See also* circle; ellipse; parabola
  history of, 657
  identification of, by equations, 703–705
  types of, 658
Conjugates, 554
Consecutive integers, 104
Consistent system, 286, 289, 309–310
Constant functions, 244–245
Constants, 3, 74
Constraints, 332–333
Contradictions, 83
Conversion, of measurements, 29–30
Coordinate plane, 166
Coordinate system, 166
Corresponding angles, 94
"CRAM" method, 80
Critical values, 639
Cubes (mathematical)
  factoring sums and differences of, 392–395
  roots, 539
Cubic equations, 611–612

D

Da Vinci, Leonardo, 495
Decimals
  equations involving, 79–83
  rational numbers as, 5
Decreasing functions, 244–245
Degree, of polynomial, 349
Denominators, rationalizing, 554–555
Dependent systems of equations, 288, 289, 309–310
Dependent variables, 224, 226
Descartes, René, 165
Differences
  of cubes, factoring of, 392–395
  of squares, factoring, 390–392
Diophantus, 109, 347
Direct variation, 505–506
Discourse on Method (Descartes), 165
Discriminant, 617–619, 666–667
Distance formula, 680–681
Distance problems, 498–499
Distributive property, 28, 55–58

I-1

Index

Dividend, 471
Division
  of complex numbers, 570–572
  with exponents, 40–43
  of fractions, 27
  of functions, 749–750
  notation, 24
  of polynomial by binomial, 475–476
  of polynomial by monomial, 470–471
  of polynomials, 471–473
  of radical expressions, 551–553
  of rational expressions, 442–443
  of real numbers, 23
  synthetic, 475–476
  zero in, 24
Division property of inequality, 127–128
Divisor, 471
Domain, of functions, 224–225, 230, 751–753

E

e, as base, 800–802
Eccentricity
  of circle, 694
  of hyperbola, 709
Elimination method, for systems of equations, 292–295, 306–309
Ellipse
  on calculator, 686
  as conic section, 658
  definition of, 684
  equation of, 704
  foci of, 684
  graphing, 684–687
Equality, properties of, 75, 77
Equations
  absolute value, 142–145
  on calculator, 819
  of circles, 682, 704
  conditional, 83
  of conic sections, 703–705
  cubic, 611–612
  definition of, 75
  of ellipses, 704
  equivalent, 76–77
  exponential
    definition of, 811
    solving, 811–815
  finding
    given parallels, 203–204
    given perpendiculars, 203–204
    given point and slope, 202
    given slope and y-intercept, 202–203
    given two points, 200–201
    of horizontal lines, 204–205
    of vertical lines, 204–205
  first-degree, 75
  of hyperbolas, 704
  with infinite solutions, 83
  of inverse functions, 762–763
  linear
    definition of, 75
    graphing, 168–171
    slope-intercept form of, 192–193
    solving, 79–83
  literal
    definition of, 87–88
    solving, 87–91
  logarithmic, solving, 786–788, 815–816
  with no solutions, 83
  nonlinear
    definition of, 713
    graphing, 174–175
    systems of, 713–715
  of parabolas, 703–704
  quadratic
    and completing the square, 597–599
    and discriminants, 617–619
    factoring, 405–410, 620–621
    of form $ax^2 + c = 0$, 592–595
    of form $a(x + b)^2 = c$, 595–597
    history of, 591
    solving, 592–599
    standard form of, 404–405, 606
  radical, 559–563
  with rational expressions, solving, 482–488, 627–632
  slope-intercept form for, 202–203
  solutions of, 76
  solving, with properties of equality, 76–79
  systems of
    on calculator, 287, 291, 292
    consistent, 286, 289, 309–310
    definition of, 285
    dependent, 288, 289, 309–310
    elimination method for, 292–295
    history of, 283
    inconsistent, 287, 289, 309–310
    linear
      on calculator, 287, 291, 292
      consistent, 286, 289
      definition of, 285
      dependent, 288, 289
      elimination method for, 292–295
      history of, 283
      inconsistent, 287, 289
      solving by graphing, 285–289
      substitution method for, 289–292, 295
      three variable, elimination method for, 306–309
    nonlinear, 713–715, 716
      on calculator, 715, 716
      solving by substitution, 713–715
      with two second-degree equations, 716
    solving by graphing, 285–289
    substitution method for solving, 289–292, 295
    three variable, elimination method for, 306–309
Equivalence property, 786, 812, 813
Equivalence relation, 240
Equivalent equations, 76–77
Exponential equations
  definition of, 811
  solving, 811–815
Exponential functions
  of base e, 800–802
  base of, 774
  on calculator, 802
  decreasing, 775–776
  definition of, 774
  exponents in, 774
  graphing, 774–776, 800–802
  increasing, 775–776
  to logarithmic, 786
Exponents
  base of, 38
  on calculator, 39
  definition of, 38
  division with, 40–43
  evaluation of, 38–39
  in exponential functions, 774
  i as base of, 572–574
  multiplication with, 40–43
  natural number, 38–39
  negative, 39–40
  power rule of, 43
  product rule for, 41
  quotient rule for, 42
  raised to powers, 43–45
  rational
    converting to radicals, 528–532
    operations with, 532–534
  and scientific notation, 45–46
  zero as, 39–40
Expressions
  algebraic, 74–75
    evaluating, 54–55
    and grouping symbols, 51–53
    and order of operations, 53–54
  radical, 527
  translation into, 91–92
  radical, 527
    addition of, 548–550
    on calculator, 541
    division of, 551–553
    multiplication of, 551–553
    reducing index of, 543–545
    similar, 549
    subtraction of, 548–550
  rational
    addition of, 449–456
    on calculator, 456
    definition of, 428–429
    division of, 442–443
    equivalent, 430–432
    fundamental rule of, 431

I-2

Index

multiplication of, 440–442
reducing, 433–436
simplification of, 443–444
solving, 627–632
solving equations with, 482–488, 627–632
standard form for, 432–433
subtraction of, 449–456
undefined, 429–430
Exterior angles, 94
Extraneous solutions, 482, 559

F

Factor theorem, 476–477
Factors and factoring
and *ac* test, 382–385
of binomials, 373
complete, 398
of difference of cubes, 392–395
of difference of two squares, 390–392
greatest common, 371–374
by grouping, 374–376
of perfect square trinomials, 388–390
of polynomials, 373, 374, 473–475
of quadratic equations, 405–410, 620–621
of sums of cubes, 392–395
of trinomials of form $ax^2 + bx + c$, 381–382
of trinomials of form $x^2 + bx + c$, 379–381
Feasible solutions, 332–333
Ferro, Scipione del, 427
First-degree equations, 75
Foci, of ellipse, 684
Formulas
definition of, 88
evaluating, 93
solving, 87–91
writing, 93
Fourth root, 539
Fractions. *See also* rational expressions
on calculator, 19, 466
complex, 460–466
definition of, 428–429
division of, 27
equations involving, 79–83
multiplication of, 23
simple, 460
Functions. *See also* relation
addition of, 749–750
on calculator, 232
combinations of, 751–753
composite, 750–751
constant, 244–245
decreasing, 244–245
definition of, 225, 232
division of, 749–750
domain of, 224–225, 230
domain of combinations of, 751–753
exponential

base of, 774
on calculator, 802
decreasing, 775–776
definition of, 774
exponents in, 774
graphing, 774–776, 800–802
increasing, 775–776
graphing, 763–765
and horizontal line test, 764
increasing, 244–245
inverse of, 761–765
linear
for data modeling, 248–252
definition of, 232
from graphs, 246–247
identifying, 243–246
logarithmic
definition of, 785
graphing, 784–785
maximum of, 754
minimum of, 754
multiplication of, 749–750
one-to-one, 764
operations with, 749–750
range of, 224–225
relation in, 224–225
subtraction of, 749–750
value of, 231–232
and vertical line test, 229

G

Gauss, Carl Friedrich, 567
Geometry, in word problems, 105–107
Goddard, Robert, 605
Golden ratio, 495
Golden rectangle, 495
Graphing
absolute value inequalities, 214–215
bar graphs, 513–514
on calculator, 631–632
of circles, 682
and distance formula, 680–681
of ellipses, 684–687
exponential functions, 774–776, 800–802
functions, 763–765
horizontal lines, 173–174
of hyperbolas, 698–703
inequalities, 124–125
inverse functions, 763–765
linear equations, 168–171
linear inequalities, 211
logarithmic functions, 784–785, 800–802
nonlinear equations, 174–175
overview of, 166–168
parabolas, 659–670, 671
second-degree inequalities, 725–727
systems of equations, 285–289, 295
systems of inequalities, 328–330
using intercepts, 171–173

using slope and point, 190–191
vertical lines, 173–174
Greatest common factor, definition of, 371, 372
Grouping, factoring by, 374–376
Grouping symbols, 51–53, 59–60

H

Hippocrates of Chios, 657
Horizontal line test, 764
Horizontal lines
finding equations of, 204–205
graphing, 173–174
slope of, 187
Hyperbola
asymptotes of, 700
on calculator, 702
as conic section, 658
definition of, 698–699, 710
eccentricity of, 709
equations of, 704
graphing, 698–703
vertices of, 699

I

i
definition of, 568
powers of, 572–574
square of, 570
Identities, 83
Imaginary numbers
definition of, 568
pure, 568
Implied constraints, 332–333
Inconsistent systems of equations, 287, 289, 309–310
Increasing functions, 244–245
Independent variables, 224, 226
Index, of radical, 527, 543–545
Inequalities
absolute value, 145–149, 214–215
addition property of, 126
compound, 129–135
definition of, 10, 123
division property of, 127–128
equivalent, 124
graphing, 124–125, 211–214
linear, 123–124
multiplication property of, 127–128
polynomial, 638–639
properties of, 125–127
quadratic, 636–638
rational, 639–642
second-degree, graphing, 725–727
solving, 125–135
subtraction property of, 126
systems of, 328–330, 726–727
on calculator, 727
graphing, 328–330
nonlinear, 726–727
from words, 135–136
Infinite intervals, 125

I-3

Index

Integers
  consecutive, 104
  definition of, 3
  word problems with, 104–105, 496
Intercepts, graphing with, 171–173, 192–193
Interior angles, 94
Intersection, of two sets, 130–131
Intervals
  bounded, 131
  unbounded, 125
Inverse
  logarithm, 797–798
  of a relation, 761–765
Inverse variation, 506–507
Investment problems, 115–116, 318–320
Irrational numbers, 6

J

Joint variation, 508

K

Key number, 382

L

Least common denominator (LCD), 451–452
Leibniz, Gottfried Wilhelm, 283
Like terms, combining, 58–59
Linear equations
  definition of, 75
  graphing, 168–171
  slope-intercept form for, 192–193
  solving, 79–83
  systems of
    on calculator, 287, 291, 292
    consistent, 286, 289
    definition of, 285
    dependent, 288, 289
    elimination method for, 292–295, 295
    history of, 283
    inconsistent, 287, 289
    solving by graphing, 285–289
    substitution method for, 289–292, 295
    three variable, elimination method for, 306–309
Linear functions
  for data modeling, 248–252
  definition of, 232
  from graphs, 246–247
  identifying, 243–246
Linear inequalities
  definition of, 123
  graphing, 124–125, 211–214
  solving, 125–129
  in two variables, 211
Linear programming, 332–333
Lines
  point-slope form for, 201
  slope-intercept form, 192–193, 202–203

Literal equations
  definition of, 87–88
  solving, 87–91
Logarithmic equations
  definition of, 815
  solving, 815–816
Logarithms
  of base e, 800–802
  on calculator, 790, 803
  changing bases of, 799–800
  common, 796–798
  converting to exponential equations, 786
  definition of, 784
  equivalence property for, 813
  graphing, 784–785
  history of, 747
  inverse, 797–798
  Napierian, 798–799
  natural, 798–799
  properties of, 788–789
  solving, 786–788
Lowest term, reducing fractions to, 433–436

M

Mach, Ernst, 548
Maclurin, Colin, 283
Mapping diagram, 224
Mathematical modeling, 248–253
Measurement conversion, 29–30
Menaechmus, 657
Mixture problems, 117–118
Modeling, mathematical, 248–253
Monomial, 349, 360–361, 470–471.
    See also polynomials
Motion problems, 116–117, 318
Multiplication
  of binomials, 362–363
  of complex numbers, 570–572
  with exponents, 40–43
  of fractions, 23
  of functions, 749–750
  of monomials by polynomials, 360–361
  by negative one, 56
  notation, 20
  of polynomials, 361–362
  of radical expressions, 551–553
  of rational expressions, 440–442
  of signed numbers, 20–23
  of sums and differences, 365
  by zero, 20
Multiplication property of inequality, 127–128
Multiplicative identity, 21, 56
Multiplicative inverse, 26

N

Napier, John, 747
Napierian logarithms, 798–799
Natural logarithms, 798–799
Natural numbers, 3, 38–39

Negative exponents, 39–40
Nonlinear equations
  definition of, 713
  graphing, 174–175
Nonlinear inequalities, systems of, 726–727
Nonlinear systems
  on calculator, 715, 716, 727
  solving by substitution, 713–715
  with two second-degree equations, 716
Nth root, 526, 538
Numbers
  additive inverse of, 7–8
  classification of, 6–7
  complex
    addition of, 569
    definition of, 568
    division of, 570–572
    history of, 567
    multiplication of, 570–572
    and quadratic formula, 610–611
    subtraction of, 569
  history of, 1
  irrational, 6
  key, 382
  natural, 3, 38–39
  perfect, 528
  prime, 371
  properties of, 28–29
  rational, 4–5
  real, 6
  signed
    addition of, 18
    multiplication of, 20–23
    subtraction of, 19–20
  whole, 3

O

One, negative, multiplication by, 56
One-to-one function, 764
Order of operations, 53–54
Ordered pairs
  on calculator, 762, 765
  graphing of, 166–168
Origin, on graph, 166

P

Parabola
  axis of, 659
  on calculator, 664, 669–670
  as conic section, 658
  equations of, 703–704
  of form $x = a(y - k)^2 + h$, 668–669
  of form $x = ay^2 + by + c$, 668–669
  of form $y = f(x) = a^2 + bx + c$, 664–668
  of form $y = f(x) = a(x - h)^2 + k$, 662–664
  of form $y = f(x) = ax^2 + k$, 659–662
  graphing, 659–670, 671
  symmetry of, 659
  vertex of, 659
Paraboloid of revolution, 678

Index

Parallel lines, 189–190, 203–204
PE(MD)(AS) method, 53
Percents, problems involving, 112–115
Perfect numbers, 528
Perfect square trinomial, 388–390
Perimeter
   definition of, 89
   formula for, 89
Perpendicular lines, 189–190, 203–204
Pi (π), on calculator, 10
Point-slope form, 201
Polynomial inequalities, 638–639
Polynomials
   and ac test, 382–385
   addition of, 352–353
   on calculator, 472
   definition of, 348–349
   degree of, 349–350
   division of, 470–473, 475–476
   evaluating, 350–351
   factoring of, 373, 398–401, 473–475
   multiplication of, 360–362
   subtraction of, 353–354
Power property, of logarithms, 788
Power rules, 43, 560
Prime numbers, 371
Product property, of logarithms, 788
Product rule, for exponents, 41
Programming, linear, 332–333
Proportions
   definition of, 488
   direct, 505–506
   inverse, 506–507
   joint, 508
   solving, 488–490
Pure imaginary numbers, 568
Pythagoras, 525
Pythagorean theorem, 411–412

Q

Quadrants, 166
Quadratic equations
   and completing the square, 597–599
   and discriminants, 617–619
   factoring, 405–410, 620–621
   of form $ax^2 + c = 0$, 592–595
   of form $a(x + b)^2 = c$, 595–597
   history of, 591
   solving, 592–599
   standard form of, 404–405, 606
Quadratic formula, 605–611
Quadratic functions, 659, 764
Quadratic inequalities, 636–638
Quotient property, of logarithms, 788
Quotient rule, for exponents, 42

R

Radical equations, 559–563
Radical expressions, 527
   addition of, 548–550
   on calculator, 541
   division of, 551–553
   multiplication of, 551–553
   reducing index of, 543–545
   similar, 549
   subtraction of, 548–550
Radical sign, 527
Radicals
   converting rational exponents to, 528–532
   overview of, 526–528
   properties of, 538–541
   simplification of, 539–545
Radicand, 527
Radius, of circle, 682–683
Range, of functions, 224–225
Rational exponents
   converting to radicals, 528–532
   operations with, 532–534
Rational expressions. See also fractions
   addition of, 449–456
   on calculator, 456
   definition of, 428–429
   division of, 442–443
   equivalent, 430–432
   fundamental rule of, 431
   multiplication of, 440–442
   reducing, 433–436
   simplification of, 443–444
   solving, 627–632
   solving equations with, 482–488, 627–632
   standard form for, 432–433
   subtraction of, 449–456
   undefined, 429–430
Rational inequalities, 639–642
Rational numbers
   as decimals, 5
   definition of, 4–5
Rationalizing denominators, 554–555
Ratios
   definition of, 488
   in proportions, 488
Real numbers, 6
Reciprocal, definition of, 26
Reflexive property, 75
Relation. See also functions
   definition of, 225
   equivalence, 240
   inverse of, 761–765
Remainder theorem, 476–477
Rhind, Henry, 73
Rhind papyrus, 73
Right angles, 106
Roots
   on calculator, 530, 534, 539
   overview of, 526–527
Roster notation, 4
RSTUV method, for word problems, 100, 102–103, 496

S

Scatter plots, 242–243, 248–251
Scientific notation, 45–46, 47
Second-degree inequalities, graphing, 725–727
Set-builder notation, 4
Sets
   intersection of two, 130–131
   notation of, 4, 131
   of real numbers, 3
   rules for, 4
   union of two, 129–130
Signed numbers
   addition of, 18
   multiplication of, 20–23
   subtraction of, 19–20
Slope
   on calculator, 192
   definition of, 185
   finding, 185–188
   graphing using, 190–191
   of horizontal line, 187
   negative, 187
   and parallel lines, 189–190
   and perpendicular lines, 189–190
   positive, 187
   of vertical line, 187
Slope-intercept form, 192–193, 202–203
Solutions
   definition of, 76
   and discriminant, 617–619
   extraneous, 482, 559
   set of, 169
Specified variables, 499–501
Square roots, 539
Squares (mathematical)
   of binomials, 364–365
   factoring difference of two, 390–392
Standard form
   for quadratic equations, 404–405, 606
   for rational expressions, 432–433
Statistical modeling, 248–253
Straight angles, 106
Substitution method
   for nonlinear systems, 713–715
   for rational expression equations, 628–632
   for system of equations, 295
Subtraction
   of complex numbers, 569
   of functions, 749–750
   of polynomials, 353–354
   of radical expressions, 548–550
   of rational expressions, 449–456
   of signed numbers, 19–20
Subtraction property of inequality, 126
Sums, of cubes, factoring of, 392–395
Supplementary angles, 106
Symmetric property, 75
Symmetry
   of parabola, 659
   y-axis, 775
Synthetic division, 475–476

Index

Systems of equations
  on calculator, 287, 291, 292
  consistent, 286, 289, 309–310
  definition of, 285
  dependent, 288, 289, 309–310
  elimination method for, 292–295
  history of, 283
  inconsistent, 287, 289, 309–310
  linear
    on calculator, 287, 291, 292
    consistent, 286, 289
    definition of, 285
    dependent, 288, 289
    elimination method for, 292–295
    history of, 283
    inconsistent, 287, 289
    solving by graphing, 285–289
    substitution method for, 289–292, 295
    three variable, elimination method for, 306–309
  nonlinear, 713–715, 716
    on calculator, 715, 716
    solving by substitution, 713–715
    with two second-degree equations, 716
  solving by graphing, 285–289
  substitution method for solving, 289–292, 295
  three variable, elimination method for, 306–309
Systems of inequalities, 328–330, 726–727
  on calculator, 727
  graphing, 328–330
  nonlinear, 726–727

T

Tallying, 1
Tartaglia, 427
Taylor, Brook, 283
Taylor series, 283
Terms
  combining like, 58–59
  definition of, 79
  degree of, 349
Transitive property, 75
Transversal, definition of, 94
Trichotomy law, 10
Trinomial. *See also* polynomials
  $ax^2 + bx + c$ form, 381–382
  definition of, 349
  perfect square, 388–390
  $x^2 + bx + c$ form, 379–381

U

Unbounded intervals, 125
Undefined rational expressions, 429–430
Union of two sets, 129–130

V

Values, critical, 639
Variables, 74, 224, 499–501
Variation
  on calculator, 509
  direct, 505–506
  inverse, 506–507
  joint, 508
Vertex
  of hyperbola, 699
  of parabola, 659
Vertical angles, 94
Vertical line test, 229
Vertical lines
  finding equations for, 204–205
  graphing, 173–174
  slope of, 187
Vieta, François, 347

W

Whole numbers, 3
Word problems
  definition of, 100
  geometry, 105–107
  integer, 104–105, 496
  investment, 115–116
  mixture, 117–118
  motion, 116–117
  percent, 112–115
  RSTUV method for, 100, 102–103
Words
  into expressions, 91–92, 100–102
  into inequalities, 135–136
Work problems, 497–498

X

X-intercept, 172

Y

Y-axis symmetry, 775
Y-intercept, 172, 202–203

Z

Zero
  in division, 24
  as exponent, 39–40
  multiplication properties of, 20